# Production Diseases in Farm Animals

# Springer Nature More Media App

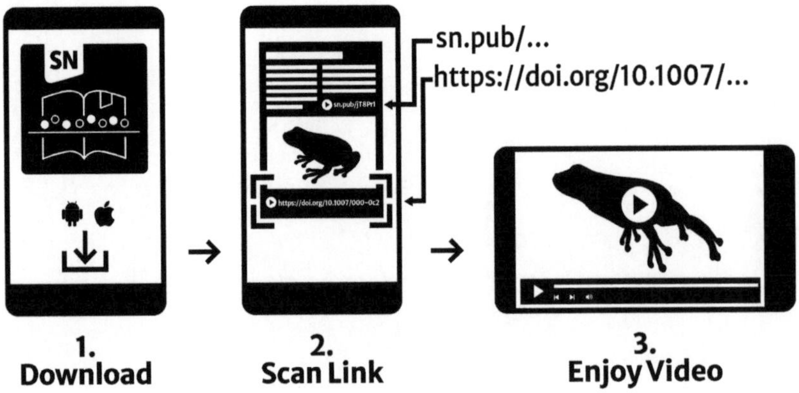


Support: customerservice@springernature.com

Josef Johann Gross
Editor

# Production Diseases in Farm Animals

## Pathophysiology, Prophylaxis and Health Management

Springer

*Editor*
Josef Johann Gross
Veterinary Physiology, Vetsuisse Faculty
University of Bern
Bern, Switzerland

This work contains media enhancements, which are displayed with a "play" icon. Material in the print book can be viewed on a mobile device by downloading the Springer Nature "More Media" app available in the major app stores. The media enhancements in the online version of the work can be accessed directly by authorized users.

ISBN 978-3-031-51787-7　　ISBN 978-3-031-51788-4 (eBook)
https://doi.org/10.1007/978-3-031-51788-4



This Springer imprint is published by the registered company Springer Nature Switzerland AG
The registered company address is: Gewerbestrasse 11, 6330 Cham, Switzerland

# Preface

The growing demand of an increasing global population for animal-derived food and products requires intensive livestock farming. During the past few decades, performance across all livestock species worldwide increased continuously. However, breeding and husbandry of farm animals are often accompanied by conflicts of interest between biological requirements of the animal and economic needs of the producer. This conflict is increasingly gaining attention not only by producers, animal scientists, and veterinarians but also by the public. An inadequate breeding strategy or animal housing and feeding may lead to production diseases, coming along impaired animal welfare, but also reduced income for the producer and reduced product quality for the consumer. Furthermore, large-scale livestock farming creates an increasing number of environmental problems which must be considered a limiting factor for animal production.

Production diseases were originally defined as "a number of metabolic disorders of increasing importance in agriculture." Over the years, this definition has expanded and covers diverse aspects of diseases and their association with performance level and management. Farm animals are permanently challenged by a variety of factors: improper nutrition, inadequate housing systems, infections, stress, etc., which affect the incidence, course, and outcome of production diseases. Despite changes and improvements in livestock farming, analytics, and veterinary care, further research is warranted in terms of gathering new information on the prevention, diagnosis, and treatment of production diseases.

This book focuses mainly on production diseases and their prevention in ruminants, swine, and poultry. Experts in different fields of animal science and veterinary medicine contribute to this book. The main topics are veterinary public health and metabolic, gastrointestinal, and respiratory health disorders. Furthermore, mammary gland health and locomotion disorders are addressed, likewise the impact of environment, behavior, genetics, and animal welfare. The major focus of the individual contributions is on an interdisciplinary view on pathophysiology, prophylaxis, and health management.

Of course, the chapters presented in this volume reflect only a small segment of current research related to production diseases in farm animals and make no claim to completeness.

Bern, Switzerland — Josef J. Gross

# Contents

# Abbreviations

| | |
|---|---|
| AA | Amino acids |
| acetyl-CoA | Acetyl-coenzyme |
| AAPP | Acute phase protein |
| BCS | Body condition score |
| BHB | Beta-hydroxybutyrate |
| C12:0 | Lauric acid |
| C14:0 | Myristic acid |
| C16:0 | Palmitic acid |
| C18:0 | Stearic acid |
| C18:1n9 | Oleic acid |
| C18:2n6 | Linoleic acid |
| CCK | Cholecystokinin |
| CLA | Conjugated linoleic acid |
| COX | Cyclooxygenase |
| CPT1 | Carnitine palmitoyltransferase 1 |
| CPT2 | Carnitine palmitoyltransferase 2 |
| DHA | Docosahexaenoic acid |
| DMI | Dry matter intake |
| EPA | Eicosapentaenoic acid |
| ETE | Eicosatrienoic acid |
| FA | Fatty acid |
| FFA | Free fatty acids |
| GH | Growth hormone |
| GIT | Gastrointestinal tract |
| GPX | Glutathione peroxidase |
| HMG-CoA | 3-Hydroxy-3-methylglutaryl-coenzyme A |
| IFN | Interferon |
| IGF-1 | Immune like growth factor |
| IL | Interleukin |
| L-FABP | Liver fatty acid binding protein |
| LPS | Lipopolysaccharide |
| NDF | Neutral detergent fiber |
| NEB | Negative energy balance |
| NEFA | Non-esterified fatty acids |
| NFC | Non-fibrous carbohydrates |
| NF-KB | Nuclear factor-K beta |
| NPB | Negative protein balance |

| | |
|---|---|
| Nrf2 | E2-related factor 2 |
| OS | Oxidative stress |
| pe | Physical effective |
| PI | Physiological imbalance |
| PIC | Proinflammatory cytokines |
| PMN | Peripheral mononuclear cells |
| PON | Paraoxonase |
| PTH | Parathyroid hormone |
| PUFA | Polyunsaturated fatty acids |
| RNS | Reactive nitrogen species |
| ROS | Reactive oxygen species |
| SARA | Subacute rumen acidosis |
| SFA | Saturated fatty acid |
| SOD | Superoxide dismutase |
| TCA | Tricarboxylic acid |
| TG | Triglycerides |
| TNF | Tumor necrosis factor |
| UFA | Unsaturated fatty acids |
| VFA | Volatile fatty acids |
| VLDL | Very low-density lipoproteins |

# Part I

# Veterinary Public Health Aspects of Production Diseases

# Epidemiology of Production Diseases

Gertraud Schüpbach-Regula

**Abstract**

Prevention and control of production diseases requires a profound knowledge on the epidemiology of the respective disease or disease complex. The objective of this chapter is therefore to introduce important epidemiological concepts and their application to production diseases. When investigating the causal factors which lead to production disease, it is important to keep in mind that, by definition, production diseases are not caused by a single pathogen or factor but rather by a combination of component causes and factors predisposing for disease. Causality should therefore be defined as a range of evidence rather than a direct relation between a factor and disease.

Methods of descriptive epidemiology are an important basis to understand disease. Frequency of disease should be expressed in relation to the population at risk, as an incidence or prevalence. Describing the distribution of disease over time and location is helpful for forming hypotheses on factors contributing to disease risk. Many epidemiological methods exist for analyzing associations between disease and risk factors. These methods range from simple descriptive measures such as relative risk and odds ratios to statistical models such as regression models and more advanced methodologies such as Bayesian methods and machine learning algorithms. For production diseases with infectious etiology, methods from infectious disease epidemiology such as SIR (susceptible-infectious-resistant) models, agent-based models, or network models can also be applied.

**Keywords**

Epidemiological methods, Causality, Prevalence, Incidence, Relative risk

## 1 Introduction

According to the World Health Organization, epidemiology is the study of the distribution and determinants of health-related states or events (including disease) and the application of this study to the control of diseases and other health problems (WHO, search reference). While many different definitions of epidemiology exist, most of them have in common that they deal with studying the distribution of health or disease in populations (Frérot et al. 2018).

Production diseases are noninfectious or infectious diseases which are caused or enhanced by management practices (Nir 2003), particularly

G. Schüpbach-Regula (✉)
Veterinary Public Health Institute, Vetsuisse Faculty
University of Bern, Bern, Switzerland
e-mail: Gertraud.Schuepbach@unibe.ch


J. J. Gross (ed.), *Production Diseases in Farm Animals*,
https://doi.org/10.1007/978-3-031-51788-4_1

in intensive production systems. Production diseases often do not result in the death of the animal but reduce the efficiency of the system (FAO 2022). This implies that production diseases are primarily a problem of animal populations and animal production systems and only secondarily problems of the individual. Understanding the epidemiology of production diseases is thus of paramount importance for investigating, preventing, and controlling production diseases.

The effects of production diseases can be manifold. While they primarily impair health of the animal populations, production diseases such as mastitis or lameness also considerably reduce animal welfare, which is causing increasing concern among consumers (Clark et al. 2016). Another important effect is the reduction of the output of production systems, for example, through decreased fertility, growth rate, or milk yield. Reduced efficiency of production systems has in turn many negative effects on economic viability of farms, the livelihood of farmers, food supply of the population, use of natural resources, and emissions of farming systems. Finally, human health, food quality, and food safety can also be affected. Examples are production diseases caused by pathogens with zoonotic potential such as *Staph. aureus* or *E. coli* and overuse of antimicrobial substances to mitigate the effects of production diseases. A large share of prophylactic and metaphylactic use of antibiotics in farm animals is attributable to production diseases (Landers et al. 2012). Also, part of the growth-promoting effect of antibiotics consists of reducing the negative effects of production diseases. Preventing production diseases is thus a prerequisite for a successful reduction of the usage of antimicrobials in animal production.

The basis for prevention is understanding the factors which cause, enable, or enhance disease. Most production diseases are not caused by a single pathogen or a single condition but rather a combination of many different factors. Therefore, a systems approach is needed which takes factors of the environment, climate, housing, husbandry, farm management, animals, and pathogens into account.

The objective of this book chapter is to introduce important epidemiological concepts and their application to production diseases. Specifically, the concept of causality, measures to describe frequency and distribution of diseases in populations, and epidemiological methods for investigating infectious and noninfectious diseases will be discussed.

## 2 Production Diseases and the Concept of Causality

Production diseases are generally complex health disorders, which are not easily understood or controlled. Mono-causal concepts of causality that postulate a direct relation between agent and disease, such as Koch's postulates, are therefore not applicable. Also, terms like sufficient and necessary cause are not very useful for describing causality of production diseases. A sufficient cause is a factor which always results in a certain effect. A necessary cause needs to be present for the effect to occur but does not always result in the effect. Instead, epidemiology uses terms like risk factors and associations to describe the factors influencing health status of animals. Another relevant causal concept uses the term component cause to describe diseases that are caused by a number of different factors (components) which only result in disease if several of them occur together. In modern epidemiology, causality is discussed as a range of evidence rather than a direct relation (Shimonovich et al. 2021). Causal models describe the relationship between different things or actions, for example, between housing system, environment, farm management, and animal health. They can be used to make predictions on the behavior of a system. For production diseases, we expect causal models to be able to predict the effect of an intervention or preventive measure (Dohoo et al. 2003). They can be visualized as causal diagrams or networks.

To understand causal factors affecting the prevalence and impact of production diseases, it is important to consider the complexity of all factors which may contribute toward disease and which are in turn often interconnected with each other:

- Factors related to the *individual animal*, such as genetic predisposition, breed, age, sex, immune status, etc.
- Factors related to the *environment and farm management*, such as climate, housing, ventilation, feeding, hygiene, biosafety, etc.
- For infectious diseases, factors related to the *infectious agent*, such as pathogenicity, virulence, contagiousness, transmissibility, immunogenicity, tenacity, resistance to treatments, etc.

While it is important for prevention and control of production diseases to understand the complexity of the system, it is often possible to substantially improve animal health by addressing one or a few of the most important factors contributing to disease. For example, measures improving milking hygiene are often successful to prevent mastitis, even though predisposing factors such as high milk yield of the animals and component causes such as the presence of *Streptococcus* spp. in the environment persist.

## 3 Descriptive Epidemiology of Production Diseases

The basis for understanding risk factors and causes of disease is a good and detailed description of disease in the population. Descriptive epidemiology answers the questions *what, who, how many, when, where, and why* (Thrusfield et al. 2018).

*What:* The first step in descriptive epidemiology is a *case definition*. A good case definition will include all relevant cases with the disease under study but exclude similar pathologies with different causalities. Case definitions can be based on clinical examination, results of diagnostic tests, characteristics of diseased animals, or a combination of these. For production diseases caused by a combination of a variety of different pathogens and predisposing factors, finding a good case definition can be quite challenging.

*Who:* Most diseases do not evenly affect all animals of a farm. Therefore, it can be helpful to characterize the individuals who are classified as diseased according to the case definition. For example, age, sex, breed/genetic background, type of use (e.g., beef or dairy), stage of production cycle (e.g., lactating/dry), level of production (e.g., milk yield), previous diseases, and immune status can be individual characteristics influencing the probability of disease.

*How many:* In epidemiology, it is not sufficient to count the number of diseased animals. This number is meaningless if we cannot relate it to a population. Five diseased animals are a very different situation in a population of 10 versus in a population of 10,000. The first rule is therefore to always get denominator data! Two measures of disease frequency are regularly used in epidemiology:

- *Prevalence* is the number of diseased animals at the time of observation divided by the total number of animals in the population.
- *Incidence* is the number of newly diseased animals over a specified period of time divided by the number of animals at risk of becoming diseased during the specified period.

Prevalence is often easier to obtain, because it can be recorded by a one-time measurement or observation. It is a useful measure for long-lasting disease (chronic states such as lameness) or serological status indicating prior contact to a pathogen. It is not well suited for short-lasting disease. For example, a one-time examination of dairy cows in a herd will detect cases of chronic mastitis more easily than cases of acute mastitis.

Incidence requires observation of the animals over a period of time to detect new cases of diseases. Incidence can be expressed as a *cumulative incidence* or an *incidence rate*. If the population is more or less stable, we can use the number of animals at the beginning of the observation period (or, alternatively, at the end, the midpoint, or the average between beginning and end of the period) as the denominator. The cumulative incidence is then the number of new cases divided by this denominator. In dynamic populations, each individual is observed over a different length of time. This can be the case if many animals are sold and purchased by a farm. In this case, the denominator is the sum of the length of time of observation for all individuals at risk of becoming diseased.

Prevalence and cumulative incidence can be expressed as a proportion or percentage, whereas incidence rate is expressed as number of diseased per number of animals and time (e.g., per 100 cow years).

*When:* The time sequence of cases can be important to understand disease. This is of particular importance for outbreaks of contagious diseases, where epidemic curves are an important tool to understand the dynamics of infection and to identify the source of infection. However, time sequence can also be relevant for noninfectious production diseases, because predisposing or causal factors may occur mainly in certain seasons or at certain times in the life of an animal.

*Where:* The location of diseased animals on the farm, or the location of farms with a high prevalence of disease in the country, is also an important information for understanding diseases. For infectious diseases, maps and geographic analyses can identify clusters of infection. For all diseases, these tools can help to identify predisposing or causal factors related to geographic factors such as climate, land use, altitude, or habitat for wildlife and potential disease vectors.

*Why:* Finally, we need to determine factors which are associated with disease. Describing associations is the first step for forming hypotheses on potential component causes or predisposing factors for the disease. However, it needs to be kept in mind that associations can also occur due to chance or due to confounding. Confounding means that a factor X can have a causal connection to a risk factor for the disease but not to the disease itself. If we only analyze factor X and disease, we will find an association between factor X and disease, which is caused by confounding. Methods to describe associations depend on the type of disease outcome to be studied. Because epidemiology often analyzes the relation between a dichotomous outcome (diseased versus healthy) and categorical risk factors, associations are often presented in 2 × 2 tables. Measures of association for this situation are:

- *Relative risk (RR):* The proportion of diseased animals (prevalence or cumulative incidence) in animals with a risk factor divided by the proportion of diseased in animals without the risk factor.
- *Odds ratio (OR)*: Odds are defined as the number of diseased divided by the number of healthy animals. The odds ratio is the odds in animals with the risk factor divided by the odds of animals without the risk factor.

If the risk (or odds) of disease are equal for animals with and without the risk factor, RR and OR are equal to 1. A RR/OR below 1 indicates a negative association, i.e., the factor is protective for the disease. Values greater than 1 indicate a positive association, i.e., a higher disease risk for animals with the risk factor. Because relative risk is more intuitive to understand for most people than odds ratio, odds ratios are often interpreted as an estimate for the relative risk. This is correct for rare diseases. However, for more frequent diseases (such as most production diseases), odds ratio and relative risk differ substantially and therefore should not be used interchangeably.

## 4 Infectious Disease Epidemiology Versus Noncontagious Diseases

Production disease is a generic term for a variety of diseases, which differ in their biological nature but are all partially caused or exacerbated by intensive production systems. They include etiologically different diseases:

- Metabolic disorders such as ketosis, ruminal acidosis, hypocalcaemia, and hypomagnesaemia
- Noninfectious diseases related to a high level of production or intensive housing conditions

such as sole ulcers in dairy cows or keel bone fractures in laying hens
- Infectious diseases caused by specific pathogens (or a specific combination of pathogens) such as *Staph. aureus* mastitis in cattle or foot-rot in sheep
- Disease complexes caused by a combination of a variety of different pathogens and predisposing factors such as bovine and porcine respiratory disease complex

This implies that for the study of production diseases, methods from infectious disease epidemiology as well as from the epidemiology of noncontagious diseases are applicable.

*Infectious disease epidemiology* studies the spread of disease in populations. In this approach, the focus is on the characteristics of the infectious agent and its interaction with the host population. Environmental risk factors enabling a more efficient disease transmission can also be considered but are typically not the focus of the analysis.

Epidemiological *field studies* are an important tool in infectious disease epidemiology. They can be used to describe disease dynamics in a well-defined population and environment. These studies have the advantage that they represent practice conditions. Examples are field studies which determine the spread of mastitis within a dairy herd (Kirkeby et al. 2019; Deng et al. 2021), studies which describe the effect of an intervention on within-herd transmission dynamics (Barlow et al. 2013), or studies which combine field data collection with modeling approaches (Dalen et al. 2018).

*Infectious disease models* combine expertise on the epidemiology of disease of many different sources, such as controlled experiments, field studies, outbreak investigation, and expert opinion. By simulating the spread of a disease in a population, the magnitude of a disease outbreak, the impact of endemic disease in a defined population, and the effectivity of intervention measures can be studied. Mathematical infectious disease models can either be deterministic, i.e., a given scenario always results in the same disease outcome, or stochastic, i.e., uncertainty and biological variability are built into the model. The classical infectious disease model is a compartmental *susceptible-infectious-recovered (SIR) model*, which uses differential equations to describe the transition between these different compartments. SIR models assume homogeneous mixing of populations, i.e., each individual has the same probability to become infected. While this is not true for most infectious diseases, good SIR models nevertheless result in accurate predictions on the average spread of disease. For subpopulations with varying disease risk, SIR models can be improved by modeling subpopulations with defined contact and transition rates between subpopulations.

With increasing processing capacity of computers becoming available, it is now possible to model the disease risk of individuals, taking individual characteristics into account which influence probability of infection. This class of infectious disease models is called *agent-based models* (Verteramo Chiu et al. 2018; Bassett et al. 2021). Because animal trade is of critical importance for the spread of infectious diseases in animals, it is often useful to specifically model the contacts between animals and farms as a network structure. These *network models* are most useful to describe and analyze between-farm transmission of disease. They are specifically useful to identify farms or animals with a high individual risk of spreading disease. Identifying these so-called super-spreaders can help to target control measures to those farms at highest risk of infecting others.

An integrated approach combining different types of models is also very useful for an improved prevention and control of infectious production diseases. For example, combining a network model with an SIR model allows for description of the effect of animal trade on disease transmission between farms (Lebl et al. 2016). Combining an SIR model with an economic model can help decision-makers to decide on whether an intervention is likely to be beneficial (Zingg et al. 2017).

*Epidemiology of noninfectious* diseases studies the risk factors and causal networks which contribute toward impaired health of the population. Classical epidemiological methods for noninfectious diseases mainly rely on the *analysis of associations and risk factors*. Descriptive epidemiology with 2 × 2 tables and graphical representation of associations is an important tool to understand associations between risk factors and disease. In further analysis, a univariable screening of potential risk factors is often followed by a variable selection procedure. *Regression models* are the methodology used most often for multivariable analysis of risk factors. Depending on the data format of the outcome, logistic regression, linear regression, ordinal logistic regression, generalized linear models, Poisson/negative binomial models, or proportional hazards models/survival analysis are applied (Dohoo et al. 2003). An important aspect to consider in these analyses is whether the assumption of independence of observations is fulfilled. Farm animals are clustered in farms. Because factors related to farm management, housing, and environment play an important role in the etiology of production diseases, animals within the same farm are more similar to each other in their health status than animals from a different farm. This clustering needs to be taken into account in regression analysis. Regression models correcting for clustering are called mixed, hierarchical, or multilevel models. Clustering can occur on more than one level. For example, if we measure mastitis once per month on the level of the individual quarter, the analysis needs to take the four levels farm-animal-quarter-monthly measure into account. Many different models exist which address this problem (Snijders and Boskers 2012).

Some more advanced methodologies for the analysis of risk factors are Bayesian approaches and machine learning. *Bayesian methods* allow for the inclusion of prior knowledge into an analysis (e.g., expert opinion, Ni et al. 2021). They can also deal with highly correlated variables and a network of direct and indirect associations and are therefore very well suited to analyze strongly interconnected component causes (Horpiencharoen et al. 2019; Hartnack et al. 2019). Results consist of a network of nodes (risk factor or disease variables) and links between the nodes, which describe dependencies between the variables of interest. Associations are corrected for the effect of other variables. Therefore, the probability of disease can be expressed as a conditional probability, taking the whole network of direct and indirect risk factors into account.

*Machine learning algorithms* can deal with large quantities of data and represent an automated way to identify the most relevant associations in a complex dataset with many variables (Ebrahimi et al. 2019; Browne et al. 2022). The strength of this approach is that the algorithms can deal with multiple interdependent variables and detect underlying correlation patterns in the data. No prior knowledge or hypothesis is needed. Thus, machine learning is very well suited for exploratory data analyses. Presently, many different algorithms, such as generalized linear model, logistic regression, naïve Bayes, deep learning, decision tree, or random forest, exist for machine learning.

## 5 Conclusions

Epidemiological concepts and methodologies are elemental for understanding, preventing, and controlling production diseases. Component causes, causal networks, and concepts that understand causality as a level of evidence rather than a direct relation are important for understanding the complex relations between factors from the host animal, pathogen, environment, farm management, and disease. Because the population perspective is of major importance for production diseases in farm animals, epidemiological methods describing the frequency and distribution of disease in populations and the factors associated with disease are important tools for their prevention and control.

## References

Barlow JW, Zadoks RN, Schukken YH (2013) Effect of lactation therapy on Staphylococcus aureus transmission dynamics in two commercial dairy herds. BMC Vet Res 9. https://doi.org/10.1186/1746-6148-9-28

Bassett J, Gethmann J, Blunk P et al (2021) Individual-based model for the control of Bovine Viral Diarrhea spread in livestock trade networks. J Theor Biol 527. https://doi.org/10.1016/j.jtbi.2021.110820

Browne N, Hudson CD, Crossley RE et al (2022) Cow- and herd-level risk factors for lameness in partly housed pasture-based dairy cows. J Dairy Sci 105:1418–1431. https://doi.org/10.3168/jds.2021-20767

Clark B, Stewart GB, Panzone LA et al (2016) A systematic review of public attitudes, perceptions and behaviours towards production diseases associated with farm animal welfare. J Agric Environ Ethics 29:455–478. https://doi.org/10.1007/s10806-016-9615-x

Dalen G, Rachah A, Nørstebø H et al (2018) Transmission dynamics of intramammary infections caused by Corynebacterium species. J Dairy Sci 101:472–479. https://doi.org/10.3168/jds.2017-13162

Deng Z, Koop G, Hogeveen H et al (2021) Transmission dynamics of Staphylococcus aureus and Streptococcus agalactiae in a Dutch dairy herd using an automatic milking system. Prev Vet Med 192. https://doi.org/10.1016/j.prevetmed.2021.105384

Dohoo L, Martin W, Stryhn H (2003) Veterinary epidemiologic research. National Library of Canada, Canada. https://projects.upei.ca/ver/

Ebrahimi M, Mohammadi-Dehcheshmeh M, Ebrahimie E, Petrovski KR (2019) Comprehensive analysis of machine learning models for prediction of sub-clinical mastitis: deep learning and gradient-boosted trees outperform other models. Comput Biol Med 114. https://doi.org/10.1016/j.compbiomed.2019.103456

FAO (2022) Animal health. In: Gateway to dairy production and products. https://www.fao.org/dairy-production-products/production/animal-health/en/. Accessed 2 June 2022

Frérot M, Lefebvre A, Aho S et al (2018) What is epidemiology? Changing definitions of epidemiology 1978-2017. PLoS One 13

Hartnack S, Odoch T, Kratzer G et al (2019) Additive Bayesian networks for antimicrobial resistance and potential risk factors in non-typhoidal Salmonella isolates from layer hens in Uganda. BMC Vet Res 15. https://doi.org/10.1186/s12917-019-1965-y

Horpiencharoen W, Thongratsakul S, Poolkhet C (2019) Risk factors of clinical mastitis and antimicrobial susceptibility test results of mastitis milk from dairy cattle in western Thailand: Bayesian network analysis. Prev Vet Med 164:49–55. https://doi.org/10.1016/j.prevetmed.2019.01.014

Kirkeby C, Zervens L, Toft N et al (2019) Transmission dynamics of Staphylococcus aureus within two Danish dairy cattle herds. J Dairy Sci 102:1428–1442. https://doi.org/10.3168/jds.2018-15106

Landers TF, Cohen B, Wittum TE, Larson EL (2012) A review of antibiotic use in food animals: perspective, policy, and potential. Public Health Rep 127:4–22. https://doi.org/10.1177/003335491212700103

Lebl K, Lentz HHK, Pinior B, Selhorst T (2016) Impact of network activity on the spread of infectious diseases through the German pig trade network. Front Vet Sci 3. https://doi.org/10.3389/fvets.2016.00048

Ni H, Klugkist I, van der Drift S et al (2021) Expert opinion as priors for random effects in Bayesian prediction models: subclinical ketosis in dairy cows as an example. PLoS One 16. https://doi.org/10.1371/journal.pone.0244752

Nir O (2003) What are production diseases, and how do we manage them? Acta Vet Scand Suppl 98:21–32

Shimonovich M, Pearce A, Thomson H et al (2021) Assessing causality in epidemiology: revisiting Bradford Hill to incorporate developments in causal thinking. Eur J Epidemiol 36:873–887. https://doi.org/10.1007/s10654-020-00703-7

Snijders T, Bosker R (2012) Multilevel analysis: an introduction to basic and advanced multilevel modeling, second edn. Sage Publishers, London etc.

Thrusfield M, Christley R, Brown H, Diggle P, French N, Howe K, Kelly L, O'Connor A, Sargeant J, Wood H (2018) Veterinary epidemiology. John Wiley & Sons Ltd. https://doi.org/10.1002/9781118280249

Verteramo Chiu LJ, Tauer LW, Al-Mamun MA et al (2018) An agent-based model evaluation of economic control strategies for paratuberculosis in a dairy herd. J Dairy Sci 101:6443–6454. https://doi.org/10.3168/jds.2017-13175

Zingg D, Steinbach S, Kuhlgatz C et al (2017) Epidemiological and economic evaluation of alternative on-farm management scenarios for ovine footrot in Switzerland. Front Vet Sci 4. https://doi.org/10.3389/fvets.2017.00070

# Societal Attitudes Towards Animal Welfare and Livestock Production Diseases

Beth Clark, Albert Boaitey, and Carmen Hubbard

**Abstract**

Farm animal health and welfare are important topics amongst global conversations and strategies surrounding food security and sustainability, including the social sustainability of food systems. Animal production systems, especially more intensive systems, are of concern to members of the public in relation to animal health and welfare and the environment. There is a large body of research demonstrating consumer and wider public concern over the welfare of farm animals, with a smaller body of research looking at production diseases in particular. This chapter looks to explore what the public know and think about production diseases. Given the limited research into this topic, the chapter first outlines the broader context of growing consumer concern over food production including animal products, in relation to food safety and farm animal health and welfare, and how concerns are addressed by government and industry. It is within this context that public attitudes towards production diseases are discussed.

**Keywords**

Public perceptions; Farm animal welfare; Animal health; Production disease

## 1 Introduction

Farm animal health and welfare (FAHW) are important topics amongst global conversations and strategies surrounding food security, sustainability and One Health (OIE 2021). Animal production systems, particularly intensive systems, are already a contentious subject (Harper and Henson 2001). There is a large and growing body of research indicating growing societal concern over farming practices (European Commission 2016; Harper and Henson 2001) in relation to several aspects of the animals' lives including specific farming practices and more generally the health and wellbeing of animals throughout their life (Clark et al. 2016). These concerns have been linked to both animal and human wellbeing (Spooner et al. 2014), with poor animal welfare being linked to animal health as well as consumer concerns over food product safety (Verbeke et al. 2010) and subsequently human health (Bennett et al. 2012). Conversely, higher welfare products are perceived as a means to guarantee safer and healthier products (Harper and Henson 2001; Lusk et al. 2007). It is recognised that poor animal welfare is a 'public bad' for humans and the environment (Cox and Bridgers 2019).

B. Clark (✉) · A. Boaitey · C. Hubbard
Centre for Rural Economy, School of Natural and Environmental Sciences, Newcastle University, Newcastle upon Tyne, UK
e-mail: Beth.Clark@newcastle.ac.uk





J. J. Gross (ed.), *Production Diseases in Farm Animals*,
https://doi.org/10.1007/978-3-031-51788-4_2

Whilst previous research has focused on perceptions of FAHW more generally, there has been less focus on the public perceptions of animal disease, in particular endemic or production diseases, such as mastitis, food pad dermatitis and lameness (Clark et al. 2016, 2017).

This chapter explores public perceptions of FAHW, focusing on production diseases. It first outlines the growth in societal concerns over food safety, farm animal welfare (FAW) and how these concerns are addressed. It then presents how these concerns and actions relate to production diseases, before concluding with a summary of recommendations and suggestions for future work.

## 2 Rise of the Concerned yet Disconnected Consumer

Lusk and Perugini (2016) describe food as one of the key concerns in society, with the public becoming increasingly concerned about several different aspects and risks within the agri-food domain, including the quality and safety of raw ingredients and the technologies and processes involved in its manufacture (Zingg et al. 2013; Frewer and Salter 2002; Frewer et al. 2016). Subsequently, the public is becoming increasingly aware and demanding of higher standards of both production and the final products (Grunert 2005). They expect food to be safe and free from contaminants, toxic substances, additives, pesticide and veterinary medicine residues including antibiotics (Scudamore 2007) and produced in a socially acceptable manner. Food safety is a long-standing and recurrent concern, particularly for European Union (EU) consumers (Eurobarometer 2019).

As a result, consumers are becoming increasingly interested in how their food is being produced, hence demanding higher product standards (Scudamore 2007) and more process-related product characteristics such as organic and welfare friendly (Grunert et al. 2004). These credence quality characteristics, i.e. those that are difficult to observe or evaluate by consumers, serve as guarantees or indicators of product safety and quality in the mind of consumers (Zingg et al. 2013). This includes FAW (Grunert et al. 2004), with consumers wanting reassurances that the food they eat is produced in an animal welfare-friendly manner (World Bank 2011). In addition, there is an increased interest in how food is produced (Grunert 2005), resulting in an increased consumer demand for greater transparency in production standards and processes. For example, the introduction of traceability systems into the beef supply chain has been motivated by human and animal health concerns (Souza Monteiro and Caswell 2004).

The public therefore represent an important user of the food chain—both as citizens and as consumers. As citizens they can petition and vote against production practices that they find unacceptable (Norwood and Lusk 2011). As consumers, they can drive demand for specific food products (Jensen 2006), expressing their preferences through their purchasing behaviour, such as by choosing labelled or assured produce, or choosing to shop specific brands or at certain retailers who advertise higher standards or quality assurances, or avoiding certain products or means of production entirely. A better understanding of public perceptions and their concerns is therefore important for providing transparency and reassurance that the food produced is safe and healthy.

At the same time there is also a growing distance and visibility between how food is produced and the consumer. This is particularly the case for animal production. It has been widely reported that the public lacks information about livestock production, particularly more intensive animal production systems (Duffy et al. [2005] in Croney and Anthony [2014]). Despite concerns over FAW and how food is produced, the public may have very little or no understanding of the technicalities of modern farming practices (Harper and Henson 2001). Thus, there is a notable disconnect between how farm animals are reared and what the public knows about the various systems of production (Spooner et al. 2014; Harper and Henson 2001; Clark et al. 2016; Hartmann and Siegrist 2019). Weible et al. (2016) highlight that research regarding consumers' or citizens' attitudes and expectations towards animal husbandry has mainly focused on animal welfare or food quality, with hardly any studies on people's opinions regarding specific animal systems of production. Therefore, it can be argued that they are unfamiliar with various issues or norms associated with FAW (Lassen et al. 2006; Te Velde et al. 2002), including disease. This unfamiliarity and lack of knowledge may create misunderstandings, and it is important to address these. It also results in the animal products of the food chain being far removed from the original animals from which these products were developed. Thus, consumers are geographically and conceptually removed from food production (Freidberg 2004). Subsequently, it is unsurprising that studies have frequently shown a lack of knowledge about modern animal production systems (Clark et al. 2016, 2019). As Sørensen et al. (2012, p. 886) point out, 'Western societies are urbanised to such a degree that only few citizens ever encounter the conditions in livestock farming or in primary processing, let alone encounter them in enough detail to be able to really distinguish the characteristics of different production systems from each other'.

## 3 Public Perceptions of Farm Animal Health and Welfare

### 3.1 Farm Animal Health and Welfare Concerns

The daily life, and so production, of farm animals was largely ignored by the public until the publication of the 1964 book *Animal Machines* by Ruth Harrison (Sayer 2013; Norwood and Lusk 2011), which brought to the public the attention of modern farming, especially the welfare of animals. Consideration of how animals are treated within different systems has remained a public concern since. Increasing discourses over animal rights and sentience have also received increasing public attention, and both the European Union (EU) and, more recently, the UK[1] have recognised animals as being sentient.

Over the last decades, animal production in particular has led to concerns with several reports (Willett et al. 2019) getting widespread media coverage (Miele and Evans 2010; Tonsor and Olynk 2011; Miele et al. 2011; Kendall et al. 2019). High-profile food scandals lead to a rise in public concern including *bovine spongiform encephalopathy* (BSE) (Frewer and Salter 2002), melamine in infant formula (Zhou and Wang 2011), the 'horsegate' scandal (Tse et al. 2016) and foot and mouth disease despite the latter not linked to food safety (Poortinga et al. 2004). These food scares and the high levels of negative media coverage surrounding these (Tonsor and Olynk 2011; Vanhonacker and Verbeke 2014; Miele et al. 2011) have further influenced public concern, with food safety being of prominent interest of EU consumers (Eurobarometer 2019).

As well as food safety, concerns also relate to FAW (Harper and Henson 2001) and the environmental impact of livestock (YouGov 2019; Grunert 2005). There is the perception that what is bad for the animal is potentially bad for the consumer (Hall and Sandilands 2007; Bennett et al. 2012). Concerns are identified from a dual perspective, i.e. both human and more altruistic (animal) perspectives (Clark et al. 2016).

The interest and concerns surrounding FAHW have resulted in a large amount of research into overall attitudes, specific elements of concern and changes in production practices and standards, either through legislative or market-based measures. Research has shown that FAHW has become a global and consistent concern with citizens and/or consumers across the world expressing their views regarding the way animals are reared (Clark et al. 2016; European Commission 2007, 2016; Eurobarometer 2005). Concerns relate to several aspects of the animals' lives including specific farming practices such as castration (Heid and Hamm 2013), transport (Miele 2011), slaughter (Vanhonacker and Verbeke 2009), animal housing systems (Tonsor et al. 2009; De Jonge and Van Trijp 2013; Jackson et al. 2022), such as the confinement of lambs in indoor pens (Coleman et al. 2016), or indoor production systems (Ventura et al. 2016). The literature highlights that modern, particularly more

[1]The Animal Welfare (Sentience) Act 2022 received the royal assent in April 2022.

intensive, indoor livestock production systems are not always viewed favourably by the public (Harper and Henson 2001). Efficiency gains through the intensification of production may result in animal rights and FAW concerns (Austin et al. 2005; World Bank 2011), in particular in relation to naturalness and humane treatment concerns (Clark et al. 2016). Modern and intensive production systems are thought to breach these concepts especially that of naturalness (Dawkins 2023).

Research would confirm that both aspects appear to be important, with naturalness considered a priority (Janssen et al. 2016). However, more recent research is challenging these aspects. For example, Cardoso et al. (2018) have traded off different attributes of farming systems or put naturalness and animal wellbeing in conflict (Robbins et al. 2018). Cardoso et al. (2018) found that US citizens would score an indoor system with fans to prevent heat stress more favourably than a pasture-based system with no shade, with the latter felt to be an indication of farmers' failure as a duty of care towards the animal, i.e. a prioritisation of welfare rather than naturalness of the animals' living conditions. Jackson et al. (2022) also found a dairy farm scenario with housing in winter and grazing in summer favourable to the public, providing both protection from inclement weather and naturalness. Naturalness would therefore appear to not compensate for poor welfare (Cardoso et al. 2018), with respondents appreciating the freedom offered to animals by outdoor systems yet placing a higher value on their welfare (Yunes et al. 2017).

More recently, a survey of the German public, compared the effect of housing conditions (straw bedding versus a slatted floor) on pig wellbeing (images of a pig looking happy or sad) (Busch et al. 2019). Findings from the study highlight that a more positively viewed animal (i.e., the happier looking pig) has a small influence on overcoming the negative expectations of more contentiously viewed surroundings, such as slatted floors (Busch et al. 2019). This reinforces the importance of both good animal welfare and the production systems animals are kept within.

An understanding of the public's concerns is therefore important. As Yunes et al. (2017) describe, societal input is important for the long-term sustainability of animal production systems. Broom (2010, p. 83) states that 'no system or procedure is sustainable if a substantial proportion of people find aspects of it now, or of its consequences in the future, morally unacceptable'. As the assessment or perception of farm animal production systems depends on the alternative or range of different production systems available at the time, the assessment of modern production systems and their acceptability may change as time progresses (Bennett et al. 2012). Thus, what was acceptable previously or is acceptable at the present time may become unacceptable in the future (The Brambell Report 1965; Broom 2014). Stakeholders' perceptions of animal production systems, including those of the public, therefore need to be taken into consideration, as they could become unsustainable if it is perceived that they pose an unacceptable social cost to FAW (Farm Animal Welfare Committee 2016; Ohl and Van Der Staay 2012).

Identifying and understanding public attitudes and concerns, including their perceptions of risks and benefits in relation to intensive farming practices and those surrounding production diseases, is important and will ensure that the processes and interventions align with the values, needs and expectations of society (Asveld et al. 2015). It also enables suitable actions to be taken where appropriate.

## 3.2 Responding to Public Concerns

Improvements to FAHW can be promoted through legislation and market-based approaches (Sørensen and Schrader 2019). Government legislation has traditionally been the main tool for ensuring and/or improving FAW standards across animal production systems, especially within the EU and UK (Bennett 1997). For example, public concerns over FAHW, production systems and practices have led to legislative initiatives and widespread regulation changes (Spooner et al.

2014). For example, public lobbying against battery cages for hens has led to 'Council directive 1999/74/EC' (1999) in the EU and Proposition 2 in the USA (Malone and Lusk 2016). However, a slowdown in the process of adoption of stricter EU FAHW regulations since the early 2000s has led to market-based solutions to facilitate improvements in FAW standards in response to public concerns. This has also been the case in the USA where FAW developments are often in response to private initiatives reacting to public awareness or changing consumer demand (Dana and Nadler 2019). It is argued that the private sector should be delivering at least acceptable or higher than minimum FAW standards (Sandøe and Christensen 2018; Grethe 2017) in order to maintain a social licence to operate[2] (Leith et al. 2014) and hence should be responding to societal concerns. This is particularly important for specific categories of consumers, such as vegans and vegetarians, who may still have an interest in how animals are reared (Bennett et al. 2012), yet do not directly express their preferences for particular production systems/practices through the market per se.

Several studies have explored if and how identified concerns might translate into a perceived consumer's responsibility through their willingness to pay (WTP) for improved FAW (Lagerkvist and Hess 2011; Clark et al. 2017; Rodrigues and Hanley 2021). Numerous studies have highlighted a strong WTP and often recommend some form of trusted labelling (Janssen et al. 2016) so as to aid consumers in making their preferred choices (Clark et al. 2016), hence providing more accurate information at the point of purchase. Despite the levels of concern reported by the public (European Commission 2016; Lusk et al. 2007), research indicates that attitudes towards FAW do not always translate into purchase decisions (Harper and Henson 2001; Toma et al. 2011), as indicated in the discrepancy between reported WTP and existing sales and market share information of higher welfare products (Baltzer 2004; Grunert 2006). Rates of vegetarian and veganism are also low, albeit growing (Bates et al. 2017) but with flexitarianism[3] more prominent (YouGov 2019).

Several reasons for this discrepancy have been suggested. In addition to the lack of information at the point of purchase and poor knowledge regarding production systems, a number of other conscious barriers to purchasing higher welfare products have been highlighted (Harper and Henson 2001), such as the lack of perceived influence their purchase decisions may have and accountability (who should be responsible for animal welfare). Consumers may not view themselves as responsible for FAHW, leaving to their own governments to deal with it (Grunert 2006). Another potential explanation of this might be that whilst individuals may be motivated to purchase higher welfare products, they may not necessarily have the means to do so with these products perceived as relatively expensive (Clark et al. 2017; Cicia and Colantuoni 2010). Whilst research so far offers a wealth of information into societal concerns surrounding FAHW more generally, how this extends to production diseases is less clear.

## 4 The Public and Production Diseases

Whilst there is a large body of research exploring public attitudes towards FAHW, attitudes towards animal disease, including production diseases, are less explored (Clark et al. 2016, 2017). The existing literature often follows disease outbreaks or interventions aimed at addressing these, hence focusing on zoonotic or epidemic disease outbreaks. This is usually the case when a specific disease hits the media headlines (Scudamore 2007).

From a public perspective, how disease relates to animal health and good welfare has been less

[2]Hampton et al. (2020) describe social license to operate (SLO) 'the community tacitly giving an industry the right to conduct its current business, particularly to exploit publicly owned resources. Hence, SLO informally reflects prevailing public values and should not be confused with regulatory licenses'.

[3]Flexitarianism is a mostly plant-based diet where people do not eat meat several days of the week.

well explored. However, as part of the Welfare Quality project,[4] Tuyttens et al. (2010) investigated the importance of 12 criteria central to FAW. These criteria developed as part of the Welfare Quality project provide an expansion of the 'five freedoms' (The Brambell Report 1965), considered as the backbone of defining FAW. Tuyttens et al. (2010) found that the 'absence of disease' was the highest rated of the criteria for both farmers and citizens in terms of importance of FAWe. Interestingly, vegetarians attributed 'absence of pain', 'absences of prolonged thirst' and 'absence of fear' as more important on average, although the 'absence of disease' still received a high score. This emphasises that disease is an important consideration of FAW and that health and welfare may not be viewed separately by stakeholders including the public. However, there is growing evidence to suggest that the public can distinguish between animal health and animal welfare (Rodrigues and Hanley 2021), emphasising the importance of addressing animal health concerns per se.

Two systematic reviews have identified gaps in relation to production diseases and public perceptions (Clark et al. 2016, 2017). The studies included in these reviews have typically explored interventions to prevent, treat or reduce the incidence of particular diseases or conditions, rather than attitudes towards the category of or individual diseases. For example, some studies have explored public preferences for interventions to reduce production diseases in farmed fish (Grimsrud et al. 2013) and to minimise footpad lesions (and associated dermatitis) in broiler chickens (Mcvittie et al. 2006) both in relation to FAHW improvements and for earlier disease detection from a food safety perspective (Doherty and Campbell 2014). Koistinen et al. (2013) examined disease and health in pigs and beef cows. Several studies have also looked to establish consumers' preferences for antibiotic-free meat and dairy products (Clark et al. 2017), with antibiotic use considered as a proxy for production diseases (Hughes and Heritage 2002). These studies found a WTP for antibiotic-free meat (Clark et al. 2017).

A large European online survey (2330 people) from the UK, Spain, Germany, Finland and Poland specifically explored public attitudes towards production diseases. The findings indicated that the public is particularly concerned about endemic livestock diseases in general (Clark et al. 2019). The research also explored the risks and benefits associated with and acceptability of production diseases and associated interventions in pig, laying hens and poultry production systems. Table 1 summarises respondents' preferences for the different interventions listed within the survey.

The most preferred interventions as identified were proactive disease prevention measures, as opposed to reactive disease mitigation strategies. This included interventions involving changes to housing design, increasing the space available for the birds and pigs and enhanced hygiene and disease prevention measures. The proactive interventions preferred also emphasise the key aspects of FAW mentioned above, humane treatment and naturalness (Clark et al. 2016), especially in relation to (perceived) naturalness.

Conversely, the least preferred interventions relating to factors known to drive public concerns and risks were also identified (Clark 2018), namely, medicine-based interventions that are associated with both humane animal care and public health concerns, associated with food safety or antimicrobial resistance (AMR). This relates to the dual perspectives of FAW, in that the benefits of livestock production systems can be perceived from both anthropocentric (human) and zoocentric (animal) perspectives. These were chosen in relation to both food safety and humane treatment reasons, again emphasising that the public cannot separate FAHW from human food safety (Clark 2018).

The findings surrounding the least acceptable interventions also reflect previous research results regarding animal vaccination and disease outbreaks. Breakwell (2003) conducted focus groups after the UK foot and mouth disease crisis of 2001, and their results suggest that there was no indication that consumers would not consume

[4]http://www.welfarequality.net/en-us/home/.

**Table 1** Ranking the acceptability of interventions to control production diseases in broiler chickens, laying hens and pigs, EU Survey, 2017

| Intervention | Broilers | Laying hens | Pigs |
|---|---|---|---|
| Enhanced hygiene and disease prevention measures | 1 | 1 | 4 |
| Providing materials and an environment where animals can perform natural behaviours | 3 | 2 | 1 |
| Housing that allows birds greater freedom to move | 2 | 3 | – |
| Efficient monitoring of pigs and pig housing conditions | – | – | 3 |
| Reducing the number of animals in a given area | 4 | 4 | 6 |
| Improvements in housing design | 5 | 5 | 6 |
| Enhanced maintenance of the quality of the bedding | 6 | 6 | – |
| Housing that protects the animals from adverse natural conditions | 7 | 7 | 9 |
| Enhanced control of air movement in the houses | 8 | 8 | 4 |
| Providing farmers with a price premium that encourages enhanced animal health | 9 | 8 | 8 |
| Adjustments to feed or diet composition | 10 | 10 | 2 |
| Changes in the amount and time of light provision | 11 | 11 | – |
| The use of vaccination | 12 | 13 | 11 |
| Adjustments to the quantity of feed available | 13 | 12 | 10 |
| Using antibiotics and medicines to treat sick animals | 14 | 14 | 12 |
| Breeding for genetically tougher or more resilient pigs | – | – | 13 |
| Use of feed supplements, e.g. probiotics | 15 | 15 | 14 |
| The preventative use of veterinary drugs including antibiotics | 16 | 16 | 15 |
| Doing nothing | 17 | 17 | 16 |

meat from vaccinated animals, when presented as an alternative to the culling strategy employed.

In another study, Coleman et al. (2016) found that the Australian public was less concerned over vaccination than in the Breakwell (2003) study, although their question was phrased around the indication of the importance of vaccination for animal health, rather than the degree of concern resulting from vaccination. However, research by Scudamore (2007) indicates that consumers had some reservations towards eating meat from animals vaccinated against foot and mouth disease (Scudamore 2007). Zingg and Siegrist (2012) found that although consumers seem to accept vaccination programmes for animals for both epidemic and zoonotic diseases, they were not willing to consume meat from animals that had been vaccinated.

The use of antibiotics and AMR were repeatedly identified as concerns both in relation to human and animal health and more generally in relation to intensive animal production systems (Clark 2018). Concerns over antibiotic usage in animal production are not new, with previous research highlighting that the public is concerned about routine antibiotic and growth hormone use in livestock (Miele et al. 2011; Bernard and Bernard 2009; Vanhonacker et al. 2010), with public concerns primarily relating to prophylactic use of antibiotics (Clark et al. 2016, 2019).

Antibiotics, and the use of other veterinary medicines, have obvious benefits to animal health (Perry et al. 2013), and the public acknowledges these benefits and accepts their use in a controlled and responsible manner, with antibiotic use for the treatment of sick animals more accepted (Clark et al. 2016). However, research would suggest that antibiotic usage and food safety are inextricably linked in the minds of consumers (Clark et al. 2019).

Healthy animals require less or no medication (Marshall and Levy 2011), and good animal welfare has a positive effect on health (Boyle and O'Driscoll 2011). At a time when antibiotic use for economic gain is coming under increased scrutiny due to antimicrobial resistance (Mateus et al. 2016) and is contested (Hughes and Heritage 2002; Chang et al. 2015), there is a need to consider the evaluation of alternatives including their public acceptability. One such example is probiotics, although these were one of the least preferred interventions (Table 1). Van Wezemael et al. (2011) have also previously identified probiotics as a concern, with over a third (37.6%) of their sample (of 2520) accepting the addition of protective bacteria to beef cattle feed and almost a quarter finding this unacceptable (24.1%).

As well as the interventions used to treat endemic or production diseases, the public also have concerns over unhealthy animals entering the food chain, usually following a disease epidemic. This has been emphasised particularly during the high-profile food scares such as BSE (Grunert et al. 2004), to a lesser extent foot and mouth disease (Breakwell 2003; Scudamore 2007) and avian influenza (Rowe et al. 2008). The literature focuses more on disease epidemics rather than on production diseases that are endemic to intensive production systems.

The food safety considerations demonstrate that the widespread concern over FAW may not be due to ethical considerations (or for more altruistic reasons) and may not be as high as initially identified (Bray and Ankeny 2017). Research has indicated that this may well be the case for animal health and livestock disease management (Clark et al. 2019). Whilst a large body of literature has found or discussed FAHW concerns in relation to altruistic concerns over animals' wellbeing (Clark et al. 2016), in some studies individuals' concerns over FAW were based on concerns for participants' own wellbeing, rather than the welfare of the animals (Clark et al. 2016, 2019). It could be argued said that this *concern* forms a 'warm glow' effect or a so-called form of 'impure altruism'(Johansson-Stenman 2006), whereby human concerns make humans feel good but do not in themselves improve the welfare conditions of the animals under consideration (Lusk and Norwood 2011). More recent UK-focused research as part of the FIELD project[5] identified that the public hold both types of concerns—about the food safety of infected meat and the wellbeing and conditions in which animals are kept that have resulted in them being 'infected' (Clark et al. 2023). This reiterates the need for food chain stakeholders to address societal concerns from both human and animal perspectives.

## 5 Conclusions

Identifying and understanding public attitudes and concerns towards FAHW including those surrounding livestock production diseases is very important and will ensure that farming practices and any disease interventions align with the values, needs and expectations of the society (Asveld et al. 2015). Conducting research to understand attitudes and perceptions, and how these change in relation to external events (e.g. outbreak diseases in livestock) and cultural shifts, is therefore an essential part of addressing societal expectations, ensuring that acceptable regulatory and production practices are in place and that industry maintains a social license to operate.

Whilst a rather limited body of literature regarding public attitudes towards production diseases exists as compared to other aspects of FAHW, the growing number of studies indicates an increase in public concerns, both in relation to the diseases themselves and preferences for the types of interventions to prevent and treat these

[5] https://field-wt.co.uk/.

diseases. Drawing on these studies, and a wider FAHW literature, it is clear that overall efforts to address FAHW are important for tackling concerns from both human and animal perspectives. Within this, naturalness, humane treatment and a preference for more proactive interventions are all important, as they consider both the health and wellbeing of animals and human health.

Identifying proactive management measures taken to mitigate production disease occurrence and severity is therefore an important part of ensuring that these measures, and any associated risk management procedures, align with societal preferences, hence highlighting the benefits of ongoing societal discussions in building consumer trust through transparency (Van Kleef et al. 2007). Best practice examples that could already be conveyed may include the work done by the UK poultry sector to reduce the amount of antibiotics used within its production systems (Griffiths 2016) and the continued and sustained efforts to reduce the risk of AMR. The role of health and disease management and control within wider quality and/or farm assurance schemes could also be better communicated. This also advocates a continual strive for improvement (rather than complacency) promoting dissemination of information about proactive approaches taken, for example, by regulatory agencies (Van Kleef et al. 2007), which acts to increase public trust.

The majority of the public cannot dissociate the risks to food safety and human health from animal health and wellbeing across various production systems, with production diseases, particularly perceived to have negative consequences for food safety (Clark et al. 2019). These are likely to have been reinforced by epidemic disease outbreaks such as the most recent avian influenza, which have received a large mass media coverage, although the risk of transmission from birds and poultry products to humans is low (Defra and APHA 2023). It is also acknowledged that the public cannot differentiate between production systems and epidemic animal diseases. The Covid-19 outbreak has also increased concerns over food safety, particularly in relation to animal-based products (Attwood and Hajat 2020), which may further serve to heighten societal concerns. Care should therefore be taken when communicating about livestock production diseases to emphasise and differentiate them from the more familiar diseases. It would also be interesting to explore further if and whether the public can distinguish between production systems and epidemic disease outbreaks.

This chapter also emphasises the need for effective communication and assurance regarding the regulations and enforcement used across the food supply chain, which are used to ensure the safety of the animal products produced. It is important to acknowledge this, as failure to do so could result in a decreased stakeholder trust and an increase in concerns and perceived risks. Research has also shown that members of the public can find it difficult to access trustworthy information and guidance to help with their food choices (Connors et al. 2022).

From an animal perspective, it is important that health and welfare is also addressed, especially with food chain stakeholders taking a more proactive approach, including communicating care.

Central to this is a need for more open discussions surrounding food production and FAHW, which is an emotive and potentially contentious topic of conversation with the public. Previous criticisms highlighted a lack of communication and transparency surrounding farming issues (Murphy-Lawless 2004), as there is insufficient information available, or (more likely) messages seem not to be reaching their intended targets. Indeed, the public are not the intended audience for most industry messages about food production (Duffy et al. (2005) in Croney and Anthony (2014)). Rather these messages are aimed at other stakeholders, such as the government and other decision-makers.

For the ordinary public to take part in shaping more socially acceptable food systems, they need to have an awareness and understanding of the issues in question as a lack of knowledge can hamper participation in decision-making (Baldy and Kruse 2019). Research has shown that members of the public are interested in learning more about farming and can handle potentially ethically difficult conversations surrounding FAHW including disease (Clark and Mahon 2023). These findings show that a more open and inclusive engagement on FAHW topics, including production diseases, would be welcomed, and this is something for researchers, industry and policy makers to consider.

## References

Asveld L, Ganzevles J, Osseweijer P (2015) Trustworthiness and responsible research and innovation: the case of the bio-economy. J Agric Environ Ethics 28:571–588

Attwood S, Hajat C (2020) How will the COVID-19 pandemic shape the future of meat consumption? Public Health Nutr 23:3116–3120

Austin EJ, Deary IJ, Edwards-Jones G, Arey D (2005) Attitudes to farm animal welfare: factor structure and personality correlates in farmers and agriculture students. J Individ Differ 26:107–120

Baldy J, Kruse S (2019) Food democracy from the top down? State-driven participation processes for local food system transformations towards sustainability. Polit Gov 7:68–80

Baltzer K (2004) Consumers' willingness to pay for food quality–the case of eggs. Food Econ Acta Agric Scand Sect C 1:78–90

Bates B, Roberts C, Lepps H, Porter L (2017) Food & You Survey Wave 4. Executive Summary Combined Report. Food Standards Agency. https://www.food.gov.uk/sites/default/files/media/document/food-and-you-w4-exec-summary.pdf. Accessed 12 Mar 2023

Bennett RM (1997) Farm animal welfare and food policy. Food Policy 22:281–288

Bennett R, Kehlbacher A, Balcombe K (2012) A method for the economic valuation of animal welfare benefits using a single welfare score. Anim Welf 21:125–130

Bernard JC, Bernard DJ (2009) What is it about organic milk? An experimental analysis. Am J Agric Econ 91:826–836

Boyle L, O'Driscoll K (2011) Animal welfare: an essential component in food safety and quality. In: Hoorfar J, Jordan K, Butler F, Prugger R (eds) Food chain integrity. Woodhead Publishing, pp 169–186

Bray HJ, Ankeny RA (2017) Happy chickens lay tastier eggs: motivations for buying free-range eggs in Australia. Anthrozoös 30:213–226

Breakwell GM (2003) Public perceptions concerning animal vaccination: a case study of foot and mouth 2001

Broom DM (2010) Animal welfare: an aspect of care, sustainability, and food quality required by the public. J Vet Med Educ 37:83–88

Broom DM (2014) Sentience and animal welfare. CABI, Wallingford

Busch G, Gauly S, Von Meyer-Höfer M, Spiller A (2019) Does picture background matter? People's evaluation of pigs in different farm settings. PLoS One 14:e0211256

Cardoso CS, Von Keyserlingk MAG, Hötzel MJ, Robbins J, Weary DM (2018) Hot and bothered: public attitudes towards heat stress and outdoor access for dairy cows. PLoS One 13:e0205352

Chang Q, Wang W, Regev-Yochay G, Lipsitch M, Hanage WP (2015) Antibiotics in agriculture and the risk to human health: how worried should we be? Evol Appl 8:240–247

Cicia G, Colantuoni F (2010) Willingness to pay for traceable meat attributes: a meta-analysis. Int J Food Syst Dyn 1:252–263

Clark B (2018) Production diseases and farm animal welfare: what do the public think? PhD thesis. Newcastle University, Newcastle upon Tyne

Clark B, Mahon N (2023) In the shoes of a farmer: (re) connecting the public with animal health and welfare in livestock production. Research for All

Clark B, Stewart GB, Panzone LA, Kyriazakis I, Frewer LJ (2016) A systematic review of public attitudes, perceptions and behaviours towards production diseases associated with farm animal welfare. J Agric Environ Ethics 29:455–478

Clark B, Stewart GB, Panzone LA, Kyriazakis I, Frewer LJ (2017) Citizens, consumers and farm animal welfare: a meta-analysis of willingness-to-pay studies. Food Policy 68:112–127

Clark B, Panzone L, Stewart G, Kyriazakis I, Niemi JK, Latvala T, Tranter R, Jones P, Frewer L (2019) Consumer attitudes towards production diseases in intensive production systems 14:e0210432

Clark B, Proctor A, Mahon N, Holloway L (2023) 'Ignorance is bliss': do the public want to know about farm animal health and welfare? J Agric Environ Ethics

Coleman G, Jongman E, Greenfield L, Hemsworth P (2016) Farmer and public attitudes toward lamb finishing systems. J Appl Anim Welf Sci 19:198–209

Connors C, Malan L, Esposito M, Madden C, Trikka N, Cohen M, Rothery F, Reynolds C, Sheppard C, Canavan S (2022) The UK Public's interests, needs and concerns around food: UK main report. https://www.food.gov.uk/sites/default/files/media/document/The%20uk%20Public%27s%20Interests%20Needs%20and%20Concerns%20around%20Food%20-%20Main%20UK%20report.pdf. Accessed 25 Feb 2023

Cox J, Bridgers J (2019) Why is animal welfare important for sustainable consumption and production? Perspectives 34. https://apo.org.au/sites/default/files/resource-files/2019-03/apo-nid225921.pdf. Accessed 30 Oct 2022

Croney C, Anthony R (2014) Food animal production, ethics, and quality assurance. Encyclopedia of food and agricultural ethics:845–853

Dana DA, Nadler J (2019) Regulation, public attitudes, and private governance. J Empir Leg Stud 16:69–93

Dawkins MS (2023) Farm animal welfare: beyond natural behavior. Science 379:326–328

De Jonge J, Van Trijp HCM (2013) The impact of broiler production system practices on consumer perceptions of animal welfare. Poult Sci 92:3080–3095

Defra & APHA (2023) Bird flu (avian influenza): latest situation in England [Online]. https://www.gov.uk/government/news/bird-flu-avian-influenza-latest-situation-in-england#:~:text=The%20Food%20Standards%20Agency%20(%20FSA,safety%20risk%20for%20UK%20consumers. Accessed 25 Feb 2023

Doherty E, Campbell D (2014) Demand for safety and regional certification of food results from Great Britain and the Republic of Ireland. Br Food J 116:676–689

Duffy R, Fearne A, Healing V (2005) Reconnection in the UK food chain: bridging the communication gap between food producers and consumers. Br Food J 107:17–33

Eurobarometer (2005) Attitudes of consumers towards the welfare of farmed animals. Special Eurobarometer 229/wave 63.2-tns Opinion & Social

Eurobarometer (2019) Food safety in the EU. http://www.efsa.europa.eu/sites/default/files/corporate_publications/files/Eurobarometer2019_Food-safety-in-the-EU_Full-report.pdf. Accessed 13 Oct 2022

European Commission (2007) Attitudes of EU citizens towards animal welfare. Special Eurobarometer 270 ed. http://ec.europa.eu/public_opinion/archives/ebs/ebs_270_en.pdf. Accessed 13 Oct 2022

European Commission (2016) Attitudes of Europeans towards animal welfare [Online]. European Union: Director Generate for Health and Social. http://ec.europa.eu/COMMFrontOffice/publicopinion/index.cfm/Survey/getSurveyDetail/instruments/SPECIAL/surveyKy/2096

Farm Animal Welfare Committee (2016) Sustainable agriculture and farm animal welfare [Online]. https://www.gov.uk/government/uploads/system/uploads/attachment_data/file/593479/Advice_about_sustainable_agriculture_and_farm_animal_welfare_-_final_2016.pdf. Accessed 21 Mar 2021

Freidberg S (2004) The ethical complex of corporate food power. Environ Plann D Soc Space 22:513–531

Frewer L, Salter B (2002) Public attitudes, scientific advice and the politics of regulatory policy: the case of BSE. Sci Public Policy 29:137–145

Frewer LJ, Fischer ARH, Brennan M, Bánáti D, Lion R, Meertens RM, Rowe G, Siegrist M, Verbeke W, Vereijken C (2016) Risk/benefit communication about food—a systematic review of the literature. Crit Rev Food Sci Nutr 56:1728–1745

Grethe H (2017) The economics of farm animal welfare. Ann Rev Resour Econ 9:75–94

Griffiths R (2016) British poultry sector reduces antibiotic use by 44%. British Poultry Council. https://britishpoultry.org.uk/british-poultry-sector-reduces-antibiotic-use-by-44/. Accessed 14 Oct 2022

Grimsrud KM, Nielsen HM, Navrud S, Olesen I (2013) Households' willingness-to-pay for improved fish welfare in breeding programs for farmed Atlantic salmon. Aquaculture 372:19–27

Grunert KG (2005) Food quality and safety: consumer perception and demand. Eur Rev Agric Econ 32:369–391

Grunert KG (2006) Future trends and consumer lifestyles with regard to meat consumption. Meat Sci 74:149–160

Grunert KG, Bredahl L, Brunsø K (2004) Consumer perception of meat quality and implications for product development in the meat sector—a review. Meat Sci 66:259–272

Hall C, Sandilands V (2007) Public attitudes to the welfare of broiler chickens. Anim Welf 16:499–512

Hampton JO, Jones B, Mcgreevy PD (2020) Social license and animal welfare: developments from the past decade in Australia. Animals (Basel) 10

Harper G, Henson S (2001) Consumer concerns about animal welfare and the impact on food choice. EU FAIR CT98-3678. Centre for Food Economics Research, The University of Reading, Reading

Hartmann C, Siegrist M (2019) Our daily meat: justification, moral evaluation and willingness to substitute. Food Qual Prefer 103799

Heid A, Hamm U (2013) Animal welfare versus food quality: factors influencing organic consumers' preferences for alternatives to piglet castration without anaesthesia. Meat Sci 95:203–211

Hughes P, Heritage J (2002) Antibiotics as animal growth promoters. https://www.fao.org/3/y5159e/y5159e08.htm. Accessed 12 Oct 2022

Jackson A, Green MJ, Kaler J (2022) Fellow cows and conflicting farmers: public perceptions of dairy farming uncovered through frame analysis. Front Vet Sci 9

Janssen M, Rödiger M, Hamm U (2016) Labels for animal husbandry systems meet consumer preferences: results from a meta-analysis of consumer studies. J Agric Environ Ethics 29:1071–1100

Jensen H (2006) Consumer issues and demand [Online]. https://www.choicesmagazine.org/2006-3/animal/2006-3-09.htm. Accessed 5 May 2022

Johansson-Stenman O (2006) Should animal welfare count? Working papers in economics, no. 197. ISSN 1403-2465

Kendall H, Clark B, Rhymer C, Kuznesof S, Hajslova J, Tomaniova M, Brereton P, Frewer L (2019) A systematic review of consumer perceptions of food fraud and authenticity: a European perspective. Trends Food Sci Technol 94:79–90

Koistinen L, Pouta E, Heikkila J, Forsman-Hugg S, Kotro J, Makela J, Niva M (2013) The impact of fat content, production methods and carbon footprint information on consumer preferences for minced meat. Food Qual Prefer 29:126–136

Lagerkvist CJ, Hess S (2011) A meta-analysis of consumer willingness to pay for farm animal welfare. Eur Rev Agric Econ 38:55–78

Lassen J, Sandøe P, Forkman B (2006) Happy pigs are dirty!–conflicting perspectives on animal welfare. Livest Sci 103:221–230

Leith P, Ogier E, Haward M (2014) Science and social license: defining environmental sustainability of Atlantic salmon aquaculture in South-Eastern Tasmania, Australia. Soc Epistemol 28:277–296

Lusk JL, Norwood FB (2011) Animal welfare economics. Appl Econ Perspect Policy 33:463–483

Lusk JL, Perugini M (2016) Editorial for special issue on "Food consumption behaviour: economic and psychological perspectives". J Econ Psychol 55:1–3

Lusk JL, Norwood FB, Prickett RW (2007) Consumer preferences for farm animal welfare: results of a nationwide telephone survey. Oklahoma State University, Department of Agricultural Economics

Malone T, Lusk JL (2016) Putting the chicken before the egg Price: an ex post analysis of California's battery cage ban. J Agric Resour Econ 41

Marshall BM, Levy SB (2011) Food animals and antimicrobials: impacts on human health. Clin Microbiol Rev 24:718–733

Mateus A, Takahashi EA, Abdelrazeg Elkholly D, Crotta M, Ekiri A, Wylie C, Guinat C, Patricio M, Marshall L, Ferreira JP, Stark K, Guitian J (2016) A systematic review to assess the significance of the food chain in the context of antimicrobial resistance (AMR) with particular reference to pork and poultry meat, dairy products, seafood and fresh produce on retail sale in the UK. https://webarchive.nationalarchives.gov.uk/ukgwa/20180411174552/https://www.food.gov.uk/sites/default/files/amr-systematic-review-final-report-2016.pdf. Accessed 22 Oct 2022

Mcvittie A, Moran D, Nevison I (2006) Public preferences for broiler chicken welfare: evidence from stated preference studies. Land Economy Paper Series, Land Economy Research Group, Scottish Agricultural College. http://www.sac.ac.uk/research/publications/lee/workingpapers. Accessed 23 June 2016

Miele M (2011) The taste of happiness: free-range chicken. Environ Plan A 43:2076–2090

Miele M, Evans A (2010) When foods become animals: ruminations on ethics and responsibility in care-full practices of consumption. Ethics Place Environ 13:171–190

Miele M, Veissier I, Evans A, Botreau R (2011) Animal welfare: establishing a dialogue between science and society. Anim Welf 20:103

Murphy-Lawless J (2004) The impact of BSE and FMD on ethics and democratic process. J Agric Environ Ethics 17:385–403

Norwood FB, Lusk JL (2011) Compassion, by the pound: the economics of farm animal welfare. Oxford University Press, Oxford

Ohl F, Van Der Staay FJ (2012) Animal welfare: at the interface between science and society. Vet J 192:13–19

OIE (2021) One health [Online]. https://www.oie.int/en/what-we-do/global-initiatives/one-health/. Accessed 14 Oct 2021

Perry BD, Grace D, Sones K (2013) Current drivers and future directions of global livestock disease dynamics. Proc Natl Acad Sci 110:20871–20877

Poortinga W, Bickerstaff K, Langford I, Niewöhner J, Pidgeon N (2004) The British 2001 foot and mouth crisis: a comparative study of public risk perceptions, trust and beliefs about government policy in two communities. J Risk Res 7:73–90

Robbins J, Franks B, Von Keyserlingk MAG (2018) 'More than a feeling': an empirical investigation of hedonistic accounts of animal welfare. PLoS One 13:e0193864

Rodrigues MS, Hanley N (2021) How much do we actually care? A study on consumer preference heterogeneity and WTP for farm animal health and welfare in the UK. Agricultural economics society 95th annual conference, Warwick

Rowe G, Hawkes G, Houghton J (2008) Initial UK public reaction to avian influenza: analysis of opinions posted on the BBC website. Health Risk Soc 10:361–384

Sandøe P, Christensen T (2018) Farm animal welfare in Europe: from legislation to labelling. Working paper. 2018. https://dyreetik.ku.dk/dokumenter/forskningsprojekter/From_legislation_to_labelling.pdf. Accessed 13 Oct 2022

Sayer K (2013) Animal machines: the public response to intensification in Great Britain, c. 1960–c. 1973. Agric Hist 87:473–501

Scudamore JM (2007) Consumer attitudes to vaccination of food-producing animals. Revue scientifique et technique (International Office of Epizootics) 26:451–459

Sørensen JT, Schrader L (2019) Labelling as a tool for improving animal welfare—the pig case. Agriculture 9:123

Sørensen BT, De Barcellos MD, Olsen NV, Verbeke W, Scholderer J (2012) Systems of attitudes towards production in the pork industry. A cross-national study. Appetite 59:885–897

Souza Monteiro DM, Caswell JA (2004) The economics of implementing traceability in beef supply chains: trends in major producing and trading countries. Department of Resource Economics Working Paper no. 2004–6, University of Massachusetts Amherst

Spooner JM, Schuppli CA, Fraser D (2014) Attitudes of Canadian citizens toward farm animal welfare: a qualitative study. Livest Sci 163:150–158

Te Velde H, Aarts N, Van Woerkum C (2002) Dealing with ambivalence: farmers' and consumers' perceptions of animal welfare in livestock breeding. J Agric Environ Ethics 15:203–219

The Brambell Report (1965) Report of the technical committee to enquire into the welfare of animals kept under intensive livestock husbandry systems, etc. [Chairman, Professor FW Rogers Brambell]. Her Majesty's Stationery Office, London

Toma L, Mcvittie A, Hubbard C, Stott AW (2011) A structural equation model of the factors influencing British consumers' behaviour toward animal welfare. J Food Prod Mark 17:261–278

Tonsor GT, Olynk NJ (2011) Impacts of animal well-being and welfare media on meat demand. J Agric Econ 62:59–72

Tonsor GT, Wolf C, Olynk N (2009) Consumer voting and demand behavior regarding swine gestation crates. Food Policy 34:492–498

Tse YK, Zhang M, Doherty B, Chappell P, Garnett P (2016) Insight from the horsemeat scandal: exploring the consumers' opinion of tweets toward Tesco. Ind Manag Data Syst 116:1178–1200

Tuyttens FAM, Vanhonacker F, Van Poucke E, Verbeke W (2010) Quantitative verification of the correspondence between the Welfare Quality® operational definition of farm animal welfare and the opinion of Flemish farmers, citizens and vegetarians. Livest Sci 131:108–114

Van Kleef E, Houghton JR, Krystallis A, Pfenning U, Rowe G, Van Dijk H, Van Der Lans IA, Frewer LJ (2007) Consumer evaluations of food risk management quality in Europe. Risk Anal 27:1565–1580

Van Wezemael L, Verbeke W, Kügler JO, Scholderer J, Ueland Ø (2011) European consumers' acceptance of beef safety-improving interventions at different stages of the beef chain: primary production, slaughtering, processing and packaging. Proc Food Syst Dyn:311–319

Vanhonacker F, Verbeke W (2009) Buying higher welfare poultry products? Profiling Flemish consumers who do and do not. Poult Sci 88:2702–2711

Vanhonacker F, Verbeke W (2014) Public and consumer policies for higher welfare food products: challenges and opportunities. J Agric Environ Ethics 27:153–171

Vanhonacker F, Van Poucke E, Tuyttens F, Verbeke W (2010) Citizens' views on farm animal welfare and related information provision: exploratory insights from Flanders, Belgium. J Agric Environ Ethics 23:551–569

Ventura BA, Von Keyserlingk MAG, Wittman H, Weary DM (2016) What difference does a visit make? Changes in animal welfare perceptions after interested citizens tour a dairy farm. PLoS One 11:e0154733

Verbeke W, Pérez-Cueto FJA, Barcellos MDD, Krystallis A, Grunert KG (2010) European citizen and consumer attitudes and preferences regarding beef and pork. Meat Sci 84:284–292

Weible D, Christoph-Schulz I, Salamon P, Zander K (2016) Citizens' perception of modern pig production in Germany: a mixed-method research approach. Br Food J 118:2014–2032

Willett W, Rockström J, Loken B, Springmann M, Lang T, Vermeulen S, Garnett T, Tilman D, Declerck F, Wood A (2019) Food in the anthropocene: the EAT–Lancet Commission on healthy diets from sustainable food systems. Lancet 393:447–492

World Bank (2011) Creating business opportunity through improved animal welfare [Online]. http://www.ifc.org/wps/wcm/connect/633e46004885558fb714f76a6515bb18/Animal%2BWelfare%2BQN.pdf?MOD=AJPERES. Accessed 13 Oct 2022

Yougov (2019) Is the future of food flexitarian? YouGov. https://yougov.co.uk/topics/society/articles-reports/2019/03/18/future-food-flexitarian. Accessed 13 Oct 2022

Yunes M, Von Keyserlingk M, Hötzel M (2017) Brazilian citizens' opinions and attitudes about farm animal production systems. Animals 7:75

Zhou Y, Wang E (2011) Urban consumers' attitudes towards the safety of milk powder after the melamine scandal in 2008 and the factors influencing the attitudes. China Agric Econ Rev 3(1) ISSN 1756-137X

Zingg A, Siegrist M (2012) People's willingness to eat meat from animals vaccinated against epidemics. Food Policy 37:226–231

Zingg A, Cousin M-E, Connor M, Siegrist M (2013) Public risk perception in the total meat supply chain. J Risk Res 16:1005–1020

# Antimicrobial Resistance in Bacterial Pathogens from Farm Animals

Ulf Magnusson, Heike Kaspar, Oskar Nilsson, and Stefan Schwarz

**Abstract**

The increase in antimicrobial resistance of bacterial pathogens threatens the health, welfare, and productivity of the food-producing animals. In order to curb the emergence, it is essential to reduce the overall use of antimicrobial agents as it is one of the main drivers in the development of resistance. One key element in reducing the use is to stop the use of antimicrobial agents for growth promotion or routine prophylaxis and instead use antimicrobial agents only for metaphylaxis and therapy of diseased animals. In addition, it is good veterinary medical practice to use narrow spectrum antimicrobial agents whenever possible as they are less prone to drive the development of resistance. Moreover, antimicrobial agents that are critically important for human medicine should be avoided whenever possible. Here, we present resistance data from Germany and Sweden for some major bacterial infections in cattle, pigs, and poultry. Although an antibiogram should be performed prior to the onset of an antimicrobial therapy, especially in herd treatment, this data may assist veterinary practitioners and clinicians to choose the appropriate antimicrobial agent in cases of empirical therapy. However, resistance patterns of defined bacterial pathogens can vary over time and by geographical location or even by herd.

**Keywords**

Antimicrobial resistance · Livestock · Bacterial pathogens · Antimicrobial consumption · Selection pressure

U. Magnusson (✉)
Department of Clinical Sciences, Swedish University of Agricultural Sciences (SLU), Uppsala, Sweden
e-mail: ulf.magnusson@slu.se

H. Kaspar
Federal Office of Consumer Protection and Food Safety (BVL), Berlin, Germany
e-mail: heike.kaspar@bvl.bund.de

O. Nilsson
Department of Animal Health and Antimicrobial Strategies, Swedish Veterinary Agency (SVA), Uppsala, Sweden
e-mail: oskar.nilsson@sva.se

S. Schwarz
Institute of Microbiology and Epizootics, Centre for Infection Medicine, School of Veterinary Medicine, Freie Universität Berlin, Berlin, Germany

Veterinary Centre for Resistance Research (TZR), School of Veterinary Medicine, Freie Universität Berlin, Berlin, Germany
e-mail: Stefan.Schwarz@fu-berlin.de



J. J. Gross (ed.), *Production Diseases in Farm Animals*,
https://doi.org/10.1007/978-3-031-51788-4_3

## 1 Introduction

The access to antimicrobial agents in veterinary medicine has been, and is, a valuable asset for improving animal health, welfare, and productivity. Thus, it is crucial to protect the efficacy of antimicrobial agents by reducing the emergence of antimicrobial resistance (AMR). As the use of antimicrobial agents drives the development and spread of AMR, the use should be as restricted and targeted as possible without jeopardizing animal health and welfare. It should also be noted that there are reports that resistant bacteria have been transmitted to humans, but to what extent that happens is largely unknown (Hoelzer et al. 2017). Needless to say, this zoonotic transmission can go both ways as well, i.e., also from humans to animals. It has been reported that a certain *Staphylococcus aureus* lineage resistant to methicillin—MRSA CC398—seems particularly prone to zoonotic transmission (Price et al. 2012). This indicates that it is not only the resistance of the bacteria, but also their zoonotic potential that must be taken into account when assessing the risk of spread of antimicrobial resistance. However, in this chapter, we will focus on antimicrobial resistance in the context of animal health. Thus, we mainly refer to the conditions in the farming systems typically seen in OECD countries for cattle, pigs, and laying hens/broiler chickens, as these species contribute to more than 90% of the global livestock mass. Moreover, we focus on the subset of antimicrobial agents that are used against bacterial infections in farm animals as these are the ones mostly used in the sector.

Firstly, we present current regulations and guidelines within different regions, countries, and livestock sectors and then the data available or not available about antibiotic use in the livestock sector. Thereafter, we present a comprehensive overview from Germany and Sweden of the occurrence of antimicrobial resistance for a set of bacteria known to cause major diseases in livestock. Finally, we suggest some measures regarding the use of antimicrobial agents in order to safeguard their efficacy.

## 2 Development of Regulatory Measures and Guidelines for Use of Antimicrobial Agents in Livestock

There are at least three areas where regulations may contribute, if effectively enforced, to a more rational and restrictive use of antimicrobial agents: banning antimicrobial agents as growth promoters, providing antimicrobial agents to producer by prescription only and reducing the incentive for veterinarians to earn money by selling antimicrobial agents. However, at acute diseases the latter should not hinder the veterinarian to provide the farmer antimicrobial drugs until it is possible to get them from a pharmacy.

Sweden was the first country in the world that banned antimicrobial agents as feed additives for growth promotion in 1986. This ban was followed by other countries in the following decades, but initially just for some certain antimicrobial agents. The European Union banned the use or did not renew the licensing of all antimicrobial growth promoters in animal feed in 2006. The USA launched a similar stop in 2017 and China in 2019. As of 2019, 70% of the WOAH (formerly OIE) member countries reported that they did not use antimicrobial agents for growth promotion in livestock (OIE 2021).

Regardless their preparation, antimicrobial pharmaceuticals are readily available over the counter without prescription in many countries worldwide. However, in the EU, there is a need to have a prescription from a veterinarian to buy antimicrobial agents since 2001 (European Parliament 2001). In Canada are medically important antimicrobials (as defined by national authorities) for veterinary use sold by prescription only since 2018 and in the USA, certain medically important antimicrobial agents requires veterinary oversight to be used in animals since 2023.

Especially in low- and middle-income countries, it is common that veterinarians earn a substantial share of their income from selling medicines, including antimicrobial agents (Magnusson et al. 2021). In some high-income countries, this is still a common practice, whereas it is not allowed in others.

In addition to the legal frameworks mentioned above, there are guidelines about how and when to treat livestock with antimicrobial agents published by producers' organizations, veterinary professional associations, or even governmental agencies. These are often science- or evidence-based and sound, but the compliance to them, or other best practices of the same quality are sometimes poor (reviewed by Magnusson 2022).

## 3 Use of Antimicrobial Agents in Livestock in Different Parts of the World

It has been estimated that more than 70% of the global antimicrobial use is in livestock and currently are the livestock-rich countries China, Brazil, and the USA, the largest consumers (Tiseo et al. 2020). The global use is increasing, primarily driven by intensification of livestock production to meet the growing demand for animal source foods in low- and middle-income countries where antimicrobial use is poorly regulated and antimicrobial agents are used medically irrationally to compensate for poor animal husbandry practices (reviewed by Magnusson et al. 2021). In Europe, there is the annual report on surveillance of veterinary antimicrobial consumption (ESVAC), where official sales and related data are presented (EMA 2021). In most European countries, there is a decreasing trend in the sales of antimicrobial agents for food-producing animals over the last years (Fig. 1). However, there are large differences among countries regarding the sales, as calculated by sales per population corrected unit (PCU): range from 2.3 mg/PCU to 393.9 mg/PCU and with a median value of 51.9 mg/PCU. Globally, it has been estimated that most antimicrobial agents per PCU are used in pigs followed by poultry and the least in cattle (Tiseo et al. 2020).

## 4 Monitoring Resistance

The Federal Office of Consumer Protection and Food Safety (BVL) collects resistance data from clinical isolates of animal-pathogenic bacteria as part of the National Resistance Monitoring of Animal Pathogenic Bacteria (GERM-Vet 2021). A dedicated sampling plan is drawn up for each study year, which gives precise instructions on how the bacterial isolates are to be collected. Corresponding clinical isolates from acutely diseased animals are sent by the laboratories to the BVL throughout the study year. In order to keep the bias as low as possible, a maximum of two bacterial strains of one bacterial species can be included in the study from one animal herd for each animal species and indication. However, these bacterial strains of one species from the same animal herd and indication must not originate from the same animal, but must be isolated independently of each other in time. Isolates from animals with an antimicrobial treatment in the immediate period before sampling shall be marked off.

The Swedish Veterinary Agency (SVA) compiles data regarding resistance among animal-pathogenic bacteria as a part of the Swedish Veterinary Resistance Monitoring (Svarm) program. Data in Svarm consist of isolates from clinical submissions of samples sent to the diagnostic lab at SVA (Swedres-Svarm 2022). For many samples, information on the indication for sampling is not available but the vast majority of submissions are likely from animals with infections. Therefore, data in Svarm may be biased toward samples from treated animals or from herds where antimicrobial treatment is common. Furthermore, in some cases there is more than one animal sampled from the same herd. Any assessments of trends are based on the assumption that these biases are inherent throughout the observation period.

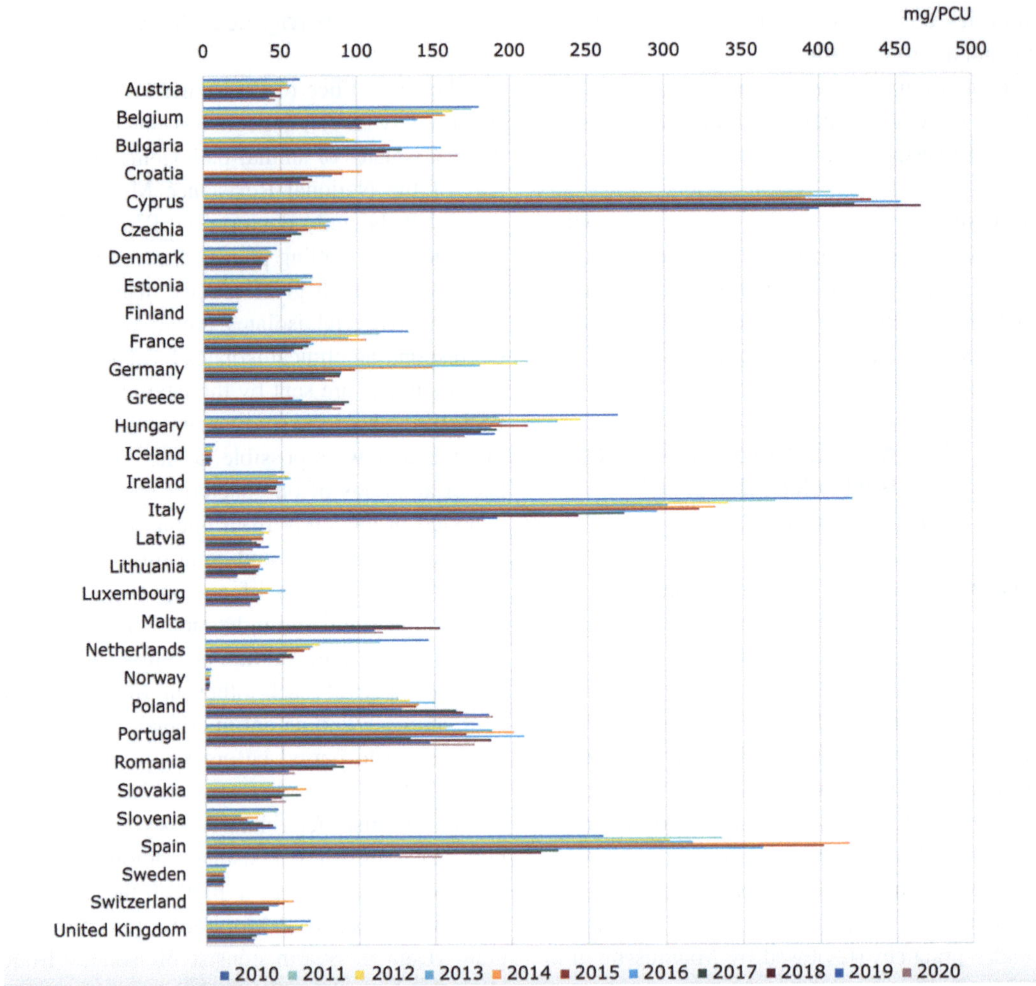


**Fig. 1** Total sales of antimicrobial veterinary medical products for food-producing animals, including horses and farmed-fish, in mg/PCU by country from 2010 to 2020 (Reproduced from EMA 2021)

## 5 Antimicrobial Resistance in Bacterial Pathogens from Cattle

### 5.1 Antimicrobial Resistance in Bovine Respiratory Tract Pathogens

Bovine respiratory disease (BRD) is a disease of particular economical relevance. Watts and Sweeney estimated global losses of the feedlot industry due to BRD of approximately $3 billion/year (Watts and Sweeney 2010). BRD is a multi-agent and multifactorial disease, in which besides bacteria and viruses also other factors play a role. Its colloquial designation, "shipping fever," refers to some of these factors that play important roles in the development of BRD. Transportation over long distances, often in combination with dehydration, starvation, overheating, chilling, or exhaustion of the animals is a stressor that supports the onset of the disease. In addition, passages through auction markets, commingling of animals from different herds, dusty environmental conditions in the feedlot, and changes in the diet represent additional stressors Initial viral infections, e.g. by parainfluenza-3 virus, bovine respiratory syncytial virus, bovine adenovirus type 3, but also bovine rhinovirus, reovirus, enterovirus, and coronavirus, often pave the way for following bacterial infections (Merck vet manual 2022). *Mannheimia haemolytica* and *Pasteurella multocida*, which represent regular inhabitants of the nasopharynx of cattle, are among the most relevant bacterial BRD pathogens Antimicrobial agents of various classes are commonly used to combat bacteria involved in BRD (Schwarz 2008; Michael et al. 2018).

### 5.1.1 *Mannheimia haemolytica*

Antimicrobial susceptibility test results of *M. haemolytica* isolates from cases of BRD in Germany, as conducted in the GE*RM*-Vet program, are available for the years 2010, 2012, 2014, 2017, and 2018 (Fig. 2a). For the classification of the minimal inhibitory concentrations into the categories susceptible, intermediate, or resistant, CLSI-approved veterinary-specific clinical breakpoints were available for penicillin, ampicillin, ceftiofur, florfenicol, tetracycline, tilmicosin, and tulathromycin (CLSI 2020). Over the years, increases in resistance were seen for ampicillin and penicillin with highest percentages of 39.0% (ampicillin) and 24.4% (penicillin) in 2018 (Fig. 2a). The higher percentages of isolates resistant to ampicillin than to penicillin are likely due to a lower clinical breakpoint for ampicillin resistance (≥0.25 mg/L) compared to that for penicillin resistance (≥1 mg/L). This is also illustrated by the fact that when 35 isolates from bovines in Sweden between 2014 and 2021 were assessed with the same breakpoint (≥1 mg/L) for both substances, the occurrence of resistance was the same for both penicillin and ampicillin (14%). In Germany, resistance to tetracycline dropped in 2018 to the lowest value of 7.3%. Furthermore, *M. haemolytica* isolates resistant to florfenicol were first seen in 2017, those resistant to tulathromycin and tilmicosin in 2014 and 2012, respectively. The percentages for all three antimicrobial agents were below 5% during the respective time periods. No isolates classified as resistant to enrofloxacin and ceftiofur were detected during the entire period 2010–2018 in the GE*RM*-Vet program. As for the isolates from bovines in Sweden, none of them showed resistance to ceftiofur, enrofloxacin, florfenicol, or tetracycline.

However, a *M. haemolytica* isolate from North America with an elevated ceftiofur MIC of 2 mg/L, which carried the extended-spectrum β-lactamase gene $bla_{\text{ROB-2}}$, has been identified (Kadlec et al. 2019). A *M. haemolytica* isolate from a BRD case in Germany was recently classified as enrofloxacin-resistant and exhibited a mutation that resulted in the amino acid exchange D87Y in the quinolone-resistance determining region of GyrA (Schink et al. 2022). In North American bovine *M. haemolytica* isolates, multiresistance-mediating integrative and conjugative elements (ICE) have been identified which carried the genes *erm*(42), *msr*(E) and *mph*(E) involved in resistance to macrolides (Michael et al. 2012). A first multiresistance ICE has recently been described in a German *M. haemolytica* isolate of cattle origin. It harbored the macrolide resistance genes *mef*(C) and *mph*(G) (Schink et al. 2022). A general overview about antimicrobial resistance genes in *M. haemolytica* isolates was published by Michael and coworkers (Michael et al. 2018).

### 5.1.2 *Pasteurella multocida*

Antimicrobial susceptibility test results of *P. multocida* isolates from cases of BRD in Germany, as conducted in the GE*RM*-Vet program, are available for the years 2011, 2013, 2014, 2016, 2017, and 2018 (Fig. 2b). From Sweden, data are available from 2005 to 2020 (Swedres-Svarm 2021). For the classification of the minimal inhibitory concentrations into the categories susceptible, intermediate, or resistant, CLSI-approved veterinary-specific clinical breakpoints were available for penicillin, ampicillin, ceftiofur, florfenicol, tetracycline, and tulathromycin (CLSI 2020). In Germany, increases in resistance over the years were seen solely for tetracycline and tulathromycin with highest percentages of 23.5% for tetracycline resistance in the years 2017 and 2018 and 14.1% for tulathromycin in 2018 (Fig. 2b). Although no veterinary-specific clinical tilmicosin breakpoints are available for bovine *P. multocida* isolates, 17/149 (11.4%) isolates exhibited tilmicosin MIC values of

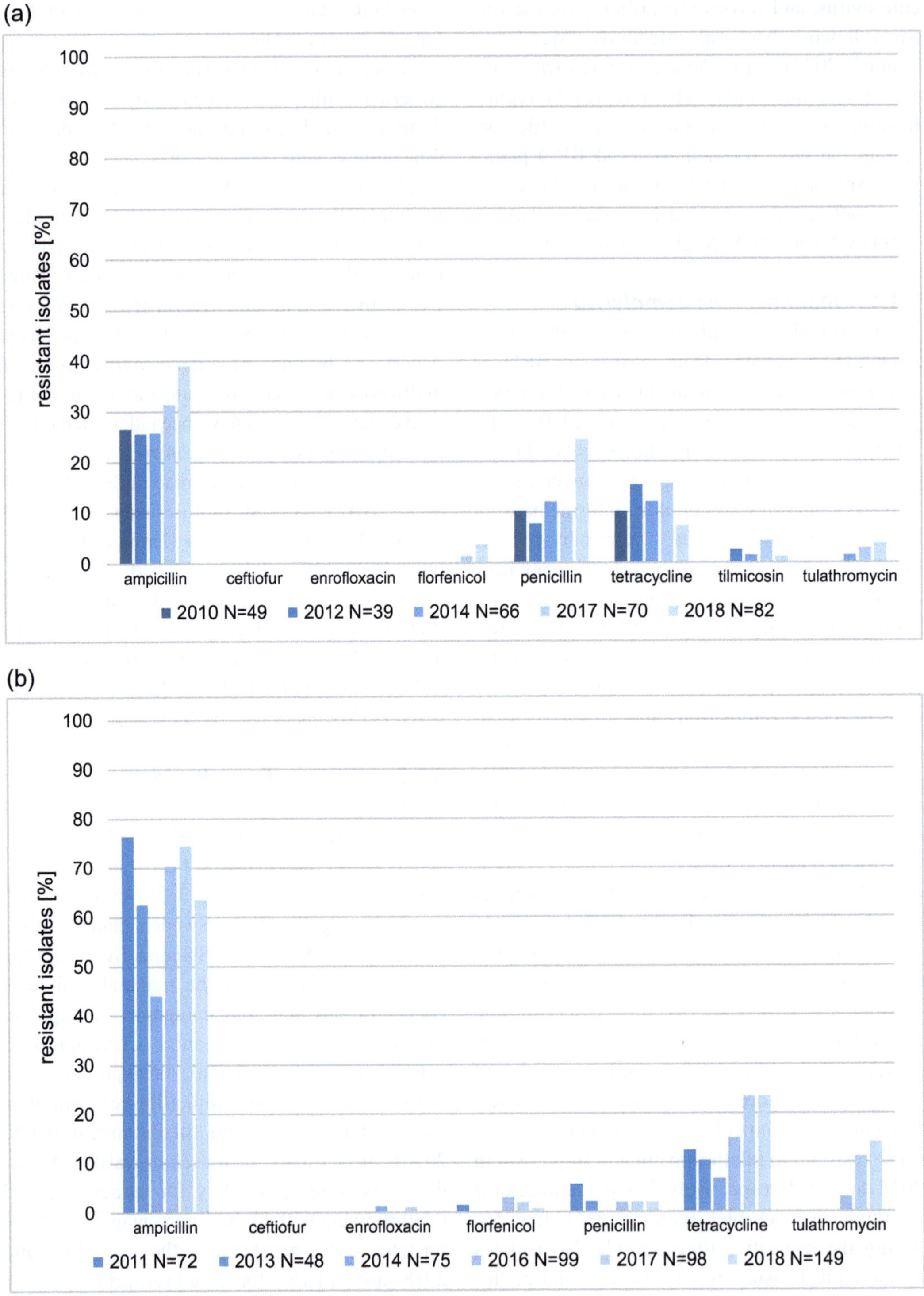


**Fig. 2** Percentages of resistance in 306 *M. haemolytica* isolates (**a**) and 541 *P. multocida* isolates (**b**), all from cases of BRD in Germany during 2010–2018 and 2011–2018, respectively

≥32 mg/L. As previously seen for bovine *M. haemolytica*, there was also a striking difference in the percentages of isolates classified as resistant to ampicillin (63.1%) and penicillin (2.0%), which again might be explained by the different CLSI-approved breakpoints for ampicillin resistance (≥0.25 mg/L) and penicillin resistance (≥1 mg/L). A look at the MIC distribution revealed that 92 of the 94 ampicillin-resistant *P. multocida* isolates had MICs of 0.25 mg/L or 0.5 mg/L. No ceftiofur-resistant isolates were detected during the entire period 2010–2018. In 2018, no enrofloxacin-resistant isolates were seen, whereas few such isolates were detected in 2014 and 2017. Florfenicol-resistant *P. multocida* isolates were detected rarely (below 5%) if at all (Fig. 2b).

Among the *P. Multocida* isolates from Sweden, antibiotic resistance was generally rare and no resistance to enrofloxacin, florfenicol, or tetracycline has been reported (Swedres-Svarm 2021; Swedres-Svarm 2022). Likewise, no resistance to ceftiofur has been detected. From 2005–2015, all isolates were also susceptible to benzyl penicillin and ampicillin but since 2016 beta-lactamase producing *P. multocida* have been isolated every year. The occurrence is still low with 6% of the isolates ($n = 397$) from 2016–2020 being resistant. However, the findings are of importance as benzyl penicillin is considered the first-choice antibiotic for pneumonia in cattle in Sweden.

Although multiresistant bovine *P. multocida* isolates were uncommon in the GE*RM*-Vet program and non-existing in Svarm, they were found frequently in North America. In these isolates, integrative and conjugative elements (ICEs) identical or related to ICE*Pmu1* have been detected. ICE*Pmu1* carries a total of 12 antimicrobial resistance genes, which include the β-lactamase gene $bla_{OXA-2}$, but also genes that confer resistance to tilmicosin and tulathromycin (*erm*(42), *msr*(E) and *mph*(E)), florfenicol (*floR*), sulfonamides (*sul2*), streptomycin (*strA*, *strB*), streptomycin and spectinomycin (*aadA25*), kanamycin and neomycin (*aphA1*), gentamicin (*aadB*), and tetracycline (*tet*(H)) (Michael et al. 2012a). As ICEs are self-transmissible elements that can move on their own from one bacterial host cell to another, they also transfer all their resistance genes in every transfer event. ICE*Pmu1* was shown to transfer even across genus boundaries and occasionally expresses its resistance genes better in *E. coli* than in *P. multocida* (Michael et al. 2012b). Most recently, a first ICE that resembled ICE*Pmu1* in its structure and carried the resistance genes *sul2*, *strA*, *strB*, *aphA1*, and *tet*(H) was detected in a multiresistant bovine *P. multocida* isolate from Germany (Schink et al. 2022). A general overview about antimicrobial resistance genes in *P. multocida* isolates was published by Michael and co-workers (Michael et al. 2018).

## 5.2 Antimicrobial Resistance in Bovine Mastitis Pathogens

Bovine mastitis is another area in veterinary medicine in which antimicrobial agents are used at high frequency. However, a comparatively small number of antimicrobial agents, mainly from the classes of penicillins, cephalosporins, macrolides, lincosamides, fluoroquinolones, and aminoglycosides are approved for use in bovine mastitis. The major bacterial pathogens involved in bovine mastitis are staphylococci, streptococci, enterococci and members of the Enterobacterales. For the classification of the minimal inhibitory concentrations into the categories susceptible, intermediate, or resistant, CLSI-approved mastitis-specific clinical breakpoints were available only for ceftiofur and pirlimycin (CLSI 2020). For the other antimicrobial agents, human-specific CLSI-approved clinical breakpoints applicable to *S. aureus* were applied (CLSI 2022).

### 5.2.1 *Staphylococcus aureus*

Antimicrobial susceptibility test results of *S. aureus* isolates from cases of bovine mastitis in Germany, as conducted in the GE*RM*-Vet program, are available for the years 2009, 2011, 2013, 2015, 2017, and 2019 (Fig. 3). In total, 1472 *S. aureus* isolates from bovine mastitis have been investigated in GE*RM*-Vet over these years.

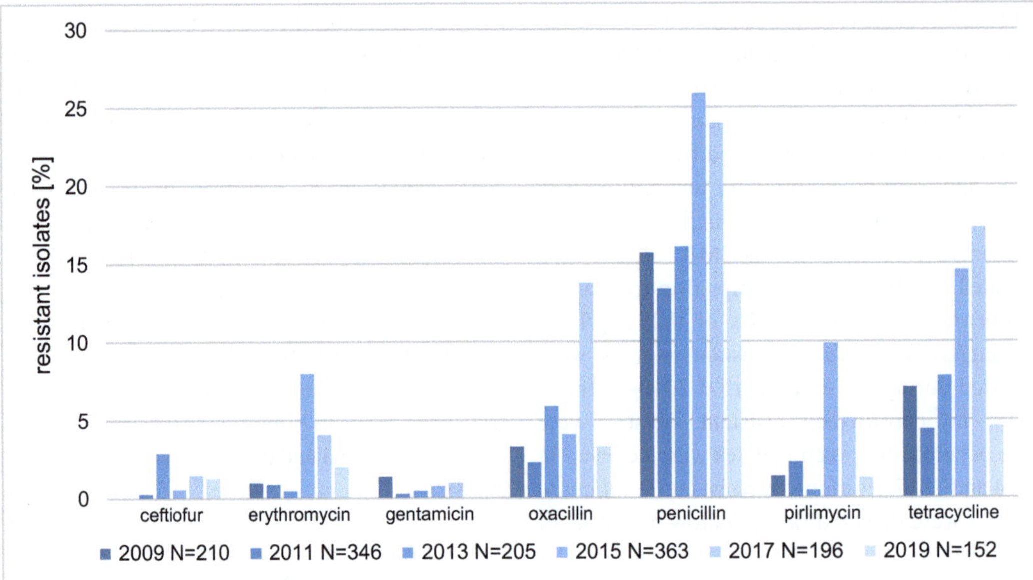


**Fig. 3** Percentages of resistance of 1472 *S. aureus* isolates from cases of bovine mastitis in Germany during 2009–2019

The percentages of resistant isolates were generally low compared to *S. aureus* isolates from other animals and disease conditions. Highest resistance rates among the isolates from 2019 were recorded for penicillin (13%) while the resistance rates for other antimicrobial agents were below 5% (Fig. 4). Solely, 5/152 (3%) isolates were resistant to oxacillin. All these MRSA isolates represented the sequence type 398, were *mecA*-positive, and displayed the *spa* types t011, t034, and t1451. MRSA of ST398 have frequently been detected in milk samples from cows suffering from clinical and subclinical mastitis in Germany (Feßler et al. 2010a) and the Netherlands (Feßler et al. 2012). In comparison with the years 2015 and 2017, the percentages of isolates resistant to penicillin, oxacillin, erythromycin, pirlimycin, and tetracycline were distinctly lower in 2019. A study on macrolide-resistant *S. aureus* from bovine mastitis identified the macrolide resistance genes *erm*(A), *erm*(B), *erm*(C) or *erm*(T) alone or in various combinations (Entorf et al. 2014).

In Sweden, the treatment guidelines for mastitis caused by *S. aureus* state that penicillin is the first-choice substance and that in case of penicillinase production measures to avoid spread should be taken. Concordantly, resistance to penicillin is rare. Between 2013 and 2018, only 3% of 227 investigated isolates were penicillinase producing (Duse et al. 2021). Furthermore, in 2020 and 2021, 527 and 605 isolates were analyzed for penicillinase production of which 1.5% and 1.2% were positive (Swedres-Svarm 2021; Swedres-Svarm 2022).

### 5.2.2 Coagulase-Negative Staphylococci (CoNS)

CoNS represent a heterogeneous group of staphylococci which are involved mainly in subclinical mastitis. In GE*RM*-Vet, CoNS from cases of bovine mastitis have been investigated in the years 2004/2005, 2009, 2015, and 2017. The major CoNS species identified among the isolates from 2017 were *S. chromogenes* ($n = 63$), *S. haemolyticus* ($n = 39$), and *S. xylosus* ($n = 27$). As seen for *S. aureus*, penicillin resistance was the most frequently detected resistance property ranging between 27% and 33% while resistance to other antimicrobial agents was below 10%. Oxacillin resistance rates were even higher than penicillin resistance rates, which, however, was

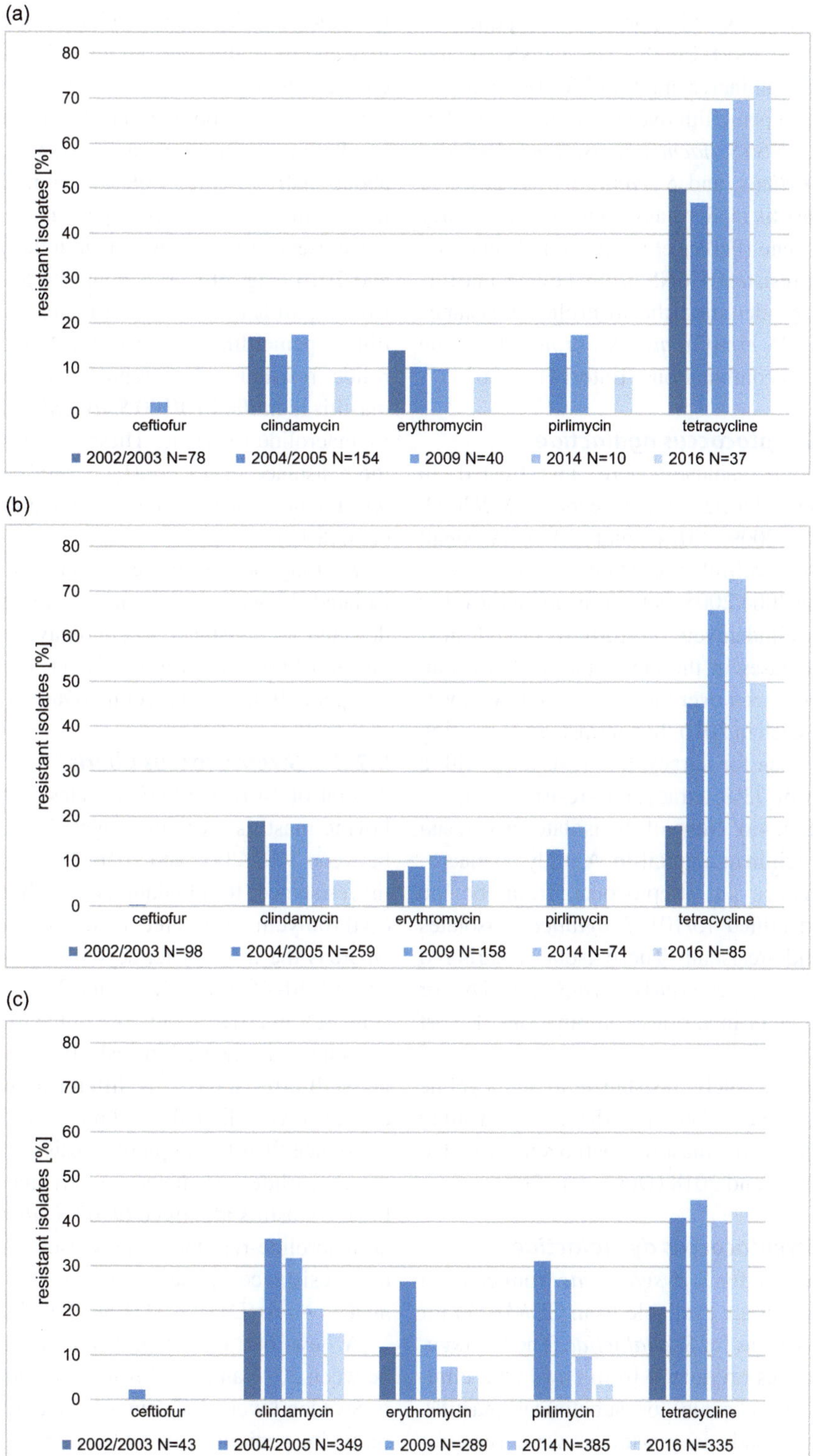


**Fig. 4** Percentages of resistance of 319 *S. agalactiae* isolates (**a**), 674 *S. dysgalactiae* isolates (**b**), and 1401 *S. uberis* isolates (**c**), all from cases of bovine mastitis in Germany during 2002/2003–2016

likely due to low CLSI-approved breakpoints for oxacillin and most CoNS. The CoNS isolates also showed an increasing trend for resistance to oxacillin and penicillin over the years. Oxacillin-resistant CoNS (*S. haemolyticus*, *S. epidermidis*, *S. saprophyticus,* and *S. capitis*) from cases of bovine mastitis in Germany were shown to carry the *mecA* gene (Feßler et al. 2010b). A study on macrolide-resistant CoNS from bovine mastitis in Germany identified the macrolide resistance genes *erm*(B), *erm*(C), *msr*(A), or *mph*(C) alone or in various combinations (Entorf et al. 2014).

### 5.2.3 *Streptococcus agalactiae*

*S. agalactiae* isolates were investigated in GE*RM*-Vet during the years 2002/2003, 2004/2005, 2009, 2014, and 2016. A small number of ceftiofur-resistant isolates were recorded in 2004/2005, but not in any of the other years. For clindamycin, erythromycin and pirlimycin decreases in the percentages of resistant isolates were seen over the years with lowest values of 8% each in 2016. In contrast, an increasing trend was seen for tetracycline resistance with a maximum of 73% tetracycline-resistant isolates in 2016 (Fig. 4a). None of the isolates was resistant to penicillin or ampicillin. A study on macrolide resistance in streptococci from bovine mastitis identified 15/101 *S. agalactiae* isolates as macrolide-resistant. These isolates harbored the resistance genes *erm*(B), *erm*(C), *mef*(A), or *msr*(D) alone or in various combinations (Entorf et al. 2016).

In Sweden solely resistance to tetracycline (two out of nine isolates was detected in a study on resistance in mastitis pathogens isolated between 2013 and 2018 (Duse et al. 2021).

### 5.2.4 *Streptococcus dysgalactiae*

Resistance data for *S. dysgalactiae* from cases of bovine mastitis are available from GE*RM*-Vet for the same years as for *S. agalactiae.* The *S. dysgalactiae* isolates originated from cows with clinical (acute or chronic) or subclinical mastitis. Decreasing trends in resistance were seen for clindamycin and erythromycin while in 2016, no pirlimycin-resistant isolates were recorded. Clindamycin and pirlimycin are both members of the lincosamides and varying resistance rates may be due to the fourfold lower human-specific clinical breakpoint of clindamycin resistance compared to the bovine mastitis-specific clinical breakpoint of pirlimycin resistance. Although about half of the isolates from 2016 was tetracycline-resistant, this percentage of resistance was distinctly lower than in the years 2009 and 2014 (Fig. 4b). As among the *S. agalactiae* isolates, all *S. dysgalactiae* isolates were susceptible to penicillin and ampicillin. A study on macrolide resistance in streptococci from bovine mastitis identified 10/100 *S. dysgalactiae* isolates as macrolide-resistant. These isolates harbored the resistance genes *erm*(B), *erm*(C), *mef*(A), or *msr*(D) alone or in various combinations (Entorf et al. 2016).

A study on resistance in mastitis pathogens isolated between 2013 and 2018 in Sweden detected no resistance to clindamycin but a few out of 120 isolates were resistant to erythromycin, penicillin, or tetracycline (Duse et al. 2021).

### 5.2.5 *Streptococcus uberis*

A total of 1401 *S. uberis* isolates from cases of bovine mastitis were investigated in GE*RM*-Vet between 2002/2003 and 2016. Decreasing trends in resistance to clindamycin, pirlimycin, and erythromycin were recorded over the years, whereas the level of tetracycline resistance was around 40–45% for the years 2004/2005–2016 (Fig. 4c). In 2016, about 40% of the isolates were classified as penicillin-resistant, and 8% as ampicillin-resistant. This difference may also be due to lower clinical breakpoints for penicillin resistance than for ampicillin resistance. A study on macrolide resistance in streptococci from bovine mastitis identified 6/102 *S. uberis* isolates as macrolide-resistant. These isolates harbored the resistance genes *erm*(B) ($n$ = 5) or *mef*(A) + *msr*(D) ($n$ = 1) (Entorf et al. 2016).

A few out of 89 *S.uberis* isolates collect between 2013 and 2018 from cows with mastitis in Sweden I were resistant to clindamycin, erythromycin, penicillin, or tetracycline (Duse et al. 2021).

### 5.2.6 *Enterococcus* spp.

It should be noted that *E. faecalis* and *E. faecium* are intrinsically resistant to cephalosporins, aminoglycosides (except high-level resistance), clindamycin, trimethoprim as well as trimethoprim-sulfamethoxazole (CLSI 2022). Susceptibility data from GE*RM*-Vet are available for 275 *Enterococcus faecalis* isolates obtained from cases of bovine mastitis in the years 2010, 2011, 2012, 2013, 2014, 2016, 2017, 2018, and 2019. Resistance to penicillin and ampicillin was recorded only in the initial years 2010 and 2011, but not in the years thereafter. Resistance to macrolides was seen during all the years with a maximum of almost 30% in 2019.

In addition, susceptibility data from GE*RM*-Vet are also available for 235 *Enterococcus faecium* isolates obtained from cases of bovine mastitis in the years 2013, 2014, 2016, 2017, 2018, and 2019. Penicillin resistance was seen at 2–11% during 2013–2018, but increased to 19% in 2019. In contrast, ampicillin resistance remained at 2% as in the years 2014–2017. Since 2017, a decreasing trend was seen for erythromycin resistance and in 2019, 14% of the *E. faecium* isolates were resistant to erythromycin, which corresponded well with the data from 2014 and 2016.

A study on 64 *E. faecalis* and 37 *E. faecium* from cases of bovine mastitis in Germany identified 86% of the *E. faecalis* isolates as tetracycline-resistant and 17% as erythromycin-resistant. High-level resistance to streptomycin and gentamicin was seen in 31% and 2%, respectively, of the isolates. Of the *E. faecium* isolates, 19% were resistant to tetracycline, 14% to erythromycin, and 6% to ampicillin while high-level streptomycin resistance was seen in a single isolate (Werner et al. 2012).

### 5.2.7 *Escherichia coli*

Among Gram-negative bacteria, *E. coli* is often found as causative agent of bovine mastitis. Antimicrobial susceptibility data for *E. coli* from bovine mastitis are available from GE*RM*-Vet for the years 2005/2006, 2010, 2012, 2014, 2016, and 2018. In total, 1901 *E. coli* isolates have been investigated over these years. In general, resistance to the tested antimicrobial agents was low at less than 20% often even less than 10% (Fig. 5a). Resistance to ampicillin was usually the most frequently observed resistance property followed by tetracycline resistance. Since 2014, a decreasing trend was seen for resistance to ampicillin, amoxicillin-clavulanic acid, gentamicin, doxycycline, tetracycline, and trimethoprim-sulfamethoxazole (Fig. 5a). Five *E. coli* isolates from 2018 were identified as producers of extended-spectrum β-lactamases (ESBLs). A study on ESBL-producing *E. coli* isolates from various food-producing animals obtained in GE*RM*-Vet during 2008–2014 identified the ESBL genes $bla_{\text{CTX-M-1}}$, $bla_{\text{CTX-M-14}}$, and $bla_{\text{CTX-M-15}}$ as the dominant ESBL genes among bovine *E. coli* including those from bovine mastitis (Michael et al. 2017). It is noteworthy that *E. coli* is intrinsically resistant to macrolides and lincosamides (CLSI 2022).

In a study on resistance in mastitis pathogens isolated between 2013 and 2018 in Sweden, resistance to the tested antimicrobial agents was less than 10% in *E. coli* isolates with resistance to ampicillin (9%) as the most common trait (Duse et al. 2021). Occurrence of resistance reported within Svarm from 2016–2020 is also low (Swedres-Svarm 2021).

### 5.2.8 *Klebsiella* spp.

Isolates of the genus *Klebsiella* have also frequently been found to be associated with bovine mastitis. In GE*RM*-Vet, a total of 580 *Klebsiella* spp. isolates from cases of bovine mastitis have been investigated during 2010–2019 (Fig. 5b). *Klebsiella* spp. are intrinsically resistant not only to macrolides and lincosamides, but also to aminopenicillins, such as ampicillin. Resistance rates for other antimicrobial agents were usually below 20% and in 2019, resistance to tetracycline was at 4%, resistance to doxycycline and trimethoprim-sulfamethoxazole at 3% each and for amoxicillin/clavulanic acid at 1%. Resistance to gentamicin was not detectable (Fig. 5b).

In a study on resistance in mastitis pathogens isolated between 2013 and 2018 in Sweden, resistance to the tested antimicrobial agents was less than 10% with resistance to tetracycline

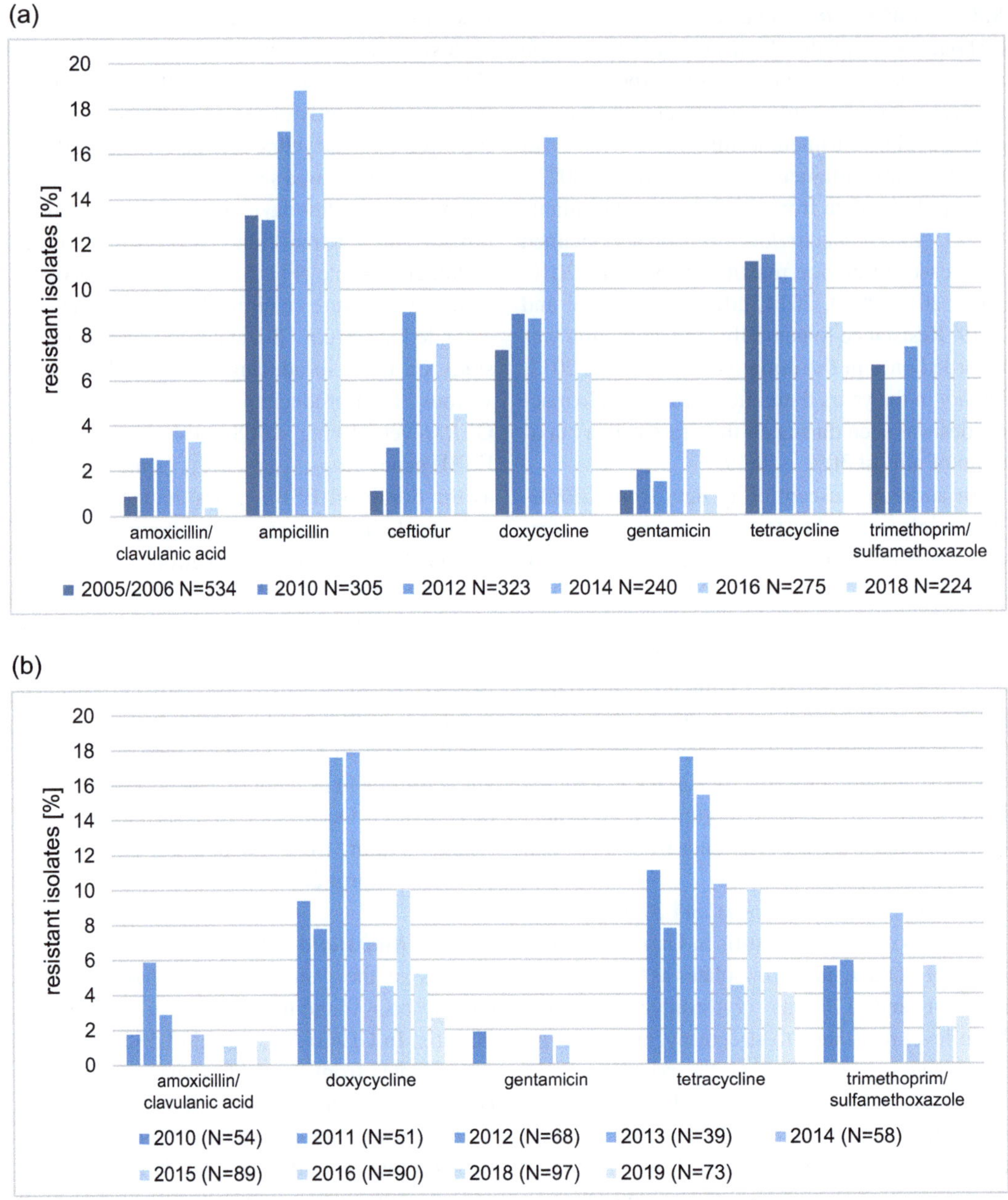


**Fig. 5** Percentages of resistance of 1901 *E. coli* isolates (**a**) and 580 *Klebsiella* spp. isolates (**b**), all from cases of bovine mastitis in Germany during 2005/2006–2018 and 2010–2019, respectively

(9%) as the most common trait (Duse et al. 2021). Occurrence of resistance reported within Svarm from 2016–2020 is also low (Swedres-Svarm 2021).

## 5.3 Antimicrobial Resistance in Bovine Enteric Pathogens

Several microorganisms play a role as enteric pathogens of cattle. Among the bacterial pathogens, *E. coli* plays a dominant role as causative agent of enteritis in calves/young cattle and adult cattle.

#### 5.3.1 *Escherichia coli* from Enteritis in Calves and Young Cattle

Antimicrobial susceptibility data were recorded for *E. coli* isolates from enteritis in calves/young cattle since 2010. The comparative analysis of 1666 *E. coli* isolates during the period 2010–2019 revealed high percentages of resistant isolates for ampicillin (66–81%), tetracycline (60–75%), trimethoprim-sulfamethoxazole (39–58%) and gentamicin (18–41%). Even the percentages of isolates resistant to amoxicillin-clavulanic acid were comparatively high (7–26%) (Fig. 6a). The percentages of ESBL-producing *E. coli* increased between 2006/2007 and 2014 from 7% to 34% whereas a decrease was seen during the years 2017–2019. In 2019, 18% of the *E. coli* isolates were ESBL-producers. Resistance is common also among *E. coli* isolates from enteritis in calves in Sweden. Occurrence presented in Svarm since 2016 has varied for ampicillin (34–58%), tetracycline (55–69%), and trimethoprim-sulfamethoxazole (10–31%). However, resistance against gentamicin has not been detected and ESBL-producing *E. coli* is detected only occasionally (Swedres-Svarm 2022).

#### 5.3.2 *Escherichia coli* from Enteritis in Adult Cattle

Distinctly fewer *E. coli* isolates from cases of enteritis in adult cattle were investigated in GE*RM*-Vet during 2015–2019 for their antimicrobial susceptibility. Among the 234 *E. coli* isolates, ampicillin resistance was the dominant resistance property followed by resistances to tetracycline and doxycycline as well as trimethoprim-sulfamethoxazole (Fig. 6b). In comparison with 2018, the percentages of resistance to all antimicrobial agents dropped except for resistance to amoxicillin-clavulanic acid which increased from 13% in 2018 to 17% in 2019. While there was a constant increase in the percentages of ESBL-producing *E. coli* between 2015 and 2018 from 12% to 23%, the percentage of ESBL-producers dropped below 10% in 2019.

## 6 Antimicrobial Resistance in Bacterial Pathogens from Pigs

### 6.1 Antimicrobial Resistance in Porcine Respiratory Tract Pathogens

The porcine respiratory disease complex (PRDC) is—similar to BRD—also a multifactorial and multiagent disease of particular economical relevance. Besides viruses, such as the porcine circovirus type 2, the porcine reproductive and respiratory syndrome virus, swine influenza virus and corona virus, various bacteria are involved in the PRDC. Among them, *Mesomycoplasma hyopneumoniae*, *Glaesserella parasuis*, *Pasteurella multocida*, *Bordetella bronchiseptica,* and *Streptococcus suis* among others play important roles (Dee 2021). In GE*RM*-Vet, antimicrobial susceptibility data are available for *P. multocida*, *B. bronchiseptica,* and *S. suis* isolates from porcine respiratory diseases.

#### 6.1.1 *Pasteurella multocida*

In total, 982 *P. multocida* isolates from respiratory tract infections of pigs in Germany were investigated for their antimicrobial susceptibility in GE*RM*-Vet between 2008 and 2019 (Fig. 7). The percentages of resistance were low—commonly below 10%. Lowest percentages of resistant isolates—if present at all—were recorded for enrofloxacin, tilmicosin, and tulathromycin. The 118 isolates investigated in 2019 comprised 29 isolates from piglets/weaners and 89 isolates from adult pigs. In 2019, resistance to all antimicrobial agents tested was below 5%, except for tetracycline resistance (8%). In Sweden, almost all isolates tested within Svarm since 2005 have been susceptible to all tested substances. A review on antimicrobial resistance in Pasteurellaceae from animals also provides data on the resistance genes so far identified in porcine *P. multocida* isolates (Michael et al. 2018).

#### 6.1.2 *Bordetella bronchiseptica*

Antimicrobial susceptibility data of *B. bronchiseptica* isolates from porcine respiratory disease

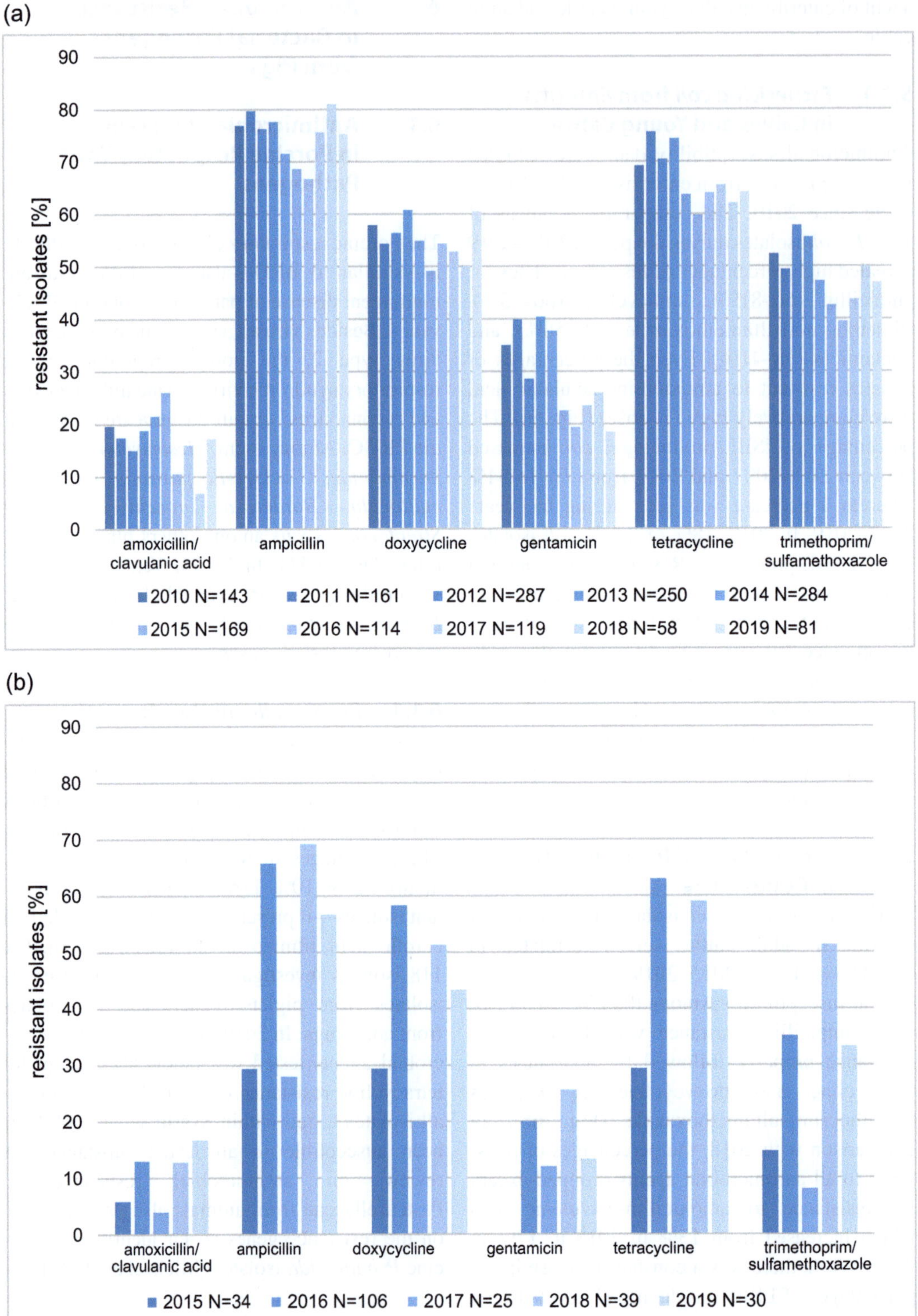


**Fig. 6** Percentages of resistance of 1666 *E. coli* isolates from enteritis of calves/young cattle (**a**) and 234 *E. coli* isolates from enteritis of adult cattle (**b**) in Germany during 2010–2019 and 2015–2019, respectively

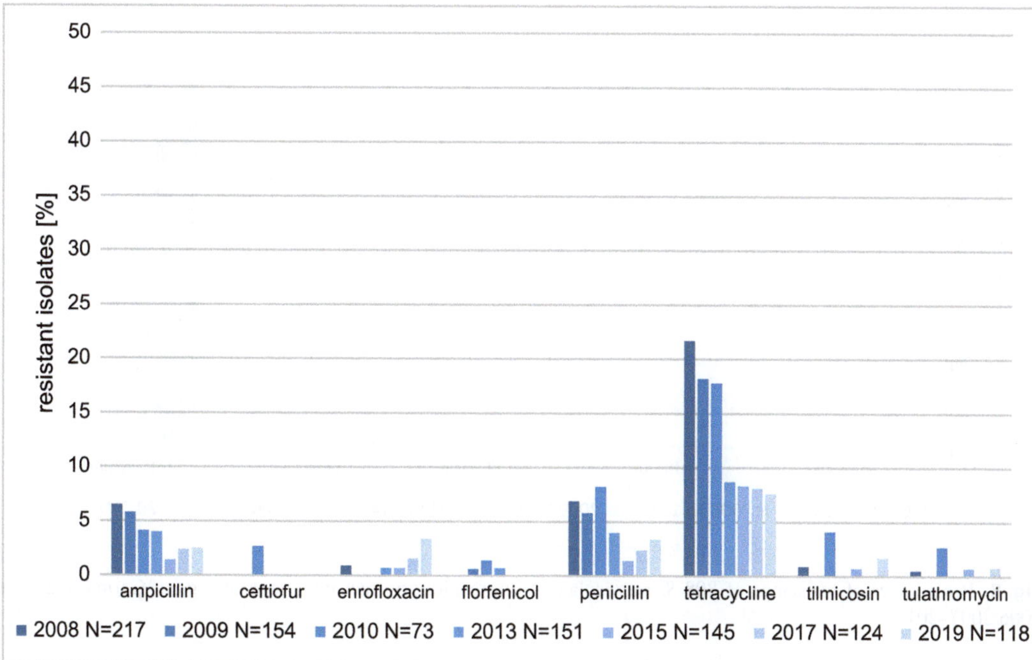


**Fig. 7** Percentages of resistance of 982 *P. multocida* isolates from respiratory tract infections of pigs in Germany during 2008–2019

are available from GE*RM*-Vet for the years 2010–2019. In total, 510 *B. bronchiseptica* isolates have been investigated. This organism is intrinsically resistant to β-lactam antimicrobials due to reduced membrane permeability and the production of a species-specific β-lactamase (Kadlec et al. 2007a). Pig-specific clinical breakpoints for clinically used antimicrobial agents are only available for florfenicol, tildipirosin, and tulathromycin. Tulathromycin resistance has rarely been detected among the isolates from GE*RM*-Vet. Solely in 2011, few isolates were classified as tulathromycin-resistant. The susceptibility data from GE*RM*-Vet showed variable resistances to florfenicol ranging between 3% and 13% over the years. Only 4% of the isolates from 2019 were florfenicol-resistant. In a previous study, the chloramphenicol/florfenicol resistance gene *floR* was identified in *B. bronchiseptica* (Kadlec et al. 2007b). A recent review provides information about the antimicrobial resistance genes so far identified in *B. bronchiseptica* (Kadlec and Schwarz 2018).

### 6.1.3 *Streptococcus suis*

In GE*RM*-Vet, a total of 808 *S. suis* isolates obtained from cases of respiratory diseases of pigs during 2006/2007 and 2017 were investigated for their susceptibility to antimicrobial agents (Fig. 8). In general, resistance to all antimicrobial agents, except tetracycline, was detected in not more than 10% of the isolates. In contrast, the tetracycline resistance rates ranged between 73% and 95%. In 2017, 82 isolates—30 from piglets, 10 from weaners, and 42 from adult pigs—were investigated for their antimicrobial susceptibility. Among these isolates, increases in resistance to ampicillin, penicillin, ceftiofur, and enrofloxacin were recorded in comparison with previous years. In Sweden the situation is similar, with high occurrence of resistance to tetracycline (67%) but low occurrence of resistance to other investigated substances among isolates investigated from 2013 to 2020. A study on three porcine *S. suis* isolates that differed in their susceptibility to penicillin revealed that the number of amino acid substitutions in the penicillin-binding proteins PBP2x and PBP2b was

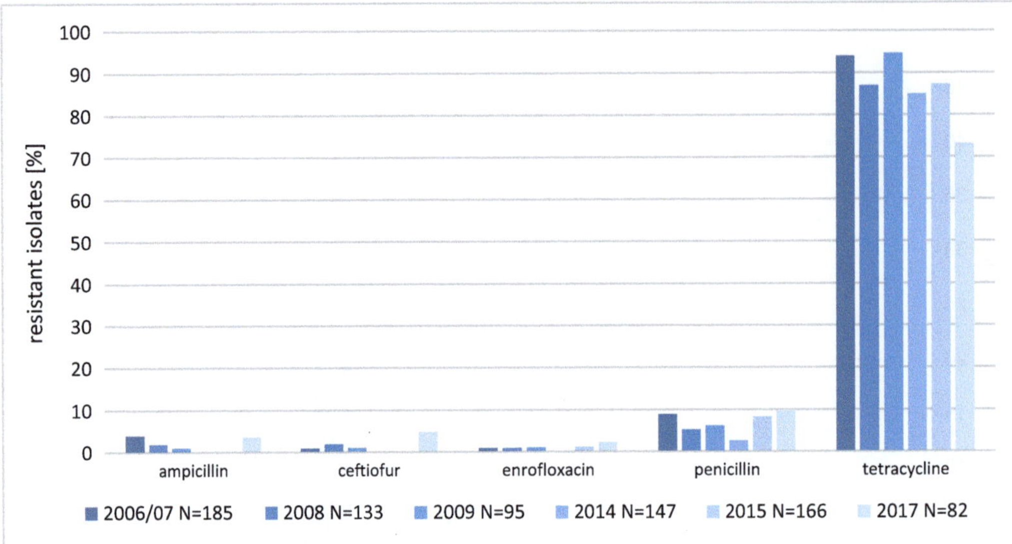


**Fig. 8** Percentages of resistance of 808 *S. suis* isolates from respiratory tract infections of pigs in Germany during 2006/2007–2017

correlated with increases in the penicillin MICs (Niemann et al. 2019). Moreover, the tetracycline resistance genes *tet*(O) and *tet*(O/W/32/O) as well as the macrolide resistance genes *erm*(B), *mef*(A) and *msr*(D) were detected in these isolates. A recent review provides information about the resistance genes so far identified in streptococci from animals, including *S. suis* from pigs (Haenni et al. 2018).

## 6.2 Antimicrobial Resistance in Porcine Enteric Pathogens

Similar to the situation among cattle, various viral and bacterial pathogens can cause enteric diseases in pigs, mainly in pigs of younger ages (piglets and weaners). Among the bacterial pathogens, *E. coli* is the most important one. Different pathotypes of *E. coli* can be involved, some of which cause enteric diseases with enterotoxemia, such as edema disease *E. coli* (EDEC), others without enterotoxemia, such as enteropathogenic (EPEC) or enterotoxigenic (ETEC) *E. coli* (Dubreuil et al. 2016; Tseng et al. 2014). Regardless of the intestinal pathogenic *E. coli* present, all of them cause diarrhea usually accompanied with an inflammation of the intestine. Although antimicrobial chemotherapy is not indicated in such disease conditions, studies on the respective *E. coli* isolates have shown that they are often resistant to multiple antimicrobial agents.

### 6.2.1 *Escherichia coli*

In 2019, a total of 298 *E. coli* isolates from pigs suffering from infections of the intestinal tract were investigated for their antimicrobial susceptibility. Among them, 161 isolates were from piglets, 33 from weaners, and 104 from fattening pigs. Comparative analysis of the isolates from the different production levels did not reveal striking differences. Thus, Fig. 9 exemplifies the situation by showing the data of the isolates from piglets over the years 2010–2019. During this time span, 1164 *E. coli* isolates from piglets were investigated for their susceptibility to antimicrobial agents. In general, decreasing trends can be seen for resistance to ampicillin, tetracycline, doxycycline, trimethoprim-sulfamethoxazole, and gentamicin (Fig. 9). Nevertheless, resistances to ampicillin and tetracycline were above 50% and those of doxycycline and trimethoprim-sulfamethoxazole above 40%. Resistances to amoxicillin/clavulanic acid and gentamicin were below 5% among the isolates from 2019. Seven

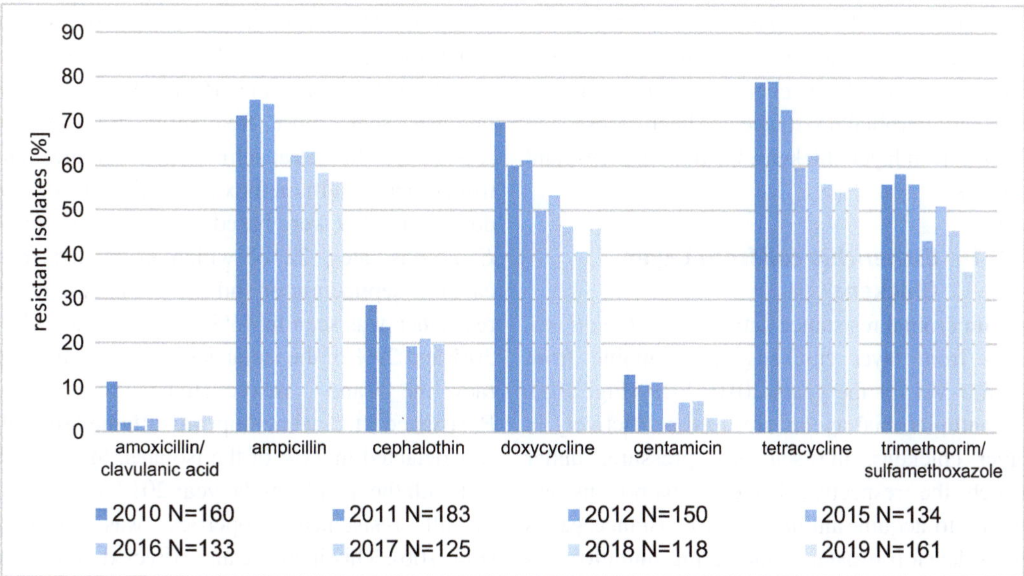


**Fig. 9** Percentages of resistance of 1164 *E. coli* isolates from intestinal infections of pigs in Germany during 2010–2019

isolates exhibited a colistin MIC of >2 mg/L and carried either the gene *mcr-1* ($n$ = 6) or *mcr-1* + *mcr-4* ($n$ = 1). The percentages of ESBL-producing *E. coli* varied between 2% and 11% during 2006 and 2019 with 6% in 2019. Thus, the percentages of ESBL-producers were distinctly lower among *E. coli* from pigs than among those from calves. The ESBL genes found among porcine *E. coli* from GE*RM*-Vet 2008–2014 comprised the resistance genes $bla_{\text{CTX-M-1}}$ in the majority of the cases, while $bla_{\text{CTX-M-15}}$, $bla_{\text{TEM-52}}$, and $bla_{\text{SHV-12}}$ were only detected in few isolates (Michael et al. 2017).

Resistance is common also among *E. coli* isolates from enteritis in pigs in Sweden. Occurrence presented in Svarm since 2016 has varied for ampicillin (24–43%), tetracycline (12–28%), and trimethoprim-sulfamethoxazole (27–40%). However, resistance against gentamicin or third generation cephalosporins (i.e., ESBL-producing *E. coli*) has not been detected (Swedres-Svarm 2022). Furthermore, only a handful of isolates with elevated MIC for colistin have been detected but no *mcr*-genes have been detected in those isolates.

## 7 Antimicrobial Resistance in Bacterial Pathogens from Poultry

### 7.1 Avian Colibacillosis

*E. coli* also plays an important role as a pathogen in poultry. Avian colibacillosis is a systemic or localized infection caused by avian pathogenic *E. coli* (APEC). The disease manifests in poultry in various ways of which acute fatal septicemia is seen most frequently. Other disease conditions may be pericarditis, airsacculitis, salpingitis, or peritonitis. Avian colibacillosis occurs worldwide and is one of the economically most important bacterial infections of poultry (Nolan 2020). APEC comprise a heterogeneous group of *E. coli*, which are often resistant not only to multiple antimicrobial agents, but also to disinfectants (Nolan 2020). Oral administration of antimicrobial agents has been shown to increase antimicrobial resistance in *E. coli* from chickens (Simoneit et al. 2015). It must be noted that cephalosporins are not approved for use in poultry. This might have an impact on the presence of ESBL-producing *E. coli*. During 2008–2014, only 20 ESBL-producing *E. coli* isolates were identified in GE*RM*-Vet, all of which harbored the resistance genes *bla*$_{CTX\text{-}M\text{-}1}$ ($n = 12$), *bla*$_{TEM\text{-}52}$ ($n = 4$) or *bla*$_{SHV\text{-}12}$ ($n = 4$) (Michael et al. 2017). In GE*RM*-Vet, antimicrobial resistance has been assessed in *E. coli* from layer chickens, broiler chickens, and turkeys.

### 7.1.1 *Escherichia coli* from Layer Chickens

Antimicrobial resistance data of 1848 *E. coli* isolates from layer chickens are available from GE*RM*-Vet for the years 2010–2019 (Fig. 10a). As antimicrobial agents are used restrictively in layer chickens, the selection pressure, under which the respective bacterial pathogens are forced to acquire antimicrobial resistance genes or to develop resistance-mediating mutations, is comparatively low. This is also reflected by the percentages of resistance shown in Fig. 10a. Over the years, resistances to amoxicillin-clavulanic acid, enrofloxacin, gentamicin, and trimethoprim-sulfamethoxazole were always below 10% if detected at all. Higher percentages were solely detected for ampicillin, tetracycline, and doxycycline. The latest data from 2019 based on the analysis of 274 isolates, 252 of which originated from cases of septicemia and sudden death. Resistance to ampicillin was seen most frequently (28%) whereas resistances to tetracycline (9%) and doxycycline (7%) also dropped below 10%. Resistances to amoxicillin-clavulanic acid, enrofloxacin, gentamicin, and trimethoprim-sulfamethoxazole were distinctly below 5%. Only a single *E. coli* isolate was multiresistant, i.e., resistance to ampicillin, tetracycline, enrofloxacin, and trimethoprim-sulfamethoxazole. In Sweden, resistance among *E. coli* from laying hens was investigated in 2017–2018 (Swedres-Svarm 2019). Only for tetracycline (13%) and ampicillin (11%), occurrence of resistance was above 10%. For the remaining investigated substances, occurrence of resistance was absent or below 5%.

### 7.1.2 *Escherichia coli* from Broiler Chickens

During 2010–2019, a total of 557 *E. coli* isolates from broiler chickens in Germany were investigated for their resistance to antimicrobial agents (Fig. 10b). In general, the percentages of resistant isolates were higher than those of the *E. coli* isolates from layer chickens. Resistance to ampicillin was seen most frequently, followed by resistances to tetracycline, doxycycline, and trimethoprim-sulfamethoxazole (Fig. 10b). The data from 2019 were based on the analysis of 70 *E. coli* isolates, 22 of which originated from cases of septicemia or sudden death. Ampicillin resistance was seen in 43% of the isolates while 30% and 29% of the isolates were resistant to tetracycline and doxycycline, respectively. Resistance to trimethoprim-sulfamethoxazole was detected in 21% of the isolates. In comparison with the data from the year 2018, an increase in all resistance properties was recorded (Fig. 10b). This increase also affected enrofloxacin, a fluoroquinolone classified by the WHO among the highest priority critically important antimicrobial agents. Here, an increase was seen from 2% in 2018 to 16% in 2019. In addition, one isolate with a colistin MIC of >2 mg/L carried the *mcr-1* gene.

### 7.1.3 *Escherichia coli* from Turkeys

Between 2010 and 2019, a total of 998 *E. coli* isolates from infections of turkeys have been investigated in GE*RM*-Vet for their antimicrobial susceptibility. Over the years, lowest percentages of resistance were found for amoxicillin, enrofloxacin, and gentamicin, whereas distinctly higher percentages of resistance were recorded for ampicillin, tetracycline, doxycycline, and trimethoprim-sulfamethoxazole (Fig. 10c). In 2019, 55 *E. coli* isolates from turkeys were analyzed. Among them 39 isolates originated from cases of septicemia. Ampicillin resistance was seen in 55% of the isolates followed by resistances to tetracycline and trimethoprim-sulfamethoxazole (18% each) and doxycycline (11%). In comparison with the year 2018, distinct increases in resistance to ampicillin (39% in 2018 and 55% in 2019) as well as trimethoprim-sulfamethoxazole (8% in 2018 and 18% in 2019) were recorded. In contrast, enrofloxacin resistance slightly decreased from 9% in 2018 to 7% in 2019. Four isolates with colistin MICs of >2 mg/L harbored the *mcr-1* gene.

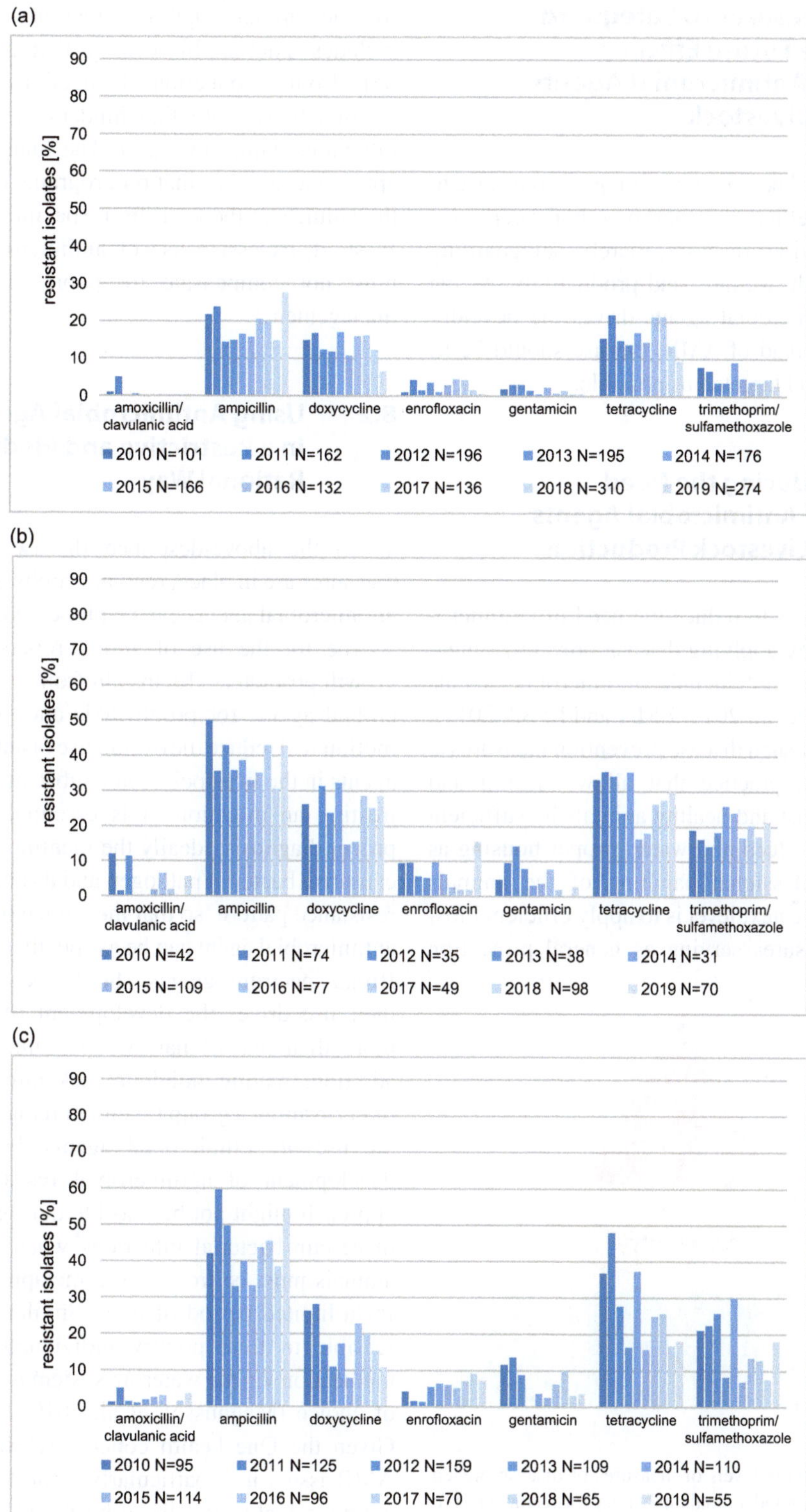


**Fig. 10** Percentages of resistance of (**a**) 1848 *E. coli* isolates from infections of layer chickens, (**b**) 557 *E. coli* from infections of broiler chickens, and (**c**) 998 *E. coli* from infections of turkeys in Germany during 2010–2019

# 8 Measures to Safeguard the Future Efficacy of Antimicrobial Agents in Livestock

As mentioned before, it is of major importance to protect the efficacy of antimicrobial agents for curing infectious diseases, thereby safeguarding animal health, welfare, and productivity. As the use of antimicrobial agents drives the development and spread of AMR, the use should be as restrictive and targeted as possible.

## 8.1 Reducing the Need for Antimicrobial Agents in Livestock Production

A key factor is to reduce the need of antimicrobial agents by applying disease preventive measures that do not include antimicrobial agents (Magnusson et al. 2019; EMA and EFSA 2017). The basis for such disease prevention measures is good animal practice that allows raising and keeping robust and healthy animals by sufficient and adequate feed and water, proper housing as well as professional handling of the animals (Fig. 11). The next level is to apply efficient biosecurity measures serving as general protection for the animals against infections. This level includes external biosecurity that protects the farm from introduction of infections as well as internal biosecurity that hinders the spread of infections within the farm. The final level is to apply relevant vaccination programs that protect the animals at the farm from specific pathogens. In short, excessive use of antimicrobial agents must not compensate for poor animal health management.

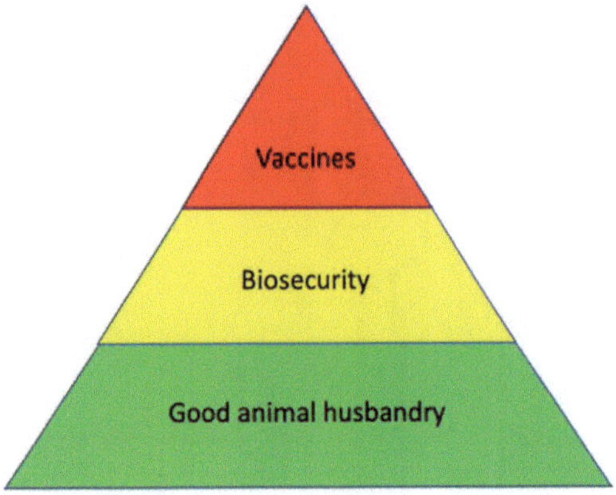


**Fig. 11** The three levels of measures of disease prevention on a farm: good animal husbandry forms the basis for disease prevention, effective biosecurity as a general protection from infectious agents, and vaccines targets specific pathogens

## 8.2 Using Antimicrobial Agents in a Restrictive and Medical Rational Way

When the above-described disease preventive measures are in place, regular prophylactic use of antimicrobial agents can be phased out. This also is true for the use of antimicrobial agents as growth promoters. To abstain from using antimicrobial agents for prophylaxis and growth promotion will reduce the overall use of antimicrobial agents in the livestock sector substantially. When treating an infection, it is essential to have a proper diagnosis, ideally the identification of the causative bacterial pathogen and its antimicrobial resistance pattern, so that the choice of a suitable antimicrobial agent can be as specific as possible. Broad-spectrum agents should be avoided as their use drives the development of resistance more than that of narrow spectrum agents. In addition, treating individual sick animals instead of performing a group treatment reduces not only the use of antimicrobial agents, but also the development of antimicrobial resistance. This approach might not be feasible during outbreaks of certain bacterial infections when a group of animals must be treated in a metaphylactic way for a limited period of time. Similarly, it is not feasible to treat poultry individually; instead, medication via the watering system is the method of choice (Magnusson et al. 2019; FAO 2020). Given the One health concept relevant for the AMR issue, it is particularly important that the antimicrobial agents listed by WHO as critically important for human medicine (WHO 2018) should not be used in animals whenever possible.

## References


CLSI (2020) Clinical and Laboratory Standards Institute. Performance standards for antimicrobial disk and dilution susceptibility tests for bacteria isolated from animals. VET01S-Ed5. CLSI, Wayne, PA

CLSI (2022) Clinical and Laboratory Standards Institute. Performance standards for antimicrobial susceptibility testing. M100 Ed 32. CLSI, Wayne

Dee SA (2021) Overview of respiratory diseases of pigs. https://www.msdvetmanual.com/respiratory-system/respiratory-diseases-of-pigs/overview-of-respiratory-diseases-of-pigs. Accessed 6 June 2022

Dubreuil JD, Isaacson RE, Schifferli DM (2016) Animal enterotoxigenic *Escherichia coli*. EcoSal Plus 7(1):https://doi.org/10.1128/ecosalplus.ESP-0006-2016

Duse A, Persson Waller K et al (2021) Microbial aetiology, antibiotic susceptibility and pathogen-specific risk factors for udder pathogens from clinical mastitis in dairy cows. Animals 11:2113

EMA (European Medicines Agency) (2021) Sales of veterinary antimicrobial agents in 31 European countries in 2019 and 2020 - trends from 2010 to 2020. Eleventh ESVAC report. https://doi.org/10.2809/636389.TC-AE-21-001-EN-N

EMA (European Medicines Agency) and EFSA (European Food Safety Authority) (2017) EMA and EFSA Joint Scientific Opinion on measures to reduce the need to use antimicrobial agents in animal husbandry in the European Union, and the resulting impacts on food safety (RONAFA). [EMA/CVMP/570771/2015]. EFSA J 15(1):4666, 245 pp. https://doi.org/10.2903/j.efsa.2017.4666

Entorf M, Feßler AT, Kadlec K et al (2014) Tylosin susceptibility of Staphylococci from bovine mastitis. Vet Microbiol 171(3–4):368–373. https://doi.org/10.1016/j.vetmic.2013.12.014

Entorf M, Feßler AT, Kaspar H et al (2016) Comparative erythromycin and tylosin susceptibility testing of streptococci from bovine mastitis. Vet Microbiol 194:36–42. https://doi.org/10.1016/j.vetmic.2015.12.003

European Parliament (2001) Directive 2001/82/EG. https://eur-lex.europa.eu/legal-content/SV/ALL/?uri=CELEX%3A32001L0082. Accessed 18 May 2022

FAO (2020) Tackling antimicrobial use and resistance in dairy cattle: lessons learned in Sweden. Rome. https://doi.org/10.4060/cb2201en. Accessed 20 May 2022

Feßler A, Scott C, Kadlec K et al (2010a) Characterization of methicillin-resistant *Staphylococcus aureus* ST398 from cases of bovine mastitis. J Antimicrob Chemother 65(4):619–625. https://doi.org/10.1093/jac/dkq021

Feßler AT, Billerbeck C, Kadlec K et al (2010b) Identification and characterization of methicillin-resistant coagulase-negative staphylococci from bovine mastitis. J Antimicrob Chemother 65(8):1576–1582. https://doi.org/10.1093/jac/dkq172

Feßler AT, Olde Riekerink RG, Rothkamp A et al (2012) Characterization of methicillin-resistant *Staphylococcus aureus* CC398 obtained from humans and animals on dairy farms. Vet Microbiol 160(1–2):77–84. https://doi.org/10.1016/j.vetmic.2012.05.005

GERMVET 2019 (2021) Resistenzsituation bei klinisch wichtigen tierpathogenen Bakterien. https://www.bvl.bund.de/SharedDocs/Berichte/07_Resistenzmonitoringstudie/Bericht_Resistenzmonitoring_2019.html. Accessed 14 June

Haenni M, Lupo A, Madec JY (2018) Antimicrobial resistance in *Streptococcus* spp. Microbiol Spectr 6(2). https://doi.org/10.1128/microbiolspec.ARBA-0008-2017

Hoelzer K, Wong N, Thomas J et al (2017) Antimicrobial drug use in food-producing animals and associated human health risks: what, and how strong, is the evidence? BMC Vet Res 13:211. https://doi.org/10.1186/s12917-017-1131-3

Kadlec K, Schwarz S (2018) Antimicrobial resistance in *Bordetella bronchiseptica*. Microbiol Spectr 6(4). https://doi.org/10.1128/microbiolspec.ARBA-0024-2017

Kadlec K, Wiegand I, Kehrenberg C et al (2007a) Studies on the mechanisms of beta-lactam resistance in *Bordetella bronchiseptica*. J Antimicrob Chemother 59(3):396–402. https://doi.org/10.1093/jac/dkl515

Kadlec K, Kehrenberg C, Schwarz S (2007b) Efflux-mediated resistance to florfenicol and/or chloramphenicol in *Bordetella bronchiseptica*: identification of a novel chloramphenicol exporter. J Antimicrob Chemother 59(2):191–196. https://doi.org/10.1093/jac/dkl498

Kadlec K, Watts JL, Schwarz S et al (2019) Plasmid-located extended-spectrum β-lactamase gene $bla_{ROB-2}$ in *Mannheimia haemolytica*. J Antimicrob Chemother 74(4):851–853. https://doi.org/10.1093/jac/dky515

Magnusson U (2022) Antimicrobial use and resistance in food-producing animals - how can we protect the efficacy of antibiotics for reproductive diseases? Reproduction in domestic animals = Zuchthygiene. Reprod Domest Anim 57(Suppl 4):13–20. https://doi.org/10.1111/rda.14170

Magnusson U, Sternberg S, Eklund G et al (2019) Prudent and efficient use of antimicrobials in pigs and poultry. FAO Animal Production and Health Manual 23. FAO, Rome http://www.fao.org/3/ca6729en/CA6729EN.pdf. Accessed 19 May 2022

Magnusson U, Moodley A, Osbjer K (2021) Antimicrobial resistance at the livestock-human interface: implications for veterinary services. Antimicrobial resistance at the livestock–human interface: implications for veterinary services. Revue Sci Tech 40(2):511–521. https://doi.org/10.20506/rst.40.2.3241

Merck vet manual (2022). https://www.merckvetmanual.com/respiratory-system/respiratory-diseases-of-cattle/bacterial-pneumonia-in-cattle. Accessed 26 Feb 2022

Michael GB, Eidam C, Kadlec K et al (2012) Increased MICs of gamithromycin and tildipirosin in the presence of the genes *erm*(42) and *msr*(E)-*mph*(E) for bovine Pasteurella multocida and Mannheimia haemolytica. J Antimicrob Chemother 67(6):1555–1557. https://doi.org/10.1093/jac/dks076

Michael GB, Kadlec K, Sweeney MT et al (2012a) ICE*Pmu1*, an integrative conjugative element (ICE) of *Pasteurella multocida*: analysis of the regions that comprise 12 antimicrobial resistance genes. J Antimicrob Chemother 67(1):84–90. https://doi.org/10.1093/jac/dkr406

Michael GB, Kadlec K, Sweeney MT et al (2012b) ICE*Pmu1*, an integrative conjugative element (ICE) of *Pasteurella multocida*: structure and transfer. J Antimicrob Chemother 67(1):91–100. https://doi.org/10.1093/jac/dkr411

Michael GB, Kaspar H, Siqueira AK et al (2017) Extended-spectrum β-lactamase (ESBL)-producing *Escherichia coli* isolates collected from diseased food-producing animals in the GERM-Vet monitoring program 2008-2014. Vet Microbiol 200:142–150. https://doi.org/10.1016/j.vetmic.2016.08.023

Michael GB, Bossé JT, Schwarz S (2018) Antimicrobial resistance in Pasteurellaceae of veterinary origin. Microbiol Spectr 6(3). https://doi.org/10.1128/microbiolspec.ARBA-0022-2017

Niemann L, Eichhorn I, Müller P et al (2019) Draft genome sequences of three porcine *Streptococcus suis* isolates which differ in their susceptibility to penicillin. Microbiol Resour Announc 8(9):e01711–e01718. https://doi.org/10.1128/MRA.01711-18

Nolan LK (2020) Colibacillosis in poultry. https://www.msdvetmanual.com/poultry/colibacillosis/colibacillosis-in-poultry. Accessed 6 June 2022

OIE (World Organisation for Animal Health) (2021) Annual report on antimicrobial agents intended for use in animals – fifth report. https://www.oie.int/app/uploads/2021/05/a-fifth-annual-report-amr.pdf. Accessed 30 March 2022

Price LB, Stegger M, Hasman H et al (2012) *Staphylococcus aureus* CC398: host adaptation and emergence of methicillin resistance in livestock. MBio 3(1):e00305–e00311. https://doi.org/10.1128/mBio.00305-11

Schink AK, Hanke D, Semmler T, et al. (2022) Novel multiresistance-mediating integrative and conjugative elements carrying unusual antimicrobial resistance genes in Mannheimia haemolytica and Pasteurella multocida. J Antimicrob Chemother. 77(7):2033-2035. doi:10.1093/jac/dkac116

Schwarz S (2008) Mechanisms of antimicrobial resistance in Pasteurellaceae. In: Kuhnert P, Christensen H (eds) Pasteurellaceae: biology, genomics and molecular aspects. Caister Academic Press, Norfolk, pp 199–228

Simoneit C, Burow E, Tenhagen BA et al (2015) Oral administration of antimicrobials increase antimicrobial resistance in E. coli from chicken--a systematic review. Prev Vet Med 118(1):1–7. https://doi.org/10.1016/j.prevetmed.2014.11.010

Swedres-Svarm 2018 (2019) Sales of antibiotics and occurrence of resistance in Sweden. Solna/Uppsala ISSN1650-6332. https://www.sva.se/media/0hihej1c/swedres-svarm-2019.pdf. Accessed 21 June 2022

Swedres-Svarm 2020 (2021) Sales of antibiotics and occurrence of resistance in Sweden. Solna/Uppsala ISSN1650-6332. https://www.sva.se/media/8d9678c390929e9/swedres_svarm_2020.pdf. Accessed 21 June 2022

Swedres-Svarm 2021 (2022) Sales of antibiotics and occurrence of resistance in Sweden. Solna/Uppsala ISSN1650-6332. https://www.sva.se/media/8da3e20fbc56137/swedres_svarm_2021.pdf. Accessed 21 June 2022

Tiseo K, Huber L, Gilbert M et al (2020) Global trends in antimicrobial use in food animals from 2017 to 2030. Antibiotics 9(12):918. https://doi.org/10.3390/antibiotics9120918

Tseng M, Fratamico PM, Manning SD et al (2014) Shiga toxin-producing *Escherichia coli* in swine: the public health perspective. Anim Health Res Rev 15(1):63–75. https://doi.org/10.1017/S1466252313000170

Watts JL, Sweeney MT (2010) Antimicrobial resistance in bovine respiratory disease pathogens: measures, trends, and impact on efficacy. Vet Clin North Am Food Anim Pract 26(1):79–88. https://doi.org/10.1016/j.cvfa.2009.10.009

Werner G, Fleige C, Fessler AT et al (2012) Improved identification including MALDI-TOF mass spectrometry analysis of group D streptococci from bovine mastitis and subsequent molecular characterization of corresponding *Enterococcus faecalis* and *Enterococcus faecium* isolates. Vet Microbiol 160(1–2, 162):–9. https://doi.org/10.1016/j.vetmic.2012.05.019

WHO (2018) Critically important antimicrobials for human medicine 6th revision 2018. https://apps.who.int/iris/bitstream/handle/10665/312266/9789241515528-eng.pdf. Accessed 20 May 2022

# Part II

# Metabolic Health Disorders

# Metabolism and Inflammation in Dairy Cows

M. Mezzetti, G. Carpenter, B. Bradford, and E. Trevisi

**Abstract**

During the transition to calving, dairy ruminants undergo physiological changes affecting several metabolic functions, including feeding behavior, nutrient flux and redistribution, hormonal patterns, and immune responses. Failure in those adaptations often induces a physiological imbalance condition, predisposing animals to developing several metabolic and infectious disorders at the onset of lactation.

The immune system exerts a pivotal role in allowing a proper adaptation of dairy ruminants to the transition period, as immune dysfunctions and chronic inflammatory conditions contribute to increasing the development of physiological imbalance condition. Several managerial practices could be adopted to improve dairy ruminants' adaptation to the transition period through improving the immune system function and mainly mitigating the inflammatory conditions. This chapter is aimed at (1) exploring physiological adaptations affecting metabolic functions of dairy ruminants while approaching calving, (2) understanding metabolic processes that are more likely affected by a physiological imbalance condition, (3) highlighting the role of the immune system in affecting a successful adaptation to the transition period, and (4) listing possible intervention aimed at improving such adaptation, mainly through modulating inflammation.

**Keywords**

Peripartum · Physiological imbalance · Immune dysfunction · Immune modulation · Dairy cows

## Acronyms

| | |
|---|---|
| AA | Amino acids |
| APP | Acute phase protein |
| BHB | Beta-hydroxybutyrate |
| C12:0 | Lauric acid |
| C14:0 | Myristic acid |
| C16:0 | Palmitic acid |

M. Mezzetti · E. Trevisi (✉)
Department of Animal Sciences, Food and Nutrition (DIANA), Facoltà di Scienze Agrarie, Alimentari e Ambientali, Università Cattolica del Sacro Cuore, Piacenza, Italy
e-mail: matteo.mezzetti@unicatt.it; erminio.trevisi@unicatt.it

G. Carpenter
Department of Animal Science, Iowa State University, Ames, IA, USA

B. Bradford
Department of Animal Science, Michigan State University, East Lansing, MI, USA
e-mail: bjbrad@msu.edu


J. J. Gross (ed.), *Production Diseases in Farm Animals*,
https://doi.org/10.1007/978-3-031-51788-4_4

| | |
|---|---|
| C18:0 | Stearic acid |
| C18:1n9 | Oleic acid |
| C18:2n6 | Linoleic acid |
| CCK | Cholecystokinin |
| COX | Cyclooxygenase |
| DHA | Docosahexaenoic acid |
| EPA | Eicosapentaenoic acid |
| ETE | Eicosatrienoic acid |
| GH | Growth hormone |
| GIT | Gastrointestinal tract |
| GPX | Glutathione peroxidase |
| IFN | Interferon |
| IGF-1 | Immune like growth factor |
| IL | Interleukin |
| LPS | Lipopolysaccharide |
| NDF | Neutral detergent fiber |
| NEB | Negative energy balance |
| NEFA | Non-esterified fatty acids |
| NFC | Non-fibrous carbohydrates |
| NF-KB | Nuclear factor-K beta |
| NPB | Negative protein balance |
| Nrf2 | E2-related factor 2 |
| OS | Oxidative stress |
| pe | Physical effective |
| PI | Physiological imbalance |
| PIC | Proinflammatory cytokines |
| PMN | Peripheral mononuclear cells |
| PON | Paraoxonase |
| PTH | Parathyroid hormone |
| PUFA | Polyunsaturated fatty acids |
| RNS | Reactive nitrogen species |
| ROS | Reactive oxygen species |
| SARA | Subacute rumen acidosis |
| SFA | Saturated fatty acid |
| SOD | Superoxide dismutase |
| TG | Triglycerides |
| TNF | Tumor necrosis factor |
| UFA | Unsaturated fatty acids |
| VFA | Volatile fatty acids |
| VLDL | Very low-density lipoproteins |

# 1 The Connection Between Metabolism, Immune Functions, and Inflammation

Sudden changes in body homeostasis are—to a certain extent—physiological during the transition to calving of dairy ruminants (Drackley 1999; Trevisi et al. 2011). Those changes affect several metabolic functions, including feeding behavior, nutrient flux and redistribution, hormonal patterns, and immune responses. Furthermore, the collective impacts of endocrine, neural, and immune signals must be integrated at the cellular level to determine impacts on metabolic pathways. These integrations occur (based on research in cell systems and model organisms) through competing stimulation and repression of specific gene transcription, protein translation, post-translational modifications, and mRNA and protein degradation. Although contribution of most of these mechanisms is poorly studied in cattle, a very active research community is increasingly demonstrating the inter-relationship among these phenomena across species. The objective of this chapter is to provide an overview of typical changes to these systems at the onset of lactation in dairy cattle and explore mechanistic links between seemingly unrelated shifts in physiology.

## 1.1 Classical View of Transition Metabolism: Nutrient Flux

### 1.1.1 Gastrointestinal Anatomy and Physiology

A massive morpho-physiological adaptation affects the whole gastrointestinal tract (GIT) of dairy ruminants while they approach calving. In the time course spanning from the 21 days before to the 21 days after calving, a substantial increase of the reticulorumen (12.25 vs 14.20 kg), liver (9.00 vs. 9.59 kg), small intestine (8.87 vs. 9.54 kg), and large intestine masses (4.87 vs. 5.94 kg) has been reported to occur in dairy cows (Reynolds et al. 2004). Furthermore, dairy cattle raised in conventional farms typically undergo three to four diet changes when transitioning from dry-off to the middle of the following lactation, implying a substantial alteration of nutrient concentrations in the diet (Table 1). Together, physical adaptations to changed conceptus size and diet changes progressively alter the feeding behavior (i.e., dry matter intake, rumination time; Fig. 1a, b) and the passage rate of digesta along the GIT of dairy cows approaching calving (Dirksen et al. 1985; Sordillo et al. 2009; Soriani et al. 2012; Calamari et al. 2014).

Altered nutrient flux resulting from these adaptations coincides with changes in the functional conformation of the GIT, as highlighted in several experiments (Dann et al. 2006; Zobel et al. 2015; Shi et al. 2020). Rumen papillae and gut microvilli decrease their length and rate of absorption after dry-off (Dijkstra et al. 2012; Dieho et al. 2016), gradually recovering their normal size and functional proprieties with the increased inclusion of fermentable carbohydrates in the early lactation diet (Dieho et al. 2016). These changes are paired with alterations in the rumen microbiota composition (Odensten et al. 2005; Calsamiglia et al. 2008; Minuti et al. 2015b), ruminal pH, and nutrient absorption efficiency (mainly for volatile fatty acids, VFA) (Hayirli et al. 2002; Grummer et al. 2004).

### 1.1.2 Energy and Protein Balance and Homeostasis Regulation

In dairy ruminants approaching calving, the nutrient redistribution shifts from homeostatic toward homeorhetic regulation. This is aimed at prioritizing nutrient requirements of the growing fetus and the mammary gland over those of the other metabolic functions. Three weeks prior to calving, nutrient requirements of the growing fetus reach maximal levels (Bell 1995; Dingwell et al. 2001) and, as soon as milk production starts, mammary gland induces a further increase in the energy, protein, and mineral requirements to start copious milk synthesis (Drackley 1999; Leroy et al. 2008). Nutrient needs cannot be satisfied from feed intake during the transition to calving, forcing dairy ruminants to rely heavily on their endogenous sources for satisfying their nutrient requirements, as reflected by body condition score trends (Fig. 1c).

**Glucose and Energy** Circulating glucose is the primary energy source for several key metabolic processes (Table 2). A ruminant's requirements for glucose are met almost exclusively through the diversion of different substrates (primarily propionate, but also amino acids [AA], glycerol, and lactate) to gluconeogenic processes occurring in the liver and, to a lesser extent, in the kidney cortex (Bell and Bauman 1997). The proportion of total glucose utilization that is insulin-dependent is only 10–30% in ruminants under normal physiological conditions (Bell and Bauman 1997). Glucose uptake by a ruminant tissues involves five classes of transporters (GLUT1, GLUT2, GLUT3, GLUT4, and GLUT5), and only one of these (GLUT4) is insulin-dependent (Sasaki 2002). GLUT4 is mainly expressed in peripheral tissues of ruminants, whose energy requirements are largely met by alternative energy sources (i.e., acetate, triglycerides [TG], non-esterified fatty acids [NEFA], and ketone bodies such as acetoacetate and 3-hydroxybutyrate) (Habel and Sundrum 2020). Conversely, tissues having a basal glucose requirement to sustain them (i.e., brain and other nervous tissue) rely mainly on insulin-independent transporters for glucose uptake (Sasaki 2002).

The homeorhetic regulation for glucose metabolism at the onset of lactation consists in a lowered expression of GLUT1 in peripheral tissues (i.e., adipose tissue) (Komatsu et al. 2005), paired with the reduction of the GLUT4 transporter's sensitivity to insulin-mediated stimuli. These changes impair the glucose uptake by peripheral tissues. At the same time, those tissues shift their metabolism from complete glucose oxidation toward lactate production, aimed at further saving circulating glucose (Habel and Sundrum 2020). Glucose uptake is then prioritized for tissues with insulin-independent transporters (i.e., GLUT1 and GLUT3), whose abundance increases substantially on the placenta and mammary gland cells. These tissues therefore become the two main sinks for circulating glucose in healthy dairy ruminants during the transition period. The most severe challenge for

**Table 1** Performances inputs, dietary recommendations, and sample diets for a medium-performance dairy cow undergoing the lactation, far-off, close-up, and fresh phases (adapted from NRC 2001)

| | Unit | Phase | | | |
|---|---|---|---|---|---|
| Item [a] | | Lactation | Far-off | Close-up [b] | Fresh |
| Inputs | | | | | |
| Days [c] | day | 90 | 240 | 270 | 11 |
| Body weight [d] | kg | 680 | 730 | 751 | 680 |
| BCS | – | 3 | 3.3 | 3.3 | 3.3 |
| Age | months | 49 | 57 | 58 | 58 |
| Milk production | kg | 35 | – | – | 35 |
| Butterfat | % | 3.5 | – | – | 3.5 |
| True protein | % | 3 | – | – | 3 |
| Lactose | % | 4.8 | – | – | 4.8 |
| Dry matter intake | kg | 23.6 | 14.4 | 13.7 | 15.6 |
| Daily weight change | kg | 0.3 | 0.67 | 0.67 | −1.6 |
| Days to gain one condition score | day | 316 | na | na | – |
| Days to lose one condition score | day | – | na | na | 55 |
| Dietary recommendations | | | | | |
| $NE_L$ | Mcal/day | 34.8 | 14 | 14.4 | 34.8 |
| $NE_L$ | Mcal/kg DM | 1.47 | 0.97 | 1.54–1.62 | 2.23 |
| MP | g/day | 2407 | 871 | 910 | 2157 |
| Diet MP | % | 10.2 | 6 | 6.6 | 13.8 |
| RDP | g/day | 2298 | 1114 | 1358 | 1634 |
| Diet RDP | % | 9.7 | 7.7 | 9.9 | 10.5 |
| RUP | g/day | 1291 | 317 | 172 | 1405 |
| Diet RUP | % | 5.5 | 2.2 | 1.3 | 9 |
| Diet NDF min | % | 25–33 | 33 | 33 | 25–33 |
| Diet ADF min | % | 17–21 | 21 | 21 | 17–21 |
| Diet NFC max | % | 36–44 | 42 | 43 | 36–44 |
| Absorbable Ca | g/day | 65 | 18.1 | 21.5 | 64 |
| Diet Ca | % | 0.61 | 0.44 | 0.45 (0.5–1.5) | 0.79 |
| Absorbable P | g/day | 56.5 | 19.9 | 20.3 | 49 |
| Diet P | % | 0.35 | 0.22 | 0.3–0.4 | 0.42 |
| Diet Mg [e] | % | 0.19 | 0.11 | 0.35–0.4 | 0.29 |
| Diet Cl | % | 0.26 | 0.13 | 0.15 (0.8–1.2) | 0.4 |
| Diet K | % | 1.04 | 0.51 | 0.52 | 1.24 |
| Diet Na | % | 0.23 | 0.1 | 0.1 | 0.34 |
| Diet S | % | 0.2 | 0.2 | 0.2 (0.3–0.4) | 0.2 |
| Diet Co | mg/kg DM | 0.11 | 0.11 | 0.11 | 0.11 |
| Diet Cu | mg/kg DM | 11 | 12 | 13 | 16 |
| Diet I | mg/kg DM | 0.5 | 0.4 | 0.4 | 0.77 |
| Diet Fe | mg/kg DM | 15 | 13 | 13 | 22 |
| Diet Mn | mg/kg DM | 14 | 16 | 18 | 21 |
| Diet Se | mg/kg DM | 0.3 | 0.3 | 0.3 | 0.3 |
| Diet Zn | mg/kg DM | 48 | 21 | 22 | 73 |
| Diet vitamin A | IU/kg DM | 3169 | 5576 | 6030 | 4795 |
| Diet vitamin D | IU/kg DM | 864 | 1520 | 1644 | 1308 |
| Diet vitamin E | IU/kg DM | 23 | 81 | 88 | 35 |
| DCAD | meq/kg DM | na | na | 10 (−75 to 0) | na |
| Sample diet (ingredients listed as kg/day DM) | | | | | |
| Corn silage, normal | – | 8.21 | – | 5.55 (5.40) | 36.44 |

(continued)

**Table 1** (continued)

| Item [a] | Unit | Phase | | | |
|---|---|---|---|---|---|
| | | Lactation | Far-off | Close-up [b] | Fresh |
| Grass silage, mid-maturity | – | – | 8.1 | 2.48 (2.42) | – |
| Legume forage silage, mid-maturity | – | 4.57 | – | – | – |
| Legume forage hay, immature | – | – | – | – | 20.17 |
| Grass hay, mid-maturity | – | 3.21 | – | – | – |
| Sugar beet pulp, dried | – | – | – | 2.15 (2.09) | – |
| Corn grain, steam flaked | – | 4.33 | – | – | 18.29 |
| Soybean meal, 48% CP | – | 1.62 | – | 0.79 (0.77) | 2.53 |
| Soybean meal, expellers | – | – | – | – | 7.65 |
| Blood meal, ring dried | – | – | – | – | 1.02 |
| Cottonseed, whole with lint | – | – | – | – | 8.41 |
| Calcium soaps of fatty acids | – | – | – | – | 0.65 |
| Vitamin and mineral premix | – | 0.49 | 0.02 | 0.43 (0.42) | 3.18 |
| Calcium carbonate | – | 0.07 | 0.46 | – | 0.56 |
| Calcium phosphate (di-) | – | – | – | 0.05 (0.03) | – |
| Calcium chloride | – | – | – | – (0.14) | – |
| Monosodium phosphate | – | 0.02 | – | – (0.07) | 0.4 |
| Sodium chloride | – | 0.011 | 5.79 | 0.03 | 0.7 |
| Magnesium oxide | – | – | – | 0.05 (0.03) | – |

[a] *BCS* body condition score, $NE_L$ net energy for lactation, *MP* metabolizable protein, *RDP* rumen degradable protein, *RUP* rumen undegradable protein, *NDF* neutral detergent fiber, *ADF* acid detergent fiber, *NFC* nonfibrous carbohydrates, *DCAD* dietary cation–anion difference (calculated as [(Na + K) – (Cl + S)]

[b] Values enclosed within brackets are referred to anionic close-up diets. na is "not available". –is "absent"

[c] Days in milk while lactating, pregnant while dry

[d] With conceptus while pregnant

[e] Provided as $MgSO_4$ or $MgCl_2$ during close-up and MgO during fresh phase

energy metabolism occurs at the onset of lactation in dairy cows. In fact, the expression of GLUT4 on lactocytes of early lactating cows decreases in favor of GLUT1, which exerts the highest affinity for glucose among this family of transporters (Komatsu et al. 2005; Mattmiller et al. 2011). The mammary glucose requirement of a dairy cow represents 80% of the whole-body supply at the onset of lactation, which is more than 2.5 times the glucose required by the gravid uterus in late pregnancy and four- to seven-fold greater than utero-fetal demand during the far-off period (Bickerstaffe et al. 1974; Bell and Bauman 1997). To face the sudden increase in glucose requirements, hepatic gluconeogenesis must abruptly double during the periparturient period, implying a proportional increase in gluconeogenic precursors required. Dietary sources (i.e., rumen-derived propionate and part of the AA absorbed along the GIT) may be inadequate to met the liver gluconeogenic capacity requirements in this phase, due to the physiological limitations hindering the GIT functions during the transition period. Thus, a negative energy balance condition (NEB) physiologically affects ruminants while adapting to the late gestation or to the early lactation phases (depending on the ruminant species considered). In Holstein dairy cows having an average milk yield of 45 kg/d, the severity of NEB has been quantified at >12 Mcal/d during the first 3 weeks of lactation (Rastani et al. 2001).

**Proteins** All the AA (except for leucine and lysine) can be utilized for gluconeogenesis, and as storage of carbohydrates in the body is limited, AA from protein turnover represent the key endogenous gluconeogenic substrate (Overton et al. 1999; Drackley et al. 2001). Additionally, AA are required for fetal growth, colostrogenesis, and milk synthesis (Table 2), boosting the mobilization of endogenous proteins and trigger-

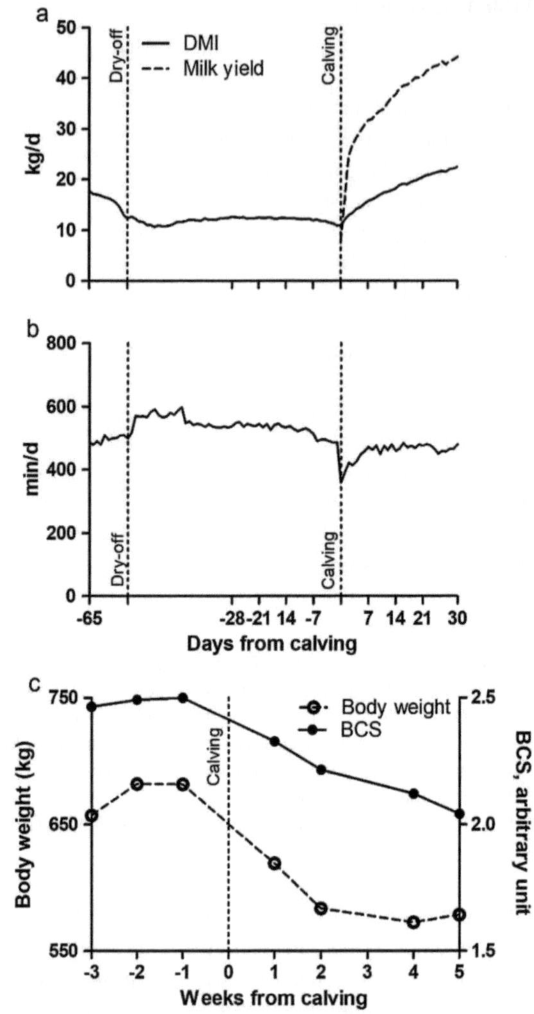


**Fig. 1** Trends of dry matter intake (DMI) and milk yield (MY) (**a**), rumination time (**b**), body weight and body condition score (BCS) (Agricultural Development and Advisory Service 1986) (**c**), detected in 41 healthy multiparous Holstein dairy cows housed in individual tied stalls under controlled environmental conditions (room temperature of 20 °C, relative humidity of 65%, 14 h of light) (Data adapted from Mezzetti et al. 2019a, b, c, 2020)

ing a negative protein balance condition (NPB) (Drackley et al. 2001; McCabe and Boerman 2020). Mobilized proteins come mostly from the skeletal muscle, where a 25% reduction in fiber diameter and a lowered anabolic activity have been reported in several ruminant species during the peripartum phase (Komaragiri and Erdman 1997; Bell and Bauman 1997; Doepel et al. 2006). Nevertheless, the skeletal muscle is low in branched chain AA, and insufficient dietary protein intake induces a massive protein mobilization to meet specific AA requirements for the gravid uterus and the mammary gland (Doepel et al. 2004; Kuhla et al. 2011). In dairy cows, the NPB condition could last up to 21 days after calving (Bell et al. 2000), and the amount of mobilized proteins ranges from 13% to 25% of total body protein (equal to up to 500 g of protein/d in an early lactating cow weighing 760 kg) depending on the protein intake (Grummer 1995; Komaragiri and Erdman 1997).

**Fats** Besides mobilization of proteins to provide gluconeogenic AA, NEB induces a massive mobilization of fatty acids from the adipose tissue. In dairy cows, up to 80 kg of fat mobilization has been documented from before calving to 12 weeks postpartum (McCabe and Boerman 2020). Within the adipocytes, TG are continuously broken down into their functional units (glycerol and NEFA) and resynthesized (Herdt

**Table 2** Nutrients and functional molecules affected by homeorhetic shift during the peripartum phase, body processes involved in increasing (+) or decreasing (−) their circulating concentrations, and main metabolic functions they are involved in

| Molecule | Body process | Function | Reference |
|---|---|---|---|
| Carbohydrates | | | |
| Glucose | Maintenance requirements (−) | Reduction of nicotinamide adenine dinucleotide phosphate through pentose phosphate pathway<br>Production of adenosine tri-phosphate | Habel and Sundrum (2020) |
| | Fetal tissues and placenta (−) | Essential for fetal growth and development. Maximal requirements are reached 3 weeks before parturition | Bell and Bauman (1997), Dingwell et al. (2001) and Marutsova and Binev (2020) |
| | Mammary gland (−) | Fundamental for mammary gland metabolism<br>Exerts an osmotic regulation on the milk volume through allowing lactose synthesis<br>Glucose-derived carbon is found in milk fat and protein | Linzell (1972), Chaiyabutr et al. (1980) and Rigout et al. (2002) |
| | Immune system (−) | Main energy source used by immune cells (drives proliferation, survival, differentiation, phagocytosis, and reactive oxygen species production in leukocytes) | Sordillo et al. (2009) and Ingvartsen and Moyes (2015) |
| Proteins | | | |
| Methionine | Fetal tissues and placenta (−) | Placental formation, fetal growth, and development | Zhou et al. (2016) |
| | Liver (−) | Methyl donor participating in the very low-density lipoproteins synthesis | Zhou et al. (2018) |
| Isoleucine, valine | Mammary gland (−) | Catabolic activities driving milk synthesis | Appuhamy et al. (2012) |
| Arginine | Mammary gland (−) | Catabolic activities driving milk synthesis | Appuhamy et al. (2012) |
| Alanine | Liver, mammary gland, immune system (−) | Together with glutamine is the main gluconeogenic amino acid | Drackley et al. (2001) |
| Leucine | Mammary gland (−) | Catabolic activities driving milk synthesis<br>Protein synthesis (mTOR regulation) | Appuhamy et al. (2012) |
| Leucine, lysine | Liver (−) | Ketogenesis | Drackley et al. (2001) |
| Isoleucine | Mammary gland (−) | Protein synthesis (mTOR regulation) | Appuhamy et al. (2012) |
| Glutamine | Liver, mammary gland, immune system (−) | Regulates gene transcription and translation in the mammary gland<br>Main gluconeogenic amino acid (together with alanine): up to 74% of circulating glutamine is completely oxidized for energy production in early lactating cows | Newsholme et al. (1985), Drackley et al. (2001) and Rezaei et al. (2016) |
| | Immune system (−) | The production of nicotinamide adenine dinucleotide phosphate during the respiratory burst is glutamine dependent | Newsholme et al. (1985) |
| Glutamate | Mammary gland (−) | Regulates gene transcription and translation | Rezaei et al. (2016) |
| Fats | | | |

(continued)

**Table 2** (continued)

| Molecule | Body process | Function | Reference |
|---|---|---|---|
| Triglycerides, non-esterified fatty acids, ketone bodies | Mammary gland (+) | Milk fat synthesis | Herdt (2000) |
| | Peripheral tissues (+) | Alternative energy sources | Herdt (2000) and Calder (2013) |
| | Immune system (+) | non esterified fatty acids are used as alternative energy sources (low efficiency) | Calder (2013) |
| Saturated fatty acids | Lipolysis (+) | C12:0; C14:0 and C16:0 are the main outcome of the lipolytic processes (>35% of total plasmatic long chain-fatty acids in dairy cows after calving) | Douglas et al. (2007) |
| Minerals | | | |
| Ca, P, Mg, K | Maintenance requirements (−) | Essential for nerves and muscle's function (direct action for Ca, P, and K and indirect action for Mg) | Goff (2006) |
| Ca, P | Mammary gland (−) | Colostrum and milk synthesis (2 g Ca and 1 g P/kg of produced milk in dairy cows) | Horst et al. (1997) and Goff (2006) |
| P | Maintenance requirements (−) | Structural component (bone and tooth, phospholipids, enzymes, DNA)<br>Muscle growth (0.3 kg P/kg body tissue in dairy cows)<br>Allows energy transportation (oxidative phosphorylation, glycolysis, glucose and triglycerides synthesis)<br>Metabolic buffer<br>Saliva (30–90 g P/day depending on chewing time and PTH status, in dairy cows) | NRC (2001) and Goff (2006) |
| P | Fetal tissues and placenta (−) | Fetal skeletal development (up to 10 g P/d depleted from the circulating pool in dairy cows) | Goff (2006) |
| K | Maintenance requirements (−) | Protein synthesis and carbohydrate metabolism (co-factor)<br>Intracellular osmotic and acid–base equilibrium<br>Permits nerves and muscle cell excitability (maintains cell membranes potential) | Goff (2006) |
| Oxidant species | | | |
| Reactive oxygen and nitrogen species | Whole body (+) (driven by fetal tissues and mammary gland metabolisms) | Produced in the mitochondria as a normal by-product of cellular respiration at the electron transport chain reaction (greatest acceleration occurs in the first 10 days of lactation in dairy cows) | Bell (1995), Bernabucci et al. (2002, 2005), Hodgkinson et al. (2007), Sordillo et al. (2007), Valko et al. (2007), Halliwell (2007), Sordillo and Aitken (2009) and Halliwell and Gutteridge (2015) |
| | Liver (+) | Produced in the peroxisome consequently to beta-oxidation (fatty acids metabolism) | Drackley (1999), Grum et al. (2002) and Schönfeld and Wojtczak (2008) |

(continued)

**Table 2** (continued)

| Molecule | Body process | Function | Reference |
|---|---|---|---|
| | Immune system (+) | During the respiratory burst, reactive oxygen species are produced by phagocytes from nicotinamide adenine dinucleotide phosphate oxidase system, nitric oxide is produced by macrophages as a cytotoxic agent and hypochlorous acid is produced by neutrophils through the myeloperoxidase activity, generating di-tyrosil residues from serum albumin.<br>During the oxylipids biosynthesis, lipid peroxide and superoxide anion by-product of the oxidative reactions of polyunsaturated fatty acids triggered by cyclooxygenase, lipoxygenase, or cytochrome P450 | Babior (1999), Dedon and Tannenbaum (2004), Finkel (2011), Raphael and Sordillo (2013), Bordignon et al. (2014), Brown and Griendling (2015) and Serhan (2017) |
| Antioxidant systems | | | |
| Exogenous | Maintenance, mammary gland, immune system (−) | High consumption rates (depletion to counteract oxidants load) compared to their low intake reduce plasma concentrations of serum-derived micronutrients (i.e., vitamins and minerals) with antioxidant properties | Bell (1995), Hodgkinson et al. (2007), Sordillo et al. (2007), Valko et al. (2007), Spears and Weiss (2008), Sordillo and Aitken (2009) and Sordillo and Mavangira (2014) |
| | Fetal tissues (−) | Increased requirements for vitamins and microminerals during growth | Spears and Weiss (2008) and Sordillo and Mavangira (2014) |
| Endogenous | Immune system, fetal tissues, and mammary gland (+) | Upregulated transcription and activity driven by increased circulating concentration of reactive oxygen and nitrogen species | Abuelo et al. (2015) |
| | Immune system (−) | Acute phase response reduces plasma cholesterol, high-density lipoproteins (which bound paraoxonase) and glutathione availability and decreases hepatic production of different carrier proteins binding antioxidant vitamins | Turk et al. (2005) and Abd Eldaim et al. (2010) |
| | Immune system (+) | Inflammation upregulates Nerf2 through the PEPK pathway, increasing the antioxidant enzymes activity (GSH-Px 1 and 2) | Gessner et al. (2013) |

2000). Adaptation to NEB decreases lipogenesis, as the low blood glucose reduces the availability of glycerol required for TG synthesis and endocrine signals drive a shift to a catabolic state. Conversely, increased lipolysis rate elevates the efflux of NEFA into blood and their availability to other tissues. Here, their usage as alternative energy sources reduces the glucose consumption. However, a large portion of mobilized NEFA is taken up by the liver and offered for two alternative metabolic fates. Firstly, liver NEFA can enter the mitochondria to be oxidized through β-oxidation and the Krebs cycle. As NEB becomes more severe, the lack of oxaloacetate (used for gluconeogenesis) induces an accumulation of acetyl-CoA in the liver. Thus, hepatocytes could be forced to switch acetyl-CoA utilization from complete (Krebs cycle) toward incomplete oxidation (ketogenesis), generating ketone bodies (i.e., acetone, acetoacetate, and beta-hydroxybutyrate [BHB]) (Ingvartsen 2006; Drackley et al. 2006). These are used as alternative energy sources in peripheral tissues. Alternatively, NEFA reaching the hepatocytes could remain in the cytosol to be re-esterified to TG. The latter could be sent back to the peripheral tissues via very low-density lipoproteins (VLDL) (Herdt 2000). The fate of the liver NEFA is related to their efficient transportation across the mitochondrial membrane which acts as a metabolic control for their diversion to the oxidation or re-esterification processes.

Besides altering the energy metabolism, lipolytic processes occurring under NEB conditions induce dramatic changes in the profile of fatty acids circulating in plasma, and these changes are rapidly reflected by the profile of fatty acids included in cell membranes. Palmitic (C16:0) and stearic acid (C18:0) are the main saturated fatty acids (SFA) present in the cell membranes of dairy ruminants (Contreras et al. 2010). Conversely, the profile of unsaturated fatty acids (UFAs) in cell membranes largely depends on dietary composition (Moallem 2018), and oleic (C18:1, n-9) and linoleic (C18:2, n-6) acids represent the two UFAs most commonly included in dairy ruminant diets (Contreras et al. 2010). In early lactation, the SFA concentrations increase in plasma as compared with mid-lactation, accounting for >35% of total plasma long chain-fatty acids after calving, while UFA account for the remainder (Drackley 1999; Douglas et al. 2007). Thus, the proportion of several SFA, including lauric (C12:0), myristic (C14:0), and palmitic (C 16:0), increases in cellular membranes (Drackley 1999; Douglas et al. 2007). Conversely, the amount of UFA, particularly polyunsaturated fatty acids (PUFA) ascribed to the n-3 series as eicosatrienoic acid (ETE; C20:3, n-3), eicosapentaenoic acid (EPA; C20:5, n-3), and docosahexaenoic acid (DHA; C22:6, n-3), decreases substantially in cellular membranes as compared with mid-lactation cows (Douglas et al. 2006; Sordillo et al. 2009).

### 1.1.3 Mineral Metabolism

Minerals are essential nutrients responsible for maintaining normal tissue functions and cellular homeostasis (Table 2). As described for energy and proteins, the metabolism of several minerals undergoes an homeorhetic regulation in dairy ruminants during their transition to lactation, aimed at prioritizing the requirements of the fetal tissues and the mammary gland over the other metabolic functions.

**Calcium** In dairy cows under homeostatic conditions, the circulating Ca pool is maintained at 3–4 g. Body Ca utilization of late gestating cows has been estimated at 18 g/d and depends on maintenance and fetal requirements (Table 2). At the onset of lactation, the galactopoietic process rapidly increases the daily Ca utilization rate to 55 g/d (Ramberg et al. 1970; Horst et al. 1997). Thus, the circulating Ca pool is rapidly exhausted by mammary gland requirements (Mulligan and Doherty 2008). When the circulating Ca concentration decreases, parathyroid glands immediately secrete parathyroid hormone (PTH) which increases the circulating Ca through combined actions on the kidneys and bones (Goff et al. 2014). For ensuring maximum efficiency in reestablishing Ca homeostasis, PTH requires blood pH to be around 7.35 (slightly below the typical homeostatic target of 7.4) and high blood Mg availability (>0.66 mmol/L in dairy cows) (Van De Braak et al. 1987; Goff et al. 1991). In these conditions, PTH immediately binds with its receptors on cell surfaces, stimulating the adenyl cyclase enzymatic complex to release cAMP, and activating metabolic functions aimed at increasing Ca in blood. At the kidney level, PTH immediately inhibits urinary Ca excretion, but in dairy cows this accounts for less than 1 g/d of Ca recovered by the circulating pool. Within 10–20 h from the decrease of the circulating Ca pool, PTH induces the kidneys to convert vitamin D to its active form (1,25-dihydroxivitamin D [1,25-(OH2)-D]). The 1,25-(OH2)-D stimulates the expression of Ca-binding proteins by the enterocytes, increasing intestinal Ca absorption within 12–24 h from the circulating Ca decrease, depending on blood pH. At the bone level, PTH stimulates Ca reabsorption from the bone fluid and induces Ca reabsorption by osteoclasts. In dairy cows, bone fluid reabsorption occurs within 6–12 h from PTH release and provides the circulating pool with a total of 9–15 g of Ca, while osteoclast reabsorption could require 16–96 h to be fully activated and allows 800–1200 g of Ca to be recovered from the bone marrow during the first 45 weeks of lactation.

**Phosphorus** Phosphorus availability undergoes a sudden decrease in dairy ruminants at their transition to calving. Phosphorus is essential for skeletal development of the growing fetus during late gestation, and milk itself drives a massive requirement at the onset of lactation (Table 2). Besides these factors, the reestablishment of Ca homeostasis contributes to lowering P availability. In fact, PTH action on parathyroids at parturition increases salivary P secretion two- to three-fold, decreasing its circulating pool (Wright et al. 1984).

**Magnesium** Despite the lack of any specific homeorhetic regulation on Mg metabolism, Mg availability during the transition period could be transiently reduced by the inadequate feed intake level as compared to body requirements, the sudden changes occurring in the diet composition and in the rumen pH. In fact, the reticulorumen is the main site for Mg absorption in adult dairy ruminants, while the small intestine is a net excretion site for this mineral (Schweigel and Martens 2000). The primary rumen mechanism for Mg absorption is a highly efficient Na-linked active transport. In dairy cows, this mechanism ensures animal survival when dietary Mg levels drop below 0.25% of diet dry matter. The secondary mechanism for Mg absorption operates when the Mg concentration in the rumen fluid is high (> 4 mM, which is equal to the rumen Mg concentration of dairy cows receiving a diet that is 0.35% Mg, Table 1). It consists of a passive transport allowing free Mg ions to flow down their concentration gradient into the extracellular fluid (Ram et al. 1998; Schweigel and Martens 2000). Primary absorption undergoes a drastic inhibition when rumen pH is alkaline (>6.5 in dairy cows), as those driven by high K or nonprotein nitrogen concentrations in the diet. Conversely, the secondary mechanism is not pH-dependent, and it allows the absorption of rumen-soluble Mg only (Goff 2004).

### 1.1.4 Redox Balance

Redox balance depends on the amount of circulating oxidant species and on the antioxidant capacity of the body to cope with their harmful effects on biological molecules (Sordillo 2016). The main physiological oxidant classes include reactive oxygen species (ROS), such as oxygen ions (i.e., superoxide anion), free radicals (i.e., hydroxyl radical), and hydroperoxides (i.e., hypochlorous acid, lipid, and hydrogen peroxides); and reactive nitrogen species (RNS; i.e., nitric oxide) (Reilly et al. 1991; Leprell et al. 2007; Cipak Gasparovic et al. 2017). These molecules are able to oxidize other molecules by reducing themselves, and they exert a pivotal role in cell communication processes, serving as messengers in signaling pathways (Abuelo et al. 2015). In the immune cells, they are involved in nuclear factor-κ beta (NF-κB) and mitogen-activated protein kinase pathways, regulating the production of cytokines, oxylipids, and immunoregulatory factors (Babior 1999; Dedon and Tannenbaum 2004; Finkel 2011; Bordignon et al. 2014; Brown and Griendling 2015; Serhan 2017). Antioxidants consist of a network of substances able to cope with oxidant species (Abuelo et al. 2015). Exogenous antioxidants include dietary molecules directly exerting an antioxidant activity (i.e., polyphenols) or serving as cofactors for endogenous antioxidant systems (vitamins and certain trace minerals, i.e., Se) (Dobbelaar et al. 2010). Endogenous antioxidants include enzymatic antioxidants (i.e., glutathione peroxidase [GPX] and superoxide dismutase [SOD]), non-enzymatic protein-antioxidants, and non-enzymatic low-molecular-weight antioxidants (Abuelo et al. 2015). Although, to the best of our knowledge, neither oxidant production nor antioxidant activities undergo a specific homeorhetic regulation during the transition to lactation, several metabolic activities occurring during this phase boost the release of oxidant species in the body (Table 2). To cope with increased oxidant species release, redox-sensitive transcription factors (i.e., E2-related factor 2 [Nrf2]) upregulate the expression of cytoprotective proteins and endogenous antioxidant systems (i.e., GPX) (Lubos et al. 2011), increasing the antioxidant capacity of the body and maintaining the redox balance homeostasis during the peripartum period.

## 1.2 Updated View of Transition Metabolism: Signaling Is a Key Factor in Lipolysis, Liver Metabolism, and Systemic Nutrient Availability

The classical view of transition cow metabolism—a paradigm in which nutrient flux in some cases strains the capacity of the primary metabolic pathway(s) related to the nutrient—has been revised in recent years. Although current scientists do not reject the discoveries made in the latter half of the twentieth century, there is a growing view that shifts in metabolic pathway flux may reflect more than a passive response to rapid changes in nutrient demand and availability. Rather, new insights into regulatory processes, involving both classic metabolic hormones and stress signaling agents, suggest that shifts in metabolic flux may be primarily due to programmed changes in enzyme activity, with substantial variability among animals. These insights are described below.

One key study that raised questions about the nutrient demand-oriented view of transition cow metabolism was carried out by Gross et al. (2011a). In this study, 50 cows were monitored through the transition to lactation, tracking energy balance as well as hormone and metabolite concentrations. On average, this cohort of cows experienced a net energy deficit of 42 MJ/d $NE_L$ over the first 3 weeks of lactation, accompanied by rapid lipolysis (indicated by plasma NEFA rising to a mean of 0.9 mM in week 2) and ketone production (plasma BHB reached a mean of 1.0 mM in week 3). The innovative aspect of this study was that the investigators then challenged the same cows with a low-energy diet and restricted feeding for 3 weeks beginning at 98 days in milk to attempt to match the NEB observed in the first weeks after calving; protein supply was also restricted to mimic postpartum NPB as well. The restriction was successful at recapitulating the NEB observed early postpartum; in fact, the deficit was greater, averaging 62 MJ/d $NE_L$ during the mid-lactation restriction. Furthermore, loss of body weight, body condition, and backfat thickness clearly showed that release of every store was, if anything, more rapid than in early lactation.

What were the metabolic impacts of this mid-lactation restriction (Gross et al. 2011a)? As might be expected, cows on the restriction treatment displayed lower plasma glucose and increased plasma NEFA and BHB concentrations. However, the limited magnitude of these changes relative to the postpartum period was remarkable. Plasma glucose declined by only 5% (compared to over 15% postpartum), whereas plasma NEFA and BHB peaked at just 0.3 mM and 0.64 mM, respectively, in mid-lactation. These responses challenge the nutrient flux-oriented explanation for metabolic disease in the transition to lactation. If the pathway bottlenecks in the bovine liver are to blame for maladaptive metabolism after calving, why were the same problems not observed with a similar nutrient imbalance in mid-lactation? Certainly, the mid-lactation cows showed a different response in terms of milk yield, with a 10% decline in milk yield and a 15% drop in milk protein yield compared to the rise in milk production during early lactation, in the face of nutrient deficiency. But the calculated energy balance accounted for this decline in output and the reduced milk yield cannot therefore account for the apparently mild metabolite shift in mid-lactation vs. early lactation energy deficit. This direct comparison study points to the possibility that metabolic control systems are inherently different in early vs. mid-lactation.

### 1.2.1 Role of Metabolic Hormones

The general metabolic responses to nutrient deficit are well defined—the animal must initiate some combination of increasing nutrient inputs (greater intake, consumption of more nutrient-dense feed, or improved digestibility), minimizing outputs (decreasing milk production or heat production), and utilizing stored nutrients (e.g., mobilization of adipose tissue and muscle). All of these processes are controlled at least in part by endocrine factors. The most obvious endocrine responses to short-term energy deficit are typically a decline in insulin secretion and increased glucagon secretion (de Boer et al. 1985). Secondary to the drop in insulin secretion, the normal insulin-like growth factor 1 (IGF-1) response to growth hormone (GH) diminishes; since IGF-1 provides a negative feedback on GH release, GH concentrations rise during period of nutrient deficit (Butler et al. 2003). Increased GH and glucagon concentrations, along with decreased insulin and IGF-1, promote a catabolic state enabling release of stored nutrients and processes like gluconeogenesis. Simultaneously, there is a substantial decline in circulating thyroid hormones, which may help to conserve energy substrates (Ehrhardt et al. 2016).

In addition to impacting concentrations of hormones from classical endocrine organs, nutrient deficiency induces endocrine responses in other organs as well. Low nutrient intake leads to diminished concentrations of gut-derived satiety factors like cholecystokinin (CCK) and gastrin (Samuelsson et al. 1996) and also promotes gastric release of the appetite-stimulating hormone ghrelin (Wertz-Lutz et al. 2006; Laeger et al. 2013). Even adipose-derived leptin, which is generally released in proportion to adipose stores, declines rapidly with short-term nutrient restriction (Gross et al. 2011b). Decreased gut peptides and leptin as well as increased ghrelin contribute to a stimulation of appetite that can help to resolve the nutrient deficit.

How do these endocrine control systems differ in early lactation vs. mid-lactation? First, changes in hormones that signal nutrient deficiency (ghrelin, GH) around feeding are far more dramatic in early lactation than in mid-lactation, even in the absence of any type of feed restriction; conversely, mid-lactation cows showed a much stronger insulin (anabolic) response to a meal (Bradford and Allen 2008). Energy status clearly impacts short-term endocrine responses to meals.

The study by Gross et al. (2011b) described above showed that mid-lactation cows subjected to NEB had much more subtle changes in endocrine profiles than they had in early lactation. Mid-lactation restriction increased GH while decreasing insulin, IGF-1, leptin, and thyroid hormones, but as with metabolite patterns, these changes were less than half the magnitude observed in early lactation. Endocrine control systems in early lactation appear to be altered more violently than can be explained based on nutrient balance alone.

### 1.2.2 Contribution of Neural Controls

Most research on the control of metabolism has traditionally focused on endocrine factors as the primary regulatory controls, but some findings in model species have provided evidence that the nervous system may be directly involved in regulating metabolic pathway flux, as well.

The most obvious point of control by the central nervous system is on feed intake. The hypothalamus is the central regulator of appetite and satiety, integrating peripheral signals related to nutrient availability, gut distention, and other factors such as immune system activation. The dominant appetite-suppressing effects of proinflammatory signaling reaching the hypothalamus may help to explain the poor feed intake of some cows in weeks after parturition (Kuhla 2020; Brown and Bradford 2021), as explained in detail in the reminder of this chapter.

Less familiar than the role of the central nervous system are the documented roles of the peripheral nervous system in regulating tissue metabolic function and endocrine signaling. One study demonstrated that systemic insulin resistance in response to glucocorticoids is dependent on intact hepatic afferent vagus nerve innervation (Bernal-Mizrachi et al. 2007). Other studies have documented the impact of signals from nerve cells on function of intestinal immune cells (Gabanyi et al. 2016) as well as diverse cell types in adipose tissue (Guilherme et al. 2019). Although little is known about ruminant-specific nervous system regulation of peripheral metabolic tissues, it is likely that similar mechanisms have a role in ruminant physiology.

## 1.3 On the Edge of Physiological Imbalance: What Is Adaptation and What Is Disease?

An exaggerated homeorhetic shift or, more likely, a maladaptation to changed metabolic fluxes and signaling changes often induces a physiological imbalance (PI) condition in dairy ruminants at their transition to lactation. Predisposing factors for developing a PI condition include severe and cumulative stressful events that occur during late pregnancy (i.e., inadequate dry-off procedures, psychologically stressful conditions resulting from overcrowding or uncomfortable equipment, heat stress, nutritional impairment, poor hygienic conditions) (Trevisi and Minuti 2018; Mezzetti et al. 2021). The establishment of PI conditions predisposes animals to developing several metabolic and infectious disorders. Thus, it is not surprising that 75% of disease incidence within dairy herds have been documented in the period extended between dry-off and the first month of lactation, with the maximum risk reported for the first 10 days after calving (Ingvartsen et al. 2003; LeBlanc et al. 2006).

### 1.3.1 Rumen-Driven PI

Physiological alterations in rumen pH as a result of changes in diet composition implemented while dairy ruminants approach calving (Table 1) require at least 4 weeks to allow the rumen milieu to adapt (Dirksen et al. 1985). Applying sudden changes in the diet composition (i.e., suddenly feeding a high starch lactation diet to animals coming from the far-off ration) could impair the animal capacity to cope with the short chain-fatty acids release in the rumen, predisposing them to developing severe pH drop and a rumen acidosis condition (Dijkstra et al. 2012). Rumen acidosis could be defined as acute or subacute (SARA) depending on the severity and duration of the pH drop and on the amount of lactic acid accumulated in the rumen fluid (Table 3). While acute acidosis primarily affects beef animals receiving high starch diets, the SARA condition has been commonly reported in dairy ruminants during early lactation (Enemark 2008). Despite that, SARA etiology has not been fully elucidated, and diet changes alone seems to be insufficient to fully account for its occurrence in dairy herds. This condition alters the rumen microbiota composition through increasing the abundance of Gram-positive bacteria and shifts the fermentation patterns in favor of propionic and butyric acids (Hibbard et al. 1995; Wang et al. 2012; Steele et al. 2015). Under this scenario, the epithelial structure of the rumen milieu could be damaged by the acidic pH, further impairing the rate of absorption of the rumen papillae and allowing the translocation of molecules with adverse effects into blood (Trevisi et al. 2014). Among these, histamine is known to act as a vasoactive compound, inducing secondary laminitis in dairy cows at the onset of lactation (Nocek 1997). Furthermore, rumen acidosis could reduce feed intake in early lactation and increase the severity of NEB conditions in this phase (Allen et al. 2009).

### 1.3.2 Energy-Driven PI

A PI condition could arise from the maladaptive response of dairy ruminants to the NEB conditions around calving. Excessive lipid deposition that occurs during the dry period and/or severe and prolonged NEB predispose the animal to developing a PI condition by increasing the magnitude and duration of lipid mobilization processes (Minuti et al. 2020). The outcome is a hypercatabolic response described as metabolic stress, characterized by a massive mobilization of NEFA. These exacerbate the severity of NEB conditions through exerting anorexic effects and further reducing feed intake (Ingvartsen and Andersen 2000; Allen et al. 2009; Sordillo and Raphael 2013; Habel and Sundrum 2020). Furthermore, during this intense lipomobilization, adipose tissue releases many signaling molecules (i.e., leptin, adiponectin, tumor necrosis factor [TNF]-α) which can affect liver metabolism (Fain 2010). This maladaptation could occur through at least two different metabolic mechanisms, predisposing dairy ruminants to a massive release of ketone bodies (Table 3) or rapid liver TG accumulation (Herdt 2000). These scenarios are further explored in the chapter by Gross et al.

**Table 3** Peripartum diseases preceded from a physiological imbalance condition, and biological fluids used to define their pathological thresholds in different ruminant species

| Disease | Biological fluid | Reference measure [a] | Animal[b] | | | Reference |
|---|---|---|---|---|---|---|
| | | | Cow | Ewe | Goat | |
| Subclinical rumen acidosis | Rumen fluid | pH (−) | 5.5–5 | | | Hibbard et al. (1995) |
| | | LA (mmol/L) | 5–10 | | | Hibbard et al. (1995) |
| Clinical rumen acidosis | Rumen fluid | pH (−) | <5 for longer than 24 h | | | |
| | | LA (mmol/L) | 11–300 | | | |
| Subclinical ketosis | Plasma | BHB (mmol/L) | >1.20 | >0.50 | >0.70 | Andrews et al. (1997), Duffield (2000) and Doré et al. (2015) |
| Clinical ketosis | Plasma | BHB (mmol/L) | >3.00 | >1.60 | >1.60 | Herdt et al. (1981), Andrews et al. (1997) and Doré et al. (2015) |
| Stage I hypocalcemia | Serum | Ca (mmol/L) | <1.92 | <1.92 | <1.92 | Oetzel (1988) and Goff (2006) |
| Stage II hypocalcemia | Serum | Ca (mmol/L) | <1.62 | <1.62 | <1.62 | Oetzel (1988) and Goff (2006) |
| Stage III hypocalcemia | Serum | Ca (mmol/L) | <0.87 | <0.87 | <0.87 | Oetzel (1988) and Goff (2006) |
| Stage I hypophosphatemia | Serum | P (mmol/L) | <1.20 | <1.20 | <1.20 | Oetzel (1988) and Goff (2006) |
| Stage II hypophosphatemia | Serum | P (mmol/L) | <0.60 | <0.60 | <0.60 | Oetzel (1988) and Goff (2006) |
| Stage I hypermagnesemia | Serum | Mg (mmol/L) | >1.23 | >1.23 | >1.23 | Oetzel (1988) |
| Stage II hypermagnesemia | Serum | Mg (mmol/L) | >2.05 | >2.05 | >2.05 | Oetzel (1988) |
| Stage I hypomagnesemia | Serum | Mg (mmol/L) | <0.78 within 12 h after calving | na | na | Goff (2006) |
| Stage II hypomagnesemia | Serum | Mg (mmol/L) | <0.66 | na | na | Goff (2006) |
| Hypokalemia | Serum | K (mmol/L) | <2.50 | na | na | Goff (2006) |

[a]Analytes and physical measures used as golden standard for disease diagnosis: *LA* lactic acid, *BHB beta-hydroxybutyrate*; [b]*na* not available

### 1.3.3 Mineral-Driven PI

A PI condition could affect the metabolism of several minerals, resulting in severe dysfunction in the cellular homeostasis of dairy ruminants at their transition to lactation. Mineral dysfunctions could reduce feed intake and productivity, increase the susceptibility of the animal to metabolic and infectious diseases, and induce nerve and muscle dysfunctions (Goff 2006). Severe dysfunctions occurring in Ca, P, Mg, and K metabolism commonly cause general weakness and impair the ability of the animal to rise to its feet, and the syndrome takes the name from the limiting mineral into the blood.

A PI condition affecting Ca metabolism commonly induces a hypocalcemic state, classified into three pathological stages depending on its severity (Table 3). These deficiencies could impair GIT motility and possibly evolve into a milk fever condition when the animal is unable to stand (Goff 1999, 2008). Pathological Ca deficiencies could occur when the reestablishment of the circulating Ca pool homeostasis is impaired. Three metabolic factors possibly contribute to impairing animals' capacity in reestablishing Ca homeostasis (Goff et al. 1991). (1) A reduced number of 1,25-(OH2)-D receptors impairs the Ca absorption at the gut level. (2) Reduced osteoclast population impairs Ca reabsorption from the bone marrow. These two factors increase the risk of developing a Ca-driven PI condition in older animals, as both the abundance of 1,25-(OH2)-D receptors on the surface of enterocytes and of osteoclasts in the bone marrow decrease substantially with aging (Goff 2008). (3) Alkaline pH in blood (higher than 7.45 in dairy cows) alters the conformation of PTH receptors, inhibiting the cAMP release consequential to the adenyl cyclase activation and preventing increased uptake of Ca from the gut and bone marrow (Bushinsky 1996). The latter condition occurs when dietary cations (i.e., K, Na, Ca, and Mg) exceed the dietary anion supply (Cl, $SO_4$, and $PO_4$) (Goff 2004). Commonly, excessive K loads in forages (i.e., higher than 1.4% on DM basis) is the main factor altering the dietary anion–cation difference and predisposing dairy cows to developing a hypocalcemic state.

Under hypocalcemia, urinary P losses rise, and the increased amount of P in saliva as a result of the PTH action could become unavailable to the animal due to the impaired GIT motility (Goff 2006). Thus, animals undergoing a hypocalcemic state are also likely to develop secondary hypophosphatemia (Table 3), contributing to impaired mobility. In some cases, the ability of the animal to rise to its feet does not recover once the Ca homeostasis is reestablished, as the circulating P pool remains low. Hypophosphatemic downer syndrome has been documented in animals receiving adequate dietary P levels, suggesting lowered P to be subsequent to the hypocalcemic state rather than the primary cause of this disease (Goff 2000, 2006). Despite this, the reason that circulating P pool remains low even after reestablished Ca homeostasis has not been fully elucidated (Goff 2006).

A PI condition affecting Mg metabolism could result in hypermagnesemic or hypomagnesemic state (Table 3). Hypermagnesemia has been commonly reported in dairy cows developing a milk fever (hypocalcemia + hypophosphatemia) state. It probably results from the combined action of a compensatory mechanism aimed at maintaining the systemic anion–cation balance when circulating P is depleted, and of an impaired renal Mg excretion driven by decreased renal blood flow during hypocalcemia (Oetzel 1988). In contrast, hypomagnesemia is commonly driven by dysfunctional rumen Mg absorption. This could occur when excessive amounts of cations (i.e., K, Na, Ca, and Mg) or nonprotein nitrogen are included in the diet, elevating ruminal pH and depressing primary Mg absorption (Goff 2004). Besides exerting deleterious effects on feed intake and depressing rumen fermentation (Ammerman et al. 1971), hypomagnesemia predisposes dairy ruminants toward secondary hypocalcemic states, named tetany. Low circulating Mg directly impairs PTH release by parathyroids and induces a dysfunctional PTH–receptor interaction (Rude et al. 1978). The latter is consequential to the co-factor role exerted by Mg during the PTH–receptor interaction and results in an alkaline-like condition blocking the mechanisms aimed at Ca recovery when Mg availability is low. Under this scenario, a hyperphosphatemic state could occur as a compensatory mechanism to preserve systemic anion—cation balance with low circulating Mg (Oetzel 1988).

A less common Ca-independent downer cause could be the hypokalemia-driven flaccid paralysis, occurring when the intra- or extracellular K concentration decreases substantially, disrupting normal cell membrane potential and impairing several metabolic functions (Table 2). In dairy cows, extracellular K ranges between 3.9 and 5.8 mmol/L, while intracellular K is 150–160 mmol/L. Factors depleting intracellular K includes prolonged fasting conditions and exaggerated renal excretion. In fasting conditions (i.e., those following a clinical ketosis status lasting longer than 5 days), the muscles release K to support normokalaemia. If this K is being rapidly excreted by the kidney, the muscle K pool will be quickly depleted, and severe hypokalemia can ensue (Goff 2006). In addition, also severe stressful conditions are signaled to cause the depletion of plasma K (Trevisi and Bertoni 2009).

### 1.3.4 Oxidative Stress: Another Form of PI

Insufficient antioxidant activity to counterbalance oxidant species load induces a metabolic stress condition named oxidative stress (OS) (Bionaz et al. 2007; Celi 2010, 2011; Celi and Gabai 2015). Several metabolic processes intersect in developing such a PI condition in dairy ruminants at their transition to lactation, both through boosting the release of oxidant species and reducing the antioxidant availability (Table 2). Recently, it has also been hypothesized that an excessive supplementation of dietary antioxidant could act as a putative cause for OS through impairing endogenous antioxidant function (Rizzo et al. 2013; Abuelo et al. 2015). In fact, high amounts of exogenous antioxidants administered through the diet might transiently lower the body oxidant load, suppressing the redox-dependent Nrf2-mediated transcription and lowering the expression of endogenous antioxidant enzymes (Abuelo et al. 2015). OS directly contributes to lipolysis through boosting the metabolic demand, acting as a putative cause for NEB-related diseases (i.e., ketosis and fatty liver) (Sordillo and Raphael 2013) and contributes to the dysfunction of the vascular endothelium in peripartum ruminants (Halliwell and Whiteman 2004; Lacetera et al. 2005; Sordillo et al. 2007). Finally, several organelles and tissues (i.e., DNA, proteins, and lipids) are damaged by uncontrolled oxidant activities occurring when the redox balance is lost (Sordillo and Aitken 2009; Sordillo 2016). These OS damages could detrimentally affect the reproductive tract, inducing ovarian diseases and reducing reproductive performances (Rizzo et al. 2012).

## 1.4 The Role of the Immune System in Metabolic Adaptation to the Transition Period

The field of endocrinology largely developed in isolation from immunology. As such, when immunologists began describing soluble factors that were released by activated immune cells, and which in turn influenced other cells, these "cytokines" were not generally categorized as endocrine factors. This helped to maintain early thinking that cytokines were primarily involved in crosstalk with the immune system rather than with parenchymal tissue. Over the past 30 years, however, these walls have largely been eliminated. Cytokines—and other signals generated by immune cells—are now known to influence essentially all organ systems.

Immune signals include a variety of soluble proteins, including cytokines and chemokines, as well as oxylipids. Non-immune cells can make small quantities of these signals as well, but systemically relevant concentrations of these signals are typically produced by immune cells.

Among the earliest studies of cytokine effects on metabolism in lactating dairy cattle was a report on TNFα. Kushibiki et al. (2003) found that subcutaneous administration of TNFα for a week caused a number of changes that mimicked profiles during the transition period, including increased plasma cortisol, GH, and NEFA, and decreased plasma triiodothyronine and IGF-1 concentrations. Some of these effects were likely mediated by a 34% decline in dry matter intake during treatment. Nonetheless, this study clearly demonstrated that immune signals can dramatically influence metabolic control systems in lactating dairy cattle. TNF-α is a proinflammatory cytokine (PIC); however, it is possible that immune signals that are less involved in inflammation may also influence metabolism of early lactating cows. Besides these direct effects sorted by immune mediators on metabolism, a wide literature documented side effects sorted by an activated immune system to have a main role in the adaptation of dairy ruminants to the peripartum challenges. In this part of the chapter, we will explore alterations affecting the innate immune system of dairy cows while approaching calving and we will describe their contribution in affecting the successful adaptation of the animal to the new lactation.

### 1.4.1 Inflammation: A Key Signaling Factor Influencing Metabolic Pathways

During infections such as mastitis or metritis, immune cells in the body recognize invading pathogens and become activated. When the infection is caused by Gram-negative bacteria, lipopolysaccharide (LPS) released by the bacteria also activates immune cells. The activation of local and systemic host defense mechanisms requires crosstalk between numerous types of immune cells, and one component of this response is inflammation. The host of signaling molecules released by activated immune cells includes inflammatory mediators such as nitric oxide, prostaglandins, and cytokines. While many of these molecules promote local inflammation and increased blood flow to the infected tissue, PICs play a key role in stimulating systemic inflammatory responses, including increased body temperature, increased heart rate, and decreased feed intake (Dantzer and Kelley 2007). Key PICs include TNF-α, interleukin (IL)-1β, and IL-6; these cytokines act through many of the same signaling cascades and often produce similar responses in cells.

One effect of PICs is to activate the acute phase response. During the acute phase, the liver shifts protean synthesis in favor of positive acute phase proteins (APP) (Cappa et al. 1989; Trevisi et al. 2005; Hotamisligil 2006; Loor et al. 2007; Bertoni et al. 2008). Positive APPs are generally found in very low abundance in the bloodstream but are greatly elevated during systemic activation of the immune system. These proteins encompasses α-globulins as serum amyloid alpha, haptoglobin, ceruloplasmin, orosomucoid, and C-reactive protein (Bertoni and Trevisi 2013) (Fleck 1989; Ceciliani et al. 2012). Synthesis of positive APPs drives severe reduction in the ability of the liver to synthesize other protein classes. The resultant is a reduced bloodstream concentration of albumin, retinol binding protein, paraoxonase (PON), and lipoproteins, so named negative APPs (Schreiber et al. 1982; Bertoni et al. 2008). The importance of APP shifts in the response to infection is somewhat unclear, but they have gained widespread acceptance as markers of inflammation.

### 1.4.2 The Innate Immune Response Around Calving: Acute vs. Chronic Inflammations and Changes in the Leukocyte Function

In ruminants, an acute systemic inflammatory state is a physiological consequence of events related to parturition (i.e., fatigue, psychological stress, tissue trauma, altered epithelial permeability, uterine infections). Damage-associated molecules, such as those that might be released by tissue traumatized by a difficult calving, can induce inflammation and sickness behavior even if there is no active infection. Likewise, translocation of pathogen-associated molecules across the GIT barrier can induce inflammation even if no viable bacteria have entered the body proper. The establishment of a certain degree of systemic inflammatory state is involved in maintaining normal body functions in this phase. For example, Nrf2 is a transcription factor aimed at counteracting proinflammatory signaling pathways (Kim et al. 2010), but its activation is also involved in upregulating several target genes with antioxidative properties. Its activation is normally driven by increased amounts of circulating oxidants (Nair et al. 2008). In dairy cows at the onset of lactation, the unfolded protein response in liver (driven by inflammation) contributes to activating Nrf2 via the PERK pathway (Cullinan and Diehl 2006; Gessner et al. 2014). Thus, inflammation-driven Nrf2 activation has been hypothesized to contribute to the ability of the animal to cope with increased amounts of oxidants produced in early lactation, preventing the establishment of an OS condition (Gessner et al. 2013; Abuelo et al. 2015).

Despite the metabolic relevance of acute inflammation in assuring successful adaptation to the transition period, dairy ruminants are prone to developing chronic inflammation in this physiological phase (Cappa et al. 1989). This is partially driven by the altered immune competence affecting innate immune cells during the transition period (Goff and Horst 1997). The impairment of the immune system is driven by a transcriptomic change in polymorphonuclear (PMN) cells between −1 and 2 weeks from calving that negatively affects viability, survival, and sensitivity to proinflammatory mediators as well as their chemotaxis, phagocytosis, respiratory burst, and antioxidant systems (Cai et al. 1994; Kimura et al. 1999; Mehrzad et al. 2001; O'Boyle et al. 2012; Crookenden et al. 2016, 2017). The resultant is a hypo-responsive state driven by a delayed migration of hypofunctional neutrophils (Sordillo 2016) and predisposing dairy ruminants at developing immune dysfunction-related diseases (i.e., retained fetal membranes, uterine infections and mastitis) (Herdt 2000; Kankofer 2001; Sordillo and Raphael 2013; Minuti et al. 2015a; Abuelo et al. 2015). The impairment of PMNs functions is accompanied by the increased responsiveness of monocytes to inflammatory stimulants in peripartum cows, resulting in greater secretion of inflammatory cytokines when stimulated (Sordillo et al. 1995). Thus, it is clear that any disease occurring during the transition period could predispose dairy ruminants at developing chronic inflammatory conditions and metabolic disorders at the onset of lactation (Trevisi et al. 2011; Trevisi and Minuti 2018). For example, mammary and uterine infections result in both local and systemic inflammation, mastitis results in release of LPS into the bloodstream and increased plasma concentrations of cytokines and positive APP (Hoeben et al. 2000), retained placenta and metritis are associated with an acute phase response, as elevated plasma haptoglobin has been documented in transition cows prior to clinical signs of metritis (Trevisi et al. 2008; Huzzey et al. 2009).

### 1.4.3 Causes of Changed Leukocytes Functions and Chronic Inflammations in Transition Cows

Proper function of the innate immune system is essential to ensure protection against pathogens and to avoid the occurrence of uncontrolled inflammation. Thus, altered leukocyte function represents a primary factor predisposing dairy ruminants to developing a chronic inflammatory state while they are approaching calving. Although there is a vast amount of scientific literature regarding this topic, the cause of changed leukocytes function in dairy ruminants approaching calving remains unclear. The sudden metabolic changes driven by physiological adaptation to lactation that occur during this phase make it challenging to identify the driving mechanism of immune alterations. It is very likely that there is not a single cause, but multiple factors are at play (Mezzetti et al. 2020), and the contribution of each of these factors is boosted under PI conditions, when the magnitude of physiological changes accompanying the parturition process exceed the control of the body mechanisms.

**Immunotolerance** Altered leukocyte function could partially arise from the normal maternal immunotolerance against the allogeneic conceptus during the late gestation period (Esposito et al. 2014). Other than fluctuations in steroid hormones, immunotolerance is also triggered by specific immune cells (i.e., M2 macrophages, γδ T-cells, and T regulatory cells) that can inhibit inflammation and T-cell responses (Majewski and Hansen 2002; Padua et al. 2005; Esposito et al. 2014). In particular, M2 macrophages are involved in inhibition of inflammatory conditions that could increase the risk of developing uterine infections (Esposito et al. 2014).

**Dysfunctional Activation of Dendritic Cells** Excessive levels of circulating PICs prior to calving could induce the establishment of chronic inflammation around calving through impairing the capacity of the adaptive immune system to intervene during inflammatory processes, delaying the recovery of normal function of tissues (Jahan et al. 2015; Trevisi et al. 2015). This could arise from the dysfunctional activation of dendritic cells driven by PIC, which is known to block the antiphlogistic role of T-lymphocytes (Joffre et al. 2009). This hypothesis is consistent with data by Nightingale et al. (2015), who found a lower production of interferon (IFN)-γ by whole blood cells after an LPS stimulation in cows having high haptoglobin levels after calving, reflecting an impairment of T-lymphocytes and adaptive immunity, and suggesting the interaction between the innate and adaptive immune systems was reduced as compared with cows having low haptoglobin levels after calving (Nightingale et al. 2015). Even though rare clinical signs are reported, several events possibly contribute to activating the immune system before calving.

**Hormonal Trends** A direct relationship with immune functions has been well documented for most of the hormones fluctuating during the peripartum phase of dairy ruminants. Steroidal changes occurring around calving (i.e., increase of estradiol and reduction of progesterone) decrease functional capabilities of lymphocytes and PMN and reduce the capacity of these cells to face bacterial infections (Roth et al. 1982; Davis 1998; Drackley et al. 2005; Lamote et al. 2006; Kelley et al. 2007; Sordillo 2016). Besides these physiological changes in hormonal patterns related to parturition, the occurrence of stressful conditions during the peripartum phase could boost the release of stress-related hormones (i.e., glucocorticoids and catecholamines). Among these, cortisol is known for impairing PMN chemotaxis and diapedesis through inducing a downregulation of selectin-L and cluster of differentiation-18 adhesion molecules (Burton et al. 1995; Drackley 1999; Drackley et al. 2005; Bertulat et al. 2013; Putman et al. 2018). Furthermore, cortisol, epinephrine, and norepinephrine are known for altering cytokines production through interactions with the NF-κB and cAMP-RBP networks (Higuchi et al. 1994; Burton et al. 1995; Drackley 1999; Padgett and Glaser 2003; Do Nascimento et al. 2004).

Cortisol is a dualistic regulator of inflammation, serving both as a pro- and anti-inflammatory mediator depending on the evolution of its circulating levels over time (Yeager et al. 2011). It could affect the hypothalamus, depressing the release of PIC (Irwin and Cole 2011). It can also increase the severity of systemic inflammations, as its release stimulates haptoglobin production in parenchymal liver cells (Putman et al. 2018). Thus, the occurrence of stressing events inducing the release of huge amounts of cortisol around calving could predispose animals at developing exaggerated inflammatory conditions through boosting the severity of the acute phase response.

Although these fluctuations occurring around calving could partly contribute to determining immune dysfunctions, hormonal trends do not overlap with the entire duration of the peripartum immune alterations. Experiments conducted on mastectomized pregnant cows highlighted that these animals face a shorter and less marked immune suppression status during calving as compared with animals with mammary glands (Nonnecke et al. 1993; Kimura et al. 1999, 2002). These results suggest that the largest cause of reduced immune competence around calving could be the homeorhetic regulation to support lactation, and that most of the hormones fluctuating during the peripartum phase (i.e., increasing catecholamine, cortisol, estrogen, and GH and decreasing IGF-1 and leptin) indirectly contribute to altering the immune functions through reducing the availability of nutrients to the immune cells (Mezzetti et al. 2020).

**Altered Nutrient Availability** Studies inducing early lactation-like NEB and NPB conditions through feed restriction on mid-lactation cows (without the endocrine alteration related to calving) revealed that lack of nutrients alone affected the transcription of genes important for immune competence in PMN. Those included the downregulation of genes encoding for defensins, cytokines, antigen presentation, respiratory burst, and inflammatory response (Moyes et al. 2010; Stevens et al. 2011; Crookenden et al. 2017). The lack of nutrients also represses the expression of genes involved in the inflammatory response and chemotaxis in PMN during intramammary infection with *Streptococcus uberis* (Moyes et al. 2010). At the same time, greater mRNA expression of anti-inflammatory genes occurred, indicating a greater likelihood of developing additional infectious diseases (Moyes et al. 2010) or in establishing chronic pathological conditions. The deficiency of energy sources, AA and minerals that are essential for immune cell function, the poor efficiency of those cells in utilizing alternative energy sources (Sordillo 2016), and the direct toxic effects of certain metabolites at high concentration could account for these results and partially explain the reduced immune competence reported during the transition to lactation (Calder 2013).

Glucose is the main energy source used by immune cells (Sordillo et al. 2009). When glucose availability is scarce, macrophages and PMN decrease their viability, proliferation, differentiation, chemotaxis, and phagocytosis (Barghouthi et al. 1995; Gamelli et al. 1996; Pithon-Curi et al. 2004). Furthermore, proliferation and secretory activities of lymphocytes are glucose-dependent, and their mitogen-induced activation is reduced under hypoglycemia (Sordillo 2016).

Glutamine has a high oxidation efficiency in immune cells. It drives several immune functions and could be used as an alternative energy source when glucose availability is low. In macrophages and PMN, phagocytosis and production of IL-1β and IL-6 are impaired when glutamine availability is low (Wallace and Keast 1992; Ogle et al. 1994; Yassad et al. 1997). Reduced glutamine availability impairs intracellular signaling through inhibiting the formation of nitric oxide by macrophages and that of superoxide anion by PMN. Finally, glutamine is an essential anabolic precursor in purine and pyrimidine synthesis, impairing lymphocytes replication when its availability is low (Newsholme et al. 1985, 1999; Pithon-Curi et al. 2004).

Leukocytes have a low efficiency in utilizing NEB-related metabolites as alternative energy sources, and both NEFA and ketones have been shown to inhibit leukocyte functions (or even compromising immune cell viability) when their concentration is relatively high (Calder 2013; Ingvartsen and Moyes 2015). Studies performed both in vitro and in vivo demonstrated that high circulating NEFA could inhibit phagocytosis and DNA synthesis in PMN, decrease diapedesis, impair secretion of IgM, TNF, and IL-6 in monocytes, and alter the proliferation of lymphocytes and their secretion of IFN-γ (Lacetera et al. 2004). High circulating ketone bodies decrease chemotaxis, respiratory burst, and superoxide anion production in macrophages and PMN (Calder et al. 1990; Hoeben et al. 1997; Sartorelli et al. 1999; Suriyasathaporn et al. 1999) and decrease blastogenesis, mitogenic responses, and IgM production in lymphocytes (Targowski and Klucinski 1983; Nonnecke et al. 1992; Takeuchi and Akira 2010). Furthermore, NEB-related metabolites induce the shift of the immune cell populations over a proinflammatory status, as widely documented by the Drackley's group after exposing dairy cows to a prolonged excess of energy during the dry period (Janovick Guretzky et al. 2007; Janovick and Drackley 2010). Cows overfeed during the dry period developed metabolic alterations supporting the establishment of a low-grade systemic inflammation in early lactation (Janovick-Guretzky et al. 2023), similar to the "metaflammation" state reported in humans which is associated with obesity and chronic diseases (Egger and Dixon 2009).

**Altered SFA/UFA Ratio in Cells Membrane** Increased SFA/UFA ratio in cell membranes due to a large mobilization of NEFA may have a role in altering the immune cell functions (Calder 2006; Raphael and Sordillo 2013). In epithelial cells, reduced UFA abundance decreases IL-8 release, impairing PMN chemotaxis and diapedesis. On the immune cell membranes, SFA can covalently modify proteins through fatty acylation, altering membrane fluidity, influencing how proteins anchor to plasma membrane, and affecting the formation of glycoproteins integral to lipid rafts. Altered lipid raft conformation inhibits CD14 functioning and LPS-induced TNF production by lymphocytes, affecting immune cells activation, antibody production, and inflammatory processes (Drackley 1999; Raphael and Sordillo 2013; Sordillo 2016). In macrophages, increased cell membrane SFA/UFA ratio decreases the secretion of IgM, IFN-γ, TNF, and IL-6 and alters the production of ROS (Lee et al. 2003, 2004; Shi et al. 2006; Scalia et al. 2006; Sordillo and Aitken 2009; Esposito et al. 2014).

In PMN, free SFA including C12:0, C14:0, and C16:0 are structurally similar to the lipid A associated with bacterial LPS, and they can activate NF-κB mediated gene expression driven by pathogen-associated molecular patterns, increasing inflammation and respiratory burst activity (Lee et al. 2003, 2004; Scalia et al. 2006; Shi et al. 2006; Sordillo et al. 2009; Esposito et al. 2014). Conversely, UFAs could act both as pro- and anti-inflammatory factors through modulating NF-κB activation (Lee et al. 2003, 2004, 2010; Scalia et al. 2006; Gorjão et al. 2006; Moallem 2018).

Reduced UFA abundance in immune cell membranes also modifies the production of oxylipids that act as immune mediators (Calder 2006; Raphael and Sordillo 2013). Pro-resolving oxylipids, primarily derived from n-3 PUFAs, are known for promoting "resolution of inflammation, clearance of microbes, reduction of pain, and tissue regeneration via novel mechanisms" (Serhan 2017). Shifted oxylipid profiles could thus predispose dairy ruminants to developing systemic inflammatory conditions. Reduced abundance of long chained n-3 PUFAs reduces the expression of mRNA for anti-inflammatory cytokines (i.e., IL-10), stimulates those of proinflammatory mediators (i.e., IL-1B, TNF, ICAM1), alters the expression of cyclooxygenase (COX)-2, boosts the phagocytosis activity and the production of ROS. In macrophages, the lack of long chained n-3 PUFAs could increase phagocytic activity and favor the maintaining of an activated immune status, as those PUFA inhibit toll like receptor (TLR) 2 and TLR 4 and reduce the expression of NF-κB (Calder et al. 1990; Lee et al. 2003, 2004). In lymphocytes, EPA and DHA increase the production of IL-10, IL-4 and TNF through inhibiting the production of prostaglandin E2, improves IFN-γ synthesis and lymphocyte proliferation (Hughes and Pinder 2000; Trebble et al. 2003; Gorjão et al. 2006; Brassard et al. 2007).

**Oxidative Stress** OS-mediated damages to DNA, proteins, and lipids can contribute to dysfunction of the immune system (Dalle-Donne et al. 2005; Sordillo et al. 2009; Celi and Gabai 2015). Proteins are highly susceptible to OS, and their oxidation alters the function of receptors, transporters, enzymes, and structural elements involved in the immune response (Celi and Gabai 2015). As a significant amount of proteins function as enzymes, the effect of oxidation on one molecule is greater than stoichiometric, as enzymes serve as catalytic mediators in cells (Dalle-Donne et al. 2005). Furthermore, oxidized proteins can generate new antigens, inducing or exacerbating the inflammatory response (Halliwell and Whiteman 2004). Oxidation of lipids is also involved in immune dysfunction, as immune cells are highly susceptible to lipid peroxide chain reaction due to the high PUFA content of their membranes (Calder 2008). Lipid peroxidation generates the reactive electrophile species malondialdehyde, which causes toxic stress in cells, altering cellular functions and signal transduction (Herdt 2000; Kankofer 2001; Lacetera et al. 2005; Del Rio et al. 2005; Sordillo et al. 2007, 2009; Sordillo and Raphael 2013; Abuelo et al. 2015).

Oxidants can also serve as messengers in a wide range of signaling pathways, including pathways related to NF-κB and mitogen-activated protein kinase, that are involved in the production of cytokines, oxylipids, and immunoregulatory factors (Finkel 2011; Brown and Griendling 2015; Serhan 2017). OS increases the proinflammatory phenotype of immune cells acting on redox-regulated proinflammatory factors (Cao et al. 2000; Weaver et al. 2001; Sordillo et al. 2009) and is also related to the upregulation of proinflammatory genes related to leukocyte extravasation (i.e., intracellular adhesion molecule-1 and vascular adhesion molecule-1) that are known to induce pathologic proinflammatory reactions (Maddox et al. 1999; Sordillo et al. 2008).

**Endotoxin Tolerance** Repeated exposure of leukocytes to microbial stressors could induce an endotoxin tolerance condition (Biswas and Lopez-Collazo 2009). This consists in a transient unresponsive state of leukocytes against endotoxin challenges, likely driven by the suppressed expression of genes related to survival as well as the upregulation of anti-inflammatory genes in these cells. Recently, these conditions have been documented in bovine leukocytes during the transition period (Filipe et al. 2021), suggesting endotoxin tolerance as a possible mechanism behind the immune dysfunctions occurring around calving. The possible causes for developing an endotoxin tolerance include several events activating the immune system before calving, or boosting the physiological inflammatory responses related to parturition during the early lactation (Mezzetti et al. 2021).

**Potential Genetic Contribution** A recent study (Milanesi et al. 2021) found the expression of two inflammatory biomarkers related to the acute phase response (PON and ceruloplasmin) to be primarily under *cis*-regulation in dairy cattle. This significant genetic association was identified in two dairy breeds (Italian Holstein and Italian Simmental), characterized by highly different selection histories, immune-metabolic status, and production performance (Bomba et al. 2015; Marras et al. 2015; Lopreiato et al. 2019). Thus, a different expression of inflammatory biomarkers driven by genotype could account for the different likelihood of developing chronic inflammation in early lactating cows (Mezzetti et al. 2021). Despite that, the relationship between the candidate causative variants identified by Milanesi et al. (2021) and animal performance during the transition period has not yet been investigated, and the contribution of genotype on the successful adaptation of dairy cows to the new lactation is still to be elucidated.

### 1.4.4 Detrimental Effects of Chronic Inflammations on Systemic Metabolism

A common denominator for developing several PI conditions during the transition period is the activation of the immune system and the development of chronic inflammations.

**Feed Intake Depression** Chronic inflammation may explain the poor dry matter intake (DMI) in many cows after calving, representing a key driver of sub-optimal transition period health. Many types of diseases induce what is known as sickness behavior, including decreased wakefulness, exploratory behavior, movement, and appetite (Johnson 1998; Bertoni 1999; Bertoni et al. 2009; Brown and Bradford 2021). The mechanisms underlying this sickness behavior are driven primarily by inflammatory signaling cascades which influence feed intake set points in the hypothalamus. Critically, because this regulatory mechanism is mediated by the host immune response to pathogens, the same response can be triggered by abiotic stressors that induce the same physiological response. In addition to the impacts of chronic inflammation on metabolic and immune function, which may promote ongoing disease challenges, the impairment of feed intake adds an additional burden to the health of the early lactation cow. Poor feed intake is the primary cause of extreme NEB, which is associated with increased risk of culling, infertility, and poor productivity. The behavioral impacts of inflammation may therefore be just as important as the direct metabolic effects.

**Increased Nutrient Deficiency** Leukocyte activation increases by two to threefold the metabolic demand for several essential nutrients by these cells (Table 2) (Pond and Newsholme 1999), further contributing to the NEB and NPB when coexisting with the metabolic adaptations occurring during the peripartum phase (Wannemacher et al. 1980; Hotamisligil 2006; Trevisi et al. 2007). Several leukocyte types (both from innate and adaptive immune systems) undergo a metabolic switch from oxidative phosphorylation toward glycolysis during activation (Kominsky et al. 2010), becoming highly glucose-dependent and releasing lactate at a rapid rate during this "oxidative burst." In fact, the use of alternative energy sources (i.e., glutamine, NEFA, and ketone bodies) reduces substantially the efficiency of leukocyte functions (Calder 1995; Castell and Newsholme 1998; Gross et al. 2018). Data obtained in dairy cows exposed to LPS quantified the glucose requirements for a fully activated immune system as 2.5 to 3.1 kg of glucose/d (>1 kg of glucose utilized within 720 min from the onset of inflammation) (Kvidera et al. 2017), nearly equal to the 2.7 kg/d required by the lactocytes of a dairy cow yielding 40 kg milk/day. To sustain their highly oxidative metabolism, activated leukocytes are known to shift the expression of their glucose transporter from insulin-dependent GLUT4 toward GLUT1 and GLUT3 (Maratou et al. 2007). In the meantime, inflammatory pathways (i.e., TLR 4 signaling) upregulate the expression of gluconeogenic enzymes in the liver (Mamedova et al. 2013). Circulating proinflammatory mediators (i.e., TNF-α) inhibit the expression of solute carrier family 4 (*SLC2A4*) on the cell surfaces, inducing a metabolic insulin resistance status aimed at prioritizing their transporters in glucose allocation (Kushibiki et al. 2001; Shi et al. 2006; Bradford et al. 2009; De Koster and Opsomer 2013; Horst et al. 2019). These mechanisms allow activated leukocytes to compete with placental tissues and lactocytes for circulating glucose around calving, further boosting the metabolic requirement for glucose in this challenging phase and increasing the risk of developing a Type-I ketosis by the animal (Herdt 2000; Mezzetti et al. 2019a; Habel and Sundrum 2020). Furthermore, the acute phase response consequential to leukocytes activation alters several metabolic pathways, including those of proteins exerting pivotal roles in lipid metabolism (i.e., VLDL, apolipoproteins) (Table 2), predisposing dairy ruminants to developing fatty liver condition and type-II ketosis (Herdt 2000). Besides that, several minerals (i.e., zinc and iron) are sequestered from circulating blood by the liver during the acute phase response (Bertoni and Trevisi 2013), predisposing dairy cows undergoing severe inflammatory conditions at developing deficiencies for these micronutrients.

**Boosted Oxidative Stress** Activated leukocytes massively increase the body oxidant load through releasing both ROS and RNS, directly contributing to the risk of developing an OS condition (Celi 2011). Furthermore, several proteins with fundamental antioxidant functions are produced by the liver (Table 2), and the occurrence of the acute phase response consequential to leukocytes activation predisposes dairy ruminants to developing OS-related diseases (Ceciliani et al. 2012).

**Altered Hormonal Patterns** Impairment of liver metabolism subsequent to the acute phase response could alter the normal hormonal patterns in the peripartum period. For instance, cortisol levels depend on the liver synthesis of a specific carrier protein, termed corticosteroid binding globulin. Under acute inflammation, reduced production of corticosteroid binding globulin triggers an initial increase in free cortisol, which results in more bioactivity (Pugeat et al. 1989). Then, negative feedback prevails and the hypophysis-pituitary-adrenal axis secretes less cortisol (Trevisi et al. 2013).

## 2 Strategies to Improve Dairy Cattle Adaptation to the Transition Period

The best strategy to ensure a "good transition" in dairy herds consists of a holistic approach to avoid stressful conditions and reducing the degree of PI during this phase, as the latter appears to be a common denominator between immune dysfunction and an increased likelihood of disease. This approach should include nutritional interventions, specific treatments aimed at modulating metabolic processes that may exceed homeorhetic control and management practices aimed at optimizing animal welfare (i.e., consistent dry-off routine, hygienic conditions, good animal well-being, and limited stress around parturition). Furthermore, researchers and practitioners should consider the adoption of predictive markers aimed at preventing pathological conditions and genetic selection criteria aimed at increasing animals' resilience (Trevisi and Minuti 2018; Halachmi et al. 2019).

### 2.1 Management of the Energy Balance Through Modulating Diet Concentration

Considerable literature exists on the best diet formulation strategies for dairy cows during the initial dry period and subsequent new lactation (Ingvartsen 2006; Douglas et al. 2006; Dann et al. 2006). These strategies focus on satisfying nutrient requirements while avoiding metabolic challenges driven by sudden changes in diet composition, as guided by nutritional models such as the INRA, NRC, and NASEM (INRA 1989; NRC 2001; NASEM 2021).

Avoiding excessive lipid deposition during the dry period and minimizing NEFA mobilization processes in early lactation is crucial to reduce NEB-related dysfunctions of leukocytes. Thus, monitoring BCS of dairy cows and applying a plan aimed at decreasing early lactation NEB are essential considerations during the transition period. Dairy cows should enter the dry period with a BCS of no more than 3.5 out of 5 (Agricultural Development and Advisory Service 1986) in order to minimize the amount of stored TG. Thus, NRC 2001 and NASEM 2022 recommended feeding a low-energy and high-fiber content diet between dry-off and 3 weeks prior to parturition (Table 1) in order to increase rumen fill and to help maximize feed intake early postpartum (Contreras et al. 2004; Esposito et al. 2014).

Diet density should be increased during the last 3 weeks prior to parturition (Table 1) to meet the growing requirements of the fetus and ensure a BCS of approximately 3.5 at calving (Overton and Waldron 2004). During this phase, it is common to increase dietary fermentable starch prior to calving and subsequent early lactation levels, as a sudden increase of starch in the diet could reduce rumen pH and result in SARA condition. Increasing non-fiber carbohydrates (NFC) or highly digestible neutral detergent fiber (NDF) levels 3 weeks prior to calving in rations with adequate physical effective (pe) NDF can improve the development of rumen papillae and ameliorate VFA absorption, minimizing the likelihood of developing SARA in early lactation (Rabelo et al. 2003). Furthermore, such a strategy increases levels of propionate, improves gluconeogenesis, and bacterial protein synthesis at rumen level (Overton and Waldron 2004), increasing glucose availability and reducing liver TG content, with a reduction of the incidence of metabolic disorders and reproductive issues in early lactation (Gong et al. 2002).

Avoiding excessive amounts of NFC and highly fermentable NDF is beneficial during the first 1–3 weeks of lactation, as one of the limiting factors for feed intake is the chemical effect of metabolites related to the oxidation of fuels in this phase, and excessive propionate productions could directly contribute to reducing DMI (Choi and Allen 1999). Conversely, rumen availability of starch, NFC, and fermentable fiber should be increased by peak lactation, as cows overcome the NEB condition. In fact, feed intake is limited by gut fill at this phase rather than chemical signaling, and increased production of propionate driven from the increased fermentable substrate increases insulin concentration and reduces lipomobilization without affecting DMI (Drackley 1999).

## 2.2 Immune-Modulatory Treatments

Recent interest in changes of the immune function of dairy cows during the peripartum period has promoted the development of specific treatments aimed at reducing their negative effects. These treatments include both dietary interventions and pharmacological products aimed at modulating immune system functions and mitigating chronic inflammatory states.

### 2.2.1 Supplementation of Nutraceuticals

There is a growing interest regarding the inclusion of nutraceuticals on the diet of dairy ruminants at their transition to calving due to their specificity in modulating metabolic pathways, allowing a fine-tuning on the metabolic processes that are more likely to exceed the control of homeorhetic mechanisms. These compounds include probiotics, prebiotics, dietary lipids, functional peptides, phytoextracts, microminerals, vitamins, and methyl donors and their beneficial effects on the transition cows' metabolism have been widely reviewed elsewhere (Bertoni et al. 2013; Bradford and Swartz 2020; Lopreiato et al. 2020a).

Immune modulation represents the novel frontier for nutraceutical products included in transition cows' diets. In this respect, acetylsalicylate is well known for its anti-inflammatory action exerted through inhibiting COX-enzymatic complex. Studies administering this nutraceutical by injection documented a mitigation of the acute phase response and an improvement of the liver function (Trevisi and Bertoni 2008). These results were paired with increased milk yield, lowered milk somatic cells count (SCC), and improved reproduction in dairy cows receiving acetylsalicylate, suggesting mitigation of the inflammatory conditions at the onset of lactation to be an effective strategy for improving performances of the animals. Successively, similar results were obtained by administering the active principle orally, confirming nutraceutical role exerted by acetylsalicylate (Trevisi and Bertoni 2009).

Despite promising results obtained in research, the animals' response threshold for most of the nutraceutical products is very low, and their direct dietary inclusion may be difficult to incorporate into the feeding routines on many commercial dairy farms. Furthermore, each active principle mentioned above affects few metabolic pathways through a highly specific mode of action, while molecules interacting with multiple functions could likely provide a greater benefit against the multifactorial challenges affecting the metabolism of transition dairy ruminants. Thus, several commercial products containing a mixture of active substances have been developed to address these limitations, allowing direct dietary inclusion at the farm level. Among these patented mixture of compounds exerting immune modulatory action, Omnigen-AF (OAF) could be included in the diet of dairy cows from dry-off to the first month of lactation, providing positive effects on leukocyte function and a consequent reduction of the incidence of infectious diseases in the transition period (Wang et al. 2007; Nace et al. 2014). Although a direct effect of OAF on the gene expression of neutrophils has been reported (Wang et al. 2004; Ortiz-Marty et al. 2013), a recent study found no effects on neutrophils diapedesis measured in vivo or on the amelioration of liver metabolism and inflammation of dairy cows receiving the feed additive (Mezzetti et al. 2019b). Nevertheless, the same study described dairy cows receiving OAF to have an increased abundance of lymphocytes at the blood level and improved leukocyte function after ex-vivo stimulation with LPS (lower lactate production and lower glucose consumption). These outcomes, paired with the improved rumination time and energy metabolism found in dairy cows receiving the additive, suggest that OAF has an indirect effect on immune function, likely mediated by positive effects on energy metabolism and feeding behavior.

### 2.2.2 Pharmacological Compounds

Besides nutritional interventions, several pharmacological compounds could be used for modulating the immune response of dairy cows during the transition period. Pegbovigrastim (Imrestor; Elanco Animal Health Greenfield, IL, USA) consists of pegylated recombinant bovine granulocytes-colony stimulating factor (PEG-rbG-CSF), aimed at stimulating the circulating numbers, and possibly function of neutrophils. Some studies using this product have demonstrated substantial increases in circulating neutrophil, monocyte, and eosinophil counts for several weeks (McDougall et al. 2017; Van Schyndel et al. 2018; Zinicola et al. 2018; Lopreiato et al. 2020b). These were paired with modest increases in per-cell function, most consistently of extracellular release of myeloperoxidase (Kimura et al. 2014; McDougall et al. 2017). Recently, Lopreiato et al. (2019) showed the effect of PEG-rbG-CSF in maintaining stable PIC levels during the first month after parturition, reflecting greater regulation of neutrophil recruitment, tracking, and maturation during the inflammatory response. However, when taken together, results of multiple studies suggest that PEG-rbG-CSF do not consistently lead to a reduction in the incidence of diseases thought to be related to neutrophil function (i.e., retained placenta, metritis, and endometritis) (LeBlanc 2014), suggesting that a massive increase in the number of neutrophils available in circulation does not correspond to fewer diseases.

Besides products specifically aimed at modulating the immune system, other pharmacological treatments exerting their primary role on modulating other metabolic processes have been recently recognized as effective immune modulators. Among these, ionophores have been classically adopted in transition cows to modulate rumen fermentation, but they have been revealed to have a potential role in mitigating immune dysfunction occurring immediately before and after parturition. Monensin is known for its effect on rumen bacteria populations, resulting in increased propionic acid and reduced lactic acid and methane production (McGuffey et al. 2001; Mammi et al. 2021), and modulation of rumen fermentation through monensin is an effective strategy to address the risk of both ketosis and acidosis in dairy cows at the onset of lactation (Duffield et al. 2008). This positive effect on rumen fermentation is reflected by trends of several plasma analytes (i.e., reduced BHB and NEFA concentrations and higher plasma glucose) in cows treated with monensin (Fiore et al. 2021; Mammi et al. 2021). A recent study (Mezzetti et al. 2019c) found reduced infiltration of T and B leukocytes in the rumen fluid from transition dairy cows that received a controlled-release capsule of monensin 21 days before the expected calving date (expected release rate, 335 mg/d for 95 days). Such an outcome suggests monensin promotes the stabilization of the rumen environment, possibly by mitigating pH fluctuations related to the adaptation of dairy cows to a lactation diet. This provides a new perspective on the utilization of ionophores to modulate the immune system of early lactating cows.

## 2.3 Management Practices to Ensure Animal Welfare

The concept of animal welfare is based on human perspectives as it is a social construct (Fraser 2008). Maintaining dairy cattle welfare in intensive systems during the transition to calving represents a significant management challenge, considering the flow of animals successively into different physical spaces. The movement of animals through transition pens should be aimed at facilitating the provision of a balanced diet to fulfill their requirements (see Table 1 and previous sections) and at allowing caregiving interventions aimed at preventing metabolic diseases. Despite that, multiple movements and treatments could inevitably induce stressful conditions on cattle, and a fine-tuning of management practices and facilities design is essential to guarantee a proper welfare status during this phase (Mills et al. 2020).

### 2.3.1 Ensuring Optimal Facilities

**Grouping Strategies** For an individual dairy cow, the transition phase often consists of four groups: dry, prefresh, maternity, and fresh (Cook 2019). The move to the prefresh pen should occur approximately 21 days before parturition. Then, cows are often moved to a contiguous maternity pen, where they should stay alone for no longer than 3 days to avoid any distress from prolonged isolation. Alternatively, some farms use a "just in time" approach, where cows are moved into a maternity pen for just a few hours prior to calving. After calving, cows may be moved to the fresh pen, where they often stay for approximately 21–30 days before including them in the lactation group.

Regrouping or inclusion of new animals requires the stabilization of social hierarchy, increasing agonistic and competitive behaviors (von Keyserlingk et al. 2008; Schirmann et al. 2011); these social interactions can have detrimental effects during the days closer to calving (i.e., reducing feed and water intake, increasing standing time and decreasing lying time) and also result in physical injuries and abortion (Cook 2019). Thus, grouping strategies should be planned carefully to avoid pointless movements and minimize social stressors in dairy cows, although direct experimental assessment did not show benefits when adequate feed bunk space was available (Silva et al. 2013). Besides planning group movements, separating primiparous animals from the herd during their first pregnancy and lactation could beneficially affect social interactions, as these animals are more susceptible to psychological stress induced by dominant subjects (Wathes et al. 2007; Neave et al. 2017).

**Design and Spaces Management** Among different aspects included in the animal welfare evaluation, a central role is revested by the suitability of the spaces provided to animals undergoing the different phases of the productive cycle. During the whole transition phase, adequate ventilation should be provided (i.e., minimum airspeed of 1 m $s^{-1}$). A soft and comfortable lying surface (i.e., dry and deep layer of shavings, straw, or sand as bedding material) has been consistently recommended for dairy cows (Fregonesi et al. 2007; Tucker et al. 2009; van Gastelen et al. 2011; Campler et al. 2014; Mondaca 2019). Particularly in the early fresh period, feed bunks with headlocks (76 cm wide) are preferred to post-and-rail design. These provide a means of animal restraint during a period with frequent caregiver examinations; however, it is important to note that lockup time should be as short as possible. They also mitigate competition while feeding, providing subordinate animals a certain level of protection (Rioja-Lang et al. 2012). Overstocking should be minimized to ensure sufficient bunk and resting space (at least 13 $m^2$ per cow). Thus, each pen should be sized using 120–150% of the average calving rate to avoid overstocking when calving rate is higher (Cook 2019).

Dry and prefresh pens can have either freestall or loose bedded pack layout. Maternity pens should guarantee the expression of innate maternal behavior through ensuring calm parturition in an appropriate environment, minimizing disturbances and risk of mistaken identity of the offspring (Rørvang et al. 2018). Good results can be obtained both with individual and group pens, separated from the rest of the herd. Maternity pens should be designed to ensure adequate space (>13 $m^2$ per cow), a soft and clean surface (deep straw bedding is preferred), easy access, and proximity to prefresh area. There is increasing interest in offering cattle options for hiding during parturition (i.e., a plywood barrier covering half of the pen) (Proudfoot 2019; Creutzinger and Proudfoot 2020; Mezzetti et al. 2021).

### 2.3.2 Dry-Off Procedures

Dry-off and subsequent remodeling and involution of the mammary gland is accompanied by alterations in the distribution of nutrients, rumen environment, and physiology (Dingwell et al. 2001; Dieho et al. 2016; Putman et al. 2018). These are typically paired with psychological challenges related to feed restrictions and to the modification of social structure after regrouping (von Keyserlingk et al. 2008). A fundamental role of dry-off in the development systemic inflammation has been demonstrated in dairy cows (Mezzetti et al. 2020; Cattaneo et al. 2021b). It has been hypothesized that inflammatory conditions occurring after dry-off could arise from the contribution of leukocytes in the involution phase of the mammary gland (Putman et al. 2018). These phenomena, paired with the immune suppressive effect of cortisol released at dry-off as a consequence of metabolic stresses (Bertulat et al. 2013), could account for the reduced responsiveness of leukocytes reported during the dry period (Kehrli et al. 1989), suggesting a potential contribution of this management practice to the etiology of immune dysfunctions affecting dairy cows around calving (Mezzetti et al. 2021). Besides limiting the detrimental effects related to cortisol release through adopting proper feeding and grouping strategies and providing suitable environments (see previous sections), several aspects should be considered to limit the severity of metabolic inflammations while drying-off dairy ruminants.

**Mammary Health** Compromised mammary health increases the severity of inflammatory conditions at milking interruption. Milk SCC effectively reflects the mammary health status of dairy cows. High milk SCC at dry-off could also possibly increase the severity of inflammatory conditions experienced by the animals around calving through boosting the likelihood of developing intramammary infections at the onset of lactation. Pantoja et al. (2009) collected three sterile quarters milk samples from dairy cows (i.e., at dry-off, at calving day, and on the first dairy herd improvement test day). Compared to normal quarters, they found quarters having SCC $\geq$ 200,000 cells/mL at both dry-off and post-calving sampling periods to be more likely infected (both by a major and by a minor pathogen) at the screening performed on the first dairy herd improvement test. Thus, SCC screening performed at dry-off could likely represent a predictive tool aimed at identifying "risky" subjects for developing new intramammary infections and chronic inflammatory states at the onset of the new lactation. Recently, differential somatic cell count (DSCC) has been developed to highlight relative abundance of PMN, lymphocytes, and macrophages populations in milk SCC, allowing a closer monitoring of the immune response occurring in the mammary gland. Thus, combining SCC and DSCC use at dry-off could allow a closer monitoring of intramammary infections occurring at dry-off, minimizing mastitis occurrence at the onset of the new lactation.

**Milk Yield Threshold to Dry-Off** Maintaining high milk yield at dry-off boosts the likelihood of developing severe inflammatory conditions after milking interruption (Mezzetti et al. 2020). These could be driven by greater amounts of milk and parenchymatic tissue being reabsorbed and by the higher risk of milk leakage and intramammary infections in animals maintaining high milk yield at the halting of milk removal (Putman et al. 2018). Thus, a milk yield of 15 kg/d has been consistently proposed as the "safety threshold" to dry-off a modern Holstein dairy cow (Vilar and Rajala-Schultz 2020). The National Mastitis Council recommends abrupt cessation of milking once the target safety threshold for milk production has been reached (National Mastitis Council 2006), stating that gradual cessation of milking can be used in combination with decreased dietary concentrate as an effective strategy for mitigating negative effects of high milk yields at dry-off (Vilar and Rajala-Schultz 2020). This practice is known for reducing intramammary pressure, stress, and discomfort after dry-off through lowering milk yield at milking interruption, accelerating mammary involution and elevating natural protective factors in mammary secretions (Vilar and Rajala-Schultz 2020).

**Antibiotic Use** Systematic antibiotic use at milking cessation (alone or combined with a teat sealant) has been widely adopted as an effective practice for reducing the likelihood of developing acute mastitis events after dry-off and for lowering the SCC and the risk of chronic intramammary infections at the onset of the new lactation (Poutrel et al. 1997; Godden et al. 2003). Raising public concerns on the extensive antibiotic use on farm animals forced dairy farmers to adopt different management practices. Dry-off of dairy ruminants is a focal point in this respect, and selective dry-off procedures basing on milk SCC have already been proposed and validated in dairy cows (Rowe et al. 2020). These could be implemented through performing a total and DSCC on single quarters of the animal before milk removal is halted, limiting antibiotic use to the animals having high risk of developing chronic intramammary infections, and orienting active principles choice on the most effective one.

### 2.3.3 Continuous Lactation or Dry Period Length

A 2-month period is the classical duration for the dry phase adopted in all the ruminant species. Extensive literature exists on the effects sorted by enlarging, shortening, or even omitting this phase on animal health and performances. Both in dairy cows and small ruminants, drying animals for longer than 60 days has been consistently documented as a predisposing factor for excessive lipid deposition and metabolic diseases development (Grum et al. 1996; Loor et al. 2006; Pires et al. 2013). Oppositely, several papers evaluated a shortened dry period as a potential strategy to increase produced milk and minimize stress induced on the animal by transition changes (Grummer and Rastani 2004). Outcomes obtained in dairy cows consistently documented that higher milk productions obtained by extreme shortening the dry phase (e.g., <30 days) were counterbalanced by a lower milk production in the subsequent lactation (Sørensen and Enevoldsen 1991; Watters et al. 2008), spanning between a 20 and 25% reduction when the dry phase is completely omitted (Grummer and Rastani 2004). This is due to the impairment of mammary renewal process when the dry phase is shortened. The effects on milk production aside, both significant shortening and omitting the dry period have been consistently indicated as an effective strategy to reduce the amount of mobilized fat at the onset of lactation, likely resulting in beneficial effects on dairy ruminants metabolism during this phase (Rastani et al. 2005; Watters et al. 2008). Furthermore, omitting dry phase in dairy cows effectively mitigated the acute phase response (as reflected by lower positive APPs concentrations in plasma), but increased plasma oxidant concentrations as compared with those observed in cows undergoing a normal dry period (Mayasari et al. 2017, 2019). In conclusion, altering the length or even by omitting the dry phase on animal health and performances are highly debated topics, with inconsistent results reported in literature. Ultimate responses on the advantages behind these management practices could be provided only by their consistent application on commercial farms, allowing a wide and long-term evaluation of the animals' responses.

## 2.4 Preventive Diagnosis

Research efforts have focused on identifying candidate biomarkers that might reflect "at risk" condition for developing PI and early lactation diseases in transitioning dairy ruminants (i.e., ketosis, mastitis, metritis, lameness, retained placenta, and milk fever) (Wankhade et al. 2017; Wisnieski et al. 2019a, b). Studies have been performed on biological fluids (i.e., urine, blood, saliva, rumen fluid) using different analytical procedures and on the main indicators related to the animal's behavior (i.e., rumination, activity, and lying times) as reflected by modern livestock precision systems.

### 2.4.1 Blood Indexes

Plasmatic trends of positive APP have been proposed as biomarkers of the severity of inflammation, as they reflect the alteration of liver functions induced by the acute phase response (Castell et al. 1989; Fleck 1989; Trevisi et al. 2012). Nevertheless, positive APP have different sensitivity to the acute phase, as they are released at different intervals from the onset of systemic inflammation. Among positive APP, haptoglobin is the most promising biomarker of inflammation in dairy cows (Bionaz et al. 2007; Huzzey et al. 2011). In fact, it has a longer plasma half-life relative to other positive APP (i.e., serum amyloid alpha), requiring 60–96 h to return to normal circulating levels after an acute phase response

(Giblett 1961). Associations of haptoglobin concentration with white blood cell's function have been investigated by Nightingale et al. (2015), who grouped 240 Holstein transition dairy cows into three classes according to their postpartum haptoglobin level: low (0–8.4 μg/mL), moderate (8.5–458 μg/mL), and high (459–1757 μg/mL). Animals with the highest haptoglobin concentrations had lower neutrophil counts in blood, likely due to the recruitment of PMN to inflamed peripheral tissues. Furthermore, PMN from cows with the highest haptoglobin concentrations produced greater amounts of TNFα, had an increased expression of L-selectin, and a greater respiratory burst activity, confirming that the level of plasma haptoglobin was related to the activation of PMN, and thus to the inflammatory condition of the animals. Despite these promising outcomes on the use of plasma haptoglobin as a diagnostic tool, several studies tried to define reference intervals for physiological trends of APPs in dairy cows around calving (Quiroz-Rocha et al. 2009; Moretti et al. 2017; Premi et al. 2021) and reported a wide individual variability for inflammation and immune functions-related analytes. This makes challenging the definition of a "safety threshold" for preventing the development of PI conditions basing on the interpretation of single plasma analytes.

Besides considering single analytes, a valuable strategy could be the calculation of indexes aggregating trends of multiple plasma analytes detected at specific time points relative to the calving date. Several promising indexes have been developed on postpartal trends of plasma analytes reflecting liver function and the acute phase response of dairy cows. Bertoni et al. (2008) calculated the liver activity index (LAI) by aggregating the albumin, cholesterol, and retinol concentrations measured 5, 15, and 30 days after calving. Similarly, Trevisi et al. (2012) calculated the liver functionality index (LFI) aggregating the albumin, cholesterol, and bilirubin concentrations measured 3 and 28 days after calving. Retrospectively dividing a group of transition dairy cows based on their liver condition (as reflected by these indexes), three studies (Bertoni et al. 2008; Trevisi et al. 2010, 2012) consistently found animals with the lowest LAI or LFI values to have the most severe inflammatory condition in early lactation (i.e., higher PICs and oxidant species concentrations in blood and higher SCC in milk). Furthermore, animals with the lowest LAI or LFI values had the lowest milk yield, worst reproductive performances, and greatest disease incidence in early lactation. This suggests LAI and LFI calculations are valuable tools for following up the adaptation of dairy cows to the new lactation (Bertoni and Trevisi 2013; Trevisi et al. 2016).

Recent research by the same group (Cattaneo et al. 2021a) provided new perspectives on the use of blood indexes as predictive biomarkers, as they found the plasma albumin/globulin ratio (AGR) of late lactating cows to reflect their adaptation to subsequent calving and lactation. Here, cows were followed from 7 days before dry-off to 28 days after calving and retrospectively divided into three groups according to their AGR before dry-off. Compared with cows having the lowest AGR (reflecting the highest inflammatory condition before dry-off), cows with the highest AGR showed lower inflammation through the dry period and up to the first month of lactation. Furthermore, these cows had a lower SCC and better productive and reproductive performances in the subsequent lactation.

### 2.4.2 Livestock Precision Systems

The availability of automatic sensors allows the detection of alterations in behavior that might reflect illness or disease risk in an animal (Weary et al. 2009; Soriani et al. 2012; Borchers et al. 2016; Hendriks et al. 2020).

**Rumination Time** Chewing and rumination times are affected by several factors (Beauchemin 2018), and their trends around calving reflect the likelihood of dairy cows developing severe inflammatory conditions or diseases in early lactation (Soriani et al. 2012; Calamari et al. 2014; Liboreiro et al. 2015). Cows with reduced rumination time before calving are known to maintain this even after calving, and these animals are more likely to develop health disorders during the peripartum phase (i.e., metabolic and digestive disorders and severe cases of mastitis and metritis) compared to cows with a greater rumination time ante-partum (Soriani et al. 2012; Stangaferro et al. 2016). Recently, Abuelo et al. (2021) extended such a relationship beyond the transition period, demonstrating that cows with the greatest reduction of rumination time at dry-off were more likely to develop ketosis or lameness in early lactation. Moreover, peripartum trends of plasma analytes from cows with a reduced rumination time during the first few days of lactation typically reflect a more marked inflammatory condition compared with those from cows having high rumination time at the onset of lactation (Soriani et al. 2012; Calamari et al. 2014). This suggests that the occurrence of severe inflammation around parturition is associated with a slower increase of rumination time after calving. Saha et al. (2019) have found that cows undergoing a SARA condition (as reflected by the increased VFAs content, altered acetate to propionate ratio and decreased pH of the rumen fluid) spent a similar amount of time ruminating during the day as compared to healthy animals, but their rumination time differed after the morning feeding.

**Rest Time** Time spent lying can predict the onset of several diseases, although this relationship has not been fully elucidated. Itle et al. (2015) reported clinically ketotic cows to have a reduced lying time in the week preceding calving. Conversely, Rodriguez-Jimenez et al. (2018) hypothesized that longer lying time (and reduced standing time) during the days leading to parturition could predispose dairy cows to postpartum ketosis through reducing DMI. Such a hypothesis is supported by the positive relationship between lying time, changes in plasma NEFA concentration, and the risk of developing ketosis (alone or paired with other diseases) detected by others during the first weeks of lactation (Kaufman et al. 2016; Piñeiro et al. 2019). In a recent paper on multiparous cows, Cattaneo et al. (2020) reported cows developing retained placenta to have increased lying time during the last 3 weeks before calving, but no alterations in prepartum lying time were found for cows developing metritis. Conversely, Neave et al. (2018) found that cows later diagnosed with metritis had reduced lying time and fewer lying bouts before calving, while Barragan et al. (2018) detected an opposite trend in prepartum lying time for primiparous and no differences for multiparous cows developing metritis.

**Body Temperature** Body temperature represents a promising indicator reflecting health status of the animals. The raise of rectal temperature above 39.5 °C in the peripartum is often associated with a reduced feed intake and rumination time and can be used as reliable alarm of the onset of illness (Wenz et al. 2011). Thus, individually monitoring rectal temperature of postpartum animals to detect acute inflammatory conditions accompanying abnormal uterine recovery (i.e., uterine infections, endometritis, metritis) has become a common practice in dairy herds (Sheldon et al. 2004). Nowadays, thermographic measurements allow a close monitoring of the body temperature of the whole herd, providing a promising tool for extending body temperature monitoring to the early detection of heat stress-associated symptoms (Liu et al. 2019). Because of the deleterious effect of heat stress on dairy ruminants transitioning to calving (Fabris et al. 2019; Dahl et al. 2020), heat abatement and monitoring is crucial.

## 2.5 Genomic Association Studies to Prevent Metabolic Dysfunctions

As the transition period severely challenges the metabolism and immunity of dairy cows, breeding cattle for lower susceptibility to disturbances could be a key step toward establishing resilient animal production systems (Berghof et al. 2019; Mezzetti et al. 2021). The existence of genetic differences in dairy cattle to common health disorders was investigated and confirmed over the years (Zwald et al. 2004). Consequently, a growing interest in including health-related traits in genetic selection programs led to the concept of a comprehensive diseases prevention strategy, accounting also for genetic information (Egger-Danner et al. 2014; Weller et al. 2017; König and May 2019). The feasibility of such kind of breeding program has been reported by scientific literature (Parker Gaddis et al. 2014; Guarini et al. 2019) and implemented with the introduction of the genomic prediction Wellness Trait Index (McNeel et al. 2017), which includes six disease resistance traits (mastitis, metritis, ketosis, retained placenta, displaced abomasum, and lameness) in estimating differences in expected lifetime profit. Moreover, in 2018, the US genetic selection index (NM$18) was updated with the insertion of the Health Trait Subindex, which accounted for six disease resistance traits (milk fever, displaced abomasum, ketosis, mastitis, metritis, and retained placenta) (Vanraden et al. 2018).

A feasible solution to improve the efficiency of current breeding programs at preventing transition cow's diseases could be understanding the genetic architecture of specific biomarkers related to metabolism, immune system, and inflammation. Indeed, the genomic prediction quality of a model strictly depends on the trait's genetic architecture (Hayes et al. 2010; Lopes et al. 2020), and a control exerted by a large number of genes with small additive effect (polygenic control) is often to take into account when dealing with complex traits—such as health-related ones—(Cole et al. 2009). Thus, identifying some intermediate phenotypes (or "endophenotypes"), which are measurable markers correlated to an illness—expected to be characterized by a simpler genetic control than the end-point phenotype—could improve our power in detecting candidate genes underlying the resistance to a disease (Glahn et al. 2014; Höglund et al. 2019; Zhang et al. 2020).

In this respect, increased evidence of the role of genotype in modulating the inflammation process has already been reported in murine models (Jofre-Monseny et al. 2008; Dose et al. 2016). Significant associations between genetic variants expressing different levels of inflammatory biomarkers and the occurrence of several diseases have been well investigated in human medicine (Prins et al. 2017; Höglund et al. 2019; Ruotsalainen et al. 2021). Thus, identifying similar associations in the genome of dairy cows has the potential to drive breeding programs aiming to improve the effectiveness of animal responses to physiological and environmental stressors (Raszek et al. 2016). This represents a promising solution against the metabolic challenges occurring during the transition period.

Recently, the genetic structure of several blood, hair, and milk biomarkers related to metabolism, immunity, inflammation, and OS response in Holstein and Simmental cows has been investigated (Milanesi et al. 2021). Specific and precise genomic regions were identified to be associated with three biomarkers levels (gamma-glutamyl transferase, PON, and ceruloplasmin), all involved in antioxidant functions and in the acute phase response (Bertoni and Trevisi 2013). Interestingly, the identified genomic regions included the genes that directly code for each protein (GGT1 and GGT5 genes on chromosome 17 for gamma-glutamyl transferase, PON1 on chromosome 4 for PON, and CP on chromosome 1 for ceruloplasmin), suggesting cis-regulation. Candidate causative variants were identified but are not yet validated.

Genetic resistance to ketosis and NEB-related diseases has already been explored, and promising results have been obtained in dairy cows. RNA-seq studies performed on dairy cows undergoing different degrees of NEB have revealed that genes involved in fat metabolism have the greatest differences in transcription levels (McCabe et al. 2012). Although transcriptional and post-transcriptional regulation of gene expression do not relate differences in RNA transcripts found in this study to differences at the genotype level, energy metabolism genes were selected as suitable targets for genetic selection in subsequent years. Polymorphisms on the gene encoding for apolipoprotein B receptor (APOBR gene on chromosome 25) have been suggested to affect the likelihood of Holstein cows developing ketosis (Tetens et al. 2015). Accordingly, that gene was found to be significantly associated with glycerophosphocholine-phosphocholine ratio in milk, which was previously proposed as a ketosis biomarker (Klein et al. 2012). It was suggested that cows with a high glycerophosphocholine-phosphocholine ratio were disposed to a greater ability in utilizing blood phosphatidylcholine as a source for milk fat synthesis. This likely reflects their lower susceptibility to developing severe lipomobilization processes that precede ketosis status. APOBR gene was suggested as promising candidate gene associated with metabolic status of dairy cattle also by Huang and colleagues (Huang et al. 2019). In this study a genome wide association study (GWAS) on ketosis resistance in Chinese Holstein cattle was performed, finding out that genes explaining largest variance portions were all involved in either insulin or lipid metabolism. Several other SNPs and genes mostly located on chromosomes 6, 14, and 20 were found to have a significant genetic association with milk BHB concentration in dairy cows (Nayeri et al. 2019). In particular, five genes (HSD17B10 and HTR2 on chromosome X; ABCA1 and ABCA2 on chromosome 8; LIPC on chromosome 10) were found to be positively related to the longitudinal concentration of BHB in Holstein milk (Leal Yepes et al. 2019).

Finally, further research is related to the resistance of infectious diseases, as the transition period alters immuno-competences. A genetic effect on this phenotype was demonstrated in Canadian Holstein cattle (Thompson-crispi et al. 2014) and subsequent studies focused on resistance to specific diseases, such as bovine respiratory disease, tuberculosis, and mastitis (reviewed in Raszek et al. 2016).

## 3 Perspectives/Conclusions

As highlighted in this chapter, metabolic background for understanding dairy cows' adaptation to the transition period challenges is very complex. Despite the massive effort made by research on this topic, the real contribution of several processes (the immune system functioning first of all) in determining the success of such adaptation has not been fully elucidated. Furthermore, the pivotal role exerted by several external factors should also be considered, as stressing conditions could detrimentally affect several metabolic processes. Besides understanding the potential interactions between metabolism and management practices, dairy farmers are interested in implementing an effective operative protocol aimed at optimizing the dairy cow's adaptation to the new lactation. A tentative for implementing such an operative guideline has been previously defined by Trevisi et al. (2011), who proposed a decalogue for practices that should be adopted between dry-off and the first month of lactation for improving transition management.

To conclude, genomics is helping in adding valuable information to the development of an efficient health management system in transition cows. Each disease is characterized by a specific genetic architecture and correlations with other traits which should be properly investigated; independent functional validation of candidate causative variants is also crucial for the implementation of breeding programs aimed at ameliorating resilience of the animals. In this respect, additional efforts in improving current disease recoding systems are required to ensure an easier estimation of genetic prediction.

## References


Abuelo A, Hernández J, Benedito JL, Castillo C (2015) The importance of the oxidative status of dairy cattle in the periparturient period: revisiting antioxidant supplementation. J Anim Physiol Anim Nutr (Berl) 99:1003–1016. https://doi.org/10.1111/jpn.12273

Abuelo A, Wisnieski L, Brown JL, Sordillo LM (2021) Rumination time around dry-off relative to the development of diseases in early-lactation cows. J Dairy Sci 104:5909–5920. https://doi.org/10.3168/jds.2020-19782

Agricultural Development and Advisory Service (1986) Condition scoring of dairy cows. In: Pubblication 612. Ministry of Agriculture, Fisheries Food, Lion House, Alhwick

Allen MS, Bradford BJ, Oba M (2009) Invited review: the hepatic oxidation theory of the control of feed intake and its application to ruminants. J Anim Sci 87:3317–3334. https://doi.org/10.2527/jas.2009-1779

Ammerman CB, Chicco CF, Moore JE et al (1971) Effect of dietary magnesium on voluntary feed intake and rumen fermentations. J Dairy Sci 54:1288–1293. https://doi.org/10.3168/jds.S0022-0302(71)86022-8

Abd Eldaim MA, Kamikawa A, Soliman MM, Ahmed MM, Okamatsu-Ogura Y, Terao A, Miyamoto T, Kimura K (2010) Retinol binding protein 4 in dairy cows: its presence in colostrum and alteration in plasma during fasting, inflammation, and the peripartum period. J Dairy Res 77:27–32. https://doi.org/10.1017/S0022029909990276

Andrews AH, Holland-Howes VE, Wilkinson JID (1997) Naturally occurring pregnancy toxaemia in the ewe and treatment with recombinant bovine somatotropin. Small Rumin Res 23:191–197. https://doi.org/10.1016/S0921-4488(96)00912-1

Appuhamy JADRN, Knoebel NA, Nayananjalie WAD, Escobar J, Hanigan MD (2012) Isoleucine and Leucine independently regulate mTOR signaling and protein synthesis in MAC-T cells and bovine mammary tissue slices. J Nutr 142:484–491. https://doi.org/10.3945/jn.111.152595

Babior BM (1999) NADPH oxidase: an update. Blood 93:1464–1476. https://doi.org/10.1182/blood.V93.5.1464

Barghouthi S, Everett KD, Speert DP (1995) Nonopsonic phagocytosis of Pseudomonas aeruginosa requires facilitated transport of D-glucose by macrophages. J Immunol 154:3420–3428

Barragan AA, Piñeiro JM, Schuenemann GM et al (2018) Assessment of daily activity patterns and biomarkers of pain, inflammation, and stress in lactating dairy cows diagnosed with clinical metritis. J Dairy Sci 101:8248–8258. https://doi.org/10.3168/jds.2018-14510

Beauchemin KA (2018) Invited review: current perspectives on eating and rumination activity in dairy cows. J Dairy Sci 101:4762–4784. https://doi.org/10.3168/jds.2017-13706

Bell AW (1995) Regulation of organic nutrient metabolism during transition from late pregnancy to early lactation. J Anim Sci 73:2804. https://doi.org/10.2527/1995.7392804x

Bell AW, Bauman DE (1997) Adaptations of glucose metabolism during pregnancy and lactation. J Mammary Gland Biol Neoplasia 2:265–278. https://doi.org/10.1023/a:1026336505343

Bell AW, Burhans WS, Overton TR (2000) Protein nutrition in late pregnancy, maternal protein reserves and lactation performance in dairy cows. Proc Nutr Soc 59:119–126. https://doi.org/10.1017/S0029665100000148

Berghof TVL, Poppe M, Mulder HA (2019) Opportunities to improve resilience in animal breeding programs. Front Genet 9:1–15. https://doi.org/10.3389/fgene.2018.00692

Bernal-Mizrachi C, Xiaozhong L, Yin L et al (2007) An afferent vagal nerve pathway links hepatic PPARα activation to glucocorticoid-induced insulin resistance and hypertension. Cell Metab 5:91–102. https://doi.org/10.1016/j.cmet.2006.12.010

Bernabucci U, Ronchi B, Lacetera N, Nardone A (2002) Markers of oxidative status in plasma and erythrocytes of transition dairy cows during hot season. J Dairy Sci 85:2173–2179. https://doi.org/10.3168/jds.S0022-0302(02)74296-3

Bernabucci U, Ronchi B, Lacetera N, Nardone A (2005) Influence of body condition score on relationships between metabolic status and oxidative stress in periparturient dairy cows. J Dairy Sci 88:2017–2026. https://doi.org/10.3168/jds.S0022-0302(05)72878-2

Bertoni G (1999) Welfare, health and management of dairy cows. In: Piva G, Bertoni G, Masoero F et al (eds) Recent progress in animal production science. FrancoAngeli, Milan, pp 59–78

Bertoni G, Trevisi E (2013) Use of the liver activity index and other metabolic variables in the assessment of metabolic health in dairy herds. Vet Clin North Am Food Anim Pract 29:413–431. https://doi.org/10.1016/j.cvfa.2013.04.004

Bertoni G, Trevisi E, Han X, Bionaz M (2008) Effects of inflammatory conditions on liver activity in puerperium period and consequences for performance in dairy cows. J Dairy Sci 91:3300–3310. https://doi.org/10.3168/jds.2008-0995

Bertoni G, Trevisi E, Lombardelli R (2009) Some new aspects of nutrition, health conditions and fertility of intensively reared dairy cows. Ital J Anim Sci 8:491–518. https://doi.org/10.4081/ijas.2009.491

Bertoni G, Grossi P, Trevisi E (2013) Use of nutraceuticals for improving animal health during the transition period of dairy cows. In: Makkar HPS (ed) Enhancing animal welfare and farmer income through strategic animal feeding. Some case studies. Food and Agriculture Organization of the United Nations (FAO), Roma, pp 79–83

Bertulat S, Fischer-Tenhagen C, Suthar V et al (2013) Measurement of fecal glucocorticoid metabolites and evaluation of udder characteristics to estimate stress after sudden dry-off in dairy cows with different milk yields. J Dairy Sci 96:3774–3787. https://doi.org/10.3168/jds.2012-6425

Bickerstaffe R, Annison EF, Linzell JL (1974) The metabolism of glucose, acetate, lipids and amino acids in lactating dairy cows. J Agric Sci 82:71–85. https://doi.org/10.1017/S0021859600050243

Bionaz M, Trevisi E, Calamari L et al (2007) Plasma paraoxonase, health, inflammatory conditions, and liver function in transition dairy cows. J Dairy Sci 90:1740–1750. https://doi.org/10.3168/jds.2006-445

Biswas SK, Lopez-Collazo E (2009) Endotoxin tolerance: new mechanisms, molecules and clinical significance. Trends Immunol 30:475–487. https://doi.org/10.1016/j.it.2009.07.009

Bomba L, Nicolazzi EL, Milanesi M et al (2015) Relative extended haplotype homozygosity signals across breeds reveal dairy and beef specific signatures of selection. Genet Sel Evol 47:25. https://doi.org/10.1186/s12711-015-0113-9

Borchers MR, Chang YM, Tsai IC et al (2016) A validation of technologies monitoring dairy cow feeding,

ruminating, and lying behaviors. J Dairy Sci 99:7458–7466. https://doi.org/10.3168/jds.2015-10843
Bordignon M, Da Dalt L, Marinelli L, Gabai G (2014) Advanced oxidation protein products are generated by bovine neutrophils and inhibit free radical production in vitro. Vet J 199:162–168. https://doi.org/10.1016/j.tvjl.2013.10.028
Bradford BJ, Allen MS (2008) Negative energy balance increases periprandial ghrelin and growth hormone concentrations in lactating dairy cows. Domest Anim Endocrinol 34:196–203
Bradford BJ, Swartz TH (2020) Review: following the smoke signals: inflammatory signaling in metabolic homeostasis and homeorhesis in dairy cattle. Animal 14:s144–s154. https://doi.org/10.1017/S1751731119003203
Bradford BJ, Mamedova LK, Minton JE et al (2009) Daily injection of tumor necrosis factor-α increases hepatic triglycerides and alters transcript abundance of metabolic genes in lactating dairy cattle. J Nutr 139:1451–1456. https://doi.org/10.3945/jn.109.108233
Brassard P, Larbi A, Grenier A et al (2007) Modulation of T-cell signalling by non-esterified fatty acids. Prostaglandins Leukot Essent Fat Acids 77:337–343. https://doi.org/10.1016/j.plefa.2007.10.025
Brown WE, Bradford BJ (2021) Invited review: mechanisms of hypophagia during disease. J Dairy Sci 104:9418–9436. https://doi.org/10.3168/jds.2021-20217
Brown DI, Griendling KK (2015) Regulation of signal transduction by reactive oxygen species in the cardiovascular system. Circ Res 116:531–549. https://doi.org/10.1161/CIRCRESAHA.116.303584
Burton JL, Kehrli ME, Kapil S, Horst RL (1995) Regulation of L-selectin and CD18 on bovine neutrophils by glucocorticoids: effects of cortisol and dexamethasone. J Leukoc Biol 57:317–325. https://doi.org/10.1002/jlb.57.2.317
Bushinsky DA (1996) Metabolic alkalosis decreases bone calcium efflux by suppressing osteoclasts and stimulating osteoblasts. Am J Physiol 271:F216–F222. https://doi.org/10.1152/ajprenal.1996.271.1.F216
Butler ST, Marr AL, Pelton SH et al (2003) Insulin restores GH responsiveness during lactation-induced negative energy balance in dairy cattle: effects on expression of IGF-I and GH receptor 1A. J Endocrinol 176:205–217. https://doi.org/10.1677/joe.0.1760205
Cai TQ, Weston PG, Lund LA et al (1994) Association between neutrophil functions and periparturient disorders in cows. Am J Vet Res 55:934–943
Calamari L, Soriani N, Panella G et al (2014) Rumination time around calving: an early signal to detect cows at greater risk of disease. J Dairy Sci 97:3635–3647. https://doi.org/10.3168/jds.2013-7709
Calder PC (1995) Fatty acids, dietary lipids and lymphocyte functions. Biochem Soc Trans 23:302–309. https://doi.org/10.1042/bst0230302
Calder PC (2006) Polyunsaturated fatty acids and inflammation. Prostaglandins Leukot Essent Fat Acids 75:197–202. https://doi.org/10.1016/j.plefa.2006.05.012
Calder PC (2008) The relationship between the fatty acid composition of immune cells and their function. Prostaglandins Leukot Essent Fat Acids 79:101–108. https://doi.org/10.1016/j.plefa.2008.09.016
Calder PC (2013) Feeding the immune system. Proc Nutr Soc 72:299–309. https://doi.org/10.1017/S0029665113001286
Calder PC, Bond JA, Harvey DJ et al (1990) Uptake and incorporation of saturated and unsaturated fatty acids into macrophage lipids and their effect upon macrophage adhesion and phagocytosis. Biochem J 269:807–814. https://doi.org/10.1042/bj2690807
Calsamiglia S, Cardozo PW, Ferret A, Bach A (2008) Changes in rumen microbial fermentation are due to a combined effect of type of diet and pH. J Anim Sci 86:702–711. https://doi.org/10.2527/jas.2007-0146
Campler M, Munksgaard L, Jensen MB et al (2014) Short communication: flooring preferences of dairy cows at calving. J Dairy Sci 97:892–896. https://doi.org/10.3168/jds.2013-7253
Cao Y-Z, Reddy CC, Sordillo LM (2000) Altered eicosanoid biosynthesis in selenium-deficient endothelial cells. Free Radic Biol Med 28:381–389. https://doi.org/10.1016/S0891-5849(99)00251-8
Cappa V, Trevisi E, Bertoni G (1989) Variazioni ematiche e produttive nel 1° mese di lattazione in bovine di allevamenti con o senza problemi "post-partum.". Zootec e Nutr Anim 15:645–660
Castell LM, Newsholme EA (1998) Glutamine and the effects of exhaustive exercise upon the immune response. Can J Physiol Pharmacol 76:524–532. https://doi.org/10.1139/y98-054
Castell JV, Gómez-Lechón MJ, David M et al (1989) Interleukin-6 is the major regulator of acute phase protein synthesis in adult human hepatocytes. FEBS Lett 242:237–239
Cattaneo L, Lopreiato V, Trevisi E, Minuti A (2020) Association of postpartum uterine diseases with lying time and metabolic profiles of multiparous Holstein dairy cows in the transition period. Vet J 263:105533. https://doi.org/10.1016/j.tvjl.2020.105533
Cattaneo L, Lopreiato V, Piccioli-Cappelli F et al (2021a) Plasma albumin-to-globulin ratio before dry-off as a possible index of inflammatory status and performance in the subsequent lactation in dairy cows. J Dairy Sci 104:8228–8242. https://doi.org/10.3168/jds.2020-19944
Cattaneo L, Mezzetti M, Lopreiato V et al (2021b) Gene network expression of whole blood leukocytes in dairy cows with different milk yield at dry-off. PLoS One 16:e0260745. https://doi.org/10.1371/journal.pone.0260745
Ceciliani F, Ceron JJ, Eckersall PD, Sauerwein H (2012) Acute phase proteins in ruminants. J Proteome 75:4207–4231. https://doi.org/10.1016/j.jprot.2012.04.004
Celi P (2010) Biomarkers of oxidative stress in ruminant medicine. Immunopharmacol Immunotoxicol 33:1–8

Celi P (2011) Oxidative stress in ruminants. In: Mandelker L, Vajdovich P (eds) Studies on veterinary medicine. Oxidative stress in applied basic research and clinical practice. Humana Press, Totowa, pp 191–231

Celi P, Gabai G (2015) Oxidant/antioxidant balance in animal nutrition and health: the role of protein oxidation. Front Vet Sci 2:48. https://doi.org/10.3389/fvets.2015.00048

Chaiyabutr N, Faulkner A, Peaker M (1980) The utilization of glucose for the synthesis of milk components in the fed and starved lactating goat in vivo. Biochem J 186:301–308. https://doi.org/10.1042/bj1860301

Choi BR, Allen MS (1999) Intake regulation by volatile fatty acids and physical fill. S Afr J Anim Sci 29:40–41

Cipak Gasparovic A, Zarkovic N, Zarkovic K et al (2017) Biomarkers of oxidative and nitro-oxidative stress: conventional and novel approaches. Br J Pharmacol 174:1771–1783. https://doi.org/10.1111/bph.13673

Cole JB, van Raden PM, O'Connell JR et al (2009) Distribution and location of genetic effects for dairy traits. J Dairy Sci 92:2931–2946. https://doi.org/10.3168/jds.2008-1762

Contreras LL, Ryan CM, Overton T (2004) Effects of dry cow grouping strategy and prepartum body condition score on performance and health of transition dairy cows. J Dairy Sci 87:517–523. https://doi.org/10.3168/jds.S0022-0302(04)73191-4

Contreras GA, O'Boyle NJ, Herdt TH, Sordillo LM (2010) Lipomobilization in periparturient dairy cows influences the composition of plasma nonesterified fatty acids and leukocyte phospholipid fatty acids. J Dairy Sci 93:2508–2516. https://doi.org/10.3168/jds.2009-2876

Cook NB (2019) Designing facilities for the adult dairy cow during the nonlactation and early lactation period. Vet Clin North Am Food Anim Pract 35:125–138. https://doi.org/10.1016/j.cvfa.2018.10.008

Creutzinger KC, Proudfoot KL (2020) Invited review: design and management of group maternity areas for dairy cows. Appl Anim Sci 36:124–132. https://doi.org/10.15232/aas.2019-01899

Crookenden MA, Heiser A, Murray A et al (2016) Parturition in dairy cows temporarily alters the expression of genes in circulating neutrophils. J Dairy Sci 99:6470–6483. https://doi.org/10.3168/jds.2015-10877

Crookenden MA, Walker CG, Heiser A et al (2017) Effects of precalving body condition and prepartum feeding level on gene expression in circulating neutrophils. J Dairy Sci 100:2310–2322. https://doi.org/10.3168/jds.2016-12105

Cullinan SB, Diehl JA (2006) Coordination of ER and oxidative stress signaling: the PERK/Nrf2 signaling pathway. Int J Biochem Cell Biol 38:317–332. https://doi.org/10.1016/j.biocel.2005.09.018

Dahl GE, Tao S, Laporta J (2020) Heat stress impacts immune status in cows across the life cycle. Front Vet Sci 7:116. https://doi.org/10.3389/fvets.2020.00116

Dalle-Donne I, Scaloni A, Giustarini D et al (2005) Proteins as biomarkers of oxidative/nitrosative stress in diseases: the contribution of redox proteomics. Mass Spectrom Rev 24:55–99. https://doi.org/10.1002/mas.20006

Dann HM, Litherland NB, Underwood JP et al (2006) Diets during far-off and close-up dry periods affect periparturient metabolism and lactation in multiparous cows. J Dairy Sci 89:3563–3577. https://doi.org/10.3168/jds.S0022-0302(06)72396-7

Dantzer R, Kelley KW (2007) Twenty years of research on cytokine-induced sickness behavior. Brain Behav Immun 21:153–160. https://doi.org/10.1016/j.bbi.2006.09.006

Davis SL (1998) Environmental modulation of the immune system via the endocrine system. Domest Anim Endocrinol 15:283–289. https://doi.org/10.1016/S0739-7240(98)00034-4

de Boer G, Trenkle A, Young JW (1985) Glucagon, insulin, growth hormone, and some blood metabolites during energy restriction ketonemia of lactating cows. J Dairy Sci 68:326–337. https://doi.org/10.3168/jds.S0022-0302(85)80829-8

De Koster JD, Opsomer G (2013) Insulin resistance in dairy cows. Vet Clin North Am Food Anim Pract 29:299–322. https://doi.org/10.1016/j.cvfa.2013.04.002

Dedon PC, Tannenbaum SR (2004) Reactive nitrogen species in the chemical biology of inflammation. Arch Biochem Biophys 423:12–22. https://doi.org/10.1016/j.abb.2003.12.017

Del Rio D, Stewart AJ, Pellegrini N (2005) A review of recent studies on malondialdehyde as toxic molecule and biological marker of oxidative stress. Nutr Metab Cardiovasc Dis 15:316–328. https://doi.org/10.1016/j.numecd.2005.05.003

Dieho K, Bannink A, Geurts IAL et al (2016) Morphological adaptation of rumen papillae during the dry period and early lactation as affected by rate of increase of concentrate allowance. J Dairy Sci 99:2339–2352. https://doi.org/10.3168/jds.2015-9837

Dijkstra J, Ellis JL, Kebreab E et al (2012) Ruminal pH regulation and nutritional consequences of low pH. Anim Feed Sci Technol 172:22–33. https://doi.org/10.1016/j.anifeedsci.2011.12.005

Dingwell RT, Kelton DF, Leslie KE, Edge VL (2001) Deciding to dry-off: does level of production matter? In: Reno NV (ed) National Mastitis Council Annual Meeting Proceedings. National Mastitis Council, Inc., Madison, pp 69–79

Dirksen GU, Liebich HG, Mayer E (1985) Adaptive changes of the ruminal mucosa and their functional and clinical significance. In: Bovine practitioner, pp 116–120

Do Nascimento CO, Hunter L, Trayhurn P (2004) Regulation of haptoglobin gene expression in 3T3-L1 adipocytes by cytokines, catecholamines, and PPARγ. Biochem Biophys Res Commun 313:702–708. https://doi.org/10.1016/j.bbrc.2003.12.008

Dobbelaar P, Bouwstra RJ, Goselink RMA et al (2010) Effects of vitamin E supplementation on and the association of body condition score with changes in peroxidative biomarkers and antioxidants around calving in dairy heifers. J Dairy Sci 93:3103–3113. https://doi.org/10.3168/jds.2009-2677

Doepel L, Pacheco D, Kennelly JJ et al (2004) Milk protein synthesis as a function of amino acid supply. J Dairy Sci 87:1279–1297. https://doi.org/10.3168/jds.S0022-0302(04)73278-6

Doepel L, Lessard M, Gagnon N et al (2006) Effect of postruminal glutamine supplementation on immune response and milk production in dairy cows. J Dairy Sci 89:3107–3121. https://doi.org/10.3168/jds.S0022-0302(06)72585-1

Doré V, Dubuc J, Bélanger AM, Buczinski S (2015) Definition of prepartum hyperketonemia in dairy goats. J Dairy Sci 98:4535–4543. https://doi.org/10.3168/jds.2014-9172

Dose J, Huebbe P, Nebel A, Rimbach G (2016) APOE genotype and stress response - a mini review. Lipids Health Dis 15:1–15. https://doi.org/10.1186/s12944-016-0288-2

Douglas GN, Overton TR, Bateman HG et al (2006) Prepartal plane of nutrition, regardless of dietary energy source, affects periparturient metabolism and dry matter intake in Holstein cows. J Dairy Sci 89:2141–2157. https://doi.org/10.3168/jds.S0022-0302(06)72285-8

Douglas GN, Rehage J, Beaulieu AD et al (2007) Prepartum nutrition alters fatty acid composition in plasma, adipose tissue, and liver lipids of periparturient dairy cows. J Dairy Sci 90:2941–2959. https://doi.org/10.3168/jds.2006-225

Drackley JK (1999) Biology of dairy cows during the transition period: the final frontier? J Dairy Sci 82:2259–2273. https://doi.org/10.3168/jds.S0022-0302(99)75474-3

Drackley JK, Overton TR, Douglas GN (2001) Adaptations of glucose and long-chain fatty acid metabolism in liver of dairy cows during the periparturient period. J Dairy Sci 84:E100–E112. https://doi.org/10.3168/jds.S0022-0302(01)70204-4

Drackley JK, Dann HM, Douglas N et al (2005) Physiological and pathological adaptations in dairy cows that may increase susceptibility to periparturient diseases and disorders. Ital J Anim Sci 4:323–344. https://doi.org/10.4081/ijas.2005.323

Drackley JK, Donkin SS, Reynolds CK (2006) Major advances in fundamental dairy cattle nutrition. J Dairy Sci 89:1324–1336. https://doi.org/10.3168/jds.S0022-0302(06)72200-7

Duffield TF (2000) Subclinical ketosis in lactating dairy cattle. Vet Clin North Am Food Anim Pract 16:231–253. https://doi.org/10.1016/S0749-0720(15)30103-1

Duffield TF, Rabiee AR, Lean IJ (2008) A meta-analysis of the impact of monensin in lactating dairy cattle. Part 3. Health and reproduction. J Dairy Sci 91:2328–2341. https://doi.org/10.3168/jds.2007-0801

Egger G, Dixon J (2009) Obesity and chronic disease: always offender or often just accomplice? Br J Nutr 102:1238–1244. https://doi.org/10.1017/S0007114509371676

Egger-Danner C, Cole JB, Pryce JE et al (2014) Invited review: overview of new traits and phenotyping strategies in dairy cattle with a focus on functional traits. Animal 9:191–207. https://doi.org/10.1017/S1751731114002614

Ehrhardt RA, Foskolos A, Giesy SL et al (2016) Increased plasma leptin attenuates adaptive metabolism in early lactating dairy cows. J Endocrinol 229:145–157. https://doi.org/10.1530/JOE-16-0031

Enemark JMD (2008) The monitoring, prevention and treatment of sub-acute ruminal acidosis (SARA): a review. Vet J 176:32–43. https://doi.org/10.1016/j.tvjl.2007.12.021

Esposito G, Irons PC, Webb EC, Chapwanya A (2014) Interactions between negative energy balance, metabolic diseases, uterine health and immune response in transition dairy cows. Anim Reprod Sci 144:60–71. https://doi.org/10.1016/j.anireprosci.2013.11.007

Fabris TF, Laporta J, Skibiel AL et al (2019) Effect of heat stress during early, late, and entire dry period on dairy cattle. J Dairy Sci 102:5647–5656. https://doi.org/10.3168/jds.2018-15721

Fain JN (2010) Release of inflammatory mediators by human adipose tissue is enhanced in obesity and primarily by the nonfat cells: a review. Mediat Inflamm 2010:1–20. https://doi.org/10.1155/2010/513948

Filipe J, Inglesi A, Amadori M et al (2021) Preliminary evidence of endotoxin tolerance in dairy cows during the transition period. Genes (Basel) 12:1801. https://doi.org/10.3390/genes12111801

Finkel T (2011) Signal transduction by reactive oxygen species. J Cell Biol 194:7. https://doi.org/10.1083/jcb.201102095

Fiore E, Perillo L, Gianesella M et al (2021) Comparison between two preventive treatments for hyperketonaemia carried out pre-partum: effects on non-esterified fatty acids, β-hydroxybutyrate and some biochemical parameters during peripartum and early lactation. J Dairy Res 88:38–44. https://doi.org/10.1017/S0022029921000108

Fleck A (1989) Clinical and nutritional aspects of changes in acute-phase proteins during inflammation. Proc Nutr Soc 48:347–354. https://doi.org/10.1079/PNS19890050

Fraser D (2008) Animal welfare and the intensification of animal production, pp 167–189

Fregonesi JA, Veira DM, von Keyserlingk MAG, Weary DM (2007) Effects of bedding quality on lying behavior of dairy cows. J Dairy Sci 90:5468–5472. https://doi.org/10.3168/jds.2007-0494

Gabanyi I, Muller PA, Feighery L et al (2016) Neuroimmune interactions drive tissue programming in intestinal macrophages. Cell 164:378–391. https://doi.org/10.1016/j.cell.2015.12.023

Gamelli RL, Liu H, He LK, Hofmann CA (1996) Augmentations of glucose uptake and glucose transporter-1 in macrophages following thermal injury and sepsis in mice. J Leukoc Biol 59:639–647

Gessner DK, Schlegel G, Keller J et al (2013) Expression of target genes of nuclear factor E2-related factor 2 in the liver of dairy cows in the transition period and at different stages of lactation. J Dairy Sci 96:1038–1043. https://doi.org/10.3168/jds.2012-5967

Gessner DK, Schlegel G, Ringseis R et al (2014) Up-regulation of endoplasmic reticulum stress induced genes of the unfolded protein response in the liver of periparturient dairy cows. BMC Vet Res 10:46. https://doi.org/10.1186/1746-6148-10-46

Giblett ER (1961) Haptoglobin: a review. Int J Transfus Med 6:513–524. https://doi.org/10.1111/j.1423-0410.1961.tb03200.x

Glahn DC, Knowles EEM, McKay DR et al (2014) Arguments for the sake of endophenotypes: examining common misconceptions about the use of endophenotypes in psychiatric genetics. Am J Med Genet Part B Neuropsychiatr Genet 165:122–130. https://doi.org/10.1002/ajmg.b.32221

Godden S, Rapnicki P, Stewart S et al (2003) Effectiveness of an internal teat seal in the prevention of new intramammary infections during the dry and early-lactation periods in dairy cows when used with a dry cow intramammary antibiotic. J Dairy Sci 86:3899–3911. https://doi.org/10.3168/jds.S0022-0302(03)73998-8

Goff JP (1999) Treatment of calcium, phosphorus, and magnesium balance disorders. Vet Clin North Am Food Anim Pract 15:619–639. https://doi.org/10.1016/S0749-0720(15)30167-5

Goff JP (2000) Pathophysiology of calcium and phosphorus disorders. Vet Clin North Am Food Anim Pract 16:319–337. https://doi.org/10.1016/S0749-0720(15)30108-0

Goff JP (2004) Macromineral disorders of the transition cow. Vet Clin North Am Food Anim Pract 20:471–494. https://doi.org/10.1016/j.cvfa.2004.06.003

Goff JP (2006) Mineral disorders of the transition period: origin and control. In: Proceedings of the World Buiatrics Congress. Nice, France

Goff JP (2008) The monitoring, prevention, and treatment of milk fever and subclinical hypocalcemia in dairy cows. Vet J 176:50–57. https://doi.org/10.1016/j.tvjl.2007.12.020

Goff JP, Horst RL (1997) Physiological changes at parturition and their relationship to metabolic disorders. J Dairy Sci 80:1260–1268

Goff JP, Reinhardt TA, Horst RL (1991) Enzymes and factors controlling vitamin D metabolism and action in normal and milk fever cows. J Dairy Sci 74:4022–4032. https://doi.org/10.3168/jds.S0022-0302(91)78597-4

Goff JP, Liesegang A, Horst RL (2014) Diet-induced pseudohypoparathyroidism: a hypocalcemia and milk fever risk factor. J Dairy Sci 97:1520–1528. https://doi.org/10.3168/jds.2013-7467

Gong JG, Lee WJ, Garnsworthy PC, Webb R (2002) Effect of dietary-induced increases in circulating insulin concentrations during the early postpartum period on reproductive function in dairy cows. Reproduction 123:419–427. https://doi.org/10.1530/rep.0.1230419

Gorjão R, Verlengia R, de Lima TM et al (2006) Effect of docosahexaenoic acid-rich fish oil supplementation on human leukocyte function. Clin Nutr 25:923–938. https://doi.org/10.1016/j.clnu.2006.03.004

Gross J, van Dorland HA, Bruckmaier RM, Schwarz FJ (2011a) Performance and metabolic profile of dairy cows during a lactational and deliberately induced negative energy balance with subsequent realimentation. J Dairy Sci 94:1820–1830

Gross J, van Dorland HA, Schwarz FJ, Bruckmaier RM (2011b) Endocrine changes and liver mRNA abundance of somatotropic axis and insulin system constituents during negative energy balance at different stages of lactation in dairy cows. J Dairy Sci 94:3484–3494. https://doi.org/10.3168/jds.2011-4251

Gross JJ, Grossen-Rösti L, Héritier R et al (2018) Inflammatory and metabolic responses to an intramammary lipopolysaccharide challenge in early lactating cows supplemented with conjugated linoleic acid. J Anim Physiol Anim Nutr (Berl) 102:e838–e848. https://doi.org/10.1111/jpn.12843

Grum DE, Drackley JK, Younker RS et al (1996) Nutrition during the dry period and hepatic lipid metabolism of periparturient dairy cows. J Dairy Sci. https://doi.org/10.3168/jds.S0022-0302(96)76553-0

Grum DE, Drackley JK, Clark JH (2002) Fatty acid metabolism in liver of dairy cows fed supplemental fat and nicotinic acid during an entire lactation. J Dairy Sci 85:3026–3034. https://doi.org/10.3168/jds.S0022-0302(02)74388-9

Grummer RR (1995) Impact of changes in organic nutrient metabolism on feeding the transition dairy cow. J Anim Sci 73:2820. https://doi.org/10.2527/1995.7392820x

Grummer RR, Rastani RR (2004) Why reevaluate dry period length? J Dairy Sci 87:E77–E85. https://doi.org/10.3168/jds.S0022-0302(04)70063-6

Grummer RR, Mashek DG, Hayirli A (2004) Dry matter intake and energy balance in the transition period. Vet Clin North Am Food Anim Pract 20:447–470. https://doi.org/10.1016/j.cvfa.2004.06.013

Guarini AR, Lourenco DAL, Brito LF et al (2019) Genetics and genomics of reproductive disorders in Canadian Holstein cattle. J Dairy Sci 102:1341–1353. https://doi.org/10.3168/jds.2018-15038

Guilherme A, Henriques F, Bedard AH, Czech MP (2019) Molecular pathways linking adipose innervation to insulin action in obesity and diabetes mellitus. Nat Rev Endocrinol 15:207–225. https://doi.org/10.1038/s41574-019-0165-y

Habel J, Sundrum A (2020) Mismatch of glucose allocation between different life functions in the transition period of dairy cows. Animals 10:1028. https://doi.org/10.3390/ani10061028

Halachmi I, Guarino M, Bewley J, Pastell M (2019) Smart animal agriculture: application of real-time sensors to improve animal well-being and production. Annu Rev Anim Biosci 7:403–425. https://doi.org/10.1146/annurev-animal-020518-114851

Halliwell B (2007) Biochemistry of oxidative stress. Biochem Soc Trans 35:1147–1150. https://doi.org/10.1042/BST0351147

Halliwell B, Whiteman M (2004) Measuring reactive species and oxidative damage in vivo and in cell culture: how should you do it and what do the results mean? Br J Pharmacol 142:231–255. https://doi.org/10.1038/sj.bjp.0705776

Halliwell B, Gutteridge JMC (2015) Free Radicals in Biology and Medicine, 5th edn. Oxford university press

Hayes BJ, Pryce J, Chamberlain AJ et al (2010) Genetic architecture of complex traits and accuracy of genomic prediction: coat colour, milk-fat percentage, and type in Holstein cattle as contrasting model traits. PLoS Genet 6:e1001139. https://doi.org/10.1371/journal.pgen.1001139

Hayirli A, Grummer RR, Nordheim EV, Crump PM (2002) Animal and dietary factors affecting feed intake during the prefresh transition period in Holsteins. J Dairy Sci 85:3430–3443. https://doi.org/10.3168/jds.S0022-0302(02)74431-7

Hendriks SJ, Phyn CVC, Huzzey JM et al (2020) Graduate student literature review: evaluating the appropriate use of wearable accelerometers in research to monitor lying behaviors of dairy cows. J Dairy Sci 103:12140–12157. https://doi.org/10.3168/jds.2019-17887

Herdt TH et al (1981) Blood concentrations of beta hydroxybutyrate in clinically normal Holstein-Friesian herds and in those with a high prevalence of clinical ketosis. Am J Vet Res 42(3):503–506

Herdt TH (2000) Ruminant adaptation to negative energy balance. Influences on the etiology of ketosis and fatty liver. Vet Clin North Am Food Anim Pract 16:215–230. https://doi.org/10.1016/S0749-0720(15)30102-X

Hibbard B, Peters JP, Chester ST et al (1995) The effect of slaframine on salivary output and subacute and acute acidosis in growing beef steers2. J Anim Sci 73:516–525. https://doi.org/10.2527/1995.732516x

Higuchi H, Katoh N, Miyamoto T et al (1994) Dexamethasone-induced haptoglogin release by calf liver parenchymal cells. Am J Vet Res 55:1080–1085

Hoeben D, Heyneman R, Burvenich C (1997) Elevated levels of β-hydroxybutyric acid in periparturient cows and in vitro effect on respiratory burst activity of bovine neutrophils. Vet Immunol Immunopathol 58:165–170. https://doi.org/10.1016/S0165-2427(97)00031-7

Hoeben D, Burvenich C, Trevisi E et al (2000) Role of endotoxin and TNF-α in the pathogenesis of experimentally induced coliform mastitis in periparturient cows. J Dairy Res 67:503–514. https://doi.org/10.1017/S0022029900004489

Hodgkinson AJ, Carpenter EA, Smith CS, Molan PC, Prosser CG (2007) Adhesion molecule expression in the bovine mammary gland. Vet Immunol Immunopathol 115:205–215. https://doi.org/10.1016/j.vetimm.2006.10.020

Höglund J, Rafati N, Rask-Andersen M et al (2019) Improved power and precision with whole genome sequencing data in genome-wide association studies of inflammatory biomarkers. Sci Rep 9:1–14. https://doi.org/10.1038/s41598-019-53111-7

Horst RL, Goff JP, Reinhardt TA (1997) Calcium and vitamin D metabolism during lactation. J Mammary Gland Biol Neoplasia. https://doi.org/10.1023/A:1026384421273

Horst EA, Kvidera SK, Dickson MJ et al (2019) Effects of continuous and increasing lipopolysaccharide infusion on basal and stimulated metabolism in lactating Holstein cows. J Dairy Sci 102:3584–3597. https://doi.org/10.3168/jds.2018-15627

Hotamisligil GS (2006) Inflammation and metabolic disorders. Nature 444:860–867. https://doi.org/10.1038/nature05485

Huang H, Cao J, Hanif Q et al (2019) Genome-wide association study identifies energy metabolism genes for resistance to ketosis in Chinese Holstein cattle. Anim Genet 50:376–380. https://doi.org/10.1111/age.12802

Hughes DA, Pinder AC (2000) N-3 polyunsaturated fatty acids inhibit the antigen-presenting function of human monocytes. Am J Clin Nutr 71:357S–360S

Huzzey JM, Duffield TF, LeBlanc SJ et al (2009) Short communication: haptoglobin as an early indicator of metritis. J Dairy Sci 92:621–625. https://doi.org/10.3168/jds.2008-1526

Huzzey JM, Nydam DV, Grant RJ, Overton TR (2011) Associations of prepartum plasma cortisol, haptoglobin, fecal cortisol metabolites, and nonesterified fatty acids with postpartum health status in Holstein dairy cows. J Dairy Sci 94:5878–5889. https://doi.org/10.3168/jds.2010-3391

Ingvartsen KL (2006) Feeding- and management-related diseases in the transition cow: physiological adaptations around calving and strategies to reduce feeding-related diseases. Anim Feed Sci Technol 126:175–213

Ingvartsen KL, Andersen JB (2000) Integration of metabolism and intake regulation: a review focusing on periparturient animals. J Dairy Sci 83:1573–1597. https://doi.org/10.3168/jds.S0022-0302(00)75029-6

Ingvartsen KL, Moyes KM (2015) Factors contributing to immunosuppression in the dairy cow during the periparturient period. Jpn J Vet Res 63:S15–S24. https://doi.org/10.14943/jjvr.63.suppl.s15

Ingvartsen K, Dewhurst R, Friggens N (2003) On the relationship between lactational performance and health: is it yield or metabolic imbalance that cause production diseases in dairy cattle? A position paper. Livest Prod Sci 83:277–308. https://doi.org/10.1016/S0301-6226(03)00110-6

INRA (1989) Ruminant nutrition: recommended allowances and feed tables. John Libbey, London

Irwin MR, Cole SW (2011) Reciprocal regulation of the neural and innate immune systems. Nat Rev Immunol 11:625–632. https://doi.org/10.1038/nri3042

Itle AJ, Huzzey JM, Weary DM, von Keyserlingk MAG (2015) Clinical ketosis and standing behavior in transition cows. J Dairy Sci 98:128–134. https://doi.org/10.3168/jds.2014-7932

Jahan N, Minuti A, Trevisi E (2015) Assessment of immune response in periparturient dairy cows using ex vivo whole blood stimulation assay with lipopolysaccharides and carrageenan skin test. Vet

Immunol Immunopathol 165:119–126. https://doi.org/10.1016/j.vetimm.2015.04.003
Janovick NA, Drackley JK (2010) Prepartum dietary management of energy intake affects postpartum intake and lactation performance by primiparous and multiparous Holstein cows. J Dairy Sci 93:3086–3102. https://doi.org/10.3168/jds.2009-2656
Janovick Guretzky NA, Dann HM, Bionaz M et al (2007) Evaluation of acute phase reactants and indices of liver function in serum from dairy cows fed different levels on energy prepartum. In: Joint annual meeting ADSA, PSA, AMPA, ASAS. S. Antonio, Texas, p 408
Janovick-Guretzky NA, Trevisi E, Bertoni G et al (2023) Prepartum plane of energy intake affects serum biomarkers for inflammation and liver function during the periparturient period. J Dairy Sci 106:168–186
Joffre O, Nolte MA, Spörri R, Sousa CRE (2009) Inflammatory signals in dendritic cell activation and the induction of adaptive immunity. Immunol Rev 227:234–247. https://doi.org/10.1111/j.1600-065X.2008.00718.x
Jofre-Monseny L, Minihane AM, Rimbach G (2008) Impact of apoE genotype on oxidative stress, inflammation and disease risk. Mol Nutr Food Res 52:131–145. https://doi.org/10.1002/mnfr.200700322
Johnson R (1998) Immune and endocrine regulation of food intake in sick animals. Domest Anim Endocrinol 15:309–319. https://doi.org/10.1016/S0739-7240(98)00031-9
Kankofer M (2001) Antioxidative defence mechanisms against reactive oxygen species in bovine retained and not-retained placenta: activity of glutathione peroxidase, glutathione transferase, catalase and superoxide dismutase. Placenta 22:466–472. https://doi.org/10.1053/plac.2001.0650
Kaufman EI, LeBlanc SJ, McBride BW et al (2016) Short communication: association of lying behavior and subclinical ketosis in transition dairy cows. J Dairy Sci 99:7473–7480. https://doi.org/10.3168/jds.2016-11185
Kehrli ME, Nonnecke BJ, Roth JA (1989) Alterations in bovine neutrophil function during the periparturient period. Am J Vet Res 50:207–214
Kelley KW, Weigent DA, Kooijman R (2007) Protein hormones and immunity. Brain Behav Immun 21:384–392. https://doi.org/10.1016/j.bbi.2006.11.010
Kim J, Cha Y-N, Surh Y-J (2010) A protective role of nuclear factor-erythroid 2-related factor-2 (Nrf2) in inflammatory disorders. Mutat Res 690:12–23. https://doi.org/10.1016/j.mrfmmm.2009.09.007
Kimura K, Goff JP, Kehrli ME (1999) Effects of the presence of the mammary gland on expression of neutrophil adhesion molecules and myeloperoxidase activity in periparturient dairy cows. J Dairy Sci 82:2385–2392. https://doi.org/10.3168/jds.S0022-0302(99)75489-5
Kimura K, Goff JP, Kehrli ME et al (2002) Effects of mastectomy on composition of peripheral blood mononuclear cell populations in periparturient dairy cows. J Dairy Sci 85:1437–1444. https://doi.org/10.3168/jds.S0022-0302(02)74211-2
Kimura K, Goff JP, Canning P et al (2014) Effect of recombinant bovine granulocyte colony-stimulating factor covalently bound to polyethylene glycol injection on neutrophil number and function in periparturient dairy cows. J Dairy Sci 97:4842–4851. https://doi.org/10.3168/jds.2013-7242
Klein M, Buttchereit N, Miemczyk S et al (2012) NMR metabolomic analysis of dairy cows reveals milk glycerophosphocholine to phosphocholine ratio as prognostic biomarker for risk of ketosis. J Proteome Res 11:1373–1381
Komaragiri MV, Erdman RA (1997) Factors affecting body tissue mobilization in early lactation dairy cows. 1. Effect of dietary protein on mobilization of body fat and protein. J Dairy Sci 80:929–937
Komatsu T, Itoh F, Kushibiki S, Hodate K (2005) Changes in gene expression of glucose transporters in lactating and nonlactating cows1. J Anim Sci 83:557–564. https://doi.org/10.2527/2005.833557x
Kominsky DJ, Campbell EL, Colgan SP (2010) Metabolic shifts in immunity and inflammation. J Immunol 184:4062–4068. https://doi.org/10.4049/jimmunol.0903002
König S, May K (2019) Invited review: phenotyping strategies and quantitative-genetic background of resistance, tolerance and resilience associated traits in dairy cattle. Animal 13:897–908. https://doi.org/10.1017/S1751731118003208
Kuhla B (2020) Review: pro-inflammatory cytokines and hypothalamic inflammation: implications for insufficient feed intake of transition dairy cows. Animal 14:s65–s77. https://doi.org/10.1017/S1751731119003124
Kuhla B, Nürnberg G, Albrecht D et al (2011) Involvement of skeletal muscle protein, glycogen, and fat metabolism in the adaptation on early lactation of dairy cows. J Proteome Res 10:4252–4262. https://doi.org/10.1021/pr200425h
Kushibiki S, Hodate K, Shingu H et al (2001) Insulin resistance induced in dairy steers by tumor necrosis factor alpha is partially reversed by 2,4-thiazolidinedione. Domest Anim Endocrinol 21:25–37. https://doi.org/10.1016/S0739-7240(01)00102-3
Kushibiki S, Hodate K, Shingu H et al (2003) Metabolic and lactational responses during recombinant bovine tumor necrosis factor-α treatment in lactating cows. J Dairy Sci 86:819–827. https://doi.org/10.3168/jds.S0022-0302(03)73664-9
Kvidera SK, Horst EA, Abuajamieh M et al (2017) Glucose requirements of an activated immune system in lactating Holstein cows. J Dairy Sci 100:2360–2374. https://doi.org/10.3168/jds.2016-12001
Lacetera N, Scalia D, Franci O et al (2004) Short communication: effects of nonesterified fatty acids on lymphocyte function in dairy heifers. J Dairy Sci 87:1012–1014. https://doi.org/10.3168/jds.S0022-0302(04)73246-4
Lacetera N, Scalia D, Bernabucci U et al (2005) Lymphocyte functions in overconditioned cows

around parturition. J Dairy Sci 88:2010–2016. https://doi.org/10.3168/jds.S0022-0302(05)72877-0

Laeger T, Sauerwein H, Tuchscherer A et al (2013) Concentrations of hormones and metabolites in cerebrospinal fluid and plasma of dairy cows during the periparturient period. J Dairy Sci 96:2883–2893

Lamote I, Meyer E, De Ketelaere A et al (2006) Influence of sex steroids on the viability and CD11b, CD18 and CD47 expression of blood neutrophils from dairy cows in the last month of gestation. Vet Res 37:61–74. https://doi.org/10.1051/vetres:2005038

Leal Yepes FA, Nydam DV, Mann S et al (2019) Longitudinal phenotypes improve genotype association for hyperketonemia in dairy cattle. Animals 9:1059. https://doi.org/10.3390/ani9121059

LeBlanc SJ (2014) Reproductive tract inflammatory disease in postpartum dairy cows. Animal 8:54–63. https://doi.org/10.1017/S1751731114000524

LeBlanc SJ, Lissemore KD, Kelton DF et al (2006) Major advances in disease prevention in dairy cattle. J Dairy Sci 89:1267–1279. https://doi.org/10.3168/jds.S0022-0302(06)72195-6

Lee JY, Plakidas A, Lee WH et al (2003) Differential modulation of Toll-like receptors by fatty acids: preferential inhibition by n-3 polyunsaturated fatty acids. J Lipid Res 44:479–486. https://doi.org/10.1194/jlr.M200361-JLR200

Lee JY, Zhao L, Youn HS et al (2004) Saturated fatty acid activates but polyunsaturated fatty acid inhibits toll-like receptor 2 dimerized with toll-like receptor 6 or 1. J Biol Chem 279:16971–16979. https://doi.org/10.1074/jbc.M312990200

Lee JY, Zhao L, Hwang DH (2010) Modulation of pattern recognition receptor-mediated inflammation and risk of chronic diseases by dietary fatty acids. Nutr Rev 68:38–61. https://doi.org/10.1111/j.1753-4887.2009.00259.x

Leprell C, Hughes L, Miller J (2007) Free radical scavengers vitamins A, C, and E plus magnesium reduce noise trauma. Free Radic Biol Med 42:1454–1463. https://doi.org/10.1016/j.freeradbiomed.2007.02.008

Leroy JLMR, Vanholder T, Van Knegsel ATM et al (2008) Nutrient prioritization in dairy cows early postpartum: mismatch between metabolism and fertility? Reprod Domest Anim 43:96–103. https://doi.org/10.1111/j.1439-0531.2008.01148.x

Liboreiro DN, Machado KS, Silva PRB et al (2015) Characterization of peripartum rumination and activity of cows diagnosed with metabolic and uterine diseases. J Dairy Sci 98:6812–6827. https://doi.org/10.3168/jds.2014-8947

Linzell JL (1972) Mechanism of secretion of the aqueous phase of milk. J Dairy Sci 55:1316–1322. https://doi.org/10.3168/jds.S0022-0302(72)85670-4

Liu J, Li L, Chen X et al (2019) Effects of heat stress on body temperature, milk production, and reproduction in dairy cows: a novel idea for monitoring and evaluation of heat stress — a review. Asian-Aust J Anim Sci 32:1332–1339. https://doi.org/10.5713/ajas.18.0743

Loor JJ, Dann HM, Guretzky NAJ et al (2006) Plane of nutrition prepartum alters hepatic gene expression and function in dairy cows as assessed by longitudinal transcript and metabolic profiling. Physiol Genomics 27:29–41. https://doi.org/10.1152/physiolgenomics.00036.2006

Loor JJ, Everts RE, Bionaz M et al (2007) Nutrition-induced ketosis alters metabolic and signaling gene networks in liver of periparturient dairy cows. Physiol Genomics 32:105–116. https://doi.org/10.1152/physiolgenomics.00188.2007

Lopes F, Rosa G, Pinedo P et al (2020) Genome-enable prediction for health traits using high-density SNP panel in US Holstein cattle. Anim Genet 51:192–199. https://doi.org/10.1111/age.12892

Lopreiato V, Vailati-Riboni M, Bellingeri A et al (2019) Inflammation and oxidative stress transcription profiles due to in vitro supply of methionine with or without choline in unstimulated blood polymorphonuclear leukocytes from lactating Holstein cows. J Dairy Sci 102:10395–10410. https://doi.org/10.3168/jds.2019-16413

Lopreiato V, Mezzetti M, Cattaneo L et al (2020a) Role of nutraceuticals during the transition period of dairy cows: a review. J Anim Sci Biotechnol 11:96. https://doi.org/10.1186/s40104-020-00501-x

Lopreiato V, Palma E, Minuti A et al (2020b) Pegbovigrastim treatment around parturition enhances postpartum immune response gene network expression of whole blood leukocytes in Holstein and Simmental cows. Animals 10:621. https://doi.org/10.3390/ani10040621

Lubos E, Loscalzo J, Handy DE (2011) Glutathione peroxidase-1 in health and disease: from molecular mechanisms to therapeutic opportunities. Antioxid Redox Signal 15:1957–1997. https://doi.org/10.1089/ars.2010.3586

Maddox JF, Aherne KM, Reddy CC, Sordillo LM (1999) Increased neutrophil adherence and adhesion molecule mRNA expression in endothelial cells during selenium deficiency. J Leukoc Biol 65:658–664. https://doi.org/10.1002/jlb.65.5.658

Majewski AC, Hansen PJ (2002) Progesterone inhibits rejection of xenogeneic transplants in the sheep uterus. Horm Res Paediatr 58:128–135. https://doi.org/10.1159/000063578

Mamedova LK, Yuan K, Laudick AN et al (2013) Toll-like receptor 4 signaling is required for induction of gluconeogenic gene expression by palmitate in human hepatic carcinoma cells. J Nutr Biochem 24:1499–1507. https://doi.org/10.1016/j.jnutbio.2012.12.009

Mammi LME, Guadagnini M, Mechor G et al (2021) The use of Monensin for ketosis prevention in dairy cows during the transition period: a systematic review. Animals 11:1988. https://doi.org/10.3390/ani11071988

Maratou E, Dimitriadis G, Kollias A et al (2007) Glucose transporter expression on the plasma membrane of resting and activated white blood cells. Eur J Clin Investig 37:282–290. https://doi.org/10.1111/j.1365-2362.2007.01786.x

Marras G, Gaspa G, Sorbolini S et al (2015) Analysis of runs of homozygosity and their relationship with inbreeding in five cattle breeds farmed in Italy. Anim Genet 46:110–121. https://doi.org/10.1111/age.12259
Mattmiller SA, Corl CM, Gandy JC et al (2011) Glucose transporter and hypoxia-associated gene expression in the mammary gland of transition dairy cattle. J Dairy Sci 94:2912–2922. https://doi.org/10.3168/jds.2010-3936
Mayasari N, Chen J, Ferrari A et al (2017) Effects of dry period length and dietary energy source on inflammatory biomarkers and oxidative stress in dairy cows. J Dairy Sci 100:4961–4975. https://doi.org/10.3168/jds.2016-11857
Mayasari N, Trevisi E, Ferrari A et al (2019) Relationship between inflammatory biomarkers and oxidative stress with uterine health in dairy cows with different dry period lengths. Transl Anim Sci 3:607–619. https://doi.org/10.1093/tas/txz040
McCabe CJ, Boerman JP (2020) Invited review: quantifying protein mobilization in dairy cows during the transition period. Appl Anim Sci 36:389–396. https://doi.org/10.15232/aas.2019-01929
McCabe M, Waters S, Morris D et al (2012) RNA-seq analysis of differential gene expression in liver from lactating dairy cows divergent in negative energy balance. BMC Genomics 13:193. https://doi.org/10.1186/1471-2164-13-193
McDougall S, LeBlanc SJ, Heiser A (2017) Effect of prepartum energy balance on neutrophil function following pegbovigrastim treatment in periparturient cows. J Dairy Sci 100:7478–7492. https://doi.org/10.3168/JDS.2017-12786
McGuffey RK, Richardson LF, Wilkinson JID (2001) Ionophores for dairy cattle: current status and future outlook. J Dairy Sci 84:E194–E203. https://doi.org/10.3168/jds.S0022-0302(01)70218-4
McNeel AK, Reiter BC, Weigel D et al (2017) Validation of genomic predictions for wellness traits in US Holstein cows. J Dairy Sci 100:9115–9124. https://doi.org/10.3168/jds.2016-12323
Marutsova VJ, Binev RG (2020) Changes in blood enzyme activities and some liver parameters in goats with subclinical ketosis. Bulg J Vet Med 23:70–79. https://doi.org/10.15547/bjvm.2175
Mehrzad J, Dosogne H, Meyer E et al (2001) Respiratory burst activity of blood and milk neutrophils in dairy cows during different stages of lactation. J Dairy Res 68:399–415. https://doi.org/10.1017/S0022029901005039
Mezzetti M, Minuti A, Piccioli-Cappelli F et al (2019a) The role of altered immune function during the dry period in promoting the development of subclinical ketosis in early lactation. J Dairy Sci 102:9241–9258. https://doi.org/10.3168/jds.2019-16497
Mezzetti M, Minuti A, Piccioli-Cappelli F et al (2019b) Administration of an immune stimulant during the transition period improved lipid metabolism and rumination without affecting inflammatory status. Animals 9:619. https://doi.org/10.3390/ani9090619
Mezzetti M, Piccioli-Cappelli F, Bani P et al (2019c) Monensin controlled-release capsule administered in late-pregnancy differentially affects rumination patterns, metabolic status, and cheese-making properties of the milk in primiparous and multiparous cows. Ital J Anim Sci 18:1271–1283. https://doi.org/10.1080/1828051X.2019.1645623
Mezzetti M, Minuti A, Piccioli-Cappelli F, Trevisi E (2020) Inflammatory status and metabolic changes at dry-off in high-yield dairy cows. Ital J Anim Sci 19:51–65. https://doi.org/10.1080/1828051X.2019.1691472
Mezzetti M, Cattaneo L, Passamonti MM et al (2021) The transition period updated: a review of the new insights into the adaptation of dairy cows to the new lactation. Dairy 2:617–636. https://doi.org/10.3390/dairy2040048
Milanesi M, Passamonti MM, Cappelli K et al (2021) Genetic regulation of biomarkers as stress proxies in dairy cows. Genes (Basel) 12:534. https://doi.org/10.3390/genes12040534
Mills KE, Weary DM, von Keyserlingk MAG (2020) Identifying barriers to successful dairy cow transition management. J Dairy Sci 103:1749–1758. https://doi.org/10.3168/jds.2018-16231
Minuti A, Bani P, Piccioli-Cappelli F et al (2015a) Metabolic and biochemical changes in plasma of the periparturient rabbit does with different litter size. Animal 9:614–621. https://doi.org/10.1017/S1751731114002675
Minuti A, Palladino A, Khan MJ et al (2015b) Abundance of ruminal bacteria, epithelial gene expression, and systemic biomarkers of metabolism and inflammation are altered during the peripartal period in dairy cows. J Dairy Sci 98:8940–8951. https://doi.org/10.3168/jds.2015-9722
Minuti A, Bionaz M, Lopreiato V et al (2020) Prepartum dietary energy intake alters adipose tissue transcriptome profiles during the periparturient period in Holstein dairy cows. J Anim Sci Biotechnol 11:1. https://doi.org/10.1186/s40104-019-0409-7
Moallem U (2018) Invited review: roles of dietary n-3 fatty acids in performance, milk fat composition, and reproductive and immune systems in dairy cattle. J Dairy Sci 101:8641–8661. https://doi.org/10.3168/jds.2018-14772
Mondaca MR (2019) Ventilation systems for adult dairy cattle. Vet Clin North Am Food Anim Pract 35:139–156. https://doi.org/10.1016/j.cvfa.2018.10.006
Moretti P, Paltrinieri S, Trevisi E et al (2017) Reference intervals for hematological and biochemical parameters, acute phase proteins and markers of oxidation in Holstein dairy cows around 3 and 30 days after calving. Res Vet Sci 114:322–331. https://doi.org/10.1016/j.rvsc.2017.06.012
Moyes KM, Drackley JK, Morin DE, Loor JJ (2010) Greater expression of TLR2, TLR4, and IL6 due to negative energy balance is associated with lower expression of HLA-DRA and HLA-A in bovine blood neutrophils after intramammary mastitis challenge

with Streptococcus uberis. Funct Integr Genomics 10:53–61. https://doi.org/10.1007/s10142-009-0154-7
Mulligan FJ, Doherty ML (2008) Production diseases of the transition cow. Vet J 176:3–9. https://doi.org/10.1016/j.tvjl.2007.12.018
Nace EL, Nickerson SC, Kautz FM et al (2014) Modulation of innate immune function and phenotype in bred dairy heifers during the periparturient period induced by feeding an immunostimulant for 60 days prior to delivery. Vet Immunol Immunopathol 161:240–250. https://doi.org/10.1016/j.vetimm.2014.08.013
Nair S, Doh ST, Chan JY et al (2008) Regulatory potential for concerted modulation of Nrf2- and Nfkb1-mediated gene expression in inflammation and carcinogenesis. Br J Cancer 99:2070–2082. https://doi.org/10.1038/sj.bjc.6604703
NASEM (2021) Nutrient requirements of dairy cattle, 8th rev edn. National Academies Press, Washington, D.C.
National Mastitis Council (2006) Dry cow therapy. NMC Factsheet. https://www.nmconline.org/wp-content/uploads/2016/09/Dry-Cow-Therapy.pdf
Nayeri S, Schenkel F, Fleming A et al (2019) Genome-wide association analysis for β-hydroxybutyrate concentration in milk in Holstein dairy cattle. BMC Genet 20:58. https://doi.org/10.1186/s12863-019-0761-9
Neave HW, Lomb J, von Keyserlingk MAG et al (2017) Parity differences in the behavior of transition dairy cows. J Dairy Sci 100:548–561. https://doi.org/10.3168/jds.2016-10987
Neave HW, Lomb J, Weary DM et al (2018) Behavioral changes before metritis diagnosis in dairy cows. J Dairy Sci 101:4388–4399. https://doi.org/10.3168/jds.2017-13078
Newsholme EA, Crabtree B, Ardawi MSM (1985) The role of high rates of glycolysis and glutamine utilization in rapidly dividing cells. Biosci Rep 5:393–400. https://doi.org/10.1007/BF01116556
Newsholme P, Curi R, Pithon-Curi TC et al (1999) Glutamine metabolism by lymphocytes, macrophages, and neutrophils: its importance in health and disease. J Nutr Biochem 10:316–324. https://doi.org/10.1016/S0955-2863(99)00022-4
Nightingale CR, Sellers MD, Ballou MA (2015) Elevated plasma haptoglobin concentrations following parturition are associated with elevated leukocyte responses and decreased subsequent reproductive efficiency in multiparous Holstein dairy cows. Vet Immunol Immunopathol 164:16–23. https://doi.org/10.1016/j.vetimm.2014.12.016
Nocek JE (1997) Bovine acidosis: implications on laminitis. J Dairy Sci 80:1005–1028. https://doi.org/10.3168/jds.S0022-0302(97)76026-0
Nonnecke BJ, Franklin ST, Young JW (1992) Effects of ketones, acetate, and glucose on in vitro immunoglobulin secretion by bovine lymphocytes. J Dairy Sci 75:982–990. https://doi.org/10.3168/jds.S0022-0302(92)77840-0
Nonnecke BJ, Reinhardt TA, Franklin ST (1993) Retinoid-induced modulation of immunoglobulin M secretion by bovine mononuclear leukocytes in vitro. J Dairy Sci 76:2175–2183. https://doi.org/10.3168/jds.S0022-0302(93)77554-2
NRC (2001) Nutrient requirement of dairy cattle, 7th ed. Washington, DC
O'Boyle NJ, Contreras GA, Mattmiller SA, Sordillo LM (2012) Changes in glucose transporter expression in monocytes of periparturient dairy cows. J Dairy Sci 95:5709–5719. https://doi.org/10.3168/jds.2012-5327
Odensten MO, Chilliard Y, Holtenius K (2005) Effects of two different feeding strategies during dry-off on metabolism in high-yielding dairy cows. J Dairy Sci 88:2072–2082. https://doi.org/10.3168/jds.S0022-0302(05)72884-8
Oetzel GK (1988) Parturient paresis and hypocalcemia in ruminant livestock. Vet Clin North Am Food Anim Pract 4:351–364
Ogle CK, Ogle JD, Mao J-X et al (1994) Effect of glutamine on phagocytosis and bacterial killing by normal and pediatric burn patient neutrophils. J Parenter Enter Nutr 18:128–133. https://doi.org/10.1177/0148607194018002128
Ortiz-Marty RJ, Forsberg NE, Chapman JD, Mullarky IK (2013) OmniGen-AF® restores GR-1, L-selectin, and RANTES expression by immunosuppressed murine PMN challenged with lipopolysaccharide in a MyD88-dependent manner. J Dairy Sci 96:397
Overton TR, Waldron MR (2004) Nutritional management of transition dairy cows: strategies to optimize metabolic health. J Dairy Sci 87:E105–E119. https://doi.org/10.3168/jds.S0022-0302(04)70066-1
Overton TR, Drackley JK, Ottemann-Abbamonte CJ et al (1999) Substrate utilization for hepatic gluconeogenesis is altered by increased glucose demand in ruminants. J Anim Sci 77:1940. https://doi.org/10.2527/1999.7771940x
Padgett DA, Glaser R (2003) How stress influences the immune response. Trends Immunol 24:444–448
Padua MB, Tekin Ş, Spencer TE, Hansen PJ (2005) Actions of progesterone on uterine immunosuppression and endometrial gland development in the uterine gland knockout (UGKO) ewe. Mol Reprod Dev 71:347–357. https://doi.org/10.1002/mrd.20301
Pantoja JCF, Hulland C, Ruegg PL (2009) Dynamics of somatic cell counts and intramammary infections across the dry period. Prev Vet Med 90:43–54. https://doi.org/10.1016/j.prevetmed.2009.03.012
Parker Gaddis KL, Cole JB, Clay JS, Maltecca C (2014) Genomic selection for producer-recorded health event data in US dairy cattle. J Dairy Sci 97:3190–3199. https://doi.org/10.3168/jds.2013-7543
Piñeiro JM, Menichetti BT, Barragan AA et al (2019) Associations of postpartum lying time with culling, milk yield, cyclicity, and reproductive performance of lactating dairy cows. J Dairy Sci 102:3362–3375. https://doi.org/10.3168/jds.2018-15387
Pires JAA, Delavaud C, Faulconnier Y et al (2013) Effects of body condition score at calving on indicators of fat and protein mobilization of periparturient Holstein-Friesian cows. J Dairy Sci. https://doi.org/10.3168/jds.2013-6801

Pithon-Curi TC, De Melo MP, Curi R (2004) Glucose and glutamine utilization by rat lymphocytes, monocytes and neutrophils in culture: a comparative study. Cell Biochem Funct 22:321–326. https://doi.org/10.1002/cbf.1109
Pond CM, Newsholme EA (1999) Coping with metabolic stress in wild and domesticated animals. BSAP Occas Publ 24:9–20. https://doi.org/10.1017/S146398150004303X
Poutrel B, de Crémoux R, Ducelliez M, Verneau D (1997) Control of intramammary infections in goats: impact on somatic cell counts. J Anim Sci 75:566. https://doi.org/10.2527/1997.752566x
Premi M, Mezzetti M, Ferronato G et al (2021) Changes of plasma analytes reflecting metabolic adaptation to the different stages of the lactation cycle in healthy multiparous Holstein dairy cows raised in high-welfare conditions. Animals 11:1714. https://doi.org/10.3390/ani11061714
Prins BP, Kuchenbaecker KB, Bao Y et al (2017) Genome-wide analysis of health-related biomarkers in the UK Household Longitudinal Study reveals novel associations. Sci Rep 7:1–9. https://doi.org/10.1038/s41598-017-10812-1
Proudfoot KL (2019) Maternal behavior and design of the maternity pen. Vet Clin North Am Food Anim Pract 35:111–124. https://doi.org/10.1016/j.cvfa.2018.10.007
Pugeat M, Bonneton A, Perrot D et al (1989) Decreased immunoreactivity and binding activity of corticosteroid-binding globulin in serum in septic shock. Clin Chem 35:1675–1679. https://doi.org/10.1093/clinchem/35.8.1675
Putman AK, Brown JL, Gandy JC et al (2018) Changes in biomarkers of nutrient metabolism, inflammation, and oxidative stress in dairy cows during the transition into the early dry period. J Dairy Sci 101:9350–9359. https://doi.org/10.3168/jds.2018-14591
Quiroz-Rocha GF, LeBlanc SJ, Duffield TF et al (2009) Reference limits for biochemical and hematological analytes of dairy cows one week before and one week after parturition. Can Vet J = La Rev Vet Can 50:383–388
Rabelo E, Rezende RL, Bertics SJ, Grummer RR (2003) Effects of transition diets varying in dietary energy density on lactation performance and ruminal parameters of dairy cows. J Dairy Sci 86:916–925. https://doi.org/10.3168/jds.S0022-0302(03)73674-1
Ram L, Schonewille JT, Martens H et al (1998) Magnesium absorption by wethers fed potassium bicarbonate in combination with different dietary magnesium concentrations. J Dairy Sci 81:2485–2492. https://doi.org/10.3168/jds.S0022-0302(98)70140-7
Ramberg CF, Mayer GP, Kronfeld DS et al (1970) Calcium kinetics in cows during late pregnancy, parturition, and early lactation. Am J Phys. https://doi.org/10.1152/ajplegacy.1970.219.5.1166
Raphael W, Sordillo LM (2013) Dietary polyunsaturated fatty acids and inflammation: the role of phospholipid biosynthesis. Int J Mol Sci 14:21167–21188. https://doi.org/10.3390/ijms141021167
Rastani RR, Andrew SM, Zinn SA, Sniffen CJ (2001) Body composition and estimated tissue energy balance in Jersey and Holstein cows during early lactation. J Dairy Sci 84:1201–1209. https://doi.org/10.3168/jds.S0022-0302(01)74581-X
Rastani RR, Grummer RR, Bertics SJ et al (2005) Reducing dry period length to simplify feeding transition cows: milk production, energy balance, and metabolic profiles. J Dairy Sci 88:1004–1014. https://doi.org/10.3168/jds.S0022-0302(05)72768-5
Raszek MM, Guan LL, Plastow GS (2016) Use of genomic tools to improve cattle health in the context of infectious diseases. Front Genet 7:30. https://doi.org/10.3389/fgene.2016.00030
Reilly PM, Schiller HJ, Bulkley GB (1991) Pharmacologic approach to tissue injury mediated by free radicals and other reactive oxygen metabolites. Am J Surg 161:488–503. https://doi.org/10.1016/0002-9610(91)91120-8
Rezaei R, Wu Z, Hou Y, Bazer FW, Wu G (2016) Amino acids and mammary gland development: nutritional implications for milk production and neonatal growth. J Anim Sci Biotechnol 7:20. https://doi.org/10.1186/s40104-016-0078-8
Reynolds CK, Dürst B, Lupoli B et al (2004) Visceral tissue mass and rumen volume in dairy cows during the transition from late gestation to early lactation. J Dairy Sci 87:961–971. https://doi.org/10.3168/jds.S0022-0302(04)73240-3
Rigout S, Lemosquet S, van Eys JE, Blum JW, Rulquin H (2002) Duodenal glucose increases glucose fluxes and lactose synthesis in grass silage-fed dairy cows. J Dairy Sci 85:595–606. https://doi.org/10.3168/jds.S0022-0302(02)74113-1
Rioja-Lang FC, Roberts DJ, Healy SD et al (2012) Dairy cow feeding space requirements assessed in a Y-maze choice test. J Dairy Sci 95:3954–3960. https://doi.org/10.3168/jds.2011-4962
Rizzo A, Roscino M, Binetti F, Sciorsci R (2012) Roles of reactive oxygen species in female reproduction. Reprod Domest Anim 47:344–352. https://doi.org/10.1111/j.1439-0531.2011.01891.x
Rizzo A, Pantaleo M, Mutinati M et al (2013) Blood and milk oxidative status after administration of different antioxidants during early postpartum in dairy cows. Res Vet Sci 95:1171–1174. https://doi.org/10.1016/j.rvsc.2013.07.016
Rodriguez-Jimenez S, Haerr KJ, Trevisi E et al (2018) Prepartal standing behavior as a parameter for early detection of postpartal subclinical ketosis associated with inflammation and liver function biomarkers in peripartal dairy cows. J Dairy Sci 101:8224–8235. https://doi.org/10.3168/jds.2017-14254
Rørvang MV, Nielsen BL, Herskin MS, Jensen MB (2018) Prepartum maternal behavior of domesticated cattle: a comparison with managed, feral, and wild ungulates. Front Vet Sci 5:45. https://doi.org/10.3389/fvets.2018.00045

Roth JA, Kaeberle ML, Hsu WH (1982) Effect of estradiol and progesterone on lymphocyte and neutrophil functions in steers. Infect Immun 35:997–1002. https://doi.org/10.1002/jemt

Rowe SM, Godden SM, Nydam DV et al (2020) Randomized controlled non-inferiority trial investigating the effect of 2 selective dry-cow therapy protocols on antibiotic use at dry-off and dry period intramammary infection dynamics. J Dairy Sci 103:6473–6492. https://doi.org/10.3168/jds.2019-17728

Rude RK, Oldham SB, Sharp CF, Singer FR (1978) Parathyroid hormone secretion in magnesium deficiency. J Clin Endocrinol Metab 47:800–806. https://doi.org/10.1210/jcem-47-4-800

Ruotsalainen SE, Partanen JJ, Cichonska A et al (2021) An expanded analysis framework for multivariate GWAS connects inflammatory biomarkers to functional variants and disease. Eur J Hum Genet 29:309–324. https://doi.org/10.1038/s41431-020-00730-8

Saha S, Gallo L, Bittante G et al (2019) A study on the effects of rumen acidity on rumination time and yield, composition, and technological properties of milk from early lactating Holstein cows. Animals 9:66. https://doi.org/10.3390/ani9020066

Samuelsson B, Uvnäs-Moberg K, Gorewit RC, Svennersten-Sjaunja K (1996) Profiles of the hormones somatostatin, gastrin, CCK, prolactin, growth hormone, oxytocin and cortisol. II. In dairy cows that are milked during food deprivation. Livest Prod Sci 46:57–64. https://doi.org/10.1016/0301-6226(96)00021-8

Sartorelli P, Paltrinieri S, Agnes F (1999) Non-specific immunity and ketone bodies. I: In vitro studies on chemotaxis and phagocytosis in ovine neutrophils. Zentralbl Veterinarmed A 46:613–619. https://doi.org/10.1046/j.1439-0442.1999.00252.x

Sasaki S (2002) Mechanism of insulin action on glucose metabolism in ruminants. Anim Sci J 73:423–433. https://doi.org/10.1046/j.1344-3941.2002.00059.x

Scalia D, Lacetera N, Bernabucci U et al (2006) In vitro effects of nonesterified fatty acids on bovine neutrophils oxidative burst and viability. J Dairy Sci 89:147–154. https://doi.org/10.3168/jds.S0022-0302(06)72078-1

Schirmann K, Chapinal N, Weary DM et al (2011) Short-term effects of regrouping on behavior of prepartum dairy cows. J Dairy Sci 94:2312–2319. https://doi.org/10.3168/jds.2010-3639

Schönfeld P, Wojtczak L (2008) Fatty acids as modulators of the cellular production of reactive oxygen species. Free Radic Biol Med 45:231–241. https://doi.org/10.1016/j.freeradbiomed.2008.04.029

Schreiber G, Howlettf G, Nagashima M et al (1982) The acute phase response of plasma protein synthesis during experimental inflammation. J Biol Chem 257:10271–10277

Schweigel M, Martens H (2000) Electrophysiologic changes in rumen epithelium in their effect on magnesium transport--a review. Berl Munch Tierarztl Wochenschr 113:97–102

Serhan CN (2017) Treating inflammation and infection in the 21st century: new hints from decoding resolution mediators and mechanisms. FASEB J 31:1273–1288

Sheldon IM, Rycroft AN, Zhou C (2004) Association between postpartum pyrexia and uterine bacterial infection in dairy cattle. Vet Rec 154:289–293. https://doi.org/10.1136/vr.154.10.289

Shi H, Kokoeva MV, Inouye K et al (2006) TLR4 links innate immunity and fatty acid–induced insulin resistance. J Clin Invest 116:3014–3025. https://doi.org/10.1172/JCI28898

Shi W, Haisan J, Inabu Y et al (2020) Effects of starch concentration of close-up diets on rumen pH and plasma metabolite responses of dairy cows to grain challenges after calving. J Dairy Sci 103:11461–11471. https://doi.org/10.3168/jds.2020-18768

Silva PRB, Moraes JGN, Mendonça LGD et al (2013) Effects of weekly regrouping of prepartum dairy cows on metabolic, health, reproductive, and productive parameters. J Dairy Sci 96:4436–4446. https://doi.org/10.3168/jds.2012-6464

Sordillo LM (2016) Nutritional strategies to optimize dairy cattle immunity. J Dairy Sci 99:4967–4982. https://doi.org/10.3168/jds.2015-10354

Sordillo LM, Aitken SL (2009) Impact of oxidative stress on the health and immune function of dairy cattle. Vet Immunol Immunopathol 128:104–109. https://doi.org/10.1016/j.vetimm.2008.10.305

Sordillo LM, Raphael W (2013) Significance of metabolic stress, lipid mobilization, and inflammation on transition cow disorders. Vet Clin North Am Food Anim Pract 29:267–278

Sordillo LM, Pighetti GM, Davis MR (1995) Enhanced production of bovine tumor necrosis factor-alpha during the periparturient period. Vet Immunol Immunopathol 49:263–270. https://doi.org/10.1016/0165-2427(95)05465-0

Sordillo LM, O'Boyle N, Gandy JC et al (2007) Shifts in thioredoxin reductase activity and oxidant status in mononuclear cells obtained from transition dairy cattle. J Dairy Sci 90:1186–1192. https://doi.org/10.3168/jds.S0022-0302(07)71605-3

Sordillo LM, Streicher KL, Mullarky IK et al (2008) Selenium inhibits 15-hydroperoxyoctadecadienoic acid-induced intracellular adhesion molecule expression in aortic endothelial cells. Free Radic Biol Med 44:34–43. https://doi.org/10.1016/j.freeradbiomed.2007.09.002

Sordillo LM, Contreras GA, Aitken SL (2009) Metabolic factors affecting the inflammatory response of periparturient dairy cows. Anim Health Res Rev 10:53–63. https://doi.org/10.1017/S1466252309990016

Sordillo LM, Mavangira V (2014) The nexus between nutrient metabolism, oxidative stress and inflammation in transition cows. Anim Prod Sci 54:1204–1214. https://doi.org/10.1071/AN14503

Sørensen JT, Enevoldsen C (1991) Effect of dry period length on milk production in subsequent lactation. J Dairy Sci 74:1277–1283. https://doi.org/10.3168/jds.S0022-0302(91)78283-0

Soriani N, Trevisi E, Calamari L (2012) Relationships between rumination time, metabolic conditions, and health status in dairy cows during the transition period. J Anim Sci 90:4544–4554. https://doi.org/10.2527/jas.2012-5064

Spears JW, Weiss WP (2008) Role of antioxidants and trace elements in health and immunity of transition dairy cows. Vet J 176:70–76. https://doi.org/10.1016/j.tvjl.2007.12.015

Stangaferro ML, Wijma R, Caixeta LS et al (2016) Use of rumination and activity monitoring for the identification of dairy cows with health disorders: Part III. Metritis. J Dairy Sci 99:7422–7433. https://doi.org/10.3168/jds.2016-11352

Steele MA, Schiestel C, AlZahal O et al (2015) The periparturient period is associated with structural and transcriptomic adaptations of rumen papillae in dairy cattle. J Dairy Sci 98:2583–2595. https://doi.org/10.3168/jds.2014-8640

Stevens MGH, Peelman LJ, De Spiegeleer B et al (2011) Differential gene expression of the toll-like receptor-4 cascade and neutrophil function in early- and mid-lactating dairy cows. J Dairy Sci 94:1277–1288. https://doi.org/10.3168/jds.2010-3563

Suriyasathaporn W, Daemen AJJM, Noordhuizen-Stassen EN et al (1999) β-Hydroxybutyrate levels in peripheral blood and ketone bodies supplemented in culture media affect the in vitro chemotaxis of bovine leukocytes. Vet Immunol Immunopathol 68:177–186. https://doi.org/10.1016/S0165-2427(99)00017-3

Takeuchi O, Akira S (2010) Pattern recognition receptors and inflammation. Cell 140:805–820. https://doi.org/10.1016/j.cell.2010.01.022

Targowski SP, Klucinski W (1983) Reduction in mitogenic response of bovine lymphocytes by ketone bodies. Am J Vet Res 44:828–830

Tetens J, Heuer C, Heyer I et al (2015) Polymorphisms within the APOBR gene are highly associated with milk levels of prognostic ketosis biomarkers in dairy cows. Physiol Genomics 47:129–137. https://doi.org/10.1152/physiolgenomics.00126.2014

Thompson-crispi KA, Sargolzaei M, Ventura R et al (2014) A genome-wide association study of immune response traits in Canadian Holstein cattle. BMC Genomics 15:1–10

Trebble TM, Wootton SA, Miles EA et al (2003) Prostaglandin E2 production and T cell function after fish-oil supplementation: response to antioxidant cosupplementation. Am J Clin Nutr 78:376–382. https://doi.org/10.1093/ajcn/78.3.376

Trevisi E, Bertoni G (2008) Attenuation with acetylsalicylate treatments of inflammatory conditions in periparturient dairy cows. In: Quinn PI (ed) Aspirin and heath research progress. Nova Science Publishers, Inc, pp 22–37

Trevisi E, Bertoni G (2009) Some physiological and biochemical methods for acute and chronic stress evaluation in dairy cows. Ital J Anim Sci 8:265–286. https://doi.org/10.4081/ijas.2009.s1.265

Trevisi E, Minuti A (2018) Assessment of the innate immune response in the periparturient cow. Res Vet Sci 116:47–54

Trevisi E, Bionaz M, Ferrari A, Bertoni G (2005) Changes of plasma proteins synthesized by the liver and of performances in periparturient dairy cows affected by inflammations. In: Greppi GF, Bonizzi L, Roncada P (eds) 40° symposium of animal production: "from genome to proteome in animal science". Lodi, pp 194–201

Trevisi E, Gubbiotti A, Bertoni G (2007) Effects of inflammation in peripartum dairy cows on milk yield, energy balance and efficiency. In: Ortigues-Marty I, Miraux N, Brand-Williams W (eds) Energy and protein metabolism and nutrition. Wageningen Academic Publishers, Wageningen, pp 395–396

Trevisi E, Ferrari AR, Bertoni G (2008) Productive and metabolic consequences induced by the retained placenta in dairy cows. Vet Res Commun 32:363–366. https://doi.org/10.1007/s11259-008-9149-4

Trevisi E, Zecconi A, Bertoni G, Piccinini R (2010) Blood and milk immune and inflammatory profiles in periparturient dairy cows showing a different liver activity index. J Dairy Res 77:310–317. https://doi.org/10.1017/S0022029910000178

Trevisi E, Amadori M, Archetti I et al (2011) Inflammatory response and acute phase proteins in the transition period of high-yielding dairy cows. In: Vea F (ed) Acute phase protein/book 2. InTech Rijeka, Croatia, pp 355–380

Trevisi E, Amadori M, Cogrossi S et al (2012) Metabolic stress and inflammatory response in high-yielding, periparturient dairy cows. Res Vet Sci 93:695–704. https://doi.org/10.1016/j.rvsc.2011.11.008

Trevisi E, Bertoni G, Lombardelli R, Minuti A (2013) Relation of inflammation and liver function with the plasma cortisol response to adrenocorticotropin in early lactating dairy cows. J Dairy Sci 96:5712–5722. https://doi.org/10.3168/jds.2012-6375

Trevisi E, Amadori M, Riva F et al (2014) Evaluation of innate immune responses in bovine forestomachs. Res Vet Sci 96:69–78. https://doi.org/10.1016/j.rvsc.2013.11.011

Trevisi E, Jahan N, Bertoni G et al (2015) Pro-inflammatory cytokine profile in dairy cows: consequences for new lactation. Ital J Anim Sci 14:3862. https://doi.org/10.4081/ijas.2015.3862

Trevisi E, Moscati L, Amadori M (2016) Disease-predicting and prognostic potential of innate immune responses to noninfectious stressors: human and animal models. In: Amadori M (ed) The innate immune response to noninfectious stressors. Elsevier, Amsterdam, pp 209–235

Tucker CB, Weary DM, von Keyserlingk MAG, Beauchemin KA (2009) Cow comfort in tie-stalls: increased depth of shavings or straw bedding increases lying time. J Dairy Sci 92:2684–2690. https://doi.org/10.3168/jds.2008-1926

Turk R, Juretić D, Gereš D, Turk N, Rekić B, Simeon-Rudolf V, Robić M, Svetina A (2005) Serum paraoxonase activity in dairy cows during pregnancy. Res Vet Sci 79:15–18. https://doi.org/10.1016/j.rvsc.2004.09.010

Valko M, Leibfritz D, Moncol J, Cronin MTD, Mazur M, Telser J (2007) Free radicals and antioxidants in normal physiological functions and human disease. Int J Biochem Cell Biol 39:44–84. https://doi.org/10.1016/j.biocel.2006.07.001

Van De Braak AE, Klooster ATH, Van't Malestein A (1987) Influence of a deficient supply of magnesium during the dry period on the rate of calcium mobilisation by dairy cows at parturition. Res Vet Sci 42:101–108. https://doi.org/10.1016/S0034-5288(18)30663-5

van Gastelen S, Westerlaan B, Houwers DJ, van Eerdenburg FJCM (2011) A study on cow comfort and risk for lameness and mastitis in relation to different types of bedding materials. J Dairy Sci 94:4878–4888. https://doi.org/10.3168/jds.2010-4019

Van Schyndel SJ, Carrier J, Bogado Pascottini O, LeBlanc SJ (2018) The effect of pegbovigrastim on circulating neutrophil count in dairy cattle: a randomized controlled trial. PLoS One 13:e0198701. https://doi.org/10.1371/journal.pone.0198701

Vanraden PM, Cole JB, Parker Gaddis KL (2018) AIP research report NM$7: net merit as a measure of lifetime profit: 2018 revision. In: AIP res. Rep. NM$7. https://aipl.arsusda.gov/reference/nmcalc-2018.htm

Vilar MJ, Rajala-Schultz PJ (2020) Dry-off and dairy cow udder health and welfare: effects of different milk cessation methods. Vet J 262:105503. https://doi.org/10.1016/j.tvjl.2020.105503

von Keyserlingk MAG, Olenick D, Weary DM (2008) Acute behavioral effects of regrouping dairy cows. J Dairy Sci 91:1011–1016. https://doi.org/10.3168/jds.2007-0532

Wallace C, Keast D (1992) Glutamine and macrophage function. Metabolism 41:1016–1020. https://doi.org/10.1016/0026-0495(92)90130-3

Wang YQ, Puntenney SB, Forsberg NE (2004) Identification of the mechanisms by which OmniGen-AF®, a nutritional supplement, augments immune function in ruminant livestock. In: Proceedings of the Western Section, American Society of Animal Science, Savoy, pp 349–352

Wang Y, Puntenney SB, Burton JL, Forsberg N (2007) Ability of a commercial feed additive to modulate expression of innate immunity in sheep immunosuppressed with dexamethasone. Animal 1:945. https://doi.org/10.1017/S1751731107000365

Wang X, Li X, Zhao C et al (2012) Correlation between composition of the bacterial community and concentration of volatile fatty acids in the rumen during the transition period and ketosis in dairy cows. Appl Environ Microbiol 78:2386–2392. https://doi.org/10.1128/AEM.07545-11

Wankhade PR, Manimaran A, Kumaresan A et al (2017) Metabolic and immunological changes in transition dairy cows: a review. Vet World 10:1367–1377. https://doi.org/10.14202/vetworld.2017.1367-1377

Wannemacher RW, Beall FA, Canonico PG et al (1980) Glucose and alanine metabolism during bacterial infections in rats and rhesus monkeys. Metabolism 29:201–212. https://doi.org/10.1016/0026-0495(80)90061-X

Wathes DC, Cheng Z, Bourne N et al (2007) Differences between primiparous and multiparous dairy cows in the inter-relationships between metabolic traits, milk yield and body condition score in the periparturient period. Domest Anim Endocrinol 33:203–225. https://doi.org/10.1016/j.domaniend.2006.05.004

Watters RD, Guenther JN, Brickner AE et al (2008) Effects of dry period length on milk production and health of dairy cattle. J Dairy Sci 91:2595–2603. https://doi.org/10.3168/jds.2007-0615

Weary DM, Huzzey JM, von Keyserlingk MAG (2009) Board-invited review: using behavior to predict and identify ill health in animals1. J Anim Sci 87:770–777. https://doi.org/10.2527/jas.2008-1297

Weaver JA, Maddox JF, Cao Y-Z et al (2001) Increased 15-HPETE production decreases prostacyclin synthase activity during oxidant stress in aortic endothelial cells. Free Radic Biol Med 30:299–308. https://doi.org/10.1016/S0891-5849(00)00466-4

Weller JI, Ezra E, Ron M (2017) Invited review: a perspective on the future of genomic selection in dairy cattle. J Dairy Sci 100:8633–8644. https://doi.org/10.3168/jds.2017-12879

Wenz JR, Moore DA, Kasimanickam R (2011) Factors associated with the rectal temperature of Holstein dairy cows during the first 10 days in milk. J Dairy Sci 94:1864–1872. https://doi.org/10.3168/jds.2010-3924

Wertz-Lutz AE, Knight TJ, Pritchard RH et al (2006) Circulating ghrelin concentrations fluctuate relative to nutritional status and influence feeding behavior in cattle. J Anim Sci 84:3285–3300. https://doi.org/10.2527/jas.2006-053

Wisnieski L, Norby B, Pierce SJ et al (2019a) Predictive models for early lactation diseases in transition dairy cattle at dry-off. Prev Vet Med 163:68–78. https://doi.org/10.1016/j.prevetmed.2018.12.014

Wisnieski L, Norby B, Pierce SJ et al (2019b) Prospects for predictive modeling of transition cow diseases. Anim Health Res Rev 20:19–30. https://doi.org/10.1017/S1466252319000112

Wright RD, Blair-West JR, Nelson JF, Tregear GW (1984) Handling of phosphate by a parotid gland (ovine). Am J Physiol 246:F916–F926. https://doi.org/10.1152/ajprenal.1984.246.6.F916

Yassad A, Lavoinne A, Bion A et al (1997) Glutamine accelerates interleukin-6 production by rat peritoneal macrophages in culture. FEBS Lett 413:81–84. https://doi.org/10.1016/S0014-5793(97)00881-8

Yeager MP, Pioli PA, Guyre PM (2011) Cortisol exerts bi-phasic regulation of inflammation in humans. Dose-Response 9:332–347. https://doi.org/10.2203/dose-response.10-013.Yeager

Zhang Q, Cai Z, Lhomme M et al (2020) Inclusion of endophenotypes in a standard GWAS facilitate a detailed mechanistic understanding of genetic elements that control blood lipid levels. Sci Rep 10:18434. https://doi.org/10.1038/s41598-020-75612-6

Zhou Z, Bulgari O, Vailati-Riboni M, Trevisi E, Ballou MA, Cardoso FC, Luchini DN, Loor JJ (2016) Rumen-protected methionine compared with rumen-protected choline improves immunometabolic status in dairy cows during the peripartal period. J Dairy Sci 99:8956–8969. https://doi.org/10.3168/jds.2016-10986

Zhou Z, Ferdous F, Montagner P, Luchini DN, Corrêa MN, Loor JJ (2018) Methionine and choline supply during the peripartal period alter polymorphonuclear leukocyte immune response and immunometabolic gene expression in Holstein cows. J Dairy Sci 101:10374–10382. https://doi.org/10.3168/jds.2018-14972

Zinicola M, Korzec H, Teixeira AGV et al (2018) Effects of pegbovigrastim administration on periparturient diseases, milk production, and reproductive performance of Holstein cows. J Dairy Sci 101:11199–11217. https://doi.org/10.3168/jds.2018-14869

Zobel G, Weary DM, Leslie KE, von Keyserlingk MAG (2015) Invited review: cessation of lactation: effects on animal welfare. J Dairy Sci 98:8263–8277. https://doi.org/10.3168/jds.2015-9617

Zwald NR, Weigel KA, Chang YM et al (2004) Genetic selection for health traits using producer-recorded data. I. Incidence rates, heritability estimates, and sire breeding values. J Dairy Sci 87:4287–4294. https://doi.org/10.3168/jds.S0022-0302(04)73573-0

# Etiology of Oxidative Stress in Dairy Cow

Loredana Basiricò, Tiziana Vetturini, and Umberto Bernabucci

**Abstract**

Farm animals are exposed to a range of stresses that negatively affect productive and reproductive performances. In dairy cows, four main groups of stress can be identified that cause disturbance in the redox homeostasis of cells and tissues and may be the etiological agent of oxidative stress. The peripartum period, heat stress, mastitis, and high stocking density participate in the induction of oxidative stress and may be considered important physiological, environmental, disease, and psychological stresses in the productive career of dairy cattle. Oxidative stress, during the productive life of dairy cow, may be the main cause of immune and inflammatory dysfunction that increases incidence of health problems in these animals. The administration of antioxidants can protect dairy cows from oxidative stress conditions, and several novel compounds are promising for future applications to counteract oxidative stress in dairy cows.

**Keywords**

Cattle · Oxidative status · Transition cow · Hot environment · Diseases · Antioxidants

L. Basiricò · T. Vetturini · U. Bernabucci (✉)
Department of Agriculture and Forest Sciences, University of Tuscia, Viterbo, Italy
e-mail: basiri@unitus.it; bernab@unitus.it

## 1 Introduction

Exposure of farm animals to a range of stresses is globally recognized to negatively affect productive and reproductive performances of animals, and dairy cows are the most affected. Inevitable or adverse environmental conditions determine the onset of stress in animals. Four main groups of stress can be identified in dairy cows: physical, chemical, physiological, and psychological. These stresses cause disturbance in the redox homeostasis of cells and tissues and may be the etiological agent of oxidative stress (Puppel et al. 2015). When the physiological balance between oxidizing and reducing species fails and the formation of oxidizing species does not allow the organism to counteract cell damage, a condition called oxidative stress is established. The imbalance results from the excessive accumulation of oxidants, a depletion of critical antioxidant defenses, or a combination of these situations. Oxidative stress is associated with an increased rate of cellular damage induced by oxidant substances. The main categories of oxidant are known as reactive oxygen species (ROS) and reactive nitrogen species (RNS). The ROS and RNS are metabolites commonly generated during the cellular normal metabolic activity and are involved in the immune response to remove pathogens (Phaniendra et al. 2015). Moreover, they play a key role in physiological processes, such as gene activation, cellular growth and


J. J. Gross (ed.), *Production Diseases in Farm Animals*,
https://doi.org/10.1007/978-3-031-51788-4_5

death, or biosynthesis of immunoregulatory substances, essential for optimizing inflammatory responses (Abuelo et al. 2015, 2019). Many of these reactive species are free radicals possessing one or more unpaired electrons and are highly reactive. The reactive species generated in cells include the superoxide anion radical ($O_2^{\bullet-}$), the hydroxyl radical ($OH^{\bullet}$), the nitric oxide radical ($NO^{\bullet}$), the lipid peroxyl radical ($LOO^{\bullet}$), and other non-radical reactive species as hydrogen peroxide ($H_2O_2$) and hypochlorous acid (HClO). Detrimental consequences occur when the rate at which reactive species are formed and eliminated is altered (Kumar et al. 2012). Most importantly, the excess of ROS production and their accumulation that exceeds the antioxidant capacity of cells lead to oxidative stress. This imbalance results in the damage of cellular macromolecules including lipids, proteins, and DNA leading to tissue damage and even cell death (Surai et al. 2019).

The free radical attack on lipids, which are important constituents of the lipid bilayer of the cellular membrane, may cause damage to membrane integrity. Polyunsaturated fatty acids are very sensitive to oxidation and peroxidation, which may be responsible to initiate chain reactions resulting in further oxidative damage. This event is associated with the appearance of alterations in fluidity and permeability of cellular membrane as well as in several physiological processes, including signaling, growth, and survival consequently of the whole cell (Ayemele et al. 2021). Moreover, the integrity of cellular compartmentalization will be damaged, leading to enzyme leakage, electrolyte diffusion, and metal and small molecule passage between cellular organelles (Behairy et al. 2020; Nemec Svete et al. 2021). The interactions between ROS and DNA are very complex; ROS activity can cause various types of damage at the level of the nucleotide chain, such as oxidation of nitrogenous bases, sugar lesions, single-strand breaks, and DNA-protein close bond (Yu et al. 2016). Results of DNA damage are aging process, genetic mutations, genome instability, carcinogenesis, and neurodegenerative disease.

Several studies have shown oxidative stress compromises health status and leads to immunosuppression and decreased productive and reproductive performance in dairy cows (Colitti et al. 2019; Putman et al. 2018; Boudjellaba et al. 2018). Oxidative stress may impair the health of dairy cows both directly and indirectly (Celi 2011).

Farm animals, like other organisms, are well prepared to neutralize oxidative damage with a network of antioxidant compounds, to maintain redox homeostasis, to increase the elimination of radicals, or to block their production. These systems work synergistically to enhance cellular defense and to combat oxidative stress. Antioxidants can be categorized in several ways. Based on their activity, they can be classified as enzymatic and nonenzymatic antioxidants or on their origins as endogenous antioxidants, synthesized in the body, and exogenous antioxidants, obtained through dietary means (Irato and Santovito 2021). Antioxidants have also been classified as water-soluble or fat-soluble molecules. The antioxidant enzymes like superoxide dismutase (SOD), glutathione peroxidase (GPx), glutathione reductase (GSR), glutathione-S-transferase, NADPH oxidase, and catalases (CAT) are macromolecules, whereas most of the remaining natural antioxidants are relatively smaller organic molecules with low molecular weights and are classified as phytochemicals and vitamins (Moussa et al. 2019). Natural nonenzymatic exogenous antioxidants are vitamins A, C, and E, flavonoids, carotenoids, plant polyphenols, theaflavin, allyl sulfides, and curcumin; nonenzymatic small endogenous antioxidants are glutathione (GSH), uric acid, melatonin, bilirubin, and polyamines (Mironczuk-Chodakowska et al. 2018; Nimse and Palb 2015). Several of these antioxidants are present in the cytosol or cytoplasmic matrix of cells since they are water-soluble, while the fat-soluble ones are found in cell membranes. The enzymatic antioxidants (Moussa et al. 2019), using cofactors such as selenium, iron, zinc, copper, and manganese, act by converting oxidized metabolic products in a multistep process to $H_2O_2$ and then to water, while nonenzymatic antioxidants intercept, preventing chain reactions, or terminate free radical chain reactions, acting on intermediate radicals and inhibiting other oxidation reactions oxidizing themselves (give up their external electron), thus behaving as reducing agents. Hence, cells protect themselves from oxidative injury by expressing genes that code for antioxidant enzymes and endogenous antioxidant to keep oxidants at inoffensive levels. Since oxidants have beneficial effects and are involved in some physiological functions, the role of antioxidant systems is not to eliminate oxidants completely but to keep them at right amount. All that underlines the importance of the antioxidant defense system in maintaining homeostasis and normal physiological processes, in preventing diseases, and in promoting immunity (Moussa et al. 2019).

## 2 Physiological Stress: Periparturient Dairy Cows

The last two months of gestation and the first months after calving in the dairy cow are known as the peripartum period and are characterized by the transition from late gestation to early lactation. In particular, the period from 3 weeks pre-calving to 3 weeks post-calving, defined by Grummer (1995) as transition period, represents one of the most stressful phases in the productive career of dairy cattle. The fetal growth, calving, and the onset of lactation determine a drastic decrease in feed intake and an increase of energy demands so the animals are subjected to important endocrine and metabolic variations. Impaired nutrient metabolism, caused by a decreased amount of glucose in the blood not rewarded by the diet, results in negative energy balance and drives the cow to use its own energy reserves. The mobilization of lipids is activated, and non-esterified fatty acids (NEFA) are released into the bloodstream and used as an energy source. Excessive lipomobilization may cause dysregulation of inflammatory and immune responses and the generation of ROS (Bernabucci et al. 2005; Lacetera et al. 2005). In addition, dairy cows are subjected to high metabolic activity, and tissues consume more oxygen through normal cellular respiration, which could lead to an overproduction of ROS. In this situation, an imbalance between ROS production and the reduced intake of dietary antioxidants, due to a reduced overall feed intake, can lead the redox balance in favor of a pro-oxidative condition of the dairy cows and therefore their risk of suffering oxidative stress (OS) (Abuelo et al. 2019; Sordillo and Aitken 2009). The susceptibility to OS of transition cows exacerbated by several stressful factors, including increased plasma NEFA concentrations, inflammatory dysregulation, and changes in immune response, predisposes animals to an increased risk of infectious and metabolic health diseases. Several studies (Table 1) indicate that high-producing dairy cows experience OS around the transition period (Bernabucci et al. 2005; Castillo et al. 2006; Sordillo and Aitken 2009). In blood samples of periparturient dairy cows, some authors have observed an increased lipid peroxidation (Bernabucci et al. 2005; Castillo et al. 2005), decreased plasma antioxidant concentrations (Bouwstra et al. 2010), low antioxidative enzyme activity (Gaal et al. 2006), high ROM and thiobarbituric acid reactive substance (TBARS), and low levels of antioxidants above all in periparturient cows with higher body condition score (BCS) (Bernabucci et al. 2005). Laubenthal et al. (2017) reported that dairy cows with a body condition score between 3.5 and 5 are more sensitive to OS and metabolic disorders during the transition period. Another determining factor is also the degree of change in BCS during delivery other than the BCS at calving. Several authors have reported that peripartum cows with higher BCS loss have higher concentrations of NEFA and higher levels of OS than those with lower BCS loss (Zachut et al. 2013; Contreras et al. 2017; Weber et al. 2013). The loss of BCS is associated with the increase of fat and protein catabolic pathways to generate energy. At the mitochondrial level, electrons derived from cellular metabolism are passed along the mitochondrial electron transport chain generating a proton gradient across the mitochondrial membrane that ultimately drives synthesis of ATP. Electrons that are transferred to the electron transport chain leak back across prematurely and can also leak back across the membrane and reduce molecular

**Table 1** Important sources of oxidative stress in dairy cows

| Stress | Conditions of oxidative stress in dairy cows | References |
|---|---|---|
| **Peripartum period** | | |
| Biological stress | ↑ lipid peroxidation, ↓ antioxidant, ↑ ROM and TBARS in blood<br>↓ plasma antioxidant<br>↓ antioxidative enzyme activity<br>↓ antioxidants synthesized in liver<br>↑ plasma oxidative status indices, ↓ hepatic Apo-B100 synthesis and secretion<br>↓ GSH, ↓ myeloperoxidase activity in polymorphonuclear neutrophils<br>↓ antioxidant activity of skeletal muscle mitochondrial | Bernabucci et al. (2005) and Castillo et al. (2005)<br>Bouwstra et al. (2010)<br>Gaal et al. (2006)<br>Abuelo et al. (2019)<br>Bernabucci et al. (2009)<br>Ingvartsen et al. (2003)<br>Bassel and Caswell (2018)<br>Favorit et al. (2021) |
| **Heat stress** | | |
| Environmental stress | ↑ ROS, ↑ lipid peroxidation, ↑ mitochondrial dysfunction<br>↑ $O_2^{\cdot -}$, ↑ hydrogen peroxide, ↑ lipid peroxidation<br>↓ antioxidant defenses | Bernabucci et al. (2002), Belhadj Slimen et al. (2014) and Zachaut et al. (2017)<br>Verma et al. (2020), Chauhan et al. (2015) and Bernabucci et al. (2002)<br>Bernabucci et al. (2002) and Li et al. (2021) |
| **Mastites** | | |
| Pathogen/Disease stress | ↓ SOD activity, ↓ catalase, ↓ GSH, ↑ lipid peroxidation in blood<br>↑ phenylpyruvic acid, ↑ xanthine/guanine ratio in presence of Streptococcus | Jhambh et al. (2013)<br>Ayemele et al. (2021) |
| **Stocking density (High)** | | |
| Psychological stress | ↑ MDA and ↑ cortisol in plasma<br>↑ MDA and ↑ cortisol in milk, ↓ TAC, ↑ MDA, ↓ SOD and GPx activity in plasma | Lin et al. (2018)<br>Lin et al. (2019) |

oxygen to produce the highly reactive anion radical superoxide. Superoxide radicals formed during the transport of electrons capture electrons from near molecules and damage the cell membrane consequently. Cell membrane fatty acids are highly susceptible to oxidation, which generates lipid radicals that can act on adjacent cell lipids, creating a positive feedback loop that can cause lipid peroxidation. Lipid peroxidation is known to be a free radical chain reaction, which forms lipid hydroperoxides and by-products (Abuelo et al. 2019; Surai et al. 2019). The latter are highly reactive and have been shown to interact with many macromolecules such as DNA (Yu et al. 2016). Mitochondrial DNA (mtDNA) is more susceptible to OS damage because it is not protected by proteins, for example, histones. Shokolenko et al. (2009) reported that OS causes degradation of mtDNA and that strand breaks and basic sites prevail over mutagenic base lesions in ROS-damaged mtDNA. Damaged mtDNA can cause a decline in mtRNA transcription and lead to dysfunction of mitochondrial biogenesis (Laubenthal et al. 2017). Probably, the intensification of metabolic activity may increase electron leakage and subsequently production of ROS (Favorit et al. 2021). Moreover, raised serum NEFA concentrations due to excessive lipomobilization can further increase OS (Sordillo and Raphael 2013). In dairy cattle tissues, adaptations in mitochondrial physiology are critical to support the metabolic changes of the transition period. Upregulation of genes and transcription factors involved in mitochondrial regulation during the transition period was observed in mammary and extramammary tissues of dairy cattle to facilitate changes in nutrient metabolism and increase bioenergetic capacity (Schaff et al. 2013; Li et al. 2020). Mitochondria produce most of the ATP (up to 90%) available for use by tissues (Hadsell et al. 2011). In the mammary gland, mitochondrial density and mtDNA copy number increase, thus allowing for greater ATP production at the beginning of lactation as mammary energy demand and milk production are increasing (Hadsell et al. 2011; Rosano and Jones 1976). Extramammary tissues such as the liver and skeletal muscle also undergo metabolic changes due to breastfeeding. During lactation, an upregulation of genes involved in mitochondrial biogenesis, an increase in ROS emission, and a steady antioxidant activity were observed in the liver (Favorit et al. 2021). Several studies suggest that reduced liver function in peripartum cows contributes to a reduction of antioxidants synthesized by the liver impacting on the antioxidant capacity of cows (Abuelo et al. 2019). Moreover, Bernabucci et al. (2009) reported that the condition of OS in early postpartum dairy cows, as indicated by plasma oxidative status indices, is related to a decrease of hepatic Apo-$B_{100}$ synthesis and secretion, which may lead to an increased lipid accumulation in liver of transition dairy cows. Conversely, in skeletal muscle, mitochondrial efficiency remains invariable during lactation but undergoes oxidative damage, which is associated with reduced antioxidant activity (Favorit et al. 2021). Therefore, a greater oxidative capacity of the mitochondria, particularly at the beginning and middle of lactation, would result in a greater emission of oxidants, culminating in oxidative damage.

Oxidative stress during the transition period may be the main cause of immune and inflammatory dysfunction in dairy cattle. The suboptimal immune responses due to various conditions like lipomobilization and hypoglycemia increased the incidence of health problems in periparturient dairy cows (Ingvartsen et al. 2003). Molecular mechanisms responsible for those changes are still not clear. The cellular redox imbalance due to the increased ROS production probably contributes to the exhaustion of antioxidant mechanisms (Sordillo et al. 2007) and to the inactivation of antioxidant enzymes induced by oxidation (Pigeolet et al. 1990) causing a defective function of neutrophils and lymphocytes (Kehrli et al. 1989a, b). Ingvartsen et al. (2003) in polymorphonuclear neutrophils observed a decrease in GSH concentrations and less effective oxidative bursts that kill pathogens linked to low glucose availability. Bassel and Caswell (2018) reported that cow neutrophils are characterized by decreased myeloperoxidase and elastase activities and an absence of α-defensins with respect to other species. This status is intrinsically associated with the increased incidence of peripartum diseases such as retained fetal membranes, elevated somatic cell counts, and mastitis (Abuelo et al. 2015, 2019). During the transition period, OS and excessive lipomobilization probably contribute to excessive and dysregulated inflammatory responses. The proinflammatory condition of periparturient dairy cow may contribute to facilitation of parturition (Van Engelen et al. 2009) and in homeorhetic adaptations to the onset of lactation (Farney et al. 2013). During inflammation, proinflammatory cytokines such as tumor necrosis factor alpha, interleukin-1, and interleukin-6 are produced in large amounts by immune cells (Bionaz et al. 2007). Proinflammatory and chemotactic cytokines play a key role in the recruitment and activation of phagocytic cells, which are main producers of ROS (Agarwal et al. 2012). Moreover, Schonfeld and Wojtczak (2008) reported that ROS may also increase inflammation by activation of nuclear factor kappa-B (NF-κB). When the peripartal inflammatory response is limited, it does not create problems, but if this response becomes excessive and permanent, it leads dairy cows to chronic inflammation, OS, and consequently infertility (Sordillo and Mavangira 2014; Dirandeh et al. 2021).

In dairy cows, OS can lead to a compromise of the reproductive status as it contributes to the onset of various reproductive disorders such as metritis, placental retention, and endometritis (Abuelo et al. 2019). Duhig et al. (2016) reported that complications of pregnancy such as spontaneous abortion and recurrent pregnancy loss develop as a consequence of OS condition. In multiparous Holstein dairy cows, with normal luteal activity, greater GPx and SOD activities and total antioxidant capacity are observed postpartum, compared with pregnant cows that lost their pregnancy at days

32 and 60 after artificial insemination. Moreover, Celi et al. (2011) reported that OS is involved in increased embryonic mortality in dairy cows. In pregnant cows, it is observed that plasma OS parameters and molecular indicators of inflammation were altered in peripheral leukocytes causing embryonic losses compared to cows with a full-term pregnancy (Dirandeh et al. 2021). Certainly, the female reproductive system of farm animals is affected by OS at several levels, from oocyte maturation to fertilization and embryo development (Talukder et al. 2017). Gilbert (2019) reported that oocytes and other follicular cells are susceptible to oxidative damage, causing the depletion of the ovarian pool of primordial follicles and damaging surviving oocytes.

Furthermore, the redox balance is essential for periparturient dairy cows. Indeed, the redox status of the animals is provided by the ratio of total oxidants to antioxidants; sometimes, the result of the separate expression of oxidants and antioxidants may lead to contradictory results regarding the redox status during the transition period. The evaluation of oxidative status of dairy cows is needed to monitor shifts in redox balance between markers of oxidative damage and antioxidant capacity using the ratio of pro- to antioxidants as proposed by Celi (2011).

Abuelo et al. (2013) studied the redox status of dairy cows during the transition period using the oxidant state index (OSi), a combined quantification of ROS and serum antioxidant capacity. Those authors demonstrated that OSi is an integrated measure of oxidants and antioxidants, able to determine the variations of redox balance more accurately in dairy cows, often not detectable by the determination of the components separately.

## 3 Environmental Stress: Heat Stress

Heat stress (HS) is recognized to be one of the most important factors to negatively affect animal well-being and dairy industries. Exposure of dairy cattle to high ambient temperatures is responsible (indirectly and directly) for the alteration of physiology, metabolism, and then productive and reproductive performances and can lead to the increase of mortality rate (Bernabucci et al. 2010; Vitali et al. 2009, 2015). Heat stress can be defined as a physiological condition when the core body temperature of a given species exceeds its range specified for normal activity, which results from a total heat load (internal production and environment) exceeding the capacity for heat dissipation, and this prompts physiological and behavioral responses to reduce the strain (Bernabucci et al. 2010).

Heat stress participates in the induction of oxidative stress. Increase of ROS and indices of lipid peroxidation and mitochondrial dysfunction have been reported in dairy cattle exposed to warm environment (Bernabucci et al. 2002; Belhadj Slimen et al. 2014; Zachaut et al. 2017). Heat stress is responsible for the increase of $O_2^{\bullet-}$, the first by-product of the one-electron reduction of oxygen (Verma et al. 2020), hydrogen peroxide (Chauhan et al. 2015), and lipid peroxidation (Bernabucci et al. 2002). Contemporarily, heat stress exposure is responsible for the decrease of antioxidant defenses (Bernabucci et al. 2002; Li et al. 2021).

The coordinate increase of erythrocyte SOD and GPx activities and SH group concentration has been considered an indirect compensatory response of cells to increased oxidant challenge during heat stress (Bernabucci et al. 2002). This was confirmed by the increased concentrations of TBARS in erythrocytes of heat-stressed cows (Bernabucci et al. 2002).

One possible cause of the increased risk of oxidative stress in heat-stressed cows is the increase of oxygen pressure of blood due to the increased respiration rate.

Some studies have demonstrated the positive effect of antioxidant supplementation in heat-stressed ruminants (Chauhan et al. 2014; Sivakumar et al. 2010) confirming the close association between heat stress and oxidative stress.

## 4 Pathogen/Disease Stress: Mastitis

The dairy industry is a major player in the agricultural economy. There are numerous problems that farmers must face to counteract the economic losses linked, above all, to infectious diseases that continue to be inherent in farms all over the world. In fact, despite the innovations related to management practices that have taken place in the past decades, mastitis remains the common enemy of all farmers.

Mastitis is a complex multifactorial disease characterized by severe inflammation of the mammary gland parenchyma that causes severe economic losses including decreased milk production and quality, premature killing, reduced conception rate, and treatment costs for dairy animals (Seegers et al. 2003; Radostits et al. 2010).

The onset of this pathology is caused by fungi, yeasts, algae and above all bacteria with a significant finding of those belonging to the genera *Escherichia*, *Streptococcus*, and *Staphylococcus*. Mastitis can be classified into clinical and subclinical. In the first, inflammation is accompanied by the development of symptoms such as swelling, redness, udder pain, and clots in milk; the second, unlike the previous one, is devoid of these symptoms, without visual changes either at a macroscopic level in the milk or on the mammary system, and is detectable only with microbiological or biochemical analyses (subclinical mastitis is generally defined using the somatic cell count; Chaneton et al. 2011). Sørensen et al. (2009) managed to correlate cases of both clinical and subclinical mastitis to harmful bacteria such as *Streptococcus dysgalactiae*, *Streptococcus uberis*, *Escherichia coli*, *Enterococcus* spp., and *Staphylococcus aureus*. The latter is mainly associated with subclinical mastitis, which often results in chronic disease, while *E. coli* induces severe acute clinical mastitis. Typical bacteria involved in breast infections activate the local immune system in several ways, which can affect the severity of the outcome. *E. coli* infection induces a rapid and massive increase in somatic cell count (SCC), while *S. aureus* infections cause a more gradual increase over a few days. Unfortunately, the pathogens involved in the inflammatory process have a lower susceptibility to antibiotics, and over 90% of cases of this disease are caused by environmental pathogens (Klaas and Zadoks 2018). Regardless of the pathogens involved, the inflammatory response leads to an increase in OS at the local level: a preliminary study found a statistically significant correlation between advanced oxidation protein products (AOPP) and the percentage of neutrophils in milk (Colitti et al. 2019). The specificity of OS and its relationship with mastitis is a complex phenomenon. If on the one hand the imbalance between oxidants and antioxidants during certain periods of the animal's life facilitates the attack by pathogens (Bernabucci et al. 2005; Kankofer 2002; Miller et al. 1993), on the other hand, the mastitis itself could induce an increase in the formation of free radicals in milk and sustain the state of OS (Gu et al. 2009), especially during the first period of lactation (Sordillo et al. 2007). In cows with clinical mastitis, a significant reduction in SOD activity, CAT, and GSH concentration was observed as well as an increase in lipid peroxidation (Jhambh et al. 2013), conditions that support OS (Weiss et al. 2004). Tong et al. (2019) identified bacterial species of the milk microbiota of cows with subclinical mastitis from *Streptococcus agalactiae* and related them to those of healthy cow milk. The drastic alteration of microbial diversity in samples from cows affected by the disease is evident by the correlation between the milk microbiota and metabolic factors. In addition to the presence of different bacterial genera, a higher level of some metabolites was measured, such as phenylpyruvic acid and the xanthine/guanine ratio. The latter together with its catalytic enzyme, xanthine oxidase, which can synthesize uric acid, can lead to an increase in ROS, $O_2^{\bullet-}$ and $H_2O_2$, all involved in the pathogenesis. Therefore, the presence of *Streptococcus spp.* is positively correlated to the xanthine/guanine ratio, favoring OS. On the contrary, the presence of D-mannose, a known regulator of microbial infections, tends to decrease in the presence of pathogenic *Streptococcus spp.* (Ayemele et al. 2021).

## 5 Psychological Stress: Stocking Density

The farming methods have an unconditional impact on the productive and reproductive performances of dairy cows. The preservation of animal welfare and the improvement of the different breeding conditions optimize the animal's behavior and reduce the incidence of OS. It is therefore important to minimize the conditions triggering the excessive production of oxidants, mainly related to the onset of inflammatory diseases, metabolic intensity, and farm stress (Celi 2011). In fact, in livestock farms where animal welfare is not fully satisfied, it is known there is a greater incidence of farm stress. This negatively affects the neuroendocrine and metabolic physiology of the animals, resulting in negative effects not only on health but also on production and reproduction (Moberg 1985). The several types of breeding have both advantages and disadvantages on the success of animal welfare. If, on the one hand, extensive farming systems must be concerned with satisfying the entire nutritional needs and climatic conditions, in indoor farms, on the other hand, the limitation of the animal's behavior may generate an imbalance in health conditions and an increase in mammary diseases and foot diseases (Arnott et al. 2016). Some conditions such as thermal stress (as reported before), overcrowding, unsuitable structures, and hierarchical competitions can induce breeding stress and consequently OS (Celi 2011). Ando et al. (1997) affirm that in case of farm stress the production of ROS increases, due both to the increased oxygen consumption necessary for the stress adaptation mechanisms and to the secondary imbalances of normal physiological processes. As well described before, since ROS are extremely reactive molecules and therefore difficult to measure, biomarkers have been used in studies as indirect indicators of the animal's state of well-being (Marco-Ramell et al. 2012). In particular, the main markers used are MDA, GSH, SOD, GPx, and AOPP.

Lin et al. (2018) carried out an experiment to verify whether the stocking density could have effects on the oxidation of milk in medium or late lactating multiparous Holstein cows. The results showed that cortisol and MDA content in plasma was lower in medium-late lactating cows with low stoking density (LSD, 75%) compared with that of medium-late lactating cows with high SD (HSD, 100%). The same research group (Lin et al. 2019) evaluated the effects of herd density on oxidative status and on mammary gland permeability in early lactating dairy cows. The experiment demonstrated that cows kept in an open barn with a rearing density of 75% (75SD) had higher concentrations of SOD with respect to a group with a density of 100% (100SD) and similar plasma activity of GPx. Other parameters were evaluated such as the level of $Na^+$, albumin, and plasmin which were lower in 75SD cows. This confirmed that the high SD is associated with higher risk of OS. Therefore, Lin et al. (2019) proposed that biomarkers of OS can be used as new and reliable indicators of animal welfare since any type of stress triggers an increase in the production of oxidants and on the contrary a decrease in the body's antioxidant resources.

## 6 Antioxidants to Counteract Oxidative Stress in Dairy Cow

The redox balance plays a fundamental role during the productive life of dairy cows, especially in periods of greatest stress and susceptibility. Physiologically, living organisms have developed a complex antioxidant defense system capable of naturally counteracting the production of ROS, continuously produced during metabolism as they are involved in key functions of the organism, as discussed in the previous paragraphs.

In recent years, identifying and protecting against free radical damage has become a very important aspect of reducing damage to cow health and consequently the management losses that occur because of related problems. Nutritional and other management strategies represent a possible short-term solution in improving cow health and well-being in critical stages.

Therefore, several strategies are needed to promote the increase in the antioxidant capacity of dairy cows so that they can counteract the increase in ROS and prevent related diseases.

The integration with vitamins and trace elements (Table 2) tries to limit the harmful consequences caused by the excessive production of ROS and oxidative stress, trying to improve the health of the animals by acting both on the antioxidant effect and, consequently, on the incidence of dairy cows' diseases (Abuelo et al. 2015).

Vitamin E is an essential and chain-breaking fat-soluble antioxidant that in lipid environments eliminates hydroperoxyl radicals (Traber and Stevens 2011), preventing their propagation in plasma lipoproteins and membranes (Traber and Atkinson 2007). The most abundant and biologically active form of vitamin E is α-tocopherol (Brigelius-Floh and Traber 1999). The reason for the decrease in plasma vitamin E at the onset of lactation is still unknown. Goff and Horst (1997) suggested that this decrease was linked to colostrum synthesis but could also be related to the decrease in the hepatic production of some proteins that carry the vitamin (Abd Eldaim et al. 2010). The incidence of intramammary infections with the consequent onset of mastitis in the transition period is associated with the decrease in the plasma concentration of vitamin E (Persson Waller et al. 2007). It has been shown that calving cows with plasma concentration of α-tocopherol below 3.0 mg/ml are at higher risk of contracting mastitis within the first 7 days of lactation compared with cows with greater vitamin E concentration (Weiss et al. 1997). Already in the late 1990s, Weiss et al. (1997) demonstrated that vitamin E and Se reduced the number of somatic cells and the severity of mastitis, apparently acting synergistically in the defense of the mammary gland. The association of vitamin E with Se showed that the supplementation of vitamin E allowed a 37% reduction in the onset of mastitis and Se reduced the duration of the symptoms of the disease (Politis 2012). Allison and Laven (2000) suggested that vitamin E may play a protective role especially against environmental breast pathogens such as *E. coli* and *S. uberis*. Another study reported that a single parenteral injection of vitamin E equal to 3000 IU/cow administered 1 week before calving did not show significant effects against the onset of mastitis. Despite some conflicting results, a meta-analysis study (Politis 2012) suggests supplementing 3000 IU/cow per day in the dry period of vitamin E for preventing mastitis.

**Table 2** Antioxidant compounds to counteract oxidative stress in dairy cow

| Antioxidants | Indicators/Activity | *In vitro/vivo* study | References |
|---|---|---|---|
| Vitamin E | ↓ milk somatic cells, ↓ clinical mastitis | *In vivo* | Smith et al. 1984 and Persson Waller et al. (2007) |
| | ↓ clinical mastitis | | Weiss et al. (1997) |
| Selenium | ↑ selenoproteins, ↓ subclinical mastitis | *In vivo* | Sordillo and Aitken (2009) |
| | ↓ clinical mastitis ↓ milk somatic cells, ↑ GPx | | Malbe et al. (1995) |
| | ↓ ROS ↑ SOD, ↑ GPS, ↑ CAT activity, ↓ MDA, ↑ plasma Se, ↑ GSH | | Gong and Xiao (2018) |
| Vitamin A | ↓ clinical mastitis | *In vivo* | LeBlanc (2004) |
| CLA | ↑ antioxidant enzymes | *In vitro* | Cantwell et al. (1999) |
| | ↑ GSH, ↑ γGLC activity, ↑ GST, SOD, GPx ↓ GR activity, ↓ ROS, ↓ TBARS | *In vitro* | Basirico et al. (2015) |
| | ↑ GSH, ↑ NADPH, ↑ γGLC activity, ↓ ROS, ↓ MDA | *In vitro* | Basiricò et al. (2016) |
| | ↓ TBARS, ↑ antioxidant response | *In vivo* | Hanschke et al. (2016) |
| Polyphenols | ↓ ROS, ↑ NFE2L2 | *In vitro* | Ma et al. (2018) |
| | ↑ GSH, ↑ γGLC activity, ↓ ROS, ↓ MDA | | Basiricò et al. (2019) |
| | ↓ ROS, ↓ MDA | | Mastrogiovanni et al. (2020) |

Among the trace elements that can be added to the animal diet to improve antioxidant defense, Se is certainly one of the major protagonists. Being the precursor of at least 25 selenoproteins (Surai 2018), it is present in a series of physiological functions related to the maintenance of redox balance and protection from OS. The selenoproteins are, in fact, involved both in the redox signaling functions and in the peroxidase and reductase activities (Zoidis et al. 2018). While the Se deficiency in monogastric animals has been solved by supplementing the diet, this deficiency is still a problem in cattle. Selenium is usually integrated in feed in the form of sodium selenite and selenate, but they are probably poorly bioavailable or adversely affected by the rumen (Galbraith et al. 2016). As a solution to the problem, organic selenium, in the form of selenomethionine (SeMet), has been reported as an additive in the feeding of dairy cattle, especially in the transition period. In fact, it has been seen that over 50% of the total Se present in feed is found in the organic form of SeMet. Probably the digestive system of animals has adapted to the supply of organic Se and SeMet which, being very similar to methionine, can be absorbed and incorporated into proteins, thus acting as an endogenous reserve of Se (Surai et al. 2018). The decrease in feed intake of dairy cows leads to a reduction in the intake of Se and consequently a reduction in the reserves of SeMet to be used in phases of stress (Surai et al. 2018; Surai 2018).

It has been estimated that the requirement of Se in dairy cows is 0.30 mg/kg dry matter (NRC 2001; NASEM 2021) which can be integrated with the diet. Furthermore, Malbe et al. (1995) compared the benefits of supplementing organic Se or selenite on 100 dairy cows. Levels of GPx present in the blood ranged from 0.22 to 3.0 μKat/g Hb for organic Se and 2.3 μKat/g Hb for selenite. It should be noted that despite the interruption of integration, the GPx continued to increase for the next 10 weeks. Another important aspect was the positive effect on breast health. Miranda et al. (2009) reported that low levels of Se and GPx activity increase OS in the mammary gland. The integration of Se leads to the formation of GPx, a selenoprotein that plays a role in the antioxidant defense process in cells by catalyzing $H_2O_2$. The increase in selenoprotein has been linked to a reduction in the frequency of subclinical mastitis in dairy cattle (Sordillo and Aitken 2009). Consistent with what has been said so far, Salman et al. (2009) have documented the effect of Se on improving the immunity of the mammary gland. A decrease in the incidence of mastitis and in milk somatic cells has been recorded (Hogan et al. 1993; Malbe et al. 1995). These results were confirmed by another study on Holstein cows fed with 0.3 mg Se/kg dry matter plus Na selenite at 0.3 mg Se/kg dry matter in their ration or without Se, during the last 4 weeks before calving. Selenium treatments improved the oxidative status inducing a decrease in the concentration of lipid peroxidation, ROS, and $H_2O_2$ as well as an increase in GPx activity (Gong and Xiao 2018). Furthermore, Surai et al. (2019) have observed the benefits of replacing selenate with organic Se, in the form of SeMet, in dairy cows as the increase in Se reserves of cows ensures the maintenance of antioxidant defenses also to the calf. Many studies clearly affirm the positive action of the Se supplementation in improving antioxidant functions and oxidative status. Given the benefits obtained from the integration of vitamins and trace elements, several studies have been dedicated to the supplementation of antioxidants. Probably Se and vitamin E are the antioxidants most included in the diets of dairy cattle.

Vitamin A, or retinol, has numerous functions that are not fully understood. Vitamin A plays a role in improving the resistance against infectious diseases, in particular mastitis, in the periparturient dairy cows (NRC 2001; NASEM 2021). An in vivo study on 1051 dairy cows showed that elevated serum retinol levels in the week before calving decreased the risk of clinical mastitis by 60% in the early lactation (LeBlanc et al. 2004). Up to now it is not clear yet what is the right amount of antioxidants to be administered in diets so that harmful consequences are not unleashed.

Among the classes of preventive molecules taken into consideration as antioxidant

(Table 2), importance was also given to the protective effects of the isomers of conjugated linoleic acid (CLA) which can increase the antioxidant enzyme defense system in rat hepatocytes (Cantwell et al. 1999). CLA-based dietary supplements are used in dairy cows mainly for their depressing effect on milk fat content but are also proposed to have antioxidant properties. Several in vitro and in vivo studies have been carried out to examine the effects of CLA on oxidative and antioxidant status of lactating dairy cows. In vitro studies carried out on a mammary epithelial cell line, BME-UV1, treated with two isomers (cis-9, trans-11 and trans-10, cis-12) of CLA showed the increased activity of GSTs, SOD, and GPx and the reduction of GR activity (Basirico et al. 2015) in treated cells. Furthermore, in vitro treatment of BME-UV1 cells with CLA isomers and other essential fatty acids (FA) induced an increase in intracellular GSH, accompanied by elevated NADPH concentrations and an increase in γ-glutamyl-cysteine ligase activity (Dipasquale et al. 2018). Cells supplemented with FA showed a reduction in intracellular levels of MDA and an antioxidant cellular response against OS induced by $H_2O_2$ (Basiricò et al. 2016). These results were confirmed by in vivo study in German Friesian cows, where supplementation of lipid-encapsulated CLA with a 12% composition of CLA trans-10, cis-12 and CLA cis-9, trans-11, resulted in an antioxidant effect in lactating cows (Hanschke et al. 2016).

The useful way to act on the health of cows is undoubtedly feeding, so it is also essential, with a view to sustainable livestock nutrition and a circular economy, to be able to use natural substances and by-products to obtain natural antioxidants. In recent years, interest has moved toward products derived from plant extracts rich in polyphenols capable of giving added value to products of animal origin, with beneficial effects on human health (Castillo et al. 2019). In this regard, an in vitro study on cultured bovine mammary epithelial cells (BMEC) investigated the cytoprotective effects of tea polyphenols (TP). The exposure of cells to different concentrations of TP after treatment with $H_2O_2$ had positive effects on cell viability and reduced the accumulation of intracellular ROS. Tea polyphenols also increased the abundance of factor related to nuclear-erythroid factor 2-like 2 (NFE2L2), an important transcription factor that acts in maintaining redox balance (Ma et al. 2018). Basiricò et al. (2019) evaluated, on BME-UV1, the antioxidant and anti-inflammatory effects of hydroxytyrosol (HTyr), a phenol found in olive oil, and (-)-epigallocatechin-3-gallate (EGCG), the major phenolic compound of green tea. The results indicated that EGCG and HTyr were able to mitigate oxidative stress and inflammatory responses, proposing that these phenolic compounds are potential natural alternatives to be used in dairy cattle as feed supplement for reducing the development of oxidative and inflammatory processes related to parturition. Another in vitro study showed the antioxidant and anti-inflammatory activity of pomegranate peel extracts, a compound rich in hydrolyzable tannins, on BME-UV1 cells (Mastrogiovanni et al. 2020). Pomegranate peel extracts induced a reduction of ROS production and a decreased MDA concentration.

## 7 Conclusions

Numerous are the stressful conditions able to generate the imbalance between increased production of free radicals and the capacity of antioxidant mechanisms to neutralize these oxidants in cells and tissues of dairy cows. The development of OS in these animals has been associated with metabolic, immune, and reproductive dysfunctions. Several studies have shown that the administration of antioxidants can protect dairy cows from OS conditions and several novel compounds are promising for future applications to counteract OS in dairy cows. However, further studies are needed to validate the use of these new substances and to obtain guidelines on the concentrations and timing of antioxidant additions to be routinely used on farms to achieve improved animal welfare and optimal production.

## References


Abd Eldaim MA, Kamikawa A, Soliman MM et al (2010) Retinol binding protein 4 in dairy cows: its presence in colostrum and alteration in plasma during fasting, inflammation, and the peripartum period. J Dairy Res 77(1):27–32. https://doi.org/10.1017/s0022029909990276

Abuelo A, Hernández J, Benedito JL et al (2013) Oxidative stress index (OSi) as a new tool to assess redox status in dairy cattle during the transition period. Animal 7(8):1374–1378. https://doi.org/10.1017/s1751731113000396

Abuelo A, Hernández J, Benedito JL et al (2015) The importance of the oxidative status of dairy cattle in the periparturient period: revisiting antioxidant supplementation. J Anim Physiol Anim Nutr 99(6):1003–1016. https://doi.org/10.1111/jpn.12273

Abuelo A, Hernández J, Benedito JL et al (2019) Redox biology in transition periods of dairy cattle: role in the health of periparturient and neonatal animals. Antioxidants 8(1):20. https://doi.org/10.3390/antiox8010020

Agarwal A, Aponte-Mellado A, Premkumar BJ et al (2012) The effects of oxidative stress on female reproduction: a review. Reprod Biol Endocrinol 10:49. https://doi.org/10.1186/1477-7827-10-49

Allison RD, Laven RA (2000) Effect of vitamin E supplementation on the health and fertility of dairy cows: a review. Vet Rec 147(25):703–708

Ando M, Katagiri K, Yamamoto S et al (1997) Age-related effects of heat stress on protective enzymes for peroxides and microsomal monooxygenase in rat liver. Environ Health Perspect 105(7):727–733. https://doi.org/10.1289/ehp.97105726

Arnott G, Ferris CP, O'Connell NE (2016) Review: welfare of dairy cows in continuously housed and pasture-based production systems. Animal 11(2):261–273. https://doi.org/10.1017/S1751731116001336

Ayemele AG, Tilahun M, Lingling S et al (2021) Oxidative stress in dairy cows: insights into the mechanistic mode of actions and mitigating strategies. Antioxidants 10(12):1918. https://doi.org/10.3390/antiox10121918

Basirico L, Morera P, Dipasquale D et al (2015) Conjugated linoleic acid isomers strongly improve the redox status of bovine mammary epithelial cells (BME-UV1). J Dairy Sci 98(10):7071–7082. https://doi.org/10.3168/jds.2015-9787

Basiricò L, Morera P, Dipasquale D et al (2016) Comparison between conjugated linoleic acid and essential fatty acids in preventing oxidative stress in bovine mammary epithelial cells. J Dairy Sci 100(3):2299–2309. https://doi.org/10.3168/jds.2016-11729

Basiricò L, Morera P, Dipasquale D et al (2019) (-)-Epigallocatechin-3-gallate and hydroxytyrosol improved antioxidative and anti-inflammatory responses in bovine mammary epithelial cells. Animal 13(12):2847–2856. https://doi.org/10.1017/s1751731119001356

Bassel LL, Caswell JF (2018) Bovine neutrophils in health and disease. Cell Tissue Res 371(3):617–637. https://doi.org/10.1007/s00441-018-2789-y

Behairy A, El-Sharkawy NI, Saber TM et al (2020) The modulatory role of vitamin C in boldenone undecylenate induced testicular oxidative damage and androgen receptor dysregulation in adult male rats. Antioxidants 9(11):1053. https://doi.org/10.3390/antiox9111053

Belhadj Slimen I, Najar T, Ghram A et al (2014) Reactive oxygen species, heat stress and oxidative-induced mitochondrial damage. A review. Int J Hyperthermia 30(7):513–523. https://doi.org/10.3109/02656736.2014.971446

Bernabucci U, Ronchi B, Lacetera N et al (2002) Markers of oxidative status in plasma and erythrocytes of transition dairy cows during hot season. J Dairy Sci 85(9):2173–2179. https://doi.org/10.3168/jds.s0022-0302(02)74296-3

Bernabucci U, Ronchi B, Lacetera N et al (2005) Influence of body condition score on relationships between metabolic status and oxidative stress in periparturient dairy cows. J Dairy Sci 88(6):2017–2026. https://doi.org/10.3168/jds.s0022-0302(05)72878-2

Bernabucci U, Basiricò L, Pirazzi D et al (2009) Liver apolipoprotein B 100 expression and secretion are down-regulated early postpartum in dairy cows. Livest Sci 125:169–176. https://doi.org/10.1016/j.livsci.2009.04.006

Bernabucci U, Lacetera N, Baumgard LH et al (2010) Metabolic and hormonal acclimation to heat stress in domesticated ruminants. Animal 4(7):1167–1183. https://doi.org/10.1017/s175173111000090x

Bionaz M, Trevisi E, Calamari L et al (2007) Plasma paraoxonase, health, inflammatory conditions, and liver function in transition dairy cows. J Dairy Sci 90(4):1740–1750. https://doi.org/10.3168/jds.2006-445

Boudjellaba S, Ainouz L, Tennah S et al (2018) Reproduction performance and blood biochemical parameters in dairy cows: relationship with oxidative stress status. Vet World 11(6):883–888. http://www.veterinaryworld.org/Vol.11/June2018/23.html#:~:text=doi%3A%2010.14202/vetworld.2018.883%2D888

Bouwstra RJ, Nielen M, Newbold JR et al (2010) Vitamin E supplementation during the dry period in dairy cattle. Part II: oxidative stress following vitamin E supplementation may increase clinical mastitis incidence postpartum. J Dairy Sci 93(12):5696–5706. https://doi.org/10.3168/jds.2010-3161

Brigelius-Floh R, Traber MG (1999) Vitamin E: function and metabolism. FASEB J 13:1145–1155. https://doi.org/10.1096/fasebj.13.10.1145

Cantwell H, Devery R, OShea M et al (1999) The effect of conjugated linoleic acid on the antioxidant enzyme defense system in rat hepatocytes. Lipids 34(8):833–839. https://doi.org/10.1007/s11745-999-0430-4

Castillo C, Hernandez J, Bravo A et al (2005) Oxidative status during late pregnancy and early lactation in dairy cows. Vet J 169(2):286–292. https://doi.org/10.1016/j.tvjl.2004.02.001

Castillo C, Hernandez J, Valverde I et al (2006) Plasma malonaldehyde (MDA) and total antioxidant status (TAS) during lactation in dairy cows. Res Vet Sci 80(2):133–139. https://doi.org/10.1016/j.rvsc.2005.06.003

Castillo C, Abuelo A, Hernández J (2019) Ruminant (bovine, caprine, and ovine) milk and meat production: the challenge of food quality and sustainability through the use of plant extracts. In: Encyclopedia of food security and sustainability. Elsevier, New York, pp 25–42

Celi P (2011) Biomarkers of oxidative stress in ruminant medicine. Immunopharmacol Immunotoxicol 33:233–240. https://doi.org/10.3109/08923973.2010.514917

Celi P, Merlo M, Da Dalt L et al (2011) Relationship between late embryonic mortality and the increase in plasma advanced oxidised protein products (AOPP) in dairy cows. Reprod Fertil Dev 23(4):527–533. https://doi.org/10.1071/rd10268

Chaneton J, Pérez Sáez M, Bussmann LE (2011) Antimicrobial activity of bovine β-lactoglobulin against mastitis-causing bacteria. J Dairy Sci 94(1):138–145. https://doi.org/10.3168/jds.2010-3319

Chauhan SS, Celi P, Leury BJ et al (2014) Dietary antioxidants at supranutritional doses improve oxidative status and reduce the negative effects of heat stress in sheep. J Anim Sci 92(8):3364–3374. https://doi.org/10.2527/jas.2014-7714

Chauhan SS, Celi P, Leury B et al (2015) Exhaled breath condensate hydrogen peroxide concentration, a novel biomarker for assessment of oxidative stress in sheep during heat stress. Anim Prod Sci 56(7):1105–1112. https://www.publish.csiro.au/an/an14070#:~:text=https%3A//doi.org/10.1071/AN14070

Colitti M, Stefanon B, Gabai G et al (2019) Oxidative stress and nutraceuticals in the modulation of the immune function: current knowledge in animals of veterinary interest. Antioxidants 8(1):28. https://doi.org/10.3390%2Fantiox8010028

Contreras GA, Strieder-Barboza C, Raphael W (2017) Adipose tissue lipolysis and remodeling during the transition period of dairy cows. J Anim Sci Biotechnol 5(8):41. https://doi.org/10.1186/s40104-017-0174-4

Dipasquale D, Basiricò L, Morera P et al (2018) Anti-inflammatory effects of conjugated linoleic acid isomers and essential fatty acids in bovine mammary epithelial cells. Animal 12(10):2018–2114. https://doi.org/10.1017/s1751731117003676

Dirandeh E, Sayyar MA, Ansari-Pirsaraei Z et al (2021) Peripheral leucocyte molecular indicators of inflammation and oxidative stress are altered in dairy cows with embryonic loss. Sci Rep 11:12771. https://doi.org/10.1038/s41598-021-91535-2

Duhig K, Chappell LC, Shennan AH (2016) Oxidative stress in pregnancy and reproduction. Obstetric Med 9(3):113–116. https://doi.org/10.1177/1753495x16648495

Farney JK, Mamedova LK, Coetzee JF et al (2013) Anti-inflammatory salicylate treatment alters the metabolic adaptations to lactation in dairy cattle. Am J Physiol Regul Integr Comp Physiol 305(2):R110–R117. https://doi.org/10.1152/ajpregu.00152.2013

Favorit V, Hood WR, Kavazis AN et al (2021) Mitochondrial bioenergetics of extramammary tissues in lactating dairy cattle. Animals 11(9):2647. https://doi.org/10.3390/ani11092647

Gaal T, Ribiczeyne-Szabo P, Stadler K et al (2006) Free radicals, lipid peroxidation and the antioxidant system in the blood of cows and newborn calves around calving. Comp Biochem Physiol B Biochem Mol Biol 143(4):391–396. https://doi.org/10.1016/j.cbpb.2005.12.014

Galbraith ML, Vorachek WR, Estill CT (2016) Rumen microorganisms decrease bioavailability of inorganic selenium supplements. Biol Trace Elem Res 171(2):338–343. https://doi.org/10.1007/s12011-015-0560-8

Gilbert RO (2019) Symposium review: mechanisms of disruption of fertility by infectious diseases of the reproductive tract. J Dairy Sci 102(4):3754–3765. https://doi.org/10.3168/jds.2018-15602

Goff JP, Horst RL (1997) Physiological changes at parturition and their relationship to metabolic disorders. J Dairy Sci 80(7):1260–1268. https://doi.org/10.3168/jds.S0022-0302(97)76055-7

Gong J, Xiao M (2018) Effect of organic selenium supplementation on selenium status, oxidative stress, and antioxidant status in selenium-adequate dairy cows during the periparturient period. Biol Trace Elem Res 186(2):430–440. https://doi.org/10.1007/s12011-018-1323-0

Grummer RR (1995) Impact of changes in organic nutrient metabolism on feeding the transition dairy cow. J Anim Sci 73(9):2820–2833. https://doi.org/10.2527/1995.7392820x

Gu B, Zhu Y, Zhu W et al (2009) Retinoid protects rats against neutrophil-induced oxidative stress in acute experimental mastitis. Int Immunopharmacol 9(2):223–229. https://doi.org/10.1016/j.intimp.2008.11.014

Hadsell DL, Olea W, Wei J et al (2011) Developmental regulation of mitochondrial biogenesis and function in the mouse mammary gland during a prolonged lactation cycle. Physiol Genom 43(6):271–285. https://doi.org/10.1152/physiolgenomics.00133.2010

Hanschke N, Kankofer M, Ruda L et al (2016) The effect of conjugated linoleic acid supplements on oxidative and antioxidative status of dairy cows. J Dairy Sci 99(10):8090–8102. https://doi.org/10.3168/jds.2015-10685

Hogan JS, Weiss WP, Smith KL (1993) Role of vitamin E and selenium in host defense against mastitis. J Dairy Sci 76(9):2795–2803. https://doi.org/10.3168/jds.s0022-0302(93)77618-3

Ingvartsen KL, Dewhurst RJ, Friggens NC (2003) On the relationship between lactational performance and health: is it yield or metabolic imbalance that cause production diseases in dairy cattle? Livest Prod Sci 83(2–3):277–308. https://doi.org/10.1016/S0301-6226(03)00110-6

Irato P, Santovito G (2021) Enzymatic and non-enzymatic molecules with antioxidant function. Antioxidants 10(4):579. https://doi.org/10.3390/antiox10040579

Jhambh R, Dimri U, Gupta VK et al (2013) Blood antioxidant profile and lipid peroxides in dairy cows with clinical mastitis. Vet World 6:271–273. https://doi.org/10.5455/vetworld.2013.271-273

Kankofer M (2002) Placental release/retention in cows and its relation to peroxidative damage of macromolecules. Reprod Domest Anim 37(1):27–30. https://doi.org/10.1046/j.1439-0531.2002.00318.x

Kehrli ME Jr, Nonnecke BJ, Roth JA (1989a) Alterations in bovine neutrophil function during the periparturient period. Am J Vet Res 50(2):207–214

Kehrli ME Jr, Nonnecke BJ, Roth JA (1989b) Alterations in bovine lymphocyte function during the periparturient period. Am J Vet Res 50(2):215–220

Klaas IC, Zadoks RN (2018) An update on environmental mastitis: challenging perceptions. Transbound Emerg Dis 65:166–185. https://doi.org/10.1111/tbed.12704

Kumar B, Manuja A, Aic P (2012) Stress and its impact on farm animals. Front Biosci 4(5):1759–1767. https://doi.org/10.2741/496

Lacetera N, Scalia D, Bernabucci U et al (2005) Lymphocyte functions in overconditioned cows around parturition. J Dairy Sci 88(6):2010–2016. https://doi.org/10.3168/jds.S0022-0302(05)72877-0

Laubenthal L, Ruda L, Sultana N et al (2017) Effect of increasing body condition on oxidative stress and mitochondrial biogenesis in subcutaneous adipose tissue depot of nonlactating dairy cows. J Dairy Sci 100(6):4976–4986. https://doi.org/10.3168/jds.2016-12356

LeBlanc SJ, Herdt TH, Seymour WM (2004) Peripartum serum vitamin E, retinol, and beta-carotene in dairy cattle and their associations with disease. J Dairy Sci 87(3):609–619. https://doi.org/10.3168/jds.s0022-0302(04)73203-8

Li Y, Ding H, Liu L et al (2020) Non-esterified fatty acid induce dairy cow hepatocytes apoptosis via the mitochondria-mediated ROS-JNK/ERK signaling pathway. Front Cell Dev Biol 8:245. https://doi.org/10.3389%2Ffcell.2020.00245

Li H, Zhang Y, Li R et al (2021) Effect of seasonal thermal stress on oxidative status, immune response and stress hormones of lactating dairy cows. Anim Nutr 7(1):216–223. https://doi.org/10.1016/j.aninu.2020.07.006

Lin S, Liu J, Wang K et al (2018) Effects of stocking density on milk fatty acids composition and oxidative stability of mid- and late-lactating dairy cows. Animal 8(9):148. https://doi.org/10.3390/ani8090148

Lin S, Liu J, Wang K et al (2019) Effects of stocking density on oxidative stress status and mammary gland permeability in early lactating dairy cows. Animal 90(7):894–902. https://doi.org/10.1111/asj.13206

Ma Y, Zhao L, Gao M et al (2018) Tea polyphenols protect bovine mammary epithelial cells from hydrogen peroxide-induced oxidative damage in vitro. J Anim Sci 96(10):4159–4172. https://doi.org/10.1093%2Fjas%2Fsky278

Malbe M, Klaassen M, Fang W et al (1995) Comparisons of selenite and selenium yeast feed supplements on Se-incorporation, mastitis and leucocyte function in Se-deficient dairy cows. Zentralbl Veterinarmed A 42(2):111–121. https://doi.org/10.1111/j.1439-0442.1995.tb00362.x

Marco-Ramell A, Arroyo L, Saco Y et al (2012) Proteomic analysis reveals oxidative stress response as the main adaptative physiological mechanism in cows under different production systems. J Proteomics 75(14):4399–4411. https://doi.org/10.1016/j.jprot.2012.04.002

Mastrogiovanni F, Bernini R, Basiricò L et al (2020) Antioxidant and anti-inflammatory effects of pomegranate peel extracts on bovine mammary epithelial cells BME-UV1. Nat Prod Res 34(10):1465–1469. https://doi.org/10.1080/14786419.2018.1508149

Miller JK, Brzezinska-Slebodzinska E, Madsen FC (1993) Oxidative stress, antioxidants, and animal function. J Dairy Sci 76(9):2812–2823. https://doi.org/10.3168/jds.s0022-0302(93)77620-1

Miranda S, Wang YJ, Purdie NG et al (2009) Selenomethionine stimulates expression of glutathione peroxidase 1 and 3 and growth of bovine mammary epithelial cells in primary culture. J Dairy Sci 92(6):2670–2683. https://doi.org/10.3168/jds.2008-1901

Mironczuk-Chodakowska I, Witkowska AM, Zujko ME (2018) Endogenous non-enzymatic antioxidants in the human body. Adv Med Sci 63(1):68–78. https://doi.org/10.1016/j.advms.2017.05.005

Moberg GP (1985) Animal stress. Am. Physiol. Soc., Bethesda

Moussa ZA, Judeh ZMAA, Ahmed SAA (2019) Nonenzymatic exogenous and endogenous antioxidants. In: Das K, Das S, Biradar MS, Bobbarala V, Tata SS (eds) Book free radical medicine and biology. https://doi.org/10.5772/intechopen.87778

NASEM (2021) National Academy of Science, Engineering and Medicine. Nutrient requirements of dairy cattle, 8th revised edition. The National Academies Press, Washington D.C. https://doi.org/10.17226/25806

Nemec Svete A, Vovk T, Bohar Topolovec M et al (2021) Effects of vitamin E and coenzyme Q10 supplementation on oxidative stress parameters in untrained leisure horses subjected to acute moderate exercise. Antioxidants 10:908. https://doi.org/10.3390/antiox10060908

Nimse SB, Palb D (2015) Free radicals, natural antioxidants, and their reaction mechanisms. RSC

Adv 5:27986–28006. https://doi.org/10.1039/C4RA13315C
NRC (2001) National Research Council. Nutrient requirements of dairy cattle, 7th revised edition. The National Academies Press, Washington, DC
Persson Waller K, Hallen Sandgren C, Emanuelson U et al (2007) Supplementation of RRR-alpha-tocopheryl acetate to periparturient dairy cows in commercial herds with high mastitis incidence. J Dairy Sci 90(8):3640–3646. https://doi.org/10.3168/jds.2006-421
Phaniendra A, Jestadi DB, Periyasamy L (2015) Free radicals: properties, sources, targets, and their implication in various diseases. Indian J Clin Biochem 30(1):11–26. https://doi.org/10.1007/s12291-014-0446-0
Pigeolet E, Corbisier P, Houbion A et al (1990) Glutathione peroxidase, superoxide dismutase, and catalase inactivation by peroxides and oxygen derived free radicals. Mech Ageing Dev 51(3):283–297. https://doi.org/10.1016/0047-6374(90)90078-t
Politis I (2012) Reevaluation of vitamin E supplementation of dairy cows: bioavailability, animal health and milk quality. Animal 6(9):1427–1434. https://doi.org/10.1017/s1751731112000225
Puppel K, Kapusta A, Kuczynska B (2015) The etiology of oxidative stress in the various species of animals, a review. J Sci Food Agric 95:2179–2184. https://doi.org/10.1002/jsfa.7015
Putman AK, Brown JL, Gandy JC et al (2018) Changes in biomarkers of nutrient metabolism, inflammation, and oxidative stress in dairy cows during the transition into the early dry period. J Dairy Sci 101(10):9350–9359. https://doi.org/10.3168/jds.2018-14591
Radostits OM, Gay CG, Hinehcliff KW et al (2010) Veterinary medicines. A textbook of the diseases of cattle, sheep, pigs, goats, and horses, 9th edn. W.B. Saunder's Publication Kent, China, p 611
Rosano TG, Jones DH (1976) Developmental changes in mitochondria during the transition into lactation in the mouse mammary gland I. Behavior on isopycnic gradient centrifugation preparation of homogenates and mitochondria from mammary glands gradient ultracentrifugation. J Cell Biol 69(3):573–580. https://doi.org/10.1083/jcb.69.3.573
Salman S, Khol-Parisini A, Schafft H et al (2009) The role of dietary selenium in bovine mammary gland health and immune function. Anim Health Res Rev 10(1):21–34. https://doi.org/10.1017/s1466252308001588
Schaff C, Borner S, Hacke S et al (2013) Increased muscle fatty acid oxidation in dairy cows with intensive body fat mobilization during early lactation. J Dairy Sci 96(10):6449–6460. https://doi.org/10.3168/jds.2013-6812
Schonfeld P, Wojtczak L (2008) Fatty acids as modulators of the cellular production of reactive oxygen species. Free Radic Biol Med 45(3):231–241. https://doi.org/10.1016/j.freeradbiomed.2008.04.029
Seegers H, Fourichon C, Beaudeau F (2003) Production effects related to mastitis and mastitis economics in dairy cattle herds. Vet Res 345(5):475–491. https://doi.org/10.1051/vetres:2003027
Shokolenko I, Venediktova N, Bochkareva A et al (2009) Oxidative stress induces degradation of mitochondrial DNA. Nucleic Acids Res 37(8):2539–2548. https://doi.org/10.1093/nar/gkp100
Sivakumar AVN, Singh G, Varshney VP (2010) Antioxidants supplementation on acid base balance during heat stress in goats. Asian-Aust J Anim Sci 23(11):1462–1468. https://doi.org/10.5713/ajas.2010.90471
Smith KL, Harrison JH, Hancock DD, Todhunter DA, Conrad HR (1984) Effect of vitamin E and selenium supplementation on incidence of clinical mastitis and duration of clinical symptoms. J Dairy Sci 67(6):1293–1300. https://doi.org/10.3168/jds.s0022-0302(84)81436-8
Sordillo LM, Aitken SL (2009) Impact of oxidative stress on the health and immune function of dairy cattle. Vet Immunol Immunopathol 128(1–3):104–109. https://doi.org/10.1016/j.vetimm.2008.10.305
Sordillo L, Mavangira V (2014) The nexus between nutrient metabolism, oxidative stress and inflammation in transition cows. Anim Prod Sci 54(9):1204–1214. https://doi.org/10.1071/AN14503
Sordillo LM, Raphael W (2013) Significance of metabolic stress, lipid mobilization, and inflammation on transition cow disorders. Vet Clin N Am Food Anim Pract 29(2):267–278. https://doi.org/10.1016/j.cvfa.2013.03.002
Sordillo LM, O'Boyle N, Gandy JC et al (2007) Shifts in thioredoxin reductase activity and oxidant status in mononuclear cells obtained from transition dairy cattle. J Dairy Sci 90(3):1186–1192. https://doi.org/10.3168/jds.S0022-0302(07)71605-3
Sørensen LP, Mark T, Madsen P et al (2009) Genetic correlations between pathogen-specific mastitis and somatic cell count in Danish Holsteins. J Dairy Sci 92(7):3457–3471. https://doi.org/10.3168/jds.2008-1870
Surai PF (2018) Selenium in poultry nutrition and health. Wageningen Academic Publishers, The Netherlands. https://doi.org/10.3920/978-90-8686-865-0
Surai PF, Kochish II, Fisinin VI et al (2018) Selenium in poultry nutrition: from sodium selenite to organic Se sources. J Poult Sci 55(2):79–93. https://doi.org/10.2141%2Fjpsa.0170132
Surai PF, Kochish II, Fisinin VI et al (2019) Revisiting oxidative stress and the use of organic selenium in dairy cow nutrition. Animals 9(7):462. https://doi.org/10.3390/ani9070462
Talukder S, Kerrisk KL, Gabai G et al (2017) Role of oxidant–antioxidant balance in reproduction of domestic animals. Anim Prod Sci 57(8):1588–1597. https://doi.org/10.1071/AN15619
Tong J, Zhang H, Zhang Y et al (2019) Microbiome and metabolome analyses of milk from dairy cows with subclinical *Streptococcus agalactiae* mastitis - potential biomarkers. Front Microbiol. https://doi.org/10.3389/fmicb.2019.02547

Traber MG, Atkinson J (2007) Vitamin E, antioxidant and nothing more. Free Radical Biol Med 43(1):4–15. https://doi.org/10.1016/j.freeradbiomed.2007.03.024

Traber MG, Stevens JF (2011) Vitamins C and E: beneficial effects from a mechanistic perspective. Free Radical Biol Med 51(5):1000–1013. https://doi.org/10.1016/j.freeradbiomed.2011.05.017

Van Engelen E, De Groot MW, Breeveld-Dwarkasing VNA et al (2009) Cervical ripening and parturition in cows are driven by a cascade of pro-inflammatory cytokines. Reprod Domest Anim 44:834–841. https://doi.org/10.1111/j.1439-0531.2008.01096.x

Verma N, Alyethodi RR, Kathuria A et al (2020) Effect of heat stress on superoxide anion production in native and crossbred cattle under in vitro whole blood culture model. J Thermal Biol 87:102457. https://doi.org/10.1016/j.jtherbio.2019.102457

Vitali A, Segnalini M, Bertocchi L et al (2009) Seasonal pattern of mortality and relationships between mortality and temperature-humidity index in dairy cows. J Dairy Sci 92(8):3781–3790. https://doi.org/10.3168/jds.2009-2127

Vitali A, Felici A, Esposito S et al (2015) The effect of heat waves on dairy cow mortality. J Dairy Sci 98(7):4572–4579. https://doi.org/10.3168/jds.2015-9331

Weber C, Hametner C, Tuchscherer A et al (2013) Variation in fat mobilization during early lactation differently affects feed intake, body condition, and lipid and glucose metabolism in high-yielding dairy cows. J Dairy Sci 96(1):165–180. https://doi.org/10.3168/jds.2012-5574

Weiss WP, Hogan JS, Todhunter DA et al (1997) Effect of vitamin E supplementation in diets with a low concentration of selenium on mammary gland health of dairy cows. J Dairy Sci 80(8):1728–1737. https://doi.org/10.3168/jds.s0022-0302(97)76105-8

Weiss WP, Hogan JS, Smith KL (2004) Changes in vitamin C concentrations in plasma and milk from dairy cows after an intramammary infusion of Escherichia coli. J Dairy Sci 87(1):32–37. https://doi.org/10.3168/jds.s0022-0302(04)73138-0

Yu Y, Cui Y, Niedernhofer LJ et al (2016) Occurrence, biological consequences, and human health relevance of oxidative stress-induced DNA damage. Chem Res Toxicol 29(12):2008–2039. https://doi.org/10.1021/acs.chemrestox.6b00265

Zachaut M, Gra G, Livshitz L et al (2017) Seasonal heat stress affects adipose tissue proteome toward enrichment of the Nrf2-mediated oxidative stress response in late-pregnant dairy cows. J Proteomics 158:52–61. https://doi.org/10.1016/j.jprot.2017.02.011

Zachut M, Honig H, Striem S et al (2013) Periparturient dairy cows do not exhibit hepatic insulin resistance, yet adipose-specific insulin resistance occurs in cows prone to high weight loss. J Dairy Sci 96(9):5656–5669. https://doi.org/10.3168/jds.2012-6142

Zoidis E, Seremelis I, Kontopoulos N et al (2018) Selenium-dependent antioxidant enzymes: actions and properties of selenoproteins. Antioxidants 7(5):66. https://doi.org/10.3390/antiox7050066

# Ketosis in Dairy Cows

Josef J. Gross

**Abstract**

Ketosis is a major production disease in early lactating dairy cows. Subclinical and clinical ketosis negatively affects animal health and performance. Monitoring of ketone body concentrations in either blood, milk, or urine is prerequisite for an early identification of cows at risk. Besides the etiology of ketosis, this chapter addresses further physiological interactions of ketone bodies with metabolic regulation of energy homeostasis and consequences for other concomitantly occurring health disorders such as fatty liver. Detection of ketone bodies in various media, treatment approaches, and risk factors and preventive measures for ketosis are summarized based on latest research findings.

**Keywords**

Ketosis · Hyperketonemia · Ketogenesis · Negative energy balance · Energy metabolism · Lactation · Dairy cow

**Supplementary Information** The online version contains supplementary material available at https://doi.org/10.1007/978-3-031-51788-4_6. The videos can be accessed individually by clicking the DOI link in the accompanying figure caption or by scanning this link with the SN More Media App.

J. J. Gross (✉)
Veterinary Physiology, Vetsuisse Faculty, University of Bern, Bern, Switzerland
e-mail: josef.gross@unibe.ch

## Abbreviations

| | |
|---|---|
| acetyl-CoA | Acetyl-coenzyme A |
| BCS | Body condition score |
| BHB | Beta-hydroxybutyrate |
| CLA | Conjugated linoleic acid |
| CPT1 | Carnitine palmitoyltransferase 1 |
| CPT2 | Carnitine palmitoyltransferase 2 |
| DMI | Dry matter intake |
| FA | Fatty acid |
| FFA | Free fatty acids |
| GH | Growth hormone |
| HMG-CoA | 3-Hydroxy-3-methylglutaryl-coenzyme A |
| IGF-1 | Insulin-like growth factor-1 |
| L-FABP | Liver fatty acid binding protein |
| NEB | Negative energy balance |
| NEFA | Nonesterified fatty acids |
| TCA | Tricarboxylic acid |
| VLDL | Very-low-density lipoprotein |

## 1 Introduction

Among the various health disorders in dairy cows, ketosis is one of the most prominent production diseases during early lactation. It results predominantly from an imbalance in energy metabolism, i.e., a negative energy balance (NEB), due to low energy intake at simultaneous high requirements. Ketosis is characterized by elevated concentrations of ketone bodies (beta-hydroxybutyrate [BHB], acetoacetate, and ace-



J. J. Gross (ed.), *Production Diseases in Farm Animals*,
https://doi.org/10.1007/978-3-031-51788-4_6

tone) and concomitantly reduced glucose concentrations in circulation (Andersson 1988; Duffield 2000; Duffield et al. 2009). Furthermore, ketonuria, inappetence, loss of body weight, decrease in milk yield, and incoordination may occur. Some animals show neural signs of either lethargy or even excitability.

Reports on prevalence rates for subclinical ketosis range from 3% to 42% in the first 2 to 3 months of lactation (Dohoo 1983; McArt et al. 2012b; Suthar et al. 2013; Brunner et al. 2019), whereas prevalence of clinical ketosis ranges at lower levels from 2% to 15% (Baird 1982; Duffield 2000). At a herd level, subclinical ketosis may affect up to 40% of cows in a herd although the incidence can be as high as 80% (McArt et al. 2013; van der Kolk et al. 2017). Both peak incidence and prevalence of subclinical ketosis occurred at 5 days in milk with values of 22% and 29%, respectively (McArt et al. 2012b). Herds with more than a 15–20% prevalence of excessively elevated concentrations of nonesterified fatty acids (NEFA) and BHB in early lactation have higher rates of subsequent health disorder (McArt et al. 2013; Raboisson et al. 2014). A review by van der Kolk et al. (2017) indicated that cows first testing subclinical ketosis positive from 3 to 5 days postpartum were 6.1 times more likely to develop a displaced abomasum than cows first testing subclinical ketosis positive at later lactational stages. Moreover, cows first testing subclinical ketosis positive from 3 to 7 days postpartum were 4.5 times more likely to be removed from the herd. Cows with ketosis often show a reduced performance in terms of milk yield and fertility (Walsh et al. 2007; Raboisson et al. 2014). Furthermore, ketosis triggers the susceptibility toward further production diseases like retained placenta, fatty liver, displaced abomasum, metritis, and lameness (Duffield et al. 2009; Ospina et al. 2010; McArt et al. 2012b). Cows diagnosed with subclinical ketosis during the first 5 days of lactation were 0.7 times as likely to conceive to first service and produced 2.2 kg less milk per day for the first 30 days in milk than cows first testing positive at 8 days in milk or later (van der Kolk et al. 2017). Each 0.1 mmol/L increase in BHB at first subclinical ketosis-positive test increased the risk of developing a displaced abomasum by a factor of 1.1, increased the risk of removal from herd by factor of 1.4, and was associated with a decrease in milk production by 0.5 kg/day for the first 30 days in milk (McArt et al. 2012b; van der Kolk et al. 2017). Besides the impairment of animal welfare, significant economic losses occur due to costs for reduced milk yield and veterinary treatments (McArt et al. 2015; Cainzos et al. 2022).

Etiology, diagnosis, and treatment of ketosis in dairy cows were subject of many scientific review papers over the last decades (e.g., Baird 1982; Gordon et al. 2013; Raboisson et al. 2014; Guliński 2021). The present chapter adds latest scientific findings to existing knowledge on ketosis, its etiology, relationships with physiological processes, and prophylactic and treatment approaches. However, it needs to be emphasized that hyperketonemia is not equivalent to ketosis. Cows with hyperketonemia do not necessarily develop ketosis, but cows with ketosis always show elevated concentrations of circulating ketone bodies.

## 2 Metabolism of Dairy Cows During Early Lactation and Etiology of Ketosis

### 2.1 Metabolic Priority of the Mammary Gland, Energy Requirements for Lactation, and Negative Energy Balance

Worldwide, yearly milk production in dairy cows increases almost linearly since many decades (FAOSTAT 2022). During the last 40 to 50 years, average milk production per cow more than doubled in many countries. Concomitantly, a considerable number of dairy cows are affected by health disorders that are primarily associated with the high performance level and associated failures of metabolic and immunological adaptation, although the relationship between the occurrence of production diseases and milk yield is not definitely clarified (Fleischer et al. 2001).

However, the selection toward higher milk yield continues worldwide (Capper et al. 2009; von Keyserlingk et al. 2013). It cannot be denied that elevated milk yields aggravated metabolic load during early lactation. Therefore, a further increase of prevalence and incidence rates of subclinical and clinical ketosis can be expected as the risk toward developing health disorders rises with the concomitantly increasing milk production. The prerequisites for the occurrence of ketosis in dairy cows are complex. The following explanations are mainly based on characteristic alterations in the carbohydrate and lipid metabolism during the transition period from late gestation to lactation. The discussions here are mainly derived from the detailed and excellent summary provided in a review paper by van der Kolk et al. (2017) but also from more recent scientific publications in this field.

In contrast to the rapidly increasing nutrient and energy demands for milk production, the concomitant feed intake rises slowly after parturition (Gross et al. 2011a), whereas toward the end of lactation and during the dry period, a risk of over-conditioning exists. Consequently, cows face a considerable energy deficiency during the first weeks of lactation until feed intake can meet the energy requirements for maintenance and lactation (Drackley 1999; Gross and Bruckmaier 2019a).

Furthermore, the onset of lactation in dairy cows is characterized by a high priority in directing nutrients to the mammary gland (Bauman and Currie 1980; Bell 1995; Bruckmaier and Gross 2017). Metabolic disorders are forced by the NEB through an inadequate feed intake around parturition, which increases only slowly, in contrast to a milk production which starts mostly at a rather high level and increases further despite energy deficiency (Grummer 1993; Ingvartsen and Andersen 2000; Kessel et al. 2008). Despite the presence of a NEB, milk production is maintained and further increases until peak lactation (Gross et al. 2011a). High milk production in the first days of lactation concomitantly implies elevated energy and nutrient requirements. However, the decline in periparturient dry matter intake (DMI) and the already low feed intake level does not cover the needs (Grummer et al. 2004; Douglas et al. 2006; Gross et al. 2011a). To compensate for the temporary NEB, cows mobilize body reserves (predominantly adipose tissue) until the peak of lactation is reached, in order to replenish and accumulate corresponding reserves again later. This mobilization pattern of body reserves is reflected by the simultaneous decrease and increase in body weight and body condition (BCS). Thus, the prevailing NEB triggering lipolysis exposes high-producing cows to metabolic stress already at the time of calving (Bell 1995; LeBlanc 2010). Consequently, predisposition to metabolic and infectious diseases is initiated at the very beginning of lactation. Additional factors such as inadequate feeding and over-conditioning during the dry period may further aggravate metabolic adaptation and favor hypocalcemia, endometritis, etc. (Douglas et al. 2006; Roche et al. 2009). Nowadays, the aim to feed animals according to their energy and nutrient requirements (e.g., Mertens and Dado 1993; Sniffen et al. 1993) throughout all stages of the gestation-lactation cycle is not always possible (Gross 2022). Owing to the limited feed intake capacity and restrictions from the cow's side (e.g., limited capacity for glucose and lipid absorption, maintenance of rumen function), the actual requirements cannot be met by feeding. Despite advances in animal nutrition and veterinary care (Drackley et al. 2005; Eastridge 2006), dairy cows are at a high risk of developing production diseases with highest incidence rates coinciding with the greatest metabolic load during the energy deficiency following parturition (Ingvartsen et al. 2003; Mulligan and Doherty 2008).

## 2.2 Metabolic Situation in Dairy Cows During Early Lactation

The beginning of lactation requires considerable metabolic adaptations to meet increased energy demands for milk production of dairy cows (Bell 1995; Rukkwamsuk et al. 1999; Vernon 2005; Wathes et al. 2012; Gross et al. 2013). Dairy cows experience an abrupt change from anabolic to catabolic metabolism when passing through a period of NEB from late gestation to early lactation (Herdt et al. 1981; Bell 1995; Rukkwamsuk et al. 1999; Gross et al. 2011a; McArt et al. 2013). The necessary metabolic adaptations to lactation are reflected by changes in a number of blood parameters. For example, the concentration of glucose decreases, while the levels of BHB and NEFA increase, concomitantly with related changes in the endocrine system. During the transition period, cows are highly susceptible to developing metabolic and related diseases such as ketosis, liver lipidosis, retained placenta, mastitis, displaced abomasum, laminitis, or reproduction problems (Goff and Horst 1997; Ingvartsen et al. 2003; Ingvartsen and Moyes 2015). An affected cow may enter a vicious circle as postpartum disorders have been observed to be interrelated and one metabolic disorder may predispose the cow to other disorders. Correa et al. (1993) observed that ketosis is a risk factor for displaced abomasum and metritis. Erb and Grohn (1988) observed that milk fever plays a central role in predisposing the cow to disorders such as ketosis, retained placenta, displaced abomasum, and metritis.

Glucose is a key nutrient and limiting source for lactation when we consider the increased specific demands for lactose and milk fat synthesis. More than 80% of the glucose turnover is required by the mammary gland for the synthesis of lactose during peak lactation (Bauman and Currie 1980). Therefore, hepatic gluconeogenesis and glycogenolysis are increased to a maximum at the start of lactation. Nevertheless, the capacity of gluconeogenesis and dietary intake of glucogenic precursors remain low during early lactation (Gross 2022). Considering the essentiality of glucose for other organs and tissues (e.g., immune system, reproductive tissues, etc.), it is not surprising that the health disorders are associated with the period of low circulating glucose concentrations in early lactation (Wankhade et al. 2017; Gross 2022). Lipids therefore become the main energy source. To cover the increased needs of fatty acids as an energy source and for milk fat synthesis, body lipid stores are mobilized (lipomobilization) and circulate in blood as NEFA, while lipogenesis is simultaneously reduced. Mobilized NEFA arrive at the liver, where they are normally burned in the presence of oxaloacetate after being activated by binding to acetyl-coenzyme A (acetyl-CoA), thereby releasing energy (Fig. 1).

Fatty acids that are not completely oxidized are resynthesized to triglycerides, bound to apolipoprotein, and secreted into the bloodstream as very-low-density lipoprotein (VLDL). However,

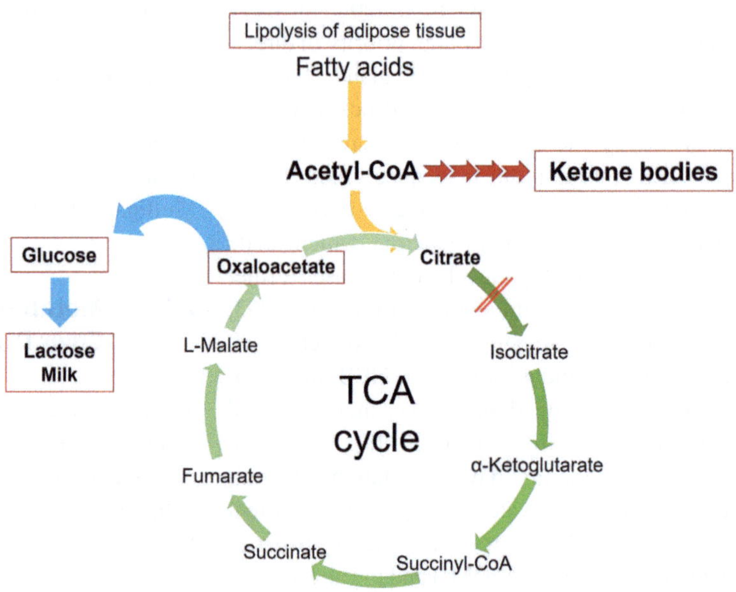


**Fig. 1** Metabolism of fatty acids in the tricarboxylic acid (TCA) cycle and relationships with ketogenesis and gluconeogenesis

the ability of the liver to export triglycerides is limited. Consequently, fatty liver can develop (Bobe et al. 2004; Gross et al. 2013; Kessler et al. 2014a; Gross et al. 2015).

## 2.3 Endocrine Control of Body Tissue Turnover and Metabolic Adaptation to Lactation

Although the metabolic load is known to be highest after parturition, the metabolic adaptations with respect to lactation already start prior to calving. Despite no obvious additional need for energy shortly before parturition, insulin and insulin-like growth factor-1 (IGF-1) concentrations already start to decline, and the elevated release of growth hormone (GH) and mobilization of adipose tissue commence (Rhoads et al. 2004). The drop of circulating thyroid hormones, leptin, adiponectin, insulin, and IGF-1 shortly before parturition enables the initiation of lipolysis while concomitantly inhibiting lipogenesis (Bell 1995; Block et al. 2001; Kuhla et al. 2016). A key mechanism of metabolic adaptation is the decrease of plasma insulin concentrations (combined with insulin resistance of peripheral tissues) and simultaneously rather high plasma concentrations of glucagon. Along with enhanced sensitivity to catecholamines, this is one of the main triggers to activate hormone-sensitive lipases in adipose tissue and hence fat mobilization (Locher et al. 2011). Low circulating insulin is involved in the uncoupling of the somatotropic axis in the liver via downregulation of the hepatic GH receptor (Kobayashi et al. 1999; Rhoads et al. 2004). Consequently, plasma GH levels increase, while plasma IGF-1 concentrations remain low (Gross et al. 2011b; Kessler et al. 2013). A direct effect of GH is the stimulation of lipolysis from adipose tissue supported by enhanced sensitivity to beta-adrenergic effects of catecholamines (Bruckmaier and Gross 2017). While plasma insulin concentrations are low, the high levels of GH and NEFA induce an additional insulin resistance in the peripheral tissues.

Glucose is predominantly derived from hepatic gluconeogenesis in ruminants. Hepatic gluconeogenesis depends on overall tricarboxylic acid (TCA) cycle activity, especially the oxaloacetate availability, and particularly on the availability of substrates, such as propionate, lactate, amino acids, and glycerol (Fig. 1). The upregulation of gluconeogenesis postpartum relies on oxaloacetate reserves and therefore reduces the availability of oxaloacetate in hepatocyte mitochondria. Consequently, the oxidation of fatty acids via the TCA cycle is limited, and an alternative metabolic pathway (besides storage of triglycerides in the liver) is activated through enhanced synthesis of ketone bodies from acetyl-CoA in the liver (Fig. 1).

## 2.4 Excessive Lipomobilization Triggers the Development of Ketosis

Elevated plasma BHB concentrations are attributed to the high mobilization of body fat store-derived fatty acids, which exceed the oxidative capacity of the liver (Bell 1995; Drackley et al. 2005). Cows face a dilemma by the competition for oxaloacetate either depleted for gluconeogenesis or required to maintain the TCA cycle for oxidation of fatty acids (Fig. 1; Aschenbach et al. 2010; White 2015). Cows with a greater milk yield are likely to have a prolonged and more severe NEB. Consequently, high-yielding cows are forced to mobilize more adipose tissue. However, if the TCA cycle gets overloaded (e.g., by capacious drainage of oxaloacetate for gluconeogenesis), the acetyl-CoA is shunted off to produce ketone bodies (acetoacetic acid, acetone, and BHB) in order to prevent cessation of the TCA cycle and accumulation of acetyl-CoA (Fig. 1; Sato et al. 1999; Sugden et al. 2001). Consequently, fatty acids that are not oxidized are converted to ketone bodies or re-esterified to triglycerides within the liver (Grummer 1993; Bobe et al. 2004). We could show a mild to severe fatty liver already within a very narrow time after calving, although NEFA concentrations and metabolic load reached highest values a few weeks later (Gross et al. 2013). The concomitant low cholesterol concentrations in early lactation are likely limiting the hepatic triglyceride export despite an upregulation of hepatic cholesterol biosynthesis (Kessler et al. 2014a). Ha et al. (2017) showed with a liver transcriptome analysis that pathways associated with cholesterol are essentially involved in adaptation processes of metabolism during the transition period. Forcing a NEB with adipose tissue mobilization at a later lactational stage, however, did not result in a considerable lipid accumulation in the liver as cholesterol concentrations were even elevated supporting hepatic triglyceride export as VLDL (Gross et al. 2013, 2015). Supplementing dairy cow diets with conjugated linoleic acids (CLA) resulted in higher plasma cholesterol and triglyceride concentrations in early lactation indicating similar beneficial effects on hepatic lipid export and improved efficiency of circulating nutrients (Galamb et al. 2017; Grossen-Rösti et al. 2018).

Ketone bodies are known to depress feed intake (Laeger et al. 2010), reduce fertility (Ospina et al. 2010; Raboisson et al. 2014), and impair immune function (Suriyasathaporn et al. 1999). Thus, excessive release of fatty acids from the adipose tissue and consequent formation of ketone bodies are detrimental, when immune competence is low, and cows in early lactation are highly susceptible to metabolic and infectious diseases anyway (Sordillo et al. 2009; Trevisi et al. 2012; Bradford et al. 2015). Despite their adverse effects, ketone bodies may be considered an energy source for certain tissues and cells (e.g., skeletal muscle, heart, and brain; Ruderman and Goodman 1973; Laffel 1999). Recently, Zarrin et al. (2014a) and Gross et al. (2018) showed that BHB is preferentially used before glucose by the immune system during inflammation. This might further support the ability of dairy cows to spare glucose. On the other hand, elevated BHB concentrations were shown to depress gluconeogenesis and thus plasma glucose concentrations independent of lactational stage (Zarrin et al. 2014b, 2017). The apparently "abundant" energy in circulation of early lactating dairy cows as signalized by elevated ketone bodies, however, paradoxically results in a reduction of feed intake (Laeger et al. 2010; Rojas-Morales et al. 2016). Plasma BHB concentrations above 1.2 to 1.4 mmol/L might already indicate subclinical ketosis (LeBlanc 2010; Suthar et al. 2013). Moreover, the risk of accompanying metabolic and infectious health disorders increases disproportionally when ketone bodies are elevated (Ospina et al. 2010; McArt et al. 2012b).

Ketogenesis is the branch of mammalian metabolism concerned with the synthesis of ketone bodies. Although ketone bodies are always present at a low level in healthy individuals, dietary manipulation and certain pathological conditions can increase the levels of these compounds in vivo (McPherson and McEneny 2012). Rather than metabolization via the TCA cycle, acetyl-CoA might also be used instead in biosynthesis of ketone bodies via acetoacetyl-CoA and 3-hydroxy-3-methylglutaryl-CoA (HMG-CoA). Ketone bodies are important vectors of energy transport from the liver to extrahepatic tissues, especially during fasting, when glucose supply is low. Blood total ketone body levels should be evaluated in parallel with blood glucose and free fatty acids (FFA; Fukao et al. 2014). It has been shown that ketone bodies inhibit protein degradation and therefore gluconeogenesis and also are able to spare glucose by inhibiting glucose utilization (Zarrin et al. 2013, 2014a, b, 2017). They also can inhibit lipolysis and function as a regulatory safety system, replacing insulin, in situations when the activity of this hormone is low, as in type I ketosis. Ketone bodies thus have important functions as substrates replacing glucose in many tissues including the immune system and also as signal substances in the regulation of energy metabolism (Holtenius and Holtenius 1996; Zarrin et al. 2014a, b). Under fasting conditions, hepatic FA-binding protein (L-FABP) contributes to hepatic long-chain FA oxidation and ketogenesis by a nontranscriptional mechanism, whereas L-FABP can activate ketogenic gene expression in fed mice. Thus, the mechanisms whereby L-FABP affects FA oxidation may vary with physiological status (Erol et al. 2004). Hepatic mRNA and protein levels of long-chain acyl-CoA synthetase were significantly higher in ketotic cows than in nonketotic cows. In contrast, mRNA levels of carnitine palmitoyltransferase (CPT) 1 and mRNA and protein levels of CPT2, long-chain acyl-CoA dehydrogenase, HMG-CoA synthase, and acetyl-CoA carboxylase were decreased in the liver of ketotic cows. Serum NEFA concentrations positively correlated with long-chain acyl-CoA synthetase protein levels and negatively correlated with protein levels of CPT2, long-chain acyl-CoA dehydrogenase, HMG-CoA synthase, and acetyl-CoA carboxylase. In addition, serum BHB concentration negatively correlated with protein levels of CPT2, long-chain acyl-CoA dehydrogenase, and HMG-CoA synthase. Overall, beta-oxidation capability of FA was reduced in the liver of ketotic compared with nonketotic cows (Murondoti et al. 2004; Carlson et al. 2007; Li et al. 2012).

## 3 Classification and Types of Ketosis

Different classification approaches are used in research and practice depending on the presence of symptoms, level of ketone bodies, time of diagnosis, or relationship with concomitant other production diseases (Duffield 2000; Zhang and Ametaj 2020).

### 3.1 Primary and Secondary Ketosis

Primary ketosis manifests in the situation of exclusive hyperketonemia (with or without clinical signs of ketosis), where other concomitant health disorders can be excluded (Baird 1982). When looking at the etiology of ketosis, primary ketosis is therefore most likely due to either excessive lipomobilization or hyperketonemia induced by ketogenic diets (e.g., butyrate in poorly fermented silage).

In contrast, secondary ketosis is the consequence of various factors resulting in lipomobilization and hyperketonemia, although physiological stage of the animal would allow for lower circulating ketone body levels. Typically, health disorders not directly related to energy metabolism (e.g., lameness, hypocalcemia, mastitis) are triggering energy-deficient situations by the reduction of appetite, feed intake, and/or forestomach and intestinal digestion. Clinical symptoms of the primary manifest disease may therefore mask the symptoms related to ketosis.

### 3.2 Types I–III of Ketosis

Another scheme for the classification of ketosis accounts for the etiology and time when hyperketonemia is detected (Holtenius and Holtenius 1996; Zhang and Ametaj 2020). **Type I** ketosis (more or less identical with primary ketosis as described above) occurs during the first weeks after parturition, where insufficient energy intake at concomitantly high milk production forces the mobilization of fat reserves. Furthermore, inadequate feeding techniques or limited feed quality may limit the energy supply. Health disorders promoting inappetence (see secondary ketosis above) result in hyperketonemia, too.

**Type II** ketosis most likely occurs immediately after calving and is closely associated with the fat cow syndrome and fatty liver disease, respectively. The reason here is over-conditioning during the dry period, which supports adipose tissue mobilization and development of steatosis already prior to parturition. Upon the diagnosis of type II ketosis, concomitantly elevated glucose and insulin concentrations may be observed (Holtenius and Holtenius 1996).

Ketogenic rations providing, e.g., butyric acid as precursor of ketone bodies are causative for **type III** ketosis. Butyric acid can be found in high concentrations in poor-quality silage (Tsunoda et al. 2017).

### 3.3 Subclinical and Clinical Ketosis

One of the most favored categorization schemes for ketosis refers to the presence, and absence, respectively, of clinical signs of ketosis.

**Subclinical ketosis (SCK)** is defined as a state with elevated ketone body concentrations (above 1.2 to 1.4 mmol/L) but without showing any clinical signs of a manifest ketosis (Duffield et al. 2009; Ospina et al. 2010; McArt et al. 2012b). However, cows with SCK may already show reduced feed intake, milk yield, or reproductive performance (Benedet et al. 2019). Depending on feeding and other accompanying circumstances, this premorbid condition can rapidly change into a clinically pronounced ketosis (**clinical ketosis, CK**). Therefore, increasing attention is paid to an early detection of (subclinical) ketosis.

## 3.4 Nervous Form of Ketosis

A prolonged exposure toward elevated ketone body concentrations can have adverse effects on neurological functions. Cows with clinical ketosis may show additional nervous signs, which often appear suddenly. The cow may walk uncoordinatedly and even seem to be blind. In milder cases, cows appear merely unenthusiastic to apathetic or somnolent. In more severe cases, cows can be either comatose (permanent lying with the head turned in toward the chest) or periodically recurring conspicuously excited. Further signs of nervous ketosis imply excessive salivation, empty chewing, wild licking or gnawing of the own skin or of objects of the environment, until wildly aggressive to raving behavior. In terms of movements, an uncoordinated gait, walking in circles or against the wall with a forward pushing, and/or stumbling gait can be indicative of nervous ketosis. On occasion, the signs may progress to resemble milk fever with paresis and collapse.

The following three video clips (Figs. 2, 3, and 4) show examples of nervous ketosis in dairy cows:

**Fig. 2** Cow showing signs of nervous ketosis: Excessive salivation and licking of surrounding objects. © med. vet. Eva Ehrnböck, Dr. med. vet. Florian Straßer, Veterinary Practice for Cattle, Hunding, Germany (▶ https://doi.org/10.1007/000-brw)

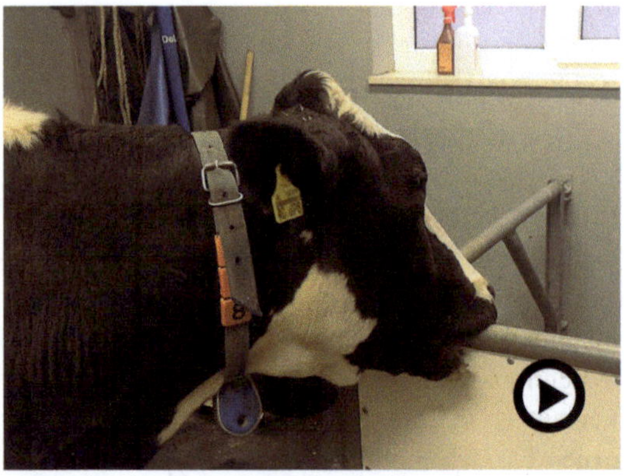

**Fig. 3** Cow with nervous ketosis licking surrounding objects. A ketone body concentration (BHB) of 3.8 mmol/L was measured in blood. © med. vet. Uwe Müller, Prof. Dr. med. vet., Alexander Starke, Dip. ECBHM, Veterinary Specialist for Cattle, Clinic for Ruminants and Swine, Faculty of Veterinary Medicine, University of Leipzig, Leipzig, Germany (▶ https://doi.org/10.1007/000-brv)

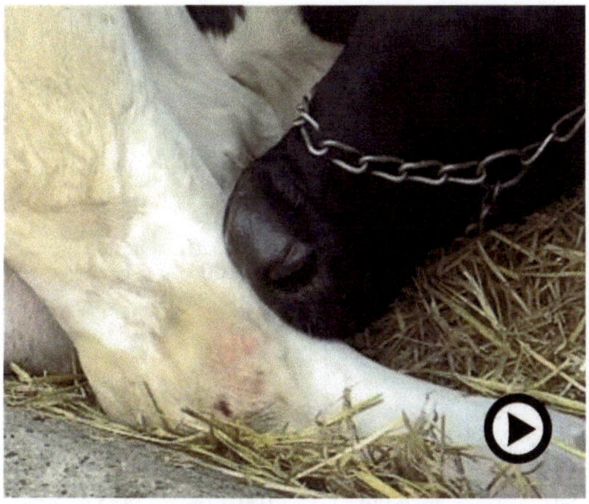

**Fig. 4** Cow with nervous ketosis: Gnawing of the own skin. © Dr. med. vet. Henning Christian Schenk, PhD Neuroscience, Dipl. ECVN, Veterinary Clinic Lüneburg, Germany; Department of Small Animal Medicine and Surgery, & Clinic for Cattle, University of Veterinary Medicine Hannover Foundation, Hannover, Germany (▶ https://doi.org/10.1007/000-brx)

## 4 Detection of Ketone Bodies and Diagnosis of Ketosis

The decline in feed intake and milk yield can be first unspecific signs of ketosis. Nowadays, sensor-based technology can early detect deviations from rumination and chewing behavior in ketotic compared to healthy dairy cows.

Ketone bodies are present in various body fluids and can be detected in blood, urine, and milk. Furthermore, acetone from breath and skin exhalations can be noticed by some persons. Although measurements in blood are considered as gold standard, a frequent sampling is not practicable at an individual animal basis. Therefore, advantages of a noninvasive and continuous measurement of ketone bodies, e.g., in milk gained attention during the last years. Furthermore, automatization of sampling can be easily implemented in dairy farms with automatic milking systems.

Independent of the media collection and analysis, it is of general importance to early and specifically detect ketone bodies prior to the manifestation of clinical signs of ketosis. This chapter does not address laboratory methods that are used in clinical laboratories. For immediate testing on-site, a variety of cow-side tests (primarily test strips) are available where ketone body concentrations can be measured in blood, milk, and/or urine. Both qualitative and quantitative tests are available. However, sensitivity and specificity for the detection of different ketone bodies depend on the media for testing (Table 1). It should be noted, however, that a high ketone body content of urine or milk does not prove the presence of ketosis. In terms of selecting lasting preventive and acute treatments, concomitant concentrations of glucose should be measured besides ketone bodies. Thus, conclusions on the present rate of gluconeogenesis and the effectiveness of additional treatments (e.g., glucogenic precursors, butaphosphan and cyanocobalamin) can be made (Gordon et al. 2017a, b).

As shown in Table 1, the preferred cow-side test to identify ketotic cows is the measurement of BHB in blood due to its high test accuracy. Different handheld devices to test BHB in blood are available, e.g., Precision Xtra, BHBCheck meter, FreeStyle Precision, etc., and were evaluated in numerous studies (Iwersen et al. 2009, 2013; Voyvoda and Erdogan 2010; Sailer et al. 2018). Milk-based tests to assess BHB (e.g., KetoTest, Ketolac, etc.) and urine-based tests to measure acetoacetate (e.g., Ketostix, etc.) are also used for the diagnosis of hyperketonemia in dairy cows. However, sensitivity and specificity of these tests are much lower compared to blood tests. Milk ketone tests are easy to use, but milk BHB concentrations are severalfold lower in milk than in blood, which makes a detection at low levels more difficult. Furthermore, BHB in milk can be affected by (sub)clinical mastitis inducing an opening of the blood-milk barrier with an increased transfer of BHB (Geishauser et al. 2000). Measurements of ketone bodies in urine, however, are mostly limited to the difficulty of urine sampling.

Milk mid-infrared spectra-based values for ketone bodies provide **indirect** estimates for ketone body concentrations. Instead, algorithms for predicting ketones refer to concomitant alterations in milk composition and milk metabolites during energy-deficient situations (e.g., milk fatty acids). Therefore, milk spectra do not provide figures on concentrations as ketones are indirectly estimated. Ketone body concentrations

**Table 1** Examples for rapid cow-side tests for the diagnosis of hyperketonemia

| Medium | Measured ketone body | Available rapid tests | BHB reference threshold in blood BHB (mmol/L) | Threshold test (mmol/L) | Sensitivity | Specificity | References |
|---|---|---|---|---|---|---|---|
| Blood | BHB | Precision Xtra, | 1.2 | 1.2 | 85–100 | 84–94 | 1, 2, 3 |
| | | BHBCheck meter, | 1.4 | 1.4 | 90–100 | 98–100 | 1, 2, 4 |
| | | FreeStyle Precision, etc. | | | | | |
| Urine | Acetoacetate | Ketostix | 1.4 | 0.5–1.5 | 67–90 | 75–97 | 4, 5, 6 |
| | | | 1.4 | 4 | 49–70 | 97–100 | 4, 5, 6 |
| | | | 1.4 | >4 | 4–44 | 100 | 4, 5 |
| Milk | BHB | KetoTest, Ketolac, etc. | 1.4 | 0.05 | 88–91 | 56–90 | 5, 6, 7 |
| | | | 1.4 | 0.1 | 73–90 | 76–96 | 4, 5, 6, 7 |
| | | | 1.4 | 0.2 | 27–59 | 90–99 | 4, 5, 6, 7 |

1 Voyvoda and Erdogan (2010)
2 Iwersen et al. (2013)
3 Sailer et al. (2018)
4 Iwersen et al. (2009)
5 Carrier et al. (2004)
6 Oetzel (2004)
7 Geishauser et al. (2000)

in milk are lower compared to respective concentrations in blood. Correlation of milk ketone concentration with ketone body concentration in blood is greater in cows with ketosis (de Roos et al. 2007; van Knegsel et al. 2010). The suitability to early detect ketosis using milk spectra is limited when metabolic load in cows is low (Caldeira et al. 2020). Furthermore, a frequent sampling during early lactation is crucial for following animals at risk. Measuring ketone bodies only at milk recording can therefore not be recommended in small herds or at an individual cow level. In larger dairy herds, however, measurements in a subpopulation of representative individuals may allow conclusions on the general metabolic situation of the entire herd. Consequently, adjustments in feeding, herd management, and preventive measures can be implemented at an earlier stage. However, milk spectra-based estimates for ketone bodies in combination with established markers for energy-deficient situations (e.g., milk fat-to-protein ratio) improve the diagnostic value of milk (Denis-Robichaud et al. 2014; Tremblay et al. 2019; Gross and Bruckmaier 2019b; Mandujano Reyes et al. 2021). More details on the suitability of milk as diagnostic medium for health disorders can be found in the chapter 24.

In contrast, in-line systems (e.g., HerdNavigator™) that measure ketone bodies **directly** (i.e., by enzymatic or colorimetric assays) and frequently during early lactation can be a valuable and supportive tool to early detect metabolic imbalances in dairy cows (Nielsen et al. 2005; Caldeira et al. 2020).

## 5 Treatment of Ketosis

The following chapter shall address treatment approaches for ketotic cows in short only. The general efforts in modern dairy cow management should be focused on preventive measures to avoid ketosis. Treatments described below are summarized from existing scientific literature and have no claim to completeness. Recommendations are not substitute for the medical treatment of diseases and cannot promise any healing success. Legal authorization of products and procedures may vary between countries and require official verification prior to their application. Besides direct interventions to alleviate symptoms of ketosis, other animal-related factors, potential health disorders, and environmental circumstances concomitantly contributing to the maintenance of the ketosis situation need to be identified. In terms of diseases such as endometritis, mastitis, abomasal displacement, or lameness, specific veterinary care is required. Figure 5 summarizes major adjusting screws where potential limitations particularly related to the TCA cycle can be positively affected.

General attention (independent from apparent signs of ketosis) should be paid to the following issues:

- Quality and access to **feed**. Roughage and concentrates should be of high quality and palatability. Animals should have free access to feed throughout the day. However, digestive physiology of ruminants must be considered (e.g., sufficient fiber and structure content; Zebeli et al. 2012). Although the negative energy balance is most severe during the first weeks of lactation, overall dry matter intake is still low and greater amounts of cereal-based concentrates supplied in only few portions may bear the risk of rumen acidosis (Humer et al. 2018; Gross 2022).
- Promotion of **gluconeogenesis** and **reduction of hepatic triglyceride accumulation**. By stimulating the appetite (by offering palatable food such as fodder beets, high-quality hay, etc.), an elevated passage rate can be realized along with a high digestibility. Consequently, dry matter and thus energy intake increase, concomitantly reducing the need of additional adipose tissue breakdown. Ketogenic feeds such as high-butyric acid silage or high-fat concentrates should be used restrictively.
- **Stimulation of forestomach digestion** (including supplementation of yeast products, or transfer of rumen fluid from clinically

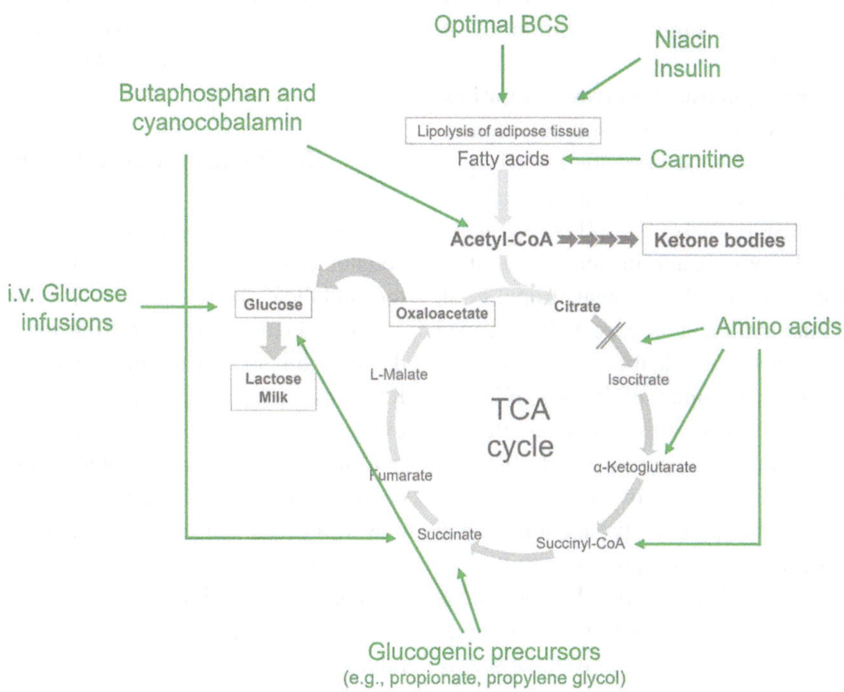


**Fig. 5** Interactions of preventive and therapeutic measures with processes of lipolysis, TCA cycle activity, ketogenesis, and gluconeogenesis

healthy donor animals). Sufficient supply of cobalt and other trace elements ensures microbial growth in the rumen. Prerequisites are furthermore simultaneously available energy and nitrogen sources from feed.

Once hyperketonemia is detected and clinical signs are present, the **reduction of excessive lipomobilization** is of priority to reduce hepatic load and further ketogenesis. Concomitant glucose concentrations are most likely at a lower level. Therefore, treatment of acute ketosis implies commonly a combination of measures listed below:

- **Glucogenic precursors** (oral treatments with sodium propionate or propylene glycol [e.g., 300 mL of propylene glycol once daily for 3 d]) can be efficacious as both are rapidly absorbed and converted to glucose in the liver (McArt et al. 2011, 2012a, b). An extended supply of propylene glycol seems to particularly benefit hyperketonemic cows with ketone body concentrations above 2 mmol/L or low circulating glucose concentrations (<2.2 mmol/L; Gordon et al. 2017a).
- **Intravenous glucose (dextrose) solutions** (500 mL of 40–50% glucose solutions administered as an i.v. bolus) provide directly glucose and may temporarily alleviate clinical signs of ketosis (Capel et al. 2021). Commonly, glucose solutions are administered in addition to a propylene glycol treatment (Capel et al. 2021). This treatment is especially intended for the use in severely ketotic cows and cows showing nervous signs of ketosis (Gordon et al. 2013). However, this beneficial effect of glucose is merely transient (Oetzel 2020). In some cases, the "surplus" of administered glucose directly boosts milk production instead of compensating for the glucose shortage of other tissues. Considering the enormous amounts of glucose required for milk production in early lactation, the intravenous supply of glucose cannot sustainably cure ketosis. Furthermore, recent studies showed potential negative effects of the sudden dextrose administration on phosphorus homeostasis (Grünberg et al. 2006; Mann et al. 2017).
- **Butaphosphan and cyanocobalamin** supports different processes related to gluconeogenesis and the TCA cycle (Nuber et al. 2016; Gordon et al. 2013, 2017b).
- Administration of **antilipolytic agents** like niacin (nicotinic acid, resp.; Niehoff et al. 2009). Insulin should be used only with concomitant administration of glucocorticoids as well as intravenous dextrose continuous drip to prevent a decline in blood glucose levels (Gordon et al. 2017b).
- **Glucocorticoids** (e.g., **dexamethasone**) and **ACTH** were commonly used to alter glucose metabolism by inhibiting the effects of insulin. Both treatments act similarly by stimulating the tricarboxylic acid cycle and gluconeogenesis from glucogenic precursors. However, the effectiveness of glucocorticoid therapy for ketosis is discussed controversially due to its side effects.
- **Ionophores** (e.g., monensin, lasalocid; Henno et al. 2021; Mammi et al. 2021).
- Reducing **milking frequency** lowers milk yield and therefore metabolic load in dairy cows (Williamson et al. 2022).

## 6 Risk Factors and Prophylaxis

For the derivation of prophylactic measures to counteract ketosis, it is crucial to know risk factors and their role in etiology of ketosis.

Undoubtedly, **over-conditioning** during the dry period is one of the most potent risk factors for ketosis (Gerloff 2000). Cows that have been overfed during dry period release their endogenous fat stores excessively rapidly during the first weeks of lactation. In addition, feeding above the cow's nutrient and energy requirements during the dry period leads to over-conditioning at calving and to depression of dry matter intake after calving. As a consequence, over-conditioned and high-producing dairy cows at calving inevitably run into a more severe negative energy balance than cows that have a normal appetite and BCS. In these animals, increased lipomobilization provides more fatty acids than can be oxidized, which further triggers ketogenesis and hepatic triglyceride accumulation.

To counteract over-conditioning, particularly in low-producing animals in advanced lactation, feeding should be moderately restrained until normal BCS is achieved. After dry-off, BCS should be maintained as losses can be detrimental for fetal and maternal health. If cows are already over-conditioned at the time of scheduled dry-off, one strategy can be continuing milking (i.e., **omitting the dry period**). Although milk yield in the next lactation period after calving is lower, the metabolic load and therefore the risk to develop ketosis are significantly lower (Chen et al. 2015, 2016; Kok et al. 2019). In a similar manner, lactation periods can be extended prior to dry-off to mitigate potential failures in BCS adjustments (Sehested et al. 2019; van Knegsel et al. 2022).

Besides maintenance of a normal body condition (Fig. 5), preventive measures for other peripartal health disorders (e.g., hypocalcemia, displacement of the abomasum, lameness, etc.) that are accompanied by reduced feed intake while energy requirements for milk production are should be promoted. As shown in the introduction, cows suffering from "primary" diseases have a severalfold elevated risk to develop ketosis (Suthar et al. 2013; Raboisson et al. 2014).

Due to the progress in genetic selection and improvements in housing and feeding conditions, milk yield increased markedly during the last decades. Greater milk production requires more energy and nutrients. Despite increased milk yields, feed intake could not be raised during the very early lactation period. Consequently, energy balance is more negative and needs to be compensated by the mobilization of endogenous body fat stores. Therefore, the risk of ketosis but also of other production diseases increases with greater performance level. However, it needs to be clearly stated that a considerable variation in metabolic adaptation exists among the population and among individuals within a herd (Kessel et al. 2008; Gross and Bruckmaier 2015). While some dairy cows fail to successfully adapt even at moderate production levels, other cows can cope very well with the imposed lactational challenges at high milk yields. Resilient cows are able to use mobilized fatty acids as well as ketone bodies as energy source. Therefore, they may spare glucose for essential metabolic processes (e.g., synthesis of lactose). Despite elevated ketone body concentrations, cows do not necessarily develop ketosis. However, there is no question that ketone bodies have a negative impact on blood glucose concentration (Zarrin et al. 2017) and immune competence (Suriyasathaporn et al. 1999). The observed genetic associations with ketone body concentrations and metabolic adaptation (Ha et al. 2015, 2017) supposed that a **genetic predisposition** toward ketosis may exist in dairy cows. **Breeding** goals nowadays try to implement further markers of metabolic health. However, phenotypic expression largely depends on the final feeding and management conditions in practice.

More recent research showed that cows starting more moderately into lactation do not necessarily have an overall lower lactation performance but benefit from an improved metabolic status (Kessler et al. 2014b, 2018). Thus, the likelihood to develop metabolic disorders might be reduced by selecting cows with a lower milk yield during the first days of lactation.

## 7 Conclusions and Outlook

Ketosis remains a major issue in dairy cows. Besides acute cases, particularly subclinical ketosis requires attention to reduce the negative impact of hyperketonemia on animal health and performance. Monitoring of ketone body concentrations in either blood, milk, or urine is prerequisite for an early identification of cows at risk. However, sensitivity and specificity of the testing devices should be as highest as possible for reliable measurements. Sensor-based technology and smart farming devices have the potential to record deviations from normal animal behavior and milk composition, thereby indicating an upcoming ketosis. Existing therapeutic approaches can only transiently alleviate clinical manifestations of ketosis. In terms of diagnosis, both glucose and BHB concentrations should be concomitantly evaluated for a promising therapeutic treatments. However, preventive strategies such as optimizing body condition score prior to dry-off and ensuring of a maximum feed intake have the greatest potential to reduce the incidence of ketosis and concomitant production diseases in early lactating dairy cows. In the near future, genetic selection targets for dairy cows may contribute to the breeding of more robust and resilient animals.

## References


Andersson L (1988) Subclinical ketosis in dairy cows. Vet Clin North Am Food Anim Pract 4:233–251. https://doi.org/10.1016/s0749-0720(15)31046-x

Aschenbach JR, Kristensen NB, Donkin SS, Hammon HM, Penner GB (2010) Gluconeogenesis in dairy cows: the secret of making sweet milk from sour dough. IUBMB Life 62:869–877. https://doi.org/10.1002/iub.400

Baird GD (1982) Primary ketosis in the high-producing dairy cow: clinical and subclinical disorders, treatment, prevention, and outlook. J Dairy Sci 65:1–10. https://doi.org/10.3168/jds.s0022-0302(82)82146-2

Bauman DE, Currie WB (1980) Partitioning of nutrients during pregnancy and lactation: a review of mechanisms involving homeostasis and homeorhesis. J Dairy Sci 63:1514–1529. https://doi.org/10.3168/jds.s0022-0302(80)83111-0

Bell AW (1995) Regulation of organic nutrient metabolism during transition from late pregnancy to early lactation. J Anim Sci 73:2804–2819. https://doi.org/10.2527/1995.7392804x

Benedet A, Manuelian CL, Zidi A, Penasa M, De Marchi M (2019) Invited review: β-hydroxybutyrate concentration in blood and milk and its associations with cow performance. Animal 13:1676–1689. https://doi.org/10.1017/S175173111900034X

Block SS, Butler WR, Ehrhardt RA, Bell AW, Van Amburgh ME, Boisclair YR (2001) Decreased concentration of plasma leptin in periparturient dairy cows is caused by negative energy balance. J Endocrinol 171:339–348. https://doi.org/10.1677/joe.0.1710339

Bobe G, Young JW, Beitz DC (2004) Invited review: pathology, etiology, prevention, and treatment of fatty liver in dairy cows. J Dairy Sci 87:3105–3124. https://doi.org/10.3168/jds.S0022-0302(04)73446-3

Bradford BJ, Yuan K, Farney JK, Mamedova LK, Carpenter AJ (2015) Invited review: inflammation during the transition to lactation: new adventures with an old flame. J Dairy Sci 98:6631–6650. https://doi.org/10.3168/jds.2015-9683

Bruckmaier RM, Gross JJ (2017) Lactational challenges in transition dairy cows. Anim Prod Sci 57:1471–1481. https://doi.org/10.1071/AN16657

Brunner N, Groeger S, Canelas Raposo J, Bruckmaier RM, Gross JJ (2019) Prevalence of subclinical ketosis and production diseases in dairy cows in Central and South America, Africa, Asia, Australia, New Zealand, and Eastern Europe. Transl Anim Sci 3:84–92. https://doi.org/10.1093/tas/txy102

Cainzos JM, Andreu-Vazquez C, Guadagnini M, Rijpert-Duvivier A, Duffield T (2022) A systematic review of the cost of ketosis in dairy cattle. J Dairy Sci 105:6175–6195. https://doi.org/10.3168/jds.2021-21539

Caldeira MO, Dan D, Neuheuser AL, Stürmlin R, Weber C, Glauser DL, Stierli M, Schuler U, Moll J, Wegmann S, Bruckmaier RM, Gross JJ (2020) Opportunities and limitations of milk mid-infrared spectra-based estimation of acetone and β-hydroxybutyrate for the prediction of metabolic stress and ketosis in dairy cows. J Dairy Res 87:196–203. https://doi.org/10.1017/S0022029920000230

Capel MB, Bach KD, Mann S, McArt JAA (2021) A randomized controlled trial to evaluate propylene glycol alone or in combination with dextrose as a treatment for hyperketonemia in dairy cows. J Dairy Sci 104:2185–2194. https://doi.org/10.3168/jds.2020-19111

Capper JL, Cady RA, Bauman DE (2009) The environmental impact of dairy production: 1944 compared with 2007. J Anim Sci 87:2160–2167. https://doi.org/10.2527/jas.2009-1781

Carlson DB, McFadden JW, D'Angelo A, Woodworth JC, Drackley JK (2007) Dietary L-carnitine affects periparturient nutrient metabolism and lactation in multiparous cows. J Dairy Sci 90:3422–3441. https://doi.org/10.3168/jds.2006-811

Carrier J, Stewart S, Godden S, Fetrow J, Rapnicki P (2004) Evaluation and use of three cowside tests for detection of subclinical ketosis in early postpartum cows. J Dairy Sci 87:3725–3735. https://doi.org/10.3168/jds.S0022-s0302(04)73511-0

Chen J, Gross JJ, van Dorland HA, Remmelink GJ, Bruckmaier RM, Kemp B, van Knegsel AT (2015) Effects of dry period length and dietary energy source on metabolic status and hepatic gene expression of dairy cows in early lactation. J Dairy Sci 98:1033–1045. https://doi.org/10.3168/jds.2014-8612

Chen J, Remmelink GJ, Gross JJ, Bruckmaier RM, Kemp B, van Knegsel ATM (2016) Effects of dry period length and dietary energy source on milk yield, energy balance, and metabolic status of dairy cows over 2 consecutive years: effects in the second year. J Dairy Sci 99:4826–4838. https://doi.org/10.3168/jds.2015-10742

Correa MT, Erb H, Scarlett J (1993) Path analysis for seven postpartum disorders of Holstein cows. J Dairy Sci 76:1305–1312. https://doi.org/10.3168/jds.S0022-0302(93)77461-5

de Roos AP, van den Bijgaart HJ, Hørlyk J, de Jong G (2007) Screening for subclinical ketosis in dairy cattle by Fourier transform infrared spectrometry. J Dairy Sci 90:1761–1766. https://doi.org/10.3168/jds.2006-203

Denis-Robichaud J, Dubuc J, Lefebvre D, DesCôteaux L (2014) Accuracy of milk ketone bodies from flow-injection analysis for the diagnosis of hyperketonemia in dairy cows. J Dairy Sci 97:3364–3370. https://doi.org/10.3168/jds.2013-6744

Dohoo IR (1983) The effects of calving to first service interval on reproductive performance in normal cows and cows with postpartal disease. Can Vet J 24:343–346

Douglas GN, Overton TR, Bateman HG 2nd, Dann HM, Drackley JK (2006) Prepartal plane of nutrition, regardless of dietary energy source, affects periparturient metabolism and dry matter intake in Holstein cows. J Dairy Sci 89:2141–2157. https://doi.org/10.3168/jds.S0022-0302(06)72285-8

Drackley JK (1999) ADSA Foundation Scholar Award. Biology of dairy cows during the transition period: the final frontier? J Dairy Sci 82:2259–2273. https://doi.org/10.3168/jds.s0022-0302(99)75474-3

Drackley JK, Dann HM, Douglas N, Janovick Guretzky NA, Litherland NB, Underwood JP, Loor JJ (2005) Physiological and pathological adaptations in dairy cows that may increase susceptibility to periparturient diseases and disorders. Ital J Anim Sci 4:323–344. https://doi.org/10.4081/ijas.2005.323

Duffield T (2000) Subclinical ketosis in lactating dairy cattle. Vet Clin North Am Food Anim Pract 16:231–253. https://doi.org/10.1016/s0749-0720(15)30103-1

Duffield TF, Lissemore KD, McBride BW, Leslie KE (2009) Impact of hyperketonemia in early lactation dairy cows on health and production. J Dairy Sci 92:571–580. https://doi.org/10.3168/jds.2008-1507

Eastridge ML (2006) Major advances in applied dairy cattle nutrition. J Dairy Sci 89:1311–1323. https://doi.org/10.3168/jds.S0022-0302(06)72199-3

Erb HN, Grohn YT (1988) Epidemiology of metabolic disorders in the periparturient dairy cow. J Dairy Sci 71:2557–2571. https://doi.org/10.3168/jds.S0022-0302(88)79845-8

Erol E, Kumar LS, Cline GW, Shulman GI, Kelly DP, Binas B (2004) Liver fatty acid binding protein is required for high rates of hepatic fatty acid oxidation but not for the action of PPARalpha in fasting mice. FASEB J 18:347–349. https://doi.org/10.1096/fj.03-0330fje

FAOSTAT (2022) Statistics Division. Food and Agriculture Organization of the United Nations. Accessed 26 Sep 2022. https://www.fao.org/faostat/en/#data

Fleischer P, Metzner M, Beyerbach M, Hoedemaker M, Klee W (2001) The relationship between milk yield and the incidence of some diseases in dairy cows. J Dairy Sci 84:2025–2035. https://doi.org/10.3168/jds.S0022-0302(01)74646-2

Fukao T, Mitchell G, Sass JO, Hori T, Orii K, Aoyama Y (2014) Ketone body metabolism and its defects. J Inherit Metab Dis 37:541–551. https://doi.org/10.1007/s10545-014-9704-9

Galamb E, Faigl V, Keresztes M, Csillik Z, Tröscher A, Elek P, Kulcsár M, Huszenicza G, Fébel H, Husvéth F (2017) Effect of pre- and post-partum supplementation with lipid-encapsulated conjugated linoleic acid on milk yield and metabolic status in multiparous high-producing dairy cows. J Anim Physiol Anim Nutr (Berl) 101:1026–1035. https://doi.org/10.1111/jpn.12544

Geishauser T, Leslie K, Tenhag J, Bashiri A (2000) Evaluation of eight cow-side ketone tests in milk for detection of subclinical ketosis in dairy cows. J Dairy Sci 83:296–299. https://doi.org/10.3168/jds.S0022-0302(00)74877-6

Gerloff BJ (2000) Dry cow management for the prevention of ketosis and fatty liver in dairy cows. Vet Clin North Am Food Anim Pract 16:283–292. https://doi.org/10.1016/s0749-0720(15)30106-7

Goff JP, Horst RL (1997) Physiological changes at parturition and their relationship to metabolic disorders. J Dairy Sci 80:1260–1268. https://doi.org/10.3168/jds.S0022-0302(97)76055-7

Gordon JL, Leblanc SJ, Duffield TF (2013) Ketosis treatment in lactating dairy cattle. Vet Clin North Am Food Anim Pract 29:433–445. https://doi.org/10.1016/j.cvfa.2013.03.001

Gordon JL, LeBlanc SJ, Kelton DF, Herdt TH, Neuder L, Duffield TF (2017a) Randomized clinical field trial on the effects of butaphosphan-cyanocobalamin and propylene glycol on ketosis resolution and milk production. J Dairy Sci 100:3912–3921. https://doi.org/10.3168/jds.2016-11926

Gordon JL, Duffield TF, Herdt TH, Kelton DF, Neuder L, LeBlanc SJ (2017b) Effects of a combination butaphosphan and cyanocobalamin product and insulin on ketosis resolution and milk production. J Dairy Sci 100:2954–2966. https://doi.org/10.3168/jds.2016-11925

Gross JJ (2022) Limiting factors for milk production in dairy cows: perspectives from physiology and nutrition. J Anim Sci 100:skac044. https://doi.org/10.1093/jas/skac044

Gross JJ, Bruckmaier RM (2015) Repeatability of metabolic responses to a nutrient deficiency in early and mid lactation and implications for robustness of dairy cows. J Dairy Sci 98:8634–8643. https://doi.org/10.3168/jds.2014-9246

Gross JJ, Bruckmaier RM (2019a) Invited review: metabolic challenges and adaptation during different functional stages of the mammary gland in dairy cows: perspectives for sustainable milk production. J Dairy Sci 102:2828–2843. https://doi.org/10.3168/jds.2018-15713

Gross JJ, Bruckmaier RM (2019b) Review: metabolic challenges in lactating dairy cows and their assessment via established and novel indicators in milk. Animal 13(S1):s75–s81. https://doi.org/10.1017/S175173111800349X

Gross J, van Dorland HA, Bruckmaier RM, Schwarz FJ (2011a) Performance and metabolic profile of dairy cows during a lactational and deliberately induced negative energy balance with subsequent realimentation. J Dairy Sci 94:1820–1830. https://doi.org/10.3168/jds.2010-3707

Gross J, van Dorland HA, Schwarz FJ, Bruckmaier RM (2011b) Endocrine changes and liver mRNA abundance of somatotropic axis and insulin system constituents during negative energy balance at different stages of lactation in dairy cows. J Dairy Sci 94:3484–3494. https://doi.org/10.3168/jds.2011-4251

Gross JJ, Schwarz FJ, Eder K, van Dorland HA, Bruckmaier RM (2013) Liver fat content and lipid metabolism in dairy cows during early lactation and during a mid-lactation feed restriction. J Dairy Sci 96:5008–5017. https://doi.org/10.3168/jds.2012-6245

Gross JJ, Kessler EC, Albrecht C, Bruckmaier RM (2015) Response of the cholesterol metabolism to a negative energy balance in dairy cows depends on the lactational stage. PLoS One 10:e0121956. https://doi.org/10.1371/journal.pone.0121956

Gross JJ, Grossen-Rösti L, Héritier R, Tröscher A, Bruckmaier RM (2018) Inflammatory and metabolic responses to an intramammary lipopolysaccharide challenge in early lactating cows supplemented with conjugated linoleic acid. J Anim Physiol Anim Nutr (Berl) 102:e838–e848. https://doi.org/10.1111/jpn.12843

Grossen-Rösti L, Kessler EC, Tröscher A, Bruckmaier RM, Gross JJ (2018) Hyperglycaemia in transition dairy cows: effects of lactational stage and conjugated linoleic acid supplementation on glucose metabolism and turnover. J Anim Physiol Anim Nutr (Berl) 102:483–494. https://doi.org/10.1111/jpn.12771

Grummer RR (1993) Etiology of lipid-related metabolic disorders in periparturient dairy cows. J Dairy Sci 76:3882–3896. https://doi.org/10.3168/jds.S0022-0302(93)77729-2

Grummer RR, Mashek DG, Hayirli A (2004) Dry matter intake and energy balance in the transition period. Vet Clin North Am Food Anim Pract 20:447–470. https://doi.org/10.1016/j.cvfa.2004.06.013

Grünberg W, Morin DE, Drackley JK, Constable PD (2006) Effect of rapid intravenous administration of 50% dextrose solution on phosphorus homeostasis in postparturient dairy cows. J Vet Intern Med 20:1471–1478. https://doi.org/10.1892/0891-6640(2006)20[1471:eoriao]2.0.co;2

Guliński P (2021) Ketone bodies - causes and effects of their increased presence in cows' body fluids: a review. Vet World 14:1492–1503. https://doi.org/10.14202/vetworld.2021.1492-1503

Ha NT, Gross JJ, van Dorland A, Tetens J, Thaller G, Schlather M, Bruckmaier R, Simianer H (2015) Gene-based mapping and pathway analysis of metabolic traits in dairy cows. PLoS One 10:e0122325. https://doi.org/10.1371/journal.pone.0122325

Ha NT, Drögemüller C, Reimer C, Schmitz-Hsu F, Bruckmaier RM, Simianer H, Gross JJ (2017) Liver transcriptome analysis reveals important factors involved in the metabolic adaptation of the transition cow. J Dairy Sci 100:9311–9323. https://doi.org/10.3168/jds.2016-12454

Henno M, Ling K, Kaart T, Ariko T, Karis P, Jaakson H, Kuusik S, Ots M (2021) Effect of monensin on milk fatty acid profile in dairy cows and on the use of fatty acids for early diagnosis of elevated blood plasma concentrations of nonesterified fatty acids and hyperketonemia. J Dairy Sci 104:10355–10362. https://doi.org/10.3168/jds.2020-20041

Herdt TH, Stevens JB, Olson WG, Larson V (1981) Blood concentrations of beta hydroxybutyrate in clinically normal Holstein-Friesian herds and in those with a high prevalence of clinical ketosis. Am J Vet Res 42:503–506

Holtenius P, Holtenius K (1996) New aspects of ketone bodies in energy metabolism of dairy cows: a review. Zentralbl Veterinarmed A 43:579–587. https://doi.org/10.1111/j.1439-0442.1996.tb00491.x

Humer E, Petri RM, Aschenbach JR, Bradford BJ, Penner GB, Tafaj M, Südekum KH, Zebeli Q (2018) Invited review: practical feeding management recommendations to mitigate the risk of subacute ruminal acidosis in dairy cattle. J Dairy Sci 101:872–888. https://doi.org/10.3168/jds.2017-13191

Ingvartsen KL, Andersen JB (2000) Integration of metabolism and intake regulation: a review focusing on periparturient animals. J Dairy Sci 83:1573–1597. https://doi.org/10.3168/jds.S0022-0302(00)75029-6

Ingvartsen KL, Moyes KM (2015) Factors contributing to immunosuppression in the dairy cow during the periparturient period. Jpn J Vet Res 63(Suppl 1):S15–S24

Ingvartsen KL, Dewhurst RJ, Friggens NC (2003) On the relationship between lactational performance and health: is it yield or metabolic imbalance that cause production diseases in dairy cattle? A position paper. Livest Prod Sci 83:277–308. https://doi.org/10.1016/S0301-6226(03)00110-6

Iwersen M, Falkenberg U, Voigtsberger R, Forderung D, Heuwieser W (2009) Evaluation of an electronic cowside test to detect subclinical ketosis in dairy cows. J Dairy Sci 92:2618–2624. https://doi.org/10.3168/jds.2008-1795

Iwersen M, Klein-Jöbstl D, Pichler M, Roland L, Fidlschuster B, Schwendenwein I, Drillich M (2013) Comparison of 2 electronic cowside tests to detect subclinical ketosis in dairy cows and the influence of the temperature and type of blood sample on the test results. J Dairy Sci 96:7719–7730. https://doi.org/10.3168/jds.2013-7121

Kessel S, Stroehl M, Meyer HH, Hiss S, Sauerwein H, Schwarz FJ, Bruckmaier RM (2008) Individual variability in physiological adaptation to metabolic stress during early lactation in dairy cows kept under equal conditions. J Anim Sci 86:2903–2912. https://doi.org/10.2527/jas.2008-1016

Kessler EC, Gross JJ, Bruckmaier RM (2013) Different adaptation of IGF-I and its IGFBPs in dairy cows during a negative energy balance in early lactation and a negative energy balance induced by feed restriction in mid lactation. Vet Med Czech 58:459–467

Kessler EC, Gross JJ, Bruckmaier RM, Albrecht C (2014a) Cholesterol metabolism, transport, and hepatic regulation in dairy cows during transition and early lactation. J Dairy Sci 97:5481–5490. https://doi.org/10.3168/jds.2014-7926

Kessler EC, Bruckmaier RM, Gross JJ (2014b) Milk production during the colostral period is not related to the later lactational performance in dairy cows. J Dairy Sci 97:2186–2192. https://doi.org/10.3168/jds.2013-7573

Kessler EC, Wall SK, Hernandez LL, Bruckmaier RM, Gross JJ (2018) Short communication: circulating serotonin is related to the metabolic status and lactational performance at the onset of lactation in dairy cows. J Dairy Sci 101:11455–11460. https://doi.org/10.3168/jds.2018-14626

Kobayashi Y, Boyd CK, Bracken CJ, Lamberson WR, Keisler DH, Lucy MC (1999) Reduced growth hormone receptor (GHR) messenger ribonucleic acid in liver of periparturient cattle is caused by a specific down-regulation of GHR 1A that is associated with decreased insulin-like growth factor I. Endocrinology 140:3947–3954. https://doi.org/10.1210/endo.140.9.7000

Kok A, Chen J, Kemp B, van Knegsel ATM (2019) Review: dry period length in dairy cows and consequences for metabolism and welfare and customised management strategies. Animal 13(S1):s42–s51. https://doi.org/10.1017/S1751731119001174

Kuhla B, Metges CC, Hammon HM (2016) Endogenous and dietary lipids influencing feed intake and energy metabolism of periparturient dairy cows. Domest Anim Endocrinol 56(Suppl):S2–S10. https://doi.org/10.1016/j.domaniend.2015.12.002

Laeger T, Metges CC, Kuhla B (2010) Role of beta-hydroxybutyric acid in the central regulation of energy balance. Appetite 54:450–455. https://doi.org/10.1016/j.appet.2010.04.005

Laffel L (1999) Ketone bodies: a review of physiology, pathophysiology and application of monitoring to diabetes. Diabetes Metab Res Rev 15:412–426. https://doi.org/10.1002/(sici)1520-7560(199911/12)15:6<412::aid-dmrr72>3.0.co;2-8

LeBlanc S (2010) Monitoring metabolic health of dairy cattle in the transition period. J Reprod Dev 56(Suppl):S29–S35. https://doi.org/10.1262/jrd.1056s29

Li P, Li XB, Fu SX, Wu CC, Wang XX, Yu GJ, Long M, Wang Z, Liu GW (2012) Alterations of fatty acid β-oxidation capability in the liver of ketotic cows. J Dairy Sci 95:1759–1766. https://doi.org/10.3168/jds.2011-4580

Locher LF, Meyer N, Weber EM, Rehage J, Meyer U, Dänicke S, Huber K (2011) Hormone-sensitive lipase protein expression and extent of phosphorylation in subcutaneous and retroperitoneal adipose tissues in the periparturient dairy cow. J Dairy Sci 94:4514–4523. https://doi.org/10.3168/jds.2011-4145

Mammi LME, Guadagnini M, Mechor G, Cainzos JM, Fusaro I, Palmonari A, Formigoni A (2021) The use of monensin for ketosis prevention in dairy cows during the transition period: a systematic review. Animals (Basel) 11:1988. https://doi.org/10.3390/ani11071988

Mandujano Reyes JF, Walleser E, Hachenberg S, Gruber S, Kammer M, Baumgartner C, Mansfeld R, Anklam K, Döpfer D (2021) Full model selection using regression trees for numeric predictions of biomarkers for metabolic challenges in dairy cows. Prev Vet Med 193:105422. https://doi.org/10.1016/j.prevetmed.2021.105422

Mann S, Yepes FAL, Behling-Kelly E, McArt JAA (2017) The effect of different treatments for early-lactation hyperketonemia on blood β-hydroxybutyrate, plasma nonesterified fatty acids, glucose, insulin, and glucagon in dairy cattle. J Dairy Sci 100:6470–6482. https://doi.org/10.3168/jds.2016-12532

McArt JA, Nydam DV, Ospina PA, Oetzel GR (2011) A field trial on the effect of propylene glycol on milk yield and resolution of ketosis in fresh cows diagnosed with subclinical ketosis. J Dairy Sci 94:6011–6020. https://doi.org/10.3168/jds.2011-4463

McArt JA, Nydam DV, Oetzel GR (2012a) A field trial on the effect of propylene glycol on displaced abomasum, removal from herd, and reproduction in fresh cows diagnosed with subclinical ketosis. J Dairy Sci 95:2505–2512. https://doi.org/10.3168/jds.2011-4908

McArt JA, Nydam DV, Oetzel GR (2012b) Epidemiology of subclinical ketosis in early lactation dairy cattle. J Dairy Sci 95:5056–5066. https://doi.org/10.3168/jds.2012-5443

McArt JA, Nydam DV, Oetzel GR, Overton TR, Ospina PA (2013) Elevated non-esterified fatty acids and β-hydroxybutyrate and their association with transition dairy cow performance. Vet J 198:560–570. https://doi.org/10.1016/j.tvjl.2013.08.011

McArt JA, Nydam DV, Overton MW (2015) Hyperketonemia in early lactation dairy cattle: a deterministic estimate of component and total cost per case. J Dairy Sci 98:2043–2054. https://doi.org/10.3168/jds.2014-8740

McPherson PA, McEneny J (2012) The biochemistry of ketogenesis and its role in weight management, neurological disease and oxidative stress. J Physiol Biochem 68:141–151. https://doi.org/10.1007/s13105-011-0112-4

Mertens DR, Dado RG (1993) System of equations for fulfilling net energy and absorbed protein requirements for milk component production. J Dairy Sci 76:3464–3478. https://doi.org/10.3168/jds.S0022-0302(93)77686-9

Mulligan FJ, Doherty ML (2008) Production diseases of the transition cow. Vet J 176:3–9. https://doi.org/10.1016/j.tvjl.2007.12.018

Murondoti A, Jorritsma R, Beynen AC, Wensing T, Geelen MJ (2004) Unrestricted feed intake during the dry period impairs the postpartum oxidation and synthesis of fatty acids in the liver of dairy cows. J Dairy Sci 87:672–679. https://doi.org/10.3168/jds.S0022-0302(04)73210-5

Niehoff ID, Hüther L, Lebzien P (2009) Niacin for dairy cattle: a review. Br J Nutr 101:5–19. https://doi.org/10.1017/S0007114508043377

Nielsen NI, Friggens NC, Chagunda MG, Ingvartsen KL (2005) Predicting risk of ketosis in dairy cows using in-line measurements of beta-hydroxybutyrate: a bio-

logical model. J Dairy Sci 88:2441–2453. https://doi.org/10.3168/jds.S0022-0302(05)72922-2
Nuber U, van Dorland HA, Bruckmaier RM (2016) Effects of butafosfan with or without cyanocobalamin on the metabolism of early lactating cows with subclinical ketosis. J Anim Physiol Anim Nutr 100:146–155. https://doi.org/10.1111/jpn.12332
Oetzel GR (2004) Monitoring and testing dairy herds for metabolic disease. Vet Clin North Am Food Anim Pract 20:651–674. https://doi.org/10.1016/j.cvfa.2004.06.006
Oetzel GR (2020) Rational treatments for ketosis in fresh cows. AABP Proc 53(1):122–126. https://doi.org/10.21423/aabppro20207983
Ospina PA, Nydam DV, Stokol T, Overton TR (2010) Evaluation of nonesterified fatty acids and beta-hydroxybutyrate in transition dairy cattle in the northeastern United States: critical thresholds for prediction of clinical diseases. J Dairy Sci 93:546–554. https://doi.org/10.3168/jds.2009-2277
Raboisson D, Mounié M, Maigné E (2014) Diseases, reproductive performance, and changes in milk production associated with subclinical ketosis in dairy cows: a meta-analysis and review. J Dairy Sci 97:7547–7563. https://doi.org/10.3168/jds.2014-8237
Rhoads RP, Kim JW, Leury BJ, Baumgard LH, Segoale N, Frank SJ, Bauman DE, Boisclair YR (2004) Insulin increases the abundance of the growth hormone receptor in liver and adipose tissue of periparturient dairy cows. J Nutr 134:1020–1027. https://doi.org/10.1093/jn/134.5.1020
Roche JR, Friggens NC, Kay JK, Fisher MW, Stafford KJ, Berry DP (2009) Invited review: body condition score and its association with dairy cow productivity, health, and welfare. J Dairy Sci 92:5769–5801. https://doi.org/10.3168/jds.2009-2431
Rojas-Morales P, Tapia E, Pedraza-Chaverri J (2016) β-Hydroxybutyrate: a signaling metabolite in starvation response? Cell Signal 28:917–923. https://doi.org/10.1016/j.cellsig.2016.04.005
Ruderman NB, Goodman MN (1973) Regulation of ketone body metabolism in skeletal muscle. Am J Physiol 224:1391–1397. https://doi.org/10.1152/ajplegacy.1973.224.6.1391
Rukkwamsuk T, Kruip TA, Wensing T (1999) Relationship between overfeeding and overconditioning in the dry period and the problems of high producing dairy cows during the postparturient period. Vet Q 21:71–77. https://doi.org/10.1080/01652176.1999.9694997
Sailer KJ, Pralle RS, Oliveira RC, Erb SJ, Oetzel GR, White HM (2018) Technical note: validation of the BHBCheck blood β-hydroxybutyrate meter as a diagnostic tool for hyperketonemia in dairy cows. J Dairy Sci 101:1524–1529. https://doi.org/10.3168/jds.2017-13583
Sato H, Matsumoto M, Hanasaka S (1999) Relations between plasma acetate, 3-hydroxybutyrate, FFA, glucose levels and energy nutrition in lactating dairy cows. J Vet Med Sci 61:447–451. https://doi.org/10.1292/jvms.61.447
Sehested J, Gaillard C, Lehmann JO, Maciel GM, Vestergaard M, Weisbjerg MR, Mogensen L, Larsen LB, Poulsen NA, Kristensen T (2019) Review: extended lactation in dairy cattle. Animal 13(S1):s65–s74. https://doi.org/10.1017/S1751731119000806
Sniffen CJ, Beverly RW, Mooney CS, Roe MB, Skidmore AL, Black JR (1993) Nutrient requirements versus supply in the dairy cow: strategies to account for variability. J Dairy Sci 76:3160–3178. https://doi.org/10.3168/jds.S0022-0302(93)77655-9
Sordillo LM, Contreras GA, Aitken SL (2009) Metabolic factors affecting the inflammatory response of periparturient dairy cows. Anim Health Res Rev 10:53–63. https://doi.org/10.1017/S1466252309990016
Sugden MC, Bulmer K, Holness MJ (2001) Fuel-sensing mechanisms integrating lipid and carbohydrate utilization. Biochem Soc Trans 29:272–278. https://doi.org/10.1042/0300-5127:0290272
Suriyasathaporn W, Daemen AJ, Noordhuizen-Stassen EN, Dieleman SJ, Nielen M, Schukken YH (1999) Beta-hydroxybutyrate levels in peripheral blood and ketone bodies supplemented in culture media affect the in vitro chemotaxis of bovine leukocytes. Vet Immunol Immunopathol 68:177–186. https://doi.org/10.1016/s0165-2427(99)00017-3
Suthar VS, Canelas-Raposo J, Deniz A, Heuwieser W (2013) Prevalence of subclinical ketosis and relationships with postpartum diseases in European dairy cows. J Dairy Sci 96:2925–2938. https://doi.org/10.3168/jds.2012-6035
Tremblay M, Kammer M, Lange H, Plattner S, Baumgartner C, Stegeman JA, Duda J, Mansfeld R, Döpfer D (2019) Prediction model optimization using full model selection with regression trees demonstrated with FTIR data from bovine milk. Prev Vet Med 163:14–23. https://doi.org/10.1016/j.prevetmed.2018.12.012
Trevisi E, Amadori M, Cogrossi S, Razzuoli E, Bertoni G (2012) Metabolic stress and inflammatory response in high-yielding, periparturient dairy cows. Res Vet Sci 93:695–704. https://doi.org/10.1016/j.rvsc.2011.11.008
Tsunoda E, Gross JJ, Kawashima C, Bruckmaier RM, Kida K, Miyamoto A (2017) Feed-derived volatile basic nitrogen increases reactive oxygen species production of blood leukocytes in lactating dairy cows. Anim Sci J 88:125–133. https://doi.org/10.1111/asj.12608
van der Kolk JH, Gross JJ, Gerber V, Bruckmaier RM (2017) Disturbed bovine mitochondrial lipid metabolism: a review. Vet Q 37:262–273. https://doi.org/10.1080/01652176.2017.1354561
van Knegsel AT, van der Drift SG, Horneman M, de Roos AP, Kemp B, Graat EA (2010) Short communication: ketone body concentration in milk determined by Fourier transform infrared spectroscopy: value for the detection of hyperketonemia in dairy cows. J Dairy Sci 93:3065–3069. https://doi.org/10.3168/jds.2009-2847

van Knegsel ATM, Burgers EEA, Ma J, Goselink RMA, Kok A (2022) Extending lactation length: consequences for cow, calf and farmer. J Anim Sci 20:skac220. https://doi.org/10.1093/jas/skac220

Vernon RG (2005) Lipid metabolism during lactation: a review of adipose tissue-liver interactions and the development of fatty liver. J Dairy Res 72:460–469. https://doi.org/10.1017/S0022029905001299

von Keyserlingk MA, Martin NP, Kebreab E, Knowlton KF, Grant RJ, Stephenson M, Sniffen CJ, Harner JP 3rd, Wright AD, Smith SI (2013) Invited review: sustainability of the US dairy industry. J Dairy Sci 96:5405–5425. https://doi.org/10.3168/jds.2012-6354

Voyvoda H, Erdogan (2010) Use of a hand-held meter for detecting subclinical ketosis in dairy cows. Res Vet Sci. 89:344–351. https://doi.org/10.1016/j.rvsc.2010.04.007

Walsh RB, Walton JS, Kelton DF, LeBlanc SJ, Leslie KE, Duffield TF (2007) The effect of subclinical ketosis in early lactation on reproductive performance of postpartum dairy cows. J Dairy Sci 90:2788–2796. https://doi.org/10.3168/jds.2006-560

Wankhade PR, Manimaran A, Kumaresan A, Jeyakumar S, Ramesha KP, Sejian V, Rajendran D, Varghese MR (2017) Metabolic and immunological changes in transition dairy cows: a review. Vet World 10:1367–1377. https://doi.org/10.14202/vetworld.2017.1367-1377

Wathes DC, Clempson AM, Pollott GE (2012) Associations between lipid metabolism and fertility in the dairy cow. Reprod Fertil Dev 25:48–61. https://doi.org/10.1071/RD12272

White HM (2015) The role of TCA cycle anaplerosis in ketosis and fatty liver in periparturient dairy cows. Animals (Basel) 5:793–802. https://doi.org/10.3390/ani5030384

Williamson M, Couto Serrenho R, McBride BW, LeBlanc SJ, DeVries TJ, Duffield TF (2022) Reducing milking frequency from twice to once daily as an adjunct treatment for ketosis in lactating dairy cows-a randomized controlled trial. J Dairy Sci 105:1402–1417. https://doi.org/10.3168/jds.2021-20551

Zarrin M, De Matteis L, Vernay MC, Wellnitz O, van Dorland HA, Bruckmaier RM (2013) Long-term elevation of β-hydroxybutyrate in dairy cows through infusion: effects on feed intake, milk production, and metabolism. J Dairy Sci 96:2960–2972. https://doi.org/10.3168/jds.2012-6224

Zarrin M, Wellnitz O, van Dorland HA, Bruckmaier RM (2014a) Induced hyperketonemia affects the mammary immune response during lipopolysaccharide challenge in dairy cows. J Dairy Sci 97:330–339. https://doi.org/10.3168/jds.2013-7222

Zarrin M, Wellnitz O, van Dorland HA, Gross JJ, Bruckmaier RM (2014b) Hyperketonemia during lipopolysaccharide-induced mastitis affects systemic and local intramammary metabolism in dairy cows. J Dairy Sci 97:3531–3541. https://doi.org/10.3168/jds.2013-7480

Zarrin M, Grossen-Rösti L, Bruckmaier RM, Gross JJ (2017) Elevation of blood β-hydroxybutyrate concentration affects glucose metabolism in dairy cows before and after parturition. J Dairy Sci 100:2323–2333. https://doi.org/10.3168/jds.2016-11714

Zebeli Q, Aschenbach JR, Tafaj M, Boguhn J, Ametaj BN, Drochner W (2012) Invited review: role of physically effective fiber and estimation of dietary fiber adequacy in high-producing dairy cattle. J Dairy Sci 95:1041–1056. https://doi.org/10.3168/jds.2011-4421

Zhang G, Ametaj BN (2020) Ketosis an old story under a new approach. Dairy 1:42–60. https://doi.org/10.3390/dairy1010005

# Part III

# Gastrointestinal Disorders

# Displacement of the Abomasum

Andrew J. Niehaus

**Abstract**

Abomasal displacements are one of the most common abdominal conditions affecting production dairy cattle. The abomasum can displace to the left side of the abdomen or dilate and displace on the right. Additionally, a right-sided displacement can turn into an abomasal volvulus which is an emergent condition. Due to their commonality, abomasal displacements pose significant economic ramifications to the dairy industry. It is a disease that likely has a genetic predisposition, but management factors play a significant role in their pathogenesis. Most simple displacements have a good to excellent prognosis for surviving, but production will likely be impacted in the affected lactation. Multiple surgical methods have been developed for correction of abomasal displacements from open surgical techniques to minimally invasive procedures. There are limitations with each technique, and none are ideal for every situation.

**Keywords**

Abomasum · Displacement · Forestomach · Dairy cattle · Surgery · Abdomen · Stomach

A. J. Niehaus (✉)
Department of Veterinary Clinical Sciences, College of Veterinary Medicine, The Ohio State University, Columbus, OH, USA
e-mail: niehaus.25@osu.edu

## 1 Introduction

The abomasum is considered the "true stomach" of the ruminant and is responsible for secretion of acid and digestive enzymes. The body of the abomasum in the adult cow is normally positioned in the right cranial abdomen and lies with its longitudinal axis against right ventral body wall (Fig. 1). The fundic part of the abomasum extends to the left side of the abdomen contacting the reticulum, ruminal atrium, and ventral sac of the rumen and is covered by the superficial wall of the greater omentum. Greater omentum extensively attaches to the greater curvature of the abomasal body and pyloric part, while the lesser omentum attaches to its lesser curvature. The orad portion of the abomasal body is situated cranially on the right ventral abdomen in a cleft formed by the rumen and the omasum, while the aborad (pyloric) portion is caudal. The large bovine omasum normally prevents the abomasum from contacting the liver. The pyloric part of the abomasum is normally situated dorsal to the abomasal body and has a cranial incline as it transitions to the duodenum (Habel and Smith 1981; Dyce et al. 2010).

A displaced abomasum (DA) can be defined as any abnormal positioning of the abomasum. It is very rare in small ruminants and camelids (Shakespeare 1997; Meymandi et al. 2008; Song et al. 2022) but is arguably the most common surgical disease of dairy cattle with significant eco-



J. J. Gross (ed.), *Production Diseases in Farm Animals*,
https://doi.org/10.1007/978-3-031-51788-4_7

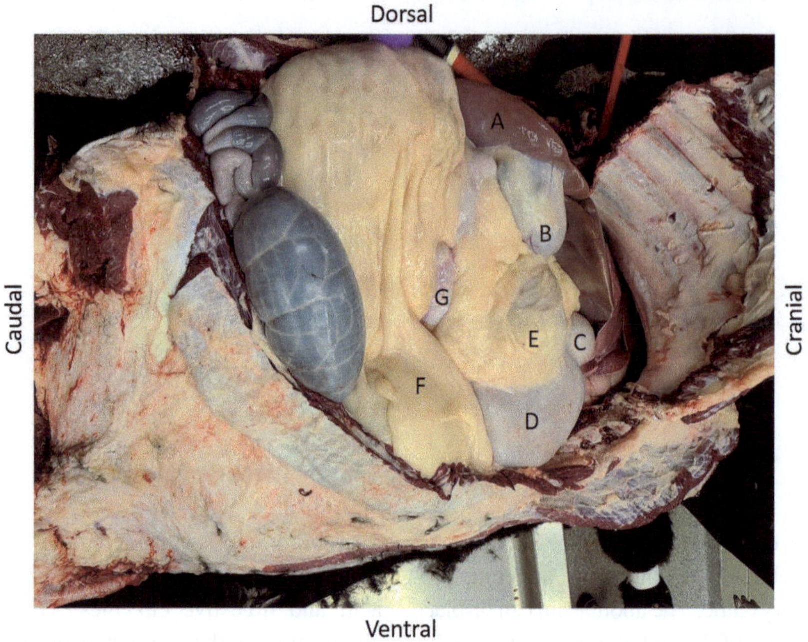


**Fig. 1** Gross postmortem examination on a cow. The abdominal viscera have been exposed by reflecting the ribs cranially. (**A**) liver, (**B**) gall bladder, (**C**) reticulum, (**D**) abomasum, (**E**) omasum covered by lesser omentum, (**F**) greater omentum, and (**G**) cranial duodenum (Photo courtesy of Dr. Joe Lozier, The Ohio State University)

nomic impacts. Losses can be experienced through decreased production, reduced reproductive efficiency, veterinary expenses, and replacement costs. Incidence of abomasal displacement varies with geographic region and operation. Reports estimate incidence of disease between 2.5% and 3.5% among US dairy cows (Pehrson and Shaver 1992; USDA 2018). A case of a left displaced abomasum is estimated to cost the producer approximately $430 for primiparous cows and $640 for multiparous cows (using 2015 market prices). This is greater on a "per cow" basis than the other common periparturient diseases. The largest portion of this cost is due to veterinary treatment (Liang et al. 2017).

Most commonly the abomasum displaces to the left of the rumen and becomes trapped between the rumen and the left body wall. This is referred to as a left displaced abomasum (LDA) and was first reported by Begg in the 1950s (Begg 1950; Begg and Whiteford 1956). Rarely, a left displaced abomasum can displace "internally" and can become entrapped between the rumen and other abdominal viscera (Pardon et al. 2012). Alternatively, the abomasum can migrate dorsally on the right side of the abdomen, referred to as a right dilated (displaced) abomasum (RDA). Early reports of RDAs emerged in the veterinary literature around 1900 with several prior to 1950 (Carougeau and Preston 1898; Lagerlöf 1925; Loje 1949). An emergent condition results if the RDA further twists around its mesenteric attachments becoming a right abomasal volvulus (RAV). Volvulus of the abomasum can also involve other structures like the omasum or reticulum (Espersen 1964; Geishauser 1995; Kumper 1995) and is associated with a poor prognosis for recovery (Kumper 1995).

## 2 Etiopathogenesis

Abomasal displacement is found with the highest prevalence in cattle from high-producing dairy herds (Constable et al. 1992a; Pentecost et al. 2014). This condition less frequently occurs in beef cattle (Oman et al. 2016) or cattle without extreme metabolic demands. Reports suggest abomasal displacement in beef cattle may occur by a different mechanism than that in dairy cattle (Ichijo et al. 2005, 2014). LDA is observed most commonly in adult cattle between 4 and 7 years (Mather and Dedrick 1966; Wallace 1975; Willeberg et al. 1982; Constable et al. 1992a), and a cow's risk increases with age (Willeberg et al. 1982), while RDAs often occur in younger animals with several reports of right-sided displacements (including abomasal volvulus) occurring in calves (Carougeau and Preston 1898; Frazee 1984). LDAs generally occur shortly after freshening with a reported mean of 14 days in milk (Mather and Dedrick 1966; Pentecost et al. 2014). Some reports show a seasonal distribution of LDAs with a higher incidence in winter months when natural forages and exercise are less (Mather and Dedrick 1966); others suggest a higher incidence in the spring (Constable et al. 1992a).

The etiology of abomasal displacements has been debated; however, most of researchers agree that the fundamental cause of abomasal displacements is altered abomasal motility (Espersen 1964; Mather and Dedrick 1966; Constable et al. 1992a; Doll et al. 2009). Abomasal atony results when the abomasum loses muscle tone and motility, possibly due to interference with vagal innervation (Itoh et al. 2011). Gas produced within the abomasum or passed from the forestomachs accumulates within the abomasum and increases its buoyancy causing it to displace (Doll et al. 2009; Itoh et al. 2011). Other abdominal viscera can entrap the abomasum preventing it from returning to its normal abdominal position resulting in a vicious cycle as outflow and motility are further compromised. Clinical signs result from decreased abomasal outflow, compromised abomasal perfusion, and the metabolic derangements that follow. Some displacements will self-correct (Varden 1979), and these are known as "floating DAs" or simply "floaters."

If the distended abomasum displaces to the left side of the abdomen, it becomes an LDA; if it remains on the right, it becomes an RDA or RAV. The exact mechanisms behind where the abomasum displaces within the abdomen (right vs. left displacement) are less clear and more widely debated, but most commonly the abomasum is pushed to the left of the rumen and becomes an LDA (Robertson 1968; Geishauser 1995).

## 2.1 Left Displaced Abomasum (LDA)

LDAs result when the abomasum, normally positioned on the ventral right body wall, slips under the rumen and becomes trapped between the rumen and the left body wall. The fundus of the abomasum normally lies to the left of midline and is slightly higher than body of the abomasum and typically retains a small gas pocket. In cases of abomasal atony, excessive gas may accumulate in this portion of the viscus and pull the greater curvature to the left. As gas continues to accumulate, the abomasum "grows" on the left side and rises dorsally (Mather and Dedrick 1966).

Left-sided displacements are most common in postpartum cows occurring within 2 months after calving (Robertson 1968; Wallace 1975; Constable et al. 1992a; Pentecost et al. 2014). Mechanical factors related to rearrangement of abdominal viscera in late pregnancy and the void created following parturition have been speculated to contribute to abomasal displacements (Robertson 1968; Constable et al. 1992a). During the last trimester of pregnancy through the first 3 months of lactation, there are significant changes in abdominal dimensions, and the vertical distance between the ventral abdomen and the descending duodenum increases (Wittek et al. 2007); the latter is thought to decrease complete abomasal emptying. Additionally, as the uterus enlarges during late pregnancy, the uterus forces the abomasum cranially and to the left (Lagerlof 1929; Robertson 1968; Wallace 1975). This is also a common time for many concurrent diseases that can affect abomasal motility. The above factors are thought to contribute to LDAs during the postpartum period.

It has also been theorized that a full rumen will block the abomasum from traveling to the left side of the abdomen (Constable et al. 1992a), and this protective blocking effect is reduced during the initial postpartum period. During gestation, the uterus slowly enlarges which inhibits normal rumen fill. This is especially prominent during late gestation when the uterus is large. During parturition, the uterus has a sudden decrease in size, and the space-occupying effect by the uterus is reduced; however, the rumen is slower to return to its pre-pregnant volume.

### 2.2 Right Displaced Abomasum (RDA)

Right-sided displacements are less common and have been found to be more random in occurrence compared to LDA. Cattle with relatively normal rumen fill have a decreased tendency to develop left-sided displacements because the full rumen acts as a barrier to the abomasum's path to the left side of the abdomen (Constable et al. 1992a); abomasal displacements in these cattle are generally confined to the right side of the abdomen. Beef cattle, males, and cattle without severe metabolic demands are not subjected to great changes in the size and fill of the rumen, and it is thought that this protects them from developing left-sided displacements; however, they have a higher frequency of right-sided displacements. If the abomasum remains on the right side, atony and subsequent dilation of the abomasum leads to an RDA or RAV. RDAs form when the abomasum flips counterclockwise as viewed from the cow's rear around an axis through the lesser omentum. The greater curvature flips outward against the right body wall and assumes a dorsally oriented position. RDAs precede RAVs; the latter forms when the right displaced abomasum turns counterclockwise as viewed from the top of the cow and twists around its mesenteric attachments.

### 2.3 Concurrent Diseases and Other Predisposing Factors

Cows with abomasal displacements frequently have concurrent systemic illnesses such as mastitis, metritis, ketosis, and hypocalcemia, and these conditions are predisposing factors for developing abomasal displacements by causing decreased feed intake, decreased rumen fill, and abomasal atony leading to left-sided displacements (Box 1). These diseases are also common during the recently fresh period, which clouds their association to abomasal displacements (Robertson 1968; Wallace 1975; Constable et al. 1992a). The association between LDAs and concurrent disease conditions may be exacerbated by the higher prevalence in older cows (Willeberg et al. 1982).

Abrupt diet changes and metabolic demands, including changes in acid/base balance after freshening (Poulsen 1976), are also theorized to predispose to development of abomasal atony and DAs. Metabolic issues like insulin resistance, hyperglycemia, and ketosis delay abomasal emptying and have been associated with abomasal displacements (Muylle et al. 1989; Kuiper 1991). However, a strong causal relationship has not been established, and others have observed decreased insulin and glucose levels in cows preceding displacement and hypothesized that changes in glucose and insulin may be the effect of the disease rather than the cause (Van Winden et al. 2003). Endotoxemia has been postulated to predispose to abomasal displacements but has not been shown to increase the incidence of displacements when studied (Kuiper 1991). Although a study looking at the effect of low serum calcium showed minimal effects on abomasal motility until the hypocalcemia was severe (Madison and Troutt 1988), it is generally agreed that cattle with hypocalcemia have decreased abomasal tone and are predisposed to abomasal displacement (Hull 1971; Massey et al. 1993).

Dietary and other management factors have been strongly correlated with abomasal displacements. High-producing dairy cows (Grymer et al. 1982), cattle in intensive production operations, and those from high-producing herds are predisposed to development of displaced abomasum. High-concentrate diets leading to volatile fatty acid (VFA) production decrease motility of the abomasum. Also a low rumen fiber content at calving, a rapid change of concentrates after calving, and social stresses in the herd at calving also predispose to abomasal displacement (Hultgran and Pehrson 1998).

**Box 1 Predisposing Factors for Abomasal Displacements**

- Abomasal atony—precede all abomasal displacements
- Excessive gas production within abomasum
- General predispositions
  - Deep abdomen
  - Female
  - Older multiparous cows (4–7 years of age)
- Factors related to parturition
  - Recent parturition/early lactation
  - Retained placenta
  - Dystocia
  - Twining
- Nutritional factors
  - High-concentrate diets (high VFA production)
  - Overfeeding in dry period
  - Low fiber intake during dry period
  - High body condition scores
- Energy and metabolic diseases
  - Ketosis
  - Insulin resistance
  - Hypocalcemia
  - Vitamin A deficiency
- Other concurrent diseases and conditions
  - Metritis
  - Mastitis
  - Lameness
- Breed predispositions (dairy cattle)
  - Holstein
  - Guernsey
  - Jersey

## 2.4 Inherited Factors

Inherited factors have been implicated to predispose certain animals and breeds to development of abomasal displacements (Parker Gaddis et al. 2014) with heritability estimates ranging from 0.03 to as high as 0.53 (Zerbin et al. 2015). Most cows with abomasal displacements come from high-producing dairy herds, and there are genetic correlations with milk production traits (Zerbin et al. 2015); however, a clinical correlation between milk production and displacements within individual cows has not been demonstrated (Rohrbach et al. 1999). Holstein-Friesian and the channel island breeds of cattle appear to be the most susceptible (Mather and Dedrick 1966; Doll et al. 2009). Cattle that are deep chested and those that have an abomasal outflow higher than the body of the abomasum are thought to be predisposed to development of displacements (Wittek et al. 2007). The position of the outflow in these individuals is thought to lead to excessive gas accumulation within the abomasum when motility is poor leading to buoyancy of the abomasum and subsequent displacement. The high percentage of Holstein cattle reported in the literature with displacements is possibly a reflection of general cattle population at the institutions reporting on DAs (Wallace 1975).

Possibly inherited differences in the intrinsic nervous system between animals predispose certain individuals to abomasal displacement. An in vitro study demonstrated that abomasal displacement is associated with malfunctions of the intrinsic nervous system of the abomasum with a predominance of inhibitory mechanisms over those that are excitatory (Geishauser et al. 1998).

# 3 Diagnosis and Clinical Signs

A thorough history and physical examination are usually all that is necessary to make an accurate diagnosis of a DA. Routine cow-side diagnostics can be used to support the diagnosis and formulate a prognosis. Examination should include a rectal examination to feel for an abomasum and rule out common competing differentials like cecal dilatation or excessive rumen gas. Checking for the presence of urine ketones should also be part of a complete physical examination on lactating dairy cows. Clinical signs associated with LDA and RDA are similar. Those associated with

RAV are more severe, often reflecting a state of shock.

Some animals with simple displacements are asymptomatic or show mild, nonspecific signs (Ingling et al. 1975). Frequently affected cattle are depressed (Espersen 1964), and most producers will generally notice that the cow drops in milk production and feed intake (Mather and Dedrick 1966; Grymer et al. 1982; Van Winden et al. 2003). Affected cows may have a preference for hay and forages over concentrate feeds. Colic is rarely a clinical sign, and cows may become dehydrated as the disease progresses (Espersen 1964). Cows that develop DAs are in a negative energy balance and have increased levels of ketones in the blood and urine (Wallace 1975). It is unclear whether the negative energy balance is a cause or effect of the DA, but a strong correlation exists between elevation of ketones and DAs; measuring blood or urine ketones can be a useful diagnostic tool (Mather and Dedrick 1966).

Physical examination of animals with simple displacements reveals a normal temperature pulse, temperature, and respiratory rate. Feces are normal to slightly loose. A sunken left paralumbar fossa with a slightly "sprung" left rib cage is a common finding in cows with an LDA (Mather and Dedrick 1966) due to an empty rumen and the mass effect of the abomasum cranially. Large LDAs may be visible as a "bubble" in the left paralumbar fossa, and rarely the abomasum is palpable per rectum (Mather and Dedrick 1966).

Auscultation of the viscus reveals "high-pitched tinkling reverberations" corresponding to borborygmal sounds (Mather and Dedrick 1966). Ruminations will be absent or diminished, and ruminal sounds will be difficult to impossible to auscultate with LDAs because the rumen is displaced medially by the abomasum. A hyperresonant "ping" will be elicited by auscultation and percussion over the displaced abomasum because of the high gas pressure within the viscus (Robertson 1965; Mather and Dedrick 1966). Outlining the margins of the "ping" will allow the examiner to determine the extent of the abomasum. It is important to keep in mind that auscultation of a "ping" is not pathognomonic for a DA; any gas-distended viscus that is in contact with the body wall may elicit a "ping" when percussed; differentials include cecal and duodenal gas for right-sided pings and rumen gas for left-sided pings. The location of the audible ping and the other physical exam findings are interpreted in relation to each other to arrive at a diagnosis. Pings from DAs are usually best auscultated on a line from the tuber coxae to the elbow. The ninth to 13th intercostal spaces (Robertson 1965) represent the cranial and caudal margins of the pings. Pings resulting from RAVs are higher pitch and larger diameter than those from RDAs or LDAs and may extend from the sixth or seventh intercostal space cranially to the right paralumbar fossa caudally (Habel and Smith 1981; Garry et al. 1988a, b; Perkins et al. 2017).

Cows with RAV are sicker, have more pain, and may have cardiovascular compromise compared to cows with simple displacements. Affected animals frequently have diarrhea and hypovolemia. A sign of abomasal mucosal compromise with ulceration, diarrhea may be fetid with melena in late stages (Espersen 1964). An abomasal volvulus will be under greater pressure than a simple displacement. Distension of the right flank may be apparent, and a large volvulus may be palpable per rectum (Espersen 1964; Garry et al. 1988a, b). A large abomasal volvulus will often push against the liver and caudal vena cava. This physical compression can lead to elevation in liver enzymes like alkaline phosphatase (ALP) and gamma-glutamyl transferase (GGT) (Constable et al. 1991).

Cows with DAs (more pronounced with RAV) will have compromise to abomasal outflow. This will lead to sequestration of HCl and other digestive enzymes in the abomasum and forestomachs. Systemic chloride decreases as the rumen chloride increases. The acid/base status shifts with the declining systemic chloride. The resultant classic metabolic profile for cattle with DAs is a hypochloremic, hypokalemic, metabolic alkalosis. The hypokalemia is due to a combination of the alkalemia, low feed intake with high amount of milk produced, hypovolemia, and hyperglycemia (Constable et al. 2013). The metabolic profile may be normal for patients with acute-onset, simple displacements, and the severity of these changes reflects the degree of impairment to abomasal outflow and abomasal perfusion. Metabolic changes will be most severe with abomasal volvulus. Cases of RAV with secondary ischemia of the abomasum combined with generalized dehydration and hypovolemic shock will have a lactic acidosis that masks the primary alkalosis (Garry et al. 1988a, b). Peritoneal fluid samples from patients with abomasal displacements had elevated lactate compared to normal cows suggesting compromised perfusion to the abomasum (Grosche et al. 2012).

An ultrasound may be used to visualize the abomasum and determine the location of the abomasum. However, ultrasound waves do not penetrate gas well, and DAs generally have a lot of gas within the lumen preventing good ultrasonic visualization of the structure.

If the veterinarian is unsure that the "ping" truly represents an LDA, the Liptak test can be used to support the diagnosis (Robertson 1965). Rarely performed, the Liptak test involves aspirating a small amount of fluid from the suspected abomasum. Measurement of the pH of the fluid can help to determine if the viscus is truly the abomasum or the rumen. Abomasal contents will be acidic with a pH between 2.0 and 4.1, while rumen contents will be closer to neutral (pH = 6.0–7.6) (Mather and Dedrick 1966). Another method to differentiate between rumen and abomasum is by forcing air through an esophageal tube. By blowing into the tube, audible bubbles will be heard by an examiner listening over the left paralumbar fossa if the rumen is in its normal anatomical position against the left body wall. If an animal has an LDA, the abomasum will push the rumen medially, and the ruminal bubbles will not be heard by external auscultation.

## 4 Medical Treatment and Prevention

Treatment of concurrent conditions, including atony of the abomasum, is an important part of treatment for abomasal displacements. Although medical therapy is seldom used alone for abomasal displacements, restoration of abomasal motility should expel gas and help it to return to its normal anatomical position (Trent 2004). Oral or systemic calcium to correct hypomotility due to hypocalcemia may be useful. Pro-motility agents such as parasympathomimetic agents can help to stimulate GI motility. Sometimes just promotion of appetite in an anorectic animal stimulates motility of the GI tract. Rumen transfaunation can help stimulate rumen motility and secondary abomasal motility. Dehydrated animals or animals that suffer from severe electrolyte imbalances may benefit from oral or systemic fluid therapy.

High-producing lactating dairy cows have a negative energy balance after freshening. Severe ketosis can ensue and lead to appetite suppression and GI atony predisposing to a DA. An abomasal displacement can further exacerbate the negative energy balance by decreasing appetite and nutrient assimilation. Mild ketosis may resolve following correction of the DA and an increase in appetite. Severe cases of ketosis should be treated. Intravenous dextrose, insulin therapy, oral niacin, and oral propylene glycol can be used in the treatment of ketosis.

Given the multifactorial nature of abomasal displacements, proper management of dairy cows, especially during the pre-fresh period and early lactation, is essential to prevent DAs. Nutrition and a properly formulated ration is likely the single most important herd-level factor, but other factors include cow comfort during the transition period, proper stocking density, adequate bunk space, access to water, proper stall design, and appropriate bedding (Caixeta et al. 2018). A consistent diet that is available to the cows at all times will reduce DAs. Feeding a total mixed ration will help to maintain dietary consistency and prevent sorting. High-fiber ingredients should be processed so that the longest particles are approximately 5 cm in length (Caixeta et al. 2018). Separating primiparous animals from multiparous animals will reduce the likelihood that primiparous animals will be displaced from the bunk (Caixeta et al. 2018). Over-conditioning of cattle, especially during the dry period, can result in increased cases of abomasal displacement (Furll et al. 1999). Regular body condition scoring of cattle and maintaining proper body condition score are important.

High-energy as well as energy-deficient diets can both predispose to DAs (Caixeta et al. 2018). Ruminal fiber helps to maintain a normal fiber mat and good rumen fill and stimulates rumen motility and salivary production; lack of adequate crude fiber in the diet is a risk factor for abomasal displacements (Hultgran and Pehrson 1998). A crude fiber content of >16% is recommended (Grymer et al. 1981). A Canadian study showed that feeding monensin 2 to 4 weeks prior to calving reduced the incidence of clinical ketosis and abomasal displacements post-calving (Duffield et al. 1999, 2002). Hypocalcemia is a risk factor for DAs, and prevention of hypocalcemia by feeding dry cows anionic diets and supplementing calcium in early lactation may be useful in some operations (Caixeta et al. 2018).

## 5 Surgical Treatment

Surgical correction of abomasal displacements is a commonly performed procedure in cattle (St Jean et al. 1987; Trent 1992; Baird and Harrison 2001). Several surgical techniques have been described, and no single technique is ideal for every situation. Abomasal displacements can be corrected with right flank, left flank, ventral midline, or paramedian techniques (Table 1). Flank approaches offer the advantage of allowing the patient to stand during the procedure and eliminate risks associated with recumbency and general anesthesia. Ventral or paramedian fixation offers the advantage of a more secure fixation and absence of a visible scar in the flank region. The best technique depends on the type of abomasal displacement, pregnancy status or stage of lactation, the use and general health of the animal, the chronicity of the displacement and presence of adhesions, economics, facilities, equipment, and surgeon preference. All of the procedures involve correction of the displacement, followed by a technique to prevent redisplacement of the abomasum. Open surgical procedures include right and left flank abomasopexy, right flank omentopexy, paramedian abomasopexy, and their combinations.

Decompression of the abomasum aids in replacing the abomasum. Decompression can be performed with a needle connected to a long section of sterile tubing. Active suction speeds decompression but is not required.

### 5.1 Right Flank Techniques

The right flank laparotomy is a common surgical approach for fixation of abomasal displacements and is the most common technique employed by the author in his practice. Right flank techniques can be used to correct all three types of abomasal displacements (RDA, LDA, RAV). Right flank techniques also offer a superior abdominal exploratory and the ability to work with other right-sided structures like the cecum, intestines, and liver.

Through a right flank, the surgeon can reach across the abdomen and under the rumen and return a left displaced abomasum to the right side (Gabel and Heath 1969). In cases of RDA, the surgeon can reach on the medial side of the abomasum and push the abomasum ventrally; recall that the abomasum flips in a counterclockwise direction as viewed from the rear of the cow, causing the greater curvature of the abomasum (which normally is positioned ventrally) to be positioned dorsally to the body of the abomasum. To correct an RDA, the author prefers placing his left arm medially to the dilated abomasum and pushing the greater curvature ventrally. Decompression of the abomasum aids in replacement. After the abomasum is decompressed, the greater curvature of the abomasum can normally be grasped which can aid in pushing the greater curvature of the abomasum ventrally. Often bub-

**Table 1** Advantages and disadvantages of common procedures used to correct abomasal displacements

| Procedure | Advantages | Disadvantages |
|---|---|---|
| Right flank omentopexy | Allows for complete abdominal exploratory and fixation of multiple types of displacements (LDA/RDA/RAV). Can be done standing with minimal assistance | May be challenging in late pregnant cattle. Omentopexy can stretch or tear; adhesion less robust than with abomasopexy |
| Right flank abomasopexy ("pyloropexy") | Same as for right flank omentopexy but stronger reported adhesion formation. Can be combined with omentopexy | Challenging in late pregnant cattle. Increased risk of compromise to abomasal wall. Suture close to pylorus can cause pyloric stricture |
| Left flank abomasopexy | Standing surgery. Easier to perform than right flank techniques in late pregnant cattle. Offers ability to work with LDA with abomasal adhesions to left body wall | Left flank does not allow for complete abdominal exploratory due to blocking effect of rumen. Requires assistance to pull suture. Can only correct LDA (not option for RDA/RAV). Abomasal fistulas possible if sutures left intact |
| Paramedian abomasopexy | Can correct all types of displacements. Superior abomasal adhesions. Cosmesis; no scar in flank region | Requires assistance to restrain cow and place on back. Possible compromise from cow in dorsal recumbency for procedure |
| Blind suture (roll and tack) | Minimally invasive procedure; abdomen not opened. Procedure fast and relatively inexpensive. Cosmesis; no scar in flank region | Does not allow for visualization of abdominal contents or abdominal exploratory. Uncertainty with blind suture placed in abdomen; penetration of incorrect structure possible. Only option for LDA (not option for RDA/RAV). Assistance to roll cow typically required. Abomasal fistula possible if sutures left intact |
| Roll and toggle | Similar to blind suture. Outcome and prognosis similar to open surgical techniques. Verification of abomasal penetration with cannula by analysis of gas/fluid. Cosmesis; no scar in flank region | Does not allow for visualization of abdominal contents or abdominal exploratory. Only option for LDA (not option for RDA/RAV). Assistance to roll cow typically required. Abomasal fistula possible if sutures left intact |
| Laparoscopic abomasopexy | Minimally invasive procedure; abdomen not opened. Laparoscopic visualization of abomasum and other abdominal viscera while sutures are placed | Generally only option for LDA. Assistance to roll cow typically required. Specialized equipment and expertise required. Abomasal fistula possible if sutures left intact |

bles of abomasal gas can be felt as it escapes the abomasum and passes through the duodenum.

To replace an RAV, the volvulus is turned back into a simple RDA by flipping the abomasum in a clockwise direction around its omental twist as viewed from the top. Then the RDA is corrected as discussed previously.

When the abomasum is replaced into its normal position, the greater curvature should be positioned along the right ventral abdomen and the pylorus situated caudodorsal to the abomasal body. A few techniques have been described for securing the abomasum and preventing redisplacement. The omentopexy is the most common. Also, suture can be used to attach the abomasum, usually close to the pylorus, to the right body wall. This technique is known as an abomasopexy, pyloroantropexy, or simply a pyloropexy (Fig. 2).

The bovine greater omentum is robust and attaches to the greater curvature of the abomasum and can be used to secure the abomasum to the right body wall. In the most simplistic omentopexy, large bites of greater omentum are secured into the incisional closure, promoting an adhesion between the omentum and the right body wall. Mattress sutures between omentum and body wall can be placed on the cranial and caudal

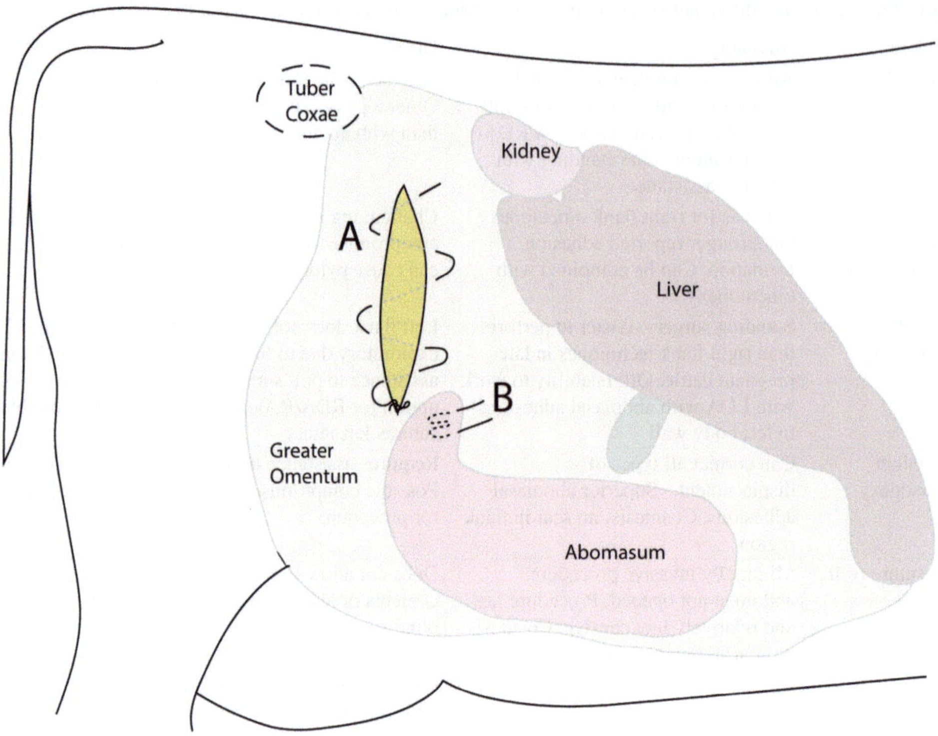


**Fig. 2** Right flank techniques for securing the abomasum following DA correction. Prior to fixation, the abomasum is pulled caudally and elevated slightly toward the flank incision. (**A**) An omentopexy is performed by incorporating large bites of omentum into the incisional closure. (**B**) An abomasopexy can provide for a more secure adhesion by suturing the seromuscular layer of the abomasum to the body wall. This is commonly called a "pyloropexy" due to the proximity of pylorus

aspects of the incision to augment the omentopexy or initially hold the omentum in place while the incision is closed.

The right flank abomasopexy is commonly called the pyloropexy because the abomasum is sutured to the right body wall approximately 3–5 cm orad to the pylorus. Passing suture directly through the pylorus should be avoided directly because of risks of pyloric inflammation and stricture. When performing a right flank abomasopexy, care should be taken to suture the seromuscular layer of the abomasum and avoid penetrating the abomasal lumen. The abomasum should be positioned so that the pylorus is near the caudal aspect of the flank incision. When performed correctly, the abomasopexy offers more security than the omentopexy because the omentum is friable tissue that can easily tear or stretch. The author advocates inclusion of an omentopexy during incisional closure when performing abomasopexy as an added level of security.

## 5.2 Left Flank Techniques

Overall, the left flank offers less versatility than the right flank because the rumen normally occupies the left hemiabdomen and blocks access to much of the other abdominal viscera. An LDA can be corrected via a left flank approach; however, the left does not allow access to correct RDA and RAV. Advantages of a left flank approach include correction of an LDA in late pregnant cows where the gravid uterus interferes with replacement of the abomasum from a right flank. Chronic LDA or LDA with concurrent abomasal ulcers can be associated with adhesions of the abomasum to the left body wall. A left flank approach is superior if adhesions need to be broken down prior to abomasal replacement. Adhesions of the abomasum are broken down with care to avoid creating disseminated intra-abdominal contamination if the adhesions are sealing perforating abomasal ulcers. Decompression of the abomasum facilitates replacement from the left flank.

One of the first techniques used to correct LDAs and prevent their reoccurrence was a rumenopexy (Mather and Dedrick 1966). This technique involved correction of the LDA from the left flank and suturing of the rumen to the ventral abdomen ablating the potential space between the rumen and ventral body wall, preventing the abomasum from passing back to the left side. This technique is not viable for right displacements and does not prevent right displacements from occurring. It is also possible that the abomasum might slip caudal to the rumenopexy site leading to reoccurrence of the LDA. The author is not aware of this procedure being performed any longer and is mentioned here for historical completeness.

The left flank abomasopexy is commonly performed by the author in cases of late pregnant cows with an LDA. Requirements include the ability to reach the right ventral abdomen from the left flank, and the this approach requires an assistant who can pull the slack out of the suture as the surgeon replaces the abomasum by pushing it under the rumen and back to the right hemiabdomen.

To perform a left flank abomasopexy, the LDA is visualized through the left flank incision, and multiple suture bites are taken through the seromuscular layer of the abomasum. Suture is placed in the greater curvature of the abomasum, and the suture ends are left long enough that they can reach the right ventral abdomen. Using suture that comes on a cassette (reel) will allow the suture to be cut extra-long. The surgeon takes each suture end and passes it through the right ventral body wall guided by a non-sterile assistant. The suture ends should exit the ventral abdomen a few centimeters caudal to the xyphoid and between midline and the right milk vein. Care should be taken to not pass suture through the milk vein. The suture ends should exit the body 2–3 cm from each other oriented in a cranial-caudal direction. The assistant applies ventral tension on the suture as the surgeon pushes the abomasum ventrally and to the right. Once the greater curvature of the abomasum is in contact with the ventral body wall, the assistant ties the suture externally. The surgeon should check and make sure that no other viscera were entrapped in the abomasopexy (Fig. 3).

A strong adhesion between the abomasum and the right ventral body wall should result. Ideally partial thickness sutures (seromuscular layer) should be placed within the abomasum, but penetration of the lumen is likely. Sutures passing from the skin surface into the lumen of the abomasum can result in an abomasal fistula, especially if the sutures are left intact. Abomasopexy sutures should be removed in 2 to 4 weeks which is sufficient time to form a functional abomasal adhesion.

A left flank omentopexy can be performed in a similar manner to the aforementioned left flank abomasopexy. Suture is passed through the greater omentum near its attachment to the greater curvature, and the ends are passed ventrally to exit the right paramedian. The suture ends are tied by a non-sterile assistant in a similar manner to the left flank abomasopexy.

### 5.3 Ventral Approaches

An abomasopexy can also be performed from a right (Trent 2004) or left (Lee et al. 2002) paramedian approach. With this approach, the cow is placed in dorsal recumbency. An incision is created parallel to and 3–4 cm from midline, extending caudally from a point 4–5 cm caudal to the xyphoid. The incision should be approximately 15 to 20 cm in length (Trent 2004).

With the more commonly employed right paramedian approach, following reposition of the abomasum, the seromuscular layer of the greater curvature of abomasum is sutured to the peritoneum and the internal rectus fascia (Trent 2004). The seromuscular layer of the abomasum may be incorporated into the incisional closure or sutured

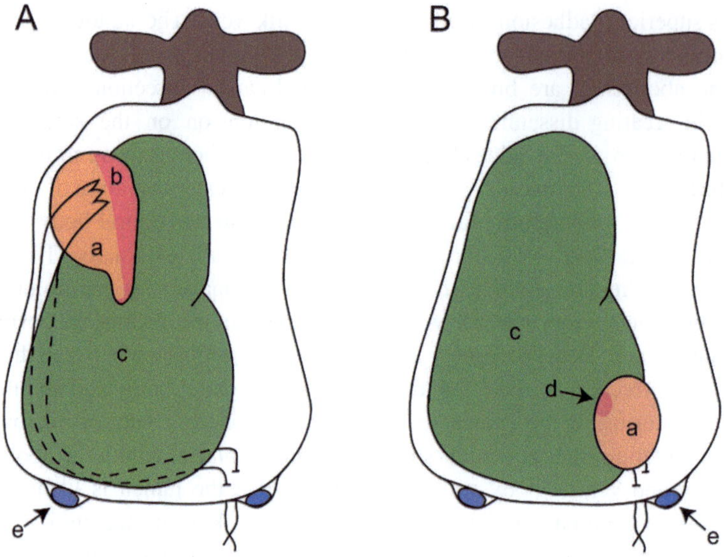


**Fig. 3** (**A**) Cross section of bovine abdomen with an LDA. A left flank abomasopexy is being performed. Suture is placed in the greater curvature of the abomasum, while the suture ends are left long and passed under the rumen to exit the ventral right paramedian. An assistant (not pictured) pulls the slack from the suture, while the surgeon pushes the abomasum down. (**B**) The abomasum is replaced and is pexied to the right ventral abdomen. (a) Abomasum covered by greater omentum, (b) portion of abomasum not covered by greater omentum, (c) rumen, (d) pyloric area of the abomasum, and (e) milk veins

to the body wall remote from the incision. Advantages of this approach include a strong adhesion of the abomasum, as well as the ability to work with unwanted adhesions. With the left paramedian abomasopexy, a 15 cm section near the reticulo-abomasal junction, along with the origin of attachment of the greater omentum, is sutured to the incision line of the internal rectus sheath (Lee et al. 2002). Complications with dorsal techniques include those associated with maintaining the cow in dorsal recumbency during surgery, abomasal fistula formation, and herniation of abomasal leaves.

## 6 Minimally Invasive Methods of DA Fixation

There are multiple minimally invasive techniques that have been described for correction of left abomasal displacements. Minimally invasive techniques are faster to perform compared to open techniques, animals generally experience a faster recovery, and they have a decreased risk of surgical complications. These techniques are less expensive and offer producers an option when the value of the animal does not warrant open surgery. Disadvantages of minimally invasive techniques include the inability to conduct an abdominal exploratory, an inability to visualize or work with other abdominal structures and they are only advocated for correction of LDAs. Minimally invasive techniques involve rolling the cow to replace the LDA. However, correcting the LDA by rolling without abomasal fixation is associated with a high rate of redisplacement (Uyanik et al. 1982), so correction via rolling is typically combined with the blind suture (Hull 1972) or toggle pin suture (TPS) (Grymer and Sterner 1982) techniques. Laparoscopic techniques have also been described which are similar in theory but offer the added advantage of visualization of the abomasum. Since their original descriptions, many practices use minimally invasive techniques as the primary technique for correction of LDAs (Newman et al. 2008).

## 6.1 Toggle Pin Suture (TPS)

The toggle pin suture technique (Grymer and Sterner 1982) has largely replaced the blind suture technique as it has an advantage of being able to verify placement of the toggle into the abomasum by examination of the gas and fluid emitted from the cannula. Today the TPS procedure is the most widely employed minimally invasive procedure for LDA correction. It does require the specialized toggle equipment (trocar/cannula and toggle sutures). The equipment is relatively inexpensive and available through Jorgensen Laboratories (JorVet™) (Fig. 4).

If placing the cow on her back will cause significant distress, sedation is recommended, but the procedure may be accomplished without sedation in quiet cattle. Two or three persons are recommended to aid in casting the cow. Adequate space is necessary in order to safely cast the cow and roll her. The cow is laid down on her right side and rolled up onto her back and slightly tilted onto her left. Auscultation is used to verify that the abomasum has been replaced by following the "ping" to the ventral abdomen. The area from the xyphoid caudally 15–20 cm and from midline to the right approximately 10 cm should be aseptically prepared. The toggle/cannula system should be assembled, and the first toggle should be placed approximately 5–7 cm to the right of midline and approximately 10–15 cm caudal to the xyphoid. Note this area should be located in a nondependent location when placing the suture due to the abomasum floating to the ventral body wall. Abomasal gas should escape the cannula after abomasal penetration. This allows the surgeon to verify correct placement by its characteristic smell or measurement of fluid pH. The bar (toggle) suture is placed through the cannula, and excessive gas should not be allowed to escape prior to placement of the second suture. A hemostat is used to secure the first suture until the second suture can be placed. The second toggle suture is placed in a similar manner 4–7 cm cranial to the first (approximately halfway between the first suture and the xyphoid). The abomasum can be allowed to deflate after placement of the second toggle suture. It is recommended to tie the sutures loose to allow for shifting of the abomasum after the cow stands. A

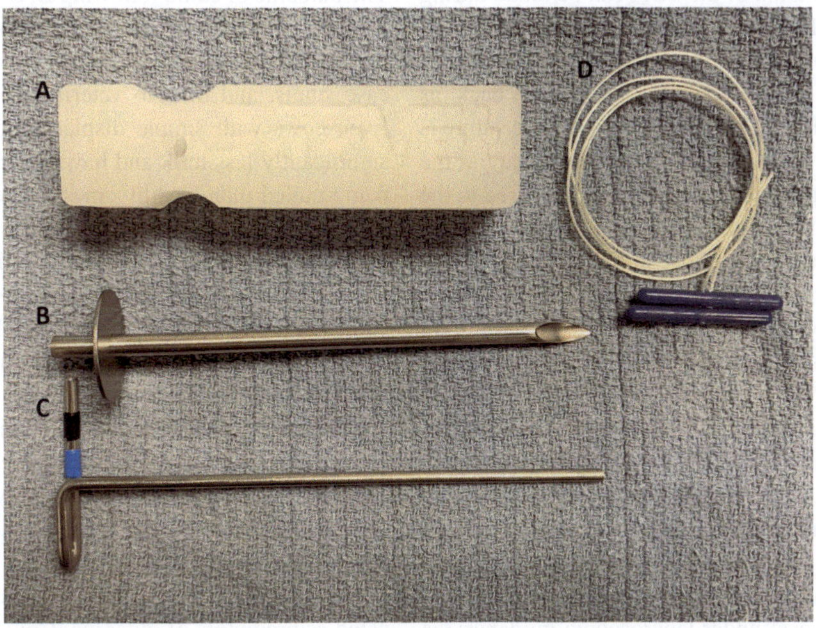


**Fig. 4** Equipment for placing a toggle pin suture. Nylon handle (**A**), trocar/cannula (**B**), suture push rod (**C**), bar/toggle suture (**D**)

stent or manufactured button can be used to distribute force of the suture and prevent it from cutting through the skin (Sterner 2002). The TPS sutures should be cut after an adhesion forms to avoid abomasal fistulation. A complete description of the procedure is given on the Grymer/Sterner toggle suture website (http://www.ldatogglesuture.com/).

## 6.2 Laparoscopic Methods

Most laparoscopic abomasopexy procedures are modifications of the TPS procedure. Advantages of the laparoscopic procedures are that they allow for visualization of the abomasum avoiding complications with misguided suture placements. They do require specialized equipment and surgical expertise in their use.

The first laparoscopic method described was the two-step laparoscopic abomasopexy developed by Janowitz (1998). The first step involves placement of a toggle suture with the cow standing via the left paralumbar fossa with laparoscopic guidance. Successful completion of the first step requires that the cow has an LDA and the confirmation of the LDA can be made during this step. The cow can also be evaluated for the presence of intra-abdominal adhesions or peritonitis. The cow is then rolled to her back as described for the TPS procedure, and the suture is recovered under laparoscopic guidance in the second step of the procedure and is secured to the ventral abdomen as for the TPS.

The one-step laparoscopic abomasopexy for LDA correction (Barisani 2004; Christiansen 2004) involves placement of the toggle suture in the abomasum under laparoscopic guidance similar to the two-step procedure. A long instrument called a spieker tool is used to direct the suture to the ventral abdomen, and a lancet contained within the spieker penetrates the ventral abdominal wall and drives the suture through.

An alternative one-step procedure developed by Newman et al. (2005) involves placement and suture retrieval under laparoscopic guidance via the right paramedian area with the cow in dorsal recumbency. The abomasum is subsequently deflated, slack removed from the suture, and the suture ends tied similar to the TPS procedure. A similar method has been successfully used to correct right abomasal displacements under laparoscopic guidance. This technique is not viable for correction of RAV (Newman et al. 2008).

A ventral laparoscopic abomasopexy (Mulon et al. 2006) where the abomasum is sutured to the ventral abdominal wall with the cow in dorsal recumbency has also been described. This technique requires greater technical expertise with laparoscopic procedures but has an advantage of being used for correction of LDA and RDA and to perform a prophylactic abomasopexy if no displacement is present (Mulon et al. 2006). Cows undergoing ventral laparoscopic abomasopexy have been shown to have a similar 305-day milk production to matched controls without displacement but did have a longer calving interval and a 1.5 times greater risk of being culled from the herd within the next 3 years compared to matched controls (Jorritsma et al. 2008).

# 7 Outcome and Prognosis

The outcome of cattle following DAs is multifactorial and depends on the type of displacement, the overall condition of the cow, concurrent disease status, and time to veterinary innervation. Even cows with simple displacements produce significantly less milk and have a greater risk of being culled than healthy cows (Melendez et al. 2017). Studies comparing the various techniques for correction have largely failed to demonstrate a superior technique, and most have similar outcomes in terms of survival in the herd and return to milk production. The prognosis for cows with simple displacements is generally good to excellent following correction, and the animal can be expected to return to productivity once the DA is corrected (Grymer et al. 1997; Meylan 1999). Cows with abomasal volvulus have a guarded to poor prognosis (Meylan 1999).

Many preoperative factors have been investigated as predictors of outcome. Cows with LDA and concurrent diarrhea had a death rate twice as great as cows that did not have diarrhea (Wallace

1975, 1976). Concurrent fatty liver disease reduced prognosis in LDA cows in another study; cows with severe fatty liver had elevated liver triglycerides and low concentrations of calcium, potassium, and magnesium and had a guarded prognosis (Kalaitzakis et al. 2010). One study showed that cows which died following DA surgery had significantly higher levels of preoperative serum glucose compared to surviving cows (Markiewicz et al. 2009). A heart rate ≥90 BPM and a serum BUN ≥10 mmol/L, a potassium of ≤3.3 mmol/L, and a chloride of ≤85 mmol/L have been shown to be indicators for a nonproductive outcome (Meylan 1999).

Most studies have found the various surgical procedures to be comparable in terms of return to function after the procedure (Gabel and Heath 1969). A study comparing right paramedian abomasopexy and right flank omentopexy for correction of LDAs showed comparable outcomes with cows undergoing abomasopexy having a slightly higher milk yield at 1 month post-surgery (Fubini et al. 1992). A study comparing the right flank omentopexy to the "pyloro-omentopexy" (right flank abomaso-omentopexy) showed to have similar outcomes in regard to herd retention and reproductive performance. However, the addition of the abomasopexy seemed to reduce the rate of recurrence of the LDA (Baird et al. 2017). The roll and toggle technique has also been shown to have comparable success rates with surgical laparotomy when performed by a veterinarian (Sterner et al. 2008).

The incidence of specific complications following abomasal fixation is low. However, a fairly newly recognized complication is a volvulus of the duodenal sigmoid flexure (Vogel et al. 2012). Cows with duodenal sigmoid flexure volvulus have a focal, dorsal right-sided "ping" and succussable fluid combined with a metabolic profile indicative of a duodenal obstruction. These animals require right-sided surgical innervation to correct the volvulus. Prognosis following correction is fair to good (Vogel et al. 2012).

Cows with RAV can have variable degrees of compromise to the abomasal wall and have a more varied outcome, doing poorer in general than cows with simple displacements (Fubini et al. 1991; Constable et al. 1992b). Cows with abomasal volvulus have been shown to have increased risk of gastric wall damage, peritonitis, and vagal indigestion (Sattler et al. 2000). Predicting the outcome, especially presurgical outcome, of cows with RAV has been a long-standing research topic. Cows with anion gaps ≥30 mEq/L (Garry et al. 1988a, b; Fubini et al. 1991) or systemic lactate ≥6 mmol/L (Boulay et al. 2014) were indicative of a poor prognosis. A prospective study of 80 cows with abomasal volvulus showed that preoperative heart rate, hydration, period of inappetence, and serum ALP were the best prognostic indicators. Postoperatively productivity was associated with the preoperative findings of a significantly lower heart rate, less dehydration, lower serum ALP, more frequent rumen contractions, and shorter periods of inappetence (Constable et al. 1991). Hemostatic dysfunction consistent with disseminated intravascular coagulation (DIC), including a prolonged thrombin time, prothrombin time, and activated partial thromboplastin time and increased fibrinogen and D-dimer along with decreased platelets, is a risk factor for mortality and was found in cows with abomasal displacements (Irmak and Turgut 2005; Sobiech et al. 2008; Maden et al. 2018). Cows requiring fluid decompression of the abomasum or those having a necrotic appearance to their mucosa had a poorer prognosis than cows that only required gas decompression (Fubini et al. 1991). Cows that have a metabolic acidosis (lower base excess) likely have abomasal necrosis and a poorer prognosis (Simpson et al. 1985; Fubini et al. 1991). Cows with RAV have higher abomasal luminal pressures compared to cows with simple displacements, and abomasal luminal pressure was significantly correlated with outcome in one study with 16 mmHg being the cutoff value that optimized the distribution of cattle into productive and nonproductive groups (Constable et al. 1992b).

Fewer studies have investigated preoperative prognostic factors for cows with LDAs than RDAs. The presence of ketosis preoperatively has been shown to have a protective effect of cows with LDAs; cows with low β-hydroxybutyrate (BHB) have a greater risk for premature removal from the herd possibly because cows with low BHB may have lower milk productions (Croushore et al. 2013; Reynen et al. 2015). Higher serum concentrations of non-esterified fatty acids and magnesium also had a protective effect against culling for cows with LDA (Reynen et al. 2015).

Much data suggests that minimally invasive techniques can be associated with a successful outcome similar to surgical techniques and can be performed faster and with less expense compared to open surgical techniques for uncomplicated LDAs (Sterner and Grymer 1982; Kelton et al. 1988; Grymer et al. 1997; Nottebrock and Frerking 1997; Seeger et al. 2006; Roy et al. 2008; Zadnik and Lombar 2011). Minimally invasive techniques may restore abomasal motility faster and result in a faster return to milk production (Wittek et al. 2009). Most complications associated with the minimally invasive techniques result from the inability to fully visualize the abomasum prior to fixation or those related to maintaining the cow in dorsal recumbency. The minimally invasive techniques involve rolling the cow to replace the LDA back to the right side, and induction of an RDA is possible following this technique; some research suggests that some form of an RDA may occur after rolling in approximately 44% of cases (Gonzalez-Martin et al. 2019). Limitations of the minimally invasive techniques include an inability to do an abdominal exploratory, correct RAVs, or correct displacements with adhesions.

## References


Baird AN, Harrison S (2001) Surgical treatment of left displaced abomasum. Compend Cont Educ Pract Vet 23(10):S102–S108

Baird AN, Mohamed A, Moore GE, Hawkins JF (2017) Comparison of omentopexy versus pyloro-omentopexy for treatment of left abomasal displacement in dairy cows: 87 cases (2001-2005). J Am Vet Med Assoc 251(10):1182–1187

Barisani C (2004) Evoluzione della tecnica di Janowitz per la risoluzione della dislocazione abomasale sinistra secondo Barisani. Summa 5:35–39

Begg H (1950) Diseases of the stomach of the adult ruminant. Vet Rec 62(51):797–808

Begg H, Whiteford WA (1956) Displacement of the abomasum in the cow. Vet Record 68:122–124

Boulay G, Francoz D, Dore E, Dufour S, Veillette M, Badillo M, Belanger AM, Buczinski S (2014) Preoperative cow-side lactatemia measurement predicts negative outcome in Holstein dairy cattle with right abomasal disorders. J Dairy Sci 97(1):212–221

Caixeta LS, Herman JA, Johnson GW, McArt JAA (2018) Herd-level monitoring and prevention of displaced abomasum in dairy cattle. Vet Clin North Am Food Anim Pract 34(1):83–99

Carougeau, Preston (1898) Torsion de la cailette chez un veau. J Med Vet 2:340

Christiansen K (2004) Laparoskopisch kontrollierte Operation des nach links verlagerten Labmagens (Janowitz-Operation) ohne Ablegen des Patienten. Tierärztliche Praxis Ausgabe G: Großtiere/Nutztiere 32(02):118–121

Constable PD, St Jean G, Hull BL, Rings DM, Hoffsis GF (1991) Preoperative prognostic indicators in cattle with abomasal volvulus. J Am Vet Med Assoc 198(12):2077–2085

Constable PD, Miller GY, Hoffsis GF, Hull BL, Rings DM (1992a) Risk factors for abomasal volvulus and left abomasal displacement in cattle. Am J Vet Res 53(7):1184–1192

Constable PD, St-Jean G, Koenig GR, Hull BL, Rings DM (1992b) Abomasal luminal pressure in cattle with abomasal volvulus or left displaced abomasum. J Am Vet Med Assoc 201(10):1564–1568

Constable P, Grunberg W, Staufenbiel R, Stampfli HR (2013) Clinicopathologic variables associated with hypokalemia in lactating dairy cows with abomasal displacement or volvulus. J Am Vet Med Assoc 242(6):826–835

Croushore WS Jr, Ospina PA, Welch DC, Zawisza DJ, Nydam DV (2013) Association between beta-hydroxybutyrate concentration at surgery for correction of left-displaced abomasum in dairy cows and removal from the herd after surgery. J Am Vet Med Assoc 243(9):1329–1333

Doll K, Sickinger M, Seeger T (2009) New aspects in the pathogenesis of abomasal displacement. Vet J 181(2):90–96

Duffield TF, Leslie KE, Sandals D, Lissemore K, McBride BW, Lumsden JH, Dick P, Bagg R (1999) Effect of a monensin-controlled release capsule on cow health and reproductive performance. J Dairy Sci 82(11):2377–2384

Duffield T, Bagg R, DesCoteaux L, Bouchard E, Brodeur M, DuTremblay D, Keefe G, LeBlanc S, Dick P (2002) Prepartum monensin for the reduction of energy associated disease in postpartum dairy cows. J Dairy Sci 85(2):397–405

Dyce KM, Sack WO, Wensing CJG (2010) The abomasum. Textbook of veterinary anatomy. Saunders Elsevier, St. Louis, pp 688–690

Espersen G (1964) Dilatation and displacement of the abomasum to the right flank, and dilatation and dislocation of the caecum. Vet Rec 76(49):1423–1428
Frazee LS (1984) Torsion of the abomasum in a one month old calf. Can Vet J 25(7):293–295
Fubini SL, Grohn YT, Smith DF (1991) Right displacement of the abomasum and abomasal volvulus in dairy cows: 458 cases (1980-1987). J Am Vet Med Assoc 198(3):460–464
Fubini SL, Ducharme NG, Erb HN, Sheils RL (1992) A comparison in 101 dairy cows of right paralumbar fossa omentopexy and right paramedian abomasopexy for treatment of left displacement of the abomasum. Can Vet J 33(5):318–324
Furll M, Dabbagh MN, Jakel L (1999) Body condition and dislocated abomasum: comparative investigations into back fat thickness and additional criteria in cattle. Dtsch Tierarztl Wochenschr 106(1):5–9
Gabel AA, Heath RB (1969) Correction and right-sided omentopexy in treatment of left-sided displacement of the abomasum in dairy cattle. J Am Vet Med Assoc 155(4):632–641
Garry F, Hull BL, Rings DM, Hoffsis G (1988a) Comparison of naturally occurring proximal duodenal obstruction and abomasal volvulus in dairy cattle. Vet Surg 17(4):226–233
Garry FB, Hull BL, Rings DM, Kersting K, Hoffsis GF (1988b) Prognostic value of anion gap calculation in cattle with abomasal volvulus: 58 cases (1980-1985). J Am Vet Med Assoc 192(8):1107–1112
Geishauser T (1995) Abomasal displacement in the bovine--a review on character, occurrence, aetiology and pathogenesis. Zentralbl Veterinarmed A 42(4):229–251
Geishauser T, Reiche D, Schemann M (1998) In vitro motility disorders associated with displaced abomasum in dairy cows. Neurogastroenterol Motil 10(5):395–401
Gonzalez-Martin JV, Perez-Villalobos N, Baumgartner W, Astiz S (2019) An investigation into the development of right displaced abomasum by rolling 268 dairy cows with left displaced abomasum. J Dairy Sci 102(12):11268–11279
Grosche A, Furll M, Wittek T (2012) Peritoneal fluid analysis in dairy cows with left displaced abomasum and abomasal volvulus. Vet Rec 170(16):413
Grymer J, Sterner KE (1982) Percutaneous fixation of left displaced abomasum, using a bar suture. J Am Vet Med Assoc 180(12):1458–1461
Grymer J, Hesselholt M, Willeberg P (1981) Feed composition and left abomasal displacement in dairy cattle. A case-control study. Nord Vet Med 33(6–8):306–309
Grymer J, Willeberg P, Hesselholt M (1982) Milk production and left displaced abomasum: cause and effect relationships. Nord Vet Med 34(11):412–415
Grymer J, Bartlett PC, Houe H, Sterner KE (1997) One-year survival of cows with left displacement of the abomasum corrected with the role-and-toggle procedure. Bov Pract:80–82
Habel RE, Smith DF (1981) Volvulus of the bovine abomasum and omasum. J Am Vet Med Assoc 179(5):447–455
Hull BL (1971) Hypocalcemia as a contributing causative factor in the development of abomasal displacement. Masters of Science, Iowa State University
Hull BL (1972) Closed suturing technique for correction of left abomasal displacement. Iowa State Univ Vet 34(3):142–144
Hultgran J, Pehrson B (1998) Risk factors of displaced abomasum in Wisconsin dairy herds. Bov Pract:56–57
Ichijo T, Matsuta K, Takahata S, Sato S (2005) Incidence and clinical findings of displaced abomasum in fattened cattle. Jpn J Vet Clin (Japan)
Ichijo T, Satoh H, Yoshida Y, Murayama I, Kikuchi T, Sato S (2014) Characteristics of fluid composition of left displaced abomasum in beef cattle fed high-starch diets. J Vet Med Sci 76(8):1165–1167
Ingling AL, Albert TF, Schueler RL (1975) Left displacement of the abomasum in a clinically normal cow. J Am Vet Med Assoc 166(6):601–603
Irmak K, Turgut K (2005) Disseminated intravascular coagulation in cattle with abomasal displacement. Vet Res Commun 29(1):61–68
Itoh M, Sasaki N, Kawamoto S, Yamada H, Inokuma H (2011) A mechanism of excessive accumulation of abomasal gas in vagotomized cattle determined using fluoroscopy. J Vet Med Sci 73(5):567–571
Janowitz H (1998) Laparoscopic reposition and fixation of the left displaced abomasum in cattle. Tierarztl Prax Ausg G Grosstiere Nutztiere 26(6):308–313
Jorritsma R, Westerlaan B, Bierma MP, Frankena K (2008) Milk yield and survival of Holstein-Friesian dairy cattle after laparoscopic correction of left-displaced abomasum. Vet Rec 162(23):743–746
Kalaitzakis E, Panousis N, Roubies N, Kaldrymidou E, Karatzias H (2010) Macromineral status of dairy cows with concurrent left abomasal displacement and fatty liver. N Z Vet J 58(6):307–311
Kelton DF, Garcia J, Guard CL, Dinsmore RP, Powers PM, Smith MC, Stehman S, Ralston N, White ME (1988) Bar suture (toggle pin) vs open surgical abomasopexy for treatment of left displaced abomasum in dairy cattle. J Am Vet Med Assoc 193(5):557–559
Kuiper R (1991) Abomasal diseases. Bov Pract:111–117
Kumper H (1995) Right displacement of the abomasum in cattle. 1. Pathogenesis, clinical course and prognosis. Tierarztl Prax 23(4):351–359
Lagerlöf N (1925) Ett fall av kongenital (?) lägeförändring och kronisk löpmagsutvidgning hos nöt med ett 50-tal sår å löpmagens slemhinna. Swedish, Skandinavisk Veterinär-tidskrift för bakteriologi, patologi samt kött- och mjölkhygien 15(1):71–98
Lagerlof NP (1929) Investigations of the topography of abdominal organs in cattle: and some clinical observations and remarks. Almqvist & Wiksells, p 253
Lee I, Yamagishi N, Oboshi K, Yamada H (2002) Left paramedian abomasopexy in cattle. J Vet Sci 3(1):59–60
Liang D, Arnold LM, Stowe CJ, Harmon RJ, Bewley JM (2017) Estimating US dairy clinical disease

costs with a stochastic simulation model. J Dairy Sci 100(2):1472–1486
Loje K (1949) Torsio abomasi in cattle with special reference to its diagnosis and treatment. Vet Rec 61:179
Maden M, Yildiz R, Col R, Arican M, Ider M, Garip M, Tras B (2018) The evaluation of hemostatic dysfunction and disseminated intravascular coagulation in dairy cows with abomasal displacement. Pol J Vet Sci 21(4):769–778
Madison JB, Troutt HF (1988) Effects of hypocalcaemia on abomasal motility. Res Vet Sci 44(2):264–266
Markiewicz H, Gehrke M, Malinowski E (2009) Selected biochemical blood compounds in cows with abomasum displacement. Pol J Vet Sci 12(4):515–518
Massey C, Wang C, Donovan G, Beede D (1993) Hypocalcemia at parturition as a risk factor for left displacement of the abomasum in dairy cows. J Am Vet Med Assoc 203(6):852–853
Mather MF, Dedrick RS (1966) Displacement of the abomasum. Cornell Vet 56(3):323–344
Melendez P, Romero C, Pithua P, Marin MP, Pinedo P, Duchens M (2017) Retrospective evaluation of milk production and culling risk following either surgical, toggle-pin suture or conservative treatment of left displaced abomasum in Chilean dairy cows. N Z Vet J 65(6):292–296
Meylan M (1999) Prognostic indicators in cattle with right-sided displacement of the abomasum and abomasal volvulus. Schweiz Arch Tierheilkd 141(9):413–418
Meymandi PA, Bigham A, Roushan GAA (2008) Scientific report: displacement of the abomasum to the left side and pyloric obstruction in a goat
Mulon PY, Babkine M, Desrochers A (2006) Ventral laparoscopic abomasopexy in 18 cattle with displaced abomasum. Vet Surg 35(4):347–355
Muylle E, Van Den Hende C, Oyaert W (1989) Spijsverteringsstoornis sen en metabole afwijkingen bij hoog productieve runderen. Mededelingen van het commite voor de studie van ziekten en pathologische afwijkingen bij het rund. I. WONL Belgie (Communications from the Committee for the Study of Diseases and Pathological Abnormalities in Bovine. I. WONL Belgium)
Newman KD, Anderson DE, Silveira F (2005) One-step laparoscopic abomasopexy for correction of left-sided displacement of the abomasum in dairy cows. J Am Vet Med Assoc 227(7):1142–1147, 1090
Newman KD, Harvey D, Roy JP (2008) Minimally invasive field abomasopexy techniques for correction and fixation of left displacement of the abomasum in dairy cows. Vet Clin North Am Food Anim Pract 24(2):359–382. viii
Nottebrock A, Frerking H (1997) Percutaneous fixation--or doubling the left-sidedness-- and righting of the displaced abomasum in milk cows. Dtsch Tierarztl Wochenschr 104(8):295–297
Oman RE, Streeter RN, Reppert EJ, Chako CZ (2016) Left displacement of the abomasum in 4 beef calves. J Vet Intern Med 30(4):1376–1380
Pardon B, Vertenten G, Cornillie P, Schauvliege S, Gasthuys F, van Loon G, Deprez P (2012) Left abomasal displacement between the uterus and rumen during bovine twin pregnancy. J Vet Sci 13(4):437–440
Parker Gaddis KL, Cole JB, Clay JS, Maltecca C (2014) Genomic selection for producer-recorded health event data in US dairy cattle. J Dairy Sci 97(5):3190–3199
Pehrson BG, Shaver RD (1992) Displaced abomasum: clinical data and effects of peripartal feeding and management on incidence. In: American Association of Bovine Practitioners Proceedings of the Annual Conference, pp 116–121
Pentecost RL, Niehaus AJ, Anderson DE, Miesner MD, Rings DM (2014) Outcome following surgical correction of abomasal displacement in lactating dairy cattle: a retrospective study of 127 cases (1999-2010). J Vet Sci Anim Husb 1(4):1
Perkins GA, Divers TJ, Smith MC, C. R.J. (2017) Examination of the surgical patient. In: Fubini SL, Ducharme NG (eds) Farm animal surgery. Elsevier, St. Louis, pp 1–22
Poulsen JS (1976) Aetiology and pathogenesis of abomasal displacement in dairy cattle. Nord Vet Med 28(6):299–303
Reynen JL, Kelton DF, LeBlanc SJ, Newby NC, Duffield TF (2015) Factors associated with survival in the herd for dairy cows following surgery to correct left displaced abomasum. J Dairy Sci 98(6):3806–3813
Robertson JM (1965) Diagnosis of left displacement of the bovine abomasum. J Am Vet Med Assoc 146:820–827
Robertson JM (1968) Left displacement of the bovine abomasum: epizootiologic factors. Am J Vet Res 29(2):421–434
Rohrbach BW, Cannedy AL, Freeman K, Slenning BD (1999) Risk factors for abomasal displacement in dairy cows. J Am Vet Med Assoc 214(11):1660–1663
Roy JP, Harvey D, Belanger AM, Buczinski S (2008) Comparison of 2-step laparoscopy-guided abomasopexy versus omentopexy via right flank laparotomy for the treatment of dairy cows with left displacement of the abomasum in on-farm settings. J Am Vet Med Assoc 232(11):1700–1706
Sattler N, Fecteau G, Helie P, Lapointe JM, Chouinard L, Babkine M, Desrochers A, Couture Y, Dubreuil P (2000) Etiology, forms, and prognosis of gastrointestinal dysfunction resembling vagal indigestion occurring after surgical correction of right abomasal displacement. Can Vet J 41(10):777–785
Seeger T, Kumper H, Failing K, Doll K (2006) Comparison of laparoscopic-guided abomasopexy versus omentopexy via right flank laparotomy for the treatment of left abomasal displacement in dairy cows. Am J Vet Res 67(3):472–478
Shakespeare A (1997) Left displaced abomasum in sheep. Vet Rec 141(20):527–528
Simpson DF, Erb HN, Smith DF (1985) Base excess as a prognostic and diagnostic indicator in cows with abomasal volvulus or right displacement of the abomasum. Am J Vet Res 46(4):796–797

Sobiech P, Radwinska J, Krystkiewicz W, Snarska A, Stopyra A (2008) Changes in the coagulation profile of cattle with left abomasal displacement. Pol J Vet Sci 11(4):301–306

Song Y, Wang Z, Li R, Hao D, Wang Z, Xia C, Bai Y (2022) Left displacement of the third gastric compartment in an alpaca: the first case report in China. BMC Vet Res 18(1):85

St Jean GD, Hull BL, Hoffsis GF, Rings DM (1987) Comparison of the different surgical techniques for correction of abomasal problems. Compend Cont Educ Pract Vet 9(11):F377–F382

Sterner KE (2002) Grymer/Sterner Toggle Suture, from http://www.ldatogglesuture.com/

Sterner K, Grymer J (1982) Closed suturing techniques using a bar-suture for correction of left displaced abomasum. Bov Pract:80–84

Sterner KE, Grymer J, Bartlett PC, Miekstyn MJ (2008) Factors influencing the survival of dairy cows after correction of left displaced abomasum. J Am Vet Med Assoc 232(10):1521–1529

Trent AM (1992) Surgery of the abomasum. Agri-Practice 13(9):12–14

Trent AM (2004) Surgery of the abomasum. In: Fubini SL, Ducharme NG (eds) Farm animal surgery. Saunders, St. Louis, pp 196–239

USDA (2018) Health and management practices on U.S. dairy operations, 2014. Fort Collins, CO, USDA–APHIS–VS–CEAH–NAHMS

Uyanik N, van Dijk S, van Beukelen P, Kuiper R, Breukink HJ (1982) Results of treatment by rotation in left-sided abomasal displacement (author's transl). Tijdschr Diergeneeskd 107(7):259–263

Van Winden SC, Jorritsma R, Muller KE, Noordhuizen JP (2003) Feed intake, milk yield, and metabolic parameters prior to left displaced abomasum in dairy cows. J Dairy Sci 86(4):1465–1471

Varden SA (1979) Displacement of the abomasum in the cow. Incidence, etiological factors and results of treatment. Nord Vet Med 31(3):106–113

Vogel SR, Nichols S, Buczinski S, Desrochers A, Babkine M, Veillette M, Francoz D, Dore E, Fecteau G, Belanger AM, Badillo M (2012) Duodenal obstruction caused by duodenal sigmoid flexure volvulus in dairy cattle: 29 cases (2006-2010). J Am Vet Med Assoc 241(5):621–625

Wallace CE (1975) Left abomasal displacement -a retrospective study of 315 cases. Bov Pract 10:50–58

Wallace CE (1976) Prognostic significance of diarrhea in cows with left displacement of the abomasum. Bov Pract:62–63

Willeberg P, Grymer J, Hesselholt M (1982) Left displacement of the abomasum: relationship to age and medical history. Nord Vet Med 34(11):404–411

Wittek T, Sen I, Constable PD (2007) Changes in abdominal dimensions during late gestation and early lactation in Holstein-Friesian heifers and cows and their relationship to left displaced abomasum. Vet Rec 161(5):155–161

Wittek T, Locher LF, Alkaassem A, Constable PD (2009) Effect of surgical correction of left displaced abomasum by means of omentopexy via right flank laparotomy or two-step laparoscopy-guided abomasopexy on postoperative abomasal emptying rate in lactating dairy cows. J Am Vet Med Assoc 234(5):652–657

Zadnik T, Lombar R (2011) Our experience with left-sided abomasal displacement correction via the roll-and-toggle-pin suture procedure according to Grymer/Sterner Model. ISRN Vet Sci 2011:572842

Zerbin I, Lehner S, Distl O (2015) Genetics of bovine abomasal displacement. Vet J 204(1):17–22

# The Gut Microbiome in Pigs and Its Impact on Animal Health

Jürgen Zentek, Wilfried Vahjen,
Łukasz Grześkowiak, Beatriz Martínez-Vallespín,
Johannes Schulze Holthausen,
and Eva-Maria Saliu

**Abstract**

The microbiome undoubtedly plays a major role in stabilizing health and ensuring the highest possible related resilience against environmental stressors and infections. Pigs have distinctly developed microbiomes depending on age, husbandry, and diet. Furthermore, the microbial composition in the digestive tract differs depending on the localization. While the stomach and small intestine are largely dominated by lactic acid bacteria and enterobacteria, the spectrum in the large intestine changes significantly. In addition, there are also important differences between the mucosa-associated and the luminal microbiomes in different intestinal sections. Animal health is supported by a balanced colonization of microorganisms in the sense of eubiosis. Disturbances of the microbial ecological system are referred to as dysbiosis and can lead to an increased incidence of diarrheal diseases in particular as a result of infection with pathogens. An important factor in the control of the microbiome is the diet, in addition to the husbandry and genetics of the animals. A balanced diet formulation is an important basis; additionally, stabilization can be achieved by using feed additives. In this context, it is important to ensure that macronutrients are within nutritional recommendations, because protein, carbohydrates, fiber, and fats can influence the microbiome. Supplementation of minerals and feed additives also plays a major role. The latter have been studied very intensively in recent years; stabilizing effects are described for probiotics, prebiotics, synbiotics, organic acids, enzymes, and plant additives. High pharmacological concentrations of zinc oxide that were commonly used in the EU have been banned since 2022, so that the importance of alternative approaches became much greater. Improved knowledge of mechanisms, signaling pathways, and molecular interactions with the digestive tract and host organism is necessary to exploit the potential of the microbiome to stabilize animal health.

**Keywords**

Piglets · Diversity · Diarrhea · Nutrition · Weaning · Eubiosis · Dysbiosis

## 1 Introduction

The composition and activity of the microbiome in the digestive tract of farm animals is influenced by the host organism and external factors

J. Zentek (✉) · W. Vahjen · Ł. Grześkowiak · B. Martínez-Vallespín · J. S. Holthausen · E.-M. Saliu
Institute of Animal Nutrition, bei Universität Berlin, Berlin, Germany
e-mail: juergen.zentek@fu-berlin.de



J. J. Gross (ed.), *Production Diseases in Farm Animals*,
https://doi.org/10.1007/978-3-031-51788-4_8

including farm environment and diet but also by antimicrobial treatments (Hellström et al. 2010; Vandeplas et al. 2010; Beyer et al. 2015; Lührmann et al. 2021; Trevisi et al. 2021). The intestinal microbiome of pigs under conventional production conditions is naturally influenced by domestication effects but also by the use of typical cereal-based diets. This leads to a disproportionate colonization with enterobacteria and lactobacilli compared to the wild boar (Ushida et al. 2016). The intestinal microbiome is not only essential for the development, resilience, and health of the host but can also modify the efficiency of nutrient use. Feed components that escape digestion in the small intestine are partially fermented in the hind gut to usable products for the host, for example, short-chain fatty acids or vitamins (Kolmeder and de Vos 2021; Martin-Gallausiaux et al. 2021). The microbiota and its metabolites are of fundamental biological importance for intestinal function, for development of the immune system, and thus finally for health of domestic animals (Stokes 2017; Chen et al. 2021; Bindari and Gerber 2022).

The gastrointestinal tract contains a complex and very metabolically active microbial community, which is found in the lumen but also associated with the outer mucus layer (Miller et al. 2021). In addition to bacteria, viruses, the phylogenetically very old archaebacteria, fungi, and protozoa colonize the digestive tract (Aluthge et al. 2019). Microbial communities are interacting with each other and with the host organism. Their compositional and metabolic profiles are strongly affected by host species and genotype (Ryu and Davenport 2022). Less intensively than the bacteria, the other representatives of the intestinal microbiota have been studied. Because of its complexity, the intestinal microbiome is often considered a metabolic organ, as it is involved in a plethora of regulatory pathways which can be useful for the host and support animal health (Chen et al. 2021; Johnson et al. 2022). The multitude of published studies on the intestinal microbiome in pigs vary with regard to methodology and animal-, diet- and husbandry-related influences. Nevertheless, a comprehensive comparison of studies in the context of a recent meta-analysis revealed commonalities (Holman et al. 2017). Therefore, it can be assumed that a core microbiome with dominant members exists; however, the composition of subdominant intestinal bacteria can vary greatly from individual to individual.

## 2 Terminology

Historically, the very unspecific and inaccurate term "intestinal flora" was used to describe the intestinal microbiota (Fuller 1992; Farré-Maduell and Casals-Pascual 2019). However, with the ever-expanding analytical spectrum of high-throughput analytical methods, often referred to as "omics- or multi-omics technologies," a differentiation and specification of terms has emerged (Marchesi and Ravel 2015; Borda-Molina et al. 2018; Tröscher-Mußotter et al. 2019; Wang et al. 2020; Dehau et al. 2022):

- Microbiota: Assemblage of all microorganisms in an environment (bacteria, *Archaea*, lower and higher eukaryotes, and viruses are also often included)
- Microbiome: Microbiota, their genomes (i.e., genes), structural elements (e.g., cell wall components), metabolites, and the surrounding environmental conditions
- Metagenome: All microbial genomes in a sample
- Metabolome: Microbial metabolites present in a sample
- Metatranscriptome: All microbial messenger ribonucleic acids (mRNA), providing information on gene expression and thus metabolic activity of the microbiome
- Metaproteome: All proteins and peptides in a sample

While cultural methods were initially used to analyze the composition of the microbiota, these have increasingly been replaced by high-throughput methods. In particular, sequencing of the 16S rRNA gene (Kim and Isaacson 2015) and, in recent years, whole-genome sequencing play an important role here. This allows in-depth insights in the microbiome and its metabolic potential. However, the use of different analytical and bioinformatics evaluation methods makes direct comparisons of studies increasingly difficult.

In the following, the terms microbiota and microbiome are predominantly used. In this context, not only precise measurements of the composition or metabolic activity of the microbiome are of interest, various indices are used to obtain information about the diversity of the microbial community in and between samples (Finotello et al. 2018):

- Alpha diversity: Average species diversity in a site in the gastrointestinal tract (diversity in a single sample)
- Beta diversity: Differentiation among sites, samples, or animals (diversity in a group of samples)
- Richness: Number of different detected species present in a sample

From the baseline measurements of the diversity of the intestinal microbiota, a number of more complex mathematical derivations have emerged; details can be found in the pertinent literature (Finotello et al. 2018). MetaGenome Rapid Annotation using Subsystem Technology (MG-RAST) and Quantitative Insights into Microbial Ecology 2 (QIIME2) are commonly used to determine domain, phylum, family, and genus levels of richness, relative abundance, and diversity. Both pipelines may result in slightly different outcomes (Lima et al. 2021).

Other terms frequently used to describe the situation of the intestinal microbiome refer to the potential health effects of its composition, although the exact definition of the normal state is lacking in both pigs and other animals:

- Eubiosis: Balance of the intestinal microbiota
- Dysbiosis: Imbalance of the intestinal microbiota

Both terms can be traced back historically, noticeably first mentioned by Elliot Furney, used in the same sense also by Elli Metchnikoff and Arthur Scheunert (Hooks and O'Malley 2017). In this context, the term pathobiome is also significant. It has also not yet been given an exact definition but is important for the basic understanding of the interactions of the microbial ecological community with their pathogenic members (Vayssier-Taussat et al. 2014).

- Pathobiome: The pathogenic agent integrated within its biotic environment

## 3 General Aspects of Bacterial Gastrointestinal Communities

Historically, bacteria were classified based on their cell wall properties in Gram-positives and Gram-negatives, which is useful for microscopical identification. Based on their electron transfer systems, bacteria can be designated as aerobic or anaerobic with facultative forms that tolerate or use oxygen in between. Bacterial counts obtained by traditional culture methods increase markedly from the stomach (0.1 mill./g) through the small intestine to the colon (up to 100 billion/g); the number of aerobes decreases, while anaerobes increase (Richards et al. 2005).

Data based on studies using laboratory culture methods underestimate the "true" numbers and taxonomic units of bacteria (Mun et al. 2021). A large proportion of the bacteria living in the digestive tract have not been cultured yet due to their unknown demands on nutrients and growth conditions and are therefore termed "unculturable." However, significant progress has been made in recent years, so that complex cultural methods ("culturomics") are complementary to molecular biological investigations and lead to an extended gain in knowledge in chickens and pigs (Crhanova et al. 2019; Fenske et al. 2020).

Differences in bacterial colonization can be observed not only between intestinal sections, but also within a particular section there are special ecological niches. The composition of mucus-associated bacteria is different from the luminal microbiota (Kelly et al. 2017), and both interact with the host and may modify mucus production and composition (Schroeder 2019; Duangnumsawang et al. 2021). The mucus-associated microbiome includes mucus-degrading specialists such as *Akkermansia muciniphila* or *Bacteroides thetaiotaomicron* (Ouwerkerk et al. 2013). Bacterial oxygen consumption along the upper intestinal tract ensures that strict anaerobic bacteria increase in lower intestinal sites. However, an increased oxygen concentration in epithelial layers due to proximity of blood vessels may stimulate growth of microaerophilic bacteria with the need for specific oxygen concentrations (Marteyn et al. 2011). The mucosa of the lower digestive tract obtains a large part of the necessary energy from the short-chain fatty acids released from fermentation by the microbiota (Roediger 1980).

Strict anaerobic bacteria are specialists that have adapted to the site-specific low oxygen tension in the intestinal tract. As a result of evolutionary adaptation pressures, anaerobic microorganisms have "learned" to design their energy metabolism in such a way that oxidative processes involving electron transfer occur without the involvement of the respiratory chain, i.e., without oxygen. The electrons generated during anaerobic fermentation processes are transported out of the bacterial cell into the surrounding medium via intermediary electron acceptors such as hydrogen, lactic acid, short-chain fatty acids, and other metabolites. For instance, the metabolic end products of electron transfer in specialized *Archaea* and sulfate reducers are methane or hydrogen sulfide (Gibson et al. 1993; Jensen and Jørgensen 1994).

## 4 Intestinal Microbiomes in Pigs

### 4.1 Microbiomes in the Stomach and Intestinal Tract

The microbiota of the digestive tract changes in terms of density and diversity from the stomach through the small intestine to the colon. In sow-milk-fed piglets, the stomach content is dominated by Firmicutes, followed by Bacteroidetes and Gram-positive microorganisms with a dominance of lactobacilli; however, the colonization pattern may change when milk replacers are given (Passlack et al. 2015; Pieper et al. 2016a; Poulsen et al. 2017). The principal dominance of lactobacilli in the stomach contents is characteristic not only in piglets but also in older pigs (Castillo et al. 2007; Eberhard et al. 2007). In suckling piglets, far more diversified colonization patterns occur in the small intestine in the aboral direction, which is continued in the cecum and colon (Zentek et al. 2013; Passlack et al. 2015; Pieper et al. 2016a; Choudhury et al. 2021b). In the jejunum of pigs 4 weeks after weaning, Firmicutes (mainly lactic acid bacteria) dominated, while other phyla played a subordinate role. There were clear effects of the feed composition on the microbial composition that vary with protein source (soybean extraction meal, rapeseed extraction meal) and grain source (rye, wheat) (Ellner et al. 2022). Examination of the intestinal sections of pigs from day 1 to day 180 revealed characteristic forms of colonization in the duodenum-jejunum with the most abundant genera represented by *Lactobacillus* and *Bacteroides* spp., while *Fusobacterium* and *Escherichia* spp. were highly abundant in the ileum shifting to higher proportions of *Prevotella* spp. in the cecum-rectum section (Liu et al. 2019b). Age-dependent changes of the intestinal microbiome in the small intestine could also be confirmed in further studies. Differences between normally developed and intrauterine growth-retarded piglets were characterized by reduced alpha diversity, a decrease in Bacteroidetes, and significant increases of Proteobacteria (Zhang et al. 2019). In the colon of 5- and 12-day-old piglets, there were no significant differences between normally developed and growth-retarded piglets, with Firmicutes and Bacteroidetes predominating in the digesta (Schulze Holthausen et al. 2022). The microbiome of the cecum and colon of 4-month-old piglets showed a comparable high diversity of bacteria and *Archaea* with highly diverse metabolic activities, which could be stimulated by the administration of fermentable carbohydrates (Metzler-Zebeli et al. 2019). It can generally be stated that site-specific microbiomes are established and that the frequently examined fecal samples do not adequately reflect the microbiome compositions and metabolic activities in the upper segments of the intestine.

## 4.2 Microbiota Associated with the Mucosa

The understanding of the mucosa-associated microbiota is fundamental to animal health and gut function due to their direct association with mucus or epithelial cells, as well as with the local immune system (Schierack et al. 2007; Duarte and Kim 2022a). Compared to studies that have addressed the fecal and luminal microbiota, work on the mucosa-associated microbiota is limited due to methodological problems. An important prerequisite for the development of a mucosa-associated microbiome is the ability of bacteria to colonize through interaction with mucosal glycans or with receptors and other structures of epithelial cells. Furthermore, those bacteria must have a relatively high tolerance to oxygen (Kelly et al. 2017). Due to this close contact, many authors assume that the mucosa-associated microbiota exerts a strong influence on the immune system and the balance of pro- and anti-inflammatory processes. Furthermore, the mucosa-associated microbiota may favorably influence resilience to pathogens and their interaction with the gut epithelium. Changes to a luminal microbiome are probably fluid, and the two compartments should not be viewed strictly separately.

After birth, there is a relatively rapid shift in the mucosa-associated microbiome, with the dynamics being much more pronounced in the posterior digestive tract than in the anterior small intestine (Liu et al. 2019a). Proteobacteria dominated in the mucosa of the duodenum and jejunum of piglets, whereas their proportion was significantly reduced in the ileum and colon, although they were still present in considerable proportions. The proximo-distal reduction in Proteobacteria was accompanied by an increase in Bacteroidetes and Firmicutes, which dominated in the colon mucosa (Kelly et al. 2017). Significant shifts in the mucosa-associated microbiota in the jejunum and colon of piglets at 7 and 27 d after weaning were detected. The mucosal microbiota was dominated by Bacteroidetes, whereas the digesta was dominated by Firmicutes. Proteobacteria were detected on the mucosa, particularly on day 7 postpartum, and their proportions differed markedly from those in the digesta (Adhikari et al. 2019). The interaction with the mucosa-associated immune system may develop through different pathways. On the one hand, it is conceivable that epithelial cells are involved and furthermore also the M cells as well as dendritic cells, which have receptors into the intestinal lumen and which function as important antigen-presenting cells. Increasingly, interactions with the innate immune system are also being studied, particularly with the various Toll-like receptors. Here, for example, lipopolysaccharides and other structures can lead to recognition of microorganisms through the host organism with the induction of corresponding signaling pathways. Piglets weaned at lower body weights showed higher alpha diversity of the mucosa-associated microbiome in the jejunum, which was accompanied by increased proinflammatory mediators (Holanda and Kim 2022). The importance of the mucosa-associated microbiome has been shown in relation to animal health. The infection with *Escherichia coli* induces an increase in the diversity and changes in the abundances of mucosa-associated microorganisms (Duarte and Kim 2022b).

## 4.3 The Preweaning Microbiome

The presence of bacteria in utero is discussed in humans and rodents; this is however controversial for the pig so far (Duarte and Kim 2022a). The microbiome of pigs is subject to lifelong changes (Wang et al. 2019) and undergoes particularly dynamic shifts in the first days to weeks of life (Petri et al. 2010; Knecht et al. 2020). In neonatal piglets, surprisingly few data are available on the microbiome of the individual sections of the digestive tract. Microbial maturation is probably also defined by correspondingly rapid development of digestive and absorptive capacities, the immune system, and metabolic traits (Zabielski et al. 2008; Juul-Madsen et al. 2010; Sinkora and Butler 2016; Metzler-Zebeli et al. 2021). Many of those processes are affected in piglets by individual birth weight. The frequency of underweight and especially the intrauterine growth-retarded piglets is correlated to high litter sizes. Underweight piglets differ in their microbiomes (D'Inca et al. 2010; Huang et al. 2019; Tang et al. 2022) and gene expression patterns in the digestive tract compared to normal-weight piglets (Villagómez-Estrada et al. 2022). Growth-retarded piglets show a markedly different colonization not only in the luminal microbiota, there are also shifts in bacteria associated with the intestinal mucosa. Furthermore, lower numbers of beneficial bacteria such as lactobacilli appear to be present (Jiang et al. 2019). Initial colonization of the gastrointestinal tract with microorganisms most likely occurs during the parturition process, with the sow's vaginal microbiota playing a major role as an inoculum (Wang et al. 2013; Chen et al. 2022). Later, bacteria are also acquired from the mother's skin, milk, feces, and the environment (Liu et al. 2019a). Coprophagy plays an important role in this context, both for the microbiome and for the long-term development of the animals (Sansom and Gleed 1981; Aviles-Rosa et al. 2019). The coprophagic transfer of bacteria from sow to piglet has a higher importance with increasing suckling time (Chen et al. 2022). Finally, reverse inoculation of bacteria back to the mother cannot be excluded, though not proven to the best of our knowledge.

Compared to older animals, newborn piglets have a greatly reduced diversity of their intestinal microbiome, which also undergoes massive changes during the first days and weeks of life (Mach et al. 2015; Bian et al. 2016; Arnaud et al. 2020; Grzeskowiak et al. 2022). While enterobacteria and clostridia dominated at the time of birth, lactobacilli subsequently develop within a few days (Petri et al. 2010). The development of the microbiome in the first phase of life is very strongly influenced by the intake of the mother's milk. The use of formula feeding also shows a high impact on the development of the preweaning microbiota (Grzeskowiak et al. 2018). Other important effects can be traced back to the early ingestion of creep feed as well as environmental factors. For instance, creep feed contains fermentable dietary fiber, nucleotides, or other dietary factors, while access to soil introduces a completely new variety of bacteria to the piglet (Vo et al. 2017; Van Hees et al. 2019; Choudhury et al. 2021a, b; Correa et al. 2021). On the other hand, dietary supplementation with glutamine had only minor effects on the composition of the intestinal microbiota in neonatal pigs (Schulze Holthausen et al. 2022). At the time of weaning, depending mainly on conventional or organic farm type in a range of 20–40 days after birth, the piglet's microbiome is still far from mature (Grzeskowiak et al. 2020). The dynamics of change are particularly high during the first few weeks of life and then weaken. In newborn piglets, *Clostridioides difficile* is an excellent example of a pathogenic bacterium that has a very narrow time window of colonization and subsequent infection (3–10 d of life) (Songer 2004). There is evidence that a maturing microbiota actively reduces *C. difficile* counts to below detection limit in the further course of the preweaning phase (Songer 2004; Grzeskowiak et al. 2016, 2019b).

## 4.4 Weaning and Postweaning Microbiomes

With increasing diversity of the microbiome and higher stabilization of intestinal function, more resilience against external influences and efficient protection of animal health are expected (Dou et al. 2017), but more knowledge is needed on this topic. The weaning period is considered a sensitive phase, since on the one hand it involves a physiological transition from the digestion of mainly milk to solid feed sources and on the other hand introduces psychological stress, as weaning usually also involves a new grouping with non-littermate piglets. It is well known that piglets behave very differently after weaning. Especially feed intake can differ considerably between individuals, and fasting periods of several days are not uncommon (Bruininx et al. 2002). Different levels of feed intake seem to have only a moderate impact on the intestinal microbiome (Rist et al. 2012). However, when piglets do not consume any feed, fasting shifts the intestinal microbiota to a significantly reduced diversity and evenness and favors Bacteroidetes and Proteobacteria instead of the usually dominating Firmicutes (Ma et al. 2022). These and other factors can contribute to health problems and predispose to diarrhea in piglets. It has been shown that piglets with diarrheal disorders can have a significantly reduced microbial alpha diversity in their intestinal microbiomes (Yang et al. 2019; Kong et al. 2022).

Bacterial pathogens interact with the commensal microbiome of the digestive tract, as they both inhabit the same ecological niche. At weaning and in the early postweaning phase of piglets, there is a massive change in microbiome composition and diversity (Mach et al. 2015; Lührmann et al. 2021; Luise et al. 2021). This can create a disposition for the proliferation of pathogens, especially enterotoxigenic *Escherichia coli* (ETEC) and Shiga toxin-producing *E. coli* (STEC), which trigger the problem of diarrhea and edema disease after weaning in commercial animal production (Fairbrother et al. 2005). A reduced bacterial diversity of the intestinal microbiome can be a prerequisite for the spread of toxinogenic strains of *E. coli*. Interestingly, the classical Henle-Koch's postulates, according to which a pathogen must be detected in the diseased organism, can be grown in pure culture, and induces disease upon experimental infection of a susceptible macroorganism, seem not to be fully applicable to non-mortal diseases. For *E. coli* infections, animal-related factors, in particular the intestinal receptor (F4) status of the piglets, as well as the so-called pathobiome play a role (Vayssier-Taussat et al. 2014). The latter states that all pathogens interact with the microbiome in its ecological environment, which is thus influenced by it. In this respect, animals can be clinically healthy despite the presence of pathogenic *E. coli* strains, while others become ill (Schierack et al. 2006, 2007).

## 4.5 Microbiomes in Maturating Pigs

With age, the microbiome of pigs becomes increasingly diversified (Luo et al. 2022). The dominant phyla seem to be comparable across all life stages and include Firmicutes, Bacteroidetes, Proteobacteria, and Actinobacteria. On further analysis, the microbiomes of neonatal piglets, suckling piglets, weanling piglets, and fattening pigs can be clearly distinguished. In addition to the persisting core microbiome, there are phases of transient colonizers that are present and may subsequently be no longer detectable (Grzeskowiak et al. 2019a; Wang et al. 2019). Interestingly, it is becoming increasingly clear that the production performance of pigs is influenced by the microbiome; in particular, effects on feed conversion could be shown (Verschuren et al. 2018). Breeding sows reach the highest life span and have a very diversified and "mature" microbiome with a large number of different bacterial species. This was recently demonstrated, where a total of nearly 2800 different species could be differentiated per sow, remarkably with a very high variability indicating substantial individual differences (Kubasova et al. 2017). In this study, it was shown that *Prevotella*, *Oscillospira*, and *Ruminococcus* spp. dominated the sow fecal samples, which were much less detectable in piglets. Under practical housing conditions, massive shifts in the microbiome of sow feces occur due to parturition or the associated feed change. Firmicutes (86%), Bacteroidetes (11%), and Actinobacteria (1%) were found in the sow feces before and after birth, Bacteroidetes and Actinobacteria decreased, and Firmicutes and Proteobacteria increased after birth (Lührmann et al. 2021). The microbiome of sows is subject to massive influences from feeding. This is partly responsible for the changes that occur between prepartum and postpartum samples. The effects of dietary fiber supplements differing in fermentability are particularly well documented (Grzeskowiak et al. 2022; Lu et al. 2022).

# 5 Gastrointestinal Microbiomes and Animal Health

## 5.1 The Basic Concept of Eubiosis and Dysbiosis

Eubiosis is defined as the endogenous and exogenous mechanisms that ensure a relatively stable state of equilibrium of the intestinal microbiota that is beneficial to the host (Trevisi et al. 2021).

Infections, inflammations, or inappropriate nutrition can trigger dysbiosis, which can be associated with clinically manifested diseases.

The intestinal microbiota is subject to numerous regulatory factors by the host but also produces factors that can regulate its host (Peng et al. 2021). Main host factors include the stomach pH barrier, which leads to a reduction and selection of ingested bacteria toward acid-tolerant species, intestinal peristalsis, the secretion of immunoactive molecules (immunoglobulins, antimicrobial peptides), the composition of the mucus layer, or secretion of digestive enzymes and bile. Microbial factors include bacteriocins or antimicrobial peptides that only act on the microbiota, as well as metabolites such as hydrogen sulfide, hydrogen peroxide, or organic acids, which exert strong effects on competing microorganisms. Moreover, regulatory bacterial peptides have been shown to directly interfere with host mucus composition (Hooper et al. 1999).

Eubiotic microbial development has advantages for the host, since, for example, vitamins are synthesized, the degradation of plant fibers to energy-delivering metabolites can only take place with the aid of microbial enzymes, and the colonization of externally ingested or pathogenic bacteria is impeded or suppressed. The so-called core microbiome, which is present in the vast majority of pigs, offers the option of positive influence on animal health (Luo et al. 2022). In the case of inappropriate feed composition, but also for disease-related reasons, dysbiotic consequences for the host include a change in the intestinal microbiome, either by a selective increase in the number of certain bacterial species or groups, increased colonization of certain parts of the gastrointestinal tract, e.g., the small intestine, or the formation of microbial metabolites. Dysbiosis is a particular problem during the weaning phase, as an abrupt change in diet overburdens the adaptive capacity of the intestinal microbiota and the host organism (Gresse et al. 2017).

## 5.2 Gastritis and Gastric Ulceration

Ulcerations of the gastric mucosa, especially of the pars esophagea, play a major role in practice and are triggered by dietary influencing factors but also by unfavorable husbandry conditions and stress in general (Doster 2000). Particular importance is attached to excessive fineness of the feed, which can also arise as a secondary effect of intensive hydrothermal treatments (Mösseler et al. 2012; Kiarie and Mills 2019). The microbiome can affect the health of the gastric mucosa. Thus, it was found that piglets with gastric ulcers have a lower colonization of lactobacilli (Almeida et al. 2018). Recent studies suggest that the pars esophagea in pigs is intensively colonized with bacteria, where Proteobacteria (56%), Firmicutes (30%), Bacteroidetes (10%), Fusobacteria (3%), and Actinobacteria (1%) dominated. In diseased pigs, the mucus-associated *Helicobacter suis* and *Fusobacterium gastrosuis* were detected to a higher extent, which may be synergistically responsible for pathogenesis (De Witte et al. 2019; Taillieu et al. 2022). Prophylaxis includes the administration of diets that are not too finely ground and the avoidance of external stress factors. To what extent further effects on the microbiome can be achieved by specific support measures is presently unclear.

## 5.3 Viral Infections and Intestinal Microbiomes

Infectious diseases are often associated with diarrhea in piglets, whereby on the one hand suckling piglets and on the other hand weaner piglets are mostly affected. Dysbiosis of the microbiome in the small intestine can promote infections with various pathogens, including viruses. The digestive tract of pigs is colonized by a vast number of viruses, including those of plant origin, and furthermore bacteriophages as well as animal pathogens (Sachsenröder et al. 2014). The role of bacteriophages has been increasingly studied in recent years, and they are important for the regulation of the composition of the microbiome itself (Wu et al. 2022). An interesting question is whether the microbiome can influence susceptibility to viral infections; however, little information is available. It was recently demonstrated for the African swine fever virus that African warthogs have a high resistance to this infection; this was partially transmissible by fecal transplantation to domestic pigs (Zhang et al. 2020). However, most studies have addressed the extent to which viral infections can lead to alterations of the microbiome. In this context, enteric viral diseases in particular play a role in pigs. Acute porcine epidemic diarrhea virus infection triggered a significant change in the microbiome of the cecum and small intestine; however, interestingly, these also turned out differently in piglets of different ages (Tan et al. 2019; Yang et al. 2020b). A clear effect on the intestinal microbiota was also observed after infection with the diarrhea-inducing porcine deltacoronavirus. In the digestive tract, there was a reduced bacterial diversity and an increase in the proportions of Fusobacteria and Proteobacteria (Li et al. 2022). Significant shifts in the intestinal microbiome had already been shown in a previous study (Li et al. 2020). Porcine reproductive and respiratory syndrome (PRRS) virus infections, which cause multiple problems on pig farms, resulted in significant shifts in intestinal microbiomes, with effects that were strain-dependent. The more virulent strain resulted in decreased colonization of *Roseburia*, *Anaerostipes*, *Butyricicoccus*, and *Prevotella* spp., which in turn was associated with increased severity of clinical signs, as well as altered immune parameters (Argüello et al. 2022). Overall, one has to consider that due to the intricate involvement of the intestinal microbiome with its host, a drastic virus-induced change in host physiology will inevitably lead to changes in its microbiome.

With regard to possible microbiome-based prevention and therapy measures, there are presently relatively few findings in pigs. Studies on piglets showed a positive effect of supplementation with a human probiotic strain *Lactobacillus rhamnosus* GG in case of infection with rotaviruses (Mao et al. 2016). Modification of the intestinal microbiome by fecal microbiota transplant in piglets reduced clinical effects of an infection model with porcine reproductive and respiratory syndrome virus (PRRSV) and porcine circovirus type 2 (PCV-2) (Niederwerder et al. 2018). Thus, stabilizing effects of the intestinal microbiota on host resilience may be a valid mechanism to reduce virus-induced damages.

## 5.4 Bacterial Infections and Microbiomes

Bacterial infections play a major role in causing intestinal diseases, especially in young pigs. In suckling piglets, diseases caused by clostridia, especially *Clostridium perfringens* type C-induced necrotizing enteritis and *C. difficile*-induced necrotizing enterocolitis, are of particular importance in early life. Mainly around weaning time, but sometimes also in very young animals, infections with pathogenic *E. coli* cause diarrhea or edema disease. Infections caused by *Salmonella* are of importance in pigs of all age groups; they can cause enteritis but are also relevant, as pigs may become carriers of this zoonotic pathogen. The causative agent of ileitis is *Lawsonia intracellularis*. Affected are predominantly weaned piglets and fattening pigs. Dysentery, caused by *Brachyspira hyodysenteriae*, represents a muco-fibrinous, diphtheroid to necrotizing typhlocolitis. Spirochete enteritis is classified as a disease on its own and is caused by *B. pilosicoli* (Ritzmann and Heinritzi 2006). Many of these pathogens are characterized by a broad variability in their biological characteristics. For example, *E. coli* strains can produce a variety of toxins and carry a number of different adhesion factors, which are relevant for the clinical effects (Nadeau et al. 2017; Dubreuil 2021).

The importance of the intestinal microbiome and possible dysbiosis is discussed in particular in connection with diarrhea disposition and the intestinal microbiome during weaning (Konstantinov et al. 2006). As already mentioned, stress factors, the change in feed composition, and possibly prolonged periods of feed refusal play a particular role here. This may result in an imbalance of the intestinal microbiome (Kim et al. 2022); when comparing healthy and diseased piglets, there were clear differences in the dominant bacterial species (Sun et al. 2019). The infection with ETEC can lead to a marked change in the intestinal microbiome of piglets, particularly in the ileum (Pollock et al. 2019). Studies with newborn piglets infected with ETEC showed that transplantation of jejunum contents from non-diseased piglets to recipient animals provides protection against diarrhea after an *E. coli* infection. Diseased animals showed lower microbial diversity in the intestinal tract, further showing a decreased incidence of Bacteroidetes and a significant increase in Firmicutes. These changes were prevented by transplantation of jejunal contents from healthy animals, clearly demonstrating that the microbiome influences disease insight (Bin et al. 2018). Interestingly, even low infection doses of pathogenic *E. coli* can lead to dysbiosis in the digestive tract of piglets, which is expressed not only in the composition but also in shifts in the bacterial metabolite spectrum (Boeckman et al. 2022). While no more detailed data are available on the importance of the microbiome for *C. perfringens* infection in young piglets, *C. difficile*-induced clinical diseases in piglets decreased in an age-dependent manner with higher diversity of the cecal microbiota (Proctor et al. 2021). As expected, infections with *Salmonella* can also lead to changes in the intestinal microbiome. *Salmonella* Typhimurium resulted in a significant shift in intestinal mucosa-associated microorganisms in piglets, and *Ruminococcaceae* and *Veillonellaceae*, which were predominantly identified with the mucosa of healthy piglets, were detected at significantly lower levels in the infected animals (Bescucci et al. 2020). It was also shown for *L. intracellularis* that infection of pigs leads to a marked change in the intestinal microbiome. Clear signs of dysbiosis were evident in both the small and large intestines, characterized in particular by an increase in potentially pathogenic bacteria such as *Collinsella*, *Campylobacter*, *Chlamydia*, and *Fusobacterium*. The changes were interpreted as suggesting that the microbiome may be conducive to the spread of *Lawsonia* but also the development of other diseases (Leite et al. 2021).

## 5.5 Mycobiomes and Microbiomes

Yeasts are in use as probiotic feed additives with some success, and the indigenous mycobiome may also contain certain species with similar effects, which may be enhanced through dietary means. For instance, in weaned piglets fed cereal-based diets, there was a marked increase in the colonization with the yeast *Kazachstania slooffiae*, which correlated with numbers of lactobacilli and enterobacteria (Urubschurov et al. 2011). Experiments with the addition of *K. slooffiae* showed that it may have a noticeable effect on the intestinal microbiome of pigs (Urubschurov et al. 2017). Recent work has shown that the mycobiome and the microbiome show some correlations in their development. *Kazachstania* was positively associated with *Lactobacillus*. In contrast, negative correlations were found between *Aspergillus* and *Butyricicoccus*, *Subdoligranulum*, and *Fusicatenibacter* spp. However, the health significance of these associations or interactions needs further elucidation (Arfken et al. 2019).

## 5.6 Animal Housing and Microbiomes

Housing influences the development of the intestinal microbiome. This is evident when comparing the composition of the microbiota in wild and domestic pigs. In extreme environments, when piglets were kept conventionally with their mothers, they had a significantly more diverse microbiota compared to piglets kept in isolators. The immune system also developed fundamentally different between the two groups (Mulder et al. 2009; Inman et al. 2010). The conditions inside a barn influence intestinal microbiomes. Thus, very clear effects of the concentrations of ammonia were shown on the microbiome of piglets. The diversity of *Bacteroidetes* strains was highest in piglets from the control group, while animals exposed to high ammonia concentrations showed increased levels of clostridia as well as *Alphaproteobacteria*, whereas *Gammaproteobacteria*, *Bacteroidia*, *Fusobacteria*, and *Actinobacteria* were significantly decreased (Li et al. 2021b). In conventional housing or an enriched housing system with straw bedding, the microbiome of sow feces changed significantly, while it was comparable in the piglets. This was attributed to the ingestion of straw by the sows (Kubasova et al. 2017). Interestingly, when sows were compared under conventional or organic housing conditions, there were no differences in fecal microbiomes or bacterial resistomes (Gerzova et al. 2015). Also, when comparing animals that were kept outdoors or indoors, there were only minor differences depending on the setup.

### 5.7 Effects of Intestinal Microbiomes on Host Metabolism

It is generally difficult to infer the effects of the intestinal microbiome on animal health in terms of causality unless gnotobiotic animals are used. There is no doubt that the functioning of the immune system and its transcriptomic profile is significantly modulated by colonization with microorganisms and their metabolic activities (Chowdhury et al. 2007; Laycock et al. 2012; Sun et al. 2018). Under conventional conditions, the microbiome is important for maintaining the barrier function of the digestive tract as well as signaling through the intestinal-liver axis; an altered microbiome as a result of heat stress can have negative repercussions on the host and especially the immune system (Ringseis and Eder 2022). Comprehensive studies show that the intestinal microbiome influences numerous physiological control processes, including behavior, appetite, hormonal control mechanisms, and physiological metabolic processes within the framework of the gut-brain axis (Torres-Fuentes et al. 2017; Wessels 2022). This is of importance in pigs, for example, as feed intake and feed conversion are influenced by the microbiome (Gardiner et al. 2020), although the complex mechanisms need to be elucidated.

## 6 Dietary Effects on the Intestinal Microbiomes

### 6.1 Fermentable Carbohydrates and Dietary Fiber

Fermentable carbohydrates can affect the microbiome and consequently animal health in pigs. Milk sugar (lactose) and milk oligosaccharides are the first dietary carbohydrates for piglets provided via colostrum and milk. Lactose is highly digestible in neonates; as lactase activity decreases, lactose is increasingly metabolized by microorganisms having a more "prebiotic" role (Zhao et al. 2021). Milk oligosaccharides and lactose in older animals are microbially fermented and represent an important regulatory factor for the intestinal microbiota with a long-term impact not only in the suckling phase but also beyond with impact on performance and gut health (Jang and Kim 2022).

Prebiotics are fermentable carbohydrates that are selectively used in feed to exert beneficial effects on the intestinal microbiota and its metabolites. These effects may be desirable and serve to promote a "desirable" microbiome of the gut. In this context, dose-response relationships are important. If the intake of fermentable carbohydrates is too high, acidic fermentation occurs in the gastrointestinal tract (Pierce et al. 2006); clinically, sour-smelling, watery diarrhea may develop.

A large body of work has shown that prebiotics, mostly fermentable short- to medium-chain carbohydrates such as fructo-oligosaccharides and inulin, have a positive effect on intestinal health and development in pigs (Passlack et al. 2015; Samolinska et al. 2018). Similar effects were induced by different types of resistant starch (Metzler-Zebeli et al. 2019; Tiwari et al. 2019; Trachsel et al. 2019), as well as by soluble and fermentable plant fibers. In general, fermentable carbohydrates promote a microbiome that is often characterized by increases of lactobacilli and bifidobacteria, while the occurrence of potentially harmful or pathogenic bacteria was reduced in most studies. Cellulose, the main building material of the plant cell wall, is metabolized by the intestinal microbiota to a much lesser extent, because it is embedded in mechanically very solid structures and therefore requires longer fermentation times.

Besides the microbiota-related effects, the physicochemical properties of fermentable fiber can exert positive effects on intestinal health in pigs in terms of a positive impact on physiological traits, as well as the gut-associated immune system (Jha et al. 2019; Li et al. 2021a). For instance, an increase in digesta viscosity can be caused by gel-forming soluble carbohydrates with high water-holding capacity, which can have a negative impact on nutrient digestion and gut health (Dikeman and Fahey 2006). In this context, it is interesting to note that not only the typical short-chain carbohydrates and fermentable fibrous feed materials but also common cereals used in feed affect the microbiome. Thus, it could be shown that the use of rye instead of wheat in weanling piglets leads to significant changes in the small and large intestinal microbiome with a massively increased fermentation activity in rye-fed piglets (Ellner et al. 2022). On the one hand, this can influence the microbiome and its metabolic activity; on the other hand, the partially fermentable carbohydrates in rye led to a reduction in the *Salmonella* load of pigs (Hankel et al. 2022). Resistant starch, occurring either in native feed materials or after thermal treatment of feed, is largely metabolized by the intestinal microbiota.

## 6.2 Protein

It is common today to use diets in ration formulation with significantly reduced protein content compared to earlier years, but supplementation with synthetic amino acids is a must in conventional farming. This has positive economic and ecological effects but also positively affects animal health in the sense that lower protein diets have a lower risk for the development of postweaning diarrhea (Pieper et al. 2016b; Gao et al. 2019). The increased influx of proteins into the posterior segments of the intestine leads to a significant change in the microbiome and its metabolic activity (Rist et al. 2013). Thus, not only excessive protein intake but also the use of protein sources with low precaecal digestibility may increase the risk (Pieper et al. 2012). Thus, inadequate protein supply is usually accompanied by dysbiosis and dysfermentation and impairment of animal health.

## 6.3 Fats and Oils

Oils and fats can also influence the intestinal microbiota, although the effects do not seem to be as pronounced as with fermentable carbohydrates or proteins. In pigs, relatively few studies exist on the effect of different lipids or dietary inclusion rates used in feeding. On the one hand, certain medium-chain fatty acids can be used as feed additives, as they exert a relatively strong effect on the microbiome in the digestive tract and are therefore of interest not only as a rapidly available energy source for pigs (Zentek et al. 2011). Long-chain fatty acids also have effects that can be positive for animal health, particularly due to interactions with the immune system, especially through n-3 fatty acids (Lauridsen 2020). Positive effects on the microbiome have been shown with linseed oil, decreasing the *Spirochaetes* and increasing the phylum Actinobacteria and genera *Blautia* and *Bifidobacterium* in colonic digesta (Che et al. 2019). When comparing different fat supplements, a correlation has been established between lower apparent digestibility of fats and effects on the microbiome (Yang et al. 2020a).

## 6.4 Minerals

It has been shown that there are interactions between the intestinal microbiome, intestinal mineral concentrations (Pieper et al. 2016a), the supply of different calcium sources (Yang et al. 2022), and especially phosphorus (Metzler-Zebeli et al. 2010). Most studies in recent years have focused particularly on phosphorus and the microbiome in pigs. On the one hand, the microbiome is interesting because of its intrinsic need for phosphorus and the interaction with different inositol phosphates (Heyer et al. 2015, 2019); on the other hand, many microorganisms build phytases that can increase the availability of the phytate-bound phosphorus. This can compensate to some extent for different phosphorus uptakes or availabilities (Reyer et al. 2021). The microbiota has a need for phosphorus for its metabolism, and in this context the ratio of calcium to phosphorus is significant (Heyer et al. 2019). Possible influences on the part of the intake of sodium and potassium as well as chloride have been little investigated. There are indications that a variation of the electrolyte ratio in the feed leads to an influence on the microbiome in the pig (Deng et al. 2020).

### 6.5 Feed Additives

Feed additives have been widely studied, and their effects on the intestinal microbiome are described in the literature. They are frequently used in practice to influence the microbiome and animal health. Based on the published work, there are clear stabilizing effects on animal health from the use of trace elements, probiotics, enzymes, organic acids, and phytogenic additives. The mechanisms of action of feed additives are presented very differently and often also contradictory. Due to the legally binding limitation of the contents of zinc and copper in feed for piglets, the stabilizing effect on health and the microbiome exerted by the so-called pharmacological doses of zinc oxide can no longer be used in Europe, but globally high contents are still contained in particular in feed for weaner piglets. The mechanisms of action have been extensively described, especially for zinc (Broom et al. 2021). Pharmacological zinc doses above 2500 mg/kg feed induce a substantial change of the microbiome (Starke et al. 2014) and interestingly result in a reduction of lactobacilli with simultaneous stabilization of enterobacterial diversity (Katouli et al. 1999). Their effect of increased diversity of enterobacteria might be that pathogenic strains have less opportunity to establish in the intestine. Disadvantageous and ultimately decisive for the prohibition of pharmacological dosages of zinc oxide as well as of high concentrations of copper sulfate in feed are not only environmental effects but also the problem of increased antimicrobial resistance in the digestive tract of pigs (Bednorz et al. 2013; Vahjen et al. 2015; Ghazisaeedi et al. 2020). Traditionally, but especially following the ban on pharmacological zinc oxide dosages in the EU, interest is focusing more on direct-fed microorganisms and probiotics, as an essential pillar of diarrhea prevention (Fuller 1989). The mechanisms of action are so far insufficiently described; some probiotics may cause a change in microbial colonization and metabolic activity in the digesta, while other probiotics seem to have more of an effect on the immune system and intermediary metabolism (Trevisi et al. 2021; Luise et al. 2022; Vasquez et al. 2022). More recent developments combine directly fed bacteria with appropriate fermentable substrates, which is called a synbiotic (Aluthge et al. 2019; Girard et al. 2021). The idea of postbiotics also represents a further development, whereby an attempt is made to harness microbial products that result from fermentation (Zamojska et al. 2021). Organic acids are used in particular for their preservative effects, but they can also lead to a reduction in diarrhea (Pluske et al. 2021). Practically relevant are on the one hand medium-chain fatty acids (Zentek et al. 2011) but then also acids and salts of short-chain fatty acids (Partanen and Mroz 1999). They have an effect on the microbiome, particularly in the upper digestive tract, and can apparently therefore reduce the downstream transfer of pathogenic bacteria to the intestine and regulate the composition of the microbiome. Enzymes support digestive processes in piglets, particularly by improving the exclusion of non-starch polysaccharides, releasing phosphorus from phytic acid, and improving protein digestibility. The effects vary depending on the matrix of the feed and the effectiveness of the enzyme supplement (Duarte et al. 2021; Petry et al. 2021; Moita et al. 2022). Newly developed feed additives from the group of plant extracts also play a major role in practical diets. In particular, essential oils but also polyphenols and other secondary ingredients can interact with the microbiome and can positively influence performance and intestinal health of piglets (Vanrolleghem et al. 2019; Rossi et al. 2020).

## 7 Conclusion

The porcine gut microbiome is clearly a particularly relevant factor for animal health. Controlling the intestinal microbial ecology is an important tool for future management. The balanced porcine gut microbiome is relevant to better control environmental stressors and infectious diseases.

With the advent of technological advances, our understanding of the gut microbiome in pigs continues to evolve rapidly. Here, holistic, meta-omics methods together with novel in vitro approaches move our focus from composition to function and from observation to mechanism, helping to understand complex interactions between the microbiome and the host. Some of these relationships have been well studied, but others require further elucidation. These host-microbiome interactions are very specific processes that can be dysregulated by external disrupters, leading to a disease state. Thus, gaining knowledge on the gut microbiome dynamics and its interplay with the host may provide new insights for maintaining pig's intestinal homeostasis and general health. It seems important to combine new bioinformatics methods that capture the complexity of the microbiome and its functional effects in upcoming studies. Future strategies to maintain animal health should focus rather on healthy gut microbiome development and its conservation. Indeed, diet is one of the most effective microbiome-modulating factors in animals. Whether diet alone can completely restore the disrupted gut homeostasis in pigs is doubtful. In case of gut dysbiosis, new conditions should be established to correct the microbiome and its interactions with the host. In the future, focus on the gut microbiome as a biomarker should allow various modulating factors, in particular nutritional measures, to be specifically selected to achieve an improvement in animal health.

## References


Adhikari B, Kim SW, Kwon YM (2019) Characterization of microbiota associated with digesta and mucosa in different regions of gastrointestinal tract of nursery pigs. Int J Mol Sci 20(7)

Almeida LR, Costa PS, Nascimento AMA, Reis MP, Barros KO, Alvim LB, Nunes ÁC, Queiroz DMM, Rocha GA, Nicoli JR, de Moura SB (2018) Porcine stomachs with and without gastric ulcer differ in Lactobacillus load and strain characteristics. Can J Microbiol 64(7):493–499

Aluthge ND, Van Sambeek DM, Carney-Hinkle EE, Li YS, Fernando SC, Burkey TE (2019) BOARD INVITED REVIEW: the pig microbiota and the potential for harnessing the power of the microbiome to improve growth and health1. J Anim Sci 97(9):3741–3757

Arfken AM, Frey JF, Ramsay TG, Summers KL (2019) Yeasts of burden: exploring the mycobiome-bacteriome of the piglet GI tract. Front Microbiol 10:2286

Argüello H, Rodríguez-Gómez IM, Sánchez-Carvajal JM, Pallares FJ, Díaz I, Cabrera-Rubio R, Crispie F, Cotter PD, Mateu E, Martín-Valls G, Carrasco L, Gómez-Laguna J (2022) Porcine reproductive and respiratory syndrome virus impacts on gut microbiome in a strain virulence-dependent fashion. Microb Biotechnol 15(3):1007–1016

Arnaud AP, Rome V, Richard M, Formal M, David-Le Gall S, Boudry G (2020) Post-natal co-development of the microbiota and gut barrier function follows different paths in the small and large intestine in piglets. Faseb J 34(1):1430–1446

Aviles-Rosa EO, Rakhshandeh A, McGlone JJ (2019) Preliminary study: depriving piglets of maternal feces for the first seven days post-partum changes piglet physiology and performance before and after weaning. Animals (Basel) 9(5)

Bednorz C, Oelgeschläger K, Kinnemann B, Hartmann S, Neumann K, Pieper R, Bethe A, Semmler T, Tedin K, Schierack P, Wieler LH, Guenther S (2013) The broader context of antibiotic resistance: zinc feed supplementation of piglets increases the proportion of multi-resistant Escherichia coli in vivo. Int J Med Microbiol 303(6–7):396–403

Bescucci DM, Moote PE, Ortega Polo R, Uwiera RRE, Inglis GD (2020) Salmonella enterica Serovar Typhimurium temporally modulates the enteric microbiota and host responses to overcome colonization resistance in swine. Appl Environ Microbiol 86(21)

Beyer A, Baumann S, Scherz G, Stahl J, von Bergen M, Friese A, Roesler U, Kietzmann M, Honscha W (2015) Effects of ceftiofur treatment on the susceptibility of commensal porcine E. coli--comparison between treated and untreated animals housed in the same stable. BMC Vet Res 11:265

Bian G, Ma S, Zhu Z, Su Y, Zoetendal EG, Mackie R, Liu J, Mu C, Huang R, Smidt H (2016) Age, introduction of solid feed and weaning are more important determinants of gut bacterial succession in piglets than breed and nursing mother as revealed by a reciprocal cross-fostering model. Environ Microbiol 18(5):1566–1577

Bin P, Tang Z, Liu S, Chen S, Xia Y, Liu J, Wu H, Zhu G (2018) Intestinal microbiota mediates enterotoxigenic Escherichia coli-induced diarrhea in piglets. BMC Vet Res 14(1):385

Bindari YR, Gerber PF (2022) Centennial review: factors affecting the chicken gastrointestinal microbial composition and their association with gut health and productive performance. Poult Sci 101(1):101612

Boeckman JX, Sprayberry S, Korn AM, Suchodolski JS, Paulk C, Genovese K, Rech RR, Giaretta PR, Blick AK, Callaway T, Gill JJ (2022) Effect of chronic and acute enterotoxigenic E. coli challenge on growth

performance, intestinal inflammation, microbiome, and metabolome of weaned piglets. Sci Rep 12(1):5024
Borda-Molina D, Seifert J, Camarinha-Silva A (2018) Current perspectives of the chicken gastrointestinal tract and its microbiome. Comput Struct Biotechnol J 16:131–139
Broom LJ, Monteiro A, Piñon A (2021) Recent advances in understanding the influence of zinc, copper, and manganese on the gastrointestinal environment of pigs and poultry. Animals (Basel) 11(5)
Bruininx EM, Binnendijk GP, van der Peet-Schwering CM, Schrama JW, den Hartog LA, Everts H, Beynen AC (2002) Effect of creep feed consumption on individual feed intake characteristics and performance of group-housed weanling pigs. J Anim Sci 80(6):1413–1418
Castillo M, Skene G, Roca M, Anguita M, Badiola I, Duncan SH, Flint HJ, Martín-Orúe SM (2007) Application of 16S rRNA gene-targeted fluorescence in situ hybridization and restriction fragment length polymorphism to study porcine microbiota along the gastrointestinal tract in response to different sources of dietary fibre. FEMS Microbiol Ecol 59(1):138–146
Che L, Zhou Q, Liu Y, Hu L, Peng X, Wu C, Zhang R, Tang J, Wu F, Fang Z, Lin Y, Xu S, Feng B, Li J, Jiang P, Wu D, Chen D (2019) Flaxseed oil supplementation improves intestinal function and immunity, associated with altered intestinal microbiome and fatty acid profile in pigs with intrauterine growth retardation. Food Funct 10(12):8149–8160
Chen S, Luo S, Yan C (2021) Gut microbiota implications for health and welfare in farm animals: a review. Animals (Basel) 12(1)
Chen W, Ma J, Jiang Y, Deng L, Lv N, Gao J, Cheng J, Liang JB, Wang Y, Lan T, Liao X, Mi J (2022) Selective maternal seeding and rearing environment from birth to weaning shape the developing piglet gut microbiome. Front Microbiol 13:795101
Choudhury R, Middelkoop A, Boekhorst J, Gerrits WJJ, Kemp B, Bolhuis JE, Kleerebezem M (2021a) Early life feeding accelerates gut microbiome maturation and suppresses acute post-weaning stress in piglets. Environ Microbiol 23(11):7201–7213
Choudhury R, Middelkoop A, de Souza JG, van Veen LA, Gerrits WJJ, Kemp B, Bolhuis JE, Kleerebezem M (2021b) Impact of early-life feeding on local intestinal microbiota and digestive system development in piglets. Sci Rep 11(1):4213
Chowdhury SR, King DE, Willing BP, Band MR, Beever JE, Lane AB, Loor JJ, Marini JC, Rund LA, Schook LB, Van Kessel AG, Gaskins HR (2007) Transcriptome profiling of the small intestinal epithelium in germfree versus conventional piglets. BMC Genomics 8:215
Correa F, Luise D, Archetti I, Bosi P, Trevisi P (2021) Investigation of early supplementation of nucleotides on the intestinal maturation of weaned piglets. Animals (Basel) 11(6)
Crhanova M, Karasova D, Juricova H, Matiasovicova J, Jahodarova E, Kubasova T, Seidlerova Z, Cizek A, Rychlik I (2019) Systematic culturomics shows that half of chicken caecal microbiota members can be grown in vitro except for two lineages of clostridiales and a single lineage of bacteroidetes. Microorganisms 7(11)
D'Inca R, Kloareg M, Gras-Le Guen C, Le Huërou-Luron I (2010) Intrauterine growth restriction modifies the developmental pattern of intestinal structure, transcriptomic profile, and bacterial colonization in neonatal pigs. J Nutr 140(5):925–931
De Witte C, Demeyere K, De Bruyckere S, Taminiau B, Daube G, Ducatelle R, Meyer E, Haesebrouck F (2019) Characterization of the non-glandular gastric region microbiota in Helicobacter suis-infected versus non-infected pigs identifies a potential role for Fusobacterium gastrosuis in gastric ulceration. Vet Res 50(1):39
Dehau T, Ducatelle R, Van Immerseel F, Goossens E (2022) Omics technologies in poultry health and productivity - part 1: current use in poultry research. Avian Pathol 51(5):407–417
Deng Q, Shao Y, Wang Q, Li J, Li Y, Ding X, Huang P, Yin J, Yang H, Yin Y (2020) Effects and interaction of dietary electrolyte balance and citric acid on the intestinal function of weaned piglets. J Anim Sci 98(5)
Dikeman CL, Fahey GC (2006) Viscosity as related to dietary fiber: a review. Crit Rev Food Sci Nutr 46(8):649–663
Doster AR (2000) Porcine gastric ulcer. Vet Clin North Am Food Anim Pract 16(1):163–174
Dou S, Gadonna-Widehem P, Rome V, Hamoudi D, Rhazi L, Lakhal L, Larcher T, Bahi-Jaber N, Pinon-Quintana A, Guyonvarch A, Huërou-Luron IL, Abdennebi-Najar L (2017) Characterisation of early-life fecal microbiota in susceptible and healthy pigs to post-weaning diarrhoea. PLoS One 12(1):e0169851
Duangnumsawang Y, Zentek J, Goodarzi Boroojeni F (2021) Development and functional properties of intestinal mucus layer in poultry. Front Immunol 12:745849
Duarte ME, Kim SW (2022a) Intestinal microbiota and its interaction to intestinal health in nursery pigs. Anim Nutr 8(1):169–184
Duarte ME, Kim SW (2022b) Significance of mucosa-associated microbiota and its impacts on intestinal health of pigs challenged with F18(+)E. coli. Pathogens 11(5)
Duarte ME, Sparks C, Kim SW (2021) Modulation of jejunal mucosa-associated microbiota in relation to intestinal health and nutrient digestibility in pigs by supplementation of β-glucanase to corn-soybean meal-based diets with xylanase. J Anim Sci 99(7)
Dubreuil JD (2021) Pig vaccination strategies based on enterotoxigenic Escherichia coli toxins. Braz J Microbiol 52(4):2499–2509
Eberhard M, Hennig U, Kuhla S, Brunner RM, Kleessen B, Metges CC (2007) Effect of inulin supplementation on selected gastric, duodenal, and caecal microbiota and short chain fatty acid pattern in growing piglets. Arch Anim Nutr 61(4):235–246

Ellner C, Wessels AG, Zentek J (2022) Effects of dietary cereal and protein source on fiber digestibility, composition, and metabolic activity of the intestinal microbiota in weaner piglets. Animals (Basel) 12(1)

Fairbrother JM, Nadeau E, Gyles CL (2005) Escherichia coli in postweaning diarrhea in pigs: an update on bacterial types, pathogenesis, and prevention strategies. Anim Health Res Rev 6(1):17–39

Farré-Maduell E, Casals-Pascual C (2019) The origins of gut microbiome research in Europe: from Escherich to Nissle. Hum Microbiome J 14:100065

Fenske GJ, Ghimire S, Antony L, Christopher-Hennings J, Scaria J (2020) Integration of culture-dependent and independent methods provides a more coherent picture of the pig gut microbiome. FEMS Microbiol Ecol 96(3)

Finotello F, Mastrorilli E, Di Camillo B (2018) Measuring the diversity of the human microbiota with targeted next-generation sequencing. Brief Bioinform 19(4):679–692

Fuller R (1989) Probiotics in man and animals. J Appl Bacteriol 66(5):365–378

Fuller R (1992) History and development of probiotics. In: Probiotics. Springer, pp 1–8

Gao J, Yin J, Xu K, Li T, Yin Y (2019) What is the impact of diet on nutritional diarrhea associated with gut microbiota in weaning piglets: a system review. Biomed Res Int 2019:6916189

Gardiner GE, Metzler-Zebeli BU, Lawlor PG (2020) Impact of intestinal microbiota on growth and feed efficiency in pigs: a review. Microorganisms 8(12)

Gerzova L, Babak V, Sedlar K, Faldynova M, Videnska P, Cejkova D, Jensen AN, Denis M, Kerouanton A, Ricci A (2015) Characterization of antibiotic resistance gene abundance and microbiota composition in feces of organic and conventional pigs from four EU countries. PLoS One 10(7):e0132892

Ghazisaeedi F, Ciesinski L, Bednorz C, Johanns V, Pieper L, Tedin K, Wieler LH, Günther S (2020) Phenotypic zinc resistance does not correlate with antimicrobial multi-resistance in fecal E. coli isolates of piglets. Gut Pathog 12:4

Gibson G, Macfarlane G, Cummings J (1993) Sulphate reducing bacteria and hydrogen metabolism in the human large intestine. Gut 34(4):437

Girard M, Tretola M, Bee G (2021) A single dose of synbiotics and vitamins at birth affects piglet microbiota before weaning and modifies post-weaning performance. Animals (Basel) 11(1)

Gresse R, Chaucheyras-Durand F, Fleury MA, Van de Wiele T, Forano E, Blanquet-Diot S (2017) Gut microbiota dysbiosis in postweaning piglets: understanding the keys to health. Trends Microbiol 25(10):851–873

Grzeskowiak L, Zentek J, Vahjen W (2016) Determination of the extent of Clostridium difficile colonisation and toxin accumulation in sows and neonatal piglets. Anaerobe 40:5–9

Grzeskowiak L, Martínez-Vallespín B, Dadi TH, Radloff J, Amasheh S, Heinsen FA, Franke A, Reinert K, Vahjen W, Zentek J, Pieper R (2018) Formula feeding predisposes neonatal piglets to clostridium difficile gut infection. J Infect Dis 217(9):1442–1452

Grzeskowiak L, Dadi TH, Zentek J, Vahjen W (2019a) Developing gut microbiota exerts colonisation resistance to clostridium (syn. Clostridioides) difficile in piglets. Microorganisms 7(8)

Grzeskowiak L, Pieper R, Huynh HA, Cutting SM, Vahjen W, Zentek J (2019b) Impact of early-life events on the susceptibility to Clostridium difficile colonisation and infection in the offspring of the pig. Gut Microbes 10(2):251–259

Grzeskowiak L, Teske J, Zentek J, Vahjen W (2020) Distinct patterns of microbial metabolic fingerprints in sows and their offspring: a pilot study. Arch Microbiol 202(3):511–517

Grzeskowiak L, Saliu EM, Martínez-Vallespín B, Wessels AG, Männer K, Vahjen W, Zentek J (2022) Fiber composition in Sows' diets modifies clostridioides difficile colonization in their offspring. Curr Microbiol 79(5):154

Hankel J, Chuppava B, Wilke V, Hartung CB, Muthukumarasamy U, Strowig T, Bach Knudsen KE, Kamphues J, Visscher C (2022) High dietary intake of rye affects porcine gut microbiota in a Salmonella Typhimurium infection study. Plants (Basel) 11(17)

Hellström S, Laukkanen R, Siekkinen KM, Ranta J, Maijala R, Korkeala H (2010) Listeria monocytogenes contamination in pork can originate from farms. J Food Prot 73(4):641–648

Heyer CM, Weiss E, Schmucker S, Rodehutscord M, Hoelzle LE, Mosenthin R, Stefanski V (2015) The impact of phosphorus on the immune system and the intestinal microbiota with special focus on the pig. Nutr Res Rev 28(1):67–82

Heyer CME, Schmucker S, Burbach K, Weiss E, Eklund M, Aumiller T, Capezzone F, Steuber J, Rodehutscord M, Hoelzle LE, Seifert J, Mosenthin R, Stefanski V (2019) Phytate degradation, intestinal microbiota, microbial metabolites and immune values are changed in growing pigs fed diets with varying calcium-phosphorus concentration and fermentable substrates. J Anim Physiol Anim Nutr (Berl) 103(4):1185–1197

Holanda DM, Kim SW (2022) Impacts of weaning weights and mycotoxin challenges on jejunal mucosa-associated microbiota, intestinal and systemic health, and growth performance of nursery pigs. J Anim Sci Biotechnol 13(1):43

Holman DB, Brunelle BW, Trachsel J, Allen HK (2017) Meta-analysis to define a core microbiota in the swine gut. mSystems 2(3)

Hooks KB, O'Malley MA (2017) Dysbiosis and its discontents. mBio 8(5):e01492–e01417

Hooper LV, Xu J, Falk PG, Midtvedt T, Gordon JI (1999) A molecular sensor that allows a gut commensal to control its nutrient foundation in a competitive ecosystem. Proc Natl Acad Sci 96(17):9833–9838

Huang S, Li N, Liu C, Li T, Wang W, Jiang L, Li Z, Han D, Tao S, Wang J (2019) Characteristics of the gut microbiota colonization, inflammatory profile, and plasma metabolome in intrauterine growth restricted

piglets during the first 12 hours after birth. J Microbiol 57(9):748–758
Inman CF, Haverson K, Konstantinov SR, Jones PH, Harris C, Smidt H, Miller B, Bailey M, Stokes C (2010) Rearing environment affects development of the immune system in neonates. Clin Exp Immunol 160(3):431–439
Jang KB, Kim SW (2022) Role of milk carbohydrates in intestinal health of nursery pigs: a review. J Anim Sci Biotechnol 13(1):6
Jensen BB, Jørgensen H (1994) Effect of dietary fiber on microbial activity and microbial gas production in various regions of the gastrointestinal tract of pigs. Appl Environ Microbiol 60(6):1897–1904
Jha R, Fouhse JM, Tiwari UP, Li L, Willing BP (2019) Dietary fiber and intestinal health of monogastric animals. Front Vet Sci 6:48
Jiang L, Feng C, Tao S, Li N, Zuo B, Han D, Wang J (2019) Maternal imprinting of the neonatal microbiota colonization in intrauterine growth restricted piglets: a review. J Anim Sci Biotechnol 10:88
Johnson AK, Rault JL, Marchant JN, Baxter EM, O'Driscoll K (2022) Improving young pig welfare on-farm: the five domains model. J Anim Sci 100(6)
Juul-Madsen HR, Jensen KH, Nielsen J, Damgaard BM (2010) Ontogeny and characterization of blood leukocyte subsets and serum proteins in piglets before and after weaning. Vet Immunol Immunopathol 133(2–4):95–108
Katouli M, Melin L, Jensen-Waern M, Wallgren P, Mollby R (1999) The effect of zinc oxide supplementation on the stability of the intestinal flora with special reference to composition of coliforms in weaned pigs. J Appl Microbiol 87(4):564–573
Kelly J, Daly K, Moran AW, Ryan S, Bravo D, Shirazi-Beechey SP (2017) Composition and diversity of mucosa-associated microbiota along the entire length of the pig gastrointestinal tract; dietary influences. Environ Microbiol 19(4):1425–1438
Kiarie EG, Mills A (2019) Role of feed processing on gut health and function in pigs and poultry: conundrum of optimal particle size and hydrothermal regimens. Front Vet Sci 6:19
Kim HB, Isaacson RE (2015) The pig gut microbial diversity: understanding the pig gut microbial ecology through the next generation high throughput sequencing. Vet Microbiol 177(3–4):242–251
Kim K, Song M, Liu Y, Ji P (2022) Enterotoxigenic Escherichia coli infection of weaned pigs: intestinal challenges and nutritional intervention to enhance disease resistance. Front Immunol 13:885253
Knecht D, Cholewińska P, Jankowska-Mąkosa A, Czyż K (2020) Development of swine's digestive tract microbiota and its relation to production indices-a review. Animals (Basel) 10(3)
Kolmeder CA, de Vos WM (2021) Roadmap to functional characterization of the human intestinal microbiota in its interaction with the host. J Pharm Biomed Anal 194:113751
Kong Q, Zhang W, An M, Kulyar MF, Shang Z, Tan Z, Xu Y, Li J, Liu S (2022) Characterization of bacterial microbiota composition in healthy and diarrheal early-weaned Tibetan piglets. Front Vet Sci 9:799862
Konstantinov SR, Awati AA, Williams BA, Miller BG, Jones P, Stokes CR, Akkermans AD, Smidt H, de Vos WM (2006) Post-natal development of the porcine microbiota composition and activities. Environ Microbiol 8(7):1191–1199
Kubasova T, Davidova-Gerzova L, Merlot E, Medvecky M, Polansky O, Gardan-Salmon D, Quesnel H, Rychlik I (2017) Housing systems influence gut microbiota composition of sows but not of their piglets. PLoS One 12(1):e0170051
Lauridsen C (2020) Effects of dietary fatty acids on gut health and function of pigs pre- and post-weaning. J Anim Sci 98(4)
Laycock G, Sait L, Inman C, Lewis M, Smidt H, van Diemen P, Jorgensen F, Stevens M, Bailey M (2012) A defined intestinal colonization microbiota for gnotobiotic pigs. Vet Immunol Immunopathol 149(3–4):216–224
Leite FL, Winfield B, Miller EA, Weber BP, Johnson TJ, Sylvia F, Vasquez E, Vannucci F, Beckler D, Isaacson RE (2021) Oral vaccination reduces the effects of Lawsonia intracellularis challenge on the swine small and large intestine microbiome. Front Vet Sci 8:692521
Li HY, Li BX, Liang QQ, Jin XH, Tang L, Ding QW, Wang ZX, Wei ZY (2020) Porcine deltacoronavirus infection alters bacterial communities in the colon and feces of neonatal piglets. Microbiologyopen 9(7):e1036
Li H, Yin J, Tan B, Chen J, Zhang H, Li Z, Ma X (2021a) Physiological function and application of dietary fiber in pig nutrition: a review. Anim Nutr 7(2):259–267
Li Y, Zhang R, Li X, Li J, Ji W, Zeng X, Bao J (2021b) Exposure to the environmental pollutant ammonia causes changes in gut microbiota and inflammatory markers in fattening pigs. Ecotoxicol Environ Saf 208:111564
Li J, Zhou J, Zhao S, Guo R, Zhong C, Xue T, Peng Q, Zhang B, Fan B, Liu C, Ni Y, Ren L, Zhu X, Li B (2022) Pathogenicity, infective dose and altered gut microbiota in piglets infected with porcine deltacoronavirus. Virology 567:26–33
Lima J, Manning T, Rutherford KM, Baima ET, Dewhurst RJ, Walsh P, Roehe R (2021) Taxonomic annotation of 16S rRNA sequences of pig intestinal samples using MG-RAST and QIIME2 generated different microbiota compositions. J Microbiol Methods 186:106235
Liu H, Zeng X, Zhang G, Hou C, Li N, Yu H, Shang L, Zhang X, Trevisi P, Yang F, Liu Z, Qiao S (2019a) Maternal milk and fecal microbes guide the spatiotemporal development of mucosa-associated microbiota and barrier function in the porcine neonatal gut. BMC Biol 17(1):106
Liu Y, Zheng Z, Yu L, Wu S, Sun L, Wu S, Xu Q, Cai S, Qin N, Bao W (2019b) Examination of the temporal and spatial dynamics of the gut microbiome in new-

born piglets reveals distinct microbial communities in six intestinal segments. Sci Rep 9(1):3453
Lu D, Pi Y, Ye H, Wu Y, Bai Y, Lian S, Han D, Ni D, Zou X, Zhao J, Zhang S, Kemp B, Soede N, Wang J (2022) Consumption of dietary fiber with different physicochemical properties during late pregnancy alters the gut microbiota and relieves constipation in sow model. Nutrients 14(12)
Lührmann A, Ovadenko K, Hellmich J, Sudendey C, Belik V, Zentek J, Vahjen W (2021) Characterization of the fecal microbiota of sows and their offspring from German commercial pig farms. PLoS One 16(8):e0256112
Luise D, Le Sciellour M, Buchet A, Resmond R, Clement C, Rossignol MN, Jardet D, Zemb O, Belloc C, Merlot E (2021) The fecal microbiota of piglets during weaning transition and its association with piglet growth across various farm environments. PLoS One 16(4):e0250655
Luise D, Bosi P, Raff L, Amatucci L, Virdis S, Trevisi P (2022) Bacillus spp. probiotic strains as a potential tool for limiting the use of antibiotics, and improving the growth and health of pigs and chickens. Front Microbiol 13:801827
Luo Y, Ren W, Smidt H, Wright AG, Yu B, Schyns G, McCormack UM, Cowieson AJ, Yu J, He J, Yan H, Wu J, Mackie RI, Chen D (2022) Dynamic distribution of gut microbiota in pigs at different growth stages: composition and contribution. Microbiol Spectr 10(3):e0068821
Ma Y, Lu C, Ji B, Qin J, Cai C, Yang Y, Zhao Y, Liang G, Guo X, Cao G, Li B, Gao P (2022) Integrated omics analysis reveals alterations in the intestinal microbiota and metabolites of piglets after starvation. Front Microbiol 13:881099
Mach N, Berri M, Estellé J, Levenez F, Lemonnier G, Denis C, Leplat JJ, Chevaleyre C, Billon Y, Doré J (2015) Early-life establishment of the swine gut microbiome and impact on host phenotypes. Environ Microbiol Rep 7(3):554–569
Mao X, Gu C, Hu H, Tang J, Chen D, Yu B, He J, Yu J, Luo J, Tian G (2016) Dietary Lactobacillus rhamnosus GG supplementation improves the mucosal barrier function in the intestine of weaned piglets challenged by porcine rotavirus. PLoS One 11(1):e0146312
Marchesi JR, Ravel J (2015) The vocabulary of microbiome research: a proposal. Microbiome 3(1):31
Marteyn B, Scorza FB, Sansonetti PJ, Tang C (2011) Breathing life into pathogens: the influence of oxygen on bacterial virulence and host responses in the gastrointestinal tract. Cell Microbiol 13(2):171–176
Martin-Gallausiaux C, Marinelli L, Blottière HM, Larraufie P, Lapaque N (2021) SCFA: mechanisms and functional importance in the gut. Proc Nutr Soc 80(1):37–49
Metzler-Zebeli BU, Vahjen W, Baumgärtel T, Rodehutscord M, Mosenthin R (2010) Ileal microbiota of growing pigs fed different dietary calcium phosphate levels and phytase content and subjected to ileal pectin infusion. J Anim Sci 88(1):147–158
Metzler-Zebeli BU, Newman MA, Grull D, Zebeli Q (2019) Functional adaptations in the cecal and colonic metagenomes associated with the consumption of transglycosylated starch in a pig model. BMC Microbiol 19(1):87
Metzler-Zebeli BU, Sener-Aydemir A, Sharma S, Lerch F (2021) Postnatal development of gut microbial activity and their importance for jejunal motility in piglets. J Anim Sci 99(7)
Miller BM, Liou MJ, Lee J-Y, Bäumler AJ (2021) The longitudinal and cross-sectional heterogeneity of the intestinal microbiota. Curr Opin Microbiol 63:221–230
Moita VHC, Duarte ME, Kim SW (2022) Functional roles of xylanase enhancing intestinal health and growth performance of nursery pigs by reducing the digesta viscosity and modulating the mucosa-associated microbiota in the jejunum. J Anim Sci 100(5)
Mösseler A, Wintermann M, Sander SJ, Kamphues J (2012) Effect of diet grinding and pelleting fed either dry or liquid feed on dry matter and pH in the stomach of pigs and the development of gastric ulcers. J Anim Sci 90(Suppl 4):343–345
Mulder IE, Schmidt B, Stokes CR, Lewis M, Bailey M, Aminov RI, Prosser JI, Gill BP, Pluske JR, Mayer C-D, Musk CC, Kelly D (2009) Environmentally-acquired bacteria influence microbial diversity and natural innate immune responses at gut surfaces. BMC Biol 7(1):79
Mun D, Kim H, Shin M, Ryu S, Song M, Oh S, Kim Y (2021) Decoding the intestinal microbiota repertoire of sow and weaned pigs using culturomic and metagenomic approaches. J Anim Sci Technol 63(6):1423
Nadeau É, Fairbrother JM, Zentek J, Bélanger L, Tremblay D, Tremblay CL, Röhe I, Vahjen W, Brunelle M, Hellmann K, Cvejić D, Brunner B, Schneider C, Bauer K, Wolf R, Hidalgo Á (2017) Efficacy of a single oral dose of a live bivalent E. coli vaccine against post-weaning diarrhea due to F4 and F18-positive enterotoxigenic E. coli. Vet J 226:32–39
Niederwerder MC, Constance LA, Rowland RRR, Abbas W, Fernando SC, Potter ML, Sheahan MA, Burkey TE, Hesse RA, Cino-Ozuna AG (2018) Fecal microbiota transplantation is associated with reduced morbidity and mortality in porcine circovirus associated disease. Front Microbiol 9:1631
Ouwerkerk JP, de Vos WM, Belzer C (2013) Glycobiome: bacteria and mucus at the epithelial interface. Best Pract Res Clin Gastroenterol 27(1):25–38
Partanen KH, Mroz Z (1999) Organic acids for performance enhancement in pig diets. Nutr Res Rev 12(1):117–145
Passlack N, Vahjen W, Zentek J (2015) Dietary inulin affects the intestinal microbiota in sows and their suckling piglets. BMC Vet Res 11:51
Peng J, Tang Y, Huang Y (2021) Gut health: the results of microbial and mucosal immune interactions in pigs. Anim Nutr 7(2):282–294

Petri D, Hill J, Van Kessel A (2010) Microbial succession in the gastrointestinal tract (GIT) of the preweaned pig. Livest Sci 133(1–3):107–109

Petry AL, Patience JF, Koester LR, Huntley NF, Bedford MR, Schmitz-Esser S (2021) Xylanase modulates the microbiota of ileal mucosa and digesta of pigs fed corn-based arabinoxylans likely through both a stimbiotic and prebiotic mechanism. PLoS One 16(1):e0246144

Pieper R, Kröger S, Richter JF, Wang J, Martin L, Bindelle J, Htoo JK, von Smolinski D, Vahjen W, Zentek J, Van Kessel AG (2012) Fermentable fiber ameliorates fermentable protein-induced changes in microbial ecology, but not the mucosal response, in the colon of piglets. J Nutr 142(4):661–667

Pieper R, Vahjen W, Zentek J (2016a) Intestinal lactose and mineral concentration affect the microbial ecophysiology along the gastrointestinal tract of formula-fed neonatal piglets. J Anim Sci 94(9):3786–3795

Pieper R, Villodre Tudela C, Taciak M, Bindelle J, Pérez JF, Zentek J (2016b) Health relevance of intestinal protein fermentation in young pigs. Anim Health Res Rev 17(2):137–147

Pierce KM, Sweeney T, Brophy PO, Callan JJ, Fitzpatrick E, McCarthy P, O'Doherty JV (2006) The effect of lactose and inulin on intestinal morphology, selected microbial populations and volatile fatty acid concentrations in the gastro-intestinal tract of the weanling pig. Anim Sci 82(3):311–318

Pluske JR, Turpin DL, Sahibzada S, Pineda L, Han Y, Collins A (2021) Impacts of feeding organic acid-based feed additives on diarrhea, performance, and fecal microbiome characteristics of pigs after weaning challenged with an enterotoxigenic strain of Escherichia coli. Transl Anim Sci 5(4):txab212

Pollock J, Hutchings MR, Hutchings KEK, Gally DL, Houdijk JGM (2019) Changes in the ileal, but not fecal, microbiome in response to increased dietary protein level and enterotoxigenic Escherichia coli exposure in pigs. Appl Environ Microbiol 85(19)

Poulsen AR, de Jonge N, Sugiharto S, Nielsen JL, Lauridsen C, Canibe N (2017) The microbial community of the gut differs between piglets fed sow milk, milk replacer or bovine colostrum. Br J Nutr 117(7):964–978

Proctor A, Cornick NA, Wang C, Mooyottu S, Arruda PA, Kobs K, Phillips GJ (2021) Neonatal piglets are protected from Clostridioides difficile infection by age-dependent increase in intestinal microbial diversity. Microbiol Spectr 9(2):e0124321

Reyer H, Sjöberg PJR, Oster M, Wubuli A, Murani E, Ponsuksili S, Wolf P, Wimmers K (2021) Mineral phosphorus supply in piglets impacts the microbial composition and phytate utilization in the large intestine. Microorganisms 9(6)

Richards J, Gong J, De Lange C (2005) The gastrointestinal microbiota and its role in monogastric nutrition and health with an emphasis on pigs: current understanding, possible modulations, and new technologies for ecological studies. Can J Anim Sci 85(4):421–435

Ringseis R, Eder K (2022) Heat stress in pigs and broilers: role of gut dysbiosis in the impairment of the gut-liver axis and restoration of these effects by probiotics, prebiotics and synbiotics. J Anim Sci Biotechnol 13(1):126

Rist VT, Eklund M, Bauer E, Sauer N, Mosenthin R (2012) Effect of feeding level on the composition of the intestinal microbiota in weaned piglets. J Anim Sci 90(Suppl 4):19–21

Rist VT, Weiss E, Eklund M, Mosenthin R (2013) Impact of dietary protein on microbiota composition and activity in the gastrointestinal tract of piglets in relation to gut health: a review. Animal 7(7):1067–1078

Ritzmann M, Heinritzi K (2006) Durchfallerkrankungen beim Schwein. Tierärztliche Praxis Ausgabe G: Großtiere/Nutztiere 34(06):411–416

Roediger W (1980) Role of anaerobic bacteria in the metabolic welfare of the colonic mucosa in man. Gut 21(9):793–798

Rossi B, Toschi A, Piva A, Grilli E (2020) Single components of botanicals and nature-identical compounds as a non-antibiotic strategy to ameliorate health status and improve performance in poultry and pigs. Nutr Res Rev 33(2):218–234

Ryu EP, Davenport ER (2022) Host genetic determinants of the microbiome across animals: from Caenorhabditis elegans to cattle. Annu Rev Anim Biosci 10:203–226

Sachsenröder J, Twardziok SO, Scheuch M, Johne R (2014) The general composition of the faecal virome of pigs depends on age, but not on feeding with a probiotic bacterium. PLoS One 9(2):e88888

Samolinska W, Kowalczuk-Vasilev E, Grela ER (2018) Comparative effect of different dietary inulin sources and probiotics on growth performance and blood characteristics in growing-finishing pigs. Arch Anim Nutr 72(5):379–395

Sansom BF, Gleed PT (1981) The ingestion of sow's faeces by suckling piglets. Br J Nutr 46(3):451–456

Schierack P, Steinrück H, Kleta S, Vahjen W (2006) Virulence factor gene profiles of Escherichia coli isolates from clinically healthy pigs. Appl Environ Microbiol 72(10):6680–6686

Schierack P, Walk N, Reiter K, Weyrauch KD, Wieler LH (2007) Composition of intestinal Enterobacteriaceae populations of healthy domestic pigs. Microbiology (Reading) 153(Pt 11):3830–3837

Schroeder BO (2019) Fight them or feed them: how the intestinal mucus layer manages the gut microbiota. Gastroenterol Rep 7(1):3–12

Schulze Holthausen J, Schregel J, Sciascia QL, Li Z, Tuchscherer A, Vahjen W, Metges CC, Zentek J (2022) Effects of oral glutamine supplementation, birthweight and age on colonic morphology and microbiome development in male suckling piglets. Microorganisms 10(10):1899

Sinkora M, Butler JE (2016) Progress in the use of swine in developmental immunology of B and T lymphocytes. Dev Comp Immunol 58:1–17

Songer JG (2004) The emergence of Clostridium difficile as a pathogen of food animals. Anim Health Res Rev 5(2):321–326

Starke IC, Pieper R, Neumann K, Zentek J, Vahjen W (2014) The impact of high dietary zinc oxide on the development of the intestinal microbiota in weaned piglets. FEMS Microbiol Ecol 87(2):416–427

Stokes CR (2017) The development and role of microbial-host interactions in gut mucosal immune development. J Anim Sci Biotechnol 8:12

Sun J, Zhong H, Du L, Li X, Ding Y, Cao H, Liu Z, Ge L (2018) Gene expression profiles of germ-free and conventional piglets from the same litter. Sci Rep 8(1):10745

Sun J, Du L, Li X, Zhong H, Ding Y, Liu Z, Ge L (2019) Identification of the core bacteria in rectums of diarrheic and non-diarrheic piglets. Sci Rep 9(1):18675

Taillieu E, Chiers K, Amorim I, Gärtner F, Maes D, Van Steenkiste C, Haesebrouck F (2022) Gastric Helicobacter species associated with dogs, cats and pigs: significance for public and animal health. Vet Res 53(1):42

Tan Z, Dong W, Ding Y, Ding X, Zhang Q, Jiang L (2019) Changes in cecal microbiota community of suckling piglets infected with porcine epidemic diarrhea virus. PLoS One 14(7):e0219868

Tang W, Zhang W, Azad MAK, Ma C, Zhu Q, Kong X (2022) Metabolome, microbiome, and gene expression alterations in the colon of newborn piglets with intrauterine growth restriction. Front Microbiol 13:989060

Tiwari UP, Singh AK, Jha R (2019) Fermentation characteristics of resistant starch, arabinoxylan, and beta-glucan and their effects on the gut microbial ecology of pigs: a review. Anim Nutr 5(3):217–226

Torres-Fuentes C, Schellekens H, Dinan TG, Cryan JF (2017) The microbiota-gut-brain axis in obesity. Lancet Gastroenterol Hepatol 2(10):747–756

Trachsel J, Briggs C, Gabler NK, Allen HK, Loving CL (2019) Dietary resistant potato starch alters intestinal microbial communities and their metabolites, and markers of immune regulation and barrier function in swine. Front Immunol 10:1381

Trevisi P, Luise D, Correa F, Bosi P (2021) Timely control of gastrointestinal eubiosis: a strategic pillar of pig health. Microorganisms 9(2)

Tröscher-Mußotter J, Tilocca B, Stefanski V, Seifert J (2019) Analysis of the bacterial and host proteins along and across the porcine gastrointestinal tract. Proteomes 7(1)

Urubschurov V, Janczyk P, Souffrant WB, Freyer G, Zeyner A (2011) Establishment of intestinal microbiota with focus on yeasts of unweaned and weaned piglets kept under different farm conditions. FEMS Microbiol Ecol 77(3):493–502

Urubschurov V, Büsing K, Freyer G, Herlemann DP, Souffrant WB, Zeyner A (2017) New insights into the role of the porcine intestinal yeast, Kazachstania slooffiae, in intestinal environment of weaned piglets. FEMS Microbiol Ecol 93(2)

Ushida K, Tsuchida S, Ogura Y, Toyoda A, Maruyama F (2016) Domestication and cereal feeding developed domestic pig-type intestinal microbiota in animals of suidae. Anim Sci J 87(6):835–841

Vahjen W, Pietruszyńska D, Starke IC, Zentek J (2015) High dietary zinc supplementation increases the occurrence of tetracycline and sulfonamide resistance genes in the intestine of weaned pigs. Gut Pathog 7:23

Van Hees HMJ, Davids M, Maes D, Millet S, Possemiers S, den Hartog LA, van Kempen T, Janssens GPJ (2019) Dietary fibre enrichment of supplemental feed modulates the development of the intestinal tract in suckling piglets. J Anim Sci Biotechnol 10:83

Vandeplas S, Dubois Dauphin R, Beckers Y, Thonart P, Théwis A (2010) Salmonella in chicken: current and developing strategies to reduce contamination at farm level. J Food Prot 73(4):774–785

Vanrolleghem W, Tanghe S, Verstringe S, Bruggeman G, Papadopoulos D, Trevisi P, Zentek J, Sarrazin S, Dewulf J (2019) Potential dietary feed additives with antibacterial effects and their impact on performance of weaned piglets: a meta-analysis. Vet J 249:24–32

Vasquez R, Oh JK, Song JH, Kang DK (2022) Gut microbiome-produced metabolites in pigs: a review on their biological functions and the influence of probiotics. J Anim Sci Technol 64(4):671–695

Vayssier-Taussat M, Albina E, Citti C, Cosson JF, Jacques M-A, Lebrun M-H, Le Loir Y, Ogliastro M, Petit M-A, Roumagnac P, Candresse T (2014) Shifting the paradigm from pathogens to pathobiome: new concepts in the light of meta-omics. Front Cell Infect Microbiol 4

Verschuren LMG, Calus MPL, Jansman AJM, Bergsma R, Knol EF, Gilbert H, Zemb O (2018) Fecal microbial composition associated with variation in feed efficiency in pigs depends on diet and sex. J Anim Sci 96(4):1405–1418

Villagómez-Estrada S, Pérez JF, Melo-Durán D, Gonzalez-Solè F, D'Angelo M, Pérez-Cano FJ, Solà-Oriol D (2022) Body weight of newborn and suckling piglets affects their intestinal gene expression. J Anim Sci 100(6)

Vo N, Tsai TC, Maxwell C, Carbonero F (2017) Early exposure to agricultural soil accelerates the maturation of the early-life pig gut microbiota. Anaerobe 45:31–39

Wang M, Radlowski EC, Monaco MH, Fahey GC Jr, Gaskins HR, Donovan SM (2013) Mode of delivery and early nutrition modulate microbial colonization and fermentation products in neonatal piglets. J Nutr 143(6):795–803

Wang X, Tsai T, Deng F, Wei X, Chai J, Knapp J, Apple J, Maxwell CV, Lee JA, Li Y, Zhao J (2019) Longitudinal investigation of the swine gut microbiome from birth to market reveals stage and growth performance associated bacteria. Microbiome 7(1):109

Wang Y, Hu Y, Liu F, Cao J, Lv N, Zhu B, Zhang G, Gao GF (2020) Integrated metagenomic and metatranscriptomic profiling reveals differentially expressed resistomes in human, chicken, and pig gut microbiomes. Environ Int 138:105649

Wessels AG (2022) Influence of the gut microbiome on feed intake of farm animals. Microorganisms 10(7):1305

Wu R, Cao Z, Jiang Y, Chen W, Sun Y, Li Q, Mi J, Deng L, Liao X, Feng Y, Lan T, Ma J (2022) Early life dynamics of ARG and MGE associated with intestinal virome in neonatal piglets. Vet Microbiol 274:109575

Yang Q, Huang X, Wang P, Yan Z, Sun W, Zhao S, Gun S (2019) Longitudinal development of the gut microbiota in healthy and diarrheic piglets induced by age-related dietary changes. Microbiologyopen 8(12):e923

Yang F, Zhang S, Tian M, Chen J, Chen F, Guan W (2020a) Different sources of high fat diet induces marked changes in gut microbiota of nursery pigs. Front Microbiol 11:859

Yang S, Li Y, Wang B, Yang N, Huang X, Chen Q, Geng S, Zhou Y, Shi H, Wang L, Brugman S, Savelkoul H, Liu G (2020b) Acute porcine epidemic diarrhea virus infection reshapes the intestinal microbiota. Virology 548:200–212

Yang A, Wang K, Peng X, Lv F, Wang Y, Cui Y, Wang Y, Qu D, Zhou J, Si H (2022) Effects of different sources of calcium in the diet on growth performance, blood metabolic parameters, and intestinal bacterial community and function of weaned piglets. Front Nutr 9:885497

Zabielski R, Godlewski MM, Guilloteau P (2008) Control of development of gastrointestinal system in neonates. J Physiol Pharmacol 59(Suppl 1):35–54

Zamojska D, Nowak A, Nowak I, Macierzyńska-Piotrowska E (2021) Probiotics and postbiotics as substitutes of antibiotics in farm animals: a review. Animals (Basel) 11(12)

Zentek J, Buchheit-Renko S, Ferrara F, Vahjen W, Van Kessel AG, Pieper R (2011) Nutritional and physiological role of medium-chain triglycerides and medium-chain fatty acids in piglets. Anim Health Res Rev 12(1):83–93

Zentek J, Ferrara F, Pieper R, Tedin L, Meyer W, Vahjen W (2013) Effects of dietary combinations of organic acids and medium chain fatty acids on the gastrointestinal microbial ecology and bacterial metabolites in the digestive tract of weaning piglets. J Anim Sci 91(7):3200–3210

Zhang W, Ma C, Xie P, Zhu Q, Wang X, Yin Y, Kong X (2019) Gut microbiota of newborn piglets with intrauterine growth restriction have lower diversity and different taxonomic abundances. J Appl Microbiol 127(2):354–369

Zhang J, Rodriguez F, Navas MJ, Costa-Hurtado M, Almagro V, Bosch-Camos L, Lopez E, Cuadrado R, Accensi F, Pina-Pedrero S, Martinez J, Correa-Fiz F (2020) Fecal microbiota transplantation from warthog to pig confirms the influence of the gut microbiota on African swine fever susceptibility. Sci Rep 10(1):17605

Zhao J, Zhang Z, Zhang S, Page G, Jaworski NW (2021) The role of lactose in weanling pig nutrition: a literature and meta-analysis review. J Anim Sci Biotechnol 12(1):10

# Physiology of Gut Water Balance and Pathomechanics of Diarrhea

Ted H. Elsasser and Shollie Faulkenberg

**Abstract**

Diarrhea is costly to animals in terms of overall health and the incurred disruption of the internal physiologic milieu. While diarrhea can be classified into a number of different categories, all have in common the development of excess water residing in the bowel, as a result of water coming from the incapacitation of regulatory mechanisms in the impacted host cells to block water loss and reclaim water from the bowel. The common mechanism underlying the vast majority of these infectious as well as noninfectious origins of diarrhea is the activation of a matrix of inflammatory processes in the epithelial cells or near-by immune and neural cells that unleash a variety of effector molecules in a proinflammatory cascade. It may be reasonable to consider novel therapy to control the progression of diarrhea and the severity of symptoms by targeting some of these inflammatory pathways to limit the extent of self-damage. This intervention is perhaps complicated by the fact that some of the observed impacts of the inflammatory process are perhaps essential to the host, especially as regards pathogen-mediated diarrhea, in its attempts to limit the extent of the disease by shedding the cells that are infected. The progression of diarrhea to a state of systemic dehydration further challenges host recovery as more and more homeostatic mechanisms that regulate processes such as acid–base balance, thermoregulation, kidney function, and cardiovascular function are all dependent on a stable internal milieu of ions, nutrients, and water. Remission of diarrhea will proceed as the integrity of the bowel is regenerated by the replacement of shed cells, the reestablishment of protein-based barrier junctions between the cells, and the reestablishment of normal homeostatic ion transport mechanisms in the gut cells.

**Keywords**

Diarrhea · Epithelial barrier · Inflammation · Ion transport · Osmotic balance · Water flux

T. H. Elsasser (✉)
Independent Animal Health Consultant,
Berwyn Heights, MD, USA

Visiting Adjunct Professor, Center for Vaccine Development and Global Health, University of Maryland School of Medicine, Baltimore, MD, USA

S. Faulkenberg
College of Veterinary Medicine, Department of Pathobiology, Auburn University, Auburn, AL, USA
e-mail: Smf0076@auburn.edu



J. J. Gross (ed.), *Production Diseases in Farm Animals*,
https://doi.org/10.1007/978-3-031-51788-4_9

# 1 Overview Perspective

## 1.1 The Generalized Impact of Diarrhea

The focus of this chapter addresses the nature of compromised gut water homeostasis in domestic food animals and consequences thereof. As a generalization, "commercial food animal production" is basically any scenario wherein animals are bred and reared for commercial sale. Nutrition and nutrient management can be the key to success or failure of production operations. For the purpose of discussion here, nutrient management is not only the fundamental aspects of feeds and feeding but also all of the inputs that collectively impact the efficiency with which those nutrients are used for a given physiological or production purpose and outcome (Jha et al. 2019). As exemplified by Makkar (2018), feed is valued as the single most costly input to animal production. For example, feed amounts to as much as 70% of the costs of raising commercial broiler chickens (El-Deek et al. 2020) and highly variable (40–70%) for cow-calf pairs. Feed and nutrient use efficiency in pork production impacts projections on sustainability in this animal production market (Shurson and Urriola 2022). There is significant fluctuation in feed ingredient commodity pricing that relates to the speculative projections of the market, and as such, the added concern of just the potential for feed cost to further increase impacts producer confidence in decision-making. Therefore, and reasonably so, issues that worsen feed conversion efficiency (i.e., amount of feed consumed to generate one unit of animal-derived product) even to a small extent, diarrhea syndromes in particular, can have a substantial impact on the profitability of the enterprise. Feed and health maintenance costs are additive (morbid animals are less efficient at converting feed to gain) and only escalate as production scenarios intensify, with a higher chance of disease outbreaks spreading more quickly within herds and flocks (Kaniyamattam et al. 2021). In addition, for the companion and sport animal owner or manager, lost efficiency due to poor nutrient use relates to several input factors ranging from aging to disease issues. This can also lead to increased veterinary expenses both to determine, for example, why high protein and energy diets might be insufficient to fully support growth in younger animals or provide maintenance nutrition adequate to maintain weight and good health or competitive performance in mature animals. A World Bank analysis of longer-term impacts on animal production costs associated with disease issues estimates that over a ten-year study period, while direct production of disease costs (i.e., mortality, morbidity, culling, treatment, depopulation, etc.) amounted to as much as 20 billion US dollars, indirect production disease costs added an additional 200 billion dollars to the realistic total. These costs were also shown to be associated directly with the challenges of zoonotic carryover, environmental concerns, and a myriad of collateral impacts as defined under the One Health concept (Machalaba et al. 2021). As summarized in an economic impact study employing time series analysis to estimate indirect disease costs in a foot and mouth disease outbreak, the authors commented on these indirect costs occurring because, in the present world scenario, "… the costs of disease do not stop at the farm-gate, within the agricultural sector, or after disease-freedom is declared" (Barratt et al. 2019). Overall, these concepts of nutrient management correlate directly with gut health in one way or another and also play a central role in the general principles impacting both human and animal welfare (Kraimi et al. 2019; Chen et al. 2021). Therefore, as diarrhea is a direct challenge to nutrient use by animals, the scale of the economic impact is tremendous.

## 1.2 Diarrhea Is Complex Because the Gut Is Complex and Diverse in Structure

Simply stated, diarrhea is largely regarded as present when a person or animal passes gut fecal material from the body as a watery and loose stool and often times more frequent in occurrence. In many situations, one of the first signs that the generalized gut health of an animal(s) is perturbed and in a state of imbalance is the observation of diarrhea in individual animals, the herd, or the flock. Depending on the animal species, food, and environment, fecal stool normally has a particular "normal" range in the ratio of solids to water and perhaps in the frequency and intensity that the digesta is moved by peristalsis through the gut and discharged as fecal material from the body. Somewhat counterintuitive but still a demonstrable fact, the very hypertonic digesta becomes essentially isotonic with the circulating blood plasma shortly after entering the bowel (at least into the proximal jejunum), and this isotonicity stems from the capacity of the duodenum and jejunum to secrete the appropriate amount of water (Fordtran et al. 1965; Field 2003) into the lumen of the bowel. This is necessary largely because the drier feed being processed in the digestive tract would otherwise be too dense to facilitate the proper mixing, dissolution/suspension of nutrients, and exposure to the absorptive surfaces of the gut, the villi, and microvilli. Further down the alimentary tract, in the ileum and colon in particular, the net transcellular flux of water is in the direction of reabsorption, and in this process, the water content of the fecal material progressively decreases as the digesta moves toward the terminal colon. Everything that in states of normal gut health determines the direction of the net flux of water into or out of the gut lumen depends on (1) the location in the gut being considered and (2) the molecular attributes of the epithelial cells and intercellular junctions that secure the integrity and relative impermeability of the villi and crypts of the cell barrier layer at that specific location. More detail on these molecular water flow processes and their regulatory components will be presented in the subsequent review sections.

It is relevant to this discussion to provide an updated review of the essential features of the gut architecture, cellular organization, and spatial and temporal considerations of cell regeneration in the gut, and the fundamental relationship between the epithelial cells in the gut boundary layer of the host and the local microbiome. This all relates back to how perturbations in the normal functioning of the gut result in diarrhea. In a general sense, most people consider the integrity of the gut and the proper functioning of the gut as a reflection of the stability and consistency of the homeostatic processes that underlie good digestion and nutrient uptake. In this view, gastrointestinal disease states can be considered a fundamental loss of that stability and function. However, there is little that is truly stable and steadfast in the gut. There has developed a popular saying about some new version of turmoil in life: "Nothing is constant but change". From a physiological perspective, it is exactly the changing, and more accurately "updating" nature of the gut architecture, components, and processes that gut health is dependent on for proper functioning, digestion, and absorption of water and nutrients. All of the dynamic changes in processes and architecture (cell and microbial turnover, cell replacement, etc.) in the gut, especially over time, constitute the true homeostatic status and adjustments needed to establish the cellular conditions that ensure proper and efficient digestion, secretion, and absorption across all of the changing dietary, health, and environmental states animals encounter on a daily basis. The obvious example of this is the accommodations and changes in villus length that occur in the small intestine segments associated with plane of nutrition (longer in the high plane of nutrition and vice versa), diet composition (fiber, protein and energy sources, ingredients with inflammatory potential), pathogen exposures (usually shorter in states of infection and inflammation), management strategies (heat stress-associated shortening, socio-behavioral stress shortening), and age or physiological state (shorter villus length with deeper crypt depth in older aged animals) (Serena et al. 2008; Nealon et al. 2017; Proszkowiec-Weglarz et al. 2020; Qu et al. 2021; Kuzmuk et al. 2005), all while ensuring the proper separation of the lumen environment from the body interior. These have the added potential for influencing susceptibility to the onset of diarrhea because of localized impacts on barrier membrane integrity.

Newer technologies, based on rapid reading and polymerase chain reaction quantification of 16s-RNA extractable from the gut microbial populations, are making more and more clear exactly what differentiates a healthy microbiome (eubiotic, eubiosis) from one associated with many gut diseases, especially diarrhea, in an unbalanced state referred to as dysbiotic/dysbiosis. In general, eubiosis is characterized by balanced relative numbers of commensal, beneficial microbes out competing pathogenic varieties; dysbiosis stems from imbalances between both beneficial and pathogenic organisms as well as disproportionate imbalances in regional commensal bacteria that can change the local microenvironments downstream, thus facilitating bacterial overgrowth or pathogen emergence (Ghoshal and Ghoshal 2017; Singh et al. 2021). It should be obvious that the environmental conditions in the gut change as digesta progresses down the alimentary tract, especially with regard to oxygen tension versus an anaerobic state; as such, it is not surprising that the upper intestine, the duodenum, and jejunum would be more highly colonized by Gram-positive bacteria while the ileum and colon are dominated by more Gram-negative organisms (Ghoshal and Ghoshal 2017). This being the case, it stands to reason that in dysbiosis, where pathogenic bacteria arise and cause gastric disturbance and diarrhea, it is likely the situation that the toxins and actions of the pathogens impacting host gut cell function and integrity follow the lines of which Gram-positive or Gram-negative organisms are present in that environment.

In the small and large intestines, the mature epithelial cells lining the gut surface in proximity to the lumen are either absorptive or secretory in function [a full description of the structure–function cell type relationships in the gut can be found in Noah et al. (2011)]. From the cellular point of view, the secretory and absorptive surfaces of the intestine are specialized epithelial surface cells of the villi and crypts. These cells are called "enterocytes" in the villus protrusions into the lumen with crypt depressions between the villi in the small intestine. In the large intestine, only the flattened epithelial "colonocytes" and crypts are present (Singhal and Shah 2020). More specifically, there are at least five different lineages of epithelial cells. Each cell type (i.e., enterocytes, goblet cells, Paneth cells, enteroendocrine, M-cells, etc.) emerges from the crypt as a fundamental stem cell under the influence of a myriad of induction factors, transcriptional factors, and signal transduction pathways (Heath 2010). These stem cells differentiate into a designated cell type as the emergent cells migrate out of the crypt and up the villus toward the apical end and the gut lumen (Heath 2010; Elmentaite et al. 2021). Interestingly, though perhaps not unexpectedly, the generation, differentiation, and abundance of epithelial cell types throughout the intestine segments differ (Zhang et al. 2019). The distribution and proportions of the cell types vary not only along the length of the gut but may also vary by species, poultry versus mammals, for example. These differences reflect, for the most part, the particular function of each specific portion of the gut for a given type of animal and its dietary needs and dietary component availability. In general, the enterocytes in the small intestine can account for more than 85% of the epithelial cells, with goblet cells accounting for 10% of the cell types; in contrast, large intestine colonocytes vary with Goblet cells where Goblet cells can be as much as 50% of the cells in a given location (Noah et al. 2011). Paneth cells are not found in the large intestine. Important epithelial cell secretions include digestive enzymes, mucin proteins/mucus, antimicrobial peptides, or gut hormones as a function of the specific differentiated cell type. Interestingly and more recently discovered, a growing list of micro-RNAs (Lee et al. 2016) have been associated to a large extent with colonocytes (which are the colonic equivalent of the small intestine enterocytes but with several basic differences in nutrient sensors, ionic transporter proteins, and some surface receptors) with the micro-RNAs capable of shaping the nature of interactions between the host and the microbiome (Liu et al. 2016; Zhao et al. 2020) (and vice versa!). From the pathology perspective, three of these (miR-1224-5P, miR-3473A, and miR-5128) have actions during colonic inflammation to increase the selective expression of the regulatory water channel protein aquaporin-8 (AQP-8) in colonocytes, thus with a great potential to influence the propensity for diarrhea to develop in the large intestine. Originally thought to be determined by programmed intrinsic host/body factors, more recently described features indicate that a large part of the differentiation stimulus, and therefore the local proportion of epithelial cell types, is greatly influenced also by cellular and lumen environmental factors such as the microbial milieu and the segment-related patterns of digesta metabolites (Arpaia et al. 2013; Dheer and Young 2021; Nakamura et al. 2021). Integrated together with the host factors, these lumen environmental factors include both nutrient factors derived from feed and the populations of the microbes that inhabit the various niches of the gut all along the tract and regionally further refine the generation of digesta metabolites. Collectively, experimental studies document that when the vital presence of these largely defensive secretions is impaired, the likelihood of diarrhea diseases emerging (as a result of dysbiosis and pathogen microbe imbalances) increases greatly since gut water balance depends on the quality (eubiosis of sound gut health) of the host–microbe–nutrient homeostatic interactions. As will be discussed, dysbiosis results in inflammation. Inflammation triggers impairment of homeostatic mechanisms, including those regulating water balance resulting in impaired absorption of water from the gut lumen back into the plasma compartment as the hallmark of diarrhea.

As a final thought on this matter, there is incredible plasticity and variability in gut function underlying sound gut health, with these attributes stemming from the integrated host/nutrient/microbe interactions. For example, feed-derived nutrients shape the presence and abundance of the various microbes. Some of the best-documented interactions at this level are (1) the observed maintenance of a favorable microbial milieu in the colon as orchestrated by the availability of the short-chain fatty acids (Louis and Flint 2009; Puertollano et al. 2014) and (2) oligosaccharide modulation of several populations of microbes in the small intestine (Cai et al. 2020) including the immunomodulatory segmented filamentous bacteria Candidatus Arthromitus (Rahimi et al. 2020). In fact, there is bi-directional communication between the microbes and host based on metabolite signaling. This involves receptor-based signaling, nutrient sensors, and neuronal activation that contribute to the homeostatic response. For example, microbes process dietary fiber, generating butyrate and propionate, which feedback on the intestinal epithelial cells. Nutrients also interact with epithelial cells, which further control microbial populations in microenvironments of the gut via diverse mechanisms ranging from the release of microbe population-modifying antimicrobial peptides to the subtle effects of micro-RNAs (Liu et al. 2016; Zhao et al. 2020). All of this can and does occur on a relatively short time scale, reflecting the fast dynamics of the gut system and ultimately the nature of modified patterns of function and distribution of surface epithelial cell types. Why is any of this important to a discussion of mechanisms underlying diarrhea? All of these factors impact the stability of the selective permeability of the single-cell thickness of the epithelial barrier between the lumen of the gut and the internal milieu. If the barrier is perturbed, diarrhea will be the result.

## 2 Gut Integrity, Semi-impermeable Barriers, and Consequences of Inflammation

The next consideration in regard to how gut architecture, or more precisely changes in gut architecture, might relate to the development of diarrhea is gut and cellular integrity and barrier function security. The ability to regulate water flux in the bowel is totally dependent on a rather continuous barrier to impede and restrict otherwise unregulated, passive water, ion, and nutrient movement into and out of the lumen. Importantly, however, a natural challenge to this integrity is continually present in states of good gut health, mainly because an epithelial cell doesn't just stay in one place in the villus once it leaves the crypt and begins its ascending migration to become part of the villus. The differentiated epithelial cells arising in the crypts of the gut soon cease to divide but continuously migrate toward the distal end of the villus. Without the need to proliferate, these cells conserve their metabolic energy for their main functions of absorption, synthesis and secretion. Along the way the cells prepare for their programmed shedding into the lumen with the induction of naturally occurring, programmed apoptotic signals (de Santa Barbara et al. 2003; Blander 2018). Well-regulated epithelial cell apoptotic (Toll-like receptor/caspase-mediated programmed cell death (Saavedra et al. 2018; Van Opdenbosch et al. 2017)) signaling is extremely important to bowel integrity. In regard to the relationship between diarrhea and apoptosis, certain gut pathogen infections that have a presentation of diarrhea cause the unregulated passage of water to channel into the hypertonic lumen. This can occur in part because in sensing the pathogen challenge, impacted host cells (enterocytes as well as immune cells) upregulate epithelial cell apoptosis in epithelial cells, perhaps prematurely shedding cells to eliminate infected cells, thus limiting the infection (Saavedra et al. 2018; Van Opdenbosch et al. 2017), or further triggering self-lytic events that unfortunately result in fragmentation of the barrier and disruption of otherwise well-regulated water and ion flow. All of the movement in this dynamic changing of position of the cell populations requires the remodeling of the protein matrix barrier junctions between the cells such that paracellular (between adjacent cells) water passage is restricted. Complicating this, the overall length of villi changes down the intestinal tract; the longest villi are most frequently observed in the duodenum and shortest in the ileum and colon (personal observations on the calf and chicken intestine). Furthermore, the villus length, and therefore the functional secretory and absorptive surface areas, is subject to change or modification, as previously mentioned, in response to such factors as the plane of nutrition, nutritional/dietary components, and presence of differing populations of commensal and pathogenic microorganisms. In most instances of gut disturbance, perturbation, or diarrhea, either from a pathogen or nonpathogen origin, inflammation inherent to almost all levels of disease largely results in the shortening of villi, loss of or reduction in absorptive capacity, and in many instances an increased rate of apoptosis and a higher rate of shedding of terminal epithelial cells. In severe instances, inflammatory processes as well as some direct effector molecules released from pathogens (i.e., toxins), induce significant increases in signaling that triggers higher levels of apoptosis, cell autophagy, and necroptosis/necrosis (Günther et al. 2013; Patankar and Becker 2020). Obviously, as cell losses occur, the basic integrity of the villi ends is compromised, as observed by the fragmentation of the villi tips or "clubbing," characterized by a blunting and rounding of the apical villus surface. The weaned piglet has been used as a model for studying the etiology of diarrhea, and data strongly implicate observable shortening of intestinal villi along with discontinuous integrity and fragmentation of the epithelial cells composing the barrier surface commensurate with the deepening of crypts as a prime factor in diarrhea-associated mortality. Interestingly, however, with nutrient uptake compromised by this loss of absorptive gut surface area, supplemental feeding, especially as sow's milk, curbed the villus-shortening impacts of the inflammatory stressors and improved outcomes (Nabuurs 1998).

This loss of villus length and structure certainly has negative consequences for the host, but more recent theories (Kapczuk et al. 2020) suggest that this process might, in fact, also be a survival mechanism associated with host attempts to shed infected and damaged cells in an attempt to clear an infection more quickly. Most often, the lateral intercellular barrier junctions deteriorate, allowing unregulated flow of water and ions between the lumen and plasma compartments, and, in total, this results in what is referred as the "leaky gut" syndrome. When this occurs, not only can excess water leave the bowel, but also gut toxins such as endogenous lipopolysaccharide/endotoxin/enterotoxins and even microorganisms can now get into the circulation and begin to compromise liver and other organ functions. This condition has been documented, for example, in association with infectious pathogens in chickens (Awad et al. 2017), heat stress, and diet/feeding changes in pigs (Cottrell et al. 2020; Mayorga et al. 2018; Pearce et al. 2013), heat and *E. coli* infection stress in calves (Opgenorth et al. 2021; He et al. 2022), and even excessive exercise in horses and dogs (Stewart et al. 2017), conditions all generating some form of inflammation. The relevance here is that circumstances that promote leaky gut and permeability-integrity issues are basically risk factors for diarrhea, especially regarding rotavirus and bacterial pathogens such as *Campylobacter* (Barmeyer et al. 2017; Gecse et al. 2012; Lobo de Sá et al. 2021; Mao et al. 2019; Yang et al. 2015) that are associated with the cellular targets many of the pathogen virulence factors and toxins interact with.

In response to sensed inflammatory conditions in the gut, several cell populations on the villus surface respond with increased caspase activity, setting into motion the pathways associated with apoptosis (Fig. 1, from Connor et al. (2013, 2016), and Elsasser et al. (2018)), and increased activity of enzymes generating excessive amounts of nitric oxide (NO) and superoxide anion (SOA). These reactive NO and SOA molecules are often generated quite locally from epithelial sources of nitric oxide synthase and/or inflammatory perturbation of mitochondria and a molecular "short-circuiting" of the electron transport chain. The result of this increase in NO and SOA can be visualized in immunofluorescence microscopy in the production of cell-specific immunostaining for elevated levels of 3′-tyrosine-nitrated proteins, reaction products of the aberrant interaction between the NO/SOA reaction product peroxynitrite ($ONOO^-$), and tyrosine residues of proteins (Elsasser et al. 2012a). Nitrated proteins, specifically the nitration of tyrosine in the 3′-position of the phenolic ring, arise from the overgeneration of intracellular levels of NO and SOA that under ordinary circumstances are sufficiently spatially distant or temporally distinct as to permit their intended intracellular signaling functions and natural decay and unlikely chance of interacting (Elsasser et al. 2008). The consequence of these protein tyrosine nitrations is that (1) functional enzymes like kinases that undergo activation through being phosphorylated at the tyrosine phenolic ring 4′-position cannot be phosphorylated due to steric and charge hindrance of a more rapidly acting nitration at the 3′-position (as we have demonstrated with the nitration of Janus Kinase-2 in the liver (Elsasser et al. 2007)), (2) changes in secondary and tertiary protein structure changing needed atomic distances and charges to perform their molecular actions efficiently, and (3) developed intracellular loss of self-recognition and sensed structural perturbations that lead to ubiquitination of the damaged proteins and delivery to proteosomes for premature dissolution (Elsasser et al. 2012b). The greatest stimulus to epithelial cell protein nitration is the upregulation of NO and SOA via proinflammatory cytokines such as tumor necrosis factor-α (TNF-α), interleukin-1, and interleukin-6 from resident and infiltrating immune cells, as well as pathogen or toxin detection via pathogen-associated molecular pattern (PAMP) receptors. The epithelial cells can respond to both the pathogen toxins directly (i.e., endotoxins via Toll-like receptor-4) or secondarily to the immune cell-released cytokines via cytokine receptors. Further details on the relationship of protein nitration to gut cell health are presented in the accompanying figure legend. Further increased nitrosative stress occurs with conditions that result in perturbations to local intestinal blood flow, low oxygen levels and increased partial pressure of carbon dioxide.

## 3 Evaluating the Causes of Diarrhea

### 3.1 Definitions and Classifications

In diarrhea, the most observable sign is significantly increased water content of the fecal material, often times accompanied by increased frequency of fecal discharge. In many situations, the intensity with which the fecal material is discharged increases for the most part with generalized effects of enterotoxins that are sensed by or have a direct impact on cells collateral to enterocytes. This results in the increased release of neuromuscular stimuli, including peptide activators

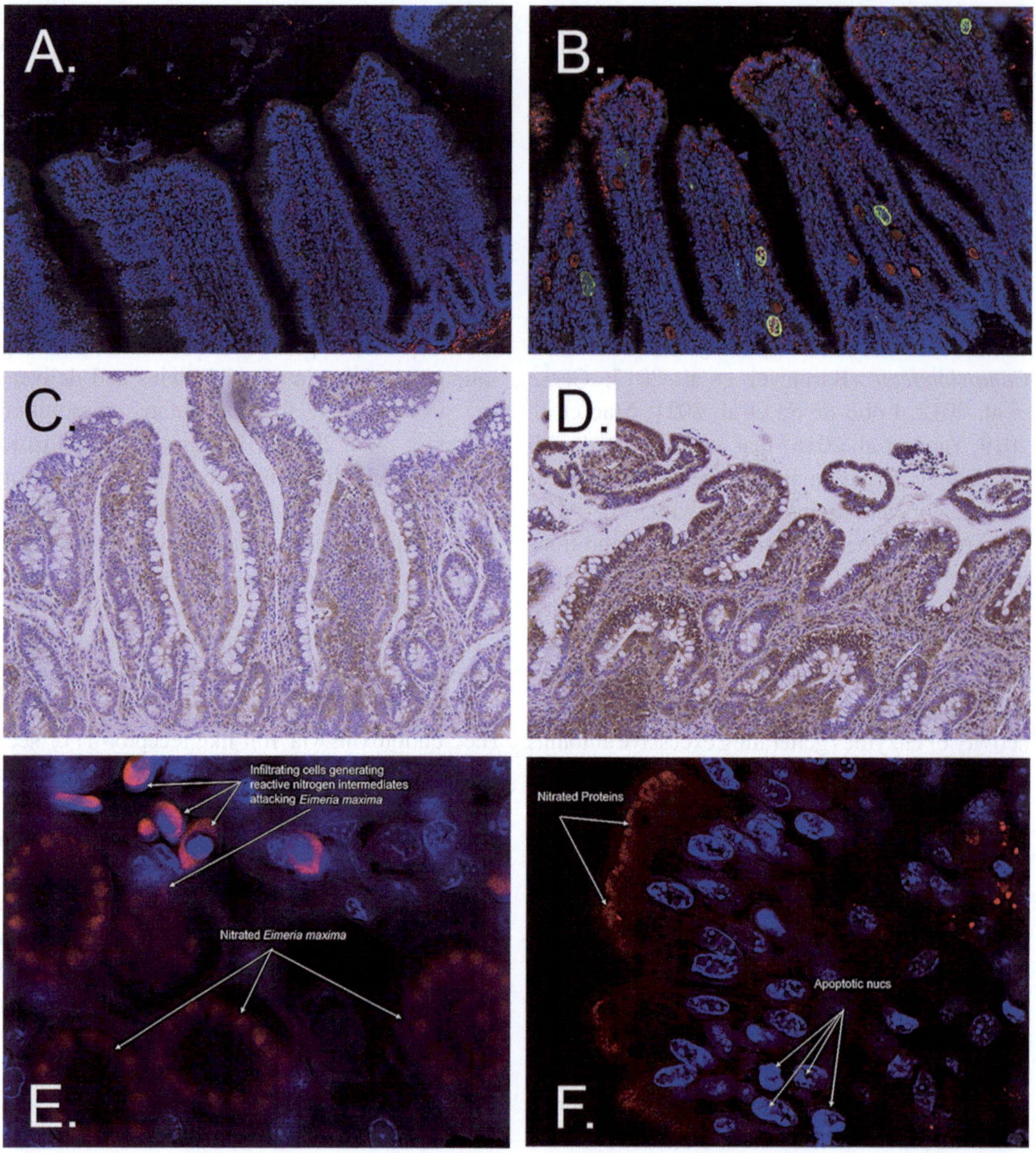


**Fig. 1** Structural impacts of intestinal infection-related inflammation in diarrhea-producing conditions. Ileal villus morphology in healthy, noninfected tissues (Panels **a** and **c**) show elongated villi and epithelial edge integrity. Tissues from animals with active infection (Panel **b**, *Eimeria maxima*-infected chicken and Panel **d**, *Eimeria bovis*-infected calf) display villi with blunted ends and disruption of the epithelial cell surface consistent with epithelial leakage. Immunofluorescence microscopy (Panels **a** and **b**) and immunohistochemistry labeling (Panels **c** and **d**) for the presence tyrosine-nitrated proteins indicate that the inflammation associated with infection increases the generation of these proteins with negative impacts on cell integrity and function. The green fluorescence labeling in Panel **b** identifies the presence of the *E. maximum* parasite. Higher magnification of immunofluorescence images of ileum from *E. maxima*-infected chicken villi reveals high levels of nitrated intracellular proteins associated with innate immune infiltrating cells, rich in nitric oxide and superoxide anion generation in their attack on the parasites as well as the parasite surface membrane, and apical surface epithelial cells (Panel **e**). Epithelial as well as nonepithelial cells in disease-perturbed villi undergo programmed death, apoptosis, with the characteristic nuclear chromatin condensation and/or nucleic acid fragmentation (Panel **f**). Blue immunostaining is associated with nuclei, and inspection of nuclei in images from coccidia-infected chickens reveals the upregulated presence of cells with apoptotic nuclei in infected animals

like vasoactive intestinal peptide (VIP) and several prostaglandins (in particular the -E1 and -F2 series) known to increase autonomic propulsive forces in the muscularis (Palsson et al. 2004; Yakabi et al. 2018; Conklin and Anuras 1996; Grabauskas et al. 2020; Hunt et al. 1975) or conversely, depletion of bioactive peptides and metabolites with actions known to calm propulsive forces such as Peptide YY and serotonin, respectively (El-Salhy et al. 2012).

In evaluating a cause for and then an appropriate treatment plan to mitigate diarrhea, it can be considered useful to employ a more standardized classification system for the disturbance. Several somewhat related rubrics are available for such. First, and perhaps the most generalized scheme, diarrhea can be classified into two general types, acute and chronic, as has been applied to diarrhea in humans (Guarino et al. 2018; Brandt et al. 2015). In general, the fundamental discriminator is the rapidity of onset and duration of symptoms. Descriptions of acute diarrhea presented by Brandt et al. (2015) stated "abrupt onset, presumably infectious etiology, potentially self-limited, with a course of less than 14 days, increased volume and/or frequency of stool, and fecal loss of nutrients (mainly water and electrolytes)." In this review, however, while consideration of the presence of infection is clearly warranted, aspects of diarrhea of noninfectious origins, such as osmotic and malabsorption sources, might also fall under part or all of Brandt et al.'s criteria. Alternatively, differentiating more clearly between acute and chronic diarrhea states, Guerino et al. further classified diarrhea into three versions: acute, lasting 1 to 7 days, prolonged, lasting 8 to 13 days, and chronic or persistent, lasting more than 14 days (Guarino et al. 2018). The history of the client or patient is an important component with respect to the classification of diarrhea. Second, Juckett and Trivedi classified watery diarrhea into three categories based on presented symptoms, duration, any laboratory-based findings, and history (Juckett and Trivedi 2011). The three varieties were defined as "osmotic (water retention due to poorly absorbed substances), secretory (reduced water absorption), and functional (hypermotility) types." Lastly, in some instances, a better discrimination between secretory and osmotic sources of diarrhea can be obtained from what is termed the fecal osmotic gap (Duncan et al. 1992; Eherer and Fordtran 1992), an interpreted score based on a relationship between serum osmolality and the fecal water $Na^+$ and $K^+$ of fecal fluid. The score is calculated as serum osmolarity $-[2 \times (\text{fecal } Na^+ + K^+ \text{ concentration})]$. Largely by definition (and in fact the *functional definition* can vary with the measured gap being between 40 and 100 mOsmol/L), diarrhea is considered osmotic in origin when the gap is positive. The most meaningful interpretation of the fecal osmotic gap, however, will take into consideration the presence or absence of carbohydrates in the fecal material in that bacterial degradation of osmotically active carbohydrates in the colon can change (decrease) the final osmotic values obtained and therefore challenge the interpretation of results and thus the course of patient treatment (Corinaldesi et al. 2012).

## 3.2 The "Big Picture" View of Generalized Causes Underlying Diarrhea

With regard to causes that directly relate to the consequences of diarrhea, several considerations apply. As will be discussed, these include (1) perturbed functional reabsorption of water somewhere in the gut, (2) potential alterations in systemic ion and electrolyte balance but especially in the small and large intestine, (3) dysbiosis and altered balance in commensal and pathogenic microbial species, (4) pathogen-released factors or -directed changes in gut epithelial and immunocyte cell signaling to maintain homeostasis, (5) overall increased inflammatory character of the bowel, (6) changed rate of passage of digesta through the bowel, (7) malabsorption syndromes indicative of nutrient uptake problems, (8) toxemias secondary to disturbed barrier membrane integrity (i.e., "leaky bowel syndrome," endotoxin transfer from the gut to the systemic circulation), and (9) potential anemia from damaged microvasculature in the villi, or hemotoxins. Diarrhea occurs as a result of one or combinations of the nine factors just mentioned that impinge negatively on several regulatory systems in the gut, and individually these regulatory processes have components of neurogenic, endocrine, and immune system inputs (Anbazhagan et al. 2018; Grenham et al. 2011; Lange et al. 2016; Viswanathan et al. 2009; Hodges and Gill 2010).

An important veterinary clinical consideration of causes for diarrhea is whether or not the trigger for the gut perturbation is from infectious or noninfectious origins. The point critical to this determination involves a further decision to pharmacologically treat with an appropriate drug or alternatively provide supportive measures to mitigate symptoms where the duration of diarrhea, either way, is self-limiting with the restitution of homeostasis of cell processes and resolved in a relatively short time. Ultimately, however, there is usually a commonality, a degree of overlap in the molecular causes for infections versus noninfectious diarrhea, and this involves the impacts on gut cell function imparted by a core of shared inflammation mediator mechanisms; the overlap can also be influenced by, and the ensuing diagnosis of cause made more complicated by, the severity of the underlying cause at the time of examination. Discussed more fully in later parts of this review, suffice it so say at this point that these inflammation mechanisms change cell function and/or integrity in conjunction with signal transduction cascades of effector molecules involved with altered gene transcription/translation, altered protein synthesis, increased protein degradation, metalloprotease release into intracellular spaces destabilizing barrier junctions between cells, and stimulated or inhibited ion flux with accompanying changes in water flux. In the case of diarrhea, this water flux leaves excessive quantities of water in the lumen of the gut.

Diarrhea can occur from nonpathogenic origins in the absence of infective pathogenic microbes such as that arising from, for example, allergy (Konturek et al. 2011; Stokes et al. 1987), food component intolerance (https://hospital.vetmed.wsu.edu/2022/01/04/diarrhea/), and diet-derived osmotic substances, usually nonmetabolizable and nonabsorbable factors that by their osmotic character would serve to pull water into or prevent water from being reabsorbed from the lumen of the gut. Such occurrences are frequently encountered when feed components in the diet are changed rapidly and time is required for the gut to adapt to new demands placed on it for digestive purposes. Common to all these causes is a water pull into the lumen of the gut at volumes not reabsorbed by properly functioning water balance mechanisms (Stokes et al. 1987). As they relate to allergy and food sensitivity, many of the effector molecules driving associated diarrhea stem from a stimulated underlying inflammatory process such as mast cell release of histamine (Brandt et al. 2003; Nguyen et al. 2021) or in conjunction with inflammatory processes that compromise otherwise normal processes through which that water would be reabsorbed (Anbazhagan et al. 2018; Barnes et al. 2000; Cave 2006; Sauter et al. 2006). It bares mentioning, perhaps even stressing, that underlying causes of diarrhea stemming from allergy, food intolerance, osmotic components, and even social and behavioral stress (Pohl et al. 2017; Taché et al. 1993; Qin et al. 2014) can generate an environment in the gut tissues that increases the probability for diarrhea and tissue pathology to develop. For example, opportunistic colonization of the different gut segments with enteric commensal nonpathogenic bacteria or resident bacteria with pathogenic potential, which is maintained in check by the balanced communications between commensal organisms, can occur when gut environmental conditions permit the translocation of microbes to different gut segments or proliferation of pathogens to a point that disease is triggered. These occurrences can change the dynamics of the homeostatic checks and balances between localized populations of microbes and can result in the deterioration of processes through which the pathogens are kept in check (Lacey et al. 2018). A good example of this occurs in cases of perturbed mesenteric blood flow (stress, trauma, toxins, heat stress, collateral parasitism?) with the development of regional or localized anoxic, hypoxia, or hypercapnia conditions that disfavor aerobic and favor anaerobic microbial function (Tripathi et al. 2019). The oxygen status of the bowel is critical to ensuring the integrity of the gut, and in most locations of the gut, the oxygen tension, $pO_2$, exists in a gradient, higher in the small intestine and near the base of the crypts and lower in the large intestine, being almost absent in the center of the lumen within the flowing digesta; in this regard, it stands to reason that different populations of microorganisms that are more or less oxygen tolerant will colonize not only different segments of the gut but also different vertical microniches from the base of the crypts to the center of the luminal contents (Singhal and Shah 2020). Obviously, this high crypt $pO_2$ is of necessity because of the high amount of ATP energy needed in the crypts to support the proliferation of the epithelial cells as they develop out of the crypt stem cells ("crypt base columnal stem cells"). Thus, if blood flow and oxygen status in the crypts are impaired, the effect can decrease the efficiency of cell regeneration in the villi or surface colonocytes as well as favor conditions that promote anaerobe survival in areas that with higher $pO_2$ would be unfavorable. Such pO2-related gut challenges with resulting barrier/permeability compromise have been identified in heat stress as well as toxic sepsis (Vargas and Marino 2016; Swank and Deitch 1996).

With the above brief introduction of the concept of "noninfectious diarrhea causes," the next important category for the cause of diarrhea is a gut infection or overpopulation of the gut with enteric pathogens largely of bacterial, viral, or parasite etiology. In their drive to survive and propagate, pathogenic organisms employ a myriad of capabilities to modify a local environment in the gut and further optimize it for their purposes (Balloux and van Dorp 2017; Casadevall and Pirofski 2019; Casadevall et al. 2011). In contrast to some viral gut infections like rotavirus that cumulatively associate with diarrhea through a large amount of induced villus epithelial cell degeneration (Dhama et al. 2009), some gut pathogens modify their needed environmental locale, the host's organ-cell network, by causing often severe modification to host cellular processes, directly or indirectly (Hodges and Gill 2010). Because of the impacts of these pathogen-facilitated cell process modifications to electrolyte and ion movement in addition to the fundamental integrity of gut epithelial cells and intercellular margins, in many instances these modifications result in what is clinically observed as diarrhea. As it relates to pathogens and diarrhea, their impact(s) on water transport affect both water flux through cells (transcellular movement) as well as between (paracellular) adjacent epithelial/enterocytes, goblet, Paneth epithelial cells (Buckley and Turner 2018; Turner 2009). Importantly, the nature of the villus and crypt architecture is based on not a single enterocyte-type of cell but a diverse collection of all the specialized epithelial cells that abut the enterocytes whose individual functional characteristics collectively established the needed barrier between the "outside" lumen of the gut and the "inside" of the body. An in-depth review of the cell- types and cell-to cell interactions in the different segments of the gut can be found elsewhere (Ogata and Itoh 2016; Pelaseyed et al. 2014; Gribble and Reimann 2019; Shashikanth et al. 2017). Gut pathogens can also influence epithelial cell function by redirecting the functionality of some neighboring resident immune cells and further causing the infiltration of the affected tissues by circulating immune cells such as neutrophils or heterophils (Barrett 2017; Qadri et al. 2004; Sørensen et al. 2013). Consequently, the inflammatory effectors and actions inflict damage to the host's epithelial cell barrier layer. Mechanistically, activation of neutrophil populations in response to pathogen detection can impart collateral damage to the epithelial cells of the gut through the release of cytokines, prostaglandins, thromboxanes, reactive oxygen (superoxide anion, hydrogen peroxide), and nitrogen (peroxynitrite) intermediates with harmful impacts not only on the pathogens but also on adjacent enterocytes, endocrine cells, goblet cells, etc. These issues will be discussed further in a subsequent section of this review. However, to best understand how diarrhea arises from both infectious and noninfectious pathogens, it is best to start with the critical control points in gut water balance through which water flux is maintained under normal states of gut health.

## 4 Physiology and Pathophysiology of Water Balance in the Gut

### 4.1 General Principles—Osmosis and Solute Gradients: The Molecular Source of Gut Water Balance

Over evolutionary spans of time, as multicellular organisms emerged and further developed into multiorgan entities, three fundamental premises have survived: (1) no matter what the organism is and even for single-celled organisms, some sort of semipermeable membrane separates an "inside" from and "outside", (2) for any requisite cellular function, some form of solute transfer into and out of cells occurs, and (3) in more complex multicellular organisms where specialized centralized tubular structures like a gut developed to facilitate those solute transfers in a manner somewhat protected from the general "outside" environment, the lumen of that tubular transport structure is still technically "outside" the organism. The essential impermeability of cell membranes functionally separating "outside" from "inside" is well characterized as a polarized continuous lipid bilayer with hydrophilic components of the lipid complexes oriented toward the water solvent polar environment and the less polar or nonpolar hydrophobic components sequestered and interacting in the interior region as formed by the interaction of the tails of the supporting fatty acids (de Mendoza and Pilon 2019; Lesieur et al. 1997; Rand 2002). For the most part, the apical brush border membranes of intestinal epithelial cells have inclusions called glycolipid rafts (cells in other organs largely have cholesterol lipid rafts) and caveolae that function as localized workbenches for specialized attributes of the membrane of a particular cell type and span the depth of the bilayer providing a signal communication route between the outside lumen and the inside of the cell (Danielsen and Hansen 2008). In addition, these lipid rafts are thought to contain structural elements that are targets for *Clostridium perfringens* iota-toxin and *Vibrio cholera* toxins to bind to, actions which initiate some of the host cellular changes and events leading to diarrhea (Danielsen and Hansen 2008; Nagahama et al. 2004).

In addition to lipid rafts and caveolae, other classes of membrane-embedded proteins and structures are very relevant to the mechanism(s) underlying diarrhea, the molecular transmembrane portals referred to as ion transporters, "water channels," and aquaporins (Almeida et al. 2005; Hooper 1999). However, in the physiological movement of water, the process is not just a simple single membrane transfer of water. Rather, the water movement into and out of cells in the gut barrier layer occurs in concert with a set of gated electrolyte and sugar transport mechanisms. For functional considerations, it is the epithelial cells/enterocytes that regulate all of this. In contrast to the polar localization of inclusions of the rafts and caveolae in the apical membrane, the ion and sugar transporters are inserted in specific regions on the basal, apical, and basolateral surfaces between cells.

From the perspective of pathology, however, while the functionality of the lipid bilayer is essential and universal across species, it is this exact evolutionarily conserved commonality that opens the door to the negative impacts of many pathogen toxins known to disrupt cellular water balance and trigger diarrhea. For example, toxins released by *E. coli*, Staphylococci, and *Serratia* are transmembrane pore-forming toxins. The toxins are secreted as monomers that, upon interaction with proteolipid and glycolipid/raft structures in cell membranes, polymerize into a β-barrel structure. This structure forms a channel ranging from 1 to more than 10 Å, differentially allowing the free unregulated passage of water and ions out of cells and into the hyperosmotic lumen. This further destabilizes both water and electrical charge homeostasis in affected cells (Geny and Popoff 2006; Bhakdi and Tranum-Jensen 1991; Roderer and Glockshuber 2017; Tilley and Saibil 2006).

With particular regard to the movement of water and nutritional/ionic/electrolyte compounds through membranes, the transmembrane movement of these components is known to be achieved through both passive and passive-facilitated energy-independent processes as well as direct and indirect energy/ATP-dependent mechanisms. Perturbation of these regulatory mechanisms by conditions like infection, inflammation, and aberrant solute (dietary components) has been shown to have a major role in the development and nature of diarrhea. Thus, in line with the overall topic, diarrhea is a matter of disrupted regulation of active and passive water balance in the gut. In an attempt to understand the ramifications of this condition, we will need to focus on the nature of what regulatory mechanisms are compromised and how or why they become compromised in different segments of the gut. The net effect of the causative agents, be they pathogens, cytokines, or proteases that result in diarrhea, is the fundamental failure to reabsorb water from the lumen back into the (circulatory) system on the serosal side of the gut. The consequences of this overt loss of water are varying levels of dehydration, electrolyte imbalance, possible issues with acid–base balance, malabsorption syndromes leading to nutrient and energy deficits, anemia, etc.

Throughout the gut, various segments impact water balance differently, with some regions like the upper small intestine (i.e., duodenum) having a net secretion of water into the lumen. Further down the small intestine (jejunum, ileum), approximately 90% of the water that would be absorbed is reabsorbed, with the remaining 10% taken up by the colon (Turvill and Farthing 1999; Ber et al. 2021; Edmonds 1981; Said 2011). In humans, for example, an estimated average of 8 to 9 liters of water are secreted into the proximal intestine, while further down into the distal gut well-controlled reabsorption ensures that only a fraction of the luminal water exits the gut as the final excreta with the higher solids content (Ma and Verkman 1999). The number of water molecules crossing a membrane barrier is so spectacularly large that any consideration of an active transport process translocating water specifically would be energetically impossible. However, this does not mean that there isn't an energetic component to water transport; rather, the energy component is mediated in terms of ATP-dependent ion movements, the flux of which contributes to the following passive movement of water. Therefore, solutes move across membranes passively following the gradients of osmotically active molecules on one side or the other of the membrane. Interestingly, though extensive research over decades has amassed a tremendous understanding of individual molecular mechanisms and mechanism components [i.e., tight junction proteins (TJ), coupled transporters, water movement channels (sodium glucose lateral transporter-1, aquaporins)] that impact water and solute movement across the gut membranes and epithelia, the consensus thoughts on exactly how water moves across the membranes, and for reabsorption in particular, are at best still only theories. These theories take these newly discovered molecular regulatory components into consideration in a logical flow scenario but still do not account for the flux of water that is measured.

Although they are theories, three proposed mechanisms dominate the conversation on how water is moved across compartments in the gut. Interestingly, all of these models are relatively old but to date still function as a fundamental framework and the basis for building upon the newer findings on the molecular mechanisms through which molecules move through and across cell membranes and organ surfaces. Briefly, these include (1) the Diamond hypothesis and its proposed lateral channel (LC) solute "standing" gradient that functions to "pull" water down the LC channel in a mathematically modeled relationship between LC length and the local osmotic milieu (Diamond and Bossert 1967), (2) the balanced osmolality model with no LC solute gradient (Mathias 1985), and (3) the tricompartment flux model (Tripathi and Boulpaep 1989; Curran and Macintosh 1962). All three model theories have common components. The most significant commonalities are that (1) water movement into and out of the gut occurs both transcellular (through the cells) and paracellular (between the cells), (2) ultimately all water movement is dependent on solute osmotic forces with an added component of hydrostatic pressure, and (3) the lateral architecture (lateral space, lateral compartment) of the adjacent cells and the stability of the tight junctions are critical to compartment separation and healthy water balance.

Osmotic particles in the gut lumen can be present from both internal cell-based ion transfers or simply from nutrient components (glucose in particular) in the digesta. In general, digesta with a fairly high-water content, such as that in the small intestine, favors complex nutrient breakdown and mixing needed for efficient absorption. Having a high-water content digesta in some gut segments increases the ease with which mechanical mixing of the chyme occurs as directed and facilitated by the coordinated smooth muscle contractions throughout the gut. This is important for mixing host-released digestive enzymes and associated transport factors like liposomes with the complex nutrients and more recently defined microbiome processing of digesta forming unique metabolites. Watery digesta also constantly presents new concentrations of released nutrients to the absorbing cells lining the gut, such as the enterocytes of the villi. However, further down the gut, the general rule is that as the digesta flows the termination of the tract, the net flux of water is toward reabsorption, mainly occurring in the large bowel and colon. Needless to say, different molecular mechanisms govern the fundamental direction of water flow into or reabsorbed from the lumen of the gut. In essence, all water transport in the gut depends on directional net fluxes in electrolyte ions, with water being passively transferred across membranes as dictated by osmotic principles and laws of mass solute flux. However, that doesn't mean that transmembrane water transfer is passive but rather means that water flux is both energy dependent and cotransport dependent. The energy component comes in where energy is needed to support the active transport of ions, and the cotransport component comes in the form of, for example, ion-glucose cotransporters—ATP-dependent $Na^+$, $K^+$ pumps in particular that facilitate movement of ions against osmotic gradients. All of this however requires an intact barrier to unrestricted fluid transfer, and this barrier takes the form of intact cell lipid bilayer membranes in individual cells and functional tight junctions between adjacent cells. It will become apparent in the subsequent section of this review that these very water flux regulatory features are the targets toward which pathogens affect diarrhea, as well as how the presence of these pathogens triggers host immune responses in the gut that result in collateral damage to the host's own gut cells.

## 4.2 Mechanisms of Gut Water Transfer and the Nature of Gut Cell Architecture

A schematic representation of epithelial cell components necessary for regulated water balance is depicted in Fig. 2. With regard to diarrhea, these same membrane components that manage water flow are, in most instances, the very targets that are acted upon by infectious pathogens like bacteria and viruses as well as inflammation mediators that trigger diarrhea. The cell regions that manage water balance are the apical/lumen surface, the basolateral surface, and the intercellular lateral space. In Fig. 2, the apical cell surface contains numerous microvilli (1) and a mucus layer (2). Microvilli further increase the surface area needed for nutrient absorption efficiency while the mucus layer performs several important functions. The microvilli establish the brush border of gut epithelial cells, the cellular region essential to digestion, nutrient absorption, raft and caveolae insertion, and ion/sugar/water transporters as previously described (Danielsen and Hansen 2008; Crawley et al. 2014). A critical component of mucus is the presence of more than 20 mucin proteins (MUC-1 through MUC-21) that are differentially expressed by enterocytes in response to multiple factors, including diet composition, the regional and microenvironmental microbial community, and generalized health status of the gut as sensed by a plethora of pathogen detection receptors (Mogensen 2009; Kawasaki and Kawai 2014) [pathogen-associated molecu-

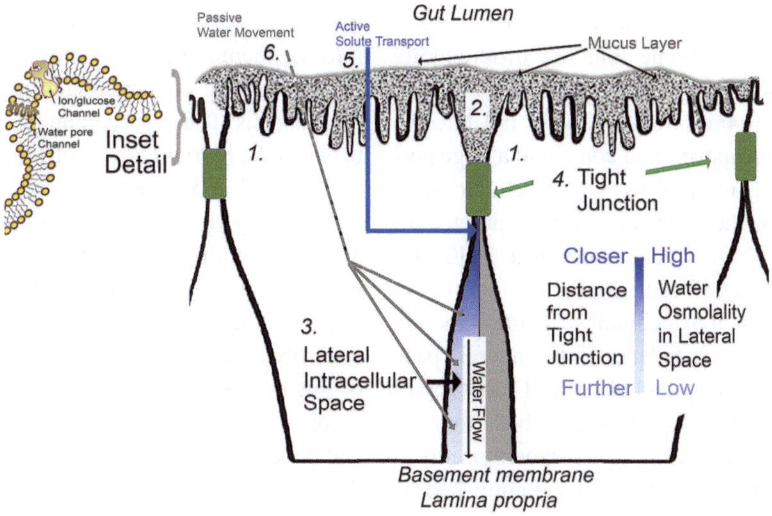


**Fig. 2** Representative generalized structure of adjacent epithelial cells on intestinal surfaces and cellular regions important to water balance. Specific elements of water regulation are indicated by the numbers 1–6; the inset detail on the left depicts an enlargement of a apical region with surface-associated ion and aquaporin water channels, the particular details of which are presented in the text

lar patterns (PAMPs) and Toll-like receptors (TLRs) 1–9, etc.]. These receptors recognize potential insults or threats to cells such as bacterial Gram-negative endotoxins and Gram-positive peptidoglycans (Ramachandran 2014), foreign nucleic acids, flagellin proteins, viral components, etc. The brush border-mucin environment is critical to the maintenance of eubiosis wherein commensal/beneficial microorganisms establish microenvironments favorable to gut stability while outcompeting populations of infectious/detrimental pathogens until the gut balance is disturbed facilitating opportunistic overpopulation by pathogens that trigger disease.

The nature and functionality of the proteins positioned in the apical and basolateral cell surfaces of the gut barrier epithelial cells are potentially the most critical attributes governing the coordinated regulation of water balance in the gut. Following inherent patterns of gene expression and protein synthesis in these gut barrier epithelial cells, these protein components are shuttled to and from the respective membranes in an intricate process of intracellular actin–microtubule filament assembly and disassembly. Interestingly, the targeted disruption of the host cell's normal shuttling of these junction proteins, as well as ion transporters, is a major mechanism by which many pathogen-derived substances, toxins, and induced actions function through as they promote cell decay and diarrhea, especially as noted for *Vibrio* and the secreted cholera toxins (Ritchie et al. 2012). As will be subsequently highlighted, a significant portion of the disruption of cell function accompanying pathogen infection, in particular, Apicomplexa (*Cryptosporidia*, *Eimeria*, *Giardia*, etc.) protozoan parasite infection that results in watery diarrhea, can be derived from the capability of these intracellular organisms to redirect the host cell's organization of transport microtubules away from host purposes to support parasite functions (Frénal et al. 2020). These parasites are capable of secreting into the epithelial cell proteins such as actin destabilizing factor (ADF) that dissolve needed actin polymerizations, thus further perturbing the host cell insertion of the membrane proteins (Wang et al. 2016; Xu et al. 2008). The apical surface contains several class-specific ion transporters that are critical to water balance in the gut. Since water flows passively via osmotic gradients, it is the ion pump channeling of electrolytes that sets up the needed gradients directing the direction of water flow. In addition, there are water channels— essentially passive pores—spanning the membrane that are differentially opened or closed as physiological needs dictate. These normal pore proteins largely fall into the category of aquaporins and constitute a large portion of what is referred to as transcellular water flux, and again these pathways can be targeted for perturbation in diarrhea (Laforenza 2012; Lv et al. 2022; Pelagalli et al. 2016).

Another essential cellular structure–function critical to gut water balance is the lateral space between adjacent epithelial cells. Data support the concept that water being absorbed from the lumen back to the plasma compartment occurs in these lateral spaces, and this is accomplished by the presence of specific ion channels embedded in the associated cell membranes. The ion pumps move electrolytes into the space, with higher solute concentrations toward the apical end and diminishing in concentration via dilution from the passive water pulled into the space, and with subsequently transferred to the plasma compartment in part from hydrostatic pressure (3) (Loo et al. 2002; Wright and Loo 2000). Located toward the distal/apical/luminal end of the lateral space, barrier junctions (4) are formed between adjacent epithelial surface cells, forming a semi-permeable/selective permeable barrier to water flux referred to as the paracellular water flux (Suzuki 2013, 2020). For the case of diarrhea, the critical feature of disruption of the barrier integrity is that water, which otherwise in states of normal physiology would have been reabsorbed unidirectionally into the plasma compartment, is lost to and remains in the lumen. However, it is important to recognize that in some normal states TJs can also pass water and solutes *toward* the lumen because some associations between various lateral space membrane protein components facilitate selective permeability in the process of gut water homeostasis (Krug et al. 2014).

During intestinal infection and inflammation, pathogen- and host immune cell-derived effector molecules, including proteases and generalized receptor-signal transduction reactions, culminate in blocking junctional barrier protein mRNA transcription and translation as well as upregulation of enzymes. Typically, under normal conditions, termination of some aspects of TJ protein into the membranes is accomplished via signaling impairment of intracellular protein transport mechanisms. The degraded integrity of the bonds between the barrier cells results in what is termed "leaky gut" commensurate with the propensity for not only diarrhea but further complications associated with toxin and bacterial translocation into the body (Groschwitz and Hogan 2009; Zuo et al. 2020). Thus, the barrier to random water, electrolyte, and nutrient movement (and in states of gut infection, pathogenic microbes and their secreted virulence factors and toxins) is fragile and delicate since it is only a single layer of epithelium immediately adjacent to the lumen (under the mucin/mucus). The efficiency of the cellular component of the gut barrier, and thus the relative impermeability of the structure to free water flux, depends on two major factors: overall cellular integrity and the relative linking associations between the various tight junction proteins and the cytoskeletal actin.

It stands to reason that host cell responses of inflammation-triggered apoptosis, necrolysis, and autophagy will destroy the affected epithelial cell. At the same time, inflammatory cytokines and immune effectors (prostaglandins and leukotrienes) and pathogen toxins and virulence factors destabilize the integrity of the barrier via induction of proteolytic–lysosome–proteosome activity that destroys the barrier protein associations and selectivity. To further elaborate on the component nature of the barrier proteins, the functionality of the barrier is formed and stabilized by a set of three major intercellular "junctions": tight junctions (TJ), adherens junctions (AH), and desmosomes (DES) (Guttman and Finlay 2009). Together, these barrier junctions regulate, among other physiological functions, the "between cell" paracellular flux of water, which is a major contributor to the water lost to the lumen as diarrhea in pathologic conditions where epithelial cells are not directly ruptured as a result of pathogen influences. In brief, these junctions are assembled from numerous protein components that are embedded between the lateral membranes of epithelial cells in specific locations between the base and the apical end of the cells (for reviews, see Guttman and Finlay (2009), Krug et al. (2014), Groschwitz and Hogan (2009), and Lee et al. (2018)). The insertion of the proteins into the membranes is regulated in part by hormone, nutrient, and commensal as well as pathogenic microorganism influences on gene expression and translation (Metzler-Zebeli et al. 2015; von Buchholz et al. 2022; Walker et al. 2015). Additionally, a variety of dynamic post-translational modifications are enzymatically made on these proteins by kinases (phosphorylations) and dephosphatases (phosphate removal) (Reiche and Huber 2020; Van Itallie and Anderson 2018).

Various protein components of the TJs consist of species- and tissue-specific claudins, occludins, junctional adhesion molecule, tricellulin, and zona occludins, the so-called ZO-proteins -1, -2, and -3. While the claudins and occludins are the functional barrier bridges of the TJs between adjacent epithelial cells, it is the ZO-proteins that bridge these proteins and further connect them to the internal actin cytoskeleton of the cell. Interestingly, TJ components are targets for infectious agents and as such are strongly associated with the conditions needed to generate diarrhea (Roxas and Viswanathan 2018). The linkage of the TJ proteins occludin and ZO-1 occurs in conjunction with the phosphorylation of specific serine and threonine residues of occludin as further modulated enzymatically by protein kinases, PI3K/Akt, and phosphatases that remove the phosphate modifications. The AH-associated proteins are located between several cell types and can be generally characterized as a set of $Ca^{++}$-dependent adhesion proteins (Guo et al. 2003). In the intestine, the major AH protein is E-cadherin, which noncovalently crosslinks itself similar proteins in an adjacent cell to strengthen cell-to-cell adhesion. In addition, another protein of this class, β-catenin, was shown to interact with E-cadherin and again stabilize cell–cell interactions in the barrier function (Guo et al. 2002). AH protein function is influenced by polyamine (derived from diet components as well as bacterial conversion and metabolism of food components) concentrations with regard to formation, interaction, and membrane insertion. Fluorescence confocal microscopy using antibodies against AH proteins is often used to visually assess aspects of the integrity of the barrier.

Desmosomal proteins assemble to form dense plaque-like adhesions within the lateral spaces between the gut surface epithelial cells called desmosomes. The desmosome plaques are, in comparison to other barrier junctions, relatively stronger and contribute resistance to physical shear stresses imposed on structures like the barrier membrane formed of epithelial cells (Capaldo et al. 2014; Holthöfer et al. 2007). Desmosomes in the gut are mainly composed of proteins called cadherins, which can be further broken down into several other proteins. The critical function of the desmosomes is to provide mechanical stability to cell–cell interfaces. As described in the gut, stretch and compression forces associated with peristalsis and distention, as food and gas pass through the digestive tract, must be accommodated. The integrity of the desmosomes may also be important to the functional lateral space to counteract any deformation effects of hydrostatic pressure or other stretch forces that could compromise the integrity of the barrier and foster water and nutrient leakage.

Active and passive facilitated transport of nutrients, mainly glucose, electrolytes $Na^+$, $K^+$, $Ca^{++}$, $Cl^-$, $HCO_3^-$, the principle osmotic agents affecting water flux, are moved across membranes in conjunction with a variety of specific transport complexes, pumps, and channels (5). A very detailed review by Rao of all of the presently defined electrolyte transporters/mechanisms to date can be found elsewhere (Rao 2019). However, where a particular ion transporter is impacted in cases of diarrhea and relevant to a particular pathogen, the mechanism(s) associated with this perturbation will be discussed in this review. The important ion/solute transporters for gut epithelial cell function and regulation of water balance are summarized in Fig. 3. It is important to note that there are additional ion transporters present in gut cells, but for the simplification of this review, the major transporters discussed will be those that are associated with documented impacts of either infectious pathogens or as a generalized site of impact of proinflammatory mediators on the transporters that when perturbed result in aberrant water loss to the lumen. As depicted in Fig. 3, there is preferential insertion of the various transporters into the cell membrane as a function of the spatial location of the membrane segment as determined by the polar orientation of the cell relative to the internal and luminal compartments.

A further discovery at Johns Hopkins University in 1992 presented the first evidence for regulated passive transcellular water channels in red blood cells, initially referred to as "CHIP-28" (Denker et al. 1988; Brown 2017) and later renamed to aquaporin (AQP-)-1. In 1995, the first report was made of an AQP channel in luminal epithelial cells of the colon (Frigeri et al. 1995), a major contribution to the understanding of gut water balance with the addition of a mechanism. This was additive to what was already known about electrolyte-nutrient-associated water flux but now with an added component of a much faster time frame. Aquaporin channels vary in their tightness or "leakiness," and this quality is determined by both the combination(s) of aquaporins present and the tissues and their functions in which they are expressed. In comparison to other tissues, the intestinal epithelial aquaporin system is relatively tight but changes dynamically in a short time. Aquaporins (element "6" in Fig. 1) are significant to gut epithelial cell water balance because, in comparison to ionic channels, aquaporin channels move larger quantities of water across membranes faster. There are 11 forms of AQPs present in the gastrointestinal tract of mammals, and of these, 7 are differentially present in the intestines. As reviewed extensively by Zhu et al. (2016), colonic epithelial cells express AQPs-1, -3, -4, and -8, while AQPs-1, -3, -7, -10, and -11 are found principally in the

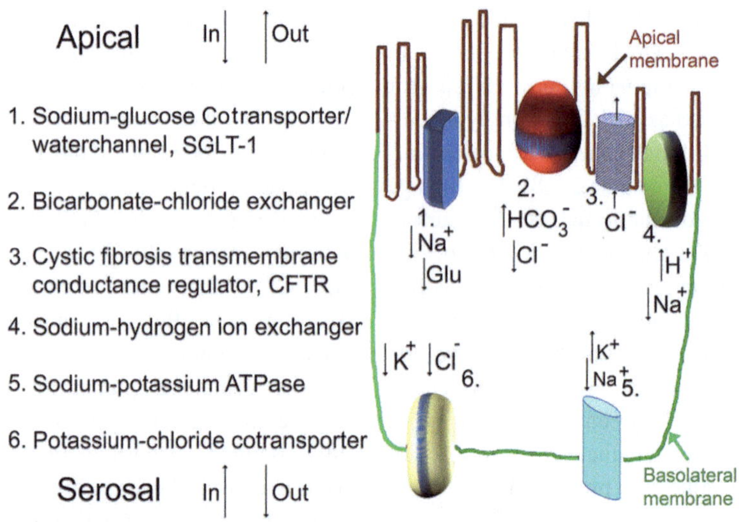


**Fig. 3** Intestinal epithelial cell localization of ion/nutrient transporters important to the homeostatic regulation of gut water balance and implicated as key targets for inflammation-associated causes of diarrhea. The six major types of ion transporters and their relative insertion profile into the epithelial cell membranes are represented by the various geometric shapes and the related numbering in the legend. The direction of ion transport is indicated by the arrows. Transporters associated with ATPase activity are energetically active, exhibiting energy-dependent characteristics that mainly affect Na+ and K+ flux. Facilitated transporters, in general but not always, move a particular ion against its concentration gradient in conjunction with another ion that follows its passive movement down a concentration gradient. Facilitated transport largely involves neutral (#1, glucose, absorption) or anionic electrolytes—$Cl^-$ and $HCO_3^-$ (#2, $Cl^-$ absorption; #3, $Cl^-$ secretion; #4, $Na^+$ absorption; #5, asymmetric 3 $Na^+$ absorption for 2 $K^+$ secretion). A difference between epithelial cells of the small and large intestine segments is that the colonocytes possess a direct $Na^+$ channel (inward-directed $Na^+$ flux) while enterocytes express the $Na^+$- glucose cotransporter SGLT-1 (Loo et al. 2002; Das et al. 2018). A $Ca^+$-activated $Cl^-$ transporter $Cl^-$ secretion) is thought to exist in enterocytes and colonocytes and colonic goblet cells in apical membranes of crypt cells

small intestine. AQPs-1, -3, -7, -8, and -11 are differentially expressed in villi and crypts as a function of gut segment (Ma and Verkman 1999). The quantity and speed of AQP-facilitated or induced diarrhea are consistent with several aspects of "non-destructive" diarrhea where there is no degradation of cell integrity as might occur with dietary osmotic diarrhea. Early stages of bowel inflammation that would be further destined to develop into severe cytokine- and infection-mediated inflammation are associated with the downregulation of intestinal AQP mRNA expression (Ma and Verkman 1999). Similarly, this can also occur with negative impacts of heat stress (Wang et al. 2015a), an issue perhaps consistent with attempts to repartition water to skin surfaces needed for evaporative cooling vis sweating (authors' speculation).

### 4.3 Molecular Targets Impacted by Inflammation and Infection/Pathogen Vectors

Solute transport structures under pathogenic conditions are mainly targeted by various enterotoxins released by pathogens, resulting in the "poisoning" and inactivation of the structures. In many instances, post-translational modification of the proteins in these structures by pathogen processes triggers further cell responses that effectively remove these structures from membranes. The modifications made to the proteins through the actions of the pathogens result in the cell sensing the self-proteins as damaged. This sensing occurs through the unfolded protein response (UPR (Elsasser et al. 2012b)) as well as the proteasome degradation pathway (PDP). It has further been demonstrated in pig enterocyte models, among others, that a portion of the dysregulation of the tight junctions via this PDP recognition mechanism is coordinated with nutrient sensing, which may further suggest that the actions of the pathogen effectors impact these tight junction proteins via simultaneous cooperating mechanisms (Li and Ajuwon 2021; Wang et al. 2015b). In the PDP, damaged proteins are tagged with poly-ubiquitin moieties that direct the proteins to the proteosome for enzymatic degradation leaving a functional gap in and destabilizing the homeostatic process. Protein nitration has been associated with activation of the UPR and premature degradation of intracellular proteins weakening cell structure and compromising function.

## 5 Focusing on the Molecular-Cellular Attributes of Inflammatory Diarrhea

### 5.1 Ion/Solute Transport Affecting Water Balance in the Gut

In the broadest sense, diarrhea occurs as a function of disruption of one or more of the principle mechanisms that under the normal state of the healthy gut (a) maintain the integrity of the epithelial boundary separating the gut lumen from the interior body and limits the free flow of water between these compartments and/or (b) changes the function of the solute and ion transporter mechanisms that facilitate the formation and maintenance of osmotic gradients which further dictating the new water flux into or out of the bowel. These mechanisms relevant to pathogen- and inflammation-associated stimuli are summarized in Fig. 3.

Inflammatory and pathogen challenges are detected in the epithelial cells through the interaction of pathogens or molecules sensed as toxic to the cells with specific Toll-like receptors (TLRs) and Pathogen Associated Molecular Patterns (PAMPs), cell membrane receptors specifically capable of detecting molecules released from or contained in pathogen microorganisms including peptidoglycans, lipopolysaccharides, double- and single-stranded nucleic acids (or fragments), viral capsid proteins, flagella proteins, and so forth. Additionally, molecules that are toxic to the cell or directed into the host cell from the pathogen disrupt intracellular signaling and functional processes, causing perturbations in the homeostatic stability and associated feedback mechanisms. The consequences of these events are the impairment of the host cell's most basic functions. These toxins can include, for example, molecules or factors that interact with the mitochondrial electron transport chain and disrupt ATP generation in the cell or toxins that mimic natural cell signaling molecules but, in fact, cause block cell regulation. Often, and in particular regard to diarrhea, these responses are brought about in the enterocyte cells themselves or on other differentiated epithelial cells in the immediate area where infiltrating immune cells such as neutrophils or heterophils are directed to the site of the insult to mitigate the challenge. In panel "A," cellular degeneration can be triggered or upregulated wherein epithelial cells die and breakdown due to one or more of recognized "cell death" pathways characterized as apoptosis, autophagy, anoikis (a subtype of apoptosis where programmed cell death occurs in enterocytes that become detached from an anchoring substrate matrix), or necrosis and lysis (Jung et al. 2016; Olkowski et al. 2008; Sherman and Petrak 2005; von Boyen et al. 2006). It is beyond the scope of this review to go into depth on each of these cell death pathway responses with respect to an infectious pathological challenge. In general, though, these pathway responses are induced in the cells in conjunction with well-defined stimuli and signal transduction processes (self-contained or transferred via infiltrating innate immune cells), many of which involve the actions of upregulated ubiquitin-proteosome trafficking, lysozyme release, protease production, and nucleic acid dysfunction. Those interested in further understanding of these concepts and principles are referred to Elkon (2018), Maelfait et al. (2020), Wu et al. (2019), and Larabi et al. (2020).

## 5.2 Examples of Pathogen, Toxin, and Inflammatory Mechanisms for Diarrhea

The following points highlight the targets for the major bacterial, viral, parasitic, toxin, or inflammatory mediator impacts on cell integrity and control of water balance that can result in diarrhea. In reviewing this information, the reader is reminded that in many instances the onset and progression of diarrhea may span more than one mechanism. For example, in infection-induced diarrhea, the early onset of diarrhea and watery stools may occur as a result of inflammation resulting from such actions as PAMP-detected pathogens or their toxins, and this may simply be a function of pathogen or toxin load or the nature of the intestinal and immune cell types responding, which may be lower or milder at the onset. Additionally, these associations might only impact electrolyte-facilitated water transfer to the bowel. Progression of pathogen replication, as well as more organisms releasing more toxins during infection, may cause a progressing elaboration of a more complete spectrum of microbe-derived disease sequelae, which then amplify host cell immune responses, leading to more cell lytic rupture and destabilization of the barrier function of the gut epithelial cell complex. The remission of diarrhea is contingent on the reestablishment of the barrier and normalization of the ion/solute fluxes. The regeneration of these factors will depend on the replenishment of healthy epithelial cell populations to migrate up the villus as well as the reestablishment of the critical barrier junctions and mucus secretion from the goblet cells. This will be a coordinate coordinated process relying on paracrine cooperation in signaling between the differentiating cells emerging from the crypts and their migration up the villi.

The incidence of diarrhea can be influenced by age and therefore in part by the level to which immune memory from previous exposures can be initiated when re-exposure to a pathogen(s) occurs. Immune memory can initiate coordinated immune responses to quickly and effectively target and eliminate the infectious pathogen. Diarrhea is probably most common in young animals and occurs opportunistically when co-stressors are present that weaken an underdeveloped immune system. A notable example of this is weaning stress, and it is recognized as a major input to piglet diarrhea-associated morbidity and mortality with the facilitation of infection of the pig gut with bacterial and viral pathogens (Nabuurs 1998; Panah et al. 2021). Even in older postweaning pigs, psychosocial/behavioral disturbances can establish stress conditions that foster opportunistic gut infection by pathogens, resulting in diarrhea-associated production losses. These stresses along with perturbed patterns of microbial colonization in the gut (microbiota dysbiosis), as well as the latent presence of coliform and rotavirus pathogens, open the door to pathogen proliferation and subsequent gut disease (Gresse et al. 2017). The other factor influencing the occurrence of diarrhea is the actual pathogens themselves. For example, the review paper of Thomson et al. (2017) presents the evidence for causative agents for pathogen-associated diarrhea in young calves. When considered as sole or co-infection agent(s), the following percent incidents were stated: *Cryptosporidium* alone—37%, *Rotavirus* alone—25%, *Cryptosporidium* co-infection—20%, *Coccidia* sole agent—8%, *E. coli* sole agent—4%, *Coronavirus* sole agent—3%, and non-Crypotosporidia co-infections—3%. In general, therefore, it must be recognized that for the most part the majority of pathogen vectors capable of causing diarrhea and gut disease are most likely already present in the gut (as refined and highly sensitive molecular detection of gut microbes by 16-S-RNA sequencing has revealed). Typically, these pathogens are maintained in check via commensal microbial dominance and host immune system inputs. This interaction between commensal and pathogenic bacteria wherein the commensals keep the eruption of pathogens into disease in check is referred to as "colonization resistance." Because so many intestinal pathogen outbreaks involve the symptom of diarrhea at some point in the eruption-disease timeframe, colonization resistance therefore can be considered a major natural form of control preventing diarrhea. A significant part of how colonization resistance actually functions basically is a matter of ecological dynamics. Commensals and pathogen compete for a given environment or niche. The major leverage that commensals have over pathogens involves preferential uptake and utilization of nutrient resources by the commensals, essentially making them unavailable for pathogen use. In addition, metabolic byproducts derived from commensal nutrient use have significant effects on repressing pathogen virulence factors; it is a complex interaction beyond the immediate scope of the preset review, but further information on the subject can be obtained from Belkaid and Hand (2014). It is the perturbations of gut microenvironmental niches unfavorable to commensal microbes that allow conditions to develop which favor the proliferation of pathogens to cause diarrhea and overt disease as commensals lose a foothold on the competitive advantage they otherwise would have over the pathogens. A point worth mentioning comes from microbiome studies of dysbiosis, gastritis, and diarrhea associated with irritable bowel syndrome (IBS) in dogs and cats (Janeczko et al. 2008; Suchodolski 2011). While there are data to indicate that dysbiosis creates conditions favorable to gut inflammation and diarrhea resulting from pathogen infection, more recent evidence further suggests that abrupt changes in otherwise nonpathogenic commensal microbes can also trigger diarrhea (Suchodolski 2011). Mechanistically, this commensal dysbiosis involvement in gut inflammation and diarrhea can originate from two sources. The first could relate to alterations in the fundamental nature of the osmotic character of the digesta where the imbalances in microbiota result in a greater abundance of osmotic matter in the bowel. Second, the stress hormone axis component of gut inflammation can trigger increases in localized, regional immune function, resulting in the generation of specific commensal microbe-binding immunoglobulins of the IgA class. In doing so, it alters the population balance of the gut, resulting in dysbiosis and therefore diarrhea (Rengarajan et al. 2020).

## 6 Attributes of Specific Pathogens and Inflammatory Mediators Associated with Diarrhea

*Diarrheagenic enteropathogenic coliform bacteria*—Example—*E. coli*

- Variability in subtype virulence factors affects the pathogenicity and is passed via horizontal gene transfer, further influencing preferred gut environmental sites for colonization.
- The nature of the virulence factors expressed determines clinical symptoms and outcomes (Gomes et al. 2016).
- A significant portion of the water loss is due to disruption of $Cl^-$ reabsorption: specific toxins and lipopolysaccharide released from a particular variant inhibit SGLT-1, causing electroneutral accumulation of NaCl in the lumen in early-stage diarrhea, inhibit H+/Na+ exchange, or disrupt actin-facilitated membrane insertion of the $Cl^-/HCO_3^-$ transporters.
- General cellular inflammatory response is triggered by TLR-4 receptor activation.
- Necrotizing variants disrupt barrier integrity via compromised function of barrier junction protein as well as necrosis/autolysis of enterocytes (Hodges and Gill 2010; Das et al. 2018; Hodges et al. 2006).

*Clostridium* sp.—Examples—*C. perfringens*; *C. difficile*

- Presents itself as several strains, each of which releases a spectrum of different toxins.
- Pathology from overpopulation/infection associates with both deranged ion transport as well as progressive necrotic/autolytic degradation of intercellular proteins and the fundamental integrity of the cell membrane.
- *C. perfringens* Types B and C—secrete β-pore-forming toxins causing unregulated transcellular water loss and when affecting cells deeper than the surface epithelium vascular compromise (blood losses, reduced blood flow, tissue anoxia, thrombus formation, and clotting).
- Type B toxin underlies fatal hemorrhagic dysentery in sheep.
- Type C promotes epithelial cell disruptive necrotic enteritis.
- Enterotoxemias in livestock.
- Early onset effects on transporter function result in $K^{-+}$ losses accompanied by influx of $Ca^{++}$, $Na^+$, and $Cl^-$ potentiating increased intracellular hydrostatic pressure, which effects subsequently combine with those of necrotic and apoptotic signaling induced junctional and cytoskeletal compromise to burst the boundary membrane killing the cell.
- Epsilon toxin (ETX) released by Types B and D—main *Clostridia enteritis* virulence factor in sheep, goats, and less so in cattle.
- Leakage into the systemic circulation triggers further multiorgan dysfunction, significant energy depletion in affected cells due to mitochondrial atresia, and collapse of ATP formation.
- *Clostridium perfringens* Necrotic Enteritis B-like toxin (NetB) is the dominant factor associated with necrotic enteritis in chickens (Awad et al. 2017). The initial association with diarrhea and intestinal water loss arises from multiple points of host cell interaction compromise, including the loss of integrity of the lamina propria, downregulation, and enzymatic degradation of barrier junction proteins. This process leads progressively toward osmotic cell lysis from the influx of $Na^+$, $Cl^-$, and $Ca^{++}$ (Navarro et al. 2018; Freedman et al. 2016; Uzal et al. 2014).

*Rotavirus*—Example—General rotavirus serotypes

- Group and sub-group defined by VP6/VP7 proteins, which differentiate the rotaviruses into seven groups (A to G, serotype determines species susceptibility) (Das et al. 2018).
- Generalized inflammatory responses triggered by TLR-3, -7, and -9 interactions with host cells (Arnold et al. 2013).
- Disruption of host cell ribosome-directed protein translation via virus-directed perturbation of protein synthesis regulation factors like eukaryotic initiation factor (eIFs) and including nonstructural protein-4 (enterotoxin NSP-4) and a host translation inhibitor nonstructural protein-3 (NSP-3) of host blocking ribosomal tRNA initiation and shuttling of mRNA from the nucleus to the cytoplasm leading to lytic barrier integrity events.

- NSP-4 disruption of SGLT1-mediated $Na^+$/glucose transport.
- $Ca^{++}$-activated proinflammatory cytokine increases in neuronal trafficking that increases $Cl^-$ secretion into the lumen (Lorrot and Vasseur 2007), disruption of junctional barrier proteins, and induction of microbial dysbiosis(Amimo et al. 2021).
- Tips of the villi are the first noted sites of viral infection, with progressive damage, villous atrophy with the fusion of some villi, and barrier integrity losses. The effects on the villi destruction are variable, ranging from slight and obscure to overt and severe. An interesting feature of these infections is that digestive intestinal disaccharidase secretion is impaired, leaving intact disaccharides with a high osmotic value in the lumen, which functions in diarrhea to retain water or the osmotic force capable of pulling water out from the intestinal/plasma compartment.

*Pestivirus*—Example—Bovine viral diarrhea virus (BVDV)

- In the model for *Pestivirus* diseases and diarrhea, BVDV-type infections are insidious, despite the name including the word "diarrhea."
- Effects of infection are largely systemic, increasing the severity of morbidity and mortality with pathogenic involvement, causing not only gastric perturbations but further complications from respiratory infection, reproductive losses, and immune dysfunction.
- Effects manifest in many domestic food, as well as wild animal species, not just cattle, as the name might imply.
- BVDV immune dysfunction/depression associates strongly with further opportunistic infection with other bacterial and viral pathogens. The actual causative nature of this viral infection in promoting diarrhea arises from (1) the virus itself and (2) the gastrointestinal population hosting other gut pathogens, which are also capable of causing diarrhea.
- Several genotypes and subgenotypes exist; isolates may vary in virulence factors/presentation.
- Persistent infection may develop in utero, resulting in persistently infected (PI) calves; this persistence in infection maintains the environmental reservoir of the virus as virions are continuously shed from infected animals.
- Mechanistically, viral infection can cause gastrointestinal lesions throughout the digestive tract, in particular the intestines.
- Pestiviral diarrhea appears to be caused by a generalized release of proinflammatory cytokines via TLR-3-mediated gene inductions and further signal transduction mechanisms involving the prostaglandin series.
- Cytokine actions impact a spectrum of cellular homeostatic mechanisms including accelerated apoptosis and cell shedding, protein degradation, signal transduction failures, and flux changes in ion pumping and pumping efficiency.
- Coincident infections with pathogens (bacterial or parasitic) possessing diarrheagenic capabilities can coincide with water transport dysfunctions and conflicts in cell barrier membrane integrity, particularly related to the infecting organism. This drives the progression toward watery diarrhea.

*Salmonella, Shigella*—Example—*Salmonella enterica/Shigella flexneri*

- Highly transmissible zoonotic organisms, usually associated with a fecal-oral route of transmission from animals to humans.
- In animals, diarrhea is associated with host susceptibility and disease dependent on the organism serovar present (Stevens et al. 2009; Wallis and Barrow 2005).
- Pathogenic varieties of organisms cause diarrhea mainly as a result of a set of enterotoxins.
- *Salmonella* infection requires introduction into an intracellular environment, and such cell invasions have been modeled in calf infections with organisms found in enterocytes,

resident macrophages, M-cells, and dendritic cells.

- In pigs, as an example, the determinate of whether an infection follows a systemic versus an enteric location is serovar-dependent.
- Multicomponent contributions to diarrheagenic effects of infection in gut water balance.
- Water balance regulator transporters are impaired as a result of mast cell histamine release, further promoting the proinflammatory cascade of cytokines through arachidonic acid-derived mediators, including cyclooxygenase-formed prostaglandins (Collins et al. 2007).
- In addition, changes in the functionality of ion transport mechanisms are perturbed in Salmonellosis, including the loss of the colonic apical surface $Na^+$ Ion channel and the change in the insertion of the $Na^+/K^+$- ATPase transporter from the basolateral to the apical surface, and a further colonic reduction in the $Cl^-/HCO3^-$ co-transporter making the lumen hypertonic (Das et al. 2018) resulting in a toxin-mediated intense watery diarrhea.
- Shiga toxins generally inhibited the absorption of luminal $Na^+$ and $Cl^-$ in the absence of stimulated $Cl^-$ secretion after binding with high affinity to apical surface membranes (Kandel et al. 1989).
- Shiga toxin is associated strongly with an increase in paracellular outward water flux coincident with a reduced intercellular expression of TJ proteins in general and ZO-1 in particular; with the inhibition of ZO-1, the coordinated crosslinking between occludin, claudin-1, and ZO-1 anchoring to the cytoskeleton was disrupted causing free water passage to the hyperosmotic lumen (Fiorentino et al. 2014).

Parasite-associated diarrhea—Examples—*Coccidia*, *Cryptosporidium*, *Giardia* (impacts considered collectively)

- Protozoan parasite infection of the gut generally results in significant changes in activation and repression of the host cell gene expression patterns (Li et al. 2019). This includes the activation of TLR-2, -4, $NF_{kappa}$ B, NO, and SOA signaling cascade in affected epithelial cells as a result of proinflammatory cytokine release from innate immune system cells. Cytokine effectors direct an acute phase response on genes controlling barrier protein and ion transport (especially the apical $Cl^-/HCO3^-$ exchanger (Kumar et al. 2019)) synthesis (downregulated), protease generation (upregulated), and capable of destabilizing TJ proteins and causing dysregulation of paracellular water flux regulation. Evidence suggests that some parasites release microRNAs (miR-181d) that conflict with defensive TLR-host response by downregulating this pathway to increase its infectivity (Feng et al. 2022). These parasites secrete into the targeted epithelial cells an actin destabilizing factor (Wang et al. 2016; Xu et al. 2008) (ADF) that in infection functions to block the intracellular migration of TJ and AH proteins as well as ion transporters as mediated by the assemblage of the intracellular actin cytoskeleton to their respective functional membranes sites perturbing the integrity and function of these regulators.
- While barrier junction protein stability is critical for the maintenance of paracellular homeostatic water balance, the initiation of pathologic events of Apicomplexa parasitism begins with the establishment of specific parasite-controlled insertion of their own tight junction proteins between themselves and the host cell to stabilize the binding and facilitate the efficient insertion of parasite genomic material into the host cell (Frénal et al. 2020).
- Infection is most common in young, immunologically immature, or naïve animals and often self-limiting, though severe enough to be an economic burden in terms of reduced growth, feed efficiency, and associated causes of increased morbidity and mortality in severe diarrhea cases.
- Parasites, especially *C. parvum* in cryptosporidiosis, are capable of initiating self-gene expression for a protein called MEDLE-2; translocation of MEDLE-2 from the parasite

directly into the cytoplasm of adjacent epithelial cells initiates the host cell endoplasmic reticulum stress response, which further triggers the unfolded protein response and proteasomal degradation of targeted host cell proteins leading to cellular dysfunction (Dumaine et al. 2021).

## References


Almeida PF, Pokorny A, Hinderliter A (2005) Thermodynamics of membrane domains. Biochim Biophys Acta 1720(1–2):1–13

Amimo JO, Raev SA, Chepngeno J, Mainga AO, Guo Y, Saif L et al (2021) Rotavirus interactions with host intestinal epithelial cells. Front Immunol 12:793841

Anbazhagan AN, Priyamvada S, Alrefai WA, Dudeja PK (2018) Pathophysiology of IBD associated diarrhea. Tissue Barriers 6(2):e1463897

Arnold MM, Sen A, Greenberg HB, Patton JT (2013) The battle between rotavirus and its host for control of the interferon signaling pathway. PLoS Pathog 9(1):e1003064

Arpaia N, Campbell C, Fan X, Dikiy S, van der Veeken J, deRoos P et al (2013) Metabolites produced by commensal bacteria promote peripheral regulatory T-cell generation. Nature 504(7480):451–455

Awad WA, Hess C, Hess M (2017) Enteric pathogens and their toxin-induced disruption of the intestinal barrier through alteration of tight junctions in chickens. Toxins (Basel) 9(2)

Balloux F, van Dorp L (2017) Q&A: what are pathogens, and what have they done to and for us? BMC Biol 15(1):91

Barmeyer C, Fromm M, Schulzke JD (2017) Active and passive involvement of claudins in the pathophysiology of intestinal inflammatory diseases. Pflugers Arch 469(1):15–26

Barnes HJ, Guy JS, Vaillancourt JP (2000) Poult enteritis complex. Rev Sci Tech 19(2):565–588

Barratt AS, Rich KM, Eze JI, Porphyre T, Gunn GJ, Stott AW (2019) Framework for estimating indirect costs in animal health using time series analysis. Front Vet Sci:6

Barrett KE (2017) Endogenous and exogenous control of gastrointestinal epithelial function: building on the legacy of Bayliss and Starling. J Physiol 595(2):423–432

Belkaid Y, Hand TW (2014) Role of the microbiota in immunity and inflammation. Cell 157(1):121–141

Ber Y, García-Lopez S, Gargallo-Puyuelo CJ, Gomollón F (2021) Small and large intestine (II): inflammatory bowel disease, short bowel syndrome, and malignant tumors of the digestive tract. Nutrients 13(7)

Bhakdi S, Tranum-Jensen J (1991) Alpha-toxin of Staphylococcus aureus. Microbiol Rev 55(4):733–751

Blander JM (2018) On cell death in the intestinal epithelium and its impact on gut homeostasis. Curr Opin Gastroenterol 34(6):413–419

Brandt EB, Strait RT, Hershko D, Wang Q, Muntel EE, Scribner TA et al (2003) Mast cells are required for experimental oral allergen-induced diarrhea. J Clin Invest 112(11):1666–1677

Brandt KG, Castro Antunes MM, Silva GA (2015) Acute diarrhea: evidence-based management. J Pediatr (Rio J) 91(6 Suppl 1):S36–S43

Brown D (2017) The discovery of water channels (aquaporins). Ann Nutr Metab 70(Suppl 1):37–42

Buckley A, Turner JR (2018) Cell biology of tight junction barrier regulation and mucosal disease. Cold Spring Harb Perspect Biol 10(1)

Cai Y, Folkerts J, Folkerts G, Maurer M, Braber S (2020) Microbiota-dependent and -independent effects of dietary fibre on human health. Br J Pharmacol 177(6):1363–1381

Capaldo CT, Farkas AE, Nusrat A (2014) Epithelial adhesive junctions. F1000Prime Rep 6:1

Casadevall A, Pirofski LA (2019) Benefits and costs of animal virulence for microbes. mBio 10(3)

Casadevall A, Fang FC, Pirofski LA (2011) Microbial virulence as an emergent property: consequences and opportunities. PLoS Pathog 7(7):e1002136

Cave NJ (2006) Hydrolyzed protein diets for dogs and cats. Vet Clin North Am Small Anim Pract 36(6):1251–1268. vi

Chen S, Luo S, Yan C (2021) Gut microbiota implications for health and welfare in farm animals: a review. Animals (Basel) 12(1)

Collins CB, McGrath J, Baird AW, Campion DP (2007) Effect of mast cell degranulation on chicken ileal ion transport in vitro. Poult Sci 86(5):843–849

Conklin JL, Anuras S (1996) Effects of prostaglandins E2 and F2 alpha on electromyogram of cat colon in vitro. Dig Dis Sci 41(4):720–726

Connor EE, Kahl S, Elsasser TH, Baldwin RL, Fayer R, Santin-Duran M et al (2013) Glucagon-like peptide 2 therapy reduces negative effects of diarrhea on calf gut. J Dairy Sci 96(3):1793–1802

Connor EE, Evock-Clover CM, Wall EH, Baldwin RL, Santin-Duran M, Elsasser TH et al (2016) Glucagon-like peptide 2 and its beneficial effects on gut function and health in production animals. Domest Anim Endocrinol 56(Suppl):S56–S65

Corinaldesi R, Stanghellini V, Barbara G, Tomassetti P, De Giorgio R (2012) Clinical approach to diarrhea. Intern Emerg Med 7(Suppl 3):S255–S262

Cottrell JJ, Furness JB, Wijesiriwardana UA, Ringuet M, Liu F, DiGiacomo K et al (2020) The effect of heat stress on respiratory alkalosis and insulin sensitivity in cinnamon supplemented pigs. Animals (Basel) 10(4)

Crawley SW, Mooseker MS, Tyska MJ (2014) Shaping the intestinal brush border. J Cell Biol 207(4):441–451

Curran PF, Macintosh JR (1962) A model system for biological water transport. Nature 193:347–348

Danielsen EM, Hansen GH (2008) Lipid raft organization and function in the small intestinal brush border. J Physiol Biochem 64(4):377–382
Das S, Jayaratne R, Barrett KE (2018) The role of ion transporters in the pathophysiology of infectious diarrhea. Cell Mol Gastroenterol Hepatol 6(1):33–45
de Mendoza D, Pilon M (2019) Control of membrane lipid homeostasis by lipid-bilayer associated sensors: a mechanism conserved from bacteria to humans. Prog Lipid Res 76:100996
de Santa Barbara P, van den Brink GR, Roberts DJ (2003) Development and differentiation of the intestinal epithelium. Cell Mol Life Sci 60(7):1322–1332
Denker BM, Smith BL, Kuhajda FP, Agre P (1988) Identification, purification, and partial characterization of a novel Mr 28,000 integral membrane protein from erythrocytes and renal tubules. J Biol Chem 263(30):15634–15642
Dhama K, Chauhan RS, Mahendran M, Malik SV (2009) Rotavirus diarrhea in bovines and other domestic animals. Vet Res Commun 33(1):1–23
Dheer R, Young VB (2021) Stem-cell-derived models: tools for studying role of microbiota in intestinal homeostasis and disease. Curr Opin Gastroenterol 37(1):15–22
Diamond JM, Bossert WH (1967) Standing-gradient osmotic flow. A mechanism for coupling of water and solute transport in epithelia. J Gen Physiol 50(8):2061–2083
Dumaine JE, Sateriale A, Gibson AR, Reddy AG, Gullicksrud JA, Hunter EN et al (2021) The enteric pathogen Cryptosporidium parvum exports proteins into the cytosol of the infected host cell. Elife:10
Duncan A, Robertson C, Russell RI (1992) The fecal osmotic gap: technical aspects regarding its calculation. J Lab Clin Med 119(4):359–363
Edmonds CJ (1981) Water and ionic transfer pathways of mammalian large intestine. Clin Sci (Lond) 61(3):257–263
Eherer AJ, Fordtran JS (1992) Fecal osmotic gap and pH in experimental diarrhea of various causes. Gastroenterology 103(2):545–551
El-Deek AA, Abdel-Wareth AAA, Osman M, El-Shafey M, Khalifah AM, Elkomy AE et al (2020) Alternative feed ingredients in the finisher diets for sustainable broiler production. Sci Rep 10(1):17743
Elkon KB (2018) Review: cell death, nucleic acids, and immunity: inflammation beyond the grave. Arthritis Rheumatol 70(6):805–816
Elmentaite R, Kumasaka N, Roberts K, Fleming A, Dann E, King HW et al (2021) Cells of the human intestinal tract mapped across space and time. Nature 597(7875):250–255
El-Salhy M, Gundersen D, Ostgaard H, Lomholt-Beck B, Hatlebakk JG, Hausken T (2012) Low densities of serotonin and peptide YY cells in the colon of patients with irritable bowel syndrome. Dig Dis Sci 57(4):873–878
Elsasser TH, Li CJ, Caperna TJ, Kahl S, Schmidt WF (2007) Growth hormone (GH)-associated nitration of Janus kinase-2 at the 1007Y-1008Y epitope impedes phosphorylation at this site: mechanism for and impact of a GH, AKT, and nitric oxide synthase axis on GH signal transduction. Endocrinology 148(8):3792–3802
Elsasser TH, Caperna TJ, Li CJ, Kahl S, Sartin JL (2008) Critical control points in the impact of the proinflammatory immune response on growth and metabolism. J Anim Sci 86(14 Suppl):E105–E125
Elsasser TH, Kahl S, Capuco AV, Schmidt W (2012a) Effects of stress on endocrine and metabolic processes and redirection: cross talk between subcellular compartments. Domest Anim Endocrinol 43(2):132–145
Elsasser TH, Li C-J, Shaffer J, Collier RJ (2012b) Effects of environment on animal health: mechanisms and regulatory inputs. In: Environmental physiology of livestock, pp 129–164
Elsasser TH, Miska K, Kahl S, Fetterer RH, Martínez Ramirez A (2018) Temporal pattern changes in duodenal protein tyrosine nitration events in response to Eimeria acervulina infection in chickens. J Anim Sci 96(6):2125–2138
Feng R, Niu Z, Zhang X, Hou W, Zhang Y, Jian F et al (2022) Cryptosporidium parvum downregulates miR-181d in HCT-8 cells via the p50-dependent TLRs/NF-κB pathway. Vet Parasitol 305:109710
Field M (2003) Intestinal ion transport and the pathophysiology of diarrhea. J Clin Invest 111(7):931–943
Fiorentino M, Levine MM, Sztein MB, Fasano A (2014) Effect of wild-type Shigella species and attenuated Shigella vaccine candidates on small intestinal barrier function, antigen trafficking, and cytokine release. PLoS One 9(1):e85211
Fordtran JS, Rector FC Jr, Ewton MF, Soter N, Kinney J (1965) Permeability characteristics of the human small intestine. J Clin Invest 44(12):1935–1944
Freedman JC, Shrestha A, McClane BA (2016) Clostridium perfringens enterotoxin: action, genetics, and translational applications. Toxins (Basel) 8(3)
Frénal K, Krishnan A, Soldati-Favre D (2020) The actomyosin systems in apicomplexa. Adv Exp Med Biol 1239:331–354
Frigeri A, Gropper MA, Turck CW, Verkman AS (1995) Immunolocalization of the mercurial-insensitive water channel and glycerol intrinsic protein in epithelial cell plasma membranes. Proc Natl Acad Sci U S A 92(10):4328–4331
Gecse K, Róka R, Séra T, Rosztóczy A, Annaházi A, Izbéki F et al (2012) Leaky gut in patients with diarrhea-predominant irritable bowel syndrome and inactive ulcerative colitis. Digestion 85(1):40–46
Geny B, Popoff MR (2006) Bacterial protein toxins and lipids: pore formation or toxin entry into cells. Biol Cell 98(11):667–678
Ghoshal UC, Ghoshal U (2017) Small intestinal bacterial overgrowth and other intestinal disorders. Gastroenterol Clin North Am 46(1):103–120
Gomes TA, Elias WP, Scaletsky IC, Guth BE, Rodrigues JF, Piazza RM et al (2016) Diarrheagenic Escherichia coli. Braz J Microbiol 47(Suppl 1):3–30

Grabauskas G, Wu X, Gao J, Li JY, Turgeon DK, Owyang C (2020) Prostaglandin E(2), produced by mast cells in colon tissues from patients with irritable bowel syndrome, contributes to visceral hypersensitivity in mice. Gastroenterology 158(8):2195–207.e6

Grenham S, Clarke G, Cryan JF, Dinan TG (2011) Brain-gut-microbe communication in health and disease. Front Physiol 2:94

Gresse R, Chaucheyras-Durand F, Fleury MA, Van de Wiele T, Forano E, Blanquet-Diot S (2017) Gut microbiota dysbiosis in postweaning piglets: understanding the keys to health. Trends Microbiol 25(10):851–873

Gribble FM, Reimann F (2019) Function and mechanisms of enteroendocrine cells and gut hormones in metabolism. Nat Rev Endocrinol 15(4):226–237

Groschwitz KR, Hogan SP (2009) Intestinal barrier function: molecular regulation and disease pathogenesis. J Allergy Clin Immunol 124(1):3–20; quiz 1-2

Guarino A, Lo Vecchio A, Dias JA, Berkley JA, Boey C, Bruzzese D et al (2018) Universal recommendations for the management of acute diarrhea in non-malnourished children. J Pediatr Gastroenterol Nutr 67(5):586–593

Günther C, Neumann H, Neurath MF, Becker C (2013) Apoptosis, necrosis and necroptosis: cell death regulation in the intestinal epithelium. Gut 62(7):1062–1071

Guo X, Rao JN, Liu L, Rizvi M, Turner DJ, Wang JY (2002) Polyamines regulate beta-catenin tyrosine phosphorylation via Ca(2+) during intestinal epithelial cell migration. Am J Physiol Cell Physiol 283(3):C722–C734

Guo X, Rao JN, Liu L, Zou TT, Turner DJ, Bass BL et al (2003) Regulation of adherens junctions and epithelial paracellular permeability: a novel function for polyamines. Am J Physiol Cell Physiol 285(5):C1174–C1187

Guttman JA, Finlay BB (2009) Tight junctions as targets of infectious agents. Biochim Biophys Acta 1788(4):832–841

He L, Wang C, Simujide H, Aricha H, Zhang J, Liu B et al (2022) Effect of early pathogenic Escherichia coli infection on the intestinal barrier and immune function in newborn calves. Front Cell Infect Microbiol 12:818276

Heath JK (2010) Transcriptional networks and signaling pathways that govern vertebrate intestinal development. Curr Top Dev Biol 90:159–192

Hodges K, Gill R (2010) Infectious diarrhea: cellular and molecular mechanisms. Gut Microbes 1(1):4–21

Hodges K, Gill R, Ramaswamy K, Dudeja PK, Hecht G (2006) Rapid activation of Na+/H+ exchange by EPEC is PKC mediated. Am J Physiol Gastrointest Liver Physiol 291(5):G959–G968

Holthöfer B, Windoffer R, Troyanovsky S, Leube RE (2007) Structure and function of desmosomes. Int Rev Cytol 264:65–163

Hooper NM (1999) Detergent-insoluble glycosphingolipid/cholesterol-rich membrane domains, lipid rafts and caveolae (review). Mol Membr Biol 16(2):145–156

Hunt RH, Dilawari JB, Misiewicz JJ (1975) The effect of intravenous prostaglandin F2 alpha and E2 on the motility of the sigmoid colon. Gut 16(1):47–49

Janeczko S, Atwater D, Bogel E, Greiter-Wilke A, Gerold A, Baumgart M et al (2008) The relationship of mucosal bacteria to duodenal histopathology, cytokine mRNA, and clinical disease activity in cats with inflammatory bowel disease. Vet Microbiol 128(1–2):178–193

Jha R, Singh AK, Yadav S, Berrocoso JFD, Mishra B (2019) Early nutrition programming (in ovo and post-hatch feeding) as a strategy to modulate gut health of poultry. Front Vet Sci 6:82

Juckett G, Trivedi R (2011) Evaluation of chronic diarrhea. Am Fam Physician 84(10):1119–1126

Jung K, Hu H, Saif LJ (2016) Porcine deltacoronavirus induces apoptosis in swine testicular and LLC porcine kidney cell lines in vitro but not in infected intestinal enterocytes in vivo. Vet Microbiol 182:57–63

Kandel G, Donohue-Rolfe A, Donowitz M, Keusch GT (1989) Pathogenesis of Shigella diarrhea. XVI. Selective targetting of Shiga toxin to villus cells of rabbit jejunum explains the effect of the toxin on intestinal electrolyte transport. J Clin Invest 84(5):1509–1517

Kaniyamattam K, Tauer LW, Gröhn YT (2021) System economic costs of antibiotic use elimination in the US beef supply chain. Front Vet Sci 8:606810

Kapczuk P, Kosik-Bogacka D, Kupnicka P, Metryka E, Simińska D, Rogulska K et al (2020) The influence of selected gastrointestinal parasites on apoptosis in intestinal epithelial cells. Biomolecules 10(5)

Kawasaki T, Kawai T (2014) Toll-like receptor signaling pathways. Front Immunol 5:461

Konturek PC, Brzozowski T, Konturek SJ (2011) Stress and the gut: pathophysiology, clinical consequences, diagnostic approach and treatment options. J Physiol Pharmacol 62(6):591–599

Kraimi N, Dawkins M, Gebhardt-Henrich SG, Velge P, Rychlik I, Volf J et al (2019) Influence of the microbiota-gut-brain axis on behavior and welfare in farm animals: a review. Physiol Behav 210:112658

Krug SM, Schulzke JD, Fromm M (2014) Tight junction, selective permeability, and related diseases. Semin Cell Dev Biol 36:166–176

Kumar A, Jayawardena D, Anbazhagan AN, Chatterjee I, Priyamvada S, Alrefai WA et al (2019) Decreased SLC26A3 expression and function in intestinal epithelial cells in response to Cryptosporidium parvum infection. Am J Physiol Cell Physiol 317(6):C1205–C1c12

Kuzmuk KN, Swanson KS, Tappenden KA, Schook LB, Fahey GC Jr (2005) Diet and age affect intestinal morphology and large bowel fermentative end-product concentrations in senior and young adult dogs. J Nutr 135(8):1940–1945

Lacey JA, Stanley D, Keyburn AL, Ford M, Chen H, Johanesen P et al (2018) Clostridium perfringens-mediated necrotic enteritis is not influenced by the pre-existing microbiota but is promoted by large changes

in the post-challenge microbiota. Vet Microbiol 227:119–126
Laforenza U (2012) Water channel proteins in the gastrointestinal tract. Mol Aspects Med 33(5–6):642–650
Lange K, Buerger M, Stallmach A, Bruns T (2016) Effects of antibiotics on gut microbiota. Dig Dis 34(3):260–268
Larabi A, Barnich N, Nguyen HTT (2020) New insights into the interplay between autophagy, gut microbiota and inflammatory responses in IBD. Autophagy 16(1):38–51
Lee J, Park EJ, Kiyono H (2016) MicroRNA-orchestrated pathophysiologic control in gut homeostasis and inflammation. BMB Rep 49(5):263–269
Lee B, Moon KM, Kim CY (2018) Tight junction in the intestinal epithelium: its association with diseases and regulation by phytochemicals. J Immunol Res 2018:2645465
Lesieur C, Vécsey-Semjén B, Abrami L, Fivaz M, Gisou van der Goot F (1997) Membrane insertion: the strategies of toxins (review). Mol Membr Biol 14(2):45–64
Li E, Ajuwon KM (2021) Mechanism of endocytic regulation of intestinal tight junction remodeling during nutrient starvation in jejunal IPEC-J2 cells. Faseb J 35(2):e21356
Li C, Yan X, Lillehoj HS, Oh S, Liu L, Sun Z et al (2019) Eimeria maxima-induced transcriptional changes in the cecal mucosa of broiler chickens. Parasit Vectors 12(1):285
Liu S, da Cunha AP, Rezende RM, Cialic R, Wei Z, Bry L et al (2016) The host shapes the gut microbiota via fecal MicroRNA. Cell Host Microbe 19(1):32–43
Lobo de Sá FD, Schulzke JD, Bücker R (2021) Diarrheal mechanisms and the role of intestinal barrier dysfunction in Campylobacter infections. Curr Top Microbiol Immunol 431:203–231
Loo DD, Wright EM, Zeuthen T (2002) Water pumps. J Physiol 542(Pt 1):53–60
Lorrot M, Vasseur M (2007) How do the rotavirus NSP4 and bacterial enterotoxins lead differently to diarrhea? Virol J 4:31
Louis P, Flint HJ (2009) Diversity, metabolism and microbial ecology of butyrate-producing bacteria from the human large intestine. FEMS Microbiol Lett 294(1):1–8
Lv H, Li Y, Xue C, Dong N, Bi C, Shan A (2022) Aquaporin: targets for dietary nutrients to regulate intestinal health. J Anim Physiol Anim Nutr (Berl) 106(1):167–180
Ma T, Verkman AS (1999) Aquaporin water channels in gastrointestinal physiology. J Physiol 517(Pt 2):317–326
Machalaba C, Raufman J, Anyamba A, Berrian AM, Berthe FCJ, Gray GC et al (2021) Applying a one health approach in global health and medicine: enhancing involvement of medical schools and global health centers. Ann Glob Health 87(1):30
Maelfait J, Liverpool L, Rehwinkel J (2020) Nucleic acid sensors and programmed cell death. J Mol Biol 432(2):552–568
Makkar HPS (2018) Review: feed demand landscape and implications of food-not feed strategy for food security and climate change. Animal 12(8):1744–1754
Mao X, Hu H, Xiao X, Chen D, Yu B, He J et al (2019) Lentinan administration relieves gut barrier dysfunction induced by rotavirus in a weaned piglet model. Food Funct 10(4):2094–2101
Mathias RT (1985) Epithelial water transport in a balanced gradient system. Biophys J 47(6):823–836
Mayorga EJ, Kvidera SK, Horst EA, Al-Qaisi M, Dickson MJ, Seibert JT et al (2018) Effects of zinc amino acid complex on biomarkers of gut integrity and metabolism during and following heat stress or feed restriction in pigs. J Anim Sci 96(10):4173–4185
Metzler-Zebeli BU, Mann E, Ertl R, Schmitz-Esser S, Wagner M, Klein D et al (2015) Dietary calcium concentration and cereals differentially affect mineral balance and tight junction proteins expression in jejunum of weaned pigs. Br J Nutr 113(7):1019–1031
Mogensen TH (2009) Pathogen recognition and inflammatory signaling in innate immune defenses. Clin Microbiol Rev 22(2):240–273. Table of Contents
Nabuurs MJ (1998) Weaning piglets as a model for studying pathophysiology of diarrhea. Vet Q 20(Suppl 3):S42–S45
Nagahama M, Yamaguchi A, Hagiyama T, Ohkubo N, Kobayashi K, Sakurai J (2004) Binding and internalization of Clostridium perfringens iota-toxin in lipid rafts. Infect Immun 72(6):3267–3275
Nakamura A, Kurihara S, Takahashi D, Ohashi W, Nakamura Y, Kimura S et al (2021) Symbiotic polyamine metabolism regulates epithelial proliferation and macrophage differentiation in the colon. Nat Commun 12(1):2105
Navarro MA, McClane BA, Uzal FA (2018) Mechanisms of action and cell death associated with Clostridium perfringens toxins. Toxins (Basel) 10(5)
Nealon NJ, Yuan L, Yang X, Ryan EP (2017) Rice bran and probiotics alter the porcine large intestine and serum metabolomes for protection against human rotavirus diarrhea. Front Microbiol 8:653
Nguyen SMT, Rupprecht CP, Haque A, Pattanaik D, Yusin J, Krishnaswamy G (2021) Mechanisms governing anaphylaxis: inflammatory cells, mediators, endothelial gap junctions and beyond. Int J Mol Sci 22(15)
Noah TK, Donahue B, Shroyer NF (2011) Intestinal development and differentiation. Exp Cell Res 317(19):2702–2710
Ogata M, Itoh T (2016) Gamma/delta intraepithelial lymphocytes in the mouse small intestine. Anat Sci Int 91(4):301–312
Olkowski AA, Wojnarowicz C, Chirino-Trejo M, Laarveld B, Sawicki G (2008) Sub-clinical necrotic enteritis in broiler chickens: novel etiological consideration based on ultra-structural and molecular changes in the intestinal tissue. Res Vet Sci 85(3):543–553
Opgenorth J, Abuajamieh M, Horst EA, Kvidera SK, Johnson JS, Mayorga EJ et al (2021) The effects of zinc amino acid complex on biomarkers of gut integ-

rity, inflammation, and metabolism in heat-stressed ruminants. J Dairy Sci 104(2):2410–2421

Palsson OS, Morteau O, Bozymski EM, Woosley JT, Sartor RB, Davies MJ et al (2004) Elevated vasoactive intestinal peptide concentrations in patients with irritable bowel syndrome. Dig Dis Sci 49(7–8):1236–1243

Panah FM, Lauridsen C, Højberg O, Nielsen TS (2021) Etiology of colitis-complex diarrhea in growing pigs: a review. Animals (Basel) 11(7)

Patankar JV, Becker C (2020) Cell death in the gut epithelium and implications for chronic inflammation. Nat Rev Gastroenterol Hepatol 17(9):543–556

Pearce SC, Mani V, Weber TE, Rhoads RP, Patience JF, Baumgard LH et al (2013) Heat stress and reduced plane of nutrition decreases intestinal integrity and function in pigs. J Anim Sci 91(11):5183–5193

Pelagalli A, Squillacioti C, Mirabella N, Meli R (2016) Aquaporins in health and disease: an overview focusing on the gut of different species. Int J Mol Sci 17(8)

Pelaseyed T, Bergström JH, Gustafsson JK, Ermund A, Birchenough GM, Schütte A et al (2014) The mucus and mucins of the goblet cells and enterocytes provide the first defense line of the gastrointestinal tract and interact with the immune system. Immunol Rev 260(1):8–20

Pohl CS, Medland JE, Mackey E, Edwards LL, Bagley KD, DeWilde MP et al (2017) Early weaning stress induces chronic functional diarrhea, intestinal barrier defects, and increased mast cell activity in a porcine model of early life adversity. Neurogastroenterol Motil 29(11)

Proszkowiec-Weglarz M, Schreier LL, Kahl S, Miska KB, Russell B, Elsasser TH (2020) Effect of delayed feeding post-hatch on expression of tight junction- and gut barrier-related genes in the small intestine of broiler chickens during neonatal development. Poult Sci 99(10):4714–4729

Puertollano E, Kolida S, Yaqoob P (2014) Biological significance of short-chain fatty acid metabolism by the intestinal microbiome. Curr Opin Clin Nutr Metab Care 17(2):139–144

Qadri F, Bhuiyan TR, Dutta KK, Raqib R, Alam MS, Alam NH et al (2004) Acute dehydrating disease caused by Vibrio cholerae serogroups O1 and O139 induce increases in innate cells and inflammatory mediators at the mucosal surface of the gut. Gut 53(1):62–69

Qin HY, Cheng CW, Tang XD, Bian ZX (2014) Impact of psychological stress on irritable bowel syndrome. World J Gastroenterol 20(39):14126–14131

Qu Y, Kahl S, Miska KB, Schreier LL, Russell B, Elsasser TH et al (2021) The effect of delayed feeding post-hatch on caeca development in broiler chickens. Br Poult Sci 62(5):731–748

Rahimi S, Kathariou S, Fletcher O, Grimes JL (2020) The effectiveness of a dietary direct-fed microbial and mannan oligosaccharide on ultrastructural changes of intestinal mucosa of turkey poults infected with Salmonella and Campylobacter. Poult Sci 99(2):1135–1149

Ramachandran G (2014) Gram-positive and gram-negative bacterial toxins in sepsis: a brief review. Virulence 5(1):213–218

Rand RP (2002) The lipid-water interface: revelations by osmotic stress. Int Rev Cytol 215:33–48

Rao MC (2019) Physiology of electrolyte transport in the gut: implications for disease. Compr Physiol 9(3):947–1023

Reiche J, Huber O (2020) Post-translational modifications of tight junction transmembrane proteins and their direct effect on barrier function. Biochim Biophys Acta Biomembr 1862(9):183330

Rengarajan S, Knoop KA, Rengarajan A, Chai JN, Grajales-Reyes JG, Samineni VK et al (2020) A potential role for stress-induced microbial alterations in IgA-associated irritable bowel syndrome with diarrhea. Cell Rep Med 1(7)

Ritchie JM, Rui H, Zhou X, Iida T, Kodoma T, Ito S et al (2012) Inflammation and disintegration of intestinal villi in an experimental model for Vibrio parahaemolyticus-induced diarrhea. PLoS Pathog 8(3):e1002593

Roderer D, Glockshuber R (2017) Assembly mechanism of the α-pore-forming toxin cytolysin A from Escherichia coli. Philos Trans R Soc Lond B Biol Sci 372(1726)

Roxas JL, Viswanathan VK (2018) Modulation of intestinal paracellular transport by bacterial pathogens. Compr Physiol 8(2):823–842

Saavedra PHV, Huang L, Ghazavi F, Kourula S, Vanden Berghe T, Takahashi N et al (2018) Apoptosis of intestinal epithelial cells restricts Clostridium difficile infection in a model of pseudomembranous colitis. Nat Commun 9(1):4846

Said HM (2011) Intestinal absorption of water-soluble vitamins in health and disease. Biochem J 437(3):357–372

Sauter SN, Benyacoub J, Allenspach K, Gaschen F, Ontsouka E, Reuteler G et al (2006) Effects of probiotic bacteria in dogs with food responsive diarrhoea treated with an elimination diet. J Anim Physiol Anim Nutr (Berl) 90(7–8):269–277

Serena A, Hedemann MS, Bach Knudsen KE (2008) Influence of dietary fiber on luminal environment and morphology in the small and large intestine of sows. J Anim Sci 86(9):2217–2227

Shashikanth N, Yeruva S, Ong M, Odenwald MA, Pavlyuk R, Turner JR (2017) Epithelial organization: the gut and beyond. Compr Physiol 7(4):1497–1518

Sherman MP, Petrak K (2005) Lactoferrin-enhanced anoikis: a defense against neonatal necrotizing enterocolitis. Med Hypotheses 65(3):478–482

Shurson GC, Urriola PE (2022) Sustainable swine feeding programs require the convergence of multiple dimensions of circular agriculture and food systems with One Health. Anim Front 12(6):30–40

Singh R, Zogg H, Wei L, Bartlett A, Ghoshal UC, Rajender S et al (2021) Gut microbial dysbiosis in the pathogenesis of gastrointestinal dysmotility and metabolic disorders. J Neurogastroenterol Motil 27(1):19–34

Singhal R, Shah YM (2020) Oxygen battle in the gut: hypoxia and hypoxia-inducible factors in metabolic and inflammatory responses in the intestine. J Biol Chem 295(30):10493–10505
Sørensen NB, Nielsen HL, Varming K, Nielsen H (2013) Neutrophil activation by Campylobacter concisus. Gut Pathog 5(1):17
Stevens MP, Humphrey TJ, Maskell DJ (2009) Molecular insights into farm animal and zoonotic Salmonella infections. Philos Trans R Soc Lond B Biol Sci 364(1530):2709–2723
Stewart AS, Pratt-Phillips S, Gonzalez LM (2017) Alterations in intestinal permeability: the role of the "leaky gut" in health and disease. J Equine Vet Sci 52:10–22
Stokes CR, Miller BG, Bailey M, Wilson AD, Bourne FJ (1987) The immune response to dietary antigens and its influence on disease susceptibility in farm animals. Vet Immunol Immunopathol 17(1–4):413–423
Suchodolski JS (2011) Companion animals symposium: microbes and gastrointestinal health of dogs and cats. J Anim Sci 89(5):1520–1530
Suzuki T (2013) Regulation of intestinal epithelial permeability by tight junctions. Cell Mol Life Sci 70(4):631–659
Suzuki T (2020) Regulation of the intestinal barrier by nutrients: the role of tight junctions. Anim Sci J 91(1):e13357
Swank GM, Deitch EA (1996) Role of the gut in multiple organ failure: bacterial translocation and permeability changes. World J Surg 20(4):411–417
Taché Y, Mönnikes H, Bonaz B, Rivier J (1993) Role of CRF in stress-related alterations of gastric and colonic motor function. Ann N Y Acad Sci 697:233–243
Thomson S, Hamilton CA, Hope JC, Katzer F, Mabbott NA, Morrison LJ et al (2017) Bovine cryptosporidiosis: impact, host-parasite interaction and control strategies. Vet Res 48(1):42
Tilley SJ, Saibil HR (2006) The mechanism of pore formation by bacterial toxins. Curr Opin Struct Biol 16(2):230–236
Tripathi S, Boulpaep EL (1989) Mechanisms of water transport by epithelial cells. Q J Exp Physiol 74(4):385–417
Tripathi A, Xu ZZ, Xue J, Poulsen O, Gonzalez A, Humphrey G et al (2019) Intermittent hypoxia and hypercapnia reproducibly change the gut microbiome and metabolome across rodent model systems. mSystems 4(2)
Turner JR (2009) Intestinal mucosal barrier function in health and disease. Nat Rev Immunol 9(11):799–809
Turvill JL, Farthing MJ (1999) Water and electrolyte absorption and secretion in the small intestine. Curr Opin Gastroenterol 15(2):108–112
Uzal FA, Freedman JC, Shrestha A, Theoret JR, Garcia J, Awad MM et al (2014) Towards an understanding of the role of Clostridium perfringens toxins in human and animal disease. Future Microbiol 9(3):361–377
Van Itallie CM, Anderson JM (2018) Phosphorylation of tight junction transmembrane proteins: many sites, much to do. Tissue Barriers 6(1):e1382671
Van Opdenbosch N, Van Gorp H, Verdonckt M, Saavedra PHV, de Vasconcelos NM, Gonçalves A et al (2017) Caspase-1 engagement and TLR-induced c-FLIP expression suppress ASC/Caspase-8-dependent apoptosis by inflammasome sensors NLRP1b and NLRC4. Cell Rep 21(12):3427–3444
Vargas N, Marino F (2016) Heat stress, gastrointestinal permeability and interleukin-6 signaling - Implications for exercise performance and fatigue. Temperature (Austin) 3(2):240–251
Viswanathan VK, Hodges K, Hecht G (2009) Enteric infection meets intestinal function: how bacterial pathogens cause diarrhoea. Nat Rev Microbiol 7(2):110–119
von Boyen GB, Steinkamp M, Geerling I, Reinshagen M, Schäfer KH, Adler G et al (2006) Proinflammatory cytokines induce neurotrophic factor expression in enteric glia: a key to the regulation of epithelial apoptosis in Crohn's disease. Inflamm Bowel Dis 12(5):346–354
von Buchholz JS, Ruhnau D, Hess C, Aschenbach JR, Hess M, Awad WA (2022) Paracellular intestinal permeability of chickens induced by DON and/or C. jejuni is associated with alterations in tight junction mRNA expression. Microb Pathog:105509
Walker MP, Evock-Clover CM, Elsasser TH, Connor EE (2015) Short communication: glucagon-like peptide-2 and coccidiosis alter tight junction gene expression in the gastrointestinal tract of dairy calves. J Dairy Sci 98(5):3432–3437
Wallis TS, Barrow PA (2005) Salmonella epidemiology and pathogenesis in food-producing animals. EcoSal Plus 1(2)
Wang YH, Liu TT, Kung WM, Chen CC, Wen YT, Lin IC et al (2015a) Expression of aquaporins in intestine after heat stroke. Int J Clin Exp Pathol 8(8):8742–8753
Wang H, Ji Y, Wu G, Sun K, Sun Y, Li W et al (2015b) l-Tryptophan activates mammalian target of rapamycin and enhances expression of tight junction proteins in intestinal porcine epithelial cells. J Nutr 145(6):1156–1162
Wang D, Naydenov NG, Feygin A, Baranwal S, Kuemmerle JF, Ivanov AI (2016) Actin-depolymerizing factor and Cofilin-1 have unique and overlapping functions in regulating intestinal epithelial junctions and mucosal inflammation. Am J Pathol 186(4):844–858
Wright EM, Loo DD (2000) Coupling between Na+, sugar, and water transport across the intestine. Ann N Y Acad Sci 915:54–66
Wu Y, Tang L, Wang B, Sun Q, Zhao P, Li W (2019) The role of autophagy in maintaining intestinal mucosal barrier. J Cell Physiol 234(11):19406–19419
Xu JH, Qin ZH, Liao YS, Xie MQ, Li AX, Cai JP (2008) Characterization and expression of an actin-

depolymerizing factor from Eimeria tenella. Parasitol Res 103(2):263–270

Yakabi S, Wang L, Karasawa H, Yuan PQ, Koike K, Yakabi K et al (2018) VIP is involved in peripheral CRF-induced stimulation of propulsive colonic motor function and diarrhea in male rats. Am J Physiol Gastrointest Liver Physiol 314(5):G610–Gg22

Yang X, Twitchell E, Li G, Wen K, Weiss M, Kocher J et al (2015) High protective efficacy of rice bran against human rotavirus diarrhea via enhancing probiotic growth, gut barrier function, and innate immunity. Sci Rep 5:15004

Zhang H, Li D, Liu L, Xu L, Zhu M, He X et al (2019) Cellular composition and differentiation signaling in chicken small intestinal epithelium. Animals (Basel) 9(11)

Zhao Y, Zeng Y, Zeng D, Wang H, Zhou M, Sun N et al (2020) Probiotics and MicroRNA: their roles in the host-microbe interactions. Front Microbiol 11:604462

Zhu C, Chen Z, Jiang Z (2016) Expression, distribution and role of aquaporin water channels in human and animal stomach and intestines. Int J Mol Sci 17(9)

Zuo L, Kuo WT, Turner JR (2020) Tight junctions as targets and effectors of mucosal immune homeostasis. Cell Mol Gastroenterol Hepatol 10(2):327–340

# Diarrhea in Calves

Ingrid Lorenz and Florian M. Trefz

**Abstract**

Calf diarrhea is a multifactorial disease. The most common pathogens involved are viruses and cryptosporidia, which cause diarrhea by damaging the intestinal mucosa. The main pathophysiological consequences are loss of fluids and electrolytes and the development of metabolic acidosis. The most important treatment measure is oral rehydration therapy, which has to be initiated as soon as diarrhea is diagnosed. Intravenous therapy is indicated if dehydration is advanced and if clinical signs indicate profound acidosis. Antimicrobial treatment is only recommended in calves with severe systemic disease. The most important factors to prevent calf diarrhea are enhanced and prolonged colostrum feeding followed by a biologically normal (intensive) milk feeding program with ad libitum or close to ad libitum feeding in the first three to four weeks of life followed by a slow reduction of feeding volumes up to weaning.

I. Lorenz (✉)
Bavarian Animal Health Service (Tiergesundheitsdienst Bayern e.V.), Poing, Germany
e-mail: Ingrid.Lorenz@tgd-bayern.de

F. M. Trefz
Clinic for Ruminants with Ambulatory and Herd Health Services at the Centre for Clinical Veterinary Medicine, Ludwig-Maximilians-Universität (LMU) Munich, Oberschleißheim, Germany
e-mail: f.trefz@med.vetmed.uni-muenchen.de

**Keywords**

Calf diarrhea · Colostrum management · Ad libitum feeding · Dehydration · Metabolic acidosis · Hyperkalemia · Oral rehydration · Intravenous treatment

## 1 Etiology and Prevention

Calf diarrhea is a classical multifactorial disease, which means that besides—generally ubiquitous—infectious agents, environmental and management factors contribute to the outbreak of disease.

*Cryptosporidium parvum* is one of the most commonly isolated pathogens in diarrheic calves; however, it can also be found in calves without diarrhea. Infection occurs via oral ingestion of oocysts from the environment or from cows that can serve as a reservoir. The organism then undergoes numerous asexual multiplication stages and a final sexual reproductive stage, where thin- and thick-walled oocysts are created. The thick-walled oocysts are excreted with the feces to contaminate the environment, whereas the thin-walled oocysts initiate an auto-infection cycle, which may prolong the disease. After excystation of the oocysts in the intestinal lumen, sporozoites damage the intestinal mucosa, mainly in the ileum, by creating parasitophorous vacuoles in the enterocytes of the villi.



J. J. Gross (ed.), *Production Diseases in Farm Animals*,
https://doi.org/10.1007/978-3-031-51788-4_10

Rotavirus causes diarrhea in calves, mainly in the first three weeks of life. After ingestion, the virus replicates in the mature absorptive and enzyme-producing enterocytes on the villi of the small intestine, causing villous atrophy.

Coronavirus is also commonly associated with diarrhea in calves. It replicates in enterocytes of the intestine, where it produces similar lesions to rotavirus but also infects the epithelial cells of the large intestine to produce atrophy of the colonic ridges.

Enterotoxic *E. coli* (ETEC) causes diarrhea in calves, mainly in the first 4 days of life. To be able to induce diarrhea, ETEC needs to attach to the intestinal wall, which it does via fimbrial antigens in calves, mainly K99 (E5). Once attached, ETEC produces heat stable toxin, which leads to secretory diarrhea (Foster and Smith 2009).

Management factors contributing to calf diarrhea generally either lower the resistance of the calf or increase the infectious pressure. Infectious pressure is regulated by means of hygiene in all areas of calf management (e.g., calving, housing, feeding). However, during the last two decades, it became increasingly clear that especially the way we feed our calves has a major impact on the resistance in artificially reared calves.

Good colostrum management is still recognized as the single most important factor in preventing calf morbidity and mortality; however, it is now known that immunoglobulins are only one of many components of colostrum that are vital for the calf's development. Other non-nutrient factors like leucocytes, hormones, and growth factors, oligosaccharides as well as microRNAs have significant effects on the development and maturation of the intestinal and systemic immune functions. They also promote the maturation and function of the intestine, thus enabling the calf to digest and absorb the nutrients provided by colostrum and milk. The improved energetic status of colostrum-fed neonates is reflected by an accelerated maturation of the somatotropic axis, which stimulates body growth and organ development. Colostrum oligosaccharides are presumed to play a major role in the development of a healthy intestinal flora. Colostrum also contains an abundance of antimicrobial components, like lactoferrin, lactoperoxidase, or lysozyme. Extended feeding of colostrum or transition milk further promotes small intestinal growth and maturation and prevents calf diarrhea. A biologically normal milk-feeding program is subsequently necessary for optimal body growth, organ development, and resistance to infectious diseases. Ad libitum or close to ad libitum feeding in the first three to four weeks of life, followed by a slow reduction of feeding volumes up to weaning, has proven to be the optimal feeding regime to prevent calves from diarrhea and other infectious diseases (Lorenz 2021).

Specific preventive measures are limited. Vaccination is the only prophylactic measure available against viral causes of diarrhea. Vaccination of the dam before calving has been used to enhance the content of rotavirus-specific antibodies in colostrum. Commercially available vaccines usually also contain coronavirus and *E. coli* F5 antigen. There is no doubt that modern vaccines are able to increase the level of specific antibodies in the serum and milk of vaccinated cows as well as in the serum of calves that have ingested colostrum from vaccinated dams. However, evidence as to clinical efficacy in naturally acquired diarrhea is either not available for all vaccines or is conflicting (Lorenz et al. 2011). Halofuginone has a demonstrated cryptosporidiostatic effect in vivo and is licensed for the prevention and treatment of cryptosporidiosis in calves in some countries. A review and meta-analysis found a positive effect for the time period where the substance was given, but not later on. On day 21, the control group had a lower infection prevalence than the halofuginone-treated group. The authors conclude that halofuginone should be used in combination with measures taken to improve calf management routines (Silverlås et al. 2009).

## 2 Pathophysiology

It has long been assumed that cryptosporidia and viral pathogens cause exclusively malabsorptive diarrhea due to villous atrophy. However, it has been shown that rotavirus causes secretory diarrhea by producing an enterotoxin-like substance, whereas cryptosporidia causes prostaglandin-mediated hypersecretion. Furthermore, due to the impairment of villous fluid absorption, continuous secretion from the crypts increases the net secretion of fluids into the intestinal lumen. ETEC, however, causes only secretory diarrhea (Foster and Smith 2009).

## 2.1 Dehydration

Depending on the type of dehydration (isotonic or hypotonic), the extracellular space either sustains the full amount of fluid losses or loses additional water to the intracellular space. The "average" diarrheic calf loses about 10% of its body weight in fluids per day additionally to maintenance requirements of 10% of body weight, but losses of over 20% have been observed. Clinical assessment of the loss of body fluid has been described many times, but studies directed at the quantification of changes occurring in relation to varying degrees of dehydration are rare. However, there have been two studies addressing this question. In general, the extent of enophthalmos was found to be the best predictor of dehydration. A study in calves with naturally acquired diarrhea has shown, that the degree of dehydration in per cent of body weight can be estimated by multiplying the enophthalmos in mm by 1.91 (Niethammer 2007). For practical purposes, it is important to note that a visible gap between the eyeball and the orbit starts to form at a degree of dehydration of about 8% of body weight. This coincides with the somewhat arbitrary, but generally accepted rule that from a degree of dehydration of 8% and greater, oral rehydration is unlikely to be successful and intravenous rehydration therapy should be initiated.

## 2.2 Metabolic Acidosis

Metabolic acidosis is a frequently occurring consequence of neonatal diarrhea in calves. The associated decrease of plasma bicarbonate concentration was traditionally related to excessive intestinal losses of electrolytes and bicarbonate ions as well as to buffering of endogenously produced acids. Dehydration resulting in a decrease of renal blood flow and therefore excretion of hydrogen ions as well as formation of L-lactate in response to concomitant tissue hypoperfusion was further considered to be a significant contributor to this condition (Glawischnig et al. 1990; Kaske 1994; Hartmann et al. 1997; Kasari 1999). An increase in plasma concentration of unmeasured anions (as indicated by an increased anion gap) has long been known as a key feature of metabolic acidosis in neonatal diarrheic calves. However, this increase could formerly not be explained by elevated L-lactate concentrations and therefore resulted in an incomplete understanding of the pathophysiology of metabolic acidosis in neonatal diarrheic calves for a long time (Kasari and Naylor 1986; Grove-White 1997; Hartmann et al. 1997; Binding et al. 2000).

In 1998, hyper-D-lactatemia was reported in markedly acidotic suckler Charolais calves with no or minimal diarrhea (Schelcher et al. 1998). Eventually, subsequent studies confirmed that increased blood D-lactate also plays an important role in the pathogenesis of metabolic acidosis in neonatal diarrheic calves and are largely responsible for the observed increase of plasma unmeasured anion concentrations (Omole et al. 2001; Ewaschuk et al. 2003; Lorenz 2004a). In neonatal diarrheic calves, increased blood D-lactate concentrations arise from excessive bacterial fermentation processes that occur in the reticulorumen or in the intestinal tract. Maldigestion and malabsorption of carbohydrates in the small intestine are considered to favor the growth of lactic acid-producing bacteria (such as lactobacilli or bifidobacteria) in the large intestine and therefore production of large amounts of D-lactate (Ewaschuk et al. 2003; Lorenz 2009; Lorenz and Gentile 2014). Also, failure of the esophageal groove reflex and subsequent intraruminal fermentation of milk represents a further mechanism as intraruminal and systemic D-lactic acidosis has also been demonstrated in calves with naturally acquired or experimentally induced ruminal drinking syndrome (Grude 2003; Gentile et al. 2004). In calves with neonatal diarrhea, the risk of developing D-lactic acidosis is age-dependent (Trefz et al. 2012b). D-Lactic acidosis is less likely to happen in the first few days of life, where enterotoxic *E. coli* are a frequent cause of diarrhea. As a consequence, calves that suffer from diarrhea within their first 4 days of life have significantly higher plasma bicarbonate concentrations and significantly lower D-lactate concentrations than older calves (Fig. 1).

Due to its bacterial origin and the assumed lack of specific metabolic pathways, it has historically been believed that D-lactate can be poorly metabolized by mammals. However, an experimental study gave evidence for the considerable capacity of D-lactate metabolization in healthy calves (Lorenz et al. 2005). In diarrheic calves with concomitant acidemia and/or dehydration, the exact mechanisms for accumulation of D-lactate in plasma are therefore not fully understood, but it is likely that D-lactate production exceeds the body's capacity for metabolization and/or excretion (Lorenz and Gentile 2014). Decreased renal perfusion in dehydrated calves, impairment of D-lactate metabolism due to acidemia, and a low intestinal pH favoring bacterial D-lactate production and absorption are considered potential contributors to hyper-D-lactatemia in diarrheic calves (Ewaschuk et al. 2004; Lorenz and Gentile 2014).

A further improved understanding of the pathophysiology of metabolic acidosis in neonatal diarrheic has been provided by a 2005 study that applied the simplified strong ion approach to acid–base balance in those animals (Constable et al. 2005). That study questioned the long-held view about the significance of fecal bicarbonate losses in the development of acidemia and metabolic acidosis and identified a strong ion acidosis due to a low plasma strong ion difference as a result of hyponatremia (with concomitant normo- or hyperchloremia) and/or an increase of unmeasured strong anions such as D-lactate as the major cause of acidemia in those animals. Dehydration and a loss of free water and a resulting nonvolatile buffer ion acidosis due to increases in serum protein concentrations was also reported as a significant, but clinically less important, contributor to acidemia (Constable et al. 2005). This and subsequent studies confirmed that increased concentrations of unmeasured strong anions (which can be quantified by calculation of the anion or strong ion gap) are quantitatively the most important contributor to the development of acidemia in those animals (Constable et al. 2005; Gomez et al. 2013; Trefz et al. 2015a).

Metabolic acidosis is of clinical relevance in neonatal diarrheic calves as it is associated with a clinical picture characterized by general weakness, depression, impairment of the ability to

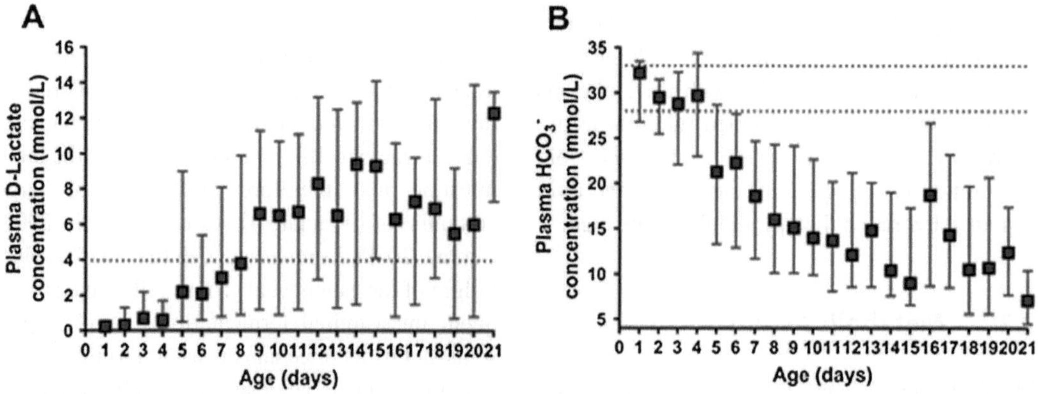


**Fig. 1** Median and interquartile ranges for plasma D-lactate (**a**) and bicarbonate concentrations (**b**) in 1400 hospitalized calves with neonatal diarrhea in relation to age. Dashed horizontal lines indicate the reference range for D-lactate (≤3.96 mmol/L) and plasma bicarbonate concentration (28–33 mmol/L), respectively

stand, ataxia, and neurologic abnormalities (Fig. 2).

Along with the D-lactate research in neonatal diarrheic calves, new light was also shed on the clinical signs of metabolic acidosis in those animals. Clinical studies have shown that in diarrheic calves with moderate to severe acidosis, variations in behavior and posture can be better explained by increases of serum D-lactate concentration than by decreases of base excess (Lorenz 2004b). Also, it was established that alterations of the palpebral reflex represent a reliable clinical sign of hyper-D-lactatemia in calves (Lorenz 2004b). Subsequent experimental studies confirmed that hyper-D-lactatemia, and not metabolic acidosis per se, is associated with changes of posture and behavior as well as alterations of the palpebral reflex (Lorenz et al. 2005; Gentile et al. 2008). D-Lactate exerts a neurotoxic effect that is most likely related to alterations of neuronal energy provision within the central nervous system (Ling et al. 2012). The clinical signs of hyper-D-lactatemia are completely reversible and resolve after normalization of plasma D-lactate concentrations in response to treatment (Lorenz and Vogt 2006; Trefz et al. 2012a). However, despite the well-proven relationship between hyper-D-lactatemia and clinical signs, clinical studies still reported profound alterations in posture and behavior in calves with neonatal diarrhea in spite of normal or only slightly elevated plasma D-lactate concentrations (Trefz et al. 2013b). This implies that there are obviously other factors aside from D-lactate that influence the clinical picture in those animals as well.

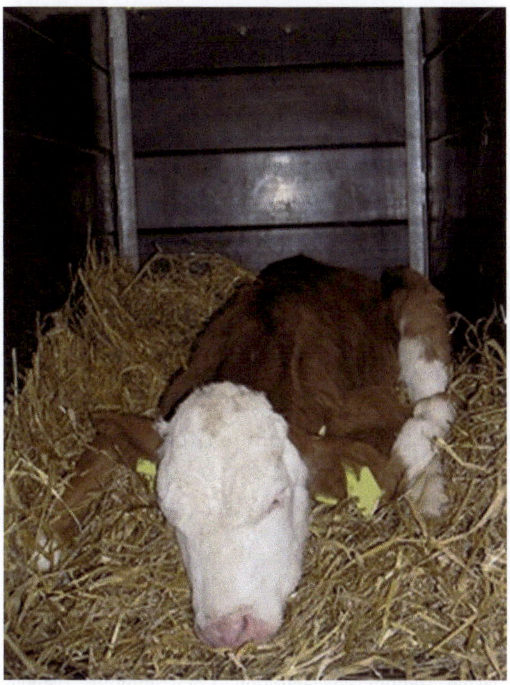

**Fig. 2** Clinical picture of marked D-lactic acidosis in a neonatal diarrheic calf

## 2.3 Electrolyte Imbalances

Electrolyte imbalances are also common in neonatal diarrheic calves and are closely linked to derangements of acid–base status. Moderate hyponatremia is a typical finding, whereas hypernatremia is less commonly seen and frequently associated with inadequate composition of oral electrolyte solutions and/or pretreatment with (hypertonic) sodium bicarbonate infusion solution in calves that do not have access to free water (Pringle and Berthiaume 1988; Trefz et al. 2017c).

Although neonatal diarrheic calves have a clear negative potassium balance due to intestinal losses, they usually have normal or even increased plasma potassium concentrations (Trefz et al. 2013a). In severely ill hospitalized diarrheic calves, hyperkalemia has a prevalence of about 30% (Trefz et al. 2013a, b). Hyperkalemia was historically attributed to an acidemia-induced dysregulation of potassium ions between the intra- and extracellular space (Lewis and Phillips 1973). However, the occurrence of this electrolyte imbalance was recently reported to be strictly dependent on the nature of an existing acidosis, with D-lactic acidosis being rarely associated with a hyperkalemic state. Consequently, an association between increased plasma potassium concentrations and decreased venous blood pH is only seen in diarrheic calves with normal D-lactate concentrations (Trefz et al. 2013a). More importantly, increased plasma potassium concentrations in neonatal diarrheic calves are most closely correlated with clinical and laboratory indices of dehydration, which strongly indicates that a decrease in glomerular filtration rate plays a central role in the development of a hyperkalemic state (Trefz et al. 2013a, b; Trefz and Lorenz 2017). Furthermore, a recent study indicated that neonatal diarrhea is not associated with a pronounced depletion of intracellular potassium stores in calves (Golbeck et al. 2018).

Clinical effects of hyperkalemia are related to impaired neuromuscular excitability, which can lead to muscle weakness and cardiac conduction abnormalities (Trefz et al. 2013b, 2018). Those effects are further exacerbated by the presence of hyponatremia and metabolic acidosis, conditions which are usually present in hyperkalemic neonatal diarrheic calves. Increased voltages of S wave, T wave, P wave, and Ta wave were identified as the first electrocardiographic changes in response to an increase of plasma potassium concentrations in neonatal diarrheic calves (Trefz et al. 2018). More pronounced hyperkalemia can also result in a loss of atrial repolarization, manifesting as the disappearance of P waves, and prolonged ventricular depolarization, which manifests as increased QRS duration, ST segment elevation, or ventricular escape rhythms and arrhythmias (Fig. 3).

Furthermore, the clinical picture of hyperkalemic diarrheic calves is characterized by profound clinical dehydration, cyanosis, and impairments of the ability to stand and affected calves are frequently presented with signs of shock. Hyperkalemia in diarrheic calves can therefore produce a clinical picture that is somehow comparable to marked D-lactic acidosis. However, a hyperkalemic state in diarrheic calves is usually accompanied by normal or only slightly elevated plasma D-lactate concentrations. Consequently, these animals usually exhibit a normal palpebral reflex, which allows to clinically distinguish between these two conditions (Trefz et al. 2013b). Impairment of ability to stand in a diarrheic calf with no clinical signs of D-lactatemia (prompt palpebral reflex), marked clinical dehydration, cardiac arrhythmias, and the finding of pale-cyanotic muzzle or mucous membranes should therefore alert the clinician to the presence of hyperkalemia. Due to its cardiotoxic effects, acute hyperkalemia in diarrheic calves should be considered as an emergency.

In critically ill neonatal diarrheic calves, hypokalemia is more commonly seen during the course of treatment than on initial presentation. An observational study in hospitalized calves indicated that the occurrence of hypokalemia was associated with a more pronounced metabolic acidosis on admission to the hospital, a lower intake of milk during hospitalization, and a longer duration of diarrhea than in calves with a normokalemic state after 24 h of treatment (Trefz et al. 2015c).

## 2.4 Hypoglycemia

Hypoglycemia has a low prevalence in hospitalized neonatal calves. In a study involving 1400 neonatal diarrheic calves, only 10% of calves had plasma glucose concentration <2.9 mmol/L (Trefz et al. 2017c). Similar findings were made in a retrospective study using a much larger data set of 10,060 neonatal hospitalized calves (Trefz et al. 2016). In that study, severe hypoglycemia (plasma glucose concentration <2 mmol/L) had a prevalence of only 6.3%. Given that calves admitted to a veterinary teaching hospital are usually preselected for severe disease, it can be assumed that this condition occurs even rarer in

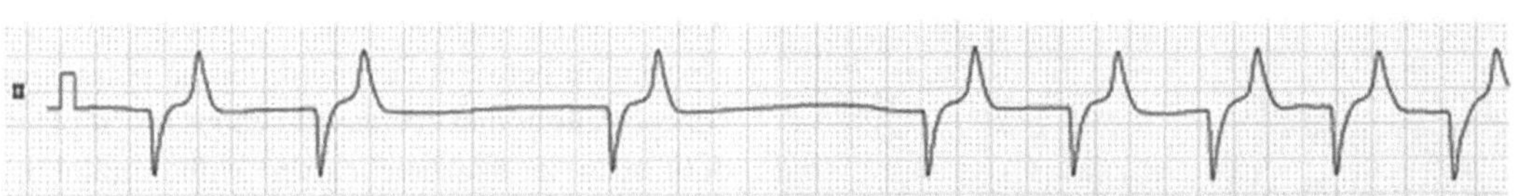

**Fig. 3** Bradycardia (ventricular rate 52/min), missing P waves, and arrhythmias in a severely hyperkalemic diarrheic calf (plasma potassium concentration 9.4 mmol/L, plasma sodium concentration 124.4 mmol/L, venous blood pH 7.081). Paper speed 25 mm/s, sensitivity 5 mm/mV

ambulatory field practice. In critically ill neonatal calves, severe hypoglycemia has been shown to be associated with serious health conditions such as hypothermia, malnutrition, systemic inflammatory response syndrome, acute abdominal emergencies including peritonitis of varied origin, and clinical or post-mortem evidence of septicemia, but not with the presence of diarrhea per se (Trefz et al. 2016). Consequently, in the aforementioned study, severe hypoglycemia was associated with a poor survival rate of 20.6%, compared to 74% of calves that had plasma glucose concentrations within the reference range (4.4–6.9 mmol/L). Detection of plasma glucose alterations requires measurement of plasma glucose concentration, which can be easily performed by means of portable point-of-care glucometers used in human medicine (Karapinar et al. 2020).

## 3 Treatment

### 3.1 Treatment of Metabolic Alterations

Oral fluid therapy by administration of electrolyte solutions is the cheapest and easiest way to replace intestinal fluid losses and to correct acid–base and electrolyte abnormalities in diarrheic calves. A suitable oral rehydration solution for use in neonatal calves should contain sodium in an amount of 90–130 mmol/L, alkalinizing agents such as bicarbonate or bicarbonate precursors that create an effective strong ion difference (buffer capacity) of 60–80 mmol/L, glucose and other components (such as glycine, acetate, or propionate) that facilitate the absorption of sodium and water, and potassium in an amount of 10–30 mmol/L in order to counterbalance a negative potassium balance (Smith and Berchtold 2014; Constable et al. 2021). The composition of an ideal oral rehydration solution differs from those used in human medicine, as neonatal diarrheic calves frequently develop hyponatremia and a more pronounced metabolic acidosis than human infants with diarrhea. However, feeding of hypertonic rehydration solutions (>600 mOsm/kg) should be avoided as these can negatively affect the abomasal emptying rate (Constable et al. 2021).

In diarrheic calves that require intravenous fluid therapy for reasons of advanced dehydration (usually defined as the presence of enophthalmos), marked depression and/or anorexia, intravenous administration of sodium bicarbonate remains the treatment of choice for correction of metabolic acidosis (Berchtold 2009). Traditionally, the amount of sodium bicarbonate needed for intravenous correction of metabolic acidosis is calculated by multiplying the existing base deficit with body mass and a factor reflecting the distribution space of sodium bicarbonate (Lorenz and Vogt 2006). Remarkably, studies found a large variability when retrospectively calculating the size of this factor. A recent study shed light on this phenomenon and revealed that the apparent bicarbonate space is increased in calves with metabolic acidosis (Trefz et al. 2021). Based on findings from this study, values for the apparent bicarbonate space of 0.68, 0.59, 0.52, and 0.46 L/kg should be used to correct the bicarbonate deficit in diarrheic calves with an initial plasma bicarbonate concentration of 4, 6, 10, and 20 mmol/L, respectively.

However, laboratory equipment is usually not available in ambulatory field practice, and several protocols and “depression scores” for diagnosing and treating metabolic acidosis in diarrheic calves have therefore been suggested (Berchtold 2009; Bellino et al. 2012). Based on own research findings, the following simple guidelines for the dosage of sodium bicarbonate can be recommended (Table 1), which are based on the clinical assessment of posture/ability to stand, degree of enophthalmos, and degree of loss of the palpebral reflex (Trefz et al. 2012a).

In the case of clinical dehydration, intravenous administration of an additional volume of a crystalloid infusion solution of 5–10 L over a period of 24 needs to be considered based on the existing fluid deficit and expected ongoing fluid losses. Slight separation of eyeball and orbit (presence of enophthalmos) is indicative of an existing fluid deficit of about 8% of body weight, a situation which is also commonly seen as the borderline to the need for intravenous fluid therapy.

Glucose should be added to the infusion solution in calves with a history of malnutrition or lack of milk intake for more than 24 h, clinical signs of sepsis, or the presence of hypothermia (Trefz et al. 2012a). A short-term correction of acute hypoglycemia can be achieved by means of an intravenous bolus injection of 0.3 g glucose per kg of body weight (Trefz et al. 2017b). In calves with (risk for) hypoglycemia, a continuous drip infusion should be carried out, and a glucose dosage of 4–6 mg/kg of body weight per minute of infusion should be administered (Scholz and Hoppe 1987; Trefz et al. 2021).

**Table 1** Clinical prediction of severity of metabolic acidosis and resulting simple guidelines for the dosage of sodium bicarbonate

| Clinical signs | Degree of acidosis (base deficit, BD) | Therapy/ required amount of sodium bicarbonate (40 kg calf) |
|---|---|---|
| Alert, standing, strong sucking reflex, no enophthalmos | No acidosis | Oral electrolyte solutions |
| Standing securely, enophthalmos | Mild (BD about 10 mmol/L) | 250 mmol (21 g) |
| Tired or listless, wobbly standing, has to be helped to stand up | Moderate (BD 10–20 mmol/L) | 500 mmol (42 g) |
| Apathetic to comatose, sternal or lateral recumbency or barely able to stand and impaired palpebral reflex | Severe (BD over 20 mmol/L) | 750 mmol (63 g) |

Both increased plasma potassium and D-lactate concentrations normalize along with the correction of dehydration and metabolic acidosis (Trefz et al. 2012a, 2015b). However, in hyperkalemic calves, stress needs to be avoided as much as possible, as major stress insults can lead to sudden death. In calves with clinical signs of hyperkalemia, treatment objectives should therefore focus on a rapid stabilization of the calf. In this context, hypertonic sodium-containing infusion solutions such as 8.4% hypertonic sodium bicarbonate or 7.2% NaCl have a sound physiological basis in the initial treatment of affected calves. These infusion solutions induce a rapid plasma volume expansion (and therefore exert a dilutional effect on plasma potassium concentration), potentially enhance the redistribution of potassium ions into the cells, and most importantly induce a rapid reversal of electrocardiographic manifestations (Fig. 4). A randomized clinical trial compared the potassium-lowering effects of hypertonic sodium chloride-, sodium bicarbonate-, and glucose-containing infusion solutions in the initial treatment of hyperkalemic diarrheic calves (Trefz et al. 2017a). The findings of this study suggested a treatment advantage of hypertonic sodium bicarbonate over the use of hypertonic saline with an identical sodium load, indicating

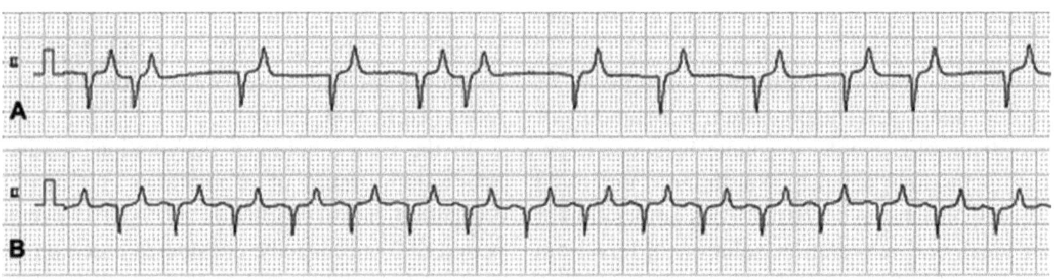


**Fig. 4** Electrocardiographic findings in a markedly hyperkalemic calf (plasma potassium concentration 8.5 mmol/L) with neonatal diarrhea and a body weight of 52 kg before (**a**) and immediately at the end (**b**) of intravenous administration of 330 mL of an 8.4% sodium bicarbonate solution over a period of five minutes. The electrocardiogram before infusion (**a**) reveals arrhythmias and missing P waves, whereas the follow-up electrocardiograms (**b**) indicate a normal sinus rhythm and return of P waves

that alkalinization is an effective potassium-lowering mechanism. Also, the administration of hypertonic glucose was inferior to the use of hypertonic bicarbonate, as it resulted in a delayed potassium-lowering effect despite marked hyperinsulinemia (Trefz et al. 2017a).

### 3.2 Antimicrobial Therapy

Antibiotics are still frequently used for the treatment of uncomplicated and undifferentiated diarrhea (etiologic agent unknown) in calves. Since indiscriminate use of antibiotics promotes the selection, and subsequent proliferation, of antibiotic-resistant strains of bacteria, this general practice has been scrutinized.

There are two potential reasons to justify antibiotic treatment in diarrheic calves: first, calves with diarrhea show small intestinal overgrowth with *E. coli* bacteria, and second, systemically sick calves with diarrhea have an increased risk of developing *E. coli* bacteremia or septicemia. Therefore, in calves with diarrhea and severe systemic involvement, antimicrobial therapy must be pondered carefully. If antimicrobial treatment is performed, it should target *E. coli* in the blood and in the small intestines.

Fecal bacterial culture and antimicrobial susceptibility testing are not useful in calves with diarrhea since fecal bacterial populations do not reflect small intestine or blood bacteria (Smith 2015). Antimicrobials should be administered to calves with diarrhea that exhibit systemic signs of disease (inappetence, dehydration, depression, or fever) or have bloody feces, which indicates a breakdown of the blood–intestine barrier. Antimicrobials used in this indication should preferably be Gram-negative in the spectrum and should be administered parenterally. Routine use of oral or injectable antibiotics cannot be recommended in calves without systemic illness (Constable 2004).

After a thorough revision of the available scientific literature (Constable 2004) and considering the subsequent developments with regard to antibiotic resistance, it is concluded that amoxicillin, ampicillin, and potentiated sulfonamides are the first choice for the treatment of diarrheic calves with systemic illness (Constable 2009). Second choices are third- and fourth-generation cephalosporines, and the last choice of antimicrobials is fluoroquinolones. Obviously, rules for the prudent use of antibiotics as well as regional legislation have to be considered by the veterinarian while choosing an appropriate antimicrobial for the treatment of systemically ill calves with diarrhea. In particular, the use of the critically important cephalosporins and fluoroquinolones should be highly exceptional.

## References


Bellino C, Arnaudo F, Biolatti C, Borrelli A, Gianella P, Maurella C, Zabaldano G, Cagnasso A, D'Angelo A (2012) Development of a diagnostic diagram for rapid field assessment of acidosis severity in diarrheic calves. J Am Vet Med Assoc 240:312–316

Berchtold J (2009) Treatment of calf diarrhea: intravenous fluid therapy. Vet Clin North Am Food Anim Pract 25:73–99, vi

Binding U, Seemann G, Klee W (2000) Untersuchungen zur Art und zur Korrektur der metabolischen Azidose bei jungen Kälbern mit Durchfall. Praktischer Tierarzt 81:314–317

Constable PD (2004) Antimicrobial use in the treatment of calf diarrhea. J Vet Intern Med 18:8–17

Constable PD (2009) Treatment of calf diarrhea: antimicrobial and ancillary treatments. Vet Clin North Am Food Anim Pract 25:101–120

Constable PD, Stampfli HR, Navetat H, Berchtold J, Schelcher F (2005) Use of a quantitative strong ion approach to determine the mechanism for acid-base abnormalities in sick calves with or without diarrhea. J Vet Intern Med 19:581–589

Constable PD, Trefz FM, Sen I, Berchtold J, Nouri M, Smith G, Grünberg W (2021) Intravenous and oral fluid therapy in neonatal calves with diarrhea or sepsis and in adult cattle. Front Vet Sci 7:603358

Ewaschuk JB, Naylor JM, Zello GA (2003) Anion gap correlates with serum D- and DL-lactate concentration in diarrheic neonatal calves. J Vet Intern Med 17:940–942

Ewaschuk JB, Naylor JM, Palmer R, Whiting SJ, Zello GA (2004) D-lactate production and excretion in diarrheic calves. J Vet Intern Med 18:744–747

Foster DM, Smith GW (2009) Pathophysiology of diarrhea in calves. Vet Clin North Am Food Anim Pract 25:13–36

Gentile A, Sconza S, Lorenz I, Otranto G, Rademacher G, Famigli-Bergamini P, Klee W (2004) D-Lactic acidosis in calves as a consequence of experimentally induced ruminal acidosis. J Vet Med A Physiol Pathol Clin Med 51:64–70

Gentile A, Lorenz I, Sconza S, Klee W (2008) Experimentally induced systemic hyperchloremic acidosis in calves. J Vet Intern Med 22:190–195

Glawischnig E, Greber N, Schlerka G (1990) Die Dauertropfinfusion bei Kälbern mit hochgradiger Azidose. Tierärztl. Umschau 45:562–569

Golbeck L, Cohrs I, Leonhard-Marek S, Grünberg W (2018) Effect of dehydration and acidemia on the potassium content of muscle tissue and erythrocytes in calves with neonatal diarrhea. J Dairy Sci 101:9339–9349

Gomez DE, Lofstedt J, Stampfli HR, Wichtel M, Muirhead T, Mcclure JT (2013) Contribution of unmeasured anions to acid-base disorders and its association with altered demeanor in 264 calves with neonatal diarrhea. J Vet Intern Med 27:1604–1612

Grove-White D (1997) Pathophysiology and treatment of metabolic acidosis in the diarrhoeic calf. Bov Pract:56–60

Grude T (2003) Concentrations of L-and D-lactate in serum, urine, and rumen liquid in calves, with special reference to "ruminal drinkers". Tierarztliche Praxis Ausgabe Grosstiere Nutztiere 31:72-+

Hartmann H, Berchtold J, Hofmann W (1997) Pathophysiologische Aspekte der Azidose bei durchfallkranken Kälbern. Tierärztliche Umschau 52:568–574

Karapinar T, Tumer KC, Buczinski S (2020) Evaluation of the Freestyle Optium Neo H point-of-care device for measuring blood glucose concentrations in sick calves. J Vet Intern Med 34:1650–1656

Kasari TR (1999) Metabolic acidosis in calves. Vet Clin North Am Food Anim Pract 15(473-86):v

Kasari TR, Naylor JM (1986) Further studies on the clinical features and clinicopathological findings of a syndrome of metabolic acidosis with minimal dehydration in neonatal calves. Can J Vet Res 50:502–508

Kaske M (1994) Pathophysiologische Aspekte der neonatalen Kälberdiarrhoe. Tierärztl Umschau 49:336–348

Lewis L, Phillips R (1973) Diarrheic induced changes in intracellular and extracellular ion concentrations in neonatal calves. Annales de recherches vétérinaires:99–111

Ling B, Peng F, Alcorn J, Lohmann K, Bandy B, Zello GA (2012) D-Lactate altered mitochondrial energy production in rat brain and heart but not liver. Nutr Metab (Lond) 9:6

Lorenz I (2004a) Influence of D-lactate on metabolic acidosis and on prognosis in neonatal calves with diarrhoea. J Vet Med A Physiol Pathol Clin Med 51:425–428

Lorenz I (2004b) Investigations on the influence of serum D-lactate levels on clinical signs in calves with metabolic acidosis. Vet J 168:323–327

Lorenz I (2009) D-Lactic acidosis in calves. Vet J 179:197–203

Lorenz I (2021) Calf health from birth to weaning - an update. Irish Vet J 74:5

Lorenz I, Gentile A (2014) D-lactic acidosis in neonatal ruminants. Vet Clin North Am Food Anim Pract 30:317–331, v

Lorenz I, Vogt S (2006) Investigations on the association of D-lactate blood concentrations with the outcome of therapy of acidosis, and with posture and demeanour in young calves with diarrhoea. J Vet Med A Physiol Pathol Clin Med 53:490–494

Lorenz I, Gentile A, Klee W (2005) Investigations of D-lactate metabolism and the clinical signs of D-lactataemia in calves. Vet Rec 156:412–415

Lorenz I, Fagan J, More SJ (2011) Calf health from birth to weaning. II. Management of diarrhoea in preweaned calves

Niethammer F (2007) Untersuchungen zur Dehydratation bei Kälbern mit akuter Diarrhoe unter Berücksichtigung ausgewählter klinischer und labordiagnostischer Parameter. Doctoral Thesis, Munich

Omole OO, Nappert G, Naylor JM, Zello GA (2001) Both L- and D-lactate contribute to metabolic acidosis in diarrheic calves. J Nutr 131:2128–2131

Pringle JK, Berthiaume LM (1988) Hypernatremia in calves. J Vet Intern Med 2:66–70

Schelcher F, Marcillaud S, Braun J, Contrepois M, Valarcher J, Navetat H (1998) Metabolic acidosis without dehydration and no or minimal diarrhoea in suckler calves is caused by hyper D-lactatemia. Proc XX World Buiatrics Congr, pp 371–374

Scholz H, Hoppe S (1987) Renale Glukoseverluste nach intravenöser Glukoseinfusion beim Kalb. Dtsch Tierärztl Wochenschr 94:473–477

Silverlås C, Björkman C, Egenvall A (2009) Systematic review and meta-analyses of the effects of halofuginone against calf cryptosporidiosis. Prev Vet Med 91:73–84

Smith GW, Berchtold J (2014) Fluid therapy in calves. Vet Clin North Am Food Anim Pract 30(409-27):vi

Smith G (2015) Antimicrobial decision making for enteric diseases of cattle. Vet Clin North Am Food Anim Pract 31:41–60

Trefz FM, Lorenz I (2017) Plasma potassium concentrations in neonatal diarrhoeic calves are correlated with serum aldosterone concentrations but not with insulin concentrations. Vet J 230:41–44

Trefz FM, Lorch A, Feist M, Sauter-Louis C, Lorenz I (2012a) Construction and validation of a decision tree for treating metabolic acidosis in calves with neonatal diarrhea. BMC Vet Res 8:238

Trefz FM, Lorch A, Feist M, Sauter-Louis C, Lorenz I (2012b) Metabolic acidosis in neonatal calf diarrhea-clinical findings and theoretical assessment of a simple treatment protocol. J Vet Intern Med 26:162–170

Trefz FM, Constable PD, Sauter-Louis C, Lorch A, Knubben-Schweizer G, Lorenz I (2013a) Hyperkalemia in neonatal diarrheic calves depends on the degree of dehydration and the cause of the metabolic acidosis but does not require the presence of acidemia. J Dairy Sci 96:7234–7244

Trefz FM, Lorch A, Feist M, Sauter-Louis C, Lorenz I (2013b) The prevalence and clinical relevance of hyperkalaemia in calves with neonatal diarrhoea. Vet J 195:350–356

Trefz FM, Constable PD, Lorenz I (2015a) Quantitative physicochemical analysis of acid-base balance and clinical utility of anion gap and strong ion gap in 806 neonatal calves with diarrhea. J Vet Intern Med 29:678–687

Trefz FM, Lorch A, Zitzl J, Kutschke A, Knubben-Schweizer G, Lorenz I (2015b) Effects of alkalinization and rehydration on plasma potassium concentrations in neonatal calves with diarrhea. J Vet Intern Med 29:696–704

Trefz FM, Lorch A, Zitzl J, Kutschke A, Knubben-Schweizer G, Lorenz I (2015c) Risk factors for the development of hypokalemia in neonatal diarrheic calves. J Vet Intern Med 29:688–695

Trefz FM, Feist M, Lorenz I (2016) Hypoglycaemia in hospitalised neonatal calves: prevalence, associated conditions and impact on prognosis. Vet J 217:103–108

Trefz FM, Constable PD, Lorenz I (2017a) Effect of intravenous small-volume hypertonic sodium bicarbonate, sodium chloride, and glucose solutions in decreasing plasma potassium concentration in hyperkalemic neonatal calves with diarrhea. J Vet Intern Med 31:907–921

Trefz FM, Lorenz I, Constable PD (2017b) Effects of profound acidemia on the dynamic glucose and insulin response and plasma potassium and phosphorus concentrations during an intravenous glucose tolerance test in neonatal calves. J Dairy Sci 100:9163–9176

Trefz FM, Lorenz I, Lorch A, Constable PD (2017c) Clinical signs, profound acidemia, hypoglycemia, and hypernatremia are predictive of mortality in 1,400 critically ill neonatal calves with diarrhea. PLoS One 12:e0182938

Trefz FM, Lorenz I, Constable PD (2018) Electrocardiographic findings in 130 hospitalized neonatal calves with diarrhea and associated potassium balance disorders. J Vet Intern Med 32:1447–1461

Trefz FM, Lorenz I, Constable PD (2021) Dependence of the apparent bicarbonate space on initial plasma bicarbonate concentration and carbon dioxide tension in neonatal calves with diarrhea, acidemia, and metabolic acidosis. J Vet Intern Med 35:644–654

# Infectious Enteric Diseases in Pigs

Mirjam Arnold, Thomas Echtermann, and Heiko Nathues

**Abstract**

Infectious diseases of the gastrointestinal tract of pigs significantly impact animal health, animal welfare and finally the economic outcome of any pig production worldwide. Preventing enteric diseases in pigs allows producers maximising the average daily weight gain and the feed conversion ratio in pigs, thereby making pig production a very meaningful process of protein refinement and source of food products. Bacterial diseases of the gastrointestinal tract of pigs often necessitate the use of antimicrobials, which can create antimicrobial resistance that threats both animals and humans. Many viral diseases of the gastrointestinal tract provoke long-lasting damage of the gut and/or the gut function, which causes poor performance and an extended period from birth to slaughter along with the need of more resources to reach slaughter weight. Therefore, farmers and veterinarians should focus on the prevention of infectious enteric diseases by means of vaccination where justified and improved internal and external biosecurity and last but not the least by optimised management of the different age and production categories of pigs.

In this chapter, bacterial, parasitical and viral diseases of pigs are described for the corresponding age category, where they have the major impact.

**Keywords**

Infectious enteric disease · Diarrhoea · Pig

## 1 Suckling Pigs

### 1.1 *Clostridium perfringens* Type A and C

#### 1.1.1 Aetiology

The gram-positive, spore-forming *Clostridium perfringens* Type C (*Cl. perfringens* Type C) occurs worldwide, and infections are reported in calves, sheep, goats, horses and humans and particularly in pigs (Niilo 1988; Songer 1996; Songer and Uzal 2005). Due to its capacity to form spores, *Cl. perfringens* Type C survives for a long time in the environment and is resistant to heat, most disinfectants and ultraviolet light. Although *Cl. perfringens* Type C is rarely found

M. Arnold · H. Nathues (✉)
Clinic for Swine, Department of Clinical Veterinary Medicine, Vetsuisse Faculty, University of Bern, Bern, Switzerland
e-mail: Mirjam.Arnold@lgl.bayern.de; heiko.nathues@unibe.ch

T. Echtermann
Division of Swine Medicine, Department for Farm Animals, Vetsuisse Faculty, University of Zurich, Zurich, Switzerland
e-mail: techtermann@vetclinics.uzh.ch


J. J. Gross (ed.), *Production Diseases in Farm Animals*,
https://doi.org/10.1007/978-3-031-51788-4_11

and in low numbers in the gut of healthy adult pigs, it is considered a primary pathogen for neonatal pigs (Songer and Uzal 2005; Posthaus et al. 2020).

Like *Cl. perfringens* Type C, *Clostridium perfringens* Type A (*Cl. perfringens* Type A) is a gram-positive, anaerobic rod with lesser potential for sporulation than *Cl. perfringens* Type C (Wendt et al. 2013; Uzal and Songer 2019). Although some field study data suggest a role of *Cl. perfringens* Type A in diarrhoea in neonatal pigs, evidence regarding causality is missing, as *Cl. perfringens* Type A is a commensal of normal gut flora in pigs. Virulence factors based on epidemiological data are still under discussion and diseases have not been reproduced experimentally. Thus, microbiological or epidemiological postulates for considering *Cl. perfringens* Type A as an infectious disease are not fulfilled so far (Bueschel et al. 2003; Chan et al. 2012; Farzan et al. 2013; Diab et al. 2016b; Uzal 2016; Uzal and Songer 2019).

A total of seven different *Cl. perfringens* types are known, differing in their toxinotyping, with *Cl. perfringens* Type C carrying the two typing toxin genes cpa (encoding for α-toxin or CPA) and the main virulence factor cpb (encoding for β-toxin or CPB) and can carry the enterotoxin (CPE) (Sayeed et al. 2008; Rood et al. 2018). In addition to these major toxins, other toxins such as cpb2 (encoding β2-toxin or CPB2) have been described (Rood et al. 2018) (Table 1).

*Cl. perfringens* Type A is missing the cpb, is carrying cpa and might also carry cpe and the minor toxin cpb2 (Rood et al. 2018).

### 1.1.2 Pathogenesis

*Cl. perfringens* Type C infection starts with an overgrowth of the pathogen in the intestine of newborn piglets which can reach $10^8$ to $10^9$ colony-forming units (CFU)/g of contents in only a few hours (Ohnuma et al. 1992). An initial damage of the small intestine epithelial cells by CPB, other toxins, enzymes and/or accumulation of metabolites is suspected, but the mechanisms are not fully understood (Sayeed et al. 2008; Posthaus et al. 2020). However, it is known that once CPB reaches the endothelial cells of the mucosa, vascular endothelial damage and widespread vascular necrosis occur (Gurtner et al. 2010; Autheman et al. 2013). This damage leads to increased vascular permeability, oedema and haemorrhage in the lamina propria of the small intestine (Box 1) (Niilo 1988; Uzal et al. 2010). The CPB toxin is trypsin sensitive, and trypsin inhibitors in colostrum protecting the immunoglobulins predispose acute necro-haemorrhagic enteritis in neonatal piglets (Sakurai and Duncan 1978; Songer and Uzal 2005; Sayeed et al. 2008). In addition to CPB, the minor toxin CPB2 is also being discussed as a virulence factor, as it is isolated from most *Cl. perfringens* Type C strains, but the exact role of this toxin has not been confirmed so far (Bueschel et al. 2003; Songer and Uzal 2005).

The pathogenesis of *Cl. perfringens* Type A enteric disease in pigs is poorly established, and although CPB2 and CPE are discussed as virulence factors, diagnosis is mostly made in neonatal diarrhoea by excluding other pathogens (Collins et al. 1989; Uzal 2016). However, this conclusion might be misleading as *Cl. perfringens* Type A is found in the intestinal flora of healthy pigs and shows marked colonisation of the stomach and small intestine of neonatal piglets within the first hours of life (Smith and Jones 1963; Farzan et al. 2013).

### 1.1.3 Clinical Signs

Transmission of *Cl. perfringens* Type C to neonatal pigs works either by contact to contaminated sow faeces, due to lateral transmission by littermates, or by exposure to spores in the environment.

**Table 1** Toxinotypes of *Clostridium perfringens* Types A and C

| | Major toxins | | | Minor toxin |
|---|---|---|---|---|
| Toxinotypes | Alpha (CPA) | Beta (CPB) | Epsilon (CPE) | Beta 2 (CPB2) |
| Type A | + | – | –/+ | –/+ |
| Type C | + | + | –/+ | –/+ |

Necro-haemorrhagic enteritis can develop within 12 hours after birth but occurs more frequently on the third to fourth day of life and is not observed in piglets older than 3 weeks (Songer and Uzal 2005; Jäggi et al. 2009; Diab et al. 2016b). The disease can occur as peracute-to-acute or subacute-to-chronic form (Høgh 1967; Jäggi et al. 2009; Posthaus et al. 2020). The prevalence in naive, non-vaccinated populations can reach up to 100% (Bergeland et al. 1966). Total herd mortality was reported usually being less than 60% but can reach up to 100% in litters of non-immune gilts (Bergeland et al. 1966). In endemically infected populations with *Cl. perfringens* Type C, peracute and acute fatal clinical disease is observed predominantly in non-immune litters of gilts (Uzal and Songer 2019).

Clinical lesions of a peracute-to-acute infection with *Cl. perfringens* Type C include reddish-brown diarrhoea containing grey shreds of tissue debris, weakness and rectal temperature of 35 °C or below. Piglets become rapidly moribund or are already found dead without clinical signs (Songer and Uzal 2005). Subacute-to-chronic diseased piglets show intermittent yellow to grey and mucoid diarrhoea and faecal staining of the perianal region and might die after several weeks (Songer and Uzal 2005).

Clinical reports regarding disease associated with *Cl. perfringens* Type A point out creamy or pasty faeces typically seen in piglets within 48 h of birth lasting for several days with lower or no mortality (Songer and Uzal 2005).

### 1.1.4 Lesions

Segmental mucosal necrosis, emphysema and massive haemorrhage in the small intestine especially in the jejunum are characteristics of a peracute-to-acute infection with *Cl. perfringens* Type C (Høgh 1969; Posthaus et al. 2020). The more chronic the infection, the less haemorrhage is present in affected intestinal segments and the more fibrinous-diphtheritic pseudomembranes can be observed (Bergeland et al. 1966; Songer and Uzal 2005). Histologically, intestinal villi are necrotic, and there is extensive acute haemorrhage in the underlying mucosa, often extending into all other layers of the intestine (Niilo 1988; Jäggi et al. 2009). In peracute-to-acute infections, numerous gram-positive bacteria and thrombosis but only few inflammatory cells can be observed, while in more subacute-to-chronic infections, especially the number of neutrophils is increased (Posthaus et al. 2020).

Necropsy of *Cl. perfringens* Type A-associated cases shows flaccid, thin-walled, gaseous or watery small intestine and pasty colonic contents (Collins et al. 1989; Songer and Uzal 2005). Microscopically, intestinal sections may have superficial villous tip necrosis, with fibrin and neutrophils, or appear normal or sometimes slightly congested (Songer and Uzal 2005).

### 1.1.5 Diagnosis

The presumptive diagnosis of necro-haemorrhagic enteritis caused by *Cl. perfringens* Type C is made by combining typical haemorrhagic diarrhoea and rapid death in neonatal piglets with gross lesions of segmental necro-haemorrhagic or fibrino-necrotic enteritis (Uzal and Songer 2019). The diagnosis is confirmed by bacteriological examination of intestinal content, faeces or mucosal swabs on, e.g. blood agar. Subsequently, detection of cpb gene or CPB toxin in isolated bacteria by means of molecular biological methods such as polymerase chain reaction (PCR), antigen immunoassays such as immunohistochemistry (IHC) or enzyme-linked immunosorbent assays (ELISA) should be performed (Box 2–4) (Møller and Ahrens 1996; Posthaus et al. 2020). In live animals, swabs and faecal samples should be collected and stored under anaerobic conditions. CPB toxin degradation can be limited by adding trypsin inhibitors. A failure in CPB toxin detection does not exclude a *Cl. perfringens* Type C infection. As *Cl. perfringens* Type C can be part of the gut flora of pigs, a detection in culture without toxinotyping is not confirming the disease in live animals (Songer and Uzal 2005).

In general, the technical diagnostic approach of *Cl perfringens* Type A follows the principles of *Cl. Perfringens* Type C. Although detection of CPB2 and CPE has been associated with disease by some authors, currently no criteria are available to establish a final diagnosis (Collins et al.

1989; Bueschel et al. 2003; Waters et al. 2003; Farzan et al. 2013; Uzal 2016; Uzal and Songer 2019).

A list of differential diagnoses for infectious causes of diarrhoea can be found in Table 2.

### 1.1.6 Treatment and Prevention

Early treatment of an acute *Cl. perfringens* Type C infection in a non-immune herd with antimicrobial metaphylactic therapy with, e.g. penicillin 15,000 UI/kg body weight intramuscularly every 24 h can reduce mortality (Springer and Selbitz 1999; Uzal and Songer 2019). Other antimicrobials of choice include aminopenicillins and spectinomycin/lincomycin, whereas resistances are described against tetracyclines, macrolides, pleuromutilins, cephalosporins and lincosamides (Rood et al. 1985; Salvarani et al. 2012; Uzal and Songer 2019). To support acutely affected piglets, they can be rehydrated orally or intraperitoneally and should be kept in a warm environment (Box 5) (Tucker et al. 2022). However, since the prognosis of acutely affected piglets is poor, focus is on prevention. Vaccination of sows protects the piglets from diseases via colostral antibodies (Kennedy et al. 1977). The basic immunisation of sows should be completed 2–3 weeks before farrowing, and vaccination should be repeated 2–3 weeks before each further farrowing (Kennedy et al. 1977). Some gilts may not develop sufficient titres of anti-CPB antibodies in their colostrum. In such cases, colostral antibody levels can be improved by optimising gilt acclimatisation with 2 weeks of isolation, followed by systematic contact to the pathogens on the farm (Kraeling and Webel 2015). In addition, gilts can be vaccinated three times, e.g. two times prior to their first insemination and a booster prior to farrowing (Richard et al. 2019a). In combination with cleaning and sanitation procedures, disinfection of the pens and vaccination of sows can reduce the infection pressure for piglets (Richard et al. 2019b; Posthaus et al. 2020). Cleaning and sanitation including changing

**Table 2** Most typical differential diagnoses for infectious causes of diarrhoea in pigs according to their age (modified (Ramirez et al. 2019))

<table>
<tr><th>1-2 days</th><th>3-4 days</th><th>5-6 days</th><th>1 week</th><th>2 weeks</th><th>3 weeks</th><th>1 month</th><th>2 months</th><th>3 months</th><th>4 months</th><th>5 months</th><th>6 months</th><th>Adults</th></tr>
<tr><td colspan="4">Clostridium difficile</td><td colspan="9"></td></tr>
<tr><td colspan="4">Clostridium perfringens Type A</td><td colspan="9"></td></tr>
<tr><td colspan="5">Clostridium perfringens Type C</td><td colspan="8"></td></tr>
<tr><td colspan="8">Escherichia coli</td><td colspan="5"></td></tr>
<tr><td colspan="8">Rotavirus</td><td colspan="5"></td></tr>
<tr><td colspan="8">Teschovirus</td><td colspan="5"></td></tr>
<tr><td colspan="13">African swine fever virus<br>Campylobacter spp.<br>Classical swine fever virus<br>Porcine delta coronavirus (PDCV, Corona virus)<br>Porcine epidemic diarrhea virus (PEDV, Coronavirus)<br>Toxoplasma gondii<br>Transmissible gastroenteritis virus (TGEV, Coronavirus)</td></tr>
<tr><td></td><td colspan="7">Cryptosporidium spp.</td><td colspan="5"></td></tr>
<tr><td></td><td colspan="5">Strongyloides spp.</td><td colspan="7"></td></tr>
<tr><td></td><td colspan="6">Cystisospora suis</td><td colspan="6"></td></tr>
<tr><td colspan="5"></td><td colspan="8">Oesophagostomum dentatum</td></tr>
<tr><td colspan="5"></td><td colspan="8">Salmonella spp.</td></tr>
<tr><td colspan="6"></td><td colspan="4">Brachyspira pilosicoli</td><td colspan="3"></td></tr>
<tr><td colspan="6"></td><td colspan="7">Ascaris suum</td></tr>
<tr><td colspan="6"></td><td colspan="7">Brachyspira (B.) hyodysenteriae/ B. suanatina / B.hampsonii</td></tr>
<tr><td colspan="6"></td><td colspan="7">Lawsonia intracellularis</td></tr>
<tr><td colspan="6"></td><td colspan="7">Porcine circovirus Type 2</td></tr>
<tr><td colspan="6"></td><td colspan="7">Trichuris suis</td></tr>
</table>

clothes between the different production areas, maintaining a strict all-in/all-out management, routinely removing sow faeces and washing the sows before entering the farrowing unit can reduce the risk of diseases (Uzal and Songer 2019). Disinfection with hydrogen peroxide, peracetic acid or natrium hypochlorite is used to inactivate spores of *Clostridium* spp. (Talukdar et al. 2017).

Although the growth of *Cl. perfringens* Type A can be limited by antimicrobial treatment and passive, colostral immunity in piglets can be increased by commercial vaccines and vaccination protocols following these for *Cl. perfringens* Type C, treatment and prevention strategies are limited to special indications, where other causes of diarrhoea are excluded (Table 2).

## 1.2 *Clostridioides difficile*

### 1.2.1 Aetiology and Public Health

*Clostridioides difficile* (*C. difficile*) is a gram-positive, strict anaerobic rod with zoonotic potential (Diab et al. 2016a). Like other *Clostridia* spp., *C. difficile* spores are environment containments that resist to most disinfectants (Worsley 1998; Uzal 2016). *C. difficile* can cause enteric diseases in humans and numerous animal species, including, but not limited to, gerbils, guinea pigs, hamsters, horses, rabbits and pigs (Diab et al. 2016a).

### 1.2.2 Pathogenesis

*C. difficile* is a commensal found in the large intestine and faeces of domestic pigs. The primary source of infection of neonatal pigs is via ingestion of spores from sow faeces or the environment. Spores survive the stomach and are activated by bile salts in the small intestine (Songer et al. 2000; Uzal and Songer 2019). *C. difficile* belongs to the natural early colonisers of the gastrointestinal tract of pigs, and up to 100% of piglets test positive within 2 days after birth, followed by a rapid decline with age. There is an increased risk of clinical disease in newborn piglets (Grześkowiak et al. 2016). The most important virulence factors of *C. difficile* are toxins A (TcdA) and B (TcdB), although in recent years, a role for an actin-specific ADP-ribosyltransferase (CDT) produced by some strains of *C. difficile* has been discussed (Cowardin et al. 2016; Lyon et al. 2016; Songer et al. 2016). These toxins, once secreted in the large intestine, leading to the loss of cell-cell contact, increased paracellular permeability of the intestinal mucosa and cell death by apoptosis or necroptosis (Voth and Ballard 2005; Awad et al. 2014).

### 1.2.3 Clinical Signs

The clinical signs of *C. difficile* are age dependent and typically observed in pigs 2–14 days of age with an average age of 5 days (Waters et al. 1998). Piglets during their first week of age show early-onset pasty to watery diarrhoea, growth retardation, dehydration, less often respiratory distress, scrotal oedema or sudden death (Diab et al. 2016a; Grześkowiak et al. 2019). Mortality in piglets can reach up to 14% (Squire et al. 2013). Preventing infection during the first 10 days of life will successfully prevent piglets from presenting clinical signs, which is related to changes in the diversity of intestinal microbiota (Proctor et al. 2021).

### 1.2.4 Lesions

Gross lesions are non-specific and include flaccid caecum and colon with pasty contents and mesocolon oedema (Keel and Songer 2006; Diab et al. 2016a; Uzal and Songer 2019). Besides mesocolon oedema, hydrothorax, hydropericardium, scrotal oedema and ascites have been described (Keel and Songer 2006). Microscopic lesions are characterised by fibrino-necrotic colitis and/or typhlocolitis including suppurative small erosions and ulcers with effusion of fibrin and neutrophils into the intestinal lumen, so-called volcano lesions (Waters et al. 1998; Keel and Songer 2006).

### 1.2.5 Diagnosis

Presumptive diagnosis of *C. difficile*-associated diseases in pigs is based on clinical history, clinical signs and gross and microscopic findings (Box 2–4) (Uzal and Songer 2019). Following the lesions, mesocolon oedema and necro-suppurative typhlocolitis are highly suggestive of *C. difficile* (Waters et al. 1998; Keel and Songer 2006; Grześkowiak et al. 2019). Since *C. difficile* can also be found in healthy pigs, pure isolation is not meaningful and of limited gain in knowledge. It must be followed by detection of TcdA and/or TcdB by PCR to confirm the diagnosis (Knight et al. 2014). Other methods for direct toxin detection include loop-mediated isothermal amplification (LAMP)-PCR, real-time PCR and enzyme immunoassays (EIA) (Knight et al. 2014; Diab et al. 2016a). Since most isolates of *C. difficile* are toxin carriers and bacterial count in the large intestine can reach $10^7$–$10^{12}$ bacterial per gram faeces, findings of toxins must be linked to clinical sings and lesions in both living and post-mortem diagnosed animals (Steele et al. 2010; Burnham and Carroll 2013; Uzal and Songer 2019). A list of differential diagnoses for infectious causes of diarrhoea by age category can be found in Table 2.

### 1.2.6 Treatment and Prevention

Infections with *C. difficile* are treated with antimicrobials including tylosin, clindamycin and tetracycline (Keessen et al. 2013). Little is known about immunoprophylaxis in pigs, but studies from other species suggest that antibodies against TcdA and TcdB prevent toxin binding in mouse and hamster models and prevent clinical diseases (Allo et al. 1979; Kink and Williams 1998; Songer et al. 2007). Disinfection with hydrogen peroxide, peracetic acid and natrium hypochlorite can be used to inactivate spores of *Clostridium* spp., and chlorine-based products are effective in 5 minutes against *C. difficile* spores (Talukdar et al. 2017; Uwamahoro et al. 2018).

## 1.3 Coronavirus Infections

### 1.3.1 Aetiology and Pathogenesis

There are six coronaviruses associated with diseases in swine (Saif et al. 2019). Four of them are associated with gastrointestinal disorders and diarrhoea, namely, transmissible gastroenteritis virus (TGEV), porcine epidemic diarrhoea virus (PEDV), porcine delta coronavirus (PDCoV) and swine acute diarrhoea syndrome virus (SAD-CoV) (Pensaert et al. 1986; Niederwerder and Hesse 2018; Wang et al. 2019; Jung et al. 2020; Turlewicz-Podbielska and Pomorska-Mól 2021). This chapter will focus on historically important and re-emerging TGEV and PEDV.

TGEV was first detected in the USA in 1946 and first reported in Belgium in 1984. It was a primary source of enteric infection and mortality in piglets around the globe from the 1960s to the 1980s (Pensaert et al. 1986). The widespread emergence of porcine respiratory coronavirus (PRCoV), which is a spike gene deletion variant of TGEV, limited the clinical impact of TGEV, leading to the fact that diarrhoea in pigs caused by TGEV is currently nearly absent (Saif et al. 2019; Turlewicz-Podbielska and Pomorska-Mól 2021).

PEDV was found on an island in the UK in 1971, and single outbreaks were reported in Mexico, Czech Republic, Hungary, Belgium, China, Korea, Italy and Thailand during the 1990s (Niederwerder and Hesse 2018). PEDV re-emerged as a major pathogen in pigs which started in China, followed by other countries in Asia and America and also Europe in the 2010s (Theuns et al. 2015; Pensaert and Martelli 2016; Niederwerder and Hesse 2018). Currently, three PEDV genotypes are differentiated on the basis of S gene (G1a PEDV = mild virulent strains from Europe and Asia; G2 PEDV = 'Original USA PEDV', highly lethal strains from Asia, presently prevalent in the USA; and G1b PEDV, which embody S-INDEL strains = mild clinical outbreaks) (Decaro and Lorusso 2020).

In addition, a recombination of TGEV and PEDV was described in European countries (Belsham et al. 2016; Boniotti et al. 2016). Infected pigs are the major source of infection. Carrier pigs harbour the virus up to 104 days and shed in faeces, milk and partly nasal secretions. At least for PEDV, airborne transmission is under discussion (Underdahl et al. 1975; Niederwerder and Hesse 2018). TGEV and PEDV can also be spread through contaminated water or food, automobiles, shoes and other inanimate objects (Underdahl et al. 1975; Niederwerder and Hesse 2018; Saif et al. 2019). Houseflies, birds and wild boars but also workers who are involved in farm activities can similarly serve as a vehicle for virus transmission to newborn pigs (Dee et al. 2016; Schumacher et al. 2016; Saif et al. 2019).

TGEV and PEDV replicate in the cytoplasmic cells of the small intestinal epithelium, leading to a destruction of intestinal villi starting within 12–18 hours post and showing a maximum effect 24–36 hours post-infection. This becomes the cause of malabsorption and maldigestion (Box 1) (Saif et al. 2019; Turlewicz-Podbielska and Pomorska-Mól 2021). The virus selectively infects and destroys the mature enterocytes lining the small intestinal villi, resulting in a significant shortening of villi, leading to a villous to crypt ratio from 7:1 to 3:1. This leads to a reduction of critical digestive enzymes, responsible for digestion of milk (Jung et al. 2020). The undigested milk increases osmolarity and results in further loss of water and electrolytes into the bowel lumen, thereby provoking diarrhoea, acidosis and severe dehydration (Thakor et al. 2022). Intestinal crypt epithelial cells remain uninfected, which is why the restoration of the villi function is rapid after recovery (Fenner 1987). It is known that the disease is more severe in younger pigs than in older ones, because of the immaturity of the immune system in piglets and the lower rate of enterocyte regeneration in younger pigs (Moon et al. 1975; Jung et al. 2015; Jung et al. 2020).

### 1.3.2 Clinical Signs

Clinical manifestations of TGEV and PEDV infections are very similar and mainly include loss of appetite, diarrhoea and vomiting, which leads to dehydration, acidosis, hyperkalaemia, weight loss, lethargy and death (Saif et al. 2019; Wang et al. 2019). Clinical signs appear 18 h–3 days after infection and usually do not last longer than 10 days (Möstl 1990). The disease is generally fatal in newborn piglets, especially those born from seronegative mothers. In older pigs, the morbidity can be high but the mortality is low (Wang et al. 2019).

Two to 7 days after infection, piglets may die, and mortality for both TGEV and certain strains of PEDV is described being up to 100% for piglets under 2 weeks or under 5 days of age, respectively (Niederwerder and Hesse 2018; Saif et al. 2019; Turlewicz-Podbielska and Pomorska-Mól 2021). In seropositive herds and animals over 5 weeks of age, mortality is generally low (Saif et al. 2019).

In seronegative herds, TGEV and PEDV quickly spread between all age categories, with a typical seasonal appearance in winter months (Song et al. 2012). Sows may show vomiting, diarrhoea, fever, anorexia and agalactiae or dysgalactia, which further contributes to piglet mortality (Saif et al. 2019; Thakor et al. 2022). Suckling pigs older than 3 weeks of age in such an epidemic situation mainly survive but remain stunted. Finishing pigs may show inappetence, diarrhoea and vomiting (Saif et al. 2019).

In endemic situations, the TGEV and PEDV spread slowly among susceptible adults with sows typically being asymptomatic and transfer passive lactogenic immunity to their offspring (Saif et al. 2019). Mild diarrhoea and mortality below 10–20% in piglets and weaners until 2 weeks post weaning is described (Xia et al. 2018). The virus persists as long as susceptible animals are in the herd, which typically is the case in large herds with frequent farrowing (Xia et al. 2018).

### 1.3.3 Lesions

Macroscopic findings are limited to the gastrointestinal tract and include a bloated stomach and fluid-filled intestine. They contain undigested milk and gas and reveal flabby consistency on palpation (Thakor et al. 2022). Damage to the villous enterocytes causes thinning of the intestinal wall, which can be seen when segments of the intestine are immersed in an isotonic buffer and examined underneath a dissecting microscope. For TGEV, villous atrophy is marked in the jejunum and to some extent in the ileum but not in the proximal duodenum (Guo et al. 2020). The mammary gland tissues and alveolar macrophages have been identified as extra-intestinal sites for replication of TGEV (Vlasova et al. 2020). In addition, PEDV-infected pigs in which vomiting was observed showed significantly lower levels of serotonin-secreting cells in the duodenum, middle jejunum, ileum and colon, suggesting that serotonin secretion may have an impact on vomiting during PEDV infection (Jung et al. 2018).

### 1.3.4 Diagnosis

Rapid diagnosis is critical for the control of these diseases to prevent them from spreading (Wang et al. 2019). Detection of viral antigen in faeces or from lesions using either immunofluorescence or immunohistochemical techniques of formalin-fixed tissue is recommended (Saif et al. 2019). In case of necropsy, animals should be freshly dead (< 30 minutes) and less than 1 week of age (lesions most severe in young animals) and have just started to show clinical signs. Ideally, four different regions in the posterior part of the small intestine each 2 cm in length are sampled and fixated in 10% buffered formalin (Saif et al. 2019).

Alternatively diagnostic methods including the detection of viral RNA and/or viral antigens via RT-PCR or real-time RT-PCR and antibodies via ELISA are possible (Wang et al. 2014; Diel et al. 2016; Saif et al. 2019). Virus isolation from faeces or small intestine content is very precise, but time consuming (Saif et al. 2019). Herd monitoring via antibody detection by ELISA for long-term neutralising antibodies is used for both TGEV and PEDV as these antibodies are present 7–8 or 10 days post-infection and persist for a minimum of 18 months or 42 days for TGEV or PEDV, respectively (Diel et al. 2016; Saif et al. 2019). A list of differential diagnoses for infectious causes of diarrhoea by age category can be found in Table 2.

### 1.3.5 Therapy and Prevention

Although TGEV and PEDV occur in pigs of all ages, piglets up to 1 week of age experience high mortality and need to be protected by transfer of maternal antibodies, especially neutralising antibodies, via colostrum and milk from immunised dams (Langel et al. 2016; Wang et al. 2019). Colostrum and milk IgA or neutralising antibody titres are better markers for lactogenic immunity than antibodies in serum (Jordan Bjustrom-Kraft et al. 2018). Pigs that have recovered from TGEV or PEDV are immune to subsequent short-term challenge with the same strain (Saif et al. 2019). At the beginning of an acute outbreak situation, in which spread of the virus within the herd is likely, two strategies are performed to reduce losses in suckling pigs: First, sows that are more than 2 weeks apart from farrowing are allowed or exposed to the pathogen, to allow maternal immunity to be built. Second, exposure of sows less than 2 weeks apart from farrowing must be strictly avoided until 3 weeks post farrowing (Saif et al. 2019). Cross-fostering with TGEV immune sows was useful to reduce losses in suckling pigs in some outbreaks in the field (Stěpánek et al. 1982). If new sows are aimed to be introduced to the herd, there should be a minimum of 4 weeks between the last signs of disease and introduction of new animals, especially in winter months (Saif et al. 2019).

A negative effect of stress (injection of corticosteroids, dexamethasone) on severity which worsened clinical signs was reported and thus should be avoided (Kwonil Jung et al. 2007). In addition, coinfections like rotavirus infection or presence of *E. coli* worsen the enteritis as well (Underdahl et al. 1975). Due to malabsorptive diarrhoea, oral administration of balanced electrolyte or glucose solutions, as well as early feed initiation, is contraindicated. Isotonic fluid may be used (Box 1 and 5). However, other supportive measurements like warm environment (min. 32 °C in the nest for young suckling pigs) and draft-free and dry environment may reduce mortality. If secondary infections occur, antimicrobial therapy is beneficial in animals 2–5 weeks of age. TGEV is stable at freezing temperatures, but labile at room or higher temperatures. Further, it is highly photosensitive and is inactivated at sunlight within a few hours. Thorough cleaning with hot water and disinfection, using sodium hydrochloride, quaternary ammonium compounds or other disinfectants, is therefore recommended to inactivate the virus (Saif et al. 2019). Live and inactivated vaccines targeting sows have been developed for TGEV and PEDV, respectively, but all of them are not sufficient to control outbreaks in seronegative farms and currently no vaccine against coronaviruses in pigs is licensed in the EU (Song et al. 2015; Saif et al. 2019).

### 1.4 Rotavirus Infections

#### 1.4.1 Aetiology and Pathogenesis

Rotaviruses are RNA viruses which can affect a variety of hosts including humans (Vlasova et al. 2017). Direct transmission from animals to humans was recently discussed (Gabbay et al. 2008; Martella et al. 2010; Wilhelm et al. 2015; Flores et al. 2021). Rotaviruses are considered ubiquitous and are shed by sows during farrowing (Benfield et al. 1982; Homwong et al. 2016). They are very stable in the environment and transmission within a farm typically occurs by faeco-oral contact (Fu and Hampson 1989; Fongaro et al. 2015; Wilhelm et al. 2015). Whether a clinical disease occurs in pigs depends especially on the age and immune status of the animals (Vlasova et al. 2017). Young suckling pigs and piglets without proper passive maternal immunity are particularly at risk for disease (Lecce and King 1978; Saif et al. 1996; Vlasova et al. 2017). Rotaviruses A, B and C are associated with diarrhoea in pigs, whereby especially Rotavirus A was considered pathogenic and prevalent (Malik et al. 2014; Marthaler et al. 2014; Amimo et al. 2015; Otto et al. 2015; Homwong et al. 2016; Theuns et al. 2016b). Today, rotaviruses B and C also appear to be relevant as a cause of enteritis in either younger, neonatal pigs or older pigs (> 3 weeks old), respectively (Marthaler et al. 2013; Theuns et al. 2016a; Theuns et al. 2016b). They increase intestinal motility, lead to jejunal and ileal villus atrophy and thus result in malabsorption and diarrhoea (Box 1) (Shepherd et al. 2019).

#### 1.4.2 Clinical Signs

The incubation period in pigs is 1–4 days (Torres-Medina and Underdahl 1980). Clinical signs consist of watery diarrhoea, vomiting, lethargy, anorexia and weight loss, especially in piglets less than 7 days of age (Lecce and King 1978; Shepherd et al. 2019). Clinical signs may persist for 1–10 days, while it is typically shorter in older animals (Torres-Medina and Underdahl 1980). Mortality is about 20% in young suckling pigs and lower in older animals (Lecce and King 1978).

#### 1.4.3 Lesions

The small and large intestine of infected pigs may be filled with watery content and the intestinal wall appears thin (Marthaler et al. 2013; Shepherd et al. 2019). The stomach can be filled with undigested milk (Neog et al. 2011). Histologically, a shortening of villi and elongation of intestinal crypts with cellular hyperplasia can be seen (Estes et al. 2001).

#### 1.4.4 Diagnosis

A quantitative reverse transcriptase PCR from faeces of typically and acutely infected piglets is the method of choice for diagnostics of disease in live animals. A multiplex real-time RT-PCR allows differentiation between rotaviruses A, B and C (Box 2–4) (Marthaler et al. 2014; Liu et al. 2019; Shepherd et al. 2019). Commercial ELISA kits, haemagglutination inhibition test, polyacrylamide gel electrophoresis and cell culture immunofluorescence are described for diagnostics as well (Shepherd et al. 2019; Memon et al. 2022). Post-mortem, intestinal tissue can be used for immunohistochemistry, in situ hybridisation or electron microscopy (Shepherd et al. 2019). The antibody status of piglets and sows can be analysed; however, serum antibodies compared to mucosal antibodies are poor indicators of protective immunity (Markowska-Daniel et al. 1996; To et al. 1998; Memon et al. 2022). A list of differential diagnoses for infectious causes of diarrhoea by age category can be found in Table 2.

#### 1.4.5 Therapy and Prevention

Symptomatic therapy can minimise diarrhoea and reduce mortality (Box 5) (Bywater and Woode 1980; Tucker et al. 2022). Due to malabsorptive diarrhoea, oral administration of balanced electrolyte or glucose solutions, as well as early feed initiation, is contraindicated. Isotonic fluid may be used (Bywater and Woode 1980; Wendt et al. 2013). In case of bacterial coinfections, antimicrobial therapy may be indicated (Shepherd et al. 2019). The key point in prevention is to ensure the supply of colostrum and thus the absorption of maternal antibodies by the piglets (Vlasova et al. 2017). Since there are many rotaviruses, also regionally different, the protection of piglets via the mother's colostrum remains the best method to protect the piglets (Saif et al. 1996). Currently, there is no vaccine approved for pigs on the European market, whereas modified live vaccines are used in the USA (Vlasova et al. 2017; Shepherd et al. 2019). Cross-protectivity between rotaviruses of different species has not been scientifically described, but multiple exposure or vaccination increases the likelihood of cross-immunity (Vlasova et al. 2017). Due to its endemic nature in pig production, elimination or eradication of rotaviruses in conventional herds is not recommended (Shepherd et al. 2019). Rotaviruses are very resistant in the environment and drying will not eliminate all rotaviruses, but chlorine-based or phenolic disinfectants can help to control the spread (Nemoto et al. 2014). Early weaning is described as a risk factor for rotavirus diarrhoea. Improving gut microbiota via pre- and probiotics has been described as beneficial (Lecce and King 1978; Vlasova et al. 2017; Kim et al. 2020; Kumar et al. 2022).

## 1.5 Coccidiosis/*Cystoisospora suis*

### 1.5.1 Aetiology and Pathogenesis

Coccidia are obligate intracellular protozoan parasites. Neonatal coccidiosis in piglets is typically caused by *Cystoisospora suis* (Lindsay et al. 2019). Regarding their life cycle, it is important that the eggs sporulate outside the animal at a temperature of 20–37 °C. These sporulated oocysts are resistant to most disinfectants (Harleman and Meyer 1984; Aliaga-Leyton et al. 2011) and are ingested by the piglets, in which they undergo endogenous development in the distal parts of the villi of the jejunum and the ileum, sometimes caecum and colon (Lindsay et al. 1980; Harleman and Meyer 1985). Un-sporulated oocysts may be excreted 5 days after the first inoculation (prepatency) (Lindsay et al. 2019; Feix et al. 2020). A list of differential diagnoses for infectious causes of diarrhoea by age category can be found in Table 2.

### 1.5.2 Clinical Signs

There is an age-related susceptibility, with piglets 1–2 days old while infected developing much more severe disease than animals which were inoculated at 2–4 weeks of age. Due to the prepatency, clinical signs usually appear in suckling pigs 7–11 days of age. Clinical signs include diarrhoea due to a reduced absorption resulting in fluid losses by the gastrointestinal tract (Box 1) (Lindsay et al. 2019). Morbidity within a litter is high, with moderate mortality. Coinfections however may lead to high mortality rates (Mengel et al. 2012; Pettersson et al. 2019).

Disease in fattening pigs is uncommon but may occur in weaners typically 4–7 days after weaning when piglets are exposed to environmental oocytes (Stuart et al. 1982). Morbidity may be high, while mortality is low (Lindsay et al. 2019). Pigs, once infected and recovered, are resistant to subsequent infections.

### 1.5.3 Lesions

Macroscopically, fibrino-necrotic membranes in the jejunum and ileum, without haemorrhages, are seen. Histologically, villous atrophy, villous fusion, crypt hyperplasia and necrotic enteritis can be found 4 days post-infection (Lindsay et al. 2019).

### 1.5.4 Diagnosis

Diarrhoea in suckling pigs starting at 7–14 days of age, not responding to antibiotic treatment, is suggestive of neonatal coccidiosis. For diagnosis, oocytes in faeces of clinically affected animals are to be shown (Lindsay et al. 2019). Therefore, faeces or mucosa should be sampled from piglets of several litters showing clinical signs for at least 2–3 days (Joachim et al. 2018). As diarrhoea may appear before oocytes are excreted, and oocyte excretion peaks 2–3 days after onset of diarrhoea, such animals should be prioritised. Further, pasty faecal samples are more likely to contain high concentrations of oocytes than liquid faeces. Sampling in weekly intervals from the second until the fourth week of life improved detection rates. Due to age resistance, detection of oocytes in older animals is of questionable clinical relevance (Stuart et al. 1982). To increase sensitivity, flotation is often combined with autofluorescence (Joachim et al. 2018).

### 1.5.5 Therapy and Prevention

Treatment of acutely diseased piglets is symptomatic by means of rehydration solutions and increasing the ambient temperature for the piglets (Box 5). Animals that are not yet infected can be treated with toltrazuril (20 mg/kg body weight) although first reports of resistance have been described (Skampardonis et al. 2010; Mengel et al. 2012; Shrestha et al. 2017; Joachim et al. 2019).

For control, hygiene and management are of highest importance. Since contaminated farrowing crates have ideal temperatures for sporulation of oocytes, cleaning with removal of all organic material directly after removal of the animals should be performed. After washing, the dry pens should be disinfected, e.g. using a cresolic agent, which is one of the few effective disinfectants (Joachim et al. 2014). However, only reduction and no elimination are possible (Lindsay et al. 2019). All-in/all-out management and washing the sows before housing them in the farrowing pen are recommended. After stocking, farrowing crates should be walked on as little as possible to minimise the spread of medium food ware. Pets should not have access to the barn and rodents should be continuously controlled (Box 6) (Lindsay et al. 2019). A partial protection of piglets against clinical disease was described, when sows were exposed to *Cystoisospora suis* 14d a.p. and passive immunity might predict severity of disease (Schwarz et al. 2013; Schwarz et al. 2014). However, further studies on this are lacking.

## 2 Weaning Pigs

### 2.1 *Escherichia coli*

#### 2.1.1 Aetiology

*Escherichia coli* (*E. coli*) is a gram-negative peritrichously flagellated bacteria of the family *Enterobacteriaceae*, and apathogenic *E. coli* are physiologically found in the gastrointestinal tract (Luppi 2017). Enterohaemorrhagic *E. coli* can be found sporadically in the faeces of healthy pigs and have zoonotic potential as they are causing bloody diarrhoea, haemorrhagic colitis or haemolytic uremic syndrome in humans (Fairbrother et al. 2005). In general, *E. coli* are the causative agents of a variety of diseases in pigs, including neonatal and postweaning diarrhoea (PWD) (Fairbrother and Nadeau 2019). In addition, *E. coli* causes septicaemia in piglets, oedema disease in weaners and mastitis and urogenital infections in sows, which will not be part of this chapter and information can be found elsewhere (Frydendahl 2002; Gerjets et al. 2011; Dubreuil et al. 2016; Fairbrother and Nadeau 2019). Neonatal-like postweaning diarrhoea caused either by enterotoxic (ETEC) or enteropathogenic (EPEC) *E. coli* results in significant welfare issues and economic losses due to increased mortality, decreased weight gain, costs of treatments, vaccination and feed supplementation in both suckling and weaned pigs, respectively (Sjölund et al. 2014; Luppi 2017). Serotyping, which is based on surface structures of the bacterial cells, allows a certain grouping of the isolates. Antigens O (component of the liposaccharide cell envelope), K (capsular polysaccharide), H (flagellar type) and F (fimbria antigen) are distinguished (Wendt et al. 2013). Today, the classification is more based on virulence genes allowing a virotyping of *E. coli* based on pathotype, adhesin and toxin and is usually run by molecular-biological methods like PCR or MALDI-TOF (Table 3) (DebRoy and Maddox 2001; Fairbrother and Nadeau 2019).

#### 2.1.2 Pathogenesis

##### 2.1.2.1 Enterotoxic (ETEC) *E. coli*

For ETEC, adhesion to the intestinal epithelium or the mucosal lining of the intestine, which is mediated by the fimbrial antigens, is an important pathogenicity factor (Luppi 2017). F4, F5, F6, F41, F4 and F18 are the most important fimbrial antigens for neonatal diarrhoea and for PWD, respectively (Wendt et al. 2013). A single strain of *E. coli* may be capable of producing several

**Table 3** Overview of adhesins and toxins of pathogenic *E. coli* associated with diarrhoea in pigs

| Pathotype | Adhesins | Toxins | Disease |
|---|---|---|---|
| ETEC[a] | F5 (K99), F6, F41<br>F4 (K88) | STa[d]<br>STa, STb, LT, EAST-1, α-haemolysin | Neonatal diarrhoea |
| | F4 (K88), AIDA[c]<br>F18, AIDA | Sta, STb, LT, EAST-1, α-haemolysin<br>STa, STb, LT, EAST-1, α-haemolysin, STx2e[e] | PWD [f] |
| EPEC[b] | Eae (intimin) | – | PWD |

[a]Enterotoxic, [b]Enteropathogenic *Escherichia coli,* [c]Adhesin involved in diffuse adherence (AIDA), [d]Heat-stable (STa/b), heat-labile (LT) and enteroaggregative *Escherichia coli* heat-stable enterotoxin (EAST1), [e] Shiga toxin 2e-producing *Escherichia coli*
[f]Post-weaning diarrhoea (PWD)

fimbriae (Fairbrother and Nadeau 2019). The corresponding receptors for the fimbriae are distributed differently in the small intestine, and receptors for F5, F6 and F41 are found in the posterior jejunum and ileum, while receptors for F4 are found in the entire jejunum and ileum (Fairbrother et al. 2005; Fairbrother and Nadeau 2019). Some pigs lack the receptors for F4, whereas the concentration of receptors for F5 decreases during the suckling period while that of receptors for F18 increases (Wendt et al. 2013). Regarding the subtypes of fimbrial antigens, F4 ac and F18 ac/ F18 ab are correlated with PWD (Francis 2002; Luppi 2017). Also, non-fimbrial adhesin identified as adhesin involved in diffuse adherence (AIDA) has been associated with ETEC strains, but their exact role in neonatal and postweaning diarrhoea has to be validated (Ravi et al. 2007; Zajacova et al. 2013). After adhesion, ETEC produces heat-labile (LT), heat-stable (STa/b) toxins and the enteroaggregative *E. coli* heat-stable enterotoxin (EAST1) (Fairbrother and Nadeau 2019). LT stimulates the production of cyclic adenosine monophosphate (AMP) via adenylate cyclase, and STa stimulates the production of cyclic guanosine monophosphate, which leads to increased secretion of $Cl^-$, $Na^+$ and $HCO_3$ along with water into the intestinal lumen and causes dehydration and acidosis (Box 1) (Johnson et al. 2009; Dubreuil et al. 2016). STb increase uptake of $Ca^{2+}$ into intestinal cells and causes low-grade villi atrophy (Wendt et al. 2013; Dubreuil et al. 2016). The role of EAST1 still needs to be further investigated (Luppi 2017).

#### 2.1.2.2 Enteropathogenic (EPEC) *E. coli*

EPEC colonise the intestinal epithelium by forming specific fimbriae (bundle-forming pili) and then attach to the villous epithelium by intimin ("attaching and effacing") inducing the destruction of the microvilli, which leads to decreased resorption and osmotic diarrhoea (Zahavi et al. 2011; Wendt et al. 2013; Cepeda-Molero et al. 2017). Compared to ETEC, EPEC is not only limited to the small intestine, as this pathogenesis can also be found in the large intestine (Fairbrother and Nadeau 2019).

### 2.1.3 Clinical Signs

Neonatal diarrhoea due to enterotoxigenic *E. coli* is observed most commonly in piglets aged from 0 to 4 days of life and is predominant in litters of gilts in endemic situations due to the potential lack of passive immunity by colostral antibodies (Dubreuil et al. 2016; Luppi 2017). Diarrhoea can start 2–3 hours after birth. A variety of piglets within litters of a farrowing batch can be affected. Various clinical signs such as mild to severe watery, white to yellow diarrhoea with a distinctive smell might be present (Wendt et al. 2013; Luppi 2017). Severe dehydration and in septicaemic cases neurological signs and apathy can occur. In some cases, only a perianal faecal staining might be observed (Wendt et al. 2013). Deaths can be seen after 2–4 days, whereby the lethality can reach up to 50%. It is high during the first week of life and 10–20% in older suckling pigs (Wendt et al. 2013; Dubreuil et al. 2016; Fairbrother and Nadeau 2019).

PWD due to *E. coli* is commonly observed during the first 2–3 weeks after weaning, but it also can be recorded at 6–8 weeks after weaning (Luppi 2017). Clinical signs include reduced feed intake and watery, grey-green over slightly pink to brown diarrhoea with a characteristic smell and generally lasting approximately 1 week (Wendt et al. 2013; Luppi 2017). Affected pigs are dehydrated and show reduced weight gain but fever is only observed occasionally. Morbidity and lethality can reach up to 50% and 10%, respectively (Wendt et al. 2013).

### 2.1.4 Lesions

Macroscopic lesions for neonatal and postweaning diarrhoea include dehydration, curdled milk or undigested feed and hyperaemic fundus of the stomach and watery contents in the dilated, oedematous and hyperaemic small intestine as well as enlarged mesenterial lymph nodes (Luppi 2017; Fairbrother and Nadeau 2019). Villous atrophy is not typical, but mild forms can be observed in few cases or in mixed infections with viruses (Fairbrother et al. 2005; Wendt et al. 2013). In uncomplicated cases and recently euthanised or dead animals, there may be no inflammatory or other pathological changes in the small intestine (Wendt et al. 2013). Microscopic findings include hyperaemia, in septicaemic cases micro thrombi in intestinal blood vessels and a high amount of bacteria observed in patches on the apical surface of villous epithelial cells in the ileum and less consistently in the jejunum (Fairbrother et al. 2005; Fairbrother and Nadeau 2019). For this reason, necropsy, both in the cases of neonatal and postweaning colibacillosis, helps the pathologist in the choice of subsequent laboratory examination (Luppi 2017).

### 2.1.5 Diagnosis

Diagnosis of ETEC or EPEC neonatal and postweaning diarrhoea is based on certain decision steps which are described by Luppi (2017) (Fig. 1) (Luppi 2017). Following the clinical presumptive diagnosis, either three to five euthanised or freshly dead pigs or faecal samples or faecal swabs of three to five acutely infected, untreated pigs with typical clinical appearance are taken for further investigation (Box 2–4). If euthanised or dead animals are available, they should have been fallen to death during the last 4–6 hours, and unopened segments of the small intestine (in particular, ileum and jejunum) and large intestine with the ends tied off are taken fresh and fixed in 10% buffered formalin for bacteriological culture and histology, respectively. Fresh samples should be stored at +4 °C and timely arrive at the laboratory. The definitive diagnosis is made by a combination of quantitative bacteriology culture; identification of virulence factors, usually by PCR; and histopathology to have an integrated interpretation of the microscopic lesions (Luppi 2017).

Growth on blood agar and lactose fermentation allow to distinguish between haemolytic and non-haemolytic colonies of *E. coli* (Fairbrother and Nadeau 2019). In general, haemolysis gives a hint for F4 in neonatal ETEC diarrhoea and F4 or F18 in PWD ETEC diarrhoea, whereas *E. coli* isolates carrying only F5, F6 and F41 are often non-haemolytic (Fairbrother et al. 2005; Fairbrother and Nadeau 2019). In cases of PWD caused by ETEC, 97.6% of isolates showed haemolysis, and the remaining 2.4% of non-haemolytic ETEC isolates shared the same virotype (Luppi et al. 2016). Cultural isolation should be followed by toxinotyping by PCR recognising genes encoding for toxins (STa, STb, LT and EAST1) and fimbriae (F4, F5, F6, F18, F41) of ETEC, for the outer membrane protein Eae or intimin in enteropathogenic *E. coli* (EPEC are available and can be used for daily routine diagnostics) (Casey and Bosworth 2009). Although PCR toxinotyping directly from faeces can have higher sensitivity in detection of fimbrial antigen, only the interpretation of PCR results combined with bacteriological findings and clinical signs confirms the diagnosis of ETEC and EPEC as causative agents (Ståhl et al. 2011; Luppi 2017).

The diagnostic approach of clinical signs, pathological lesions, culture and toxinotyping

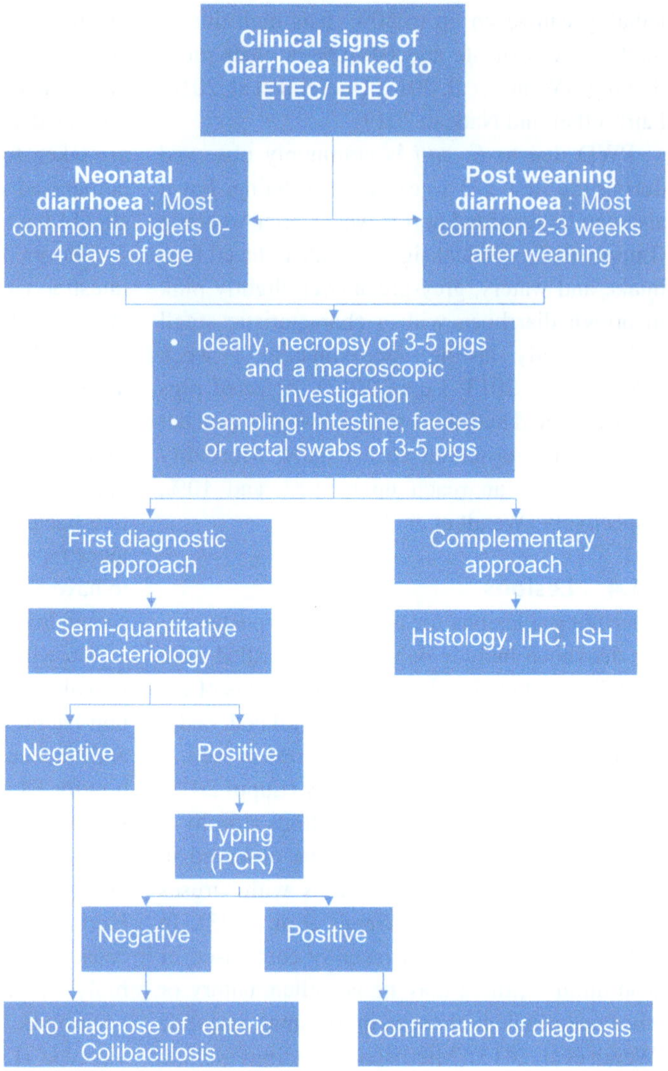


**Fig. 1** Diagnostic approach of enteric colibacillosis in neonatal and postweaning pigs. Modified after (Luppi 2017). *IHC* immunohistochemistry, *FISH* fluorescence in situ hybridisation, *PCR* polymerase chain reaction

can be completed by detection of *E. coli* in the intestine by immunohistochemistry and fluorescence in situ hybridisation (Fairbrother and Nadeau 2019). A list of differential diagnoses for infectious causes of diarrhoea by age category can be found in Table 2.

### 2.1.6 Treatment and Prevention

Treatment of acutely infected pigs with ETEC or EPEC is comprised of antimicrobial treatment of the pathogen, limitation of dehydration and management changes such as keeping piglets and weaners in a warm and clean environment (Box 5). Regarding antimicrobial treatments, general considerations include oral or parenteral treatments for both neonatal and postweaning diarrhoea (Luppi 2017). Especially in severely infected animals, the ability of water/milk and/or feed intake is doubtful and individual parenteral or oral treatment is indicated (Box 7). Amoxicillin/clavulanic acid, apramycin, gentamicin, neomycin, colistin and trimethoprim-sulphonamide are all antimicrobials generally effective for the treatment as well as metaphylaxis of ETEC infections (Luppi 2017; Cutler et al. 2020; Laird et al. 2021). However, due to the fact that resistance to *E. coli* against various antimicrobial classes are present and increasing globally, decision should be based on antimicrobial resistance testing in the particular cases (Kempf et al. 2013; Abraham et al. 2014; Luppi et al. 2015; Brand et al. 2017; Rosager et al. 2017; Jiang et al. 2019; Laird et al. 2021). Besides antimicrobials, other strategies to decrease the amount of ETEC and EPEC include the use of bacteriophages, improvement of the intestinal microbiota by various pre- and probiotics or the use of zinc oxide; however, the latter is banned in the European countries for treatment usage, since increased zinc and also combined multidrug resistance have been described (CVMP n.d.; Jamalludeen et al. 2009; Yazdankhah et al. 2014; Vahjen et al. 2015; Angelakis 2017; Rhouma et al. 2017; Luise et al. 2019a). Rehydration with solution including electrolytes and glucose can be performed either intraperitoneally or orally for piglets and weaners, respectively (Box 5) (Wendt et al. 2013; Fairbrother and Nadeau 2019; Tucker et al. 2022). If group treatment is chosen, drinkers and cups should be replaced and cleaned routinely during the day (Wendt et al. 2013). Correct environmental temperature of 30–34 °C and 26–28 °C for piglets and weaners, respectively, will improve recovery and decrease severity of disease (Le Dividich and Herpin 1994; Fairbrother and Nadeau 2019).

Vaccination is the most important aspect of prevention, and several strategies including using live and inactivated vaccines have been described for *E. coli* in pigs (Melkebeek et al. 2013). Regarding neonatal diarrhoea, passive immunity of the piglets via colostral and/or lactogenic antibodies induced by vaccination of the sow is the most applied and successful vaccination strategy (Melkebeek et al. 2013; Dubreuil 2021). For basic immunisation, gilts are vaccinated 3 and 5–6 weeks prior to their first farrowing followed by sow vaccination 3 weeks prior to each following farrowing (Wendt et al. 2013). As lactogenic immunity might not protect beyond weaning, passive immunity via sow's milk is more difficult, and although there are some vaccine candidates, there is no overall evidence that passive immunity protects against PWD (Melkebeek et al. 2013; Ruan and Zhang 2013; Fairbrother et al. 2017; Ntakiyisumba et al. 2022). There are several oral and parenteral live attenuated and inactivated vaccines described for the usage in piglets, which induce an active, mucosal immunity protecting against PWD (Melkebeek et al. 2013; Zhang 2014; Fairbrother et al. 2017; Ntakiyisumba et al. 2022). Besides vaccination, porcine susceptibility or resistance to F4ac and F18ac ETEC in genes is another important approach in preventing neonatal and postweaning diarrhoea caused by *E. coli* in pigs (Zhao et al. 2014; Luise et al. 2019b). Genetically, some piglets lack the fimbriae receptor genes and thus are resistant to ETEC infections. This can be used as a selection criterion in breeding pigs (van den Broeck et al. 1999; Fairbrother and Nadeau 2019).

## 2.2 Porcine Circovirus Type 2 (PCV2)

### 2.2.1 Aetiology and Pathogenesis

Porcine circovirus type 2 (PCV2) is a DNA virus ubiquitous in all pig-keeping countries (Segalés et al. 2005; Opriessnig and Langohr 2013). Within the group of PCV2, eight genotypes (a–h) are currently known, and several "genotype-shifts" regarding the most prevalent genotype causing disease in a given pig populations have been described (Dupont et al. 2008; Franzo and Segalés 2020; Link et al. 2021). Non-porcine species are, with the exception of mice and rats, not susceptible to PCV2 infections, but may serve as potential vectors within herds (Lőrincz et al. 2010; Segalés et al. 2019). PCV2 is mainly transmitted via oronasal exposure but can be found in nearly all secretions and excretions of pigs (Opriessnig et al. 2007; Segalés 2012).

PCV2 is recognised as one of the most important viruses causing severe economic impact in the swine industry worldwide. (Armstrong and Bishop 2004; Segalés et al. 2005; Madec et al. 2008). Many diseases are associated with PCV2 and are summarised under the term PCV2-associated diseases (PCV2-AD) (Table 4) (Opriessnig et al. 2007; Segalés 2012). Diseases affecting the gastrointestinal tract are PCV2-systemic disease (PCV2-SD) and PCV2-enteric disease (PCV2-ED) (Segalés 2012). Both forms

**Table 4** Overview of PCV2-associated diseases (PCV2-AD) (Segalés 2012)

| |
|---|
| • PCV2 systemic disease (PCV2-SD): Is characterised by multiple organic systems involved and includes post-weaning multisystemic wasting syndrome |
| • PCV2 enteric or lung disease[a] (PCV2-ED/PCV2-LD): Only one organ—and its lymphatic system—is involved |
| • PCV2 reproductive disease (PCV2-RD) |
| • Subclinical infection (PCV2-SI) |
| • Porcine dermatitis and nephropathy syndrome (PDNS) |

[a]As PCV2-ED/PCV2-LD is nowadays detected rarely, a combination with PCV2-SD is controversially discussed (Baró et al. 2015; Franzo and Segalés 2020)

may be caused by PCV2a, b and d and are associated with diarrhoea (Kim et al. 2004; Segalés et al. 2004; Opriessnig et al. 2007). Whether PCV2-ED nowadays exists independently or rather, as a part of the systemic disease is discussed (Segalés 2012; Baró et al. 2015; Franzo and Segalés 2020).

Like all PCV-associated diseases, enteritis caused by PCV2 is a multifactorial disease in which PCV2 infection is the necessary factor (Segalés et al. 2019). Although PCV2-SD has been reproduced under experimental conditions, a definitive, repeatable disease model is lacking, and both infectious factors like infectious doses and/or other pathogens and non-infectious factors such as immune status and genetic seem to play a role in the pathogenesis of PCV2 (Opriessnig et al. 2006; Tomás et al. 2008; Grau-Roma et al. 2011; Rose et al. 2012).

### 2.2.2 Clinical Signs

Among many other symptoms such as wasting, growth retardation, enlargement of lymph nodes and respiratory distress, PCV2 can also cause diarrhoea, especially in 4–11-week-old pigs. Such systemic disease including wasting after weaning is typically called post-weaning multisystemic wasting syndrome (PMWS) (Segalés et al. 2019). Morbidity and mortality are commonly described with 4–30% and 4–20%, respectively, and can be decreased by vaccination (Segalés 2012, 2015). The symptoms are nonspecific and are usually accompanied by enlargement of the regional lymph nodes (Segalés et al. 2019). Less frequently than PCV2-SD, exclusively gastrointestinal symptoms and lymphatic organs of the gastrointestinal tract are involved, thus called PCV2-ED.

### 2.2.3 Lesions

The most striking general macroscopic changes of PCV2-SD are non-collapsed, tan-mottled lungs and enlargement of lymph nodes (mainly inguinal, submandibular, mesenteric and mediastinal) (Rosell et al. 1999; Segalés et al. 2004). Other gross lesions associated with PCV2-SD include reduction or increase in liver size, with orange-yellow discoloration in those cases where icterus is present and kidneys with multiple pale foci of variable diameter (Rosell et al. 2000; Segalés et al. 2004). Characteristic histopathological lymphoid lesions in PCV2-SD-affected pigs include lymphocyte depletion with infiltration by large histiocytic cells and giant multinucleate cells (Rosell et al. 1999).

Both PVC2-SD and PCV2-ED enteric lesions such as catarrhal enteritis/colitis, fibrinous lesions, granulomatous inflammation and other lesions such as haemorrhages or ulceration are described (Segalés 2012; Baró et al. 2015). Microscopic findings including lymphocyte depletion in Peyer's patches and in case of PCV2-SD also other lymphoid tissues (Segalés 2012).

### 2.2.4 Diagnosis

PCV2-SD can only be diagnosed in dead animals, fulfilling all the following three aspects (Segalés et al. 2005):

1. Typical clinical signs like growth retardation and wasting and enlargement of lymph nodes.
2. Moderate to severe characteristic, histopathological lesions in lymphoid tissue.
3. Moderate to high amount of PCV2 within lesions of affected pigs. High viral load in quantitative PCR expressed as genome/ml serum, plasma or tissue (Brunborg et al. 2004; Segalés 2012). In situ hybridisation and immunohistochemistry are thereby the most widely used methods.

Following these aspects, a PCV2-ED is diagnosed when (Segalés 2012) the following are present:

1. Pigs suffer from diarrhoea.
2. Granulomatous enteritis and lymphocyte depletion are observed microscopically in Peyer's patches, but not in other lymphoid tissues.
3. Moderate to high amount of PCV2 is detected in intestinal mucosa or Peyer's patches without detection or lesions in other lymphoid tissues. As quantitative PCR is not suitable in this particular case, in situ hybridisation and immunohistochemistry are used.

A list of differential diagnoses for infectious causes of diarrhoea by age category can be found in Table 2.

### 2.2.5 Therapy and Prevention

Once animals are diseased, only symptomatic therapy can be applied (Box 5).

Prevention is of major importance and vaccination is a cornerstone in preventing the occurrence or at least reducing the severity of PCV2-SD or PCV2-ED. Vaccination of suckling pigs against PCV2 should be routinely performed in all herds (Segalés et al. 2009; Young et al. 2011; Franzo and Segalés 2020). Vaccination of gilts may be indicated in some herds to further improve maternally derived immunity in piglets and to prevent reproductive disorders associated with PCV2 (Segalés et al. 2019). It should be noted that the efficiency of the PVC2a vaccines against other genotypes of PCV2 is under discussion (Rose et al. 2016; Opriessnig et al. 2017; Franzo and Segalés 2020).

Besides vaccination, several management measures to lower the impact of PCV2 infections should be applied. The most famous list of measures has been published as "Madec's 20-point plan" (Table 5) (Madec et al. 1999, 2000; Segalés et al. 2005).

**Table 5** Management measures to lower the impact of PCV2 on farm ("Madec's 20-point plan")

| Production phase | Management measures |
|---|---|
| Farrowing pen | 1. Empty the trough by cleaning and disinfecting between batches (use all-in/all-out)<br>2. Clean the sows and treat them against parasites before farrowing<br>3. Limit adoptions (cross-fostering) in the farrowing pen to only those that are really necessary and only in the first days of life |
| Transition | 4. Use small pens (<13 animals), with solid dividers<br>5. Empty the trough, clean and disinfect, use all-in/all-out<br>6. Suitable animal density of three piglets/m$^2$<br>7. Guarantee at least 7 cm of eating space per piglet<br>8. Guarantee good air quality (NH3 <10 ppm, $CO_2$ <0.15%)<br>9. Guarantee a suitable room temperature<br>10. Avoid mixing batches |
| Fattening | 11. Use small pens, with solid dividers<br>12. Empty the trough, clean and disinfect, use all-in/all-out<br>13. Avoid mixing animals from different transition pens<br>14. Avoid mixing animals from different barns<br>15. Suitable animal density of >0.75 m$^2$/pig<br>16. Guarantee good air quality |
| Others | 17. Assure suitable vaccine programme<br>18. Assure suitable animal flow between buildings<br>19. Strict hygiene in castration, injections, etc.<br>20. Assure a fast separation of sick animals; place them in a hospital pen. Remove dead animals |

Slightly modified from Madec et al. (1999, 2000)

## 3 Fattening Pigs

### 3.1 *Brachyspira* spp.

#### 3.1.1 Aetiology

*Brachyspira (B.)* spp. are gram-negative, obligate anaerobic bacteria typically infecting the large intestine of their host. Currently, there are seven *Brachyspira* spp. described to naturally infect pigs. These are divided into weakly and strong beta-haemolytic *Brachyspira* spp. Strong beta-haemolytic *Brachyspira* spp. are *B. hyodysenteriae, B. suanatina* and *B. hampsonii*. All three may act as the causative agents of swine dysentery (SD), a disease characterised by mucohaemorrhagic diarrhoea. While *B. hyodysenteriae* has been known for a long time and is widespread worldwide, the latter two have been discovered only in the last decades in much fewer countries. *B. suanatina*, for example, is only described in Scandinavian countries and Germany so far. *B. hampsonii*, on the other hand, has already been described in Belgium, Canada, Germany, the UK and the USA (Atyeo et al. 1999; Chander et al. 2012; Rubin et al. 2013; Mahu et al. 2014; Rohde et al. 2014; Mushtaq et al. 2015; Burrough 2017). Although all three are classified as strong beta-haemolytic, some isolates of *B. hyodysenteriae* with less haemolysis and reduced or even absence of clinical signs in case of infection have been described. Therefore, the presence of weakly virulent or avirulent strains of *B. hyodysenteriae* is discussed. However, also herd-specific co-factors may influence the severity of clinical signs (Mahu et al. 2016; Card et al. 2019; Joerling et al. 2020)..

Weakly haemolytic *Brachyspira* spp. infecting pigs include *B. pilosicoli, B. murdochii, B. intermedia* and *B. innocence*. Thereby, *B. pilosicoli* is the causative agent of porcine intestinal spirochaetosis (PIS) or porcine colonic spirochaetosis (PCS), a disease that is accompanied by diarrhoea typically without blood or mucus added. *B. murdochii* and *B. intermedia* are mild or non-pathogenic, while *B. innocence* is described as non-pathogenic (Hampson et al. 2015, 2019).

Beside pigs, also other animals such as rodents or birds can carry *Brachyspira* spp. Especially *B. pilosicoli* is known for its ability to infect several mammalian species including humans, as well as a variety of birds (Jansson et al. 2004; Backhans et al. 2010; F. Zeeh et al. 2018b). Once excreted, *Brachyspira* spp. can survive in the environment for a long time. Thus, live *B. hyodysenteriae* and *B. pilosicoli* were still detected in pig faeces after 112 and 119 days, respectively (Boye et al. 2001). High ambient temperatures of 25 °C or more drastically shorten their survival time (Chia and Taylor 1978).

#### 3.1.2 Pathogenesis

Infection typically happens by oral ingestion of contaminated faeces. *Brachyspira* spp. survive the acid environment of the stomach and ultimately reach their target habitat, which is the large intestine of the pig. In the caecum and colon, they move through the mucus into the crypts and attach to the cells. *B. pilosicoli* typically attaches only onto mature apical enterocytes between, and not onto, immature cells deeper within the crypts. Subsequently, *Brachyspira* spp. stimulate and increase the output of mucus. Thereby, the mucous layer become disorganised; however, the mechanism of tissue destruction is not fully understood yet. Secondary infections with bacteria or protozoa such as *Balantidium coli* may contribute to the damage. Colonic malabsorption due to failure of epithelial transport of sodium and chloride from the gut lumen to the blood results in diarrhoea in affected animals (Box 1) (Wilcock and Olander 1979; Schmall et al. 1983; Trott et al. 1996). The bacteria are shed via faeces already 1–4 days before clinical signs appear, and shedding may last for several weeks, even if animals are asymptomatic again (Costa et al. 2014; Massacci et al. 2018). Pigs that have once recover from an infection show less or no clinical signs for up to 17 weeks after initial inoculation when they get re-infected (Olson 1974; Joens et al. 1979).

#### 3.1.3 Clinical Signs

SD is mainly observed in grower and finishing pigs and less frequently in younger animals or adults. Typical clinical sign is diarrhoea with mucohaemorrhagic admixtures. Growth performance in affected animals is reduced. Increased internal body temperatures may occur. In acute outbreaks, about 90% of the animals may be affected with mortality rates up to 30% if no treatment occurs. Sudden deaths may occur, while most animals recover after several weeks (Burrough 2017; Hampson et al. 2019). Different factors such as stress, composition of the diet, group size and pigs' body weight can influence the course of SD (Hansen et al. 2011; Alvarez-Ordóez et al. 2013). If no continuous treatment or eradication is performed, the disease becomes endemic in herds and clinical signs typically reoccur in cyclic intervals within the herd (Patterson et al. 2013).

Clinical signs of PIS are mainly seen soon after weaning but can occur as well in finishers and more rarely in the breeding stock. The most typical clinical sign is liquid to pasty diarrhoea, mainly without blood or mucoid admixtures. Diarrhoea is usually self-limiting and last for about 2–14 days. Up to 100% of the pigs in the age group may become infected. However, diarrhoea is only observed in 17–67% of the cases. Colitis, on the other hand, was described in up to 100% of infected pigs, meaning animals were also affected without clinical diarrhoea. Besides diarrhoea, reduced feed conversion and depressed growth rates may be observed while mortality is low (Trott et al. 1996; Duhamel 2001; Hampson 2018).

### 3.1.4 Lesions

Lesions due to *Brachyspira* spp. infection are mainly located in the large intestine. In case of *B. hyodysenteriae*, such lesions can be expected, when faeces contains $10^5$ colony-forming units of per gram mucosa or more (Wilcock and Olander 1979). Typical for SD, the spiral and descending colon, along with the caecum, is frequently affected. Hyperaemia and oedema of the large intestinal wall, mesentery and mesenteric lymph nodes in combination with oedematous mucosa covered with mucus, fibrin and flecks of blood are hallmarks of SD. Over time, exudation of fibrin becomes more severe and pseudomembranes, membranes out of fibrin, mucus and blood may be seen. In chronic forms, fibrinonecrotic lesions become visible. Histologically, elongation of intestinal crypts, thickening of the mucosa, goblet cell hyperplasia with plenty of mucus secretion, superficial necrosis (no deep ulceration) and neutrophilic infiltration of the lamina propria are the most typical changes (Olson 1974; Wilcock and Olander 1979; B. L. Wilberts et al. 2014a).

Gross lesions of PIS/PIC resemble those of SD, but typically less pronounced. The caecum and colon are most commonly affected. Watery, partly mucoid, content in flaccid intestine with thin walls has been described. Mesenteric lymph nodes and lymphoid follicles may be enlarged, mesocolonic oedema may be present and mucosal erosions may be seen. Histologically, signs of inflammation (typhlocolitis) and elongated crypts filled with mucous or cells may be present. Attachment of spirochetes end-on forming a "false brush border" is pathognomonic for PIS/PIC; however, it is not always present (Girard et al. 1995; Trott et al. 1996; Duhamel 2001).

### 3.1.5 Diagnosis

Currently, SD is defined as a combination of detection of a strong beta-haemolytic *Brachyspira* spp. from faeces or colonic mucosa of pigs with typical clinical signs and/or pathological lesions (Burrough 2017). PIS/PCS diagnosis is defined as the confirmation of *B. pilosicoli* in pigs with typical clinical signs and lesions. Clinically, normal pigs may shed *B. pilosicoli*, which is why detection of the pathogen within faeces alone does not confirm PIC/PCS (Hampson et al. 2019).

In order to obtain such a definitive diagnosis, several protocols exist. The first one is to detect *Brachyspira* spp. or specific genome fragments of the bacteria in living pigs out of native rectal faeces or faecal swabs, whereby these not pretreated pigs show typical clinical signs (Box 2–4). Animals with such clinical signs can be expected to excrete $10^8$–$10^9$ bacterial cells per g faeces, while asymptomatic animals are only periodically excreting detectable levels of $>10^3$ bacterial cells per g of faeces (Songer and Harris 1978; Hampson et al. 2019). To identify the *Brachyspira* spp. within the sample, two methods have prevailed in diagnostics in the field. One is the direct detection of different *Brachyspira* spp. by PCR from the sample. This procedure has the advantage of being fast and low-cost intensive. Alternatively, a selective culture for *Brachyspira* spp. can be performed. It is important to note that the anaerobic *Brachyspira* spp. arrive in the laboratory alive, if culture is aimed at (Box 3). This method has the advantage of being considered more sensitive and enables the preparation of an antibiogram. In addition, the risk of overlooking the more rare or atypical *Brachyspira* spp. isolates, for which no suitable primer may be included in the PCR, is reduced. The disadvantage is that cultivation takes 6–10 days, as *Brachyspira* spp. grow slowly. Subsequently, a differentiation of the different *Brachyspira* spp. is necessary, e.g. by means of PCR, MALDI-TOF or biochemical tests, since *Brachyspira* spp. cannot be distinguished on the agar plate. The procedure thus takes longer than direct PCR and is associated with higher costs (Råsbäck et al. 2006; Burrough et al. 2012; Patterson et al. 2013; Burrough 2017; Warneke 2017).

Alternatively, samples may be collected from freshly dead or euthanised animals and examined by the tests already mentioned above. As sampling material from such animals, mucosal scrapings from the caecum and/or colon are recommended (Box 8) (Burrough et al. 2012; Friederike Zeeh et al. 2018a). Patho-histological examination via histological examination of haematoxylin and eosin or Warthin-Starry silver-stained sections of the caecum and colon is another alternative. Immunohistochemical staining or in situ hybridisation can help to differentiate the different *Brachyspira* spp. (Wilberts et al. 2015; Burrough 2017). A list of differential diagnoses for infectious causes of diarrhoea by age category can be found in Table 2.

Due to their low sensitivity, oral fluid samples are not suitable for diagnostic purposes and may only be used during acute outbreaks and on herd level. The same applies to serology. Since it is currently almost impossible to distinguish between the different *Brachyspira* spp. and antibodies do not correlate with protection against SD, serology has not been used in routine diagnostics but rather monitoring to date. Analysis of meat juice from slaughterhouses via ELISA is currently investigated but not established in diagnostics so far (Diarra et al. 1995; Song et al. 2012).

### 3.1.6 Therapy and Prevention

Only few antimicrobials are suitable for the treatment of animals diseased with *Brachyspira* spp. These include pleuromutilins (tiamulin, valnemulin), macrolides (tylosin, tylvalosin), lincosamides (lincomycin) and tetracyclines. Pleuromutilins are thereby described as the most effective with one study showing resolved clinical disease of SD within 24 h after initiation of oral treatment via drinking water and stop of shedding within 72 h, respectively (B. L. Wilberts et al. 2014b). However, many countries report resistance or even multi-drug resistances for pleuromutilins (Karlsson et al. 2004; Pringle et al. 2004; Šperling and Smola 2011; Rugna et al. 2015; Card et al. 2018; García-Martín et al. 2018; FINRES-Vet 2019; Arnold et al. 2022). Therefore, an antibiogram prior to treatment as well as an adequate administration in appropriate doses is highly recommended (Table 6). If oral therapy is used, animals with reduced water or feed intake should receive additional parenteral therapy to achieve adequate drug levels (Box 7).

Treatment of pigs with antimicrobials alone does not eliminate the pathogen in the group of animals, which is why, if susceptible animals are still in the herd or are newly introduced, a recurrence of the symptoms in these susceptible animals has to be expected (Patterson et al. 2013; Neirynck et al. 2020). Several management factors can also positively influence the severity and course of the disease. This includes all-in/all-out management with separation of different age categories and cleaning (high pressure cleaner, hot water) and disinfection, e.g. with oxygen compounds at 4% for 10 minutes or quaternary ammonium compounds, sodium hypochlorite or organic acids, between batches. Moving affected batches after treatment to clean buildings may help as well. Bedding must be changed, and equipment used in affected batches should be cleaned and disinfected. Further, disinfectant

**Table 6** Dosage and duration of some antimicrobials for treatment of swine dysentery

| Antimicrobial | Dosage and route of administration[a] | Duration |
|---|---|---|
| Tiamulin | 10 mg/kg BW, i.m.<br>≥8 mg/kg BW, p.o. in drinking water<br>100–180 ppm, p.o. in feed | 1–3 days<br>5–7 days<br>10–14 days |
| Valnemulin | 3–4 mg/kg BW, p.o. in feed | 1–4 weeks |
| Tylosin | 10–12.5 mg/kg BW, i.m.<br>5–25 mg /kg BW, p.o. in drinking water | 2x daily for 3 days<br>5–7 days |
| Tylvalosin | 4.25 mg/kg BW, p.o. in feed | 10–14 days |
| Lincomycin | 11–15 mg/kg BW, i.m.<br>8–10 mg/kg BW, p.o. in drinking water<br>100 g/ton feed | 3–7 days<br>5–10 days<br>3 weeks |

*mg* milligram, *BW* bodyweight, *i.m.* intramuscular, *p.o.* per os
[a]Is a combination of product information and publications (Alvarez-Ordóez et al. 2013; Hampson et al. 2019)

footbaths and change of clothing between batches or age groups may be necessary. Since *Brachyspira* spp. have been detected in rodents, professional rodent control should be implemented (Box 6). Minimising contact with birds and other potential vectors would also be desirable, but is often difficult, especially in herds with outdoor access. At the animal level, possible stressors should be avoided. These include crowding, transportation, pig handling or dietary changes. However, the composition of the diet can influence the course of the disease (Alvarez-Ordóez et al. 2013; Burrough 2017; Zeeh et al. 2020; Alarcón et al. 2021). For SD, it has been shown that highly digestible diets (cooked rice, milk by-products, animal protein), fine grounded feed, feed high in inulin, linoleic acid feed, steam flaked corn or sorghum, sweet lupines, barley, acid herbs and conjugated linolic acid were likely protecting against or delaying the onset of SD. On the other hand, insoluble fibre or hindgut fermentation, distiller dried grains with soluble soybean meal, wheat and high concentration of non-polysaccharides were associated with more severe or earlier onset of SD (Pluske et al. 1996; Durmic et al. 2002; Jacobson et al. 2003; Lindecrona et al. 2003; Thomson 2009; Bailey L Wilberts et al. 2014; Helm et al. 2020, 2021).

In order to reduce the use of antimicrobials in the long term and optimise the health of the animals, eradication of *B. hyodysenteriae* from the herd has been described and should be considered in herds with a low risk of reintroduction. Simultaneous eradication of *B. pilosicoli* with the same methods seems less successful (Neirynck et al. 2020). Eradication can be performed by either total depopulation and repopulation with negative animals or by partial depopulation. In the latter case, the number of animals in the stock is reduced as far as possible. All remaining animals in the herd are treated with suitable antimicrobials (antibiogram) for at least 2–3 weeks. In addition, rodent control in the herd is intensified as well as strict cleaning and disinfection using all-in/ all-out management including emptying the manure and disinfecting the residual manure, e.g. with hydrogen cyanamide. Eradication exclusively with antimicrobials or partial depopulation is not very promising. Since the survival time of *B. hyodysenteriae* in the environment is reduced at warmer temperatures, eradication should only be carried out in the warm summer months (Burrough 2017; Cadetg et al. 2019; Hampson et al. 2019; Neirynck et al. 2020).

Research is being conducted into the production and use of vaccines against *B. hyodysenteriae*. However, to the knowledge of the authors, commercial vaccines are not in use in most pig-keeping countries yet (Diego et al. 1995; La et al. 2004; Mahu et al. 2017).

## 3.2 Lawsonia intracellularis

### 3.2.1 Aetiology

*Lawsonia* (L.) *intracellularis* is a microaerophilic, gram-negative, obligate intracellular bacterium and the causative agent of porcine proliferative enteropathy (PPE) also known as ileitis. The bacterium can be found in pigs throughout the world with varying prevalence. The prevalence in European herds with a history of diarrhoea has averaged over 80% of positive pig farms (Arnold et al. 2019). Similarly, sometimes even higher prevalence is reported worldwide (Marsteller et al. 2003; Holyoake et al. 2010; Wu et al. 2014). The number of positive weaner or fattener pigs within such herds often averages about 25–30% (Stege et al. 2000; Arnold et al. 2019). Consequently, the spread of the pathogen can be described as ubiquitous. Other animal species may be infected, diseased and/or shed *L. intracellularis* as well. These include, among others, mice, rats, cats, foxes, dogs and wild boars (Landsverk 1973; Husnik et al. 2003; Dezorzova-Tomanova et al. 2006; Collins et al. 2011; Hwang et al. 2017). Experimentally, it has been shown that mice can become infected with *L. intracellularis* by oral ingestion of infected pig faeces, and that infection of pigs by oral ingestion of mice faeces is possible as well (Gabardo et al. 2017). The likelihood of such an infection as well as the role of other animal species as potential vectors within and between pig herds requires further investigation (Collins 2013; Hwang et al. 2017). From current knowledge, an infection of humans seems unlikely (Jacobson et al. 2007).

Transmission of *L. intracellularis* mainly occurs by faeco-oral route. Infected animals start shedding the pathogen about 1 week post-infection with their faeces. Highest shedding rates have been described about 3 weeks post-infection while shedding continues, sometimes intermittently, for approximately 4 weeks and in rare cases up to 12 weeks (Fig. 2) (Smith and Mcorist 1997; Guedes and Gebhart 2003; Collins and Love 2007). Furthermore, not only field but also vaccine strains can be excreted. For a conventional attenuated live vaccine, excretions of up to 9 weeks has been described in individual cases (Guedes and Gebhart 2003). While in challenge studies a number of up to $5.4 \times 10^{10}$ organism of *L. intracellularis* per pig is usually sufficient for infection, excretion levels of up to $7 \times 10^{8}$ *L. intracellularis*/gram of faeces have been described in clinical cases. The infectious dose is rather low (McOrist et al. 1993; Smith and Mcorist 1997; Guedes and Gebhart 2003; Guedes et al. 2003; Macintyre et al. 2003). In addition, *L. intracellularis* can remain viable in cell-free suspension for up to 6 days at 5 °C and even 2 weeks within faeces at temperatures between 5 and 15 °C (Collins et al. 2000). Thus, both infected animals and contaminated environment and objects can contribute to the spread of the pathogen. In cases of uncomplicated PPE, recovery can be expected approximately 4–5 weeks after infection (Kroll et al. 2005; Collins and Love 2007). Pigs that recovered from a previous infection with *L. intracellularis* are protected against re-colonisation and clinical disease for up to 10 weeks (Collins and Love 2007; Riber et al. 2011, 2014). Such a complete protection against re-infection could not be achieved by vaccination so far, although clinical faecal scores may be lowered, weight gain improved and shedding and pathological lesions reduced (Guedes and Gebhart 2003; Jacobs et al. 2020). Antibodies in pigs challenged with pathogenic *L. intracellularis* isolates can be first detected 2–3 weeks and last for 10–13 weeks post exposure, respectively (Guedes et al. 2002; Guedes and Gebhart 2003; Collins and Love 2007; Riber et al. 2011). In case of vaccination, seroconversion was described 4 weeks after vaccination when using an inactivated intradermal vaccine (Jacobs et al. 2020). Using an attenuated live vaccine, no systemic antibodies but rather local immune response is reported (Kroll et al. 2004). Maternal antibodies may be detectable in piglets of up to 5 weeks of age, although they do not seem to fully protect against infection (Guedes et al. 2002; Jensen et al. 2005).

### 3.2.2 Pathogenesis

Twelve hours post-infection, single *L. intracellularis* can be found in the cytoplasm and lamina propria of mature enterocytes on the tips of villi in the small intestine. From 5 to 28 days post-infection, persistent infection of poorly differentiated epithelial cells occurs. The majority of bacteria and lesions can thereby typically be seen within the lower jejunum and ileum although the colon may be also involved. *L. intracellularis*

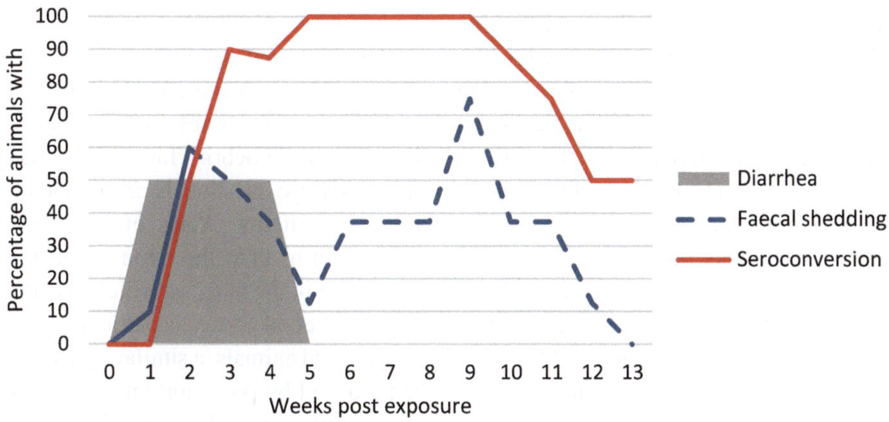


**Fig. 2** Percentage of weaner pigs ($n = 10$) showing diarrhoea, faecal shedding and/or seroconversion after experimental infection with *Lawsonia intracellularis* (Guedes and Gebhart 2003). From week 4 onwards, only eight weaners are remaining. Faecal shedding (blue line) tested by PCR—with broken line due to intermittent shedding. Seroconversion (red line) analysed using immunoperoxidase monolayer assay serology. Grey area, occurrence of diarrhoea

proliferates in the apical cytoplasm of intestinal crypt epithelial cells and causes cells to proliferate without maturation. This process is called adenomatous proliferation, the hallmark of PPE (Boutrup et al. 2010). Such changes are likely involved in reduced nutrient absorption as well as losses of protein and amino acid into the lumen, which may lead to later clinical signs (Gogolewski et al. 1991; Vannucci and Gebhart 2014; Visscher et al. 2018). Intestinal bleedings due to leakage from the capillary bed in connection with degeneration and desquamation of epithelial cells are described (Hampson et al. 2019).

### 3.2.3 Clinical Signs

Gnotobiotic pigs do not show PPE when infected with pure culture of *L. intracellularis* or filtrates of affected mucosa. Conventional pigs, on the other hand, develop lesions of PPE when fed with intestinal mucosa of infected pigs or culture of the organism (McOrist and Lawson 1989; McOrist et al. 1993). In conventional pigs, three different forms of PE are most commonly observed: acute, chronic and subclinical.

The acute form, also called porcine haemorrhagic enteropathy (PHE), is most commonly seen in fattening pigs or gilts of 4–12 months of age. It is characterised by sudden deaths without prior signs of disease, or diarrhoea with bloody admixtures, which can subsequently lead to death of the animals. Due to blood loss, anaemia and circulatory disorders can be observed, which are reflected in pallor, apathy or decreased feed intake in affected animals. Rarely, fever up to 40 °C and associated reproductive disorders have been described in pregnant animals. Depending on the herd structure, an outbreak may persist for several weeks (Yates et al. 1979; Mcorist et al. 1999; Guedes et al. 2002). Morbidity in naive animals may be high, with lethality of up to 50%.

The chronic form, also called porcine intestinal adenomatosis (PIA), is mainly seen in nursery and fattening pigs of 6–20 weeks of age. One typical clinical sign is pasty to watery diarrhoea without blood or mucus admixtures. Diarrhoea is typically associated with excretion levels of up to $1.8 \times 10^8$ *L. intracellularis*/g faeces, while subclinically infected animals showed significantly less excretion of about $7 \times 10^6$ *L. intracellularis*/g faeces (Collins and Barchia 2014). Further, an increase in shedding rates from $10^6$ *L. intracellularis*/g faeces onwards could statistically be associated with a decrease in average daily weight gain. Moreover, an increase in shedding rate of one $\log_{10}$ unit increased the odds for low growth rates by two times (Pedersen et al.

2012a; Johansen et al. 2013; Collins and Barchia 2014). Other clinical signs in animals with PIA may include reduced feed conversion, wasting and minor disturbance of general state. Fever and sudden deaths are rarely reported for the chronic form (Gogolewski et al. 1991; McOrist et al. 1996). In uncomplicated cases, recovery can be expected about 5 weeks post-infection. Rarely, complications such as necrotising enteritis (NE) or regional ileitis (RI) may be seen. These forms appear to be transitional or complicated forms of PIA or PHE and can only be distinguished from them via pathological examinations.

Subclinical infection, meaning neither diarrhoea nor other clinical signs in the presence of the pathogen, is very common in pigs (Brandt 2008; Arnold et al. 2019). However, histological lesions may be present, while macroscopic changes can rarely be seen (Jacobson et al. 2003; Brandt 2008). Excretion of the pathogen is anyway possible; thus, simple detection of *L. intracellularis* in faeces is not necessarily associated with clinical signs, and quantitative analysis should be performed for diagnostic purposes (McOrist and Lawson 1989; McOrist et al. 1993; Stege et al. 2004; Vannucci and Gebhart 2014).

### 3.2.4 Lesions

Five days after infection, *L. intracellularis* antigen may be detected via immunohistochemistry. Lesions suggestive for PPE can be expected from 11 to 24 days after infection (Guedes et al. 2017). Such lesions are most typically seen in the terminal ileum as well as the distal jejunum. More rarely, lesions can be detected in the caecum and colon, which, if present, typically appear 1–2 weeks later than those of the small intestine. Inoculation with $3.7 \times 10^6$ organisms per animal led to lesions in more than 60% of pigs, while such changes could only be seen in less than 5% of pigs inoculated with $4 \times 10^5$ organisms (McOrist et al. 1993). Further, excretion levels in pigs with PPE lesions (median excretion 5.92 $\log_{10}$ bacteria/g faeces) were higher than those without such lesions (median excretion <3.3 $\log_{10}$ bacteria/g faeces) (Pedersen et al. 2012a, 2012b). Proliferation of immature epithelial cells, which leads to a thickening of the intestinal mucosa, is the hallmark of PPE. In case of PHE, in addition to the thickened mucosa and serosa oedema, blood or blood clots are present in the lumen of the ileum, sometimes in combination with fibrino-necrotic debris. Tarry blood may be visible in the large intestine. Erosions and ulcerations are less common (Yates et al. 1979). In the chronic form, PIA, the strong proliferation and thickening of the mucosa, which visually appears like the cerebral cortex, is particularly striking. In freshly dead animals, a similar optical image may be produced by post-mortem contractions of the tunica muscularis. However, these are typically dissipative and usually involve several areas of the intestine.

Histologically, the proliferation of immature crypt epithelial cells forming hyperplastic and dysplastic crypts, loss of villi and reduction of goblet cells are the most remarkable lesions. Fibrino-haemorrhagic lumina clots in case of PHE and fibrino-necrotic mucosal exudate in PIA and NE may be seen (Lomax et al. 1982; Jensen et al. 1997; Guedes et al. 2017).

### 3.2.5 Diagnosis

PPE is diagnosed by the clinical picture in combination with *L. intracellularis* in characteristic pathological lesion or quantitative detection of the pathogen from intestinal samples or faeces from affected animals. Immunohistochemical (IHC) staining and immunofluorescence (IF) or less common in situ hybridisation may be used to confirm the pathogen within the lesions. IHC and IF are considered the most reliable tests for post-mortem diagnosis. Warthin-Starry silver staining is a cost-effective method that is used in identifying the intracellular curved bacteria, but it does not allow specific identification of *L. intracellularis*. Further, multiple segments should be analysed, when using histology, due to segmental distribution of PPE (Guedes and Gebhart 2003; Ladinig et al. 2009; Campillo et al. 2021).

For diagnosis in living animals, polymerase chain reaction (PCR) is a widely used, rapid and accurate method. PCR can be applied on faecal samples or intestinal samples (Box 3). Due to the ubiquitous distribution of the pathogen, the use of quantitative or semi-quantitative PCR is highly recommended. Quantification can support a conclusion regarding the severity of the disease. Ct values of <20 were significantly linked to presence of pathological lesions, while a Ct values >30 strongly indicated their absence (Burrough et al. 2015). Furthermore, shedding rates of $1.8 \times 10^8$ *L. intracellularis*/g faeces were associated with diarrhoea, while sub-clinically infected animals showed significantly less excretion of about $7 \times 10^6$ *L. intracellularis*/g faeces or less (Collins and Barchia 2014). To allow this interpretation of laboratory results, samples must not be pooled. Moreover, shedding might occur intermittently, a reason why multiple animals should be sampled. Cultivation and thus the preparation of an antibiogram is only possible in very few laboratories worldwide and is not available in routine diagnostics yet (Campillo et al. 2021). Antibodies against *L. intracellularis* in serum can be detected from about 2 weeks post-infection onwards. The presence of antibodies may be due to maternally derived antibodies, vaccination and/or infection. Therefore, a meaningful interpretation for diagnosis is very difficult, especially since the proportion of seropositive animals increases with age due to the ubiquitous occurrence of the pathogen (Arnold et al. 2019). A list of differential diagnoses for infectious causes of diarrhoea by age category can be found in Table 2.

### 3.2.6 Therapy and Prevention

Antimicrobials used successfully as therapy of PPE include lincosamides (lincomycin), macrolides (tylosin, tylvalosin), pleuromutilin (tiamulin, valnemulin) and tetracyclines. Macrolides and pleuromutilin were described as particularly effective (Schwartz et al. 1999; Walter et al. 2001; Pommier et al. 2008; Yeh et al. 2011). Macrolides, however, are classified by the WHO and OIE in the class of critically important antimicrobials and should therefore be used very restrictively. Pleuromutilins are classified as less critical by both organisations; however, in some countries, they are one of the few agents available for antimicrobial control of other pathogens such as *Brachyspira* spp. (World-Health-Organization 2019; World-organisation-for-animal-health 2021). Consequently, these antimicrobials should only be used in appropriate dosage and duration (usually 10–21 days) after a confirmed diagnosis (Box 7). Since the cultivation of *L. intracellularis* is still not routinely offered, the preparation of an antibiogram is currently not possible for routine diagnostics.

In addition to, or instead of, antimicrobial therapy, many other therapeutic and supportive measures are available (Box 5). To limit pathogen pressure and spread, separation of age categories, respecting rules of all-in/all-out, small pens of less than 30 animals per pen, routine rodent control, slatted floor of more than 78% of total pen surface, use of non-pelleted diets, rodent control and strict hygiene with cleaning and disinfection of pens and equipment are recommended (Bronsvoort et al. 2001; Stege et al. 2001; Mølbak et al. 2008; Bae et al. 2013; Dors et al. 2015; Resende et al. 2015; Arnold et al. 2021). For example, quaternary ammonium compounds (3.3% cetrimide) are suitable for disinfection. Little less but still good susceptibility in laboratory trials was seen using 1% povidone-iodine and 1% potassium peroxygen-sulphate. Less good or varying results per strain were determined for 0.33% phenolic mixtures and sodium hypochlorite (0.001%), while no reduction in growth was seen for hydrogen peroxide/peracetic acids (Collins et al. 2000). In addition to management and hygiene measures, commercial vaccines are available for prophylaxis to reduce clinical and pathological lesions and economic losses. Thereby, pigs can be vaccinated intramuscularly or intradermally, with a whole-cell bacteria vaccine, or orally, using an attenuated live vaccine from 3 weeks of age onwards. When using the attenuated live vaccine, antimicrobials should not be administered 3 days before and after vaccination (Park et al. 2013; Roerink et al. 2018; Jacobs et al. 2020). Eradication attempts via medication and partial depopulation were not successful so far.

### 3.3 *Salmonella* spp.

#### 3.3.1 Aetiology

*Salmonella (S.)* spp. are facultative intracellular, facultative anaerobe, gram-negative bacteria that can infect a wide range of hosts. These include pigs and humans, as well as nearly all vertebrates in which searches have been conducted. Thus, ingestion of contaminated food from carrier animals as well as direct contact to clinically sick or sub-clinically infected pigs may be a threat to human health (Ellin Doyle et al. 2009; Schroeder et al. 2015). Typical *Salmonella* spp. that have adapted to pigs are *S.* Choleraesuis and *S.* Typhisuis. Also frequently described in pigs, but with a broad host range, are *S.* Typhimurium, *S.* I1,4, [5],12:i: and *S.* Derby (EFSA 2009; Bernad-Roche et al. 2021). In Europe, about 29.0% of all pig holdings and in the USA 52.6% of all herds were *Salmonella* spp. positive (EFSA 2009, 2011; National-animal-health-monitoring-System 2009). Transmission is mainly horizontal by the faeco-oral route, but also via oropharyngeal secretions, meaning nose-to-nose contact, and vertical, meaning dam to offspring. Airborne transmission over short distance is described in other animals and confirmed in pig farms as a source of contamination in the environment (Wathes et al. 1988; Fedorka-Cray et al. 1995; Proux et al. 2001). In addition, also food, bedding, clothings or vectors such as birds, rodents, insects or pets may be involved in the transmission of the pathogen (Jones 2011). If transmission occurs by animal or vector contact, the disease spreads from pen to pen. In case of transmission by feed or contaminated environment, several pens may be suddenly involved. Intermittent shedding can be detected few days after experimental inoculation and may last over 4–5 months or even longer (Wood et al. 1989; Fedorka-Cray et al. 1994). Once excreted, *Salmonella* spp. remain viable and infectious for multiple months in soil and water, where it develops biofilms, and on surfaces, plants, slurry and multiple other environmental places (Davies and Wray 1996; Baloda et al. 2001; Jensen et al. 2006; Bondo et al. 2016). Stressors like mixing, high pig density, transportation, disease, antimicrobial therapy and food deprivation may increase faecal shedding as well as susceptibility of exposed pigs.

#### 3.3.2 Pathogenesis

Many factors such as serotype, virulence, host resistance, route and quantity of affective dosage, interference with intestinal flora, gastric pH and others have an impact on the course of infection. Having successfully passed the stomach, the M-cells of Payer's patches in the jejunum and ileum are major entry points into the submucosa. *Salmonella* spp. may survive in macrophages and neutrophils and also infect mesenteric lymph nodes. Inflammatory mediators as well as endotoxins are discussed to be associated with microvascular damage including thrombosis in the lamina propria and submucosa leading to ischaemia and probably necrosis Decreased sodium resorption and increased chloride secretion due to enterotoxins cause diarrhoea in the pigs with salmonellosis. Endotoxemia is likely involved in fever, inflammation and systemic signs and lesions (Reed et al. 1985; Roof et al. 1992; Meyerholz and Stabel 2003; Schauser et al. 2004).

#### 3.3.3 Clinical Signs

While infection may occur in all age categories, clinical signs are mainly seen in nursery and fattening pigs. However, naive sows as well as piglets without lactogenic protection sometimes show clinical signs (Wilcock et al. 1976; Van Der Wolf et al. 2001; Veterinary Laboratories Agencys 2009; Pires et al. 2014). The disease with clinical signs caused by *Salmonella* spp. is called salmonellosis. The detection of *Salmonella* spp. from an animal that does not show clinical signs is called *Salmonella* infection and is particularly relevant for food safety. The infective dose to cause salmonellosis in pigs under field conditions is unclear. However experimental disease could be produced using $10^{8-11}$ organisms/pig (Gray et al. 1996). The clinical picture of salmonellosis differs per serovar.

Infections with *S.* Choleraesuis may cause systemic disease due to sepsis with lethargy, high fever of more than 40.5 °C, abortion, coughing, cyanosis, reluctance to move, icterus and infrequently nervous signs. Case fatality may be high, with morbidity typically less than 10.0%. Diarrhoea occurs, if at all, only in the later course of the disease after a few days. During the acute phase, $10^6$ *S.* Choleraesuis/gram faeces may be shed (Smith and Jones 1967; Gray et al. 1995, 1996).

Infections with *S.* Typhimurium typically cause clinical signs in the gastrointestinal tract including watery diarrhoea, which may contain blood and necrotic material. Dehydration, wasting, inappetence, fever and more rarely rectum strictures, due to ulcerative proctitis leading to distended abdomen and difficulties in defecation, may be seen as well. Shedding of $10^7$ S. Typhimurium/g faeces is described. With excretion at this level, visible lesions in the intestines can be assumed. The morbidity is high, while the mortality caused by hypokalaemia and dehydration is rather low. Most pigs completely recover within some weeks, with individuals remaining carriers and continuing shedding the pathogen over months. Disease is mainly seen in stables with poor hygiene or stocked naive pigs (Gutzmann et al. 1976; Wilcock et al. 1976; Wilcock and Olander 1977).

Other *Salmonella* spp. are associated with chronic diarrhoea and wasting with caseous lesions like *S.* Typhisuis, with post-weaning diarrhoea and mild lesions due to *S.* Heidelberg or suppurative meningitis in suckling pigs due to *S.* Dublin and S. Enteritidis.

### 3.3.4 Lesions

In animals suffering from salmonellosis caused by *S.* Choleraesuis, septicaemia lesions in a variety of organs can be observed. This includes enlarged, moist and congested gastrohepatic and mesenteric lymph nodes, splenomegaly, hepatomegaly with foci of 1–2 mm of parenchymal necrosis (=paratyphoid nodules), oedematous and thickened wall of the gallbladder, acute interstitial pneumonia, icterus, dermal cyanosis and enterocolitis similar to the enteric infection with *S.* Typhimurium. Histologically, a necrotising and histiocytic vasculitis may be seen. In the gut, fibrino-necrotic exudate may form so-called button ulcers in the large intestine. Splenomegaly together with hepatomegaly and lymphomegaly is very suggestive of septicaemia by *S.* Choleraesuis (Sherman 1972; Fedorka-Cray et al. 2000; Savic et al. 2021).

The enteric forms are characterised by fibrino-necrotic enteritis or typhlocolitis characterised by thickened oedematous walls, red mucosa, erosions and ulcers covered with fibrino-necrotic debris (button ulcers) mainly in the ileum, caecum and spiral colon. Oedema in the mesocolon may be seen. Ileocaecal lymph nodes may be enlarged, congested and moist (lymphadenopathy). In the case of *S.* Typhisuis infection, lymph nodes with caseous abscess and chronic ulcerative colitis may be observed (Fedorka-Cray et al. 2000).

### 3.3.5 Diagnosis

Ideally, diagnosis of salmonellosis is made by detection of *Salmonella* spp. using culture from faeces or organ material of clinically sick animals in the presence of corresponding lesions (Andrews and Ryan 2015). The detection of the pathogen from the environment cannot be used for the diagnosis of disease. In case of *S.* Choleraesuis septicaemia, culture from the lung, liver or spleen from euthanised or fresh dead animals (<2 h) is recommended. Intestinal content or faeces should never be used for isolation and diagnostic of *S.* Choleraesuis in pigs with acute septicaemia.

For enterocolitis due to *S.* Typhimurium, a pool of ileum, colon and ileocaecal lymph nodes for culture is recommended. Alternatively, tonsils and the caecal wall may be sampled as well. In live animals, large quantities of faeces (10 g) or pharyngeal tonsil scrapings should be used for enrichment and culture (Box 2–4). Rectal swabs should be avoided as well as direct PCR due to lower sensitivity compared to culture. Detection of antibodies in oral fluids does not correlate well with presence of serum antibodies or shedding and lacks sensitivity as well as specificity (De Lucia et al. 2020, 2021). Serology or meat juice analysis at the slaughterhouse using enzyme-linked immunosorbent assays (ELISA) for herd screenings may be used for monitoring but likewise have little relevance in regard to carrier or shedding status (Wilhelm et al. 2007; Szabó et al. 2008; Alban et al. 2010). A list of differential diagnoses for infectious causes of diarrhoea by age category can be found in Table 2.

### 3.3.6 Therapy and Prevention

Antibiotic usage in cases of enteric salmonellosis is highly controversial. Unlike in human medicine, antibiotic treatment does not seem to increase or prolong disease or shedding in pigs. However, significantly reduced duration or magnitude of shedding is neither uniformly reported and no benefit on mortality is described (Finlayson and Barnum 1973; Gutzmann et al. 1976; Jones et al. 1983; Sirinavin and Garner 1999). Only in case of septicaemic salmonellosis, antibiotics are described to possibly reduce clinical signs, mortality and faecal shedding of S. Choleraesuis (Jacks et al. 1981). Aminoglycosides, amoxicillin-clavulanic acid, fluoroquinolones, trimethoprim-sulphonamides and others are reported to be effective in vitro. In contrast, increasing resistance in more than 80% of isolates analysed could be seen for ampicillin, chloramphenicol, streptomycin, sulphonamides and tetracyclines, thus displaying the ACSSuT phenotype (Sisak et al. 2006; Teng et al. 2022). Resistance was more often detected in *Salmonella* spp. isolated from nursery than grow-finishing pigs or sows. Therefore, antimicrobial susceptibility test is always recommended (Rosengren et al. 2008). Anti-inflammatory drugs such as flunixin or meloxicam can be used to combat fever. Since stressors like mixing of pigs, transportation, occurrence of other disease, high pig density, antimicrobial therapy or food/water deprivation can increase shedding of the pathogen, such stressors must be avoided as much as possible (Chalias et al. 2022). In addition to avoiding these stressors, other management measures can contribute to the reduction of *Salmonella* spp. within the herd. These include, for example, the usage of fermented liquid feeding or mash/meal feed, instead of pelleted feed, increasing microbial digestion of starch and addition of bacteriophages or organic acids (e.g. formic or lactic acid of 0.6%) (Larsen et al. 1993; Lebel et al. 2017; Zimmermann et al. 2019). Further, hygiene measures including cleaning and disinfection, all-in/all-out, uni-directed animal flow, no mixing of age categories, reduced stocking density, slatted floor, rodent control (Box 6) and separation of pigs and wild/other domestic animals are important to prevent the spread of *Salmonella* spp. (Fosse et al. 2009; EFSA 2011; Casanova-Higes et al. 2016). In some countries, commercial vaccines against *Salmonella* spp. are licensed. Live attenuated and inactivated vaccines have been developed, with the former being excreted after vaccination, thus warranting time between vaccination and slaughter. Furthermore, no antimicrobials in the days before and after vaccination with an attenuated live vaccine should be used (de la Cruz et al. 2017; van der Wolf et al. 2021).

Such measure, which consists of the prevention of the entry of the infection to the herd, prevention of survival and spread within the herd and reduction of infection, is known as pre-harvest control and is particularly important in food safety (World-Health-Organization 2002). To further limit the risk of *Salmonella*-associated food poisoning in humans, lairage at the slaughterhouse should be as short as possible and contamination and cross-contamination of the carcasses with faeces have to be prevented (Morgan et al. 1987; Hurd et al. 2001; Swanenburg et al. 2001; Wales and Davies 2017). No differentiation can be made between vaccinated and naturally infected animals.

## 3.4 *Trichuris suis*

### 3.4.1 Aetiology and Life Cycle

The whipworm *Trichuris suis* is one of the two nematodes described in the large intestine of pigs. For its life cycle, *Trichuris suis* inhabits mainly the caecum and colon of pigs without the need for an intermediate host. In rare cases, *Trichuris suis* may infect and reproduce in humans as well (Turck et al. 2019). Affected animals pass non-infectious lemon-shaped eggs via faeces into the environment, where they develop to the infective $L_1$ stage within 3–8 weeks, depending on environmental conditions. The infective $L_1$ stage is highly resistant and can remain viable for several years. Once ingested by a pig, the larvae hatch and penetrate the mucosa in the distal ileum, caecum and colon. Subsequently, they undergo four moults causing tissue damage, mucosal oedema, haemorrhages and ulceration. During 4–5 weeks, they reach the adult stage $L_5$. The adult worms reach a size of up to 8 cm in length, with 2/3 typically stitched into the intestinal mucosa and do not cause further lesions. Sporadic shedding of eggs starts about 6–8 weeks after infection with a rapid decrease within 4–5 weeks. Six to eight eggs per gram faeces per adult female at the peak of egg production have been described. The lifespan of adult worms is 4–5 months (Beer 1973; Murrell 1986; Pedersen and Saeed 2001; Kringel and Roepstorff 2006; Pittman et al. 2010).

### 3.4.2 Clinical Signs

Under experimental conditions, continuous re-infection over 4 weeks using up to 4200 infective eggs or single infections using 5000 infective eggs caused infection and shedding in weaner pigs with depressed growth rate being the only clinical sign observed. Massively affected pigs may show bloody and mucoid diarrhoea, anorexia, anaemia, dehydration, poorer feed conversion, reduced daily weight gains and rarely rectal prolapse. Affected pigs, typically weaner or fattener, recover within few weeks, while mortality is low. Acquired immunity with nearly sterile immunity after previous infection is described and could explain the reduced number of adults being affected (Pedersen and Saeed 2000, 2001; Kringel and Roepstorff 2006; Nejsum et al. 2009; Pittman et al. 2010).

### 3.4.3 Lesions

Macroscopically, watery to bloody mucoid faeces may be observed in the large intestine. In severely infected animals, ulceration of the mucosa, mucosal oedema and haemorrhages may be seen. At close inspection, the anterior part of the adult worms in the mucosa may be visible. Microscopically, typhlitis and colitis with inflammatory nodules around the adult worms as well as larvae and/or adults may be seen within the mucosa (Pittman et al. 2010; Zimmermann et al. 2019).

### 3.4.4 Diagnosis

Flotation from native rectal faecal samples is the diagnostic tool of choice in life animals (Box 2–4). Due to the long prepatent period and relatively low and periodic shedding, multiple samplings from several animals over time are recommended. False-negative results may occur due to periodic shedding, low egg production and some pigs being worm positive but do not shed at all. Eggs present a very typical lemon-shaped form and are easy to distinguish from others. At necropsy, adult whipworms may be seen but are easily overlooked. Mucosal scrapings or histology may be used for diagnosis as well (Pedersen and Saeed 2001; Nejsum et al. 2009; Pittman et al. 2010). Antibodies can be detected from 3–4 and peaking at about 6–8 weeks post inoculation. Serum antibody levels remain elevated for at least 6 weeks longer (Hill et al. 1997; Pedersen and Saeed 2001; Nejsum et al. 2009). A list of differential diagnoses for infectious causes of diarrhoea by age category can be found in Table 2.

### 3.4.5 Therapy and Prevention

Oral treatment with organophosphate compounds or benzimidazoles (e.g. fenbendazole or flubendazole) can control the infection with *Trichuris suis* in pigs. However, the varying responsiveness of larval and adult stages with the high resistance of $L_1$ stage in the environment over several years makes eradication in herds very difficult and nearly impossible in pasture. Cleaning and disinfecting using a chlorocresol in combination with an all-in/all-out management can reduce the number of worms in the herd (Thienpont et al. 1982; Rossignol and Maisonneuve 1984; Zimmermann et al. 2019). An increase in severity and shortening of pre-patency period in pigs with iron deficiency seem to occur (Pedersen et al. 2001). Further, a negative impact of low protein, low energy or high non-fermentable carbohydrate diets (low inulin) on the pigs' ability to resist an *Trichuris suis* infection is discussed (Pedersen et al. 2002; Krag et al. 2006).

**Box 1 Three Major Pathogenic Mechanisms for Diarrhoea**

Hypersecretion: Secretion by the mucosa exceeds the absorptive capacity of the intestine. An example is colibacillosis, which causes hypersecretion due to toxins. Diarrhoea is characterised by being isotonic, alkaline and rich in electrolytes.

- Malabsorption: Often due to the destruction of villous' absorptive cells. The osmotically active material in the intestine thus causes fluid to flow into the intestine causing diarrhoea. Fermentation causes an acid pH. An example is an infection with rota- or coronaviruses as well as Cystoisospora suis infection.
- Increased intestinal permeability: Inflammatory or necrotising lesions in the intestine can result in the passage of fluid exudates into the lumen and may cause a reduction in the absorptive capacity of the intestine. An example is a C. perfringens Type C or Strongyloides infection.

**Box 2 How to Select Animals for Diagnostics in Infectious Disease**

1. Clinical signs: Exclusively sample animals that show clinical signs typical for the problem, e.g. diarrhoea, fever, etc.
2. Acute: Sample animals with acute or subacute disease. Pathogen isolation out of animals chronically diseased is less often successful.
3. No pre-treatment: Sample animals not pre-treated with antimicrobials (antibiotics, zinc oxide). Previous antimicrobial treatments reduce the likelihood of detecting bacterial pathogens as the cause of disease.

**Box 3 General Remarks on Faecal Sampling and Sampling Material**

- Take samples exclusively from the rectum of the animals and not from the floor.
- If you want to perform a polymerase chain reaction (PCR), take native faeces or swabs without medium. Some media may inhibit the PCR.
- Native faeces have the advantage over swabs that repeated testing is possible due to the larger amount of material.
- If you want to perform culture for anaerobic bacteria like Brachyspira spp. or Clostridia spp., take faecal swabs with medium or tubes fully filled with faeces (without air) to ensure survival of the pathogen. This does not apply to PCR, as the pathogen does not need to be alive.

**Box 4 Appropriate Sample Size for the Detection of a Pathogen**

It often takes several samples to identify a pathogen as the cause of a disease. However, as laboratory tests cost money, a minimum number of samples should be used to obtain the maximum reliable result.

The following should be taken into account regarding the sample size:

- At least three samples should always be taken! Like this, random findings, not associate with the cause of the disease, can be interpreted correctly.
- Pooling of faecal samples is not recommended. Faeces are very inhomogeneous and therefore difficult to mix. Pooling can lead to dilution, which is why the pathogen may no longer be detectable.

If you want to calculate the sample size needed, there is an easy rule of thumb. It is called the "rule of three":

$$3\,/\,\text{expected prevalence} * 100 = \text{Number of samples needed}$$

Example: 20% of 100 piglets suffer from diarrhoea

$$3\,/\,20 * 100 = 15$$

Interpretation: Randomly, 15 out of the 100 piglets need to be sampled, to detect the pathogen. By selecting only animals with clinical signs, the expected prevalence among the sampled animals can be increased, thus reducing the number of samples needed.

**Box 5 Supportive Measurements in Case of Diarrhoea**

Animals may lose a lot of fluid due to diarrhoea. Ensure adequate water and electrolyte intake by the following:

- In suckling pigs: Control if sow has enough teats per piglet and let them drink (e.g. no post-partum dysgalactia syndrome).
- In weaner/fattener: Control the number of animals per drinker and appropriate flow rate (max 12 animals in dry feed per drinker).
- Compensate electrolyte and fluid loss by additional offering of oral rehydration solution, e.g. with conventional solutions, conventional sugar-containing drinks, self-mixed solutions using 5 g salt, 50 g glucose to 1 L warm water or conventional ones.
  - Do not give oral glucose in case of malabsorptive diarrhoea.
- In severe cases of dehydration in suckling piglets, administration of 5–20% glucose intraperitoneally may be indicated.
- Animals use a lot of energy to maintain their body temperature; increase environmental temperature (in nest area of suckling piglets up to 35 °C in their first week of life), e.g. by covering the nest area, more bedding and heating lamps.
- Provide a draft-free, dry, and clean environment to rest for sick animals.

**Box 6 Rodent Control**

Specialised companies should carry out rodent control; nevertheless, there are additional things that the farmer can also consider, to make the farm less attractive to rodents:

- Eliminate loose materials (machinery, old feed bags, etc.) that the rodent can hide, or build its nest, within a 30 m radius of the building.
- Remove or cut long grass regularly within a 30 m radius of the building.
- Make access to food and water sources more difficult to rodents.
  - Mice can go through openings 6–8 mm in diameter.
- Dispose of animal carcasses regularly and seal them in tight containers.

**Box 7 General Remark on Oral Treatment of Pigs**

Oral treatment in pigs is done via feed or water. The following should be taken into account:

- Sick animals reduce the amount of food they consume. This must be taken into account when dosing.
- Food consumption is reduced before water consumption. Consequently, animals eating less may still drink normally.
- Very sick animals also reduce their water intake. Once again, this must be taken into account when dosing.

Consequently, parenteral treatment of animal with reduced general state and thus assumed reduced feed or water intake is recommended.

**Box 8 How to Do a Mucosal Smear**

The intestinal mucosa is scraped with a scalpel using enough pressure to dislodge villi. The scraping should be prepared as a smear on a glass microscope slide for microscopic examination or put into a vessel for further investigation such as polymerase chain reaction.

## References


Abraham S, Trott DJ, Jordan D, Gordon DM, Groves MD, Fairbrother JM, Smith MG, Zhang R, Chapman TA (2014) Phylogenetic and molecular insights into the evolution of multidrug-resistant porcine enterotoxigenic Escherichia coli in Australia. Int J Antimicrob Agents 44(2):105–111

Alarcón LV, Alberto AA, Mateu E (2021) Biosecurity in pig farms: a review. Porcine Health Manag 7. https://doi.org/10.1186/S40813-020-00181-Z

Alban L, Barfod K, Petersen JV, Dahl J, Ajufo JC, Sandø G, Krog HH, Aabo S (2010) Description of extended pre-harvest pig salmonella surveillance-and-control programme and its estimated effect on food safety related to pork. Zoonoses Public Health 6:1. https://doi.org/10.1111/j.1863-2378.2010.01367.x

Aliaga-Leyton A, Webster E, Friendship R, Dewey C, Vilaça K, Peregrine AS (2011) An observational study on the prevalence and impact of Isospora suis in suckling piglets in southwestern Ontario, and risk factors for shedding oocysts. Can Vet J 52(2):184

Allo M, Silva J, Fekety R, Rifkin GD, Waskin H (1979) Prevention of clindamycin-induced colitis in hamsters by Clostridium sordellii antitoxin. Gastroenterology 76(2):351–355

Alvarez-Ordóez A, Martínez-Lobo F, Arguello H, Carvajal A, Rubio P (2013) Swine dysentery: aetiology, pathogenicity, determinants of transmission and the fight against the disease. Int J Environ Res Public Health 10:1927–1947. https://doi.org/10.3390/ijerph10051927

Amimo JO, Junga JO, Ogara WO, Vlasova AN, Njahira MN, Maina S, Okoth EA, Bishop RP, Saif LJ, Djikeng A (2015) Detection and genetic characterization of porcine group A rotaviruses in asymptomatic pigs in smallholder farms in East Africa: predominance of P [8] genotype resembling human strains. Vet Microbiol 175(2–4):195–210

Andrews JR, Ryan ET (2015) Diagnostics for invasive Salmonella infections: current challenges and future directions. Vaccine 33:C8–C15. https://doi.org/10.1016/j.vaccine.2015.02.030

Angelakis E (2017) Weight gain by gut microbiota manipulation in productive animals. Microb Pathog 106:162–170

Armstrong D, Bishop SC (2004) Does genetics or litter effect influence mortality in PMWS. Mortality 13:9

Arnold M, Crienen A, Swam H, Von Berg S, Jolie R, Nathues H (2019) Prevalence of Lawsonia intracellularis in pig herds in different European countries. Porcine Health Manag 5. https://doi.org/10.1186/s40813-019-0137-6

Arnold M, Crienen A, Swam H, von Berg S, Jolie R, Nathues H (2021) Correlation of Lawsonia intracellularis positivity in quantitative PCR and herd factors in European pig herds. Porcine Health Manag. 7:13. https://doi.org/10.1186/s40813-021-00192-4

Arnold M, Schmitt S, Collaud A, Rossano A, Hübschke E, Zeeh F, Nathues H, Perreten V (2022) Distribution, genetic heterogeneity, and antimicrobial susceptibility of Brachyspira pilosicoli in Swiss pig herds. Vet Microbiol 269:109421. https://doi.org/10.1016/J.VETMIC.2022.109421

Atyeo RF, Oxberry SL, Hampson DJ (1999) Analysis of Serpulina hyodysenteriae strain variation and its molecular epidemiology using pulsed-field gel electrophoresis. Epidemiol Infect 123(1):133–138

Autheman D, Wyder M, Popoff M, D'Herde K, Christen S, Posthaus H (2013) Clostridium perfringens beta-toxin induces necrostatin-inhibitable, calpain-dependent necrosis in primary porcine endothelial cells (in eng). PLoS One 8(5):e64644

Awad MM, Johanesen PA, Carter GP, Rose E, Lyras D (2014) Clostridium difficile virulence factors: insights into an anaerobic spore-forming pathogen (in eng). Gut Microbes 5(5):579–593

Backhans A, Johansson K-E, Fellström C (2010) Phenotypic and molecular characterization of Brachyspira spp. isolated from wild rodents. Environ Microbiol Rep 2:720–727. https://doi.org/10.1111/j.1758-2229.2010.00165.x

Bae JK, Wieland B, Sait M, Longbottom D, Smith DGE, Alarcon P, Wheelhouse N (2013) Risk factors associated with Lawsonia intracellularis in English pig farms. Vet J 197:707–711. https://doi.org/10.1016/j.tvjl.2013.03.020

Baloda SB, Christensen L, Trajcevska S (2001) Persistence of a Salmonella enterica Serovar Typhimurium DT12 Clone in a Piggery and in Agricultural Soil Amended with Salmonella-Contaminated Slurry. Appl Environ Microbiol 67:2859–2862. https://doi.org/10.1128/AEM.67.6.2859-2862.2001

Baró J, Segalés J, Martínez J (2015) Porcine circovirus type 2 (PCV2) enteric disease: an independent condition or part of the systemic disease? (in eng). Vet Microbiol 176(1–2):83–87

Beer RJS (1973) Morphological descriptions of the egg and larval stages of Trichuris suis Schrank, 1788. Parasitology 67:263–278. https://doi.org/10.1017/S0031182000046503

Belsham GJ, Rasmussen TB, Normann P, Vaclavek P, Strandbygaard B, Bøtner A (2016) Characterization of a novel chimeric swine enteric coronavirus from diseased pigs in Central Eastern Europe in 2016 (in eng). Transbound Emerg Dis 63(6):595–601

Benfield DA, Stotz I, Moore R, McADARAGH JP (1982) Shedding of rotavirus in feces of sows before and after farrowing. J Clin Microbiol 16(1):186–190

Bergeland ME, Dermody TA, Sorensen DK (eds) (1966) Porcine enteritis due to Clostridium perfringens type CI Epizootiology and diagnosis

Bernad-Roche M, Casanova-Higes A, Marín-Alcalá CM, Cebollada-Solanas A, Mainar-Jaime RC (2021) Salmonella infection in nursery piglets and its role in the spread of salmonellosis to further production periods. Pathogens 10(2):123. https://doi.org/10.3390/pathogens10020123

Bjustrom-Kraft J, Woodard K, Giménez-Lirola L, Setness B, Ji J, Lasley P, Nelson E, Zhang J, Baum D, Gauger P, Main R, Zimmerman J (2018) Serum and mammary secretion antibody responses in porcine epidemic diarrhea-immune gilts following porcine epidemic diarrhea vaccination. J Swine Health Prod 26(1):34–40

Bondo KJ, Pearl DL, Janecko N, Boerlin P, Reid-Smith RJ, Parmley J, Jardine CM (2016) Impact of season, demographic and environmental factors on Salmonella occurrence in raccoons (procyon lotor) from swine farms and conservation areas in Southern Ontario. PLoS One 11:161497. https://doi.org/10.1371/journal.pone.0161497

Boniotti MB, Papetti A, Lavazza A, Alborali G, Sozzi E, Chiapponi C, Faccini S, Bonilauri P, Cordioli P, Marthaler D (2016) Porcine epidemic Diarrhea virus and discovery of a recombinant swine enteric coronavirus, Italy (in eng). Emerg Infect Dis 22(1):83–87

Boutrup TS, Boesen HT, Boye M, Agerholm JS, Jensen TK (2010) Early pathogenesis in porcine proliferative enteropathy caused by Lawsonia intracellularis. J Comp Pathol 143:101–109. https://doi.org/10.1016/j.jcpa.2010.01.006

Boye M, Baloda SB, Leser TD, Møller K (2001) Survival of Brachyspira hyodysenteriae and B. pilosicoli in terrestrial microcosms. Vet Microbiol 81:33–40. https://doi.org/10.1016/S0378-1135(01)00328-5

Brand P, Gobeli S, Perreten V (2017) Pathotyping and antibiotic resistance of porcine enterovirulent Escherichia coli strains from Switzerland (2014–2015) (in eng). Schweizer Archiv fur Tierheilkunde 159(7):373–380

Brandt D (2008) Untersuchungen zum klinischen Verlauf einer Lawsonia-intracellularis-Infektion bei Schweinen. Tierärztliche Hochschule Hannover

van den Broeck W, Cox E, Goddeeris BM (1999) Receptor-dependent immune responses in pigs after oral immunization with F4 fimbriae (in eng). Infect Immun 67(2):520–526

Bronsvoort M, Norby B, Bane DP, Gardner IA (2001) Management factors associated with seropositivity to Lawsonia intracellularis in US swine herds. J Swine Health Prod 9:285–290

Brunborg IM, Moldal T, Jonassen CM (2004) Quantitation of porcine circovirus type 2 isolated from serum/plasma and tissue samples of healthy pigs and pigs with postweaning multisystemic wasting syndrome using a TaqMan-based real-time PCR. J Virol Methods 122(2):171–178

Bueschel DM, Helen Jost B, Billington SJ, Trinh HT, Glenn Songer J (2003) Prevalence of cpb2, encoding beta2 toxin, in Clostridium perfringens field isolates: correlation of genotype with phenotype (in eng). Vet Microbiol 94(2):121–129

Burnham C-AD, Carroll KC (2013) Diagnosis of Clostridium difficile infection: an ongoing conundrum for clinicians and for clinical laboratories (in eng). Clin Microbiol Rev 26(3):604–630

Burrough ER (2017) Swine dysentery: etiopathogenesis and diagnosis of a reemerging disease. Vet Pathol 54:22–31. https://doi.org/10.1177/0300985816653795

Burrough ER, Strait EL, Kinyon JM, Bower LP, Madson DM, Wilberts BL, Schwartz KJ, Frana TS, Songer JG (2012) Comparative virulence of clinical Brachyspira spp. isolates in inoculated pigs. J Vet Diagn Invest 24:1025–1034. https://doi.org/10.1177/1040638712457927

Burrough ER, Rotolo ML, Gauger PC, Madson DM, Schwartz KJ (2015) Correlation of Lawsonia intracellularis semi-quantitative fecal polymerase chain reaction assay results with the presence of histologic lesions of proliferative enteropathy and positive immunohistochemical staining. J Swine Health Prod 23(4):204–207

Bywater RJ, Woode GN (1980) Oral fluid replacement by a glucose glycine electrolyte formulation in E. coli and rotavirus diarrhoea in pigs. Vet Rec 106(4):75–78

Cadetg RSS, Vidondo B, Nathues H, Schüpbach G, Zeeh F (2019) Retrospective study on the eradication of Swine Dysentery (Brachyspira hyodysenteriae) in Switzerland. Schweiz Arch Tierheilkd 161:217–230. https://doi.org/10.17236/sat00202

Campillo M, Smith SH, Gally DL, Opriessnig T (2021) Review of methods for the detection of Lawsonia intracellularis infection in pigs. J Vet Diagn Invest 33(4):621–631. https://doi.org/10.1177/10406387211003551

Card RM, Stubberfield E, Rogers J, Nunez-Garcia J, Ellis RJ, AbuOun M, Strugnell B, Teale C, Williamson S, Anjum MF (2018) Identification of a new antimicrobial resistance gene provides fresh insights into pleuromutilin resistance in Brachyspira hyodysenteriae, aetiological agent of swine dysentery. Front Microbiol 9:1183. https://doi.org/10.3389/fmicb.2018.01183

Card RM, La T, Burrough ER, Ellis RJ, Nunez-Garcia J, Thomson JR, Mahu M, Phillips ND, Hampson DJ, Rohde J, Tucker AW (2019) Weakly haemolytic variants of Brachyspira hyodysenteriae newly emerged in Europe belong to a distinct subclade with unique genetic properties. Vet Res 50:21. https://doi.org/10.1186/s13567-019-0639-x

Casanova-Higes A, Andr Es-Barranco S, Mainar-Jaime RC (2016) Influence of On-farm pig Salmonella status on Salmonella Shedding at Slaughter. https://doi.org/10.1111/zph.12301

Casey TA, Bosworth BT (2009) Design and evaluation of a multiplex polymerase chain reaction assay for the simultaneous identification of genes for nine different virulence factors associated with Escherichia coli that cause diarrhea and edema disease in swine. J Vet Diagn Investig 21(1):25–30

Cepeda-Molero M, Berger CN, Walsham ADS, Ellis SJ, Wemyss-Holden S, Schüller S, Frankel G, Fernández LÁ (2017) Attaching and effacing (A/E) lesion formation by enteropathogenic E. coli on human intestinal mucosa is dependent on non-LEE effectors (in eng). PLoS Pathog 13(10):e1006706

Chalias A, Grispoldi L, Cenci Goga B (2022) A risk assessment model for Salmonella spp. in swine carcasses. EFSA J 20:e200405. https://doi.org/10.2903/J.EFSA.2022.E200405

Chan G, Farzan A, Soltes G, Nicholson VM, Pei Y, Friendship R, Prescott JF (2012) The epidemiology of Clostridium perfringens type A on Ontario swine farms, with special reference to cpb2-positive isolates (in eng). BMC Vet Res 8:156

Chander Y, Primus A, Oliveira S, Gebhart CJ (2012) Phenotypic and molecular characterization of a novel strongly hemolytic Brachyspira species, provisionally designated "Brachyspira hampsonii". J Vet Diagn Invest 24:903–910. https://doi.org/10.1177/1040638712456975

Chia SP, Taylor DJ (1978) Factors affecting the survival of Treponema hyodysenteriae in dysenteric pig faeces. Vet Rec 103:68–70. https://doi.org/10.1136/VR.103.4.68

Collins A (2013) Advances in ileitis control, diagnosis, epidemiology and the economic impacts of disease in commercial pig herds. Agriculture 3:536–555. https://doi.org/10.3390/agriculture3030536

Collins AM, Barchia IM (2014) The critical threshold of Lawsonia intracellularis in pig faeces that causes reduced average daily weight gains in experimentally challenged pigs. Vet Microbiol 168:455–458. https://doi.org/10.1016/j.vetmic.2013.12.003

Collins AM, Love RJ (2007) Re-challenge of pigs following recovery from proliferative enteropathy. Vet Microbiol 120:381–386. https://doi.org/10.1016/J.VETMIC.2006.11.004

Collins JE, Bergeland ME, Bouley D, Ducommun AL, Francis DH, Yeske P (1989) Diarrhea associated with Clostridium perfringens type A enterotoxin in neonatal pigs. J Vet Diagn Investig 1(4):351–353

Collins A, Love RJ, Pozo J, Smith SH, McOrist S, Ecvp D (2000) Studies on the ex vivo survival of Lawsonia intracellularis. Swine Health and Production

Collins AM, Fell S, Pearson H, Toribio J-A (2011) Colonisation and shedding of Lawsonia intracellularis in experimentally inoculated rodents and in wild rodents on pig farms. Vet Microbiol 150:384–388. https://doi.org/10.1016/j.vetmic.2011.01.020

Costa MO, Hill JE, Fernando C, Lemieux HD, Detmer SE, Rubin JE, Harding JCS (2014) Confirmation that "Brachyspira hampsonii" clade I (Canadian strain 30599) causes mucohemorrhagic diarrhea and colitis in experimentally infected pigs. BMC Vet Res 10:1–12. https://doi.org/10.1186/1746-6148-10-129/FIGURES/6

Cowardin CA, Buonomo EL, Saleh MM, Wilson MG, Burgess SL, Kuehne SA, Schwan C, Eichhoff AM, Koch-Nolte F, Lyras D, Aktories K, Minton NP, Petri WA (2016) The binary toxin CDT enhances Clostridium difficile virulence by suppressing protective colonic eosinophilia (in eng). Nat Microbiol 1(8):16108

de la Cruz ML, Conrado I, Nault A, Perez A, Dominguez L, Alvarez J (2017) Vaccination as a control strategy against Salmonella infection in pigs: a systematic review and meta-analysis of the literature. Res Vet Sci 114:86–94. https://doi.org/10.1016/J.RVSC.2017.03.005

Cutler R, Gleeson B, Page S, Norris J, Browning G (2020) Antimicrobial prescribing guidelines for pigs. Aust Vet J 98(4):105–134

CVMP (n.d.) Opinions on veterinary medicinal products. European Medicines Agency. http://www.ema.europa.eu/ (accessed 11.10.2022)

Davies RH, Wray C (1996) Seasonal Variations in the Isolation of Salmonella typhimurium, Salmonella enteritidis, Bacillus cereus and Clostridium perfringens from Environmental Samples. J Veterinary Med Ser B 43:119–127. https://doi.org/10.1111/J.1439-0450.1996.TB00295.X

De Lucia A, Cawthraw S, Davies R, Smith RP, Bianco C, Ostanello F, Martelli F (2020) Correlation of anti-Salmonella antibodies between serum and saliva samples collected from finisher pigs. Front Vet Sci 6:489. https://doi.org/10.3389/FVETS.2019.00489/BIBTEX

De Lucia A, Cawthraw SA, Smith RP, Davies R, Bianco C, Ostanello F, Martelli F (2021) Pilot investigation of anti-salmonella antibodies in oral fluids from salmonella typhimurium vaccinated and unvaccinated swine herds. Animals 11. https://doi.org/10.3390/ani11082408

DebRoy C, Maddox CW (2001) Identification of virulence attributes of gastrointestinal Escherichia coli isolates of veterinary significance (in eng). Anim Health Res Rev 2(2):129–140

Decaro N, Lorusso A (2020) Novel human coronavirus (SARS-CoV-2): a lesson from animal coronaviruses. Vet Microbiol 244:108693

Dee S, Neill C, Singrey A, Clement T, Cochrane R, Jones C, Patterson G, Spronk G, Christopher-Hennings J, Nelson E (2016) Modeling the transboundary risk of feed ingredients contaminated with porcine epidemic diarrhea virus (in eng). BMC Vet Res 12:51

Dezorzova-Tomanova K, Smola J, Trcka I, Lamka J, Pavlik I (2006) Detection of Lawsonia intracellularis in wild boar and fallow deer bred in one game enclosure in the Czech Republic. J Vet Med Ser B Infect Dis Vet Public Heal 53:42–44. https://doi.org/10.1111/j.1439-0450.2006.00909.x

Diab SS, Uzal FA, Songer JG (2016a) Diseases produced by Clostridium difficile. Clostridial Dis Anim 177

Diab SS, Uzal FA, Songer JG, Prescott JF, Popoff (2016b) Diseases produced by Clostridium perfringens type C. Clostridial Dis Anim 143155

Diarra AT, Achacha M, Mittal KR (1995) Evaluation of different serological tests for detection of antibodies against Serpulina hyodysenteriae in pig sera. Comp Immunol Microbiol Infect Dis 18:215–221. https://doi.org/10.1016/0147-9571(95)00002-P

Diego R, Lanza I, Carvajal A, Rubio P, Cármenes P (1995) Serpulina hyodysenteriae challenge of fattening pigs vaccinated with an adjuvanted bivalent bacterin against swine dysentery. Vaccine 13:663–667. https://doi.org/10.1016/0264-410X(94)00018-I

Diel DG, Lawson S, Okda F, Singrey A, Clement T, Fernandes M, Christopher-Hennings J, Nelson EA (2016) Porcine epidemic diarrhea virus: an overview of current virological and serological diagnostic methods. Virus Res 226:60–70

Dors A, Pomorska-Mól M, Czyżewska E, Wasyl D, Pejsak Z (2015) Prevalence and risk factors for Lawsonia intracellularis, Brachyspira hyodysenteriae and Salmonella spp. in finishing pigs in Polish farrow-to-finish swine herds. Pol J Vet Sci 18:825–831. https://doi.org/10.1515/pjvs-2015-0107

Dubreuil JD (2021) Pig vaccination strategies based on enterotoxigenic Escherichia coli toxins (in eng). Braz J Microbiol [publication of the Brazilian Society for Microbiology] 52(4):2499–2509

Dubreuil JD, Isaacson RE, Schifferli DM (2016) Animal Enterotoxigenic Escherichia coli (in eng). EcoSal Plus 7(1)

Duhamel GE (2001) Comparative pathology and pathogenesis of naturally acquired and experimentally induced colonic spirochetosis. Anim Health Res Rev 2:3–18. https://doi.org/10.1079/ahrr200128

Dupont K, Nielsen EO, Baekbo P, Larsen L (2008) Genomic analysis of PCV2 isolates from Danish archives and a current PMWS case–control study supports a shift in genotypes with time. Vet Microbiol 128(1–2):56–64

Durmic Z, Pethick DW, Mullan BP, Accioly JM, Schulze H, Hampson DJ (2002) Evaluation of large-intestinal parameters associated with dietary treatments designed to reduce the occurrence of swine dysentery. Br J Nutr 88:159–169. https://doi.org/10.1079/bjnbjn2002607

EFSA (2009) Analysis of the baseline survey on the prevalence of Salmonella in holdings with breeding pigs, in the EU, 2008, Part A Salmonella prevalence estimates, EFSA Journal. https://doi.org/10.2903/j.efsa.2009.1377

EFSA (2011) Analysis of the baseline survey on the prevalence of Salmonella in holdings with breeding pigs, in the EU, 2008 – Part B: Analysis of factors associated with Salmonella pen positivity, EFSA Journal. Wiley-Blackwell. https://doi.org/10.2903/J.EFSA.2011.2329

Ellin Doyle M, Kaspar C, Archer J, Klos R (2009) White paper on human illness caused by salmonella from all food and non-food vectors

Estes MK, Kang G, Zeng CQ, Crawford SE, Ciarlet M (2001) Pathogenesis of rotavirus gastroenteritis (in eng). Novartis Found Symp 238:82–96. discussion 96–100

Fairbrother JM, Nadeau É (2019) Colibacillosis. In: Zimmerman JJ (ed) Diseases of swine, 11th edn. Wiley-Blackwell, Hoboken, NJ, pp 807–834

Fairbrother JM, Nadeau E, Gyles CL (2005) Escherichia coli in postweaning diarrhea in pigs: an update on bacterial types, pathogenesis, and prevention strategies (in eng). Anim Health Res Rev 6(1):17–39

Fairbrother JM, Nadeau É, Bélanger L, Tremblay C-L, Tremblay D, Brunelle M, Wolf R, Hellmann K, Hidalgo Á (2017) Immunogenicity and protective efficacy of a single-dose live non-pathogenic Escherichia coli oral vaccine against F4-positive enterotoxigenic Escherichia coli challenge in pigs (in eng). Vaccine 35(2):353–360

Farzan A, Kircanski J, DeLay J, Soltes G, Songer JG, Friendship R, Prescott JF (2013) An investigation into the association between cpb2-encoding Clostridium perfringens type A and diarrhea in neonatal piglets (in eng). Can J Vet Res 77(1):45–53

Fedorka-Cray PJ, Whipp SC, Isaacson RE, Nord N, Lager K (1994) Transmission of Salmonella typhimurium to swine. Vet Microbiol 41:333–344. https://doi.org/10.1016/0378-1135(94)90029-9

Fedorka-Cray PJ, Kelley LC, Stabel TJ, Gray JT, Laufer JA (1995) Alternate routes of invasion may affect pathogenesis of Salmonella typhimurium in swine.

Infect Immun 63:2658–2664. https://doi.org/10.1128/IAI.63.7.2658-2664.1995

Fedorka-Cray PJ, Gray JT, Wray C (2000) Salmonella infections in pigs. In: Salmonella in domestic animals, pp 191–208. https://doi.org/10.1111/j.1751-0813.2000.tb11907.x

Feix AS, Cruz-Bustos T, Ruttkowski B, Joachim A (2020) Characterization of Cystoisospora suis sexual stages in vitro (in eng). Parasit Vectors 13(1):143

Fenner FJ (1987) Veterinary virology (in eng). Elsevier Science, Burlington, p 1156

Finlayson M, Barnum DA (1973) The effect of chlortetracycline feed additive on experimental salmonella infection of swine and antibiotic resistance transfer. Can J Comp Med 37:139–146

FINRES-Vet (2019) Finnish veterinary antimicrobial resistance monitoring and consumption of antimicrobial agents, Finnish Food Authority, Helsinki, Finland

Flores PS, Costa FB, Amorim AR, Mendes GS, Rojas M, Santos N (2021) Rotavirus A, C, and H in Brazilian pigs: potential for zoonotic transmission of RVA. J Vet Diagn Investig 33(1):129–135

Fongaro G, Padilha J, Schissi CD, Nascimento MA, Bampi GB, Viancelli A, Barardi CR (2015) Human and animal enteric virus in groundwater from deep wells, and recreational and network water. Environ Sci Pollut Res 22(24):20060–20066

Fosse J, Seegers H, Magras C (2009) Prevalence and risk factors for bacterial food-borne zoonotic hazards in slaughter pigs: a review. Zoonoses Public Health 56:429–454. https://doi.org/10.1111/j.1863-2378.2008.01185.x

Francis DH (2002) Enterotoxigenic Escherichia coli infection in pigs and its diagnosis. J Swine Health Prod 10(4):171–175

Franzo G, Segalés J (2020) Porcine circovirus 2 genotypes, immunity and vaccines: multiple genotypes but one single serotype. Pathogens 9:1–12. https://doi.org/10.3390/pathogens9121049

Frydendahl K (2002) Prevalence of serogroups and virulence genes in Escherichia coli associated with postweaning diarrhoea and edema disease in pigs and a comparison of diagnostic approaches. Vet Microbiol 85(2):169–182

Fu ZF, Hampson DJ (1989) Natural transmission of group A rotavirus within a pig population. Res Vet Sci 46(3):312–317

Gabardo M d P, Sato JPH, Daniel AG d S, Andrade MR, Pereira CER, Rezende TP, Otoni LVA, Rezende LA, Guedes RMC (2017) Evaluation of the involvement of mice (Mus musculus) in the epidemiology of porcine proliferative enteropathy. Vet Microbiol 205:75–79. https://doi.org/10.1016/J.VETMIC.2017.05.009

Gabbay YB, Borges AA, Oliveira DS, Linhares AC, Mascarenhas JDP, Barardi CRM, Simões CMO, Wang Y, Glass RI, Jiang B (2008) Evidence for zoonotic transmission of group C rotaviruses among children in Belém, Brazil. J Med Virol 80(9):1666–1674

García-Martín AB, Perreten V, Rossano A, Schmitt S, Nathues H, Zeeh F (2018) Predominance of a macrolide-lincosamide-resistant Brachyspira hyodysenteriae of sequence type 196 in Swiss pig herds. Vet Microbiol 226:97–102. https://doi.org/10.1016/j.vetmic.2018.10.007

Gerjets I, Traulsen I, Reiners K, Kemper N (2011) Comparison of virulence gene profiles of Escherichia coli isolates from sows with coliform mastitis and healthy sows (in eng). Vet Microbiol 152(3–4):361–367

Girard C, Lemarchand T, Higgins R (1995) Porcine colonic spirochetosis: a retrospective study of eleven cases. Can Vet J

Gogolewski RP, Cook RW, Batterham ES (1991) Suboptimal growth associated with porcine intestinal adenomatosis in pigs in nutritional studies. Aust Vet J 68:406–408. https://doi.org/10.1111/j.1751-0813.1991.tb03114.x

Grau-Roma L, Fraile L, Segalés J (2011) Recent advances in the epidemiology, diagnosis and control of diseases caused by porcine circovirus type 2. Vet J 187(1):23–32

Gray JT, Fedorka-Cray PJ, Stabel TJ, Ackermann MR (1995) Influence of inoculation route on the carrier state of Salmonella choleraesuis in swine. Vet Microbiol 47:43–59. https://doi.org/10.1016/0378-1135(95)00060-N

Gray JT, Fedorka-Cray PJ, Stabel TJ, Kramer TT (1996) Natural transmission of Salmonella choleraesuis in swine. Appl Environ Microbiol 62:141–146. https://doi.org/10.1128/aem.62.1.141-146.1996

Grześkowiak Ł, Zentek J, Vahjen W (2016) Physical pretreatment improves efficient DNA extraction and qPCR sensitivity from Clostridium difficile spores in faecal swine specimens. Curr Microbiol 73(5):727–731

Grześkowiak ŁM, Pieper R, Huynh HA, Cutting SM, Vahjen W, Zentek J (2019) Impact of early-life events on the susceptibility to Clostridium difficile colonisation and infection in the offspring of the pig. Gut Microbes 10(2):251–259

Guedes RMC, Gebhart CJ (2003) Onset and duration of fecal shedding, cell-mediated and humoral immune responses in pigs after challenge with a pathogenic isolate or attenuated vaccine strain of Lawsonia intracellularis. Vet Microbiol 91:135–145. https://doi.org/10.1016/S0378-1135(02)00301-2

Guedes RMC, Gebhart CJ, Armbruster GA, Roggow BD (2002) Serologic follow-up of a repopulated swine herd after an outbreak of proliferative hemorrhagic enteropathy. Can J Vet Res 66:258–263

Guedes RM, Winkelman NL, Gebhart CJ (2003) Relationship between the severity of porcine proliferative enteropathy and the infectious dose of Lawsonia intracellularis. Vet Rec 153:432–433

Guedes RMC, Machuca MA, Quiroga MA, Pereira CER, Resende TP, Gebhart CJ (2017) Lawsonia intracellularis in Pigs: progression of lesions and involvement of apoptosis. Vet Pathol 54(4):620–628. https://doi.org/10.1177/0300985817698206

Guo R, Fan B, Chang X, Zhou J, Zhao Y, Shi D, Yu Z, He K, Li B (2020) Characterization and evaluation of the pathogenicity of a natural recombinant transmissible gastroenteritis virus in China. Virology 545:24–32

Gurtner, C., Popescu, F., Wyder, M., Sutter, E., Zeeh, F., Frey, J., Schubert, C. von, Posthaus, H., 2010. Rapid cytopathic effects of Clostridium perfringens beta-toxin on porcine endothelial cells. Infect Immun 78 (7), 2966–2973

Gutzmann F, Layton H, Simkins K, Jarolmen H (1976) Influence of antibiotic-supplemented feed on occurrence and persistence of Salmonella typhimurium in experimentally infected swine - PubMed. Am J Vet Res 37:649–655

Hampson DJ (2018) The spirochete brachyspira pilosicoli, enteric pathogen of animals and humans. Clin Microbiol Rev 31. https://doi.org/10.1128/CMR.00087-17

Hampson D, La T, Phillips ND (2015) Emergence of Brachyspira species and strains: reinforcing the need for surveillance. Porcine Health Manag https://doi.org/10.1186/s40813-015-0002-1

Hampson DJ, Burrough ER, Zimmermann J, Karriker L, Ramirez A, Schwartz K, Stevenson G, Zhang J (2019) Swine dysentery and brachyspiral colitis. In: Diseases of swine, 11th ed. pp 951–970

Hansen CF, Hernández A, Mansfield J, Hidalgo Á, La T, Phillips ND, Hampson DJ, Pluske JR (2011) A high dietary concentration of inulin is necessary to reduce the incidence of swine dysentery in pigs experimentally challenged with Brachyspira hyodysenteriae. Br J Nutr 106:1506–1513. https://doi.org/10.1017/S000711451100208X

Harleman JH, Meyer RC (1984) Life cycle of Isospora suis in gnotobiotic and conventionalized piglets. Vet Parasitol 17(1):27–39

Harleman JH, Meyer RC (1985) Pathogenicity of Isospora suis in gnotobiotic and conventionalised piglets. Vet Rec 116(21):561–565

Helm ET, Lin SJ, Gabler NK, Burrough ER (2020) Brachyspira hyodysenteriae infection reduces digestive function but not intestinal integrity in growing pigs while disease onset can be mitigated by reducing insoluble fiber. Front Vet Sci 7:587926. https://doi.org/10.3389/fvets.2020.587926

Helm ET, Gabler NK, Burrough ER (2021) Highly fermentable fiber alters fecal microbiota and mitigates swine dysentery induced by brachyspira hyodysenteriae. Animals 11:1–20. https://doi.org/10.3390/ani11020396

Hill DE, Romanowski RD, Urban JF (1997) A Trichuris specific diagnostic antigen from culture fluids of Trichuris suis adult worms. Vet Parasitol 68:91–102. https://doi.org/10.1016/S0304-4017(96)01055-2

Høgh P (1967) Necrotizing infectious enteritis in piglets, caused by clostridium perfringens type C: II. Incidence and clinical features. Acta Vet Scand 8(4):301

Høgh P (1969) Necrotizing infectious enteritis in piglets, caused by Clostridium perfringens type C. IV Bacteriological diagnosis. Acta Vet Scand 10(1):84–100

Holyoake PK, Emery D, Gonsalves J, Donahoo M, Collins A (2010) Prevalence of antibodies to Lawsonia intracellularis in pig herds in Australia. Aust Vet J 88:186–188. https://doi.org/10.1111/j.1751-0813.2010.00558.x

Homwong N, Diaz A, Rossow S, Ciarlet M, Marthaler D (2016) Three-level mixed-effects logistic regression analysis reveals complex epidemiology of swine rotaviruses in diagnostic samples from North America. PLoS One 11(5):e0154734

Hurd HS, Gailey JK, McKean JD, Rostagno MH (2001) Rapid infection in market-weight swine following exposure to a Salmonella Typhimurium-contaminated environment H. AJVR 62:1194–1197

Husnik R, Klimes J, Tomanova K, Smola J, Halouzka R, Tichy F, Brazdil J (2003) Lawsonia intracellularis in a dog with inflammatory bowel disease. Vet Med Czech

Hwang J-M, Seo M-J, Yeh J-Y (2017) Lawsonia intracellularis in the feces of wild rodents and stray cats captured around equine farms. BMC Vet Res 13:233. https://doi.org/10.1186/s12917-017-1155-8

Jacks TM, Welter CJ, Fitzgerald GR, Miller BM (1981) Cephamycin C treatment of induced swine salmonellosis. Antimicrob Agents Chemother 19:562–566. https://doi.org/10.1128/AAC.19.4.562

Jacobs AAC, Harks F, Pauwels R, Cao Q, Holtslag H, Pel S, Segers RPAM (2020) Efficacy of a novel intradermal Lawsonia intracellularis vaccine in pigs against experimental infection and under field conditions. Porcine Health Manag 6:25. https://doi.org/10.1186/s40813-020-00164-0

Jacobson M, Hard Af Segerstad C, Gunnarsson A, Fellström C, De Verdier Klingenberg K, Wallgren P, Jensen-Waern M (2003) Diarrhoea in the growing pig – a comparison of clinical, morphological and microbial findings between animals from good and poor performance herds. Res Vet Sci 74:163–169. https://doi.org/10.1016/S0034-5288(02)00187-X

Jacobson M, Råsbäck T, Flöistrup H, Benz M, Braun-Fahrländer C, Riedler J, Schram-Bijkerk D, Fellström C (2007) Survey on the occurrence of Brachyspira species and Lawsonia intracellularis in children living on pig farms. Epidemiol Infect 135:1043–1045. https://doi.org/10.1017/S0950268806007606

Jäggi M, Wollschlager N, Abril C, Albini S, Brachelente C, Wyder M, Posthaus H (2009) Retrospective study on necrotizing enteritis in piglets in Switzerland. Schweiz Arch Tierheilkd 151:369–375

Jamalludeen N, Johnson RP, Shewen PE, Gyles CL (2009) Evaluation of bacteriophages for prevention and treatment of diarrhea due to experimental enterotoxigenic Escherichia coli O149 infection of pigs. Vet Microbiol 136(1–2):135–141

Jansson DS, Johansson KE, Olofsson T, Råsbäck T, Vågsholm I, Pettersson B, Gunnarsson A, Fellström C (2004) Brachyspira hyodysenteriae and other strongly β-haemolytic and indole-positive spirochaetes isolated from mallards (Anas platyrhynchos). J Med Microbiol Microbiol Soc:293–300. https://doi.org/10.1099/jmm.0.05488-0

Jensen TK, Møller K, Leser TD, Jorsal SE (1997) Comparison of histology, immunohistochemistry and polymerase chain reaction for detection of Lawsonia

intracellularis in natural proliferative enteropathy. Eur J Vet Pathol 3:115–123
Jensen TK, Vigre H, Sørensen V, Møller K (2005) Naturally acquired Lawsonia intracellularis infection in pigs studied from weaning to slaughter by indirect immunofluorescence antibody test and polymerase chain reaction on faeces. Res Vet Sci 79:93–98. https://doi.org/10.1016/j.rvsc.2004.08.001
Jensen AN, Dalsgaard A, Stockmarr A, Nielsen EM, Baggesen DL (2006) Survival and transmission of Salmonella enterica serovar typhimurium in an outdoor organic pig farming environment. Appl Environ Microbiol 72:1833–1842. https://doi.org/10.1128/AEM.72.3.1833-1842.2006
Jiang F, Wu Z, Zheng Y, Frana TS, Sahin O, Zhang Q, Li G (2019) Genotypes and antimicrobial susceptibility profiles of hemolytic Escherichia coli from diarrheic piglets. Foodborne Pathog Dis 16(2):94–103
Joachim A, Schwarz L, Hinney B, Ruttkowski B, Vogl C, Mundt H-C (2014) Which factors influence the outcome of experimental infection with Cystoisospora suis? Parasitol Res 113(5):1863–1873
Joachim A, Ruttkowski B, Sperling D (2018) Detection of Cystoisospora suis in faeces of suckling piglets – when and how? A comparison of methods. Porcine Health Manag 4(1):20
Joachim A, Guerra N, Hinney B, Hodžić A, Karembe H, Shrestha A, Sperling D (2019) Efficacy of injectable toltrazuril-iron combination product and oral toltrazuril against early experimental infection of suckling piglets with Cystoisospora suis (in eng). Parasit Vectors 12(1):272
Joens LA, Harris DL, Baum DH (1979) Immunity to Swine dysentery in recovered pigs. Am J Vet Res 40:1352–1354
Joerling J, Willems H, Ewers C, Herbst W (2020) Differential expression of hemolysin genes in weakly and strongly hemolytic Brachyspira hyodysenteriae strains. BMC Vet Res 16. https://doi.org/10.1186/s12917-020-02385-5
Johansen M, Nielsen M, Dahl J, Svensmark B, Bækbo P, Kristensen CS, Hjulsager CK, Jensen TK, Ståhl M, Larsen LE, Angen Ø (2013) Investigation of the association of growth rate in grower-finishing pigs with the quantification of Lawsonia intracellularis and porcine circovirus type 2. Prev Vet Med 108:63–72. https://doi.org/10.1016/j.prevetmed.2012.07.004
Johnson AM, Kaushik RS, Francis DH, Fleckenstein JM, Hardwidge PR (2009) Heat-labile enterotoxin promotes Escherichia coli adherence to intestinal epithelial cells. J Bacteriol 191(1):178–186
Jones FT (2011) A review of practical Salmonella control measures in animal feed. J Appl Poult Res 20:102–113. https://doi.org/10.3382/japr.2010-00281
Jones FT, Langlois BE, Cromwell GL, Hays VW (1983) Effect of feeding chlortetracycline or virginiamycin on shedding of salmonellae from experimentally-infected swine. J Anim Sci 57:279–285. https://doi.org/10.2527/jas1983.572279x
Jung K, Alekseev KP, Zhang X, Cheon D-S, Vlasova AN, Saif LJ (2007) Altered pathogenesis of porcine respiratory coronavirus in pigs due to immunosuppressive effects of dexamethasone: implications for corticosteroid use in treatment of severe acute respiratory syndrome coronavirus. J Virol 81(24):13681–13693
Jung K, Annamalai T, Lu Z, Saif LJ (2015) Comparative pathogenesis of US porcine epidemic diarrhea virus (PEDV) strain PC21A in conventional 9-day-old nursing piglets vs. 26-day-old weaned pigs (in eng). Vet Microbiol 178(1–2):31–40
Jung K, Miyazaki A, Saif LJ (2018) Immunohistochemical detection of the vomiting-inducing monoamine neurotransmitter serotonin and enterochromaffin cells in the intestines of conventional or gnotobiotic (Gn) pigs infected with porcine epidemic diarrhea virus (PEDV) and serum cytokine responses of Gn pigs to acute PEDV infection. Res Vet Sci 119:99–108
Jung K, Saif LJ, Wang Q (2020) Porcine epidemic diarrhea virus (PEDV): an update on etiology, transmission, pathogenesis, and prevention and control (in eng). Virus Res 286:198045
Karlsson M, Fellström C, Johansson K-E, Franklin A (2004) Antimicrobial resistance in Brachyspira pilosicoli with special reference to point mutations in the 23S rRNA gene associated with macrolide and lincosamide resistance. Microb Drug Resist. https://doi.org/10.1089/mdr.2004.10.204
Keel MK, Songer JG (2006) The comparative pathology of Clostridium difficile-associated disease. Vet Pathol 43(3):225–240
Keessen EC, Hensgens MP, Spigaglia P, Barbanti F, Sanders IM, Kuijper EJ, Lipman LJ (2013) Antimicrobial susceptibility profiles of human and piglet Clostridium difficile PCR-ribotype 078 (in eng). Antimicrob Resist Infect Control 2:14
Kempf I, Fleury MA, Drider D, Bruneau M, Sanders P, Chauvin C, Madec J-Y, Jouy E (2013) What do we know about resistance to colistin in Enterobacteriaceae in avian and pig production in Europe? Int J Antimicrob Agents 42(5):379–383
Kennedy KK, Norris SJ, Beckenhauer WH, Hogg A (1977) Vaccination of pregnant sows with Clostridium perfringens type C toxoid. Vet Med Small Anim Clin 72(6):1047–1049
Kim J, Ha Y, Jung K, Choi C, Chae C (2004) Enteritis associated with porcine circovirus 2 in pigs. Can J Vet Res 68(3):218
Kim AH, Hogarty MP, Harris VC, Baldridge MT (2020) The complex interactions between rotavirus and the gut microbiota (in eng). Front Cell Infect Microbiol 10:586751
Kink JA, Williams JA (1998) Antibodies to recombinant Clostridium difficile toxins A and B are an effective treatment and prevent relapse of C. difficile-associated disease in a hamster model of infection. Infect Immun 66(5):2018–2025
Knight DR, Squire MM, Riley TV (2014) Laboratory detection of Clostridium difficile in piglets in Australia (in eng). J Clin Microbiol 52(11):3856–3862

Kraeling RR, Webel SK (2015) Current strategies for reproductive management of gilts and sows in North America. J Anim Sci Biotechnol 6(1):1–14

Krag L, Thomsen LE, Iburg T (2006) Pathology of Trichuris suis infection in pigs fed an inulin-and a non-inulin-containing diet. J Vet Med 53:405–409

Kringel H, Roepstorff A (2006) Trichuris suis population dynamics following a primary experimental infection. Vet Parasitol 139:132–139. https://doi.org/10.1016/j.vetpar.2006.03.002

Kroll JJ, Roof MB, McOrist S (2004) Evaluation of protective immunity in pigs following oral administration of an avirulent live vaccine of Lawsonia intracellularis. Am J Vet Res 65:559–565. https://doi.org/10.2460/ajvr.2004.65.559

Kroll JJ, Roof MB, Hoffman LJ, Dickson JS, Harris DLH (2005) Proliferative enteropathy: a global enteric disease of pigs caused by Lawsonia intracellularis. Anim Sci Publ https://doi.org/10.1079/AHR2005109

Kumar D, Shepherd FK, Springer NL, Mwangi W, Marthaler DG (2022) Rotavirus infection in swine: genotypic diversity, immune responses, and role of gut microbiome in rotavirus immunity. Pathogens 11(10):1078

La T, Phillips ND, Reichel MP, Hampson DJ (2004) Protection of pigs from swine dysentery by vaccination with recombinant BmpB, a 29.7 kDa outer-membrane lipoprotein of Brachyspira hyodysenteriae. Vet Microbiol 102:97–109. https://doi.org/10.1016/j.vetmic.2004.06.004

Ladinig A, Sommerfeld-Stur I, Weissenböck H (2009) Comparative evaluation of diagnostic methods for Lawsonia intracellularis infection in pigs, with emphasis on cases lacking characteristic lesions. J Comp Pathol 140:140–148. https://doi.org/10.1016/j.jcpa.2008.09.011

Laird TJ, Abraham S, Jordan D, Pluske JR, Hampson DJ, Trott DJ, O'Dea M (2021) Porcine enterotoxigenic Escherichia coli: antimicrobial resistance and development of microbial-based alternative control strategies (in eng). Vet Microbiol 258:109117

Landsverk T (1973) Intestinal adenomatosis in a blue fox (Alopex lagopus). Equine Vet J 565

Langel SN, Paim FC, Lager KM, Vlasova AN, Saif LJ (2016) Lactogenic immunity and vaccines for porcine epidemic diarrhea virus (PEDV): historical and current concepts. Virus Res 226:93–107

Larsen GJ, Rolow AM, Nelson CE (1993) Research note: the effect of organic acids on Salmonella contamination originating from mouse fecal pellets, Poultry science. Poultry Science Association Inc. https://doi.org/10.3382/ps.0721797

Le Dividich J, Herpin P (1994) Effects of climatic conditions on the performance, metabolism and health status of weaned piglets: a review. Livest Prod Sci 38(2):79–90

Lebel P, Letellier A, Longpré J, Laplante B, Yergeau E, Fravalo P (2017) Feed presentation options in Swine early fattening mitigates Salmonella shedding and specifically modulates the faecal microbiota. J Appl Microbiol 122:30–39. https://doi.org/10.1111/JAM.13305

Lecce JG, King MW (1978) Role of rotavirus (reo-like) in weanling diarrhea of pigs. J Clin Microbiol 8(4):454–458

Lindecrona RH, Jensen TK, Jensen BB, Leser TD, Jiufeng W, Møller K (2003) The influence of diet on the development of swine dysentery upon experimental infection. Anim Sci 76:81–87. https://doi.org/10.1017/S1357729800053340

Lindsay DS, Stuart BP, Wheat BE, Ernst JV (1980) Endogenous development of the swine coccidium, Isospora suis Biester 1934. J Parasitol:771–779

Lindsay DS, Dubey JP, Santín-Durán M (2019) Coccidia and other protozoa. In: Zimmerman JJ (ed) Diseases of swine, 11th edn. Wiley-Blackwell, Hoboken, NJ, pp 1015–1027

Link EK, Eddicks M, Nan L, Ritzmann M, Sutter G, Fux R (2021) Discriminating the eight genotypes of the porcine circovirus type 2 with TaqMan-based real-time PCR. Virol J 18(1):1–12

Liu G, Jiang Y, Opriessnig T, Gu K, Zhang H, Yang Z (2019) Detection and differentiation of five diarrhea related pig viruses utilizing a multiplex PCR assay. J Virol Methods 263:32–37

Lomax LG, Glock RD, Hogan JE (1982) Experimentally induced porcine proliferative enteritis in specific-pathogen-free pigs. Am J Vet Res 43:1615–1621

Lőrincz M, Cságola A, Biksi I, Szeredi L, Dán Á, Tuboly T (2010) Detection of porcine circovirus in rodents. Acta Vet Hung 58(2):265–268

Luise D, Bertocchi M, Motta V, Salvarani C, Bosi P, Luppi A, Fanelli F, Mazzoni M, Archetti I, Maiorano G (2019a) Bacillus sp. probiotic supplementation diminish the Escherichia coli F4ac infection in susceptible weaned pigs by influencing the intestinal immune response, intestinal microbiota and blood metabolomics. J Anim Sci Biotechnol 10(1):1–16

Luise D, Lauridsen C, Bosi P, Trevisi P (2019b) Methodology and application of Escherichia coli F4 and F18 encoding infection models in post-weaning pigs. J Anim Sci Biotechnol 10(1):53

Luppi A (2017) Swine enteric colibacillosis: diagnosis, therapy and antimicrobial resistance. Porcine Health Manag 3(1):16

Luppi A, Bonilauri P, Dottori M, Gherpelli Y, Biasi G, Merialdi G, Maioli G, Martelli P (2015) Antimicrobial resistance of F4+ Escherichia coli isolated from swine in Italy. Transbound Emerg Dis 62(1):67–71

Luppi A, Gibellini M, Gin T, Vangroenweghe F, Vandenbroucke V, Bauerfeind R, Bonilauri P, Labarque G, Hidalgo Á (2016) Prevalence of virulence factors in enterotoxigenic Escherichia coli isolated from pigs with post-weaning diarrhoea in Europe. Porcine Health Manag 2(1):1–6

Lyon SA, Hutton ML, Rood JI, Cheung JK, Lyras D (2016) CdtR regulates TcdA and TcdB production in Clostridium difficile. PLoS Pathog 12(7):e1005758

Macintyre N, Smith DGE, Shaw DJ, Thomson JR, Rhind SM (2003) Immunopathogenesis of experimentally induced proliferative enteropathy in pigs. Vet Pathol

Madec F, Eveno E, Morvan P, Hamon L, Albina E, Truong C, Hutet E, Cariolet R, Arnauld C, Croix L, Jestin A (1999) La Maladie de l ' Amaigrissement du Porcelet ( MAP ) en France 1 – Aspects descriptifs , impact en élevage. Journées Rech Porc en Fr 31:347–354

Madec F, Eveno E, Morvan P, Hamon L, Blanchard P, Cariolet R, Amenna N, Morvan H, Truong C, Mahé D, Albina E, Jestin A (2000) Post-weaning multisystemic wasting syndrome (PMWS) in pigs in France: clinical observations from follow-up studies on affected farms. Livest Prod Sci 63:223–233. https://doi.org/10.1016/S0301-6226(99)00138-4

Madec F, Rose N, Grasland B, Cariolet R, Jestin A (2008) Post-weaning multisystemic wasting syndrome and other PCV2-related problems in pigs: a 12-year experience. Transbound Emerg Dis 55(7):273–283

Mahu M, De Jong E, De Pauw N, Maele LV, Vandenbroucke V, Vandersmissen T, Miry C, Pasmans F, Haesebrouck F, Martel A, Boyen F, Maele LV (2014) First isolation of "Brachyspira hampsonii" from pigs in Europe. Vet Rec 174. https://doi.org/10.1136/vr.101868

Mahu M, De Pauw N, Maele LV, Verlinden M, Boyen F, Ducatelle R, Haesebrouck F, Martel A, Pasmans F (2016) Variation in hemolytic activity of Brachyspira hyodysenteriae strains from pigs. Vet Res 47:66. https://doi.org/10.1186/s13567-016-0353-x

Mahu M, Boyen F, Canessa S, Zavala Marchan J, Haesebrouck F, Martel A, Pasmans F (2017) An avirulent Brachyspira hyodysenteriae strain elicits intestinal IgA and slows down spread of swine dysentery. Vet Res 48:59. https://doi.org/10.1186/s13567-017-0465-y

Malik YS, Kumar N, Sharma K, Sircar S, Dhama K, Bora DP, Dutta TK, Prasad M, Tiwari AK (2014) Rotavirus diarrhea in piglets: a review on epidemiology, genetic diversity and zoonotic risks. Indian J Anim Sci 84(10):1035–1042

Markowska-Daniel I, Winiarczyk S, Gradzki Z, Pejsak Z (1996) Evaluation of different methods (ELISA, IF, EM, PAGE) for the diagnosis of rotavirus infection in piglets. Comp Immunol Microbiol Infect Dis 19(3):219–232

Marsteller TA, Abvp D, Armbruster G, Bane DP, Gebhart CJ, Muller R, Weatherford J, Thacker B (2003) Monitoring the prevalence of Lawsonia intracellularis IgG antibodies using serial sampling in growing and breeding swine herds. J Swine Health Prod 1111:127127–127130

Martella V, Bányai K, Matthijnssens J, Buonavoglia C, Ciarlet M (2010) Zoonotic aspects of rotaviruses. Vet Microbiol 140(3–4):246–255

Marthaler D, Rossow K, Culhane M, Collins J, Goyal S, Ciarlet M, Matthijnssens J (2013) Identification, phylogenetic analysis and classification of porcine group C rotavirus VP7 sequences from the United States and Canada. Virology 446(1–2):189–198

Marthaler D, Homwong N, Rossow K, Culhane M, Goyal S, Collins J, Matthijnssens J, Ciarlet M (2014) Rapid detection and high occurrence of porcine rotavirus A, B, and C by RT-qPCR in diagnostic samples. J Virol Methods 209:30–34

Massacci FR, De Luca S, Cucco L, Tentellini M, Perreten V, Pezzotti G, Magistrali CF (2018) Multiresistant Brachyspira hyodysenteriae shedding by pigs during the fattening period. Vet Rec 183:264. https://doi.org/10.1136/VR.104886

McOrist M, Lawson GH (1989) Reproduction of proliferative enteritis in gnotobiotic pigs. Res Vet Sci 46:27–33

McOrist S, Jasni S, Mackie RA, Macintyre N, Neef N, Lawson GHK (1993) Reproduction of porcine proliferative enteropathy with pure cultures of ileal symbiont intracellularis. Infect Immun 61(10):4286–4292

McOrist S, Smith SH, Shearn MF, Carr MM, Miller DJ (1996) Treatment and prevention of porcine proliferative enteropathy with oral tiamulin. Vet Rec 139:651–658

Mcorist S, Smith SH, Klein T (1999) Monitored control programme for proliferative enteropathy on British pig farms the effect of control programmes on proliferative enteropathy and its causative agent (Lawsonia intracellularis) was investigated on four farrow-to-finish pig farms in Britain. Vet Rec 144:202–204. https://doi.org/10.1136/vr.144.8.202

Melkebeek V, Goddeeris BM, Cox E (2013) ETEC vaccination in pigs. Vet Immunol Immunopathol 152(1–2):37–42

Memon AM, Chen F, Khan SB, Guo X, Khan R, Khan FA, Zhu Y, He Q (2022) Development and evaluation of polyclonal antibodies based antigen capture ELISA for detection of porcine rotavirus. Anim Biotechnol:1–8

Mengel H, Kruger M, Kruger MU, Westphal B, Swidsinski A, Schwarz S, Mundt H-C, Dittmar K, Daugschies A (2012) Necrotic enteritis due to simultaneous infection with Isospora suis and clostridia in newborn piglets and its prevention by early treatment with toltrazuril. Parasitol Res 110(4):1347–1355

Meyerholz DK, Stabel TJ (2003) Comparison of early ileal invasion by Salmonella enterica serovars choleraesuis and typhimurium. Vet Pathol 40(4):371–375

Mølbak L, Johnsen K, Boye M, Jensen TK, Johansen M, Møller K, Leser TD (2008) The microbiota of pigs influenced by diet texture and severity of Lawsonia intracellularis infection. Vet Microbiol 128:96–107. https://doi.org/10.1016/J.VETMIC.2007.09.012

Møller K, Ahrens P (1996) Comparison of toxicity neutralization-, ELISA-and PCR tests for typing of clostridium perfringens and detection of the enterotoxin gene by PCR. Anaerobe 2(2):103–110

Moon HW, Kemeny LJ, Lambert G, Stark SL, Booth GD (1975) Age-dependent resistance to transmissible gastroenteritis of Swine: III. Effects of epithelial cell kinetics on coronavirus production and on atrophy of intestinal Villi. Vet Pathol 12(5–6):434–445

Morgan IR, Krautil FL, Craven JA (1987) Effect of time in lairage on caecal and carcass salmonella contamina-

tion of slaughter pigs. Epidem Inf 98:323–330. https://doi.org/10.1017/S0950268800062075
Möstl K (1990) Coronaviridae, pathogenetic and clinical aspects: an update (in eng). Comp Immunol Microbiol Infect Dis 13(4):169–180
Murrell KD (1986) Epidemiology, pathogenesis, and control of major swine helminth parasites. Vet Clin North Am Food Anim Pract 2:439–454. https://doi.org/10.1016/S0749-0720(15)31255-X
Mushtaq M, Zubair S, Råsbäck T, Bongcam-Rudloff E, Jansson DS (2015) Brachyspira suanatina sp. nov., an enteropathogenic intestinal spirochaete isolated from pigs and mallards: genomic and phenotypic characteristics. BMC Microbiol 15:1–13. https://doi.org/10.1186/S12866-015-0537-Y/FIGURES/4
National-animal-health-monitoring-System (2009) Salmonella on U.S. Swine sites-prevalence and antimicrobial susceptibility
Neirynck W, Boyen F, Chantziaras I, Vandersmissen T, Vyt P, Haesebrouck F, Dewulf J, Maes D (2020) Implementation and evaluation of different eradication strategies for Brachyspira hyodysenteriae. Porcine Health Manag 6:27. https://doi.org/10.1186/s40813-020-00162-2
Nejsum P, Thamsborg SM, Petersen HH, Kringel H, Fredholm M, Roepstorff A (2009) Population dynamics of Trichuris suis in trickle-infected pigs. Parasitology 136:691–679. https://doi.org/10.1017/S0031182009005976
Nemoto M, Bannai H, Tsujimura K, Yamanaka T, Kondo T (2014) Virucidal effect of commercially available disinfectants on equine group A rotavirus. J Vet Med Sci 76(7):1061–1063
Neog BK, Barman NN, Bora DP, Dey SC, Chakraborty A (2011) Experimental infection of pigs with group A rotavirus and enterotoxigenic Escherichia coli in India: gross, histopathological and immunopathological study. Vet Ital 47(2):117–128
Niederwerder MC, Hesse RA (2018) Swine enteric coronavirus disease: a review of 4 years with porcine epidemic diarrhoea virus and porcine deltacoronavirus in the United States and Canada. Transbound Emerg Dis 65(3):660–675
Niilo L (1988) Clostridium perfringens type C enterotoxemia. Can Vet J 29(8):658
Ntakiyisumba E, Lee S, Won G (2022) Evidence-based approaches for determining effective target antigens to develop vaccines against post-weaning diarrhea caused by enterotoxigenic Escherichia coli in pigs: a systematic review and network meta-analysis. Animals 12(16):2136
Ohnuma Y, Kondo H, Saino H, Taguchi M, Ohno A, Matsuda T (1992) Necrotic enteritis due to Clostridium perfringens type C in newborn piglets. J Jpn Vet Med Assoc (Japan)
Olson LD (1974) Clinical and pathological observations on the experimental passage of swine dysentery. Can J Comp Med Rev Can Med Comp 38:7–13
Opriessnig T, Langohr I (2013) Current state of knowledge on porcine circovirus type 2-associated lesions (in eng). Vet Pathol 50(1):23–38
Opriessnig T, Fenaux M, Thomas P, Hoogland MJ, Rothschild MF, Meng XJ, Halbur PG (2006) Evidence of breed-dependent differences in susceptibility to porcine circovirus type-2-associated disease and lesions. Vet Pathol 43(3):281–293
Opriessnig T, Meng X-J, Halbur PG (2007) Porcine circovirus type 2–associated disease: update on current terminology, clinical manifestations, pathogenesis, diagnosis, and intervention strategies. J Vet Diagn Investig 19(6):591–615
Opriessnig T, Xiao C-T, Halbur PG, Gerber PF, Matzinger SR, Meng X-J (2017) A commercial porcine circovirus (PCV) type 2a-based vaccine reduces PCV2d viremia and shedding and prevents PCV2d transmission to naive pigs under experimental conditions. Vaccine 35(2):248–254
Otto PH, Rosenhain S, Elschner MC, Hotzel H, Machnowska P, Trojnar E, Hoffmann K, Johne R (2015) Detection of rotavirus species A, B and C in domestic mammalian animals with diarrhoea and genotyping of bovine species A rotavirus strains. Vet Microbiol 179(3–4):168–176
Park S, Lee J-B, Kim K-J, Oh Y-S, Kim M-O, Oh Y-R, Hwang M-A, Lee J-A, Lee S-W (2013) Efficacy of a commercial live attenuated Lawsonia intracellularis vaccine in a large scale field trial in Korea. Clin Exp Vaccine Res 2:135. https://doi.org/10.7774/CEVR.2013.2.2.135
Patterson AH, Rubin JE, Fernando C, Costa MO, Harding JCS, Hill JE (2013) Fecal shedding of Brachyspira spp. on a farrow-to-finish swine farm with a clinical history of "Brachyspira hampsonii"-associated colitis. BMC Vet Res 9:1–10. https://doi.org/10.1186/1746-6148-9-137/FIGURES/4
Pedersen S, Saeed I (2000) Experimental infection of pigs with three dose levels of Trichuris suis. Parasite 7:275–281. https://doi.org/10.1051/parasite/2000074275
Pedersen S, Saeed I (2001) Acquired immunity to Trichuris suis infection in pigs. Parasitology 123:95–101. https://doi.org/10.1017/S0031182001007934
Pedersen S, Saeed I, Friis H, Michaelsen KF (2001) Effect of iron deficiency on Trichuris suis and Ascaris suum infections in pigs. Parasitology 122:589–598. https://doi.org/10.1017/S0031182001007582
Pedersen S, Saeed I, Michaelsen KF, Friis H, Murrell KD (2002) Impact of protein energy malnutrition on Trichuris suis infection in pigs concomitantly infected with Ascaris suum. Parasitology 124:561–568. https://doi.org/10.1017/S0031182002001592
Pedersen KS, Skrubel R, Stege H, Angen Ø, Ståhl M, Hjulsager C, Larsen LE, Nielsen JP (2012a) Association between average daily gain, faecal dry matter content and concentration of Lawsonia intracellularis in faeces. Acta Vet Scand 54:58. https://doi.org/10.1186/1751-0147-54-58
Pedersen KS, Ståhl M, Maurício R, Guedes C, Angen Ø, Nielsen JP, Jensen TK (2012b) Association between

faecal load of lawsonia intracellularis and pathological findings of proliferative enteropathy in pigs with diarrhoea. BMC Vet Res 8:198. https://doi.org/10.1186/1746-6148-8-198

Pensaert MB, Martelli P (2016) Porcine epidemic diarrhea: a retrospect from Europe and matters of debate. Virus Res 226:1–6

Pensaert M, Callebaut P, Vergote J (1986) Isolation of a porcine respiratory, non-enteric coronavirus related to transmissible gastroenteritis. Vet Q 8(3):257–261

Pettersson E, Hestad S, Möttus I, Skiöldebrand E, Wallgren P (2019) Rotavirus and Cystoisospora suis in piglets during the suckling and early post weaning period, in systems with solid floors and age segregated rearing. Porcine Health Manag 5(1):7

Pires AFA, Funk JA, Bolin C (2014) Risk factors associated with persistence of Salmonella shedding in finishing pigs. Prev Vet Med 116:120–128. https://doi.org/10.1016/j.prevetmed.2014.06.009

Pittman JS, Shepherd G, Thacker BJ, Myers GH (2010) Trichuris suis in finishing pigs: case report and review. J Swine Health Prod 18:306–3013

Pluske JR, Siba PM, Pethick DW, Durmic Z, Mullah BP, Hampson DJ (1996) The incidence of swine dysentery in pigs can be reduced by feeding diets that limit the amount of fermentable substrate entering the large intestine. Am Inst Nutr

Pommier P, Keita A, Pagot E, Duran O, Cloet PR (2008) Comparison of Tylvalosin with Tylosin for the control of subclinical ileitis in swine. Rev Med Vet (Toulouse) 159:579–582

Posthaus H, Kittl S, Tarek B, Bruggisser J (2020) Clostridium perfringens type C necrotic enteritis in pigs: diagnosis, pathogenesis, and prevention (in eng). J Vet Diagn Invest 32(2):203–212

Pringle M, Poehlsgaard J, Vester B, Long KS (2004) Mutations in ribosomal protein L3 and 23S ribosomal RNA at the peptidyl transferase centre are associated with reduced susceptibility to tiamulin in Brachyspira spp. isolates. Mol Microbiol 54:40. https://doi.org/10.1111/j.1365-2958.2004.04373.x

Proctor A, Cornick NA, Wang C, Mooyottu S, Arruda PA, Kobs K, Phillips GJ (2021) Neonatal piglets are protected from clostridioides difficile infection by age-dependent increase in intestinal microbial diversity (in eng). Microbiol Spectr 9(2):e0124321

Proux K, Cariolet R, Fravalo P, Houdayer C, Keranflech A, Madec F (2001) Contamination of pigs by nose-to-nose contact or airborne transmission of Salmonella Typhimurium. Vet Res 32:591–600. https://doi.org/10.1051/vetres:2001148

Ramirez A, Zimmermann JJ, Karriker L, Schwartz K, Stevenson G, Zhang J (2019) Differential diagnosis of diseases. In: Diseases of swine 11th ed, pp 59–74

Råsbäck T, Fellström C, Gunnarsson A, Aspán A (2006) Comparison of culture and biochemical tests with PCR for detection of Brachyspira hyodysenteriae and Brachyspira pilosicoli. J Microbiol Methods 66:347–353. https://doi.org/10.1016/j.mimet.2005.12.008

Ravi M, Ngeleka M, Kim S-H, Gyles C, Berthiaume F, Mourez M, Middleton D, Simko E (2007) Contribution of AIDA-I to the pathogenicity of a porcine diarrheagenic Escherichia coli and to intestinal colonization through biofilm formation in pigs. Vet Microbiol 120(3–4):308–319

Reed WM, Olander HJ, Thacker HL (1985) Studies on the pathogenesis of Salmonella heidelberg infection in weanling pigs. Am J Vet Res 46:2300–2310

Resende TP, Pereira CER, Gabardo MdP, Haddad JPA, Lobato ZIP, Guedes RMC (2015) Serological profile, seroprevalence and risk factors related to Lawsonia intracellularis infection in swine herds from Minas Gerais State, Brazil. BMC Vet Res 11:1–6. https://doi.org/10.1186/s12917-015-0618-z

Rhouma M, Fairbrother JM, Beaudry F, Letellier A (2017) Post weaning diarrhea in pigs: risk factors and non-colistin-based control strategies. Acta Vet Scand 59(1):1–19

Riber U, Cordes H, Boutrup TS, Jensen TK, Heegaard PMH, Jungersen G (2011) Primary infection protects pigs against re-infection with Lawsonia intracellularis in experimental challenge studies. Vet Microbiol 149:406–414. https://doi.org/10.1016/j.vetmic.2010.11.028

Riber U, Heegaard PMH, Cordes H, Ståhl M, Jensen TK, Jungersen G (2014) Vaccination of pigs with attenuated Lawsonia intracellularis induced acute phase protein responses and primed cell-mediated immunity without reduction in bacterial shedding after challenge. Vaccine 33:156–162. https://doi.org/10.1016/j.vaccine.2014.10.084

Richard OK, Grahofer A, Nathues H, Posthaus H (2019a) Vaccination against Clostridium perfringens type C enteritis in pigs: a field study using an adapted vaccination scheme (in eng). Porcine Health Manag 5:20

Richard OK, Springer S, Finzel J, Theuß T, Wyder M, Vidondo B, Posthaus H (2019b) Application of an endothelial cell culture assay for the detection of neutralizing anti-clostridium perfringens beta-toxin antibodies in a porcine vaccination Trial (in eng). Toxins 11(4):225

Roerink F, Morgan CL, Knetter SM, Passat MH, Archibald AL, Ait-Ali T, Strait EL (2018) A novel inactivated vaccine against Lawsonia intracellularis induces rapid induction of humoral immunity, reduction of bacterial shedding and provides robust gut barrier function. Vaccine 36:1500–1508. https://doi.org/10.1016/j.vaccine.2017.12.049

Rohde J, Habighorst-Blome K, Seehusen F (2014) "Brachyspira hampsonii" clade I isolated from Belgian pigs imported to Germany. Vet Microbiol 168:432–435. https://doi.org/10.1016/J.VETMIC.2013.11.016

Rood JI, Buddle JR, Wales AJ, Sidhu R (1985) The occurrence of antibiotic resistance in Clostridium perfringens from pigs (in eng). Aust Vet J 62(8):276–279

Rood JI, Adams V, Lacey J, Lyras D, McClane BA, Melville SB, Moore RJ, Popoff MR, Sarker MR, Songer JG (2018) Expansion of the Clostridium perfringens toxin-based typing scheme. Anaerobe 53:5–10

Roof MB, Kramer TT, Kunesh JP, Roth JA (1992) In vivo isolation of Salmonella choleraesuis from porcine neutrophils. Am J Vet Res 53:1333–1336

Rosager WN, Peter NJ, Lind JSE, Svend H, Matthew D, Steen PK (2017) Comparison of antimicrobial resistance in E. coli isolated from rectal and floor samples in pens with diarrhoeic nursery pigs in Denmark. Prev Vet Med 147:42–49

Rose N, Opriessnig T, Grasland B, Jestin A (2012) Epidemiology and transmission of porcine circovirus type 2 (PCV2). Virus Res 164(1–2):78–89

Rose N, Andraud M, Bigault L, Jestin A, Grasland B (2016) A commercial PCV2a-based vaccine significantly reduces PCV2b transmission in experimental conditions. Vaccine 34(33):3738–3745

Rosell C, Segalés J, Plana-Duran J, Balasch M, Rodrıguez-Arrioja GM, Kennedy S, Allan GM, McNeilly F, Latimer KS, Domingo M (1999) Pathological, immunohistochemical, and in-situ hybridization studies of natural cases of postweaning multisystemic wasting syndrome (PMWS) in pigs. J Comp Pathol 120(1):59–78

Rosell C, Segalés J, Domingo M (2000) Hepatitis and staging of hepatic damage in pigs naturally infected with porcine circovirus type 2. Vet Pathol 37(6):687–692

Rosengren LB, Waldner CL, Reid-Smith RJ, Checkley SL, McFall ME, Raji'c A (2008) Antimicrobial resistance of fecal Salmonella spp. isolated from all phases of pig production in 20 herds in Alberta and Saskatchewan. Can J Vet Res 72:151–159

Rossignol JF, Maisonneuve H (1984) Benzimidazoles in the treatment of trichuriasis: a review. Ann Trop Med Parasitol 78:135–144. https://doi.org/10.1080/00034983.1984.11811787

Ruan X, Zhang W (2013) Oral immunization of a live attenuated Escherichia coli strain expressing a holotoxin-structured adhesin–toxoid fusion (1FaeG-FedF-LTA2: 5LTB) protected young pigs against enterotoxigenic E. coli (ETEC) infection. Vaccine 31(11):1458–1463

Rubin JE, Costa MO, Hill JE, Kittrell HE, Fernando C, Huang Y, O'connor B, Harding JCS (2013) Reproduction of Mucohaemorrhagic Diarrhea and Colitis Indistinguishable from Swine Dysentery following Experimental Inoculation with "'Brachyspira hampsonii'" Strain 30446. PLoS One 8:e57146. https://doi.org/10.1371/journal.pone.0057146

Rugna G, Bonilauri P, Carra E, Bergamini F, Luppi A, Gherpelli Y, Magistrali CF, Nigrelli A, Alborali GL, Martelli P, La T, Hampson DJ, Merialdi G (2015) Sequence types and pleuromutilin susceptibility of Brachyspira hyodysenteriae isolates from Italian pigs with swine dysentery: 2003–2012. Vet J 203:115–119. https://doi.org/10.1016/j.tvjl.2014.10.033

Saif LJ, Ward L, Yuan L, Rosen BI, To, T.L. (1996) The gnotobiotic piglet as a model for studies of disease pathogenesis and immunity to human rotaviruses. Viral Gastroenteritis, pp 153–161

Saif LJ, Wang Q, Vlasova AN, Jung K, Xiao S (2019) Coronaviruses. In: Zimmerman JJ (ed) Diseases of swine, 11th edn. Wiley-Blackwell, Hoboken, NJ, pp 488–523

Sakurai J, Duncan CL (1978) Some properties of beta-toxin produced by Clostridium perfringens type C. Infect Immun 21(2):678–680

Salvarani FM, Silveira Silva RO, Pires PS, da Costa Cruz Júnior EC, Albefaro IS, de Carvalho Guedes RM, Faria Lobato FC (2012) Antimicrobial susceptibility of Clostridium perfringens isolated from piglets with or without diarrhea in Brazil (in eng). Brazil J Microbiol 43(3):1030–1033

Savic B, Zdravkovic N, Radanovic O, Jezdimirovic N, Kureljusic B, Stevancevic O (2021) A Salmonella enterica subspecies enterica serovar Choleraesuis outbreak in weaned piglets in Serbia: clinical signs, pathologic changes, and microbiologic features. J Vet Diagn Invest 33:993–996. https://doi.org/10.1177/10406387211025507

Sayeed S, Uzal FA, Fisher DJ, Saputo J, Vidal JE, Chen Y, Gupta P, Rood JI, McClane BA (2008) Beta toxin is essential for the intestinal virulence of Clostridium perfringens type C disease isolate CN3685 in a rabbit ileal loop model. Mol Microbiol 67(1):15–30

Schauser K, Olsen JE, Larsson LI (2004) Immunocytochemical studies of Salmonella Typhimurium invasion of porcine jejunal epithelial cells. J Med Microbiol 53:691–695. https://doi.org/10.1099/jmm.0.45582-0

Schmall LM, Argenzio RA, Whipp SC (1983) Pathophysiologic features of swine dysentery: cyclic nucleotide-independent production of diarrhea – PubMed. Am J Vet Res 44:1309–1316

Schroeder S, Harries M, Prager R, Höfig A, Ahrens B, Hoffmann L, Rabsch W, Mertens E, Rimek AD (2015) A prolonged outbreak of Salmonella Infantis associated with pork products in central Germany, April–October 2013. Epidemiol Infect 144:1429–1439. https://doi.org/10.1017/S0950268815002629

Schumacher LL, Woodworth JC, Jones CK, Chen Q, Zhang J, Gauger PC, Stark CR, Main RG, Hesse RA, Tokach MD, Dritz SS (2016) Evaluation of the minimum infectious dose of porcine epidemic diarrhea virus in virus-inoculated feed (in eng). Am J Vet Res 77(10):1108–1113

Schwartz K, Knittel J, Roof M, Anderson M (1999) Effect of oral tiamulin on the development of porcine proliferative enteropathy in a pure-culture challenge model

Schwarz L, Joachim A, Worliczek HL (2013) Transfer of Cystoisospora suis-specific colostral antibodies and their correlation with the course of neonatal porcine cystoisosporosis (in eng). Vet Parasitol 197(3–4):487–497

Schwarz L, Worliczek HL, Winkler M, Joachim A (2014) Superinfection of sows with Cystoisospora suis ante partum leads to a milder course of cystoisosporosis in suckling piglets. Vet Parasitol 204(3–4):158–168

Segalés J (2012) Porcine circovirus type 2 (PCV2) infections: clinical signs, pathology and laboratory diagnosis. Virus Res 164(1–2):10–19

Segalés J (2015) Best practice and future challenges for vaccination against porcine circovirus type 2. Expert Rev Vaccines 14(3):473–487

Segalés J, Rosell C, Domingo M (2004) Pathological findings associated with naturally acquired porcine circovirus type 2 associated disease. Vet Microbiol 98(2):137–149

Segalés J, Allan GM, Domingo M (2005) Porcine circovirus diseases. Anim Health Res Rev 6(2):119–142

Segalés J, Urniza A, Alegre A, Bru T, Crisci E, Nofrarías M, López-Soria S, Balasch M, Sibila M, Xu Z (2009) A genetically engineered chimeric vaccine against porcine circovirus type 2 (PCV2) improves clinical, pathological and virological outcomes in postweaning multisystemic wasting syndrome affected farms. Vaccine 27(52):7313–7321

Segalés J, Allan GM, Domingo M (2019) Circoviruses. In: Zimmerman JJ (ed) Diseases of swine, 11th edn. Wiley-Blackwell, Hoboken, NJ, pp 473–487

Shepherd FK, Freeman MJ, Culhane MR, Marthaler DG (2019) Reoviruses (Rotaviruses and Reoviruses). In: Zimmerman JJ (ed) Diseases of swine, 11th edn. Wiley-Blackwell, Hoboken, NJ, pp 715–727

Sherman KC (1972) Pathologic and hematologic changes observed in swine inoculated with Salmonella choleraesuis var kunzendorf and its endotoxin

Shrestha A, Freudenschuss B, Jansen R, Hinney B, Ruttkowski B, Joachim A (2017) Experimentally confirmed toltrazuril resistance in a field isolate of Cystoisospora suis (in eng). Parasit Vectors 10(1):317

Sirinavin S, Garner P (1999) Antibiotics for treating salmonella gut infections. Cochrane Database Syst Rev 1. https://doi.org/10.1002/14651858.CD001167

Sisak F, Havlickova H, Hradecka H, Rychlik I, Kolackova I, Karpiskova R (2006) Antibiotic resistance of Salmonella spp. isolates from pigs in the Czech Republic. Vet Med (Praha) 51:303–310. https://doi.org/10.17221/5550-VETMED

Sjölund M, Zoric M, Wallgren P (2014) Financial impact on pig production: III. Gastrointestinal disorders

Skampardonis V, Sotiraki S, Kostoulas P, Leontides L (2010) Effect of toltrazuril treatment in nursing piglets naturally infected with Isospora suis. Vet Parasitol 172(1–2):46–52

Smith HW, Jones JE (1963) Observations on the alimentary tract and its bacterial flora in healthy and diseased pigs. J Pathol Bacteriol 86(2):387–412

Smith HW, Jones JE (1967) Observations on experimental oral infection with Salmonella dublin in calves and Salmonella choleraesuis in pigs. J Pathol Bacteriol 93:141–156. https://doi.org/10.1002/path.1700930114

Smith SH, Mcorist S (1997) Development of persistent intestinal infection and excretion of Lawsonia intracellularis by piglets. Res Vet Sci

Song Y, Frey B, Hampson DJ (2012) The use of ELISAs for monitoring exposure of pig herds to Brachyspira hyodysenteriae. BMC Vet Res 8:1–8. https://doi.org/10.1186/1746-6148-8-6/TABLES/3

Song D, Moon H, Kang B (2015) Porcine epidemic diarrhea: a review of current epidemiology and available vaccines (in eng). Clin Exp Vaccine Res 4(2):166–176

Songer JG (1996) Clostridial enteric diseases of domestic animals. Clin Microbiol Rev 9(2):216–234

Songer JG, Harris DL (1978) Transmission of swine dysentery by carrier pigs. Am J Vet Res 39:913–916

Songer JG, Uzal FA (2005) Clostridial enteric infections in pigs (in eng). J Vet Diagn Invest 17(6):528–536

Songer JG, Post KW, Larson DJ, Jost BH, Glock RD (2000) Infection of neonatal swine with Clostridium difficile. J Swine Health Prod 8(4):185–189

Songer JG, Jones R, Anderson MA, Barbara AJ, Post KW, Trinh HT (2007) Prevention of porcine Clostridium difficile-associated disease by competitive exclusion with nontoxigenic organisms (in eng). Vet Microbiol 124(3–4):358–361

Songer JG, Harmon AE, Keel MK (2016) Toxins of Clostridium difficile. Clostridial Diseas Anim:61–70

Šperling D, Smola J (2011) Characterisation of multiresistant Brachyspira hyodysenteriae isolates from Czech pig farms. Vet Rec 168:215. https://doi.org/10.1136/vr.c4247

Springer S, Selbitz H-J (1999) The control of necrotic enteritis in sucking piglets by means of a Clostridium perfringens toxoid vaccine. FEMS Immunol Med Microbiol 24(3):333–336

Squire MM, Carter GP, Mackin KE, Chakravorty A, Norén T, Elliott B, Lyras D, Riley TV (2013) Novel molecular type of Clostridium difficile in neonatal pigs. Western Austr Emerg Infect Diseas 19(5):790

Ståhl, M., Kokotovic, B., Hjulsager, C.K., Breum, S.Ø., Angen, Ø., 2011. The use of quantitative PCR for identification and quantification of Brachyspira pilosicoli, Lawsonia intracellularis and Escherichia coli fimbrial types F4 and F18 in pig feces. Vet Microbiol 151 (3–4), 307–314

Steele J, Feng H, Parry N, Tzipori S (2010) Piglet models of acute or chronic Clostridium difficile illness (in eng). J Infect Dis 201(3):428–434

Stege H, Jensen TK, Mùller K, Bñkbo P, Jorsal SE (2000) Prevalence of intestinal pathogens in Danish finishing pig herds. Prev Vet Med 46:279–292. https://doi.org/10.1016/s0167-5877(00)00148-3

Stege H, Jensen TK, Mùller K, Bñkbo P, Jorsal SE (2001) Risk factors for intestinal pathogens in Danish finishing pig herds. Prev Vet Med 20:153–164

Stege H, Jensen TK, Møller K, Vestergaard K, Baekbo P, Jorsal SE (2004) Infection dynamics of Lawsonia intracellularis in pig herds. Vet Microbiol 104:197–206. https://doi.org/10.1016/j.vetmic.2004.09.015

Stěpánek J, Menšík J, Franz J, Krejčí J (1982) Preparation of hyperimmune cow colostrum whey and its use in the protection of piglets against transmissible gastroenteritis (in eng). Acta Vet Brno 51(1–4):99–108

Stuart BP, Gosser HS, Allen CB, Bedell DM (1982) Coccidiosis in swine: dose and age response to Isospora suis. Can J Comp Med 46(3):317

Swanenburg M, Urlings HAP, Keuzenkamp DA, Snijders JMA (2001) Salmonella in the Lairage of Pig Slaughterhouses. J Food Prot 64:12–16

Szabó I, Scherer K, Roesler U, Appel B, Nöckler K, Hensel A (2008) Comparative examination and validation of ELISA test systems for Salmonella typhimurium diagnosis of slaughtering pigs. Int J Food Microbiol 124:65–69. https://doi.org/10.1016/j.ijfoodmicro.2008.02.022

Talukdar PK, Udompijitkul P, Hossain A, Sarker MR (2017) Inactivation strategies for clostridium perfringens spores and vegetative cells (in eng). Appl Environ Microbiol 83(1)

Teng KTY, Aerts M, Jaspers S, Ugarte-Ruiz M, Moreno MA, Saez JL, Collado S, de Frutos C, Dominguez L, Alvarez J (2022) Patterns of antimicrobial resistance in Salmonella isolates from fattening pigs in Spain. BMC Vet Res 18:1–17. https://doi.org/10.1186/S12917-022-03377-3/FIGURES/12

Thakor JC, Dinesh M, Manikandan R, Bindu S, Sahoo M, Sahoo D, Dhawan M, Pandey MK, Tiwari R, Emran TB, Dhama K, Chaicumpa W (2022) Swine coronaviruses (SCoVs) and their emerging threats to swine population, inter-species transmission, exploring the susceptibility of pigs for SARS-CoV-2 and zoonotic concerns (in eng). Vet Q 42(1):125–147

Theuns S, Conceição-Neto N, Christiaens I, Zeller M, Desmarets LMB, Roukaerts IDM, Acar DD, Heylen E, Matthijnssens J, Nauwynck HJ (2015) Complete genome sequence of a porcine epidemic diarrhea virus from a novel outbreak in Belgium, January 2015. Genome Announc 3(3):e00506–e00515

Theuns S, Conceição-Neto N, Zeller M, Heylen E, Roukaerts IDM, Desmarets LMB, van Ranst M, Nauwynck HJ, Matthijnssens J (2016a) Characterization of a genetically heterogeneous porcine rotavirus C, and other viruses present in the fecal virome of a non-diarrheic Belgian piglet. Infect Genet Evol 43:135–145

Theuns S, Vyt P, Desmarets LM, Roukaerts ID, Heylen E, Zeller M, Matthijnssens J, Nauwynck HJ (2016b) Presence and characterization of pig group A and C rotaviruses in feces of Belgian diarrheic suckling piglets. Virus Res 213:172–183

Thienpont D, Vanparijs O, Hermans L, De Roose P (1982) Treatment of Trichuris suis infections in pigs with flubendazole. Vet Rec 110:517–520. https://doi.org/10.1136/VR.110.22.517

Thomson J (2009) Feed-associated colitis of growing pigs and its interaction with enteric infections Colite nutricional em suínos de crescimento e sua interação com infecções entéricas. Acta Sci Vet 37:1–9

To TL, Ward L, Yuan L, Saif LJ (1998) Serum and intestinal isotype antibody responses and correlates of protective immunity to human rotavirus in a gnotobiotic pig model of disease. J Gen Virol 79(11):2661–2672

Tomás A, Fernandes LT, Valero O, Segalés J (2008) A meta-analysis on experimental infections with porcine circovirus type 2 (PCV2). Vet Microbiol 132(3–4):260–273

Torres-Medina A, Underdahl NR (1980) Scanning electron microscopy of intestine of gnotobiotic piglets infected with porcine rotavirus. Can J Comp Med 44(4):403

Trott DJ, Huxtable CR, Hampson DJ (1996) Experimental infection of newly weaned pigs with human and porcine strains of Serpulina pilosicoli. Infect Immun 64:4648–4654. https://doi.org/10.1128/IAI.64.11.4648-4654.1996

Tucker BS, Petrovski KR, Kirkwood RN (2022) Neonatal piglet temperature changes: effect of intraperitoneal warm saline injection (in eng). Animals 12(10)

Turck D, Castenmiller J, De Henauw S, Hirsch-Ernst KI, Kearney J, Maciuk A, Mangelsdorf I, McArdle HJ, Naska A, Pelaez C, Pentieva K, Siani A, Thies F, Tsabouri S, Vinceti M, Cubadda F, Engel KH, Frenzel T, Heinonen M, Marchelli R, Neuhäuser-Berthold M, Pöting A, Poulsen M, Sanz Y, Schlatter JR, van Loveren H, Fernandez Dumont A, Gelbmann W, Knutsen HK (2019) Scientific opinion on the safety of viable embryonated eggs of the whipworm Trichuris suis as a novel food pursuant to Regulation (EU) 2015/2283. EFSA J 17:e05777. https://doi.org/10.2903/J.EFSA.2019.5777

Turlewicz-Podbielska H, Pomorska-Mól M (2021) Porcine coronaviruses: overview of the state of the art (in eng). Virol Sin 36(5):833–851

Underdahl NR, Mebus CA, Torres-Medina A (1975) Recovery of transmissible gastroenteritis virus from chronically infected experimental pigs (in eng). Am J Vet Res 36(10):1473–1476

Uwamahoro MC, Massicotte R, Hurtubise Y, Gagné-Bourque F, Mafu AA, Yahia L (2018) Evaluating the sporicidal activity of disinfectants against Clostridium difficile and Bacillus amyloliquefaciens spores by using the improved methods based on ASTM E2197-11 (in eng). Front Public Health 6:18

Uzal FA (2016) Clostridial diseases of animals (in eng), 1st edn. Wiley, Hoboken, p 1427

Uzal FA, Songer JG (2019) Clostridial diseases. In: Zimmerman JJ (ed) Diseases of swine, 11th edn. Wiley-Blackwell, Hoboken, NJ, pp 792–806

Uzal FA, Vidal JE, McClane BA, Gurjar AA (2010) Clostridium perfringens toxins involved in mammalian veterinary diseases. Open Toxinol J 2:24

Vahjen W, Pietruszyńska D, Starke IC, Zentek J (2015) High dietary zinc supplementation increases the occurrence of tetracycline and sulfonamide resistance genes in the intestine of weaned pigs. Gut Pathog 7(1):1–5

Van Der Wolf PJ, Vercammen TJM, Geene JJ, Van Exsel ACA, Peperkamp NHMT, Voets MT, Zeeuwen AAPA (2001) Salmonella Typhimurium DT104 septicaemia with meningitis in neonatal piglets. Vet Q 23:199–201. https://doi.org/10.1080/01652176.2001.9695113

Vannucci FA, Gebhart CJ (2014) Recent advances in understanding the pathogenesis of Lawsonia intracellularis infections. Vet Pathol 51(2):465–477. https://doi.org/10.1177/0300985813520249

Veterinary Laboratories Agencys (2009) Salmonella typhimurium causes outbreaks of disease in pigs.

Vet Rec 165:585–588. https://doi.org/10.1136/vr.165.20.585
Visscher C, Mischok J, Sander S, Schmicke M, Peitzmeier E-U, Von Dem Busche I, Rohn K, Kamphues J (2018) Nutrient digestibility, organ morphometry and performance in vaccinated or non- vaccinated Lawsonia intracellularis infected piglets. BMC Vet Res 14. https://doi.org/10.1186/s12917-018-1662-2
Vlasova AN, Amimo JO, Saif LJ (2017) Porcine rotaviruses: epidemiology, immune responses and control strategies (in eng). Viruses 9(3)
Vlasova AN, Wang Q, Jung K, Langel SN, Malik YS, Saif LJ (2020) Porcine coronaviruses. In: Malik YS, Singh RK, Yadav MP (eds) Emerging and transboundary animal viruses. Springer, Singapore, pp 79–110
Voth DE, Ballard JD (2005) Clostridium difficile toxins: mechanism of action and role in disease (in eng). Clin Microbiol Rev 18(2):247–263
Wales AD, Davies RH (2017) Salmonella vaccination in pigs: a review. Zoonoses Public Health 64:1–13. https://doi.org/10.1111/ZPH.12256
Walter D, Knittel J, Schwartz K, Kroll J, Roof M (2001) Treatment and control of porcine proliferative enteropathy using different tiamulin delivery methods. J Swine Health Prod
Wang L, Zhang Y, Byrum B (2014) Development and evaluation of a duplex real-time RT-PCR for detection and differentiation of virulent and variant strains of porcine epidemic diarrhea viruses from the United States. J Virol Methods 207:154–157
Wang Q, Vlasova AN, Kenney SP, Saif LJ (2019) Emerging and re-emerging coronaviruses in pigs (in eng). Curr Opin Virol 34:39–49
Warneke H (2017) Methods for improving diagnostic techniques used for the identification and isolation of Brachyspira species from swine. Grad. Theses Diss. Iowa State University, Digital Repository, Ames. https://doi.org/10.31274/etd-180810-5071
Waters EH, Orr JP, Clark EG, Schaufele CM (1998) Typhlocolitis caused by Clostridium difficile in suckling piglets. J Vet Diagn Investig 10(1):104–108
Waters M, Savoie A, Garmory HS, Bueschel D, Popoff MR, Songer JG, Titball RW, McClane BA, Sarker MR (2003) Genotyping and phenotyping of beta2-toxigenic Clostridium perfringens fecal isolates associated with gastrointestinal diseases in piglets. J Clin Microbiol 41(8):3584–3591
Wathes CM, Zaidan W, Pearson GR, Hinton M, Todd N (1988) Aerosol infection of calves and mice with Salmonella typhimurium. Vet Rec 123:590–594
Wendt, et al (2013) Erkrankungen des Verdauungstraktes. Eugen Ulmer KG, Diagnostik und Gesundheitsmanagement im Schweinebestand – Band 1
Wilberts BL, Arruda PH, Kinyon JM, Frana TS, Wang C, Magstadt DR, Madson DM, Patience JF, Burrough ER (2014) Investigation of the impact of increased dietary insoluble fiber through the feeding of Distillers Dried Grains with Solubles (DDGS) on the incidence and severity of brachyspira-associated colitis in pigs. PLoS One 9. https://doi.org/10.1371/journal.pone
Wilberts BL, Arruda PH, Kinyon JM, Madson DM, Frana TS, Burrough ER (2014a) Comparison of session severity, distribution, and colonic mucin expression in pigs with acute swine dysentery following oral inoculation with "Brachyspira hampsonii" or Brachyspira hyodysenteriae. Vet Pathol 51:1096–1108. https://doi.org/10.1177/0300985813516646
Wilberts BL, Arruda PH, Warneke HL, Erlandson KR, Hammer JM, Burrough ER (2014b) Cessation of clinical disease and spirochete shedding after tiamulin treatment in pigs experimentally infected with "Brachyspira hampsonii". Res Vet Sci 97:341–347. https://doi.org/10.1016/J.RVSC.2014.08.004
Wilberts BL, Warneke HL, Bower LP, Kinyon JM, Burrough ER (2015) Comparison of culture, polymerase chain reaction, and fluorescent in situ hybridization for detection of Brachyspira hyodysenteriae and "Brachyspira hampsonii" in pig feces. J Vet Diagn Invest 27:41–46. https://doi.org/10.1177/1040638714563064
Wilcock BP, Olander HJ (1977) The pathogenesis of porcine rectal stricture. Vet Pathol 4:36–40
Wilcock BP, Olander HJ (1979) Studies on the pathogenesis of swine dysentery. Vet Pathol 16:567–573. https://doi.org/10.1177/030098587901600509
Wilcock BP, Armstrong CH, Olonder HJ (1976) The significance of the serotype in the clinical and pathological features of naturally occurring porcine salmonellosis. Can J Comp Med Rev Can Med Comp 40:80–88
Wilhelm E, Hilbert F, Paulsen P, Smulders FJM, Rossmanith W (2007) Salmonella diagnosis in pig production: methodological problems in monitoring the prevalence in pigs and pork. J Food Prot 70:1246–1248. https://doi.org/10.4315/0362-028X-70.5.1246
Wilhelm B, Waddell L, Greig J, Rajić A, Houde A, McEwen SA (2015) A scoping review of the evidence for public health risks of three emerging potentially zoonotic viruses: hepatitis E virus, norovirus, and rotavirus. Prev Vet Med 119(1–2):61–79
van der Wolf P, Meijerink M, Libbrecht E, Tacken G, Gijsen E, Lillie-Jaschniski K, Schüller V (2021) Salmonella Typhimurium environmental reduction in a farrow-to-finish pig herd using a live attenuated Salmonella Typhimurium vaccine. Porcine Health Manag 7:1–17. https://doi.org/10.1186/S40813-021-00222-1/FIGURES/3
Wood RL, Pospischil A, Rose R (1989) Distribution of persistent Salmonella typhimurium infection in internal organs of swine. Am J V 50:1015–1021
World-Health-Organization (2002) Pre-harvest food safety Report of a WHO consultation with the participation of the Food and Agriculture Organization of the United Nations and the Office International des Epizooties

World-Health-Organization (2019) Critically important antimicrobials for human medicine-6th revision. Geneva

World-organisation-for-animal-health (2021) Oie list of antimicrobial agents of veterinary importance

Worsley MA (1998) Infection control and prevention of Clostridium difficile infection. J Antimicrob Chemother 41(suppl_3):59–66

Wu Z, Ling Y, Tian D, Pan Q, Heegaard PMH, He C (2014) Seroprevalence of Lawsonia intracellularis antibodies in intensive pig farms in China. BMC Vet Res 10:100. https://doi.org/10.1186/1746-6148-10-100

Xia L, Yang Y, Wang J, Jing Y, Yang Q (2018) Impact of TGEV infection on the pig small intestine (in eng). Virol J 15(1):102

Yates WDG, Clark EG, Osborne AD, Enweani CC, Radostits OM, Th Eede A (1979) Proliferative hemorrhagic enteropathy in swine: an outbreak and review of the literature. Can Vet J 20:261–268

Yazdankhah S, Rudi K, Bernhoft A (2014) Zinc and copper in animal feed – development of resistance and coresistance to antimicrobial agents in bacteria of animal origin (in eng). Microb Ecol Health Dis 25

Yeh J-Y, Lee J-H, Yeh H-R, Kim A, Lee JY, Hwang J-M, Kang B-K, Kim J-M, Choi I-S, Lee J-B (2011) Antimicrobial susceptibility testing of two Lawsonia intracellularis isolates associated with proliferative hemorrhagic enteropathy and porcine intestinal adenomatosis in South Korea. Antimicrob Agents Chemother 55:4451–4453. https://doi.org/10.1128/AAC.00408-11

Young MG, Cunningham GL, Sanford SE (2011) Circovirus vaccination in pigs with subclinical porcine circovirus type 2 infection complicated by ileitis. J Swine Health Prod 19(3):175–180

Zahavi EE, Lieberman JA, Donnenberg MS, Nitzan M, Baruch K, Rosenshine I, Turner JR, Melamed-Book N, Feinstein N, Zlotkin-Rivkin E, Aroeti B (2011) Bundle-forming pilus retraction enhances enteropathogenic Escherichia coli infectivity (in eng). Mol Biol Cell 22(14):2436–2447

Zajacova ZS, Faldyna M, Kulich P, Kummer V, Maskova J, Alexa P (2013) Experimental infection of gnotobiotic piglets with Escherichia coli strains positive for EAST1 and AIDA. Vet Immunol Immunopathol 152(1–2):176–182

Zeeh F, De Luca S, Nicholson P, Grützner N, Nathues C, Perreten V, Nathues H (2018a) Brachyspira hyodysenteriae detection in the large intestine of slaughtered pigs. J Vet Diagn Invest 30:56–63. https://doi.org/10.1177/1040638717722816

Zeeh F, Klausmann S, Masserey Y, Nathues H, Perreten V, Rohde J (2018b) Isolation of Brachyspira hyodysenteriae from a crow (Corvus corone) in close proximity to commercial pigs. Vet J 236:111–112. https://doi.org/10.1016/j.tvjl.2018.05.002

Zeeh F, Vidondo B, Nathues H (2020) Risk factors for the infection with Brachyspira hyodysenteriae in pig herds. Prev Vet Med 174. https://doi.org/10.1016/j.prevetmed.2019.104819

Zhang W (2014) Progress and challenges in vaccine development against Enterotoxigenic Escherichia Coli (ETEC)-Associated porcine PostWeaning Diarrhoea (PWD). J Vet Med Res

Zhao Q, Liu Y, Dong W, Zhu S, Huo Y, Wu S, Bao W (2014) Genetic variations of TAP1 gene exon 3 affects gene expression and Escherichia coli F18 resistance in piglets. Int J Mol Sci 15(6):11161–11171

Zimmermann J, Karriker L, Ramirez A, Schwartz K, Stevenson G, Zhang J (2019) Diseases of swine, 11th ed

# Part IV

# Respiratory Diseases

# Bovine Respiratory Disease

Martin Kaske, Petra Reinhold, Bart Pardon, and Jade Bokma

**Abstract**

Bovine respiratory disease (BRD) represents one of the most important diseases of rearing and veal calves with enormous impact on animal welfare and the economic outcome of the dairy and beef industry. As a typical factorial disease, interactions between viral and bacterial infectious agents, host cells, abiotic factors (such as stocking density, ventilation, and air quality), and management factors (such as supply of colostrum and all in-all out) are decisive for the pathogenesis and the incidence on the farms. This review deals with the etiology of BRD, clinical signs, diagnostic approaches, therapeutical regimes, and strategies for prevention.

**Keywords**

Calf · Pneumonia · Multifactorial disease · Diagnostics · Therapy · Antibiotics · Prevention · Vaccination

## 1 Introduction

Predominantly in the first 6 months of life, calves suffer frequently from respiratory tract infections which are summarized as bovine respiratory disease (BRD; syn.: undifferentiated respiratory disease, enzootic bronchopneumonia, undifferentiated fever). The morbidity especially in groups of young calves after transport to the fattener unit is typically high and may reach up to 100%. Massive suffering of affected animals indicates the relevance of BRD with regard to animal welfare. Moreover, BRD causes high economic losses due to treatment costs, reduced growth during the acute disease, and even more important, long-lasting effects of pneumonia on subsequent weight gain and carcass quality (Jourquin et al. 2022). Not at least, treatments of calves suffering from or at high risk of BRD require substantial amounts of antimicrobial drugs. For example, systematic evaluations of veterinary use of antibiotics in Switzerland on a national level revealed that more than one third of all administrations of antibiotics in cattle was due to BRD in 2021 (ISABV 2021).

M. Kaske (✉)
Swiss Bovine Health Service, University Zürich, Zürich, Switzerland
e-mail: martin.kaske@kgd-ssv.ch

P. Reinhold
Institute of Molecular Pathogenesis, Friedrich-Loeffler-Institut, Jena, Germany
e-mail: petra.elke.reinhold@gmail.com

B. Pardon · J. Bokma
Department of Internal Medicine, Reproduction and Population Medicine, Faculty of Veterinary Medicine, Ghent University, Merelbeke, Belgium
e-mail: bart.pardon@ugent.be; jade.bokma@ugent.be



J. J. Gross (ed.), *Production Diseases in Farm Animals*,
https://doi.org/10.1007/978-3-031-51788-4_12

Since decades, the issue of BRD has bothered both veterinarians and farmers. Thus, not surprisingly, tremendous efforts have been made by researchers all over the world to achieve progress: the functional anatomy of the bovine lung was examined intensively, and the viral and bacterial pathogens, some of which being ubiquitous, were identified and characterized and extraordinarily detailed. Schemes for clinical diagnostics and for sonographic diagnostics have been suggested and found to be valid. Therapy concepts were developed, refined, specified, and adapted. Vaccines have been developed and promoted. However, the result of all these activities is quite sobering: in fact, morbidity and mortality on an international scale are not lower today than they were 20 or 30 years ago (Smith et al. 2020).

We want to solve this apparent contradiction within the framework of this review and present the current state of knowledge regarding the aspects mentioned.

## 2 Species-Specific Characteristics of the Bovine Lung

The bovine lung itself is highly susceptible to pulmonary dysfunctions. As a result of anatomical and physiological peculiarities, pneumonia usually leads to severe disorders of gas exchange. Particularly important aspects are the following:

- The bronchial tree of ruminants (sheep, cattle, goat) and pigs differs from those of other species because the right cranial lobe bronchus arises directly from the right lateral side of the trachea rather than from the mainstream bronchus.
- Anatomy of the bovine lung is characterized by extensive connective tissue septa and a high degree of lobulation and sublobulation (Hare 1955; McLaughlin et al. 1961). As a result, limited interdependence exists between lobules. Consequently, in calves suffering from pneumonia, consolidated lobules may be adjacent to almost completely healthy lobules.
- Distal airways in the bovine lung are predominantly terminal bronchioles. Collateral airways are not present at all in bovine or porcine lungs (Mitzner 1991). The lack of collateral ventilation limits the inflammatory process to a lobule. However, in species with collateral airways (e.g., humans, dogs, sheep, horses), collateral ventilation may serve partially to bypass airway obstruction. Since pigs and cattle lack pathways for collateral airflow, atelectasis easily occurs in their lungs. Consequently, ventilatory asynchronisms or regional inhomogeneities in alveolar ventilation showing both atelectatic and hyperinflated areas within the same lung develop easily.
- In relation to body weight, the lung volume of calves is markedly smaller (38 mL $kg^{-1}$) compared to other animal species (pony, 56 mL $kg^{-1}$; goat, 64 mL $kg^{-1}$; Gehr et al. 1981; Weibel et al. 1987). Due to a small alveolar surface per kg body weight and because the bovine alveolus is reported to have low numbers of pulmonary capillaries per alveolar section, lungs of calves have a small physiological gaseous exchange capacity in relation to the basal oxygen needs. To meet the metabolic oxygen requirements, more basal ventilation is necessary compared to other species. The tidal volume of cattle even at rest ventilates around 30% of the lung volume which is twice as high as in dogs and horses (Veit and Farrell 1978). Consequently, even normal breathing at rest leads to a relatively higher exposition of the bovine lung to agents present in the ambient air (dust, bacteria, viruses, etc.) compared to other species.
- Pulmonary vasculature in the bovine lung is characterized by a thick medial layer (*Tunica muscularis*) that is predominantly present in small pulmonary arteries (Kay 1983). The thickness of the medial smooth muscle layer of the small pulmonary vessels is correlated with the susceptibility of pulmonary arteries to react on vasoconstrictive stimuli and differs between species (Tucker et al. 1975; Robinson 1982). One of the most important stimuli to increase pulmonary vascular resistance is alveolar hypoxia (hypoxic pulmonary vasoconstriction). Accordingly, pulmonary hyper-

tension develops easily in the bovine lung in any case of oxygen deficiency in the alveolar space (e.g., alveolar hypoventilation). In chronic cases, this even may lead to right heart hypertrophy and/or right ventricular failure (*Cor pulmonale*).

- Calves might be more susceptible to lung diseases than adult cattle because the functional maturation of the bovine lung lasts for approximately the first year of life or until a body weight of about 300 kg, respectively (Lekeux 1993).

# 3 Etiology

## 3.1 Infectious Agents

BRD can be caused by a primary viral infection of the upper respiratory tract along with mild clinical signs or, less frequently, by a primary bacterial infection. Well-known agents are bovine herpesvirus-1 (BHV-1), bovine viral diarrhea virus (BVDV), bovine respiratory syncytial virus (BRSV), bovine coronavirus (BCV), and parainfluenza type 3 (PI-3) (Griffin et al. 2010; Grissett et al. 2015). The bacteria most frequently involved in BRD are *Mannheimia (M.) haemolytica*, *Pasteurella (P.) multocida*, *Histophilus somni*, and *Mycoplasma* (*M.*) species, particularly *M. bovis* (Assié et al. 2009; Schönecker et al. 2020). Several of these common bacteria are found also in the respiratory tract of clinically healthy calves (Deepak et al. 2021). The drastic augmentation of bacteria seems to be favored by viral-induced release of mucins and loss of cilia activity of epithelial cells (Mata et al. 2012; Kirchhoff et al. 2014) and the immune suppression induced by viral infections (Woolums 2015). Moreover, BRS virus enhances the adherence of *P. multocida* to bovine lower respiratory tract epithelial cells (Sudaryatma et al. 2020). In addition, different bacteria may act in synergism with others. For example, infections with *M. haemolytica* are promoted and exacerbated by *M. bovis* (Ridley and Hateley 2018).

The isolation of infectious agents in a representative subsample of calves with clinical signs may be expensive but can provide important information for prevention and control strategies on the farm. In addition, laboratory diagnosis is recommended to monitor the development of resistance rates on farm or regional level, to maximize the probability of successful treatments, and to minimize the proportion of relapses. However, as for any diagnostic approach, the overriding principle must be considered that a diagnostic effort is only meaningful if the result of the diagnostic approach leads to a consequence. In that way, the results cannot be used for the respective individual calf, as the incubation for identification and antimicrobial susceptibility testing takes at least 2 days. An immediate treatment of calves with clinical signs of BRD, however, is imperative because most conspicuous calves have lung lesions several days before clinical signs are obvious (Cuevas-Gómez et al. 2021). Moreover, in farms getting batches of calves from numerous origins, it is challenging to get a representative sample to identify the decisive pathogen for this group (see Sect. 3.1.2).

### 3.1.1 Sampling Techniques

When facing respiratory disease in calves, different sampling techniques can be performed by the veterinarian. Most often, taking the limitations of on-farm sampling into account, a deep nasopharyngeal swab (DNS), a non-endoscopic bronchoalveolar lavage (nBAL) or a transtracheal wash (TTW) is executed (Pardon and Buczinski 2020). Sometimes pathogens can also be detected in swabs of conjunctival fluid from the medial eye corner (Penterman et al. 2022). Due to their docile nature, calves tolerate these procedures without sedation. However, the stress and/or pain related to different sampling methods has not been elucidated until now.

Briefly, before a DNS or nBAL is performed, the nostrils are rinsed with alcohol (70–90%). To perform the DNS, a swab is introduced via the nasal cavity to sample a surface abrasion of the nasopharyngeal mucosa including epithelial cells and mucus. In case of a nBAL, a sterilized catheter (diameter 2–4 mm) is introduced in the ventral nasal meatus, introduced carefully into the larynx and trachea until the wedge position in a bronchus is reached. Subsequently, 20–60 mL of saline is injected via the catheter and immediately aspirated while gently retracting the catheter several millimeters to allow air to recirculate and negative pressure to relieve creating suction (Van Driessche et al. 2017). About half of the instilled volume is retrieved. The bronchoalveolar lavage fluid (BALf) should be foamy as evidence of mixture with surfactant. In many cases of respiratory tract infection, BALf looks cloudy and may contain pus particles.

For a TTW, a local anesthetic should be applied after the skin is shaved and sterilized. Subsequently, a human intravenous catheter (Centracath 75, Vygon) or a Large Animal Trans-Tracheal Wash Kit (MILA international) can be used to push through the skin of the lower segment of the trachea between two tracheal rings into the tracheal lumen. Next, a volume of saline is injected (typically 60 mL). The saline washes the tracheal bifurcation and can be aspirated for further analysis (Timsit et al. 2013; Pardon and Buczinski 2020). Compared to nBAL, a larger volume of saline needs to be injected and a much smaller volume is retrieved (typically 2–10 mL).

The abovementioned techniques all have their (dis)advantages considering interpretation. In airways of healthy calves, primary pathogens (e.g., BRSV, BHV-1, PI-3, *M. bovis*) should not be present. Depending on their virulence, these pathogenic germs can, either as a sole agent or in combination with other agents, provoke clinical disease. Thus, if a primary agent is found and the animal is ill, the respective agent plays most likely a significant role. Moreover, for the interpretation of test results, it is relevant whether a bacterium was isolated from the upper respiratory tract (DNS) or from the lower respiratory tract (the larger bronchial airways and alveoli, nBAL) (Chase and Kaushik 2019). Certain opportunistic pathogens (e.g., *Pasteurellaceae*) can naturally exist in the nasopharyngeal mucosa as part of the normal flora without being associated with pneumonia.

The diagnostic method used for the analysis of the samples should also be taken into account. For example, the sound interpretation of a positive culture result requires a differentiation between contamination, colonization, and infection. Contamination can be assumed in case of the presence, usually in low numbers, of bacteria in a sample that are not expected to be present. Colonization corresponds to the presence of a microorganism with growth and multiplication, but without inflammatory reaction, immune response, or clinical signs. If an inflammation of the mucosa occurs concomitant with the isolation of a high number of bacteria from a site of the respiratory tract, these agents probably caused an infection (Pardon and Buczinski 2020). Thus, when these pathogens are found in large quantities in the nBAL fluid of a clinically ill calf, it is most likely that they are involved in bronchopneumonia. In this case, antimicrobial therapy becomes necessary and antimicrobial susceptibility testing may be appropriate.

There are varying opinions regarding the relevance of pathogens isolated from TTW. Some argue that pathogens from the lung are transferred by cilia toward the bifurcation indicating the microbiological milieu in lower airways and the periphery of the lung. Others refer on the possibility of aspirating mucus originating from the nasopharynx or indicating a bacterial tracheitis without the need for antimicrobial therapy (Pardon and Buczinski 2020). Additionally, cultures obtained through DNS tend to yield more polymicrobial results (approximately 80%) than those from nBAL (~20%), making interpretation more challenging (Van Driessche et al. 2017). When comparing identifications from DNS, TTW, and BAL fluid, a very good agreement for bacteria (*P. multocida*, *M. haemolytica*, *M. bovis*) was shown. However, for BRSV and BCV, only a good agreement between TTW and BAL fluid was observed, whereas DNS showed less agreement with both sampling methods from the lower respiratory tract (Doyle et al. 2017). Another study showed that *H. somni* was more likely to be isolated from BAL fluid than from DNS due to easy overgrowth of these bacteria in polymicrobial DNS samples (Van Driessche et al. 2017).

### 3.1.2 Sampling Strategy

For routine sampling on farms, nBAL is recommended as it takes only a few minutes and meaningful results can be expected. Calves in the first days of the disease, preferably not previously treated with antimicrobials and not displaying severe respiratory signs, should be chosen. By sampling in the acute phase of the disease, the probability of detecting a primary virus is higher as shedding of the primary pathogen may cease soon while clinical signs are still detectable. The absence or presence of fever is not a reliable predictor for the probability to find the primary or secondary pathogen (Grissett et al. 2015).

Taking a nBAL of five calves out of a larger group offers, on the one hand, the option to detect reliably the primary agent with PCR as a very sensitive technique from a pooled sample. On the other hand, separate bacterial cultures from each calf allow to detect in addition opportunistic infections and provide results of antimicrobial susceptibility testing of respective pathogens (Pardon and Buczinski 2020).

## 3.2 Abiotic Factors

BRD is a classical factorial disease. Infectious agents are involved, but some of them (e.g., *Pasteurellaceae*, BCV) are found almost ubiquitous on dairy farms and in beef herds. They may infect calves without any clinical consequences. Massive occurrence of BRD arises only if abiotic risk factors and inappropriate management procedures become relevant.

### 3.2.1 Closed Herds

On beef farms and dairy farms without purchase of calves or cows ("closed herds"), severe problems with BRD are rare. However, animal health services are contacted occasionally by dairy farmers with profound problems related to BRD. At the date of the farm inspection, one scenario is marked by several older, more or less emaciated, calves with distinct expiratory dyspnea suffering obviously from severe chronic pneumonia (Fig. 1). The owners claim that a variety of different antibiotics has already been administered in the past weeks with unsatisfying results—and they expect that a different compound recommended now will be the game changer. Unfortunately, clinical signs and sonographic findings indicate clearly that the prognosis is grave for such patients. Instead, the risk factors for forthcoming calves must be eliminated. These become obvious when considering in detail treatment recordings of the farm and the anamnesis of respective calves. While the calves had been delivered on time and appeared vital after birth, many of them suffered already from diarrhea in their first 2 weeks of life and revealed some respiratory signs later on, which were not taken seriously by the personnel.

Clearly, on such farms, the colostrum management has to be taken into account as a risk factor of pivotal importance (Godden et al. 2019). Further factors impeding the immune system of the calves must be considered (e.g., insufficient supply of trace elements and vitamins to pregnant dams, subclinical or clinical anemia). Moreover, all factors are decisive that increase the infectious pressure and/or weaken the calf's natural defenses (stressors in the broadest sense, in particular as a result of restrictive energy supply, poor hygiene, heat stress, cold and drafty environment, early dehorning) (Svensson et al. 2003; Lundborg et al. 2005). Not at least, arbitrary and untidy treatments of diseased calves provoke relapses and reduce the chance of a complete recovery. On such farms, standard operating procedures (SOP) have to be conceived to minimize systematically the impact of the most important risk factors.

### 3.2.2 Fattening Operations

The prevalence and incidence of BRD on fattening units is closely linked to the structure of the different production systems which have developed over the past decades. In fact, in Europe, the majority of fattening units, whether for veal calves, starters, heifers, steers, or bulls, buy batches of calves (20–150 heads) from traders. These acquire the calves at an age of 2–4 weeks with a body weight of 50–75 kg from dairy farmers. After a transport on trucks, calves reach marketing centers where calves are sorted according to quality, genetics, and weight. Thereafter, the regrouped batches of calves are transported to the fattening units. Obviously, risk factors for subsequent diseases are the durations and conditions of

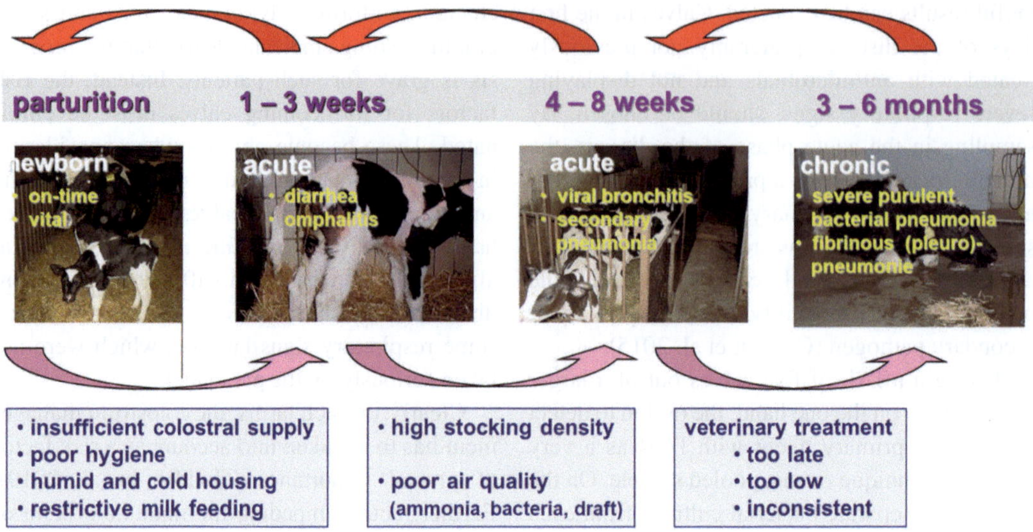


**Fig. 1** Abiotic and management-related risk factors on dairy farms faced with BRD as a herd problem

transport and regrouping, the intense contact between calves of a variety of different farms ("commingling"), and the conditions after arrival on the fattening unit with respect to group size, temperature conditions, stocking density, and access to feed (Brscic et al. 2012; Lava et al. 2016; Schnyder et al. 2019).

The risk for subsequent diseases of traded calves significantly depends on the constitution prior to transport, which is determined by colostrum supply and feeding intensity, housing, and hygiene on the dairy farm (Maccari et al. 2015; Marcato et al. 2022). According to the low demand of the market especially for male calves of dairy breeds, their sales value is minimal and decreased over the last years (Wilson et al. 2020a, b). Therefore, these so-called surplus calves may receive inadequate care on dairy farms promoting a poor constitution when sold to traders (Creutzinger et al. 2021). Moreover, the dairy farmers are interested to market the calves as early as possible. However, the ability of the immune system to combat unknown antigens and to produce endogenous antibodies increases only gradually during the first weeks of life (Hassig et al. 2007)—particularly in the high percentage of hypogammaglobulinemic calves due to insufficient intake of colostrum directly after parturition (Reschke et al. 2017). Not surprisingly, the trading of unweaned young, undernourished calves susceptible to infections due to a compromised immune system and a poor constitution enhances the risk of a high morbidity and mortality in the first weeks on the fattening unit considerably (Renaud et al. 2018a, b; Masmeijer et al. 2019; Fig. 2).

## 4 Diagnostics

The appropriate therapy of diseased calves is essential to minimize suffering and fatalities. When asked to handle a new batch of calves within the first days after arrival on a fattening unit, the veterinarian often finds different states of the disease in a group of calves: several of the animals may be healthy, while others are moderately ill (i.e., primarily after a viral infection leading to tracheitis, bronchiolitis, and sometimes interstitial pneumonia), others are in an advanced state (i.e., after a secondary bacterial infection took place leading to purulent bronchopneumonia), and some animals may be even already moribund (i.e., suffering from a fibrinous (pleuro)pneumonia). It has to be the aim of a cost-effective intervention to avoid dispensable treatments (by medicating either healthy or moribund calves) and to choose an optimal therapeu-

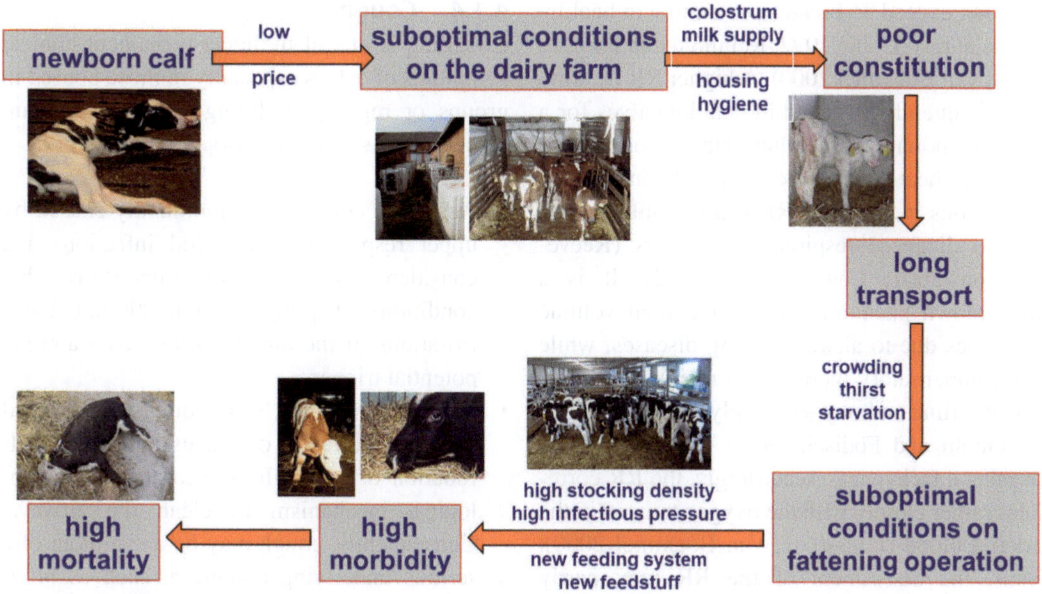


**Fig. 2** Abiotic and management-related risk factors on fattening units faced with BRD as a herd problem

tic regime for as many animals as possible. This requires a classification of the stages of the disease based on assessment of clinical signs and preferably including a sonographic examination.

## 4.1 Clinical Signs

### 4.1.1 General Behavior

Important information can already be gained by observing the animals in the pen from a distance. A substandard body weight, a rough hair coat, and a poor nutritional status can stand for chronic problems of the respective calves. The distribution of lying calves within the barn may indicate ventilation problems.

After arrival of a group of calves on a farm, the assessment of their health status is of pivotal importance. Firstly, healthy calves are vivid, alert, and difficult to catch. In contrast, acutely diseased calves appear to be tired and exhausted. Frequently, they lay separated from the rest of the group and leaning against the cold outer wall of the barn. In undisturbed animals, the facial expression of the calf may point on stress, pain, and discomfort. The features of a pain face have been described for cows (Gleerup et al. 2015) and adapted to calves (Tschoner 2021): ears either tense and backward or low ("lamb ears"), eyes with a tense stare or withdrawn appearance, tension of the muscles above the eyes ('furrow lines'), tension of the facial muscles on the side of the head, strained or dilatated nostrils, and increased tonus of the lips. However, the behavior of even severely diseased animals is often comparatively inconspicuous—phylogenetically, calves as prey animals have to hide any physical problems as long as possible to avoid attracting predators. Thus, it is—even for experienced personnel—challenging to distinguish between calves suffering from BRD and those experiencing distress after transportation, processing, and regrouping. Further clinical signs must be taken into account.

### 4.1.2 Respiratory Dynamics

In cattle, the tidal volume of spontaneous breathing at rest is lower compared to other species with the same body mass (7.1 ± 1.0 ml/kg BW *vs.*10.6 ± 1.6 ml/kg in horses; Gallivan et al. 1989). Thus, the physiological respiratory rate (RR) is higher in cattle than in horses, and higher in calves compared to adult cattle to fulfill metabolic oxygen requirements. Nevertheless, the RR does not exceed 36 breaths per minute in healthy calves at rest (Collie 1992; Reinhold and Födisch 1993; Van Boemmel 2000). A higher RR is, however, frequently observed in healthy calves for a certain amount of time after drinking or playing and in heat-stressed calves. Having these restrictions in mind, the RR is a valuable parameter to diagnose respiratory disorders (Reeve-Johnson 2001; Lowie et al. 2022). It is a well-known phenomenon that the tidal volume decreases due to airway or lung diseases, while RR compensatory increases to maintain ventilation per time and oxygen supply to the organism (Reinhold and Födisch 1993). Any hypoxemia triggers a tachypnea. Accordingly, the RR correlates rather closely with the oxygen tension in the arterial blood ($r = -0.61$; Van Boemmel 2000). Thus, the assessment of the RR may partly replace a blood gas analysis in practice.

The assessment of RR should be done from a distance without any disturbance of the animal by counting the abdominal deflections during the active inspiration process. The simpler it is to count the deflections, the higher is the breathing intensity, i.e., more labor is expended for breathing which may indicate a respiratory problem. If it is, on the other hand, not possible to count the deflections from the distance due to minimal movements of the flank, a dyspnea is improbable.

### 4.1.3 Body Temperature

Rectal body temperature should always be measured in animals that show signs of exhaustion and increased RR. As a biphasic rise of body temperature (≥39.5 °C) is typical for an acute infection, sometimes a physiological temperature (38.5–39.4 °C) is found together with clear symptoms indicating pulmonal dysfunction. In addition, a normal or even subnormal rectal temperature is frequently found in calves suffering from chronic irreversible bronchopneumonia—quite often accompanied by an expiratory dyspnea and bradypnea. If a veterinarian has to assess a group of calves at risk of BRD, it is useful to measure rectal temperature either in all calves or in a representative sample of the group to get an overview about the situation.

### 4.1.4 Cough

Coughing is one of the most prominent clinical symptoms of calves especially in those housed in groups or regrouped. Being a quite important symptom, it is generally adopted that:

- Acute dry cough is predominantly caused by upper respiratory tract viral infections and considered to be self-limited under favorable conditions (Dicpinigaitis et al. 2009), but also irritations of the airways (dust, cold air) are potential triggers.
- A "productive" or "wet" cough is associated with expectoration of mucus indicating malfunction of mucociliary clearance, a physiological mechanism to clean the airways. Consequently, cough may be a symptom of a serious underlying condition, such as bronchopneumonia or acute pulmonary edema. In such cases, coughing is very painful, and patients may try to suppress coughing (Grief and Loza 2018).

In calves, spontaneous and induced cough is significantly associated with lung consolidations assessed sonographically (Lowie et al. 2022). Fever (≥39.4 °C) was found to be the second crucial factor indicating pulmonary inflammation. Unfortunately, the diagnostic accuracy of cough as a single clinical sign is too low to detect animals with pneumonia reliably. The rather low sensitivity of 0.37 of cough to detect pneumonia (Lowie et al. 2022) may be explained by the varying and undefined duration of the disease in a group of calves: while some, probably most, calves cough due to a simple bronchitis or environmental induced irritation of the upper airways without any pathomorphological changes in lung parenchyma, in some calves suffering from an advanced stage of the disease, cough is caused by an inflammation of either airways or lung tissue, i.e., bronchopneumonia. Not surprisingly, the significance of coughing to detect bronchopneumonia differs between calves of different breeds, ages, and production types. With respect to lung tissue, thoracic ultrasound is recommended to come easily and quickly to a valid diagnosis. This technique, however, is not suitable to identify irritation or inflammation of larger airways (i.e., tracheitis, bronchitis) which may induce cough as well.

In conclusion, a high proportion of coughing calves in a group indicates clearly that the group is running out of the track and the risk of severe BRD is significantly increased. In fact, technical solutions to assess frequency of coughing in a group of calves are on the way (Vandermeulen et al. 2016).

#### 4.1.5 Nasal Discharge and Epiphora

An assessment of nasal discharge may help to differentiate between viral infections (increased amount of serous discharge) and secondary bacterial infections (mucoid-purulent discharge). Unfortunately, the absence of purulent discharge does not exclude the possibility of bacterial infections as calves will remove nasal discharge by licking the nostrils with their tongue as long as the general condition is reasonably good.

Epiphora may be indicative for an early infection caused by various viruses and bacteria but is also observed in calves exposed to high ammonia concentrations and/or dust as well as patients suffering from a conjunctivitis due to other reasons.

#### 4.1.6 Airway Auscultation

On the one hand, findings from auscultation of calves suffering from BRD were found to be inferior indicators of specific respiratory problems and had, moreover, a high inter-rater difference (Buczinski et al. 2014; Pardon et al. 2019). On the other hand, if performed properly, there remain two aspects of considerable diagnostic value:

Firstly, the auscultation one hand below the larynx (i.e., laryngotracheal) allows (i) to estimate the involvement of the upper airway system (indicated by loud hissing in- and expiratory sounds) and to estimate the respiratory ratio, i.e., the ratio between the duration of inspiration and expiration. The respiratory ratio should be around 1:1.2 in normal calves. An extended duration of the inspiration indicates inspiratory dyspnea. An inspiratory dyspnea mainly develops as a consequence of a reduced expansibility (i.e., a reduced compliance) of the lung. Thus, the respiratory effort for inspiration has to be increased. An inspiratory dyspnea has a better prognosis compared to an expiratory dyspnea (i.e., respiratory ratio > 1.5). In most calves suffering from expiratory dyspnea as indicated by a markedly prolonged expiration compared to inspiration, peripheral bronchoconstriction is present that causes high airway resistance and often leads to an obstructive emphysema (increased intrathoracic gas volume). Accordingly, the calf has to squeeze the air out of the lung actively by contraction of the abdominal oblique musculature (abdominal breathing type to overcome high airway resistance), whereas physiological expiration would be a passive process in healthy animals. The prognosis of calves suffering from expiratory dyspnea is poor – not at least due to the irreversibility of interstitial and alveolar emphysema as a consequence of trapped air due to collapsed peripheral airways during expiration.

Secondly, the quality of auscultation sound of the anteroventral aspects of the lung (i.e., trachea-bronchial auscultation) allows to roughly estimate the extent of consolidation of lung parenchyma. Ventilated, air-filled parenchyma acts sound-absorbing (Kotlikoff and Gillespie 1983). Thus, louder and fissing sounds indicate accumulation of exudate, alveolar edema, and atelectasis. This consolidation is found always at first and most extensively in the anteroventral parts of the lung which seem to be of major importance for the ventilation process. Adventitious sounds may be audible in some calves; late-inspiratory crackles are considered as typical for trapped air due to obstructions of the peripheral airways ending up in an obstructive emphysema (Kotlikoff and Gillespie 1984). Breathing sounds above the caudal lobes of the lung remain mostly moderate even in calves in an advanced state of disease (Van Boemmel 2000).

### 4.2 Imaging Techniques

#### 4.2.1 Thoracic Ultrasonography

Thoracic ultrasonography (TUS) using preferably a 5–7.5 MHz linear scanner has emerged as a recent game changer for managing (antimicrobial) treatment in calves with respiratory disease. TUS allows a quantitative estimation of subpleural alterations in calves suffering from respiratory disease. Sound waves are nearly totally reflected from ventilated lung tissue. Therefore, reverberation artifacts are typical for normal lung tissue (Braun et al. 1996). Radiating artifacts (syn.: comet tail artifacts; Thickman et al. 1983) represent a specific reverberation artifact distal to a hyperechogenic structure. They are generated by an irregular surface of the pleura which may be due to slight aggregations within the lung parenchyma or an irregular visceral surface. Radiating artifacts are typically found in patients suffering from interstitial pneumonia; such artifacts are found in the anteroventral parts of the lung in about 20% of all clinically healthy calves (Rocker 2001). Consolidations due to catarrhal, purulent bronchopneumonia and/or abscesses can be detected reliably by sonography. TUS has shown a high sensitivity (80–94%), high specificity (94–100%), and high correlation ($r = 0.92$) to postmortem findings (Ollivett et al. 2015). False-negative results may be due to the alterations which do not reach the surface of the pleura. In addition, in a clinical study carried out with 81 calves, significant correlations were found between the extent of ultrasonographically detectable consolidations and clinical score ($r = 0.67$) as well as results of arterial blood gas analysis ($pO_2$, $SatO_2$, and $AaDO_2$ with $r = 0.62$–0.65), respectively (Rocker 2001). TUS offers accurate detection of (sub)clinical pneumonia, enabling their differentiation from upper respiratory tract infections (URTI) on farm (Ollivett and Buczinski 2016; Jourquin et al. 2023). It has a great advantage in comparison to the traditional BRD scoring cards (e.g., Wisconsin respiratory scoring chart) since a significant number of animals with ultrasound-confirmed pneumonia display no clinical signs of the disease (25–87%) but could benefit from antimicrobial treatment (Binversie et al. 2020; van Leenen et al. 2020a; Jourquin et al. 2022). To facilitate the practical application of TUS, different techniques have been developed to allow quick scanning of calves within a limited timeframe. Examples of these techniques are the UGent quick thoracic ultrasound (qTUS), the focused lung ultrasound (FLUS), and scanning of the cranial part of the thorax (Pravettoni et al. 2021; Masset et al. 2022; Jourquin et al. 2023).

In case of qTUS (Jourquin et al. 2023), the whole lung is scanned in one movement (or two in a larger calf) both on the right and left side. During the scan, as soon as any abnormality (e.g., comet tail artifact, pleural irregularity, consolidation) is found, the probe is halted at that position and gently moved cranio-caudally and back to expand the area of visualization. Case definitions of pneumonia have been based on consolidation depth: those of more than 1 cm were defined as lobular pneumonia and those of more than 3 cm as lobar pneumonia. Comet tail artifacts, smaller consolidations, and pleural effusion should also be noted.

While TUS is a suitable tool to detect structural changes in lung tissue, it does not allow to identify functional disturbances in the lung diminishing gas exchange, for example, bronchoconstriction, ventilatory inhomogeneities, or ventilation-perfusion asynchronisms (Reinhold et al. 2002).

### 4.2.2 X-Ray

Chest radiography is potentially qualified for the diagnostic imaging of lobar pneumonia, larger abscesses, and severe interstitial infiltration (Lakritz et al. 1993; Jones et al. 1998). Pneumonia can be recognized in calves by radiography with a sensitivity of 77% and a specificity of 91% (Schneider 1995; Rabeling et al. 1998). Sensitivity and specificity of radiography and TUS are almost similar (Berman et al. 2021). Thus, clinically, the easier use and accessibility of TUS compared with radiography support recommending its utilization as a first-line test to detect active pneumonia in calves.

## 4.3 Blood Parameters

### 4.3.1 Blood Gas Analysis

The efficiency of pulmonary gas exchange can be assessed measuring partial pressures of oxygen and carbon dioxide in the arterialized blood along with pH. In calves, the *A. auricularis* is the most common localization to collect arterial blood. To get valid results, blood must be collected from physically unstressed animals preventing stress-induced hyperventilation.

Any disturbances in alveolar ventilation, diffusion, or lung perfusion restrict oxygen transport from alveoli into the blood. Since cattle lung tends to develop inhomogeneities in alveolar ventilation, and pulmonary vessels are usually involved in infectious lung diseases, a marked mismatching between ventilation and perfusion has to be taken into account in any case of bovine pneumonia. In consequence, gas exchange disturbances may occur for both $O_2$ and $CO_2$. With respect to oxygen, lung function disorders result in hypoxemia, an increased difference in alveolar-arterial oxygen tension ($AaDO_2$), and an increased pulmonary shunt. Elimination of $CO_2$, however, is only limited by impaired alveolar ventilation. Due to its high solubility, diffusion of $CO_2$ from blood into the alveolar space occurs much faster in comparison to the diffusion of $O_2$ in the opposite direction. Following this, $CO_2$ accumulates to measurable hypercapnia only in the case of alveolar hypoventilation or in the presence of ventilation-perfusion mismatching. Since pneumonia in calves is not primarily an obstructive airway disease, a pronounced hypercapnia is an exception even in calves suffering from severe respiratory disease. In contrast, pulmonary dysfunctions are dominated by an impairment of diffusion and lung perfusion in the presence of pneumonia.

The $AaDO_2$ indicates problems in diffusion or perfusion or a mismatching between ventilation and perfusion. It seems to be one of the most sensitive parameters (Lekeux et al. 2002) because it already increases significantly during the incubation period in febrile calves without dyspnea (Van Boemmel 2000). Another sensitive quantitative parameter is the arterial partial pressure of oxygen ($PaO_2$): hypoxemia is evident in calves with inspiratory dyspnea and even more pronounced in calves with expiratory dyspnea. The virtual plateau of the sigmoid oxyhemoglobin dissociation curve above a $PaO_2$ of roughly 70 mm Hg (9.3 kPa) (Clerbaux et al. 1993) explains that the arterial hemoglobin-oxygen saturation ($SatO_2$) is above 90% even in calves with moderate hypoxemia. A $SatO_2$ of less than 90% represents, on the other hand, a severe disturbance of gas exchange.

### 4.3.2 Pulse Oximetry

$SatO_2$ can be assessed noninvasively using a pulse oximeter (Coghe et al. 1999; Uystepruyst et al. 2000), and it has been suggested as a parameter to estimate the severity of a respiratory disease in the field (Lekeux et al. 2002). However, fixing simple human finger pulse oximeters at the ear, tongue, scrotum, vulva, or nasal septum is hampered by defence movements of the calf. In addition, any generalized disease inducing a disturbance of peripheral microcirculation will reduce the significance of the value measured. There are two more restrictions assessing $SatO_2$: (i) Areas of dark pigmented skin are not suitable for any measurements with red or infrared light which is the measurement principle of pulse oximetry. (ii) Any situation of poor peripheral blood circulation (e.g., reduced blood pressure, shock, disturbed microcirculation) will lead to false results not corresponding to arterial $SO_2$.

### 4.3.3 Lactate

Plasma lactate concentration can be easily determined even in venous blood by a portable analyzer and represents another simple parameter which may be helpful in pneumonic calves (Coghe et al. 2000). An increased concentration of >4 mmol/L has been suggested as a reliable prognostic indicator for mortality within 24 hours in Belgian Blue calves suffering from severe interstitial pneumonia and massive edema (Coghe et al. 2000). However, as a single indicator, lactate should not be used due to a variety of other causes for a rise. Moreover, in calves with high lactate concentrations, the accompanying clinical symptoms are so stunning that further laboratory analyses are normally not needed.

### 4.3.4 Acute Phase Proteins

Various attempts have been made to assess the clinical status of calves suffering from BRD by analyzing specific acute phase proteins (APP) in blood as inflammatory markers (Saco and Bassols 2023). Mainly, haptoglobin (Hp), lipopolysaccharide binding protein (LBP), and serum amyloid A (SAA) have been studied. In particular, LBP assessed in the blood of cattle turned out to be a highly sensitive biomarker associated with infections caused by gram-negative bacteria (Schroedl et al. 2001; Ostermann et al. 2013). Together with fever, APP appear to be good indicators of disease (e.g., Idoate et al. 2015). However, a systematic review of the diagnostic accuracy of Hp, SAA, and fibrinogen versus clinical reference standards for the diagnosis of BRD failed to confirm the utility of APP in the diagnosis of BRD—mainly to the difficulties to standardize the analyses and the considerable impact of age, sex, breed, and type of housing on serum concentrations (Abdallah et al. 2016). In general, APP are nonspecific markers for a systemic acute phase response. Thus, the diagnostic value with respect to a certain disease will always be limited. Even if the sensitivity would be high, the specificity remains low (Schroedl et al. 2016).

## 4.4 Pathomorphology

### 4.4.1 Biopsy

Lung biopsy in ruminants for the diagnosis of respiratory disease can be made using a Tru-Cut needle or the Bard Biopty system (Sydler et al. 2004). Nevertheless, biopsies are not recommended as the comparison of histological and microbiological results from biopsies with results from postmortem findings did not reveal satisfying correlations (sensitivity 0.64 and 0.35, respectively; Klein 1996). This is likely due to the strict segmentation of the lung parenchyma in the bovine, allowing the existence of inflammatory altered and unaltered tissue in adjacent segments even in cases of severe pneumonia.

### 4.4.2 Necropsy

A postmortem examination may be useful to obtain the distribution and type of pathomorphological alterations, being either pathognomonic for specific infectious agents or providing samples for respective tests. To get valid results:

- A systematic procedure and knowledge of the anatomical features is indispensable (Nietfeld 2010)
- The animals examined must be representative for the herd problem
- More than one calf has to be examined to avoid unrepresentative and therefore misleading results
- Cadavers in advanced autolysis and freeze-thawed calves are ineligible
- Probability of success to detect infectious agents is maximal within the first 4 days after onset of clinical symptoms (Caswell et al. 2012)
- Calves should be preferentially untreated

In practice, these preconditions can be complied only on rare occasions. If isolation of bacteria is desired, samples of consolidated lung tissue (> 4 cm in diameter) between abnormal and normal parenchyma should be taken. Old lesions and abscesses are not useful as secondary pyogenic bacteria (*Trueperella pyogenes*, *Bibersteinia trehalosi*, *Fusobacterium*) are likely to be found whose detection only distracts from the main, primary pathogen. Samples for histopathology and immunohistochemistry (roughly 1 cm thick) have to be fixed in formalin with a 1:10 ratio of tissue to formalin (Caswell et al. 2012).

## 4.5 Experimental Techniques

### 4.5.1 Pulmonary Function Testing

To assess possible respiratory malfunctions, the basic physiological functions of the lung ensuring gas exchange need to be examined, i.e., ventilation, perfusion, diffusion, and distribution. Pulmonary function tests (PFT) should be noninvasive and applicable to conscious spontaneously breathing animals. Several PFT techniques originally developed for human medicine have been evaluated in veterinary medicine and were found most suitable for calves with body weights comparable to those of adult humans (less than about

150 kg). The calf to be examined usually wears a tightly fitting facemask (according to the size of the animal's head, thus ensuring low dead space) to connect the particular PFT system.

Ventilation of the lung can be described by variables of the spontaneous breathing pattern (i.e., RR, tidal volume (Vt), volume of minute ventilation, inspiratory time, expiratory time) and variables of respiratory mechanics (i.e., airway resistance and compliance of the lung). All these variables can be assessed by the impulse oscillometry system (IOS) as one of the best validated methods in calves (Reinhold et al. 1998a, b). The technical design of this device has been described elsewhere (Smith et al. 2005). Impulse oscillometry has been successfully implemented in studies assessing pulmonary dysfunction in calves with spontaneous or experimentally induced respiratory infections (Jaeger et al. 2007; Ostermann et al. 2014; Lohr et al. 2016). Under field conditions, it was proved to be useful in assessing therapeutic effects in calves suffering from pneumonia (Lawnick et al. 2005).

Variables of the spontaneous breathing pattern are very sensitive to respiratory disorders although they are not very specific in identifying pulmonary dysfunctions. Electronic spirometry is a simple and suitable method for assessing ventilatory variables of spontaneous breathing. Again, a face mask is mandatory to connect the spirometer in front of the animal's head. To measure RR even in freely moving animals, a sensor was designed for cattle analyzing the pressure difference between inhaled and exhaled breath in one nostril (Strutzke et al. 2019). Whether this RR sensor is suitable to measure Vt as well is currently under investigation (Dißmann et al. 2023).

Further noninvasive PFT techniques that have been successfully evaluated in calves under experimental conditions are capnovolumetry and re-breathing methods. By analyzing the shape of exhaled $CO_2$ in relation to the expiratory volume per breath (capnogram), dead space volumes and inhomogeneities in lung ventilation can be identified. Volumetric capnography was found to be sensitive enough to detect even clinically inapparent obstructive lung infections in calves (Reinhold et al. 2007). The re-breathing technique using carrier gases measures the functional residual capacity (FRC) using helium (He) dilution as well as the transfer factor for carbon monoxide ($TL_{CO}$) which is also called diffusion capacity of the lung. These variables have also been carefully assessed in healthy calves and in calves with respiratory infections (Kneucker et al. 2008).

Using any pulmonary function tests in calves, one must be aware that the lung is still in a process of growing and maturation, and individual peculiarities of the respiratory system in fast growing animals do exist. Consequently, repeated measurements within one individual animal will allow the assessment of disease-related changes or therapeutic intervention in a much better way than comparing PFT results of different animals.

### 4.5.2 Exhaled Biomarkers

Using biomarkers in exhaled breath for diagnostic purposes appears promising because the method of collection is completely noninvasive and repeatable and does not require patient cooperation. There are three different strategies to identify breath-borne biomarkers: (i) exhaled breath condensate (EBC), (ii) exhaled particles (Pex), and (iii) analysis of exhaled volatile organic compounds (VOC). So far, EBC and VOC analyses have been applied to ruminants including calves.

The term EBC describes a variety of samples obtained by cooling the exhaled breath. Based on the hypothesis that aerosols present in exhaled breath contain a range of compounds reflecting their concentrations within the extracellular epithelial lining fluid in the peripheral airways, biochemical analysis of EBC has been evaluated as a diagnostic tool in calves with respiratory infections. First, promising results obtained under standardized experimental conditions indicated that, for example, the mediator of inflammation leukotriene B4 (LTB4) measured in EBC might have a high potential for diagnostics of BRD (Reinhold et al. 2000; Reinhold and Knobloch 2010). Further studies, however, revealed only a limited benefit of this technique in the field (Burciaga-Robles et al. 2009). Taking the current knowledge into account, the latter is not surprising. Profound EBC research in both human and veterinary medicine revealed that highly variable results reported in literature are mainly caused by methodological heterogeneity between different research groups. Thus, in the future, a much better standardization of collection, processing, and analysis of EBC is necessary. Most likely, different standardization steps need to be adjusted to each potential biomarker of interest also taking the numbers of exhaled particles into account (Horvath et al. 2005; Reinhold et al. 2020).

Breath gas analysis offers another new research area using exhaled VOC as indicators of physiological conditions or as biomarkers of disease. A technical setup for breath analysis in bovines has been proved to be applicable to calves as well as to dairy cows under field conditions (Küntzel et al. 2018; Oertel et al. 2018). With respect to infectious disease in cattle, an explorative study evaluated the potential of VOC to discriminate between cows with and without infections with *Mycobacterium avium* subsp. *paratuberculosis* (Weber et al. 2021). The diagnostic potential of VOC analysis in BRD-related infections remains to be elucidated.

Taking the current knowledge about analysis of exhaled breath in ruminants into account, the detection of breath-borne biomarkers linked to BRD could be a challenge of the future. Infectious processes within the airways and the lung may generate diagnostically relevant compounds at the side of infection; thus, the chance of using exhaled breath for diagnosing pulmonary infections or inflammation is quite high. However, controlled collection of different fractions of exhaled breath (i.e., dead space gas vs. alveolar gas, particles, etc.) as well as elimination of confounding compounds derived from both metabolism and eructation is mandatory and challenging (Reinhold et al. 2020).

## 5 Therapy

Self cure of animals with clinical signs of BRD may occur. However, the proportion of uncomplicated disease courses without treatment varies considerably depending on the infectious pressure, virulence of pathogens and environmental conditions. Thus, in the vast majority of cases, a systematic treatment of diseased animals is essential to minimize suffering and to avoid economic losses. Basically, all treatments must be carried out in accordance with the national regulations and restrictions.

Treatment of calves suffering from BRD consists of antimicrobials, anti-inflammatory drugs and supportive pharmaceuticals. Clearly, the success of any therapeutic interventions depends on a correct diagnosis. A scheme for grading the severity of disease has been suggested and was found to be useful although not validated completely (Table 1).

There is no effective therapy for virus-related diseases which depict the onset of BRD in most cases. If the environmental conditions for affected calves would be optimal, the probability for a complete recovery even without any treatment would be high. However, having in mind the suboptimal frame conditions on fattening units, the application of antibiotics is often essential to minimize the risk of a serious secondary bacterial disease.

With early detection and immediate start of treatment with an effective antibiotic, up to 90% of all treated animals experience a significant and sustained improvement of the clinical signs (Welling et al. 2020).

### 5.1 Antimicrobials

In general, antimicrobials should be used systematically, targeted and evidence-based. The most important aspects of antimicrobial therapy are the following:

- Currently, it has been recommended to start a treatment as early as possible after the onset of fever and respiratory signs. Based on recent studies, the decision to treat should be preferentially made based on sonographically proven presence of lung consolidations (Van Leenen et al. 2020a; Lowie et al. 2022). Clearly, as soon as massive pathomorphological remodeling processes have developed in

**Table 1** Different stages of BRD based on clinical symptoms, rectal temperature, oxygen saturation in arterial blood ($SatO_2$), and ventilatory reserve (as determined experimentally by comparing the tidal volume before and after intravenous injection of lobeline [0.25 mg/kg body weight]) and consequences for therapeutical interventions; modified from Coghe et al. (2000) and Lekeux et al. (2002)

| Severity of disease | Clinical signs | Rectal temperature [°C] | $SatO_2$ [%] | Ventilatory reserve [%] | Therapy |
|---|---|---|---|---|---|
| Healthy subclinical | Alert behavior<br>Normal feed intake<br>Respiratory rate <40 $min^{-1}$ | <39.5 | >95 | Almost 100% | No treatment/metaphylactic antibiotics |
| Compensated | Tired/exhausted behavior<br>Reduced appetite<br>Tachypnea<br>Inspiratory dyspnea | ≥39.5 | >95 | Reduced by up to 50% | Antibiotics |
| Non-compensated | Markedly reduced general condition<br>Inappetence<br>Mixed dyspnea<br>Tachypnea or bradypnea | ≥39.5 or <39.5 | 85–95 | No ventilatory reserves | Antibiotics<br>NSAID/SAID<br>Secretolytics<br>Bronchodilators<br>Antioxidants |
| Irreversible | Markedly reduced general condition<br>Inappetence<br>Expiratory dyspnea | Usually <39.5 | <85 | No ventilatory reserves | Therapy futile<br>Grave prognosis |

lung tissue, the prognosis becomes much less favorable. Having in mind the normally unavailable option to isolate acutely ill calves from the group and thereby the high infection pressure in pens with a high stocking density, the metaphylactic treatment of all animals in the group is often considered to be mandatory to avoid an escalating illness. Practitioners consider a proportion of 10-20% diseased calves in a group as a useful threshold to treat the whole group instead of single animals. As indicated by an impressive systematic meta-analysis of 58 publications summarizing 169 randomly controlled clinical trials, antimicrobial group medications demonstrated moderate yet highly variable relative risk reductions in BRD morbidity (Baptiste and Kyvsgaard 2017); treatments with critical broad-spectrum antibiotics or combinations were found to be the most effective.

- In the context of group medications, it should be considered that the success of a simultaneous vaccination may be biased by the antibiotic compound. Antibiotics exhibit immunomodulatory properties (Tauber and Nau 2008). For example, macrolides affect the functionality of macrophages and neutrophilic granulocytes (Fischer et al. 2013; Scott et al. 2018). Several studies have been conducted with rodents and broilers (Lynn and Pulendran 2018); respective results from calves are missing.
- In principle, bacteriostatic antibiotics should only be used in immunocompetent patients. A high percentage of calves arriving at the fattening unit received, however, little or no colostrum on the dairy farm (Reschke et al. 2017), was fed insufficient amounts of milk causing a poor constitution, and is highly stressed after transport and processing (Bolton and von Keyserlingk 2021). Accordingly, these calves should not be considered as immunocompetent. This favors in severely, acutely ill calves the application of bactericidal antibiotics which are also effective for

immunocompromised patients. Respective compounds are ß-lactams, aminoglycosides, and fluoroquinolones. Tetracyclines and phenicols act essentially bacteriostatically but suggested to be bactericidal in higher concentrations. Sulfonamides as competitive inhibitors of bacterial dihydrofolate synthesis are bacteriostatic compounds, but in combination with trimethoprim ("potentiated sulfonamides") bactericidal.

- The efficacy of antibiotics depends on their pharmacokinetics and pharmacodynamics (PK/PD). For the different classes of antibiotics, pharmacokinetic parameters are related to microbiologic parameters in different ways (Eyler and Shvets 2019; Fig. 3):
  - Most classes of antibiotics (such as ß-lactams, phenicols, tetracyclines, sulfonamides, and macrolides) act time-dependently, i.e., their concentration should be maintained as long as possible above the minimum inhibitory concentration (MIC) of the bacterium in focus. The optimal concentration which should be achieved is two to four times higher than the MIC; higher concentrations do not result in greater kill of bacteria (Pieralli et al. 2016).
  - In contrast, the effect of concentration-dependent antibiotics (such as aminoglycosides, fluoroquinolones, and polypeptides) depends on the maximal concentration achieved ($C_{max}$), i.e., the higher the $C_{max}$/MIC (aminoglycosides) or area under the curve (AUC)/MIC (fluoroquinolones), the better the efficacy of treatment. If these antibiotics are used, a high initial dose is decisive to achieve a $C_{max}$/MIC of 10 at best (Gorham et al. 2022). Typical for concentration-dependent antibiotics is their post-antibiotic effect (PAE). This phenomenon refers to the time interval when the growth of the target bacteria is suppressed although the concentration of the compound fell below the MIC. For concentration-dependent antimicrobials, it is valid that the greater the quotient of $C_{max}$/MIC, the longer is the PAE.
- In respect to the duration of antibiotic therapy, a duration of 7–15 days gained acceptance over decades. Interestingly, meaningful evidence-based studies supporting this concept are mainly missing. Nevertheless, it became a dogma in human as well as veteri-

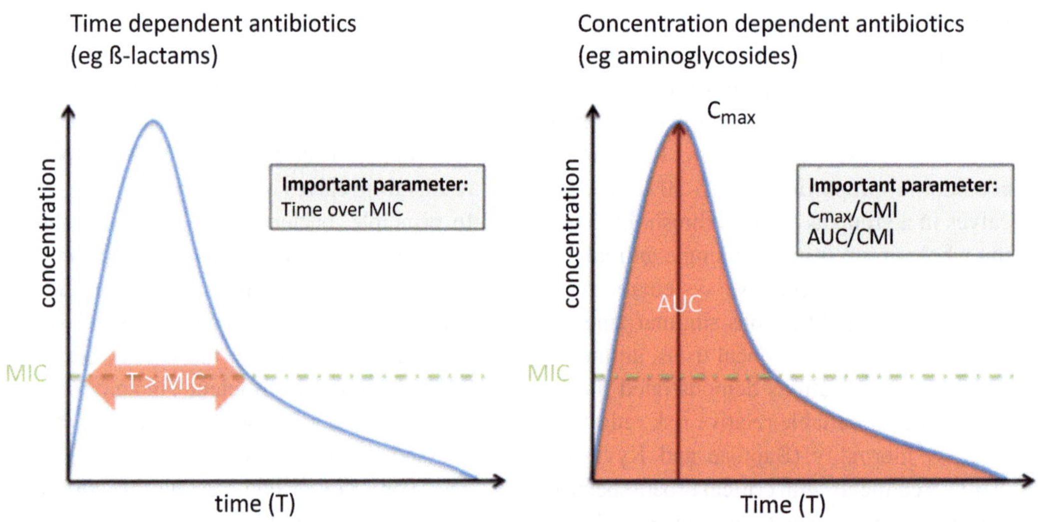


**Fig. 3** Antimicrobial substances are classified based on their killing characteristics in time-dependent and concentration-dependent antibiotics. *MIC* minimum inhibitory concentration, *AUC* area under the curve (www.futurelearn.com/info/courses/antimicrobial-stewardship/0/steps/7565)

nary medicine that high concentrations of the antibiotic in blood are decisive even several days after fever and clinical symptoms have subsided. Moreover, it had been argued that patients must complete a long course of therapy irrespective of clinical symptoms to prevent antimicrobial resistance (AMR). In fact, the opposite is true: numerous studies have shown that, in contrary, longer courses of therapy result in more emergence of AMR as natural selection of resistant bacteria is favored by a longer time of exposition (Spellberg 2016; Chotiprasitsakul et al. 2018; Li et al. 2022). Accordingly, the mantra "The shorter – the better" has been pushed in the last years in human medicine. For example, in adults with pneumonia, an amoxicillin treatment for 3 days was found to have similar clinical outcomes as the application of amoxicillin for 8 days (El Moussaoui et al. 2006; Hanretty and Gallagher 2018). In a meta-analysis evaluating the effect of long versus short period of antimicrobial treatment across 13 studies with 1727 patients suffering from common infections including pneumonia, no significant differences were found in respect to clinical efficacy, microbiologic cure, mortality, and relapses (Royer et al. 2018). Although results from human medicine cannot be transferred directly to the procedures in veterinary medicine, this issue should be taken more into consideration. By using TUS, treatment results and cure rate can easily be followed up and therefore aid in the duration of therapy (Jourquin et al. 2022). Empirically, under various conditions, good success has been achieved with shorter periods of treatment (4–6 days) compared to the traditional treatment interval of 7–15 days.

- Several studies and empirical experience indicate that AMR is not a serious problem in *M. haemolytica* and *Histophilus somni*; these bacteria are susceptible to a variety of widely used antibiotics such as ß-lactams (Schönecker et al. 2020). Also, *P. multocida* can be erased effectively in many cases (Lüthi et al. 2021). However, in production systems that commingle extensively calves from different dairy farms, AMR in *Pasteurellaceae* is frequent (Catry et al. 2016). The intense use of mainly oral metaphylactic antimicrobial therapy likely plays a key role for these high levels of AMR.
- Serious is the situation with respect to *Mycoplasma bovis*:
  - As *Mycoplasma* spp. lack a cell wall around their cell membranes, any antimicrobial classes whose impact on bacterial growth is achieved by inhibition of cell wall synthesis (i.e., ß-lactams, polymyxins) are ineffective per se.
  - Since mycoplasmas do not synthesize folic acid, there is also an intrinsic resistance to sulfonamides and trimethoprim (Valentine-King et al. 2019).
  - In the past, macrolides, tetracyclines, aminoglycosides, phenicols, pleuromutilins, and fluoroquinolones have proven to be effective agents against *M. bovis*. In recent years, however, significantly increasing MIC have been detected for almost all active substances worldwide, and thus increasing resistance rates (Gautier-Bouchardon et al. 2014; Heuvelink et al. 2016; Lysnyansky and Ayling 2016). A multinational study with 232 *M. bovis* isolates from France, Great Britain, Hungary, Italy, and Spain showed that the MIC values for enrofloxacin, tylosin, florfenicol, and oxytetracycline in 2014–2016 were more than doubled compared to 2010–2012 (Klein et al. 2019).
  - Comparison of florfenicol and oxytetracycline in Belgian calves with field infections revealed faster healing and lower antibiotic use in animals receiving florfenicol (Jourquin et al. 2022).
- In contrast to ß-lactams and sulfonamides, some antibiotics (in particular macrolides and, to a lesser extent, phenicols and tetracyclines) accumulate specifically in inflamed tissue, thereby achieving a particularly long active substance level in the lung after a single application.
- The oral administration of antibiotics via the milk feeder ("oral medication") is potentially

possible and effective but is associated with several problems: febrile calves with acute BRD are often weak and more or less inappetent due to their poor general condition. Accordingly, the most severely affected animals may receive the least amount of the drug. In these calves, maintaining therapeutically effective concentrations of the antibiotic over a sufficiently long period of time is difficult to ensure with oral medication. Moreover, the risk of AMR appears to be increased by oral administration of antibiotics compared to parenteral application; individual animal treatments were not associated with increased proportions of cattle with resistant *E. coli* isolates preslaughter (Stabler et al. 1982; Checkley et al. 2010).

- Careful documentation of each group and individual treatment is crucial on fattening units to evaluate systematically (at least twice per year) the efficacy of specific protocols by assessing the proportion of treated calves, repeatedly treated calves, relapses, fatalities, and condemned lungs after slaughter.

## 5.2 Anti-inflammatory Drugs

Cattle are considered to be the so-called fibrinous type of inflammation, since in this species, a pronounced fibrinous exudation occurs due to bacterial- or viral-induced damages of endothelia (Gruber et al. 1965). Thus, fatal excessive exudations of fibrin are common and fibrinous inflammations are seen more frequently than in horses (Meurer and Wolf 2007). Thus, not surprisingly, fibrinous pleuropneumonia is a dreaded final stage of BRD with an almost grave prognosis. The high risk of exaggerated inflammation processes in cattle is the reason that the application of anti-inflammatory drugs is recommended in severely diseased calves.

### 5.2.1 Nonsteroidal Anti-inflammatory Drugs (NSAID)

A variety of NSAIDs is available for use in calves suffering from BRD. A systematic review concerning the efficacy revealed that the administration of NSAIDs as an ancillary treatment cause a more rapid decrease in rectal temperature of the calves and may decrease lung lesions without beneficial effects on clinical signs at the end of the treatment (Francoz et al. 2012). Thus, patients benefit from the typical antipyretic, anti-inflammatory, and antiphlogistic action of these drugs. As a consequence, the general condition improves. The effects of NSAIDs are primarily due to an inhibition of the cyclooxygenase (COX) being a key enzyme in the synthesis of prostaglandins (PG). While the iso-enzyme COX-1 inhibits synthesis of cytoprotective PGI, the COX-2 impairs synthesis of proinflammatory PGE. Only meloxicam is considered as a preferential inhibitor of COX-2, while most traditional NSAIDs used in cattle medicine inhibit non-selectively both COX-1 and COX-2. Thus, a repeated or overdosed application of NSAIDs may cause dreaded abomasal ulcerations. NSAIDs are, however, unlikely to be the main factors for abomasal damages in veal calves (Bus et al. 2019).

Unfortunately, to the best of authors' knowledge, no systematic comparison has been performed to evaluate which of the NSAIDs is best suitable. Clearly, the half-life varies between flunixin (1.1 mg/kg i.v./s.c.), meloxicam (0.5 mg/kg i.v./s.c.), ketoprofen (3 mg/kg i.v./i.m.), tolfenamic acid (2 mg/kg s.c.), and carprofen (1.4 mg/kg i.v./s.c.). An older study in acutely febrile feedlot cattle revealed that a single subcutaneous dose of meloxicam was clinically as effective as three consecutive daily intravenous doses of flunixin meglumine both in combination with oxytetracycline (OTC) (Friton et al. 2004). In calves suffering from BRD, a combination of meloxicam with OTC was found to be superior compared to OTC alone and also to the combination of OTC with flumethasone (Bednarek et al. 2003).

### 5.2.2 Steroidal Anti-inflammatory Drugs (SAID)

Glucocorticoids exhibit a pronounced and lasting anti-inflammatory effect and inhibit the release of endogenous substances causing bronchoconstriction. They impair prostaglandin synthesis by inhibiting the key enzyme phosphodiesterase I. In cattle medicine, dexamethasone is by far the most used SAID, while others like flumethasone are not licensed in most European countries. The application of dexamethasone is quite popular with veterinarians and farmers also due to the significant improvement of general condition. Unfortunately, SAID have also an immunosuppressive effect, particularly if used in a high dose. Generally, they should not be used in calves suffering from BRD without applying simultaneously an antibiotic expected to be effective against the bacteria involved. Clearly, glucocorticoids should not be used routinely. However, they are the veterinarian's choice in individual cases for severely acutely ill calves with severe dyspnea. In such cases, a high dose of dexamethasone (0.2–0.5 mg/kg BW i.v.) may be clinically successful.

## 5.3 Supportive Therapy

Having in mind that increased amounts of viscous mucus are found in the airways of diseased calves, the administration of expectorant drugs (secretolytics) appears to be appropriate. Bromhexine as the prevailing drug converts mucosal mucoproteins into more fluid breakdown products. Thereby, it becomes easier for the animal to cough up the mucus. Mucolytics also lead to an accumulation of antibiotics in the lungs (Deretic and Timmins 2019). On the one hand, they can be injected, and on the other hand, they can also be given orally. Mixing the expectorant into the milk replacer of groups at high risk for BRD for several days was found to be helpful to reduce the incidence of clinical disease (Schmidt et al. 1998).

The positive clinical effects of sympathomimetics in critically ill calves suffering from BRD are due to a bronchodilatation and stimulation of the cilia in the airways to clear mucus. Clenbuterol (0.8 µg/kg BW i.v./i.m.) is the prevailing drug but prohibited for food-producing animals in many countries due to its systemic anabolic side effects. Also inhalation of an aerosolised parasympathicolyticum was found to improve respiratory functions (Genicot et al. 1995).

Preliminary studies indicate that histamine receptor antagonists may enhance the efficacy of antibiotics considerably (Bruer et al. 2019). In fact, based on clinical experience, the application of chlorphenamine (1 mg/kg slowly i.v.) in seriously ill calves seems to be a promising tool.

Medicinal plants bear a potential as additional treatment for calves suffering from BRD. Due to the versatile effects of their plant specific multicomponent compositions, medicinal plants can potentially act as "multi-target drugs." Based on a systematic evaluation of 418 references, *Echinacea purpurea* (L.) Moench, *Thymus vulgaris* L., and *Althea officinalis* L. were the most promising candidates for respiratory disease (Ayrle et al. 2016). In humans, oral doses of certain essential oils reduce symptoms of acute respiratory infections of viral origin (Prall et al. 2020).

The administration of fat-soluble vitamins (particularly vitamin E) can indirectly support the body's immune system and is particularly indicated for heavily stressed calves (Elam 2007).

All patients with alveolar hypoxia must cope pulmonary hypertension in poorly ventilated lung areas. Accordingly, the right heart has to pump excessively against the increased blood pressure in the pulmonary vessels. However, until now, there are no therapeutical approaches to support the heart pharmacologically as the relevance of this aspect for the prognosis is unclear.

For neonatal calves suffering from a respiratory distress syndrome, the direct administration of oxygen via nasal insufflation represents a simple method to improve blood gas variables. An oxygen flow rate of 3-6 L/min has been recommended (Bleul et al. 2008). For calves with BRD, this approach is only conceivable in exceptional cases.

## 6 Prevention

Several concepts have been developed to minimize the impact of the well-known risk factors for BRD on dairy farms as well as on fattening units.

### 6.1 Preconditioning

The constitution of calves at delivery on the fattening unit has an enormous impact on subsequent risk to suffer from BRD. On the dairy farm, supply of colostrum, milk feeding intensity, appropriate housing, and hygiene as well as vaccination are considered to maximize the robustness of calves (Abdallah et al. 2022; Marcato et al. 2022) which can be defined as the ability of the animal to cope with environmental challenges and to bounce back rapidly when challenges occur (Colditz and Hine 2016). Upstream measures to improve the health status prior to transport to the veal farm are known as "preconditioning" and were found to be very effective (Hilton 2015; Renaud and Pardon 2022). Remarkably, the body weight upon arrival is not a reliable predictor of robustness per se but has to be considered in combination with other factors, such as age, breed, and health status (Renaud et al. 2018b).

### 6.2 Stress

Any reduction or removal of stressors will have direct effects on morbidity and mortality of BRD. The interplay between host-pathogen interactions and stress has been underestimated for a long time. During a stress response, glucocorticoids, catecholamines, and neuroendocrine factors are released into the circulation of the host. Direct effects of these stress-related hormones on the immune system of the host are well known. For example, stress is known to cause a shift from T helper 1-mediated cellular immunity toward T helper 2-mediated humoral immunity which can affect the course of an infection and/or the susceptibility to a microorganism.

The research field exploring the interactions between microbiology and neurophysiology (named "microbial endocrinology") has elucidated that bacteria can recognize changing concentrations of stress-related hormones in the host ("quorum sensing"; Everest 2007; Mukherjee and Bassler 2019) and will respond directly to stress-related host signals. Catecholamines (i.e., adrenalin, noradrenalin) can alter metabolism and growth, motility, biofilm formation, and/or virulence of pathogens and commensal bacteria. Also mediated by catecholamines, iron will be mobilized from iron-binding proteins of the host (e.g., ferritin, transferrin), and this iron can be used by bacteria for metabolism and growth (Verbrugghe et al. 2012).

These insights in the role of neuroendocrine hormones regarding the susceptibility of mammalian hosts to a bacterial infection are the most relevant because they help to explain an increasing growth of germs already present in the host (silent carriers) and/or outbreaks of clinically relevant infections by facultative pathogens after events stressing the animals (e.g., by transport or re-grouping).

### 6.3 Vaccination

Commercially produced vaccines against BRD have been available since more than 40 years. Since then, vaccination of calves against BRD has been recommend as an effective preventive tool by scientists, veterinarians, and consultants. However, generally acknowledged recommendations regarding the vaccine (inactivated vs. attenuated), route of administration (intranasal vs. parenteral), booster interval, and time point of administration (directly after arrival vs. days/weeks thereafter) do not exist. In monofactorial infectious diseases, a robust protection of almost all vaccinated animals in response to the respective infectious agent can be expected. BRD, however, represents a typical multifactorial disease. Not surprisingly, the specific effects of vaccination on morbidity, mortality, proportion of relapses, return on investment (ROI), and protection rate of vaccinated animals vary in case of factorial diseases tremendously depending on seasons, farms, breeds, and frame conditions. The evidence of vaccination efficacy against BRD is inconsistent in literature as documented in several excellent reviews (Theurer et al. 2015; Ellis 2017; Chamorro and Palomares 2020). Clearly, no vaccination protocol conveys an overall convincing solution of BRD on fattening units. Nevertheless, some statements appear to be valid:

- Even newborn calves are able to produce an immunological response after getting in touch with an antigen; this is true both for agammaglobulinemic calves and for calves after appropriate supply with colostrum.
- The more similar the interaction between the immune system and antigen of the vaccine, the better is the success of the vaccination; i.e., it seems to be advantageous if the entry site of the antigen of the vaccine corresponds to the entrance of the natural infectious agent; in fact, intranasal application of modified live vaccines (MLV) seems to be more effective in naïve calves than parenteral application of a vaccine.
- As a rule of thumb, MLV provoke a more intensive immune response (primarily TH1 response and cell-mediated immunity) compared to inactivated vaccines (primarily TH2 response; production of IL-4, IL-6, and IL-10 favoring the differentiation of B-cells to antibody-producing plasma cells while inhibiting the generation of cytotoxic T-cells). This general statement is supported only in part based on a systematic evaluation of published studies. In fact, the difference in clinical efficiency is not convincing comparing parenterally administered live and inactivated vaccines. Remarkably, a booster vaccination was mandatory in case of using inactivated vaccines but only in roughly every second study where MLV was administered.
- Maternal antibodies may interfere with the antigen of the vaccine but do not necessarily inactivate it completely; as long as the vaccine is administered directly on the mucosa, the interaction is expected to be less distinct or missing.
- A reliable and robust protection of the vaccinated subject requires time (roughly 2 weeks after the first vaccination) and—predominantly in case of inactivated vaccines—a booster vaccination.
- An effective immune response presupposes a responsive immune system of a healthy organism, i.e., all vaccines are designated for prophylactic use. It is, however, common knowledge that this precondition is not fulfilled under field conditions having in mind that many calves arriving on the fattener unit are stressed.

During the last years, the immunomodulatory effect of vaccines in addition to the adaptive immune response has been a matter of debate. The relevance of nonspecific effects (Aaby et al. 2020) for vaccinations against BRD is, however, simply unknown until now. Moreover, interactions between stress responsiveness, temperament, and immunity in cattle have been demonstrated (Burdick et al. 2011), but it has still to be proven that immune responses after a vaccination are impaired in more temperamental animals.

## 6.4 Transport

As insufficient colostral supply and underweight increase disease susceptibility after transport (Masmeijer et al. 2019), only well-nourished calves should be transported. Moreover, educing the duration of the transport, optimizing the conditions before, during, and after the transport, avoiding hunger, thirst, and thermal stress and ensuring a gentle handling during processing and regrouping are key factors to reduce the risk of subsequent diseases after delivery on the fattening unit (Zulkifli et al. 2019). It is given that this is usually not achieved. Roadknight et al. (2021) reported that mortality of young calf rates during transport, including calves that were euthanized directly afterward, ranges from 0.06 to 0.7%—which is roughly tenfold higher than mortality rates reported for healthy adult cattle (0.009 to 0.01%; Warren et al. 2010; González et al. 2012). By far more important is the post-transport mortality within 2–4 weeks after transport. Values around 4% have been reported (Pempek et al. 2017; Wilson et al. 2020a, b) which are high compared with preweaning mortality rates in non-transported calves. However, this statement has to be qualified as the mortality on the fattening unit results from a variety of different factors.

## 6.5 Barn Climate

Different housing systems for calves became established in practise whose design and ventilation system affect significantly the health status of the animals (Mahendran et al. 2023). Clearly, barn climate plays a crucial role for respiratory health. The main risk factors are cold stress and accumulation of barn air pollutants. Cold stress results from the combination of cold environmental temperature and airspeed, and is aggravated by hampered insulation of the animal by poor body condition, hair clipping or humidification of the skin. Air pollutants originate from the animals or other substances within the barn and accumulate due to too low barn ventilation rates. The best-known air pollutants for calves are ammonia, hydrogen sulfide, bacterial air count, particulate matter and endotoxin levels (Van Leenen et al. 2021). Within the same barn with multiple microclimates, parts can be exposed to cold stress and others to accumulation of air pollutants. Obviously, the solution for cold stress (i. e., less ventilation) stands opposite to the solution for accumulation of air pollutants (i. e., more ventilation).

### 6.5.1 Temperature

Outdoor conditions can be favorable to avoid BRD in veal calves (Becker et al. 2020). However, young calves have a significant need for warmth according to their zone of thermoneutrality (Borderas et al. 2009) and are highly sensitive to sudden changes in temperature. An experimental study showed that the majority of calves adapted to 18 °C ambient temperature suffered from massive respiratory problems after staying for only 4 hours at 5 °C in a climate chamber (Reinhold and Elmer 2002). Accordingly, if calves are unloaded in the cold season after a more or less long transport and placed in a chilly barn, an initially catarrhalic bronchitis may develop favoring a subsequent pneumonia. In pig industry, preheating the barn before the arrival of new batches of piglets is common practice. A similar procedure would alleviate the adaptation of calves to the new housing situation on the fattening unit in winter.

Likewise, in outdoor barns, the pronounced temperature fluctuations between day- and nighttime being typical for early spring and late autumn are particularly problematic for calves.

### 6.5.2 Ammonia

Especially in indoor housing systems for young calves, the accumulation of noxious gases and moisture represent risk factors for BRD. Ammonia plays a key role; as calves are kept mainly on straw, hydrogen sulfide emissions are of minor importance.

Urea excreted with the urine is rapidly transformed by bacterial urease to ammonia; the degradation of the organically bound nitrogen in the feces is much slower. Especially in deep litter pens, ammonia accumulates directly above the straw. The nostrils of lying calves are less than 30 cm above the ground—thus, the continuously released ammonia is constantly inhaled. Ammonia irritates the epithelial cells in the respiratory tract because ammonia reacts with the moisture of the water-saturated air in the airways to the aggressive alkaline solution $NH_4OH$. The first sign of airway irritation is coughing.

Ammonia concentrations around 6 ppm are already perceived as a characteristic pungent ammonia smell and may harm the mucous membranes if exposed to prolonged exposure (Van Leenen et al. 2020b). Decisive factors for the ammonia concentrations in the lying area of calves are the following:

- The stocking density of the pen; in case of outdoor housing without a ventilation system, at least 3 $m^2$ (better 4 $m^2$) should be available per head as overstocking is detrimental to calf health.

- Ammonia-producing bacteria rely on warmth and moisture. Thus, the manure should be cleaned out frequently (every 2 weeks) and lying surfaces must be kept as dry as possible by spreading straw daily.
- An effective ventilation system is needed especially in indoor systems to remove ammonia as well as humidity—calves easily release 10 L of water per day via urine, feces, and expired air. Properly ventilated calf barns should enable 4–8 air exchanges per hour in winter, 10–20 air exchanges in spring and fall, and 40 air exchanges in summer (Nordlund and Halbach 2019). However, ventilation is only effective if fresh air is delivered from outside of the barn; recirculating air from inside the calf area is ineffective in removing moisture and ammonia. Moreover, the air exchange should not be limited to the upper air layers but has to reach the place directly above the lying area. To achieve high air exchange rates, fans are not recommended because calves are extremely vulnerable to a high velocity of air (>0.25 m/sec in the cold season; Van Leenen et al. 2020b) and draught (Lundborg et al. 2005). Any solutions need to be barn specific and accounting for the microclimate within a barn. Guiding plates, wind breaking nets, spaceboards, ventilators, positive pressure tubes, additional straw bedding and winter feeding strategies are all possible adjustments available for the veterinarian or ventilation expert to decide their usefulness for a given barn.

#### 6.5.3 Humidity and Dust

Excessive humidity is frequently a problem both in closed barns with poor ventilation and in outdoor housing systems during the cold, wet season. A relative humidity of 50–70% is recommended.

Dust irritates the epithelial cells of the respiratory tract and acts as a vehicle for pathogens. Functioning ventilation of the barn area is crucial for minimizing the dust exposure of the calves, but affects also the bacterial exposure of the air. Based on a study investigating air quality in 13 naturally ventilated calf barns during winter, Lago et al. (2006) demonstrated a highly significant association between pen bacterial counts in the air and prevalence of respiratory disease.

### 6.6 Group Size

Group size is another critical factor: the risk of disease transmission decreases with fewer animals kept together in a group. It is, however, almost impossible to define an upper threshold for group size because such a value depends on further factors: these are, in particular, the weight and age range of the animals in a group, the number of farms of origin, the proportion of animals vaccinated against BRD on the dairy farm, air quality in the barn, and the stocking density (Breen et al. 2023). Beside these factors, it has to be emphasized that the assessment of the health status and the separation of calves for clinical examination as well as treatments is easier in smaller pens with few calves. Many fatteners, on the other hand, prefer larger groups—not at least to use automatic calf feeder stations to their full capacity.

## 7 Future Perspectives

The prevalence of BRD as a typical multifactorial disease on a specific farm (Maier et al. 2020) can be exemplified by a chain (Fig. 4). The stronger this chain, the lower the prevalence. Each link of the chain corresponds to a specific risk factor such as constitution of the calf, supply with colostrum, body weight, duration and conditions of transport, air quality on the fattening unit, group size, etc. The crucial point is that a chain is only as strong as the weakest link in the chain. Thus, improving the rigidity of one link of the chain (e.g., by improving air quality in the pen of the fattening unit) will not markedly reduce BRD prevalence if the calves were highly stressed by a long transport—corresponding in this example to the weakest link of the chain and determining thereby the stability of the whole chain. The con-

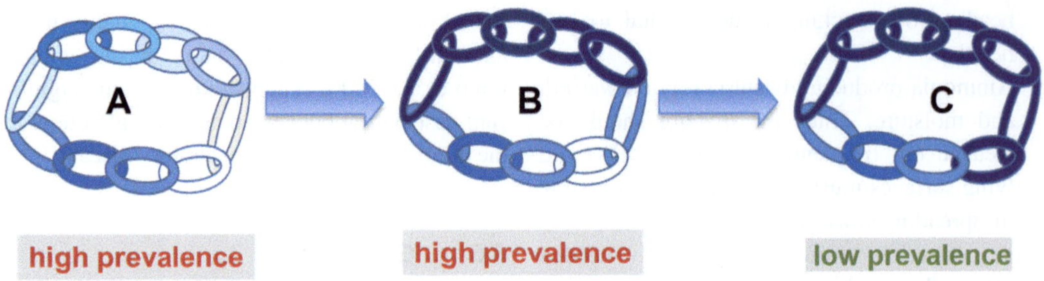


**Fig. 4** The problem of dealing with multifactorial diseases on farms: each chain link represents an infectious or noninfectious factor affecting the prevalence of the disease. The lighter the chain link, the more important is this factor for the prevalence. (**a**) Farm with many important risk factors resulting in a high prevalence; (**b**) After improvement of several factors, prevalence remains high because one crucial factor—and thus the weakest link of the chain—was not taken into account; (**c**) Significant reduction of prevalence by implementing an overall concept that takes all important factors into account

sequence is that an overall concept is required to minimize the prevalence of BRD considering all risk factors involved.

On dairy farms where female calves are raised for replacement, herd problems related to BRD are rare (see Sect. 3.2.1) as long as common rules of good agricultural practice are heeded. In fact, BRD is first a problem for farmers getting batches of young calves from traders for further fattening either as veal calves or for beef production. Unfortunately, many different parties are involved in the veal calf and beef industry (i.e., dairy farmers, traders, marketing organizations, fatteners, veterinarians, processing operations, retailers, regulatory authorities, consumer associations). Each stakeholder has its own specific interests which partially conflict with those of other parties. For example, most dairy farmers are interested to dispose the calves at the youngest possible age investing as little effort as possible, while fatteners would benefit from older, preferably vaccinated, calves with superior constitution. Therefore, a production system in the veal calf industry became established over the past decades which ignores even the simplest basics of external and internal biosecurity (Shortall et al. 2017). This is the reason why all progress of research related to diagnosis, therapy, and preventive tools in the last years did not result in a significant reduction of the problem on the farms. As long as a high morbidity and mortality could be curtailed by excessive use of antibiotics, the frame conditions of the production system remained. Nowadays, the reduction of use of antibiotics in farm animals becomes more and more an issue in face of the tremendous risks of AMR for human health.

Clearly, a turnaround in respect to our understanding of a sustainable production of veal and beef is mandatory giving animal welfare and animal health a high priority. This requires a coordinated strategy with the participation of all stakeholders which should touch upon and improve the different links of the chain. Such a concerted initiative would open up a perspective to minimize the problem of BRD in a really sustainable and long-term way.

## References

Aaby P, Benn CS, Flanagan KL, Klein SL, Kollmann TR, Lynn DJ, Shann F (2020) The non-specific and sex-differential effect of vaccines. Nature Rev Immunol 20:464–470

Abdallah A, Hewson J, Francoz D, Selim H, Buczinski S (2016) Systematic review of the diagnostic accuracy of haptoglobin, serum amyloid a, and fibrinogen versus clinical reference standards for the diagnosis of bovine respiratory disease. J Vet Intern Med 30:1356–1368. https://doi.org/10.1111/jvim.13975

Abdallah A, Francoz D, Berman J, Dufour S, Buczinski S (2022) Association between transfer of passive immunity and health disorders in multisource commingled dairy calves raised for veal or other purposes: systematic review and meta-analysis. J Dairy Sci 105:8371–8386. https://doi.org/10.3168/jds.2021-21671

Assié S, Seegers H, Makoschey B, Désiré-Bousquié L, Bareille N (2009) Exposure to pathogens and incidence of respiratory disease in young bulls on their arrival at fattening operations in France. Vet Rec 165:195–199. https://doi.org/10.1136/vr.165.7.195

Ayrle H, Mevissen M, Kaske M, Nathues H, Gruetzner N, Melzig M, Walkenhorst M (2016) Medicinal plants – prophylactic and therapeutic options for gastrointestinal and respiratory diseases in calves and piglets? A systematic review. BMC Vet Res 12:89. https://doi.org/10.1186/s12917-016-0714-8

Baptiste KE, Kyvsgaard NC (2017) Do antimicrobial mass medications work? A systematic review and meta-analysis of randomized clinical trials investigating antimicrobial prophylaxis or metaphylaxis against naturally occurring bovine respiratory disease. Pathogens Dis 75:ftx083. https://doi.org/10.1093/femspd/ftx083

Becker J, Schüpbach-Regula G, Steiner A, Perreten V, Wüthrich D, Hausherr A, Meylan M (2020) Effects of the novel concept 'outdoor veal calf' on antimicrobial use, mortality and weight gain in Switzerland. Prev Vet Med 176:104907. https://doi.org/10.1016/j.prevetmed.2020.104907

Bednarek D, Zdzisinska B, Kondracki M, Kandefer-Szerszen M (2003) Effect of steroidal and non-steroidal anti-inflammatory drugs in combination with long-acting oxytetracycline on non-specific immunity of calves suffering from enzootic bronchopneumonia. Vet Microbiol 96:53–67

Berman J, Masseau I, Fecteau G, Buczinski S, Francoz D (2021) Comparison of thoracic ultrasonography and thoracic radiography to detect active infectious bronchopneumonia in hospitalized dairy calves. J Vet Intern Med 35:2058–2068. https://doi.org/10.1111/jvim.16157

Bleul UT, Bircher BM, Kähn WK (2008) Effect of intranasal oxygen administration on blood gas variables and outcome in neonatal calves with respiratory distress syndrome: 20 cases (2004-2006). J Am Vet Med Assoc 233:289–293. https://doi.org/10.2460/javma.233.2.289

Binversie ES, Ruegg PL, Combs DK, Ollivett TL (2020) Randomized clinical trial to assess the effect of antibiotic therapy on health and growth of preweaned dairy calves diagnosed with respiratory disease using respiratory scoring and lung ultrasound. J Dairy Sci 103:11723–11735. https://doi.org/10.3168/jds.2019-18044

Bolton SE, von Keyserlingk MAG (2021) The dispensable surplus dairy calf: is this issue a "wicked problem" and where do we go from here? Front Vet Sci 8:660934. https://doi.org/10.3389/fvets.2021.660934

Borderas F, de Pasille AMB, Rushen J (2009) Temperature preferences and feed level of the newborn dairy calf. Appl Anim Behav Sci 120:56–61. https://doi.org/10.1016/j.applanim.2009.04.010

Braun U, Sicher D, Pusterla N (1996) Ultrasonography of the lungs, and mediastinum in healthy cows. Am J Vet Res 57:432–438

Breen MJ, Williams DR, Abdelfattah EM, Karle BM, Byrne BA, Lehenbauer TW, Aly SS (2023) Effect of group housing of preweaned dairy calves: health and fecal commensal antimicrobial resistance outcomes. Antibiotics 12:1019. https://doi.org/10.3390/antibiotics12061019

Brscic M, Leruste H, Heutinck LF, Bokkers EA, Wolthuis-Fillerup M, Stockhofe N, Gottardo F, Lensink BJ, Cozzi G, Van Reenen CG (2012) Prevalence of respiratory disorders in veal calves and potential risk factors. J Dairy Sci 95:2753–2764. https://doi.org/10.3168/jds.2011-4699

Bruer GG, Hagedorn P, Kietzmann M, Tohamy AF, Filor V, Schultz E, Mielke-Kuschow S, Meissner J (2019) Histamine H1 receptor antagonists enhance the efficacy of antibacterials against Escherichia coli. BMC Vet Res 15:55. https://doi.org/10.1186/s12917-019-1797-9

Buczinski S, Forté G, Francoz D, Bélanger AM (2014) Comparison of thoracic auscultation, clinical score, and ultrasonography as indicators of bovine respiratory disease in preweaned dairy calves. J Vet Intern Med 28:234–242. https://doi.org/10.1111/jvim.12251

Burciaga-Robles LO, Holland BP, Step DL, Krehbiel CR, McMillen GL, Richards CJ, Sims LE, Jeffers JD, Nomjou K, McCann PJ (2009) Evaluation of breath biomarkers and serum haptoglobin concentration for diagnosis of bovine respiratory disease in heifers newly arrived at a feedlot. Am J Vet Res 70:1291–1298. https://doi.org/10.2460/ajvr.70.10.1291

Burdick NC, Randel RD, Carroll JA, Welsh TH (2011) Interactions between temperament, stress, and immune function in cattle. Int J Zool 2011. https://doi.org/10.1155/2011/373197

Bus JD, Stockhofe N, Webb LE (2019) Abomasal damage in veal calves. J Dairy Sci 102:943–960. https://doi.org/10.3168/jds.2018-15292

Caswell JL, Hewson J, Slavic D, DeLay J, Bateman K (2012) Laboratory and postmortem diagnosis of bovine respiratory disease. Vet Clin Food Anim 28:419–441. https://doi.org/10.1016/j.cvfa.2012.07.004

Catry B, Dewulf J, Maes D, Pardon B, Callens B, Vanrobaeys M, Opsomer G, de Kruif A, Haesebrouck F (2016) Effect of antimicrobial consumption and production type on antibacterial resistance in the bovine respiratory and digestive tract. PLoS One 11:e0146488. https://doi.org/10.1371/journal.pone.0146488

Chamorro MF, Palomares RA (2020) Bovine respiratory disease vaccination against viral pathogens: modified-live versus inactivated antigen vaccines, intranasal versus parenteral, what is the evidence? Vet Clin North Am Food Anim Pract 36:461–472. https://doi.org/10.1016/j.cvfa.2020.03.006

Chase C, Kaushik RS (2019) Mucosal immune system of cattle: all immune responses begin here. Vet Clin N Am, Food Anim 35:431–451. https://doi.org/10.1016/j.cvfa.2019.08.006

Checkley SL, Campbell JR, Chirino-Trejo M, Janzen ED, Waldner CL (2010) Associations between antimicrobial use and the prevalence of antimicrobial resistance in fecal *Escherichia coli* from feedlot cattle in western Canada. Can Vet J 51:853–861

Chotiprasitsakul D, Han JH, Cosgrove SE, Harris AD, Lautenbach E, Conley AT, Tolomeo P, Wise J, Tamma PD (2018) Comparing the outcomes of adults with *Enterobacteriaceae* bacteremia receiving short-course versus prolonged-course antibiotic therapy in a multicenter, propensity score-matched cohort. Clin Infect Dis 66:172–177. https://doi.org/10.1093/cid/cix767

Clerbaux T, Gustin P, Detry B, Cao ML, Frans A (1993) Comparative study of the oxyhaemoglobin dissociation curve of four mammals: man, dog, horse and cattle. Comp Biochem Physiol A 106:687–694

Coghe J, Uystepruyst C, Bureau F, Lekeux P (1999) Non-invasive assessment of arterial haemoglobin oxygen saturation in cattle by pulse oximetry. Vet Rec 145:666–669

Coghe J, Uystepruyst CH, Bureau F, Detilleux J, Art T (2000) Lekeux P (2000): Validation and prognostic value of plasma lactate measurement in bovine respiratory disease. Vet J 160:139–146. https://doi.org/10.1053/tvjl.2000.0487

Colditz IG, Hine BC (2016) Resilience in farm animals: biology, management, breeding and implications for animal welfare. Anim Prod Sci 56:1961–1983. https://doi.org/10.1071/AN15297

Collie DDS (1992) Pulmonary function changes and clinical findings associated with chronic respiratory disease in calves. Br Vet 148:33–40

Creutzinger K, Pempek J, Habing G, Proudfoot K, Locke S, Wilson D, Renaud D (2021) Perspectives on the management of surplus dairy calves in the United States and Canada. Front Vet Sci 8:661453. https://doi.org/10.3389/fvets.2021.661453

Cuevas-Gómez I, McGee M, Sánchez JM, O'Riordan E, Byrne N, McDaneld T, Earey B (2021) Association between clinical respiratory signs, lung lesions detected by thoracic ultrasonography and growth performance in pre-weaned dairy calves. Irish Vet J 74:7. https://doi.org/10.1186/s13620-021-00187-1

Deepak ASS, Love WJ, Blanchard PC, Crossley B, Van Eenennaam AL, Lehenbauer TW (2021) Etiology and risk factors for bovine respiratory disease in pre-weaned calves on California dairies and calf ranches. Prev Vet Med 197:105506. https://doi.org/10.1016/j.prevetmed.2021.105506

Deretic V, Timmins GS (2019) Enhancement of lung levels of antibiotics by ambroxol and bromhexine. Expert Opin Drug Metab Toxicol 15:213–218. https://doi.org/10.1080/17425255.2019.1578748

Dicpinigaitis PV, Colice GL, Goolsby MJ, Rogg GI, Spector SL, Winther B (2009) Acute cough: a diagnostic and therapeutic challenge. Cough 5:11. https://doi.org/10.1186/1745-9974-5-11

Dißmann L, Reinhold P, Smith HJ, Amon T, Sergeeva A, Hoffmann G (2023) Evaluation of a respiration rate sensor for recording tidal volume in calves under field conditions. Sensors 23:4683. https://doi.org/10.3390/s23104683

Doyle D, Credille B, Lehenbauer TW, Berghaus R, Aly SS, Champagne J, Blanchard P, Crossley B, Berghaus R, Cochran S, Woolums A (2017) Agreement among 4 sampling methods to identify respiratory pathogens in dairy calves with acute bovine respiratory disease. J Vet Intern Med 31:954–959. https://doi.org/10.1111/jvim.14683

El Moussaoui R, de Bourgie CAJM, van den Broek P, Hustinx WN, Bresser P, van den Berk GEL, Krouwels FH, Bonten MJM, Weenink C, Bossuyt PMM, Speelman P, Opmeer BC, Prins JM (2006) Effectiveness of discontinuing antibiotic treatment after three days versus eight days in mild to moderate-severe community acquired pneumonia: randomised, double blind study. BMC 332:355

Elam NA (2007) Impact of Vitamin E supplementation on newly received calves: a review and meta-analysis. Prof Anim Sci 23:455–458

Ellis JA (2017) How efficious are vaccines against bovine respiratory syncytial virus in cattle? Vet Microbiol 206:59–68. https://doi.org/10.1016/j.vetmic.2016.11.030

Everest P (2007) Stress and bacteria: microbial endocrinology. Gut 56:1037–1038. https://doi.org/10.1136/gut.2006.117150

Eyler RF, Shvets K (2019) Clinical pharmacology of antibiotics. CASN 14:1080–1090. https://doi.org/10.2215/CJN.08140718

Fischer CD, Beatty JK, Duquette SC, Morck DW, Lucas MJ, Buret AG (2013) Direct and indirect anti-inflammatory effects of tulathromycin in bovine macrophages: inhibition of CXCL-8 secretion, induction of apoptosis, and promotion of efferocytosis. Antimicrob Agents Chemother 57:1385–1393

Francoz D, Buczinski S, Apley M (2012) Evidence related to the use of ancillary drugs in bovine respiratory disease (anti-inflammatory and others): are they justified or not? Vet Clin N Am Food Anim 28:23–38. https://doi.org/10.1016/j.cvfa.2011.12.003

Friton GM, Cajal C, Ramirez-Romero R, Kleemann R (2004) Clinical efficacy of meloxicam (Metacam) and flunixin (Finadyne) as adjuncts to antibacterial treatment of respiratory disease in fattening cattle. Berl Münch Tierärztl Wschr 117:304–309

Gallivan GJ, McDonell WN, Forrest JB (1989) Comparative pulmonary mechanics in the horse and the cow. Res Vet Sci 46:322–330. https://doi.org/10.1016/S0034-5288(18)31174-3

Gautier-Bouchardon AV, Ferré S, Le Grand D, Paoli A, Gay E, Poumarat F (2014) Overall decrease in the susceptibility of *Mycoplasma bovis* to antimicrobials over the past 30 years in France. PLoS One 9(2):e87672

Genicot B, Close R, Lindsey JK, Lekeux P (1995) Pulmonary function changes induced by three regimens of bronchodilating agents in calves with acute respiratory distress syndrome. Vet Rec 137:183–186. https://doi.org/10.1136/vr.137.8.183

Gehr P, Mwani DK, Ammann A, Malloiy GMO, Taylor CR, Weibel ER (1981) Design of the mammalian respiratory system. V. Scaling morphometric pulmonary diffusing capacity to body mass: wild and domestic mammals. Resp Physiol 44:61–86

Gleerup KB, Andersen PH, Munksgaard L, Forkman B (2015) Pain evaluation in dairy cattle. Appl Anim Behav Sci 171:25–32

Godden SM, Lombard JE, Woolums AR (2019) Colostrum management for dairy calves. Vet Clin North Am, Food Anim 35:535–556. https://doi.org/10.1016/j.cvfa.2019.07.005

González LA, Schwartzkopf-Genswein KS, Bryan M, Silasi R, Brown F (2012) Relationships between transport conditions and welfare outcomes during commercial long haul transport of cattle in North America. J Anim Sci 90:3640–3651. https://doi.org/10.2527/jas.2011-4796

Gorham J, Taccone FS, Hites M (2022) Ensuring target concentrations of antibiotics in critically ill patients through dose adjustment. Expert Opin Drug Metab Toxicol 18:177–187. https://doi.org/10.1080/17425255.2022.2056012

Grief SN, Loza JK (2018) Guidelines for the evaluation and treatment of pneumonia. Prim Care 45:485–503

Griffin D, Chengappa MM, Kuszak J, McVey DS (2010) Bacterial pathogens of the bovine respiratory disease complex. Vet Clin North Am Food Anim Pract 26:381–394

Grissett GP, White BJ, Larson RL (2015) Structured literature review of responses of cattle to viral and bacterial pathogens causing bovine respiratory disease complex. J Vet Intern Med 29:770–780

Gruber C, Messow C, Stöber M (1965) Histological studies of healing processes of surgical wounds in cattle. Tierärztl Wschr 72:49–53

Hanretty A, Gallagher JC (2018) Shortened courses of antibiotics for bacterial infections: a systematic review of randomized controlled trials. Pharmacotherapy 38:674–687. https://doi.org/10.1002/phar.2118

Hare WC (1955) The broncho-pulmonary segments in the sheep. J Anat 89(1955):387–402

Hassig M, Stadler T, Lutz H (2007) Transition from maternal to endogenous antibodies in newborn calves. Vet Rec 160:234–235. https://doi.org/10.1136/vr.160.7.234

Heuvelink A, Reugebrink C, Mars J (2016) Antimicrobial susceptibility of *Mycoplasma bovis* isolates from veal calves and dairy cattle in the Netherlands. Vet Microbiol 189:1–7

Hilton WM (2015) Management of preconditioned calves and impacts of preconditioning. Vet Clin N Am Food Anim Pract 31:197–207. https://doi.org/10.1016/j.cvfa.2015.03.002

Horvath I, Hunt J, Barnes PJ, Alving K, Antczak A, Baraldi E, Becher G, van Beurden WJ, Corradi M, Dekhuijzen R, Dweik RA, Dwyer T, Effros R, Erzurum S, Gaston B, Gessner C, Greening A, Ho LP, Hohlfeld J, Jobsis Q, Laskowski D, Loukides S, Marlin D, Montuschi P, Olin AC, Redington AE, Reinhold P, van Rensen EL, Rubinstein I, Silkoff P, Toren K, Vass G, Vogelberg C, Wirtz H, (ATS/ERS Task Force on Exhaled Breath Condensate) (2005) Exhaled breath condensate: methodological recommendations and unresolved questions. Eur Resp J 26:523–548

Idoate I, Ley BV, Schultz L, Heller M (2015) Acute phase proteins in naturally occurring respiratory disease of feedlot cattle. Vet Immun Immunopath 163:221–226

ISABV (2021) Prescriptions of antibiotics for animals in Switzerland. www.blv.admin.ch/blv/de/home/tiere/tierarzneimittel/antibiotika.html

Jaeger J, Liebler-Tenorio E, Kirschvink N, Sachse K, Reinhold P (2007) A clinically silent respiratory infection with *Chlamydophila* spp. in calves is associated with airway obstruction and pulmonary inflammation. Vet Res 38:711–728

Jones GF, Feeney DA, Mews C (1998) Comparison of radiographic and necropsy findings of lung lesions in calves after challenge exposure with Pasteurella multocida. Am J Vet Res 59:1108–1112

Jourquin S, Bokma J, De Cremer L, van Leenen K, Vereecke N, Pardon B (2022) Randomized field trial comparing the efficacy of florfenicol and oxytetracycline in a natural outbreak of calf pneumonia using lung reaeration as a cure criterion. J Vet Intern Med 36:820–828

Jourquin S, Lowie T, Debruyne F, Chantillon L, Vereecke N, Boyen F, Boone R, Bokma J, Pardon B (2023) Dynamics of subclinical pneumonia in male dairy calves in relation to antimicrobial therapy and production outcomes. J Dairy Sci 106:676–689. https://doi.org/10.3168/jds.2022-22212

Kay JM (1983) Pulmonary vasculature and nerves. Comparative morphological features of the pulmonary vasculature in mammals. Am Rev Resp Dis 128:53–57

Kirchhoff J, Uhlenbruck S, Goris K, Keil GM, Herrler G (2014) Three viruses of the bovine respiratory disease complex apply different strategies to initiate infection. Vet Res 45, 20; www.veterinaryresearch.org/content/45/1/20

Klein S (1996) Collection and examination of lung biopsies from calves with bronchopneumonia. Thesis, Univ Vet Med Hannover, Germany

Klein U, de Jong A, Youala M, El Garch F, Stevenin C, Moyaert H, Rose M, Catania S, Gyuranecz M, Pridmore A, Ayling RD (2019) New antimicrobial susceptibility data from monitoring of Mycoplasma bovis isolated in Europe. Vet Microbiol 238:108432. https://doi.org/10.1016/j.vetmic.2019.108432

Kneucker A, Langenberg A, Smith HJ, Reinhold P (2008) Functional residual capacity and lung diffusion capacity in calves with chronic airflow obstruction caused by persistent chlamydial infection. Eur Respir J 32(Suppl):52

Kotlikoff MI, Gillespie JR (1983) Lung sounds in veterinary medicine. Part I. Terminology and mechanisms of sound production. Compend Contin Educ Pract Vet 5:634–639

Kotlikoff MI, Gillespie JR (1984) Lung sounds in veterinary medicine. Part II. Deriving clinical information from lung sounds. Compend Contin Educ Pract Vet 6:462–467

Küntzel A, Oertel P, Trefz P, Miekisch W, Schubert J, Köhler H, Reinhold P (2018) Animal science meets agricultural practice: preliminary results of an innovative technical approach for exhaled breath analysis in cattle under field conditions. Berl Münch Tierärztl Wochenschr 131:444–452. https://doi.org/10.2376/0005-9366-17101

Lago A, McGuirk SM, Bennett TB, Cook NB, Nordlund KV (2006) Calf respiratory disease and pen microenvironments in naturally ventilated calf barns in winter. J Dairy Sci 89:4014–4025

Lakritz J, Wilson WD, Berry CR, Schrenzel MD, Carlson GP, Madigan JE (1993) Bronchointerstitial pneumonia and respiratory distress in young horses: clinical, clinicopathologic, radiographic, and pathological findings in 23 cases (1984-1989). J Vet Intern Med 7:277–288

Lava M, Pardon B, Schüpbach-Regula G, Keckeis K, Deprez P, Steiner A, Meylan M (2016) Effect of calf purchase and other herd-level risk factors on mortality, unwanted early slaugther, and use of antimicrobial group treatments in Swiss veal calf operations. Prev Vet Med 126:81–88

Lawnick U, Reinhold P, Doll K (2005) The therapeutic effect of inhalation of hypertonic saline solution in calves with bronchopneumonia. Tierärztl Prax 33:373–382

Lekeux P (1993) Pulmonary function in healthy, exercising and diseased animals. Flemish Vet J, Special issue, Gent

Lekeux P, Borceux JP, Boutet P, Bureau F, Coghe J, Uys-Tepruyst C (2002) : Recent advances in bovinepneumology. In : Kaske M, Scholz ZH, Höltershinken M (eds.) : Recent Developments and Perspectives in BovineMedicine. 12th World Buiatrics Congress: 18-23 August 2002, Hannover, 144–149.

Li Q, Zhou Q, Florez ID, Mathew JL, Shang L, Zhang G, Tian X, Fu Z, Liu E, Luo Z, Chen Y (2022) Short-course vs long-course antibiotic therapy for children with nonsevere community-acquired pneumonia: a systematic review and meta-analysis. JAMA Pediatr 176:1199–1207. https://doi.org/10.1001/jamapediatrics.2022.4123

Lohr M, Prohl A, Ostermann C, Diller R, Greub G, Reinhold P (2016) Effect of *Parachlamydia acanthamoebae* on pulmonary function parameters in a bovine respiratory model. Vet J 213:9–15

Lowie T, van Leenen K, Jourquin S, Pas ML, Bokma J, Pardon B (2022) Differences in the association of cough and other clinical signs with ultrasonographic lung consolidation in dairy, veal, and beef calves. J Dairy Sci 105:6111–6124. https://doi.org/10.3168/jds.2021-21570

Lundborg GK, Svensson EC, Oltenacu PA (2005) Herd-level risk factors for infectious diseases in Swedish dairy calves aged 0–90 days. Prev Vet Med 68:123–143

Lynn DJ, Pulendran B (2018) The potential of the microbiota to influence vaccine responses. J Leukoc Biol 103:225–231. https://doi.org/10.1189/jlb.5MR0617-216R

Lüthi K, Bähler C, Overesch G, Kaske M (2021) Laboratory diagnostic examinations as part of «Pathocalf» in case of herd problems on Swiss calf rearing and fattening farms. Schweiz Arch Tierheilk 163:493–503

Lysnyansky I, Ayling RD (2016) Mycoplasma bovis: mechanisms of resistance and trends in antimicrobial susceptibility. Front Microbiol 7:595. https://doi.org/10.3389/fmicb.2016.00595

Mahendran SA, Blackie N, Wathes DC, Booth RE (2023) Comparison of environment quality measurements between 3 types of calf housing in the United Kingdom. J Dairy Sci 106:2461–2474. https://doi.org/10.3168/jds.2022-22613

Maccari P, Wiedemann S, Kunz HJ, Piechotta M, Sanftleben P, Kaske M (2015) Effects of two different rearing protocoils for Holstein bull calves in the first 3 weeks of life on health status, metabolism and subsequent performance. J Anim Physiol Anim Nutr 99:737–746. https://doi.org/10.1111/jpn.12241

Maier GU, Love WJ, Karle BM, Dubrovsky SA, Williams DR, Champagne JD, Anderson RJ, Rowe JD, Lehenbauer TW, Van Eenennaam AL, Aly SS (2020) A novel risk assessment tool for bovine respiratory disease in preweaned dairy calves. J Dairy Sci 103:9301–9317

Masmeijer C, Devriendt B, Rogge T, van Leenen K, De Cremer L, Van Ranst B, Deprez P, Cox E, Pardon B (2019) Randomized field trial on the effects of body weight and shorttransport on stress and immune variables in 2- to 4-week-olddairy calves. J Vet Intern Med 33:1514–1529. https://doi.org/10.1111/jvim.15482

Marcato F, van den Brand H, Kemp B, Engel B, Schnabel SK, Hoorweg FA, Wolthuis-Fillerup M, van Reenen K (2022) Effects of transport age and calf and maternal characteristics on health and performance of veal calves. J Dairy Sci 105:1452–1468. https://doi.org/10.3168/jds.2021-2063

Masset N, Assié S, Herman N, Jozan T, Herry V (2022) Ultrasonography of the cranial part of the thorax is a quick and sensitive technique to detect lung consolidation in veal calves. Vet Med Sci 8:1229–1239

Mata M, Sarrion I, Armengot M, Carda C, Martinez I (2012) Respiratory syncytial virus inhibits ciliagenesis in differentiated normal human bronchial epithelial cells: effectiveness of N-acetylcysteine. PLoS One 7(10):e48037

McLaughlin RF, Tyler WS, Canada RO (1961) A study of subgross pulmonary anatomy in various mammals. Am J Anat 108:149–165

Meurer DG, Wolf S (2007): Formen der Entzündung. In: Meurer DG, Wolf S (eds) Allgemeine Pathologie: Kompendium für die Veterinärmedizin, 2. Aufl., Verlag Schattauer, Stuttgart, S. 183–222

Mitzner W (1991) Collateral ventilation. In: Crystal RG, West JB (eds) The Lung. Raven Press, New York, pp 1053–1063

Mukherjee S, Bassler BL (2019) Bacterial quorum sensing in complex and dynamically changing environments. Nat Rev Microbiol 17:371–382. https://doi.org/10.1038/s41579-019-0186-5

Nietfeld JC (2010) Field necropsy techniques and proper specimen submission for investigation of emerging infectious diseases of food animals. Vet Clin North Am Food Anim Pract 26:1–13

Nordlund KV, Halbach CE (2019) Calf barn design to optimize health and ease of management. Vet Clin North Am Food Anim Pract 35:29–45

Oertel P, Küntzel A, Reinhold P, Köhler H, Kolb J, Schubert J, Miekisch W (2018) Continuous real-time breath analysis in ruminants: effect of eructation on exhaled VOC profiles. J Breath Res 12:036014. https://doi.org/10.1088/1752-7163/aabdaf

Ollivett TL, Buczinski S (2016) On-farm use of ultrasonography for bovine respiratory disease. Vet Clin North Am Food Anim Pract 32:19–35. https://doi.org/10.1016/j.cvfa.2015.09.00

Ollivett TL, Caswell JL, Nydam DV, Duffield T, Leslie KE, Hewson J, Kelton D (2015) Thoracic ultrasonography and bronchoalveolar lavage fluid analysis in Holstein calves affected with subclinical lung lesions. J Vet Intern Med 29:1728–1734. https://doi.org/10.1111/jvim.13605

Ostermann C, Ruttger A, Schubert E, Schroedl W, Sachse K, Reinhold P (2013) Infection, di-sease, and transmission dynamics in calves after experimental and natural challenge with a bovine *Chlamydia psittaci* isolate. PLoS One 8:e64066

Ostermann C, Linde S, Siegling-Vlitakis C, Reinhold P (2014) Evaluation of pulmonary dysfunctions and acid–base imbalances induced by *Chlamydia psittaci* in a bovine model of respiratory infection. Multidiscip Resp Med 9:10

Pardon B, Buczinski S (2020) Bovine respiratory disease diagnosis: what progress has been made in infectious diagnosis? Vet Clin North Am Food Anim Pract 36:425–444

Pardon B, Buczinski S, Deprez PR (2019) Accuracy and inter-rater reliability of lung auscultation by bovine practitioners when compared with ultrasonographic findings. Vet Rec 185:109. https://doi.org/10.1136/vr.105238

Pempek J, Trearchis D, Masterson M, Habing G, Proudfoot K (2017) Veal calf health on the day of arrival at growers in Ohio. J Anim Sci 95:3863–3872. https://doi.org/10.2527/jas2017.1642

Penterman PM, Holzhauer M, van Engelen E, Smits D, Velthuis AGJ (2022) Dynamics of Mycoplasma dairy herds during acute clinical outbreaks. Vet J 283–284:105841. https://doi.org/10.1016/j.tvjl.2022.105841

Pieralli F, Mancini A, Crociani A (2016) Appropriate antibiotic therapy in critically ill patients. Ital J Med 10:275–281

Prall S, Bowles EJ, Bennett K, Cooke CG, Agnew T, Steel A, Hausser T (2020) Effects of essential oils on symptoms and course (duration and severity) of viral respiratory infections in humans: a rapid review. Adv Integr Med 7:218–221. https://doi.org/10.1016/j.aimed.2020.07.005

Pravettoni D, Buczinski S, Sala G, Ferrulli V, Bianchi F, Boccardo A (2021) Diagnostic accuracy of focused lung ultrasonography as a rapid method for the diagnosis of respiratory disease in dairy calves. J Dairy Sci 104:4929–4935. https://doi.org/10.3168/jds.2020-19377

Rabeling B, Rehage J, Döpfer D, Scholz H (1998) Ultrasonographic findings in calves with respiratory disease. Vet Rec 143:468–471

Reeve-Johnson (2001) Relationships between clinical and pathological signs of disease in calves infected with *Mannheimia* (*Pasteurella*) *haemolytica* type A1. Vet Rec 149:549–552. https://doi.org/10.1136/vr.149.18.549

Reinhold P, Elmer S (2002) Consequences of short term fluctuations of the environmental temperatures in calves. Part 2: Effects on the health status of animals within three weeks after exposure. Dtsch Tierärztl Wschr 109:193–200

Reinhold P, Födisch G (1993) Lungenfunktionsdiagnostik bei gesunden und an Pneumonie erkrankten Kälbern. Mh Vet Med 48:113–117

Reinhold P, Knobloch H (2010) Exhaled breath condensate: lessons learned from veterinary medicine. J Breath Res 4:017001. https://doi.org/10.1088/1752-7155/4/1/017001

Reinhold P, Smith HJ, Langenberg A, Lekeux P (1998a) Measurement of respiratory impedance in healthy calves using the impulse oscillation technique – physiological and methodological aspects. Vet J 155:27–38

Reinhold P, Smith HJ, Close R, Genicot B, Lekeux P (1998b) Validation of impulse oscillometry in Friesian and Blue Belgian calves with respect to changes in extrathoracic upper airway resistance. Res Vet Sci 65:93–101

Reinhold P, Becher G, Rothe M (2000) Evaluation of the measurement of leukotriene B4 concentrations in exhaled condensate as a noninvasive method for assessing mediators of inflammation in the lungs of calves. Am J Vet Res 61:742–749. https://doi.org/10.2460/ajvr.2000.61.742

Reinhold P, Rabeling B, Günther H, Schimmel D (2002) Comparative evaluation of ultrasonography and lung function testing with the clinical signs and pathology of calves inoculated experimentally with *Pasteurella multocida*. Vet Rec 150:109–114

Reinhold P, Reissig S, Jaeger J, Langenberg A, Smith HJ (2007) Capnovolumetry is a useful technique to detect peripheral airway obstruction. Eur Respir J 30, Suppl 51, 8s (233)

Reinhold PE, Gierschner P, Küntzel A, Schubert JK, Miekisch W, Köhler HU (2020) Ruminants. In: Beauchamp J, Davis C, Pleil J (eds) Breathborne biomarkers and the human volatilome, 2nd edn. Elsevier, pp 441–460. ISBN: 9780128199671

Renaud D, Pardon B (2022) Preparing male dairy calves for the veal and dairy beef industry. Vet Clin Food Anim 38:77–92. https://doi.org/10.1016/j.cvfa.2021.11.006

Renaud DL, Kelton DF, LeBlanc SJ, Haley DB, Duffield DF (2018a) Calf management risk factors on dairy farms associated with male calf mortality on veal farms. J Dairy Sci 101:1785–1794. https://doi.org/10.3168/jds.2017-13578

Renaud DL, Duffield TF, LeBlanc SJ, Ferguson S, Haley DB, Kelton DF (2018b) Risk factors associated with mortality at a milk-fed veal calf facility: a prospective cohort study. J Dairy Sci 101:2659–2668. https://doi.org/10.3168/jds.2017-13581

Reschke C, Schelling E, Michel A, Remy-Wohlfender F, Meylan M (2017) Factors associated with colostrum quality and effects on serum gamma globulin concentrations of calves in Swiss dairy herds. J Vet Intern Med 31:1563–1571

Ridley A, Hateley G (2018) Mycoplasma bovis investigations in cattle. Vet Rec 183:256–258

Roadknight N, Mansell P, Jongman E, Courtman N, Fisher A (2021) The welfare of young calves transported by road. J Dairy Sci 104:6343–6357. https://doi.org/10.3168/jds.2020-19346

Robinson NE (1982) Some functional consequences of species differences in lung anatomy. Adv Vet Sci Comp Med 26:1–33

Rocker A (2001) Ultrasonographic findings in the lung of clinically healthy and pneumonic calves: correlations to clinical, haematological and impulsoscillometric results. Thesis, Univ Vet Med Hannover, Germany

Royer S, DeMerle KM, Dickson RP, Prescott HC (2018) Shorter versus longer courses of antibiotics for tinection in hospitalized patients: a systematic review and meta-analysis. J Hosp Med 13:336–342. https://doi.org/10.12788/jhm2905

Saco Y, Bassols A (2023) Acute phase proteins in cattle and swine: a review. Vet Clin Pathol 52(Suppl. 1):50–63

Schmidt H, Philipp H, Hamel U, Quirke JF (1998) Investigations on the treatment of bovines suffering from acute respiratory disease with Bisolvon in combination with either enrofloxacin, cefquinome, ceftiofur or florfenicol. Tierärztl Prax 26(G):127–132

Schneider B (1995) Comparative sonographic and radiological examination of lung diseases in calves. Thesis, Univ Vet Med Hannover, Germany

Schnyder P, Schönecker L, Schüpbach-Regula G, Meylan M (2019) Effects of management practices, animal transport and barn climate on animal health and antimicrobial use in Swiss veal calf operations. Prev Vet Med 167:146–157

Schönecker L, Schnyder P, Schüpbach-Regula G, Meylan M, Overesch G (2020) Prevalence and antimicrobial resistance of opportunistic pathogens associated with bovine respiratory disease isolated from nasopharyngeal swabs of veal calves in Switzerland. Prev Vet Med 185:105182. https://doi.org/10.1016/j.prevetmed.2020.105182

Schroedl W, Fuerll B, Reinhold P, Krueger M, Schuett C (2001) A novel acute phase marker in cattle: lipopolysaccharide binding protein (LBP). J Endotoxin Res 7:49–52

Schroedl W, Büchler R, Wendler S, Reinhold P, Muckova P, Reindl J, Rhode H (2016) Acute phase proteins as promising biomarkers: perspectives and limitations for human and veterinary medicine. Proteomics Clin Appl 10:1077–1092

Scott NA, Andrusaite A, Andersen P, Lawson M, Alcon-Giner C, Leclaire C, Caim S, Le Gall G, Shaw T, Connolly JPR (2018) Antibiotics induce sustained dysregulation of intestinal T cell immunity by perturbing macrophage homeostasis. Sci Transl Med 10:464

Shortall O, Green M, Brennan M, Wapenaar W, Kaler J (2017) Exploring expert opinion on the practicality and effectiveness of biosecurity measures on dairy farms in the United Kingdom using choice modeling. J Dairy Sci 100:2225–2239. https://doi.org/10.3168/jds.2016-11435

Smith H, Reinhold P, Goldman M (2005) Forced oscillation technique and impulse oscillometry. Eur Respir Monogr 31:72–105

Smith RA, Step DL, Woolumns AR (2020) Bovine respiratory disease: looking back and looking forward – what do we see? Vet Clin North Am Food Anim Pract 36:239–251

Spellberg B (2016) The new antibiotic mantra – "shorter is better". JAMA Intern Med 176:1254–1255

Stabler SL, Fagerberg DJ, Quarles CL (1982) Effects of oral and injectable tetracyclines on bacterial drug resistance in feedlot cattle. Am J Vet Res 43:1763–1766

Strutzke S, Fiske D, Hoffmann G, Ammon C, Heuwieser W, Amon T (2019) Technical note: development of a noninvasive respiration rate sensor for cattle. J Dairy Sci 102:690–695

Sudaryatma PE, Saito A, Mekata H, Kubo M, Fahkrajang W, Mazimpaka E, Okabayashi T (2020) Bovine respiratory syncytial virus enhances the adherence of Pasteurella multocida to bovine lower respiratory tract epithelial cells by upregulating the platelet-activating factor receptor. Front Microbiol 11:1676. https://doi.org/10.3389/fmicb.2020.01676

Svensson EC, Lundborg GK, Emanuelson U, Olsson SO (2003) Morbidity in Swedish dairy calves from birth to 90 days of age and individual calf-level risk factors for infectious diseases. Prev Vet Med 68:123–143

Sydler T, Braun U, Estermann U, Pospischil A (2004) A comparison of biopsy and post-mortem findings in the lungs of healthy cows. J Vet Med A 51:184–187

Tauber SC, Nau R (2008) Immunomodulatory properties of antibiotics. Curr Mol Pharmacol 1:68–79

Theurer ME, Larson RL, White BJ (2015) Systematic review and meta-analysis of the effectiveness of commercially available vaccines against bovine herpesvirus, bovine viral diarrhea virus, bovine respiratory syncytial virus, and parainfluenza type 3 virus for mitigation of bovine respiratory disease complex in cattle. J Am Vet Med Assoc 246:126–142. https://doi.org/10.2460/javma.246.1.126

Thickman DI, Ziskin MC, Goldenberg NJ, Linder BE (1983) Clinical manifestations of the comet tail artifact. J Ultrasound Med 2:225–230. https://doi.org/10.7863/jum.1983.2.5.225

Timsit E, Christensen H, Bareille N, Seegers H, Bisgaard M, Assié S (2013) Transmission dynamics of *Mannheimia haemolytica* in newly-received beef bulls at fattening operations. Vet Microbiol 161:295–304. https://doi.org/10.1016/j.vetmic.2012.07.044

Tschoner T (2021) Methods for pain assessment in calves and their use for the evaluation of pain during different procedures — a review. Animals 2021(11):1235. https://doi.org/10.3390/ani11051235

Tucker A, McMurtry IF, Reeves JT, Alexander AF, Will DH, Grover RF (1975) Lung vascular smooth muscle as a determinant of pulmonary hypertension at high altitude. Am J Phys 228:762–767. https://doi.org/10.1152/ajplegacy.1975.228.3.762

Uystepruyst C, Reinhold P, Coghe J, Bureau F, Lekeux P (2000) Mechanics of the respiratory system in healthy newborn calves using impulse oscillometry. Res Vet Sci 68:47–55. https://doi.org/10.1053/rvsc.1999.0337

Valentine-King MA, Cisneros K, James MO, Huigens RW, Brown MB (2019) Turning the tide against antibiotic resistance by evaluating novel, halogenated phenazine, quinoline, and NH125 compounds against Ureaplasma species clinical isolates and Mycoplasma type strains. Antimicrob Agents Chemother 63:e02265–e02218. https://doi.org/10.1128/AAC.02265-18

Van Boemmel C (2000) Examination of calves with enzootic bronchopneumonia: clinical, laboratory diagnostic and impulse oscilloresistometric results. Thesis, Univ Vet Med Hannover, Germany

Van Driessche L, Valgaeren BR, Gille L, Boyen F, Ducatelle R, Haesebrouck F, Deprez P, Pardon B (2017) A deep nasopharyngeal swab versus nonendoscopic bronchoalveolar lavage for isolation of bacterial pathogens from preweaned calves with respiratory disease. J Vet Intern Med 31:946–953

Van Leenen K, Van Driessche L, De Cremer L, Masmeijer C, Boyen F, Deprez P, Pardon B (2020a) Comparison of bronchoalveolar lavage fluid bacteriology and cytology in calves classified based on combined clinical scoring and lung ultrasonography. Prev Vet Med 176:104901. https://doi.org/10.1016/j.prevetmed.2020.104901

Van Leenen K, Jouret J, Demeyer P, Van Driessche L, De Cremer L, Masmeijer C, Boyen F, Deprez P, Pardon B (2020b) Associations of barn air quality parameters with ultrasonographic lung lesions, airway inflammation and infection in group-housed calves. Prev Vet Med 181:105056. https://doi.org/10.1016/j.prevetmed.2020.105056

van Leenen K, Jouret J, Demeyer P, Vermeir P, Leenknecht D, Van Driessche L, De Cremer L, Masmeijer C, Boyen F, Deprez P, Cox E, Devriendt B, Pardon B (2021) Particulate matter and airborne endotoxin concentration in calf barns and their association with lung consolidation, inflammation, and infection. J Dairy Sci 104:5932–5947. https://doi.org/10.3168/jds.2020-18981

Vandermeulen J, Bahr C, Johnston D, Earley B, Tullo E, Fontana I, Guarino M, Exadaktylos V, Berckmans D (2016) Early recognition of bovine respiratory disease in calves using automated continuous monitoring of cough sounds. Comput Electron Agric 129:15–26. https://doi.org/10.1016/j.compag.2016.07.014

Veit HP, Farrell RL (1978) The anatomy and physiology of the bovine respiratory system relating to pulmonary disease. Cornell Vet 68:555–581

Verbrugghe E, Boyen F, Gaastra W, Bekhuis L, Leyman B, Van Parys A, Haesebrouck F, Pasmans F (2012) The complex interplay between stress and bacterial infections in animals. Vet Microbiol 155:115–127. https://doi.org/10.1016/j.vetmic.2011.09.012

Warren LA, Mandell IB, Bateman KG (2010) An audit of transport conditions and arrival status of slaughter cattle shipped by road at an Ontario processor. Can J Anim Sci 90:159–167. https://doi.org/10.4141/CJAS09068

Weber M, Gierschner P, Klassen A, Kasbohm E, Schubert JK, Miekisch W, Reinhold P, Köhler H (2021) Detection of paratuberculosis in dairy herds by analyzing the scent of feces, alveolar gas, and stable air. Molecules 26:2854. https://doi.org/10.3390/molecules26102854

Weibel ER, Marques LB, Constantinopol M, Doffey F, Gehr P, Taylor CR (1987) Adaptive variation in the mammalian respiratory system in relation to energetic demand. VI. The pulmonary gas exchanger. Resp Physiol 69:81–100

Welling V, Lundeheim N, Bengtsson B (2020) A pilot study in Sweden on efficacy of benzylpenicillin, oxytetracycline, and florfenicol in treatment of acute undifferentiated respiratory disease in calves. Antibiotics 9:736. https://doi.org/10.3390/antibiotics9110736

Wilson DJ, Canning D, Giacomazzi T, Keels K, Lothrop R, Renaud DL, Sillett N, Taylor D, Van Huigenbos H, Wynands B, Zuest D, Fraser D (2020a) Hot topic: health and welfare challenges in the marketing of male dairy calves – findings and consensus of an expert consultation. J Dairy Sci 103:11628–11635. https://doi.org/10.3168/jds.2020-18438

Wilson DJ, Stojkov J, Renaud DL, Fraser D (2020b) Risk factors for poor health outcomes for male calves undergoing transportation in western Canada. Can Vet J 61:1265–1272

Woolums AR (2015) The bronchopneumonias (respiratory disease complex of cattle, sheep, and goats). In: Smith BL (ed) Large animal internal medicine, 5th edn. Mosby Elsevier, St. Louis, MO, pp 584–603

Zulkifli I, Abubakar AA, Sazili AQ, Goh YM, Imlan JC, Kaka U, Sabow AB, Awad EA, Othman AH, Raghazali R, Phillips CJC, Quaza Nizamuddin HN, Mitin H (2019) The effects of sea and road transport on physiological and electroencephalographic responses in Brahman crossbred heifers. Animals 9:199. 10.3390/ani9050199. PMID: 31035550; PMCID: PMC6563091

# Respiratory Diseases in Pigs

Maria Pieters and Albert Canturri

**Abstract**

Respiratory diseases cause significant issues in pigs worldwide. Welfare concerns, disease burden, decreased production, and low worker morale can all be effects of swine respiratory diseases. Various infectious and noninfectious agents, alone or in combination, can induce or predispose to respiratory disease in pigs. Viruses and bacteria are the main types of infectious agents affecting the respiratory tract of pigs. Infections of polymicrobial origin are usually termed porcine respiratory disease complex and group disease manifestations of multiviral, multibacterial, or mixed nature. In this chapter, the main diseases affecting the respiratory tract of pigs are briefly described, making emphasis on the basics of etiology, epidemiology, clinical presentation, lesions, treatment, and control for each agent. Viruses such as porcine circovirus type 2, porcine reproductive and respiratory syndrome virus, and influenza A virus and bacteria, namely, *Mycoplasma hyopneumoniae*, *Actinobacillus pleuropneumoniae*, *Pasteurella multocida*, and *Bordetella bronchiseptica*, are depicted given their importance in swine health. Other infectious agents and factors potentially affecting the respiratory system of pigs are listed, and readers are encouraged to dig deeper into the understanding of the processes that drive the lack of health of this important body system.

**Keywords**

Swine · Respiratory disease · Virus · Bacteria · Lung · Cough · Lesions · Pneumonia · Disease transmission · Pathogen

## 1 Introduction

Respiratory diseases can significantly affect the welfare and production of farmed animals, especially pigs. The diagnosis, prevention, treatment, and control of respiratory diseases has improved dramatically in the most recent decades. However, pathologies of the respiratory tract continue to be one of the most prevalent health issues in porcine production worldwide. The prevailing presence of respiratory pathologies in pigs in the field is evidenced in various forms, from increased disease burden, decreased animal production, diagnostic laboratory case submissions, and carcass condemnation at slaughter, among others. Thus, the prevention and control of respiratory diseases in pigs is of paramount importance. Commonly, strategies to avoid or to manage the effects caused

M. Pieters (✉) · A. Canturri
Veterinary Population Medicine Department, Veterinary Diagnostic Laboratory & Swine Disease Eradication Center, College of Veterinary Medicine, University of Minnesota, St. Paul, MN, USA
e-mail: piet0094@umn.edu; cantu0028@umn.edu



J. J. Gross (ed.), *Production Diseases in Farm Animals*,
https://doi.org/10.1007/978-3-031-51788-4_13

by respiratory infections vary depending on the disease-causing agent(s). Overall, optimization of management practices that can lead to decreased pathogen transmission, preventative measures that can aid escaping infection, and treatments, either palliative or curative, are put in place alone or in combination, to decrease the problems caused by respiratory diseases. Therefore, often times, the control of respiratory diseases in pigs includes antibiotic administration. Several different compounds with distinct modes of action can be used to treat respiratory disease in pigs. Although the beneficial effects of antimicrobials in the control and prevention of disease are widely accepted, the development of antimicrobial resistances is currently of major concern, both in veterinary and human medicine. Regulations are being adopted at local, national, and international levels, aiming at promoting antibiotic stewardship and judicious use.

The development of respiratory disease in pigs can have a multifactorial origin, many times resulting from the combination of the effects of both primary and opportunistic pathogens, the environment, and management practices. Thus, the ultimate disease control strategy is the elimination of pathogens from the herd, when feasible, and the implementation of strict biosecurity measures. Raising healthy pigs is not only a matter of animal welfare, sustainable food animal production, and responsible practice, but also beneficial from the economic standpoint. The morale and well-being of pig farmers, caretakers, and veterinary practitioners are also essentially influenced by rearing pigs that are free of respiratory disease.

## 2 Polymicrobial Diseases

### 2.1 Porcine Respiratory Disease Complex

In the respiratory tract of pigs, simultaneous infections by more than one type of pathogen, such as viruses, bacteria, and parasites, are much more common than infections with a single pathogen. In the mixed infection scenarios (i.e., co-infection), respiratory signs can be caused by one or more primary pathogens. The disease-causing agents harbor virulence factors that enable them to overcome the host immune defenses and to cause disease by themselves, together with other secondary pathogens, or with opportunistic commensal microorganisms that thrive as a consequence of the impaired immunity. Porcine respiratory disease complex (PRDC) was the term composed to describe such co-infections in pigs, clinically characterized by slow growth, decreased feed efficiency, lethargy, anorexia, fever, cough, and dyspnea, and in which multiple primary and secondary pathogens are detected by laboratory diagnostic testing (Thacker et al. 2001). Several studies have evaluated the effect that multiple infections, including viral-viral, viral-bacterial, and bacterial-bacterial infections, have on the respiratory system of pigs, evidencing that in numerous cases, synergistic interactions between pathogens occur and lead to an increase in the severity and duration of respiratory disease (Saade et al. 2020).

## 3 Viral Diseases

### 3.1 Porcine Circovirus Type 2

Porcine circoviruses are nonencapsulated DNA viruses, extremely small in size, and with a high mutation capacity. In pigs, four different types of porcine circoviruses have been described (Opriessnig et al. 2020). Porcine circovirus type 1 (PCV1) was the first discovered, in 1974, as a human cell line contaminant. The virus infects pigs in moderate to low prevalence worldwide, but has no pathogenic potential. Porcine circovirus type 2 (PCV2) was discovered in the late 1990s and is the most widely known, as it has been, and continues to be, associated with high prevalence and severe disease if not controlled. Recently, two other circoviruses, porcine circovirus type 3 (PCV3) and type 4 (PCV4), have been discovered via metagenomic sequencing, in 2016 and 2019, respectively, and much remains to be known about prevalence and pathogenicity in pigs of these two viruses. However, PCV3 has been associated with reproductive failure and multisystemic inflammation, although only the latter has been reproduced experimentally (Tan et al. 2021). Lastly, PCV4 has been detected in pigs with respiratory and enteric disease, although its clinical significance and pathogenesis warrants further investigation (Zhang et al. 2020). In this chapter, only PCV2 will be described since it is the viral type most associated with respiratory disease in pigs.

PCV2 is ubiquitous in both domestic and wild pigs worldwide. The principal route of transmission between an infected and a naïve pig is horizontal through the oronasal route, although the virus is excreted in secretions from different organ systems, namely, saliva, urine, feces, nasal and ocular secretions, milk, and semen. Transplacental transmission has also been demonstrated experimentally. In the field, most pigs become infected at 4–11 weeks of age and viral DNA can be detected in serum up to 22 weeks (Segales et al. 2019).

PCV2 is considered an immunosuppressive virus as it attacks cells of the immune system, mainly cells from the lymphocytic and monocytic lineages, thus promoting and increasing the severity of co-infections with other pathogens. Infection with the virus triggers different clinical manifestations, being the most important the systemic disease (PCV2-SD, formerly known as porcine multisystemic wasting syndrome (PMWS)), in which the respiratory disease is included.

It is important to note that most infections with PCV2 are subclinical in nature, and only a proportion of infected pigs develop PCV2-SD, usually between 8 and 16 weeks of age. The disease is characterized by failure to thrive and wasting with variable respiratory distress, diarrhea, and jaundice.

At necropsy, macroscopic characteristic lesions include enlargement of lymph nodes and interstitial pneumonia, while microscopically the most distinctive lesion is a moderate to severe lymphocytic depletion and substitution by histiocytes and giant multinucleated cells in the lymph nodes and, in lesser extent, in the lungs. Histiocytes occasionally contain eosinophilic, botryoid viral inclusions.

PCV2 can have a high economic impact at the farm if uncontrolled, and thus the use of commercial vaccines, which are very successful at protecting pigs from clinical disease, is generalized in swine production worldwide. Immunity against PCV2, either natural or artificial through vaccination, is not sterilizing, meaning that although vaccination will reduce infection pressure and the appearance of clinical disease and lesions, the virus will not be eliminated from the farm and pigs will continue to get infected if there is virus circulation (Karuppannan and Opriessnig 2017). It has been described that vaccination, besides preventing clinical disease, also helps improving production parameters in subclinically infected pigs (Franzo and Segalés 2020).

### 3.2 Porcine Reproductive and Respiratory Syndrome Virus

Porcine reproductive and respiratory syndrome virus (PRRSV) is a small, enveloped RNA virus, first isolated in the early 1990s in Europe (Wensvoort et al. 1991) and North America (Collins et al. 1992) from pigs with respiratory disease signs and reproductive failures. By means of phylogenetic and antigenic analyses, the two strains were considered to belong to two different genotypes of the virus, the European or type 1 genotype (PRRSV-1) and the American or type 2 genotype (PRRSV-2) (Murtaugh et al. 1995). Nowadays, the genetic diversity between the two genotypes is so remarkable that PRRSV-1 and PRRSV-2 are classified as two distinct viral species within the *Arteriviridae* family. Indeed, the high mutation rates for this RNA virus, and perhaps recombination between viral strains in co-infections, have led to the emergence of novel genetic variants that have been further classified in linages and sub-linages that circulate at different rates over time (Paploski et al. 2021). Importantly, different viral strains have been associated with increased severity of clinical signs and high mortality, although a clear terminology to refer to those strains has not been uniformly used (Ruedas-Torres et al. 2021).

Porcine reproductive and respiratory virus is present in most countries with relevant porcine production, although with notable exceptions such as Brazil, Norway, and Switzerland, among others.

The main mode of transmission of PRRSV is direct through the nasal, oral, or coital route, as infected pigs shed virus in oral and nasal secretions, urine, semen, mammary secretions, and occasionally feces (Pileri and Mateu 2016). Interestingly, the parenteral route of exposure requires the lowest minimum dose to infect pigs, and $\leq$ 20 PRRSV particles inoculated intramuscularly are sufficient to cause infection (Yoon et al. 1999). This route is particularly important when considering even standard husbandry practices, such as ear notching, tail docking, and castration, or the administration of medications and vaccines. The virus is also transmitted transplacentally (i.e., vertically) from viremic sows to fetuses, especially in the last trimester of pregnancy.

Indirect transmission of PRRSV has been demonstrated in a wide array of mechanical vectors, as the virus is moderately stable outside the host and can contaminate the environment. Aerosol transmission has also been experimentally demonstrated, although its efficiency may be dependent on the PRRSV strain involved and, in the case of long-distance airborne transmission, on environmental and meteorological factors (Dee et al. 2010). Additionally, the role of arthropods as mechanical vectors of the virus has also been studied (Otake et al. 2003; Pitkin et al. 2009).

Within a given pig herd, PRRSV tends to become endemic and circulates indefinitely between two general subpopulations of pigs: pigs that are chronically infected with the virus and newly susceptible pigs that enter the herd either through birth or through lateral introduction, most likely as replacement gilts in a sow farm.

In general, PRRSV infection can be classified as a chronic persistent infection, as there is continuous viral replication in specific target cells and the host immune response is inefficient, resulting in a slow clearance of the virus over time. This lengthy infection period allows for the identification of three different stages following PRRSV exposure: acute infection, persistence, and extinction (Lunney et al. 2016).

Acute infection is characterized by rapid localization to primary sites of replication, which contain replication-permissive cells characterized by the presence of the cell receptor CD163. These permissive cells are primarily pulmonary alveolar and intravascular macrophages in the lungs and monocyte-derived macrophages in lymphoid tissues. After pig exposure to the virus, viremia develops rapidly, and is usually detectable within 6–48 hours, while viral loads peak by 4–14 days post-infection, depending on the virulence of the strain. Respiratory clinical disease is mostly observed at this stage, with anorexia, lethargy, dyspnea, and variable coughing caused by diffuse interstitial pneumonia. Clinical signs and lesions are influenced by the virulence of the strain and the presence of concurrent infections, as PRRSV predisposes to a higher incidence of secondary endemic diseases, such as streptococcal meningitis or Glasser's disease, which significantly contribute to the increase in mortality observed in the growth-finishing period (Brockmeier et al. 2017).

Viral titers in serum decrease rapidly after peaking, and when viremia ends, the persistence stage starts. Persistence is defined by the absence of clinical signs and a progressive reduction of viral replication in lymphoid tissues and shedding. When viral shedding finally ceases, the extinction phase begins, in which there is presence of residual virus in lymphoid tissues that has not been cleared by the immune system. The ability to define the transition between one stage and the next is heavily dependent on diagnostics; thus, the correct choice of the clinical specimen and the diagnostic test in each stage is essential (Henao-Diaz et al. 2020).

Preventing the introduction of PRRSV into a negative herd, or a new strain into a positive herd, is probably the most efficient control measure for the disease. The implementation of strict biosecurity to prevent introduction of the virus, directed at restricting and controlling both swine and staff movements, is considered critical. Examples include testing the incoming semen and gilts, disinfection protocols for transport vehicles, or sanitation standards for farm personnel, such as shower-in farms for personnel, or UV light chambers for incoming supplies. Despite the high costs associated with it, air filtration has been implemented in regions with high density of swine farms, helping at significantly decreasing the risk of introduction of novel PRRSV into large sow farms (Alonso et al. 2013).

In the event of an outbreak, specific treatments aiming to combat PRRSV infection are limited, and thus, most efforts are directed at reducing the clinical signs by administering anti-inflammatory drugs and/or antibiotics to control secondary infections. Additionally, the use of modified live virus vaccines can reduce production losses and the shedding of wild-type virus (Linhares et al. 2012). Thus, regardless of the herd status or the absence of clinical signs, PRRSV surveillance based on routine testing is recommended.

### 3.3 Swine Influenza Virus

Influenza viruses are included in the family *Orthomyxoviridae* and, although genera A, B, C, and D have been identified in pigs, only influenza A virus (IAV) is clinically significant for porcine production worldwide. Influenza A virus possesses eight segments of negative-sense RNA coated with nucleoprotein and altogether delimited by a lipid envelope, a feature that makes the virus susceptible to most detergents used in livestock farms. The segmented structure of IAV genome permits that, when two distinct IAVs co-infect the same host cell, the RNA segments exchange or reassort between the two viruses during viral replication, resulting in two antigenically novel viruses. This process, termed antigenic shift or genetic reassortment, is one of the main sources of genetic variability for IAVs. Antigenic drift, a different source of genetic variability, is characterized by the gradual accumulation of mutation during RNA replication due to the error-prone nature of the viral RNA-dependent RNA polymerase (Flint et al. 2015). The constant evolution of IAVs has led to the generation of several antigenic subtypes that are defined by the major surface glycoproteins, hemagglutinin (H) and neuraminidase (N), which project from the surface of the viral envelope. There are 16 Hs and 9 Ns that are distinguished antigenically and genetically, although in pigs there are only two types of Hs, H1 and H3. The combination of H and N in a given virus is what characterizes the viral subtype, for example, H1N1 or H3N2.

Pigs are regarded as a "mixing vessel" for influenza viruses, as their respiratory tract has cell receptors for both avian and human influenza viruses and thus, novel reassortant viruses can be generated when viruses of different origins infect a single porcine respiratory epithelial cell (Ma 2020). This particular aspect of IAV epidemiology is critical for public health, as human infections with swine IAVs of various subtypes have been reported. In 2009, the appearance of an H1N1 originated from the reassortment between North American and Eurasian linages of swine IAV infected millions of humans in many parts of the world, resulting in a pandemic (Garten et al. 2009). A posteriori, this same subtype, termed H1N1pdm09, infected swine populations via humans, highlighting the high frequency of bidirectional, human-to-swine transmission of IAVs in pig farms (Lopez-Moreno et al. 2022).

In pig populations, IAV is rapidly transmitted via direct contact through oronasal secretions. Aerosolized virus and mechanical vectorization of the virus through fomites and staff movements within a farm can be also significant routes of transmission leading to outbreaks.

Pig oronasal exposure to IAV results in an acute infection of the respiratory tract. The virus binds to and replicates in epithelial cells expressing sialic acid receptors in the upper and especially in the lower respiratory tract, with bronchial and bronchiolar epithelial cells being the main cellular targets. Consequently, viral circulation and shedding occurs exclusively via the respiratory route and there is no viremia. The acute, transient nature of the infection is exemplified by the fact that the virus is cleared from the respiratory system within 7 days after experimental inoculation. Accordingly, the onset of clinical signs is sudden after an incubation period or 1–3 days and characterized, at first, by high fever, anorexia, and lack of natural activity, resulting from the rapid production of cytokines by the host, and then followed by sneezing and coughing after direct epithelial cell damage. Morbidity is high but mortality is very low in high health scenarios without co-infections, and rapid recovery occurs 5–7 days after the onset.

Lesions are restricted to the respiratory tract, with mild catarrhal rhinitis with epithelial cell degeneration and exfoliation and, more commonly, multifocal broncho-interstitial pneumonia with bronchiolar necrosis. This specific type of pneumonia is macroscopically characterized by multifocal areas of consolidation and reddening, sharply demarcated, following a lobular pattern (also called "chessboard pattern") mainly in the apical and medial lung lobes. Microscopically, the affected areas show necrosis of the epithelial cells lining the bronchioles, which become obstructed by necrotic epithelial cells and neutrophils. The alveolar septa surrounding the affected airways are infiltrated and expanded by mononuclear cells (Caswell and Williams 2016).

Vaccination is the most common strategy to control IAV infections in pig farms. Most commercial vaccines are inactivated whole virus vaccines, and thus, there is no production of mucosal IgA antibodies, which are essential for virus blockage in the upper respiratory tract. Inactivated whole virus vaccines induce serum antibodies to the viral H, and because H antibodies are subtype and strain specific (i.e., there is no serologic cross-reaction between H subtypes), the inactivated vaccine strains should match the circulating strains in order to induce protection (Salvesen and Whitelaw 2021). Other ways of controlling IAV infections at the herd level are by means of internal and external biosecurity, preventing the introduction of positive gilts, limiting the movement of farmworkers between rooms, and implementing preventive protocols targeting the transmission from humans to pigs.

## 4 Bacterial Diseases

### 4.1 *Mycoplasma hyopneumoniae*

*Mycoplasma hyopneumoniae* (*M. hyopneumoniae*) is arguably the most important respiratory bacterium affecting pigs. Although several *Mycoplasma* species have been identified and isolated from swine, only *M. hyopneumoniae* has been shown to be a primary respiratory pathogen, causing enzootic pneumonia, and being one of the main contributors of PRDC. In fact, *M. hyopneumoniae* is usually termed as the "door opener" to most other important respiratory pathogens in pigs. Mare and Switzer (1965) and Goodwin et al. (1965) identified *M. hyopneumoniae* simultaneously. The bacterium has a small size and set of genes and lacks a cell wall. The agent is difficult to isolate in the laboratory from clinical specimens. Strains of different virulence have been described in the literature for *M. hyopneumoniae*. Research studies have shown that single pigs can be infected with different variants of *M. hyopneumoniae* at the same time, and it has been suggested that disease severity is influenced by the number of variants involved in the infection (Michiels et al. 2017). Various strains of *M. hyopneumoniae* have been sequenced to date; nevertheless, genetic markers of virulence have not been described.

The disease caused by *M. hyopneumoniae* is chronic in nature and tends to become endemic in farms, hence the name, enzootic pneumonia. Infections with *M. hyopneumoniae* have been reported in most pig-producing areas around the world, although a small number of countries indicate freedom of disease in commercial production, following intensive disease eradication programs (Yeske et al. 2020). Pigs of all ages are deemed susceptible to infection with *M. hyopneumoniae*. Newborn piglets appear to be born free of the bacterium, but can be exposed by the dam at an early age, during the lactation period (Sibila et al. 2007). Pathogen transmission from the dam to the piglet and subsequent transmission among litter and pen mates seems to be the most likely way in which pigs are infected and the disease becomes endemic.

Infections with *M. hyopneumoniae* are slow in onset, with an incubation period of approximately 15 days in experimental conditions, which could be longer, or go unnoticed at first in field settings. Once infected, pigs may show persistence and potentially shed the bacterium for extended periods of up to 8 months (Pieters et al. 2009). The long bacterial persistence in infected hosts, combined with the constant influx of incoming pigs into a herd, which is common in swine operations, are the crucial pieces that help create and maintain endemicity (Pieters et al. 2020).

The pathogenesis of *M. hyopneumoniae* has not been fully described to date. However, inhalation of infected aerosols seems to be the most likely way in which pigs are initially exposed to this pathogen (Pieters and Maes 2019). The bacterium expresses various adhesion proteins, which participate in the attachment to cilia in the airways. To date, no toxins have been identified to be involved in *M. hyopneumoniae* that contribute to disease causation. It is generally accepted that *M. hyopneumoniae* does not cause direct cell or tissue damage and that an indirect effect of the host immune response is responsible for the marked reaction that characterizes the infection.

Pigs infected with *M. hyopneumoniae* exhibit a temporary dry cough, which is the only true clinical sign associated with infection. Production parameters, such as average daily gain and days to market weight, will be affected, especially in endemically infected herds. Although mortality is not usually increased with *M. hyopneumoniae* infections, morbidity can be high. Gross lesions in lungs of infected pigs are characterized by purplish-dark areas of consolidation, mainly in the tip of the apical, medial, and accessory lobes. Assessment of lung lesions at the abattoir is frequently employed to monitor *M. hyopneumoniae* disease at the herd level (Maes et al. 2023). Microscopically, although lesions are not pathognomonic, they feature marked lymphocyte infiltration in perivascular and peribronchiolar areas (Pieters and Maes 2019).

Control of *M. hyopneumoniae* infection is usually attempted with the application of management practices that can help decrease pathogen transmission and disease development, such as gilt acclimation (Pieters and Fano 2016), all-in/all-out, proper pig density, and ventilation (Marco et al. 2020). Administration of vaccines (mainly bacterins) can help reduce transmission of the bacterium (Garza-Moreno et al. 2018; Betlach et al. 2021b) and the burden and detrimental effects of the disease, although infection will not be prevented. Infections caused by *M. hyopneumoniae* can be treated with antibiotics. However, it is important to keep into account that the bacterium is naturally resistant to antimicrobial compounds that have an effect on the cell wall, such as β-lactams. Treatment of *M. hyopneumoniae* outbreaks can be accomplished with antimicrobials like macrolides, pleuromutilins, lincosamides, or fluoroquinolones (Gautier-Bouchardon 2020), although clearance of the pathogen will not be achieved using this approach (Betlach et al. 2021a). Treatments should be applied under strict veterinary oversight and careful consideration should be given to the pig's age and production stage, among others.

Eradication of *M. hyopneumoniae* from swine herds, production systems, and countries has been reported with the use of different methods, from depopulation and repopulation to partial depopulation and a combination of herd closure, vaccination, and treatments (Yeske et al. 2020).

## 4.2 *Actinobacillus pleuropneumoniae*

*Actinobacillus pleuropneumoniae* (*A. pleuropneumoniae*) is a small Gram-negative, encapsulated coccobacillus that affects swine of different ages, although outbreaks are usually observed in finishing pig (Gottschalk and Broes 2019). The disease condition caused by *A. pleuropneumoniae* is known as porcine pleuropneumonia. Bacterial growth of *A. pleuropneumoniae* in laboratory conditions can be difficult based on the bacterium's requirements. However, supplementation of blood agar plates with nicotinamide adenine dinucleotide (NAD) increases the likelihood of culture and isolation (Pohl et al. 1983). To date, two bacterial biovars (I and II) and 16 serotypes have been identified for *A. pleuropneumoniae*, some of which are avirulent. Biovar I and biovar II differ by the NAD requirement for growth. Serotypes are based on the bacterial capsule, which confers resistance from the first phagocytosis onslaught (Rioux et al. 2000). The virulence of *A. pleuropneumoniae* can be classified as low, moderate, or high and is mainly serotype and co-infection dependent. Other virulence factors of *A. pleuropneumoniae* include LPS endotoxin, hemolysins/cytotoxins (Apx I–IV), and outer membrane proteins, among others.

Infections with *A. pleuropneumoniae* have been reported worldwide. However, the incidence and prevalence of the infection can vary dramatically depending on the country or region. The bacterium can be harbored in the tonsils of healthy hosts (carriers) and can be transmitted, mainly nose-to-nose, to other pigs in the population, which can develop disease quickly post exposure. Three forms of the disease have been described: peracute, acute, and chronic. High fever and anorexia characterize the peracute form, with skin cyanosis evidenced in affected pigs, which collapse and foamy oral and nasal discharges tinged with blood are observed (Gottschalk and Broes 2019). In the acute form, morbidity as well as mortality can be high, with evident clinical signs, most pigs in the group affected, and pathogen spread that does not seem to be associated with specific management factors. The acute form of disease is usually observed in new introductions of the pathogen, especially in high health status herds. In the chronic form, which appears to be associated with specific serotypes, clinical signs may not be present and mortality is low (2–4%), although morbidity can remain above 50%. In these cases, growth reduction and production cost can still inflict sufficient damage in the herd.

A myriad of clinical signs can be observed in herds affected with *A. pleuropneumoniae*, including, but not limited to, fever, some coughing, labored breathing, nose bleeding, and death. At the lung level, involvement can be between 50 and 90%, with notable congestion, focal necrosis, friable tissue, fibrinous pleuritis, and accumulation of pleural fluid that have been reported, which can be summarized as fibrinohemorrhagic and necrotizing pneumonia (Gottschalk and Broes 2019).

Pig treatment with antibiotics can help in controlling a clinical outbreak. However, the speed of disease development can be faster than that of diagnosis, and a proportion of the affected population may pass the effective treatment point (Desrosiers 2004). It is highly recommended to obtain a bacterial culture and perform antibiotic susceptibility testing prior to treatment administration, as drug activity can vary based on serotypes. Penicillins, especially injectable, oxytetracyclines, lincosamides, fluroquinolones, and pleuromutilins show good activity against *A. pleuropneumoniae* (Gottschalk and Broes 2019). Pig vaccination for *A. pleuropneumoniae* seems to confer a protection, in the form of reduced clinical signs and lung lesions (Del Pozo Sacristan et al. 2014), that is serotype-specific.

Disease eradication can be attempted using various methods, with depopulation and repopulation with *A. pleuropneumoniae*-free pigs being highly effective. However, other forms of disease elimination can be successful for this pathogen, most of which are directed at either identifying or excluding carriers pigs from the population in question.

## 4.3 *Pasteurella multocida*

*Pasteurella multocida* (*P. multocida*) is a Gram-negative bacterium that can be easily grown in the laboratory and causes respiratory disease in pigs, such as pneumonic pasteurellosis and progressive atrophic rhinitis, alone or when in combination, with other bacterial agents. Five distinct capsular types of the bacterium are recognized, types A, B, D, E, and F (Rimler and Rhoades 1987).

The non-toxigenic type A *P. multocida* appears to be mainly associated with the lower respiratory tract, especially lungs, and causes fibrinous pleuritis that is associated with pneumonia. Commonly, a co-infection between *P. multocida* and *M. hyopneumoniae* is reported in veterinary diagnostic laboratories. The transmission of *P. multocida* among pigs appears to be nose-to-nose via aerosols between pen mates and from the sow to the piglet (Register and Brockmeier 2019).

Pigs affected with pneumonia are mostly grower-finishers, which show obvious respiratory distress, fever, cough, abdominal breathing, and emaciation in older animals. In some cases, infected pigs die extremely emaciated. In modern swine production, infections with *P. multocida* do not seem to represent a common respiratory issue in most large commercial systems. Gross lung lesions may resemble enzootic pneumonia, as anteroventral pneumonia is observed. Nevertheless, signs of pneumonia may not always be present. Pleuritis and lungs with a thin, translucent, and dry aspect are common, with adhesions to the thoracic cavity. In mild cases, interlobular adhesion is present. Less commonly, small to medium (2–10 mm) abscesses, mostly in anteroventral sites, can be observed at necropsy.

Antibiotic treatment with ampicillin, ceftiofur, enrofloxacin, and tulathromycin is preferred for pneumonic infections with *P. multocida* (Register and Brockmeier 2019). To date, no vaccines for *P. multocida* are generalized in the market, and autogenous vaccines may play a role to avoid disease outbreaks in specific geographical locations or production systems.

The toxigenic, type D *P. multocida* mainly causes progressive atrophic rhinitis, and its pathogenesis depends on the co-infection with *Bordetella bronchiseptica* (*B. bronchiseptica*). It has been shown that infection with *P. multocida* is facilitated by the initial colonization of the nasal cavity with *B. bronchiseptica* (Dugal et al. 1992). A protein toxin produced by *P. multocida* is crucial for development of progressive atrophic rhinitis by causing shortening of the snout via turbinate atrophy (Il'ina and Zasukhin 1975). At the same time and via mechanisms not completely known—although systemic effects of the *P. multocida* toxin have been suggested—progressive atrophic rhinitis slows pig growth and decreases feed efficiency. Certain pig management practices and environmental factors have also been suggested to contribute to the development to progressive atrophic rhinitis. Rutter et al. (1984) have suggested disease severity to be dependent on infection age, with older pigs exhibiting mild clinical presentation, but not lack of disease expression.

Young pigs will sneeze, snuffle, and snort, and nasal and ocular discharge may be observed as well. Snout deformation and shortening of the upper jaw are common, and considered a hallmark of infection, although it may not be evident in all pigs in an outbreak (Register and Brockmeier 2019).

Macroscopic lesions associated with progressive atrophic rhinitis include variable degrees of atrophy of the dorsal and ventral turbinates, leading to deviations of the nasal septum (Register and Brockmeier 2019). Microscopically, the bony trabeculae of the nasal turbinates are replaced with fibrous connective tissue (Pedersen and Elling 1984). At different stages of disease, various degrees of inflammation and repair are observed histologically.

As with most respiratory diseases, a combination of various strategies, including management, vaccination, and medication, can be utilized for effective control. Proper pig density, all-in/all-out, hygiene, and ventilation are a must to minimize pathogen transmission. Vaccination of the sow herd and maintaining a slightly mature parity structure can aid with prevention of disease development in piglets via transfer of a solid maternal immunity. Vaccine products for prevention of progressive atrophic rhinitis should be directed against *P. multocida* as well as *B. bronchiseptica*. Vaccination with a toxoid product for the *P. multocida* toxin may not prevent infection, but will prevent lesion development. Medication of the sow with antimicrobial compounds such as sulfonamides and tetracyclines is generally effective at decreasing the bacteria that are shed and potentially transmitted to piglets. However, bacterial isolation and antimicrobial susceptibility testing are strongly recommended for the application of antibiotic treatment (Register and Brockmeier 2019). Feed or water medication of growing pigs with tetracyclines, tilmicosin, and trimethoprim/sulfa can be used in the case of postweaning infections (Register and Brockmeier 2019). Ultimately, depopulation and repopulation can be utilized to eradicate the disease. Strict biosecurity measures and monitoring should be in place to maintain the health status of the herd.

### 4.4 *Bordetella bronchiseptica*

The Gram-negative bacterium *B. bronchiseptica* is a pathogen that can affect various animal species, including humans. In pigs, *B. bronchiseptica* is the causative agent of bronchopneumonia and nonprogressive atrophic rhinitis. The wide range of virulence observed with distinct strains of this bacterium suggest that different virulent factors play a role in disease development. Bacterial isolation of *B. bronchiseptica* can be obtained from samples of diseased as well as apparently healthy pigs. Susceptibility to the bacterium does not seem to be age-dependent and infections have been reported worldwide.

Transmission of *B. bronchiseptica* within pigs is mainly airborne, through droplets in close contact and aerosols at short distances. Entry of *B. bronchiseptica* into a herd can be through carrier pigs (Smith et al. 1982), hosts other than swine (Le Moine et al. 1987), or potentially insects (Beatson 1972). Moreover, the bacterium survives well in the environment.

Disease causation is associated with phenotypic modulation, in which certain virulence factors are expressed based on the change of temperature in the environment where the bacterium is located. Moreover, toxin expression contributes to the progression of disease (Brockmeier et al. 2019). Co-infection of *B. bronchiseptica* and other disease-causing agents is well documented, but perhaps the most important disease interaction occurs with *P. multocida* in the form of progressive atrophic rhinitis (see above), as colonization of the upper respiratory tract by *P. multocida* results from predisposition with *B. bronchiseptica*.

High morbidity and low mortality are characteristics of infections with *B. bronchiseptica* in pigs, although mortality can be significant in complicated cases in piglets with necrohemorrhagic bronchopneumonia (Brockmeier and Lager 2002). Typically, sneezing, coughing, and inflammation of the lachrymal duct are observed in *B. bronchiseptica* infection that is not complicated with other pathogens. In these cases, clinical signs appear only days after colonization and are present for a short number of weeks, even if pigs are still colonized (Brockmeier et al. 2019).

Nonprogressive atrophic rhinitis, caused by *B. bronchiseptica*, is characterized by atrophy of the nasal cavity, including the production of nasal exudate and mild to moderate atrophy of the nasal turbinates (Duncan et al. 1966), which is reversible in nature and most times may be unnoticed. On the other hand, necrohemorrhagic bronchopneumonia in piglets is acute, and gross lung lesions will consist of hemorrhagic consolidated areas and localized fibrinous pleuritis. Under the microscope, hemorrhage, necrosis, and concentration of inflammatory infiltrates are observed (Duncan et al. 1966).

Antibiotic treatments directed at respiratory disease in pigs are generally effective against *B. bronchiseptica*, with reported exceptions with ampicillin and cephalosporins (Zhao et al. 2011). Vaccines for prevention of disease with *B. bronchiseptica* are formulated either alone or in combination with toxigenic *P. multocida*, as protection from development of progressive atrophic rhinitis is highly desirable at herd level. Sow vaccination, especially pre-farrow, is generally practiced as a way to prevent pathogen transmission and increase immunity transfer to piglets (Riising et al. 2002).

## 5 Other Bacterial Agents Causing Respiratory Disease in Pigs

A number of bacterial agents have been identified and isolated in the respiratory tract of pigs. However, their pathogenic potential has not been documented for all. Notably, respiratory diseases in pigs have been reported when infection with agents like *A. suis* (pleuropneumonia), *Glaesserella parasuis* (polyserositis), *M. hyorhinis* (pneumonia and polyserositis), and *Chlamydia suis* (pneumonia) occurs.

## 6 Environmental and Management-Related Respiratory Diseases

Respiratory disease in pigs can also develop by the effect of nonpathogenic factors, which are usually related to the environment and/or management practices. Of special interest is the control of factors that can induce or potentiate disease in an indirect manner, by increasing pathogen transmission or promoting disease manifestation. Management strategies such as all-in/all-out, adequate ventilation, proper stocking density, and gilt acclimation can help reduce issues associated with disease in intensive production systems (Marco et al. 2020). Additionally, optimizing farm's climatic conditions, improving air quality, and decreasing air contaminants contribute to overall respiratory health. For instance, avoiding irritation and/or toxicity of the respiratory airways is of paramount importance in close facilities housing large numbers of pigs. Ammonia, hydrogen sulfide, and anhydrous ammonia are examples of gases that need to be monitored, as excessive concentrations can be detrimental and in some cases lethal to pigs (Ensley and Radke 2019).

## References

Alonso C, Murtaugh MP, Dee SA, Davies PR (2013) Prev Vet Med 112(1–2):109–117

Beatson SH (1972) Lancet 1:425–427

Betlach AM, Baumert D, Utrera V, Galina Pantoja L, Pieters M (2021a) Vet Microbiol 262:109222

Betlach A, Fano E, VanderWaal K, Pieters M (2021b) Vaccine. 39(4):767–774

Brockmeier SL, Lager KM (2002) Vet Microbiol 89:267–275

Brockmeier SL, Loving CL, Palmer MV, Spear A, Nicholson TL, Faaberg KS, Lager KM (2017) Vet Microbiol 203:6–17

Brockmeier SL, Register KB, Nicholson TL, Loving CL (2019) In: Zimmerman JJ, Karriker LA, Ramirez A (eds) Diseases of swine, 11th edn. Wiley, pp 767–777

Caswell JL, Williams KJ (2016) Respiratory system. In: Maxie MG (ed) Jubb, Kennedy and Palmer's pathology of domestic animals, 6th edn. Elsevier, pp 533–535

Collins JE, Benfield DA, Christianson WT, Harris L, Hennings JC, Shaw DP, Goyal SM, McCullough S, Morrison RB, Joo HS, Gorcyca D, Chladek D (1992) J Vet Diagn Investig 4(2):117–126

Dee S, Otake S, Deen J (2010) Virus Res 154(1–2):177–184

Del Pozo SR, Michiels A et al (2014) Vet Rec 174:302

Desrosiers R (2004) Epidemiology, diagnosis and control of swine diseases: Howard Dunne memorial lecture. Proc Am Assn Swine Vet 9:37

Dugal F, Bélanger M, Jacques M (1992) Can J Vet Res 56:260–264

Duncan JR, Ramsey FK, Switzer WP (1966) Am J Vet Res 27:467–472

Ensley SM, Radke S (2019) In: Zimmerman JJ, Karriker LA, Ramirez A (eds) Diseases of swine, 11th edn. Wiley, pp 1072–1087

Flint J, Racaniello VR, Rall GF, Skalka AM (2015) Principles of virology, 4th edn. ASM Press

Franzo G, Segalés J (2020) Pathogens 9(12):1–12

Garten RJ, Davis CT, Russell CA, Shu B, Lindstrom S, Balish A, Sessions WM, Xu X, Skepner E, Deyde V, Okomo-Adhiambo M, Gubareva L, Barnes J, Smith CB, Emery SL, Hillman MJ, Rivailler P, Smagala J, de Graaf M et al (2009) Science (New York, NY) 325(5937):197–201

Garza-Moreno L, Segalés J, Pieters M, Romagosa A, Sibila M (2018) Acclimation strategies in gilts to control Mycoplasma hyopneumoniae infection. Vet Microbiol 219:23–29. https://doi.org/10.1016/j.vetmic.2018.04.005

Gautier-Bouchardon AV (2020) In: Mycoplasmas in Swine (344 pp). Edition 1. Acco. ISBN 9789463797962

Goodwin RFW, Pomeroy AP, Whittlestone P (1965) Vet Rec 77:1247–1249

Gottschalk M, Broes A (2019) In: Zimmerman JJ, Karriker LA, Ramirez A (eds) Diseases of swine, 11th edn. Wiley, pp 749–766

Henao-Diaz A, Ji J, Giménez-Lirola L, Baum DH, Zimmerman J (2020) Res Vet Sci 131(February):173–176

Il'ina ZM, Zasukhin MI (1975) Sibirskii Nauchno-Issledovatel'skii Veterinaryni Institut. Omsk 25:76

Karuppannan AK, Opriessnig T (2017) Viruses 9(5):1–15

Le Moine V, Vannier P, Jestin A (1987) Prev Vet Med 4:399–408

Linhares DCL, Cano JP, Wetzell T, Nerem J, Torremorell M, Dee SA (2012) Vaccine 30(2):407–413

Lopez-Moreno G, Davies P, Yang M, Culhane MR, Corzo CA, Li C, Rendahl A, Torremorell M (2022) Zoonoses and Public Health

Lunney JK, Fang Y, Ladinig A, Chen N, Li Y, Rowland B, Renukaradhya GJ (2016) Ann Rev Anim Biosci 4:129–154

Ma W (2020) Virus Res 288:198118

Maes D, Sibila M, Pieters M, Haesebrouck F, Segales J, Oliveira LG (2023) Vet Res 54(1):8

Marco E, Yeske P, Pieters M (2020) In: Mycoplasmas in Swine (344 pp). Edition 1. Acco. ISBN 9789463797962

Mare CJ, Switzer WP (1965) Vet Med 60:841–846

Michiels A, Vranckx K, Piepers S et al (2017) Vet Res 48:2

Murtaugh MP, Elam MR, Kakach LT (1995) Arch Virol 140(8):1451–1460

Opriessnig T, Karuppannan AK, Castro AMMG, Xiao CT (2020) Virus Res 286(June):198044

Otake S, Dee SA, Rossow KD, Moon RD, Trincado C, Pijoan C (2003) Vet Rec 152(3):73–76

Paploski IAD, Pamornchainavakul N, Makau DN, Rovira A, Corzo CA, Schroeder DC, Cheeran MC-J, Doeschl-Wilson A, Kao RR, Lycett S, VanderWaal K (2021) Vaccine 9(6)

Pedersen KB, Elling F (1984) Comp Pathol 94:203–214

Pieters M, Fano E (2016) Vet Rec 178:122–123

Pieters MG, Maes D (2019) In: Diseases of swine (pp. 863–883), 11th ed. Blackwell, Hoboken, NJ: Wiley

Pieters M, Pijoan C, Fano E et al (2009) Vet Microbiol 134:261–266
Pieters M, Betlach A, Nathues H, Maes D (2020) In: Mycoplasmas in Swine (344 pp). Edition 1. Acco. ISBN 9789463797962
Pileri E, Mateu E (2016) Vet Res 47(1):1–13
Pitkin A, Deen J, Otake S, Moon R, Dee S (2009) Can J Vet Res 73(2):91–96
Pohl S, Bertschinger H, Frederiksen W et al (1983) Inst J Syst Bacteriol 33:510–514
Register KB, Brockmeier SL (2019) In: Zimmerman JJ, Karriker LA, Ramirez A (eds) Diseases of swine, 11th edn. Wiley, pp 884–897
Riising HJ, van Empel P, Witvliet M (2002) Vet Rec 150:569–571
Rimler RB, Rhoades KR (1987) J Clin Microbiol 25:615–618
Rioux S, Galarneau C, Harel J et al (2000) Microb Pathog 28:279–289
Ruedas-Torres I, Rodríguez-Gómez IM, Sánchez-Carvajal JM, Larenas-Muñoz F, Pallarés FJ, Carrasco L, Gómez-Laguna J (2021) Vet Microbiol 260
Rutter JM, Taylor RJ, Crighton WG et al (1984) Vet Rec 115:615–619
Saade G, Deblanc C, Bougon J, Marois-Créhan C, Fablet C, Auray G, Belloc C, Leblanc-Maridor M, Gagnon CA, Zhu J, Gottschalk M, Summerfield A, Simon G, Bertho N, Meurens F (2020) Vet Res 51(1):80
Salvesen HA, Whitelaw CBA (2021) Porcine Health Manag 7(1):1–17
Segales J, Allan GM, Domingo M (2019) Circoviruses. In: Zimmerman JJ, Karriker LA, Ramirez A (eds) Diseases of swine, 11th edn. Wiley, pp 473–487
Sibila M, Nofrarias M, Lopez-Soria S et al (2007) Vet Microbiol 121:352–356
Smith IM, Giles CJ, Baskerville AJ (1982) Vet Rec 110:488–494
Tan CY, Lin C-N, Ooi PT (2021) Transbound Emerg Dis 68(6):2915–2935
Thacker EL, Thacker BJ, Janke BH (2001) J Clin Microbiol 39(7):2525–2530
Wensvoort G, Terpstra C, Pol JM, ter Laak EA, Bloemraad M, de Kluyver EP, Kragten C, van Buiten L, den Besten A, Wagenaar F (1991) Vet Q 13(3):121–130
Yeske P, Betlach A, Holst S, Pieters M (2020) In: Mycoplasmas in swine (344 pp). Edition 1. Acco. ISBN 9789463797962
Yoon KJ, Zimmerman JJ, Chang CC, Cancel-Tirado S, Harmon KM, McGinley MJ (1999) Vet Res 30(6):629–638
Zhang H-H, Hu W-Q, Li J-Y, Liu T-N, Zhou J-Y, Opriessnig T, Xiao C-T (2020) Transbound Emerg Dis 67(3):1057–1061
Zhao Z, Wang C, Xue Y et al (2011) Vet J 188:377–340

# Part V

# Mammary Gland Health and Lactation

# Postpartum Dysgalactia Syndrome in Sows: A Review

Dominiek Maes and Chantal Farmer

**Abstract**

Postpartum dysgalactia syndrome (PDS) in sows is characterized by inadequate and insufficient colostrum and milk production during the first days after farrowing. PDS occurs worldwide and incurs major financial losses to affected pig herds. The pathophysiology is multifactorial and different possible pathways related to involvement of endotoxins, stress, and inappropriate body condition have been proposed. Diagnosis of PDS can be established based on local or general signs in the sows, and increased incidence of poor performance and mortality in piglets. Many different risk factors are associated with PDS: sow-related factors (e.g., genetics, parity, body condition, and constipation), nutrition (e.g., feeding strategy, feed composition, antioxidants, and contaminants), housing and climate conditions, and management practices (e.g., farrowing induction and supervision). Treatment of sows mostly includes non-steroidal anti-inflammatory drugs, oxytocin, and eventually antimicrobials. Piglets should receive an alternative source of energy and/or be transferred to another sow. Control of PDS largely relies on decreasing the risk factors.

**Keywords**

Agalactia · Colostrum yield · Milk yield · Parturition · Swine

## 1 Introduction

Postpartum or periparturient dysgalactia syndrome (PDS) in sows is characterized by inadequate and insufficient colostrum and milk production during the first days after farrowing. PDS occurs worldwide and incurs major financial losses to affected pig herds, as the disease decreases the survival and performance of piglets and leads to a higher sow replacement rate. Piglets that ingested insufficient colostrum and milk suffer from reduced performance, not only in the farrowing unit (Decaluwé et al. 2014a), but also post-weaning until the end of the fattening period (Declerck et al. 2015, 2016a). Using stochastic modelling, Niemi et al. (2017) estimated the losses due to PDS to be €300–€470 per affected sow. Costs due to increased labor for care and treatment of affected sows must also be accounted for.

D. Maes
Unit of Porcine Health Management, Faculty of Veterinary Medicine, Ghent University, Merelbeke, Belgium
e-mail: Dominiek.Maes@UGent.be

C. Farmer (✉)
Agriculture and Agri-Food Canada, Sherbrooke R & D Centre, Sherbrooke, QC, Canada
e-mail: chantal.farmer@AGR.GC.CA



J. J. Gross (ed.), *Production Diseases in Farm Animals*,
https://doi.org/10.1007/978-3-031-51788-4_14

The word "peripartal" or "periparturient" is constructed from the Greek prefix "peri-" which can be translated as "around" and the scientific term "parturition" meaning "giving birth." Although the term peripartal period is used by several authors, a clear definition of the time period it covers is usually not mentioned. In general, it relates to the period starting 7–5 days prior to parturition, which coincides with the transfer from the gestation to the farrowing unit of the sow (Papadopoulos et al. 2009), and ends several days to one week postpartum.

Many names to describe problems with reduced milk production in the postpartum sow have been used in the literature. These include mastitis-metritis-agalactia (MMA), chronic mastitis, agalactia complex, and lactation failure (Gerjets and Kemper 2009). The name MMA has been commonly used in the older literature and describes a condition where mastitis is the central symptom, with infected glands showing signs of inflammation. However, the name MMA is misleading because in many instances there is no metritis or complete agalactia, and the role of clinical mastitis is often not clear (Kaiser et al. 2020). Mastitis-metritis-agalactia is therefore currently considered as a subtype of PDS. The pathophysiology of PDS is complex and is not yet fully elucidated. It is generally accepted to be a multifactorial disease in which many different risk factors related to animal characteristics, nutrition, management, and housing can play a role (Martineau 2005). As the pathophysiology is not clear and there are numerous potential risk factors, the implementation of appropriate therapeutic and preventive measures in affected farms is a challenge.

The present chapter aims to review the current knowledge of PDS in sows. Emphasis is placed on the pathophysiology, clinical signs, occurrence, risk factors, treatment, and control measures.

## 2 Physiology of Lactation in the Sow

The lactation process in sows is under hormonal control. The combination of the decrease in concentrations of progesterone and the increase in prolactin concentrations in late gestation triggers lactogenesis. Prolactin is most important for its effects on sow lactation. An early study indicated that serum concentrations of prolactin are lower 41–58 hours postpartum in sows with lactation failure compared with unaffected sows (Threlfall et al. 1974). Peter et al. (1985) corroborated this by showing that inhibition of uterine secretion of prostaglandin-F2α after delivery could lower prolactin secretion and play a part in the onset of PDS. It was later demonstrated that prolactin is essential for mammary development taking place from day 90 of gestation to farrowing (Farmer and Petitclerc 2003), for the onset and also for the maintenance of lactation (Farmer et al. 1998b). When the prepartum peak in prolactin is inhibited, the onset of lactation is abolished leading to complete agalactia (Farmer et al. 1998b). There is also a positive relation between colostrum yield in sows and circulating concentrations of prolactin 30–40 hours prepartum (Foisnet et al. 2010), and the prolactin to progesterone ratio 2 days prepartum (Loisel et al. 2015). Prolactin is also important during established lactation. When prolactin concentrations are drastically decreased (from ≥100 ng/mL to ≤10 ng/mL) during any of a 4-week lactation, milk yield is inhibited (Farmer et al. 1998b). Rydhmer et al. (2001) also reported lower prolactin concentrations in sows with disturbed milk production on day 28 of lactation.

Recent results showed a pronounced drop in circulating concentrations of triiodothyronine (T3) from 30 days before farrowing to 3 days postpartum in sows affected with PDS compared to unaffected sows (Patra et al. 2021). These authors suggested that such a rapid decrease in T3 could potentially serve as a metabolic marker to predict PDS in sows. Nevertheless, more studies are needed to substantiate this.

## 3 Pathophysiology of Postpartum Dysgalactia Syndrome

The pathophysiology of PDS is multifactorial and there are at least three possible pathways that can impact lactogenesis and galactopoiesis (Martineau et al. 2012). The first one is via endotoxins that affect the innate immune system. It is likely due to health conditions such as constipation, cystitis, mastitis, and endometritis. The second pathway is due to stress that can arise from housing, environmental, or management conditions. The third pathway is related to body condition, namely the "fat sow syndrome" (FSS) or the "over-muscled sow syndrome" (OMSS). All these pathways will potentially lead to PDS via impaired mammary development, problems at farrowing, and metabolism or hormonal imbalances during the colostrum or milk production period. In turn, these conditions will cause dishomeorhesis and/or behavioral problems at farrowing and/or nursing, which could then induce PDS. Each of these aspects will be covered in detail in the following sections of this chapter.

## 4 Clinical Signs

Hypogalactia within the first 3 days postpartum is the primary clinical sign of PDS. However, this is not always easy to observe in the sow during the initial stages of the disease (Klopfenstein et al. 1997). Careful observation of the piglets is likely a better way to detect the problem. Piglets of problem litters are hungry, restless, fight more and for longer periods, appear thin, lose weight, and often remain close to the sow between suckling occasions. This proximity between sow and piglets increases the risk for piglet mortality. Litters from PDS sows also show an increased incidence of diarrhea requiring antimicrobial treatment, and piglet weights within a litter become heterogeneous. Weaning weights are also decreased.

Affected sows might have increased rectal temperature. An arbitrary threshold value of 39.5 °C in the period from 12 to 24 hours postpartum has been proposed to classify sows as having fever (Kemper and Gerjets 2009). However, the threshold value used to define increased rectal temperature as fever is not straightforward because healthy sows also have increased rectal temperatures in the periparturient period and because rectal temperatures may vary considerably between sows and within sows over time, depending on different factors such as metabolic status, parity, and circadian rhythm (Messias de Braganca et al. 1998; Claeyé et al. 2015) (Fig. 1). Decaluwé et al. (2013) observed an increase in rectal temperature of 0.8 °C, namely from 38.1 °C the day before farrowing to 38.9 °C the day after farrowing. In other studies (Messias de Braganca et al. 1998) more pronounced increases in rectal temperature were reported. The physiological increase in rectal temperature during the periparturient period can be explained by the greater metabolism for milk production, the release of pro-inflammatory cytokines in response to tissue damage caused by parturition, and possibly also by the higher ambient temperatures in the farrowing compared to the gestation environment (Pomorska-Mól et al. 2020). Nevertheless, there appears to be some form of physiological hyperthermia in PDS sows during the periparturient period (Klopfenstein et al. 1997).

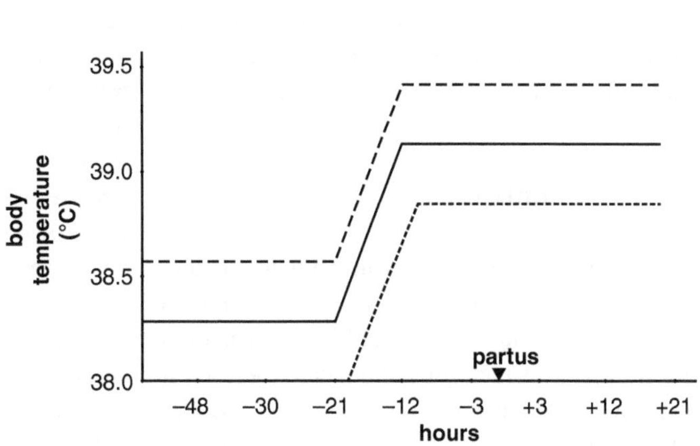


**Fig. 1** Physiological increase of body temperature in sows around parturition. The space between both dotted lines indicates the possible variation between sows (adapted from Martineau et al. 2012)

**Table 1** Clinical signs associated with postpartum dysgalactia syndrome in sows (adapted from Martineau et al. 1992)

| | Signs |
|---|---|
| Sows | • Local: absent, mastitis, vaginal discharge.<br>• General: absent, fever (>39.5 °C), prostration, anorexia. |
| Piglets | • < 1 w: mortality ↑, diarrhea, heterogeneity of weight among litter.<br>• > 1 w: heterogeneity of weight among litter, low weaning weight. |
| Herd | • Decreased number of piglets/sow/year.<br>• Consequences of variable and low piglet weaning weights.<br>• Variation in the number of affected sows and intensity of signs.<br>• Decreased sow longevity. |

Body temperature should, however, not be used as a single tool in the clinical evaluation of sows with PDS (Klopfenstein et al. 1997). Other symptoms may also be present, such as anorexia or reduced appetite, depression, and constipation. In the case of mastitis, affected glands may show signs of inflammation, including edema, swelling and redness, and may feel warm upon palpation, which can be painful. Posterior glands are generally more affected than anterior ones (Bostedt et al. 1998; Baer and Bilkei 2005), likely because of the higher risk for fecal contamination. However, other studies showed no difference in the incidence of mastitis related to anatomical location of the gland (Gerjets et al. 2008). Edema without signs of mastitis can be present as well, especially in primiparous sows (Nachreiner et al. 1971). In the case of hypogalactia without mastitis, the udder tends to be limp, the mammary glands are not filled with milk and palpation is not painful.

It should be remembered that not all sows exhibit the same range or intensity of clinical signs and that the number of affected sows may vary within a herd. An overview of early and late signs of PDS that may be present in sows, piglets, and at the herd level is shown in Table 1.

## 5 Diagnosis

Studies were carried out to investigate whether sows that will develop PDS can be identified prior to farrowing. This would be most helpful for preventative measures that cannot be used if detection is done once clinical signs, such as reduced milk intake by piglets, are present. To this end, different potential biomarkers, such as white blood cell types and cytokine responses, were examined in sows during the periparturient period. Kaiser et al. (2018) suggested that white blood cells, neutrophil and lymphocyte counts, and serum concentrations of tumor necrosis factor alpha (TNF-α), interleukin (IL) 6, haptoglobin, and iron differed between affected and unaffected sows and may potentially serve as tools for the detection of PDS. On the other hand, inflammatory changes were also found in unaffected sows rendering the diagnostic value of measuring inflammatory markers unclear. Such detection methods definitely need further investigation.

Combining measurements of inflammatory markers with assessment of clinical and behavioral parameters may improve identification of sows at risk of developing PDS. Before parturition, of the white blood cell counts measured, only lymphocytes differed between affected and unaffected sows. Wierzchosławski et al. (2018) found a strong correlation between selected acute-phase proteins and PDS severity in the periparturient period. Pomorska-Mól et al. (2020) demonstrated highly dynamic changes in concentrations of different cytokines in sows postpartum. They reported that in healthy and PDS-affected sows concentrations of IL-6 and TNF-α in serum were very variable, whereas those of IL-8 and IL-10 were relatively stable in healthy sows but highly variable in PDS-affected sows. The authors concluded that these cytokines cannot be useful early diagnostic markers for sows with PDS. Also, Stiehler et al. (2016) reported that haptoglobin and C-reactive protein cannot be used as markers for detection of PDS, mainly due to inherently large variations in values of healthy sows. In most cases, concentrations of such biomarkers in serum show a large between animal variation due to factors such as reproductive state and parity (Verheyen et al. 2007), thereby hampering their predictive value for PDS. In addition, blood sampling in sows at the end of gestation is difficult to do and will lead to stress.

Hoy (2006) reported that PDS has a negative impact on the subsequent reproductive performance of sows. In two German pig herds suffering from PDS, the following parameters were affected by PDS (compared with non-PDS sows): frequency of late oestrus (5.9% versus 3.4%), return to oestrus after insemination (21.7% versus 16.1%), frequency of abortion (2.3% versus 0.9%), and sow mortality (4.1% versus 1.2%). These authors showed that the negative effects of PDS were present in primiparous as well as in multiparous sows.

## 6 Occurrence

The prevalence of PDS, either at the animal or herd level, depends on the criteria used to assess the occurrence and the severity of the problem. Signs and criteria used to define PDS vary largely between studies (reviewed by Papadopoulos 2008). Jorsal (1986) defined PDS when one or more of the following signs were present: agalactia, anorexia, constipation, vaginal discharge, inflammation of mammary gland, or a rectal temperature higher than 39.8 °C. In contrast, Persson et al. (1996) considered sows with PDS based on a milk somatic cell count (SCC) greater than 10 million/mL, even if rectal temperature was normal (<39.5 °C) and no clinical mastitis was present. Bäckström et al. (1984) defined sows affected by PDS as sows with presence of starving piglets and one or more of the clinical signs (anorexia, fever, depression, signs of mastitis, or vaginal discharge) within 3 days after parturition. Van Gelder and Bilkei (2005) classified sows as hypogalactic, agalactic without clinical mastitis, and agalactic with clinical mastitis using an individual sow score based on severity of fever, milk production, degree of disturbed appetite, and assessment of the condition of newborn piglets. Hirsch et al. (2003) reported a detailed scoring system and classified sows as affected by PDS when there was at least a slight disturbance of their general demeanor, a reduction in feed intake by up to one-third, either pathological vaginal discharge (even only drops) and/or barely perceptible inflammation of mammary glands and a body temperature equal or higher than 39.5 °C.

The large variation in criteria used for diagnosis of PDS makes it difficult to directly compare prevalence figures between studies. Bäckström et al. (1984) reported MMA in 6.9% of 16,405 farrowings recorded over one year in 31 swine herds in Illinois, with in-herd prevalence ranging from 1.1% to 37.2%. The mean prevalence was 13% in multiparous sows and 4.2% in primiparous sows. Threlfall and Martin (1973) studied 27,656 farrowings in the state of Missouri and found that 13% of the sows were affected by MMA. The range of incidence of MMA in Swedish herds varied from 5.5% in small herds to 10.3% in large herds (Bäckström et al. 1982). In Denmark, Larsen and Thorup (2006) used a definition of PDS as one or more of the following: loss of appetite, abnormal mammary glands (redness, swelling), or rectal temperature > 39.4 °C, and found an occurrence of 32.5% on the first day post-farrowing, 31.5% on the second day, and 10.1% on the third day. In a survey by Papadopoulos et al. (2010) including 110 pig herds, 34% of the farms reported to have experienced PDS problems during the year preceding the study. This high percentage can be explained by the fact that herds with minor problems were considered as PDS herds. A Swiss study showed a maximum occurrence of PDS in herds of 75%, with the average being around 37% (Pendl et al. 2017). Angjelovski et al. (2019) reported a 23.3% occurrence of PDS in sows examined the day after farrowing, and the average occurrence between farms ranged from 15% to 38%.

# 7 Risk Factors for Postpartum Dysgalactia Syndrome

Because PDS is a multifactorial disease, many different risk factors are involved (Kemper 2020). The identification and correction of risk factors are key for reducing the incidence of PDS problems in herds. The risk factors will be arbitrarily classified into sow-related factors (Table 2), nutrition and housing (Table 3), management (Table 4), and infections or other disease conditions (Table 5).

**Table 2** Sow-related factors increasing the risk for postpartum dysgalactia syndrome

| Risk factor | References |
|---|---|
| Genetic predisposition → resistance to infections (heritability 0.02–0.20) | Awad Masalmeh et al. (1990), Lingaas and Ronningen (1991), Preissler et al. (2012) |
| Parity: | |
| – Increased in young sows (parity 1,2)<br>– Increased in older sows (parity >4) | Bostedt et al. (1998), Hoy (2006), Gerjets et al. (2011)<br>Baer and Bilkei (2005) |
| Body condition and conformation: | |
| – Fat sow syndrome<br>– Overmuscled sow syndrome | Martineau and Klopfenstein (1996), Oliviero et al. (2009)<br>Solignac et al. (2010) |
| PDS in previous lactation | Hultén et al. (2003) |
| Long gestation (>116 days) | Awad Masalmeh et al. (1990) |
| Litter data: | |
| – More piglets born alive (PDS sows had 1.1 piglets more than non-PDS sows) | Bostedt et al. (1998) |
| – >13 piglets born alive (odds ratio = 1.65)<br>– One (odds ratio = 1.04) or > 1 (odds ratio = 1.45) stillborn piglets | Gerjets et al. (2011)<br>Gerjets et al. (2011) |
| Dystocia and long farrowing duration (>3 hours) | Bostedt et al. (1998) |
| Constipation | Hermansson et al. (1978); Martineau et al. (1992); Bostedt et al. (1998); Oliviero et al. (2009); Jenny et al. (2015) |

**Table 3** Nutritional factors and housing conditions increasing the risk for postpartum dysgalactia syndrome in sows

| Risk factor | References |
|---|---|
| Feeding strategy: | |
| – Feeding ad libitum the first days after farrowing (versus restricted feeding)<br>– Feeding ad libitum after farrowing (versus one day before and on the day of farrowing)<br>– Rapid changes in nutrition | Papadopoulos et al. (2010)<br>Neil et al. (1996)<br>Plonait and Bickhart (1997) |
| Feed composition: | |
| – Lack of crude fiber in the diet (<7–8%)<br>– High n-6:n-3 ratio (10:1 versus 2:1) of fat in the diet the last week before farrowing<br>– Not adding whey to the lactation feed of sows | Pendl et al. (2017)<br>Papadopoulos et al. (2009)<br>Hultén et al. (2004) |
| Antioxidants in feed: | |
| Low vitamin E level (16 or 33 IU/kg versus 66 IU/kg dietary level) | Mahan (1991) |
| Feed contaminants: | |
| Ergot contamination in feed | Kopinski et al. (2007, 2008) |
| Housing: | |
| – Farrowing crates and lack of exercise (versus loose housing)<br>– No enrichment in the periparturient period<br>– Crate width of 60 cm (versus 67 cm)<br>– Solid floor in farrowing pens (versus slatted floor)<br>– Contaminated bedding material in farrowing house | Oliviero et al. (2008, 2010)<br>Oliviero et al. (2008, 2010)<br>Cariolet (1991)<br>Hultén et al. (2004)<br>Gerjets and Kemper (2009) |
| Ambient temperature: | |
| – High ambient temperatures (indirect effect via decreased feed intake)<br>– High ambient temperatures (direct effect)<br>– Overheating of mammary glands | Quiniou and Noblet (1999)<br>Messias de Braganca et al. (1998)<br>Muirhead and Alexander (1997) |

## 7.1 Sow-Related Factors

### 7.1.1 Genetics

Information on the role of genetics on the occurrence of PDS is limited. Different studies have shown heritability values for PDS ranging from 0.02 to 0.20 (Awad Masalmeh et al. 1990; Lingaas and Ronningen 1991; Preissler et al.

**Table 4** Management factors increasing the risk for postpartum dysgalactia syndrome in sows

| Risk factor | References |
|---|---|
| Transfer of sows from gestation to farrowing unit: | |
| – Abrupt change from pasture gestation to crates a few days before farrowing<br>– 4 days before expected farrowing (versus 7 days or earlier) | Bäckström et al. (1984)<br>Papadopoulos et al. (2010) |
| Poor hygiene: | |
| – Not using disinfectants between batches in the farrowing house<br>– Not using disinfectants in the breeding area<br>– Soiled feeding troughs during lactation | Hultén et al. (2004)<br>Hultén et al. (2004)<br>Jenny et al. (2015) |
| Farrowing induction: | |
| – Increased risk (odds ratio = 4.80) for PDS in farms practicing farrowing induction (versus farms not practicing induction)<br>– Decreased risk for PDS<br>– No effect on PDS | Papadopoulos et al. (2010)<br>Holtz et al. (1983); Bostedt et al. (1998)<br>Hansen (1979); Gerjets et al. (2011) |
| Farrowing supervision: | |
| – Increased risk for PDS<br>– Decreased risk for PDS | Jorsal (1986); Gerjets et al. (2011)<br>Papadopoulos et al. (2010) |
| Prostaglandins postpartum: | |
| – No effect on PDS<br>– Decreased risk for PDS | Vanderhaeghe et al. (2008)<br>Waldmann and Heide (1996) |
| Cutting or grinding the canine teeth of piglets shortly after birth | Hultén et al. (2004) |

**Table 5** Infections and disease conditions increasing the risk for postpartum dysgalactia syndrome in sows

| Risk factor | Reference |
|---|---|
| *Mycoplasma suis* infection | Strait et al. (2012) |
| Urinary tract infections | Petersen (1983) |
| Teat wounds at weaning | Hultén et al. (2003) |
| Sow lameness | Jenny et al. (2015) |

2012), indicating a genetic predisposition. Preissler et al. (2013) reported that there is no clear case for a single causative gene or genomic region for PDS, but rather several chromosomal regions that are contributing causative genes. These authors were able to detect PDS-associated regions on porcine chromosomes 13, 15, and 17 and suggested that this opens opportunities for genetic improvement in animal breeding even with low heritable traits. Kemper and Gerjets (2009) found that the bacterial microbiota of the udder in PDS-affected sows was largely similar to that of healthy sows. They therefore suggested that developing PDS was largely dependent on the individual immune resistance of the sow, rather than on the presence of specific pathogens.

### 7.1.2 Parity

Most studies have reported a higher risk for PDS in primiparous or low parity sows (first and second parity) (Bostedt et al. 1998; Hoy 2006; Gerjets et al. 2011). However, in one study (Baer and Bilkei 2005) sows of higher parity (>4) had increased risk for PDS. The reason for greater risk in young sows is not clear. It might be due to the less mature immunity in these sows compared with older sows, or to their more pronounced physiological hyperthermia (Klopfenstein et al. 1997; Gerjets et al. 2008) which could lead to misinterpretations, or to the fact that young sows with PDS problems are prematurely removed from the farm.

### 7.1.3 Body Condition and Conformation

Two conditions that have been recognized as risk factors for PDS include the "fat sow syndrome" (FSS) and the "over-muscled sow syndrome" (OMSS). Fat sows at farrowing have a greater risk of periparturient hypophagia, prolonged farrowing, PDS (Oliviero et al. 2009), and they also have a lower colostrum yield than non-fat sows (Decaluwé et al. 2014b). Van der Peet-Schwering et al. (2004) showed that increasing the dietary energy in late pregnancy can negatively affect feed intake during early lactation due to a reduction in glucose tolerance and insulin resistance. Sows with high backfat also have greater leptin concentrations, which is associated with reduced feed intake (Barb et al. 2008; Cools et al. 2013). A high backfat thickness will, however, not always lead to severe PDS problems (Maes et al. 2004; Cools et al. 2014). The absence of individual feeding during gestation often creates more variable body weights, with dominant sows being overweight and submissive sows being underweight (Martineau 1990).

The "over-muscled sow syndrome" (OMSS) is an emerging syndrome that has been reported mainly in modern hyperprolific sows (Solignac et al. 2010). It is likely a consequence of genetic improvement and hyperprolificacy (Solignac et al. 2010). Such sows are leaner, thereby increasing the risk of catabolism before farrowing. This leanness may be due to the increased weight gain of individual fetuses during the last 10 days of gestation, i.e., 50 g/d versus 4–5 g/d (Ji et al. 2005), with faster protein deposition (Whittemore 1998). In addition to fetal growth, mammary glands also grow significantly during the last third of gestation (Sorensen et al. 2002). In lean sows, both conditions may contribute to a catabolic state before or around farrowing, which may lead to lower milk production (Van den Brand and Kemp 2006). Increased concentrations of blood non-esterified fatty acids (NEFAs) are an indicator of a catabolic state (Messias de Bragança and Prunier 1999). Le Cozler et al. (1998) and Oliviero et al. (2009) found that circulating concentrations of NEFAs increased rapidly a few days before farrowing, reaching a peak on the day of parturition. A catabolic state can also be identified by greater circulating urea and creatinine concentrations (Mitchell and Scholz 2001).

### 7.1.4 Previous Occurrence of PDS, Gestation, and Farrowing Data

Hultén et al. (2003) reported that sows which suffered from PDS in the previous lactation were at higher risk of subsequently developing PDS. Awad Masalmeh et al. (1990) reported more problems with PDS in sows with gestation lengths higher than 116 days. Whether this association is also valid in current hyperprolific sows remains to be investigated. Differences in litter data at farrowing were noted between affected and non-affected sows. An increase in the number of piglets born alive and one or more stillborn piglets were both associated with a higher risk for PDS problems (Gerjets et al. (2011). Similar results were reported by Bostedt et al. (1998). However, Van Gelder and Bilkei (2005) did not find differences in the number of stillborn piglets between healthy and PDS-affected sows.

A long farrowing duration and dystocia are associated with a higher risk for contamination and introduction of pathogens in the urogenital tract. Infections of the urogenital tract may lead to endotoxin production and subsequent PDS. Therefore, all factors that may slow down the farrowing process (van Dijk et al. 2005) also increase the risk for PDS (Bostedt et al. 1998).

### 7.1.5 Constipation

Sow constipation and endotoxemia are considered to be risk factors for PDS. As sows approach farrowing, water absorption in the intestine increases due to the fluid needed for milk production. This, in turn, alters the consistency of feces. The threshold between "normal dry feces" and constipation is however not clear (Oliviero et al. 2009). Hermansson et al. (1978) reported constipation to occur in approximately 25% of the cases of PDS in sows.

Low-fiber diets and low feed intake in late gestation are well known to increase the risk of constipation and the absorption of bacterial endotoxins, leading to PDS (Hermansson et al. 1978; Martineau et al. 1992; Persson et al. 1996; Jenny et al. 2015). The beneficial effects of adding extra fibers to the diet on lactation performance depend on many factors such as the type of fiber (fermentable or inert carbohydrates) and the composition and fiber levels in the basic diet of the sow (Guillou et al. 2016). Further research on the inclusion level of fibers and the effects of different fiber types (soluble versus insoluble, degree of fermentability) in sow diets is needed. Laxative products can also be used to prevent constipation. Pearodwong et al. (2016) noted that constipation in sows raised in hot climates is not only a problem around farrowing, but also during gestation. Constipation slightly increased the incidence of PDS. These authors showed that a commercial laxative (combination of magnesium, choline, betaine, methionine, sorbitol, lysine, and the herb extracts MoreMeta® from More Standing Enterprise Co., Ltd., Yunlin, Taiwan) given as top-dressing on the feed of gestating sows reduced the proportion of sows with severe constipation. On the other hand, providing the laxative to periparturient sows did not reduce the incidence of constipation or PDS.

In addition, insufficient water intake of sows in the peripartal period contributes to the risk of constipation and lactation problems (Fraser and Phillips 1989). Jenny et al. (2015) reported that a low flow rate of the drinking nipples, i.e., less than 2 lit/min is a risk factor for PDS.

## 7.2 Nutrition

### 7.2.1 Feeding Strategy

The results of different feeding strategies in the peripartal period on sow PDS occurrence are not univocal.

According to Papadopoulos et al. (2010), ad libitum feeding during the first days of lactation increased the odds for PDS 3.15 times compared to sows fed restrictedly. Neil et al. (1996) showed that 16% of the sows that were allowed feed ad libitum before or on the day of farrowing became agalactic, whereas 31% became agalactic when the sows were allowed feed ad libitum after the day of farrowing. The precise underlying mechanism for the higher PDS risk of ad libitum feeding at or shortly after farrowing is not clear. It might increase intestinal microbial growth, thereby increasing bacterial toxin production in the gut.

However, Moser et al. (1987) reported no adverse effects (e.g., preweaning pig weights and survival of piglets) in sows that were allowed ad libitum feed consumption within 16 h postpartum compared to sows that received a restricted amount of feed, starting with 0.45 kg on day 0 and increasing to 0.91 kg each day reaching ad libitum by day 6 of lactation. Decaluwé et al. (2014b) found that feeding sows a transition diet (meal), 4.5 kg/day versus 1.5 kg/day, from day 108 of gestation until day 3 of lactation resulted in higher colostrum yield. Cools et al. (2014) compared ad libitum feeding of a lactation diet from day 106 of gestation onward until day 7 of lactation with a commonly used restricted feeding scheme during this period (decreasing until farrowing-increasing after farrowing until reaching ad libitum levels at day 7). The sows that were fed restrictedly consumed 47% of the amount of feed of the ad libitum group during that period. The authors found that, on the premise that back fat of sows is beneath 22 mm at late gestation, ad libitum feeding limited excessive loss of sow body reserves, and improved litter growth and weaning weight.

The conflicting results between studies (restricted versus ad libitum feeding) may be attributed to differences in the feed composition, sow body condition, experimental design, and last but not least the genetics of the sows. It can be expected that results from older studies might not be applicable anymore to highly prolific sows currently used in many commercial pig farms.

### 7.2.2 Feed Composition

Feed composition and, more specifically, the type of dietary fat was shown to be important for the immune status of peripartal sows. Papadopoulos et al. (2009) reported that a lactation diet with a low *n*-6:*n*-3 ratio (2:1) fed to sows as of 8 days before farrowing improved their feed intake during the first days postpartum and was associated with a better metabolic change and inflammatory profile (less insulin resistance, lower leptin concentrations, lower rectal temperature), when compared to a diet with a high *n*-6:*n*-3 ratio (10:1) and/or administered as of 3 days before farrowing. Providing feed mixed with whey (compared to dry feed) to sows throughout lactation was associated with a decreased risk for PDS (Hultén et al. 2004). The preventive effect of whey may have been caused by a shift in the bacterial population in the environment because of changes in the sows' fecal microbiota, feed spillage, or both. Yu et al. (2020) showed that the addition of *Bacillus licheniformis* fermented products to the feed of sows throughout lactation was associated with a significantly higher piglet weaning weight (6.61 versus 7.22 kg), as well as changes in milk composition (lower fat content, higher protein content) and in fecal microbiota (increased abundance of the beneficial genus (*Eubacterium*) coprostanoligenes). Böhmer et al. (2006) reported that diets containing probiotics (*Enterococcus faecium*) from day 90 of pregnancy onward and around parturition resulted in a tendency toward slightly lower temperatures in the probiotic group on day 1 and 2 post-farrowing in sows (by 0.1–0.2 °C on average) and to higher feed intake during lactation. Further research is definitely needed to determine the true potential and possible modes of action of such dietary treatments on the occurrence of PDS.

It is well known that nutrition plays a key role in the hypothalamo–hypophysial–gonadal axis (Messias de Bragança and Prunier 1999; Barb et al. 2008; Farmer and Quesnel 2009). The latter studies emphasized the importance of appropriate nutritional management and energy status to support the endocrine system and their influence on lactogenesis and, more specifically, colostrum composition.

### 7.2.3 Antioxidants

It has been suggested that diets deficient in selenium and vitamin E increase the risk for lactational problems because these micronutrients have antioxidant properties that can alleviate the effects of endotoxins (Elmore and Martin 1986). An early study indicated that a low level of vitamin E (16 or 33 IU/kg versus 66 IU/kg) could be a risk factor for PDS (Mahan 1991), but Mahan et al. (2000) later reported that increasing dietary levels of vitamin E from 30 to 60 IU did not decrease the prevalence of MMA at parturition. In another study, injections of vitamin E (400 IU) and selenium (3 mg) to sows on days 30, 60, and 100 of gestation, while fed normal levels of these nutrients, increased the survival rate of piglets but did not affect litter weights at weaning (Chavez and Patton 1986).

### 7.2.4 Contaminants

Mycotoxins are often reported by swine veterinarians as a risk factor for PDS. However, only ergot derivatives of *Claviceps purpurea* are currently well recognized as being associated with decreased milk production (Kopinski et al. 2007), largely due to their inhibitory effect on prolactin secretion when fed in high amounts. Sows fed 1.5% ergot for 6–10 days preceding farrowing produced no milk, whereas ergot inclusions of 0.6–1.2% caused less problems in milk release and neonatal piglet mortality (Kopinski et al. 2007). When 3% ergot was added to the feed of lactating sows, milk production stopped in 2 of the 12 sows and was reduced in the others (Kopinski et al. 2008). Sow plasma prolactin concentrations were reduced after feeding ergot for 7 days, therefore indicating that the negative effects of ergot toxins are mediated by suppressed prolactin secretion.

## 7.3 Housing and Climate Conditions

### 7.3.1 Housing and Hygiene

Housing conditions may influence the farrowing process, the welfare of sows, and also the hygienic conditions. Oliviero et al. (2008) showed that the farrowing process lasts longer when sows are housed in crates with no bedding material (311 min) compared to sows housed in farrowing pens enriched with straw bedding (218 min). Such findings imply a potential indirect effect of housing conditions on PDS via farrowing duration. A low level of activity of sows in the peripartal period or the impossibility of moving around, as is the case when they are housed in crates, can also increase the risk for constipation (Fraser and Phillips 1989). Providing adequate enrichment material for nest-building can mitigate the negative effects of crate housing and reduce the incidence of PDS (Oliviero et al. 2008, 2010). Cariolet (1991) found that the incidence of PDS was greater in sows housed in narrower crates (60 cm) compared with wider crates (67 cm).

The use of slatted floors in farrowing pens, compared with solid floors, was associated with a decreased risk of chronic mastitis in sows (Hultén et al. 2004). This might have been due to the more hygienic conditions with slatted floors (Munsterhjelm et al. 2008). Hultén et al. (2004)

also reported that the use of disinfectants between batches decreased the incidence of PDS. Jenny et al. (2015) found that soiled troughs in the peripartal period increased the risk for PDS. Improving the level of hygiene of the farrowing unit was also associated with lower piglet preweaning mortality (Ravel et al. 1996).

When used as bedding material, wood shavings may contain pathogenic *Klebsiella pneumoniae* bacteria which could lead to environmental contamination, infection of the mammary glands, and an increased risk for PDS (Gerjets and Kemper 2009).

### 7.3.2 Stable Climate and Heat Stress

In a recent study using 50 sows, even though not significant, the occurrence of PDS was higher (37.5% versus 26.9%) in the winter than summer months (Patra et al. 2021). This was linked to lower postpartum concentrations of triiodothyronine indicating metabolic dysfunction. On the other hand, during established lactation, high ambient temperature or heat stress results in decreased sow feed intake, subsequently leading to decreased milk production (Quiniou and Noblet 1999). Other studies reported also a direct effect of high ambient temperatures on milk production, independent of the reduction in feed intake (Messias de Bragança and Prunier 1999). Temperature control in the farrowing room is therefore a crucial factor for the performance of the sow during the entire lactation. These studies mainly relate to the entire lactation and not specifically to the peripartal period. Newborn piglets require a localized warm environment because the thermoneutral zone of sows is much lower than that for piglets. It is generally recommended to maintain room temperature warm (20–22 °C) for 2–3 days after farrowing to favor piglet survival. After this crucial period, room temperature can be gradually decreased to attain 18 °C or even lower on day 10 of lactation as long as the piglets have a draft-free heated creep area (Farmer et al. 1998a).

Overheating of the mammary glands by inappropriate placement of heat lamps can decrease milk production (Muirhead and Alexander 1997). Supplementary heat should be directed away from the sow and her udder and be removed when not needed to favor the sow's well-being and maximize her milk production.

## 7.4 Management

### 7.4.1 Timing of Moving Sows to the Farrowing Unit

An abrupt change of environment from group housing in gestation to restraint in crates only a few days before farrowing was related to a higher incidence of PDS (Bäckström et al. 1984). Papadopoulos et al. (2010) also found that moving pregnant sows to the farrowing unit 4 days before expected farrowing, instead of 7 days or earlier, increased the risk for PDS (odds ratio = 6.2). An adaptation period to the new farrowing-lactation environment would appear to be beneficial. However, not all studies showed consistent results. Klopfenstein et al. (1995) did not find a lower prevalence of problem litters when sows were allowed to adapt to the farrowing house for more than a week before parturition, compared to only a few days.

### 7.4.2 Farrowing Induction

There are conflicting results on the effect of farrowing induction on the incidence of PDS. In some herds, with a significant percentage of gilts and sows showing PDS, induction of parturition with prostaglandins was effective in reducing the incidence of PDS (Holtz et al. 1983; Bostedt et al. 1998), whereas no influence of birth induction on PDS was found in other studies (Hansen 1979; Gerjets et al. 2011). In some studies, farrowing induction was even identified as a risk factor for PDS (odds ratio 4.8) (Papadopoulos et al. 2010) or for a lower colostrum yield when performed on day 109 of gestation (Devillers et al. 2007). The timing of induction as well as other factors (product, dose, characteristics of the treated sow, farm management) may explain the conflicting results. Prostaglandins could be effective in treating PDS caused by delayed lactogenesis because incomplete luteolysis of corpora lutea can lead to high progesterone concentrations that could potentially inhibit lactogenesis.

However, farrowing induction at 113 days of pregnancy did not modify the chronology of endocrine changes taking place during the peripartum period, nor did it affect colostrum yield and colostral IgG concentrations (Foisnet et al. 2010), albeit some hematological variables of piglets were modified.

### 7.4.3 Farrowing Supervision and Birth Assistance

Supervision and assisting sows exhibiting dystocia may help reduce the occurrence of PDS (Bäckström et al. 1984). Papadopoulos et al. (2010) reported that frequent farrowing supervision and neonatal care given by the stockperson, compared with no farrowing supervision, decreased the risk of PDS. Other studies, however, reported that obstetric aid by either herd managers or veterinarians increased the sow's risk of acquiring early lactation problems and increased the incidence of postpartum vulvar discharges and endometritis (Jorsal 1986; Gerjets et al. 2011). Differences between studies may depend on the criteria used to define dystocia and PDS (mastitis or dysgalactia). It is also possible that the sows requiring manual birth assistance are at a higher risk for developing PDS compared to sows not requiring birth assistance, irrespective of whether birth assistance is practiced. Nevertheless, when there are clear indications of dystocia, manual assistance should be done for animal health and welfare reasons. If obstetric intervention is practiced, it should be done properly and according to strict hygiene conditions (Bostedt et al. 1998).

### 7.4.4 Prostaglandin Administration Postpartum

Prostaglandins are luteolytic agents causing a prepartum decline in progesterone and the release of relaxin from the corpora lutea. Postpartum administration of prostaglandins can have a beneficial effect on uterine involution and prevention of endometritis (Waldmann and Heide 1996). Vanderhaeghe et al. (2008), however, did not observe beneficial effects of administering a prostaglandin F2α analogue to sows within 24–48 hours after farrowing on preweaning mortality and litter average daily gain.

### 7.4.5 Cutting the Teeth of Piglets

Cutting the canine and lateral incisor teeth of piglets was identified as a risk factor for PDS (Hultén et al. (2004). This is somewhat surprising because cutting teeth might decrease the risk of teat wounds in the sow. The authors explained that teeth cutting could have led to pulpitis in the piglets, and bacteria involved in pulpitis might have been transferred to the mammary glands of the sows during suckling and have caused mastitis, infections, and disease conditions.

In the case of mastitis, the bacteria most frequently involved belong to the class of coliforms, covering the genera *Escherichia*, *Enterobacter*, *Citrobacter*, and *Klebsiella*. Yet, staphylococci and streptococci have also been isolated from sow milk (Hirsch et al. 2003; Kemper and Gerjets 2009). These bacteria may originate from the intestinal microbiota of the sow, the environment, or the oral microbiota of the piglets. Such bacteria are not specific pathogens but are commonly found in the environment and are also present in non-PDS sows. Therefore, their presence alone cannot be considered as problematic. Strait et al. (2012) reported a case of PDS with sudden onset of extreme dysgalactia in gilts and sows in a 1000-head herd due to *Mycoplasma suis* infection. Signs of dysgalactia were identified in sows as of one day after parturition and lasted 4–6 days, resulting in an increased piglet preweaning mortality of (from 6 to 18%). Sows were neither off-feed nor had fever. Udders were not inflamed or congested. Treatment with chlortetracycline in the feed resulted in a near complete absence of dysgalactia in sows that farrowed the weeks after treatment. The authors reported that the impact of *M. suis* on the health of lactating sows is likely underestimated.

Sows with urinary tract infections were identified as being at greater risk for PDS (Petersen 1983). Hultén et al. (2003) also reported that the number of teat wounds at weaning was positively associated with the occurrence of PDS, suggesting that skin lesions caused by the piglets or the housing conditions in the farrowing house might predispose to PDS. The same was found for sow lameness, whereby a higher incidence of PDS occurred in farms with a high prevalence of lameness (Jenny et al. 2015). Sows with locomotory problems lay down more frequently and are possibly subjected to less hygienic conditions, therefore increasing the risk for inflammation of the mammary glands. Lame sows are also at greater risk for constipation because of lower water and feed intakes and because they are less active.

## 8 Treatment

In herds suffering from PDS, both affected sows and their offspring need to be treated.

### 8.1 Sow Treatment

Affected sows need to be treated as soon as possible. Treatment mostly includes non-steroidal anti-inflammatory drugs (NSAID), as well as oxytocin to stimulate milk letdown. In cases of complete agalactia pharmacological agents to increase prolactin synthesis may be required. Antimicrobials may be warranted depending on individual situations. An overview of studies that have assessed treatment strategies for sows with PDS is shown in Table 6.

#### 8.1.1 Non-steroidal Anti-inflammatory Drugs

Non-steroidal anti-inflammatory drugs have an anti-inflammatory, analgesic, and antipyretic action via inhibition of cyclooxygenase (COX) activity, and thus prostaglandin synthesis. They are used in sows with PDS to reduce fever and alleviate the effects of inflammation and endotoxemia. Drugs used include flunixin (2 mg/kg) (Stringer 1989; Hirsch et al. 2003), tolfenamic acid (2–4 mg/kg) (Tummaruk and Pearodwong 2015), ketoprofen (1 mg/kg) (Claeyé et al. 2015), and meloxicam (0.4 mg/kg) (Hirsch et al. 2003; Hoy and Friton 2005; Schoos et al. 2020). Treatment strategies often consist of a single administration on the day of parturition and sometimes a second treatment 1 or 2 days later. The benefits of NSAID treatment, e.g., better performance of the piglets, usually outweigh the costs (Hirsch et al. 2003). It is likely important to prevent rather than to stop or decrease the cascade of inflammatory processes that lead to PDS (Schoos et al. 2020). The last authors therefore tested the administration of either meloxicam or paracetamol orally in a prophylactic way for 7 days, starting 2 days prior to expected farrowing. Paracetamol is not an NSAID, but an NSAID-like drug with central antipyretic and analgesic action and without peripheral anti-inflammatory action. Due to its strong antipyretic

**Table 6** Overview of efficient treatment strategies for sows affected by PDS

| Source | Anti-inflammatory drugs | Oxytocin | Antibiotics | Other |
|---|---|---|---|---|
| Persson et al. (1989) | Corticosteroids | Yes | Broad spectrum | |
| Stringer (1989) | Flunixin | Yes | Trimethoprim-sulfadoxine | |
| Heinritzi and Hagn (1999) | | | Cefquinome, amoxycillin | |
| Hirsch et al. (2003) | Flunixin, meloxicam | Yes | Enrofloxacin | |
| Baer and Bilkei (2005) | | Yes | Type not specified | Prostaglandins |
| Hoy and Friton (2005) | Meloxicam | | Cefquinome | |
| Claeyé et al. (2015) | Ketoprofen | | | |
| Tummaruk and Pearodwong (2015) | Tolfenamic acid | | Enrofloxacin | Combination product (iron, cobalt, vit B3, vit B12) |
| Schoos et al. (2020) | Meloxicam, paracetamol | | | |

action, it is recommended to treat animals having fever (Schoos et al. 2019). The authors were, however, unable to show beneficial effects of both products on both health and performance of sows and their offspring.

### 8.1.2 Oxytocin

The administration of oxytocin in clinical cases of PDS is recommended to stimulate milk let-down and minimize the consequences of PDS. Oxytocin may be injected intravenously at 5–10 IU/sow 4–5 times at 2–3 hourly intervals (Martineau 2005). There should be at least 30 minutes between two subsequent injections. In most cases, 5 IU is enough to induce milk let-down. When using this dosage, repeated administrations do not cause adverse effects as the half-life of oxytocin is short, namely approximately 6 minutes. Parenteral administration of synthetic oxytocin is a very efficient way to trigger milk ejection. Intramuscular administration is most common, but the intravenous route, e.g., in the ear vein, is also possible, yet the dosage needs to be lowered. Overuse of oxytocin (higher dosage, more administrations in a shorter time) may result in poor herd performance (Ravel et al. 1996) and increased somatic cell counts in milk (Garst et al. 1999).

### 8.1.3 Increasing Prolactin Concentrations

Dusza et al. (1991) were the first to inject recombinant porcine prolactin into lactating sows and demonstrated that it could stimulate milk yield in sows with PDS. The increase in prolactin was rapid, with concentrations going from 0.2 to 20 ng/mL in 15 minutes. Recombinant porcine prolactin is, however, not currently available so that pharmacological agents must be used to increase prolactin concentrations. Furthermore, there is no commercial kit to measure prolactin concentrations in swine and an in-house radioimmunoassay must be used. The regulation of prolactin secretion is largely under the negative control of dopamine. Hence, dopamine agonists (such as ergocriptine or bromocriptine) decrease prolactin concentrations whereas dopamine antagonists (such as domperidone and metoclopramide) enhance its concentrations. However, not all dopamine antagonists seem to have an effect.

The opioidergic system is thought to affect prolactin secretion in sows. Endogenous opioid peptides stimulate prolactin secretion by suppressing the tuberoinfundibular dopaminergic neurons (Barb et al. 1987), and the opioid antagonist naloxone can induce a marked suppression of plasma prolactin concentrations in lactating sows (Mattioli et al. 1986). A dose as small as 0.5 mg/kg was in fact sufficient to suppress prolactin concentrations for an hour in lactating sows (Armstrong et al. 1986) and intravenous injections of 0.25 mg/kg of naloxone decreased prolactin on day 21 of lactation (Armstrong et al. 1990). De Rensis et al. (1993) observed a reduction in plasma prolactin after naloxone treatment on day 10 of lactation but not after chronic naloxone treatment during the first 78 hours postpartum. It was also demonstrated that it takes 4–5 hours for prolactin concentrations to increase to pretreatment values after cessation of naloxone infusion (Armstrong et al. 1988a). On the other hand, morphine was reported to suppress prolactin concentrations in lactating sows (Armstrong et al. 1988b; De Rensis et al. 1999).

### 8.1.4 Antimicrobials

Antimicrobial treatment is only recommended if generalized signs of PDS are present. Antimicrobials should not be administered based on increased rectal temperature only, as sow body temperature is highly variable and physiological hyperthermia may occur in the periparturient period. The selection of an antibiotic should be based on its spectrum of activity against those bacterial organisms identified or thought to be responsible for the problem. Antimicrobials must also reach sufficient concentrations in the mammary gland so that pharmacokinetics should be considered. National formularies have been developed in many countries to guide veterinarians in the selection of a proper antimicrobial for specific disease conditions, including PDS. Examples of antimicrobials that can be used to treat PDS include potentiated sulfonamides, tetracyclines, lincospectin, amoxycillin, and the combination of penicillin and neomycin. Fluoroquinolones and cephalosporins are also effective. However, within the scope of responsible use of antimicrobials and the prevention of antimicrobial resistance development, their use is discouraged because these molecules are critically important in the treatment of sick humans. Nevertheless, a survey done by De Briyne et al. (2014) in Europe showed that antibiotics that are critical for the treatment of humans, e.g., fluoroquinolones, were often prescribed for PDS problems.

#### 8.1.5 Supportive Treatment

Supportive treatment, mainly to alleviate dehydration, may be recommended in severe cases of PDS. Sows with endotoxemia suffer from dehydration (Reiner et al. 2009). If the sow is not drinking then fluids can be administered intravenously or eventually via the rectum. Daily fluid requirement is considered to be approximately 7% of the body weight.

### 8.2 Treatment of Piglets

The first step in treating piglets is to provide an alternative source of energy to colostrum or milk (Declerck et al. 2016b) and/or to transfer piglets to another healthy sow. Piglets with higher birth weights can be cross-fostered with greater success because they generally have more energy reserves. Supplementing the water with electrolytes and glucose may help to prevent dehydration and promote survival of piglets. Intraperitoneal infusion of approximately 15 mL of 5% glucose can temporarily alleviate dehydration and starvation so that piglets may be able to find a teat and start suckling. Piglets may also be raised with a combination of milk replacers and highly digestible creep feed. However, immunoglobulins included in some milk replacers are unlikely to contain antibodies specific for a farm and cannot be considered as a replacement for colostrum. The decreased energy intake of piglets from affected sows will predispose to hypothermia and providing additional heat therefore becomes of utmost importance.

## 9 Control and Prevention

Since PDS is a multifactorial disease, it is important to control or prevent the many risk factors discussed above from being present on a farm. Vaccination against PDS is not widely practiced. The *E. coli* strains causing neonatal diarrhea are usually not detected in sows with PDS. Gerjets and Kemper (2009) reported that after natural infection with PDS mammary glands did not develop resistance to subsequent infection. Therefore, the effects of vaccines against *E. coli* with regard to PDS are questionable. More research on pathogen-host interaction and efficacy studies are needed to assess whether vaccination against PDS could be a solution in the future. As mentioned above, induction of parturition with PGF2α can be effective in reducing the incidence of PDS, although conflicting results have been obtained. Pendl et al. (2017) reported that the most effective control measures to fight PDS were the use of a transition feed with soluble fiber (e.g., sugar beet pulp and wheat bran) and sufficient crude fiber levels (7–8%) prepartum to avoid constipation, careful and early PDS diagnostic and treatment with NSAIDs and oxytocin in case of clinical signs.

## 10 Conclusions

Many sow herds suffer from problems with PDS and incur major financial losses. PDS is a syndrome with a complex pathophysiology and many risk factors have been identified. As PDS causes major health and welfare problems, affected sows and their piglets need to be treated as soon as possible. The control of PDS largely relies on decreasing the risk factors, such as optimization of feeding, housing, climate, and last but not least management practices. Extra care for the sow and the piglets is particularly important in herds with hyperprolific sows and larger litter sizes. Further research is warranted to better understand the pathophysiology of PDS, the influence of genetics, and to further explore control measures.

## References


Angjelovski B, Radeski M, Djadjovski I, Mitrov D, Bojkovski J, Adamov N, Dovenski T (2019) Prevalence and clinical signs of postpartum dysgalactia syndrome at the first year after farrowing in farmed sows in the Republic of Macedonia. Mac Vet Rev 42:79–86. https://doi.org/10.2478/macvetrev-2019-0014

Armstrong JD, Kraeling RR, Britt JH (1986) Effect of bolus administration of naloxone on secretion of LH and prolactin in lactating sows. J Anim Sci 63(Suppl. 1):332

Armstrong JD, Kraeling RR, Britt JH (1988a) Effects of naloxone or transient weaning on secretion of LH and prolactin in lactating sows. J Reprod Fert 83:301–308

Armstrong JD, Kraeling RR, Britt JH (1988b) Morphine suppresses luteinizing hormone concentrations in transiently weaned sows and delays onset of estrus after weaning. J Anim Sci 66:2216–2222

Armstrong JD, Esbenshade KL, Coffey MT, Heimer E, Campbell R, Mowles T, Felix A (1990) Opioid control of growth hormone in the suckled sow is primarily mediated through growth hormone releasing factor. Domest Anim Endocrinol 7:191–198

Awad Masalmeh M, Baumgartner W, Passering A, Silber R, Hinterdorfer F (1990) Bacteriological studies in sows with puerperal mastitis in different herds in Austria. Tierärztl Umschau 45:526–535

Bäckström L, Connors J, Price W, Larson R, Morko A (1982) Mastitis-metritis-agalactia (MMA) in the sow: a field survey of MMA and other farrowing disorders under different gestation and farrowing housing conditions. Proc Int Pig Vet Soc Congr 7:175

Bäckström L, Morkoc AC, Connor J, Larson R, Price W (1984) Clinical study of mastitis-metritis-agalactia in sows in Illinois. J Am Vet Med Assoc 185:70–73

Baer C, Bilkei G (2005) Ultrasonographic and gross pathological findings in the mammary glands of weaned sows having suffered recidiving mastitis metritis agalactia. Reprod Dom Anim 40:544–547

Barb CR, Kraeling RR, Rampacek GB, Whisnant CS (1987) Opioid inhibition of luteinizing hormone secretion in the postpartum lactating sow. Biol Reprod 35:368–371

Barb C, Hausman G, Lents C (2008) Energy metabolism and leptin: effects on neuroendocrine regulation of reproduction in the gilt and sow. Reprod Dom Anim 43(Suppl 2):324–330

Böhmer B, Kramer W, Roth-Maier D (2006) Dietary probiotic supplementation and resulting effects on performance, health status and microbial characteristics of primiparous sows. J Anim Physiol Anim Nutr 90:309–315

Bostedt H, Maier G, Herfen K, Hospen R (1998) Clinical examinations of gilts with puerperal septicemia and toxemia. Tierärztl Praxis 26:332–338

Cariolet R (1991) Etude des postures en phase the repos chez des truies bloques en gestation: Relation entre la largeur de la stalle et la pathologie de la mise bas. Journ Rech Porcine France 23:189–194

Chavez ER, Patton KL (1986) Response to injectable Selenium and vitamin E on reproductive performance of sows receiving a standard commercial diet. Can J Anim Sci 66:1065–1074

Claeyé E, Beek J, Meyns T, Maes D (2015) Effect of ketoprofen treatment for the prevention of postpartum dysgalactia syndrome in sows. Vl Diergeneesk Tijdschr 84:127–132

Cools A, Maes D, Decaluwé R, Buyse J, van Kempen TA, Janssens GP (2013) Peripartal changes in orexigenic and anorexigenic hormones in relation to back fat thickness and peripartal feeding strategy of sows. Domest Anim Endocrinol 45:22–27

Cools A, Maes D, Decaluwé R, Buyse J, van Kempen TA, Liesegang A, Janssens GP (2014) Ad libitum feeding during the peripartal period affects body condition, reproduction results and metabolism of sows. Anim Reprod Sci 145:130–140

De Briyne N, Atkinson J, Pokludová L, Borriello SP (2014) Antibiotics used most commonly to treat animals in Europe. Vet Rec. https://doi.org/10.1136/vr.102462

Decaluwé R, Maes D, Declerck I, Cools A, Wuyts B, De Smet S, Janssens GPJ (2013) Changes in back fat thickness during late gestation predicts colostrum yield in sows. Animal 7:1999–2007

Decaluwé R, Maes D, Cools A, Piepers S, Janssens GPJ (2014a) Piglets'colostrum intake associates with daily weight gain and survival until weaning. Livest Sci 162:185–192

Decaluwé R, Maes D, Cools A, Wuyts B, De Smet S, Marescau B, De Deyn PP, Janssens GPJ (2014b) Effect of peripartal feeding strategy on colostrum yield and composition in sows. J Anim Sci 92:3557–3567

Declerck I, Dewulf J, Piepers S, Decaluwé R, Maes D (2015) Sow and litter factors influencing colostrum yield and nutritional composition. J Anim Sci 93:1309–1317

Declerck I, Dewulf J, Sarrazin S, Maes D (2016a) Long-term effects of colostrum intake in piglet mortality and performance. J Anim Sci 94:1–11

Declerck I, Dewulf J, Decaluwé R, Maes D (2016b) Effects of energy supplementation to neonatal (very) low birth weight piglets on mortality, weaning weight, daily weight gain and colostrum intake. Livest Sci 183:48–53

Devillers N, Farmer C, Le Dividich J, Prunier A (2007) Variability of colostrum yield and colostrum intake in pigs. Animal 1:1033–1041

van Dijk AJ, van Rens BT, van der Lende T, Taverne M (2005) Factors affecting duration of the expulsive stage of parturition and piglet birth intervals in sows with uncomplicated, spontaneous farrowings. Theriogenology 64:1573–1590

Dusza L, Sobczak J, Jana B, Mudza A, Bluj W (1991) Using biolactin 2 (purified porcine prolactin) to stimulate lactation in sows. Medycyna Wet 47:418–421

Elmore RG, Martin CE (1986) Mammary glands. In: Leman AD, Straw BE, Glock WL, Mengeling WL, Penny HC, Scholl E (eds) Diseases of swine, 6th edn. Iowa State University Press, Ames

Farmer C, Petitclerc D (2003) Specific window of prolactin inhibition in late gestation decreases mammary parenchymal tissue development in gilts. J Anim Sci 81:1823–1829

Farmer C, Quesnel H (2009) Nutritional, hormonal and environmental effects on colostrum in sows. J Anim Sci 87:56–64

Farmer C, Robert S, Choinière Y (1998a) Reducing ambient temperature in farrowing houses with a new controlled-environment system. Can J Anim Sci 78:23–28

Farmer C, Robert S, Rushen J (1998b) Bromocriptine given orally to periparturient or lactating sows inhibits milk production. J Anim Sci 76:750–757

Foisnet A, Farmer C, David C, Quesnel H (2010) Altrenogest treatment during late pregnancy did not reduce colostrum yield in primiparous sows. J Anim Sci 88:1684–1693

Fraser D, Phillips PA (1989) Lethargy and low water intake by sows during early lactation: A cause of low piglet weight gains and survival? Appl Anim Behav Sci 24:13–22

Garst AS, Ball SF, Williams BL, Wood CM, Knight JW, Moll HD, Aardema CH, Gwazdauskas FC (1999) Influence of pig substitution on milk yield, litter weights, and milk composition of machine milked sows. J Anim Sci 77:1624–1630

Gerjets I, Kemper N (2009) Coliform mastitis in sows: a review. JSAP 17:97–105

Gerjets I, Kruse S, Krieter J, Kemper N (2008) Diagnosis of MMA affected sows: bacteriological differentiation, temperature measurement and water intake. Proc Int Vet Pig Soc Congr. Durban, South Africa, 236

Gerjets I, Traulsen I, Reiners K, Kemper N (2011) Assessing individual sow risk factors for coliform mastitis: a case-control study. Prev Vet Med 100:248–251

Guillou D, de Grave X, van der Aar PJ, Molist F (2016) Influence of fiber type and content and amino acid levels in the sow lactation diet on farrowing process, sow health and piglet vitality. J Anim Sci 94(Suppl 3):146–149

Hansen LH (1979) Reproductive efficiency and incidence of MMA after controlled farrowing using a prostaglandin analogue, cloprostenol. Nord Vet Med 31:122–128

Heinritzi K, Hagn J (1999) Effectiveness and disposition of the newly developed cephalosporin cefquinome in puerperal septicemia and toxemia in gilts. Tierärztl Prax 27:114–121

Hermansson I, Einarsson S, Larson K, Backstrom L (1978) On the agalactia post partum in the sow. A clinical study. Nord Vet Med 30:465–473

Hirsch AC, Philipp H, Kleemann R (2003) Investigation on the efficacy of meloxicam in sows with mastitis-metritis-agalactia syndrome. J Vet Pharm Ther 26:355–360

Holtz W, Hartmann FJ, Welp C (1983) Induction of parturition in swine with prostaglandin analogs and oxytocin. Theriogenology 19:583–592

Hoy S (2006) The impact of puerperal diseases in sows on their fertility and health up to next farrowing. Anim Sci 82:701–704

Hoy S, Friton GM (2005) Effects of meloxicam treatment on performance of sows with MMA syndrome. Prakt Tierarzt 86:568–576

Hultén F, Persson A, Eliasson-Selling L, Heldmer E, Lindberg M, Sjogren U, Kugelberg C, Ehlorsson CJ (2003) Clinical characteristics, prevalence, influence on sow performance, and assessment of sow-related risk factors for granulomatous mastitis in sows. Am J Vet Res 64:463–469

Hultén F, Persson A, Eliasson-Selling L, Heldmer E, Lindberg M, Sjögren U, Kugelberg C, Ehlorsson CJ (2004) Evaluation of environmental and management-related risk factors associated with chronic mastitis in sows. Am J Vet Res 65:1398–1403

Jenny B, Vidondo B, Pendl W, Kümmerlen D, Sidler X (2015) Evaluation of risk factors for mastitis-metritis-agalactia in pig farms in Switzerland. Schweiz Arch Tierheilk 157:689–696. https://doi.org/10.17236/sat00047

Ji F, Wu G, Blanton JR, Kim SW (2005) Changes in weight and composition in various tissues of pregnant gilts and their nutritional implications. J Anim Sci 83:366–375

Jorsal SE (1986) Epidemiology of the MMA-syndrome: a field survey in Danish sow herds. Proc Congr Int Vet Pig Soc 93

Kaiser M, Jacobson M, Andersen PH, Bakbo P, Ceron JJ, Dahl J, Escribano D, Jacobsen S (2018) Inflammatory markers before and after farrowing in healthy sows and in sows affected with postpartum dysgalactia syndrome. BMC Vet Res 14:83. https://doi.org/10.1186/s12917-018-1382-7

Kaiser M, Jacobson M, Bækbo P, Dahl J, Jacobsen S, Guo YZ, Larsen T, Andersen PH (2020) Lack of evidence of mastitis as a causal factor for postpartum dysgalactia syndrome in sows. Transl Anim Sci 4:250–263. https://doi.org/10.1093/tas/txz159

Kemper N (2020) Update on postpartum dysgalactia syndrome in sows. J Anim Sci 98(suppl 1):S117–S125

Kemper N, Gerjets I (2009) Bacteria in milk from anterior and posterior mammary glands in sows affected and unaffected by postpartum dysgalactia syndrome (PPDS). Acta Vet Scand 51:26–33

Klopfenstein C, D'Allaire S, Martineau GP (1995) Effect of adaptation to the farrowing crate on water intake of sows. Livest Prod Sci 43:243–252

Klopfenstein C, Bigras Poulin M, Martineau GP (1997) Reliability of sow 'fever' as an indicator of lactation problems in swine neonatology. Journ Rech Porcine France 29:53–58

Kopinski J, Blaney B, Downing J, McVeigh JF, Murray SA (2007) Feeding sorghum ergot (*Claviceps africana*) to sows before farrowing inhibits milk production. Aust Vet J 85:169–176

Kopinski J, Blaney B, Murray SA, Downing J (2008) Effect of feeding sorghum ergot (*Claviceps Africana*) to sows during mid-lactation on plasma prolactin and litter performance. J Anim Physiol Anim Nutr 92:554–561
Larsen I, Thorup F (2006) The diagnosis of MMA. In: Proc Int Vet Pig Soc Congr, Copenhagen, Denmark, O.52-02:256
Le Cozler Y, David C, Beaumal V, Johansen S, Dourmad J-Y (1998) Effect of the feeding level during rearing on performance of Large White gilts. Part 2: Effect on metabolite profiles during gestation and lactation, and on glucose tolerance. Reprod Nutr Dev 38:377–390
Lingaas F, Ronningen K (1991) Epidemiological and genetical studies in Norwegian pig herds. V Estimates of heritability and phenotypic correlations of the most common diseases in Norwegian pigs. Acta Vet Scand 32:115–122
Loisel F, Farmer C, van Hees H, Quesnel H (2015) Relative prolactin-to-progesterone concentrations around farrowing influence colostrum yield in primiparous sows. Domest Anim Endocrinol 53:35–41
Maes D, Janssens G, Delputte P, Lammertyn A, de Kruif A (2004) Backfat measurements in sows from three commercial pig herds: relationship with reproductive efficiency and correlation with visual body condition scores. Livest Prod Sci 91:57–67
Mahan D (1991) Assessment of the influence of dietary vitamin E on sows and offspring in three parities: reproductive performance, tissue tocopherol, and effects on progeny. J Anim Sci 69:2904–2917
Mahan DC, Kim YY, Stuart RL (2000) Effect of vitamin E sources (RRR- or all-rac-alpha-tocopheryl acetate) and levels on sow reproductive performance, serum, tissue, and milk alpha-tocopherol contents over a five-parity period, and the effects on the progeny. J Anim Sci 78:110–119
Martineau GP (1990) Body building syndrome in sows. Proc Annu Meet Am Assoc Swine Pract:345–348
Martineau GP (2005) Postpartum dysgalactia syndrome and mastitis in sows. In: Kahn CM (ed) Reproduction. The Merck veterinary manual, 9th edn. Merck Co, Whitehouse Station, NJ, pp 1134–1137
Martineau GP, Klopfenstein C (1996) Body building syndromes in sows (BBS) thin sow syndrome, fat sow syndrome, accordeon sow syndrome. Journ Rech Porcine France 28:331–338
Martineau GP, Smith B, Doize B (1992) Pathogenesis, prevention and treatment of lactational insufficiency in sows. Vet Clin N Am Food Anim Pract 8:661–683
Martineau GP, Farmer C, Peltoniemi O (2012) Mammary System. In: Zimmerman JJ, Karriker LA, Ramirez A, Schwartz KJ, Stevenson GW (eds) Diseases of swine, 10th edn. Wiley-Blackwell, West Sussex, pp 270–293
Mattioli M, Conte F, Galeati G, Seren E (1986) Effect of naloxone on plasma concentrations of prolactin and LH in lactating sows. J Reprod Fert 76:167–173
Messias de Bragança M, Prunier A (1999) Effects of low feed intake and hot environment on plasma profiles of glucose, nonesterified fatty acids, insulin, glucagon, and IGF-I in lactating sows. Domest Anim Endocrinol 16:89–101
Messias de Braganca M, Mounier AM, Prunier A (1998) Does feed restriction mimic the effects of increased ambient temperature in lactating sows? J Anim Sci 76:2017–2024
Mitchell AD, Scholz AM (2001) Techniques for measuring body composition of swine. In: Lewis AL, Southern L (eds) Swine nutrition. CRC Press, New York, pp 32–37
Moser RL, Cornelius SG, Pettigrew JE, Hanke H, Heeg TR, Miller KP (1987) Influence of postpartum feeding method on performance of the lactating sow. Livest Prod Sci 16:91–99
Muirhead M, Alexander T (1997) Mastitis, Lactation failure. In: Muirhead MR, Alexander TJL (eds) Managing pig health and the treatment of disease. 5M Enterprises Limited, pp 236–239
Munsterhjelm C, Valros A, Heinonen M, Hälli O, Peltoniemi O (2008) Housing during early pregnancy affects fertility and behaviour of sows. Reprod Domest Anim 43:584–591
Nachreiner RF, Ginther OJ, Ribelin WE, Carlson IH (1971) Pathologic and endocriniologic changes associated with porcine agalactia. Am J Vet Res 32:1065–1075
Neil M, Ogle B, Annér K (1996) A two-diet system and *ad libitum* lactation feeding of the sow 1. Sow performance. Anim Sci 62:337–347
Niemi J, Bergman P, Ovaska S, Sevón-Aimonen ML, Heinonen M (2017) Modeling the costs of postpartum dysgalactia syndrome and locomotory disorders on sow productivity and replacement. Front Vet Sci 4:181. https://doi.org/10.3389/fvets.2017.00181
Oliviero C, Heinonen M, Valros A, Hälli O, Peltoniemi O (2008) Effect of the environment on the physiology of the sow during late pregnancy, farrowing and early lactation. Anim Reprod Sci 105:365–377
Oliviero C, Kokkonen T, Heinonen M, Sankari S, Peltoniemi O (2009) Feeding sows with high fiber diet around farrowing and early lactation: impact on intestinal activity, energy balance related parameters and litter performance. Res Vet Sci 86:314–319
Oliviero C, Heinonen M, Valros A, Peltoniemi O (2010) Environmental and sow-related factors affecting the duration of farrowing. Anim Reprod Sci 119:85–91
Papadopoulos GA (2008) Lactation physiology in sows: impact of feeding strategies and risk factors for postpartum dysgalactia syndrome. PhD Thesis, Ghent University, pp 229
Papadopoulos G, Maes D, Van Weyenberg S, van Kempen T, Buyse J, Janssens GPJ (2009) Peripartal feeding strategy with different n-6:n-3 ratios in sows: effects on sow's performance, inflammatory and periparturient metabolic parameters. Br J Nutr 101:348–357
Papadopoulos G, Vanderhaeghe C, Janssens GPJ, Dewulf J, Maes D (2010) Risk factors associated with postpartum dysgalactia syndrome. Vet J 184:167–171
Patra MK, De UK, Kent Y, Rungsung S, Krishnaswamy N, Deka BC (2021) Influence of seasonal variation

on post-farrowing dysgalactia syndrome (PFDS) and serum biochemistry profiles in the periparturient sow. Trop Anim Health Prod 53:346. https://doi.org/10.1007/s11250-021-02793-1

Pearodwong P, Muns R, Tummaruk P (2016) Prevalence of constipation and its influence on post-parturient disorders in tropical sows. Trop Anim Health Prod 48:525–531

van der Peet-Schwering CMC, Kemp B, Binnendijk GP, den Hartog L, Vereijken PFG, Verstegen MWA (2004) Effects of additional starch or fat in late-gestating high nonstarch polysaccharide diets on litter performance anal glucose tolerance in sows. J Anim Sci 82:2964–2971

Pendl W, Jenny B, Torgerson P, Spring P, Kümmerlen D, Sidler X (2017) Effect of herd health management on the prevalence of postpartum dysgalaktie syndrome (PPDS) and the treatment incidence. Schweiz Arch Tierheilkd 159:109–116. https://doi.org/10.17236/sat00105

Persson A, Pedersen E, Gorensen L, Kuhl W (1989) A long-term study on the health status and performance of sows on different feed allowances during late pregnancy. I. Clinical observations, with special reference to agalactia post-partum. Acta Vet Scand 30:9–17

Persson A, Pedersen E, Morner A, Kuhl W (1996) A long-term study on the health status and performance of sows on different feed allowances during late pregnancy. III. Escherichia coli and other bacteria, total cell content, polymorphonuclear leucocytes and pH in colostrum and milk during the first 3 weeks of lactation. Acta Vet Scand 37:293–313

Peter AT, Huether P, Doble E, Liptrap RM (1985) Prostaglandin F2α and prolactin in experimental hypogalactia in sows. Res Vet Sci 39:222–229

Petersen B (1983) Methods of early recognition of puerperal and fertility disorders in the sow. Livest Prod Sci 10:253–264

Plonait H, Bickhart K (1997) Birth, puerperium and perinatal losses. In: Plonait H, Bickhart H (eds) Handbook of pig diseases. Berlin-Hamburg, Parey, pp 472–512

Pomorska-Mól M, Wierzchosławski K, Włodarek J, Gogulski M, Pejsak Z (2020) Dynamics of pro- and anti-inflammatory cytokine changes in serum and assessment of their diagnostic utility during lactation impairment in pigs. Res Vet Sci 128:9–15. https://doi.org/10.1016/j.rvsc.2019.10.020

Preissler R, Hinrichs D, Reiners K, Looft H, Kemper N (2012) Estimation of variance components for postpartum dysgalactia syndrome in sows. J Anim Breed Genet 129:98–102

Preissler R, Tetens J, Reiners K, Looft H, Kemper N (2013) A genome-wide association study to detect genetic variation for postpartum dysgalactia syndrome in five commercial pig breeding lines. Anim Genet 44:502–508. https://doi.org/10.1111/age.12047

Quiniou N, Noblet J (1999) Influence of high ambient temperatures on performance of multiparous lactating sows. J Anim Sci 77:2124–2134

Ravel A, D'Allaire S, Bigras PM (1996) Influence of management, housing and personality of the stockperson on preweaning performances on independent and integrated swine farms in Québec. Prev Vet Med 29:37–57

Reiner G, Hertampf B, Richard HR (2009) Dysgalactia in the sow postpartum – a review with special emphasis on pathogenesis. Tierarztl Prax 37:305–318

de Rensis F, Cosgrove JR, Foxcroft GR (1993) Luteinizing hormone and prolactin responses to naloxone vary with stage of lactation in the sow. Biol Reprod 48:970–976

de Rensis F, Cosgrove JR, Willis S, Hofäcker S, Foxcroft GR (1999) Ontogeny of the opioidergic regulation of LH and prolactin secretion in lactating sows: interaction between suckling and morphine administration. J Reprod Fert 116:243–251

Rydhmer L, Johansson K, Eliasson-Selling L, Persson A (2001) Clinical and genetic studies of disturbed milk production in sows. Acta Agric Scand Sect A Anim Sci 51:1–6

Schoos A, Devreese M, Maes D (2019) Use of non-steroidal anti-inflammatory drugs in porcine health management. Vet Rec 185(6):172. https://doi.org/10.1136/vr.105170

Schoos A, Chantziaras I, Vandenabeele J, Biebaut E, Meyer E, Cools A, Devreese M, Maes D (2020) Prophylactic use of meloxicam and paracetamol in peripartal sows in a farm with postpartum dysgalactia syndrome. Front Vet Sci 7. https://doi.org/10.3389/fvets.2020.603719

Solignac T, Keita A, Pagot E, Martineau GP (2010) The over-muscled sow syndrome: a new emerging syndrome in a hyperprolific sow herds. Proc Congr Int Pig Vet Soc, Preliminary observations on farrowing duration. In, p 124

Sorensen MT, Sejrsen K, Purup S (2002) Mammary gland development in gilts. Livest Prod Sci 75:143–148

Stiehler T, Heuwieser W, Pfützner A, Burfeind O (2016) Serum haptoglobin and C-reactive protein concentration in relation to rectal and vaginal temperature of early post-partum sows. Theriogenology 86:862–867

Strait E, Hawkins P, Wilson W (2012) Dysgalactia associated with Mycoplasma suis infection in a sow herd. J Am Vet Med Assoc 241:1666–1667

Stringer C (1989) Management of lactation failure in a sow. Can Vet J 30:904–925

Threlfall WR, Martin CE (1973) Swine agalactia in Missouri. Vet Med Small Anim Clin 68:423–426

Threlfall WR, Dale HE, Martin CE (1974) Serum and adenohypophyseal concentrations of prolactin in sows affected with agalactia. Am J Vet Res 35:313–315

Tummaruk P, Pearodwong P (2015) Postparturient disorders and backfat loss in tropical sows associated with parity, farrowing duration and type of antibiotic. Trop Anim Health Prod 47:1457–1464

Van den Brand H, Kemp B (2006) Dietary fat and reproduction in the post-partum sow. In: Proc 7th International Conference on Pig Reproduction, Netherlands, pp 177–189

Van Gelder KN, Bilkei G (2005) The course of acute-phase proteins and serum cortisol in mastitis metritis

agalactia (MMA) of the sow and sow performance. Tijdsch Diergen 130:38–41
Vanderhaeghe C, Dewulf J, Daems A, Van Soom A, de Kruif A, Maes D (2008) Influence of postpartum cloprostenol treatment in sows on subsequent reproductive performance under field conditions. Reprod Domest Anim 43:484–489
Verheyen A, Maes D, Mateusen B, Deprez P, Janssens G, Lange L, Counotte G (2007) Serum biochemical reference values for gestating and lactating sows. Vet J 174:92–98
Waldmann KH, Heide J (1996) Investigations on uterine contraction post-partum and effects of different utero-tonicia in puerperal sows. In: Proc. 14th International Pig Veterinary Society Congress, Bologna, Italy, 614
Whittemore C (1998) The science and practice of pig production. 2nd ed Blackwell Science, London. pp 91–130 and 421–454
Wierzchosławski K, Kwit K, Pejsak Z, Pomorska-Mól M (2018) Selected serum acute phase proteins in peripartum sows and evaluation of their diagnostic usefulness. Anim Reprod Sci 191:44–55
Yu YH, Hsu TY, Chen WJ, Horng YB, Cheng YH (2020) The effect of Bacillus licheniformis-fermented products and postpartum dysgalactia syndrome on litter performance traits, milk composition, and fecal microbiota in sows. Animals 10:2044

# Mastitis in Dairy Cows

Pamela L. Ruegg and Pamela R. F. Adkins

**Abstract**

Mastitis is the most frequent bacterial disease of adult cows and has numerous detrimental impacts on animal health and welfare, productivity, milk quality, and sustainability. Treatment of mastitis is the most common reason that antibiotics are administered to mature cattle, and prevention of mastitis requires daily attention to small management details. Mastitis occurs when intramammary infection is established by a variety of pathogens most of which are bacteria. Exposure to potential pathogens occurs through contact with organisms shed by infected cows (contagious transmission) or by contact with opportunistic organisms found in specific environmental niches. Depending on the magnitude of the inflammatory response to intramammary infection, mastitis can present in either a subclinical or clinical state. Treatment for mastitis is based on presentation and evidence-based protocols should be based on knowledge of etiology, which can be determined using a variety of diagnostic tests. While effective treatments exist, most efforts to control mastitis should focus on preventing infection by using well-defined best management practices that reduce exposure to common mastitis causing organisms.

**Keywords**

Dairy cows · Mastitis · Udder health · Somatic cell count · Clinical mastitis · Treatment · Bacteria

## 1 Introduction and Definitions

High-quality milk is produced from healthy cows and is an important source of many nutrients that are necessary to meet human dietary requirements. Efficient production of milk is dependent on prevention of diseases that can impact milk production and reduce milk quality. Mastitis is defined as inflammation of the mammary gland and is almost always caused by intramammary bacterial infection, but yeasts, algae, and other pathogenic organisms are also possible agents. On most dairy farms, mastitis is the most frequently diagnosed disease of mature dairy cows (Goncalves et al. 2022; USDA 2018), and treatment and prevention of mastitis are the most

P. L. Ruegg (✉)
Large Animal Clinical Sciences, College of Veterinary Medicine, Michigan State University, E. Lansing, MI, USA
e-mail: plruegg@msu.edu

P. R. F. Adkins
Food Animal Medicine and Surgery, College of Veterinary Medicine, University of Missouri, Columbia, MO, USA
e-mail: adkinsp@missouri.edu



J. J. Gross (ed.), *Production Diseases in Farm Animals*,
https://doi.org/10.1007/978-3-031-51788-4_15

common reason that antibiotics are used (Pol and Ruegg 2007; Saini et al. 2012; De Campos et al. 2021; Stevens et al. 2016; Gonzalez Pereyra et al. 2015; Kuipers et al. 2016). While other bacterial diseases such as pneumonia or metritis also reduce animal well-being and the efficiency of milk production, inflammatory responses to intramammary infections can damage milk secretory cells resulting in reduced milk yields (Heikkila et al. 2018) and impaired milk quality that decreases the processing value of milk (Bobbo et al. 2017; Goncalves et al. 2020; Pegolo et al. 2022). At the cow level, reduced milk yield is the greatest component of economic losses caused by mastitis, but mastitis is also associated with increased culling (Hertl et al. 2011; Heikkila et al. 2012; Cha et al. 2013), increased costs for labor, and reduced reproductive performance (Schrick et al. 2001; Santos et al. 2004; Ahmadzadeh et al. 2009; Fuenzalida et al. 2015). At the herd level, farmers incur considerable costs of implementing preventive practices (Aghamohammadi et al. 2018), which if not successful may result in further losses. Herds that maintain a high prevalence of subclinically infected cows may produce milk with bulk tank somatic cell count (SCC) levels that exceed market expectations and incur penalties or in extreme cases lose market access.

Intramammary infection (IMI) can be caused by a wide variety of bacteria, but most IMIs are caused by a smaller group of well-characterized pathogens. Historically, *Streptococcus agalactiae* was the most common agent but highly adopted, effective control practices have almost eradicated this organism, and it is rarely a cause of mastitis on most modern dairy farms in developed dairy regions (Makovec and Ruegg 2003). Common etiologies of bovine mastitis include *Staphylococcus aureus*, several non-*aureus* Staphylococci, and *Streptococci* spp., including *Str. uberis* and *Str. dysgalactiae*, as well as a variety of Gram-positive cocci such as *Lactococcus* spp. and *Enterococcus* spp. (Oliveira et al. 2013; Barkema et al. 2005; De Buck et al. 2021; Gao et al. 2017; Levison et al. 2016; Rowe et al. 2021a). In addition, *Escherichia coli*, *Klebsiella* spp., *Mycoplasma* spp., (especially *M. bovis*), and nonbacterial pathogens such as *Prototheca bovis* are frequently identified etiologies (Ruegg 2017b; Ruegg and Erskine 2020).

Mastitis pathogens are often classified as "contagious" or "opportunistic or environmental" based on the most likely route of exposure. Transmission of "contagious" pathogens usually occurs when the teats of healthy cows are exposed to bacteria that originated from infected udders of other cows. Cows that are infected with bacteria that cause long-term subclinical infections (such as *Staph. aureus* and *M. bovis)* frequently shed bacteria in normal appearing milk droplets that contaminate milking inflations and other surfaces that contact teats of healthy cows, thus providing repeated opportunities for exposure. In contrast, opportunistic (such as non-*aureus* Staphylococci (NAS)) and environmental (e.g., *E. coli* or *Klebsiell*a spp.) infections occur when teats are exposed to bacteria that are commonly found on teat skin or in the environment of the cows. Classifications based on transmission are not absolute and there is a spectrum of contagiousness based on the duration of IMI and the number of bacteria that are typically shed from an infected mammary gland. For example, *Streptococcus uberis* and *Strep. dysgalactiae* can both be found in many bedding sources used on dairy farms (Patel et al. 2019; Rowbotham and Ruegg 2016b), and thus infections from environmental sources are common. However, after IMI is established, *Streptococci* spp. can cause long-term subclinical infections and be transmitted in a contagious manner; thus, transmission of IMI lies on a spectrum and multiple routes of exposure are possible (Fig. 1).

The presentation of mastitis is characterized by the magnitude of the inflammatory response after IMI is established. Subclinical mastitis occurs when the immune response to IMI causes increased number of inflammatory cells in milk but there are no visible changes in the milk, in the udder, or systemically. Quantification of somatic cells in milk is the most common indicator of subclinical mastitis, and while thresholds vary, a SCC of >200,000 cells/mL is commonly used to define subclinical mastitis. Subclinical mastitis is common and often affects 15–20% of cows that

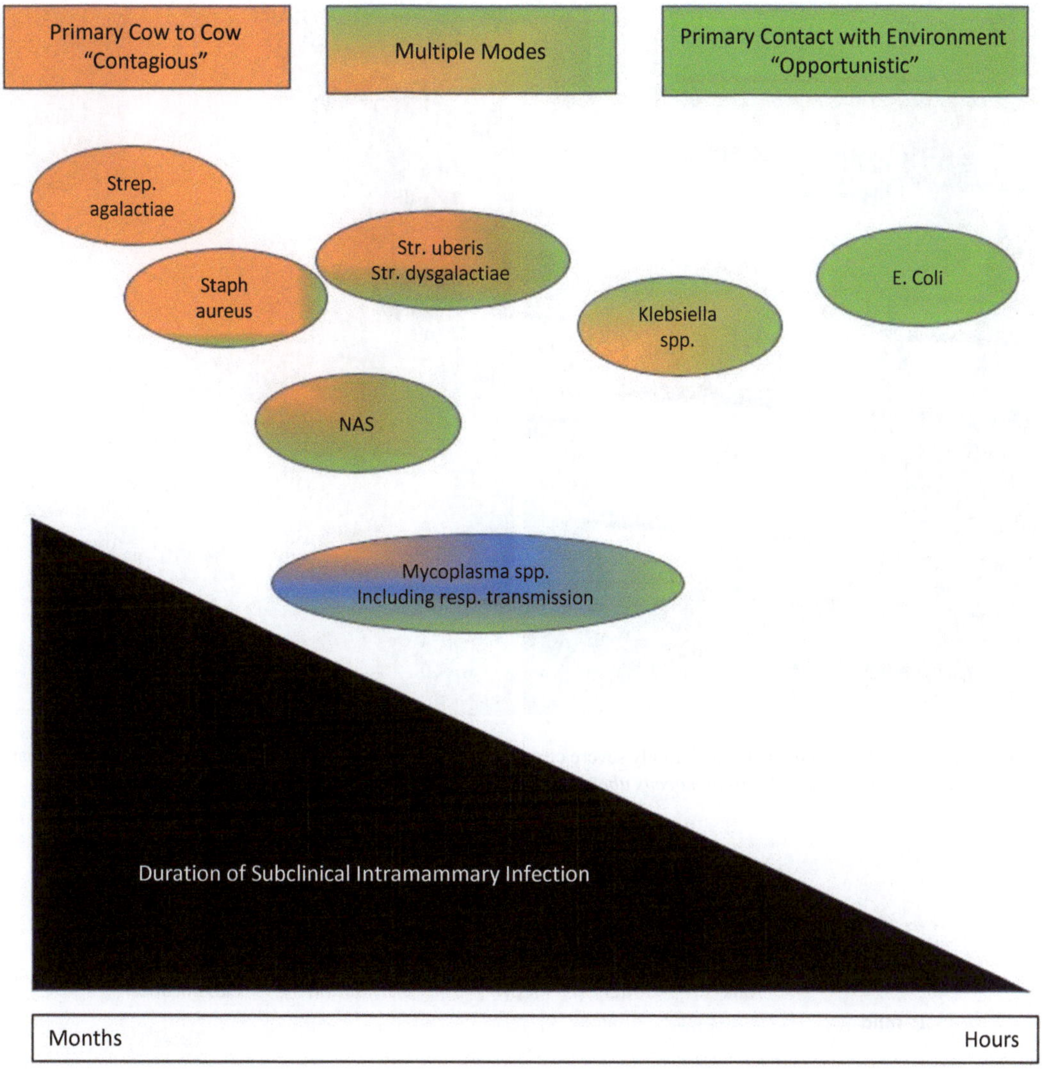


**Fig. 1** Spectrum of transmission of common mastitis pathogens ranging from primarily contagious (cow to cow) to primarily opportunistic (environmental)

are tested using monthly composite SCC values. Clinical mastitis is diagnosed based on observation of abnormal milk with or without secondary clinical signs (Fig. 2). On average, about 20–25% of cows are recorded to have at least one case of clinical mastitis each year, but there is tremendous variation in incidence among herds (USDA 2016; Goncalves et al. 2022). Fortunately, most clinical mastitis is localized to the udder and few cows develop severe systemic disease.

Severity scoring is useful to monitor clinical mastitis (Wenz et al. 2006) and a simple 3-point scale can be used to assess detection intensity (Fig. 3; (Pinzon-Sanchez and Ruegg 2011). Non-severe clinical mastitis consists of cases that present with only abnormal milk (severity score 1), or abnormal milk accompanied by localized mammary gland inflammation (severity score 2), while severe clinical mastitis is defined when abnormal milk is accompanied by fever, anorexia, rumen stasis, or other systemic signs. Using this scoring system on cases occurring on multiple herds has demonstrated that severe clinical mastitis accounts for about 15% of cases, while non-

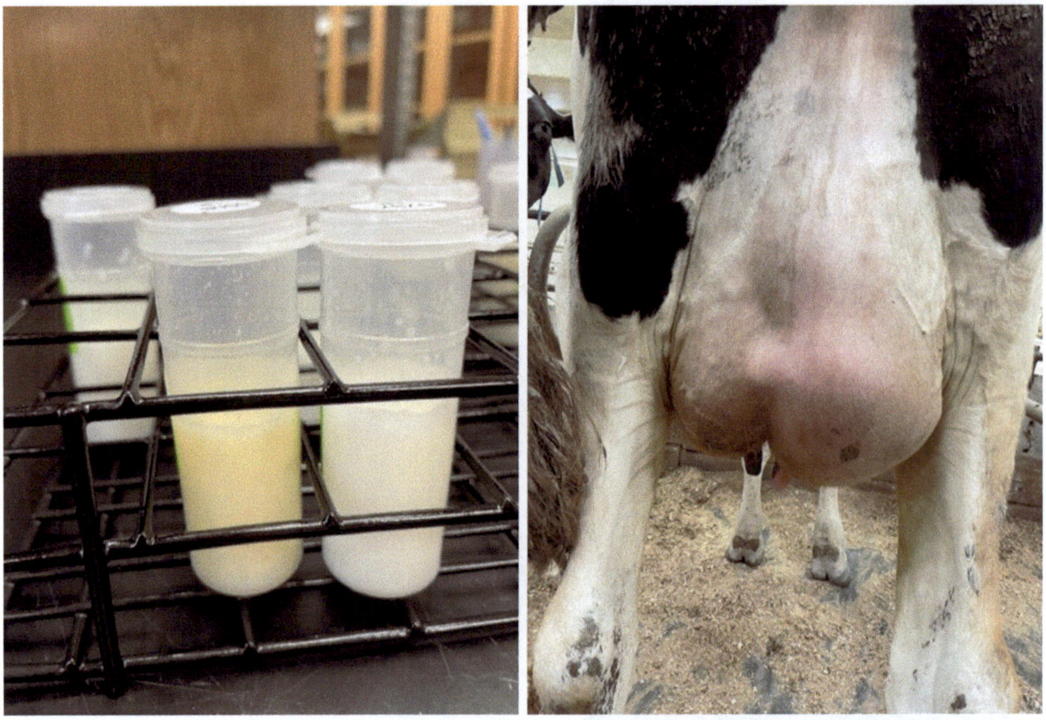

**Fig. 2** Mastitic milk and cow with moderately severe clinical mastitis (visible inflammation in the right rear mammary gland) caused by infection with *Streptococcus uberis*

| Severity Score | Level of Clinical Signs | | | | |
|---|---|---|---|---|---|
| | Milk | Udder | Cow | Expected distribution | Classification |
| 1: mild | Abnormal | Normal | Normal | 50% of cases of clinical mastitis | NON-SEVERE CLINICAL CASE |
| 2: moderate | Abnormal | Abnormal | Normal | 35% of cases of clinical mastitis | |
| 3: severe | Abnormal | Normal or Abnormal | Abnormal | <15% of cases of clinical mastitis | SEVERE CLINICAL CASE |

**Fig. 3** Classification of clinical mastitis by severity (from Pinzon-Sanchez and Ruegg 2011)

severe mastitis accounts for the remaining 85% (distributed as 50% and 35% severity scores 1 and 2, respectively) (Pinzon-Sanchez and Ruegg 2011; Oliveira et al. 2013; Tomazi et al. 2018b).

Global trends of consolidation and specialization in farming have dramatically changed herd management, but mastitis continues to be a significant problem that impacts animal well-being and sustainability of dairy farms. Throughout the world, dairy cattle are raised and managed under extensive (pasture-based) or intensive (confinement) management systems. Regardless of the management system, an increasingly specialized labor force is implementing standardized management practices that are focused on reducing exposure to mastitis pathogens in the cow's environment, during milking and throughout periods of reduced immune capacity such as the dry and early transition periods. Considerable progress in controlling mastitis has occurred but a thorough understanding of this disease is necessary to effectively reduce its impact.

## 2 Intramammary Infection

The vast majority of intramammary infections occur when a bacterial agent passes through the teat canal. After entry, the organism multiplies in the teat and gland cistern and then progresses dorsally to the milk-producing tissues. In lactating cows, infectious agents can breach the teat canal during milking or between milkings. Because the teat end is the most common entry point for infectious agents, teat health, structure, and cleanliness are vital considerations when determining risk factors for intramammary infections. Exposure and cleanliness considerations include infectious agents remaining in the milking equipment from previously milked cows and environmental exposures post milking during times of teat canal dilation.

The process of IMI is impacted by characteristics of the teat. For example, teat-end hyperkeratosis impacts the teat structure and is a risk factor for both clinical and subclinical mastitis (Guarin et al. 2017b). Teat-end hyperkeratosis is characterized by hyperplasia of the keratin layer of the teat orifice in response to chronic stimuli. Teat-end hyperkeratosis is associated with many factors, including teat shape, teat position, stage of lactation, parity, peak milk production, and milking machine-related issues. At the herd level, factors that impact teat-end hyperkeratosis include teat-end vacuum, pulsation system, pulsation cycle, liner characteristics, and overmilking. Teat-end hyperkeratosis is thought be a risk for mastitis through impairment of the teat canal and an increase in teat-end contamination by harboring bacteria on the teat end (Pantoja et al. 2020).

Another factor that can impact teat structure is udder edema. Udder edema is the accumulation of lymphatic fluid in and around the interstitial spaces of the mammary gland. During pregnancy, fetal pressure in the pelvic region disrupts blood and lymphatic circulation, resulting in fluid buildup in the udder tissues. While udder edema is not the result of an infectious condition, it can be detrimental to the structural integrity and shape of the udder and teats, resulting in an increased risk of mastitis. Beyond structural changes, the impairment of local blood circulation may prevent a proper immune response in the udder tissues. Udder edema becomes less severe and less prevalent as animals increase in parity. Udder edema, teat edema, blood in the milk, and milk leakage at the time of calving are risk factors for clinical mastitis within the first few weeks of the first calving. Udder edema can be reduced through adjusting nutrition and possibly through genetic selection. Dietary considerations should include adjusting for changes in dry matter intake, protein requirements, and consideration of anionic salts (Okkema and Grandin 2021).

Teat-end colonization or contamination with bacterial species has been shown to be associated with IMI. As compared to non-colonized teat ends, teat ends that are colonized with *Staph. aureus* have about eight times greater risk of developing IMI caused by *Staph. aureus*. Similarly, in herds that have not eradicated *Strep. agalactiae*, cows with teat skin contaminated with *Strep. agalactiae* have almost four times greater risk of IMI caused by this pathogen (Svennesen et al. 2019). Both contagious and opportunistic pathogens that colonize teat skin are associated with increased risk of IMI. Teats are in direct contact with bacteria found in lying areas, and bacterial counts of bedding are positively associated with odds of acquiring an IMI (Rowe et al. 2019).

Nonlactating cows and heifers are also at risk of developing IMI. Dairy heifers as young as

8 months of age may develop mastitis which are often caused by teat skin colonizers, such as NAS. However, dirty udders and poor hygiene in the calving area are associated with increased exposure to bacteria, resulting in increased prevalence of mastitis. In heifers, poor hygiene before calving is associated with higher odds of environmental pathogens being identified as causes of IMI shortly after parturition (Piepers et al. 2011). Flies can initiate small wounds on the distal teat skin which can become colonized by bacteria, allowing for an optimal position to invade the teat canal (Owens et al. 1998). Herds using some sort of fly control have a lower percentage of heifers with mastitis than those that do not use fly control.

## 3 Immune Responses to Infection

Anatomical barriers of the teat provide the primary defense against infection and maintenance of these barriers is vital to prevent IMI. Physical barriers include the smooth muscle sphincter that surrounds the teat canal which functions to prevent escape of milk and prevents entry of bacteria through a tight closure. Healthy teat skin is coated with fatty acids that help slow bacterial growth. Additionally, teat duct epithelium produces keratin, a waxy material that lines the teat canal. Keratin works to trap invading infectious agents, hindering migration into the gland, and contains antimicrobial agents that assist in combating invading bacteria.

The cellular components of the innate immunity of the mammary gland play a critical role in defense against pathogens that inevitably elude anatomical barriers of the teat. The healthy mammary gland has a normal population of leukocytes that help respond to infectious agents. The proportion of each cell type fluctuates but includes mostly macrophages and lymphocytes, with fewer numbers of polymorphonuclear leukocytes (PMNs). Once a stimulus is identified, macrophages and PMNs respond by secreting cytokines and chemokines. These secretions result in the recruitment of circulating leukocytes to the site of the stimulus, which can be identified by an increase in SCC. Alongside leukocytes, mammary epithelial and stromal cells can work to increase recruitment of leukocytes. Mammary epithelial cells contribute to the immune defense by expression of pathogen recognition receptors and produce cytokines when exposed to infectious agents (Gunther et al. 2011). Stromal fibroblasts and endothelial cells can also influence inflammation by regulating mammary epithelial cell function and altering the initiation, development, and resolution of inflammation (Ryman et al. 2015). The first leukocytes to be recruited are primarily PMNs. While the proportion of PMNs is only 1–11% of cells in a healthy quarter, the immune response results in an increase up to 90% or more in quarters with an IMI. The PMNs are activated by proinflammatory cytokines and release oxidizing agents and cytotoxic enzymes. After elimination of the infectious agent, macrophages clear damaged and dead cells and release anti-inflammatory cytokines to restore homeostasis.

During inflammation, the blood-milk barrier (BMB) integrity is impaired, which facilitates movement of immune cells and antibodies from the blood to the milk. In a normal gland, high concentrations of milk proteins, lactose, and potassium and low concentrations of sodium, chloride, plasma proteins, antibodies, and leukocytes can only be achieved by an intact BMB. Disruptions to the BMB during inflammation can be induced by the pathogen or the inflammatory response of the host (Bruckmaier and Wellnitz 2017). As compared to Gram-positive pathogens, IMI caused by Gram-negative pathogens result in more pronounced changes in blood constituents in milk, indicating a leaky BMB (Khatun et al. 2019; Wellnitz et al. 2013). The pathogen determines if the leaky BMB is caused solely by tight junction opening or if epithelial cells are also destroyed. Gram-negative bacteria, specifically lipopolysaccharides (LPS), cause considerable cell damage, as shown by the release of the enzyme lactate dehydrogenase (LDH) from degraded epithelial cells. Conversely, lipoteichoic acid (LTA) from Gram-positive pathogens seems to only induce opening of the epithelial tight junctions and not cause considerable cell damage (Wellnitz et al. 2016).

The outcome of any IMI is dependent on several host factors including immune status, genetics, parity, and stage of lactation. Immune status can be influenced by many factors, including stress and nutritional status. Beyond the host factors, pathogen-specific factors also play an important role in the pathophysiology of IMI. Immune responses vary depending on the number of colonies of bacteria as well as bacterial species and strain. In general, after a bacterial invasion, the acute phase of the immune response attempts to resolve the IMI immediately and is either successful (resulting in spontaneous bacteriological clearance) or unsuccessful (resulting in ongoing IMI or progression in severity). In the case of spontaneous clearance, bacteria are destroyed, and the recruitment of PMNs results in a transient IMI. When elimination is unsuccessful, the IMI may result in an aggressive immune response that developed into clinical mastitis or a less aggressive immune response resulting in a persistent subclinical infection (Leelahapongsathon et al. 2020).

## 4 Consequences of Infection

When evaluating the impact of specific pathogens on milk quality, no difference is noted based on IMI by contagious, environmental, or opportunistic organisms. In one study that primarily included IMI caused by contagious pathogens as well as milk from culture-negative samples, the greatest impairment of milk quantity and quality was found for pathogen groups that stimulated the greatest SCC (Bobbo et al. 2017). This supports the idea that effects of mastitis on milk quantity and quality are related to the inflammation rather than the infection itself.

Inflammation impacts quality of both milk and cheese products. High SCC pasteurized, homogenized 2% milk stored at 5 °C was found to have elevated lipolysis and proteolysis, resulting in bitterness and rancidity and an overall low-quality rating (Ma et al. 2000). Cheese made from milk with high SCC has a compromised sensory quality as well as a longer rennet coagulation time and a weak coagulation, which increases moisture content and results in an overall lower cheese yield (Politis 1988). Quality of milk products is impacted by inflammation because of the marked changes in the levels of nearly all major and minor constituents of the milk. The compositional changes reflect the damaged caused to the cells, resulting in impaired synthetic and secretory activity, and damage caused to the BMB. Some examples of changes that occur include a decrease in casein, fat, and lactose levels. The damage to the BMB results in an increase in the influx of ions and blood proteins, such as albumin and immunoglobulins, from the blood to the milk. Ion changes that are most notable include elevations in sodium and chloride and decreased levels of potassium (Kitchen 1981).

Beyond changes in milk quality, inflammatory responses to IMI can result in significant decrease in milk production due to impaired synthetic ability of the secretory tissue. Parity, absolute SCC value, and duration of subclinical mastitis all affect the degree of milk loss. Although reducing SCC can help producers recover income through increased milk production, not all milk loss is recoverable because not all cows can return to an SCC level less than 100,000 cells/ml, the threshold at which milk losses start to accrue (Chen et al. 2021). Therefore, prevention is key to preventing production losses associated with clinical and subclinical mastitis.

While the focus of the impacts of mastitis is often on the infected quarter and it is generally considered that each quarter functions as an independent unit, clinical mastitis in one quarter impacts the overall immune status of the mammary gland. Milk from mammary quarters adjacent to a single quarter affected with clinical mastitis have altered fat, total protein, lactose, solids-to-fat ratio, and chloride levels as well as an increase in SCC (Paixao et al. 2017). Acute but transient inflammation in a gland is sufficient to disrupt the mammary transcriptome and lactation in all quarters. The upregulation of anti-inflammatory genes in neighboring glands suggests that signaling may occur through inflammatory mediators (Shangraw et al. 2021).

Beyond the impact on milk quality and quantity, clinical mastitis can reduce cow welfare. Animals with severe clinical mastitis show signs of discomfort, such as abnormal posture and decreased social interactions. While animals with mild or moderate cases of clinical mastitis may not show as obvious signs of discomfort, some of these animals will have altered stance, heart rate, respiratory rate, and rectal temperature. In the days before diagnosis of clinical mastitis, cows can be more restless, noted by increased kicking at milking, and have a reduced feed intake. Cows with mastitis spend less time lying, consume less feed, have reduced ruminations or cud chewing, have reduced self-grooming, spend more time standing in idle, and produce less milk compared to healthy cows. Decreased lying time is particularly concerning, because lying is a high priority behavior that typically increases as part of the adaptive response to overcome disease. Together, these behavioral and physiologic changes indicate pain. To improve animal welfare, use of approved analgesics to mitigate the painful component of mastitis is often advised (Petersson-Wolfe et al. 2018).

## 5 Pathogen-Specific Characteristics

### 5.1 Gram-Positive Organisms

*Streptococci and streptococcal-like organisms.* The group streptococci and streptococci-like organisms (SSLO) are a heterogenous group of Gram-positive, catalase-negative cocci that are frequently isolated from the mammary gland of dairy cows. The term SSLO is applied to all colonies that grow on modified Edwards agar (Sawant et al. 2002). Many of the organisms within this group have been well established to be associated with mastitis, and others have only recently been identified from bovine milk samples using newer molecular methods of speciation. Organisms that are included in this category are *Streptococcus agalactiae*, *Streptococcus dysgalactiae*, *Streptococcus uberis*, *Lactococcus garvieae*, *Lactococcus lactis*, *Enterococcus faecalis*, *Enterococcus hirae*, *Aerococcus viridans*, and more. While we do not know the impact of all these organisms, late-lactation IMIs caused by *Streptococcus*, *Lactococcus*, and *Enterococcus* species are associated with increases in post-calving clinical and subclinical mastitis. In contrast, IMIs caused by *Aerococcus* species have not been associated with any post-calving udder health measures, potentially indicating this species is less pathogenic than other species within the SSLO group (Rowe et al. 2021a). When characterized as a group, the use of manure solid bedding is associated with higher SSLO counts in bulk tank milk and is associated with worse udder health (Patel et al. 2019). Large numbers of SSLO can be recovered from many different bedding types as well as on the teats of cows from many different bedding types (Patel et al. 2019; Rowbotham and Ruegg 2016a).

Among this group of organisms, the streptococcal species are the most characterized. *Streptococcus agalactiae* is regarded as a highly contagious mastitis pathogen (Fig. 1). The main reservoir of this organism is the mammary gland of infected cows with contagious transmission occurring primarily when teats are exposed to infected milk during the milking process. Recent studies have occasionally identified *S. agalactiae* in fecal and environmental samples (Jorgensen et al. 2016), but this is a minor route of potential exposure. Infections with *Strep. agalactiae* are often chronic and subclinical, resulting in a high SCC and decreased milk production. While infections often have a low rate of spontaneous cure, antibiotic treatment is highly efficacious. Before antibiotics were commonly used to treat mastitis, IMI caused by *Strep. agalactiae* was the most common cause of mastitis on many dairy farms throughout the world (Ruegg 2017a). Improved hygiene during milking routines, use of post-milking teat dip, and blanket dry cow antibiotic therapy have all proven to be useful control measures, and this organism is rarely isolated from IMI in developed dairy regions. However, in recent years, the prevalence of *Strep. agalactiae* has gradually increased in some countries, which has coincided with changes in herd management, including fewer herds, increased herd size, introduction of automated milking systems, and a reduction in the use of blanket dry cow therapy.

*Streptococcus uberis* is a pathogen with multiple environmental reservoirs, including soil and bedding material. The organism has been isolated from cow skin, lips, tonsils, respiratory tract, rumen, teat skin, mammary gland, and feces. It is commonly isolated from pastures as well as bedding used in confinement operations. While the environment is considered an important source of initial exposure, cow-to-cow transmission may occur (Fig. 1). Sources of *S. uberis* are often areas of high cow traffic, including holding pens for the milking parlor, drinking troughs, and passageways to the pasture. Multiple strains of *Strep. uberis* are often identified and both transient and persistent infections have been documented. *Streptococcus uberis* is associated with both clinical and subclinical mastitis. The organism can produce biofilms and capsules, invade mammary cells, resist phagocytosis, and withstand intracellular destruction by leukocytes. All these factors help aid in persistence, leading to chronic infections.

*Streptococcus dysgalactiae* is associated with clinical mastitis, subclinical mastitis with prolonged elevation in SCC, and decreased milk yield. Like *Strep. uberis*, *Strep. dysgalactiae* has multiple modes of transmission including cross-contamination during milking as well as exposure from environmental reservoirs (Fig. 1). Several extramammary reservoirs have been identified, including cattle tonsils, mouth, vagina, and wounds. The organism has also been isolated from flies collected from the farm environment. It is thought that cow skin is a main reservoir and environmental samples test positive due to cattle shedding bacteria into the environment, which then plays a role in the transmission. Free-stall housing has been found to be associated with a higher incidence of *Strep. dysgalactiae*, potentially related to more direct and indirect contact with other cows. *Strep. dysgalactiae* IMI may establish during the dry period or in heifers before calving (Smistad et al. 2022).

Isolates of the *Lactococcus* genus are often misidentified as enterococci or streptococci, which has hindered the understanding of their significance and has potentially led to underreporting. The genus has been labeled as an emerging mastitis pathogen; however, it is unknown if emergence is related to management changes or is a result of improved identification using molecular methods. The two main *Lactococcus* species that have been associated with bovine IMI include *Lactococcus lactis* ssp. *lactis* and *Lactococcus garvieae*, with *Lacto. lactis* often identified as the dominant species compared to *Lacto. garvieae*. *Lactococcus* species are commonly isolated from clinical mastitis, subclinical mastitis, and environmental samples. Odds of SCC resolution and bacteriologic cure are thought to be greater for cows infected with *Strep. dysgalactiae* or *Strep. uberis* compared to *Lacto. lactis* or *Lacto. garvieae* (Scillieri Smith et al. 2020). Based on high diversity among strain types, it is suspected that *Lactococcus* species originate from multiple reservoirs (Rodrigues et al. 2016).

Enterococci are a normal part of the gut flora. The genus is known for a high level of resistance to many antimicrobials through both intrinsic and acquired mechanisms. Species that have been identified among milk samples have included *Entero. faecalis*, *Entero. faecium*, *Entero. hirae*, *Entero. casseliflavus*, *Entero. durans*, and *Entero. avium*. *Aerococcus viridans* has been identified in many environments, including hospitals, seawater, and animal farms. *Aero. viridans* has been increasingly isolated from bovine milk samples with subclinical and clinical mastitis, although its pathogenic potential is not completely understood. When evaluated in a mouse model, considerable pathogenic variation was identified among isolates that originated from dairy cows (Liu et al. 2019).

*Staphylococci* spp. Staphylococci are among the most commonly isolated bacteria from bovine IMI. There are more than 50 different staphylococcal species that have been associated with bovine mastitis. Staphylococci are Gram-positive spherical organisms that form grapelike clusters when visualized under the microscope. *Staph. aureus* is the most recognized among the staphylococcal species, primarily due to the well-established characteristics of causing chronic mastitis. The main reservoir of *Staph. aureus* is infected mammary glands, and this agent is most associated with subclinical mastitis. Most infections are chronic, frequently persisting over the subsequent lactations. Transmission within a herd is influenced by differences in contagiousness among strains and variation in herd management practices (Rainard et al. 2018). Many *Staph. aureus* strains have been associated with bovine mastitis, and most herds have a predominant strain type but not all strains result in cow-to-cow transmission (Zadoks et al. 2011). Studies conducted in Europe using a simple strain typing method have identified various bovine genotypes with different virulence and pathogenicity factors. Genotype B and genotype C are the primary genotypes identified in some countries. *Staph. aureus* genotype B is primarily found in Central Europe and is highly contagious and pathogenic, causing significant herd problems with up to 87% cow-level prevalence. In contrast, genotype C and other genotypes are more sporadic and non-contagious and therefore considered less problematic (Leuenberger et al. 2019). In addition to subclinical mastitis, *Staph. aureus* can also cause clinical mastitis which is usually mild of moderate, but severe cases can occur and in some instances may result in ischemic gangrenous necrosis. The severity of disease is likely determined by a combination of strain virulence and the host condition.

In recent years, NAS species have emerged as the most frequently isolated bacterial group associated with IMI, and more than 20 different NAS species have been isolated from bovine milk samples. The relevance of NAS for udder health and milk yield is not completely clear. As a group, NAS are considered minor pathogens; however, species and strain variations have been identified. In most studies, *Staph. chromogenes* is the most prevalent NAS species (De Buck et al. 2021). Some NAS species, such as *Staph. chromogenes* and *Staph. simulans*, are associated with increased SCC and persistent infection (Fry et al. 2014; Supre et al. 2011). Non-*aureus* Staphylococci are ubiquitous on dairy farms and have been isolated from quarter-milk samples, bulk tank samples, teat ends, bedding, feces, and the environment (De Visscher et al. 2017; Wuytack et al. 2020; Piessens et al. 2011). Some NAS species seem to favor a particular niche. For example, *Staph. chromogenes* and *Staph. epidermidis* are primarily isolated from milk, while *Staph. equorum* and *Staph. devriesei* are predominately found in the environment (Piessens et al. 2011). Teat skin colonization has been found to be a risk factor for IMI for some NAS species, including *Staph. haemolyticus* and *Staph. chromogens* (Adkins et al. 2018; Leroy et al. 2015). Recent studies have identified that while NAS species can cause persistent IMI and cause a minor elevation in SCC, they may not be associated with changes in milk yield, at least in dairy heifers (Valckenier et al. 2021). Additionally, some have identified *Staph. chromogenes* to be protective against IMI with other pathogens. In one study, when *Staph. chromogenes* was found on teat ends of prepartum heifers, it was protective against a SCC of ≥200,000 cells/ml in the first few days of lactation (De Vliegher et al. 2003). Some have proposed these protective effects could be due to bacteriocin production. In vitro experiments have identified that some *Staph. chromogenes* isolates can inhibit growth of *Staph. aureus*, *Strep. dysgalactiae*, and *Strep. uberis* (De Vliegher et al. 2004). These finding need further evaluation and ongoing research is evaluating potential negative or positive impacts of NAS IMI on udder health.

Epidemiologic investigations of pathogens associated with IMI in heifers and early lactation of primiparous cows have identified NAS as the most common organisms. Risk factors for NAS IMI in heifers include poor heifer hygiene, teat dipping prior to calving, and udder clipping prior to calving. The prevalence of *Staph. aureus* is generally less in heifers compared to NAS species; however, it is still a major concern due to difficultly controlling these infections. While the major reservoir of *Staph. aureus* is the infected mammary gland of lactating cows, the organism can also be identified on teat skin, the vagina, bedding, feedstuffs, air, and equipment. Risk factors for *Staph. aureus* IMI in heifers include a higher average herd milk SCC and contact with lactating cows before calving. Additionally, horn flies (*Hydrotaea irritans*) likely play an important role in transmission of *Staph. aureus* to unbred and pregnant heifers. Application of fly control strategies prior to calving is associated with a reduced risk for *Staph. aureus* IMI in heifers (Piepers et al. 2011).

***Other Gram-Positive Organisms*** Other Gram-positive organisms include *Bacillus* species, *Corynebacterium* species, and *Trueperella pyogenes*. Bacillus species are aerobic, sporulating, Gram-positive, rod-shaped bacteria. These organisms can be identified from many sources, including soil, water, dust, air, feces, and on vegetation. The spores are resistant to heat and chemical disinfectants. *Bacillus cereus* and *Bacillus licheniformis* are important to consider as spoilage organisms in pasteurized milk and cream. *Bacillus cereus* has been identified from mammary glands with elevated SCC and is considered an uncommon cause of mastitis. Cases have been linked to intramammary infusions, chaff bedding, or heated brewers' grains. Some cases have resulted in acute gangrenous mastitis and were associated with acute death with signs of toxemia. In contrast, a few colonies of *Bacillus* species on blood agar plates are generally the result of contamination, and greater number of colonies ($\geq$ 500 CFU/ml) should be used to classify an IMI with *Bacillus* species. *Bacillus* species should be considered a cause of an IMI if cultured sequentially in pure growth with concurrent elevations in SCC or clinical signs.

*Corynebacterium* are Gram-positive and mostly aerobic, slow growing, bacilli, or club-shaped bacteria. *Corynebacterium bovis* is considered a contagious organism that commonly contaminates teat skin and is the most characterized of the *Corynebacterium* species. Other *Corynebacterium* species that have been identified among milk samples have included *C. xerosis*, *C. amycolatum*, *C. lipophiloflavum*, *C. variabile*, *C. casei*, and others. It has been reported that *C. bovis* only colonizes the teat canal of dairy cows; however, others have identified *C. bovis* in the teat cistern, gland cistern, and mammary parenchyma (Benites et al. 2003). Subclinical mastitis associated with *C. bovis* can increase SCC, decrease lactose, and decrease solids' nonfat content compared to healthy quarters. However, subclinical mastitis associated with other *Corynebacterium* species has no effect on SCC, milk yield, or composition (Goncalves et al. 2016).

*Trueperella pyogenes* is a nonmotile, facultative, anaerobic, slow-growing Gram-positive bacterium that appears as coccoid rods or coccobacilli. The organism is frequently referred to as "summer mastitis" or "dry cow mastitis" as infections often occur and with more severe clinical signs during the dry period or when cows are on pasture. Intramammary infections caused by *T. pyogenes* are associated with high somatic cell counts, significant losses in milk yield, and a high percentage of nonfunctional quarters. *Trueperella pyogenes* is often associated with teat lesions. The organism is frequently identified with other bacterial species when isolated from milk samples. The organism is more common in cows during early lactation compared to later lactation stages. It is more common in cows in loose housing compared to tie stalls and also common in cows that had previous cases of clinical mastitis during the current lactation (Duse et al. 2021).

## 5.2 Gram-Negative Organisms

Many Gram-negative bacteria can be found in dairy farm environments, and frequent teat-end exposure can result in opportunistic intramammary infections. Gram-negative genera including *Serratia*, *Proteus*, *Enterobacter*, and *Raoultella* have been recovered from both clinical and subclinical cases of mastitis, but *E. coli* and *Klebsiella pneumoniae* are the most common Gram-negative bacteria associated with bovine mastitis.

***Escherichia coli*** *E. coli* is a Gram-negative bacterial pathogen that is commonly found in the dairy farm environment. Most *E. coli* are commensal organisms that originate in the digestive tract of farm animals and contaminate the environment through fecal shedding. Many types of bedding used in confinement dairy operations support bacterial growth, and this is especially true for organic sources of bedding such as manure solids. *E. coli* can be found in large numbers in many organic bedding sources and easily transferred to teat skin, increasing the risk of IMI (Rowbotham and Ruegg 2016b). *E. coli* is one of the most common bacteria recovered from milk samples obtained from quarters of cows affected with clinical mastitis and often accounts for at least 25% of all cases of clinical mastitis occurring in cows managed using confinement systems (Oliveira et al. 2013). The outer membranes of Gram-negative bacteria contain LPS that is a potent stimulator of the immune response, and about 30% of clinical cases of mastitis that have viable *E. coli* recovered from quarter-milk samples may be classified as severe (Oliveira et al. 2013). Activation of the immune response and progression of a clinical case from mild to severe signs is dose-dependent based on magnitude of LPS (Schukken et al. 2012). Exposure to *E. coli* is ubiquitous on many dairy farms, and immune suppression associated with the postpartum period or heat stress is a risk factor for development of more severe disease. Severe cases of clinical mastitis caused by *E. coli* may result in death or failure to return to productivity, so prompt treatment of severe cases is a priority. In most instances, however, a rapid immune response to IMI by *E. coli* is effective in achieving spontaneous bacteriological clearance, and antimicrobial therapy is not necessary for most non-severe cases of clinical mastitis caused by this pathogen (Fuenzalida and Ruegg 2019a; Pyorala et al. 1994; Suojala et al. 2013). Few cases of subclinical mastitis are attributable to *E. coli* because the majority of IMI caused by this pathogen are of short duration and are successfully eliminated within 7–28 d (Pyorala et al. 1994; Todhunter et al. 1991).

***Klebsiella* spp.** *Klebsiella* spp. are known to be shed in feces of cattle (Munoz et al. 2006), and both *Klebsiella oxytoca* and *K. pneumoniae* have been frequently recovered from milk samples of cows affected with clinical and subclinical mastitis (Schukken et al. 2012; Fuenzalida and Ruegg 2020). Among Gram-negative mastitis pathogens, *Klebsiella* spp. are the second most common cause of mastitis, but they usually account for only 5–10% of all clinical cases (Oliveira et al. 2013). In contrast to IMI caused by *E. coli*, the duration of infection is much longer for *Klebsiella* spp., and some cases result in persistent subclinical infections (Schukken et al. 2012). As compared to *E. coli*, IMI caused by *Klebsiella pneumoniae* have a lower rate of spontaneous bacteriological clearance, and many have intrinsic or acquired resistance to β-lactam antimicrobials (Fuenzalida and Ruegg 2019a). While most *Klebsiella* IMI are opportunistic and are acquired from an environmental source, the longer duration of IMI and persistent subclinical infections can contribute to contagious transmission (Munoz et al. 2007). Mastitis caused by *Klebsiella* spp. can remain in a subclinical state or progress to clinical signs of any severity. As compared to IMI caused by *E. coli*, more severe responses to *Klebsiella* infections have been noted, and the duration and magnitude of milk loss after infection often result in increased culling of cows that develop clinical mastitis (Schukken et al. 2012).

## 5.3 Mycoplasma

There are over 100 known *Mycoplasma* species, and this genus is associated with several diseases in cattle, including mastitis, arthritis, pneumonia, otitis media, and reproductive disorders. *Mycoplasma* species are known for their lack of a cell wall and for having a smaller genetic code compared to most bacteria. Their simple genomes and fastidious growth requirements can be associated with their slow replication and difficult identification, often leading to underdiagnosis. The identification of *Mycoplasma* species as a cause of mastitis is highly relevant due to its highly contagious nature and association with clinical, subclinical, and chronic IMI (Parker et al. 2018; Fox 2012). *Mycoplasma* can be transmitted by contact with milk from infected cows and contaminated milking equipment. Unique among the mastitis-associated organisms, *Mycoplasma* can also be transmitted through direct contact with respiratory secretions. *Mycoplasma* IMI can affect more than one quarter, causing significant decreases in milk production, and do not respond to antimicrobial therapy (Nicholas et al. 2016). While the presence of *Mycoplasma* in the milk is an indication of infection, animals can be asymptomatic carriers and carry *Mycoplasma* species in other body sites without showing any clinical signs. While *Mycoplasma bovis* is the predominant *Mycoplasma* species identified from milk samples, at least 11 species have been identified as associated with bovine mastitis, including *Mycoplasma bovigenitalium*, *Mycoplasma canadense*, *Mycoplasma californicum*, and *Mycoplasma alkalescens* (Gioia et al. 2021).

Risk factors for *M. bovis* have been identified and include large herd size, purchased cattle, the use of a breeding bull, or contaminated semen used for insemination (Fujimoto et al. 2020; Gille et al. 2018; Haapala et al. 2018). In a naïve herd, *M. bovis* can spread quickly among all age groups and young stock are a reservoir. Respiratory secretions via aerosols and nose-to-nose contact with infected animals play a major role in spread among a positive farm (Haapala et al. 2021). Feeding waste milk or colostrum increases the risk of transmission to the rest of the herd and can be associated with otitis in calves. The lack of a separate sick or hospital pen can also contribute to the spread of the organism. Some environmental sources likely play a role in transmission, and *Mycoplasma* species have been isolated from stored recycled sand bedding and dry lots. *Mycoplasma* species can survive in recycled sand bedding for 8 months with moisture likely contributing to organism survival and replication (Justice-Allen et al. 2010). *Mycoplasma* mastitis has also been diagnosed in peripubertal calves and represent a threat to the herd health, potentially acting as a source of introduction of the organism to the lactating herd.

***Other Organisms*** *Prototheca* are algae that reside in environments that are rich in organic matter and moisture (Janosi et al. 2001). Several *Prototheca* species can be recovered from feces of cattle and other mammals (including pigs and rodents) as well as from environmental sources including mud, water, and soil (Anderson and Walker 1988; Pore and Shahan 1988; Pore et al. 1987). While *P. blaschkeae* and *Prototheca wickerhamii* have been associated with bovine mastitis (Ito et al. 2011), *P. bovis* is the main species associated with herd outbreaks of *Prototheca*-related mastitis (Moller et al. 2007; Ricchi et al. 2010). Intramammary infection with *Prototheca bovis* often results in persistent subclinical infections that are resistant to antimicrobial therapy and can persist throughout the dry period and be transmitted in a contagious manner. Diagnosis of mastitis caused by *Prototheca bovis* is enhanced by using selective agar, and segregation and culling are frequently recommended to control transmission.

*Yeasts* are additional organisms that are commonly found in the dairy farm environment and occasionally result in opportunistic IMI (Scaccabarozzi et al. 2008). Non-hygienic administration of IMM products is a well-known risk factor for IMI caused by these organisms. Antimicrobials approved for use in dairy cattle are not effective for yeast infections, and infusion of intramammary antibiotics may exacerbate yeast infections. In many instances, IMI caused by yeast will spontaneously resolve after several weeks, but cows affected with severe or recurrent clinical mastitis, reduced milk production, or persistently high SCC attributable to yeast should be culled.

## 5.4 Detection and Diagnostics

***Subclinical mastitis*** Many tests have been evaluated for detection of subclinical mastitis, with evaluation of SCC being the most used method due to cost and availability. Bulk tank SCC values are usually available on every shipment of milk and provide farm level data that roughly correlates with overall udder health among the lactating herd. This data is used by processors to assess quality of the milk that they are purchasing and is often the starting point for producers to recognize when a subclinical mastitis problem exists. At the cow level, composite milk SCC can be obtained by the producer using Dairy Herd Improvement testing or similar testing strategies. Cow-level data can be used to predict overall mammary health. Because this data is often collected once monthly, the longitudinal data can easily be used to determine number of cows with chronically elevated SCC, those with newly elevated SCC, and those that have recently had a reduction in SCC.

The most accurate relationship between IMI and SCC is at the quarter level. Healthy quarters have a mean SCC of approximately 70,000 cells/ml, and SCC of 200,000 cells/mL or greater is often used as a threshold to define infected quarters (Schukken et al. 2003). While these cut points have been established and are frequently used, the diagnostic sensitivity of quarter-level SCC for subclinical mastitis detection is imperfect and depends on the associated organism. For example, NAS species can often be associated with a mild elevation in SCC that may not elevate about 200,000 cells/ml (Fry et al. 2014). Previous work has reported a sensitivity of 0.53 for quarter-level SCC when using a threshold of 100,000 cells/ml (Middleton et al. 2004). Beyond the impact of each specific organism, other limitations of SCC include false-positive results related to SCC remaining elevated for some time after an organism has been eliminated from the mammary gland (Adkins and Middleton 2018).

Other methods to diagnose mammary gland inflammation are available and include lactate dehydrogenase (LDH), *N*-acetyl-β-D-glucosaminidase (NAGase), milk amyloid A, electrical conductivity (EC), and others. Some of these methods, such as EC and LDH, are used by automatic milking systems (AMS). Cow-side tests are also available and have the potential to provide timely data for the producer to make decisions. Available cow-side tests include the California Mastitis Test (CMT), the Wisconsin Mastitis Test (WMT), the Somaticell test, esterase tests, and milk pH. Although commercially available, not all these tests have proven to be useful. Milk pH has been shown to perform poorly when used to diagnose an IMI at dry-off or calving and is therefore considered not clinically useful (Kandeel et al. 2019). Comparisons have been done to evaluate the CMT, the cow-side WMT, and three esterase tests. Overall, the CMT provided the most accurate, practical, and least cost for on-farm screening to predict subclinical mastitis (Kandeel et al. 2017). The CMT is a rapid, inexpensive, semiquantitative cow-side test that has been used for over 60 years as an indirect indicator of the level of SCC in milk. The CMT works by using a reagent that causes cell membrane disruption. Free DNA then reacts with the test agent to form a gel. The amount of gelling that occurs is then scored as negative, trace, 1, 2, or 3. While commonly used for the reasons stated, this test is imperfect when evaluating sensitivity and specificity of identifying an IMI compared to milk culture. When evaluating early lactation cows, the CMT has a sensitivity of 0–52% and a specificity of 68–100%, depending on the cut points used. Some sensitivity can be improved when interpreted at the cow level, ranging from 0 to 72%, again depending on the cut points applied; however, this results in a decline in specificity, ranging from 25 to 100% (Godden et al. 2017). The poor sensitivity of the CMT test can result in many infected quarters going undetected.

***Clinical mastitis*** Clinical mastitis is diagnosed based on visible changes in milk including flakes, clots, and/or watery milk. Visual examination of foremilk immediately before attachment of teat cups allows for easy detection of abnormal milk. While visual detection is the primary means of detecting clinical mastitis, this is not possible on herds using AMS, and development of automated mastitis detection systems has become important. Detection of abnormal milk in AMS herds is done using a variety of sensor systems based on proprietary algorithms. Sensor systems have included EC, color, SCC, temperature, milk yield, and enzyme detection (Penry 2018). In addition to sensors that aim to measure mastitis indicators, many other sensor systems measure other characteristics that are affected when cows develop mastitis, such as activity or behavior monitors, rumen activity, temperature sensors, rumen pH sensors, and body weight measuring systems. Detection performance of these systems and algorithms varies considerably and predictions improve with usage of multiple measurements. To date, the perfect mastitis alert systems have not been identified (Rutten et al. 2013).

***Intramammary Infection*** Diagnosis of an IMI is performed to confirm a bacterial etiology and used to select appropriate treatment protocols or initiate preventive measures. Bacteriologic culture and polymerase chain reaction (PCR) are the most common methods used to diagnose etiology. Bacterial culture of milk is generally inexpensive and simple to perform (NMC 2017). Many mastitis pathogens are grown under aerobic conditions using blood-based agar. However, some organisms require specific growth conditions and media, including *Mycoplasma* species. While culture is simple, consistent procedures and definitions are needed to establish standardized diagnoses. Aseptic sample collection is vital to obtain usable data. Organisms that can cause mastitis are prevalent in the dairy environment and therefore contamination can cause confusion and misdiagnosis. If a sample is classified as contaminated, it is prudent to not make any diagnosis, but rather recommend resampling. Recommendations for defining contamination have been standardized at the quarter level, with ≥3 colony types generally being considered a contaminated sample (NMC 2017). When using culture techniques to diagnose IMI, sample storage and handling are also important. The impact of storage and freezing is variable, depending on the organisms. Some organisms, such as *E. coli*, *Trueperella pyogenes*, *Nocardia*, and *Mycoplasma* species, do not always survive extended storage periods. For *Mycoplasma* species specifically, it has been recommended that fresh samples should be used to improve detection, and if frozen samples must be used, the storage time should be minimized (Biddle et al. 2004). Conversely, long-term freezing of milk samples does not affect the viability of Gram-positive cocci. Freezing of milk samples can increase the likelihood of detecting some staphylococcal species and *Strep. agalactiae*. This may occur due to the act of freezing and thawing resulting in a disruption in the bacterial clusters.

Composite milk samples, including milk from all four quarters, are often used to diagnose organisms associated with subclinical mastitis but have a low sensitivity that is related to dilutional effects. Quarter-level milk samples are recommended to diagnose IMI. Even when using quarter-level milk samples, the major limitation of culture is a low sensitivity. Therefore, the general recommendation to consider a quarter as culture positive is to use a 100 CFU/ml cut point, with the exception of NAS, in which it is recommended to use a 200 CFU/ml cut point (Dohoo et al. 2011). That said, the cut points used to define an IMI on a specific farm can vary depending on the goals of the producer.

Selective media or secondary methods that utilize phenotypic characterization can often be used to easily categorize organisms into broad groups. These methods are often used for on-farm culture programs and can accurately categorization into groups such as Gram-negative or Gram-positive. These broad groupings are useful to make treatment decisions for clinical mastitis. If genus- or species-level identification is needed, usage of phenotypic testing to differentiate organisms is not always accurate, and advanced molecular techniques are recommended. Advanced techniques include polymerase chain reaction (PCR) for a specific target using the isolated organism or partial sequencing analysis of housekeeping genes, such as 16S rRNA, as a universal target. More recently, matrix-assisted laser desorption/ionization time-of-flight (MALDI-TOF) mass spectrometry has been used more widely to identify mastitis-associated organisms. MALDI-TOF is a high-throughput technology that uses a protein signature of the organism and a reference database to determine bacterial species. The method has been validated as accurate for many mastitis pathogens but relies on a database which currently contains about >450 bacterial species. One limitation of MALDI-TOF is that the technique cannot be used to identify bacterial species directly from milk. Traditional bacterial culture is required and therefore the limitations of culture cannot be overcome using this method.

Beyond bacterial culture, polymerase chain reaction (PCR) is commonly used to diagnose mastitis pathogens. The first commercially available real-time multiplex PCR tests for quarter-based identification of mammary pathogens was the PathoProof Mastitis PCR Assay (Thermo Scientific) that was launched in 2008 and is now available in many parts of the world. In comparison to culture-based methods, PCR performed on aseptically collected milk samples is faster and can provide user-independent identification of bacteria. Commercially available PCR assays can be used to diagnose milk samples containing preservatives and samples that are shipped at room temperature. The PCR testing has been found to provide positive results for many samples that were culture negative based on traditional methods. The detection of bacterial cells that were dead in the mammary gland is one limitation of this method. For example, in one study in which traditional culture and PCR were used to evaluate elimination of experimentally induced bacterial IMI, researchers identified that PCR results were positive much longer than traditional culture (Hiitio et al. 2018). Because of this, results should be interpreted with caution if using this modality to determine treatment plans. If using PCR to assess for bacteriological cure, it is recommended to wait 2–3 weeks after the onset of mastitis or after the quarter-level milk SCC has normalized (Hiitio et al. 2018). Furthermore, interpretation of results has limitations, especially when defining contaminated samples. Commercially available PCR testing is generally done using quantitative methods and the target bacterial species must meet a specific threshold to be counted as positive; however, that does not rule out contamination completely. One final limitation to consider when using PCR methods is that PCR can only identify targeted organisms. While commonly identified organisms are available using the commercially available kits, it is possible that organisms can go undetected if the target range is not inclusive.

## 5.5 Treatment of Mastitis

***Clinical Mastitis*** The appearance of clinical signs and the necessity of discarding abnormal milk result in strong motivation for farmers to effectively treat clinical mastitis (Valeeva et al. 2007). Most cases of CM are non-severe and are treated symptomatically using intramammary antimicrobials (Oliveira and Ruegg 2014), but severe cases may be treated using combinations of intramammary and systemic antimicrobials, supportive therapy, and/or nonsteroidal anti-inflammatories. Most cases of clinical mastitis are either mild or moderate, and the majority of cases that present with severe symptoms are caused by Gram-negative organisms (Oliveira et al. 2013). Microbiological examination of milk collected from cases of clinical mastitis results in approximately 30% Gram-positive bacteria, 30% Gram-negative bacteria, 30% culture-negative bacteria, and 10% other pathogens. The most common pathogens include *Escherichia coli* and environmental streptococci (Oliveira et al. 2013).

Knowledge of etiology of clinical mastitis allows for selective treatment decisions that reserve antimicrobial therapy for cases that will benefit. Research has suggested that antimicrobials are not necessary for the treatment of non-severe cases of mastitis caused by *E. coli* (Suojala et al. 2013; Fuenzalida and Ruegg 2019a) or cases that are microbiologically negative when detected (Fuenzalida and Ruegg 2019b). Antimicrobial treatment is considered beneficial for cases of clinical mastitis associated with environmental streptococci. Selective treatment of clinical mastitis represents a tremendous opportunity to reduce antimicrobial usage on commercial dairy farms without sacrificing efficacy or long-term health, welfare, or production of the cow. Antimicrobial treatment protocols can easily be established based on culture results through usage of a lab near the farm or on-farm culture systems. The use of on-farm culture systems to guide treatment reduces antimicrobial use and decreases milk withhold times without impacting clinical cure, bacterial cure risk, new intramammary infection risk, or treatment failure risk (Lago et al. 2011). Culture-based selective treatment protocols are based on targeting antimicrobial usage for Gram-positive cases while allowing for spontaneous cure of culture-negative and Gram-negative cases although treatment decisions should include cow-level variables associated with the ability to mount an effective immune response (Ruegg 2017c). Research has indicated that intramammary antimicrobial treatments of culture-negative non-severe cases of clinical mastitis are not expected to improve economically relevant outcomes and there is little financial or welfare benefit (Fuenzalida and Ruegg 2019a). Targeted antimicrobial usage allows for selective treatment of Gram-positive organisms and allows usage of noncritically important antimicrobials as the first-line therapy.

Systemic antimicrobials are most often suggested in cases of severe clinical mastitis. In one study, systemic antimicrobial administration reduced the proportion of severe clinical mastitis caused by coliforms that resulted in cow death or culling. This outcome is likely associated with amelioration of bacteremia-related pathogenesis (Erskine et al. 2002). Bacteremia has been found to develop in approximately 32% of cows with acute coliform mastitis, is more common among cows with severe systemic disease, and has an impact on outcome of the animal (Wenz et al. 2001). Therefore, consideration of systemic antimicrobials should be considered in these cases.

A systematic review of efficacy of non-antimicrobial therapies for treatment of clinical mastitis has been conducted (Francoz et al. 2017). The treatments included anti-inflammatory drugs, oxytocin with or without frequent milk out, biologics, homeopathy, botanicals, probiotics, and alternative products. This study concluded that evidence-based recommendations could not be given for use of alternative or non-antimicrobial conventional treatments. While use of anti-inflammatories did not impact changes in milk associated with clinical mastitis, they did have a positive impact on that cow as these drugs were associated with decreased body temperature. The study concluded that probiotics and oxytocin with or without frequent milk out should not be recommended. Another study evaluated use of nonsteroidal anti-inflammatory drug (NSAID) for cases of clinical mastitis to determine economic benefits and determined that cows treated with an NSAID had better reproductive performance based on a shorter calving to conception interval, a shorter intercalving interval, and fewer inseminations per conception compared to the non-NSAID group. Additionally, the NSAID-treated group had a lower culling rate, resulting in lower replacement costs. Overall, the use of NSAIDs was determined to be cost-effective and the authors recommended NSAID based on welfare benefits (Van Soest et al. 2018).

***Subclinical Mastitis*** Milk from quarters affected with subclinical mastitis may contain millions of somatic cells, but appear visually normal. Normal appearing milk from subclinically infected cows has economic value because when comingled with milk from healthy cows, it may be legally sold for processing into dairy products. Thus, the value of antibiotic treatment of cows affected with subclinical mastitis is based on the marginal difference between the cost of discarded milk (due to treatment) and potential benefits of treatment. If a treatment is more efficacious than expected spontaneous cure, potential benefits include increased milk yield, improved milk quality, reduced culling, and reduced transmission to other cows. However, economic models have not demonstrated cost-effectiveness of treating most cases of subclinical mastitis during lactation (Sandgren et al. 2008; Swinkels et al. 2005; Steeneveld et al. 2007). Subclinical mastitis is often caused by Gram-positive bacteria including *Staph. aureus*, NAS, and SSLO. Efficacy of antimicrobial therapy for chronic infections caused by *Staph. aureus* is very low; thus, treatment during lactation is not usually recommended for these infections (Barkema et al. 2006). In contrast, treatment efficacy for subclinical mastitis caused by *Strep. agalactiae* is high (Ruegg 2017a), and treatment of quarters infected with *Strep. agalactiae* is one of the few instances when the benefits of antimicrobial therapy during lactation exceed potential losses (Keefe 2012). With a few exceptions, due to the cost of milk discard, most subclinical infections are treated at dry-off using long-acting intramammary antibiotics.

***Dry Cow Therapy*** The purpose of antibiotic dry cow therapy (DCT) is to treat existing subclinical mastitis during a period when milk discard is not required as well as help prevent development of new IMI during the high-risk period surrounding mammary gland involution. Administration of IMM antibiotics in every quarter of all cows at the end of each lactation is often termed "blanket dry cow therapy" and is highly effective in reducing subclinical infections caused by Gram-positive cocci. Blanket DCT is part of the National Mastitis Council 5-point mastitis control plan that was originally developed to reduce the prevalence of cows infected with *Strep. agalactiae* and *Staph. aureus* (NMC 2016). Blanket DCT has been a highly adopted and efficacious practice in N. America (USDA 2016) and has contributed to near eradication of mastitis caused by *Strep. agalactiae* and considerable reduction in the prevalence of *Staph aureus* (Makovec and Ruegg 2003). This practice continues to be an integral part of mastitis control programs for herds that have higher bulk tank SCC values (>250,000 cells/mL) that reflect ineffective control in reducing prevalence of subclinically infected cows and for herds that lack recording systems to monitor mastitis.

Some regions (including the Nordic countries) have historically and successfully used "selective dry cow therapy" based on treating only cows with confirmed bacteriological infections (Rajala-Schultz et al. 2019). On many modern dairy farms, successful adoption of best management practices for udder health has greatly reduced the prevalence of subclinically infected cows and interest in selective DCT has grown. Most selective DCT programs use IMM antibiotics to treat cows suspected of having IMI and use internal or external teat sealants (without administration of antibiotics) to prevent new infections in healthy cows. A variety of methods have been successfully used to identify cows that require antimicrobials, including culture based, PCR, and algorithms based on historical review of records (Cameron et al. 2014; Scherpenzeel et al. 2014; Rowe et al. 2020b) Rowe et al. 2020a, b; McDougall et al. 2022). A recent study reported use of an algorithm using cow history to identify cows for IMM antibiotics or those that are eligible to receive only internal sealants (Rowe et al. 2020a). Cows that had at least one case of CM or at least one SCC >200,000 cells/mL were given IMM antimicrobials at dry-off, while cows that did not meet those criteria received only internal sealants. The researchers reported a 55% reduction in antimicrobial usage without negative effects on udder health or productivity. Other methods to identify cows that require antibiotics to treat existing IMI include use of CMT testing and culture or PCR testing of composite or quarter-milk samples (Swinkels et al. 2021; Rindsig et al. 1979; Kabera et al. 2020). Several economic models have demonstrated financial advantages by use of selective DCT for farmers managing herds with characteristics that indicate adoption of effective mastitis control procedures (Hommels et al. 2021; Rowe et al. 2021b).

Blanket DCT is a highly effective strategy to treat quarters infected with subclinical mastitis and remains a recommended practice for herds with a high prevalence of cows infected with contagious pathogens such as *Strep. agalactiae* or *Staph. aureus*. However, many modern dairy farms have successfully controlled these pathogens, have excellent detection and recording systems for both subclinical and clinical mastitis, and provide clean and dry areas for dry cows and at calving. These herds typically have bulk tank SCC consistently <200,000 cells/mL and an incidence of CM of <25 cases per 100 cows per year. In such herds, selective DCT can be effectively implemented and result in both reductions in antimicrobial usage and cost savings for the dairy farm.

## 6 Prevention of Mastitis

Development of mastitis causes reduced animal well-being, reduced milk quality, and considerable financial losses that decrease dairy farm sustainability. Effective mastitis control is based on prevention of new IMI by adoption of well-known, best management practices that reduce teat-end exposure to potential pathogens (NMC 2016). Prevention of mastitis emphasizes reducing exposure to mastitis pathogens by providing a clean and dry environment to all groups of cows and by ensuring that cows are milked gently using practices that minimize exposure to contagious mastitis pathogens.

Prevention of IMI caused by pathogens characterized as "contagious" has been well known for decades (Dodd et al. 1964) and is based on reducing exposure to bacteria that contaminate teat contact surfaces after being shed in milk of infected cows. Implementation of the "5-point plan" promoted by the National Mastitis Council (NMC 2016) has been very successful in control of pathogens (such as *Staph. aureus* and *Strep. agalactiae*) that cause persistent subclinical infection that can spread from cow to cow (especially during the milking process). The five points include (1) post-milking application of effective teat disinfectants to all quarters of all cows at every milking, (2) detection and recording of all cases of clinical mastitis, (3) identification and culling of all chronically infected cows, (4) administration of efficacious dry cow antibiotic therapy to all quarters of all cows at the end of every lactation, and (5) performing regular milking machine maintenance to ensure stable teat-end vacuum and gentle milking to ensure teat end health. Very high adoption of the 5-point plan has been demonstrated in herds that produce high-quality milk with >98% of large Wisconsin dairy farms indicating adoption of pre- and post-dipping teat sanitation (Rowbotham 2015) and > 90% of US dairy herds using blanket dry cow therapy (USDA 2016). These practices have contributed to great reductions in the prevalence of contagious mastitis pathogens (Makovec and Ruegg 2003) and considerable improvements in bulk tank SCC with overall US bulk tank SCC values dropping from 317,000 cells/mL in 1998 to 179,000 cell/mL in 2021 (https://queries.uscdcb.com/publish/dhi.cfm). Similar trends have been seen throughout the world in regions with modern dairy industries. Ensuring successful implementation of the 5-point plan remains the basis for initial mastitis control programs on all farms.

As dairy farms have consolidated and increased in size, opportunities for exposure to opportunistic, environmental pathogens have also increased, and many herds with lower bulk tank SCC continue to experience problems with clinical mastitis that is caused by environmental pathogens (USDA 2016; Tomazi et al. 2018a; Gao et al. 2017). Reducing teat-end exposure of highly susceptible cows to environmental bacteria is the key principle for controlling environmental pathogens. On many farms, environmental niches of bacteria such as moisture, mud, and manure contribute to exposure to potential pathogens outside of the milking center, and attention must be paid to decreasing exposure to risk factors associated with these pathogens. Regardless of management style (confinement or extensive), emphasis must be placed on maintaining clean and dry udders. Simple methods to score and assess udder hygiene have been developed (Schreiner and Ruegg 2002), and higher udder hygiene scores have been demonstrated to harbor more bacteria (Fig. 4; (Munoz et al. 2008, Guarin et al. 2017a)) and are associated with greater risks of IMI (Schreiner and Ruegg 2003). When cows are moved, to prevent splashing of manure on udders and lower legs, cows should be handled gently and allowed to walk at their natural pace, which is approximately 1.3–1.5 m/s (5.0 km/h; (Chapinal et al. 2009).

In confinement operations, selection of bedding used in stalls is an important choice, as different types of bedding have differing abilities to support growth of bacteria which present differing risks for development of IMI (Rowbotham and Ruegg 2016b; Patel et al. 2019; Hogan et al. 1989; Frechette et al. 2021). In general, as compared to sand bedding, organic bedding (especially manure-based materials) contains more bacteria and has been shown to be associated with dirtier udders, increased counts of coliforms and SSLO in bulk tank milk, and poorer udder health (Patel et al. 2019). When organic bedding sources are used, maintaining low moisture content is a key factor that can reduce bacterial

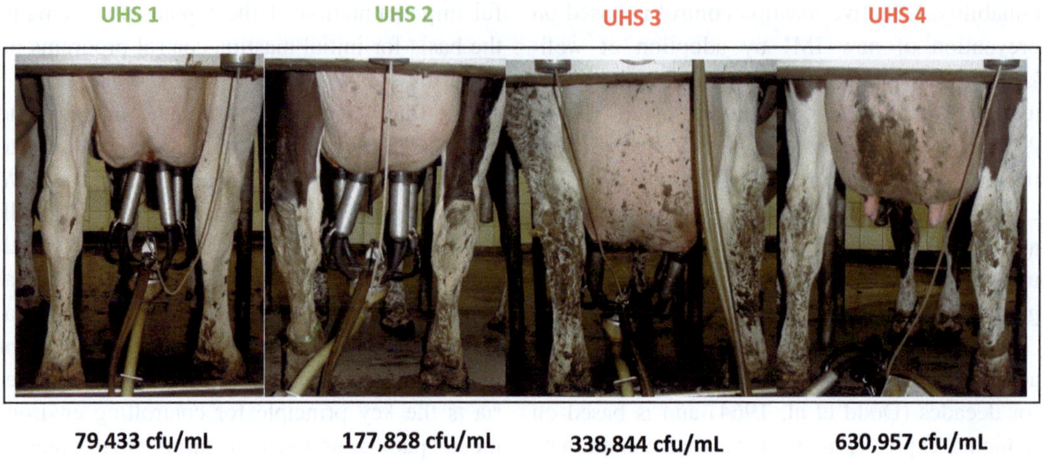


**Fig. 4** Number of bacteria found in teat skin swabs based on udder hygiene score (UHS) (cfu/mL; from (Guarin et al. 2017a))

growth and risk of IMI (Ward et al. 2002). It is especially important to provide clean and dry areas for high-risk cows (peri- and post-parturient cows) as physiological immune suppression that typically occurs during this time reduces their ability to respond to bacterial challenge at the teat end.

While teats often become exposed to mastitis pathogens in their environment, use of a consistent milking routine that includes both pre- and post-milking teat sanitation can greatly reduce risk of IMI. Teat cups should be applied to visibly clean well-stimulated teats to allow for rapid milk removal without damage to sensitive teat skin or teat sphincters. Standardized milking routines are commonly and effectively used throughout the world. Typical components of an effective milking routine include: (1) *Use of pre-milking teat sanitation*. The purpose of pre-milking teat sanitation is to remove bacteria that have accumulated on teat skin in the interval since the last milking. Numerous researchers have demonstrated efficacy in reducing the number of bacteria on teat skin with effective pre-milking sanitation (Pankey et al. 1987; Baumberger et al. 2016). Effective pre-dipping reduces bacterial counts of milk, enhancing milk safety (Galton et al. 1986). For effective reduction in bacterial numbers, teat dips should be properly formulated, stored in clean containers, completely applied to debris free teats, and allowed sufficient time (usually at least 30 seconds) for action before removal. (2) *Removal and examination of foremilk*. Manual removal of foremilk is the most effective method of pre-milking stimulation of milk letdown, which is necessary for efficient milking. Failure to properly stimulate cows before attachment of the milking unit can result in delayed milk ejection and reduced milk yields (Moore-Foster et al. 2019; Erskine et al. 2019). Removal and examination of foremilk is also a sanitary step that is required by many public health authorities. This procedure is necessary to ensure that abnormal milk produced by quarters affected with clinical mastitis is not included in milk intended for human consumption. About 50% of cases of clinical mastitis present solely with abnormal milk (no visible changes in the udder or cow (Oliveira et al. 2013)); thus, exclusion of this step from the milking routine results in under-detection of clinical mastitis. When milking is performed manually, three to four streams of foremilk from each quarter should be expressed and examined by the milking technician. To reduce risk of transferring bacteria among cows, it is important that milking technician wear disposable gloves as failure to use gloves has been documented as a risk factor associated with increased mastitis (Dufour et al. 2012). When cows are milked using automatic milking systems, lists of potentially mastitic cows are generated using indirect measures such as conductivity, activity, lactose, and others and these should be individually examined. (3) *Effective drying of teats.* This process is an important pre-milking step that greatly reduces bacterial contamination of teat skin (Galton et al. 1986, 1988), thus both improving milk quality and reducing risk of IMI. Teats should be dried using a single cloth or paper towel for each cow before the unit is attached. (4) *Efficient attachment of the milking unit.* Milking routines should be designed to attach the milking unit to well-stimulated cows that have achieved milk letdown. While cows vary in their need for physical stimulation based on stage of lactation or temperament, most cows benefit from the stimulation associated with removal of foremilk. To allow time for the physiological processes associated with milk letdown, the time between stimulation and unit attachment should be at least 60 seconds and not exceed 2.5–3 minutes. Milking units should be attached promptly to minimize air leakage into the system and removed when milk flow drops below about 500–1000 mL/minute. Higher removal thresholds may be recommended for higher producing cows that are milked multiple times per day. (5) *Post-milking management.* Application of post-milking teat disinfectants to teat skin is one of the most highly adopted and effective methods to reduce new IMI. Studies conducted in different regions and time periods have consistently demonstrated that effective post-milking teat disinfection reduces new IMI by 50% (Wesen and Schultz 1970; Williamson and Lacy-Hulbert 2013). While almost all herds practice post-milking teat dipping, the effectiveness is highly influenced by herd-level factors such as selection of the dip, method of application, and ability to fully cover teat skin.

In general, mastitis is a common disease that occurs when bacterial exposure at the teat end exceeds anatomical and immunological defenses of the cow to allow IMI to be established. Development of mastitis is based on (1) host (cow) factors associated with the ability to contact and penetrate teat sphincters and the ability of the cow to mount a rapid and effective immune response; (2) agent (usually bacterial) factors associated with virulence, pathogenicity, and host adaptation; and (3) environmental factors (moisture and organic matter that enhance bacterial growth) associated with the ability of bacteria to grow in areas that cows contact. Mastitis control is based on minimizing exposure of cow teats to bacteria and encompasses all aspects of dairy farm management. Housing areas must be designed to meet physiological needs of the cows and ensure dry resting areas, milking centers must be well maintained and designed to ensure that the milking process is hygienic, and farm workers must be well trained in how to handle cows gently. The most challenging aspect of mastitis control is that it is a never-ending task.

## References


Adkins PRF, Middleton JR (2018) Methods for diagnosing mastitis. Vet Clin North Am Food Anim Pract 34:479–491

Adkins PRF, Dufour S, Spain JN, Calcutt MJ, Reilly TJ, Stewart GC, Middleton JR (2018) Molecular characterization of non-aureus Staphylococcus spp. from heifer intramammary infections and body sites. J Dairy Sci 101:5388–5403

Aghamohammadi M, Haine D, Kelton DF, Barkema HW, Hogeveen H, Keefe GP, Dufour S (2018) Herd-level mastitis-associated costs on canadian dairy farms. Front Vet Sci 5:100

Ahmadzadeh A, Frago F, Shaffii B, Dalton JC, Price WJ, Mcguire MA (2009) Effect of clinical mastitis and other diseases on reproductive performance of Holstein cows. Anim Reprod Sci 112:273–282

Anderson KL, Walker RL (1988) Sources of Prototheca spp in a dairy herd environment. J Am Vet Med Assoc 193:553–556

Barkema HW, Riekerink RGMO, Zadoks RN, Schukken YH (2005) Epidemiology of mastitis: changes in distribution of pathogens, bulk milk somatic cell count and preventative practices in the last decade. J Dairy Sci 88:1–1

Barkema HW, Schukken YH, Zadoks RN (2006) Invited review: the role of cow, pathogen, and treatment regimen in the therapeutic success of bovine Staphylococcus aureus mastitis. J Dairy Sci 89:1877–1895

Baumberger C, Guarin JF, Ruegg PL (2016) Effect of 2 different premilking teat sanitation routines on reduction of bacterial counts on teat skin of cows on commercial dairy farms. J Dairy Sci 99:2915–2929

Benites NR, Melville PA, Costa EO (2003) Evaluation of the microbiological status of milk and various struc-

tures in mammary glands from naturally infected dairy cows. Trop Anim Health Prod 35:301–307
Biddle MK, Fox LK, Hancock DD, Gaskins CT, Evans MA (2004) Effects of storage time and thawing methods on the recovery of Mycoplasma species in milk samples from cows with intramammary infections. J Dairy Sci 87:933–936
Bobbo T, Ruegg PL, Stocco G, Fiore E, Gianesella M, Morgante M, Pasotto D, Bittante G, Cecchinato A (2017) Associations between pathogen-specific cases of subclinical mastitis and milk yield, quality, protein composition, and cheese-making traits in dairy cows. J Dairy Sci 100:4868–4883
Bruckmaier RM, Wellnitz O (2017) Triennial Lactation Symposium/Bolfa: pathogen-specific immune response and changes in the blood-milk barrier of the bovine mammary gland. J Anim Sci 95:5720–5728
Cameron M, Mckenna SL, Macdonald KA, Dohoo IR, Roy JP, Keefe GP (2014) Evaluation of selective dry cow treatment following on-farm culture: risk of post-calving intramammary infection and clinical mastitis in the subsequent lactation. J Dairy Sci 97:270–284
Cha E, Hertl JA, Schukken YH, Tauer LW, Welcome FL, Grohn YT (2013) The effect of repeated episodes of bacteria-specific clinical mastitis on mortality and culling in Holstein dairy cows. J Dairy Sci 96:4993–5007
Chapinal N, De Passille AM, Weary DM, Von Keyserlingk MA, Rushen J (2009) Using gait score, walking speed, and lying behavior to detect hoof lesions in dairy cows. J Dairy Sci 92:4365–4374
Chen H, Weersink A, Kelton D, Von Massow M (2021) Estimating milk loss based on somatic cell count at the cow and herd level. J Dairy Sci 104:7919–7931
De Buck J, Ha V, Naushad S, Nobrega DB, Luby C, middleton, J. R., De Vliegher, S. & Barkema, H. W. (2021) Non-aureus Staphylococci and Bovine Udder Health: current Understanding and Knowledge Gaps. Front Vet Sci 8:658031
De Campos JL, Kates A, Steinberger A, Sethi A, Suen G, Shutske J, Safdar N, Goldberg T, Ruegg PL (2021) Quantification of antimicrobial usage in adult cows and preweaned calves on 40 large Wisconsin dairy farms using dose-based and mass-based metrics. J Dairy Sci 104:4727–4745
De Visscher A, Piepers S, Haesebrouck F, Supre K, De Vliegher S (2017) Coagulase-negative Staphylococcus species in bulk milk: prevalence, distribution, and associated subgroup- and species-specific risk factors. J Dairy Sci 100:629–642
De Vliegher S, Laevens H, Devriese LA, Opsomer G, Leroy JL, Barkema HW, De Kruif A (2003) Prepartum teat apex colonization with Staphylococcus chromogenes in dairy heifers is associated with low somatic cell count in early lactation. Vet Microbiol 92:245–252
De Vliegher S, Opsomer G, Vanrolleghem A, Devriese LA, Sampimon OC, Sol J, Barkema HW, Haesebrouck F, De Kruif A (2004) In vitro growth inhibition of major mastitis pathogens by Staphylococcus chromogenes originating from teat apices of dairy heifers. Vet Microbiol 101:215–221
Dodd FH, Neave FK, Kingwill RG (1964) Control of udder infection by management. J Dairy Sci 47:1109–1114
Dohoo IR, Smith J, Andersen S, Kelton DF, Godden S, Mastitis Research Workers C (2011) Diagnosing intramammary infections: evaluation of definitions based on a single milk sample. J Dairy Sci 94:250–261
Dufour S, Dohoo IR, Barkema HW, Descoteaux L, Devries TJ, Reyher KK, Roy JP, Scholl DT (2012) Manageable risk factors associated with the lactational incidence, elimination, and prevalence of Staphylococcus aureus intramammary infections in dairy cows. J Dairy Sci 95:1283–1300
Duse A, Persson-Waller K, Pedersen K (2021) Microbial aetiology, antibiotic susceptibility and pathogen-specific risk factors for udder pathogens from clinical mastitis in dairy cows. Animals (Basel), 11
Erskine RJ, Bartlett PC, Vanlente JL, Phipps CR (2002) Efficacy of systemic ceftiofur as a therapy for severe clinical mastitis in dairy cattle. J Dairy Sci 85:2571–2575
Erskine RJ, Norby B, Neuder LM, Thomson RS (2019) Decreased milk yield is associated with delayed milk ejection. J Dairy Sci 102:6477–6484
Fox LK (2012) Mycoplasma mastitis: causes, transmission, and control. Vet Clin North Am Food Anim Pract 28:225–237
Francoz D, Wellemans V, Dupre JP, Roy JP, Labelle F, Lacasse P, Dufour S (2017) Invited review: a systematic review and qualitative analysis of treatments other than conventional antimicrobials for clinical mastitis in dairy cows. J Dairy Sci 100:7751–7770
Frechette A, Fecteau G, Cote C, Dufour S (2021) Clinical mastitis incidence in dairy cows housed on recycled manure solids bedding: a canadian cohort study. Front Vet Sci 8:742868
Fry PR, Middleton JR, Dufour S, Perry J, Scholl D, Dohoo I (2014) Association of coagulase-negative staphylococcal species, mammary quarter milk somatic cell count, and persistence of intramammary infection in dairy cattle. J Dairy Sci 97:4876–4885
Fuenzalida MJ, Ruegg PL (2019a) Negatively controlled, randomized clinical trial to evaluate intramammary treatment of nonsevere, gram-negative clinical mastitis. J Dairy Sci 102:5438–5457
Fuenzalida MJ, Ruegg PL (2019b) Negatively controlled, randomized clinical trial to evaluate use of intramammary ceftiofur for treatment of nonsevere culture-negative clinical mastitis. J Dairy Sci 102:3321–3338
Fuenzalida MJ, Ruegg PL (2020) Molecular epidemiology of nonsevere clinical mastitis caused by Klebsiella pneumoniae occurring in cows on 2 Wisconsin dairy farms. J Dairy Sci 103:3479–3492
Fuenzalida MJ, Fricke PM, Ruegg PL (2015) The association between occurrence and severity of subclinical and clinical mastitis on pregnancies per artificial insemination at first service of Holstein cows. J Dairy Sci 98:3791–3805
Fujimoto Y, Ito H, Higuchi H, Ohno H, Makita K (2020) A case-control study of herd- and cow-level risk factors associated with an outbreak of

Mycoplasma mastitis in Nemuro, Japan. Prev Vet Med 177:104946

Galton DM, Petersson LG, Merrill WG (1986) Effects of premilking udder preparation practices on bacterial counts in milk and on teats. J Dairy Sci 69:260–266

Galton DM, Peterson LG, Merrill WG (1988) Evaluation of udder preparations on intramammary infections. J Dairy Sci 71:1417–1421

Gao J, Barkema HW, Zhang L, Liu G, Deng Z, Cai L, Shan R, Zhang S, Zou J, Kastelic JP, Han B (2017) Incidence of clinical mastitis and distribution of pathogens on large Chinese dairy farms. J Dairy Sci 100:4797–4806

Gille L, Callens J, Supre K, Boyen F, Haesebrouck F, Van Driessche L, Van Leenen K, Deprez P, Pardon B (2018) Use of a breeding bull and absence of a calving pen as risk factors for the presence of Mycoplasma bovis in dairy herds. J Dairy Sci 101:8284–8290

Gioia G, Addis MF, Santisteban C, Gross B, Nydam DV, Sipka AS, Virkler PD, Watters RD, Wieland M, Zurakowski MJ, Moroni P (2021) Mycoplasma species isolated from bovine milk collected from US dairy herds between 2016 and 2019. J Dairy Sci 104:4813–4821

Godden SM, Royster E, Timmerman J, Rapnicki P, Green H (2017) Evaluation of an automated milk leukocyte differential test and the California Mastitis Test for detecting intramammary infection in early- and late-lactation quarters and cows. J Dairy Sci 100:6527–6544

Goncalves JL, Tomazi T, Barreiro JR, Beuron DC, Arcari MA, Lee SH, Martins CM, Araujo Junior JP, Dos Santos MV (2016) Effects of bovine subclinical mastitis caused by Corynebacterium spp. on somatic cell count, milk yield and composition by comparing contralateral quarters. Vet J 209:87–92

Goncalves JL, Kamphuis C, Vernooij H, Araujo JP Jr, Grenfell RC, Juliano L, Anderson KL, Hogeveen H, Dos Santos MV (2020) Pathogen effects on milk yield and composition in chronic subclinical mastitis in dairy cows. Vet J 262:105473

Goncalves JL, Leite De Campos J, Steinberger A, Safdar N, Sethi A, Shutske J, Rodriguez Z, Suen G, Goldberg T, Ruegg PL (2022). Incidence and treatments of mastitis and other selected bovine diseases on 37 large Wisconsin dairy farms. Pathogens 11:1282

Gonzalez Pereyra V, Pol M, Pastorino F, Herrero A (2015) Quantification of antimicrobial usage in dairy cows and preweaned calves in Argentina. Prev Vet Med 122:273–279

Guarin JF, Baumberger C, Ruegg PL (2017a) Anatomical characteristics of teats and premilking bacterial counts of teat skin swabs of primiparous cows exposed to different types of bedding. J Dairy Sci 100:1436–1444

Guarin JF, Paixao MG, Ruegg PL (2017b) Association of anatomical characteristics of teats with quarter-level somatic cell count. J Dairy Sci 100

Gunther J, Esch K, Poschadel N, Petzl W, Zerbe H, Mitterhuemer S, Blum H, Seyfert HM (2011) Comparative kinetics of Escherichia coli- and Staphylococcus aureus-specific activation of key immune pathways in mammary epithelial cells demonstrates that S. aureus elicits a delayed response dominated by interleukin-6 (IL-6) but not by IL-1A or tumor necrosis factor alpha. Infect Immun 79:695–707

Haapala V, Pohjanvirta T, Vahanikkila N, Halkilahti J, Simonen H, Pelkonen S, Soveri T, Simojoki H, Autio T (2018) Semen as a source of Mycoplasma bovis mastitis in dairy herds. Vet Microbiol 216:60–66

Haapala V, Vahanikkila N, Kulkas L, Tuunainen E, Pohjanvirta T, Autio T, Pelkonen S, Soveri T, Simojoki H (2021) Mycoplasma bovis infection in dairy herds-Risk factors and effect of control measures. J Dairy Sci 104:2254–2265

Heikkila AM, Nousiainen JI, Pyorala S (2012) Costs of clinical mastitis with special reference to premature culling. J Dairy Sci 95:139–150

Heikkila AM, Liski E, Pyorala S, Taponen S (2018) Pathogen-specific production losses in bovine mastitis. J Dairy Sci 101:9493–9504

Hertl JA, Schukken YH, Bar D, Bennett GJ, González RN, Rauch BJ, Welcome FL, Tauer LW, Gröhn YT (2011) The effect of recurrent episodes of clinical mastitis caused by gram-positive and gram-negative bacteria and other organisms on mortality and culling in Holstein dairy cows. J Dairy Sci 94:4863–4877

Hiitio H, Pyorala S, Taponen S, Rajala-Schultz P, Simojoki H (2018) Elimination of experimentally induced bovine intramammary infection assessed by multiplex real-time PCR and bacterial culture. J Dairy Sci 101:5267–5276

Hogan JS, Smith KL, Hoblet KH, Todhunter DA, Schoenberger PS, Hueston WD, Pritchard DE, Bowman GL, Heider LE, Brockett BL et al (1989) Bacterial counts in bedding materials used on nine commercial dairies. J Dairy Sci 72:250–258

Hommels NMC, Ferreira FC, Van Den Borne BHP, Hogeveen H (2021) Antibiotic use and potential economic impact of implementing selective dry cow therapy in large US dairies. J Dairy Sci 104:8931–8946

Ito T, Kano R, Sobukawa H, Ogawa J, Honda Y, Hosoi Y, Shibuya H, Sato T, Hasegawa A, Kamata H (2011) Experimental infection of bovine mammary gland with prototheca zopfii genotype 1. J Vet Med Sci 73:117–119

Janosi S, Ratz F, Szigeti G, Kulcsar M, Kerenyi J, Lauko T, Katona F, Huszenicza G (2001) Review of the microbiological, pathological, and clinical aspects of bovine mastitis caused by the alga Prototheca zopfii. Vet Q 23:58–61

Jorgensen HJ, Nordstoga AB, Sviland S, Zadoks RN, Solverod L, Kvitle B, Mork T (2016) Streptococcus agalactiae in the environment of bovine dairy herds – rewriting the textbooks? Vet Microbiol 184:64–72

Justice-Allen A, Trujillo J, Corbett R, Harding R, Goodell G, Wilson D (2010) Survival and replication of Mycoplasma species in recycled bedding sand and association with mastitis on dairy farms in Utah. J Dairy Sci 93:192–202

Kabera F, Dufour S, Keefe G, Cameron M, Roy JP (2020) Evaluation of quarter-based selective dry cow therapy using Petrifilm on-farm milk culture: a randomized controlled trial. J Dairy Sci 103:7276–7287

Kandeel SA, Megahed AA, Constable PD (2017) Comparison of six on-farm tests to estimate somatic cell count at dry-off in dairy cattle. American College of Veterinary Internal Medicine Forum, 2017 National Harbor, MD 1358

Kandeel SA, Megahed AA, Ebeid MH, Constable PD (2019) Ability of milk pH to predict subclinical mastitis and intramammary infection in quarters from lactating dairy cattle. J Dairy Sci 102:1417–1427

Keefe G (2012) Update on Control of Staphylococcus aureus and Streptococcus agalactiae for Management of Mastitis. Vet Clin North Am Food Anim Pract 28:203

Khatun M, Bruckmaier RM, Thomson PC, House J, Garcia SC (2019) Suitability of somatic cell count, electrical conductivity, and lactate dehydrogenase activity in foremilk before versus after alveolar milk ejection for mastitis detection. J Dairy Sci 102:9200–9212

Kitchen BJ (1981) Review of the progress of dairy science: bovine mastitis: milk compositional changes and related diagnostic tests. J Dairy Res 48:167–188

Kuipers A, Koops WJ, Wemmenhove H (2016) Antibiotic use in dairy herds in the Netherlands from 2005 to 2012. J Dairy Sci 99:1632–1648

Lago A, Godden SM, Bey R, Ruegg PL, Leslie K (2011) The selective treatment of clinical mastitis based on on-farm culture results: I. Effects on antibiotic use, milk withholding time, and short-term clinical and bacteriological outcomes. J Dairy Sci 94:4441–4456

Leelahapongsathon K, Schukken YH, Srithanasuwan A, Suriyasathaporn W (2020) Molecular epidemiology of Streptococcus uberis intramammary infections: persistent and transient patterns of infection in a dairy herd. J Dairy Sci 103:3565–3576

Leroy F, Van Coillie E, Braem G, Piessens V, Verbist B, De Vuyst L, De Vliegher S (2015) Short communication: subtyping of Staphylococcus haemolyticus isolates from milk and corresponding teat apices to verify the potential teat-skin origin of intramammary infections in dairy cows. J Dairy Sci 98:7893–7898

Leuenberger A, Sartori C, Boss R, Resch G, Oechslin F, Steiner A, Moreillon P, Graber HU (2019) Genotypes of Staphylococcus aureus: on-farm epidemiology and the consequences for prevention of intramammary infections. J Dairy Sci 102:3295–3309

Levison LJ, Miller-Chuson EK, Tucker AL, Bergeron R, Leslie KE, Barkema H, Devries TJ (2016) Incidence rate of pathogen-specific clinical mastitis on conventional and organic Canadian dairy farms. J Dairy Sci 99:1341–1350

Liu G, Yin J, Han B, Barkema HW, Shahid M, De Buck J, Cobo ER, Kastelic JP, Gao J (2019) Adherent/invasive capacities of bovine-associated Aerococcus viridans contribute to pathogenesis of acute mastitis in a murine model. Vet Microbiol 230:202–211

Ma Y, Ryan C, Barbano DM, Galton DM, Rudan MA, Boor KJ (2000) Effects of somatic cell count on quality and shelf-life of pasteurized fluid milk. J Dairy Sci 83:264–274

McDougall S, Williamson J, Lacy-Hulbert J (2022) Bacteriological outcomes following random allocation to quarter-level selection based on California Mastitis Test score or cow-level allocation based on somatic cell count for dry cow therapy. J Dairy Sci 105(3):2453–2472. https://doi.org/10.3168/jds.2021-21020

Makovec JA, Ruegg PL (2003) Results of milk samples submitted for microbiological examination in Wisconsin from 1994 to 2001. J Dairy Sci 86:3466–3472

Middleton JR, Hardin D, Steevens B, Randle R, Tyler JW (2004) Use of somatic cell counts and California mastitis test results from individual quarter milk samples to detect subclinical intramammary infection in dairy cattle from a herd with a high bulk tank somatic cell count. J Am Vet Med Assoc 224:419–423

Moller A, Truyen U, Roesler U (2007) Prototheca zopfii genotype 2: the causative agent of bovine protothecal mastitis? Vet Microbiol 120:370–374

Moore-Foster R, Norby B, Schewe RL, Thomson R, Bartlett PC, Erskine RJ (2019) Herd-level variables associated with premilking stimulation time in Michigan dairy herds. J Dairy Sci 102:2544–2550

Munoz MA, Ahlstrom C, Rauch BJ, Zadoks RN (2006) Fecal shedding of Klebsiella pneumoniae by dairy cows. J Dairy Sci 89:3425–3430

Munoz MA, Welcome FL, Schukken YH, Zadoks RN (2007) Molecular epidemiology of two Klebsiella pneumoniae mastitis outbreaks on a dairy farm in New York State. J Clin Microbiol 45:3964–3971

Munoz MA, Bennett GJ, Ahlstrom C, Griffiths HM, Schukken YH, Zadoks RN (2008) Cleanliness scores as indicator of Klebsiella exposure in dairy cows. J Dairy Sci 91:3908–3916

Nicholas RA, Fox LK, Lysnyansky I (2016) Mycoplasma mastitis in cattle: to cull or not to cull. Vet J 216:142–147

NMC (2016) Current concepts of bovine mastitis. 5th ed. New Prague, MN

NMC (2017) Laboratory Handbook on Bovine Mastitis, New Prague, MN, USA, NMC

Okkema C, Grandin T (2021) Graduate student literature review: udder edema in dairy cattle-A possible emerging animal welfare issue. J Dairy Sci 104:7334–7341

Oliveira L, Ruegg PL (2014) Treatments of clinical mastitis occurring in cows on 51 large dairy herds in Wisconsin. J Dairy Sci 97:5426–5436

Oliveira L, Hulland C, Ruegg PL (2013) Characterization of clinical mastitis occurring in cows on 50 large dairy herds in Wisconsin. J Dairy Sci 96:7538–7549

Owens WE, Oliver SP, Gillespie BE, Ray CH, Nickerson SC (1998) Role of horn flies (Haematobia irritans) in Staphylococcus aureus-induced mastitis in dairy heifers. Am J Vet Res 59:1122–1124

Paixao MG, Abreu LR, Richert R, Ruegg PL (2017) Milk composition and health status from mammary gland quarters adjacent to glands affected with naturally occurring clinical mastitis. J Dairy Sci 100:7522–7533
Pankey JW, Wildman EE, Drechsler PA, Hogan JS (1987) Field trial evaluation of premilking teat disinfection. J Dairy Sci 70:867–872
Pantoja JCF, Correia LBN, Rossi RS, Latosinski GS (2020) Association between teat-end hyperkeratosis and mastitis in dairy cows: a systematic review. J Dairy Sci 103:1843–1855
Parker AM, Sheehy PA, Hazelton MS, Bosward KL, House JK (2018) A review of mycoplasma diagnostics in cattle. J Vet Intern Med 32:1241–1252
Patel K, Godden SM, Royster E, Crooker BA, Timmerman J, Fox L (2019) Relationships among bedding materials, bedding bacteria counts, udder hygiene, milk quality, and udder health in US dairy herds. J Dairy Sci 102:10213–10234
Pegolo S, Tessari R, Bisutti V, Vanzin A, Giannuzzi D, Gianesella M, Lisuzzo A, Fiore E, Barberio A, Schiavon E, Trevisi E, Piccioli Cappelli F, Gallo L, Ruegg P, Negrini R, Cecchinato A (2022) Quarter-level analyses of the associations among subclinical intramammary infection and milk quality, udder health, and cheesemaking traits in Holstein cows. J Dairy Sci 105:3490–3507
Penry JF (2018) Mastitis control in automatic milking systems. Vet Clin North Am Food Anim Pract 34:439–456
Petersson-Wolfe CS, Leslie KE, Swartz TH (2018) An update on the effect of clinical mastitis on the welfare of dairy cows and potential therapies. Vet Clin North Am Food Anim Pract 34:525–535
Piepers S, Peeters K, Opsomer G, Barkema HW, Frankena K, De Vliegher S (2011) Pathogen group specific risk factors at herd, heifer and quarter levels for intramammary infections in early lactating dairy heifers. Prev Vet Med 99:91–101
Piessens V, Van Coillie E, Verbist B, Supre K, Braem G, Van Nuffel A, De Vuyst L, Heyndrickx M, De Vliegher S (2011) Distribution of coagulase-negative Staphylococcus species from milk and environment of dairy cows differs between herds. J Dairy Sci 94:2933–2944
Pinzon-Sanchez C, Ruegg PL (2011) Risk factors associated with short-term post-treatment outcomes of clinical mastitis. J Dairy Sci 94:3397–3410
Pol M, Ruegg PL (2007) Treatment practices and quantification of antimicrobial drug usage in conventional and organic dairy farms in Wisconsin. J Dairy Sci 90:249–261
Politis IAKFN-K-H (1988) Association between somatic cell counts of milk and cheese yielding capacity. J Dairy Sci 71:1720–1727
Pore RS, Shahan TA (1988) Prototheca zopfii: natural, transient, occurrence in pigs and rats. Mycopathologia 101:85–88
Pore RS, Shahan TA, Pore MD, Blauwiekel R (1987) Occurrence of Prototheca zopfii, a mastitis pathogen, in milk. Vet Microbiol 15:315–323
Pyorala S, Kaartinen L, Kack H, Rainio V (1994) Efficacy of two therapy regimens for treatment of experimentally induced Escherichia coli mastitis in cows. J Dairy Sci 77:453–461
Rainard P, Foucras G, Fitzgerald JR, Watts JL, Koop G, Middleton JR (2018) Knowledge gaps and research priorities in Staphylococcus aureus mastitis control. Transbound Emerg Dis 65(Suppl 1):149–165
Rajala-Schultz P, Persson Waller K, Halasa T, Nodtvedt A (2019) Selective approach to dry cow therapy. Vet Rec 184:29–30
Ricchi M, Goretti M, Branda E, Cammi G, Garbarino CA, Turchetti B, Moroni P, Arrigoni N, Buzzini P (2010) Molecular characterization of Prototheca strains isolated from Italian dairy herds. J Dairy Sci 93:4625–4631
Rindsig RB, Rodewald RG, Smith AR, Thomsen NK, Spahr SL (1979) Mastitis history, California Mastitis Test, and somatic cell counts for identifying cows for treatment in a selective dry cow therapy program. J Dairy Sci 62:1335–1339
Rodrigues MX, Lima SF, Higgins CH, Canniatti-Brazaca SG, Bicalho RC (2016) The Lactococcus genus as a potential emerging mastitis pathogen group: a report on an outbreak investigation. J Dairy Sci 99:9864–9874
Rowbotham RF (2015) Associations between dairy cow bedding and milk quality. PhD, University of WI
Rowbotham RF, Ruegg PL (2016a) Associations of selected bedding types with incidence rates of subclinical and clinical mastitis in primiparous Holstein dairy cows. J Dairy Sci 99:4707–4717
Rowbotham RF, Ruegg PL (2016b) Bacterial counts on teat skin and in new sand, recycled sand, and recycled manure solids used as bedding in freestalls. J Dairy Sci 99:6594–6608
Rowe SM, Godden SM, Royster E, Timmerman J, Crooker BA, Boyle M (2019) Cross-sectional study of the relationships among bedding materials, bedding bacteria counts, and intramammary infection in late-lactation dairy cows. J Dairy Sci 102:11384–11400
Rowe SM, Godden SM, Nydam DV, Gorden PJ, Lago A, Vasquez AK, Royster E, Timmerman J, Thomas MJ (2020a) Randomized controlled non-inferiority trial investigating the effect of 2 selective dry-cow therapy protocols on antibiotic use at dry-off and dry period intramammary infection dynamics. J Dairy Sci 103:6473–6492
Rowe SM, Godden SM, Nydam DV, Gorden PJ, Lago A, Vasquez AK, Royster E, Timmerman J, Thomas MJ (2020b) Randomized controlled trial investigating the effect of 2 selective dry-cow therapy protocols on udder health and performance in the subsequent lactation. J Dairy Sci 103:6493–6503
Rowe SM, Godden SM, Royster E, Timmerman J, Boyle M (2021a) Postcalving udder health and productivity in cows approaching dry-off with intramammary infections caused by non-aureus Staphylococcus,

Aerococcus, Enterococcus, Lactococcus, and Streptococcus species. J Dairy Sci 104:6061–6079
Rowe SM, Nydam DV, Godden SM, Gorden PJ, Lago A, Vasquez AK, Royster E, Timmerman J, Thomas MJ, Lynch RA (2021b) Partial budget analysis of culture- and algorithm-guided selective dry cow therapy. J Dairy Sci 104:5652–5664
Ruegg PL (2017a) A 100-year review: mastitis detection, management, and prevention. J Dairy Sci 100:10381–10397
Ruegg PL (2017b) Mastitis, detection, diagnosis and management. J Dairy Sci
Ruegg PL (2017c) Practical approaches to mastitis therapy on large dairy herds. In: Hogan J (ed) Large dairy herd management, 3rd edn. Am. Dairy Sci. Assoc, Champaign, IL
Ruegg PL, Erskine RD (2020) Mammary gland health and disorders. In: Smith BP (ed) Large animal veterinary internal medicine, 6th edn. Elsevier, New York
Rutten CJ, Velthuis AGJ, Steeneveld W, Hogeveen H (2013) Invited review: sensors to support health management on dairy farms. J Dairy Sci 96:1928–1952
Ryman VE, Packiriswamy N, Sordillo LM (2015) Role of endothelial cells in bovine mammary gland health and disease. Anim Health Res Rev 16:135–149
Saini V, Mcclure JT, Leger D, Dufour S, Sheldon AG, Scholl DT, Barkema HW (2012) Antimicrobial use on Canadian dairy farms. J Dairy Sci 95:1209–1221
Sandgren CH, Waller KP, Emanuelson U (2008) Therapeutic effects of systemic or intramammary antimicrobial treatment of bovine subclinical mastitis during lactation. Vet J 175:108–117
Santos JEP, Cerri RLA, Ballou MA, Higginbotham GE, Kirk JH (2004) Effect of timing of first clinical mastitis occurrence on lactational and reproductive performance of Holstein dairy cows. Anim Reprod Sci 80:31–45
Sawant AA, Pillai SR, Jayarao BM (2002) Evaluation of five selective media for isolation of catalase-negative gram-positive cocci from bulk tank milk. J Dairy Sci 85:1127–1132
Scaccabarozzi L, Turchetti B, Buzzini P, Pisoni G, Bertocchi L, Arrigoni N, Boettcher P, Bronzo V, Moroni P (2008) Short communication: isolation of Prototheca species strains from environmental sources in dairy herds. J Dairy Sci 91:3474–3477
Scherpenzeel CGM, Den Uijl IEM, Van Schaik G, Riekerink RGMO, Lam TJGM (2014) Evaluation of the use of dry cow antibiotics in low somatic count cows. J Dairy Sci 97:3606–3614
Schreiner DA, Ruegg PL (2002) Effects of tail docking on milk quality and cow cleanliness. J Dairy Sci 85:2503–2511
Schreiner DA, Ruegg PL (2003) Relationship between udder and leg hygiene scores and subclinical mastitis. J Dairy Sci 86:3460–3465
Schrick FN, Hockett ME, Saxton AM, Lewis MJ, Dowlen HH, Oliver SP (2001) Influence of subclinical mastitis during early lactation on reproductive parameters. J Dairy Sci 84:1407–1412
Schukken YH, Wilson DJ, Welcome F, Garrison-Tikofsky L, Gonzalez RN (2003) Monitoring udder health and milk quality using somatic cell counts. Vet Res 34:579–596
Schukken Y, Chuff M, Moroni P, Gurjar A, Santisteban C, Welcome F, Zadoks R (2012) The "other" gram-negative bacteria in mastitis: Klebsiella, serratia, and more. Vet Clin North Am Food Anim Pract 28:239–256
Scillieri Smith JC, Moroni P, Santisteban CG, Rauch BJ, Ospina PA, Nydam DV (2020) Distribution of Lactococcus spp. in New York State dairy farms and the association of somatic cell count resolution and bacteriological cure in clinical mastitis samples. J Dairy Sci 103:1785–1794
Shangraw EM, Rodrigues RO, Choudhary RK, Zhao FQ, Mcfadden TB (2021) Hypogalactia in mammary quarters adjacent to lipopolysaccharide-infused quarters is associated with transcriptional changes in immune genes. J Dairy Sci 104:9276–9286
Smistad M, Kaspersen H, Franklin-Alming FV, Wolff C, Solverod L, Porcellato D, Trettenes E, Jorgensen HJ (2022) Streptococcus dysgalactiae ssp. dysgalactiae in Norwegian bovine dairy herds: risk factors, sources, and genomic diversity. J Dairy Sci 105:3574–3587
Steeneveld W, Swinkels J, Hogeveen H (2007) Stochastic modelling to assess economic effects of treatment of chronic subclinical mastitis caused by Streptococcus uberis. J Dairy Res 74:459–467
Stevens M, Piepers S, Supre K, Dewulf J, De Vliegher S (2016) Quantification of antimicrobial consumption in adult cattle on dairy herds in Flanders, Belgium, and associations with udder health, milk quality, and production performance. J Dairy Sci 99:2118–2130
Suojala L, Kaartinen L, Pyorala S (2013) Treatment for bovine Escherichia coli mastitis – an evidence-based approach. J Vet Pharmacol Ther 36:521–531
Supre K, Haesebrouck F, Zadoks RN, Vaneechoutte M, Piepers S, De Vliegher S (2011) Some coagulase-negative Staphylococcus species affect udder health more than others. J Dairy Sci 94:2329–2340
Svennesen L, Nielsen SS, Mahmmod YS, Kromker V, Pedersen K, Klaas IC (2019) Association between teat skin colonization and intramammary infection with Staphylococcus aureus and Streptococcus agalactiae in herds with automatic milking systems. J Dairy Sci 102:629–639
Swinkels JM, Rooijendijk JG, Zadoks RN, Hogeveen H (2005) Use of partial budgeting to determine the economic benefits of antibiotic treatment of chronic subclinical mastitis caused by Streptococcus uberis or Streptococcus dysgalactiae. J Dairy Res 72:75–85
Swinkels JM, Leach KA, Breen JE, Payne B, White V, Green MJ, Bradley AJ (2021) Randomized controlled field trial comparing quarter and cow level selective dry cow treatment using the California Mastitis Test. J Dairy Sci 104:9063–9081
Todhunter DA, Smith KL, Hogan JS, Schoenberger PS (1991) Gram-negative bacterial infections of the mammary gland in cows. Am J Vet Res 52:184–188

Tomazi T, Ferreira GC, Orsi AM, Goncalves JL, Ospina PA, Nydam DV, Moroni P, Dos Santos MV (2018a) Association of herd-level risk factors and incidence rate of clinical mastitis in 20 Brazilian dairy herds. Prev Vet Med 161:9–18

Tomazi T, Ferreira GC, Orsi AM, Gonçalves JL, Ospina PA, Nydam DV, Moroni P, Dos Santos MV (2018b) Association of herd-level risk factors and incidence rate of clinical mastitis in 20 Brazilian dairy herds. Prev Vet Med 161:9–18

USDA (2016) Dairy 2014, Milk quality, milking procedures and mastitis in the United States, 2014. In: USDA–APHIS–VS–CEAH–NAHMS (ed.). Fort Collins, CO

USDA (2018) Health and Management Practices on U.S. Dairy Operations, 2014. In: USDA-APHIS-VS, C. (ed.). Fort Collins, CO

Valckenier D, Piepers S, Schukken YH, De Visscher A, Boyen F, Haesebrouck F, De Vliegher S (2021) Longitudinal study on the effects of intramammary infection with non-aureus staphylococci on udder health and milk production in dairy heifers. J Dairy Sci 104:899–914

Valeeva NI, Lam TJ, Hogeveen H (2007) Motivation of dairy farmers to improve mastitis management. J Dairy Sci 90:4466–4477

Van Soest FJS, Abbeloos E, Mcdougall S, Hogeveen H (2018) Addition of meloxicam to the treatment of bovine clinical mastitis results in a net economic benefit to the dairy farmer. J Dairy Sci 101:3387–3397

Ward WR, Hughes JW, Faull WB, Cripps PJ, Sutherland JP, Sutherst JE (2002) Observational study of temperature, moisture, pH and bacteria in straw bedding, and faecal consistency, cleanliness and mastitis in cows in four dairy herds. Vet Rec 151:199–206

Wellnitz O, Arnold ET, Lehmann M, Bruckmaier RM (2013) Short communication: differential immunoglobulin transfer during mastitis challenge by pathogen-specific components. J Dairy Sci 96:1681–1684

Wellnitz O, Zbinden C, Huang X, Bruckmaier RM (2016) Short communication: differential loss of bovine mammary epithelial barrier integrity in response to lipopolysaccharide and lipoteichoic acid. J Dairy Sci 99:4851–4856

Wenz JR, Barrington GM, Garry FB, Mcsweeney KD, Dinsmore RP, Goodell G, Callan RJ (2001) Bacteremia associated with naturally occurring acute coliform mastitis in dairy cows. J Am Vet Med Assoc 219:976–981

Wenz JR, Garry FB, Barrington GM (2006) Comparison of disease severity scoring systems for dairy cattle with acute coliform mastitis. J Am Vet Med Assoc 229:259–262

Wesen DP, Schultz LH (1970) Effectiveness of a post-milking teat dip in preventing new udder infections. J Dairy Sci 53:1391–1403

Williamson JH, Lacy-Hulbert SJ (2013) Effect of disinfecting teats post-milking or pre-and post-milking on intramammary infection and somatic cell count. N Z Vet J 61:262–268

Wuytack A, De Visscher A, Piepers S, Boyen F, Haesebrouck F, De Vliegher S (2020) Distribution of non-aureus staphylococci from quarter milk, teat apices, and rectal feces of dairy cows, and their virulence potential. J Dairy Sci 103:10658–10675

Zadoks RN, Middleton JR, Mcdougall S, Katholm J, Schukken YH (2011) Molecular epidemiology of mastitis pathogens of dairy cattle and comparative relevance to humans. J Mammary Gland Biol Neoplasia 16:357–372

# Disturbed Milk Ejection at Machine Milking and the Issues Related to Exogenous Oxytocin in Dairy Cows

Rupert M. Bruckmaier

**Abstract**

Disturbed milk ejection (DME) can occur under various conditions in practical dairy farms. As a consequence udder emptying at milking is incomplete. If milk ejection is completely lacking, only the cisternal milk fraction is available, i.e., only 0–20% of the milk stored in the udder. The disturbance is due to an insufficient release of oxytocin in response to tactile udder stimulation. In all known situations in practical dairy farming, exogenous oxytocin even at a low dosage is suitable to overcome the disturbance. However, due to the short half-life of oxytocin this treatment is only effective for a short time, i.e., for one milking event. A sustainable therapy to activate endogenous oxytocin release in response to teat stimulation is currently not yet available. In some cases, albeit not always, vaginal instead of teat stimulation leads to a sufficient oxytocin release to induce milk ejection. DME can occur in primiparous cows during the first weeks postpartum, in unfamiliar surroundings (e.g., after the changeover to a new milking system) or during peak estrus. Also in housing systems where dairy cows are kept in contact with their calves DME at machine milking has been observed in some farms. Except for clinical mastitis or inadequate milking machine settings causing pain to the cow disturbed milk ejection can mostly not be linked to obviously observed stress. Also, typical stress-related hormones such as adrenaline have not been elevated at disturbed milk ejection, and exogenous cortisol or catecholamines do not inhibit the release of oxytocin during milking. Endogenous opioids may be involved in the disturbance as beta-endorphin has been elevated in some studies during DME, and exogenous morphine caused an inhibition of oxytocin release during machine milking. DME is mostly treated by oxytocin injection at each milking. However, all commercially available oxytocin preparations can be used only at a massive overdose causing circulating blood concentrations far beyond the physiological range. Whereas endogenously released oxytocin decreases immediately to baseline after milking, the injected oxytocin remains often elevated for several hours after injection. The consequence is the development of a reduced sensitivity of the mammary gland to oxytocin within a few days of repeated treatments. Even if the oxytocin release from the pituitary is normal, the high dosage oxytocin treatment cannot be stopped before the next dry off. Therefore, exogenous oxytocin treatment needs to be recommended at a dosage as low as possible and for the shortest possible

R. M. Bruckmaier (✉)
Veterinary Physiology, Vetsuisse Faculty, University of Bern, Bern, Switzerland
e-mail: rupert.bruckmaier@unibe.ch





J. J. Gross (ed.), *Production Diseases in Farm Animals*,
https://doi.org/10.1007/978-3-031-51788-4_16

period. The i.v. injection of 0.5–1 i.u. of oxytocin instead of the often used 20–50 i.u. is sufficient to empty the udder at machine milking. This is only possible if the oxytocin product is diluted in physiological saline to achieve a volume which can be injected from a practical point of view, i.e., to inject several ml instead of μl. Besides the reduced milk harvest with a direct impact on the farm income, periods of low or no milk flow during milking cause an increased mechanical load on the teat tissue through high vacuum both at the teat end and at the mouthpiece chamber of the liner. This in turn increases the risk of hyperkeratosis and increased teat tip callosity, and hence susceptibility to mastitis. Studies to detect a genetic background of the likelihood of disturbed milk ejection were without unambiguous results until today. It remains unclear why only some individual animals show disturbed milk ejection.

**Keywords**

Dairy cow · Disturbed milk ejection · Oxytocin · Catecholamines · Opioid peptides

## 1 Physiological Regulation of Milk Ejection

Only the milk that is present in the cisternal cavities of the four-quarter glands and teats are available for harvest by the calf or the milking machine. In dairy cows, the cisternal fraction amounts to 0–20% of the total milk, with the lowest amounts in late lactation and at short intervals between milkings. The main portion of milk, the alveolar milk, is stored in the alveolar compartment of each quarter and must be actively shifted into the cistern via milk ejection to become available for removal (Bruckmaier and Blum 1998). Milk ejection occurs in response to the release of the neuropeptide hormone oxytocin from the posterior pituitary. Its release is induced via a neuroendocrine reflex by tactile stimulation of the mammary gland or the genital tract. Other types of stimulation such as acoustic or visual stimuli do obviously exist in some species but have not been observed in dairy cows. Elevated plasma oxytocin causes the contraction of smooth muscle cells surrounding the alveoli of the secretory tissue, the myoepithelial cells, and consequently, the alveolar milk is shifted through the milk duct system into the cisterns. Depending on the type and intensity of stimulation oxytocin increases from a concentration below 5 pg/ml to levels of 10–100 pg/ml. It has been demonstrated that the milk ejection response is not related to the dimension of oxytocin release. As long as a threshold of approximately 10 pg/ml of plasma oxytocin is surmounted milk ejection to a maximum intracisternal milk pressure is occurring. Consequently, there are no special requirements for the intensity of stimulation of the mammary gland to induce a sufficient oxytocin release for maximum milk ejection. Only extremely high oxytocin blood concentrations, only achieved by exogenous oxytocin at a high supraphysiological dosage, can induce a more efficient milk ejection to remove also the so-called residual milk, a portion of 5–20% of the total milk which would remain in the udder of cows under conditions of a physiological oxytocin release (Bruckmaier and Blum 1998).

## 2 Prestimulation and Continuous Milk Ejection

Ideally, milk ejection is initiated by prestimulation before teat cup attachment (Bruckmaier and Blum 1998). This leads to an increase of intracisternal milk pressure in the teat as a precondition for perfect positioning of the teat in the teat cup liner. If the milking cluster is attached without prestimulation a transient milking on empty teats can occur between the removal of the cisternal milk and the start of milk ejection. Because no milk is transported during this period the vacuum at the teat end and in the liner mouthpiece is high up to the full system vacuum, and consequently the increased mechanical impact on the teat tissue causes stretching of the teats in the liner and in the mouthpiece chamber (Holst et al. 2021) or even a reverse pressure gradient inside the empty teat when the liner opens after compression of the teat (Rasmussen et al. 1994). Besides the high mechanical impact on the teat tissue, this can partially close the connection between the mammary gland milk ducts and the teat cistern and may cause a reduced milk flow or even an incomplete udder emptying (Bruckmaier and Blum 1996). However, prestimulation does not guarantee that the whole alveolar milk is transferred into the cisternal cavities. It has been demonstrated that milk ejection continues throughout milking or suckling while simultaneously milk removal provides space in the cistern for the continuously ejected milk (Bruckmaier et al. 1994). The short half-life of oxytocin in blood circulation of only 2–3 min causes an almost immediate cessation of milk ejection when the respective tactile stimulus is terminated (Belo and Bruckmaier 2010). Thus, during suckling only the amount of milk which is required by the offspring is ejected. During machine milking, the stimulatory effect of the cyclic teat cup liner movement causes a continuous release of oxytocin throughout milking until the udder is empty (Bruckmaier et al. 1994).

## 3 Lag Time to Induce Milk Ejection

Independent of lactational stage or season, elevated blood concentrations of oxytocin in the jugular vein occur within 15–30 s after the start of a tactile teat stimulation. However, the lag time from the start of stimulation until the commencement of alveolar milk in the cistern varies widely. It has been demonstrated that the degree of alveolar filling determines the response time of alveolar contraction and milk shift into the cistern after oxytocin release. Alveolar milk appears in the cistern after the start of tactile stimulation at about 40–50 s in full udders, and the reaction to teat stimulation may take up to 3 min in udders with a very low degree of filling (Bruckmaier and Hilger 2001). This variation is caused by the different grades of myoepithelial contraction which is necessary until milk is pressed out of the alveoli. In dairy practice, different degrees of udder filling are caused by the stage of lactation or by the milking frequency. Because of the often longer milking interval during the night, the degree of udder filling is often lower at evening milkings compared to morning milkings in conventional milking systems. Due to irregular intervals between milkings the degree of udder filling and hence lag time from the start of mechanical teat stimulation until milk ejection can considerably vary between individual milking in automatic milking systems (Dzidic et al. 2004). A prestimulation prior to milking to induce milk ejection before the attachment of the milking machine can be permanent, but can also be split into a short tactile stimulation (minimum 15 s), followed by a latency period of up to 45 s until milk ejection occurs (Kaskous and Bruckmaier 2011; Tuor et al. 2022).

## 4 Disturbed Milk Ejection

Disturbed milk ejection (DME) occurs most frequently in primiparous cows during the first weeks postpartum, in unfamiliar surroundings (e.g., after the introduction of a new milking system such as from conventional to automatic milking), or for one or two milkings during estrus (Bruckmaier and Blum 1998). Occasionally, disturbed milk ejection occurs also in multiparous cows at the start of lactation, or without an obvious reason suddenly during ongoing lactation.

## 5 The Role of the Sympathetic Nervous System and Adrenaline for Milk Removal

Catecholamines such as adrenaline and noradrenaline have often been considered to be the stress-related hormones that cause disturbed milk ejection. Indeed, experiments performed during the twentieth century could stop milk ejection in cows by the administration of adrenaline or xenobiotic alpha-adrenergic agonists at a supraphysiological dosage (Barowicz 1979; Bruckmaier et al. 1991, 1997). However, newer experiments showed that catecholamines rather increase than inhibit the release of oxytocin in response to teat stimulation (Bruckmaier et al. 1997). Contrary to spontaneously disturbed milk ejection, exogenous oxytocin is not able to overcome the disturbance if catecholamines had been administered before (Bruckmaier et al. 1997). It was shown that adrenergic receptors are only present in the teats and in the tissue around the larger milk ducts of the mammary gland, but not in the secretory parenchyma (Hammon et al. 1994; Inderwies et al. 2003a). Consequently, the administration of alpha-adrenergic agonists caused a massive longitudinal contraction of the teats, and a contraction of the smooth muscles which have been detected in the larger milk ducts, thus terminating the transfer of milk from the alveolar tissue into the cisternal cavities (Bruckmaier et al. 1997). It can be speculated that such a mechanism to immediately stop milk ejection during suckling in an acute fight-and-flight situation has developed in evolution of wild animals. However, such an acute stress situation does not occur in practical dairy farming, and it has never been observed when individual cows express DME. Therefore, a role of adrenaline and other catecholamines either as systemically acting hormones or locally acting neurotransmitters of the sympathetic nervous system in the inhibition of milk ejection is very unlikely.

However, the adrenergic receptors of the smooth muscles of the milk duct system influence the transfer of milk from the secretory tissue into the cisternal cavities and hence overall milkability (Roets et al. 1989; Inderwies et al. 2003a, b). The tone of the smooth muscles of the milk ducts, teat walls, and teat canal were shown to be under control by the sympathetic nervous system (Lefcourt 1982). Changes in the teat sphincter tone were shown to barely cause changes in milk flow out of the cistern induced by the vacuum during machine milking (Bruckmaier et al. 1997). However, a low sympathetic tone of the teat sphincter muscles increases the likelihood of milk leakage ("incontinentia lactis") which can be particularly high shortly before milking, e.g., when the vacuum pump is switched on or other signals indicate that milking will start shortly. The leaking milk is only cisternal milk and is not related to oxytocin release and milk ejection (Rovai et al. 2007).

## 6 Lack of Oxytocin Release and Disturbed Milk Ejection

Disturbed milk ejection in practical dairy farming has been shown to be regulated at the level of the central nervous system and is expressed via reduced or totally lacking release of oxytocin in response to tactile teat stimulation before and during machine milking. A field study in Switzerland which included >67,000 cows showed that about 4% of the dairy cow population had bad milkability, in most cases caused by DME (Belo et al. 2009). The percentage of affected animals was similar in Holstein, Brown Swiss, and Simmental x Red Holstein breeds. Any relationship of DME with genetic strains of the animals, or with special housing and milking conditions was not found (Belo et al. 2009). In a representative sample, the reason for the bad milkability was experimentally demonstrated to be due to insufficient oxytocin release in 69% of the cows under suspicion (Belo et al. 2009).

While it was demonstrated that adrenaline is not the cause of DME in practical farming it remains unclear which factors are controlling the inhibition of oxytocin release as the main cause of disturbed milk ejection at machine milking. Neither the administration of cortisol nor catecholamines caused an inhibition of oxytocin release in dairy cows (Bruckmaier et al. 1997; Mayer and Lefcourt 1987). Some results indicate that endogenous opioid peptides, often released concomitantly with cortisol, are involved in the inhibition (Bruckmaier and Blum 1998). It has been shown that morphine inhibits the release of oxytocin during milking, and, in contrast, the opioid antagonist naloxone increases the milking-related release of oxytocin in cows with non-disturbed milk ejection (Tancin et al. 2003). Naloxone had, however, no clear effect on the milking-related release of oxytocin in cows with disturbed milk ejection (Macuhova et al. 2002). Disturbed milk ejection is mostly treated with exogenous oxytocin injected before each milking (details see next paragraph). An alternative method which represents a very strong stimulus to release oxytocin and thus often works in most cases of DME is vaginal or cervical tactile stimulation immediately before milking. This method may be used in a tie stall system but is obviously not suitable in most parlors or even robotic milking systems.

## 7 Use of Exogenous Oxytocin

Most of the animals with disturbed milk ejection are treated with oxytocin from the syringe before each milking (Belo et al. 2009). Besides that, oxytocin injection before milking is sometimes used in mastitis therapy to activate the transfer of immunoglobulins and other immune-active components from blood into milk (Wall et al. 2016) or to increase the degree of udder emptying and—as a consequence—to increase milk production (Nostrand et al. 1991).

Oxytocin is injected either intramuscularly (i.m.) or intravenously (i.v.), mostly at a dosage that results in plasma concentrations far beyond the physiological level. While oxytocin levels after the end of milking and cessation of oxytocin release from the pituitary drops to baseline within minutes due to the short half-life of 2–3 min (Belo and Bruckmaier 2010) oxytocin remains elevated for hours, especially after i.m. injection because of its continuous release from the muscle tissue into blood circulation (Macuhova et al. 2004). Besides a complete udder emptying including residual milk at milking this causes a continuation of milk ejections during the first hours after milking, and an early accumulation of cisternal milk which would start only at 4–5 hours after previous milking without the impact of elevated oxytocin (Macuhova et al. 2004). Both, more efficient udder emptying, and continued milk ejection during the time between milkings may be the basis for the increased milk production.

Often oxytocin from the syringe is used chronically at least for a period of time. This treatment has several consequences which need to be considered. Very obvious is the effect that cows become afraid of voluntarily entering the parlor because they expect the injection. Thus, it was observed in farms with frequent oxytocin use that many cows had to be driven into the parlor and that many animals started to defecate while present in the parlor. Because milk ejection and milk removal are sensitive to stress reactions it is obvious that the treatment does not support the required well-being and calmness during milking (Bruckmaier 2013).

Besides the observed behavioral changes of the animals there are also physiological effects of chronic oxytocin treatment. Already after 1 week of i.m. oxytocin injection (30 I.U.) to cows with normal milk ejection before the start of the treatment led to a dramatic reduction of udder emptying if oxytocin injection was not applied (Bruckmaier 2003), i.e., cows become addicted to the treatment, and it takes about 2–3 days until milk removal normalizes without the use of oxytocin (Bruckmaier and Macuhova, unpublished observations). Because a transient only partial udder emptying causes an increased risk for intramammary infection, and an irreversible decrease of milk production, the treatment can basically not be stopped before the next drying off.

Unexpectedly, the treatment with exogenous oxytocin increased the milking-related release of oxytocin from the pituitary, obviously based on a positive feedback mechanism of oxytocin on its own release (Macuhova et al. 2004). Despite a normal oxytocin release after cessation of the treatment, there was a pre-term stop of milk ejection during milking which led to the observed incomplete udder emptying (Macuhova et al. 2004). Most likely a desensitization of the oxytocin receptors of the myoepithelial cells is responsible for this effect. Thus, the chronic use of oxytocin at a high dosage should be avoided whenever possible.

We could demonstrate that much lower amounts of oxytocin injected i.v. are capable to induce complete milk ejection in animals with DME (Belo and Bruckmaier 2010). The required threshold of oxytocin to induce milk ejection throughout the entire milking process can easily be reached by dosages <1 I.U. of oxytocin. The needed dosage depends on the time needed to empty the udder, i.e., on the milk yield and anatomical milkability of the animal. The response phase was 2–3 min, and 3–5 min at oxytocin dosages of 0.2 I.U. and 0.5 I.U., respectively (Belo and Bruckmaier 2010). By these treatments or even up to 1 I.U., oxytocin plasma concentrations decreased to baseline within 20 min, and an effect of the above-described type of addiction to the treatment is most likely avoided. Because commercial oxytocin products are only available at concentrations of 10 I.U./ml or even 20 I.U./ml, the product must be diluted in sterile physiological saline solution, e.g., 1:100. Doing so, 0.2 I.U. or 0.5 I.U. can be injected in 2 ml or 5 ml of sterile solution, respectively (Belo and Bruckmaier 2010).

In conclusion, oxytocin from the syringe should only be used if necessary, and an alternative treatment by vaginal stimulation should be considered if suitable for the farmer.

## 8 Outlook

Disturbed milk ejection occurs at a constantly low frequency in individual cows. Genetic effects could not be confirmed and the pathophysiological mechanisms, obviously located in the central nervous system, are still unknown and almost impossible to be clarified in dairy cows. Therefore, it is rather unlikely that a sustainable therapy will be available in the near future. However, the use of endogenous oxytocin is a measure to obtain the milk that is stored in the udder. If oxytocin is used at the lowest possible dosage, and as short as possible effects of addiction can be widely avoided.

## References

Barowicz T (1979) Inhibitory effect of adrenaline on oxytocin release in the ewe during the milk-ejection reflex. J Dairy Res 46:41–46

Belo C, Bruckmaier RM (2010) Suitability of low dosage oxytocin treatment to induce milk ejection in dairy cows. J Dairy Sci 93:63–69

Belo C, Schlegel S, Moll J, Möstl E, Bruckmaier RM (2009) Milk ejection disorders in Swiss dairy cows: a field study. J Dairy Res 76:222–228

Bruckmaier RM (2003) Chronic oxytocin treatment causes reduced milk ejection in dairy cows. J Dairy Res 70:123–126

Bruckmaier RM (2013) Oxytocin from the pituitary or from the syringe: importance and consequences for machine milking in dairy cows. In: NMC Annual Meeting 2013. San Diego. 27.–29.01.2013

Bruckmaier RM, Blum JW (1996) Simultaneous recording of oxytocin release, milk ejection and milk flow during milking of dairy cows with and without prestimulation. J Dairy Res 63:201–208

Bruckmaier RM, Blum JW (1998) Oxytocin release and milk removal in ruminants (Review). J Dairy Sci 81:939–949

Bruckmaier RM, Hilger M (2001) Milk ejection in dairy cows at different degrees of udder filling. J Dairy Res 68:369–376

Bruckmaier RM, Mayer H, Schams D (1991) Effects of alpha- and beta-adrenergic agonists on intramammary pressure and milkflow in dairy cows. J Dairy Res 58:411–419

Bruckmaier RM, Schams D, Blum JW (1994) Continuously elevated concentrations of oxytocin during milking are necessary for complete milk removal in dairy cows. J Dairy Res 61:323–334

Bruckmaier RM, Wellnitz O, Blum JW (1997) Inhibition of milk ejection in cows by oxytocin receptor blockade, α-adrenergic receptor stimulation and in unfamiliar surroundings. J Dairy Res 64:315–325

Dzidic A, Macuhova J, Bruckmaier RM (2004) Effects of cleaning duration and water temperature on oxytocin release and milk removal in an automatic milking system. J Dairy Sci 87:4163–4169

Hammon HM, Bruckmaier RM, Honegger UE, Blum JW (1994) Distribution and density of alpha- and beta-adrenergic receptor binding sites in the bovine mammary gland. J Dairy Res 61:47–57

Holst GE, Adrion F, Umstätter C, Bruckmaier RM (2021) Type of teat cup liner and cluster ventilation affect vacuum conditions in the liner and milking performance in dairy cows. J Dairy Sci 104:4775–4786

Inderwies T, Pfaffl MW, Meyer HHD, Blum JW, Bruckmaier RM (2003a) Detection and quantification of mRNA expression of α- and β-adrenergic receptor subtypes in the mammary gland of dairy cows. Domest Anim Endocrinol 24:123–135

Inderwies T, Pfaffl MW, Bruckmaier RM (2003b) Milking characteristics and their relation to adrenergic receptor mRNA expression and ligand binding in the mammary gland of dairy cows. Domest Anim Endocrinol 25:275–286

Kaskous S, Bruckmaier RM (2011) Best combination of pre-stimulation and latency period duration before cluster attachment for efficient oxytocin release and milk ejection in cows with low to high udder-filling level. J Dairy Res 78:97–104

Lefcourt A (1982) Effect of teat stimulation on sympathetic tone in bovine mammary gland. J Dairy Sci 65:2317–2322

Macuhova J, Tancin V, Kraetzl WD, Meyer HHD, Bruckmaier RM (2002) Inhibition of oxytocin release during repeated milking in unfamiliar surroundings: importance of opioids and adrenal cortex sensitivity. J Dairy Res 69:63–73

Macuhova J, Tancin V, Bruckmaier RM (2004) Effects of oxytocin administration on oxytocin release and milk ejection. J Dairy Sci 87:1236–1244

Mayer HK, Lefcourt AM (1987) Failure of cortisol injected prior to milking to inhibit milk ejection in dairy cattle. J Dairy Res 54:173–177

Nostrand SD, Galton DM, Erb HN, Bauman DE (1991) Effects of daily exogenous oxytocin on lactation milk yield and composition. J Dairy Sci 74:2119–2127

Rasmussen MD, Frimer ES, Decker EL (1994) Reverse pressure gradients across the teat canal related to machine milking. J Dairy Sci 77:984–993

Roets E, Vandeputte-Van Messom G, Burvenich C, Peeters G (1989) Relationship between numbers of alpha 2- and beta 2-adrenoceptors in teat tissue and blood cells and milkability of primiparous cows. J Dairy Sci 72:3304–3313

Rovai M, Kollmann MT, Bruckmaier RM (2007) Incontinentia lactis: physiology and anatomy conducive to milk leakage in dairy cows. J Dairy Sci 90:682–690

Tancin V, Schams D, Kraetzl WD (2003) The effect of morphine and naloxone on the release of prolactin during machine milking in dairy cows. J Dairy Res 2003(70):277–282

Tuor M, Jenni B, Wellnitz O, Bruckmaier RM (2022) Reduced liner-open phase and vacuum instead of prestimulation increase parlor efficiency in dairy cows. J Dairy Sci 105:1533–1541

Wall SK, Wellnitz O, Hernández-Castellano LE, Ahmadpour A, Bruckmaier RM (2016) Supraphysiological oxytocin increases the transfer of immunoglobulins and other blood components to milk during lipopolysaccharide- and lipoteichoic acid-induced mastitis in dairy cows. J Dairy Sci 99:9165–9173

# Part VI

# Locomotion Disorders

# Lameness in Cattle

Nick J. Bell, Sara I. L. Pedersen, Laura V. Randall, John G. Remnant, and James P. Wilson

**Abstract**

Despite being of major importance for cattle health, welfare and productivity, lameness remains a highly prevalent condition, particularly in dairy cattle. Lesions of the distal limb (claw and surrounding skin) are the predominant cause of lameness and can be broadly summarised as non-infectious or infectious; with claw horn lesions and digital dermatitis, respectively, being the most common. Substantial knowledge gaps still exist relating to the pathogenesis of claw horn lesions. The body of evidence relating to risk factors and treatment have shown a history of previous lameness is an important contributor to future lameness. Early detection and prompt effective treatment appear key to improving treatment outcomes and reducing the risk of repeat lameness. Digital dermatitis is a contagious cause of lameness. Controlling environmental risk factors, when combined with footbathing and treatment of lesions appears effective in managing digital dermatitis. Genetic markers have been identified which offers potential benefits to control through genetic selection.

Addressing lameness can be challenging due to its multifactorial and complex nature. Therefore, it is important to take a strategic approach that includes the whole farm team (farm staff, vet, hoof trimmer and nutritionist). Relevant data should first be reviewed to assist in the identification of key lameness risk factors followed by the prioritisation of the most important changes required for success.

**Keywords**

Cattle · Dairy · Beef · Lameness · Claw horn lesions · Sole lesion · White line lesion · Sole ulcer · Digital dermatitis · Upper limb lameness

N. J. Bell · L. V. Randall (✉)
School of Veterinary Medicine and Science, University of Nottingham, Sutton Bonington Campus, Leicestershire, UK
e-mail: nick.bell@nottingham.ac.uk
e-mail: laura.randall@nottingham.ac.uk

S. I. L. Pedersen
Farm Dynamics Ltd,
Cowbridge, Vale of Glamorgan, Wales, UK
e-mail: sara@farmdynamics.co.uk

J. G. Remnant
Faculty of Veterinary Medicine, University of Calgary, Calgary, AB, Canada
e-mail: john.remnant@ucalgary.ca

J. P. Wilson
Department of Livestock and One Health, Institute of Infection, Veterinary and Ecological Sciences, University of Liverpool, Leahurst Campus, Chester, UK
e-mail: J.P.Wilson@liverpool.ac.uk



J. J. Gross (ed.), *Production Diseases in Farm Animals*,
https://doi.org/10.1007/978-3-031-51788-4_17

# 1 Introduction

## 1.1 General Overview

Lameness is a common and painful condition that has a detrimental impact on animal welfare, economic viability and sustainability of the cattle industry. Commonly defined in the literature as 'impaired mobility' or an 'abnormal gait or stance', lameness is not a disease but a clinical syndrome with many underlying causes. The aim of this chapter is to provide an overview of the diagnosis, treatment and management of lameness in cattle based on the latest evidence-base. Due to the weight of literature involving the study of lameness in the dairy sector, the chapter will focus predominantly on dairy cattle, whilst also considering similarities and differences with lameness in the beef sector.

### 1.1.1 Overview of Current Lameness Levels and Its Impact

Lameness presents a substantial challenge to the sustainability of dairy and beef systems worldwide. In dairy cattle, it causes financial losses through depression of milk production, reduced reproductive efficiency and increased culling risk (Huxley 2013), and greatly impacts on cow welfare (Whay and Shearer 2017). The economic significance of lameness in beef cattle is also recognised due to similar effects: longer times to finishing due to reduced weight gains (Davis-Unger et al. 2017), poorer breeding bull soundness assessments (Boakari et al. 2022), reduced fertility in cows (Fjeldaas et al. 2007) and increased risk of culling (Hird et al. 1991). These reduced efficiencies present a major barrier in terms of sustainability and the aim of the agricultural industry to reduce its environmental impact.

The current mean prevalence of lameness in dairy cattle is estimated at ~30% in the UK (Griffiths et al. 2018; Randall et al. 2019), between ~22 and 55% in the USA (Cook 2003; Espejo et al. 2006; von Keyserlingk et al. 2012), and ~19% in Australia (Ranjbar et al. 2016). There is less well-defined data for beef cattle, however, unpublished UK data suggests that it is higher in suckler cows in comparison to finishing cattle (14.2% vs. 8.3%; Tunstall 2020). The reported prevalence in both the dairy and the beef sectors of the industry means that a substantial reduction in lameness remains a key goal for the global cattle industry.

### 1.1.2 Measuring Lameness; Mobility/Locomotion Scoring, and Lesion Recording

Lameness can be measured through the use of mobility/locomotion scoring which is commonly utilised to determine lameness prevalence at a specific time point and to screen for new cases (active surveillance as opposed to passive surveillance when cows are spotted during normal husbandry activities). The use of trimming or treatment data can be used to provide the incidence of lesions over a specific timeframe.

Cross-sectional studies have demonstrated associations between lesions causing lameness and lameness behaviours. A UK study reported various behaviours were associated with presence of lesions (O'Callaghan et al. 2003). A similar study was conducted on sole ulcers with a similar finding (Flower and Weary 2006). The relationship between digital dermatitis and locomotion score is far more variable (Stokes 2011). In a cross-sectional survey of primiparous heifers, there was a linear relationship between the odds of finding a painful foot lesion and locomotion score, but not other non-painful lesions. However, there appears to be a low association between locomotion score and sole haemorrhage (Capion et al. 2009), likely to reflect the time delay between a bruising event affecting the sole corium and the appearance of sole haemorrhagic pigment on the sole surface many weeks later. There have been few studies relating serial locomotion scores with lesions. Chapinal et al (2009) found changes in locomotion score at least 4 weeks prior to the diagnosis of sole ulcer and Groenevelt et al. (2014) demonstrated cows detected with early onset lameness had a higher incidence of diagnosed sole bruising and digital dermatitis than control cows which had a longer duration of lameness and more severe lesions such as sole ulcers.

### 1.1.3 Incidence; Lesion Recording

Accurate assessment of lesion incidence requires active surveillance combined with consistent and accurate recording. The broad range of terms that are used to describe lesions can introduce a lack of consistency when benchmarking between herds or comparing results across published studies. In addition, there is a tendency towards under-recording of lameness treatments by producers, which can lead to inaccurate estimations of incidence (Main et al. 2003).

The 'ICAR Claw Health Atlas' is an international collaboration from the International Committee for Animal Recording (ICAR) that provides descriptions and images of claw and skin lesions and aims to standardise coding for application of data recording (ICAR 2020). The terminology within this chapter will broadly be used according to ICAR descriptors.

### 1.1.4 Prevalence; Visual Assessment Versus Automated Methods/Technologies

Visual assessment is the most widely used method for determining lameness prevalence whereby a mobility/locomotion score based on gait is assigned to an individual animal. A number of visual assessment scoring systems have been reported in the research literature for both dairy (Sprecher et al. 1997; Reader et al. 2011; Manson and Leaver 1988; Tranter and Morris 1991) and beef (Terrel 2016; Tunstall et al. 2020) cattle, which have formed the basis of scoring systems used on commercial farms for monitoring lameness. Adoption of single scoring systems in a standardised way helps in benchmarking and monitoring lameness across herds within countries. For example, in the UK the AHDB 4 point (0–3) (AHDB 2020) system has been adopted as an industry standard for dairy cattle, with the Register of Mobility scorers (RoMS) providing a register of independent mobility scorers trained in using this system.

Identification of cows likely to benefit from treatment using locomotion or mobility scoring reduces the prevalence of chronically lame cows (Groenevelt et al. 2014). The scoring approach identifies lame cows earlier (Leach et al. 2012) and allows the implementation of prompt and effective treatment (Thomas et al. 2015). There is a growing evidence-base surrounding the benefits of early detection in both promoting a better recovery and reducing chronic or repeat cases (Groenevelt et al. 2014; Thomas et al. 2015, 2016).

Automated methods are increasingly being developed and becoming commercially available. The predominant technologies used in commercially available automated lameness detection include sensors (e.g. accelerometers, load cells, and pressure sensors) and vision technologies (Bell and Walker 2016; Van Nuffel et al. 2015). As the research relating to these technologies becomes available their potential in lameness control programmes can be fully realised, with the capability for standardised, daily identification of lame cows being a major advantage over intermittent and error-prone mobility scoring by individual observers.

### 1.1.5 Location of Lameness and General Anatomical Considerations

Conditions causing pain resulting in lameness can occur at any level of the fore or hindlimb and occasionally higher in the animal's musculoskeletal system, e.g. the vertebrae. In cattle, the hind limbs are most commonly involved in lameness with approximately 85–90% occurring in the foot (Shearer et al. 2012; Murray et al. 1996; Barker et al. 2007) and predominantly the lateral claws (Murray et al. 1996; Barker 2007). Lesions that cause lameness can also occur in a variety of structures, including soft tissues, bone, and claw horn.

In dairy cattle, the primary causes of lameness are consistently reported to be claw horn lesions irrespective of geographical location (Capion et al. 2009; Chesterton 1989; Tadich et al. 2010; Barker et al. 2009). In comparison, there are considerably fewer reports on the main lesions causing lameness in beef cattle. Where studies exist these report varying findings predominantly based on age and management system.

In the following sections, we will review the claw horn (claw disorders), soft tissues including the skin (non-claw disorders) and upper limb causes of lameness.

# 2 Claw Disorders

There is a wide range of claw disorders reported in cattle. In this section, we focus on the most common lesions causing lameness in dairy cattle, their associated risk factors and the evidence-based surrounding treatment. The similarities and differences between dairy and beef cattle are considered at the end of the section.

## 2.1 Dairy Cattle

Summary

**Terminology**

- ICAR is an international collaboration providing standardised descriptions and coding for lesion data recording.
- Sole lesions commonly causing lameness can include sole haemorrhage (bruising) or ulceration.
- White line lesions commonly causing lameness can include; white line haemorrhage (bruising), separation, or abscessation.
- Other disorders include sand cracks, double sole, horn fissures, abnormal claw shape, toe necrosis, toe and heel ulcers, and thin soles.
- Laminitis is not considered an appropriate term to use to describe claw horn lesions.

**Risk factors**

- Can be broadly categorised into animal-based and environmental risk factors.
- Risk factors for the most common claw horn lesions (sole lesions and white line lesions) are broadly similar, although sole lesions are generally more associated with factors increasing standing times whilst white line lesions with flooring surfaces and cattle handling.
- Previous lameness history is an important risk factor and suggests prevention should remain the main focus.

**Treatment**

- Early detection and prompt effective treatment (EDPET) has been demonstrated to improve lameness outcomes for the most common claw horn lesions.
- Evidence from randomised clinical trials support the use of NSAIDs and hoof blocks in the treatment of lameness at an early stage.

### 2.1.1 Terminology

#### 2.1.1.1 Terminology to Describe Common Claw Horn Lesions

The claw horn lesions most commonly causing lameness include sole lesions (sole haemorrhage/bruising through to sole ulceration) and white line lesions (haemorrhage/bruising through to separation, fissures or abscessation). Other disorders of the claw horn include sand cracks, double sole, horn fissures (axial, horizontal and vertical), abnormal claw shape (asymmetric, dorsal concave, scissor and corkscrew), toe and bulb ulcer, toe necrosis and thin soles, as detailed in the ICAR Claw Health Atlas (ICAR 2020). Although not all lesions are associated with lameness and even lesions that are widely acknowledged to cause lameness can be present in cows with a normal gait.

#### 2.1.1.2 Laminitis

Historically, the term 'laminitis' or 'subclinical laminitis' has been used in the literature to describe claw horn lesions, however, the use of this terminology has been increasingly questioned (Randall et al. 2018a; Laven 2006). This is due to a lack of evidence existing to support the hypothesis that the classical laminitis pathways, extrapolated from the described pathogenesis for equine laminitis involving a nutritional aetiology of carbohydrate overload, is involved in the development of claw horn lesions (Randall et al. 2018a). A small number of experimental oligofructose (a readily available carbohydrate) overload models have been able to induce acute laminitic-type clinical signs, such as lameness and joint effusion alongside profound ruminal acidosis (Danscher et al. 2009; Thoefner et al. 2004; Li et al. 2021). However, the evidence to demonstrate that a non-acute lowered rumen pH (or sub-acute ruminal acidosis) has a causal relationship with development of claw horn lesions has yet to be reported. Since other potential causal pathways for the development of claw horn lesions, which have a far greater evidence-base, have become more widely known and understood the use of the terminology 'laminitis' in reference to claw horn lesions is considered inappropriate and should be reserved for referring specifically to inflammation of the laminae.

### 2.1.2 Anatomical Considerations

The distal phalanx (pedal bone) is suspended in the hoof capsule by the suspensory apparatus by strong collagenous attachments and supported in place by the digital cushion. The digital cushion is composed of three cylinders of fat which extend from the heel towards the toe proximally for approximately two-thirds the length of the sole. The function of both structures is to reduce the force exerted on the horn producing sole corium during both locomotion and standing. The germinal epithelium producing sole corium is composed of keratinocytes and is highly vascularised and therefore vulnerable to both trauma and inflammation. Recent research has highlighted how the dynamic nature of the digital cushion and distal phalanx and that when these are compromised it increases the risk of claw horn lesion development. The peri-parturient period is a particularly high-risk time due to biomechanical and histopathological changes (e.g. upregulated matrix metalloproteinases) (Tarlton et al. 2002) and body condition changes (Machado et al. 2010; Randall et al. 2015). Where claw horn lesions occur, they have been found to be more commonly found in the lateral hind claws compared with other claws (Tarlton and Webster 2002; Winckler and Margerison 2012).

### 2.1.3 Risk Factors for Common Claw Horn Lesions

Risk factors relating to the most common claw horn lesions (sole lesions and white line lesions) can be broadly categorised into animal-based and environmental. Animal-based risk factors relate to those associated with the structure and function of the claw whilst environmental risk factors are those which operate beyond the animal and in the case of claw horn lesions are predominantly factors that challenge the internal structures of the hoof capsule. There are both animal-based and environmental risk factors that are common to the two main types of claw horn lesions (sole lesions and white line lesions). Within sole lesions, as it is currently considered there is a continuum of increased severity from sole haemorrhage through to sole ulceration (and the equivalent with white line lesions), the risk factors for sole haemorrhage and sole ulceration are considered the same. Claw horn lesions are often described as 'pressure lesions' based on the evidence that a primary pathogenic pathway resulting in these lesions is increased pressure within the hoof capsule, ultimately causing damage to the sensitive tissues (corium) from which the claw horn develops. Therefore, factors that increase the pressure within the hoof horn capsule are considered risk factors for claw horn lesions as described below. White line lesions are often considered in situations where there are increased dynamic forces on the foot, for example sharp turns or poor cow flow, although the evidence to demonstrate this association is still lacking.

#### 2.1.3.1 Animal-Based Risk Factors

Body condition score (BCS) has been commonly reported as a risk factor for lameness (assessed by locomotion scoring) as well as specifically for claw horn lesions (Randall et al. 2015; Green et al. 2014). A meta-analyses of the literature relating to risk factors for lameness reported a body condition score of <2.5/5 was associated with an increased odds of lameness (Oehm et al. 2019). This BCS has been consistently found across studies to be a threshold below which there is an increased risk of lameness and claw horn lesions. Although the underlying pathological pathways have yet to be fully elucidated and understood, a number of hypotheses have been proposed and explored. A low body condition has been reported to be associated with a reduced thickness of the digital cushion, a capsule of adipose tissue that sits beneath the third phalanx and is thought to provide a force-dissipating effect (Bicalho et al. 2009). One hypothesis is that with a thinning of the digital cushion with a loss in BCS this protective force-dissipating effect reduces and therefore increases the risk of claw horn lesions. Whilst a thinner digital cushion thickness has been found to increase the risk of future claw horn lesions, a change in thickness (thinning) was not found to, challenging the concept that loss of BCS is the primary cause of decreased digital cushion thickness (Newsome et al. 2017a, b). In fact, other factors such as previous lameness events and genetics may be more important in the role of digital cushion thickness in the development of claw horn lesions (Wilson et al. 2021). Milk yield has been widely reported to be associated with lameness and claw horn lesions, with higher-yielding cows being at increased risk of lameness, as well as lameness having a negative impact on milk yield (Green et al. 2002). The underlying pathogenic pathways relating to this association between milk yield and lameness are potentially complex and are still poorly understood. For example, claw horn lesions are most severe and/or prevalent approximately 10–20 weeks after calving (Offer et al. 2000) and it may be that prior reduction in yield could be undiagnosed claw horn lesions. Another hypothesis is that as milk yield increases to peak lactation a concurrent decrease in BCS is observed which contributes to claw horn lesion risk by the pathways described above. Alternatively other direct or indirect metabolic, immune or behavioural pathways may be present. Body weight, although often raised as a risk factor for lameness has conflicting evidence relating to its association with claw horn lesions. In general, it appears that low body weight in mature dairy cows is associated with an increased risk of lameness, which is potentially due to social factors (Randall et al. 2015; Norring et al. 2014). The calving period has also been highlighted as a potentially important time period in terms of its influence on the risk of lameness and claw horn lesions. Hormonal and metabolic changes that occur have been suggested to contribute towards increased laxity of the connective tissues that suspend the pedal bone within the hoof capsule (suspensory apparatus). With increased laxity, the resulting instability of the pedal bone may cause inflammation and cellular damage to the corium leading to development of claw horn lesions (Tarlton et al. 2002; Webster 2001; Knott et al. 2007). It has been demonstrated that the use of non-steroidal anti-inflammatory drugs (NSAIDs) at calving can reduce lameness prevalence when compared with conventional best practice treatment of lesions (treatment trim +/– block and NSAID), which may also support the hypothesis that inflammation occurring during the calving period is contributing towards the risk of claw horn lesions (Wilson et al. 2022).

A clear and important risk factor for claw horn lesions is previous lameness history (Randall et al. 2018b). Numerous studies have reported a positive association between previous lameness and future lameness risk, including risk of claw horn lesions (Newsome et al. 2017a; Hirst et al. 2002). Lameness risk also increases with age and supports the hypothesis that over time long-term pathological changes associated with the occurrence of claw horn lesions lead to an increased risk of future lesions and lameness (Newsome et al. 2016). It is therefore key in lameness management that the focus is placed on prevention of lameness in the first place.

#### 2.1.3.2 Environmental Risk Factors

Environmental risk factors for claw horn lesions are primarily related to how the environment influences the biomechanical forces within the claw due to pressures at the claw-environment surface interface. The surface the cow is standing, walking or rising from is therefore a potentially important aspect as well as the duration of time spent on the surface (Randall et al. 2018a). There is conflicting evidence in relation to the direct associations between different floor surfaces (e.g. solid vs. slatted concrete alleys and rubber vs. concrete) and claw horn lesions. However, evidence from a randomised controlled trial appears to provide support for the concept that flooring surface is important around the transition period in dairy heifers and that acclimatisation of youngstock to the environment before calving is important for risk of claw horn lesions after calving (Bergsten et al. 2015). Well-designed intervention studies are still needed to really understand the influences of different flooring surfaces (including abrasiveness, coefficients of friction, compressibility) on the risk of claw horn lesions. Track surface has been demonstrated to have an association with lameness in general, including track cover and maintenance (Chesterton 1989; Burow et al. 2014), stones in paddock gateways and slats on cow tracks near the collecting yard (Browne et al. 2022). Likewise, the use of roads or concrete tracks has been identified as a risk factor for sole ulcers (Barker et al. 2009). Increased standing time is a widely accepted risk factor for lameness in general and evidence supports an association between increased standing times/standing half-in-half-out of cubicles and claw horn lesions (Galindo and Broom 2000; Proudfoot et al. 2010).

A summary of evidence-based risk factors for claw horn lesions in dairy herds is provided in Table 1.

**Table 1** Summary of key risk factors supported by peer-reviewed literature in dairy herds for the most common types of claw horn lesions

| | Animal-based risk factors | Environmental risk factors |
|---|---|---|
| Sole lesions (haemorrhage through to ulceration) | Previous lameness/claw horn lesion<br>Calving effect; changes to the suspensory apparatus<br>Low BCS or loss after calving (potentially due to reduced digital cushion thickness)<br>Increasing parity | Lack of acclimatisation to environment, e.g. concrete floors or cubicles<br>Increased standing time/ poor cubicle comfort<br>Use of roads or concrete tracks |
| White line lesions | Increasing parity<br>Increasing herd size | Solid grooved concrete floors in yards or alleys or slatted concrete alleys<br>Use of tracks; long distances, poor track quality (e.g. stones) |

### 2.1.4 Treatment of Claw Horn Lesions

#### 2.1.4.1 Approach to Treating Claw Horn Lesions and Underpinning Evidence

The evidence-base suggests the best practice approach to claw horn lesion treatment is a therapeutic trim, application of a hoof block to a healthy contralateral claw, and a 3-day course of the NSAID ketoprofen (Thomas et al. 2015). However, the success of the treatment is dependent on the severity of the lesion and the duration of lameness before intervention (Thomas et al. 2016; Miguel-Pacheco et al. 2017). The skill and experience of the operator are also likely important in determining success. Approach to treating the individual lame cow including technique is covered in more detail later in this chapter.

Key success factors underpinning the treatment of claw horn lesions (primarily sole bruising, sole ulceration and white line disease) are focussed on the early detection and prompt effective treatment protocol (EDPET) principle (Pedersen and Wilson 2021; Bell et al. 2009). Early detection relies on identifying animals with an early-stage disease (e.g. sole bruising), where evidence suggests the prognosis is much more positive in terms of response to treatment (Leach et al. 2012; Thomas et al. 2016). Fortnightly mobility scoring has been demonstrated to be effective at identifying early cases of lameness (Groenevelt et al. 2014) and is the currently recommended industry standard. Prompt treatment is then required within 24–48 h of identification to promote good welfare and ensure optimal outcomes.

The evidence-base behind what constitutes prompt effective treatment has developed further in recent years to encompass the use of NSAIDs, the administration of an appropriate therapeutic trim, and the application of a hoof block (Thomas et al. 2015). The use of NSAIDs has been proven to have significant positive impacts in the treatment of claw horn lesions and in the management of pain associated with these lesions. This two-part effect gives substantial weight to their usage as part of a lameness treatment and management protocol. Further to this, the application of hoof blocks (appropriate to the system in which the cow is managed) is critical in improving the immediate welfare of the animal (by alleviating weight from the diseased claw), but also improves the chance of recovery.

The EDPET principle has the potential to minimise long-term anatomic change to the distal phalanx and digital cushion associated with claw horn lesions (Wilson et al. 2021; Newsome et al. 2016). This is thought to be through reducing the exposure to the inflammation associated with claw horn lesions, which is understood to drive these anatomical changes, and also contributes towards the rationale for the use of NSAIDs being beneficial. Managing the potential for pathological change to anatomy is vital in the management of the animal's long-term prognosis. The degree to which new bone is formed on the distal phalanx of the extent to which the digital cushion is metabolised or scarred can set precedent for the remainder of the animal's life. If these changes occur, then it is likely the germinal epithelium will have a higher degree of pressure exerted on it, meaning that the animal is permanently more susceptible to claw horn lesions (Newsome et al. 2016).

## 2.2 Beef Cattle

Due to the anatomy of the bovine limb, the lesions described in beef cattle are broadly the same as those in dairy cattle and therefore a similar approach to prevention is warranted. Due to differences in management and temperament between dairy and beef cattle, the latter tend to be more prone to traumatic injuries, including pedal bone fractures.

There is very limited data on the prevalence of claw horn lesions in beef cattle and where these exist, they report varied outcomes depending on management system and age of animal. A recent UK study indicated that white line disease and overgrown claws were the main cause of lameness (Tunstall 2020) whereas a Canadian study reported corkscrew claws as the main disorder found (Newcomer and Chamorro 2016). Sand cracks are reported more frequently in beef cattle, with more than 20% of mature cattle in western Canada reported to be affected, however, this did not adversely impact production (Goonewardene and Hand 1995).

A specific condition predominantly seen in feedlot cattle aged 10–12 months is toe tip necrosis, accounting for approximately 16% of all lameness treatment (Jelinski et al. 2016). This usually occurs a few weeks following weaning and transportation and is related to bacterial ingress through a defect in the apical white line (Gyan et al. 2015). This then leads to ascending infection of the distal phalanx with necrosis of the tip of the pedal bone. Mostly seen in the lateral hind claws it causes lameness and reduced growth rates, negatively impacting productivity on feedlots (Paetsch 2014). Preventive measures are centred around reducing wear of the horn at the apex of the toe through attention to flooring surfaces and cattle handling during transportation.

Laminitis is commonly referred to as a cause of lameness in beef systems, especially in finishing systems where diets are predominantly fed high energy and protein diets. As with dairy cattle, there has been some focus on the causative pathways between diet and laminitis (as defined by inflammation and degeneration of the laminae). Whilst an acute laminitis as a sequalae to a short episode of grain overload is often reported, this is commonly based on the clinical signs of lameness or changes in hoof conformation. Recent research has sought to offer a hypothesis for why growing cattle on an excessively high energy and protein diet are predisposed to laminitis, however, diagnosis was based on changes in hoof conformation rather than histological evaluation (Bäßler et al. 2021). Where histological evidence has been considered, low-grade inflammatory changes have been reported although there was no detrimental effect on laminar architecture (Tavares et al. 2019). Therefore, as in dairy cattle, the link between diet, laminitis and lameness needs further investigation.

## 3 Non-claw Disorders Causing Lameness Affecting the Skin of the Foot


**Summary**

**Terminology**

- Digital Dermatitis is the term used to describe the most common non-claw (infectious) lesion causing lameness in dairy cattle.

**Risk factors**

- Digital Dermatitis is contagious and therefore environmental risk factors are important including floor and foot hygiene, and potential spread via fomites.
- Genetic markers for DD have been identified, suggesting genetic selection may be beneficial.

**Treatment/control**

- Endemic infection of a herd with DD is unlikely to be eradicated and instead, the focus is on control via footbathing and treatment.
- Intensive screening of the herd can be used to detect individual cases and treat them with a licensed topical antibacterial agent.
- The evidence-base for preventative foot bathing regimes for DD remains surprisingly weak and conflicting. Daily footbathing is recommended for milking cows and formalin and copper sulphate have most evidence for efficacy, although there are safety and legislative considerations associated with these.


### 3.1 Non-claw Disorders Causing Lameness

Most lesions causing lameness within non-claw horn structures are generally microbial infections. Some may be *contagious* with close contact responsible for infections passing from the lesions on infected to uninfected animals (e.g. spirochaetal Treponemes causing digital dermatitis). Others are not contagious, with infections arising from *opportunistic* bacteria ubiquitous in the environment (e.g. *Fusobacterium necrophorum* causing Foul-in-the-foot). Occasionally, outbreaks of infections involving opportunistic bacteria can have the appearance of a contagious outbreak when risk factors are sudden onset, severe and the environmental contamination with bacteria is high, e.g. dermatophilosis. Secondary infections are common when a mix of bacteria in the environment opportunistically invade a wound or lesion, e.g. complicated sole ulcer, white line abscess, hock lesion or tetanus. Whilst sole ulcers may be infected but are not considered infectious conditions. Secondary infections are covered in more detail elsewhere.

There are several distinct lameness conditions associated with specific pathogens affecting non-claw tissues of the foot (below the fetlock) (Table 2). These mostly affect the epidermis, dermis and subcutis. There are rare or sporadic skin conditions that are not commonly associated with lameness which include chorioptic mange, bovine viral diarrhoea and photosensitisation and readers are directed to general cattle texts for these conditions.

Of the conditions listed in Table 1, digital dermatitis and Foul-in-the-foot are the most common and important causes of lameness in cattle and will be examined in the next section.

### 3.2 Digital Dermatitis

Digital dermatitis is one of the commonest causes of lameness affecting dairy cattle worldwide. There are reports of this condition from all the major dairy producing countries with surveys

**Table 2** Non-claw conditions causing lameness affecting the skin of the foot

| Condition | Alternative terms (those not listed by ICAR in italics) | Type of condition | Primary aetiology | Treatment |
|---|---|---|---|---|
| Digital dermatitis | Mortellaro, *hairy heel warts* | Contagious skin infection | *Treponema phagedenis*, *T. medium* and *T. pedis* (Evans et al. 2009) and a mix of other bacterial species which appear to alter the microbiome to allow invasion or persistence (Zinicola et al. 2015) | Topical antibacterial treatments and non-steroidal anti-inflammatory drugs |
| Interdigital phlegmon | Foul-in-the-foot, Interdigital necrobacillosis, Foot rot, *Foul* | Opportunistic infection | *Fusobacterium necrophorum* biotypes A, B and AB (Emery et al. 1985) | Systemic and topical antimicrobial treatments and non-steroidal anti-inflammatory drugs |
| *Dermatophilosis* | *Mud fever* | Opportunistic infection | *Dermatophilus congolensis* is ubiquitous but appears in outbreaks when conditions are wet and muddy underfoot for a number of weeks | Remove scabbed and crusted lesions, apply topical and systemic antimicrobial treatment |
| Interdigital hyperplasia | Tyloma, fibroma, corns, interdigital growths | Prone to secondary infection | Dermal mass between the digits, which can be genetic but is often secondary to digital dermatitis or foul-in-the-foot between the claws | Treat any underlying infection. Trim to prevent claw pinching soft tissues. Non-steroidal anti-inflammatory drugs. Careful surgical removal in severe cases |
| Swelling of the coronet and/or bulb | *Deep digital sepsis, retrobulbar abscess* | Secondary infection | Secondary infection of an interdigital lesion, sole ulcer or heel ulcer (usually). Occasionally infections track up the leg under the skin, causing cellulitis | Claw amputation or euthanasia. Surgical drainage is possible but rarely performed for welfare and economic reasons |
| Wounds and lacerations | Skin punctures and penetrations | Secondary infection | Many different sources of injury which tend to be farm and circumstance specific. One common cause of injury is constriction injury from bandages | Clean and apply topical antibacterial treatment. Keep wound clean and dry |

(continued)

**Table 2** (continued)

| Condition | Alternative terms (those not listed by ICAR in italics) | Type of condition | Primary aetiology | Treatment |
|---|---|---|---|---|
| Chemical burn | None | Secondary infection | The most common cause of chemical burn is through formalin foot bathing when there has been mis-dosing of formalin | Cease foot bathing using problem chemical. apply topical antibacterial treatment and analgesia as necessary. Resume foot baths using a cleaning agent and milder biocide |
| Systemic disease presenting with lameness in foot | Various | Varied systemic infectious on non-infectious diseases may affect the skin of the foot with lameness as the primary presenting sign. For example, foot-and-mouth disease virus is a contagious viral infection. Viral vesicles leading to erosions on the skin around hoof | Varied, depending on the systemic disease. The reader is directed to broader cattle medicine texts for reference. As an example, Foot-and-mouth disease virus infections are endemic in many countries with outbreaks occurring infrequently and sporadically in countries free from infection | Varied, depending on the systemic disease. The reader is directed to broader cattle medicine texts for reference. As an example, country-specific statutory controls are in place for FMD (national eradication schemes). Outside of statutory controls, supportive therapy |

describing prevalence over 40% in some instances (Bell 2006; Vink 2006). Whilst it is not reported to affect beef cattle as commonly as dairy, reports suggest it is becoming an increasing problem (Sullivan et al. 2013), particularly in mixed farming systems.

Digital dermatitis is a poly-Treponemal infection. Furthermore, it appears there are significant changes to the skin microbiome which facilitates Treponemal invasion and persistence of lesions, with over 1000 microbial species identified in lesions, making digital dermatitis truly polymicrobial.

Many infections are observed to go through distinct pathological stages described in the M-score (Dopfer et al. 1997; Berry et al. 2012) which classically starts with epidermal erosion or ulceration, and progresses to hyperkeratosis and/or 'proliferation' (gross dermal and epidermal thickening) (Table 3). A number of diagrammatics have been used to describe the stages through which lesions occur as shown in Fig. 1.

### 3.3 Risks for Digital Dermatitis and Routes of Spread

Few dairy herds remain uninfected. Digital dermatitis is usually acquired through infected livestock (dairy cows, beef animals, bulls or sheep), although spread is thought to be possible on fomites, e.g. contaminated foot trimming equipment or contaminated tractor scraper.

Once herds are endemically infected, transmission is believed to occur from infected to uninfected animals via contaminated foot prints primarily (Bell 2017). However, foot-trimming knives and gloves after handling digital dermatitis and contagious ovine digital dermatitis (the ovine equivalent infection) have been found to be culture positive (Angell et al. 2017; Sullivan et al. 2014). Further work went on to show a range of disinfectants could remove viable Trepomenes (Gillespie et al. 2020). Survival at low pH (<5) and high pH (>8.5) was very short-lived meaning a range of acid and alkali biocides may be effective (Bell 2017).

**Table 3** The M-score for digital dermatitis. Note the rulers used to scale lesions had 5 mm gradations

| M-score | Description (adapted from Berry et al. 2012 and Kofler et al. 2020) | Photograph |
|---|---|---|
| M0 (M5) | Apparently healthy-looking skin (macroscopically). Note this does not mean uninfected | 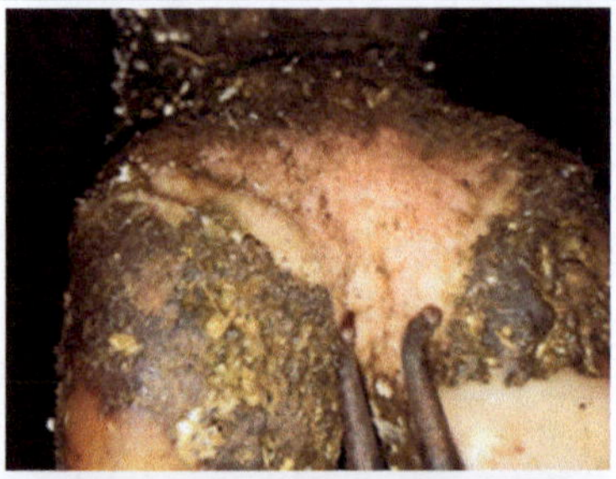 |
| M1 | Active lesion. Area of eroded or ulcerated skin less than 2 cm in diameter | 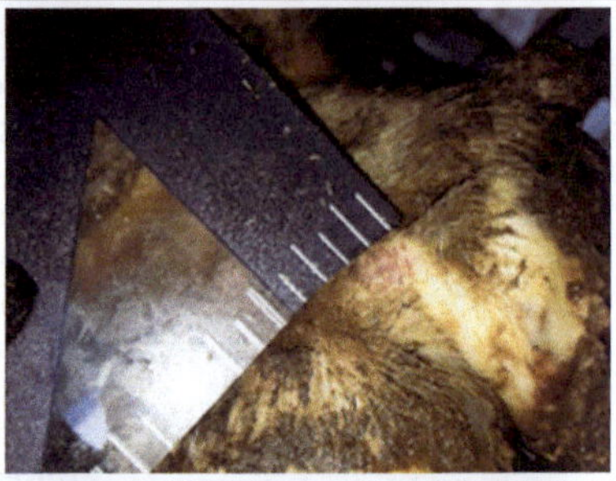 |
| M2 | Active lesion. Area of eroded or ulcerated skin 2 cm or more in diameter. Sometimes these may be swollen with grossly thickened dermis, which are then given the term 'proliferative' | 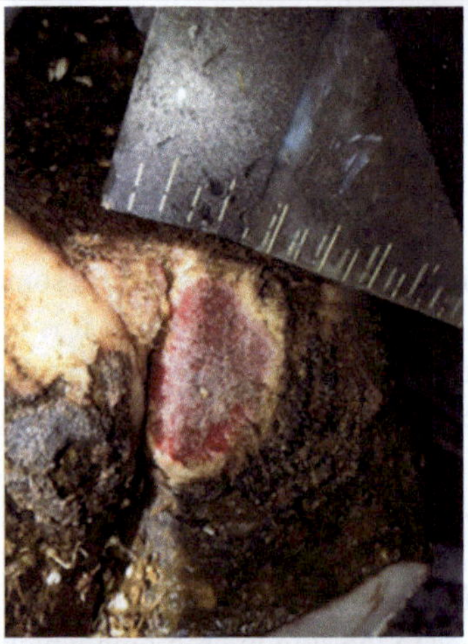 |
| M3 | Healing lesion. Flat, scabbed, brown appearance | 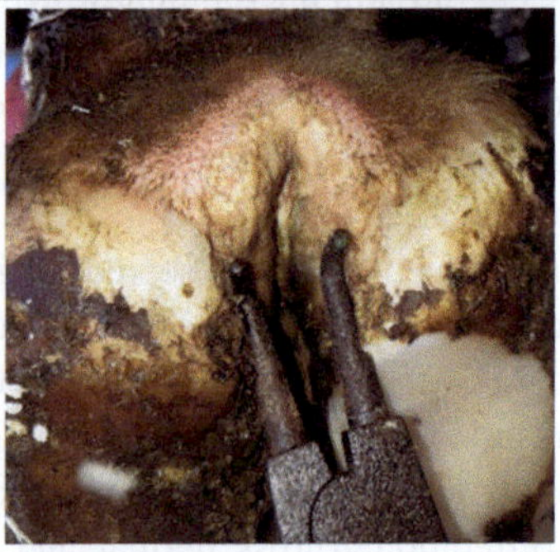 |

(continued)

**Table 3** (continued)

| M-score | Description (adapted from Berry et al. 2012 and Kofler et al. 2020) | Photograph |
|---|---|---|
| M4 | Chronic lesion. Distinctive dyskeratosis and hyperkeratosis give this stage a warty appearance with strands of keratin having a 'Medussa-head' in extreme cases | 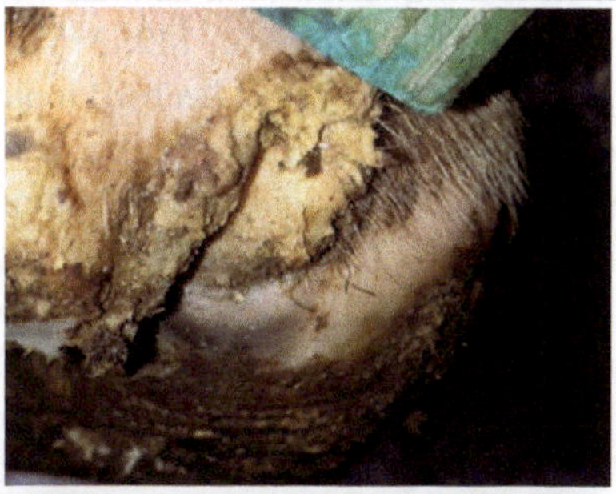 |
| M4.1 | Reactivating lesion. Area of ulceration less than 2 cm diameter present within the M4 lesion. Once ulceration exceeds 2 cm it is reclassified as M2 | 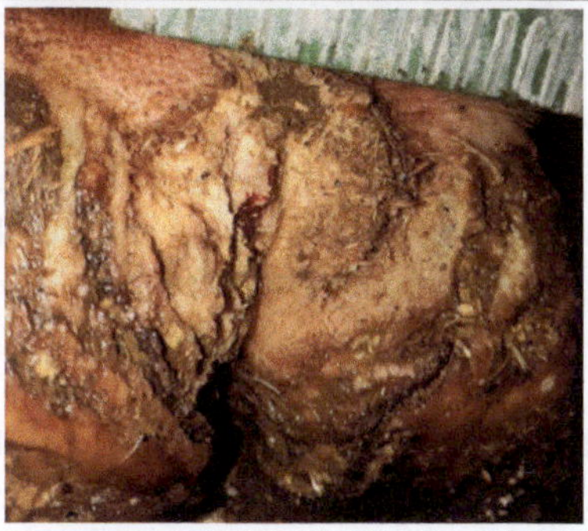 |

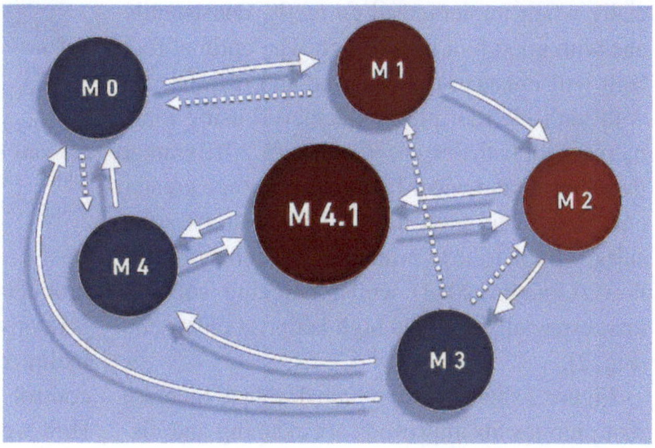


**Fig. 1** Possible M-score transitions of digital dermatitis lesions, modified from Kofler et al. (2020). Solid arrows indicate very likely transitions, dotted arrows indicate less likely transitions, red circles represent active stages and blue circles represent chronic stages including healthy skin (M0 stage)

## 3.4 Prevention and Control of Digital Dermatitis

Once endemically infected there does not appear to be a proven means of eradication. One Israeli report described eradication early in an outbreak and this involved an intensive treatment combined with a foot bathing protocol (Yeruham and Perl 1998). No further lesions were reported at follow-up in the subsequent 12 months although modern diagnostic tests (Vink 2006; Bell 2017) were not available or applied.

In the face of an outbreak in an endemically infected herds, similar principles of treatment and footbathing are applied. The intensive screening of the herd with treatment of individual animals has been used in a protocol (Pedersen 2019) where cows are screened in the parlour at milking using the 'bright light with a sharp water jet' approach (Rodriguez-Lainz et al. 1998) to assist in visualising lesions. All animals with active or reactivating lesions (M1, M2, M4.1) are segregated and treated with a licensed topical antibacterial agent. Treatments are repeated for 3 days or according to treatment protocol. Foot bathing is then used to prevent further reactivation of M4 lesions. A similar approach can be taken to control in beef herds, however, additional handling is required as lesions must be treated in an adequate foot trimming crush. In this treatment approach, it is important to prevent re-introduction of active lesions from youngstock and dry cows. These groups are commonly infected, with inadequate control using footbaths and therefore a worthwhile area to review in any infected herd struggling with control.

The evidence-base for preventative foot bathing regimes remains surprisingly weak and conflicting (Bell et al. 2014). In spite of the weak literature, empirical evidence farm data would support the daily foot disinfection regime as extremely effective (Bell and Main 2011), especially when implemented correctly, consistently and with good foot bath design. The optimal foot bath will enhance the cleaning and disinfection with multiple foot plunges achieved using long, narrow foot bath designs (Cook et al. 2012) rather than shorter and wider baths to encourage greater cow flow. This bath design also features solid sides which follow principles to enhance cow flow (Grandin 2014) whilst reducing solution loss, especially with the high kerbs on either end (Fig. 2).

**Fig. 2** Foot bath design to increase foot plunges for improved foot cleaning and disinfection. Note the large (28 cm) kerb, narrow (60 cm) foot well and solid sides which slope up to 90 cm wide passage at cow pelvis height. Also note the clean slatted area on either end

Figure 2 shows the foot bath design to increase foot plunges for improved foot cleaning and disinfection. Note the large (28 cm) kerb, narrow (60 cm) foot well and solid sides which slope up to 90 cm wide passage at cow pelvis height. Also, note the clean slatted area either end of the bath to ensure feet remain clean for greater biocide contact time after the cow has left the bath.

For endemically infected herds the main principles of prevention and control involve preventing new infections in youngstock through biocontainment, and prevent already infected animals from infecting others using foot baths. Whilst there are many commercial footbathing products available, the evidence-base in this area is relatively sparse, with copper sulphate and formalin remaining the most effective products for control (Bell et al. 2014; Jacobs et al. 2019). However, there are concerns rather the environmental impact of copper sulphate use and formalin is a probable carcinogen, thus must be handled safely. These concerns have led to restrictions on their use in some countries.

With regard to frequency of footbathing, daily regimes using 3% formalin appear effective for milking cows. For youngstock and dry cows where feet may be cleaner and drier, then twice weekly regimes appear sufficiently efficacious and practically achievable. Footbathing frequency in beef herds is largely dependent on the scale of the problem and careful attention needs to be paid to positioning the footbath to make cattle handling as practical as possible.

At the individual animal level, there appears to be a strong genetic predilection to digital dermatitis which can be explained with genetic markers and genomics. This makes the concept of genetic selection look very promising in comparison to vaccine development which has thus far yielded little success (Ertze et al. 2006). Currently, there is one genetic index for digital dermatitis produced by AHDB (UK).

# 4 Practical Approaches to Lameness

## 4.1 Equipment for the Treatment of Lame Cattle

### 4.1.1 Trimming Equipment

Hoof trimming is a routine procedure which often carries more risk to operators and cattle than any other activity on farm. Therefore, hoof trimming and lame cow treatment should only be undertaken by someone deemed safe to work by holding recognised competency qualifications. Part of the training should cover the safe use of equipment, equipment maintenance and the important personal protective equipment (PPE). Ideally, first aid at work training should also be undertaken by someone on site.

### 4.1.2 Personal Protective Equipment

Hoof trimming is one of the more hazardous procedures for farm staff, veterinarians and hoof trimmers, therefore, ensuring the right PPE is available and used is obvious and important. Every trimming site should have a risk assessment and sensible mitigation for identified risks.

Figure 3 shows the available PPE used by many professional hoof trimmers and should be considered for everyone treating lame cows or preventively trimming. Some items are considered essential whilst others are more important in situations where throughput is high or specialist equipment is used.

It can be argued that the equipment should be similar regardless of the role but the risks to farm staff may be reduced if knives and nippers are solely used (Table 4). There may be instances

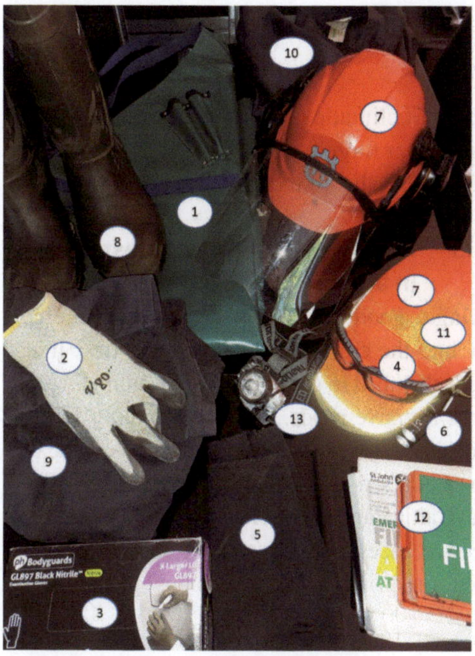

1. Apron (with knife holders)
2. Anti-cut gloves
3. Nitrile gloves over the top of anti-cut gloves
4. Eye protection
5. Wrist guards
6. Ear protection
7. Bump hat or combined hard hat with ear and face protection
8. Steel toe-capped boots
9. Overalls
10. Waterproofs
11. High visibility clothing
12. First aid kit and training
13. Adequate lighting

**Fig. 3** Personal protective equipment for use whilst trimming

**Table 4** List of PPE used by farm staff and professional contract trimmers

| | Personal protective equipment item | Essential for farm staff performing knife and nipper trimming only? | Additionally, if a professional contract trimmer using rotary cutting blades |
|---|---|---|---|
| 1 | Apron with knife holder | ✓ | ✓ |
| 2 | Anti-cut gloves | ✓ | ✓ |
| 3 | Tough nitrile gloves over the top for biosecurity | | ✓ |
| 4 | Eye protection | | ✓ |
| 5 | Wrist guards | ✓ | ✓ |
| 6 | Ear protection (to filter loud sounds but to permit some sound perception) | If trim crush is loud | ✓ |
| 7 | Bump hat or combined face, ear and head protection | Ideally yes | ✓ |
| 8 | Steel toe cap boots | ✓ | ✓ |
| 9 | Overalls to reduce cut risk and reduce zoonotic risk | ✓ | ✓ |
| 10 | Waterproofs to reduce cut risk and reduce zoonotic risk | | ✓ |
| 11 | High visibility clothing (Hi viz) | ✓ | ✓ |
| 12 | First aid kit with mobile phone | ✓ | ✓ |
| 13 | Good lighting | ✓ | ✓ |
| 14 | Safe and calm handling system and crush for secure restraint | ✓ | ✓ |
| 15 | Risk assessment and competency qualification | ✓ | ✓ |

**Fig. 4** Full PPE worn is important but can be too heavy for hot climates

where hot climatic conditions mean it is safer to wear lighter more breathable clothing rather than the full PPE shown in Fig. 4.

### 4.1.3 Handling System and Restraint in a Safe Crush

Having a good handling system will ensure cow flow into the crush is calm and stress free. It also avoids the need to get in behind the animal to push or use aversive prods. The best handling systems used either a race or a Bud Box (https://arec.tennessee.edu/wp-content/uploads/sites/17/2020/09/AlleyCattleHandlingSSHandout1.pdf). The most effective races are curved and sheeted (Fig. 5) based on principles and designs published by

**Fig. 5** Curved race with sheeted sides for efficient loading of the crush

Grandin (2014). Additional planning and care are required when dealing with beef cattle or those of a temperamental nature.

Once in the crush, the cow needs to be effectively restrained. The more secure the restraint the less likely the cow will struggle. Most modern crushes use a double bellyband system, then lift the cow (which the authors observe will further immobilise the cow) and then allow two limbs to be securely lifted to keep the operator safe from rear foot kicks whilst trimming forefeet (Fig. 5). Every crush will operate slightly differently and operators should have training specific to their crush. In Fig. 5, note the use of PPE, safe routine of cables and battery-operated grinder which has no cables to be caught, particularly important when a right-handed trimmer is working on a left forefoot.

### 4.1.4 Hoof Trimming Equipment

Like PPE, there is trimming equipment which is deemed essential and other items helpful for speed and efficiency. For preventive trimming the basic equipment involves knives (left- and right-handed), nippers and a specialist trimming crush (Fig. 6). Rotary cutting blades can make trimming quicker and easier, but in inexperienced hands they become inherently riskier for operator and cow. For lame cow treatment, a variety of blocks with glue will be needed, licensed treatments, hoof testers (pain testers), spreading pliers and a range of other items shown in Fig. 7, in addition to the basic knives and nippers.

It is useful to have a means of cleaning feet before trimming. This can be done with running water or disinfectant (provided the crush base has grip) or sawdust when there is fresh slurry easily removed. Maintaining good hygiene during trimming is a basic principle useful to prevent spread of disease or transference of zoonotic infections.

### 4.1.5 Maintenance of Equipment

After treating a cow with suspected infection, it is good practice to clean and disinfect equipment

1. PPE
2. Disinfectant
3. Safe routine of cables
4. Battery-operated grinder
5. Knife holders
6. Double bellyband, hind leg lifted to prevent kicks
7. Good lighting and visibility

**Fig. 6** Left forefoot being trimmed whilst the cow has a left hindlimb securely raised and two bellybands in place

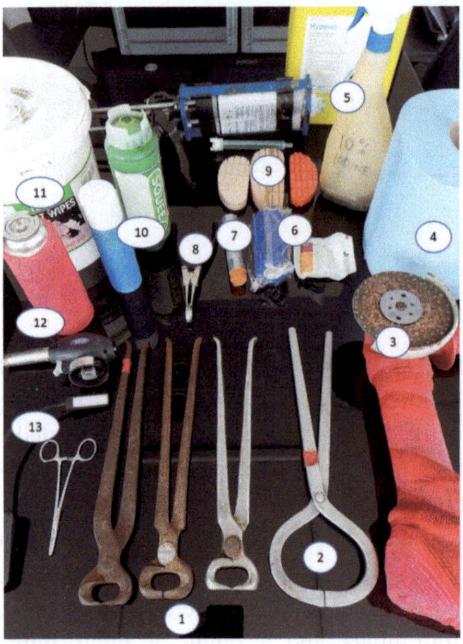

1. Range of nippers – sharp, old blunt for block removal, half-round
2. Hoof testers
3. Abrasive wheel for preparing sole to receive glue-on block
4. Paper towel
5. Disinfectant
6. Dressing and bright bandage
7. Colour crayon
8. Spreading pliers
9. Range of blocks and suitable glue
10. Licensed antibacterial treatment
11. Medicated wipes
12. Burner/heat gun
13. Forceps

**Fig. 7** A range of equipment used for the treatment of lame cows. Note there are some injectable treatments such as non-steroidal anti-inflammatory drugs (NSAIDS) and injectable antimicrobials which are not included in the picture

with an approved disinfectant. All equipment should be clean *and* disinfected if transported between herds to minimise the risk of transferring disease and zoonosis. Knives should be maintained in a very sharp condition and sharpened or replaced immediately if blunt. A rotary sharpening wheel, cloth/felt wheel and abrasive buffer paste is the most efficient way of sharpening knives (Fig. 8). Grinders need regular maintenance checks for safety also.

## 4.2 Approach to Treating the Individual Lame Cow

### 4.2.1 General Considerations

When presented with a lame animal there are several steps that should be taken before the examination. A short history to understand the nature and chronicity of the problem is important. Cattle that present as acutely lame or new cases of lameness may be managed differently from those that

1. Hot soapy detergent and disinfectant to clean then disinfect equipment
2. Rubber abrasive wheel
3. Cloth wheel for buffer paste
4. Safe storage of sharpened knives
5. Buffer paste (green for steel)
6. Safety goggles (bifocals to allow inspection of blade edge)
7. Bench for safe fixing of rotary bench grinder

**Fig. 8** Equipment needed for basic maintenance of knives

have recurrent or chronic lameness. Whilst not carried out commonly by veterinarians, many individual cows will be presented for routine preventive trimming without being identified as lame. Further history should then be obtained depending on the nature of the case, for example an adult dairy cow compared to a fattening beef calf. Once a history has been obtained, where at all possible, the animal should be observed in its environment and walking. This enables the clinician to identify an affected limb or limbs, as well as assess the severity of lameness. It also allows the clinician to look for signs of other systemic or neurological diseases that may be contributing to the condition. The animal should then be restrained as described previously.

Most cases of lameness in adult cattle will be related to the foot. Where no other clear cause of lameness has been identified on the initial history and examination, the foot of the affected limb should be examined. Whilst the use of hoof knives will often be needed to allow good observation, the authors prefer to refer to these as an 'inspection' rather than a 'trim'. Trimming may not be necessary, and even cause harm in some cases, if a large amount of claw horn is removed. If trimming is required, a full functional trim should be carried out before any further exploration.

### 4.2.2 Case Selection

On some occasions the case will be presented for examination by the owner, on other occasions it may be necessary for an advisor to select individuals for treatment directly, or to advise the owner on case selection. Any animal exhibiting signs of lameness should be examined by a suitably trained person at the earliest opportunity. Where resources or time are limited in can be tempting to focus attention on the most severely affected individual animals. It is essential that these severely, and often chronically, affected individuals receive treatment to relieve pain and discomfort. However, focusing attention on only those chronically affected individuals can lead to delayed treatment of milder new cases. Early detection and prompt effective treatment of new cases is a fundamental aspect of lameness control. Where at all possible, treatment of both groups should be implemented to avoid new cases progressing and enlarging the population of chronically lame cases.

### 4.2.3 Examination

An individual animal identified as lame should be examined using the equipment described previously at the earliest opportunity. Where there is any suspicion of a systemic cause of concurrent systemic disease a full clinical examination should be carried out. As the animal is examined, and before it is restrained, careful observation should allow the identification of the lame limb(s). Unless suffering from severe systemic disease, all lame cattle should have the relevant foot examined and all feet inspected. The foot of the affected limb should be cleaned and inspected. The limb should be observed and palpated for heat, pain and swelling, particularly in the area proximal to the coronary band. The skin above the foot and in the interdigital space should also be examined for signs of discolouration, infection or foreign bodies. Following this observation, the foot can be trimmed according to the steps described below to identify the presence of any claw horn lesions. It is important to make a distinction between an 'inspection' and a therapeutic foot trim. In the absence of lameness (for example, preventive (routine) inspection), or in the absence of any lesions, extensive removal of horn in an exploratory way is rarely warranted. Sometimes removing just a thin layer of sole horn is all that is required. Where deeper or unseen lesions are suspected, the use of hoof testers applied to the claw is preferable to exploration with a hoof knife. Where deeper lesions, such as early sole haemorrhage, may be present a standard approach to trimming as described below, should allow both identification of the lesion as well as some relief to the cow.

### 4.2.4 Approach to Therapeutic Hoof Trimming

A standard approach to foot trimming should allow the identification of claw horn lesions, whilst restoring the foot to normal balance in cases of overgrowth, as well as relieving pressure at the sole ulcer site, all whilst preserving weight-bearing surfaces. The same approach can be adopted for routine preventive trimming, inspection trims, and therapeutic trimming. Over trimming can be as greater threat to cow foot health than under trimming, therefore care should be taken and appropriate training received. Table 5 outlines an approach to trimming the dairy cow's foot based on the Five-Step Method for trimming hind feet (as adapted from Toussaint Raven (1989) and Mahendran and Bell (2015)).

**Table 5** The five-step method for trimming hind feet

| | Steps | Aim | Actions |
|---|---|---|---|
| Preventive | 1 (2 parts) | Transfer weight bearing to the toe region by steepening the foot angle to 48–55° where possible | Measure from where the claw capsule is firm at the coronary band and trim the medial claw to a length suitable for size, age and foot health |
| | | | Trim the sole flat so that it is minimum 5–7 mm thick at the toe, whilst sparing the heel |
| | 2 (2 parts) | Achieve 50:50 weight bearing between the claws | Match the length of the lateral claw to that of the medial claw (unless the medial claw is short) |
| | | | Trim the sole so that it is level at the toe and balanced at the heel with medial claw |
| | 3 | Reduce weight bearing at the sole ulcer site | Model out horn from the central third of the claws so that it slopes towards the interdigital space. A deeper, wider model is made in the lateral claw |
| Therapeutic | 4 | Reduce weight bearing on a diseased lateral claw | Create a height difference between the two claws by applying a block and/or trimming the back two-thirds of the lateral claw |
| | 5 | Remove loose horn and thin hard edges to assist in healing | Remove loose horn from the heels where there is heel horn erosion present. Remove loose horn and thin hard edges around lesions |

Adapted from Toussaint Raven (1989) and Mahendran and Bell (2015)

### 4.2.5 Treatment of Lesions

Once the foot has been examined and trimmed as described above and the lesion causing lameness identified, treatment is then required according to best practice. The evidence-base behind the current best practice treatment recommendations for the most common lesions causing lameness and summarised in Table 6.

Treatment of claw horn lesions commonly involves application of a hoof block to create sufficient rest for the affected claw (Step 4 of the Five-Step Method). A block should only be applied to a healthy claw with no pain or clinical lesion (hoof testers are useful here to check for underlying lesions). If both claws have lesions present which prevent the application of a hoof block, then it is recommended that the animal is housed in a straw yard near the milking parlour to aid recovery. The choice of block will depend upon the lesion present (type and severity) and the environment in which the animal is managed. For example, non-wearing blocks would be preferable in sand-housed herds and the use of thicker blocks for toe lesions. Blocks should be positioned to provide stability to the sound claw (i.e. placed back from the toe tip), and create optimal height difference (e.g. by banking the axial face of the block higher than the abaxial).

### 4.2.6 Aftercare

Follow-up is essential for all cases of lameness, especially those where the lesions were severe and/or a block was applied. Typically, this should take place between 3 and 6 weeks post-treatment, and should comprise a foot inspection to monitor healing, remove any abnormal hoof growth and check positioning and wear of the block. If the lesion has healed and no pain is present, then the block can be removed, and the animal is monitored through regular observation.

## 4.3 Necrotic Hoof Lesions, e.g. Wall Ulcers, Toe Necrosis and Joint Sepsis

Where treatment of claw horn lesions has not been prompt, there is the potential for secondary infection of exposed corium, especially where digital dermatitis is present on the farm. These lesions can become necrotic with infection and necrosis of deeper soft tissue structures and in some cases the pedal bone. Joint sepsis can also occur when there is ascending infection through the soft tissue structures of the foot. In these cases, lameness is often severe and requires veterinary intervention with extensive debridement

**Table 6** Summary of best practice treatment recommendations for the most common lesions causing lameness in dairy cattle

| Condition | Treatment |
|---|---|
| Digital dermatitis | Clean and debride lesions and administer a licensed topical antibiotic such as oxytetracycline spray or licensed biocide daily for at least 3 days. A systemic non-steroidal anti-inflammatory can be provided for analgesia |
| Interdigital phlegmon (foul) | The interdigital space needs careful inspection for foreign bodies. Clean, flush and debride any wound in the interdigital space. Systemic treatment with a licensed antibiotic is warranted, such as amoxicillin daily for 3 days. A systemic non-steroidal anti-inflammatory drug should be provided for analgesia |
| Sole haemorrhage/sole ulcer | A therapeutic trim should be undertaken including careful removal of underrun horn back to the margin of attachment and application of a hoof block to the unaffected claw. A systemic non-steroidal anti-inflammatory drug should be provided for analgesia. In early cases, the best outcomes are achieved with a combination of both a block and a non-steroidal anti-inflammatory drug for 3 days. Topical antimicrobial should be applied to the exposed corium |
| White line disease | A therapeutic trim should be undertaken including careful removal of underrun horn back to the margin of attachment and application of a hoof block to the unaffected claw. Sometimes debridement is best achieved using local anesthesia. The affected area should be allowed to drain. A systemic non-steroidal anti-inflammatory drug should be provided for analgesia. In early cases, the best outcomes are achieved with a combination of both a block and a non-steroidal anti-inflammatory drug for 3 days |

of the necrotic tissue under regional anesthesia or digit amputation in the case of join sepsis. These cases require a more intensive follow-up to manage pain, welfare of the animal and ensure that healing is progressing.

## 4.4 Approach to Investigating and Preventing Lameness in a Herd

Lameness can be challenging to overcome, particularly due to its multifactorial nature and the fact there is no one solution for all farms. Adopting a strategic approach to ensure that the main risks are identified and targeted is essential for success. There are a number of lameness management strategies adopted worldwide, but fundamentally they are all based on the following principles:

1. *Assessment of lameness prevalence and main lesions causing lameness*

   An initial assessment of prevalence can be determined through a mobility or locomotion score which also forms a benchmark against which the impact of interventions can be assessed. It can be particularly useful to assess lameness prevalence by parity and days in milk, and seasonality. Where hoof trimming or lameness treatment data is accurate and complete, this can be used to identify trends or patterns which may indicate specific risk periods, and the effectiveness of interventions (e.g. routine hoof trimming). It is important to highlight the impact of repeat cases of lameness on overall lameness incidence and therefore 'new' and 'repeat' cases should be clearly defined to avoid chronic cows distorting the data.
2. *Risk assessment to identify farm-specific risk factors*

   A full farm risk assessment encompassing all aspects of lameness management is important and there are a number of control programmes that provide resources to support this, such as the AHDB Healthy Feet Programme in the UK. Risk factors can broadly be divided up into key areas which have more or less of an impact on different types of lesion. For example, reducing infection pressure can be considered to be most important where digital dermatitis is the main lesion causing lameness whereas cow comfort would be the focus if sole bruising/ulcers were the predominant cause.

   Regardless of the underlying lesions causing lameness in a herd, a key component of all successful control programmes is Early Detection and Prompt Effective Treatment (see Sect. 2). Part of the risk assessment should involve assessing hoof trimming skill level on farms so that there is someone competent to both detect early cases of lameness and treat them appropriately. The gold standard in dairy herds of fortnightly mobility scoring also has additional benefits alongside early detection of lameness. Collation of mobility score data allows improved monitoring of lameness onset, alongside recovery rates and the identification of chronically lame cows (that can be identified either for more regular inspection or as voluntary culls). There are a number of available software programmes that can assist with data collection and analysis.
3. *Action plan*

   The creation of a farm-specific action plan underpins any control programme. Actions must be targeted, practical and achievable. The most successful action plans are those that are facilitated by the whole farm team and specific tasks attributed to specific team members to create ownership. Providing a timeframe during which the action should be implemented or completed can also help with motivation. Regularly reviewing and updating the action plan alongside assessing progress through mobility/locomotion scoring and record analysis ensures continued momentum and progress. This approach can enable risk factors for the predominant causes of lameness on individual farms to be assessed and prevention strategies specific to those risk factors to be discussed.
4. *The importance of the team approach*

   The key to successfully managing lameness involves the farmer, vet and hoof trimmer all working together collaboratively. Recent research has indicated that the importance of this working relationship is recognised, however, it is one that is not well established on many farms (Wynands et al. 2021). The foundation of an effective working relationship is a common understanding of each person's role and perspectives on lameness in the herd. Communication between all parties is also crucial to ensure that everyone is aware of changes being implemented and challenges that may arise.

Regular communication between the hoof trimmer and vet can also help identify any unexpected changes, for example a sudden rise in digital dermatitis, so that this can be followed up with the farmer and any necessary changes implemented at an early stage. Trimming and treatment protocols should be shared, with regular feedback through mobility/locomotion scores and record analysis used to monitor outcomes. By keeping all parties informed, it ensures that there is a single, common approach with a shared ownership of the lameness challenge on a farm resulting in a more successful outcome.

## 5 Proximal Limb Lameness (Originating Above the Foot)

### 5.1 Introduction

The most common causes of lameness in cattle originate in the foot. However, there are sporadic causes of lameness arising from disease further up the limbs although these are rarely an important cause of lameness at a herd level. It is also important to acknowledge that lameness originating from the foot may occur concurrently with other lesions. A good example of this is the presence of 'hock lesions' (Tarsal dermatitis, bursitis, or cellulitis) (Kester et al. 2014). Hock lesions, particularly when they are infected, may result in lameness. However, hock lesions result from abrasive, unyielding lying surfaces. This makes them a concurrent condition due to a common risk factor for claw horn disease. A full examination of the feet of a lame animal should be carried out, even in the presence of an obvious hock lesion.

Lameness may be a symptom of systemic disease, with or without other signs, or lameness may result from traumatic or acquired injury of the limbs. There are also congenital causes of lameness which may present in early life. Whilst an exhaustive list of every condition of cattle that might lead to an abnormality of gait is beyond the scope of this chapter, here we summarise some key conditions in which some degree of lameness is likely to be the major presenting sign. These can be grouped into conditions affecting the muscles and nerves, and those affecting the bones and joints.

### 5.2 Disorders Affecting the Nerves and Muscles

Traumatic damage to nerves and muscles is a common complication of parturition and recumbency in adult cattle (Huxley 2006). Whatever the initiating cause of recumbency the weight of cattle recumbent for long periods of time is likely to lead to compartment syndrome in the underlying muscles and compression damage to superficial peripheral nerves. Commonly affected nerves include the radial, peroneal and tibial nerves. Damage to the sciatic nerve following poor intramuscular injection technique, and damage to the obturator nerve after parturition may also occur. Muscle damage resulting from recumbency is likely to present as prolonged recumbency and an inability to stand after the initiating cause has been treated. The exact presentation of traumatic nerve damage will depend on the nerve affected. In all cases, effective treatment requires good nursing, and administration of non-steroidal anti-inflammatory medication. Euthanasia may be required in cases where response to treatment is poor.

Primary disease of the muscles may result from trace element, or mineral imbalances. A relatively common example is nutritional muscular dystrophy, or 'White Muscle Disease' in calves. This may occur congenitally in neonates where it primarily affects cardiac muscle, or in older calves where the skeletal muscle is affected. The condition is caused by a deficiency of Selenium, Vitamin E or both. Cases of White

Muscle Disease present with generalised muscle weakness due to muscle degeneration caused by a lack of antioxidants. Treatment is with rest and supplementary selenium.

There are other systemic conditions that may result in generalised muscle or nerve dysfunction. The toxins produced by the clostridial bacteria causing Tetanus and Botulism lead to spastic and flaccid paralysis, respectively (Rings 2004). In the early stages of disease, these cases may present with an abnormal gait due to generalised weakness in the case of botulism, or in the case of tetanus, limb stiffness. A variety of toxins may also act on the muscles or neurological system resulting in lameness. Central nervous system disease may also present with abnormal gait as the major sign. Conditions like congenital cerebellar hypoplasia caused by in utero infection with Bovine Viral Diarrhoea Virus will lead to ataxia and a wide-based stance. Central nervous system disease may also affect adults, conditions such as Bovine Spongiform Encephalopathy result in a range of central nervous signs, but a characteristic 'high-stepping' gait is reported.

A characteristic gait is also associated with Bovine Spastic Paresis (De Vlamynck et al. 2014). This developmental disease, most common in the Belgian Blue, leads to contraction of the gastrocnemius muscle when standing or walking. The result is that when attempting to walk there are repetitive contractions of the gastrocnemius muscle which pull the hind limb straight and caudally. The quadriceps muscles can also be affected in some animals. The aetiology is not understood, treatment involves partial tibial neurectomy when the gastrocnemius muscle is affected.

Tendon disease may result in abnormalities of gait. Contracted tendons in neonatal calves may result in persistent flexion of the forelimbs. This will sometimes respond to conservative treatment using splints, physiotherapy and corrective shoes. Tendon disease in adult cattle is more likely to result from trauma, with rupture of the gastrocnemius tendon and peroneus tertius muscle most reported (Anderson and Jean 1996).

## 5.3 Disorders Affecting the Joints and Skeleton

Traumatic damage to the skeleton and joints are the most obvious bony cause of lameness in cattle. Fractures can occur at any time, with the prognosis for recovery decreasing with age and with more proximal fractures (St. Jean and Anderson 2014). Fractures in the carpal region of the calf caused by poor application of obstetric ropes will usually respond to casting and pain relief. Long bone fractures in older, heavier animals can be more challenging to treat and euthanasia is invariably the best course of action. Identification of a fracture in a recumbent adult can be challenging due to the overlying muscle mass. Careful examination, including the use of a stethoscope to auscultate for crepitus can be helpful. A trans-rectal examination can also help identify damage to the pelvis. The overall demeanour of a recumbent cow can also be a useful indicator, with animals with fractures tending to show signs of pain and elevated heart rates in the absence of other obvious causes.

Traumatic luxation of the joints is also possible in cattle, and again diagnosis in adults is complicated by the degree of overlying muscle mass. Coxo-femoral luxation is the most frequently reported. A diagnosis can be made by comparing the length of legs, careful palpation of the joint externally and palpation of the inside of the pelvis per rectum to identify the displaced femoral head. When no damage to the acetabulum or pelvis is identified, both closed and open reduction techniques have been reported as being successful. Prompt treatment and age appears to be an important factor in success.

As well as luxation, joints can be a site of infection. Septic arthritis is particularly common in calves. Resulting in acute onset, severe lameness with swelling, it is easily confused with a fracture by the inexperienced clinician. The presence of pyrexia, concurrent omphalitis and examination of joint fluid may all aid a diagnosis. Infection is almost always haematogenous and so identifying a wound is rarely possible or necessary. Treatment is with systemic antibiotics and analgesia. Flushing of the joint may also be necessary in some cases. Prevention, if numerous cases occur, should focus on good colostrum management, navel treatment and hygiene of the calf environment.

Older cattle, often those kept as pets or high genetic merit pedigree animals, are also susceptible to degenerative joint disease as in many other species. Management of osteoarthritis in these animals can be challenging due to a lack of licensed medications, but where possible reducing exercise and providing analgesia may control clinical signs. Similarly, in younger animals, osteochondrosis is reported. Thought to be particularly common in feedlot cattle, osteochondrosis most commonly presents as a mild lameness with some swelling in the affected joint—most frequently the stifle or the hock.

Joint and bone disease may also result from in appropriate nutrition. Rickets is reported in cattle and thought to be associated with inadequate phosphorus intake or vitamin D deficiency. Rickets results in widening of the growth plates and can lead to swollen joints and pathological fracture in cattle. Pathological fractures have also been reported because of copper deficiency in cattle. Lameness may also result from a number of toxicities, for example fluorosis results in excessive exostosis formation and may lead to reduced joint function in cattle.

### 5.4 Lameness Originating Above the Foot: Summary

Whilst most cattle presenting with lameness will have foot lesions, lameness originating above the foot does occur and can result from systemic disease as well as trauma. Careful examination of the whole animal at the time of presentation, alongside examination of the feet, should enable a diagnosis to be made.

## 6 Conclusion

Lameness is a prevalent condition experienced by cattle worldwide with major implications from a health, welfare and economic perspective. This chapter has aimed to provide an evidence-based overview of the most common disorders affecting the claw and skin of the bovine foot including a summary of key risk factors and approaches to managing and treating lameness in beef and dairy cattle. Claw horn lesions remain a challenging cause of lameness due to knowledge gaps that still exist in relation to pathogenesis. The use of fortnightly mobility scoring offers an opportunity to reduce these lesions through early detection and prompt effective treatment and thereby reducing lameness recurrence and chronicity. Digital dermatitis is the most common infectious cause of lameness and should be considered a high priority on dairy farms. Appropriate management with footbathing regimes and treatment strategies will reduce transmission and chronicity respectively. Managing lameness in herds effectively involves a team approach and making effective use of available lameness data. As technologies increasingly become available that produce data relating to cow mobility (e.g. vision-based technologies for locomotion scoring) this may offer new opportunities to approach lameness management in more targeted and effective ways.

## References

AHDB (2020) Agriculture and Horticulture Development Board (AHDB) Technical Information; Mobility Scoring [cited 30 Apr 2020]. Available from: https://dairy.ahdb.org.uk/technical-information/animal-health-welfare/lameness/husbandry-prevention/mobility-scoring/#.XqrlhWhKjIU

Anderson DE, Jean GS (1996) Diagnosis and management of tendon disorders in cattle. Vet Clin North Am Food Anim Pract 12(1):85–116

Angell JW et al (2017) Survival of contagious ovine digital dermatitis (CODD)-associated treponemes on disposable gloves after handling CODD-affected feet. Vet Rec 181(4):89–89

Barker ZE (2007) Epidemiology of lameness in dairy cows. Department of Biological Sciences, University of Warwick

Barker ZE et al (2007) Management factors associated with impaired locomotion in dairy cows in England and Wales. J Dairy Sci 90(7):3270–3277

Barker ZE et al (2009) Risk factors for increased rates of sole ulcers, white line disease, and digital dermatitis in dairy cattle from twenty-seven farms in England and Wales. J Dairy Sci 92(5):1971–1978

Bäßler SC et al (2021) Association between alterations in plasma metabolome profiles and laminitis in intensively finished Holstein bulls in a randomized controlled study. Sci Rep 11(1):12735

Bell NJ (2006) The alleviation and prevention of lameness in dairy heifers. University of Bristol

Bell J (2017) Identifying infection reservoirs of digital dermatitis in dairy cattle, in Institute of Infection, Veterinary and Ecological studies. University of Liverpool

Bell NJ, Main DCJ (2011) Engaging farm staff in a lameness control programme. UK Vet Livestock 16(3):38

Bell NJ, Walker AN (2016) Technology: are we any closer to automated lameness detection or automated mobility scoring? Livestock 21(3):152–159

Bell NJ et al (2009) The development, implementation and testing of a lameness control programme based on HACCP principles and designed for heifers on dairy farms. Vet J 180(2):178–188

Bell NJ et al (2014) Disinfectant footbathing agents for the control of bovine digital dermatitis in dairy cattle. Livestock 19(1):6–13

Bergsten C, Telezhenko E, Ventorp M (2015) Influence of soft or hard floors before and after first calving on dairy heifer locomotion, claw and leg health. Animals 5(3):662–686

Berry SL et al (2012) Long-term observations on the dynamics of bovine digital dermatitis lesions on a California dairy after topical treatment with lincomycin HCl (Special issue: Bovine lameness). Vet J 193(3):654–658

Bicalho RC, Machado VS, Caixeta LS (2009) Lameness in dairy cattle: A debilitating disease or a disease of debilitated cattle? A cross-sectional study of lameness prevalence and thickness of the digital cushion. J Dairy Sci 92(7):3175–3184

Boakari YL et al (2022) Effect of lameness on breeding soundness examination results of beef bulls. Theriogenology 185:134–139

Browne N et al (2022) Cow- and herd-level risk factors for lameness in partly housed pasture-based dairy cows. J Dairy Sci 105(2):1418–1431

Burow E et al (2014) Track way distance and cover as risk factors for lameness in Danish dairy cows. Prev Vet Med 113(4):625–628

Capion N, Thamsborg SM, Enevoldsen C (2009) Prevalence and severity of foot lesions in Danish Holstein heifers through first lactation. Vet J 182(1):50–58

Chapinal N, de Passille AM, Weary DM, von Keyserlingk MAG, Rushen J (2009) Using gait score, walking speed, and lying behavior to detect hoof lesions in dairy cows. J Dairy Sci 92:4365–4374. https://doi.org/10.3168/jds.2009-2115

Chesterton N (1989) Environmental and behavioural factors affecting the prevalence of foot lameness in New Zealand dairy herds—a case-control study. N Z Vet J 37:135–142

Cook NB (2003) Prevalence of lameness among dairy cattle in Wisconsin as a function of housing type and stall surface. J Am Vet Med Assoc 223(9):1324–1328

Cook NB et al (2012) Observations on the design and use of footbaths for the control of infectious hoof disease in dairy cattle (Special issue: Bovine lameness). Vet J 193(3):669–673

Danscher AM et al (2009) Oligofructose overload induces lameness in cattle. J Dairy Sci 92(2):607–616

Davis-Unger J et al (2017) Economic impacts of lameness in feedlot cattle. Transl Anim Sci 1(4):467–479

De Vlamynck C, Pille F, Vlaminck L (2014) Bovine spastic paresis: Current knowledge and scientific voids. Vet J 202(2):229–235

Dopfer D et al (1997) Histological and bacteriological evaluation of digital dermatitis in cattle, with special reference to spirochaetes and Campylobacter faecalis. Vet Rec 140:620–623

Emery DL et al (1985) Cultural characteristics and virulence of strains of Fusobacterium necrophorum isolated from the feet of cattle and sheep. Aust Vet J 62(2):43–46

Ertze RA et al (2006) Field evaluation of prophylactic and therapeutic effects of a vaccine against (papillomatous) digital dermatitis in dairy cattle on two California dairies. Bovine Pract 40(2):76–82

Espejo LA, Endres MI, Salfer JA (2006) Prevalence of lameness in high-producing Holstein cows housed in freestall barns in Minnesota. J Dairy Sci 89:3052–3058

Evans NJ et al (2009) Treponema pedis sp. nov., a spirochaete isolated from bovine digital dermatitis lesions. Int J Syst Evol Microbiol 59(Pt 5):987–991

Fjeldaas T et al (2007) Claw and limb disorders in 12 Norwegian beef-cow herds. Acta Vet Scand 49(1):24

Flower FC, Weary DM (2006) Effect of hoof pathologies on subjective assessments of dairy cow gait. J Dairy Sci 89(1):139–146

Galindo F, Broom DM (2000) The relationships between social behaviour of dairy cows and the occurrence of lameness in three herds. Res Vet Sci 69(1):75–79

Gillespie AV et al (2020) Removal of bovine digital dermatitis-associated treponemes from hoof knives after foot-trimming: a disinfection field study. BMC Vet Res 16(1):330

Goonewardene LA, Hand RK (1995) A study of hoof cracks in grazing cattle—association of age, weight and fatness. Can J Anim Sci 75(1):25–29

Grandin T (2014) Improving welfare and reducing stress on animals at slaughter plants. In: Grandin T (ed) Livestock handling and transport. CAB International, Wallingford, UK, pp 421–450

Green LE et al (2002) The impact of clinical lameness on the milk yield of dairy cows. J Dairy Sci 85(9):2250–2256

Green LE et al (2014) Temporal associations between low body condition, lameness and milk yield in a UK dairy herd. Prev Vet Med 113(1):63–71

Griffiths BE, Grove White D, Oikonomou G (2018) A cross-sectional study into the prevalence of dairy cattle lameness and associated herd-level risk factors in England and Wales. Front Vet Sci 5:65–65

Groenevelt M et al (2014) Measuring the response to therapeutic foot trimming in dairy cows with fortnightly lameness scoring. Vet J 201(3):283–288

Gyan LA et al (2015) The lesions of toe tip necrosis in southern Alberta feedlot cattle provide insight into the pathogenesis of the disease. Can Vet J 56(11):1134–1139

Hird DW et al (1991) Expenditures for veterinary services and other costs of disease and disease prevention in 57 California beef herds in the National Animal Health Monitoring System (1988-1989). J Am Vet Med Assoc 198(4):554–558

Hirst WM et al (2002) A mixed-effects time-to-event analysis of the relationship between first-lactation lameness and subsequent lameness in dairy cows in the UK. Prev Vet Med 54(3):191–201

Huxley J (2006) Assessment and management of the recumbent cow. In Pract 28(4):176–184

Huxley JN (2013) Impact of lameness and claw lesions in cows on health and production. Livest Sci 156(1–3):64–70

ICAR (2020) ICAR Claw Health Atlas. ICAR Working Group on Functional Traits (ICAR WGFT) and International Claw Health Experts (eds) ICAR, Rome, Italy

Jacobs C et al (2019) Effect of footbath protocols for prevention and treatment of digital dermatitis in dairy cattle: A systematic review and network meta-analysis. Prev Vet Med 164:56–71

Jelinski M et al (2016) Epidemiology of toe tip necrosis syndrome (TTNS) of North American feedlot cattle. Can Vet J 57(8):829–834

Kester E, Holzhauer M, Frankena K (2014) A descriptive review of the prevalence and risk factors of hock lesions in dairy cows. Vet J 202(2):222–228

Knott L et al (2007) Effects of housing, parturition and diet change on the biochemistry and biomechanics of the support structures of the hoof of dairy heifers. Vet J 174(2):277–287

Kofler JA et al (2020) Appendix 1: Digital dermatitis stages (M-stages). In: ICAR Claw Health Atlas

Laven R (2006) Nutrition and lameness in dairy cattle—a different perspective? UK Vet Livestock 11(2):1–3

Leach KA et al (2012) The effects of early treatment for hindlimb lameness in dairy cows on four commercial UK farms (Special issue: Bovine lameness). Vet J 193(3):626–632

Li YP et al (2021) Oral oligofructose challenge reduces expression of glucose transport-1 and 5′-adenosine monophosphate-activated protein kinase in lamellar wall of Holstein heifer claw. Res Vet Sci 141:42–47

Machado VS et al (2010) The effect of claw horn disruption lesions and body condition score at dry-off on survivability, reproductive performance, and milk production in the subsequent lactation. J Dairy Sci 93(9):4071–4078

Mahendran S, Bell N (2015) Lameness in cattle 2. Managing claw health through appropriate trimming techniques. In Practice. https://doi.org/10.1136/inp.h2011

Main DCJ et al (2003) Effect of the RSPCA Freedom Food scheme on the welfare of dairy cattle. Vet Rec 153(8):227–231

Manson FJ, Leaver JD (1988) The influence of concentrate amount on locomotion and clinical lameness in dairy cattle. Anim Prod 47(2):185–190

Miguel-Pacheco GG et al (2017) Effect of claw horn lesion type and severity at the time of treatment on outcome of lameness in dairy cows. Vet J 225:16–22

Murray RD et al (1996) Epidemiology of lameness in dairy cattle: description and analysis of foot lesions. Vet Rec 138(24):586–591

Newcomer BW, Chamorro MF (2016) Distribution of lameness lesions in beef cattle: A retrospective analysis of 745 cases. Can Vet J 57(4):401–406

Newsome R et al (2016) Linking bone development on the caudal aspect of the distal phalanx with lameness during life. J Dairy Sci 99:4512–4525

Newsome R et al (2017a) A prospective cohort study of digital cushion and corium thickness. Part 2: does thinning of the digital cushion and corium lead to lameness and claw horn disruption lesions? J Dairy Sci 100(6):4759–4771

Newsome R et al (2017b) A prospective cohort study of the digital cushion and corium. Part 1: associations with body condition, lesion incidence and proximity to calving. J Dairy Sci 100(6):4745–4758

Norring M et al (2014) Short communication: Lameness impairs feeding behavior of dairy cows. J Dairy Sci 97(7):4317–4321

O'Callaghan KA et al (2003) Subjective and objective assessment of pain and discomfort due to lameness in dairy cattle. Anim Welfare 12(4):605–610

Oehm AW et al (2019) A systematic review and meta-analyses of risk factors associated with lameness in dairy cows. BMC Vet Res 15(1):346

Offer JE, McNulty D, Logue DN (2000) Observations of lameness, hoof conformation and development of lesions in dairy cattle over four lactations. Vet Rec 147(4):105–109

Paetsch CD (2014) Epidemiology of toe tip necrosis syndrome in Western Canadian Feedlot Cattle. Department of Large Animal Clinical Sciences, University of Saskatchewan

Pedersen S (2019) Digital dermatitis control in the dairy herd: incorporating the 'blitz' treatment approach. Livestock 24(3):130–135

Pedersen S, Wilson J (2021) Early detection and prompt effective treatment of lameness in dairy cattle. Livestock 26(3):115–121

Proudfoot KL, Weary DM, von Keyserlingk MA (2010) Behavior during transition differs for cows diagnosed with claw horn lesions in mid lactation. J Dairy Sci 93(9):3970–3978

Raven ET (1989) Cattle footcare and claw trimming. Farming Press Books

Randall LV et al (2015) Low body condition predisposes cattle to lameness: An 8-year study of one dairy herd. J Dairy Sci 98(6):3766–3777
Randall LV, Green MJ, Huxley JN (2018a) Use of statistical modelling to investigate the pathogenesis of claw horn disruption lesions in dairy cattle. Vet J 238:41–48
Randall LV et al (2018b) The contribution of previous lameness events and body condition score to the occurrence of lameness in dairy herds: A study of 2 herds. J Dairy Sci 101(2):1311–1324
Randall LV et al (2019) Lameness prevalence in a random sample of UK dairy herds. Vet Rec 184(11)
Ranjbar S et al (2016) Identifying risk factors associated with lameness in pasture-based dairy herds. J Dairy Sci 99(9):7495–7505
Reader JD et al (2011) Effect of mobility score on milk yield and activity in dairy cattle. J Dairy Sci 94(10):5045–5052
Rings DM (2004) Clostridial disease associated with neurologic signs: tetanus, botulism, and enterotoxemia. Vet Clin Food Anim Pract 20(2):379–391
Rodriguez-Lainz A et al (1998) Papillomatous digital dermatitis in Chilean dairies and evaluation of a screening method. Prev Vet Med 37(1):197–207
Shearer JK, Van Amstel SR, Brodersen BW (2012) Clinical diagnosis of foot and leg lameness in cattle (Special issue: Diagnostic pathology). Vet Clin North Am Food Anim Pract 28(3):535–556
Sprecher DJ, Hostetler DE, Kaneene JB (1997) A lameness scoring system that uses posture and gait to predict dairy cattle reproductive performance. Theriogenology 47(6):1179–1187
St. Jean G, Anderson DE (2014) Decision analysis for fracture management in cattle. Vet Clin Food Anim Pract 30(1):1–10
Stokes JE (2011) Investigating novel and existing methods of preventing, detecting and treating digital dermatitis in dairy cows. Department of Clinical Veterinary Science, University of Bristol
Sullivan LE et al (2013) Digital dermatitis in beef cattle. Vet Rec 173(23):582–582
Sullivan LE et al (2014) Presence of digital dermatitis treponemes on cattle and sheep hoof trimming equipment. Vet Rec 175(8):201–201
Tadich N, Flor E, Green L (2010) Associations between hoof lesions and locomotion score in 1098 unsound dairy cows. Vet J 184(1):60–65
Tarlton JF, Webster AJF (2002) A biochemical and biomechanical basis for the pathogenesis of claw horn lesions. In: Proceedings of the 12th international symposium on lameness in ruminants, Orlando, Florida, USA, 9th–13th January, pp 395–398
Tarlton JF et al (2002) Biomechanical and histopathological changes in the support structures of bovine hooves around the time of first calving. Vet J 163(2):196–204
Tavares NC et al (2019) Impact of high-energy diets on the rumen environment and digital cushion in confined cattle. Pesquisa Veterinária Brasileira 39:970–977
Terrel SP (2016) Feedlot lameness: Industry perceptions, locomotion scoring, and lameness morbidity and association of locomotion score and diagnosis with case outcome in beef cattle in Great Plains feedlots. Department of Diagnostic Medicine/Pathobiology, Kansas State University
Thoefner MB et al (2004) Acute bovine laminitis: a new induction model using alimentary oligofructose overload. J Dairy Sci 87(9):2932–2940
Thomas HJ et al (2015) Evaluation of treatments for claw horn lesions in dairy cows in a randomized controlled trial. J Dairy Sci 98(7):4477–4486
Thomas HJ et al (2016) Recovery of chronically lame dairy cows following treatment for claw horn lesions: a randomised controlled trial. Vet Rec 178(5):116–116
Tranter WP, Morris RS (1991) A case study of lameness in three dairy herds. N Z Vet J 39(3):88–96
Tunstall J (2020) Lameness in beef cattle: establishing a knowledge base. University of Liverpool
Tunstall J et al (2020) Reliability of a beef cattle locomotion scoring system for use in clinical practice. Vet Rec 187(8):319–319
Van Nuffel A et al (2015) Lameness detection in dairy cows: Part 2. Use of sensors to automatically register changes in locomotion or behavior. Animals 5(3):861–885
Vink WD (2006) Investigating the epidemiology of bovine digital dermatitis: causality, transmission and infection dynamics. University of Liverpool
von Keyserlingk MAG et al (2012) Benchmarking cow comfort on North American freestall dairies: lameness, leg injuries, lying time, facility design, and management for high-producing Holstein dairy cows. J Dairy Sci 95(12):7399–7408
Webster AJ (2001) Effects of housing and two forage diets on the development of claw horn lesions in dairy cows at first calving and in first lactation. Vet J 162(1):56–65
Whay HR, Shearer JK (2017) The impact of lameness on welfare of the dairy cow. Vet Clin North Am Food Anim Pract 33(2):153–164
Wilson J et al (2021) A history of lameness and low body condition score is associated with a reduced digital cushion volume, measured by magnetic resonance imaging, in dairy cattle. J Dairy Sci 104(6):7026–7038
Wilson JP et al (2022) Effects of routine treatment with nonsteroidal anti-inflammatory drugs at calving and when lame on the future probability of lameness and culling in dairy cows: A randomized controlled trial. J Dairy Sci 105(7):6041–6054
Winckler B, Margerison JK (2012) Mechanical properties of the bovine claw horn during lactation. J Dairy Sci 95:1714–1728
Wynands EM et al (2021) Dairy farmer, hoof trimmer, and veterinarian perceptions of barriers and roles in lameness management. J Dairy Sci 104(11):11889–11903
Yeruham I, Perl S (1998) Clinical aspects of an outbreak of papillomatous digital dermatitis in a dairy cattle herd. J S Afr Vet Assoc 69(3):112–115
Zinicola M et al (2015) Altered microbiomes in bovine digital dermatitis lesions, and the gut as a pathogen reservoir. PLoS One 10(3):e0120504

# Lameness in Pigs

Mari Heinonen, Liesbet Pluym, Dominiek Maes,
Kristin Olstad, and Mate Zoric

**Abstract**

Lameness is a frequently encountered clinical manifestation of different locomotory disorders affecting all age groups of modern pig herds. It results in decreased animal welfare and health and economic losses to pork producers (Anil et al. J Am Vet Med Assoc 226:956–961, 2005; Heinonen et al. Livest Sci 156:2–9, 2013). Lameness is a multifactorial condition with infectious, nutritional, housing, managemental, and/or genetic components. It is of great importance to detect the lame pigs, diagnose the cause of lameness, treat them efficiently, and move them to a sick pen whenever needed. In addition, culling should be considered without delay to prevent unnecessary suffering of the animals (Heinonen et al. Vet Rec 159:383–387, 2006). The final aim is to prevent lameness efficiently. This chapter elaborates on current knowledge about lameness in pigs, with emphasis on the etiology, diagnosis, treatment, and prevention.

**Keywords**

Lameness · Pig · Risk factors · Infectious · Noninfectious · Clinical examination · Scoring · Treatment · Prevention

M. Heinonen (✉)
Department of Production Animal Medicine, Faculty of Veterinary Medicine, University of Helsinki, Saarentaus, Finland

Research Centre for Animal Welfare, University of Helsinki, Helsinki, Finland
e-mail: mari.heinonen@helsinki.fi

L. Pluym
Belpork vzw, Brussels, Belgium
e-mail: liesbet.pluym@belpork.be

D. Maes
Unit of Porcine Health Management, Faculty of Veterinary Medicine, Ghent University, Merelbeke, Belgium
e-mail: dominiek.maes@ugent.be

K. Olstad
Department of Companion Animal Clinical Sciences, Faculty of Veterinary Medicine, Norwegian University of Life Sciences, Ås, Norway
e-mail: kristin.olstad@nmbu.no

M. Zoric
Department of Animal Health and Antimicrobial Strategies, Swedish Veterinary Agency, Uppsala, Sweden
e-mail: mate.zoric@sva.se

## 1 Anatomy and Physiology of Pigs: Short Introduction

Lameness can be defined as a deviation from normal gait or posture while a normal degree of consciousness is displayed (Wells 1984; Straw et al. 1999; Beusker 2007). The severity varies greatly, and lameness can manifest as an asymmetrical limb movement, an altered or shortened stride, a reduced ability or inability to bear weight, and even total recumbency (Straw et al. 1999; Maas



J. J. Gross (ed.), *Production Diseases in Farm Animals*,
https://doi.org/10.1007/978-3-031-51788-4_18

2009). Lameness may develop when the functional integration of the nervous system, bones, muscles, cartilages, tendons, joints, and ligaments is disturbed. Disorders can be located in the limbs as well as in the trunk of the animal and may include painful lesions as well as mechanical defects (Wells 1984; Smith 1988; Beusker 2007). A proper understanding of lameness requires knowledge of the gait of healthy pigs and insight into the anatomy and physiology of the locomotory system.

The walking gait of a pig is characterized by a symmetrical four-beat gait with an alternating two- and three-limb support phase and even strides. Pigs walk mainly in a diagonal sequence in which the hind foot touches down slightly later than the contralateral fore foot. Fore limbs have a longer stance time and a lower swing/stance time ratio than hind limbs. Yet, stride elevation is higher in hind limbs (Thorup et al. 2007; von Wachenfelt et al. 2008). While standing, pigs show a merely straight back and stand squarely on all four legs. The body weight of a standing pig is not evenly distributed between and within feet: feet at the fore limbs carry the highest load, i.e., 57% of the total body weight (Pluym et al. 2013a). Within feet, lateral claws carry 78% of the body weight mostly by the heel bulb, followed by the junction between the heel bulb and the abaxial wall. Within the medial claw, the tip of the toe is the most loaded area (Webb 1984).

The pig's body contains about 216 bones (Fig. 1). Bones are not static but highly metabolically active during skeletal growth as well as during bone modeling and remodeling, the latter occurring throughout life.

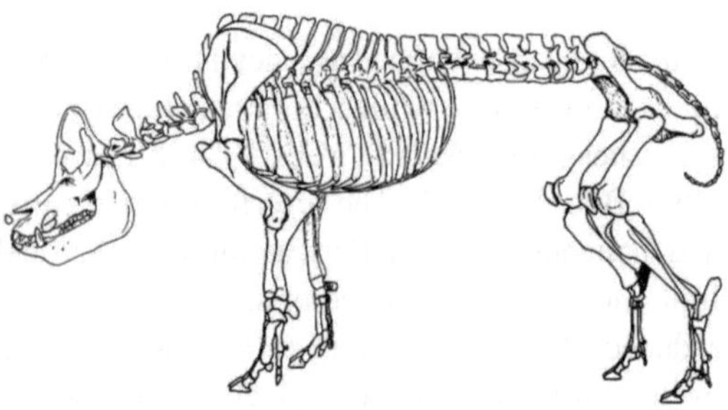

**Fig. 1** Lateral view on the skeleton of a pig (source: Prof. Dr. Pieter Cornillie, Morphology Lab, Ghent University)

The bones of skeletally immature pigs contain some anatomical components in addition to the ones found in bones of skeletally mature pigs, mainly the specialized form of cartilage known as growth cartilage and its blood supply (Carlson et al. 1991; Ytrehus et al. 2004b, 2007). The bones of skeletally immature pigs are also engaged in an additional physiological process of growth by endochondral ossification, defined as the growth of a cartilage scaffold that is progressively replaced by bone. The fact that skeletally immature pigs differ slightly in anatomy and physiology from mature pigs means that they are susceptible to different diseases and may respond differently to disease agents. This is reflected by the variation in etiology and pathogenesis of locomotory diseases, as described in the following chapters.

Specialized growth cartilage is present at two sites on either end of long bones: the sub-articular, epiphyseal growth cartilage and the metaphyseal growth plate, or physis (Carlson et al. 1991). The physis is predominantly responsible for growth in terms of increase in bone length, whereas the epiphyseal growth cartilage is responsible for shaping the articulating bone ends. Both physeal (Hill et al. 1985) and epiphyseal growth cartilage (Carlson et al. 1991; Ytrehus et al. 2004c, 2007) contain a temporary blood supply that runs within so-called cartilage canals. In epiphyseal growth cartilage, the blood supply is organized as anatomical end arteries (Ytrehus et al. 2004c, 2007), whereas in the physis, organization varies with age (Hill et al. 1985; Wormstrand et al. 2021). In both cases, the cartilage canals are regularly spaced and separated by an extra-cellular matrix, rendering chondrocytes vulnerable in the case of failure because there is limited potential for collateral supply (Carlson et al. 1991). As the pig approaches skeletal maturity, the blood supply regresses through physiological chondrification and incorporation into the advancing ossification front (Ytrehus et al. 2004c). Studies in horses have shown that regression follows an age-dependent pattern for different joints and regions (Carlson et al. 1991; Olstad et al. 2008b). The proliferation of chondrocytes slows down to such an extent that all growth cartilage eventually becomes replaced by bone, at which point the growth plates close and skeletal maturity is reached. Closure time depends on the plates and bones considered. One source (Zeder et al. 2015) states that all the major growth plates are closed by 60 months/5 years.

The process of bone modeling occurs in response to physiological influences and mechanical forces. The structure of bones is gradually adjusted to the forces they encounter through bone formation and resorption. During bone modeling, formation and resorption are not tightly coupled. The homeostatic process of bone remodeling throughout life aims to renew bone in order to maintain bone quality and mineral homeostasis. Bone remodeling is a cyclical process with four components: activation of (pre) osteoclasts, resorption of bone, transition from bone resorption to formation, and formation of bone mediated by osteoblasts (Clarke 2008).

Joints create the connection between bones, transmit the mechanical load between the bones they connect, and allow movement. They are the most loaded parts of the locomotory system (Done et al. 2012). A joint consists of articular cartilage that covers the bones it connects. Joint cartilage includes a layer of articular cartilage, beneath which, in growing animals, is the subarticular (epiphyseal) growth cartilage. The joint capsule is a membrane that surrounds the entire circumference of the end of the bones involved in the joint. The synovial membrane that secretes the synovial fluid covers the inner surface of the joint capsule. Synovial fluid acts as a lubricant and normally appears as clear to straw-colored viscous fluid. Joints have been suggested to be highly susceptible to infection after local trauma or in the case of septicemia (Done et al. 2012).

The pig is an even-toed ungulate with four digits on every limb, two principal ones (III, medial and IV, lateral) and two accessory ones (dewclaws, II and V). In contrast to cattle, the accessory digits have a full complement of phalanges, and they are attached to the distal row of the carpal and tarsal bones (Sack 1982). Dewclaws have been regarded as non-wearing accessory digits, which have no load-carrying capacity, or only some on soft underground (Sack 1982). Papadopoulos et al. (2021), however, suggested a positive association between the dewclaws and the principal claws: dewclaw length is approximately 2 cm shorter than the dorsal length of the main claws. They defined the dorsal claw length as the linear distance along the dorsal wall from just below the coronary band to the weight-bearing border (Fig. 2).

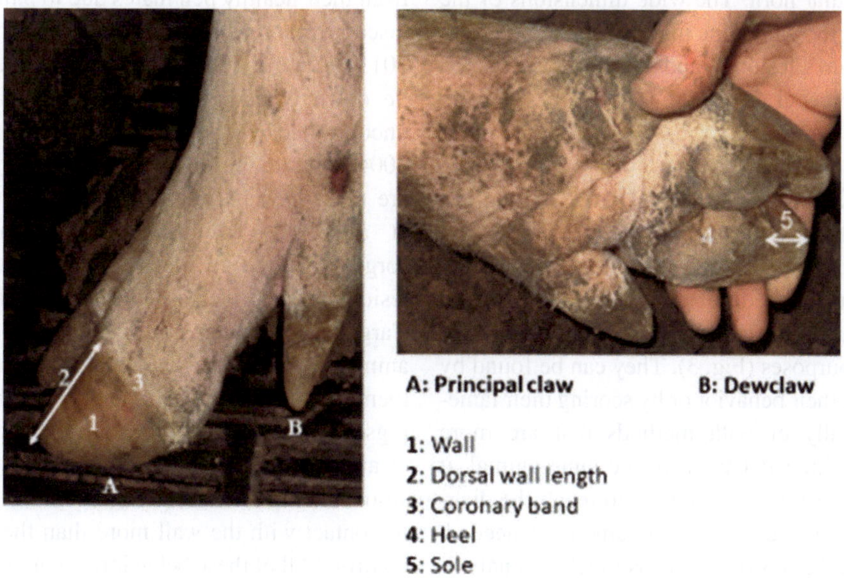


**Fig. 2** Lateral and palmar view on the principal claws (A) and dewclaws (B) of the pig with different distinguishable regions (photo: L. Pluym)

The pig's claw (Fig. 2) consists of a horn capsule covering the skeleton (distal phalanx, the distal part of the middle phalanx, and the distal sesamoid bone) and soft tissue at the end of the limb (Sack 1982; Mülling 2000). Different regions can be distinguished: a hard wall, hard sole, soft heel bulb (with cushioning function), and a non-pigmented junction between the wall and sole, the so-called white line (Ossent 2010). The hoof capsule is in fact skin that is extensively modified. The epidermis is a strongly cornified epithelium, and the dermis underneath is modified to dermal papillae or lamellae (only at the level of the wall) (Mülling 2000). Claw horn production is the result of proliferation, keratinization, and cornification (programmed cell death) of the epidermal cells at the level of the claw (i.e., keratinocytes) (Tomlinson et al. 2004). In areas with dermal lamellae, the epidermis forms horn lamellae, while tubular horn is formed at the level of the dermal papillae. A sufficient supply of nutrients, oxygen, and hormones to the metabolic active keratinocytes within the avascular epidermis essentially depends on the dense vascular system within the dermis and the process of diffusion within the epidermis (Mülling 2000; Tomlinson et al. 2004). Claw horn strength is directly related to the dimensions of the tubules in the tubular horn. The wide dimensions of the tubules in the white line make this site more susceptible to lesions and bacterial invasion (Tomlinson et al. 2004).

## 2 Clinical Examination and Diagnosis of Lame Pigs

Lame animals need to be examined for the need of treatments, for herd health evaluations, or for research purposes (Fig. 3). They can be found by observing their behavior or by scoring their lameness visually or with methods that are more objective. After detection of the lame animal, it should be examined clinically to reach the diagnosis and enable the right treatment if needed. Lameness diagnostics may also include analysis of diagnostic samples.

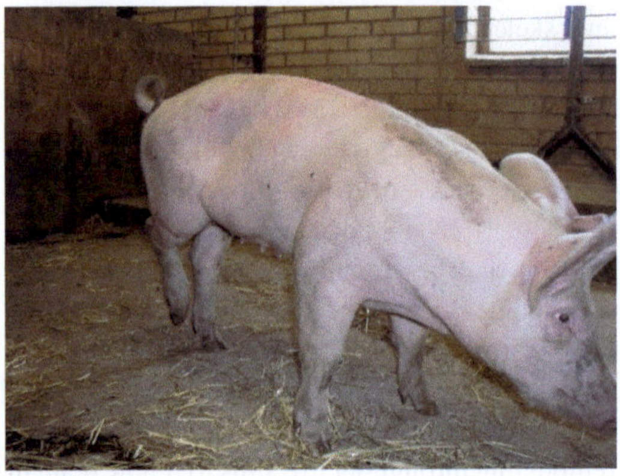

**Fig. 3** Lameness in a pig (photo: Carl-Johan Ehlorsson)

### 2.1 Pig Behavior

Foot and skin lesions can contribute to lameness either by causing pain from the injury itself or by acting as an entrance for infections that may spread to joints through bacteremia and thereby induce arthritis and pain (Gardner and Hird 1994; Zoric et al. 2004, 2016). The pig has the sufficient cognitive and emotional capacity to experience negative affective states such as pain and not just show reflexive avoidance behavior in response to tissue damage (Herskin and Di Giminiani 2018). Lame pigs behave differently from their healthy pen mates due to pain, general discomfort, or sickness behavior (Heinonen et al. 2013). Lame sows have a shorter latency time to lie down (Calderón Díaz et al. 2014) and an uncontrolled lying-down behavior (Bonde et al. 2004), and they use longer time to stand up and are more hesitant during standing up (Grégoire et al. 2013) compared to non-lame sows. Jørgensen (2000) reported that sows with claw lesions showed increased stepping of hind legs. Cargill and Buddle (2001) suggested that lame animals can be best detected by first chasing them up standing and then examining the first 20 pigs, which sit down. In a study by Ala-Kurikka et al. (2017), sows with non-resolved lameness stood and moved less, lied down more, and were in contact with the wall more than their healthy controls. All of these behavioral changes are used in finding possible lame animals for individual clinical examination. Pigs have a short neck and stiff movements, and that is why their lameness evaluation is challenging (Main et al. 2000). Mild lameness is often undetected, even though it would be very important to detect and treat the lame sows early (Alawneh et al. 2012). Possibly, in the future, remote monitoring technology applications will be developed to solve this problem.

## 2.2 Lameness Scoring in Pigs

The degree of lameness is usually evaluated visually by observing the gait of the pigs. Several different visual scoring systems have been used (reviewed by Nalon et al. 2013), but they have not always been found reliable because of high subjectivity and variable skills of the scorer (Main et al. 2000; D'Eath 2012). The scores usually include four to six levels (Main et al. 2000, Geverink et al. 2006; Karlen et al. 2007; Kilbride et al. 2009a, b; Mustonen et al. 2011; D'Eath 2012; Grégoire et al. 2013), but even ten different (Dewey et al. 1993) or only three levels have been used (Welfare Quality 2009). Visual scoring requires substantial training (Petersen et al. 2004; Abell et al. 2014). However, it is often the only available, useful, quick, and affordable way to quantify lameness (Nalon et al. 2013). If pigs are moved to walk on solid floors for lameness scoring, as they should, the ease and cheapness are lost.

More objective methods have been developed to assess the pig lameness (Table 1), but they have been used so far mostly in research. Their use typically takes a lot of time due to complex installations, motion captures and/or initial data processing (Nalon et al. 2013; Stavrakakis et al. 2015). Further validation is still needed for most automatic lameness systems before they can be used in pig herds (Ala-Kurikka 2020). The Sow Stance Information System based on force plates in a sow feeder (Pluym et al. 2013a, b) has also shown to be usable in a commercial setting (Briene et al. 2021).

**Table 1** Selected objective methods in evaluation of lameness in pigs

| Method | Parameters | Age group | Reference |
|---|---|---|---|
| Force plates | Weight distribution of each foot | Sows | Sun et al. (2011), Pluym et al. (2013b), Karriker et al. (2013), Mohling et al. (2014), Conte et al. (2015), Briene et al. (2021) |
| Pressure algometry | Mechanical nociceptive threshold in kg of force | Sows | Tapper et al. (2013) |
| Pressure mat | Weight distribution, asymmetry | Sows | Meijer et al. (2014) |
| Gait analysis walkway | Stride time, stride length, maximum pressure, stance time | Sows | Mohling et al. (2014), Karriker et al. (2013) |
| Kinematics, motion capture | Vertical head displacement, asymmetric stride, phase timing, walking speed, stance time, swing time, foot height, stride length, angle and amplitude of joint | Sows, pre-breeding gilts, growing pigs | Stavrakakis et al. (2014, 2015), Grégoire et al. (2013), Conte et al. (2015) |
| Positioning and acceleration, ear sensors in a group of sows | Path length walked, number of areas visited, variance of accelerations | Sows | Traulsen et al. (2016) |
| Accelerometers | Time standing/lying, stepping behavior (number of steps/min) | Sows | Grégoire et al. (2013), Scheel et al. (2017) |
| Footprint (image) analysis | Stride length, contralateral distance, ipsilateral distance, diagonal distance | Sows | Grégoire et al. (2013) |
| Infrared thermography | Thermal images, temperature | Sows | Amezcua et al. (2014) |

## 2.3 Clinical Examination of Lame Pigs

It is not easy to examine lame pigs: their large joints are covered with muscles, they do not like to be touched and they are not used to be tied and walked with humans in a controlled manner. Clinicians examining other animal species, pet pigs, or pigs for research can use diagnostic methods that are not available in swine practice, such as radiography, ultrasound imaging, and nerve blocks. In addition, the value of an individual pig is usually so low that clinical samples such as joint fluid for bacterial culture or blood for hematological analysis are seldom used in confirming the diagnosis of a single pig. Diagnosis is mostly based on visual inspection, palpation, and measuring of rectal temperature.

For proper diagnosis, a thorough clinical examination and collection of anamnestic information are needed, for example, age, morbidity, mortality, environment, length of lameness period, response to earlier treatments, feeding, appetite, vaccinations, and herd history. Depending on the case, all age categories of pigs on the farm may need to be examined, not just those with the most severe signs. The clinical examination includes first the inspection of the pig in detail: posture, symmetry, appearance, body condition, cleanliness, and skin wounds. High rectal temperature may reveal infections. Through examination and palpation of the legs, hoofs and joints are used in finding swellings, heat, pain, or any abnormalities, and in helping to localize the affected area. It is important to investigate if the locomotory disorder is acute or chronic and if more than one leg is affected.

Observation of the claws can be performed while the sow is lying down (e.g., in the farrowing crate). If the sow is standing, her lame legs should be lifted up and claws examined, which is difficult in adult animals. It is easy to lift a piglet (Fig. 4), but two to four persons are needed to lift the feet of adult pigs for proper examination, see Figs. 5 and 6. One person holds the sow standing still with a snout snare. One person is needed to handle a front foot, but the sow needs help with balance when her hind foot is lifted. Two people hold a broomstick under the hind leg in the groin area while they lift the leg, and then a third person can inspect the hind claws of an adult pig. In addition, technical devices exist to facilitate claw scoring such as the FeetFirst© chute from Zinpro Corporation and the Mobile Claw Scoring Device as developed by van Riet et al. (2019). The latter is based on sliding the device under a standing sow and taking video images for further analysis. Future will show if this device will be used in practice.

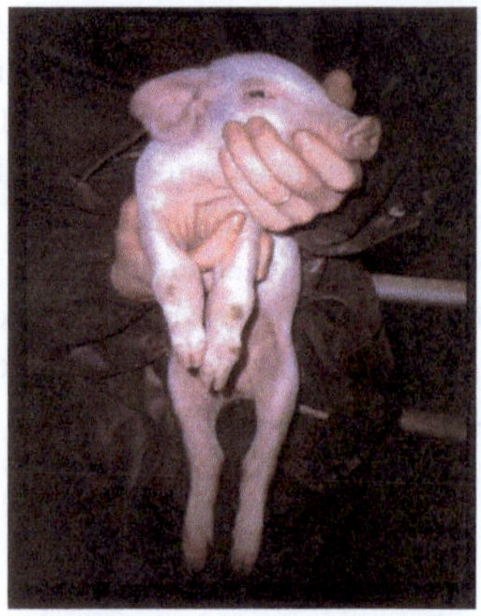

**Fig. 4** Joint swelling and lameness are the most prominent clinical signs of infectious arthritis in a piglet (photo: Magdalena Jacobson)

## 2.4 Assessment of Pig Claws

This part concentrates mostly on the procedures performed in sows, but the knowledge can also be utilized for other age groups. Assessment of the claws is an important part of the clinical examination of lame sows to reach a proper diagnosis. Although not all claw injuries are associated with lameness, they may indicate an underlying problem and may lead to lameness, reduced reproductive performance, and early culling. Therefore, claw lesion assessment can also be done to monitor the occurrence of claw lesions on the farm.

Diagnosis can be impaired by the fact that the severity of the lesions is not always commensurate with the impact on the sow's welfare. Internal damage can be more severe than expected based on the visual assessment (Kilbride et al. 2009a,

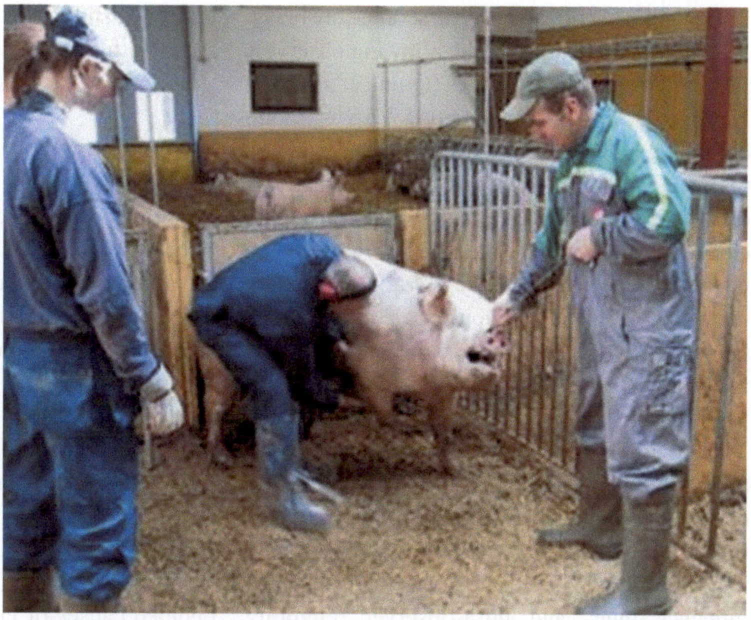

**Fig. 5** Two persons are needed in a thorough examination of front feet (photo: Sylvia Persson)

**Fig. 6** When the hind foot of an adult pig is examined, one person is needed to hold the animal and two to lift the leg (photo: Sylvia Persson)

b). On the contrary, severe claw lesions in appearance may be painful only for a short time or not at all, while they persist for a long time since it takes time for new horn to grow (Calderón Díaz et al. 2013, 2014). An ideal method would allow the detection of claw injuries before they become relevant, i.e., when they may become a risk of developing lameness, reduced feed intake, and impaired reproductive performance. Unfortunately, such a detection method does not exist at present, and this threshold has not been defined for claw lesions in sows.

Mapping of claw injuries and scoring of their severity is mainly done visually with either a numerical rating scale (NRS) or a visual analog scale (VAS). Various NRS use different numbers of ordered categories and VAS score scores claw lesions on a continuous scale, allowing observers to record more subtle changes (Quinn et al. 2007; Nalon et al. 2014). In practice, a VAS is a long horizontal line anchored by the minimum and maximum score at each end. The severity of a claw lesion type is determined by measuring the distance from 0 mm. A modified scale in which the thresholds of a numerical rating scale are indicated on the continuous scale, a so-called tagged VAS (tVAS), which can help observers make more consistent choices (Nalon et al. 2014; van Riet et al. 2018).

The scorer, the scale used, the scoring conditions, and claw cleanliness influence the outcome of scoring. Training and experience of the scorer, the type and severity of the claw lesion, the scoring condition, and the scoring scale influence inter- and intra-observer reliability (D'Eath 2012; Nalon et al. 2014; van Riet et al. 2020). Van Riet et al. (2020) compared the sow chute and the Mobile Claw Scoring Device and concluded that both conditions cannot be used interchangeably, and the choice will depend on the possibilities on-farm. Claw cleanliness clearly affects the outcome of claw lesion scoring. If claws are soiled with manure, the visibility of lesions is reduced (van Riet et al. 2020). This may lead to the detection of the injuries at a severe stage when the animals are already lame and treatment is unrewarding. Detailed claw scoring in live sows is difficult to be performed accurately and objectively and therefore some researchers have used post-mortem examination of culled sows in their claw lesion prevalence studies (see later in this chapter, epidemiology of claw lesions).

Claw conformation can be assessed by means of different claw dimensions using a digital caliper (van Riet et al. 2018). Jørgensen (2000) reported an association between claw lesions and buck-kneed forelegs, upright pasterns, steep hock joints, turn out of hind legs, standing under position on hind legs, stiff movements, swaying hindquarters, goose-stepping hind legs, tendency to slip, and lameness. However, most of these clinical signs have been associated with other leg disorders as well (such as osteochondrosis and arthrosis) and therefore have no specific diagnostic value for claw lesion diagnosis.

## 2.5 Sampling of Lame Pigs

Many pathological conditions can be diagnosed only in necropsy (Nalon et al. 2013). A conclusive diagnosis of lameness can be made through necropsy, including histopathological and microbiological examination of representative samples of lame, untreated pigs. If a pig herd suffers from therapy failure, pigs should be sent for necropsy to establish the etiological diagnosis, including antimicrobial resistance investigation. In practice, this is easy for piglets because of their small size. Some studies have collected culled sows for pathological-anatomical diagnosis and found detailed causes of lameness (Dewey et al. 1993; Kirk et al. 2005; Sanz et al. 2007; Engblom et al. 2008; Ala-Kurikka et al. 2019). These studies do not include information about the clinical signs of the culled animals, but they describe the common pathological-anatomical diagnosis. However, one should bear in mind that the study animals are not always randomly selected and therefore the results may be partly misleading. Actually, the combination of clinical and post-mortem examinations is a very good method for diagnosing the cause of lameness in a pig (Dewey et al. 1993; Ala-Kurikka 2020). However, the combination is not often used in scientific literature, possibly due to practical constraints. Dewey et al. (1993) recommended that herd lameness problems can be diagnosed by performing post-mortem examinations on a representative sample of sows culled for lameness during 6–12 months. However, this approach is time-consuming and expensive, and autolysis should be prevented. The microbial cause of lameness in piglets varies, and laboratory diagnostic methods are therefore valuable (Zoric et al. 2008, 2009).

If arthritis is suspected, aseptically obtained joint fluid can be used in the detection of pathogens, antimicrobial resistance, and antimicrobial resistance genes (Oh et al. 2020). This can be done on the living animal under anesthesia (Sack 1982) but, for the welfare of the animal and for practical reasons, it is usually done during necropsy (Gomes Neto et al. 2012). In both cases, a sterile technique is critical. If performed during necropsy, it has been recommended to obtain the sample from the medial side of the joint as this area is less covered by muscle tissue. After the skin is removed, an incision with a sterile scalpel blade is made through the joint capsule following the outline of the bones. By applying the leverage, the joint is exposed and samples can be collected. In addition, sterilization of the tissue before incision by the use of, e.g., a small propane flame has been suggested to reduce bacterial contamination (Gomes Neto et al. 2012).

In the case of arthrosis, usually necropsy is needed until a specific diagnosis can be made. If the synovial fluid is not turbid or fetid and surface cartilage lesions (e.g., fibrillation, tears, erosion, ulcers) are present, the condition is likely to be osteochondrosis/arthrosis (Rowles 2001).

Reference values for synovial fluid (Canning et al. 2018) and blood hematological analyses (e.g., Klem et al. 2010) have been established, but they are used only in research or for valuable individuals. Acute phase protein concentrations (Heinonen et al. 2006) have been found to be higher in lame than in non-lame animals, but this method is too unspecific to be used in detailed lameness diagnosis. In growing and finishing phases, some infectious agents are associated with lameness occurrence. Collection of oral fluids with soft rope has been used in the detection of *Mycoplasma hyorhinis* and *Mycoplasma hyosynoviae* connected with lameness in a group of growing pigs (Pillman et al. 2019). The same pathogens were detected also in the tonsillae of the piglets later developing lameness (Roos et al. 2019).

## 3 Lameness in Suckling Piglets and Weaned Pigs

Lameness in piglets and growing pigs cause illthrift and contribute to losses in the form of dead pigs, decreased growth, and increased use of antimicrobials and labor. Arthritis is the most common cause for individual antimicrobial treatment in piglets (Zoric et al. 2003; Zoric 2008).

### 3.1 Predisposing Factors, Prevalence, and Effects of Lameness in Piglets and Weaned Pigs

Lameness in suckling piglets causes significant problems in farrowing units. Abrasions, wounds, and necrosis in the skin or on claws (Figs. 7 and 8) and accessory digits are very common in newborn piglets (Dewey 2006). Risk factors relate to housing conditions, especially floor type, nutrition, and genetics (Baxter et al. 2011; Le et al. 2015). In addition to that, the passive immunity obtained from colostrum is very important to suckling piglets (Oliviero et al. 2019). Skin lesions in piglets are presumably mainly a result of contact with the floor, especially during suckling (Mouttotou and Green 1999; Westin and Algers 2006; Zoric et al. 2008, 2009). The total time spent suckling at 3–6 days of age and the total time spent lying near the sow at 3 and 4 days of age are positively correlated to the development of skin abrasions (Mouttotou and Green 1999).

Zoric et al. (2003) studied piglets up to 9 weeks of age in a research station. Out of 9411 piglets born alive, 9.8% were treated for lame-

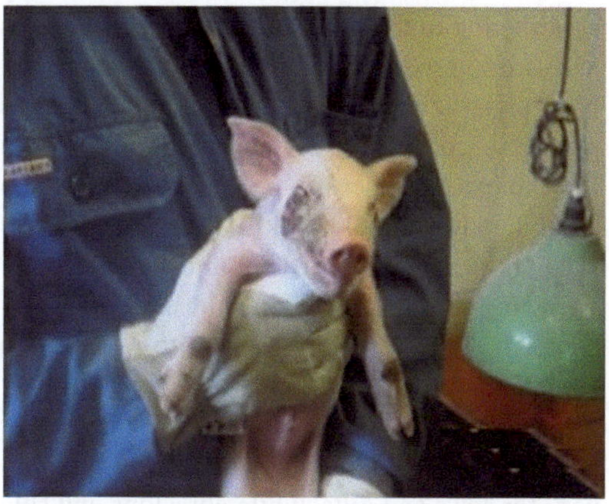

**Fig. 7** Abrasions, wounds, and necrosis in the skin of a suckling piglet (photo: Mate Zoric)

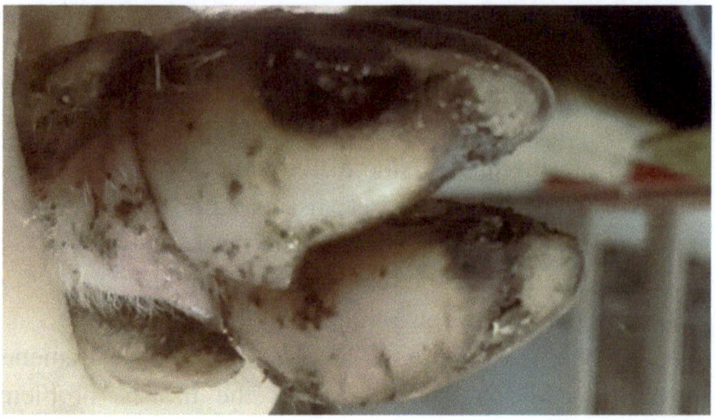

**Fig. 8** Abrasions, wounds, and necrosis on the claws of a piglet (photo: Marie Sjölund)

**Table 2** Mean weight of piglets treated for lameness during suckling compared to non-lame piglets (Zoric et al. 2017). All weights are presented as live weights, except for the weight at slaughter, which represents the carcass weight

| Age | Weight of non-lame piglets, kg | Weight of piglets treated for arthritis during suckling, kg | *p*-value |
|---|---|---|---|
| Birth | 1.5 | 1.5 | n.s. |
| 5 weeks | 11.5 | 10.2 | <0.001 |
| 9 weeks | 28.9 | 26.5 | <0.001 |
| At slaughter | 83.1 | 80.8 | <0.001 |

ness. Around 75% of the lameness cases were observed in piglets less than 3 weeks of age, and the incidence risk for lameness decreased from 2.7% during the first week of life to 0.3% after weaning. Litters with 12 or more piglets had a higher incidence of lameness, but no difference between the sexes was seen.

The housing conditions in farrowing houses present a dilemma because the needs of the sow differ from those of the piglets (Roach 1981). Concrete floor can be rough and abrasive, thereby contributing to the removal of horn from hooves and skin , especially from the limbs in newly born piglets, which may cause acute lameness (Barnett et al. 2001). Piglets prefer floors with low abrasive properties (Clark 1983; Nilsson 1988; Lewis et al. 2005). Still, floor abrasiveness is a major cause of superficial injury in piglets in farrowing pens (Clark 1985; Lium et al. 2002). In contrast, the sow requires access to an abrasive surface in order to prevent hoof overgrowth. Attempts to provide a non-abrasive surface for piglets have resulted in floors that were too slippery for sows and gilts, while abrasions and sole bruising in piglets decreased (Barnett et al. 2001). In some other studies, doubling the amount of chopped straw prevented the development of skin lesions and sole bruising to some extent (Zoric et al. 2009, 2016).

Zoric et al. (2017) analyzed arthritis and growth of pigs with a documented pedigree from birth to slaughter in a farm with a floor adapted especially to piglets. Altogether, 9569 piglets were alive at 9 weeks of age and 208 (2.2%) of these had been treated for arthritis. Of these 208 pigs, three (1.4%) were treated again during the fattening period, compared to 0.3% of those pigs remaining healthy. At slaughter, 1.0% of the lame pigs and 0.3% of the healthy ones were recorded to have arthritis in meat inspection. The mean birth weight (1.5 kg) did not differ between the groups, but the weight at 5 weeks, the weight at 9 weeks, and the carcass weight at slaughter were 1.3, 2.4 and 2.3 kg lower, respectively, in pigs having been treated for lameness during suckling ($p < 0.001$) (Table 2). Piglets treated for arthritis grew slower than healthy pigs. In addition, the results indicated a greater risk of becoming lame during the fattening period if treated for lameness during suckling (OR = 4.7). However, the results also suggested that the negative impact of arthritis can be reduced by proper floor surfaces since the incidence of pre-weaning arthritis was lower than the national average (2% vs. 10%, OR = 0.2).

Lameness in suckling piglets is often associated with skin lesions, often bilateral and most commonly observed as abrasions over the carpal joints. Such lesions are present already on day three; they increase in magnitude until day ten and thereafter decline. Hocks, face, abdomen, and tails are affected in a similar way, but at lower magnitudes, and these lesions usually vanish on day 17 (Zoric et al. 2008, 2009, 2010). Foot and skin lesions can contribute to lameness in two ways, either due to pain induced by the injury itself or by acting as an entrance for infections that spread to joints through bacteremia and thereby induce arthritis and pain (Gardner and Hird 1994; Zoric et al. 2004, 2016).

Infectious arthritis affecting single joints has been most commonly observed in piglets less than 3 weeks old. Zoric and Wallgren (2014) analyzed the localization of affected joints in lame piglets until 5 weeks of age. In total, 6.1% out of 6780 liveborn piglets were diagnosed with lameness and 91% of these diagnoses took place during the first 3 weeks of life. At 5 and 9 weeks of age, healthy piglets were 1.3 and 1.8 kg heavier than piglets recorded as having been lame. In lame animals, one, two, and three clinically affected joints were observed in 91.5%, 7.5%, and 1% of piglets, respectively. The distribution of affected joints is shown in Figs. 9 and 10. In intensive pig production, the weight of the weaned piglet has a significant influence on lifetime performance (Declerck et al. 2016; Zoric et al. 2017).

During the first week of life, the piglets spend most of their time lying in the nest, and it is obvious that their immediate environment during this period plays a primary role in the appearance and development of leg injuries. Although most commonly observed in back hoof joints, followed by elbows, front hoof joints, and hocks, arthritis was fairly evenly distributed between joints (Zoric and Wallgren 2014). This suggests a septicemic spread of the pathogens associated with lameness.

## 3.2 Infectious Causes of Lameness in Piglets and Weaned Pigs

Piglets can be infected with different pathogens, either originating from the sow and/or the environment. They may enter the bloodstream via skin wounds, the navel, or the tonsils (Gottschalk and Segura 2019). In routine diagnosis, septicemic polyarthritis can rarely be distinguished from joint infections caused by scrub and hoof injuries (Zoric 2008; Zoric et al. 2009). In piglets, usually hemolytic streptococci dominate, but also oppor-

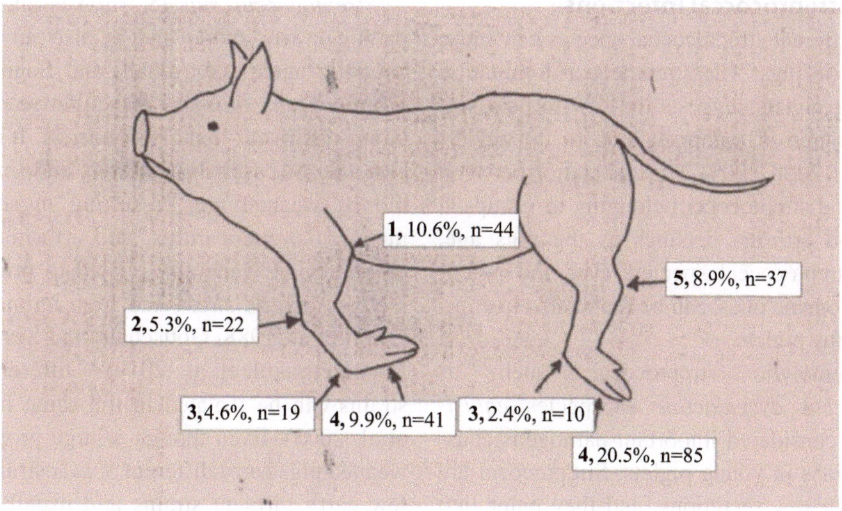


**Fig. 9** The prevalence of joints affected with infectious arthritis on the left side of the piglets (Zoric and Wallgren 2014)

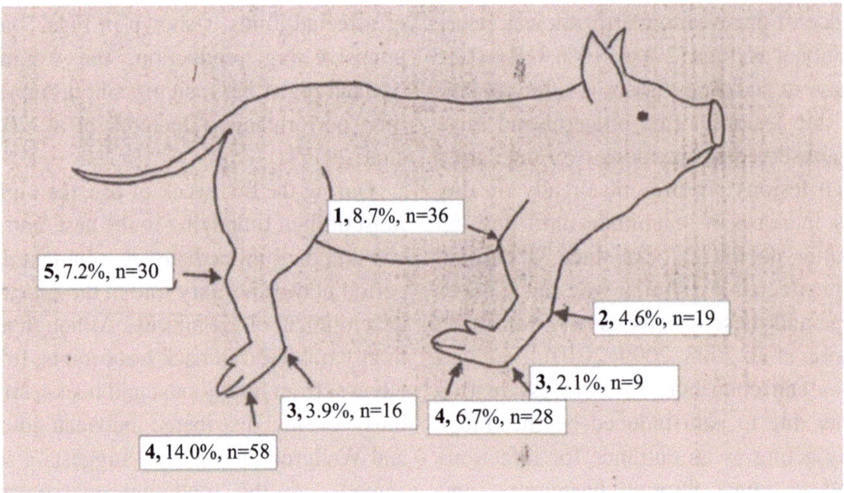


**Fig. 10** The prevalence of joints affected with infectious arthritis on the right side of the piglets (Zoric and Wallgren 2014)

tunistic bacteria such as staphylococci, *Trueperella pyogenes*, and *Escherichia coli* are frequently diagnosed. Among weaned piglets, arthritis is more typically associated with microbes such as *Glaesserella parasuis*, *Streptococcus suis*, *Erysipelothrix rhusiopathiae*, *Mycoplasma hyorhinis*, or/and *Mycoplasma hyosynoviae* (Dewey 2006; Done et al. 2012). However, the latter pathogen mainly affects growing pigs, usually during the fattening pig period.

### 3.2.1 Streptococcal Infections

Several different streptococcal species can cause infections in pigs. The streptococci domination of sucking piglets suggests the sow to be a significant source of pathogen load to the piglets (Zoric et al. 2008, 2009). The association between arthritis and streptococci belonging to groups C and L and arthritis declines as the pigs age. Several streptococcal species, even the potentially pathogenic ones, can be found also in clinically healthy piglets.

Beta-hemolytic streptococci, such as *Streptococcus dysgalactiae* subspecies *equisimilis*, are considered important pathogens causing infections in young piglets. Streptococci are common in sow secretions, and they enter into the piglets through their umbilicus, tonsils, or skin wounds. Usually pigs between 1 and 3 weeks of age are affected. The organisms settle in joints and cause arthritis or polyarthritis and affected animals become lame and their joints become swollen. In addition, they may suffer from inappetence, fever, and apathy. At necropsy, in early cases, one can see swollen and hyperemic synovial membranes, periarticular edema, and turbid synovial fluid. In later stages, necrotic articular cartilage may also become visible, along with fibrosis and abscesses in periarticular tissue (Gottschalk and Segura 2019).

*Streptococcus suis* (*S. suis*) is an important pathogen worldwide and is also an emerging zoonotic agent (Gottschalk and Segura 2019). *Streptococcus suis* is a facultative anaerobic Gram-positive α-hemolytic coccus. It can cause systemic diseases in newborns and, more commonly, weaned pigs, resulting in septicemia, meningitis, endocarditis, and arthritis. Disease outbreaks are generally related to predisposing factors such as stress and high animal density (Touil et al. 1988, Gottschalk and Segura 2019; Yongkiettrakul et al. 2019). Different *S. suis* strains can be isolated in the same herd (Vela et al. 2003). Even though a large proportion of weaned pigs carry different *S. suis* strains, only a few carry virulent strains and usually a single strain causes most diseases in a herd (Marois et al. 2007). The disease is confirmed by culturing and identifying the pathogen in tissues of the affected animals. Serotyping *S. suis* isolates helps indicate if the disease is caused by the most common serotype or if the serotype-causing diseases in a herd has changed. Serotype 2 is considered to be the predominant and most virulent serotype in many countries, but also other serotypes occur commonly (Gottschalk and Segura 2019). The pathogen has a coat of capsular polysaccharides to protect it against the immune system of the host, but this virulence factor is not the only one (Segura et al. 2017). It is difficult to control through vaccination, even though both autogenous and commercial vaccines have been used with inconsistent results (Segura 2015).

### 3.2.2 *Glaesserella parasuis*

Glässer's disease is caused by *Glaesserella parasuis* (*G. parasuis*), which is considered to belong also to the normal respiratory microbiota of pigs. The bacterium is found in pig herds all over the world, even in herds with good health, but there is usually a balance between infection pressure and immunity. The piglets are colonized soon after birth from their dams (Aragon et al. 2019). Even though several different strains can be found in a herd at the same time, usually one strain is associated with an outbreak of the disease (Rafiee et al. 2000). The immunity of the piglets, other diseases present in the herd, stress conditions (e.g., transport), the virulence of the strain involved, and genetic resistance have an effect on the severity of the disease (Aragon et al. 2019). This pathogen can act as a primary or secondary pathogen. The disease is seen mainly in pigs of 4–8 weeks old. The bacterium can cause peracute and acute lameness, depression, high fever, abdominal breathing, warm swollen joints, unwillingness to stand or move, tremors, paralysis, and sudden death (Aragon et al. 2019). Lesions include fibrinous and fibrinopurulent polyserositis, polyarthritis, and meningitis. *G. parasuis* can be detected in the lesions of affected tissues by IHC, in situ hybridization, or PCR (Aragon et al. 2019). Pigs recovering from the acute phase may become neglected and develop lameness. Vaccines prevent Glässer's disease and can be used until a balance between infection and immunity occurs or in non-immune herds when selling animals to herds where *G. parasuis* occurs (Aragon et al. 2019; Evira 2018).

### 3.2.3 Mycoplasma Infections

*Mycoplasma hyorhinis* (*M. hyorhinis*) is found in swine populations worldwide and also as a commensal in the upper respiratory tract of pigs. It usually affects pigs younger than 10 weeks of age (Heinritzi 2006; Pieters and Maes 2019) and is often associated with the decline of passive immunity (Roos and Spear 1973) and the purchase of breeding animals (Done et al. 2012). Older pigs in the herd or dams transmit the pathogen to the piglets. The process during the infection or the virulence factors are not well known. Symptoms may include moderately hot and swollen joints, polyserositis, abdominal breathing, and unwillingness to move. After about 2 weeks, clinical signs subside in most animals. In an experimental study, mortality was not found to be high (Gimenez-Lirola et al. 2019). In cases where *M. hyorhinis* causes arthritis, swollen joints and lameness may persist for months. At necropsy, the synovia is swollen and hyperemic. Infection in the joint can lead to adhesions, capsular fibrosis, and articular erosions (Pieters and Maes 2019). *M. hyorhinis* arthritis is usually confirmed by culture and/or PCR of swabs collected from joint cavities (Pieters and Maes 2019). Diagnosis of the pathogen can also be reached by detecting the characteristic microscopic lesions in synovial and serosal membranes (mono- and polymorphonuclear cells) (Gomes Neto et al. 2012) and by using in situ hybridization. Measures to control the infections caused by *M. hyorhinis* are difficult to implement because there is a lack of understanding about the epidemiology and strain virulence of this highly prevalent pathogen. Autogenous vaccines against *M. hyorhinis* may be of help (Pieters and Maes 2019).

The other mycoplasma causing arthritis, *Mycoplasma hyosynoviae*, is more common in the finishing stage, and it is discussed in detail later in the chapter.

#### 3.2.4 *Erysipelothrix rhusiopathiae*

Infections with *E. rhusiopathiae* in pigs can occur in an acute form characterized by high fever, rhomboid skin lesions, and possible arthritis, and in a chronic form that is mainly associated with chronic arthritis (Opriessnig and Coutinho 2019). Because this pathogen causes more lameness problems in finishing pigs, it is discussed in that part of the chapter.

### 3.3 Treatment of Lame Piglets and Weaned Pigs

Most lameness treatments are given parenterally to piglets because lameness usually affects single animals. Only a few studies exist about pharmacokinetics of different medicines in pigs. Due attention must be paid to the correct dosage of parenterally administered medicines in newborn piglets: overdoses and concentrated preparations must be avoided (Evira 2018). In terms of treatment time, the variables that affect the length of treatment have not been defined. Responses to different types of antimicrobial drug infections vary, and clinical experience is important in assessing response to treatment. For acute infections, improvement should be obvious within 2 days. If no response is seen at that time, both diagnosis and treatment should be reconsidered. Treatment of acute infections should be continued for at least 2 days after clinical resolution. In case of chronic infections, euthanasia of the piglet should be considered. Treatment with antimicrobials should be started as early as possible. Else, the infection may have developed into a chronic state and treatment might be ineffective.

For suckling piglets and weaned pigs, the first treatment choice is usually benzylpenicillin because many pathogens causing arthritis in young pigs are penicillin sensitive (Zoric et al. 2009; Sjölund et al. 2018). Tylosin or amoxicillin are possible alternatives. Trimethoprim-sulfa is effective against *Escherichia coli* isolated in pigs with arthritis (Zoric et al. 2009). Antimicrobial sensitivity testing is recommended, especially in the case of failure in therapy. Beta-lactam antimicrobials are not effective against mycoplasmas, and in these cases, recommended antimicrobials include tylosin, tiamulin, and lincomycin (Pieters and Maes 2019). High doses of antimicrobials should be administered parenterally as soon as clinical signs appear, and all pigs in the affected group should be treated (Desrosiers et al. 1986).

Prompt treatment with antimicrobials is required to achieve a positive treatment effect in lame piglets due to arthritis (Dewey 2006). Treatment of inflammation and pain management is important because of pig welfare. Different non-steroidal anti-inflammatory drugs are available for use in pigs, but specific indications are lacking (Schoos et al. 2019). Zoric et al. (2012) compared the clinical effects of a concurrent treatment of lame piglets with ketoprofen and with penicillin and ketoprofen. The occurrence of lameness in 6780 piglets was registered from birth until 5 weeks of age, and the clinical efficacy of the treatments was assessed daily. Lameness was diagnosed in 6.1% of the piglets, mostly (91%) during the first 3 weeks of life. Both treatment strategies improved the clinical status from day to day (Table 3), and the clinical response did not differ between the two treatment groups. Non-lame piglets were 1.3 and 2.5 kg heavier than piglets treated for lameness at 5 and 9 weeks of age, respectively. There was no difference between the body weights of the piglets. However, the NSAID most likely improved the welfare of the lame piglets. It is notable that decreasing of the pain due to arthritis did not lead to injurious use of affected joints. The study also showed that about half of the lame piglets treated with penicillin only or with penicillin and ketoprofen were still clearly lame 5 days after initiating the treatment and that the lame piglets had reduced weight gain. Possibly, the treatments were not sufficient or started too late.

### 3.4 Prevention of Lameness in Piglets and Weaned Pigs

Lameness in suckling piglets and weaned pigs can be prevented most of all by improving environmental conditions. In addition, good passive immunity obtained by ingesting a sufficient

**Table 3** Number of pigs with daily clinical scores of lame piglets treated for lameness with penicillin for 5 days ($n = 208$) or with penicillin for 5 days and ketoprofen with 3 days ($n = 207$)

| | Treated with penicillin | | | Treated with penicillin and ketoprofen | | |
|---|---|---|---|---|---|---|
| | Clinical score 0–1 | 2–3 | Difference between days, $p$-value | Clinical score 0–1 | 2–3 | Difference between days, $p$-value |
| Day 1 | 0 | 52–156 | – | 0 | 49–158 | – |
| Day 2 | 0–7 | 79–114 | D1/D2 ** | 2–8 | 71–120 | D1/D2 ** |
| Day 3 | 2–32 | 85–77 | D2/D3 *** | 4–36 | 85–77 | D2/D3 *** |
| Day 4 | 12–51 | 73–53 | D3/D4 *** | 15–53 | 76–49 | D3/D4 *** |
| Day 5 | 34–58 | 66–36 | D4/D5 ** | 38–63 | 57–40 | D4/D5 ** |

Lameness and general condition were scored clinically as 0 (good), 1 (almost good), 2 (not good), or 3 (not good at all) (Zoric et al. 2012). ** $p < 0.01$, *** $p < 0.001$. The treatment group did not have an effect on the development of the score

amount of colostrum is of utmost importance for the piglets (Oliviero et al. 2019). Adding peat on the floor for the first 3 weeks after farrowing and taping the front knees of the piglets may be of help. Proper maintenance of the floors is a relevant measure. A successful concept to reduce abrasions, sole bruisings, and lameness is to increase the amount of chopped straw during the first days of life and again after weaning. In addition, during the first week of life, the lying nest of the piglets should be warm enough (32–33 °C). A proper hygiene program is important, and it requires all-in, all-out production in farrowing and weaning rooms. In continuous production, rooms cannot be properly cleaned, disinfected, and dried. It is also important to avoid regrouping of pigs to reduce fighting and aggression to minimize stress and injuries (Coutellier et al. 2007).

## 4 Lameness in Finishing Pigs and Breeding Pigs

Lameness is a frequent health and welfare problem in finishing pigs. It also causes great economic losses. Lameness has accounted for 11–15% of the recorded clinical signs in finishing pigs, being the second most common clinical sign in Danish finishing herds (Christensen et al. 1994; Petersen et al. 2008). Similarly, arthritis, together with locomotory problems, was the second most frequent disease treated in Finnish finishing pigs (Heinonen et al. 2001), and the main reasons for antimicrobial treatments of finishing pigs were musculoskeletal diseases (Stygar et al. 2020). However, lameness is just a clinical sign, and its etiology may differ considerably.

Lameness in finishing and breeding age pigs can be caused by infectious agents (bacteria or viruses) or by noninfectious causes (trauma, fractures, muscle disorders, osteochondrosis, and degenerative joint diseases). In many cases of lameness in single pigs, the exact diagnosis remains unknown because individual pigs have low monetary value and the expenses for microbial investigations or other diagnostic procedures are too high. Therefore, usually clinical diagnosis alone is used when the choice of therapy and prophylaxis is determined. Only when several finishing pigs become affected, necropsies or additional diagnostic investigations are performed to reach the diagnosis.

### 4.1 Predisposing Factors for Lameness in Finishing Pigs

Plenty of studies have reported risk factors for lameness of sows, with emphasis on the association between claw health and either nutrition or flooring (Pluym et al. 2013b). However, much less has been published about risk factors of lameness in finishing pigs. The lameness prevalences vary in different housing systems, but the associations are not always easy to interpret because of the many factors having an effect at the same time. Some of the studies about finishing pigs focused on joint lesions in the slaughterhouse or on farm, whereas others focused on lameness in live pigs, which makes it difficult to make proper comparisons between the studies.

The most important predisposing factor for lameness in finishing pigs and especially for claw lesions is poor housing. Housing conditions can be very variable, and therefore the needed sample size of different housing conditions becomes large and difficult to study. In addition, different characteristics of the housing likely have an interacting effect, which further complicates the analysis and the interpretation of the results.

Usually, soft bedding has been considered to be beneficial for the pig welfare. However, the research results have not always been consistent. For example, de Koning et al. (2014) did not find an age-dependent effect of floor type on the prevalence of osteochondrosis in gilts reared on different floor surfaces. However, they found that gilts kept on a floor with wood shavings had greater odds of being affected with severe osteochondrosis lesions compared with those kept on the partly slatted floor (twisted metal slats and epoxy-coated concrete). In other studies, fully slatted floors were found to be a risk factor for lameness in finishing pigs compared to deep bedding (Kilbride et al. 2009b; Scott et al. 2006) or to solid concrete floors with sparse bedding (Kilbride et al. 2009b).

The floor surface should not be too slippery or too rough, and the in-between spaces of the slats should be suitable for the age of the pigs (Mouttotou et al. 1999; EU legislative 2008; Kilbride et al. 2009a, b). The front legs carry a higher load and have been observed to slip more both forward and backward when the pig walks (Applegate et al. 1988). This may have an effect on the development of lesions in front feet. Wet floors may have an effect on slipping and softening the sole, increasing the risk of erosions, especially when the surface is rough. Soft floors prevent the hoof from wearing but increase the risk of claw overgrowth, which may cause lameness. However, finishing pigs are not usually affected by claw overgrowth because they are slaughtered at a young age.

In a study investigating claws of finishing pigs in the slaughterhouse, pigs kept on bedded, partially slatted, or totally slatted floors had different kinds of claw lesions (Mouttotou et al. 1999). No ideal floor surface was found for pigs because the development of claw lesions in finishing pigs is complex and multifactorial. Another group found that finishing pigs raised on straw bedding had more severe toe erosions, whereas the ones raised on fully slatted floors had more severe sole and heel erosions (Scott et al. 2006). Heinonen et al. (2007) investigated factors associated with arthritis detected at slaughter and recorded information about the housing conditions, management, feeding and hygiene practices, health and disease prevention in the farrowing, weaning, and finishing units of integrated pig herds. The herds with high clinical sign scores, including all disease signs of lactating sows, piglets less than 1 week old, and finishing pigs in the herds, had greater chance of being a case herd with high arthritis prevalence in slaughterhouses. The case herds were generally larger, aimed at higher efficiency and tended to have housing and management practices such as the use of slatted floors and handling of manure as slurry. In addition, the case herds were more likely to wash and disinfect the pens, and they invested more in hospital pens.

The results have not always been easy to interpret. For example, Krieter et al. (2004) found that growing-finishing pigs kept on a bedded multi-surface system with outdoor run had a higher risk of lameness than pigs kept on fully slatted pens. The authors discussed that the outdoor area may have caused the lameness. However, this nicely brings out the fact that different studies are quite difficult to compare because of the very different housing conditions and the major differences in the study design.

Greater leg lesion scores (wounds or swellings on legs) and lameness scores were found in finishing pigs housed in large groups (108 pigs) compared with the ones in small groups (18 pigs) in a study of Street and Gonyou (2008). They also found an interaction between group size and space allowance for lameness. Pigs were more susceptible to lameness in large groups with restricted space allowance. In large groups, pigs might fight more, especially if their area to move is restricted. Their unconfirmed hypothesis was, however, that pigs in large groups had more space for moving around and therefore were more likely to get their claws caught in slats or run into other pigs and get hurt.

## 4.2 Infectious Causes of Lameness of Finishing Pigs

The most common infectious lameness diagnosis in finishing pigs is arthritis caused by different pathogens, the two most important ones for this age group being *E. rhusiopathiae* and *M. hyosynoviae*. Other possible pathogens are presented here shortly and in more detail in the part discussing lameness in younger pigs.

### 4.2.1 *Erysipelotrix rhusiopathiae*

*E. rhusiopathiae* is ubiquitous, and the domestic pig is an important reservoir (Opriessnig and Coutinho 2019). The presence of neuraminidase and hyaluronidase has been suggested to be the virulence factors of this organism (Wang et al. 2010). The virulence of *E. rhusiopathiae* strains is known to vary considerably. *E. rhusiopathiae* can be found in many animal species and in water, soil, and slurry. About 30–50% of clinically healthy pigs harbor the organism in their tonsils and other lymphoid tissues. Carriers and acutely infected pigs excrete the organism in the urine, feces, saliva, and nasal mucus (Opriessnig and Coutinho 2019). The organism can remain viable for 12 days in direct sunlight and for months in carcasses (Reboli and Farrar 1989).

In acute infections, cutaneous lesions and septicemia are the most prominent findings; however, the infection may also pass subclinically. Later, chronic proliferative lesions may develop in the joints of some affected pigs, resulting in lameness about 3 weeks after the initial outbreak (Opriessnig and Coutinho 2019). Affected animals show varying degrees of lameness, joint stiffness, and inappetence, and their hocks, stifles, or carpal joints can be enlarged. The joint swellings are firm and not warm. Morbidity and mortality vary in herds due to the different immune statuses of the pigs and the occurrence of other infections in the herd (Opriessnig and Coutinho 2019).

Chronic lesions of erysipelas include chronic arthritis involving several joints. The synovial membrane proliferates, and the capsule is often hyperemic. In addition, the proliferation and erosion of the articular cartilage leading to fibrosis, ankyloses, and spondylitis can be seen (Opriessnig and Coutinho 2019). Isolation of the organism from tissues or fluids from affected animals gives a definite laboratory diagnosis. However, in chronic cases, it may be difficult to isolate the organism (Friis et al. 1992). A variety of tests can be used to diagnose *E. rhusiopathiae*: direct isolation, enrichment, fluorescent antibody assay, immunohistochemistry, PCR, and serology (Opriessnig and Coutinho 2019).

Protection against *E. rhusiopathiae* is achieved through regular vaccinations of all sows in a herd. Because of short-lasting immunity, vaccinations need to be repeated every 6 months. Vaccinations are effective in the prevention of the acute form, but they may not be as effective in the prevention of the chronic form. Vaccination of pigs to prevent erysipelas in fattening pigs is rare.

### 4.2.2 *Mycoplasma hyosynoviae*

*M. hyosynoviae* causes non-purulent arthritis in finishing pigs and young breeding animals older than 10 weeks. Outbreaks with morbidity rates between 10 and 50% have been reported (Kobisch and Friis 1996). Suckling piglets are usually protected against *M. hyosynoviae* arthritis by passive immunity obtained from the sow (Lauritsen et al. 2017), and most of the pigs are infected after weaning or during the finishing stage (Hagedorn-Olsen et al. 1999b; Geudeke et al. 2016). Okholm Nielsen et al. (2005) studied the epidemiology of the disease and found that the most critical period for the spread of *M. hyosynoviae* infection in the herds is the age between 4 and 12 weeks. In addition, they concluded that other herd or host factors than virulence or antibody levels are more likely to be the triggers of clinical arthritis due to *M. hyosynoviae*. Affected pigs probably remain infected (Okholm Nielsen et al. 2005). The pathogen resides in the tonsils, nasal cavity, and airways of the pigs and occasionally invades systemically (Friis et al. 1991; Hagedorn-Olsen et al. 1999b; Okholm Nielsen 2000). The clinical outcome at the pen level has not been associated with the detection of *M. hyosynoviae* in oral fluids or tonsils of the finishing pigs (Pillman et al. 2019). The organism reaches the synovia of the joints through the blood stream and causes edema in the synovial membrane and synovial villi, and the amount of synovial fluid increases (Hagedorn-Olsen et al. 1999a). A bacteremia of about 1 week precedes the joint infection (Hagedorn-Olsen et al. 1999b). Acute signs are usually observed 7–10 days after an experimental inoculation (Gomes Neto et al. 2012). According to Gomes Neto et al. (2016), this took 1–15 days, depending on the route of inoculation. The diseased pigs may become severely lame, showing a dog-sitting position for several hours, but also only mild lameness can be seen (Okholm Nielsen 2000). The stifle and elbow joints are primarily affected. The pigs may have a slight rise in body temperature. The affected animal usually recovers, and the condition does not become chronic (Okholm Nielsen 2000).

Gross pathological changes are highly suggestive of infection of *M. hyosynoviae* (Hagedorn-Olsen et al. 1999a) but must be confirmed by showing the presence of the mycoplasma. *M. hyosynoviae* should be suspected when synovial membranes are dark, proliferative, or hyperemic, and the synovial is brown or yellow and serofibrinous or serosanguinous (Gomes Neto et al. 2012). Okholm Nielsen et al. (2001) studied arthritis in growing-finishing pigs in nine Danish herds 1–98 days before slaughter. They obtained synovial fluid from the joints of lame and non-lame pigs. *M. hyosynoviae* was the only pathogen isolated in 20% of the lame pigs and in 8% of the non-lame pigs. They showed that *M. hyosynoviae* infection can also be present in pig joints without lameness. Some of the affected pigs (5%) needed to be euthanized due to lameness, but others recovered after treatment and were slaughtered. The agent is often present in the joints at low concentrations (Kobisch and Friis 1996). Culturing is not always successful, and the diagnosis is often reached by clinical signs and the absence of other pathogens causing arthritis or by PCR tests (Friis et al. 1992; Gomes Neto et al. 2015). No commercial vaccine is available to prevent mycoplasmal arthritis (Pieters and Maes 2019).

### 4.2.3 *Streptococcus suis* and *Glaesserella parasuis*

*S. suis* infections are known to occur worldwide as epidemic outbreaks of meningitis, septicemia, and arthritis. This pathogen has a predilection for pigs in the preweaning and immediate postweaning periods, but also finishing pigs can be affected (Williamson 2018). *G. parasuis* is another bacterium affecting mostly young pigs in endemically infected herds, but having a possibility to cause infections also in finishing pigs. Both of these pathogens have been discussed in more detail in Sect. 3.

### 4.2.4 Other Bacteria

Many other opportunistic bacteria cause septicemia after entering the body of the pig, often through skin wounds or other routes, and localize in joints, meninges, serosa surfaces, lungs, and other organs (Madson et al. 2019). In finishing pigs, tail biting is a common risk factor for joint infection. Pathogens such as *Actinobacillus pleuropneumoniae*, *Trueperella pyogenes*, *Brucella suis*, *E. coli*, *M. hyorhinis*, *Salmonella* spp., and pyogenic *Streptococcus* spp. may cause arthritis in growing pigs (Okholm Nielsen 2000). Affected animals are lame, and they may have fever and inappetence. In addition, they may have swollen joints, inflamed synovial membranes, fibrinous periarthritis, and the joint fluid can be purulent. In addition, due to the septic nature of the condition, abscesses can be found in these animals. Usually, the most often affected joints are the hock, stifle, carpus, elbow, and hip (Madson et al. 2019). Joint swellings are sometime difficult to be recognized clinically because some of these joints are largely covered by muscle layers.

It is difficult to find scientific articles about the abovementioned pathogens taken from the joints of live animals. Old studies report pathogens found in affected joints in slaughterhouse material, which have revealed *E. rhusiopathiae*, streptococci, staphylococci, and mycoplasmas in arthritic joints, but the proportions present vary considerably. Hariharan et al. (1992) could isolate the following bacteria in the synovial fluid of 153 finishing pigs with swollen joints and enlarged accompanying lymph nodes: *E. rhusiopathiae* (45%), *S. suis* (16%), *Trueperella pyogenes* (10%), *Mycoplasma* spp. (7%), *Staphylococcus* spp. (7%), *Streptococcus* spp. (6%), and organisms of uncertain significance (7%). They did not isolate any bacteria in the joints of 80 healthy control animals. Turner et al. (1991) found that 6.1% of joints of 192 carcasses of finishing pigs had arthritis, and they isolated either *Staphylococcus aureus* or *Streptococcus* spp. from these joints. Buttenschon et al. (1995) studied non-purulent arthritis in slaughter pigs. They found that more than 70% of the joint samples were sterile. *E. rhusiopathiae* was isolated in 9% and *M. hyosynoviae* in 12–14% of the pigs, and the following pathogens from a few animals: *Staphylococcus* sp., hemolytic *Streptococcus* sp., and hemolytic coliforms.

### 4.2.5 Infectious Contagious Viral Diseases

Also, contagious viral infections should be considered in the differential diagnoses of lameness. These diseases include Foot and Mouth Disease (picornavirus), Seneca-virus (picornavirus), Vesicular Exanthema (calicivirus), Swine Vesicular Disease (enterovirus), and Vesicular Stomatitis (rhabdovirus) (Ramirez 2019). Even though these infections are rare, they may affect the whole herd, whenever suspected. They cause similar clinical signs, the lesions are indistinguishable from each other, and therefore the laboratory diagnosis is always needed. These diseases usually cause high fever at the beginning of the disease and then the formation of vesicles in and around the mouth and/or feet. Affected pigs are reluctant to stand and walk, and they lose their appetite. The details of these diseases are not included in this chapter.

## 4.3 Noninfectious Lameness in Finishing Pigs

### 4.3.1 Osteochondrosis in Finishing Pigs

Osteochondrosis (OC) is defined as a focal disturbance (delay) in endochondral ossification (Reiland 1978a). Osteochondrosis can occur in sub-articular, epiphyseal growth cartilage and metaphyseal growth plates or physes (see Sect. 1 of this chapter) in both the appendicular and axial skeleton, but this section will focus on articular OC in limbs.

OC occurs as a consequence of the failure of the temporary, end-arterial blood supply to growth cartilage (Carlson et al. 1991; Ytrehus et al. 2004a) (Fig. 11). Vascular failure leads to ischemic necrosis of chondrocytes at an intermediate depth of growth cartilage, outside diffusion distance from alternative sources, known as osteochondrosis latens or an infarct (Carlson et al. 1991; Ytrehus et al. 2004a, b, c) (Fig. 11). With time, the ossification front advances to surround the infarct, at which stage it causes the focal delay in endochondral ossification that is the definition of osteochondrosis, also referred to as osteochondrosis manifesta (Ytrehus et al. 2004b) (Figs. 12 and 13).

OC lesions can resolve spontaneously before joint-specific age thresholds (Olstad et al. 2014a)

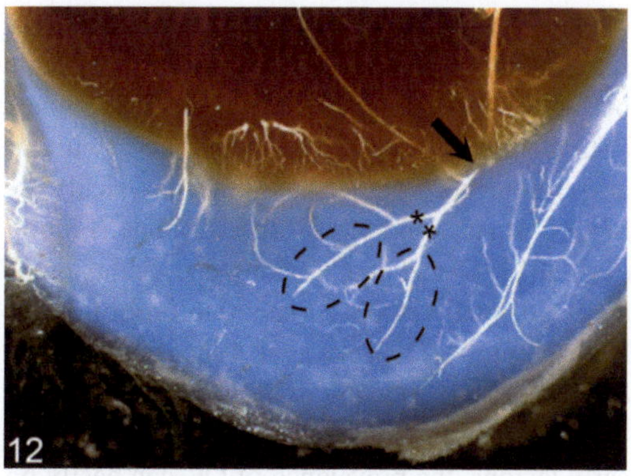


**Fig. 11** In osteochondrosis, failure of the temporary, end-arterial blood supply (white, barium-filled vessels) to growth cartilage leads to ischemic necrosis at intermediate depth of growth cartilage (dashed circles). Reprinted with permission (Olstad et al. 2015)

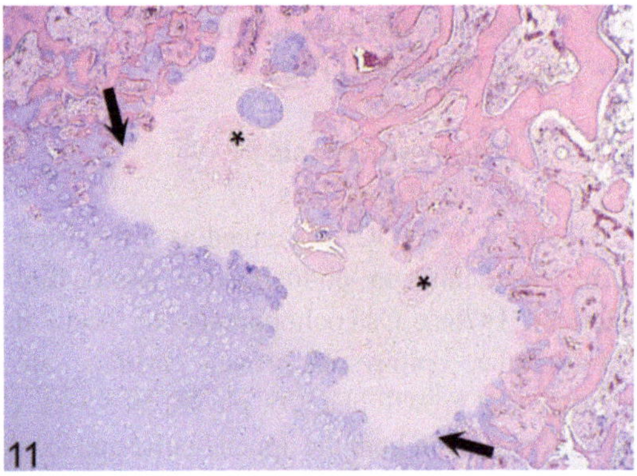


**Fig. 12** With time, the ossification front advances to surround the failed vessels (asterisks) and area of chondronecrosis (between arrows), which then causes a delay in endochondral ossification. Reprinted with permission (Olstad et al. 2015)

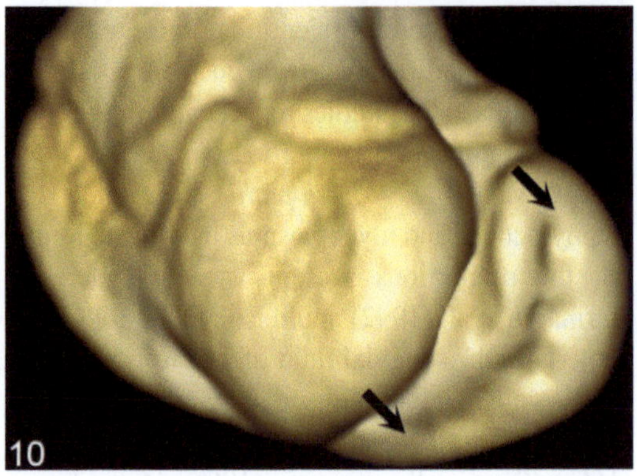


**Fig. 13** The delay in ossification is detectable in CT scans, typically as multi-lobulated or "stair-step" defects (between arrows) reflecting the configuration of the failed, branching blood vessels, as seen in Fig. 11 (between arrows). Three-dimensional rendering of bone density only; cartilage and soft tissues not shown. Reprinted with permission (Olstad et al. 2015)

or progress to pathologic fracture known as osteochondrosis dissecans (OCD), pseudo- or true subchondral bone cysts (Olstad et al. 2014a, b). Several factors have been implicated in the pathogenesis of OC such as heritability, rapid growth, over-feeding or lack of certain micronutrients in the diet, exercise, and trauma (Ytrehus et al. 2007; Olstad et al. 2015). These factors seem capable of modifying the outcome of the disease process. Whether they are capable of causing the failure of blood supply that triggers the pathogenesis in the first place, i.e., whether they are true etiological factors, remains to be confirmed (Olstad et al. 2015). The best-documented etiological factors are failure of vessels at the point where they traverse junctions between growth cartilage and other tissues (Ytrehus et al. 2004c; Olstad et al. 2008c) without any concurrent traumatic or other physical disruption of surrounding cartilage or bone (Finnøy et al. 2017; Olstad et al. 2008a), and occlusion of cartilage canal vessels by septic thrombi following bacteremia (Denecke et al. 1986; Wormstrand et al. 2018).

When using techniques capable of detecting early lesions, the prevalence of OC can easily be 100% Aasmundstad et al. 2013). OC is therefore probably present to some extent in any herd, but the condition does not necessarily cause clinical signs. Osteochondrosis latens and manifesta lesions are generally believed to represent subclinical stages, whereas OCD and cysts that communicate with the joint are considered clinical stages of the disease. When OC does cause clinical signs, articular OC has been associated with joint effusion and various manifestations of pain (Reiland 1978b).

OC is a condition that occurs multi-focally at predilection sites, and in the majority of cases, it affects more than one joint simultaneously (Olstad et al. 2014a). Thus, whether an individual pig shows lameness depends on the overall distribution and severity of OC in that particular pig. Pigs with symmetrical lesions may not show overt lameness on one limb, or lameness may change from one limb to the other depending on which joint feels worst on the day. It is a useful diagnostic aid to examine posture, as affected pigs will often stand on their toes or rocked back on their heels, with their fore- and/or hind limbs tucked under (Fig. 14), presumably to offload OC lesions in weight-bearing regions of the elbow and stifle joints (Reiland 1978b). The lameness associated with OC was originally referred to as "leg weakness," and a proportion of pigs with OC will prefer to adopt a dog-sitting position or lie down altogether. If further diagnostic work-up is indicated, live pigs can be examined using ultrasound or radiography in the field, computed tomography (CT; Fig. 15) (Olstad et al. 2014b), or magnetic resonance imaging (MRI) (Toth

**Fig. 14** Pigs affected with osteochondrosis may stand with their forelegs tucked under presumably to offload lesions in weight-bearing regions of the elbow joints (photo: Mari Heinonen)

et al. 2013) in hospital, screening, or research settings. More commonly, a definitive diagnosis is made on post-mortem examination (Reiland 1978a) through macroscopic inspection of intact joints and slabs sawed through multiple bones from slaughtered pigs. At post-mortem examination, it is also possible to collect and process slabs for histological evaluation, where standard morphological and Gram-staining enables definitive differentiation between aseptic and septic vascular failure (Wormstrand et al. 2018).

Pigs with subclinical OC lesions do not require any treatment. Pigs that show only mild signs can be monitored in the hope that their lesions will resolve (Olstad et al. 2014a) and may be put on restricted feeding in a sick pen to encourage reso-

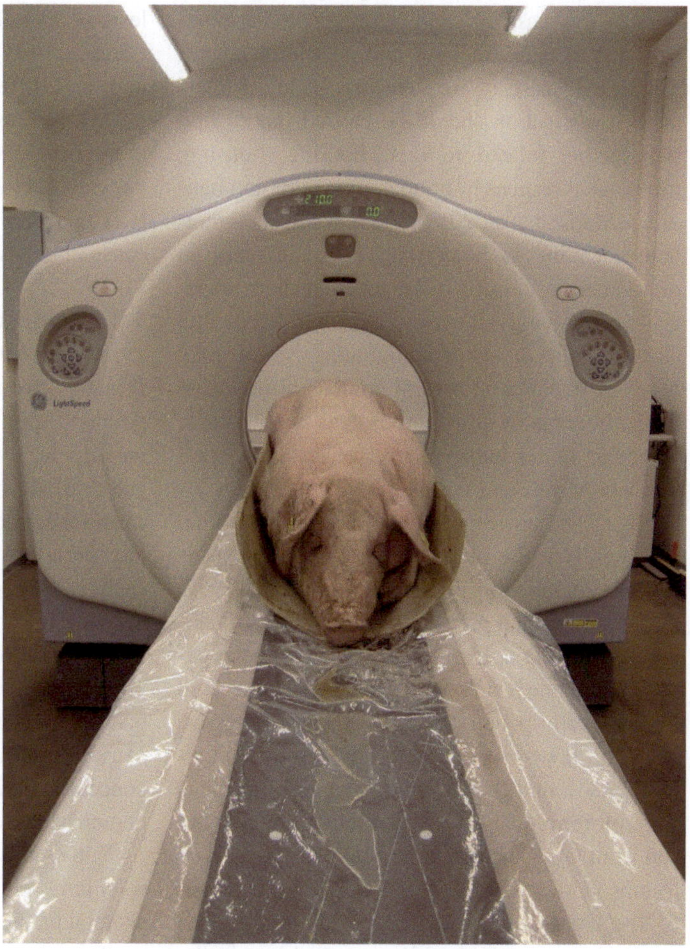

**Fig. 15** Entire live pigs can be scanned under sedation using computed tomography. Their joints are examined for screening and research purposes (photo: courtesy of Norsvin SA)

lution, if feasible. For pigs showing marked signs like overt lameness, dog-sitting, or prolonged recumbency, treatment is mandatory. In a commercial setting, the treatment of individual pigs may not be financially viable, necessitating premature culling of the pig instead. One of the main reasons why OC must be diagnosed and controlled, even if in most cases it causes only mild or no clinical signs, is that it is associated with significant losses due to premature culling. OC can also progress to irreversible, debilitating osteoarthritis in older pigs (Grøndalen 1974; Reiland 1978b), meaning the condition is important for the durability of breeding sows (Heinonen et al. 2006; Jensen et al. 2010).

### 4.3.2 Trauma in Finishing Pigs

Various traumas can cause acute, sporadic lameness due to bone fractures and muscle or tendon ruptures. In case of frequent prevalence, a thorough investigation should be carried out to find risk factors related to diet, flooring, housing, mechanical sources of trauma, transport, and handling (Madson et al. 2019). Fractures in slaughterhouse leading to hemorrhage may increase in prevalence in case of disbalances in diets, especially concerning mineral composition. Accidental electrocution or outdoor lightning strike may result in fractures in several animals at the same time affecting mostly lumbosacral and thoracic vertebrae, femur, humerus, neck of scapula, or pelvis (Van Alstine and Widmer 2003).

Pig claws are especially sensitive to trauma, and therefore the prevalence of claw lesions is common in finishing pigs. Mouttotou et al. (1997, 1999) published a prevalence of 93.8% of finishing pigs having at least one lesion (heel corrugation, white line lesion, wall separation, false sand crack, wall penetration, overgrown hooves, unequal claw size) in their slaughterhouse studies. They found that lateral digits and hind legs were more frequently affected than medial digits and front legs. The two most commonly observed lesions were sole erosion and white line lesions.

The etiology of different claw lesions has mostly been investigated in sows and less in the finishing pigs. Mouttotou et al. (1999) reported that floor type had an effect on claw lesions. Finishing pigs housed on bedded floors had less sole erosions, heel erosions and heel flaps and more white line lesions, false sand cracks, wall separations, and toe erosions than pigs kept on bare solid concrete floors. Partially slatted floors were associated with an increased prevalence of heel erosions, heel flaps, white line lesions, wall separations, and false sand cracks, and totally slatted floors were associated with an increased prevalence of sole erosions and heel flaps.

In a study of Okholm Nielsen et al. (2001), claw lesions were found in 63.2% of growing-finishing pigs. Most of the lesions (48.3% of the animals) were cracks in the soft heel and 14.9% of the pigs had deep cracks in the wall, the white line, or the soft heel.

Claw and soft heel lesions may cause lameness, and especially if they get infected, the lesions may become very painful (Gjein and Larssen 1995). The association between lameness and different kinds of claw lesions cannot always be found. In sows, lesions such as cracks in the heel or white line have not correlated with lameness of the same leg (Kroneman et al. 1993). It is therefore also possible that not all lesions cause lameness in finishing pigs.

### 4.3.3 Muscle Disorders in Finishing Pigs

Primary myopathies are rare nowadays because porcine stress syndrome (PSS) leading to PSE meat (pale, soft exudative) has mostly been mitigated (Madson et al. 2019). This stress syndrome causes the sudden death of pigs who have a hereditary predisposition, which is triggered by stress such as transport or excitement. Back muscle necrosis is considered to be one type of the syndrome, and it is found mostly in pigs heavier than 50 kg. Affected pigs have swollen, warm, and painful back muscles and difficulties in moving. Muscle enzyme CPK (creatine phosphokinase) concentration is high. The back muscle is dark and hemorrhagic when cut open (Madson et al. 2019).

Nutritional myopathies include vitamin E and selenium deficiency leading to generalized skeletal muscle myopathy but more commonly to mulberry heart disease or hepatosis dietetica. The problem may arise when diets low in selenium/vitamin E are fed to growing pigs or if factors in diets such as mycotoxins destroy vitamin E in the diet. Pigs are not very susceptible to nutritional muscular dystrophy, but in case it is seen, usually a pig of 50–60 kg is ataxic, walks stiffly, and is weak, depressed, and anorexic. The condition may lead to paralysis and death. The muscle is edematous with white streaks (Madson et al. 2019).

## 4.4 Treatment of Lameness in Finishing Pigs

Neither controlled trials nor evidence-based guidelines can be found about the treatment of lameness in sows (Ala-Kurikka 2020), and the same applies to finishing and breeding age pigs. In case of lameness suspected to be of nutritional or environmental origin, the first action is to correct the diet or environment. In these cases, the affected animals cannot totally be cured, but the condition will not appear in new, younger groups of pigs.

The decisions of treatment mostly rely on clinical experience, and the difficulty in reaching a proper diagnosis complicates the situation. Therefore, the treatment of single lame finishing pigs is usually symptomatic, aiming to relieve the symptoms without targeting the underlying cause of the disease. Usually, the treatment includes antimicrobials, if any kind of bacterial infection is suspected and pain and inflammation mitigation. The third part of the treatment is supportive care of the affected animal (Table 4).

It is very important to try to reach a good clinical diagnosis. If the condition causing lameness in the pig is detected late and the disease is already in severe or chronic stage, the pig needs to be culled rather than treated. In addition, economic considerations, long withdrawal times of some medications, or lack of effective treatment of the affected individual may lead to euthanasia rather than medication.

The microbes causing lameness should be treated with antimicrobials ideally based on culture and antimicrobial sensitivity testing. A sampling of individual lame pigs is seldom performed, and therefore the selection of the drug is usually made based on general knowledge and experience. In herds with frequent problems of infectious lameness, it is recommended to take samples from affected animals in the early stages of the disease to obtain information about the pathogens causing the lameness and their antimicrobial sensitivity. Penicillin and other beta-lactam antimicrobials are usually considered as drugs of choice. They are active against most of the pathogens causing arthritis in finishing pigs as discussed earlier in this chapter. Of these pathogens, only mycoplasmas are resistant to beta-lactam antimicrobials, and some studies have shown the efficacy of macrolides, lincosamides, or tetracyclines in their treatment (Burch and Goodwin 1984; Hannan et al. 1997a, b; Karriker et al. 2019). The duration of treatment in pigs has not been adequately defined in pigs. For acute infections, response to treatment is usually observed within 2 days, and in case no response is observed, diagnosis and treatment should be reconsidered (Karriker et al. 2019). However, very often the common practice is to treat the pigs only 1–7 days (Ala-Kurikka 2020).

Because scientific literature about the treatment of infectious arthritis in finishing pigs is

**Table 4** Main items in the treatment of lame pigs

| Cause of lameness | Medical treatment | Other treatments | Supportive care | Other considerations |
|---|---|---|---|---|
| Nutritional | | Correction of the diet | Use of proper sick pen, if needed to ensure that the lame pig can eat, drink, and rest without problems | In severe cases, consider euthanasia |
| Environmental, managemental | | Improvement of environment and/or management | | |
| Infective, bacterial | Antimicrobials, NSAIDs | | | |
| Infective, viral | NSAIDs | | | |
| Noninfective | NSAIDs | | | |

NSAIDs, non-steroidal anti-inflammatory drugs

lacking, we need to refer to other animal species. In cattle, successful therapy of infectious arthritis has included early administration of appropriate antimicrobial agents and mechanical removal of deleterious material from the joint and continuation of the therapy long enough to complete the elimination of the infectious organisms (Trent and Plumb 1991). If debilitating damage has occurred, therapy should stop the progression of infection and allow pain-free function of the limb. In pigs, any kind of joint lavage is extremely difficult to be performed under farm conditions and in addition to this, very expensive. Horses with septic arthritis are recommended to be treated systemically 10–14 days, and the affected joints lavaged (Haerdi-Landerer et al. 2010). In humans, several weeks of antimicrobial treatment is recommended together with drainage of the affected joint (Smith et al. 2006). The antimicrobial medication given to pigs may be too short to resolve the problem, but on the other hand, long treatment duration is usually not economically possible or discouraged because of antimicrobial resistance. However, clinical experience has shown that also these short treatments for lame pigs can be effective, especially if the treatment is initiated early enough. More research is definitely needed about the different aspects of antimicrobial treatment of infectious lameness in pigs.

Currently, pain mitigation is becoming more important in different disease conditions and in pigs, both in the attitudes of veterinarians and farmers (Ison and Rutherford 2014). The use of three NSAIDs has been reviewed in connection to treating lameness in sows and/or finishing pigs: flunixin-meglumine, ketoprofen, and meloxicam (Schoos et al. 2019). In most of the studies, the use of NSAID has resulted in a reduction of lameness. In one report, clear lameness due to osteochondrosis of a boar was alleviated with NSAID (Oomah 2008). However, potentially pain is not treated because of difficulties in pain recognition (Flecknell 2008). In addition, as long as we do not make proper diagnosis early enough, the treatment is likely to be applied in the chronic phase of the condition (Schoos et al. 2019). We lack lameness studies, where different NSAIDs would have been used for finishing pigs with and without antimicrobial agents.

In horses, treatment of osteochondrosis would initially be conservative by restricting diet and exercise whilst waiting until the age thresholds for spontaneous resolution have passed, after which time surgical removal of any persistent OCD fragments cures the problem. It is probable that small OC lesions may go undetected and resolve in pig populations all the time. In the case of clinical lesions, signs might be either very mild, not necessitating any treatment, or else treatment might not be economically justified, resulting in (premature) culling or slaughtering of the animal. It is more a question of deciding whether signs are so mild that pigs may live on, or must be sent to slaughter, rather than surgery. Thus, prevention is more important than cure.

The third cornerstone of treatment is the proper supportive care of the affected animal, which is sometimes neglected (Ala-Kurikka 2020). Each herd should have enough proper sick pens, in which lame finishing pigs can be moved. In these pens, the pigs should have the possibility to eat, drink, and rest without problems.

## 4.5 Prevention of Lameness in Finishing Pigs

Because the most important predisposing factor for lameness in finishing pigs is poor housing and management, their correction is the key to preventing lameness. More details are given in the chapter discussing the predisposing factors. In addition, proper nutrition and healthcare, including vaccination programs whenever available for different pathogens, are needed in lameness prevention.

The aim of prevention of osteochondrosis-type problems is to reduce the prevalence of that portion of OC that is due to heritably predisposed vascular failure (Olstad et al. 2015) through genetic selection. In pigs, the generation of breeding values still includes an element of phenotypic scoring (Grøndalen 1974; Reiland 1978a). Historically, prospective breeding boars received a phenotypic OC score inferred from the scoring of slaughtered half-siblings and pedigree analyses. Since 2008 in Norway and more recently in Canada, prospective breeding boars have been sedated to undergo whole-body CT scanning (Fig. 15), enabling phenotyping for OC and other traits in the still-alive boar itself. At present, OC is scored manually in CT scans at medial and lateral sites in the elbow and stifle (Aasmundstad et al. 2013), but manual scoring is likely to be superseded by automated diagnosis via artificial intelligence (Kongsro et al. 2017), enabling scoring of more sites and joints per pig. Breeding values are then generated by coupling the phenotypic scoring with genetic information from pedigree analyses and single nucleotide polymorphism (SNP) chip analyses of the genome (Grindflek et al. 2014), and collectively, these techniques constitute powerful tools for maintaining and improving pig health and welfare.

# 5 Lameness in Sows

Since 2013, group housing of gestating sows has been obligatory in all member states of the European Union. Although the ban on the confinement of sows was decided on welfare grounds, group housing of pregnant sows also presented welfare-related shortcomings, including an increased risk for lameness development (Anil et al. 2005; Calderón Díaz et al. 2014).

The list of differential diagnoses of lameness in sows is extensive. All components of the musculoskeletal system can be involved. However, in several studies, joint disorders (arthritis and osteochondrosis/arthrosis) and claw lesions were diagnosed to be the principal causes of lameness and were associated with culling or euthanasia in sows (Dewey et al. 1993; Kirk et al. 2005; Heinonen et al. 2006; Sanz et al. 2007; Engblom et al. 2008).

## 5.1 Arthritis in Sows

### 5.1.1 Etiology and Diagnosis of Arthritis in Sows

Infectious arthritis is commonly associated with bacteria. These may access the synovial membrane through the bloodstream, direct entrance into the joint after trauma, or an expansion of a nearby infection (e.g., infected claw lesions) (Rowles 2001; Crenshaw 2006; Done et al. 2012). Bacteria isolated from arthritic joints in pigs include *Erysipelothrix rhusiopathiae*, *Streptococcus suis*, *Glaesserella parasuis*, *Actinobacillus suis*, *Trueperella pyogenes*, *Staphylococcus* species, *Salmonella* Choleraesuis, *Mycoplasma hyorhinis*, and *Mycoplasma hyosynoviae* (Rowles 2001; Crenshaw 2006; Gomes Neto et al. 2012). *Trueperella pyogenes* has been found to be the most common bacterium in arthritis of sows, followed by hemolytic *Streptococcus* spp. and *Staphylococcus aureus* (Kirk et al. 2005).

The clinical signs of arthritis are an acute refusal to put weight on the leg, a swelling of the joint that is warm and painful when palpated, as well as fever and anorexia (Hill et al. 1986). Although clinical examination may indicate arthritis, specific diagnosis usually requires post-mortem examination and laboratory analyses, including histopathology and bacterial culture.

### 5.1.2 Epidemiology and Impact of Arthritis on Sow Welfare and Production

Arthritis may be a cause of severe lameness in sows. In a Canadian study, 22% (11/50) of the lame sows were diagnosed with arthritis after post-mortem examination, with arthritis being the second most important cause of lameness. Heinonen et al. (2006) diagnosed arthritis clinically in 8.8% (5/57) of the lame sows included in their study. In addition to the impact on welfare, arthritis can also lower herd performance and financial return. Arthritis has been reported to be the most important reason for euthanasia of sows (Kirk et al. 2005; Sanz et al. 2007; Engblom et al. 2008; Ala-Kurikka 2020) (Table 5).

### 5.1.3 Treatment and Prevention of Arthritis in Sows

Many infections are chronic by the time they are detected and therefore they respond poorly to treatment (Rowles 2001; Dewey 2006). Especially in case of severe lameness, euthanasia should considered as an alternative to treatment

**Table 5** The prevalence of primary diagnosis after post-mortem examination in euthanized sows/gilts related to the locomotory system. *N*: total number of euthanized sows/gilts examined

| | *N* | Osteochondrosis | | Arthrosis | | Fracture | | Arthritis | | Osteomyelitis (vertebral column) | | Osteomyelitis (other locations) | | Abscess in the spinal cord | | Other[a] | |
|---|---|---|---|---|---|---|---|---|---|---|---|---|---|---|---|---|---|
| | | *n* | % | *n* | % | *n* | % | *n* | % | *n* | % | *n* | % | *n* | % | *n* | % |
| Kirk et al. (2005) | 172 | | | 15 | 8.7 | 27[b] | 15.7 | 41 | 23.8 | 19 | 11 | 12 | 6.9 | | | 9 | 5.2 |
| Sanz et al. (2007) | 49 | 3 | 6.1 | | | | | 18 | 36.7 | | | | | | | | |
| Engblom et al. (2008) | 79 | 13 | 16.4 | | | 10 | 12.6 | 35 | 44.3 | | | | | 7 | 8.8 | | |
| Ala-Kurikka (2020) | 38 | | | | | 4 | 10.5 | 6 | 15.8 | 2 | 5.3 | | | | | | |

[a]Claw lesions, rupture of ligament, etc.
[b]Including 13 cases of epiphysiolysis

(Rowles 2001). Because infectious arthritis is mostly caused by bacteria, the use of antimicrobials at an early stage of the disease is recommended. If the causative bacterium has been isolated from the joints, the choice of antimicrobial can be based on the results of an antibiogram. If an antimicrobial therapy is chosen as the only intervention, even 7 days of systemic treatment will presumably not be sufficient (Ala-Kurikka 2020). Nevertheless, treatment courses for sows suffering from arthritis are often shorter in practice. Besides an antimicrobial therapy, painkillers should be used for the welfare of the sow. Although the use of non-steroid anti-inflammatory drugs (NSAID) has been reported to be effective mainly in noninfectious locomotor disorders in pigs (Friton et al. 2003; Mustonen et al. 2011), their use may also be beneficial in the case of infectious arthritis (Zoric et al. 2016). Wound management and joint lavage have been proposed as part of the treatment of infectious arthritis (Anderson and St. Jean 2012). For valuable breeding stock animals, this can be considered, but for most sows in commercial farms, this is not financially viable.

## 5.2 Arthrosis in Sows

### 5.2.1 Definition and Diagnosis of Arthrosis in Sows

Arthrosis is a noninfectious degenerative joint disease. Secondary arthrosis, also called secondary degenerative joint disease, is a progression of OC, which in turn is a local disturbance of endochondral ossification and the most common disorder found in articular growth cartilage (Nakano and Aherne 1993; Palmer 1993). Secondary arthrosis develops in a late stage of OC and is characterized by erosion of articular cartilage with exposure and excessive mineralization of subchondral bone, calcification of the synovial membrane, and formation of osteophytes (Crenshaw 2006). Non-OC-related arthrosis is called primary arthrosis (Palmer 1993). It is a degenerative joint disease, in which the articular cartilage surface is first affected. Primary arthrosis is characterized by fibrillation, surface tears, and erosion of articular cartilage covering unaffected or eburnated subchondral bone (Nakano and Aherne 1993; Palmer 1993).

Diagnosis of non-inflammatory degenerative joint disorders is difficult, especially in a living sow (Heinonen et al. 2006). The bilaterally symmetrical presence of lesions can induce lameness shifting between limbs (Rowles 2001). Gait and posture variables have been found to reflect joint lesions. Turned-out hind legs and a stiff movement in rear legs have been associated with osteochondrotic and arthrotic lesions in knee joints, while forelegs that are turned out and a stiff movement of the front and rear legs are highly indicative of lesions in the elbow joint (Jørgensen 2000; Kirk et al. 2008). Nevertheless, in most cases, a post-mortem examination will be required to find the cartilage lesions.

### 5.2.2 Epidemiology of Arthrosis in Sows

Studies focusing on arthrosis in sows are rare and often do not distinguish between the two types of arthrosis. In addition, the terms OC and arthrosis are often used ambiguously. In literature, articular cartilage changes found in sows have been defined either as OC or arthrosis and in some cases both terms are combined. Also, the method of diagnosis has a great impact on the number of affected animals found in different studies. This makes it difficult to interpret prevalence figures.

Osteochondrotic lesions or arthrosis are quite prevalent post-mortem findings in sows (Table 6). Both can be the primary reason for culling or euthanasia. In euthanized sows, arthrosis was found as a primary post-mortem diagnosis in 8% (Kirk et al. 2005) and OC in 14% (Engblom et al. 2008). Furthermore, they are an important secondary finding. Kirk et al. (2005) found arthrosis as a secondary diagnosis after postmortem examination in 88% of euthanized sows and 93% of sows that died.

Arthrosis is frequently bilateral and symmetrical, and it is found in sows most often in the elbow and stifle joint (Grøndalen 1974; Jørgensen 2000; Ytrehus et al. 2007; Kirk et al. 2008). Grøndalen (1974) reported a high incidence also in the lumbar region. The medial condyle of the

**Table 6** Prevalence (percentage) of osteochondrosis (OC) and arthrosis as primary or secondary diagnostic findings in culled/euthanized sows

| | Osteochondrosis | Arthrosis as primary diagnosis | Arthrosis as secondary diagnosis |
|---|---|---|---|
| Nakano and Aherne (1993)[a] | | 100[f] | |
| Dewey et al. (1993)[b] | 34[g] | 12 | 8 |
| Kirk et al. (2005)[c] | 49[h] | 8 | 88 |
| Sanz et al. (2007)[d] | 3 | | |
| Engblom et al. (2008)[e] | 13.5 | | |

[a]19 sows from one farm culled for impaired reproductive performance
[b]50 sows from ten farms culled for lameness
[c]172 sows from ten farms, reason for euthanasia not reported
[d]107 sows from one farm euthanized for several reasons
[e]96 sows from one herd euthanized for several reasons
[f] It was not reported whether arthrosis was the primary cause or only a secondary finding
[g] Primary cause for culling due to lameness. In 8% of all investigated lame sows, OC was found as a secondary lesion. In 94% of all examined sows, microscopic lesions of OC were seen
[h] Out of the 35 sows that were histologically examined

humerus, the anconeal process of the ulna, and the medial condyle of the femur are usually the most severely affected areas (Grøndalen 1974; Jørgensen 2000; Ytrehus et al. 2007, Kirk et al. 2008).

Severe lesions in the stifle joint have been reported to be more common in animals less than 1.5 years old. The stifle joint has been suggested to be more resistant to severe lesions once the most active growth period is over. In contrast, Grøndalen (1974) found a frequent, severe degree of arthrosis in the elbow joint in sows of different ages ranging from 9 months to 4 years. Kirk et al. (2008) reported that the elbow joint was more affected in sows compared to the stifle joint.

### 5.2.3 Impact of Arthrosis on Sow Welfare and Production

Osteochondrosis/arthrosis can be the primary cause of lameness and subsequently a reason for early culling and euthanasia. Dewey et al. (1993) reported arthrosis to be the primary cause in 12% and OC in 34% of the lame sows in their study. Heinonen et al. (2006) diagnosed clinically OC/arthrosis in 4.3% of lame sows and gilts. In a study by Engblom et al. (2008), 62% of the sows diagnosed with OC showed clinical lameness. However, the high prevalence of OC/arthrosis as a secondary diagnosis (8–93%) at post-mortem examination of sows also indicates that the presence of lesions does not always have a clinical relevance. Severe degree of arthrosis rarely occurs in the stifle joint of pigs more than 1.5 years old, whereas severe lesions in the elbow joint are found in all age groups up to 4 years. This may suggest that injuries in the knee might cause severe discomfort to the animal and consequently result in early culling or euthanasia, while injuries in the elbow joint might cause less discomfort (Grøndalen 1974).

### 5.2.4 Predisposing Factors of Arthrosis in Sows

OC has been suggested to be a multifactorial disease in which rapid growth, heredity, anatomic characteristics, trauma, nutrition, and a defect in vascular supply to the epiphysial cartilage may play a role (Ytrehus et al. 2007). Favored by intensive selection, pigs today grow to a body weight of over 100 kg within only 5–6 months. This rapid weight gain has been suggested to impair the soundness of legs and claws (Sack 1982; Kroneman et al. 1993). However, high growth rates may not play a direct role in the etiology of OC (Ytrehus et al. 2007). The pathogenesis of OC in finishing pigs is discussed earlier in this chapter. Pathogenesis of primary arthrosis in sows is still not well understood. Nakano and Aherne (1993) suggested a limited exercise and consequent restricted joint motion as a risk factor. With the transition to group housing in

Europe, it could be assumed that primary arthrosis in sows is rather rare today.

#### 5.2.5 Treatment and Prevention of Arthrosis in Sows

Arthrosis is a degenerative joint disease most often not responsive to treatment (Rowles 2001). Treatments, such as surgery, applied to horses and humans are often neither practical nor financially justifiable for pigs. For those cases that are accompanied by pain and discomfort, a symptomatic treatment with NSAID drugs is advised, and the lame animal should be moved to a proper sick pen to recover until sent to slaughter.

Prevention of secondary arthrosis in sows requires the prevention of OC development starting from skeletally immature pigs (see this chapter earlier, prevention in finishing pigs). Although rapid growth, heredity, conformation, trauma, and nutritional deficiencies have been suggested to play a role, only heredity and conformation have been well supported by scientific literature (Ytrehus et al. 2007; Belkova and Rozkot 2022).

### 5.3 Claw Lesions in Sows

#### 5.3.1 Epidemiology of Claw Lesions in Sows

Claw lesions are highly prevalent in sows and may affect almost every sow (Table 7). Different types of claw lesions can be distinguished with bruising/hemorrhage, heel erosions and overgrowth, cracks at the heel-sole junction, white line cracks, horizontal and vertical wall cracks, and overgrown claws and dewclaws being most commonly identified and investigated (Brooks et al. 1977; Anil et al. 2005; Engblom et al. 2008; Ossent 2010).

Interpreting the results of studies about the occurrence and severity of claw lesions should be done with caution because of the differing predisposing factors, the variety of scoring scales, and the different time points of scoring, i.e., in vivo or post-mortem. Post-mortem examination enables a more thorough assessment of the claws. However, sows are often euthanized or culled because of specific disorders or diseases, which may bias the results found on the prevalence and severity of claw lesions.

#### 5.3.2 Predisposing Factors for Claw Lesions in Sows

Claw lesions have been reported to be more prevalent and severe in group housed compared to individually stalled sows (Gjein and Larssen 1995; Anil et al. 2005, 2007). The effect of housing differs according to the type of claw lesion considered. Calderón Díaz et al. (2014) found that group-housed sows had a reduced risk of white line lesions, horizontal wall cracks, and dewclaw lesions but an increased risk of heel erosion and overgrowth. The degree and type of activity of the sow, particularly the ability to move around, the amount of aggressive interactions, and the possibility of avoiding confrontation can influence the type and severity of claw lesions. Fighting and aggressive behavior at the time of mixing in order to establish a new hierarchical ranking (i.e., in the case of dynamic groups) and at feeding time, especially in feeding systems associated with the highly competitive situation at feeding, increase the risk for claw lesion development (Kroneman et al. 1993; Anil et al. 2005). Similarly, group housing in pens with electronic sow feeders is associated with a higher proportion of claw lesions (Anil et al. 2007) and risk of leg injuries (Cador et al. 2014) compared to crate housing in which the crates can be accessed freely by the sows (Cador et al. 2014).

Knowledge about the predisposing factors for claw lesions in group-housed sows constitutes the basis for proper control and preventive measures. Development of claw lesions is determined by an interaction between internal (nutrition), external (housing conditions and management), and sow-related factors (claw conformation, weight distribution, genetic line, and parity). A disturbance in one or more of these factors may provoke trauma, inflammation, and/or inferior horn quality, which can result in the development of claw lesions.

**Table 7** The claw lesion prevalence/incidence (percentage) as found in different studies performed between 1993 and 2021, including information on the scoring method and time, housing conditions, and sample size

| | Global | Claw region | | | | | Overgrown (dew)claws | Scoring in vivo (1) or post-mortem (2) | Reason for culling, euthanasia or dead | Scoring method | Housing during gestation | Time at scoring | Sample size |
|---|---|---|---|---|---|---|---|---|---|---|---|---|---|
| | | Heel | Toe/sole | Heel–sole junction | Wall | White line | | | | | | | |
| Dewey et al. (1993) | 42[a] | | | | | | | 2 | Lameness | Gross examination | Not reported | | 50 |
| Gjein and Larssen (1995) | | 50–66[b] | 2–5 | 12–15 | 47–60 | 41–69 | | 1 | | Hind feet, Visual, 5-point scale (1–5) | Group housing[c] | | |
| Gjein and Larssen (1995) | | 30–40[d] | 1–4 | 5–9 | 28–33 | 32–43 | | 1 | | Hind feet, Visual, 5-point scale (1–5) | Confined | | |
| Gjein and Larssen (1995) | | 77[e] | 12 | 26 | 80 | 63 | | 2 | | Hind feet, Visual, 5-point scale (1–5) | Group housing[c] | | |
| Gjein and Larssen (1995) | | 53[f] | 7 | 7 | 51 | 46 | | 2 | | Hind feet, Visual, 5-point scale (1–5) | Confined | | |
| Pluym et al. (2011) | 99 | 93 | | | 52 | | 39 | 1 | | Hind feet, visual, 4-point scale (1–4) | Group housing[g] | Early lactation | 421 |
| Kirk et al. (2005) | 5/>84[i] | 74–84 | 75–77 | | 44–49 | 63–65 | | 2 | Several | Scoring not reported (absence vs. presence) | Both group and confined[h] | | 265 |
| Knauer et al. (2007) | 86.4 | 32.9/67.5[j] | 22.6/18.1[j] | | | | 3.5/21.1[j] | 2 | Several | Visual, 5-point scale and presence/absence | Not reported | | 3158 |

| | | | | | | | | | | | | | |
|---|---|---|---|---|---|---|---|---|---|---|---|---|---|
| Anil et al. (2005, 2007) | 96.2[l] | 86.4 | 45 | 66.3 | 88.6 | 60.9 | | 1 | | Visual, 5-point scale (0–4) | Both group and confined[k] | After moving to farrowing crates | 184 |
| Kilbride (2008) | 70.7 | 6.5/17.5[m] | 14.6/25.2[m] | 14.1/33.3[m] | 4/9[m] | 0.3/5.0[m] | 2.3/12.0[m] | 1 | | Left hind feet, visual, 4-point scale (0–3) | Group housing[n] | In the farrowing crates | 397 (gilts), 2005 (sows) |
| Calderón Díaz et al. (2013) | | 24.8 | | 8.6–11.3[o] | 12.8–20.7[o] | 9.8–10.7[o] | 7.5–19.5[o] | 1 | | Hind feet, Visual, sow chute, 4-point scale (0–3) | Group housing[p] | At insemination and between 50 and 70 days of gestation | 133 |
| Calderón Díaz et al. (2013) | | 23.3–24[o] | | 32.3 | 16.5–23.3[o] | 21.5–30.8[o] | 16.5–29.7[o] | 1 | | Hind feet, Visual, sow chute, 4-point scale (0–3) | Group housing[q] | At insemination and between 50 and 70 days of gestation | 133 |
| Pluym et al. (2011) | 99 | 93 | | | 52 | | 38.5 | 1 | | Hind feet, visual, 4-point scale (1–4) | Group housing | In farrowing crate during lactation | 421 |
| Pluym et al. (2013a, b) | | 96 | | 17 | 24 | 29 | 55 | 1 | | Hind feet, visual, 4-point scale (1–4) | Group housing | In farrowing crate during lactation | 381 |

(continued)

**Table 7** (continued)

| | Global | Claw region | | | | | Overgrown (dew)claws | Scoring in vivo (1) or post-mortem (2) | Reason for culling, euthanasia or dead | Scoring method | Housing during gestation | Time at scoring | Sample size |
|---|---|---|---|---|---|---|---|---|---|---|---|---|---|
| | | Heel | Toe/sole | Heel–sole junction | Wall | White line | | | | | | | |
| Sasaki et al. (2015) | 98.2 | 19.1–76.4[r] | 0.9–4.5[r] | 28.2–83.6[r] | 1.8–3.6[r] | 3.6–26.4[r] | | 1 | | Hind feet, visual, 5-point scale (0–4) | Confined | In farrowing crate or gestation stall | 110 |
| Bos et al. (2016) | 84.6/94.8[s] | | | | | | | 1 | | Visual, sow chute, tagged analog scale of 160 mm | Group housing[t] | At d50 of gestation and in farrowing crate at the end of lactation | 126 |
| Papadopoulos et al. (2021) | 62.5–96.5[u] | 35.9–52.6[u] | 21.9–87.7[u] | | 35.9–96.5[u] | 31.2–82.5[u] | | 2 | Several | Front feet, Visual, 3-point scale (0–2) | Group housing[v] | | 185 |
| Papadopoulos et al. (2021) | 76.6–98.2[u] | 60.9–83.6[u] | 60.9–96.5[u] | | 59.4–100[u] | 68.7–98.2[u] | | 2 | Several | Hind feet, Visual, 3-point scale (0–2) | Group housing on fully slates concrete[v] | | 185 |

[a] Diagnosed as "claw lesions"; primary diagnosis in 20% of the lame sows, secondary diagnosis in 22% of the lame sows; [b]Including heel cracks, erosion, and bruising but excluding heel overgrowth (prevalence proportion of 30–51%); [c]15 herds on slatted floors, 3 herds with deep litter; [d]Including heel cracks, erosion, and bruising but excluding heel overgrowth (prevalence proportion of 27–33%); [e]Including heel cracks, erosion, and bruising but excluding heel overgrowth (prevalence proportion of 45%); [f]Including heel cracks, erosion, and bruising but excluding heel overgrowth (prevalence proportion of 25%); [g]Four herds with free-access stalls (stable groups) on partially slatted floors); 4 herds in dynamic groups with electronic sow feeder, one of these used deep straw bedding in the lying area); [h]On 8 herds, sows were loose housed, and in two herds, sows were tethered; in 5 herds, deep bedding (straw) was used during gestation; [i]In 5%, claw lesions were the primary diagnosis for euthanasia; claw lesions were found as a secondary diagnosis in at least 84% of all sows; [j]Frequency of claw lesions were scored for front and rear feet separately. In the table, frequency is presented as front feet/rear feat; [k]In this study, 184 sows were included, 102 were housed in group pens with electronic sow feeders, and others were housed in stalls during gestation; [l]In total, 3.8% of all sows had no claw lesions, 9% of the sows in stalls had no claw lesions whereas there was no sow without claw lesions in group pens; [m]The percentage of each type of claw lesions is presented for gilts and sows separately: gilts/sows; [n]Group housing mainly on solid floors with deep bedding; [o]Gilts were followed during two parities; variation in claw lesion prevalence is due to differences between both parities. The highest percentages are found during second parity except for white line and wall lesions in sows housed on rubber slats; [p]Group housing on slatted concrete floors; [q]Group housing on slatted rubber floors; [r]The lateral and medial claw of each hind limb was scored separately. The variation in prevalence is due to a difference in the percentage of claw lesions between the lateral and medial claws of the left and right hind limb. For all claw regions, the lowest percentage was found on the medial claws; [s]The prevalence of claw lesions was significantly higher at the end of lactation (compared to day 50 of gestation); [t]Group housing with electronic sow feeder; [u]Claw lesion prevalence was measured on three different herds. Variation described is due to variation between herds; [v]Group housing on fully slatted concrete

#### 5.3.2.1 Claw Conformation and Weight Distribution

Claw lesions are more prevalent and severe on the lateral compared to the medial claws, a difference that is more pronounced on the hind feet compared to the front feet (Gjein and Larssen 1995; Anil et al. 2005, 2007; Knauer et al. 2007; Pluym et al. 2011; Lisgara et al. 2015b; Sasaki et al. 2015). Inequality in claw size may partly explain the difference in susceptibility to lesion development. Lateral claws are bigger compared to medial claws, possibly indicating that weight load, and as such mechanical impact of locomotion, is higher on lateral claws of the rear feet (Kornegay et al. 1990; Gjein and Larssen 1995; Bradley et al. 2009; Van Amstel and Doherty 2010; Sasaki et al. 2015; Papadopoulos et al. 2021). A total of 75% of the weight of the pig is born by the lateral claws (Webb 1984), and furthermore, claw size discrepancy increases as the sow ages (Bradley et al. 2009; Papadopoulos et al. 2021). The more pronounced the difference in claw size is, the more lesions are found in the area of the lateral claw (Kornegay et al. 1990).

In addition, abnormal claw and dewclaw growth may contribute to the susceptibility of claw lesion development. In the case of overgrown dewclaws, an area of contact with the ground may appear and influence the distribution of the weight-bearing force applied to the sow's feet, finally affecting its walking pattern (Papadopoulos et al. 2021). Overgrowth of the claws in general may influence the weight distribution pattern between and within claws. In addition, claw overgrowth has been associated with a decline in mechanical efficiency, expressed as Young's modulus, a measure of horn stiffness, which in turn has also been associated with an increased prevalence of wall lesions (Chalvatzi et al. 2021). Overgrown claws are associated with a greater lesion score for almost all types of claw lesions (Calderón Díaz et al. 2015; Sasaki et al. 2015; Papadopoulos et al. 2021). It also predisposes the sow to lameness development (Jørgensen 2000) and can even impair sow performance (Fitzgerald et al. 2012). Besides a genetic predisposition (Quintanilla et al. 2006; Fernàndez de sevilla et al. 2009a), claw overgrowth may naturally occur over time as horn growth may exceed the rate of wear, even if sows are housed on concrete flooring (Geyer 1979; Van Amstel and Doherty 2010). As the sow ages, an increase in toe length and prevalence of claw overgrowth has been reported in several studies (Dewey et al. 1993; Bradley et al. 2009; Fernàndez de sevilla et al. 2009a; Pluym et al. 2011; Lisgara et al. 2015b).

#### 5.3.2.2 Genetic Lines

Sows and boars from specific genetic lines may be more prone to claw lesions. Kilbride (2008) reported that pigmented breeds such as Duroc, Pietrain, and Hampshire are less likely to develop claw lesions, probably because of tougher claw horns. The prevalence of uneven hooves was significantly higher in Landrace boars compared to Yorkshire and Duroc ones, according to a report of Jørgensen and Vestergaard (1990). In a study by Fernàndez de Sevilla et al. (2009a), abnormal claw horn growth tended to affect the Landrace breed more severely than that of the Large White. The prevalence of abnormal claw horn growth at the end of the growing period was 36.5% and 28.2% for Landrace and Large White breeds, respectively. Landrace boars showed a higher likelihood of heel overgrowth and erosion, heel-sole cracks, and horizontal wall cracks but had a lower likelihood of overgrown claws compared to Duroc boars in a study by Wang et al. (2008). In the same study, the probability of horizontal wall cracks, heel-sole cracks, and dewclaw abnormalities was higher in Yorkshire than in Duroc Boars. Papadopoulos et al. (2021) investigated several claw and dewclaw length measurements and their associations with claw lesions among sows and found variation among the genetic lines studied (i.e., Pig Improvement Company (PIC), Danbred, and Topigs). Mechanical efficiency, expressed as the Young's modulus, a measure of stiffness, varied as well among those breeds with the highest values found in Topigs, followed by Danbred and PIC, in decreasing order (Chalvatzi et al. 2021).

This variation among genetic lines should be interpreted with caution. Other factors may influence claw integrity and the development of claw lesions, which may have differed between the above-mentioned studies.

#### 5.3.2.3 Parity

The odds of having white line lesions were significantly higher in sows with less than five parities compared to the older ones (Anil et al. 2007). In the same study, white line lesions were also associated with lameness. The association between white line lesions and young sows might be because older sows with white line lesions had already been culled. Lisgara et al. (2015b) found heel lesions and overgrown claws and dewclaws to be more likely seen in sows with more than three parities. These sows have a high body weight, increasing the pressure on the claws per unit area (Anil et al. 2007; Knauer et al. 2007), and exhibit a higher claw size discrepancy, leading to an increase in weight-bearing properties of the lateral claws. The highest load within the lateral claw found at the level of the heel (Webb 1984) might explain the increase in heel lesions as the sows age. A parity effect for claw measurements and claw overgrowth was reported earlier (Dewey et al. 1993; Bradley et al. 2009; Fernàndez de sevilla et al. 2009a; Pluym et al. 2011). Claw horn growth and wear in sows have been estimated to be 6.3 mm and 5.1 mm per month, respectively (Van Amstel and Doherty 2010). Although the horn growth rate has been suggested to diminish with age, the disbalance between horn growth and wear rate may indicate that claw and dewclaw overgrowth is to be expected with increasing age. In contrast, Anil et al. (2005) and Kilbride (2008) could not find any association between claw lesions and parity. As horn tissue in pigs grows faster than that of other animals, abnormal claw horn growth may develop faster, but claw lesions, especially mild ones, may heal more quickly and therefore may not be carried over to subsequent parities (Geyer 1979). In addition, claw trimming may contribute to improved claw lesion scores in successive parities (Pluym et al. 2013a, b). Moreover, the different findings for the association between parity and claw lesion prevalence may be biased by the culling management in the herds included in the different studies.

#### 5.3.2.4 Flooring

Three main floor types are used in indoor housing: solid, partially slatted, and fully slatted floors. A different amount of bedding can be used on partially slatted and solid concrete floors.

With respect to the risk for claw lesion development, solid concrete floors with deep straw bedding were reported to be the best and solid floors without bedding the worst (Jørgensen 2003; Cador et al. 2014). Although floors covered with deep bedding seem to be preferred, claw lesion development is not entirely prevented (Jørgensen 2003; Cador et al. 2014). Deep bedding reduces the risk of overburdening the claw, giving rise to less heel lesions (Kilbride 2008). Toe erosions, overgrown claws, and heel-sole cracks, however, may be more prevalent on straw bedding, likely due to the lack of natural wearing (Kilbride 2008; Cador et al. 2014). Rubber mats can be used as an alternative to deep litter. The benefits in terms of claw lesion development are threefold. First of all, as on deep litter, the impact load on claws is lower as the area of contact between the claw and the floor on mats is larger and pressure distribution is improved (Carvalho et al. 2009). Second, the cushioning effect of rubber mats may improve blood circulation in the dermis of the claw, essential for claw horn integrity. Lastly, textured rubber mats may provide more traction and reduce slipping (Boyle et al. 1998). The latter, however, only applies if the mats are not soiled; otherwise, the number of claw lesions may even increase (Calderón Díaz et al. 2013). Rubber mats may be associated with insufficient wear of claw horn, resulting in a higher prevalence of overgrown claws and heel-sole cracks (Calderón Díaz et al. 2013).

On slatted floors, the more limited contact area between the claw and the floor increases the pressure applied per unit area of the claw. Mainly, the heel bulb will be affected as this region of the claw is subject to the highest load. Heel flap development is associated with bruising and hemorrhage in the soft heel. Kilbride (2008) reported an increased risk for heel flaps on slatted floors compared to solid floors with deep bedding. The impact of slatted floors on the risk for claw lesions (particularly wall cracks) may be higher for heavier sows as the pressure per unit area of the claw is greater (Anil et al. 2007; Knauer et al. 2007; Kilbride 2008; Calderón Díaz et al. 2014). The effect of slatted floors on claw lesion development has been found both during gestation and lactation (Calderón Díaz et al. 2014). Webb (1984) defined an equation, based on the pig's live weight, to derive the maximum safe void percentage, i.e., the ratio of the area of holes per unit area of floor to unit area. For sows, this would suggest a maximum void percentage of 50%, above which a detrimental impact on the health of the heel region occurs, increasing the risk for heel overgrowth and erosions.

On solid floors, the pressure on the claws per unit area is lower, and claws and dewclaws cannot get stuck and tear off. However, urine and feces may stay on the solid floor, making the floor more slippery and the claw horn weaker (Sobestiansky et al. 1999; Gregory 2004), and increase the risk for infections (Gjein and Larssen 1995).

Although claw lesion development in outdoor-housed sows could be expected to be low, Kilbride (2008) reported a high prevalence (57%) of claw lesions in outdoor-housed sows and gilts, which was similar to that of the sows housed indoors (68%). Wet conditions and stony paddocks were suggested as possible reasons for this finding.

The influence of the material on claw lesion development is not clear due to limited investigations. Calderón Díaz et al. (2014) reported a higher risk of heel overgrowth/erosion and heel-sole cracks when sows were housed in the farrowing crates on slatted steel compared to cast iron. The higher void ratios and the more abrasive properties of steel slats compared to cast iron were suggested as possible reasons for the difference. An old study investigated the effect of steel, plastic, concrete, and aluminum slats on the occurrence of wall cracks and foot pad lesions in finishing pigs and found out that the results could be extrapolated to breeding pigs. Pigs housed on concrete slats were demonstrated to have the best foot pad condition score (i.e., the percentage of the volar surface affected by lesions) but showed the highest prevalence of wall cracks (Newton et al. 1980).

Four main quality factors contribute to a floor's total injury potential with respect to claw lesion development: slipperiness, abrasiveness, hardness, and roughness. Slippery floors and slipping has been associated with an increased risk of claw lesions (Kroneman et al. 1993; Kilbride 2008). The higher risk of hoof cracks on concrete slats compared to plastic or aluminum slats (Newton et al. 1980) and on steel slats compared to cast iron (Calderón Díaz et al. 2014) suggest wall cracks to be associated with abrasive floors (Calderón Díaz et al. 2014). Due to extensive wearing, the lateral wall horn becomes thinner and straightened, and this mechanically worn wall horn might be more susceptible to splitting. A minimum of abrasiveness, however, is necessary to prevent hoof overgrowth. On soft surfaces, e.g., soil or rubber mats, the area of contact between the claw and the floor is higher, thus resulting in reducing overloading of the claws, which might in turn prevent claw lesion development (Carvalho et al. 2009). Rough concrete, worn or uneven slats, and sharp protruding objects may also impair claw health. An increased risk of toe erosions has been associated with sharp slat edges and sharp protrusions at joints between surfaces on the farrowing pen floor (Kilbride 2008). De Belie (1997) found that concrete slats could already show degradation within 5 years, giving rise to increased surface roughness, enlarged gaps, and animal injuries.

Hygiene is, without doubt, of utmost importance. Dirty and wet floors can reduce claw horn strength, making claws more vulnerable to damage (Sobestiansky et al. 1999, Gregory 2004). Toe erosions in lactating sows (Kilbride 2008) and heel overgrowth, considered as a chronic reaction (Gjein and Larssen 1995), have been associated with bad floor hygiene. The white line is the weakest part of the claw horn capsule (Tomlinson et al. 2004). Poor floor hygiene results in the sow standing in feces and urine, which may manifest damage to this vulnerable claw area (Cador et al. 2014; Calderón Díaz et al. 2014). Dirty and wet floors will also increase the risk of claw infections. Sows with severe claw lesions seem to be more susceptible to claw infections compared to sows with minor claw lesions (Gjein and Larssen 1995). If floors are covered with feces and urine, the relative risk for claw lesions to be infected is quadrupled (Gjein and Larssen 1995). Hygiene is also important when using deep litter or rubber mats. Damp straw, soiled with feces and urine, will soften the hooves and make them more prone to lesion development and infection. On rubber mats, poor hygiene will additionally increase slipperiness. On soiled mats, the number of claw lesions may even increase (Calderón Díaz et al. 2013).

#### 5.3.2.5 Nutrition

Susceptibility to claw lesions also depends on claw horn strength, which is associated with multiple factors, including genetics, housing conditions, and nutrition (Torrison 2010). Functional integrity of claw horn highly depends on an appropriate supply of nutrients such as proteins, sulfur-containing amino acids (Cysteïne), hormones, and, considered to be the main factors, vitamins (A, D, E, and biotin) and minerals (Zn, Cu, Mn, Se, Ca, Cr, I, Mo) (Tomlinson et al. 2004; van Riet et al. 2013). Deficiencies may occur because of inadequate vascular supply or insufficient availability due to reduced dietary intake or low bioavailability. Late gestation and lactation are associated with a high nutritional demand, especially in high-producing or young (growing) sows and gilts. Any reduction in nutrient supply therefore may result in compromised claw horn quality (Anil 2011; Pluym et al. 2011).

Bioavailability is especially of importance to trace minerals. It has been reported that the bioavailability of organic trace minerals (i.e., a mineral joined with, e.g., a protein or an amino acid) is higher compared to inorganic trace minerals, and this may be even increased if a combination of organic trace minerals is used in the diet (Anil 2011; van Riet et al. 2013). It has been shown that partly replacing inorganic with organic trace minerals (Zn, Cu, Mn) in the diet could decrease both claw lesion incidence and severity and more specifically severity of heel, sole (Lisgara et al. 2016), and wall lesions (Anil 2011; Lisgara et al. 2016). Yet, the effect was less pronounced on the rear claws and lateral heel bulb (Lisgara et al. 2016). Trace minerals do not seem to affect claw and dewclaw overgrowth (Lisgara et al. 2016). The extracellular matrix secreted in the end phase of the horn production process is lipid rich and a major contributor to the mechanical strength of claw horn (Tomlinson et al. 2004). The impact of fatty acids, which can be influenced by the diet, on claw integrity in sows, however, remains unclear (van Riet et al. 2013).

Sows with a large body weight have an increased risk of developing especially heel lesions, probably because the pressure per unit area is higher (Anil et al. 2007; Knauer et al. 2007). In contrast, Kilbride (2008) found a trend for an increased prevalence of claw lesions in sows with a low body condition score. Nutritional deficiencies may have increased the risk of claw lesions, or the discomfort related to the claw lesions may have led to weight loss.

### 5.3.3 Impact of Claw Lesions on Sow Welfare and Production

The prevalence of claw lesions and the extent to which they are associated with lameness and sow performance varies considerably in published literature, including both sow and farm-specific parameters.

Although claw lesions have been considered an important cause of lameness in sows (Dewey et al. 1993; Heinonen et al. 2006), in most studies the association between lameness and claw lesions remains restricted to only a few types of claw lesions. Lameness has been associated with heel lesions (Cador et al. 2014), overgrown claws (Lisgara et al. 2015b), dewclaw lesions (Pluym et al. 2011; Cador et al. 2014; Lisgara et al. 2015b), and white line lesions (Anil et al. 2007; Lisgara et al. 2015b). Overgrown (dew)claws can get stuck between slats, crack, or even tear off (Pluym et al. 2011; Cador et al. 2014), whereas severe white line lesions may be a predilection site for bacterial invasion, especially on dirty wet floors (Cador et al. 2014, Calderón Díaz et al. 2014). Some studies have found no association between lameness and claw lesions (Gjein and Larssen 1995, Calderón Díaz et al. 2014). Claw infections, on the other hand, are clearly considered a major cause and risk for lameness in sows (Dewey et al. 1993; Gjein and Larssen 1995).

Claw lesions also have a detrimental effect on sow herd performance. They were one of the primary reasons for euthanasia or premature removal from the farm in several studies (Kirk et al. 2005; Engblom et al. 2008; Pluym et al. 2013a, b). Abnormal hoof growth has been shown to affect sow longevity in Duroc, Landrace, and Large White sows (Fernàndez de Sevilla et al. 2008, 2009b).

Combinations of claw lesions may negatively affect reproduction parameters such as the number of live-born piglets (Lisgara et al. 2015a), weaned piglets (Fitzgerald et al. 2012; Lisgara et al. 2015a), adjusted litter body weight at weaning (Fitzgerald et al. 2012), and wean-to-first service interval (Lisgara et al. 2015a) and increase the odds of having stillborn pigs (Pluym et al. 2013a, b).

### 5.3.4 Treatment of Claw Lesions in Sows

The first questions to be asked are whether the animal warrants treatment, whether it can be cured, or whether culling or euthanasia is preferred (Rowles 2001).

The dermis is a well-drained tissue in the claw that contains various nerve fibers (Tomlinson et al. 2004). Severe lesions might induce pain when they affect the dermis or when they have been an entrance for infection and inflammation of the tissues have occurred. Treatment with an NSAID is needed, if observed claw lesions are suspected to be detrimental to the welfare of the sow because of physical reduced mobility, pain, and general discomfort. The use of NSAID can be effective for pain mitigation in lame sows (Friton et al. 2003; Pairis-Garcia et al. 2015; Conte et al. 2015).

Infected claw lesions are often detected in the chronic stage, and prognosis is poor. Treatment is also challenging because of the difficulty in ensuring adequate local treatment. Ideally, infected lesions should be cleaned, debrided, and disinfected followed by an antimicrobial (topical and general) and NSAID therapy (Martineau and Morvan 2010). Penicillin is commonly recommended (Rowles 2001). In non-responsive cases, amputation of the claw has been suggested (Anderson and St. Jean 2012), but in practice this procedure is seldom carried out.

Supplementing complexed trace minerals (Zn, Mn, Cu) has been reported to contribute to the heeling of claw lesions, especially for heel and wall lesions, among group-housed sows (Anil 2011).

Claw trimming in the case of digital overgrowth is necessary as overgrown claws have been associated with lameness (Jørgensen 2000), an increased prevalence of almost all types of claw lesions (Papadopoulos et al. 2021), and a negative impact on sow performance (Fitzgerald et al. 2012). Moreover, long dewclaws can be stuck in the slots of slatted floors and may be ripped off, which is very painful for the animal (Pluym et al. 2011; Cador et al. 2014). Claw trimming has also been reported as part of the treatment strategy for pododermatitis (Martineau and Morvan 2010).

### 5.3.5 Prevention of Claw Lesions in Sows

Because treatment of claw-related lameness is frequently unrewarding, there is merit in working toward prevention and management. Factors affecting the prevalence and severity of claw lesions and infections are related to claw conformation, housing, nutrition, and management. They are the key to reducing claw lesions and related lameness.

Claw size discrepancy and claw overgrowth are both hereditary. Heritability values vary from 0.01 to 0.48 for uneven claws and from 0.24 to 0.38 for claw overgrowth, depending on the breed and statistical method used (Quintanilla et al. 2006; Fan et al. 2009; Fernàndez de sevilla et al. 2009a). Given the low to moderate heritability values and the fact that conformation traits such as uneven toes are highly polygenic, genomic selection may benefit progress in conformation traits. Standardization in methods for phenotyping conformation traits, identifying genetic correlations with economically important traits, and taking into account that conformation may evolve over a sow's reproductive life (Fernàndez de sevilla et al. 2008) are essential actions in achieving the right (genetic) strategy (Trenhaile Grannemann 2021).

It is impossible to define an ideal floor (Kroneman et al. 1993). Group housing on deep straw litter appeared to be the best solution to reduce the incidence and severity of claw lesions (Jørgensen 2003; Cador et al. 2014) although toe erosion and claw overgrowth may still be frequent (Kilbride 2008). The use of a non-bedded dunging area may help compensate the lack of natural wearing on the straw-bedded lying area. Use of a pitch of the solid floor toward the dunging area may help to drain urine. To keep the dunging area as dry and clean as possible, a slatted floor or scraper might be used (Altena and Vermeer 2004). However, maintaining a dry and clean floor is a challenge on solid floors. Even outdoor housing does not appear to be the ideal environment to minimize the risk of claw lesions. Wet conditions and stony paddocks may be risk factors in outdoor housing (Kilbride 2008).

Clean and dry floors moreover increase claw cleanliness, which improves (early) detection of claw lesions (van Riet et al. 2020) and consequently allows measures to be taken at an early stage to prevent claw lesions from developing into more severe lesions, lameness, and even premature culling.

Good-quality flooring, without protruding sharp objects, neither rough nor slippery and without crumbled slat edges and enlarged gap width, may help reduce the development of claw lesions (Kroneman et al. 1993). In slatted floors, the void ratio is also important. For sows, a maximum void percentage of 50% could be suggested based on the equation of Webb (1984). A number of suggestions related to sledge design and maximum slot width have been made in studies conducted almost 40 years ago, and we lack recent investigations.

The use of barriers may reduce aggression in group-housed sows (Arey and Edwards 1998). As the amount of aggressive interactions and the possibility to avoid confrontation can influence the type and severity of claw lesions (Kroneman et al. 1993; Anil et al. 2005), the use of barriers and increasing lying comfort may help prevent claw lesion development.

A sufficient dietary supply of bioavailable nutrients, especially during periods of high demand or low dietary intake, such as late gestation and lactation, is an important element in the prevention of claw lesions and consequently sow lameness (van Riet et al. 2013). Supplementing complexed trace minerals (Zn, Mn, Cu) can contribute to both the prevention and the healing of claw lesions, especially for heel and wall lesions (Anil 2011).

Claw trimming is an essential part of the treatment and prevention of claw lesions and associated lameness. Claw trimming can improve locomotion by inducing more comfort in gait as well as stability of the sow by reducing the risk for slipping (Tinkle et al. 2017, 2020). It may also contribute to improved claw lesion scores in successive parities (Pluym et al. 2013a, b). Functional claw trimming is recommended (i.e., trimming and shaping the full claw, balancing the sole, re-establishing even weight distribution within and between claws) although high labor inputs, training, and special equipment are needed. If functional trimming is not feasible, trimming the dewclaws may be a valuable alternative. By reducing the length of dewclaws, interference of the heel contact by the dewclaws is avoided, making stride faster and reducing the chance of dewclaws being caught, torn, or broken. Trimming too short and drawing blood should be avoided at all times. Trimming by only cutting the toe has not been recommended (Tinkle et al. 2020).

## References

Aasmundstad T, Kongsro J, Wetten M, Dolvik NI, Vangen O (2013) Osteochondrosis in pigs diagnosed with computed tomography: heritabilities and genetic correlations to weight gain in specific age intervals. Animal 7:1576–1582

Abell CE, Johnson AK, Karriker LA, Rothschild MF, Hoff SJ, Sun G, Fitzgerald RF, Stalder KJ (2014) Using classification trees to detect induced sow lameness with a transient model. Animal 8(6):1000–1009

Ala-Kurikka E (2020) Studies on sow lameness and related pathological findings. PhD thesis. University of Helsinki. http://urn.fi/URN:ISBN:978-951-51-6535-0

Ala-Kurikka E, Heinonen M, Mustonen K, Peltoniemi O, Raekallio M, Vainio O, Valros A (2017) Behavior changes associated with lameness in sows. Appl Anim Behav Sci 193:15–20

Ala-Kurikka E, Munsterhjelm C, Bergman P, Laine T, Pekkarinen H, Peltoniemi O, Valros A, Heinonen M (2019) Pathological findings in spontaneously dead an euthanised sows—a descriptive study. Porcine Health Manag 5:25

Alawneh JI, Laven RA, Stevenson MA (2012) Interval between detection of lameness by locomotion scoring and treatment for lameness: A survival analysis. Vet J 193(3):622–625

Altena H, Vermeer HM (2004) Straw housing of pregnant sows in large groups: bottlenecks and solutions. Animal Sciences Group Wageningen UR. Available via https://library.wur.nl/WebQuery/wurpubs/fulltext/23939. Accessed 11 Feb 2022

Amezcua R, Walsh S, Luimes PH, Friendship RM (2014) Infrared thermography to evaluate lameness in pregnant sows. Can Vet J 55:268–272

Anderson DE, St. Jean G (2012) Anesthesia and surgical procedures in swine. In: Zimmerman JJ, Karriker LA, Ramirez A, Schwartz KJ, Stevenson GW (eds) Diseases of swine, 10th edn. Wiley-Blackwell, Ames, IA, pp 119–140

Anil S (2011) Epidemiology of lameness in breeding female pigs. Dissertation, University of Minnesota

Anil S, Anil L, Deen J (2005) Evaluation of patterns of removal and associations among culling because of lameness and sow productivity traits in swine breeding herds. J Am Vet Med Assoc 226:956–961

Anil SS, Anil L, Deen J, Baidoo SK, Walker RD (2007) Factors associated with claw lesions in gestating sows. J Swine Health Prod 15(2):78–83

Applegate AL, Curtis SE, Groppel JL, JM MF, Widowski TM (1988) Footing and gait of pigs on different concrete surfaces. J Anim Sci 66:334–341

Aragon V, Segalés J, Oliveira S (2019) Glässer's disease. In: Zimmerman JJ, Karriker LA, Ramirez A, Schwartz KJ, Stevenson GW, Zhang J (eds) Diseases of swine, 11th edn. Wiley, Hoboken, NJ, pp 844–853

Arey DS, Edwards SA (1998) Factors influencing aggression between sows after mixing and the consequences for welfare and production. Livest Prod Sci 56(1):61–70

Barnett JL, Hemsworth PH, Cronin GM, Jongman EC, Hutson GD (2001) A review of the welfare issues for sows and piglets in relation to housing. Aust J Agric Res 52(1):1–28. https://doi.org/10.1071/AR00057

Baxter EM, Lawrence AB, Edwards SA (2011) Alternative farrowing systems: design criteria for farrowing systems based on the biological needs of sows and piglets. Animal 5(4):580–600. https://doi.org/10.1017/S1751731110002272

Belkova J, Rozkot M (2022) Gilt rearing impacts on sow performance and longevity—a review. J Swine Health Prod 30(1):10–16

Beusker N (2007) Welfare of dairy cows: lameness in cattle—a literature review. Dissertation, University of Hannover. https://d-nb.info/98788509X/34

Bonde M, Rousing T, Badsberg JH, Sørensen JT (2004) Associations between lying-down behaviour problems and body condition, limb disorders and skin lesions of lactating sows housed in farrowing crates in commercial sow herds. Livest Prod Sci 87:179–187

Bos EJ, van Riet MM, Maes D, Millet S, Ampe B, Janssens GPJ, Tuyttens FAM (2016) Effect of rubber flooring on group-housed sows' gait and claw and skin lesions. J Anim Sci 94(5):2086–2096

Boyle LA, Leonard FC, Lynch PB, Brophy P (1998) The effect of mats on the welfare of sows and piglets in the farrowing house. Anim Welfare 9:39–48

Bradley CL, Frank JW, Maxwell CV, Johnson ZB, Wilson ME, Ward TL (2009) The effects of parity, time and claw location on different claw measurements in the University of Arkansas sow herd over an 18-month period of time. Ark Anim Sci Dept Rep 574:94–100

Briene P, Szczodry O, De Geest P, van Weyenberg S, Van Nuffel A, Vangeyte J, Millet S, Ampe B, Tuyttens FA, Maselyne J (2021) Testing the potential of the Sow Stance Information System (SowSIS) based on a force plate system built into an electronic sow feeder for on-farm automatic lameness detection in breeding sows. Biosyst Eng 204:270–282

Brooks PH, Smith DA, Irwin VCR (1977) Biotin-supplementation of diets; the incidence of foot lesions, and the reproductive performance of sows. Vet Rec 101(3):46–50

Burch DGS, Goodwin RFW (1984) Use of tiamulin in a herd of pigs seriously affected with Mycoplasma hyosynoviae arthritis. Vet Rec 8:594–595

Buttenschon J, Svensmark B, Kyrval J (1995) Non-purulent arthritis in Danish slaughter pigs. I. A study of field cases. Zentralbl Veterinarmed A 42(10):633–641

Cador C, Pol F, Hamoniaux M, Dorenlor V, Eveno E, Guyomarc'h C, Rose N (2014) Risk factors associated with leg disorders of gestating sows in different group-housing systems: a cross-sectional study in 108 farrow-to-finish farms in France. Prev Vet Med 116(1–2):102–110

Calderón Díaz JC, Fahey AG, KilBride AL, Green LE, Boyle LA (2013) Longitudinal study of the effect of rubber slat mats on locomotory ability, body, limb and claw lesions, and dirtiness of group housed sows. J Anim Sci 91(8):3940–3954

Calderón Díaz JA, Fahey AG, Boyle LA (2014) Effects of gestation housing system and floor type during lactation on locomotory ability; body, limb, and claw lesions; and lying-down behavior of lactating sows. J Am Sci 92:1673–1683

Calderón Díaz JAC, Stienezen IM, Leonard FC, Boyle LA (2015) The effect of overgrown claws on behaviour and claw abnormalities of sows in farrowing crates. Appl Anim Behav Sci 166:44–51

Canning P, Viall A, O'Brien K, Madson D, Skoland K, Krull A, Linhares D, Gauger P, Ramirez A, Karriker L (2018) Determination of reference intervals for fluid analysis and cytologic evaluation variables in synovial fluid samples obtained from carpal and tarsal joints in commercial nonlame growing swine. Am J Vet Res 79:858–866

Cargill C, Buddle R (2001) Leg conformation and herd lameness. In: Cranwell PD (ed) Manipulating pig production VIII. APSA, Werribee, VIC.. 312 pages. ISBN: 0 957 7226-1-3

Carlson CS, Meuten DJ, Richardson DC (1991) Ischemic necrosis of cartilage in spontaneous and experimental lesions of osteochondrosis. J Orthop 9(3):317–329

Carvalho VC, de Alencar NI, Neto MM, de Souza SRL (2009) Measurement of pig claw pressure distribution. Biosyst Eng 103(3):357–363

Chalvatzi S, Papadopoulos GA, Kroustallas F, Cernat M, Skampardonis V, Marouda C, Fotiadou V, Psychas V, Poutahidis T, Leontides L, Fortomaris P (2021) Claw characteristics of culled sows from three farrow-to-finish Greek farms. Part 2: Mechanical indices of hoof horn and their associations with length measurements and lesion scores. Vet Sci 8(9):175–185

Christensen J, Ellegaard B, Kirkegaard Petersen B, Willeberg P, Mousing J (1994) Pig health and production surveillance in Denmark: sampling design, data recording, and measures of disease frequency. Prev Vet Med 20:47–61

Clark M (1983) Biomechanical aspects of piglet foot development. PhD thesis. University of Aberdeen, Aberdeen, Scotland

Clark M (1985) Farrowing pen floor abrasiveness measured using a rubber-block drag test. Farm Build Progr 80:29–32

Clarke B (2008) Normal bone anatomy and physiology. Clin J Am Soc Nephrol 3:131–139

Conte S, Bergeron R, Gonyou H, Brown J, Rioja-Lang FC, Connor ML, Devillers N (2015) Use of an analgesic to identify pain-related indicators of lameness in sows. Livest Sci 180:203–208

Coutellier L, Arnould C, Boissy A, Orgeur P, Prunier A, Veissier I, Meunier-Salaün MC (2007) Pig's responses to repeated social regrouping and relocation during the growing-finishing period. App Anim Behav Sci 105:102–114

Crenshaw TD (2006) Arthritis or OCD-identification and prevention. Adv Pork Prod 17:199–208

D'Eath RB (2012) Repeated locomotion scoring of a sow herd to measure lameness: consistency over time, the effect of sow characteristics and inter-observer reliability. Anim Welfare 21:219–231

De Belie N (1997) A survey on concrete floors in pig houses and their degradation. J Agric Eng Res 66:151–156

Declerck I, Dewulf J, Sarrazin S, Maes D (2016) Long-term effects of colostrum intake on piglet mortality and performance. J Anim Sci 94:1633–1643

de Koning DB, van Grevenhof EM, Laurenssen BF, van Weeren PR, Hazeleger W, Kemp B (2014) The influence of floor type before and after 10 weeks of age on osteochondrosis in growing gilts. J Anim Sci 92(8):3338–3347

Denecke R, Trautwein G, Kaup FJ (1986) The role of cartilage canals in the pathogenesis of experimentally induced polyarthritis. Rheumatol Int 6:239–243

Desrosiers R, Phaneuf JB, Broes A (1986) An outbreak of atypical Glasser's disease in Quebec. In: Proceedings of the 9th International Pig Veterinary Society congress, July 15–18, 1986, Barcelona, Spain, 9, 277

Dewey CE (2006) Disease of the nervous and locomotors systems. In: Straw BE, Zimmerman JJ, Allaire SD, Taylor DJ (eds) Diseases of swine, 9th edn. Iowa State University Press, Ames, IA, pp 87–111

Dewey CE, Friendship RM, Wilson MR (1993) Clinical and postmortem examination of sows culled for lameness. Can Vet J 34:555–556

Done S, Williamson SM, Strugnell BW (2012) Nervous and locomotor systems. In: Zimmerman JJ, Karriker LA, Ramirez A, Schwartz KJ, Stevenson GW (eds) Diseases of swine, 10th edn. Wiley-Blackwell, Ames, IA, pp 294–328

Engblom L, Eliasson-Selling L, Lundeheim N, Belak K, Andersson K, Dalin AM (2008) Post mortem findings in sows and gilts euthanized or found dead in a large Swedish herd. Acta Vet Scand 50:25

EU legislative (2008) Council Directive 2008/120/EC of 18 December 2008 laying down minimum standards for the protection of pigs. https://eur-lex.europa.eu/LexUriServ/LexUriServ.do?uri=OJ:L:2009:047:0005:0013:en:PDF

Evira, The Finnish Food Safety Authority and the Faculty of Veterinary Medicine at the University of Helsinki (2018) Recommendations for the use of antimicrobials in the treatment of the most significant infectious and contagious diseases in animals. https://www.ruokavirasto.fi/globalassets/viljelijat/elaintenpito/elainten-laakitseminen/hallittu_laakekekaytto/mikrobilaakekaytonperiaatteet/mikrobilaakkeiden_kayttosuositukset_en.pdf. Accessed 13 Jan 2022

Fan B, Onteru SK, Mote BE, Serenius T, Stalder KJ, Rothschild MF (2009) Large-scale association study for structural soundness and leg locomotion traits in the pig. Genet Sel Evol 41(1):1–9

Fernàndez de Sevilla XF, Fàbrega E, Tibau J, Casellas J (2008) Effect of leg conformation on survivability of Duroc, Landrace, and Large White sows. J Anim Sci 86(9):2392–2400

Fernàndez de Sevilla XF, Fabrega E, Tibau J, Casellas J (2009a) Genetic background and phenotypic characterization over two farrowings of leg conformation defects in Landrace and Large White sows. J Anim Sci 87(5):1606–1612

Fernàndez de Sevilla XF, Fàbrega E, Tibau J, Casellas J (2009b) Competing risk analyses of longevity in Duroc sows with a special emphasis on leg conformation. Animal 3(3):446–453

Finnøy A, Olstad K, Lilledahl MB (2017) Non-linear optical microscopy of cartilage canals in the distal femur of young pigs may reveal the cause of articular osteochondrosis. BMC Vet Res 13:270

Fitzgerald RF, Stalder KJ, Karriker LA, Sadler LJ, Hill HT, Kaisand J, Johnson AK (2012) The effect of hoof abnormalities on sow behavior and performance. Livest Sci 145(1–3):230–238

Flecknell P (2008) Analgesia from a veterinary perspective. Br J Anaesth 101(1):121–124

Friis NF, Ahrens P, Larsen H (1991) Mycoplasma hyosynoviae isolation from the upper respiratory tract and tonsils of pigs. Acta Vet Scand 32(4):425–429

Friis NF, Hansen KK, Schirmer AL, Aabo S (1992) Mycoplasma hyosynoviae in joints with arthritis in abattoir baconers. Acta Vet Scand 33(3):205–210

Friton GM, Philipp H, Schneider T, Kleemann R (2003) Investigation on the clinical efficacy and safety of meloxicam (Metacam) in the treatment of non-infectious locomotor disorders in pigs. Berl Munch Tierarztl Wochenschr 116(9–10):421–426

Gardner IA, Hird DW (1994) Risk factors for development of foot abscess in neonatal pigs. J Am Vet Med Assoc 204:1062–1067

Geudeke T, Junker K, Greijdanus S (2016) Mycoplasma hyosynoviae infections in finisher pigs with arthritis. In: Proceedings of the 7th ESPHM, Dublin, Ireland, 290

Geverink NA, Meuleman M, Van Nuffel A, van Steenbergen L, Hautekiet V, Vermeulen K, Lammens V, Geers R, van Reenen C, Tuyttens F (2006) Repeatability of lameness score measured on farm. Welfare Qual Rep 10:73–78

Geyer H (1979) Morphologie und Wachstum der Schweineklaue. Dissertation, University of Zurich

Gimenez-Lirola LG, Meiroz-De-Souza-Almeida H, Magtoto RL, McDaniel AJ, Merodio MM, MatiasFerreyra FS, Poonsuk K, Gatto IRH, Baum DH, Ross RF, Arruda PHE, Schwartz KJ, Zimmerman JJ, Derscheid RJ, Arruda BL (2019) Early detection and differential serodiagnosis of Mycoplasma hyorhinis and Mycoplasma hyosynoviae infections under experimental conditions. PLoS One 14:e0223459

Gjein H, Larssen RB (1995) The effect of claw lesions and claw infections on lameness in loose housing of pregnant sows. Acta Vet Scand 36(4):451–459

Gomes Neto JC, Gauger PC, Strait EL, Boyes N, Madson M, Schwartz KJ (2012) Mycoplasma associated arthritis: Critical points for diagnosis. J Swine Health Prod 20:82–86

Gomes Neto JC, Bower L, Erickson BZ, Wang C, Raymond M, Strait EL (2015) Quantitative real-time polymerase chain reaction for detecting Mycoplasma hyosynoviae and Mycoplasma hyorhinis in pen-based oral, tonsillar, and nasal fluids. J Vet Sci 16(2):195–201

Gomes Neto JC, Raymond M, Bower L, Ramirez A, Madson DM, Strait EL, Rosey EL, Rapp-Gabrielson VJ (2016) Two clinical isolates of Mycoplasma hyosynoviae pattern of lameness and pathogen detection in experimentally challenged pigs. J Vet Sci 17:489–496

Gottschalk M, Segura M. (2019) Streptococcosis. In: Zimmerman JJ, Karriker LA, Ramirez A, Schwartz KJ, Stevenson GW, Zhang J (eds) Diseases of swine, 11th edn. Wiley, Hoboken, NJ p 934-950

Grégoire J, Bergeron R, D'Allaire S, Meunier-Salaün M-C, Devillers N (2013) Assessment of lameness in sows using gait, footprints, postural behaviour and foot lesion analysis. Animal 7:1163–1173

Gregory NG (2004) Swelling of cattle heel horn by urine. Aust Vet J 82:161–163

Grindflek E, Hamland H, Aasmundstad T (2014) Genome-wide association study for conformation traits and osteochondrosis in pigs. In: 10th world congress on genetics applied to livestock production, Vancouver, Canada

Grøndalen T (1974) Osteochondrosis, arthrosis and leg weakness in pigs. Nord Vet Med 26:534–537

Haerdi-Landerer MC, Habermacher J, Wenger B, Suter MM, Steiner A (2010) Slow release antibiotics for treatment of septic arthritis in large animals. Vet J 184:14–20

Hagedorn-Olsen T, Basse A, Jensen TK, Nielsen NC (1999a) Gross and histopathological findings in synovial membranes of pigs with experimentally induced Mycoplasma hyosynoviae arthritis. APMIS 107(2):201–210

Hagedorn-Olsen T, Nielsen NC, Friis NF, Nielsen J (1999b) Progression of Mycoplasma hyosynoviae infection in three pig herds. Development of tonsillar carrier state, arthritis and antibodies in serum and synovial fluid in pigs from birth to slaughter. Zentralbl Veterinarmed A 46(9):555–564

Hannan PC, Windsor GD, de Jong A, Schmeer N, Stegemann M (1997a) Comparative susceptibilities of various animal-pathogenic mycoplasmas to fluoroquinolones. Antimicrob Agents Chemother 41(9):2037–2040

Hannan PCT, Windsor HM, Ripley PH (1997b) In vitro susceptibilities of recent field isolates of Mycoplasma hyopneumoniae and Mycoplasma hyosynoviae to valnemulin (Econor®), tiamulin and enrofloxacin and the in vitro development of resistance to certain antimicrobial agents in Mycoplasma hyopneumoniae. Res Vet Sci 63(2):157–160

Hariharan H, MacDonald J, Carnat B, Bryenton J, Heaney S (1992) An investigation of bacterial causes of arthritis in slaughter hogs. J Vet Diagn Invest 4(1):28–30

Heinonen M, Hämeenoja P, Saloniemi H, Tuovinen V (2001) Diagnoses and treatments in health-classified fattening herds rearing pigs all in-all out. Acta Vet Scand 42(3):365–375

Heinonen M, Oravainen J, Orro T, Seppä-Lassila L, Ala-Kurikka E, Virolainen J, Tast A, Peltoniemi O (2006) Lameness and fertility of sows and gilts in randomly selected loose-housed herds in Finland. Vet Rec 159:383–387

Heinonen M, Hakala S, Hämeenoja P, Murro A, Kokkonen T, Levonen K, Peltoniemi OAT (2007) Case-control study of factors associated with arthritis detected at slaughter in pigs from 49 farms. Vet Rec 160:573–578

Heinonen M, Peltoniemi O, Valros A (2013) Impact of lameness and claw lesions in sows on welfare, health and production. Livest Sci 156:2–9

Heinritzi K (2006) Mycoplasmen-polyserositis. In: Heinritzi K, Gindele HR, Reiner G, Schnurrbusch U (eds) Schweinekrankheiten. Ulmer, Stuttgart

Herskin MS, Di Giminiani P (2018) 11—Pain in pigs: characterisation, mechanisms and indicators. In: Špinka M (ed) Advances in pig welfare. Woodhead

Publishing series in food science, technology and nutrition, London, pp 325–355
Hill MA, Ruth GR, Hilley HD, Torrison JL, Bagent JK, Leman AD (1985) Dyschondroplasias of growth cartilages (osteochondrosis) in crossbred commercial pigs at one and 15 days of age: radiological, angiomicrographical and histological findings. Vet Rec 116(2):40–47
Hill MA, Hilley HD, Penny RHC (1986) Skeletal system. In: Leman AD, Glock RD, Mengeling WL, Penny RHC, Scholl E, Straw BE (eds) Diseases of swine, 6th edn. Iowa State University Press, Ames, IA, pp 183–197
Ison SH, Rutherford KM (2014) Attitudes of farmers and veterinarians towards pain and the use of pain relief in pigs. Vet J 202(3):622–627
Jensen TB, Bonde MK, Kongsted AG, Toft N, Sorensen JT (2010) The interrelationships between clinical signs and their effect on involuntary culling among pregnant sows in group-housing systems. Animal 4(11):1922–1928
Jørgensen B (2000) Osteochondrosis / osteoarthrosis and claw disorders in sows, associated with leg weakness. Acta Vet Scand 41:123–138
Jørgensen B (2003) Influence of floor type and stocking density on leg weakness, osteochondrosis and claw disorders in slaughter pigs. Anim Sci 77(3):439–449
Jørgensen B, Vestergaard T (1990) Genetics of leg weakness in boars at the Danish pig breeding stations. Acta Agriculturae Scandinavica 40(1):59–69
Karlen GAM, Hemsworth PH, Gonyou HW, Fabrega E, Strom AD, Smits RJ (2007) The welfare of gestating sows in conventional stalls and large groups on deep litter. Appl Anim Behav Sci 105:87–101
Karriker LA, Abell CE, Pairis-Garcia MD, Holt WA, Sun G, Coetzee JF, Johnson AK, Hoff SJ, Stalder KJ (2013) Validation of a lameness model in sows using physiological and mechanical measurements. J Anim Sci 91:130–136
Karriker LA, Coetzee JF, Friendship RM, Apley MD (2019) Drug pharmacology, therapy and prophylaxis. In: Zimmerman J, Karriker LA, Ramirez A, Schwartz KJ, Stevenson GW, Zhang J (eds) Diseases of swine, 11th edn. Wiley, Hoboken, NJ, pp 158–170
Kilbride A (2008) An epidemiological study of foot, limb and body lesions and lameness in pigs. Dissertation, University of Warwick
Kilbride AL, Gillman CE, Green LE (2009a) A cross-sectional study of the prevalence of lameness in finishing pigs, gilts and pregnant sows and associations with limb lesions and floor types on commercial farms in England. Anim Welfare 18:215–224
Kilbride A, Gillman C, Ossent P, Green L (2009b) Impact of flooring on the health and welfare of pigs. In Pract 31:390–395
Kirk RK, Jørgensen B, Jensen HE (2008) The impact of elbow and knee joint lesions on abnormal gait and posture of sows. Acta Vet Scand 50:5
Kirk RK, Svensmakr B, Ellegaard LP, Jensen HE (2005) Locomotive disorders associated with sow mortality in Danish pig herds. J Vet Med A Physiol Pathol Clin Med 52:423–428
Klem TB, Bleken E, Morberg H, Thoresen SI, Framstad T (2010) Hematologic and biochemical reference intervals for Norwegian crossbreed grower pigs. Vet Clin Pathol 39(2):221–226
Knauer M, Stalder KJ, Karriker L, Baas TJ, Johnson C, Serenius T, Layman L, McKean JD (2007) A descriptive survey of lesions from cull sows harvested at two Midwestern US facilities. Prev Vet Med 82(3–4):198–212
Kobisch M, Friis NF (1996) Swine mycoplasmoses. Rev Sci Tech 15(4):1569–1605
Kongsro J, Gangsei LE, Karlsson-Drangsholt TM, Grindflek E (2017) Genetic parameters of in vivo primal cuts and body composition (PigAtlas) in pigs measured by computed tomography (CT). Transl Anim Sci 1(4):599–606
Kornegay ET, Bryant KL, Notter DR (1990) Toe lesion development in gilts and sows housed in confinement as influenced by toe size and toe location. Appl Agric Res 5(4):327–334
Krieter J, Schnider R, Tölle KH (2004) Health conditions of growing-finishing pigs in fully-slatted pens and multi-surface systems. Dtsch Tierarztl Wochenschr 111(12):462–466
Kroneman A, Vellenga L, van der Wilt FJ, Vermeer HM (1993) Field research on veterinary problems in group-housed sows—a survey of lameness. Zentralbl Veterinarmed A 40(9–10):704–712
Lauritsen KT, Hagedorn-Olsen T, Jungersen G, Riber U, Stryhn H, Friis NF, Lind P, Kristensen B (2017) Transfer of maternal immunity to piglets is involved in early protection against Mycoplasma hyosynoviae infection. Vet Immunol Immunopathol 183:22–30
Le HT, Nilsson K, Norberg E, Lundeheim N (2015) Genetic association between leg conformation in young pigs and sow reproduction. Livest Sci 178:9–17. https://doi.org/10.1016/j.livsci.2015.05.025
Lewis E, Boyle LA, O'Doherty JV, Brophy P, Lynch PB (2005) The effect of floor type in farrowing crates on piglet welfare. Ir J Agric Food Res 44:69–81
Lisgara M, Skampardonis V, Angelidou E, Kouroupides S, Leontides L (2015a) Associations between claw lesions and reproductive performance of sows in three Greek herds. Veterinarni Medicina 60(8):415–422
Lisgara M, Skampardonis V, Kouroupides S, Leontides L (2015b) Hoof lesions and lameness in sows in three Greek swine herds. J Swine Health Prod 23(5):244–251
Lisgara M, Skampardonis V, Leontides L (2016) Effect of diet supplementation with chelated zinc, copper and manganese on hoof lesions of loose housed sows. Porcine Health Manag 2(1):1–9
Lium B, Egeli AK, Ytrehus B, Grondalen T, Gjein H (2002) Joint and foot diseases of pigs. Nor Vet Tidsskr 114:209–217
Maas J (2009) Musculoskeletal abnormalities. In: Smith BP (ed) Large animal internal medicine, 4th edn. Mosby, St Louis, MO, pp 217–231

Madson DM, Arruda PHE, Arruda BL (2019) Nervous and locomotor system. In: Zimmerman JJ, Karriker LA, Ramirez A, Schwartz KJ, Stevenson GW, Zhang J (eds) Diseases of swine, 11th edn. Wiley, Hoboken, NJ, pp 339–372

Main DC, Clegg J, Spatz A, Green LE (2000) Repeatability of a lameness scoring system for finishing pigs. Vet Rec 11:574–576

Marois C, Le Devendec L, Gottschalk M, Kobisch M (2007) Detection and molecular typing of *Streptococcus suis* in tonsils from live pigs in France. Can J Vet Res 71(1):14–22

Martineau G-P, Morvan H (2010) La Pododermatite. In: Martineau G-P, Morvan H (eds) Maladies d'élevage des porcs, 2nd edn. France Agricole, Paris, pp 316–320

Meijer E, Oosterlinck M, van Nes A, Back W, van der Staay FJ (2014) Pressure mat analysis of naturally occurring lameness in young pigs after weaning. BMC Vet Res 20(10):193

Mohling CM, Johnson AK, Coetzee JF, Karriker LA, Abell CE, Millman ST, Stalder KJ (2014) Kinematics as objective tools to evaluate lameness phases in multiparous sows. Livest Sci 165:120–128

Mouttotou N, Green LE (1999) Incidence of foot and skin lesions in nursing piglets and their association with behavioural activities. Vet Rec 145:160–165. https://doi.org/10.1136/vr.145.6.160

Mouttotou N, Hatchell FM, Lundervold M, Green LE (1997) Prevalence and distribution of foot lesions in finishing pigs in south-west England. Vet Rec 141(5):115–120

Mouttotou N, Hatchell FM, Green LE (1999) Foot lesions in finishing pigs and their associations with the type of floor. Vet Rec 144(23):629–632

Mülling CK (2000) Three-dimensional appearance of bovine epidermal keratinocytes in different stages of differentiation revealed by cell maceration and scanning electron microscopic investigation. Folia Morphologica 59(4):239–246

Mustonen K, Ala-Kurikka E, Orro T, Peltoniemi O, Raekallio M, Vainio O, Heinonen M (2011) Oral ketoprofen is effective in the treatment of non-infectious lameness in sows. Vet J 190:55–59

Nakano T, Aherne FX (1993) Articular cartilage lesions in female breeding swine. Can J Anim Sci 73(4):1005–1008

Nalon E, Conte S, Maes D, Tuyttens FAM, Devillers N (2013) Assessment of lameness and claw lesions in sows. Livest Sci 156:10–23

Nalon E, Maes D, Van Dongen S, van Riet MM, Janssens GPJ, Millet S, Tuyttens FAM (2014) Comparison of the inter- and intra-observer repeatability of three gait-scoring scales for sows. Animal 8:650–659

Newton GL, Booram CV, Hale OM, Mullinix BG Jr (1980) Effect of four types of floor slats on certain feet characteristics and performance of swine. J Anim Sci 50:7–20

Nielsen EO, Nielsen NC, Friis NF (2001) Mycoplasma hyosynoviae arthritis in grower-finisher pigs. J Vet Med A Physiol Pathol Clin Med 48(8):475–486

Nilsson C (1988) Floors in animal houses. PhD thesis. Swedish University of Animal Sciences, Lund, Sweden

Oh SI, Kim JW, Kim J, So B, Kim B, Kim HY (2020) Molecular subtyping and antimicrobial susceptibility of *Streptococcus dysgalactiae* subspecies equisimilis isolates from clinically diseased pigs. J Vet Sci 21(4):e57

Okholm Nielsen E (2000) Lameness in swine. A field study of aetiology and epidemiology of lameness in swine- with special reference to Mycoplasma hyosynoviae infections in growing-finishing pigs. Dissertation, Royal Veterinary and Agricultural University, Denmark

Okholm Nielsen EO, Lauritsen KT, Friis NF, Enøe C, Hagedorn-Olsen T, Jungersen G (2005) Use of a novel serum ELISA method and the tonsil-carrier state for evaluation of Mycoplasma hyosynoviae distributions in pig herds with or without clinical arthritis. Vet Microbiol 111(1–2):41–50

Oliviero C, Junnikkala S, Peltoniemi O (2019) The challenge of large litters on the immune system of the sow and the piglets. Reprod Domest Anim 54(Suppl 3):12–21. https://doi.org/10.1111/rda.13463

Olstad K, Cnudde V, Masschaele B, Thomassen R, Dolvik NI (2008a) Micro-computed tomography of early lesions of osteochondrosis in the tarsus of foals. Bone 43:574–583

Olstad K, Ytrehus B, Ekman S, Carlson CS, Dolvik NI (2008b) Epiphyseal cartilage canal blood supply to the distal femur of foals. Equine Vet J 40:433–439

Olstad K, Ytrehus B, Ekman S, Carlson CS, Dolvik NI (2008c) Epiphyseal cartilage canal blood supply to the tarsus of foals and relationship to osteochondrosis. Equine Vet J 40:30–39

Olstad K, Kongsro J, Grindflek E, Dolvik NI (2014a) Consequences of the natural course of articular osteochondrosis in pigs for the suitability of computed tomography as a screening tool. BMC Vet Res 10(1):212

Olstad K, Kongsro J, Grindflek E, Dolvik NI (2014b) Ossification defects detected in CT scans represent early osteochondrosis in the distal femur of piglets. J Orthop Res 32:1014–1023

Olstad K, Ekman S, Carlson CS (2015) An update on the pathogenesis of osteochondrosis. Vet Pathol 52:785–802

Oomah K (2008) Palliative therapy of osteochondrosis dessicans in a Duroc boar. Can Vet J 49:82–83

Opriessnig T, Coutinho TA (2019) Erysipelas. In: Zimmerman JJ, Karriker LA, Ramirez A, Schwartz KJ, Stevenson GW, Zhang J (eds) Diseases of swine, 11th edn. Wiley, Hoboken, NJ, pp 835–843

Ossent P (2010) An introduction to sow lamenes, claw lesions and pathogenesis theories. Zinpro Corporation, Eden Prairie, MN

Pairis-Garcia MD, Johnson AK, Abell CA, Coetzee JF, Karriker LA, Millman ST, Stalder KJ (2015) Measuring the efficacy of flunixin meglumine and meloxicam for lame sows using a GAITFour pressure

mat and an embedded microcomputer-based force plate system. J Anim Sci 93(5):2100–2110
Palmer N (1993) Bones and joints. In: Jubb KVF, Kennedy PC, Palmer N (eds) Pathology of domestic animals, 4th edn. Academic, San Diego, CA, pp 1–137
Papadopoulos GA, Chalvatzi S, Kroustallas F, Skampardonis V, Cernat M, Marouda C, Psychas V, Poutahidis T, Leontides L, Fortomaris P (2021) Claw characteristics of culled sows from three farrow-to-finish Greek farms. Part 1: Claw length measurements, lesion scores and their association. Vet Sci 8(7):126
Petersen HH, Enøe C, Nielsen EO (2004) Observer agreement on pen level prevalence of clinical signs in finishing pigs. Prev Vet Med 64:147–156
Petersen HH, Nielsen EO, Hassing AG, Ersbøll AK, Nielsen JP (2008) Prevalence of clinical signs of disease in Danish finisher pigs. Vet Rec 62(12):377–382
Pieters MG, Maes D (2019) Mycoplasmosis. In: Zimmerman JJ, Karriker LA, Ramirez A, Schwartz KJ, Stevenson GW, Zhang J (eds) Diseases of swine, 11th edn. Wiley, Hoboken, NJ, pp 863–883
Pillman D, Surendran Nair M, Schwartz J, Pieters M (2019) Detection of Mycoplasma hyorhinis and Mycoplasma hyosynoviae in oral fluids and correlation with pig lameness scores. Vet Microbiol 239:108448
Pluym L, Van Nuffel A, Dewulf J, Cools A, Vangroenweghe F, Van Hoorebeke S, Maes D (2011) Prevalence and risk factors of claw lesions and lameness in pregnant sows in two types of group housing. Veterinarni Medicina 56(3):101–109
Pluym LM, Maes D, Vangeyte J, Mertens K, Baert J, Van Weyewnberg S, Millet S, Van Nuffel A (2013a) Development of a system for automatic measurements of force and visual stance variables for objective lameness detection in sows: SowSIS. Biosyst Eng 116(1):64–74
Pluym L, Van Nuffel A, Maes D (2013b) Treatment and prevention of lameness with special emphasis on claw disorders in group-housed sows. Livest Sci 156:36–43
Quinn MM, Keuler NS, Lu YAN, Faria ML, Muir P, Markel MD (2007) Evaluation of agreement between numerical rating scales, visual analogue scoring scales, and force plate gait analysis in dogs. Vet Surg 36(4):360–367
Quintanilla R, Varona L, Noguera JL (2006) Testing genetic determinism in rate of hoof growth in pigs using Bayes factors. Livest Sci 105(1–3):50–56
Rafiee M, Bara M, Stephens CP, Blackall PJ (2000) Application of ERIC-PCR for the comparison of isolates of Haemophilus parasuis. Aust Vet J 78(12):846–849. https://doi.org/10.1111/j.1751-0813.2000.tb10507.x
Ramirez A (2019) Differential diagnosis of diseases. In: Zimmerman JJ, Karriker LA, Ramirez A, Schwartz KJ, Stevenson GW, Zhang J (eds) Diseases of swine, 11th edn. Wiley, Hoboken, NJ, pp 59–74
Reboli AC, Farrar WE (1989) Erysipelothrix rhusiopathiae: an occupational pathogen. Clin Microbiol Rev 2(4):354–359. https://doi.org/10.1128/CMR.2.4.354
Reiland S (1978a) Morphology of osteochondrosis and sequelae in pigs. Acta Radiol Suppl 358:45–90
Reiland S (1978b) Pathology of so-called leg weakness in the pig. Acta Radiol Suppl 358:23–44
Roach B (1981) Sussing out pig comfort. Pig Farm 29:47–48
Roos RF, Spear MI (1973) Role of the sow as a reservoir of infection for infection for *Mycoplasma hyosynoviae*. Am J Vet Res 34:373–378
Roos LR, Surendran Nair M, Rendahl AK, Pieters M (2019) Mycoplasma hyorhinis and Mycoplasma hyosynoviae dual detection patterns in dams and piglets. PLoS One 14(1):e0209975
Rowles C (2001) Sow lameness. J Swine Health Prod 9(3):130–131
Sack WO (1982) Essentials of pig anatomy. Veterinary Textbooks, New York
Sanz M, Roberts JD, Perfumo J, Alvarez RM, Donovan T, Almond GW (2007) Assessment of sow mortality in a large herd. J Swine Health Prod 15(1):30–36
Sasaki Y, Ushijima R, Sueyoshi M (2015) Field study of hind limb claw lesions and claw measures in sows. Anim Sci J 86(3):351–357
Scheel C, Traulsen I, Auer W, Müller K, Stamer E, Krieter J (2017) Detecting lameness in sows from ear tag-sampled acceleration data using wavelets. Animal 11:2076–2083
Schoos A, Devreese M, Maes DGD (2019) Use of non-steroidal anti-inflammatory drugs in porcine health management. Vet Rec 185(6):172
Scott K, Chennells DJ, Campbell FM, Hunt B, Armstrong D, Taylor L, Gill BP, Edwards SA (2006) The welfare of finishing pigs in two contrasting housing systems: Fully-slatted versus straw-bedded accommodation. Livest Sci 103(1–2):104–115
Segura M (2015) *Streptococcus suis* vaccines: candidate antigens and progress. Expert Rev Vaccines 14(12):1587–1608
Segura M, Fittipaldi N, Calzas C, Gottschalk M (2017) Critical *Streptococcus suis* virulence factors: Are they all really critical? Trends Microbiol 25(7):585–599
Sjölund M, Lindberg M, Pringle M, Harbom B, Unnerstad HE (2018) Occurrence, serotype and antibiotic susceptibility of *Streptococcus suis* isolates from healthy and diseased Swedish pigs. In: Proceedings of the 25th International Pig Veterinary Society congress, June 11–14, 2018, Chongqing, China, II-037, 672
Smith B (1988) Lameness in pigs associated with foot and limb disorders. In Pract 10:113–117
Smith JW, Chalupa P, Shabaz Hasan M (2006) Infectious arthritis: clinical features, laboratory findings and treatment. Clin Microbiol Infect 12(4):309–314
Sobestiansky J, Barcellos D, Moraes N, Carvalho LF, Oliveira S (1999) Clínica e Patologia suína. Universidade Federal de Goiás, Goiania
Stavrakakis S, Guy JH, Warlow OME, Johnson GR, Edwards SA (2014) Walking kinematics of growing pigs associated with differences in musculoskeletal conformation, subjective gait score and osteochondrosis. Livest Sci 165:104–113

Stavrakakis S, Guy JH, Syranidis I, Johnson GR, Edwards SA (2015) Pre-clinical and clinical walking kinematics in female breeding pigs with lameness: A nested case-control cohort study. Vet J 205:38–43

Straw BE, Dewey CE, Wilson MR (1999) Differential diagnosis of swine diseases. In: Straw BE, D'Allaire S, Mengeling WL, Taylor DJ (eds) Diseases of swine, 8th edn. Wiley-Blackwell, Hoboken, NJ, pp 41–86

Street BR, Gonyou HW (2008) Effects of housing finishing pigs in two group sizes and at two floor space allocations on production, health, behavior, and physiological variables. J Anim Sci 86(4):982–991

Stygar AH, Chantziaras I, Toppari I, Maes D, Niemi JK (2020) High biosecurity and welfare standards in fattening pig farms are associated with reduced antimicrobial use. Animal 14(10):2178–2186

Sun G, Fitzgerald RF, Stalder KJ, Karriker LA, Johnson AK, Hoff SJ (2011) Development of an embedded microcomputer-based force plate system for measuring sow weight distribution and detection of lameness. Appl Eng Agric 27:475–482

Tapper KR, Johnson AK, Karriker LA, Stadler KJ, Parsons RL, Wang C, Millman ST (2013) Pressure algometry and thermal sensitivity for assessing pain sensitivity and effects of flunixin meglumine and sodium salicylate in a transient lameness model in sows. Livest Sci 157:245–253

Thorup VM, Tøgersen FA, Jørgensen B, Jensen BR (2007) Biomechanical gait analysis of pigs walking on solid concrete floor. Animal 1:708–715

Tinkle AK, Duberstein KJ, Wilson ME, Parsley MA, Beckman MK, Torrison J, Azain MJ, Dove CR (2017) Functional claw trimming improves the gait and locomotion of sows. Livest Sci 195:53–57

Tinkle AK, Wilson ME, Torrison JL, Parsley MA, Duberstein KJ, Azain MJ, Dove CR (2020) Comparison of blunt versus functional claw trimming effects on sow gait. J Swine Health Prod 28(3):118–123

Tomlinson DJ, Mülling CH, Fakler TM (2004) Invited review: formation of keratins in the bovine claw: roles of hormones, minerals, and vitamins in functional claw integrity. J Dairy Sci 87(4):797–809

Torrison J (2010) Sow claw lesion pathology. Paper presented at the second FeetFirst® sow lameness symposium, Minneapolis, 31 August–2 September

Toth F, Nissi MJ, Zhang J, Benson M, Schmitter S, Ellermann JM, Carlson CS (2013) Histological confirmation and biological significance of cartilage canals demonstrated using high field MRI in swine at predilection sites of osteochondrosis. J Orthop Res 31:2006–2012

Touil F, Higgins R, Nadeau M (1988) Isolation of Streptococcus suis from diseased pigs in Canada. Can Vet Microbiol 17:171–177

Traulsen I, Breitenberger S, Auer W, Stamer E, Muller K, Krieter J (2016) Automatic detection of lameness in gestating group-housed sows using positioning and acceleration measurements. Animal 10(6):970–977

Trenhaile Grannemann MD (2021) Quantification and repeated measurements of conformation traits in replacement females to optimize sow longevity. Dissertation, University of Nebraska

Trent AM, Plumb D (1991) Treatment infectious arthritis and osteomyelitis. Vet Clin North Am Food Anim Pract 7(3):747–778

Turner GV, Collett MG, Veary CM, Kruger C (1991) Arthritis in slaughter pigs. J S Afr Vet Assoc 62(3):107–109

Van Alstine WG, Widmer WR (2003) Lightning injury in an outdoor swine herd. J Vet Diagn Invest 15:289–291

Van Amstel S, Doherty T (2010) Claw horn growth and wear rates, toe length, and claw size in commercial pigs: A pilot study. J Swine Health Prod 18(5):239–243

Van Riet MM, Millet S, Aluwé M, Janssens GPJ (2013) Impact of nutrition on lameness and claw health in sows. Livest Sci 156(1–3):24–35

Van Riet MM, Bos EJ, Ampe B, Bikker P, Vanhauteghem D, Van Bockstaele F, Cornillie P, Van Den Broeck W, Du Laing G, Maes D, Tuyttens FAM, Janssens GPJ, Millet S (2018) Long-term impact of zinc supplementation in sows: Impact on claw quality. J Swine Health Prod 26(1):10–24

Van Riet MM, Vangeyte J, Janssens GP, Ampe B, Nalon E, Bos EJ, Pluym L, Tuyttens FAM, Maes D, Millet S (2019) On-farm claw scoring in sows using a novel mobile device. Sensors 19(6):1473–1487

Van Riet MM, Janssens GP, Ampe B, Nalon E, Bos EJ, Pluym L, Vangeyte J, Tuyttens FAM, Maes D, Millet S (2020) Factors influencing claw lesion scoring in sows. Prev Vet Med 175:104859

Vela A, Goyache J, Tarradas C, Luque I, Mateos A, Moreno MA, Borge C, Perea JA, Domínguez L, Fernández-Garayzábal JF (2003) Analysis of genetic diversity of *Streptococcus suis* clinical isolates from pigs in Spain by pulsed-field gel electrophoresis. Clin Vet Microbiol 41:2498–2502

von Wachenfelt H, Pinzke S, Nilsson C, Olsson O, Ehlorsson CJ (2008) Gait analysis of unprovoked pig gait on clean and fouled concrete surfaces. Biosyst Eng 101:376–382

Wang Q, Chang B, Riley TV (2010) *Erysipelothrix rhusiopathiae*. Vet Microbiol 140:405–417

Wang C, Li JL, Wei HK, Zhou YF, Tan JJ, Sun HQ, Jiang SW, Peng J (2008) Analysis of influencing factors of boar claw lesions and lameness. An Sci J 89:802–809

Webb NG (1984) Compressive stresses on, and the strength of the inner and outer digits of pigs' feet, and the implications for injury and floor design. J Agric Eng Res 30:71–80

Welfare Quality® (2009) Welfare quality assessment protocol for pigs (sows, and piglets, growing and finishing pigs). Welfare Quality Consortium, Lelystad, The Netherlands

Wells GAH (1984) Locomotor disorders of the pig. In Pract 6:43–53

Westin R, Algers B (2006) Effects of farrowing pen design on health and behavior of the sow and piglets at farrowing and lactation. Sv Vet Tidn 8–9:21–27

Williamson S (2018) *Streptococcus suis* disease in pigs. Vet Rec 183(13):408–410

Wormstrand B, Ostevik L, Ekman S, Olstad K (2018) Septic arthritis/osteomyelitis may lead to osteochondrosis-like lesions in foals. Vet Pathol 55:693–702

Wormstrand BH, Fjordbakk CT, Griffiths DJ, Lykkjen S, Olstad K (2021) Development of the blood supply to the growth cartilage of the medial femoral condyle of foals. Equine Vet J 53(1):134–142

Yongkiettrakul S, Maneerat K, Arechanajan B, Malila Y, Srimanote P, Gottschalk M, Visessanguan W (2019) Antimicrobial susceptibility of *Streptococcus suis* isolated from diseased pigs, asymptomatic pigs, and human patients in Thailand. BMC Vet Res 15:5

Ytrehus B, Andreas Haga H, Mellum CN, Mathisen L, Carlson CS, Ekman S, Teige J, Reinholt FP (2004a) Experimental ischemia of porcine growth cartilage produces lesions of osteochondrosis. J Orthop Res 22:1201–1209

Ytrehus B, Carlson CS, Lundeheim N, Mathisen L, Reinholt FP, Teige J, Ekman S (2004b) Vascularisation and osteochondrosis of the epiphyseal growth cartilage of the distal femur in pigs—development with age, growth rate, weight and joint shape. Bone 34(3):454–465

Ytrehus B, Ekman S, Carlson CS, Teige J, Reinholt FP (2004c) Focal changes in blood supply during normal epiphyseal growth are central in the pathogenesis of osteochondrosis in pigs. Bone 35:1294–1306

Ytrehus B, Carlson CS, Ekman S (2007) Etiology and pathogenesis of osteochondrosis. Vet Pathol 44(4):429–448

Zeder MA, Lemoine X, Payne S (2015) A new system for computing long-bone fusion age profiles in *Sus scrofa*. J Archaeol Sci 55:135–150

Zoric M (2008) Lameness in piglets. PhD thesis. Acta Universitatis Agriculturae Sueciae, 2008:49, Faculty of Veterinary Medicine and Animal Science, Swedish University of Agricultural Sciences (SLU), Uppsala, Sweden. http://pub.epsilon.slu.se/1862/. Accessed 03 Feb 2022

Zoric M, Wallgren P (2014) Lameness in piglets. Which joints are affected? In: Proceedings of the 6th European symposium of porcine health management, May 07–09, 2014, Sorrento, Italy, pp 24, 142

Zoric M, Stern S, Lundeheim N, Wallgren P (2003) Four-year study of lameness in piglets at a research station. Vet Rec 153:323–328

Zoric M, Sjölund M, Persson M, Nilsson E, Lundeheim N, Wallgren P (2004) Lameness in piglets. Abrasions in nursing piglets and transfer of protection towards infections with Streptococci from sow to offspring. J Vet Med B 51(6):278–284. https://doi.org/10.1111/j.1439-0450.2004.00777.x

Zoric M, Nilsson E, Mattsson S, Lundeheim N, Wallgren P (2008) Abrasions and lameness in piglets born in different farrowing systems with different types of floor. Acta Vet Scand 50:37. https://doi.org/10.1186/1751-0147-50-37

Zoric M, Nilsson E, Lundeheim N, Wallgren P (2009) Incidence of lameness and abrasions in piglets in identical farrowing pens with four different types of floor. Acta Vet Scand 51:23. https://doi.org/10.1186/1751-0147-51-23

Zoric M, Norling E, Schmidt U, Wallgren P (2012) Aspects on completion of antibiotics with NSAIDs when treating lame piglets. In: Proceedings of the 22nd International Pig Veterinary Society congress, June 10–13, 2012, Jeju, Korea, I, 218

Zoric M, Schmidt U, Wallenbeck A, Wallgren P (2016) Lameness in piglets—should pain killers be included at treatment? Porcine Health Manag 2:8. https://doi.org/10.1186/s40813-016-0022-5

Zoric M, Wallenbeck A, Schmidt U, Österberg J, Wallgren P (2017) Relation between treatment for arthritis during suckling and performance the during fattening period. In: Proceedings of the 9th European symposium of porcine health management, May 03–05, 2017, Prague, Czech Republic, MIS-006, 374

# Part VII

# Environment, Behavior, Genetics, and Animal Welfare

# Associations of Behavior and Production Disease in Dairy Cattle

Meagan T. M. King and Trevor J. DeVries

**Abstract**

Dairy cattle need to devote sufficient time each day to eating, drinking, ruminating, and resting to maintain good health and production. There is mounting empirical evidence that monitoring of behavior can be used to predict and identify health and welfare concerns in dairy cattle. Interest in the association between dairy cow behavior and illness is growing, particularly with an increasing availability of commercially available equipment that is readily available for the automated monitoring of dairy cow behavior. These technologies provide ready access to vast quantities of accurate and repeatable measures of behavior. Thus, with such technologies, there are new opportunities for the automation detection (through identification and prediction) of health concerns in dairy cattle, potentially allowing for better prevention and treatment protocols to be implemented. This chapter reviews how various cow behaviors, in particular, eating, resting, and ruminating, have been associated with various production diseases, as well as reviews some of the technologies available and their application, both current and future, that may be used to improve cow health and welfare, at a cow and herd level.

**Keywords**

Behavior · Health · Welfare · Dairy cow · Technology · Automation · Rumination · Lying · Eating

## 1 Introduction

Dairy cattle have a number of things they need to accomplish each day to maintain good health and production. At minimum, these activities that make up their time budget include eating, drinking, ruminating, and resting. In modern production systems, lactating dairy cows will typically spend 3–5 h/day eating, 0.5 h/day drinking, 10–13 h/day lying down, 7–10 h/day ruminating, and, if milked in a parlor, up to 2.5–3.5 h/day waiting for and being milked (DeVries 2017). Even though every 24-h day should be enough time to allow cows to do these things, we know that any impingement of the cow's ability to devote her time to those activities may have negative consequences.

Lying down is a high-priority behavior for dairy cattle; thus, insufficient time spent lying

M. T. M. King
Department of Animal Science, University of Manitoba, Winnipeg, MB, Canada
e-mail: meagan.king@umanitoba.ca

T. J. DeVries (✉)
Department of Animal Biosciences, University of Guelph, Guelph, ON, Canada
e-mail: tdevries@uoguelph.ca



J. J. Gross (ed.), *Production Diseases in Farm Animals*,
https://doi.org/10.1007/978-3-031-51788-4_19

can be a challenge for their welfare (Tucker et al. 2021). Dairy cows are motivated to spend approximately half of their day lying down; Jensen et al. (2005) demonstrated that cows have an inelastic demand for about 12–13 h/day of rest. Other researchers have shown that when opportunities to perform behaviors are restricted, lying behavior takes precedence over eating and social behavior (Munksgaard et al. 2005). Adequate lying time has not only been linked to ensuring good milk production (Grant 2004), but prevention of cows spending too much time standing has also been linked to prevention of hoof pathologies (Proudfoot et al. 2010) and resultant lameness (Tucker et al. 2021). In fact, factors that are linked to encouraging resting time in dairy cows, such as larger, less-restrictive stalls, and use of well-maintained, deep bedding have been linked to a lower prevalence of lameness (Chapinal et al. 2013). Thus, anything that limits the ability of cows to devote the time they need and want to spend lying down may have negative consequences.

The eating behavior of dairy cattle is an equally important factor to consider, as it directly relates to their total dry matter intake (DMI), as well as to their rumen health and digestion (DeVries 2019). For production animals, feed intake needs to be sufficient to not only match maintenance requirements but also needs for growth and production. The feed intake of a dairy cow is simply a function of her eating behavior; that is the total dry matter intake (DMI; kg/day) of a cow is the result of the number of meals consumed daily (#/day) and the size of those meals (kg/meal). Similarly, the DMI can be expressed as a function of the total time a cow spends feeding per day (min/day) multiplied by the rate (kg dry matter[DM]min) at which she consumes that feed. Thus, if a cow is to consume more feed, she needs to adjust some aspect of her feeding behavior. Johnston and DeVries (2018) demonstrated that gains in DMI may be more consistent by getting cows to spend more time feeding at the bunk, broken up into smaller and more frequent meals. Thus, maximizing time and opportunities to eat, to ensure high levels of DMI, is critical. This is particularly important for post-partum cows, as insufficient DMI after calving may result in non-adaptive negative-energy balance, which may lead to subclinical ketosis (SCK) and further related negative consequences for health, production, and reproduction. Maximizing time spent feeding at the bunk, in smaller meals, is also important for keeping the rumen stable, by avoiding large post-prandial drops in rumen pH associated with large meals and resultant risk of sub-acute ruminal acidosis (SARA) (Krause and Oetzel 2006). That type of eating behavior is also more conducive to improved feed efficiency (Ben Meir et al. 2018).

Rumination behavior is highly important for dairy cattle (Beauchemin 2018). Dairy cattle rely on the process of rumination to fully digest their food. Secondary mastication and salivation during rumination serves to assist in the breakdown of particles, which not only also for greater microbial activity, thus increasing the rate of fermentation (Welch 1982), but also helps assist in passage of material from the rumen. Thus, rumination also contributes to the ability of cows to maximize their DMI. Rumination also serves to stimulate saliva production and, therefore, assists in rumen buffering and maintenance of a stable rumen environment (Beauchemin 1991). While rumination time is largely dictated by the type and amount of feed consumed, factors that influence the daily activity patterns of cows also have the potential to influence rumination. Dairy cows typically ruminate in a diurnal pattern during the time periods when the animal is at rest (lying down) and not active (feeding, milking). As such, most rumination activity occurs at night, with other major bouts of rumination occurring during the middle of the day in-between other periods of activity (DeVries et al. 2009). Position while ruminating may also be important; McWilliams et al. (2022) recently demonstrated that cows who are more likely to ruminate while lying down spend more time ruminating and lying, and tend to consume more total dry matter and produce milk with greater component yields. Thus, any disruption to a cow's normal rest time, due to other factors (for example, poor stall comfort or availability, increased need to walk, and activity related to social agitation) may result in a decrease in rumination time and also negatively affect intake and production.

Dairy cattle are social animals and, thus, part of their time budget would also be spent engaged in social activity. Some of this activity may be affiliative (e.g., allogrooming), while it may also be agonistic (e.g., competition for resources). Social environment, including social regrouping, may be stressful (von Keyserlingk et al. 2008; Padua et al. 2022), cause changes in eating and resting behavior (DeVries 2019; Tucker et al. 2021), and also be linked to risk of illness (Proudfoot et al. 2012). Furthermore, changes in social behavior may be an indicator of impaired health (Galindo and Broom 2000; Weary et al. 2009; Sepulveda-Varas et al. 2016).

There is mounting empirical evidence that measures of behavior can be used to predict and identify health and welfare concerns in dairy cattle. Interest in the association between dairy cow behavior and illness is growing, particularly with an increasing availability of commercially available equipment that is readily available for the automated monitoring of dairy cow behavior. These technologies provide ready access to vast quantities of accurate and repeatable measures of behavior. Thus, with such technologies, there are new opportunities for the automation detection (through identification and prediction) of health concerns in dairy cattle, potentially allowing for better prevention and treatment protocols to be implemented. In this chapter, we review how various cow behaviors, in particular, eating, resting, and ruminating, have been associated with various production diseases, as well as review some of the technologies available and their application, both current and future, that may be used to improve cow's health and welfare.

## 2 Effects of Health Disorders on Behavior and Production

Regardless of the type of health disorder, there are generally negative consequences for production, behavior, and the overall welfare of dairy cattle. Further, more severe and clinical cases of disease cause obvious and outward signs of sickness, whereas subclinical diseases are those which present without symptoms. However, with modern precision technologies, we can monitor cow behavior and detect symptoms that appear to be subclinical to the naked eye. General trends for cow activity, rumination time, and milk yield are consistent between studies, whether comparing sick and healthy animals or looking within affected cows relative to diagnosis day. Despite large variations between cows, there are also certain variables (particularly rumination time) which consistently deviate at least 1 day before other variables (milk yield) and may act as early indicators of disease (King et al. 2017b). There are also differences across diseases such that acute health disorders (i.e., DA and mastitis) are associated with deviations from those cows' baseline automated milking systems (AMS) data, whereas more chronic disorders (i.e., SCK and lameness) are associated with significant, but subtle, longer-term changes in milk production and behavior (King et al. 2018). In the next several sections, we highlight the changes in commonly measured behavior and production variables that occur prior to the diagnosis of various forms of production disease in dairy cattle.

### 2.1 Lameness and Associated Disorders

The impacts of lameness and associated hoof and leg disorders have been the subject of great interest in the dairy industry and scientific community due to the negative effects on cow longevity, productivity, farm profitability, and public perception. When comparing lame and sound cows, most studies (Table 1) report a net loss of milk yield associated with lameness and related disorders, ranging from 1.6 to 10 kg/day (Bach et al. 2007; Van Hertem et al. 2013; King et al. 2017a). However, Steensels et al. (2017) reported only numerically lower milk yield for lame cows and Deming et al. (2013a) found no linear association between milk yield and gait score, but the majority of cows had locomotion score of 2 or 3 out of 5 in that study. In a more complex, longitudinal analysis of the impact of hoof horn lesions on

**Table 1** Associations of lameness with milk yield, rumination time, and other variables in conventional and AMS herds

| Study | Comparison(s) | Change(s) associated with lameness | No. farms | No. cows | Milking system |
|---|---|---|---|---|---|
| 1 | • Lame vs. sound | • Milk yield ↓ | 1 | 3623 | Conventional |
| 2 | • Lame vs. sound | • Milk yield ↓ | 1 | 163 | Conventional |
| 3 | • Lame vs. sound | • Rumination time ↓ | 1 | 16 | Conventional |
| 4 | • Lame vs. sound | • Rumination time: no association<br>• Time spent walking ↓ | 1 | 59 | Conventional |
| 5 | • Lame vs. sound | • Rumination time ↓<br>• AMS visits ↓ overall and overnight<br>• Refusals and fetches: no significance | 1 | 150 | AMS |
| 6 | • Lame vs. sound | • Activity in early morning/evening ↑<br>• Activity unstable throughout the day<br>• Teat cup attachment speeds unstable | 1 | 88 | AMS |
| 7 | • Lame (most cows were moderately lame) vs. sound | • Milk yield, frequency, voluntary milkings, and refusals ↓<br>• Activity ratio night:day ↑<br>• Rumination time and activity: no association | 41 | 1184 | AMS |
| 8 | • Moderately lame vs. sound | • Activity ↓ 1st hour after feed delivery or pushup<br>• Rumination time: no association | 17 | 389 | Conventional |
| 9 | • Lame vs. sound<br>• Within lame cows<br>• Relative to diagnosis | • Milk yield ↓<br>• Activity ratio night:day ↑<br>• Rumination time overnight ↓ | 1 | 118 | Conventional |
| 10 | • Lame vs. sound and within lame relative to diagnosis | • Rumination time ↓ | 1 | 198 | Conventional |
| 11 | • Lame vs. other vs. healthy by DIM | • Rumination time ↓<br>• Activity ↓<br>• Milk yield: no association | 1 | 704 | Conventional |
| 12 | • High vs. low frequency visitors | • NRS ↓ for cows visiting more AMS frequently | 8 | 256 | AMS |
| 13 | • Linear models with NRS 1–5 | • Milk yield ↓<br>• Total milking frequency and voluntary milkings ↓ | 1 | 120 | AMS |
| 14 | • Linear models with NRS 1–5 | • Milking frequency ↓<br>• Milk yield: no association | 1 | 90 | AMS |

AMS, automated milking systems; NRS, numerical rating score

Referenced articles: 1. Bicalho et al. (2008); 2. Singh et al. (2011); 3. Almeida et al. (2008); 4. Walker et al. (2008); 5. Miguel-Pacheco et al. (2014); 6. Garcia et al. (2014); 7. King et al. (2017a); 8. Weigele et al. (2018); 9. Van Hertem et al. (2013); 10. Paudyal et al. (2017); 11. Steensels et al. (2017); 12. Borderas et al. (2008); 13. Bach et al. (2007); 14. Deming et al. (2013a) milk yield, Bicalho et al. (2008) found that when controlling for early-lactation milk yield (when lame cows produced 3.2 kg/day more milk than control cows in the first 3 weeks), lame cows in fact produced 1.0 kg/day less milk than control cows throughout lactation. They then matched 603 lame cows with 603 sound cows, again accounting for early-lactation milk yield, and found that lame cows produced 1.4 kg/day less milk than control cows (Bicalho et al. 2008). Therefore, milk yield may be greater in cows about to become lame (1.1 kg/day more milk before cows were diagnosed with lameness), but once diagnosed, their production drops to that of an average cow (Green et al. 2002), and it is important to consider the previous milk yield and lactation potential of a cow when assessing her current milk production. When examining changes within cows leading up to diagnosis, milk yield can begin to decline anywhere from 1 to 14 days prior to diagnosis of lameness, foot and hock lesions, hoof treatments, and abscesses up to a total of 4 kg (Bareille et al. 2003; Lukas et al. 2009; Van Hertem et al. 2013; King et al. 2017b). However, in a later study by King et al. (2018), there were no deviations within lame cows' milk yield, only differences from healthy, non-lame cows. There may also be changes in milk temperature or conductivity that correspond with lameness, but results vary across studies (Lukas et al. 2009; King et al. 2017b, 2018). Therefore, examining milk yield alone or even other milk characteristics may not be a reliable indicator of hoof health or lameness (Table 1).

Lame cows may be less active or spend less time ruminating than healthy individuals, but the results are more inconsistent across studies (Table 1). In terms of lying time and the daily number of bouts, the literature is consistent in finding that lame cows spend more time lying down with fewer bouts per day (Ito et al. 2010; Deming et al. 2013a, b; Westin et al. 2016; King et al. 2017a; Weigele et al. 2018). Regarding activity, as measured by neck collars or assessed visually, overall daily activity as well as overnight activity can be lower for lame cows (Walker et al. 2008; Van Hertem et al. 2013; Garcia et al. 2014; Steensels et al. 2017; King et al. 2017a). In contrast, King et al. (2017a) saw no differences in daily activity between lame and sound individuals. Regarding rumination time, while some researchers report reductions of 10% or approximately 40–50 min/day (Almeida et al. 2008; Van Hertem et al. 2013; Paudyal et al. 2017; Steensels et al. 2017), others detected no differences in rumination between lame and sound cows (Walker et al. 2008; King et al. 2017a). Interestingly, Weigele et al. (2018) reported that eating time and the number of eating chews (jaw movements) were reduced in moderately lame compared with non-lame cows. Within-day variation may also be a useful variable to monitor, as Van Hertem et al. (2013) reported lame cows to have greater daytime rumination, lower nighttime rumination, and a lower ratio of night: day rumination compared to sound cows. Within cow rumination time may also decline leading up to diagnosis of lameness and hoof disorders, by 40–50 min in total across a few days (Paudyal et al. 2017) to a few weeks beforehand while cows spend proportionately more time ruminating overnight (King et al. 2017b). Those authors also saw declining body weight, and slowly declining milking frequency by 0.05 milkings/day resulting in a total reduction of approximately 0.6 daily milkings during the 2 weeks before diagnosis (King et al. 2017b). Finally, Veissier et al. (2017) noted that the circadian activity of cows varied 1–2 days prior to diagnosis of lameness.

In herds using automated or robotic milking systems, dairy producers can also monitor the frequency at which cows visit the milking unit. Researchers have noted that lame cows visit the milking unit less often each day (0.3 or 0.4 times less per day), and they do so less overnight (Bach et al. 2007; Borderas et al. 2008; Miguel-Pacheco et al. 2014; King et al. 2017a). Although Miguel-Pacheco et al. (2014) did not observe any differences in frequencies of refused milkings or fetched milkings, they had a smaller sample size than King et al. (2017a) who did detect fewer refused milkings (–0.2/day) for lame cows and increased odds of fetching cows for milking.

**Table 2** Associations of illness with variables of interest in conventional and AMS herds

| Study | Comparison(s) | Change(s) associated with illness | No. farms | No. cows | Milking system |
|---|---|---|---|---|---|
| 1 | • Within cow relative to diagnosis | • Milk yield ↓ | 1 | 551 | Conventional |
| 2 | • Within cow relative to diagnosis | • Milk yield ↓<br>• Milk conductivity ↑ | 1 | 587 | Conventional |
| 3 | • Within cow relative to diagnosis | • Milk yield, frequency, rumination time, activity ↓ | 1 | 57 | AMS |
| 4 | • Within cow relative to diagnosis | • Activity patterns (circadian rhythm) varied 1–2 days before mastitis and lameness | 1 | 350 | AMS |
| 5 | • Within cow relative to diagnosis<br>• Healthy vs. sick | • Rumination time ↓ | 1 | 198 | Conventional |
| 6 | • Within cow relative to diagnosis<br>• Healthy vs. sick | • Milk yield, frequency, rumination time, activity ↓<br>• Milk conductivity ↑ and temperature was variable | 9 | 605 | AMS |
| 7 | • Within cow relative to diagnosis<br>• Healthy vs. sick by DIM | • Milk yield ↓<br>• Activity (steps/h) ↓ | 3 | 1445 | Conventional |
| 8 | • Within cow relative to diagnosis<br>• Healthy vs. sick by DIM | • Rumination time and activity ↓<br>• No differences within cow | 1 | 704 | Conventional |
| 9 | • RP, metritis, SCK vs. healthy by DIM | • Rumination time and activity pre-calving: no association<br>• Rumination time and activity post-calving ↓ | 1 | 296 | Conventional |
| 10 | • Deviations around diagnosis of SCK and MET (day –3 to +3 vs. day –6 to –4 and day +4 to +6) | • Rumination time ↓ with SCK (PP and MP cows)<br>• BW ↓ with metritis (PP cows only)<br>• Rumination time ↓ with metritis (MP cows only) | ,3 | SCK = 20, MET = 84 | Conventional |
| 11 | • Metritis vs. healthy by DIM | • Milk yield ↓ | 1 | 62 | Conventional |
| 12 | • Metritis vs. healthy by DIM | • Milk yield ↓ for MP cows, but not PP cows | 1 | 84 | Conventional |
| 13 | • SHC vs. healthy by DIM | • Milk yield ↑ during week 2–4 of lactation | 1 | 30 | Conventional |
| 14 | • Relative to mastitis induction | • Daily rumination time: no association<br>• 2-h rumination time intervals ↓ 4 and 8 h after<br>• Activity (steps/day) ↑ | 1 | 6 | Conventional tie-stall |
| 15 | • Relative to mastitis induction | • Milk yield ↓<br>• Rumination time ↓ | 1 | 20 | Conventional tie-stall |
| 16 | • Mastitis vs. healthy after diagnosis | • Activity (steps/h) ↓ | 1 | 60 | Conventional |
| 17 | • Relative to acidosis induction | • Rumination time ↓ | 1 | 8 | Conventional tie-stall |
| 18 | • SCK vs. healthy by DIM | • Milk yield/daily change: no association | 1 | 62 | Conventional |

| | | | | | |
|---|---|---|---|---|---|
| 19 | • SCK and SCK+ vs. healthy by DIM | • Rumination time pre- and post-calving for SCK+ cows ↓ | 4 | 202 cows | Conventional |
| 20 | • SCK vs. healthy by DIM | • Rumination time pre-calving ↓<br>• Rumination time post-calving: no association | 1 | 80 | Conventional |
| 21 | • Ketosis, mastitis, DA, metritis vs. healthy relative to day of diagnosis (from day –5 to +5) | • Rumination time ↓ for all<br>• Activity ↓ for all<br>• Milk yield ↓ all except metritis | 1 | 1121 | Conventional |

AMS, automated milking systems. Free-stall housing unless specified otherwise; DD, digestive disorders; SCK, subclinical ketosis; SCK+, SCK plus another health disorder; SCH, subclinical hypocalcemia; RP, retained placenta; DA, displaced abomasum; MET, metritis

1. Bareille et al. (2003); 2. Lukas et al. (2009); 3. King et al. (2017b); 4. Veissier et al. (2017); 5. Paudyal et al. (2017); 6. King et al. (2018); 7. Edwards and Tozer (2004); 8. Steensels et al. (2017); 9. Liboreiro et al. (2015); 10. Gáspárdy et al. (2014); 11. Huzzey et al. (2007); 12. Wittrock et al. (2011); 13. Jawor et al. (2012); 14. Siivonen et al. (2011); 15. Fogsgaard et al. (2012); 16. Fogsgaard et al. (2015); 17. DeVries et al. (2009); 18. Goldhawk et al. (2009); 19. Kaufman et al. (2016); 20. Schirmann et al. (2016); 21. Stangaferro et al. (2016a, b, c)

## 2.2 Reproductive and Calving-Related Disorders

Reductions in milk yield are also commonly seen with metritis, whether it is puerperal or chronic metritis (Bareille et al. 2003; Lukas et al. 2009; Table 2). Other changes associated with metritis include reductions in activity, rumination time, and body weight (Huzzey et al. 2007; Stangaferro et al. 2016c; King et al. 2017b; Paudyal et al. 2017). Compared to healthy cows, those with retained placenta and/or metritis were less active and ruminated less after calving (Liboreiro et al. 2015). This can be related to the cows' feeding behavior, as cows diagnosed with metritis have been reported to consume less feed in fewer meals (Neave et al. 2018). Those cows were also replaced more often by others at the feed bunk (Neave et al. 2018) and were less likely to displace others (Huzzey et al. 2007). Cows with metritis spent more time lying down in fewer bouts than healthy cows (Neave et al. 2018). However, Barragan et al. (2018) found that lying time was only different between primiparous cows with and without clinical metritis, and there were no such effects for multiparous cows. However, when comparing healthy cows to those with severe endometritis (purulent vaginal discharge assessed by scoring a sample of vaginal discharge 28–35 DIM), very few differences were detected in terms of milk yield, rumination time, activity, or other milk-related variables (King et al. 2018). Therefore, as noted by Stangaferro et al. (2016c), using rumination time and activity as indicators of metritis is more effective for severe cases and not necessarily mild cases of metritis. It is interesting to note that other behaviors may be useful to detect disease events, like metritis. For example, Mandel et al. (2017) demonstrated that the use of an automated cow brush was lower in metritic cows as compared to their healthy counterparts. Those authors concluded that health events may be more efficiently detected with the on-farm monitoring of low-resilience behaviors, such as grooming, in combination with other systems that monitor basic core behaviors (e.g., activity and rumination).

## 2.3 Mastitis

As expected, mastitis also causes reductions in milk yield (Bareille et al. 2003; Table 2) due to changes in feeding behavior such that sick cows replace others less often at the feed bin (Sepulveda-Varas et al. 2016) and they spend much less time feeding (Fogsgaard et al. 2012) and ruminating (Siivonen et al. 2011; 2–7 h/day less in Paudyal et al. 2017). Mastitis also causes reductions in daily activity (Paudyal et al. 2017) and the distribution of activity throughout the day (Veissier et al. 2017), such that mastitis cows are less active than healthy cows (King et al. 2018). What is unique about mastitis in terms of lying behavior is that infected cows spend less time lying down and spend less time lying on the side of the infected quarter (Siivonen et al. 2011; Fogsgaard et al. 2012). In herds with automated milking systems, prior to diagnosis of mastitis, those cows exhibited sharp drops in milk yield, rumination time, and milking frequency, whereas milk conductivity rose sharply (King et al. 2018). Milk conductivity is often seen to increase rapidly with the onset of mastitis (Lukas et al. 2009) due to increased permeability of mammary secretory cells and leakage of ions into the milk.

## 2.4 Digestive and Metabolic Disorders

Digestive and metabolic disorders include diarrhea, ketosis, acidosis, displaced abomasum, milk fever, and "going off feed" in general. Some studies have grouped these disorders together to determine that rumination time and activity begin to decline 1–9 days prior to their diagnosis (Edwards and Tozer 2004; Stangaferro et al. 2016a; Table 2). Milk yield decreases for up to 5 days before diagnosis of digestive disorders and their reoccurrence (Bareille et al. 2003). Looking specifically at sub-acute ruminal acidosis, DeVries et al. (2009) found that rumination time declined by >1.5 h/day on the day of induction. Prior to displaced abomasum (DA), cows showed a drastic reduction in activity, rumination time, milk yield, and milk conductivity, which all remained lower compared to healthy cows, as were their milking frequency and supplement intake (King et al. 2017b). Lukas et al. (2009) also noted changes in milk conductivity prior to milk fever, ketosis, DA, and cows going off feed.

Not only does clinical disease affect cow behavior and productivity, but subclinical disease can have substantial consequences as well (Table 2). For example, subclinical ketosis (SCK) reduces milk yield, rumination time, body weight, and activity (King et al. 2017b, 2018). Before calving, cows later diagnosed with SCK displaced others less often, consumed less feed, and visited the feeder less often for less time overall than healthy cows (Goldhawk et al. 2009). Some researchers documented changes in lying time (Itle et al. 2015) and rumination time (Kaufman et al. 2016; King et al. 2018). However, others found no effect of SCK on lying time, number of bouts, or bout length (Kaufman et al. 2016), milk yield (Goldhawk et al. 2009), or pre-calving rumination time (Liboreiro et al. 2015). The latter research group did record post-calving reductions of up to 80 min/day rumination time and 110 daily activity units, whereas Schirmann et al. (2016) only observed reductions in rumination time before calving.

With regard to hypocalcemia, on the day of calving as well as the day before, cows with clinical hypocalcemia (milk fever) spent more time lying down and were less active than healthy cows and those with subclinical hypocalcemia (Hendriks et al. 2020). There was also, in that research, a relationship between the relative change in daily steps the week before calving with the concentration of calcium in the blood, demonstrating that a greater decline in activity was associated with more severe hypocalcemia. Liboreiro et al. (2015) did not observe this relationship with subclinical hypocalcemia, but they did report lower rumination time on the day before calving (Table 2). Finally, cows with subclinical hypocalcemia have been shown to produce more milk in weeks 2–4 in lactation, but do not necessarily consume more water or feed (Jawor et al. 2012). However, in that study those cows did consume more dry matter intake in the 2 weeks before calving and in the second week after calving. Cows with subclinical hypocalcemia also stood up for 2.6 h longer the day before calving, but then spent 2.7 h less time standing the day after calving.

### 2.5 Respiratory Disorders

Prior to diagnosis with pneumonia, cows produce less milk, ruminate less, weigh less, and are less active than their baseline King et al. (2017b; Table 2). Lukas et al. (2009) also observed declining milk yield as well as increasing conductivity before diagnoses of pneumonia. Pneumonia is more commonly an issue with younger animals, who may also show changes in behavior. Calves with bovine respiratory disease (BRD) and those who relapsed exhibited depressed solid feed intake and longer lying times than healthy calves and those who recovered (Cantor and Costa 2022; Cantor et al. 2022). However, in another study, only calves with clinical BRD drank their milk more slowly than those with subclinical or no BRD and so perhaps milk-feeding behavior is not a useful indicator of subclinical BRD (Cramer et al. 2020). Larger differences were seen between healthy calves and those with respiratory disease and/or scours by Duthie et al. (2021), who reported longer lying times, and shorter and less frequent milk feeding times for sick calves.

## 3 Using Technology to Detect Disease Events

Farmers can observe and monitor cow behavior in numerous ways to assess the health of their cows. In fact, they have been doing that for many years—simply by observing the basic behaviors of their cows, including eating, lying, and ruminating. So much so that there have even been recommendations made for such observations. As example, many have suggested that when at least 40% of cows in a herd are ruminating at any given time that is indicative of good rumen function. DeVries et al. (2009) provided some empirical evidence supporting this guideline, specifically in a group of healthy cows, on average, 40% were ruminating at times of the day when rumination behavior was expected. Alternatively, they demonstrated that 10% fewer cows were ruminating during these peak rumination times when a ruminal acidosis event was occurring. However, DeVries et al. (2009) suggested that one may not be able to detect suboptimal rumen function (i.e., a herd-level acidosis event) using this criterion if only a single observation of a herd was undertaken. Rather, they calculated based on variability observed, that to detect a herd-level ruminal acidosis event, greater than 48 individual observations would be required to accurately estimate the percentage of cows ruminating within a herd. Such a task would be onerous, and potentially inaccurate if conducted visually; however, new monitoring technologies that allow for the objective, repeatable, and automated capture of various behaviors (including rumination, activity, lying, and locomotion) make this type of behavioral monitoring feasible. However, with all these data being collected comes the need to transform it into timely, useful, reliable, and actionable information for producers. Therefore, not only do the monitors themselves need to be validated to measure what they claim to measure, but also the models and algorithms making predictions must be validated for accuracy and timeliness. This will help to reduce the number of excessive alerts (false positives) as well as the number of sick animals predicted as being healthy (false negatives).

Mastitis alerts can include predictions based on teat attachment speed, lactate dehydrogenase, milk yield, cell count, conductivity, and temperature. Researchers can predict mastitis diagnosis using deviations from 5-day moving averages and standard deviations of milk yield and conductivity (Miekley et al. 2013), online somatic cell counters (Dalen et al. 2019), and even a combination of milk yield, electrical conductivity, average milk flow rate, occurrence of incompletely milked quarters in each milking session, MY per hour, and conductivity per hour between successive milking sessions (Khatun et al. 2018). Deep learning models (e.g., neural network models) have, in particular, shown promise for developing mastitis probability indexes that are reliable and timely for disease identification (e.g., Naqvi et al. 2021). Ketosis can be assessed by monitoring milk BHB. Reproductive status can be assessed by looking at milk progesterone and activity. Numerous health disorders can be flagged by monitoring changes in body weight, rumination time, and milk yield. For example, Stangaferro et al. (2016a, b, c) assessed the validity of a health index score, composed of rumination and activity data during a 5-day period before diagnosis, in a conventional herd. Steensels et al. (2016) used a decision tree model to detect ketosis and metritis by considering rumination time and activity (measured by neck collar monitors), as well as milk yield, milk slope, and current BW relative to BW at calving. Regarding lameness detection, Garcia et al. (2014) created models using activity (measured by neck collar monitors), milk flow, and teat cup attachment data from 1 ± week of lameness diagnosis and de Mol et al. (2013) combined several activity variables (i.e., number of steps negatively associated with lameness, and lying behavior data with various associations, measured by leg band accelerometers), milk yield per day (≥3 negative outliers/week), and concentrate leftovers per day (positively associated with lameness). Van Hertem et al. (2013) also created a lameness detection model, which used milk yield, activity, and rumination data from 4 to 7 days before diagnosis. Other researchers have incorporated similar variables into lameness detection models, such as BW, activity, milk order, walking speed, standing bouts, lying time, and limb weight ratio during milking (Kamphuis et al. 2013; Beer et al. 2016; Nechanitzky et al. 2016). However, most of the studies mentioned above excluded fresh cows, cows affected by health disorders other than lameness, or cows with mild/moderate cases of illness/lameness, which would reduce their false positive rates.

Adoption of these technologies by farmers has been reported to depend on five key factors: (1) relative advantage (to what they replace), (2) compatibility (with needs, experiences, values, current systems), (3) complexity (perceived difficulty to implement and understand), (4) trialability (ability to test it out), and (5) observability (how visible the results are to others; Stone 2020). It is also crucial that farmers trust the information being provided by technologies and their associated alerts. In one study (Eckelkamp and Bewley 2020), out of >24,000 alerts sent to farmers on 4 farms, only 8% of the alerts were doubted by producers and 55% of alerts were perceived to be true, especially if they were for fresh or early-lactation cows. Producers also responded more to alerts for eating time, activity, or a combination of multiple behaviors. However, there was a lot of variation between farms in terms of which cows the producers would visually assess (2–45%, average 21%). Over time and with a greater number of alerts per day, producers were more likely to completely ignore alerts (Eckelkamp and Bewley 2020). Therefore, these alerts must provide reliable and actionable information to producers while being compatible and integrated with their existing software in order to be installed and truly useful for improving animal health management.

## 4 Conclusions and Future Directions

The dairy cattle industry has seen an increasing availability of equipment that is readily available for the automation of management tasks, including milking and feeding, as well as the precision monitoring of dairy cow behavior. Such automation not only has the ability to improve production and time efficiency on farm, but also increases our ability to monitor individual cows (DeVries 2019).

Much recent research has been focused on the use of individual behavioral monitoring to detect health disorders, both in occurrence and in advance of clinical symptoms. This, in turn, has the potential to allow producers to identify and implement prevention and treatment protocols at earlier time points. Many different individual cow behaviors can now be collected automatically on farm, including feeding, rumination, lying/standing, and milking behavior. Thus, much research has been done in recent times to investigate the association these behaviors have with various health issues, including metabolic and infectious transition disease, lameness, and mastitis (King and DeVries 2018). Much progress has been made in this area, in particular, through a combination of use of multiple behavioral inputs, as well as other cow-level data (production, age, size) that may be collected in real time, to improve our predictive ability of health events. The predictive abilities of using such information are currently being strengthened through investigation of more advanced data modelling techniques (e.g., machine learning).

Despite great potential for the use of automated behavioral monitoring for early disease detection, the fact of the matter is that to make significant improvements in animal welfare as well as farm economic sustainability, our focus on disease control in dairy cattle must be more centered on disease prevention rather than detection. To that end, there is a great opportunity to use behavioral monitoring to inform management decisions on farms, particularly those that are going to reduce the incidence of disease, and thus improve overall herd health, productivity, and efficiency. For example, by monitoring behavior (e.g., eating, resting, and rumination), both at an individual and herd level, changes in response to nutritional and management alternations can be detected, and subsequent decisions can be made to ensure continued good behavior, and thus maintain cow health, productivity, and efficiency.

In conclusion, it is anticipated that through automated precision monitoring of dairy cattle behavior, dairy cattle producers will not only be able to identify and treat production diseases earlier, but also make more timely decisions for altering and adjusting nutritional programs, housing, and management, both at a cow and herd level, in effort to reduce the risk of production disease, and then also as result improve welfare, production, and efficiency.

## References

Almeida PE, Weber PSD, Burton JL, Zanella AJ (2008) Depressed DHEA and increased sickness response behaviors in lame dairy cows with inflammatory foot

lesions. Domest Anim Endocrinol 34:89–99. https://doi.org/10.1016/j.domaniend.2006.11.006
Bach A, Dinarés M, Devant M, Carré X (2007) Associations between lameness and production, feeding and milking attendance of Holstein cows milked with an automatic milking system. J Dairy Res 74:40–46. https://doi.org/10.1017/S0022029906002184
Bareille N, Beaudeau F, Billon S, Robert A, Faverdin P (2003) Effects of health disorders on feed intake and milk production in dairy cows. Livest Prod Sci 83:53–62. https://doi.org/10.1016/S0301-6226(03)00040-X
Barragan AA, Piñeiro JM, Schuenemann GM, Rajala-Schultz PJ, Sanders DE, Lakritz J, Bas S (2018) Assessment of daily activity patterns and biomarkers of pain, inflammation, and stress in lactating dairy cows diagnosed with clinical metritis. J Dairy Sci 101:8248–8258. https://doi.org/10.3168/jds.2018-14510
Beauchemin KA (1991) Effects of dietary neutral detergent fiber concentration and alfalfa hay quality on chewing, rumen function, and milk production of dairy cows. J Dairy Sci 74:3140–3151. https://doi.org/10.3168/jds.S0022-0302(91)78499-3
Beauchemin KA (2018) Invited review: Current perspectives on eating and rumination activity in dairy cows. J Dairy Sci 101:4762–4784. https://doi.org/10.3168/jds.2017-13706
Beer G, Alsaaod M, Starke A, Schuepbach-Regula G, Mülller H, Kohler P, Steiner A (2016) Use of extended characteristics of locomotion and feeding behavior for automated identification of lame dairy cows. PLoS One 11:1–19. https://doi.org/10.1371/journal.pone.0155796
Ben Meir YA, Nikbachat M, Fortnik Y, Jacoby S, Levit H, Adin G, Cohen Zinder M, Shabtay A, Gershon E, Zachut M, Mabjeesh SJ, Halachmi I, Miron J (2018) Eating behavior, milk production, rumination, and digestibility characteristics of high- and low-efficiency lactating cows fed a low-roughage diet. J Dairy Sci 101:10973–10984. https://doi.org/10.3168/jds.2018-14684
Bicalho RC, Warnick LD, Guard CL (2008) Strategies to analyze milk losses caused by diseases with potential incidence throughout the lactation: a lameness example. J Dairy Sci 91:2653–2661. https://doi.org/10.3168/jds.2007-0744
Borderas TF, Fournier A, Rushen J, de Passillé AMB (2008) Effect of lameness on dairy cows' visits to automatic milking systems. Can J Anim Sci 88:1–8. https://doi.org/10.4141/CJAS07014
Cantor MC, Costa JHC (2022) Daily behavioral measures recorded by precision technology devices may indicate bovine respiratory disease status in preweaned dairy calves. J Dairy Sci 96:318–328. https://doi.org/10.3168/jds.2021-20798
Cantor MC, Renaud DL, Neave HW, Costa JHC (2022) Feeding behavior and activity levels are associated with recovery status in dairy calves treated with antimicrobials for Bovine Respiratory Disease. Sci Rep 12:4854. https://doi.org/10.1038/s41598-022-08131-1
Chapinal N, Barrientos AK, von Keyserlingk MAG, Galo E, Weary DM (2013) Herd-level risk factors for lameness in freestall farms in the northeastern United States and California. J Dairy Sci 96:318–328. https://doi.org/10.3168/jds.2012-5940
Cramer C, Proudfoot K, Ollivett T (2020) Automated feeding behaviors associated with subclinical respiratory disease in preweaned dairy calves. Animals 10:988. https://doi.org/10.3390/ani10060988
Dalen G, Rachah A, Nørstebø H, Schukken YH, Reksen O (2019) The detection of intramammary infections using online somatic cell counts. J Dairy Sci 102:5419–5429. https://doi.org/10.3168/jds.2018-15295
Deming JA, Bergeron R, Leslie KE, DeVries TJ (2013a) Associations of cow-level factors, frequency of feed delivery, and standing and lying behaviour of dairy cows milked in an automatic system. Can J Anim Sci 93:427–433. https://doi.org/10.4141/CJAS2013-055
Deming JA, Bergeron R, Leslie KE, DeVries TJ (2013b) Associations of housing, management, milking activity, and standing and lying behavior of dairy cows milked in automatic systems. J Dairy Sci 96:344–351. https://doi.org/10.3168/jds.2012-5985
de Mol RM, André G, Bleumer EJB, van der Werf JTN, de Haas Y, van Reenen CG (2013) Applicability of day-to-day variation in behavior for the automated detection of lameness in dairy cows. J Dairy Sci 96:3703–3712. https://doi.org/10.3168/jds.2012-6305
DeVries TJ (2017) Fresh Cows – Management for best behavior. Proceedings of the 13th Western Dairy Management Conference. Reno, Nevada, pp 154–167
DeVries TJ (2019) Feeding behavior, feed space, and bunk design and management for adult dairy cattle. Vet Clin North Am Food Anim Pract 35:61–76. https://doi.org/10.1016/j.cvfa.2018.10.003
DeVries TJ, Beauchemin KA, Dohme F, Schwartzkopf-Genswein KS (2009) Repeated ruminal acidosis challenges in lactating dairy cows at high and low risk for developing acidosis: Feeding, ruminating, and lying behavior. J Dairy Sci 92:5067–5078. https://doi.org/10.3168/jds.2009-2102
Duthie CA, Bowen JM, Bell DJ, Miller GA, Mason C, Haskell MJ (2021) Feeding behaviour and activity as early indicators of disease in pre-weaned dairy calves. Animal 15:100150. https://doi.org/10.1016/j.animal.2020.100150
Eckelkamp EA, Bewley JM (2020) On-farm use of disease alerts generated by precision dairy technology. J Dairy Sci 103:1566–1582. https://doi.org/10.3168/jds.2019-16888
Edwards JL, Tozer PR (2004) Using activity and milk yield as predictors of fresh cow disorders. J Dairy Sci 87:524–531. https://doi.org/10.3168/jds.S0022-0302(04)73192-6
Fogsgaard KK, Røntved CM, Sørensen P, Herskin MS (2012) Sickness behavior in dairy cows during Escherichia coli mastitis. J Dairy Sci 95:630–638. https://doi.org/10.3168/jds.2011-4350
Fogsgaard KK, Bennedsgaard TW, Herskin MS (2015) Behavioral changes in freestall-housed dairy cows

with naturally occurring clinical mastitis. J Dairy Sci 98:1730–1738. https://doi.org/10.3168/jds.2014-8347
Galindo F, Broom GM (2000) The relationships between social behaviour of dairy cows and the occurrence of lameness in three herds. Res Vet Sci 69:75–79. https://doi.org/10.1053/rvsc.2000.0391
Garcia E, Klaas I, Amigo JM, Bro R, Enevoldsen C (2014) Lameness detection challenges in automated milking systems addressed with partial least squares discriminant analysis. J Dairy Sci 97:7476–7486. https://doi.org/10.3168/jds.2014-7982
Gáspárdy A, Efrat G, Bajcsy AC, Fekete SG (2014) Electronic monitoring of rumination activity as an indicator of health status and production traits in high-yielding dairy cows. Acta Vet Hung 62:452–462. https://doi.org/10.1556/AVet.2014.026
Goldhawk C, Chapinal N, Veira DM, Weary DM, von Keyserlingk MAG (2009) Prepartum feeding behavior is an early indicator of subclinical ketosis. J Dairy Sci 92:4971–4977. https://doi.org/10.3168/jds.2009-2242
Grant RJ (2004) Incorporating dairy cow behavior into management tools. In: Proceedings of Cornell nutrition conference for feed manufacturers, October 19–21, Syracuse, NY, pp 65–76
Green LE, Hedges VJ, Schukken YH, Blowey RW, Packington AJ (2002) The impact of clinical lameness on the milk yield of dairy cows. J Dairy Sci 85:2250–2256. https://doi.org/10.3168/jds.S0022-0302(02)74304-X
Hendriks SJ, Huzzey JM et al (2020) Associations between lying behavior and activity and hypocalcemia in grazing dairy cows during the transition period. J Dairy Sci 103:10530–10546. https://doi.org/10.3168/jds.2019-18111
Huzzey JM, Veira DM, Weary DM, von Keyserlingk MAG (2007) Prepartum behavior and dry matter intake identify dairy cows at risk for metritis. J Dairy Sci 90:3220–3233. https://doi.org/10.3168/jds.S0022-0302(06)72075-6
Itle AJ, Huzzey JM, Weary DM, von Keyserlingk MAG (2015) Clinical ketosis and standing behavior in transition cows. J Dairy Sci 98:128–134. https://doi.org/10.3168/jds.2014-7932
Ito K, von Keyserlingk MAG, LeBlanc SJ, Weary DM (2010) Lying behavior as an indicator of lameness in dairy cows. J Dairy Sci 93:3553–3560. https://doi.org/10.3168/jds.2009-2951
Jawor PE, Huzzey JM, LeBlanc SJ, von Keyserlingk MA (2012) Associations of subclinical hypocalcemia at calving with milk yield, and feeding, drinking, and standing behaviors around parturition in Holstein cows. J Dairy Sci 95:1240–1248. https://doi.org/10.3168/jds.2011-4586
Jensen MB, Pedersen LJ, Munksgaard L (2005) The effect of reward duration on demand functions for rest in dairy heifers and lying requirements as measured by demand functions. Appl Anim Behav Sci 90:207–217
Johnston C, DeVries TJ (2018) Associations of feeding behavior and milk production in dairy cows. J Dairy Sci 101:3367–3373. https://doi.org/10.3168/jds.2017-13743
Kamphuis C, Frank E, Burke JK, Verkerk GA, Jago JG (2013) Applying additive logistic regression to data derived from sensors monitoring behavioral and physiological characteristics of dairy cows to detect lameness. J Dairy Sci 96:7043–7053. https://doi.org/10.3168/jds.2013-6993
Kaufman E, LeBlanc SJ, McBride BW, Duffield TF, DeVries TJ (2016) Association of rumination time with subclinical ketosis in transition dairy cows. J Dairy Sci 99:5604–5618. https://doi.org/10.3168/jds.2015-10509
Khatun M, Thomson PC, Kerrisk KL, Lyons NA, Clark CEF, Molfino J, García SC (2018) Development of a new clinical mastitis detection method for automatic milking systems. J Dairy Sci 101:9385–9395. https://doi.org/10.3168/jds.2017-14310
King MTM, DeVries TJ (2018) Graduate Student Literature Review: Detecting health disorders using data from automatic milking systems and associated technologies. J Dairy Sci 101:8605–8614. https://doi.org/10.3168/jds.2018-14521
King MTM, LeBlanc SJ, Pajor EA, DeVries TJ (2017a) Cow-level associations of lameness, behavior, and milk yield of cows milked in automated systems. J Dairy Sci 100:4818–4828. https://doi.org/10.3168/jds.2016-12281
King MTM, Dancy KM, LeBlanc SJ, Pajor EA, DeVries TJ (2017b) Deviations in behavior and productivity data before diagnosis of health disorders in cows milked with an automated system. J Dairy Sci 100:8358–8371. https://doi.org/10.3168/jds.2017-12723
King MTM, Dancy KM, LeBlanc SJ, Pajor EA, Devries TJ (2018) Behavior and productivity of cows milked in automated systems before diagnosis of health disorders in early lactation. J Dairy Sci 101:4343–4356. https://doi.org/10.3168/jds.2017-13686
Krause KM, Oetzel GR (2006) Understanding and preventing subacute ruminal acidosis in dairy herds: A review. Anim Feed Sci Technol 126:215–236
Liboreiro DN, Machado SK, Silva PRB, Maturana MM, Nishimura TK, Brandão AP, Endres MI, Chebel RC (2015) Characterization of peripartum rumination and activity of cows diagnosed with metabolic and uterine diseases. J Dairy Sci 98:6812–6827. https://doi.org/10.3168/jds.2014-8947
Lukas JM, Reneau JK, Wallace R, Hawkins D, Munoz-Zanz C (2009) A novel method of analyzing daily milk production and electrical conductivity to predict disease onset. J Dairy Sci 92:5964–5976. https://doi.org/10.3168/jds.2009-2066
Mandel R, Nicol CJ, Whay HR, Klement E (2017) *Short communication:* Detection and monitoring of metritis in dairy cows using an automated grooming device. J Dairy Sci 100:5724–5728. https://doi.org/10.3168/jds.2016-12201
McWilliams CJ, Schwanke AJ, DeVries TJ (2022) Is greater milk production associated with dairy cows who have a greater probability of ruminating while

lying down? JDS Commun 3:66–71. https://doi.org/10.3168/jdsc.2021-0159

Miekley B, Traulsen I, Krieter J (2013) Mastitis detection in dairy cows: The application of support vector machines. J Agric Sci 151:889–897. https://doi.org/10.1017/S0021859613000178

Miguel-Pacheco GG, Kaler J, Remnant J, Cheyne L, Abbott C, French AP, Pridmore TP, Huxley JN (2014) Behavioural changes in dairy cows with lameness in an automatic milking system. Appl Anim Behav Sci 150:1–8. https://doi.org/10.1016/j.applanim.2013.11.003

Munksgaard L, Jensen MB, Pedersen LJ, Hansen SW, Matthews L (2005) Quantifying behavioural priorities—Effects of time constraints on behaviour of dairy cows, *Bos taurus*. Appl Anim Behav Sci 92:3–14

Naqvi SA, King MTM, Matson RD, DeVries TJ, Deardon R, Barkema H (2021) Data considerations for developing deep learning models for dairy applications: a simulation study on mastitis detection. Comput Electron Agric 192:106895. https://doi.org/10.1016/j.compag.2021.106618

Neave HW, Lomb J, Weary DM, LeBlanc SJ, Huzzey JM, von Keyserlingk MAG (2018) Behavioral changes before metritis diagnosis in dairy cows. J Dairy Sci 101:4388–4399. https://doi.org/10.3168/jds.2017-13078

Nechanitzky K, Starke A, Vidondo B, Müller H, Reckardt M, Friedli K, Steiner A (2016) Analysis of behavioral changes in dairy cows associated with claw horn lesions. J Dairy Sci 99:2904–2914. https://doi.org/10.3168/jds.2015-10109

Padua FH, Dancy KM, Bergeron R, DeVries TJ (2022) Impact of a dietary citrus extract on the behavior and production of lactating dairy cows following regrouping: a preliminary study. Front Vet Sci 8:773399. https://doi.org/10.3389/fvets.2021.773399

Paudyal S, Maunsell FP, Richeson JT, Risco CA, Donovan DA, Pinedo PJ (2017) Rumination time and monitoring of health disorders during early lactation. Animal 12:1484–1492. https://doi.org/10.1017/S1751731117002932

Proudfoot KL, Weary DM, von Keyserlingk MAG (2010) Behavior during transition differs for cows diagnosed with claw horn lesions in mid lactation. J Dairy Sci 93:3970–3978

Proudfoot KL, Weary DM, von Keyserlingk MAG (2012) Linking the social environment to illness in farm animals. Appl Anim Behav Sci 138:203–215. https://doi.org/10.1016/j.applanim.2012.02.008

Schirmann K, Weary DM, Heuwieser W, Chapinal N, Cerri RLA, von Keyserlingk MAG (2016) Short communication: Rumination and feeding behaviors differ between healthy and sick dairy cows during the transition period. J Dairy Sci 99:9917–9924. https://doi.org/10.3168/jds.2015-10548

Sepulveda-Varas P, Proudfoot KL, Weary DM, von Keyserlingk MAG (2016) Changes in behavior of dairy cows with clinical mastitis. Appl Anim Behav Sci 175:8–13

Siivonen J, Taponen S, Hovinen M, Pastell M, Lensink BJ, Pyörälä S, Hänninen L, Fogsgaard KK, Røntved CM, Sørensen P, Herskin MS, Haley DB, Rushen J, de Passillé AM (2011) Impact of acute clinical mastitis on cow behaviour. Can J Anim Sci 80:257–263. https://doi.org/10.4141/A99-084

Singh Y, Lathwal SS, Chakravarty AK, Gupta AK, Mohanty TK, Raja TV, Dangi RL, Roy BK (2011) Effect of lameness (hoof disorders) on productivity of Karan Fries crossbred cows. Anim Sci J 82:169–174. https://doi.org/10.1111/j.1740-0929.2010.00800.x

Stangaferro M, Wijma R, Quinteros C, Medrano M, Masello M, Giordano J (2016a) Use of a rumination and activity monitoring for the identification of dairy cows with health disorders: Part II. Mastitis. J Dairy Sci 99:7411–7420. https://doi.org/10.3168/jds.2016-10908

Stangaferro M, Wijma R, Quinteros C, Medrano M, Masello M, Giordano J (2016b) Use of a rumination and activity monitoring for the identification of dairy cows with health disorders: Part III. Metritis. J Dairy Sci 99:7422–7433. https://doi.org/10.3168/jds.2016-10908

Stangaferro M, Wijma R, Quinteros C, Medrano M, Masello M, Giordano J (2016c) Use of a rumination and activity monitoring for the identification of dairy cows with health disorders: Part I. Metabolic and digestive disorders. J Dairy Sci 99:7395–7410. https://doi.org/10.3168/jds.2016-10908

Steensels M, Antler A, Bahr C, Berckmans D, Halachmi I (2016) A decision-tree model to detect post-calving diseases based on rumination, activity, milk yield, BW and voluntary visits to the milking robot. Animal 10:1493–1500. https://doi.org/10.1017/S1751731116000744

Steensels M, Maltz E, Bahr C, Berckmans D, Antler A, Halachmi I (2017) Towards practical application of sensors for monitoring animal health: the effect of post-calving health problems on rumination duration, activity and milk yield. J Dairy Res 84:132–138. https://doi.org/10.1017/S0022029917000176

Stone A (2020) Symposium review: The most important factors affecting adoption of precision dairy monitoring technologies. J Dairy Sci 103:5740–5745. https://doi.org/10.3168/jds.2019-17148

Tucker CB, Jensen MB, de Passillé AM, Hänninen L, Rushen J (2021) Invited review: Lying time and the welfare of dairy cows. J Dairy Sci 104:20–46. https://doi.org/10.3168/jds.2019-18074

Van Hertem T, Maltz E, Antler A, Romanini CEB, Viazzi S, Bahr C, Schlageter-Tello A, Lokhorst C, Berckmans D, Halachmi I (2013) Lameness detection based on multivariate continuous sensing of milk yield, rumination, and neck activity. J Dairy Sci 96:4286–4298. https://doi.org/10.3168/jds.2012-6188

Veissier I, Mialon MM, Sloth KH (2017) Short communication: Early modification of the circadian organization of cow activity in relation to disease or estrus. J Dairy Sci 100:3969–3974. https://doi.org/10.3168/jds.2016-11853

von Keyserlingk MAG, Olineck D, Weary DM (2008) Acute behavioral effects of regrouping dairy cows. J Dairy Sci 91:1011–1016. https://doi.org/10.3168/jds.2007-0532

Walker SL, Smith RF, Routly JE, Jones DN, Morris MJ, Dobson H (2008) Lameness, activity time-budgets, and estrus expression in dairy cattle. J Dairy Sci 91:4552–4559. https://doi.org/10.3168/jds.2008-1048

Weary DM, Huzzey JM, von Keyserlingk MAG (2009) Board-invited review: Using behavior to predict and identify ill health in animals. J Anim Sci 87:770–777. https://doi.org/10.2527/jas.2008-1297

Weigele HC, Gygax L, Steiner A, Wechsler B, Burla JB (2018) Moderate lameness leads to marked behavioral changes in dairy cows. J Dairy Sci 101:2370–2382. https://doi.org/10.3168/jds.2017-13120

Welch JG (1982) Rumination, particle size reduction and passage from the rumen. J Anim Sci 54:885–894

Westin R, Vaughan A, de Passillé AM, DeVries TJ, Pajor EA, Pellerin D, Siegford JM, Vasseur E, Rushen J (2016) Lying times of lactating cows on dairy farms with automatic milking systems and the relation to lameness, leg lesions, and body condition score. J Dairy Sci 99:551–561. https://doi.org/10.3168/jds.2015-9737

Wittrock JM, Proudfoot KL, Weary DM, von Keyserlingk MAG (2011) Short communication: Metritis affects milk production and cull rate of Holstein multiparous and primiparous dairy cows differently. J Dairy Sci 94:2408–2412. https://doi.org/10.3168/jds.2010-3697

# Heat Stress in Lactating and Non-lactating Dairy Cows

Jimena Laporta and Amy L. Skibiel

**Abstract**

The negative consequences of heat stress on the dairy industry are substantial. Elevated ambient temperatures lead to a cascade of detrimental effects, including declines in productivity and elevated morbidity and mortality at various life stages, including pre- and post-weaning, during the dry period, and during lactation. The impacts of elevated ambient temperatures leading to heat stress in lactating dairy cows have been well-established. These include changes in autonomic and behavioral thermoregulation, reductions in dry matter intake, impaired immunity and health, altered mammary gene expression, and reduced milk production. More recently, heat stress implications on non-lactating cattle, for example, the dry dairy cow, have begun to be documented. Heat-stressed dry cows have altered immune function, metabolism, mammary development, and milk production in the subsequent lactation, even after the heat stress insult has been removed. Further, there is substantial evidence of maladaptive responses arising from late-gestation intrauterine hyperthermia, affecting multiple tissues and organs, and persisting as lifelong deficiencies lowering dairy cattle survival and productive potential. As predicted, climate change trends will continue to affect dairy cattle. There is an urgent need to develop practical strategies to effectively identify dairy cattle at risk of heat stress and offer interventions tailored to their age and physiological status to prevent the onset of heat stress in these animals.

**Keywords**

Hyperthermia · Climate change · Dry cows · Dairy calves · Gestation

## 1 Introduction

From a livestock production perspective, the primary function of the dairy cow is to produce high-quality and high-quantity milk with optimum efficiency. Although this goal has been targeted primarily through advancements in cattle genetic selection and the implementation of nutritional strategies, there still lies a necessity to provide optimal management of various environmental and physiological stressors, in order

J. Laporta (✉)
Department of Animal and Dairy Sciences, University of Wisconsin-Madison, Madison, WI, USA
e-mail: jlaporta@wisc.edu

A. L. Skibiel
Department of Animal, Veterinary and Food Sciences, University of Idaho, Moscow, ID, USA
e-mail: askibiel@uidaho.edu



J. J. Gross (ed.), *Production Diseases in Farm Animals*,
https://doi.org/10.1007/978-3-031-51788-4_20

to maximize metabolic and mammary gland production potential. A stressor can be defined as an external event or condition that sets a strain on a biological system. Animals respond to stress by expending energy to remove or lessen the impact of the stressor, increasing maintenance requirements, and subsequently decreasing production (Collier et al. 2017). Stressors affecting dairy cattle may include, but are not limited to challenges of the immune system (e.g., pathogen exposure, parturition), housing and cow comfort (e.g., over-crowding), social dominance and competition, and thermal environment (e.g., cold or heat). According to Collier et al. (2017), the thermal environment is "the largest single stressor in animal production" due to its ability to impact animals in different ways directly or indirectly, including alterations in food availability and quality, changes in pathogen loads, immune suppression, health, and animal performance.

Rapid changes in environmental temperature and extreme weather events may be some of the largest stressors currently facing animal agriculture. Rising environmental temperature is challenging the capacity of homeotherm animals to effectively acclimate to their changing environment. Cows have an optimal thermoneutral zone at ambient temperatures ranging between 5 and 22–25 °C, where minimal and constant energy maintains normal body temperature (38.0–39.3 °C). Physiological heat stress occurs when cattle are pushed beyond the upper limit of their thermoneutral zone through exposure to elevated ambient temperature and/or direct solar radiation, resulting in an increase in their core body temperature past the equilibrium of heat load and their ability for effective heat dissipation. Maintaining thermoneutrality is crucial to reach optimal performance in dairy cattle (West 2003). Animals respond to heat stress by coordinating behavioral, metabolic, and physiological changes from the subcellular to the whole animal level (Collier and Gebremedhin 2015). The increase in global average surface temperature between 0.4 and 4.8 °C projected for the year 2100 (IPCC 2014) poses a threat to production of high-quality and high-quantity milk with optimum efficiency. It is predicted that cows raised in regions traditionally characterized by "less extreme" climate conditions will also experience temperatures beyond their comfort zone which will be particularly damaging for high-yielding dairy cows (Kadzere et al. 2002; Collier et al. 2006a).

The impacts of elevated ambient temperatures leading to heat stress in lactating dairy cows have been well-established (Bernabucci et al. 2010) and include changes in autonomic and behavioral thermoregulation, reductions in dry matter intake, impaired immunity and health, altered mammary gene expression, and reduced milk production. More recently, heat stress implications on non-lactating cattle including the dry dairy cow, her gestating offspring (Tao et al. 2019), and dairy youngstock, such as the pre-weaned calf (Dado-Senn et al. 2020a, b) and pre-pubertal heifers (Wang et al. 2020), have begun to be documented. While recent reports describe short-term heat acclimations in these groups of cattle, much remains unknown regarding the whole-body and productive responses to heat stress, and the long-term consequences of heat exposure. If thermoneutrality is not maintained, short- and long-term consequences on the productive capacity of these animals (i.e., growth, reproduction, and lactation) may arise. There is a need to develop effective strategies and practical tools to identify dairy calves and non-lactating cows under heat stress and develop tailored interventions to prevent the onset of heat stress in these animals (Ferreira et al. 2016).

This chapter will cover how predicted climate change trends will affect dairy cattle, and our current understanding of heat stress as a biological stressor on physiological, productive, and reproductive outcomes in lactating dairy cattle. Further detail will be covered on the carryover effects of heat stress in non-lactating pregnant cows and their progeny. Last, the current knowledge of methods for effective heat stress detection in lactating cows, dry cows, and calves will be discussed.

## 2 Climate Change and the Dairy Industry

Earth's temperature has been increasing at an alarming rate over the past several decades, largely driven by anthropomorphic activities, such as increases in greenhouse gas emissions and deforestation, with relatively smaller contributions from natural phenomena, such as changes in solar activity and volcanic eruptions (Stern and Kaufmann 2014; Ara Begum et al. 2022). Since 1880, the global surface temperature has increased by roughly 0.85 °C, with 9 of the 10 warmest years on record occurring in the last decade (IPCC 2014; NOAA 2022). Further increases in the global average temperature are predicted; during the twenty-first-century surface temperature is expected to increase by 1.5–2 °C if greenhouse gas emissions are not curtailed in the near future (Ara Begum et al. 2022). Recent spatiotemporal modeling of land-surface temperatures and gridded meteorological data estimates that over the next 30 years almost a third of the US will experience a heat index above 52 °C (Wilson et al. 2022). Extreme heat can have varied and catastrophic impacts on plants and animals from changes in phenology (e.g., seasonal breeding and growth), to changes in physiology (e.g., altered metabolism and tissue composition), reduced productivity, and increased risk of morbidity and mortality (Hughes 2000; Vitali et al. 2015; Cox et al. 2016a). Additionally, more frequent and intense natural disasters, such as wildfires and droughts, are associated with global warming, and can exacerbate the adverse effects of thermal stress on animal performance and survival (Cox et al. 2016b; Pace 2022).

Dairy cattle are particularly susceptible to hyperthermia. The main dairy cattle breeds, such as Holstein-Friesian and Jersey cattle, originated in the temperate climates of northern Europe and are not heat-adapted (Mansfield 1985; Huson et al. 2020). Moreover, modern dairy cows have high metabolic rates, which are associated with low thermal tolerance. Driven by advancements in genetic technologies and selective breeding programs, improvements in nutrition, and better management practices, the modern dairy cow produces approximately 60% more milk than dairy cows did 70 years ago (Baumgard et al. 2017; USDA 2020). A staggering 10% increase in milk production has occurred just over the past 10 years, with dairy cows in 2021 producing an average of almost 11,000 kg of milk per cow (USDA-NASS 2022). Increased milk production necessitates greater metabolic activity; for example, cows producing around 30 kg/day of milk have more than a 15% increase in metabolic rate compared to cows producing only 18 kg/day of milk (Purwanto et al. 1990). However, heat production also increases with metabolic rate. A lactating cow producing 30 kg/day of milk will produce twice as much heat than the maintenance heat production of a non-lactating cow (Wolfenson and Roth 2018). Thus, high-producing dairy cows are particularly sensitive to heat stress as genetic selection for higher milk yield (i.e., greater metabolic activity) is tied to a lower capacity for heat tolerance (Kadzere et al. 2002).

The consequences of increased ambient temperature and humidity on the dairy industry are substantial. Hyperthermic cows exhibit a shift in physiological priorities to promote thermoregulation and increase the likelihood of survival at the expense of other energetically demanding processes, such as reproduction, immune function, and milk production (Baumgard and Rhoads 2013). As a consequence, heat-stressed cows have reduced fertility, increased disease risk, and lower milk yield relative to thermoneutral cows. Heat stress in lactating cows can decrease milk yield by up to 40%, which is only partially attributed to reduced dry matter intake (Rhoads et al. 2009; Wheelock et al. 2010). In addition to impacts on the lactating cow, heat stress experienced during the dry period can have immediate and carry-over effects into the next lactation, even after abatement of the thermal insult (Tao and Dahl 2013). Heat-stressed dry cows have altered immune function, metabolism, mammary development, and milk production in the subsequent lactation (do Amaral et al. 2011; Tao et al. 2011; Skibiel et al. 2018b; Dado-Senn et al. 2019). Furthermore, fetal exposure to hyperthermic conditions in utero impacts future behavior, thermoregulation, mammary development, and lactation performance, (Skibiel et al. 2018a; Davidson et al. 2022; Dado-Senn et al. 2022a, b). Together, declines in productivity and elevated morbidity and mortality associated with heat stress at various life stages in dairy cows are estimated to cost the US dairy industry more than $2 billion annually (St-Pierre et al. 2003; Ferreira et al. 2016; Laporta et al. 2020).

## 3 Environmental and Animal-Based Indicators of Heat Stress in Dairy Cattle

Historically, the onset of heat stress in lactating cows has been associated with the initiation of a decline in productive parameters (i.e., milk production). However, this is an indicator of the cow's response to the heat strain, and therefore a consequence of the damage that already occurred rather than an early indicator of the risk of heat stress. As for non-lactating dry cows and youngstock, non-productive indicators of heat stress must be assessed, as there is no direct milk output response. Efforts from different research groups have focused on identifying sensitive physiological and non-invasive variables that manifest in the early stages that can be used as indicators of heat stress onset. Heat stress indicators can be classified as *animal-based* (i.e., related to behavior, physiology, or performance) or *environmental* (i.e., related to the thermal environment the animal experiences, such as the ambient temperature, relative humidity, wind speed, and the calculated temperature–humidity index, *THI*). Animal-based indicators reflect the state of the animal in a particular environment. Two animal-based parameters that have been widely used as physiological indicators of heat stress are respiration frequency and rectal (body) temperature of the animals (Hoffmann et al. 2020). Initial detection of heat stress (i.e., early onset) can be done using segmented regression models to establish environmental breakpoints at which abrupt changes in animal physiological responses (i.e., respiratory frequency, rectal temperature) begin to occur (Brügemann et al. 2012). Presumably, these breakpoints will vary depending on the physiological stage and age of the animal.

***Lactating Cows*** Temperature–humidity index breakpoints have been widely evaluated in lactating dairy cattle to detect declining milk yield, rumination, or conception rates, or increases in thermoregulatory responses under heat stress. Temperature–humidity thresholds between 68 and 72 have been well-recognized for observed decreases in milk production in lactating cows in a subtropical (Ravagnolo et al. 2000) or arid (Zimbelman et al. 2009) environment in the US. Thresholds for lactating, high-yielding dairy cows in a moderate climatic zone for respiratory frequency in standing cows (THI = 70) and lying cows (THI = 65) and for rectal temperature (THI = 70) in standing cows have been reported (Pinto et al. 2020). However, rumination time begins to decrease at THIs as low as 52 for lactating cows in Central Europe (Müschner-Siemens et al. 2020). Considerable variability exists regarding THI thresholds at which lactating cows become heat stressed, depending on the type of climate, cow-related factors (i.e., parity, lactation stage, and milk yield), and the physiological (animal) response variable evaluated. However, these studies are crucial to developing management plans to initiate heat mitigation in high-yielding dairy cows. An important consideration is that lactating cows will begin to experience physiological adaptations to heat stress when exposed to THIs as low as 52.

***Dry Cows*** Although dry cows are more thermotolerant relative to lactating cows due to their non-lactating status and lower metabolic heat production (Hahn 1999), they carry a calf growing at an exponential rate during the last trimester of gestation (Bauman and Bruce Currie 1980), and are still susceptible to changes in environmental temperature. Indeed, exposing dry cows to environmental heat stress during the dry period (i.e., 6–8 week non-lactating period between two consecutive lactations) impairs their milk production in the subsequent lactation by 5 kg/day and has carry-over effects on the offspring postnatally (Tao and Dahl 2013; Dahl et al. 2017). Environmental breakpoints at which respiration frequency or rectal temperature began to abruptly rise in cooled or heat-stressed (i.e., with access or not to fans and water soakers, respectively) dry cows housed in a free-stall barn in a subtropical US climate were evaluated (Ouellet et al. 2021). Dry cows begin to display signs of heat stress (increased rectal temperature and respiration frequency) at a THI of 77 compared to dry cows that were cooled during the entire dry period. Although respiration frequency began to abruptly rise at a THI of 75 in cooled dry cows, the overall respiration frequency of that group never rose above 60 breaths per minute across a wide range of THIs. Research in both dry and lactating cows suggests that 60 breaths/minute is the target benchmark at which these cows begin to experience heat stress (Toledo et al. 2020). Based on these studies, in a subtropical climate, dry cows should be closely monitored when or before THI reaches 75 to avoid further heat stress-related impairments during the dry period and the subsequent lactation. No thresholds have been established for dry cows under different climates.

***Pre-weaned Calves*** While a comprehensive body of research has been developed to understand heat stress responses, detection, and mitigation in mature dairy cows, much remains to be understood about the impact of heat stress on the growing dairy calf. Pre-weaned calves are often not considered for heat abatement due to their larger surface/mass ratio and smaller heat loads (Collier et al. 1982a). Strong positive correlations between environmental measures such as dry bulb temperature and THI, and physiological variables such as respiration frequency, rectal temperature, and skin temperature have been established in dairy calves (Spain and Spiers 1996; Kovács et al. 2018). The THI threshold at which animal-based indicators start to change varies depending on the parameter evaluated and the thermal environment of the calves. Group-housed Holstein calves raised in a subtropical US climate under the shade of an open-sided barn with access to continuous cooling by fans (average wind speed 2 m/s) had higher breakpoints compared with those that did not have access to fans (Dado-Senn et al. 2020a). More specifically, calves exposed to chronic heat stress had a THI breakpoint of 67 for rectal temperature, and THI breakpoint of 65 for respiration frequency. For every unit increase in THI above these thresholds, rectal temperature and respiratory frequency begin to rise at a rate of 0.03 °C and 2.5 breath/min. Calves with continuous access to fans during summer had THI threshold of 69 for respiration frequency, with 1.4 breath/min increase for every unit increase in THI. Milk replacer intake started to abruptly decrease at a THI of 82 at a rate of 0.29 L/day for every unit increase in THI, and no significant breakpoints (i.e., significant change in response) were detected for milk replacer intake and rectal temperature in calves with access to fans.

In a temperate Midwestern US climate with calves housed outdoor in individual hutches, calf respiration frequency, and rectal temperate begin to rise when THI reaches 69 or when dry bulb temperature reaches 21–21.5 °C. Under this climate, ambient (dry bulb) temperature was a more accurate environmental indicator instead of THI (Dado-Senn et al. 2024). An important consideration is that identifying heat stress in youngstock can be challenging, as it is not practical or common to monitor calf productive and vital responses on-farm. These environmental thresholds associated with thermal discomfort tailored for different climates will aid in the early detection and monitoring of heat stress to prompt further monitoring and the development of mitigation strategies in pre-weaned dairy calves.

## 4 Effects of Heat Stress in Lactating Dairy Cows

***Health and Immune Function*** It is evident from the literature that the incidence of diseases, such as mastitis and metritis, increase with warmer temperatures. Many studies have documented seasonal variability in milk somatic cell count (SCC) and mastitis cases, whereby SCC and mastitis rates are highest in the warm summer months (MacLeod et al. 1954; Bernabucci et al. 2015; Norman et al. 2019). Across a 12-year timespan and 600 Holstein dairy herds, SCC was positively associated with THI (Hammami et al. 2013). A similar trend was found with mastitis cases when analyzing data across 15 large dairy herds in a 3-year period (Gernand et al. 2019). Mastitis reduces milk yield as mounting an immune response is energetically demanding. It has been estimated that immune activation in the lactating dairy cow requires 1 kg of glucose in a 12-h period, concomitant with an 80% drop in milk yield and an 11% decrease in milk lactose content (Kvidera et al. 2017). Heat stress also appears to contribute to transmissibility of mastitis pathogens. In a longitudinal study across five dairy herds in Germany, Hamel et al. (2021) found that pathogen shedding from infected mammary glands was increased when THI exceeded 60, particularly for yeasts and *Strep. uberis*. Pathogen shedding contributes to disease transmission and increases the risk of infection across the herd. Although the efficacy of active cooling on pathogen shedding has not yet been evaluated, providing evaporative cooling devices, and shade can reduce the incidence of mastitis on farms (Thatcher et al. 1974; Roman-Ponce et al. 1977; Safa et al. 2018).

The number of retained placenta and metritis cases also increases with elevated summer THI, which has consequences for the cow's future reproductive health and performance (Santos and Ribeiro 2014). In a herd with more than 500 fresh cows, 18% were diagnosed with retained placenta in summer in contrast to only 7% in the spring and winter (Ahmadi and Mirzaei 2006). Additionally, DuBois and Williams (1980) reported a 24% increase in the incidence of metritis and retained placenta in the warm summer months. These periparturient diseases are more likely to occur when gestation length is truncated (Wagner et al. 1974), which is common among dairy cows that are heat stressed during gestation (Monteiro et al. 2014; Fabris et al. 2019). One possible explanation for the increased disease incidence with thermal stress is the greater proclivity for pathogen survival and proliferation in hot and humid environments (Bernabucci et al. 2010). However, while certain mastitis-causing pathogens, such as *Strep. uberis*, are more prevalent in summer, others, such as *Staph. aureus*, *E. coli*, and *Strep. dysgalactiae* are more likely to cause intramammary infections in winter (Olde Riekerink et al. 2007). Furthermore, risk of metritis infection still increases under hyperthermia regardless of vaginal bacterial load (Molinari et al. 2019, 2022a). Together, these studies point to endogenous host factors that contribute to elevated disease risk with heat stress.

A number of empirical studies have demonstrated that heat stress has immunosuppressive effects on both the innate and adaptive arms of the immune system. A proteomics analysis revealed that the plasma complement system, part of the innate immune system that plays a role in the defense against infection, was suppressed in mid-lactation cows in summer versus spring (Min et al. 2016). Total circulating leukocyte count was lower in cows during periods of elevated THI and in cows under fluctuating temperature conditions (e.g., 32 °C during the day and 21 °C at night) (Paape et al. 1973; Zhang et al. 2014). Moreover, leukocyte migration to the udder was reduced in heat-stressed cows relative to thermoneutral herdmates after intramammary injection of oyster glycogen as a chemotactic agent (Elvinger et al. 1992). Decreases in circulating lymphocytes, lymphocyte to neutrophil ratios, lipopolysaccharide binding protein, and serum amyloid A are also observed with heat stress, indicative of a depressed immune response (Aggarwal and Upadhyay 2013; Muñoz et al. 2014; Abeyta et al. 2023). In corroboration, thermal stress reduced inflammatory cytokines, including TNFα, IL-1α, and IL-2, and immunoglobulins (e.g., IgG IgM, and IgA) in early lactation cows compared to cooled cows, indicating reduced systemic inflammation and adaptive immune function (Safa et al. 2018). This is problematic because acute, controlled inflammation during the periparturient period supports metabolic and endocrinological adaptations for transition to copious milk production (Bradford et al.

2015). Similar to early lactation cows, in mid-lactation heat-stressed cows, TNFα was lower in circulation relative to cows actively cooled, however, TNFα and IL-10 production was elevated in the heat-stressed group in response to a lipopolysaccharide challenge (Marins et al. 2021). Zhang et al. (2014) also found that mid-lactation cows had greater circulating TNFα and IL-10 concentrations with increasing THI. These studies suggest that lactation stage and pathogen exposure contribute to divergent immune responses under heat stress conditions.

Heat stress also alters immune cell function and viability. In one study, peripheral blood mononuclear cells (PBMC) were isolated from the blood of cows calving in summer and spring and stimulated in vitro with mitogens. PBMC isolated from cows calving in summer had reduced DNA synthesis but increased IgM secretion in response to the mitogen relative to PBMC isolated from cows calving in spring (Lacetera et al. 2005). In corroboration, incubating PBMC at a high temperature in vitro increases expression of heat shock protein 72, which has cytoprotective effects, but decreases DNA synthesis, indicative of reduced proliferative capacity of the cells (Lacetera et al. 2006). Additionally, lymphocytes cultured at elevated temperatures of 42–45 °C and have lower proliferation after mitogen stimulation and reduced cell viability relative to cells incubated at 38.5 °C (Kamwanja et al. 1994; Lacetera et al. 2006). Finally, phagocytic capacity and oxidative burst, a measure of bacterial killing ability, were both reduced in polymorphonuclear cells incubated at 41 °C compared to those incubated at 39 °C (Lecchi et al. 2016). Thus, heat stress appears to enhance pathogenesis by impairing immune cell function and survival and through depression of systemic immune responses.

***Productivity*** One of the most widely studied and economically important impacts of heat stress in dairy cattle is the decline in milk production. Milk losses between 11 and 35% have been reported (Spiers et al. 2004; Kim et al. 2010), with values varying dependent on stage of lactation. Heat stress has the greatest impact on milk production in mid-lactation cows relative to other lactation stages. In studies utilizing environmental chambers to induce heat stress, milk yield was reduced by 14–29% in early lactation cows (Baumgard et al. 2011; Lamp et al. 2015), by 27–35% in mid-lactation cows (Spiers et al. 2004; Rhoads et al. 2009, 2011), and by 11% in late lactation cows (Kim et al. 2010). Induction of heat stress through application of electric heating blankets similarly decreased milk yield of mid-lactation cows by 21–27% (Al-Qaisi et al. 2019, 2020). Contingent on stage of lactation, heat-stressed cows have declines of 9–13 kg of milk per day relative to cows with access to shade and actively cooled with water sprinklers and fans (Tao et al. 2018; Martínez et al. 2021).

Changes in milk composition associated with thermal conditions are equivocal. Several studies have described lower milk fat content with elevated THI in the summer months (Bouraoui et al. 2002; Heck et al. 2009; Bernabucci et al. 2015) whereas others have reported either no seasonal differences or increases in milk fat with increasing THI (Smith et al. 2013; Hammami et al. 2015). Similarly, in controlled heat stress experiments, no change (Flamenbaum et al. 1995; Rhoads et al. 2009; Al-Qaisi et al. 2019), a decrease (Moody et al. 1971), or an increase (Garner et al. 2016; Al-Qaisi et al. 2020) in milk fat percentage was found in heat-stressed cows relative to cooled or thermoneutral controls. Associations between milk protein and lactose content and thermal conditions are also discrepant among studies investigating seasonal differences in milk composition and those experimentally inducing or mitigating heat stress (Tarazón-Herrera et al. 1999; Bouraoui et al. 2002; Heck et al. 2009; Rhoads et al. 2011; Bernabucci et al. 2015; Hammami et al. 2015; Gao et al. 2017; Al-Qaisi et al. 2019; Al-Qaisi et al. 2020). Inconsistencies among studies may be attributed to lactation stage, differences in experimental approaches, varied climatic experiences based on geographical location, and the duration and severity of heat stress.

Dynamic structural and physiological changes occurring within tissues and across the body contribute to heat stress-induced lactation depression in dairy cows. Decreased dry matter intake is a hallmark of heat stress and has substantial impacts on milk yield. Dry matter intake can decline by as much as 30% in heat-stressed dairy cows (Collier et al. 2008). Feed intake, and consequently milk production, drops in lactating cows when ambient temperature exceeds 25 °C, likely an adaptation to reduce heat production from digestive processes and metabolism, which are increased during lactation (Renaudeau et al. 2012, Collier et al. 2017). A series of studies utilizing a pair-feeding approach have dissociated direct effects of the thermal environment from indirect effects through reduced feed intake. Pair-feeding trials typically consist of an initial period where all cows are housed under thermoneutral conditions and are fed ad libitum to obtain baseline feed intake data. In the following period, cows are either heat stressed or maintained under thermoneutral conditions and pair fed an amount that matches the same percent reduction in feed intake that the heat group experiences between periods. Employing this approach, it was found that 35–50% of milk loss during heat stress is attributable to decreased DMI (Rhoads et al. 2009; Wheelock et al. 2010).

The remaining variation in milk yield declines with heat stress can be attributed to nutrient intake-independent alterations in tissue perfusion, metabolism, mammary development, and cellular dynamics in the mammary gland. Both hepatic portal plasma flow and mammary blood flow have been found to decrease with heat stress, which reduces nutrient availability for milk biosynthesis in the mammary gland (McGuire et al. 1989; Lough et al. 1990; Rodrigues et al. 2019). Furthermore, heat-stress effects on macronutrient metabolism in lactating dairy cows are well-established (Baumgard and Rhoads 2013; Ríus 2019). Typically, lactating dairy cows exhibit a homeorhetic reduction in glucose use by peripheral tissues and shunt glucose preferentially to the mammary gland for milk lactose synthesis (Bauman and Bruce Currie 1980). However, several lines of evidence indicate that heat-stressed cows are unable to effectively deploy this glucose-sparing strategy. For example, heat-stressed cows have a greater glucose disposal rate, retain insulin sensitivity following a glucose tolerance test, and have a similar response to glucose after epinephrine challenge relative to their pair-fed thermoneutral counterparts, despite having lower milk lactose content (Wheelock 2006; Baumgard et al. 2011). These results indicate peripheral uptake and use of glucose at the expense of milk synthesis in heat-stressed cows. Indirect calorimetry further supports this work, demonstrating that lactating heat-stressed cows rely more heavily on carbohydrate oxidation than lipid oxidation relative to pair-fed thermoneutral controls (Lamp et al. 2015). Furthermore, despite depressed feed intake in heat-stressed cows, adipose tissue is resistant to adrenergic signaling as indicated by lower circulating NEFA concentrations following an epinephrine challenge in heat-stressed compared to pair-fed thermoneutral cows (Baumgard et al. 2011). Quantification of blood and ruminal metabolites and transcriptomics and proteomics analyses of multiple tissues have also revealed altered protein metabolism in heat-stressed lactating cows, including greater tissue protein catabolism, increased peripheral demand for amino acids, and impaired absorption of amino acids in the gastrointestinal tract (Lamp et al. 2015; Gao et al. 2017; Ma et al. 2019; Ríus 2019).

Within the mammary gland, heat stress derails mammary development and impairs mammary nutrient uptake and milk synthesis. Through RNA sequencing of the mammary gland, Gao et al. (2019) found that several genes involved in nutrient transport across the mammary epithelium, milk protein synthesis, and mammary glucose and fatty acid metabolism were downregulated in heat-stressed cows compared to pair-fed thermoneutral controls. In corroboration, trans-mammary disappearance of glucose was reduced in mid-lactation cows heat stressed for 14 days compared to ad libitum-fed thermoneutral cows and pair-fed thermoneutral cows (Rodrigues et al. 2019). Furthermore, expression of proteins involved in metabolic processes and milk synthesis was reduced in cultured bovine mammary epithelial cells (bMEC) 12 h after acute exposure to elevated temperature (Li et al. 2017). Heat stress also affects mammary epithelial cell structure, growth, and viability. Beginning around 8 h after incubation at 42 °C, bMEC grown in a collagen matrix had lower DNA content and stunted ductal branching patterns relative to those grown at 37 °C (Collier et al. 2006b). Incubating bMEC for just 30 min–1 h induces an apoptosis cascade and results in downregulation of genes involved in cytoskeletal integrity, cell cycle progression and differentiation, and milk synthesis (Collier et al. 2006b; Du et al. 2008; Li et al. 2015, 2017). Together, these studies indicate that even short-duration exposures to high temperatures can impede mammary function and promote mammary epithelial cell death.

***Fertility*** Many aspects of the cow's reproductive physiology, including follicular and oocyte development and competence, ovulation, conception, and early embryo development are impacted by thermal stress, leading to reduced reproductive performance. In a multi-year study in Hawaii, Holstein cows had the lowest conception rate in years with the highest summer THI (Ingraham et al. 1976). Across multiple Holstein dairy herds in Israel, pregnancy per AI insemination was only 20% in the hot summer months compared to almost 40% in the winter (Flamenbaum and Galon 2010). Analysis of records from more than 20,000 Holstein cows across 15 dairy farms in Thuringia, Germany revealed that natural inseminations of cows decreased by up to 28% when THI was greater than 57, although mild improvements were found in cows with estrous synchronization compared to those without. The same study reported that the number of pregnancies per artificial insemination decreased when THI exceeded 65 (Gernand et al. 2019). Estrous synchronization may increase conception rates in warm climates by eliminating the need for or improving estrus detection and thereby facilitating appropriately timed insemination (Roth 2008; Carvalho et al. 2018). Further, active cooling of cows, such as with fans and water, can increase conception rates (Flamenbaum and Galon 2010), although cooling should encompass the entire reproductive period from follicular maturation to early embryo development. In a study where cows were only cooled from 1 day prior to expected estrus until 8 days post-estrus, cooled cows had similar conception rates relative to non-cooled cows (Her et al. 1988).

A rich body of literature implicates impaired folliculogenesis in the reduced fertility of heat-stressed cows. Heat stress depresses follicular dominance, which increases the number of large follicles in the ovary (Badinga et al. 1993; Roth et al. 2000). Holstein heifers that were experimentally heat stressed by lack of access to water sprinklers and cooled ventilation, had 53% more large follicles over 10 mm in diameter compared to the cooled group with access to shade and cooling devices (Wolfenson et al. 1995). Furthermore, heat stress delays the atrophy of subordinate follicles and prolongs the period of preovulatory follicle dominance, which is associated with reduced fertility (Wolfenson et al. 1995; Wilson et al. 1998; Roth et al. 2000). Reduced follicular steroidogenesis, as indicated by low androstenedione and estradiol production, and low aromatase activity in follicular cells and follicular fluid, also contributes to impaired follicular dynamics and reduced fertility under heat stress conditions (Badinga et al. 1993; Wolfenson et al. 1995). Thecal and granulosa cells harvested from dominant follicles of cows in the summer had lower androstenedione and estradiol production, respectively, relative to follicles collected from cows in the winter (Wolfenson et al. 1997). Similarly, androstenedione production and estradiol concentration were reduced in thecal and granulosa cells, respectively, from medium-sized follicles collected from cows experimentally heat stressed 20 days prior to follicle collection compared to follicles harvested from thermoneutral control cows (Roth et al. 2001). Impaired steroidogenesis diminishes physical signs of estrus making estrus detection difficult, shortens estrus duration and intensity, and affects ovulation through attenuation of the LH surge (Gwazdauskas et al. 1981; Roth 2008).

In addition to impacts of the thermal environment on follicular development, heat stress impairs oocyte maturation and early embryo development. Maturation of oocytes occurs through the release from meiosis I arrest, allowing the cells to proceed through subsequent meiotic division, and is necessary for ovulation and fertilization to occur (Jamnongjit and Hammes 2005). Culturing oocytes under elevated temperatures of up to 41 °C for the first 12 h of the maturation process resulted in cytoskeletal rearrangement indicating suspension of the cells in meiosis I stages, which has implications for future fertilization and development, while oocytes maturing at a lower temperature proceeded to metaphase II (Roth and Hansen 2005). Heat stress also reduces the quality of fertilized oocytes. Oocytes harvested from cows during hot months showed delayed cleavage at the two- and four-cell stages and were less likely to develop to the blastocyst stage relative to oocytes collected from cows in cool months (Al-Katanani et al. 2002; Aroyo et al. 2007). Similarly, exposing oocytes to elevated incubation temperatures in vitro inhibits cell cleavage and blastocyst development (Edwards and Hansen 1997; Roth and Hansen 2004). Furthermore, heat stress occurring between estrus and insemination results in a greater proportion of abnormal and non-viable embryos (Putney et al. 1988; Ealy et al. 1993). Effects of heat stress on embryo viability are most evident during the first 7 days of development when the embryo is particularly sensitive to thermal perturbations (Hansen 2019). Exposure of cows to elevated temperatures 1 day after estrus, but not on subsequent days, decreases embryo survival on day 8 (Ealy et al. 1993). Applying thermal stress to fertilized oocytes in vitro ceased development of two-cell embryos but had smaller effects on four- and eight-cell embryos, and little effect on morulae (Edwards and Hansen 1997). These studies indicate that heat stress has numerous and varied effects on reproductive processes from follicular development through early embryogenesis.

## 5 Short- and Long-Term Effects of Dry Period Heat Stress on the Dam

***Health and Immune Function*** During the dry period, the mammary gland undergoes involution and redevelopment processes in preparation for the subsequent lactation (Hurley and Loor 2011). The late dry period coincides with the last trimester of gestation and the first half of the physiologically demanding transition from a non-lactating to a lactating state, which requires numerous metabolic adjustments to promote and support the biosynthesis of milk (Horst et al. 2005). Consequently, the transition period is also a time when diseases and metabolic disorders are likely to manifest (Goff and Horst 1997), particularly when cows are simultaneously exposed to a suboptimal thermal environment. Indeed, an analysis of health records collated from more than 2500 cows on commercial farms in Florida revealed that the number of mastitis and respiratory disease cases were higher among cows that were dry during the hot summer months relative to cows dried off in winter (Thompson and Dahl 2012).

Heat stress during the dry period alters both the immediate innate and adaptive immune responses and has carry-over effects on immune function in the subsequent lactation, contributing to disease and poor lactation performance. For example, PBMC collected from dry cows that were heat stressed during the dry period had greater *TNFα* mRNA expression relative to cows that were actively cooled during the dry period but there was no persistent effect on cytokine production in the subsequent lactation when all cows were cooled (Tao et al. 2013). In contrast, after stimulation with a mitogen, TNFα production was lower in PBMC harvested from dry cows that were heat stressed compared to those that were cooled during the dry period (do Amaral et al. 2010). In a follow-up study, dry cows were inoculated with an ovalbumin antigen at dry off, 32 days before expected calving, and on the calving date (do Amaral et al. 2011). Heat-stressed cows had lower anti-ovalbumin IgG production after the second inoculation relative to the cooled dry cows. The expression of immune- and inflammatory-related genes was also upregulated in mammary gland tissue of dry heat-stressed relative to cooled cows (Dado-Senn et al. 2018). These studies indicate immediate effects of the thermal environment on innate and adaptive immunity in dry dairy cows.

Immune dysregulation stemming from dry period heat stress continues into the subsequent lactation, even when heat stress is mitigated after calving. In one study, cows were either heat stressed or actively cooled for the entire 60 day dry period and all cows were subsequently cooled after parturition. In vitro stimulation of whole blood collected from the cows at 21 days postpartum increased the production of the cytokines, IL-1β, IL-10, and the chemokine, MIP-1α, in the heat-stressed group (Molinari et al. 2022b). Furthermore, inducing an intramammary infection (IMI) with *Strep. uberis* 5 days postpartum, reduced total leukocyte and neutrophil count and increased *IL-10* mRNA expression in neutrophils from cows heat stressed only in the dry period relative to cooled dry cows (Thompson et al. 2014). In the same study, toll-like receptor 2 expression in neutrophils collected after IMI was lower in the heat stressed compared to the group cooled during the dry period. These changes are indicative of an altered inflamatory state that reduces the capacity of immune cells to respond to invading pathogens. Furthermore, heat stress during the dry period increases *IL-8* expression in PBMC, decreases phagocytic and bacterial killing capacity in neutrophils, and depresses lymphocyte proliferation in the subsequent lactation (do Amaral et al. 2010, 2011; Tao et al. 2013). Phosphoproteomics and proteomics analysis of mammary gland samples collected during lactation corroborate previous empirical work, where multiple proteins and phosphoproteins involved in immune function and inflammatory pathways were largely more abundant in cows that were heat stressed during the previous dry period (Skibiel et al. 2022). These studies suggest that heat stress disrupts normal immune function both during the thermal insult and for months after heat mitigation.

***Productivity*** Heat stress during the dry period not only exerts carry-over effects on immune function and health but also on milk production in the subsequent lactation. For example, cows dried off in summer produced a total of 550 kg less milk in the next lactation than cows that were dried off in other seasons (Thompson and Dahl 2012). Studies employing an experimental approach where cows either did or did not have access to evaporative cooling devices, such as water soakers and fans, during the dry period have reported declines in milk yield of 3–8 kg/day during the subsequent lactation of cows heat stressed when dry, even though all cows were cooled after calving (do Amaral et al. 2009; Tao et al. 2011; Dado-Senn et al. 2019; Fabris et al. 2019). Interestingly, the effects of dry period heat stress on subsequent milk production is similar regardless of when the thermal insult occurs during the dry period. Cows that were heat stressed in only the first half or only the second half of the dry period still had 3–4 kg/day lower milk yield in the next lactation relative to cows that were cooled during the entire dry period (Fabris et al. 2019). No differences in milk macronutrient composition in the subsequent lactation were detected between cows that were cooled or heat stressed during the entire dry period in the majority of published studies (Wolfenson et al. 1988; Avendaño-Reyes et al. 2006; Adin et al. 2009; do Amaral et al. 2009; Tao et al. 2012b; Thompson et al. 2014; Fabris et al. 2019). However, there are some discrepancies; do Amaral et al. (2011) found that milk fat percentage was 0.5% lower among cows heat stressed when dry and Tao et al. (Tao et al. 2011) reported that milk protein content was 0.14% higher in cows heat stressed versus cooled during the dry period.

The causal mechanisms contributing to declines in milk yield after dry period heat stress are complex and multifactorial. Using a pair-feeding approach, it was estimated that up to 60% of the variation in milk yield after dry period heat stress may be attributable to reduced dry matter intake pre-calving (Seyed Almoosavi et al. 2021). Similar to lactating cows, dry matter intake decreases in cows heat stressed during the dry period, although the extent of the decline is less severe than in lactating cows (up to 12% for cows heat stressed during the dry period versus up to 30% in lactating heat-stressed cows) (Collier et al. 2008; Tao and Dahl 2013; Ouellet et al. 2021). However, despite reduced DMI, heat stress during the dry period appears to have little effect on systemic metabolism pre-calving, as circulating glucose, NEFA, BHB, and insulin concentrations are similar between heat-stressed and cooled dry cows (do Amaral et al. 2011; Tao et al. 2012b). For example, there were no pre-calving differences in insulin response to an insulin challenge or glucose tolerance test (Tao et al. 2012b). Moreover, in the same study, NEFA response to an insulin challenge (IC) and glucose clearance after a glucose tolerance test (GTT) were similar between cows that were heat-stressed or cooled while dry. However, targeted quantitation of gene and protein expression suggest that cows heat stressed during the dry period have altered tissue-specific metabolism, including lower fatty acid oxidation and protein catabolism in skeletal muscle and greater fatty acid synthesis in the mammary gland, which may affect mammary development and colostrogenesis through altered nutrient availability and cellular processes in the mammary gland (Adin et al. 2009; Koch et al. 2016).

In contrast to pre-calving, metabolic function post-calving is substantially influenced by heat stress in the previous dry period. Despite consuming as much or more feed in early lactation, cows heat stressed during the dry period produce significantly less milk than cooled cows (do Amaral et al. 2009; Tao et al. 2011). Cows heat stressed when dry have lower plasma concentrations of BHB, NEFA, and insulin, a potent antilipolytic hormone, and higher circulating glucose in early lactation relative to cooled dry cows (do Amaral et al. 2009; Tao et al. 2012b). Furthermore, in early lactation, cows heat stressed during the previous dry period have greater glucose responses after GTT and IC, and a blunted NEFA response to IC (Tao et al. 2012b). Congruent with these results, expression of hepatic proteins involved in lipid and glucose metabolism were altered in early lactation after dry period heat stress (Skibiel et al. 2018b). For example, phosphoenolpyruvate carboxykinase, a rate-limiting enzyme in gluconeogenesis, was less abundant, whereas the lipid binding protein, apolipoprotein A-IV, was more abundant in the liver of post-parturient cows that were heat stressed when dry. Together, this empirical work suggests that dry period heat stress has carry-over effects on nutrient metabolism such that dry period heat-stressed cows have greater peripheral insulin sensitivity and glucose use, lower lipid mobilization from adipose tissue, and lower reliance on fatty acid oxidation for energy post-calving.

In addition to alterations in systemic and tissue-specific metabolism, dry period heat stress impedes mammary involution and development. The dry non-lactating period between subsequent lactations improves milk yield in the following lactation through the removal of senescent mammary epithelial cells which are replaced with new cells capable of producing milk (Capuco et al. 1997). The dry period consists of an early involution phase and a later redevelopment phase that are characterized by relative changes in the rates of cell death processes, including apoptosis and autophagy, and cell proliferation (Capuco and Akers 1999; Zarzynska and Motyl 2008). The rate of cellular death exceeds proliferation in the early dry period whereas the reverse occurs in the late, redevelopment stage of the dry period. These cellular processes are disrupted with dry period heat stress. For example, cows cooled when dry had a temporal increase in the expression of proteins involved in autophagy until 7 days after dry-off whereas dry heat-stressed cows had a relatively consistent expression of the same proteins across the dry period (Wohlgemuth et al. 2016). Global RNA sequencing of the mammary gland also revealed aberrations in the normal mammary involution process with genes involved in cell death and ductal growth differing between dry heat-stressed and cooled cows (Dado-Senn et al. 2018). Similarly, expression of autophagic and pro-apoptotic genes was downregulated in dry heat-stressed cows along with a decrease in the proportion of cells undergoing apoptosis

(Fabris et al. 2020). As a result, dry heat-stressed cows have more alveoli present in the glands during the early dry period. Additionally, heat-stressed dry cows have lower mammary cell proliferation during the redevelopment phase of the dry period relative to cooled dry cows (Tao et al. 2011). Impairments to normal mammary development from heat stress during the dry period persist into the next lactation when mammary gland growth continues through further increases in cell proliferation (Akers 2002). The mammary glands of lactating cows that were heat-stressed only when dry had fewer alveoli and more connective tissue than cooled cows, but this was not attributed to changes in cell turnover as the proportion of differentiating and apoptotic mammary epithelial cells was similar between treatment groups (Dado-Senn et al. 2019).

Alterations in the hormonal milieu associated with thermal stress may contribute to impaired cellular turnover in the mammary gland with implications for mammary growth and milk yield. Prolactin (PRL) is a mammogenic and lactogenic hormone that stimulates mammary growth, functional differentiation of the mammary secretory cells, and expression of milk synthesis genes (Akers et al. 1981). Although plasma PRL concentrations increase with heat stress during the dry period, PRL-receptor expression decreases in multiple tissues (Collier et al. 1982b; do Amaral et al. 2010, 2011), suggesting the potential for reduced PRL action on target tissues. However, in an experiment where mammary tissue was collected from dry heat-stressed or cooled cows, no differences in expression of PRL signaling genes were observed between treatment groups. Besides PRL, estrogens and progesterone stimulate mammary ductal branching and alveolar growth (Akers 1985). Both progesterone and placental estrone sulfate production are lower in dry heat-stressed cows, which may also play a role in stunted mammary growth. Further studies are warranted to elucidate putative endocrine mechanisms that may contribute to thermally-induced disruptions to mammary development.

## 6 Fetal Programming Effects of Heat Stress on the Postnatal Offspring

The conditions experienced during early-life developmental windows, including the prenatal intrauterine environment, can profoundly influence fetal growth and development (i.e., organ size, organ structure, and cellular function) and exert long-term effects on postnatal disease, survivability, resilience, and productivity (Van Eetvelde and Opsomer 2017; Reynolds et al. 2019). This phenomenon, referred to as fetal programming, was first described by Barker and colleagues through epidemiological evidence of fetal origins to adult diseases in humans (Barker 1998). Several factors can influence the fetal developmental trajectory including maternal age, under- or over-nutrition, toxin exposure, health status, and social, physiological, or thermal stress. If the fetus develops under stressful intrauterine conditions, the offspring is at increased risk of numerous chronic pathologic conditions across their lifetime including growth abnormalities, metabolic syndrome, reduced life expectancy, and reproductive, immune, behavioral, or cognitive dysfunction (Reynolds et al. 2019). Initially demonstrated in humans, this concept extends to livestock species. The current knowledge on the programming effects of maternal insults including over- and under-nutrition, heat stress, and inflammation in ruminant livestock species has been recently reviewed (Vautier and Cadaret 2022).

The placenta is the interface between the developing fetus and the mother. The developing fetus responds to stimuli or challenges imposed on the mother and conveyed by the placenta causing changes in gene expression due to epigenetic modifications. These modifications (i.e., heritable changes in gene expression beyond alteration of genetic code; including DNA methylation, histone modifications, and non-coding RNAs) can influence organ development, structure, and/or function. Epigenetic shifts arising in utero can be stable and ultimately exert long-term effects on the postnatal offspring phenotype (Reynolds and Caton 2012).

A developing fetus is sensitive to changes in temperature and is impacted by changes in the mother's thermal environment (Laburn et al. 2000). Maternal hyperthermia exposes the developing fetus to temperatures above euthermic levels leading to intrauterine (prenatal) hyperthermia (IUHT). Heat exchange between the fetus and the mother occurs primarily through the fetal-placental circulation, which accounts for more than 80% of total fetal heat loss (Laburn et al. 2000; Asakura 2004). In goats and sheep, fetal-maternal temperature difference has been shown to be 0.4–0.6 °C, yet under acute heat stress conditions experienced by the dam, this difference decreases quite rapidly to 0.2–0.5 °C (Faurie et al. 2001; Laburn et al. 2002). Despite the lack of direct evidence in the bovine, these studies demonstrate that alterations in maternal core temperature (i.e., exposure to heat stress conditions) can impact fetal temperature in ruminants. Failure to maintain thermoneutrality during late gestation in dairy cows exerts short- and long-term carry-over developmental adaptations on the offspring's growth, metabolism, and organ structure and function. Fetal hyperthermia can also lead to intrauterine growth restriction and programing of cell differentiation of organs and tissues that play crucial roles in future lactational outcomes (i.e., mammary gland development), heath and survival (i.e., immune organs), adaptation to physiological stress (i.e., hypothalamic-pituitary-adrenal axis), and thermoregulation (i.e., sweat gland morphology).

***Calf Growth and Organ Weights*** Holstein dairy calves experiencing IUHT for the last 60 days of gestation weigh less, are shorter in stature, and have reduced head circumference at birth, compared to calves gestated by dams with access to cooling devices to maintain thermoneutrality (i.e., in utero thermoneutral, IUTN; Dado-Senn et al. 2021). Specific growth impairments triggered by IUHT (i.e., body weight and stature) are not transient and manifest 1 year after birth, despite only being exposed to hyperthermia while developing in utero (Monteiro et al. 2016). Two studies reported that IUHT dairy calves have impairments not only at the whole-body level but also at the organ level. Organs involved in metabolism and immunity (i.e., thymus, spleen, heart, and liver) weighed less at birth in IUHT bull calves, compared to organs harvested from IUTN bulls (Ahmed et al. 2021). Another study in dairy heifers reported reduced thymus, thyroid gland, spleen, and heart organ weights at birth in IUHT heifers relative to organs harvested from IUTN heifers (Dado-Senn et al. 2021). In the same study, IUHT heifers had larger adrenal glands and kidneys and reduced ovary weights compared to IUTN counterparts at 2 months of age and smaller mammary glands at both birth and weaning. Table 1 summarizes dairy calves' whole body and organ weights impacted by IUHT. Similar alterations in whole-body growth and organ weights have been observed in fetuses of chronically heat-stressed ewes, which are 20% smaller than thermoneutral counterparts with livers, hearts, and brains of the heat-stressed fetuses disproportionate in size relative to overall body weight (Dreiling et al. 1991; Limesand et al. 2018). The impact of IUHT in the postnatal swine includes reductions in birth weight, body weight gain, and alterations to adipose tissue accretion rates (Johnson et al. 2015a).

***Thermoregulation*** Thermoregulation is a physiological mechanism by which mammals maintain core body temperature within a tightly controlled range independent of the external temperature. Physiological changes in the developing fetus are triggered by hyperthermic conditions and these adaptations can be long-lasting and confer differential acclimation capacity to the changing environment later in life. A variety of phenotypic responses arising from IUHT have been observed in different species, ranging from increased heat susceptibility to improvements in thermotolerance. Dairy heifers exposed to IUHT have greater basal postnatal body temperature compared to their IUTN counterparts (Laporta et al. 2017; Davidson et al. 2021). This pre-programmed set-point shift in postnatal basal core body temperature has also been reported in

**Table 1** Growth, stature, and organ weights of Holstein dairy calves born to actively cooled cows (in utero thermoneutral, IUTN) or born to heat-stressed cows (in utero heat stressed, IUHT)

| | Birth | Weaning | *p*-Value | Citation |
|---|---|---|---|---|
| *Growth parameter (n = 28/group)* | | | | |
| Birth weight (kg) | −2.4 | – | 0.03 | (1) |
| Weaning weight (Kg) | – | −6.6 | 0.02 | (1) |
| Hip height (cm) | −2.72 | −1.6 | <0.09 | (1) |
| Withers height (cm) | −3.7 | −2.3 | <0.003 | (1) |
| Body length (cm) | −1.5 | −2.4 | <0.07 | (1) |
| Chest girth (cm) | −4.9 | −3.2 | <0.001 | (1) |
| Waist girth (cm) | −4.6 | −3.8 | <0.01 | (1) |
| Head circumference (cm) | −10.6 | −1.6 | <0.01 | (1) |
| *Organ/tissue (n = 8/group)* | | | | |
| Thymus (g)[a] | −20.1 | 0 | 0.10 | (1, 3) |
| Liver (g) | −155 | 0 | 0.09 | (3) |
| Heart (g) | −35 | 0 | 0.01 | (1, 3) |
| Spleen (g)[a] | −10 | 0 | 0.02 | (1, 3) |
| Adrenal gland (g)[a] | 0 | +0.35 | 0.04 | (1) |
| Thyroid gland | −2.7 | 0 | 0.07 | (1) |
| Mammary gland (g)[a] | −17.6 | −54.3 | <0.05 | (1) |
| *Parenchyma* (g)[a] | −0.17 | 0 | 0.01 | (2) |
| *Fat pad* (g) | −15.8 | −42.1 | <0.03 | (2) |
| Ovaries (g)[a] | 0 | −1.45 | 0.04 | (1) |
| Kidney (g)[a] | 0 | +39.2 | 0.05 | (1) |
| Heart (g) | −34 | 0 | 0.01 | (1) |

Positive (+) and negative (–) indicate heavier or lighter organ/structure in IUHT ($n$ = 28/treatment) relative to IUTN ($n$ = 29/treatment) at birth (day 0) and weaning (day 63), least square means, and $p$-values adapted from (1) Dado-Senn et al. (2021), (2) Dado-Senn et al. (2022b), (3) Ahmed et al. (2021). $n$ = 8 per group for organs/tissues and $n$ = 28 for growth parameters. Gastrointestinal tract, rumen, liver, and pancreas organ weights were not different between IUTN and IUHT heifers either at birth or 63 days of age

[a]Organ/tissue statistically significant when expressed also as a g/kg body weight

IUHT pigs (Johnson et al. 2015b) whereas in other non-mammalian species, such as poultry, prenatal hyperthermia seems to exert the opposite effect (Piestun et al. 2008). Skin characteristics play important functions in thermoregulation. The number and distribution of skin glands play a role in the adaptive ability of the skin to compensate for heat stress (Raghav et al. 2021). Histological skin properties and skin gland histomorphology of IUHT heifers evaluated at different postnatal life stages support the notion that hyperthermia triggers long-lasting hair and skin adaptations (Davidson et al. 2022). At birth, IUHT heifers had reduced skin thickness and had more distant sweat glands to the skin surface relative to IUTN. As the IUHT animal matures, their skin has smaller and fewer sweat glands which are localized further from the skin surface. These adaptations to the skin sweat gland number, size, and localization were not transient and remained evident 1 year after birth. In utero hyperthermia influences skin gland composition and possibly function postnatally.

***Immunity and Survival*** Heat stress profoundly alters the immune response of dairy cattle, from the prenatal stage through lactation. Growing evidence suggests that the effects of IUHT can persist even after the insult is removed postnatally. Dairy calves gestated under IUHT conditions in late gestation have impaired passive and adaptive cell-mediated immunity through the first 2 months of life. Intrauterine heat stress impairs passive transfer of colostral IgG leading to a 10–15% reduction in the apparent efficiency of IgG absorption, compared with calves born to cooled dams and receiving their dam's colostrum

(Tao et al. 2012a; Laporta et al. 2017). It is known that direct heat stress exposure can compromise intestinal integrity in cows and pigs (Pearce et al. 2013; Koch et al. 2019), however, the impact of IUHT on the offspring's intestinal integrity is less documented. Dairy calves exposed to IUHT have more apoptotic enterocytes in the jejunum compared to IUTN counterparts, suggesting that the process of gut closure is accelerated by IUHT and possibly prevents complete IgG transfer from colostrum within the first hours after birth.

Dairy calves gestated under IUHT for 60 days of gestation have lower circulating lymphocytes, and higher circulating acute phase proteins, relative to IUTN calves (Skibiel et al. 2017). Peripheral blood mononuclear cells isolated from IUHT calves incubated in vitro with a mitogen have lower proliferation rates compared with cells from IUTN (Tao et al. 2012a; Monteiro et al. 2014). An important consideration is the duration and intensity of the intrauterine heat insult (i.e., acute vs. chronic exposure). Nevertheless, an acute 7-day exposure is sufficient to trigger immune function alterations in the postnatal calf by modifying T and B cell functions (Merlot et al. 2008) and leading to reductions in circulating lymphocytes, TNFα and toll-like receptor 2 (Strong et al. 2015). Alterations to early-life immunocompetence might lead to long-term consequences ultimately reducing that animal's survival and productive life. Early-life losses of IUHT heifers and significant reductions in the length of productive life and overall lifespan of IUHT dairy cows have been documented (Laporta et al. 2020; Kipp et al. 2021).

***Mammary Gland Development and Milk Production*** Reductions in milk yield of cows born to a dam heat-stressed in late gestation (i.e., exposed to elevated THI for 3–8 weeks) have been reported (Monteiro et al. 2016; Kipp et al. 2021). Another study reported that the average dairy cow in the US born to a heat-stressed dam produces less milk for three consecutive lactations, when compared with a cow born to a dam cooled during late gestation (Laporta et al. 2020).

The mammary gland undergoes phases of expansion and regression across the animal's lifetime to promote lactation (Akers 2017). Although most of its functional growth occurs postnatally (i.e., during gestation and lactation), the mammary gland starts developing in utero and, like other developing tissues and organs, may undergo fetal programming (Laporta et al. 2022). Recent literature has demonstrated that the detriment in milk synthetic capacity of dairy cows that experienced IUHT might arise from aberrant alterations in the developmental trajectory of the fetal mammary gland which are evident at birth and weaning and persist through the first lactation. More specifically, the mammary parenchyma of IUHT heifers is smaller at birth with stunted ductal development and reduced cellular proliferation at birth and after weaning (Dado-Senn et al. 2022b). Impairments in early mammary growth and development limit later mammary epithelial development and milk synthetic capacity, impacting lactation performance at maturity (Skibiel et al. 2018a). The economic repercussions on the US dairy sector arising from raising heifers that do not reach the first lactation, and from heifers that reach the first lactation with less productive udders were estimated at around $500 million annually (Laporta et al. 2020).

Mechanistically, heat stress in pregnant cows might induce epigenetic modifications in the fetus and possibly the developing germ cells, leading to phenotypic carry-over effects in the progeny. Studies evaluating the multigenerational and transgenerational phenotypic effects of dairy heifers gestated under heat stress have been reported also in the $F_2$ ($F_2$; born to IUHT $F_1$) and the subsequent $F_3$ female progeny (Laporta et al. 2020; Weller et al. 2021). Although a subset of differentially methylated genes in the mammary tissue has been reported in their first lactation of $F_1$ IUHT (Skibiel et al. 2018a), no definitive controlled studies have demonstrated transgenerational inheritance of IUHT epigenetic marks in dairy cows.

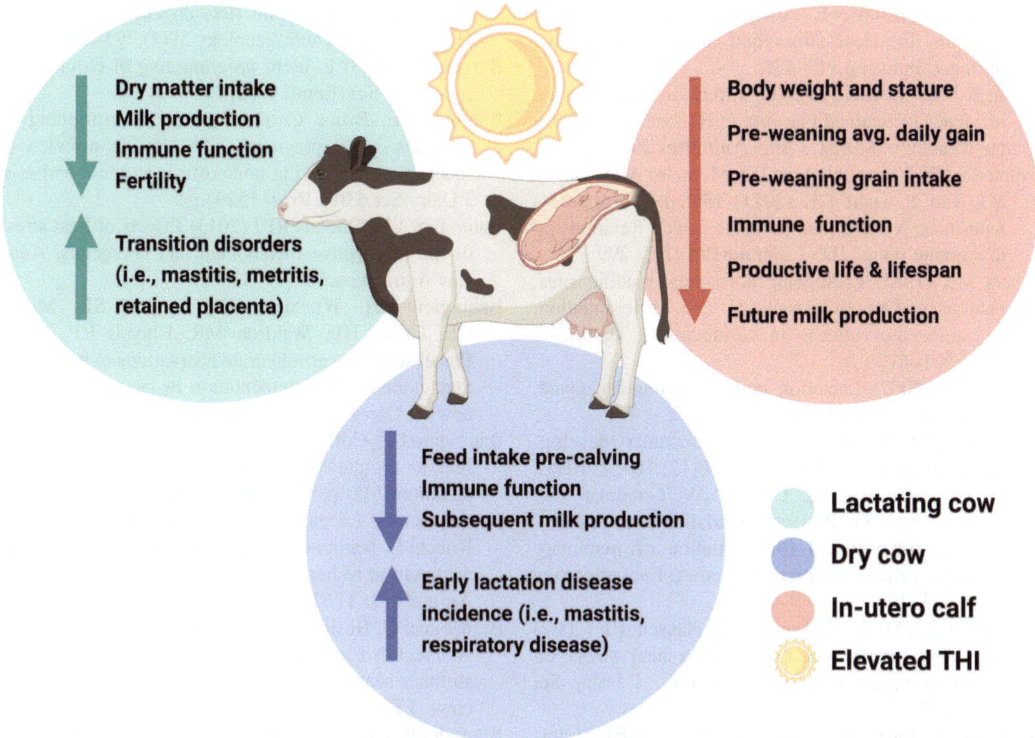


**Fig. 1** Phenotypic consequences of heat stress in dairy cattle. *Green* (top-left): dairy cows exposed to heat stress during lactation, *blue* (bottom): dairy cows exposed to heat stress during the dry period (non-lactating 6–8-week period pre-calving); *red* (top-right): postnatal consequences of in utero heat stress (last 6–8 weeks of gestation). THI: temperature-humidity Index. THI breakpoints at which lactating dairy cows begin to experience physiological and productive consequences of heat stress range from 52 to 72 (i.e., changes in rumination and milk production; Ravagnolo et al. 2000, Zimbelman et al. 2009; Müschner-Siemens et al. 2020, Pinto et al. 2020), while for dry (pregnant) cows THI breakpoints of 75 have been reported (Ouellet et al. 2021)

## 7 Summary and Conclusions

The consequences of increased ambient temperature and humidity on the dairy industry are substantial. Elevated ambient temperatures lead to a cascade of detrimental effects including declines in productivity and elevated morbidity and mortality at various life stages including pre- and post-weaning, during the dry period, and during lactation. Further, there is substantial evidence of maladaptive responses arising from intrauterine hyperthermia that are multifactorial, affect multiple tissues and organs, and persist as lifelong deficiencies, lowering the survival and productive potential of dairy cattle. Figure 1 summarizes the main physiological and productive outcomes impacted by environmental heat stress in dairy cattle.

## References

Abeyta MA, Al-Qaisi M, Horst EA, Mayorga EJ, Rodriguez-Jimenez S, Goetz BM, Carta S, Tucker H, Baumgard LH (2023) Effects of dietary antioxidant supplementation on metabolism and inflammatory biomarkers in heat-stressed dairy cows. J Dairy Sci 106(2):1441–1452

Adin G, Gelman A, Solomon R, Flamenbaum I, Nikbachat M, Yosef E, Zenou A, Shamay A, Feuermann Y, Mabjeesh SJ, Miron J (2009) Effects of cooling dry cows under heat load conditions on mammary gland enzymatic activity, intake of food and water, and performance during the dry period and after parturition. Livest Sci 124(1):189–195

Aggarwal A, Upadhyay R (2013) Heat stress and immune function. In: Heat stress and animal productivity. Springer, India, pp 113–136

Ahmadi MR, Mirzaei A (2006) Effect of heat stress on incidence of retained placenta in Holstein cows at dry hot weather of Shiraz. J Appl Anim Res 29(1):23–24

Ahmed BMS, Younas U, Asar TO, Monteiro APA, Hayen MJ, Tao S, Dahl GE (2021) Maternal heat stress reduces body and organ growth in calves: Relationship to immune status. JDS Commun 2(5):295–299

Akers MR (1985) Lactogenic hormones: binding sites, mammary Growth, secretory cell differentiation, and milk biosynthesis in ruminants. J Dairy Sci 68(2):501–519

Akers RM (2002) Lactation and the mammary gland. Blackwell, Ames, IA

Akers RM (2017) A 100-year review: mammary development and lactation. J Dairy Sci 100(12):10332–10352

Akers RM, Bauman DE, Capuco AV, Goodman GT, Tucker HA (1981) Prolactin regulation of milk secretion and biochemical differentiation of mammary epithelial cells in periparturient cows. Endocrinology 109(1):23–30

Al-Katanani YM, Paula-Lopes FF, Hansen PJ (2002) Effect of season and exposure to heat stress on oocyte competence in Holstein cows. J Dairy Sci 85(2):390–396

Al-Qaisi M, Horst EA, Kvidera SK, Mayorga EJ, Timms LL, Baumgard LH (2019) Developing a heat stress model in dairy cows using an electric heat blanket. J Dairy Sci 102:684–689

Al-Qaisi M, Mayorga EJ, Horst EA, Kvidera SK, McCarthy CS, Abeyta MA, Goetz BM, Ramirez-Ramirez HA, Timms LL, Baumgard LH (2020) Validating a heat stress model: the effects of an electric heat blanket and nutritional plane on lactating dairy cows. J Dairy Sci 103:5550–5560

Ara Begum R, Lempert R, Ali E, Benjaminsen TA, Bernauer T, Cramer W, Cui X, Mach K, Nagy G, Stenseth NC, Sukumar R, Wester P (2022) Point of departure and key concepts. In: Pörtner H-O, Roberts DC, Tignor M, Poloczanska ES, Mintenbeck K, Alegría A, Craig M, Langsdorf S, Löschke S, Möller V, Okem A, Rama B (eds) Climate change 2022: impacts, adaptation and vulnerability. Contribution of Working Group II to the sixth assessment report of the Intergovernmental Panel on Climate Change. Cambridge University Press, Cambridge, pp 121–196

Aroyo A, Yavin S, Roth Z, Arav A (2007) Hindering of cleavage timing in bovine parthenotes during the hot season. Reprod Fertil Dev 19

Asakura H (2004) Fetal and neonatal thermoregulation. J Nippon Med Sch 71(6):360–370

Avendaño-Reyes L, Alvarez-Valenzuala FD, Correa-Calderón A, Saucedo-Quintero JS, Robinson PH, Fadel JG (2006) Effect of cooling Holstein cows during the dry period on postpartum performance under heat stress conditions. Livest Sci 105:198–206

Badinga L, Thatcher WW, Diaz T, Drost M, Wolfenson D (1993) Effect of environmental heat stress on follicular development and steroidogenesis in lactating Holstein cows. Theriogenology 39(4):797–810

Barker DJ (1998) In utero programming of chronic disease. Clin Sci (Lond) 95(2):115–128

Bauman DE, Bruce Currie W (1980) Partitioning of nutrients during pregnancy and lactation: a review of mechanisms involving homeostasis and homeorhesis. J Dairy Sci 63(9):1514–1529

Baumgard LH, Rhoads RPJ (2013) Effects of heat stress on postabsorptive metabolism and energetics. Annu Rev Anim Biosci 1:311–337

Baumgard LH, Wheelock JB, Sanders SR, Moore CE, Green HB, Waldron MR, Rhoads RP (2011) Postabsorptive carbohydrate adaptations to heat stress and monensin supplementation in lactating Holstein cows. J Dairy Sci 94(11):5620–5633

Baumgard LH, Collier RJ, Bauman DE (2017) A 100-year review: regulation of nutrient partitioning to support lactation. J Dairy Sci 100(12):10353–10366

Bernabucci U, Lacetera N, Baumgard LH, Rhoads RP, Ronchi B, Nardone A (2010) Metabolic and hormonal acclimation to heat stress in domesticated ruminants. Animal 4(7):1167–1183

Bernabucci U, Basiricò L, Morera P, Dipasquale D, Vitali A, Piccioli Cappelli F, Calamari L (2015) Effect of summer season on milk protein fractions in Holstein cows. J Dairy Sci 98(3):1815–1827

Bouraoui R, Lahmar M, Majdoub A, Djemali M, Belyea R (2002) The relationship of temperature-humidity index with milk production of dairy cows in a Mediterranean climate. Anim Res 51(6):479–491

Bradford BJ, Yuan K, Farney JK, Mamedova LK, Carpenter AJ (2015) Invited review: Inflammation during the transition to lactation: New adventures with an old flame. J Dairy Sci 98(10):6631–6650

Brügemann K, Gernand E, König von Borstel U, König S (2012) Defining and evaluating heat stress thresholds in different dairy cow production systems. Arch Anim Breed 55(1):13–24

Capuco AV, Akers RM (1999) Mammary involution in dairy animals. J Mammary Gland Biol Neoplasia 4(2):137–144

Capuco AV, Akers RM, Smith JJ (1997) Mammary growth in Holstein cows during the dry period: quantification of nucleic acids and histology. J Dairy Sci 80:477–487

Carvalho PD, Santos VG, Giordano JO, Wiltbank MC, Fricke PM (2018) Development of fertility programs to achieve high 21-day pregnancy rates in high-producing dairy cows. Theriogenology 114:165–172

Collier RJ, Gebremedhin KG (2015) Thermal biology of domestic animals. Annu Rev Anim Biosci 3:513–532

Collier RJ, Beede DK, Thatcher WW, Israel LA, Wilcox CJ (1982a) Influences of environment and its modification on dairy animal health and production. J Dairy Sci 65(11):2213–2227

Collier RJ, Doelger SG, Head HH, Thatcher WW, Wilcox CJ (1982b) Effects of heat stress during pregnancy on maternal hormone concentrations, calf birth weight and postpartum milk yield of Holstein cows. J Anim Sci 54:309–319

Collier RJ, Dahl GE, VanBaale MJ (2006a) Major advances associated with environmental effects on dairy cattle. J Dairy Sci 89(4):1244–1253

Collier RJ, Stiening CM, Pollard BC, VanBaale MJ, Baumgard LH, Gentry PC, Coussens PM (2006b) Use of gene expression microarrays for evaluating environmental stress tolerance at the cellular level in cattle. J Anim Sci 84(E. Suppl):E1–E13

Collier RJ, Collier JL, Rhoads RP, Baumgard LH (2008) Genes involved in the bovine heat stress response. J Dairy Sci 91:445–454

Collier RJ, Renquist BJ, Xiao Y (2017) A 100-year review: stress physiology including heat stress. J Dairy Sci 100(12):10367–10380

Cox B, Gasparrini A, Catry B, Delcloo A, Bijnens E, Vangronsveld J, Nawrot TS (2016a) Mortality related to cold and heat. What do we learn from dairy cattle? Environ Res 149:231–238

Cox B, Gasparrini A, Catry B, Fierens F, Vangronsveld J, Nawrot TS (2016b) Ambient air pollution-related mortality in dairy cattle: does it corroborate human findings? Epidemiology 27(6):779–786

Dado-Senn B, Skibiel AL, Fabris TF, Zhang Y, Dahl GE, Peñagaricano F, Laporta J (2018) RNA-Seq reveals novel genes and pathways involved in bovine mammary involution during the dry period and under environmental heat stress. Sci Rep 8:11096

Dado-Senn B, Skibiel AL, Fabris TF, Dahl GE, Laporta J (2019) Dry period heat stress induces microstructural changes in the lactating mammary gland. PLoS One 19(9):e0222120

Dado-Senn B, Ouellet V, Dahl GE, Laporta J (2020a) Methods for assessing heat stress in preweaned dairy calves exposed to chronic heat stress or continuous cooling. J Dairy Sci 103(9):8587–8600

Dado-Senn B, Vega Acosta L, Torres Rivera M, Field SL, Marrero MG, Davidson BD, Tao S, Fabris TF, Ortiz-Colón G, Dahl GE, Laporta J (2020b) Pre- and postnatal heat stress abatement affects dairy calf thermoregulation and performance. J Dairy Sci 103(5):4822–4837

Dado-Senn B, Field SL, Davidson BD, Casarotto LT, Marrero MG, Ouellet V, Cunha F, Sacher MA, Rice CL, Maunsell FP, Dahl GE, Laporta J (2021) Late-gestation in utero heat stress limits dairy heifer early-life growth and organ development. Front Anim Sci. https://doi.org/10.3389/fanim.2021.750390

Dado-Senn B, Gingerich KN, Horvath KC, Field SL, Marrero MG, Maunsell F, Miller-Cushon EK, Laporta J (2022a) Early-life heat stress exposure impacts dairy calf feeding and thermoregulatory behavior. JDS Commun 3(1):38–43

Dado-Senn BM, Field SL, Davidson BD, Dahl GE, Laporta J (2022b) In utero hyperthermia in late gestation derails dairy calf early-life mammary development. J Anim Sci 100(10):skac186. https://doi.org/10.1093/jas/skac186

Dado-Senn B, Van Os J, Dorea J, Laporta J (2024) Actively ventilating calf hutches using solar-powered fans: Effects on hutch microclimate and calf thermoregulation. JDS Comm. 5(1):61–66

Dahl GE, Tao S, Laporta J (2017) Triennial lactation symposium/BOLFA: Late gestation heat stress of dairy cattle programs dam and daughter milk production. J Anim Sci 95(12):5701–5710

Davidson BD, Dado-Senn B, Ouellet V, Dahl GE, Laporta J (2021) Effect of late-gestation heat stress in nulliparous heifers on postnatal growth, passive transfer of immunoglobulin G, and thermoregulation of their calves. JDS Commun 2(3):165–169

Davidson BD, Sarlo Davila KM, Mateescu RG, Dahl GE, Laporta J (2022) Effect of in utero exposure to hyperthermia on postnatal hair length, skin morphology, and thermoregulatory responses. J Dairy Sci 105(11):8898–8910

do Amaral BC, Connor EE, Tao S, Hayen J, Bubolz J, Dahl GE (2009) Heat-stress abatement during the dry period: does cooling improve transition into lactation? J Dairy Sci 92:5988–5999

do Amaral BC, Connor EE, Tao S, Hayen J, Bubolz J, Dahl GE (2010) Heat stress abatement during the dry period influences prolactin signaling in lymphocytes. Domest Anim Endocrinol 38:38–45

do Amaral BC, Connor EE, Tao S, Hayen J, Bubolz JW, Dahl GE (2011) Heat stress abatement during the dry period influences metabolic gene expression and improves immune status in the transition period of dairy cows. J Dairy Sci 94:86–96

Dreiling CE, Carman FS 3rd, Brown DE (1991) Maternal endocrine and fetal metabolic responses to heat stress. J Dairy Sci 74(1):312–327. https://doi.org/10.3168/jds.S0022-0302(91)78175-7

Du J, Di H-S, Guo L, Li Z-H, Wang G-L (2008) Hyperthermia causes bovine mammary epithelial cell death by a mitochondrial-induced pathway. J Therm Biol 33:37–47

DuBois PR, Williams DJ (1980) Increased incidence of retained placenta associated with heat stress in dairy cows. Theriogenology 13(2):115–121

Ealy AD, Drost M, Hansen PJ (1993) Developmental changes in embryonic resistance to adverse effects of maternal heat stress in cows. J Dairy Sci 76(10):2899–2905

Edwards JL, Hansen PJ (1997) Differential responses of bovine oocytes and preimplantation embryos to heat shock. Mol Reprod Dev 46(2):138–145

Elvinger F, Natzke RP, Hansen PJ (1992) Interactions of heat stress and bovine somatotropin affecting physiology and immunology of lactating cows. J Dairy Sci 75(2):449–462

Fabris TF, Laporta J, Skibiel AL, Corrá FN, Senn DB, Wohlgemuth SE, Dahl GE (2019) Effect of heat stress during early, late, and entire dry period on dairy cattle. J Dairy Sci 102:5647–5656

Fabris TF, Laporta J, Skibiel AL, Dado-Senn B, Wohlgemuth SE, Dahl GE (2020) Effect of heat stress during the early and late dry period on mammary gland development of Holstein dairy cattle. J Dairy Sci 103(9):8576–8586

Faurie AS, Mitchell D, Laburn HP (2001) Feto-maternal relationships in goats during heat and cold exposure. Exp Physiol 86(2):199–204

Ferreira FC, Gennari RS, Dahl GE, De Vries A (2016) Economic feasibility of cooling dry cows across the United States. J Dairy Sci 99(12):9931–9941

Flamenbaum I, Galon N (2010) Management of heat stress to improve fertility in dairy cows in Israel. J Reprod Dev 56(Suppl):S36–S41

Flamenbaum I, Wolfenson D, Kunz PL, Maman M, Berman A (1995) Interactions between body condition at calving and cooling of dairy cows during lactation in summer. J Dairy Sci 78(10):2221–2229

Gao ST, Guo J, Quan SY, Nan XM, Fernandez MVS, Baumgard LH, Bu DP (2017) The effects of heat stress on protein metabolism in lactating Holstein cows. J Dairy Sci 100(6):5040–5049

Gao ST, Ma L, Zhou Z, Zhou ZK, Baumgard LH, Jiang D, Bionaz M, Bu DP (2019) Heat stress negatively affects the transcriptome related to overall metabolism and milk protein synthesis in mammary tissue of midlactating dairy cows. Physiol Genomics 51(8):400–409

Garner JB, Douglas ML, Williams SRO, Wales WJ, Marett LC, Nguyen TTT, Reich CM, Hayes BJ (2016) Genomic selection improves heat tolerance in dairy cattle. Sci Rep 6(1):34114

Gernand E, König S, Kipp C (2019) Influence of on-farm measurements for heat stress indicators on dairy cow productivity, female fertility, and health. J Dairy Sci 102(7):6660–6671

Goff JP, Horst RL (1997) Physiological changes at parturition and their relationship to metabolic disorders. J Dairy Sci 80:1260–1268

Gwazdauskas FC, Thatcher WW, Kiddy CA, Paape MJ, Wilcox CJ (1981) Hormonal patterns during heat stress following PGF(2)alpha-tham salt induced luteal regression in heifers. Theriogenology 16(3):271–285

Hahn GL (1999) Dynamic responses of cattle to thermal heat loads. J Anim Sci 77(Suppl 2):10–20

Hamel J, Zhang Y, Wente N, Krömker V (2021) Heat stress and cow factors affect bacteria shedding pattern from naturally infected mammary gland quarters in dairy cattle. J Dairy Sci 104(1):786–794

Hammami H, Bormann J, M'Hamdi N, Montaldo HH, Gengler N (2013) Evaluation of heat stress effects on production traits and somatic cell score of Holsteins in a temperate environment. J Dairy Sci 96(3):1844–1855

Hammami H, Vandenplas J, Vanrobays ML, Rekik B, Bastin C, Gengler N (2015) Genetic analysis of heat stress effects on yield traits, udder health, and fatty acids of Walloon Holstein cows. J Dairy Sci 98(7):4956–4968

Hansen PJ (2019) Reproductive physiology of the heat-stressed dairy cow: implications for fertility and assisted reproduction. Anim Reprod 16(3):497–507

Heck JML, van Valenberg HJF, Dijkstra J, van Hooijdonk ACM (2009) Seasonal variation in the Dutch bovine raw milk composition. J Dairy Sci 92(10):4745–4755

Her E, Wolfenson D, Flamenbaum I, Folman Y, Kaim M, Berman A (1988) Thermal, productive, and reproductive responses of high yielding cows exposed to short-term cooling in summer. J Dairy Sci 71(4):1085–1092

Hoffmann G, Herbut P, Pinto S, Heinicke J, Kuhla B, Amon T (2020) Animal-related, non-invasive indicators for determining heat stress in dairy cows. Biosyst Eng 199:83–96

Horst RL, Goff JP, Reinhardt TA (2005) Adapting to the transition between gestation and lactation: differences between rat, human, and dairy cow. J Mammary Gland Biol Neoplasia 10(2):141–156

Hughes L (2000) Biological consequences of global warming: is the signal already apparent? Trends Ecol Evol 15(2):56–61

Hurley WL, Loor JJ (2011) Mammary gland: growth, development and involution. In: Fuquay JW (ed) Encyclopedia of dairy sciences, 2nd edn. Academic, San Diego, CA, pp 338–345

Huson HJ, Sonstegard TS, Godfrey J, Hambrook D, Wolfe C, Wiggans G, Blackburn H, VanTassell CP (2020) A genetic investigation of Island Jersey Cattle, the foundation of the Jersey breed: Comparing population structure and selection to Guernsey, Holstein, and United States Jersey Cattle. Front Genet 11

Ingraham RH, Stanley RW, Wagner WC (1976) Relationship of temperature and humidity to conception rate of Holstein cows in Hawaii. J Dairy Sci 59(12):2086–2090

IPCC (2014) Climate change 2014: synthesis report. In: Core Writing Team, Pachauri RK, Meyer LA (eds) Contribution of Working Groups I, II and III of the fifth assessment report of the Intergovernmental Panel on Climate Change. IPCC, Geneva, Switzerland, p 151

Jamnongjit M, Hammes SR (2005) Oocyte maturation: the coming of age of a germ cell. Semin Reprod Med 23:234–241

Johnson JS, Sanz-Fernandez MV, Patience JF, Ross JW, Gabler NK, Lucy MC, Safranski TJ, Rhoads RP, Baumgard LH (2015a) Effects of in utero heat stress on postnatal body composition in pigs: II. Finishing Phase. J Anim Sci. 93:82–92. https://doi.org/10.2527/jas.2014-8355

Johnson JS, Sanz Fernandez MV, Seibert JT, Ross JW, Lucy MC, Safranski TJ, Elsasser TH, Kahl S, Rhoads RP, Baumgard LH (2015b) In utero heat stress increases postnatal core body temperature in pigs. J Anim Sci 93(9):4312–4322

Kadzere CT, Murphy MR, Silanikove N, Maltz E (2002) Heat stress in lactating dairy cows: a review. Livest Prod Sci 77(1):59–91

Kamwanja LA, Chase CC Jr, Gutierrez JA, Guerriero V Jr, Olson TA, Hammond AC, Hansen PJ (1994) Responses of bovine lymphocytes to heat shock as modified by breed and antioxidant status. J Anim Sci 72(2):438–444

Kim KH, Kim D-H, Oh Y-K, Lee S-S, Lee H-J, Kim D-W, Seol Y-J, Kimura N (2010) Productivity and energy partition of late lactation dairy cows during heat exposure. Anim Sci J 81(1):58–62

Kipp C, Brügemann K, Zieger P, Mütze K, Möcklinghoff-Wicke S, König S, Halli K (2021) Across-generation effects of maternal heat stress during late gestation on production, female fertility and longevity traits in dairy cows. J Dairy Res 88(2):147–153

Koch F, Lamp O, Eslamizad M, Weitzel J, Kuhla B (2016) Metabolic response to heat stress in late-pregnant and early lactation dairy cows: implications to liver-muscle crosstalk. PLoS One 11(8):e0160912

Koch F, Thom U, Albrecht E, Weikard R, Nolte W, Kuhla B, Kuehn C (2019) Heat stress directly impairs gut integrity and recruits distinct immune cell populations into the bovine intestine. Proc Natl Acad Sci U S A 116(21):10333–10338

Kovács L, Kézér FL, Ruff F, Jurkovich V, Szenci O (2018) Assessment of heat stress in 7-week old dairy calves with non-invasive physiological parameters in different thermal environments. PLoS One 13(7):e0200622

Kvidera SK, Horst EA, Abuajamieh M, Mayorga EJ, Fernandez MVS, Baumgard LH (2017) Glucose requirements of an activated immune system in lactating Holstein cows. J Dairy Sci 100(3):2360–2374

Laburn H, Faurie A, Mitchell D (2000) The thermal physiology of the ruminant fetus. CABI International, Wallingford

Laburn HP, Faurie A, Goelst K, Mitchell D (2002) Effects on fetal and maternal body temperatures of exposure of pregnant ewes to heat, cold, and exercise. J Appl Physiol 92(2):802–808

Lacetera N, Bernabucci U, Scalia D, Ronchi B, Kuzminsky G, Nardone A (2005) Lymphocyte functions in dairy cows in hot environment. Int J Biometeorol 50(2):105–110

Lacetera N, Bernabucci U, Scalia D, Basiricò L, Morera P, Nardone A (2006) Heat stress elicits different responses in peripheral blood mononuclear cells from Brown Swiss and Holstein Cows. J Dairy Sci 89(12):4606–4612

Lamp O, Derno M, Otten W, Mielenz M, Nurnberg G, Kuhla B (2015) Metabolic heat stress adaptation in transition cows: differences in macronutrient oxidation between late-gestating and early-lactating German Holstein dairy cows. PLoS One 10(5):e0125264

Laporta J, Fabris TF, Skibiel AL, Powell JL, Hayen MJ, Horvath K, Miller-Cushon EK, Dahl GE (2017) In utero exposure to heat stress during late gestation has prolonged effects on the activity patterns and growth of dairy calves. J Dairy Sci 100(4):2976–2984

Laporta J, Ferreira FC, Ouellet V, Dado-Senn B, Almeida AK, De Vries A, Dahl GE (2020) Late-gestation heat stress impairs daughter and granddaughter lifetime performance. J Dairy Sci 103(8):7555–7568

Laporta J, Dado-Senn B, Skibiel AL (2022) Late gestation hyperthermia: epigenetic programming of daughter's mammary development and function. Domest Anim Endocrinol 78:106681

Lecchi C, Rota N, Vitali A, Ceciliani F, Lacetera N (2016) In vitro assessment of the effects of temperature on phagocytosis, reactive oxygen species production and apoptosis in bovine polymorphonuclear cells. Vet Immunol Immunopathol 182:89–94

Li L, Sun Y, Li X, Luo M, Wang G (2015) The global effect of heat on gene expression in cultured bovine mammary epithelial cells. Cell Stress Chaperones 20:381–389

Li L, Wang Y, Li C, Wang G (2017) Proteomic analysis to unravel the effect of heat stress on gene expression and milk synthesis in bovine mammary epithelial cells. Anim Sci J 88:2090–2099

Limesand SW, Camacho LE, Kelly AC, Antolic AT (2018) Impact of thermal stress on placental function and fetal physiology. Anim Reprod 15(Suppl 1):886–898. https://doi.org/10.21451/1984-3143-AR2018-0056

Lough DS, Beede DL, Wilcox CJ (1990) Effects of feed intake and thermal stress on mammary blood flow and other physiological measurements in lactating dairy cows. J Dairy Sci 73:325–332

Ma L, Yang Y, Zhao X, Wang F, Gao S, Bu D (2019) Heat stress induces proteomic changes in the liver and mammary tissue of dairy cows independent of feed intake: An iTRAQ study. PLoS One 14(1):e0209182

MacLeod P, Anderson EO, Plastridge WN (1954) Cell counts of platform samples of herd milk. J Dairy Sci 37(8):919–923

Mansfield RH (1985) Progress of the breed. The history of U.S. Holsteins. Holstein-Friesian World, Sandy Creek, NY

Marins TN, Gao J, Yang Q, Binda RM, Pessoa CMB, Orellana Rivas RM, Garrick M, Melo VHLR, Chen Y-C, Bernard JK, Garcia M, Chapman JD, Kirk DJ, Tao S (2021) Impact of heat stress and a feed supplement on hormonal and inflammatory responses of dairy cows. J Dairy Sci 104(7):8276–8289

Martínez RS, Palladino RA, Banchero G, Fernández-Martín R, Nanni M, Juliano N, Iorio J, La Manna A (2021) Providing heat-stress abatement to late-lactation Holstein cows affects hormones, metabolite blood profiles, and hepatic gene expression but not productive responses. Appl Anim Sci 37(4):490–503

McGuire MA, Beede DK, DeLorenzo MA, Wilcox CJ, Huntington GB, Reynolds CK, Collier RJ (1989) Effects of thermal stress and level of feed intake on portal plasma flow and net fluxes of metabolites in lactating Holstein cows. J Anim Sci 67(4):1050–1060

Merlot E, Couret D, Otten W (2008) Prenatal stress, fetal imprinting and immunity. Brain Behav Immun 22(1):42–51

Min L, Cheng J, Zhao S, Tian H, Zhang Y, Li S, Yang H, Zheng N, Wang J (2016) Plasma-based proteomics reveals immune response, complement and coagulation cascades pathway shifts in heat-stressed lactating dairy cows. J Proteomics 146:99–108

Molinari PCC, Sheldon IM, Dahl GE, Bromfield JJ (2019) Regardless of bacterial load in the vagina, heat stress increases uterine disease incidence and severity in the dairy cow. J Dairy Sci 102:306–307

Molinari PCC, Dahl GE, Sheldon IM, Bromfield JJ (2022a) Effect of calving season on metritis incidence and bacterial content of the vagina in dairy cows. Theriogenology 191:67–76

Molinari PCC, Davidson BD, Laporta J, Dahl GE, Sheldon IM, Bromfield JJ (2022b) Prepartum heat stress in dairy cows increases postpartum inflammatory responses in blood of lactating dairy cows. J Dairy Sci 106(2):1464–1474
Monteiro AP, Tao S, Thompson IM, Dahl GE (2014) Effect of heat stress during late gestation on immune function and growth performance of calves: isolation of altered colostral and calf factors. J Dairy Sci 97(10):6426–6439
Monteiro APA, Tao S, Thompson IMT, Dahl GE (2016) In utero heat stress decreases calf survival and performance through the first lactation. J Dairy Sci 99(10):8443–8450
Moody EG, Van Soest PJ, McDowell RE, Ford GL (1971) Effect of high temperature and dietary fat on milk fatty acids. J Dairy Sci 54(10):1457–1460
Muñoz G, Rondelli F, Maiztegui L, Gherardi S, Tolini F, Amelong J, Fernández G, Coronel A (2014) Effects of heat waves on lactating Argentinian Holstein cows in the dairy module. Cienc Vet 16:113–127
Müschner-Siemens T, Hoffmann G, Ammon C, Amon T (2020) Daily rumination time of lactating dairy cows under heat stress conditions. J Therm Biol 88:102484
NOAA (2022) Global climate report—annual 2021
Norman HD, Guinan FL, Walton LM, Dürr J (2019) Somatic cell counts of milk from dairy herd improvement herds during 2019. Council on Dairy Cattle Breeding, Beltsville, MD
Olde Riekerink RGM, Barkema HW, Stryhn H (2007) The effect of season on somatic cell count and the incidence of clinical mastitis. J Dairy Sci 90(4):1704–1715
Ouellet V, Toledo IM, Dado-Senn B, Dahl GE, Laporta J (2021) Critical temperature-humidity index thresholds for dry cows in a subtropical climate. Front Anim Sci 2:706636
Paape MJ, Schultze WD, Miller RH, Smith JW (1973) Thermal stress and circulating erythrocytes, leucocytes, and milk somatic cells. J Dairy Sci 56(1):84–91
Pace A (2022) Effects of wildfire smoke PM2.5 on preweaned Holstein dairy calves. J Dairy Sci 105(Suppl. 1):113
Pearce SC, Mani V, Weber TE, Rhoads RP, Patience JF, Baumgard LH, Gabler NK (2013) Heat stress and reduced plane of nutrition decreases intestinal integrity and function in pigs. J Anim Sci 91(11):5183–5193
Piestun Y, Shinder D, Ruzal M, Halevy O, Brake J, Yahav S (2008) Thermal manipulations during broiler embryogenesis: effect on the acquisition of thermotolerance. Poult Sci 87(8):1516–1525
Pinto S, Hoffmann G, Ammon C, Amon T (2020) Critical THI thresholds based on the physiological parameters of lactating dairy cows. J Therm Biol 88:102523
Purwanto BP, Abo Y, Sakamoto R, Furumoto F, Yamamoto S (1990) Diurnal patterns of heat production and heart rate under thermoneutral conditions in Holstein Friesian cows differing in milk production. J Agric Sci 114(2):139–142
Putney DJ, Drost M, Thatcher WW (1988) Embryonic development in superovulated dairy cattle exposed to elevated ambient temperatures between days 1 to 7 post insemination. Theriogenology 30(2):195–209
Raghav S, Uppal V, Gupta A (2021) Comparative study on distribution of sebaceous and sweat glands in skin of different domestic animals. Indian J Anim Res 56:1356–1360
Ravagnolo O, Misztal I, Hoogenboom G (2000) Genetic component of heat stress in dairy cattle, development of heat index function. J Dairy Sci 83(9):2120–2125
Renaudeau D, Collin A, Yahav S, de Basilio V, Gourdine JL, Collier RJ (2012) Adaptation to hot climate and strategies to alleviate heat stress in livestock production. Animal 6(5):707–728
Reynolds LP, Caton JS (2012) Role of the pre- and post-natal environment in developmental programming of health and productivity. Mol Cell Endocrinol 354(1–2):54–59
Reynolds LP, Borowicz PP, Caton JS, Crouse MS, Dahlen CR, Ward AK (2019) Developmental programming of fetal growth and development. Vet Clin North Am Food Anim Pract 35(2):229–247
Rhoads ML, Rhoads RP, VanBaale MJ, Collier RJ, Sanders SR, Weber WJ, Crooker BA, Baumgard LH (2009) Effects of heat stress and plane of nutrition on lactating Holstein cows: Production, metabolism, and aspects of circulating somatotropin. J Dairy Sci 92:1986–1997
Rhoads RP, La Noce AJ, Wheelock JB, Baumgard LH (2011) Alterations in expression of gluconeogenic genes during heat stress and exogenous bovine somatotropin administration. J Dairy Sci 94:1917–1921
Ríus AG (2019) Invited Review: Adaptations of protein and amino acid metabolism to heat stress in dairy cows and other livestock species. Appl Anim Sci 35(1):39–48
Rodrigues O, Scaliante JR Jr, Shangraw EM, Hirtz LK, Adkins PRF, McFadden TB (2019) Heat stress reduces total mammary blood flow and transmammary disappearance of metabolites in lactating dairy cows. J Dairy Sci 102(Suppl. 1):153
Roman-Ponce H, Thatcher WW, Buffington DE, Wilcox CJ, Van Horn HH (1977) Physiological and production responses of dairy cattle to a shade structure in a subtropical environment. J Dairy Sci 60(3):424–430
Roth Z (2008) Heat stress, the follicle, and its enclosed oocyte: mechanisms and potential strategies to improve fertility in dairy cows. Reprod Domest Anim 43(s2):238–244
Roth Z, Hansen PJ (2004) Sphingosine 1-phosphate protects bovine oocytes from heat shock during maturation. Biol Reprod 71(6):2072–2078
Roth Z, Hansen PJ (2005) Disruption of nuclear maturation and rearrangement of cytoskeletal elements in bovine oocytes exposed to heat shock during maturation. Reproduction 129(2):235–244
Roth Z, Meidan R, Braw-Tal R, Wolfenson D (2000) Immediate and delayed effects of heat stress on follicular development and its association with plasma FSH and inhibin concentration in cows. J Reprod Fertil 120(1):83–90

Roth Z, Meidan R, Shaham-Albalancy A, Braw-Tal R, Wolfenson D (2001) Delayed effect of heat stress on steroid production in medium-sized and preovulatory bovine follicles. Reproduction 121(5):745–751

Safa S, Kargar S, Moghaddam GA, Ciliberti MG, Caroprese M (2018) Heat stress abatement during the postpartum period: effects on whole lactation milk yield, indicators of metabolic status, inflammatory cytokines, and biomarkers of the oxidative stress. J Anim Sci 97(1):122–132

Santos JEP, Ribeiro ES (2014) Impact of animal health on reproduction of dairy cows. Anim Reprod 11:254–269

Seyed Almoosavi SMM, Ghoorchi T, Naserian AA, Khanaki H, Drackley JK, Ghaffari MH (2021) Effects of late-gestation heat stress independent of reduced feed intake on colostrum, metabolism at calving, and milk yield in early lactation of dairy cows. J Dairy Sci 104(2):1744–1758

Skibiel AL, Fabris TF, Corrá FN, Torres YM, McLean DJ, Chapman JD, Kirk DJ, Dahl GE, Laporta J (2017) Effects of feeding an immunomodulatory supplement to heat-stressed or actively cooled cows during late gestation on postnatal immunity, health, and growth of calves. J Dairy Sci 100(9):7659–7668

Skibiel AL, Dado-Senn B, Fabris TF, Dahl GE, Laporta J (2018a) In utero exposure to thermal stress has long-term effects on mammary gland microstructure and function in dairy cattle. PLoS One 13(10):e0206046–e0206046

Skibiel AL, Zachut M, do Amaral BC, Levin Y, Dahl GE (2018b) Liver proteomic analysis of postpartum Holstein cows exposed to heat stress or cooling conditions during the dry period. J Dairy Sci 101:705–716

Skibiel AL, Koh J, Zhu N, Zhu F, Yoo M-J, Laporta J (2022) Carry-over effects of dry period heat stress on the mammary gland proteome and phosphoproteome in the subsequent lactation of dairy cows. Sci Rep 12(1):6637

Smith DL, Smith T, Rude BJ, Ward SH (2013) Short communication: Comparison of the effects of heat stress on milk and component yields and somatic cell score in Holstein and Jersey cows. J Dairy Sci 96(5):3028–3033

Spain JN, Spiers DE (1996) Effects of supplemental shade on thermoregulatory response of calves to heat challenge in a hutch environment. J Dairy Sci 79(4):639–646

Spiers DE, Spain JN, Sampson JD, Rhoads RP (2004) Use of physiological parameters to predict milk yield and feed intake in heat-stressed dairy cows. J Therm Biol 29(7):759–764

Stern DI, Kaufmann RK (2014) Anthropogenic and natural causes of climate change. Clim Change 122(1):257–269

St-Pierre NR, Cobanov B, Schnitkey G (2003) Economic losses from heat stress by U.S. livestock industries. J Dairy Sci 86(E. Suppl):E52–E77

Strong RA, Silva EB, Cheng HW, Eicher SD (2015) Acute brief heat stress in late gestation alters neonatal calf innate immune functions. J Dairy Sci 98(11):7771–7783

Tao S, Dahl GE (2013) Invited review: heat stress effects during late gestation on dry cows and their calves. J Dairy Sci 96(7):4079–4093

Tao S, Bubolz JW, do Amaral BC, Thompson IM, Hayen MJ, Johnson SE, Dahl GE (2011) Effect of heat stress during the dry period on mammary gland development. J Dairy Sci 94:5976–5986

Tao S, Monteiro AP, Thompson IM, Hayen MJ, Dahl GE (2012a) Effect of late-gestation maternal heat stress on growth and immune function of dairy calves. J Dairy Sci 95(12):7128–7136

Tao S, Thompson APA, Monteiro MJ, Hayen LJ, Young, and G. E. Dahl. (2012b) Effect of cooling heat-stressed dairy cows during the dry period on insulin response. J Dairy Sci 95:5035–5046

Tao S, Connor EE, Bubolz JW, Thompson IM, do Amaral BC, Hayen MJ, Dahl GE (2013) Effect of heat stress during the dry period on gene expression in mammary tissue and peripheral blood mononuclear cells. J Dairy Sci 96:378–383

Tao S, Orellana RM, Weng X, Marins TN, Dahl GE, Bernard JK (2018) Symposium review: The influences of heat stress on bovine mammary gland function1. J Dairy Sci 101(6):5642–5654

Tao S, Dahl GE, Laporta J, Bernard JK, Orellana Rivas RM, Marins TN (2019) Physiology symposium: Effects of heat stress during late gestation on the dam and its calf12. J Anim Sci 97(5):2245–2257

Tarazón-Herrera M, Huber JT, Santos J, Mena H, Nusso L, Nussio C (1999) Effects of bovine somatotropin and evaporative cooling plus shade on lactation performance of cows during summer heat stress. J Dairy Sci 82(11):2352–2357

Thatcher WW, Gwazdauskas FC, Wilcox CJ, Toms J, Head HH, Buffington DE, Fredricksson WB (1974) Milking performance and reproductive efficiency of dairy cows in an environmentally controlled structure. J Dairy Sci 57(3):304–307

Thompson IM, Dahl GE (2012) Dry-period seasonal effects on the subsequent lactation. Prof Anim Sci 28(6):628–631

Thompson IMT, Tao S, Monteiro APA, Jeong KC, Dahl GE (2014) Effect of cooling during the dry period on immune response after Streptococcus uberis intramammary infection challenge of dairy cows. J Dairy Sci 97(12):7426–7436

Toledo IM, Fabris TF, Tao S, Dahl GE (2020) When do dry cows get heat stressed? Correlations of rectal temperature, respiration rate, and performance. JDS Commun 1(1):21–24

USDA (2020) Milk cows and production by state and region. USDA Economic Research Service, USDA National Agricultural Statistics Service

USDA-NASS (2022) Milk: Production per cow by year, US. USDA Economic Research Service, USDA National Agricultural Statistics Service

Van Eetvelde M, Opsomer G (2017) Innovative look at dairy heifer rearing: Effect of prenatal and post-natal environment on later performance. Reprod Domest Anim 52(Suppl 3):30–36

Vautier AN, Cadaret CN (2022) Long-term consequences of adaptive fetal programming in ruminant livestock. Front Anim Sci 3

Vitali A, Felici A, Esposito S, Bernabucci U, Bertocchi L, Maresca C, Nardone A, Lacetera N (2015) The effect of heat waves on dairy cow mortality. J Dairy Sci 98(7):4572–4579

Wagner WC, Willham RL, Evans LE (1974) Controlled parturition in cattle. J Anim Sci 38(3):485–489

Wang J, Li J, Wang F, Xiao J, Wang Y, Yang H, Li S, Cao Z (2020) Heat stress on calves and heifers: a review. J Anim Sci Biotechnol 11:79

Weller JI, Ezra E, Gershoni M (2021) Broad phenotypic impact of the effects of transgenerational heat stress in dairy cattle: a study of four consecutive generations. Genet Sel Evol 53(1):69

West JW (2003) Effects of heat-stress on production in dairy cattle. J Dairy Sci 86(6):2131–2144

Wheelock JB (2006) Effects of heat stress and rbST on production parameters and glucose homeostasis. J Dairy Sci 89(1):290–291

Wheelock JB, Rhoads RP, VanBaale MJ, Sanders SR, Baumgard LH (2010) Effects of heat stress on energetic metabolism in lactating Holstein cows. J Dairy Sci 93:644–655

Wilson SJ, Marion RS, Spain JN, Spiers DE, Keisler DH, Lucy MC (1998) Effects of controlled heat stress on ovarian function of dairy cattle. 1. Lactating cows. J Dairy Sci 81(8):2124–2131

Wilson B, Porter JR, Kearns EJ, Hoffman JS, Shu E, Lai K, Bauer M, Pope M (2022) High-resolution estimation of monthly air temperature from joint modeling of in situ measurements and gridded temperature data. Climate. https://doi.org/10.3390/cli10030047

Wohlgemuth SE, Ramirez-Lee Y, Tao S, Monteiro APA, Ahmed BM, Dahl GE (2016) Effect of heat stress on markers of autophagy in the mammary gland during the dry period. J Dairy Sci 99:4875–4880

Wolfenson D, Roth Z (2018) Impact of heat stress on cow reproduction and fertility. Anim Front 9(1):32–38

Wolfenson D, Flamenbaum I, Berman A (1988) Dry period heat stress relief effects on prepartum progesterone, calf birth weight, and milk production. J Dairy Sci 71(3):809–818

Wolfenson D, Thatcher WW, Badinga L, Savio JD, Meidan R, Lew BJ, Braw-tal R, Berman A (1995) Effect of heat stress on follicular development during the estrous cycle in lactating dairy cattle. Biol Reprod 52(5):1106–1113

Wolfenson D, Lew BJ, Thatcher WW, Graber Y, Meidan R (1997) Seasonal and acute heat stress effects on steroid production by dominant follicles in cows. Anim Reprod Sci 47(1):9–19

Zarzynska J, Motyl T (2008) Apoptosis and autophagy in involuting bovine mammary gland. J Physiol Pharmacol 59(Supplement 9):275–288

Zhang FJ, Weng XG, Wang JF, Zhou D, Zhang W, Zhai CC, Hou YX, Zhu YH (2014) Effects of temperature–humidity index and chromium supplementation on antioxidant capacity, heat shock protein 72, and cytokine responses of lactating cows. J Anim Sci 92(7):3026–3034

Zimbelman RB, Rhoads RP, Rhoads ML, Duff GC, Baumgard LH, Collier RJ (2009) A re-evaluation of the impact of temperature humidity index (THI) and black globe humidity index (BGHI) on milk production in high producing dairy cows. In: Proceedings of the 24th annual southwest nutrition and management conference, Tempe, AZ

# Feather Pecking Behavior in Laying Hens: Challenges in Management and Breeding

Nicole Kemper and Jens Tetens

**Abstract**

Feather pecking is a global problem in laying hen flocks, causing massive impairments of animal welfare and economic losses. The propensity to perform this unwanted behavior is determined by intrinsic factors, which are partly under genetic control. The motivation is likely linked to explorative behavior and foraging and mechanistically, GABAergig and serotonergic neurotransmission can be assumed to play a central role. The expression of the actual behavior is triggered by various extrinsic factors and together, the knowledge about intrinsic and extrinsic factors and their interplay has important implications for management and breeding.

**Keyword**

Animal welfare · Beak trimming · Cannibalism · Feather pecking · Fear · Enrichment material · Hyperactivity · GABA

N. Kemper (✉)
Institute for Animal Hygiene, Animal Welfare and Farm Animal Behaviour (ITTN), University of Veterinary Medicine Hannover, Foundation, Hannover, Germany
e-mail: nicole.kemper@tiho-hannover.de

J. Tetens (✉)
Department of Animal Sciences, Georg-August-University, Göttingen, Germany
e-mail: jens.tetens@uni-goettingen.de

## 1 Introduction

Feather pecking (FP) is a long-standing problem in commercial layer flocks worldwide, which occurs in all husbandry systems (Rodenburg et al. 2012). It massively impairs animal welfare and causes substantial economic losses, which are largely attributable to reduced feed efficiency in denuded birds (Blokhuis et al. 2007). There are two different forms of FP to be distinguished: Gentle and severe FP, which both are clearly not a form of aggression (Rodenburg et al. 2013; Savory 1995) (Fig. 1).

Gentle pecking is directed toward the tips of the feathers and mainly occurs in young birds. It does usually not cause feather damage or loss and the recipient mostly ignores it. Severe pecking is more forceful and involves grasping and pulling out the feathers, which are frequently eaten. Thus, it causes damage and loss of feather cover and the recipient usually tries to escape the situation. The severe form can culminate in the occurrence of tissue pecking and cannibalism (Savory 1995), but it is questionable, whether gentle FP may be a precursor or predictor of severe FP. Most studies that have dealt with that question come to the conclusion that it is probably not (Rodenburg et al. 2013), while there seems to be a correlation between the two behaviors in birds of the same age (Newberry et al. 2007; Rodenburg et al. 2004a).


J. J. Gross (ed.), *Production Diseases in Farm Animals*,
https://doi.org/10.1007/978-3-031-51788-4_21

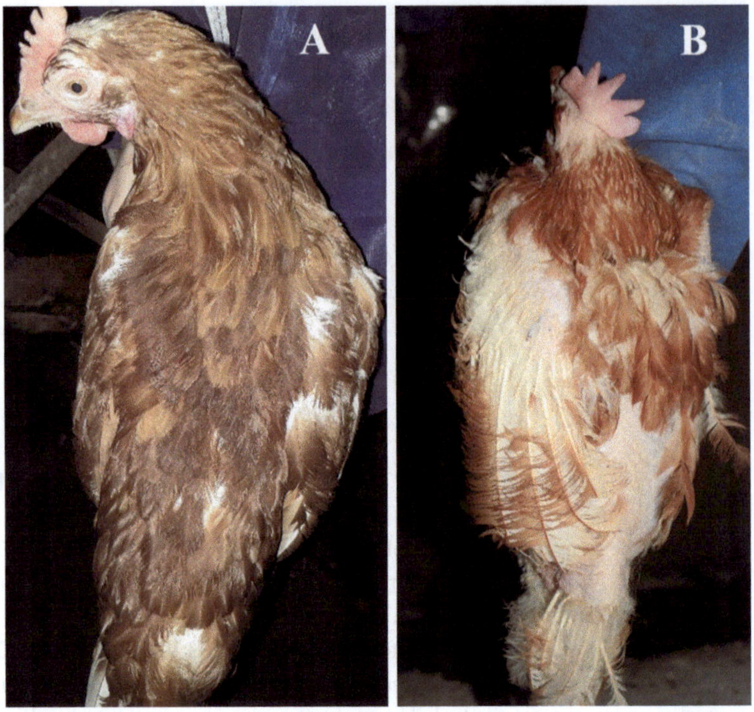


**Fig. 1** Examples of slight (**a**) and heavy (**b**) feather cover damage in victims of feather pecking in laying hens (pictures: Birgit Spindler, ITTN Hannover)

Apart from these general differences between gentle and severe pecking, birds also widely differ in terms of frequency and intensity of pecking. Several studies have identified a subgroup of extreme peckers among birds performing severe FP (Labouriau et al. 2009; Piepho et al. 2017; Wechsler et al. 1998). Iffland et al. (2020a) estimated the posterior probability of belonging to this group. They found that members of the extreme subgroup came in the mood for FP more than twice as often as other peckers, and showed a nine times higher pecking frequency during these periods.

The propensity to perform FP is a complex trait. It is partly under genetic control and influenced by a wide range of environmental factors such as stocking density, lack of foraging, nutrition, or light intensity (Nicol et al. 2013), as will be laid out in more detail later in this chapter. Further, it is a socially affected trait and individual roles of performers or victims might be rather stable once established (Bennewitz et al. 2014). This complexity has various implications for the prevention and control of FP in the different housing systems. Until recently, beak trimming was the most common measure to prevent FP, but this practice is a mutilation causing pain and discomfort, which is heavily criticized due to animal welfare reasons. While some Scandinavian countries have already outlawed beak trimming decades ago, more and more countries followed or will follow this example. Notably, in some countries, egg producers decided to ban beak trimming ahead of legislation. The German auditing system KAT (Association for Controlled Alternative Animal Husbandry) for instance prohibited the restocking of hens with trimmed beaks in 2017, which took effect for about 80 million hens in Europe (Preisinger 2018). It may be argued that such restrictions were overdue, but they still come with substantial challenges. Egg producers have to be prepared and suitable management strategies have to be in place to avoid injurious pecking behavior. It is generally accepted that this is possible, but it is complex and requires advanced management skills (van Niekerk 2017). Hence, the reduction of FP by means of selective breeding would be a highly appreciated solution. It might, however, be hard to achieve, mainly because FP is notoriously hard to measure.

Overall, FP has remained a substantial challenge to farmers and breeders with a strong impact on animal welfare and economic outcomes. In this chapter, we will review the major risk factors for FP along with suitable management strategies to control this unwanted behavior, and we will reflect on the consequences of beak trimming. Further, we will summarize the genetics of FP, take a glance at the etiology as deduced from various functional and omics-experiments, and try to point out possible implications for management and breeding.

## 2 Animal Welfare Impacts

There are two main aspects to consider when looking at potential animal welfare impairments in the context of FP. First and obviously, the direct effects of FP on the hens have to be pointed out. Secondly, in relation to prevention methods still common worldwide, the practice of beak trimming and its possible impacts on animal welfare have to be discussed.

The direct effects of FP are not limited to the "victim animals," meaning the animals being pecked, only, but also the actively pecking animals. Even though these pecking offenders do not suffer from obvious damages, active FP is always an indicator of substantial deficiencies in animal management and husbandry. Actual FP events are influenced by extrinsic and intrinsic factors. As illustrated in Fig. 2, each animal has an intrinsic threshold to respond to external pressures, which is determined by genetics and health status for instance. Moreover, each hen has an intrinsic, individual motivation to perform exploratory and foraging behavior, potentially resulting in FP. To successfully prevent FP, both extrinsic and intrinsic factors have to be kept on an optimal level by implementing best practices, and thus, avoiding stress to the laying hens. Keeping laying hens with intact beaks without any signs of FP in the flock is always a true indication that high standards are being achieved.

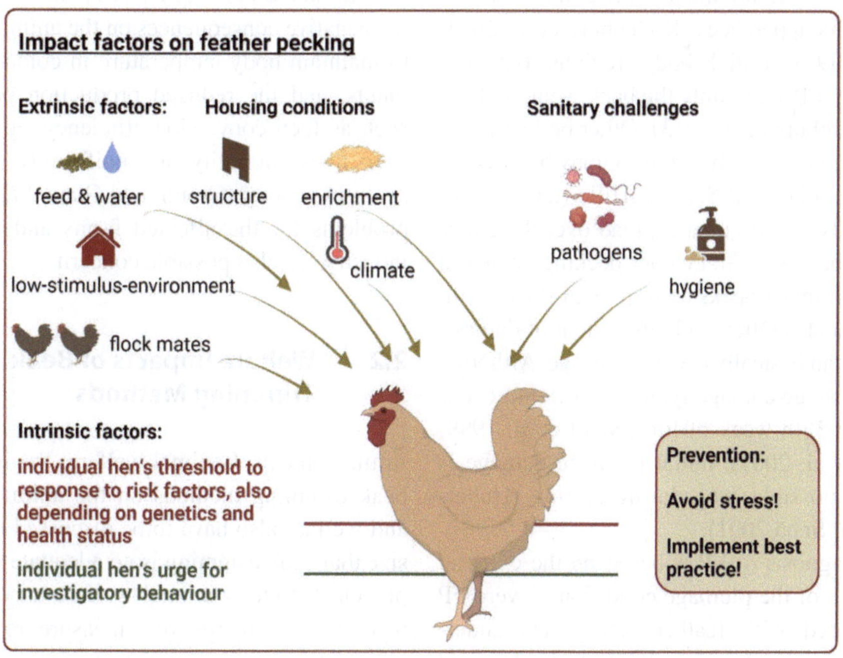


**Fig. 2** Scheme of potential extrinsic and intrinsic factors influencing feather pecking events (picture: N. Kemper, created with BioRender.com)

## 2.1 Feather Pecking and Its Link to Cannibalism: How Welfare Is Affected

Feather pecking can, but does not necessarily, occur before or together with cannibalism in affected flocks. Feather pecking and cannibalism are often thrown into the same pot, but from a behavioral point of view, they are basically two different, non-aggressively motivated behavioral disorders (Savory 1995). Although feather pecking can turn into cannibalism, the behaviors can also be observed independently of each other. Both have in common that once several animals show these behaviors, usually other animals are animated to perform FP and/or cannibalism, too, until the whole flock is affected, if no counter measures are implemented.

Feather pecking, in particular the severe form, is defined as the pecking of feathers and the pulling out and eating of feather parts or whole feathers by a conspecific (Rodenburg et al. 2013). As outlined above, this has to be differentiated from aggressive interactions to build or maintain the rank order, which are mainly directed against the head and its appendices (Rodenburg et al. 2013; Savory 1995). Initial body regions typically affected by FP are mainly the back, vent, and tail area (Rodenburg et al. 2013). Other body regions such as the laying belly and neck may be affected aswell (Spindler and Schulze-Hillert 2013). Not infrequently, a subsequent spread over the entire body can be observed. Feather pecking occurs in flocks with intact beaks as well as in flocks with trimmed beaks. Often, only the extent of damage differs. It can basically occur at any age. Although severe plumage damage typically does not occur before the 38th week of life (Nicol et al. 1999; Pötzsch et al. 2001), initial plumage damage is generally possible even during rearing (Huber-Eicher and Sebö 2001).

The diagnosis of FP is based on the external appearance of the plumage condition. Severe FP is associated with feather eating (Harlander-Matauschek and Häusler 2009; Ramadan and Borell 2008) and, therefore, when entering the barn, it is often noticeable in affected flocks that no feathers can be found in the litter in the scratching area as a result of "feather hunger/feather eating." This circumstance was also described by McKeegan and Savory (1999).

The animal welfare relevance of FP is related to substantial feather losses and plumage damages that can be followed by cannibalism. Cannibalism or cannibalistic pecking is defined as pecking and pulling at the skin and underlying tissue until bloody wounds appear. Cannibalistic pecking has severe welfare impacts and can even lead to the death of the animal. It is often a consequence of FP, as shown in several studies indicating a higher risk for following skin damages and cannibalism in hens with severe FP (Cloutier et al. 2000; Lambton et al. 2015). However, cannibalism can also appear independently as toe or vent cannibalism (Newberry 2004; Rodenburg et al. 2009).

Cannibalism and FP are often linked, but even FP alone represents an animal welfare issue. Pain, fear, and wounds can be caused by the forceful removal of feathers and the pecks on the skin. The direct impacts of FP on animal welfare have recently been reviewed comprehensively by Cronin and Glatz (2021). They also pointed out the negative consequences on the animal's ability to maintain body temperature in colder environments, and the reduced production parameters such as feed conversion efficiency, egg production, flock mortality and uniformity, and farm profit in general (Cronin and Glatz 2021). Image problems for the affected farms and the entire industry are also possible concerns.

## 2.2 Welfare Impacts of Beak Trimming Methods

In the context of animal welfare, the impacts of beak trimming methods on the animals' health and welfare also have to be discussed to emphasize that beak trimming is no adequate method to prevent FP under animal welfare aspects. Beak trimming is an effective measure to maintain intact plumage condition and prevent cannibalism (Hartcher et al. 2015a, b), but is a mutilation that simply masks the underlying problems.

To trim the beak, the most common method worldwide is to shorten the tip of the beak using a hot plate (Glatz 2000). In this procedure, a trained person holds the tip of the upper and lower beak to a hot plate of about 950 °F, which shortens and cauterizes the tip at the same time (Glatz and Hinch 2008). This technique can be used for chicks up to 10 days old. Another method that can already be used for day-old chicks in the hatchery is the largely standardizable and automatic technique using a high-energy infrared beam (Poultry Service Processor Nova-Tech; PSP; Nova-Tech Engineering of Willmar, Minnesota, USA). This technique has been used for turkeys for some time now. In addition to the possibility of precise and more uniform shortening of the beak tip compared with the usual method using a hot plate, it offers the advantage that no open wounds occur (Marchant-Forde et al. 2008). The risk of infection is thus minimized and the beak tip detaches after about 10–14 days. Despite the mentioned advantages of the infrared method, various scientific studies indicate that both methods are nevertheless associated with acute pain (Janczak and Riber 2015).

Thus, with both methods, morphological and histologically, sometimes serious changes in the tissue can be detected. With the infrared method, within the first 2 weeks of life a complete necrosis of the beak tip with an infiltration of inflammatory cells was detected histologically (Spindler et al. 2016). Likewise, reduced longitudinal growth was morphologically documented (Marchant-Forde et al. 2008). In addition, both procedures showed an influence on the behavior of the affected animals (Gentle et al. 1997; Marchant-Forde et al. 2008).

Both techniques resulted in reduced feed intake in the animals during the first weeks of life compared to animals whose beaks were not clipped (Marchant-Forde et al. 2008). In the case of shortening by infrared, this persisted even until the 18th week of life (Honaker and Ruszler 2004). The animals also rested more in a standing position (Marchant-Forde et al. 2008), and showed increased grooming behavior (Marchant-Forde et al. 2008; Sandilands and Savory 2002) along with increased pecking activity in the litter (Sandilands and Savory 2002). Associated with reduced feed intake, reduced body weight development was also observed (Henderson et al. 2009; Marchant-Forde et al. 2008), but this has not been demonstrated in other studies (Dennis et al. 2009; Sandilands and Savory 2002). Chronic effects were reported, particularly when shortening the tip of the beak in older animals (over 8 days old) using a hot plate (Sandilands and Savory 2002). Evidence of neuromas, as an indication of possible chronic pain, was present in 10-week-old animals, but not in 70-week-old animals, whose beaks had already been shortened in the hatchery using a hot plate (Lunam et al. 1996). The occurrence of neuroma was strongly dependent on the extent of shortening (Lunam et al. 1996). If shortening of the beak tip with infrared leads to persistent pain and an impairment of the animals has been discussed controversially so far (Spindler et al. 2016).

Summarizing the negative impacts of beak trimming on the hens' welfare, it is obvious that all efforts should be taken to keep the herd with untrimmed beaks, as it is already legally required in several countries, such as Norway, Sweden, and Austria. In a number of European countries, beak trimming is about to be banned (Jung and Knierim 2019). However, also the discussion about the ban on beak trimming is not led without animal welfare concerns. Once FP occurs within a flock, the affected animals are often pecked so intensively by their flock mates that bald spots appear in the plumage, resulting in completely non-feathered body regions. The extent of the damage can become multiple if the beak is intact.

## 3 Risk Factors for Feather Pecking: What Triggers Are Known?

Refraining from beak trimming is certainly one of the biggest challenges in today's laying hen management. To minimize the risk of FP in untrimmed herds, knowledge of potential risk factors and prevention strategies is key. However, the situation is complicated by the fact that the causes for the occurrence of FP are so diverse. A simple solution to prevent FP definitely does not exist, and farm-specific strategies have to be developed for each individual case. In this regard, it has to be stressed that FP is not an exclusive phenomenon of today's intensive laying hen husbandry. It was described as a bad habit of feather-eating as early as at the beginning of the twentieth century, where the difficulties to get the animals out of the habit were described in a German reference book on domestic poultry (Blancke and Klefner 1908). Feather pecking is observed in all husbandry forms, from cages, including furnished cages, over aviaries, and free-range farms to organic farms (Åkerfeldt et al. 2021; Gunnarsson et al. 1999; Habig and Distl 2013; Sherwin et al. 2010).

Even though the knowledge of potential risk factors is of special importance in untrimmed herds, it is also essential for managing beak-trimmed herds in an optimal way. Potential external and internal factors, also known as triggers, range from genetic disposition and various inadequate housing and management conditions to disturbances in exploratory and foraging behavior. The number of extrinsic factors described in scientific literature so far is nearly uncountable. Examples are listed in Table 1. Jung and Knierim (2018) identified a list of 62 factors. In addition, various intrinsic factors can play a role, which will be described later in this chapter in the context of the genetics of FP.

Most of the external risk factors are related to inadequate housing and management conditions both in the rearing and in the subsequent production phase. Although FP often manifests later in the laying period, the rearing conditions show a significant impact (Mels et al. 2022). In the whole life of a hen, the interplay of living conditions and the hen itself determines its behavior. As most important factors contributing to these living conditions, particularly the composition and structure of the feed (van Krimpen et al. 2005), litter condition (de Haas et al. 2014a; Schreiter et al. 2019), insufficient climate conditions (van Niekerk 2019), space allowance and stocking density (Zepp et al. 2018), and the intensity and quality of light (Shi et al. 2019) show strong effects on FP. All these factors can serve as stressors to the animals (Riedstra and Groothuis 2002), and excluding one of these stressors alone can reduce the risk for FP, but not prevent it entirely. To really succeed in prevention, all potential stressors have to be taken into account, balanced management plans have to be established, and in case of imbalances, immediate countermeasures have to be initiated. One factor of utmost importance, however, which is often not mentioned, is the role of humans as caretakers and farm staff. Not only the keeper of the animals, but to a larger extent the persons taking care of the flocks and having direct, daily contact with the animals have to be qualified and possess appropriate knowledge to immediately detect any irregularities in the herd. Only then, they are able to define suitable counter measures and act accordingly. The special demands of today's high-performing animals regarding feed, health, and general management have to be fulfilled to minimize the risk of FP as far as possible. This holds especially true in herds with hens with untrimmed beaks. Moreover,

**Table 1** Examples of described extrinsic factors triggering feather pecking events in laying hens in the laying period

| Risk factor | Literature |
|---|---|
| Preventing the access to litter (restriction in foraging material) | Blokhuis 1986 |
| Lack of environmental stimulation | Brunberg et al. 2016 |
| Large group size | de Haas et al. 2014b |
| High fear of humans | de Haas et al. 2014b; Mels et al. 2022 |
| Aviary housing compared to floor housing | de Haas et al. 2014b; Mels et al. 2022 |
| Standard management compared to adjusted management (including preventive measures) | de Haas et al. 2014b |
| Feather pecking in the rearing phase | Hartcher et al. 2015b |
| Lack of UV light | Prescott et al. 2003 |
| No or late access to perches | Gunnarsson et al. 2000; Lambton et al. 2010 |
| Low barn climate quality | Lambton et al. 2010 |
| Early onset of laying | Pötzsch et al. 2001 |

the farm personnel has to be trained in applying and coordinating prevention measures in advance. This requires not only experience but also a constant education to put the latest findings from science into practice.

## 4 Preventive Management Strategies

In a flock, the occurrence of FP is always an indicator that something went wrong before. Therefore, prevention has to be implemented long before any alterations in the hens' feathers can be detected. For prevention, a whole series of factors have to be taken into account to keep the animals with as little stress as possible. A number of recommendations to realize prevention, especially in hens with untrimmed beaks, have been published over the recent years (Jung and Knierim 2018). Examples are good-practice guides such as FeatherWel (University of Bristol 2013) and LayWel (Blokhuis et al. 2007). It is essential that all preventive measures are not limited to the laying phase, but have to be implemented in the rearing phase already to reach best effects. Besides the regular preventive measures, which should be continuously provided, farms should also have an emergency plan at hand, to calm the situation as quickly as possible in the event of FP problems (Spindler et al. 2016).

### 4.1 Prevention of FP in the Rearing Phase

The rearing phase lays the foundation for the hens' future performance. To successfully prevent FP over lifetime, following recommendations should be implemented as early as possible. They do not claim to be exhaustive but represent a field-tested approach based on extensive, long-term experience, and current scientific knowledge (Spindler and Schulze-Hillert 2013; Spindler et al. 2016).

The barn should be structured not only by providing different areas but also by offering three-dimensional structures. Elevated perches, available for all chicks from the first day of life, are very beneficial. Access to the scratching area with litter should be realized as soon as possible. The litter should consist of loose and dry material that attracts the chicks to peck and scratch. If necessary, the litter has to be fluffed-up. To satisfy the exploratory and foraging behavior, the pullets should be offered manipulable and changeable enrichment material as early as possible (de Haas et al. 2014b). Enrichment material has to be of good hygienic quality, to not negatively affect animal health and food safety. Picking blocks and bales of hay (alfalfa) or straw are proven to be well-accepted materials (Schreiter et al. 2019). Supplementation of wheat grains to the litter keeps the animals busy. It was shown that the constant provision of manipulable and changeable material in addition to the litter significantly reduces the risk of FP (Blokhuis and van der Haar 1992). Enrichment material should be replaced regularly to maintain its attractiveness.

Lighting (daylight and/or flicker-free artificial lighting) should be adjusted to 20 lx at the animals' eye level, and light cones should be avoided.

A twilight phase (30–45 min) is recommended to attract the animals to the facility and avoid stress. The provision of red light has an effect on the occurrence of FP, as it masks skin damages and blood, and it is often applied to control FP (Mohammed et al. 2010), but not recommended as a standard measure (Nicol et al. 2013).

Good housing climate should be ensured via sufficient ventilation, prevention of drafts, and avoidance of high ammonia levels (>10 ppm).

As FP is related to foraging behavior, and because inadequate nutrition represents a risk factor for FP, special emphasis has to be laid on the provision of feed of proper quality and quantity. A good body weight development has to be ensured and should be controlled by regularly weighing.

A selective feed intake can be prevented by a coarsely ground homogeneous structure of the feed (Desbruslais et al. 2021; Mens et al. 2020). In addition to an adequate supply of energy and nutrients, an increased crude fiber content of 5–6% is recommended to promote the development of the gastrointestinal tract and the feed intake capacity (van Krimpen et al. 2005). The feed intake capacity can also be influenced positively by the provision of insoluble grit, which also stimulates the development of the gizzard.

With regard to health management, all necessary vaccinations should be given, but be completed at least 14 days before the change of housing after the end of rearing to reduce stress.

Naturally, in rearing, the farm staff should handle the animals in a competent and caring way with close observation and intensive contact. For instance, as a result of the plumage change in the 15th week of life, numerous feathers can be found on the floor of the barn in a well-developed flock. If no feathers are found, this may be a sign of feather eating, and thus a first alarm signal for a deficiency or stress situation.

The re-housing process of the pullets on the laying farm has to take place by a smooth transition (Drake et al. 2010). This can only be achieved by intensive consultation between the laying hen farmer and the breeder. Prior to re-housing on the laying farm, the pullet flock should be optimally feathered and have a uniformity (= balance) of at least 80% in terms of weight (Du et al. 2022). The optimal time depends on the development of the pullets and is usually between the 17th and 18th week of life. A detailed re-housing protocol and precise agreements between the rearing and laying farms regarding the number of animals, light, feed, and vaccination management help to keep the process as stress-free as possible for the animals and thus guarantee them the best possible development (Spindler and Schulze-Hillert 2013).

It is recommended to carry out an incoming inspection of the pullets. The aim is to transfer a uniformly developed flock with impeccable plumage, intense comb color, and comb size as well as sufficient weight and good uniformity. To ensure that the hens find food and water quickly in their new environment, it is advisable to place them close to feed and watering facilities. In this way, feed and water are immediately available. Incentives for feed and water intake can be created by, for example, starting the feed chain several times and, in the case of nipple drinkers, adjusting the pressure so that droplets of water emerge. Documentation and control of feed and water consumption is strongly recommended to detect possible problems immediately.

The selected light intensity must be based on the rearing conditions and allow the newly housed hens to orient themselves without problems (Nicol et al. 2013). Particularly abrupt differences in light intensity and quality between rearing and production houses must be avoided. The barn has to be littered from the beginning. To avoid panic later on, the animals should be accustomed to various farm-specific work processes and noises already in the first days.

## 4.2 Prevention of FP in the Laying Phase

In principle, all recommendations for the rearing phase can similarly be transferred to the laying phase. Depending on the husbandry system, raised perches should be offered on several levels. These allow the hens to rest undisturbed, especially at night, and offer escape and evasion possibilities during the day. Access to the scratching area has to be ensured as soon as possible after re-housing. A litter height of 1–2 cm is sufficient at the beginning. The material, preferably manipulable, loose, and dry substrate, should be suitable for natural behavior such as pecking, scratching, and dust bathing. Fresh material should be added if necessary.

As in pullet rearing, manipulable and changeable enrichment material in addition to the litter should be provided to satisfy the hens' exploratory and foraging behavior (de Haas et al. 2014b).

Weight checks (by hand or with automatic animal scales) are recommended to monitor the development of the herd. Intensive animal observation (e.g., behavior, plumage, health status, and fecal condition) is required for the immediate detection of any problems and the initiation of countermeasures. The same applies to the recording, documentation, and evaluation of essential performance parameters (including feed and water consumption, laying performance, and mortality).

A good stable climate has to be ensured, characterized by no drafts, an appropriate barn temperature (16–18 °C), 60–70% relative humidity, low dust concentration, and compliance with the limits for various harmful gases in the barn air ($NH_3$ < 10 ppm, $CO_2$ < 3000 ppm) (Spindler and Schulze-Hillert 2013).

In the feed, the nutrient requirements have to be met continuously, because fluctuations lead to behavioral disturbances. The nutrient requirements of the respective breeding line (genetics) have been taken into account, following the recommendations of the breeding companies. After completion of the pre-feeding phase, 3-phase feeding should be carried out. The timing of the feed change is determined by performance and calcium requirements, less by age (Du et al. 2022). It is important to avoid abrupt feed changes (structure, components, content). When changing phases, it is best to blend feed so that a slow transition takes place. Vital, well-developed hens are a basic prerequisite for successfully managing hens without FP, especially in non-beak-trimmed flocks.

## 4.3 Countermeasures: What Has Proven Effects in Case of FP Events?

At the first sign of severe FP, countermeasures should be taken immediately to calm the situation as quickly as possible and control the problem. In addition, the search for possible triggers that may have contributed to the occurrence of FP is crucial. Only if these are identified and subsequently removed, the situation in affected herds can be stabilized in the long term.

In case of FP events, research into the causes includes, among other things, an examination of the housing environment, in particular, the climate in the barn and the light, access to manipulable material, and the quality of litter. Animal health and the feed provided have to be checked. Especially if problems occur shortly after the use of a new feed batch (approx. 2–4 days later), the feed formulation and feed structure have to be controlled.

Measures that have proven to be suitable and practical to calm flocks with FP problems so far are listed in Table 2. A good practical guide is provided by van Niekerk et al. (2017). The earlier they are applied, the better. Depending on the situation, they can be applied alone or in combination. Under certain circumstances, measures may only be carried out after veterinary indication.

All measures are only effective if implemented in time, i.e., before the FP event affects the whole flock. Therefore, early warning signs of FP have to be recognized via intensive supervision of the flock by experienced persons. This should include precise animal observation and documentation as well as evaluation of the collected data to be able to react as quickly as possible to the first signs of FP. In addition, on most farms, there is an obvious need for optimization with regard to best possible husbandry conditions and management, by means of an individual farm analysis of weak points. This also includes an intensive training and consultation not only of the animal owners

**Table 2** Countermeasures with proven beneficial impacts in cases of feather pecking events

| Measure | Literature |
|---|---|
| Immediate offer of additional enrichment material | Schreiter et al. 2019; van Staaveren et al. 2021 |
| Intensification of animal care | Spindler et al. 2016 |
| Administration of salt (NaCl) or magnesium (Mg) via the drinking water | van Niekerk et al. 2017 |
| Supplementation of essential amino acids in the feed. For example, dosing milk powder (up to 2%) "on top" of the feed for a few days | Kjaer and Bessei 2013; van Niekerk et al. 2017 |
| Preventing direct sunlight from entering the barn through windows, air inlet flaps, ventilation ducts, etc. | Spindler and Schulze-Hillert 2013 |
| A reduction of the light intensity in the barn in 10% steps every 3 days to a maximum of 30–40% of the original light intensity. However, the light intensity should not be reduced too much | van Niekerk et al. 2017 |
| A change of the lighting to red light | Mohammed et al. 2010 |

but above all, of the persons who have direct contact with the animals and take over the herd care. It has been shown that farmers' attitude is essential to prevent feather pecking (Mels et al. 2022). Especially in the rearing phase, the staff should be aware that regular animal contact reduces stress in the animals. Intensive expert care and support, however, inevitably means an additional personnel expense on the farms as well as an additional financial burden.

## 5 The Genetics of Feather Pecking

The genetic analysis of feather pecking is notoriously difficult for several reasons. Individual records are needed for each bird, which makes phenotyping a laborious task. Although tremendous progress has been made in the field of sensor technologies and computer vision (for a review see Ellen et al. 2019), there is currently no automated solution in place to accurately determine FP behavior in individual birds. Thus, phenotyping requires direct observation by trained persons able to differentiate between FP and other bird-to-bird interactions or agonistic behavior in a setting where birds can be easily identified. Typically, FP is recorded as the number of bouts during a particular time span. This count data heavily departs from a Gaussian distribution and the length of the observation period is of fundamental importance. A too-short period will result in imprecise records with many zero observations; a longer observation period is more labor intensive. A suitable recording length depends also on the general FP activity in the population, which can be challenged, e.g., by a light stimulus. In any case, the heritabilities for different observation periods have to be estimated.

### 5.1 Heritability

Early evidence for the heritable nature of FP comes from the observation of clear line differences (Hughes and Duncan 1972; Richter 1954). Since then, a number of studies applying different approaches were conducted with the aim of estimating genetic parameters for FP. From the ten studies summarized in Table 3, it can be seen that the heritability estimates vary widely depending on various factors. Besides the line or breed and the study design, the age of the hens and the statistical model as well as the duration of the examination period and the exact trait definition do affect the estimates. As a rule of thumb, however, the heritability of delivering feather pecks could be quantified around 0.10–0.20 at the age of 30 weeks.

The probably most striking demonstration of the heritability of FP comes from a selection experiment started at the Danish Institute of Animal Science in Foulum and continued at the Institute of Animal Science of the University of Hohenheim. The selection experiment was started in 1996 from a White Leghorn base population established in 1970 as a control line in a crossbreeding experiment (Grams et al. 2015b; Kjaer et al. 2001; Liljedahl et al. 1979; Su et al. 2006). The lines were divergently selected for high (HFP) and low FP (LFP) based on estimated breeding values for a total of 16 generations as of 2021. There was an immediate selection response in the first three generations and all subsequent generations showed a clear phenotypic distinction. In generation 15, approximately 0.22 FP bouts per bird and hour were recorded in the LFP line, while this number was 1.47 in the HFP line. This means an almost sevenfold difference in FP activity between the lines. Furthermore, almost half of the HFP birds were ever recorded as performers, while only roughly 10% of the LFP hens delivered pecks (Iffland et al. 2021). Mainly for this reason, the selection lines have gained much importance in FP research. Once established, the experimental populations are easily accessible in a research context and reliably show FP behavior at a reasonable frequency. This, however, comes at the price of questionable transferability into practice. Alleles might have been lost due to drift in the small populations. Further, correlated selection responses were observed in other economically important traits (Su et al. 2006) reflecting complex genetic interrelationships. Thus, it is unclear, whether the selection lines genetically

**Table 3** Heritabilities for feather pecking and related traits from ten different studies. The studies are listed in chronological order

| Reference | Breed/line | Traits and heritabilities ($h^2$) | Remarks |
|---|---|---|---|
| Cuthbertson 1980 | White Leghorn | Classification of birds as peckers / pecked: 0.09–0.56 | Estimates depend on classification<br>Done in very young birds |
| Bessei 1984 | Rhode Island Red and Sussex | Frequency of pecking: 0.07 | Done in young hens |
| Kjaer and Sørensen 1997 | White Leghorn | Pecks/bouts at 6 weeks: 0.05 / 0.13<br>38 weeks: 0.14 / 0.13<br>69 weeks: 0.38 / 0.35 | Very low phenotypic, but high genetic correlation between ages |
| Kjaer et al. 2001 | White Leghorn selection lines | Number of FP bouts per hour: 0.20 | Most birds observed in weeks 26–34 |
| Rodenburg et al. 2003 | $F_2$ design from production lines differing in FP | Gentle pecking at…<br>6 weeks: 0.08–0.12<br>30 weeks: 0.15–0.16<br>Severe pecking at…<br>6 weeks: 0.00–0.02<br>30 weeks: 0.06–0.07 | Similar $h^2$ for pecks and bouts<br>Also estimated $h^2$ for open-field test traits (0.10–0.49) |
| Bennewitz et al. 2014 | $F_2$ design from selection lines | FP delivered: 0.11<br>Aggressive pecks delivered: 0.10<br>Feathers eaten: 0.57 | $h^2$ on link scale from GLMM (overdispersed Poisson model)<br>High genetic correlation between FP and aggressive pecks delivered ($r_g = 0.81$) / feathers eaten ($r_g = 0.73$) |
| Grams et al. 2015a | White Leghorn selection lines | FP bouts…<br>…in high FP line: 0.15<br>…in low FP line: 0.01 | GLMM assuming a Poisson distribution outperformed LMM<br>Birds observed at ~30 weeks |
| Grams et al. 2015b | $F_2$ design from selection lines | FP delivered: 0.27<br>Aggressive pecks delivered: 0.14 | $h^2$ on observed scale from LMM<br>Also analyzed fear traits and found moderate genetic correlations between tonic immobility in juvenile and FP in adults |
| Lutz et al. 2016 | $F_2$ design from selection lines | FP delivered: 0.20<br>Feather eating: 0.36<br>General locomotor activity: 0.29 | Structural equation models were applied<br>Found genetic correlations between FP and general locomotor activity ($r_g = 0.04$) / feathers eaten ($r_g = 0.17$) |
| Iffland et al. 2020a | $F_2$ design from selection lines | Posterior probability of belonging to the subgroup of extreme peckers (pEFP): 0.35 | Fitted 2 negative binomial distributions to the data |

still reflect current commercial genetics. A validation in these populations is pending for many findings discovered in the selection lines.

## 5.2 Interrelationships with Other Behavioral Traits and Performance

FP shows a complex phenotypic and genetic relationship with other behavioral traits as well as performance traits such as body weight or laid eggs. These interrelationships have been considered in various ethological, functional, and genetic studies, mainly with two aims. On the one hand, mechanistic models have been developed based on these studies to explain the etiology of feather pecking. On the other hand, genetically correlated traits, which are easier to measure than FP itself, may serve as indicator traits in practical breeding.

**Foraging, Activity, Exploration, and Fear**

Access to foraging material reduces FP (Hughes and Duncan 1972; Lambton et al. 2010; Nicol et al. 2001) and there is a negative relationship between foraging activity and the frequency of FP (Blokhuis and Arkes 1984; Huber-Eicher and

Wechsler 1997, 1998). Thus, FP was considered as redirected foraging behavior (Rodenburg et al. 2013). This is consistent with the contribution of the serotonergic and dopaminergic systems to the development of FP (for a comprehensive review see de Haas and van der Eijk 2018) as these neurotransmitters are involved in the regulation of feed intake. In fact, the administration of a drug that acts as a serotonin autoreceptor agonist and postsynaptic receptor antagonist can simultaneously increase foraging and FP in chicks (van Hierden et al. 2004).

FP is also particularly related to eating feathers (de Haas et al. 2014b; Savory and Mann 1997) and feather consumption during the rearing period may increase FP in adult hens (McKeegan and Savory 1999, 2001; Ramadan and Borell 2008). In fact, "Feathers eaten" is a heritable trait and shows a considerable genetic correlation with FP (Bennewitz et al. 2014; Lutz et al. 2016, Table 1). In addition, relatively high recursive effects of feather eating on FP were obtained from structural equation models, which implies that an increase in feather eating leads to increased FP (Lutz et al. 2016). The reason for this connection could be related to the finding that feeding on feathers leads to changes in the gut microbiota composition (Meyer et al. 2012), as will be further discussed later in this chapter. This is plausible, because gut microbiota composition and metabolites do affect peripheral and central serotonin and could influence FP via that route.

The hypothesis of FP being a redirected foraging behavior should, however, not be mixed up with the latter, as in the first case, foraging and FP can be regarded as "co-regulated" by the serotonergic system. In the second case, if it holds true, feather eating would have a causative relationship with FP mediated by changes in the gut microbiota and consequently the serotonergic system.

Chickens selected for high FP are very active in behavioral test settings (Kops et al. 2017; van der Eijk et al. 2018) and exhibit high home-pen locomotor activity (Kjaer 2009). Thus, a hyperactivity disorder model for FP was proposed (Kjaer 2009), probably mainly influenced by the dopaminergic system (Kjaer et al. 2015). This is supported by results from structural equation modelling demonstrating positive recursive effects of general locomotor activity on FP (Lutz et al. 2016).

Actually, the redirected foraging hypothesis and the hyperactivity disorder model are related to a large extent. It can be argued that it might be the intrinsic exploratory motivation that leads to higher locomotor and foraging activity and ultimately FP (Kjaer 2009; Newberry et al. 2007; Rudkin 2021). It was also proposed that FP is a form of social exploration (Riedstra and Groothuis 2002). Assuming a role of exploratory motivation in FP puts it into a context with fear as these behaviors are reversely related. Both can be assessed by open-field locomotor activity, which is positively related to exploration, but negatively to fear. This led to the assumption that open-field response in young birds could predict FP in adult hens (Rodenburg et al. 2004b). Overall, quite a few studies thus considered the phenotypic and genetic interrelation between fear / exploration, FP, and various ethological tests.

**Open-Field Test** Studies on the relation between open-field activity and FP are not completely consistent. While some authors reported a positive phenotypic relationship (Jensen et al. 2005; Rodenburg et al. 2004b), others found no relationship or even the opposite (Grams et al. 2015b; Jones et al. 1995), which might partly be explained by different breeds/strains and study designs. Overall, a positive relationship between open-field activity and FP can be assumed. Reported genetic correlations between open-field activity and FP measured at the same age range from 0.08 to 0.62 (Grams et al. 2015b; Rodenburg et al. 2004b). Notably, a negative genetic correlation of −0.65 was reported between open-field activity measured in young birds and FP in adult hens (Rodenburg et al. 2004b).

**Tonic Immobility Test** Vestergaard et al. (1993) found a positive relationship between fear as measured by a tonic immobility test and individual FP activity, while Rodenburg et al. (2010) on the other hand did not find differences between FP selection lines. However, similar to open-field activity, a genetic correlation of 0.27 between the outcome of a tonic immobility test in young birds and FP in adult hens was reported (Grams et al. 2015b).

In summary, FP is linked to high activity and increased exploratory motivation. There is a genetic interrelation between fear as measured by open field and tonic immobility test and FP, also between different ages. And although studies are not completely consistent, it seems possible to use activity and fear-related traits in young birds to breed for reduced FP in adults.

**Aggression**

FP is clearly distinct from aggression (Rodenburg et al. 2013; Savory 1995), but the two behaviors are not unrelated (McKeegan and Savory 1999). And although the motivation behind FP and aggressive pecking are different, selection for high FP also increases aggressive behavior (Bessei et al. 2013; Iffland et al. 2021). This is well reflected by genetic correlations between FP and aggressive pecks delivered ranging from 0.2 to 0.81 (Bennewitz et al. 2014; Grams et al. 2015b).

**Performance**

Based on the observation of an increased mortality due to cannibalism in commercial lines selected for high egg production and low body weight, Kjaer and Mench (2003) hypothesized that direct selection for FP may have an influence on production traits. In the fifth generation of the aforementioned selection lines, correlated selection responses were indeed evident for several production traits (Su et al. 2006). Egg number and egg mass were higher in the LFP line, while egg weight, albumen height, shell thickness, and yolk proportion were higher in HFP birds. In addition, LFP animals showed better feed efficiency, partly because of lower maintenance requirements due to better feather cover. In contrast, Bennewitz et al. (2014) estimated a positive genetic correlation of 0.50 between the delivery of FP and egg number in an $F_2$ population established from the tenth generation of the selection lines. This is basically in line with Buitenhuis et al. (2004), who found significant phenotypic correlations between gentle FP and egg weight. A notable finding in this regard is the clear difference between conventional layer strains and dual-purpose hens, which have been reported to show much less pecking behavior (Giersberg et al. 2017, 2019). It is unclear, however, whether this has directly to do something with performance or might be just a line difference.

Overall, there is a clear relationship between performance and FP, and although available studies cannot be directly compared, the interrelation might be more complex than expected. If direct selection for low FP will be performed, attention has to be paid to this circumstance as negative consequences for egg production might otherwise result.

## 5.3 Mapping Causative Genome Regions and Genes

The propensity to perform FP is a complex trait and many loci with presumably small effects can be expected to contribute to its variation. Thus, sufficiently large mapping populations are required, which need to be accurately phenotyped and genotyped with a sufficiently dense marker panel. This is often prohibitive, even in a research context, and the utilization of commercial flocks for this purpose is hampered by the fact that time and intensity of FP episodes cannot be predicted. It could thus be advantageous to make use of previously established QTL mapping resource populations, which are usually phenotyped in depth. However, such designs were often established from founder lines or breeds that do not reflect commercial layer genetics (see, e.g., Jensen et al. 2005, Table 4) Thus, many mapping studies were conducted in the aforementioned FP selection lines, which combine the advantages of predicable occurrence of FP and relatedness to commercial white layer genetics. After 16 generations of selection, however, the latter statement does not necessarily hold true anymore as outlined earlier in the chapter.

When using the FP selection lines, two mapping strategies are applicable. Either one could search for selection signatures within or between lines, or one could adopt GWAS-based mapping approaches, possibly in $F_2$ designs established from the selection lines. All of these approaches have been applied in various studies (Table 4).

A major challenge in finding meaningful selection signatures is the separation of real selection effects from genetic drift. Grams et al.

**Table 4** Overview of available mapping studies applying either classical QTL mapping, GWAS, or selection signature approaches. The studies are listed in chronological order (Partly adapted from Bennewitz and Tetens 2021)

| Reference | Study design | Trait(s) | QTL/Key findings |
|---|---|---|---|
| Buitenhuis et al. 2003 | $F_2$ design from commercial lines unselected for FP<br>Interval mapping with 180 microsatellites<br>$N = 630$ $F_2$ hens | Gentle FP<br>Severe FP<br>Aggressive pecking | Genome-wide significant QTL for severe FP on GGA2<br>Various chromosome-wide significant/suggestive QTL on 8 different chromosomes |
| Jensen et al. 2005 | $F_2$ cross from red jungle fowl and White Leghorn<br>Interval mapping with 105 markers<br>N = 762 $F_2$ birds (both sexes) | Severe FP (binary coding) | Only suggestive QTL on GGA 3<br>$F_2$ feather peckers more active<br>Different resource allocation pattern suggested |
| Biscarini et al. 2010 | Study across 9 lines (White Leghorn / Rhode Island Red)<br>GWAS with 1 k SNPs<br>$N = 662$ hens | Plumage condition (modelled with direct and associative effects) | Fifty-seven significantly associated SNPs for the associative effect<br>Suggested candidate genes: Serotonin receptor 2C gene (*HTR2C*) on GGA4 and some immune-related genes |
| Grams et al. 2015a | FP selection lines<br>$F_{ST}$-based mapping of selection signatures with 33 k SNPs<br>N = 41 HFP and 34 LFP hens | – | Thirteen significant SNP clusters with most significant SNPs on GGA 3 and 4 |
| Lutz et al. 2017 | $F_2$ design from FP selection lines<br>GWAS and meta-analysis in conjunction with data from Grams et al. (2015a)<br>Prioritization of candidate genes based on expression data (Wysocki et al. 2013)<br>$N = 817$ $F_2$ hens | FP delivered<br>Aggressive pecks delivered/received | Thirteen associated clusters for FP delivered<br>Suggested candidate genes: *SLC12A9* (involved in monoamine signaling) and *GNG2* (involved in GABAergic signaling) |
| Iffland et al. 2020b | FP selection lines<br>Mapping of selection signatures between ($F_{ST}$) and within (iHS) lines + GWAS using 30 k SNPs<br>N = 270 HFP and 219 LFP hens | FP delivered<br>Posterior probability to be an extreme pecker (pEFP) | Selective sweep on GGA4 in LFP line<br>Genome-wide significant QTL for FP delivered and pEFP on GGA1<br>Positional candidate genes pointing to GABAergic signaling |
| Falker-Gieske et al. 2020b | FP selection lines and $F_2$-cross thereof<br>Imputation from SNP chip to sequence level and GWAS for the lines (4.3 mio variants) and the cross (6.7 mio variants) separately, combination in a meta-analysis<br>Downstream enrichment and pathway analyses<br>N = 489 hens of the lines and 817 $F_2$ hens<br>Forty-nine founder animals from both designs sequenced | FP delivered and pEFP | Genome-wide significant associations for FP delivered on GGA2, 4, 7, and 13 with the major peak on GGA7<br>Confirmation of the QTL on GGA1 from (Iffland et al. 2020b, 2021)<br>Suggested candidate genes (*SPATS2L, ZEB2, KCHN8, ZEB2*, and *MRPL13*) previously implicated in psychiatric and neurological disorders as well as nicotine treatment |

(continued)

**Table 4** (continued)

| Reference | Study design | Trait(s) | QTL/Key findings |
|---|---|---|---|
| Iffland et al. 2021 | GWAS (same design as in Iffland et al. 2020b) | Aggressive pecks delivered<br>Threats delivered<br>FP index (FP delivered minus FP received)<br>Activity | QTL for…<br>Aggressive pecks delivered on GGA6<br>Activity on GGA18<br>FP index on GGA1 (supporting the findings of Iffland et al. 2020b) |
| Falker-Gieske et al. 2022 | FP selection lines<br>Methodology identical to Falker-Gieske et al. (2020b)<br>GWAS in lines particularly focused on GABA receptors based on results from transcriptomic studies | FP delivered | Various associations with GABA receptor subunits including the QTL on GGA1 (Iffland et al. 2020b, 2021) |

(2015a) took advantage of the fact that the complete history of the experimental lines is known and modelled drift stochastically. In an $F_{ST}$-based approach, which is capable of detecting signs of selection between but not within lines, they found several regions with a particular differentiation between HFP and LFP with most significant SNPs being located on GGA3 and 4. Later, Iffland et al. (2020b) extended the approach and along with the inter-population $F_{ST}$ approach also applied an intrapopulation approach based on integrated haplotype scores. This resulted in the detection of a single significant selection signature in GGA4 in the LFP line. Except for the common finding of a region under selection on GGA4, the studies yielded no convincing genome regions with a substantial contribution to trait variation. This might simply reflect the complex nature of FP as well as the population history with relatively few generations of selection in comparatively small populations.

The studies applying QTL mapping or GWAS-based approaches are more diverse in terms of mapping strategy and outcome. As can be seen from Table 2, the earliest studies applied microsatellite-based QTL mapping approaches (Buitenhuis et al. 2003; Jensen et al. 2005), which naturally yield large QTL confidence intervals. Later studies used SNPs / SNPs chips with an increasing density from 1 k (Biscarini et al. 2010) to roughly 30 k (Iffland et al. 2020b, 2021; Lutz et al. 2017). Another study used imputed genome-level genotypes for several millions of markers in the selection lines as well as an $F_2$-cross (Falker-Gieske et al. 2020b, 2022). In all these studies, linked or associated genome regions were mapped to various chromosomes with apparently little overlap (Table 4). There seems to be, however, a common picture that can be derived.

In concordance with functional studies, genes involved in serotonergic signaling were pinpointed (Biscarini et al. 2010; Lutz et al. 2017). A candidate gene study also implicated dopaminergic signaling in FP (Flisikowski et al. 2009).

Several results pointed to GABAergic signaling / GABA receptors (Falker-Gieske et al. 2020b, 2022; Iffland et al. 2020b, 2021), which was also implicated in cannibalism and survival time (Brinker et al. 2018).

The immune system was implicated (Biscarini et al. 2010; Falker-Gieske et al. 2020b).

In their sequence-level study, Falker-Gieske et al. (2020b) found several candidate genes previously reported to be involved in human psychiatric disorders, mainly schizophrenia.

In summary, the available mapping studies clearly reflect the complex nature of the trait, but some common findings point to mechanisms possibly involved in the propensity to perform FP. It may be argued that most studies were conducted in the same selection lines, but there is also an overlap across populations, especially regarding GABAergig and serotonergic signaling.

## 5.4 Functional and Omics-Studies

In order to understand the genetic causes predisposing hens to show FP behavior, attempts to map causative genes as described in the previous section can only be successful to a certain extent. Downstream analyses of GWAS results such as gene set enrichment approaches can point to particular pathways. To gain a deeper understanding, however, transcriptomic and functional studies are necessary, which can then in turn reveal pathways and genes regulating the changes at these levels. Together, these types of studies can possibly also lead to new breeding tools such as genetic markers or endophenotypes suitable to serve as indicator traits. The following section summarizes the current main areas of functional and omics research.

### The Functional Role of Serotonin and Dopamine

The serotonergic system affects a wide range of physiological processes and partly also regulates dopamine signaling. It is related to diverse psychopathological behaviors in various species and both, serotonin and dopamine have been shown to influence FP (Kops et al. 2014, 2017; Rodenburg et al. 2013; van Hierden et al. 2004).

In one striking example, it was, e.g. shown that a somatodendritic serotonin autoreceptor agonist and postsynaptic serotonin receptor antagonist can decrease serotonin turnover, in particular, brain areas and increase foraging and FP behavior, which confirms that FP is triggered by low serotonin neurotransmission (van Hierden et al. 2004). Dopamine activity in high FP birds is reduced in brain areas related to motivation and reward as well goal-directed behavior (de Haas and van der Eijk 2018). Thus, is clear that serotonin and dopamine are involved in FP, but the exact mechanism is complex and possibly heterogeneous (for a comprehensive review see de Haas and van der Eijk 2018).

### The Role of the Gut Microbiota

Composition and metabolism of the gut microbiota do affect behavior and mental health (Dinan et al. 2015). This is also assumed for FP, because gut microbiota interacts with both the peripheral and the central serotonergic system (de Haas and van der Eijk 2018). Actually, high and low FP genotypes from selection lines can be distinguished according to their gut microbiota composition (Birkl et al. 2018; van der Eijk et al. 2019). This does, however, not answer the question, whether this is a consequence or the cause of FP. The fact that the inclusion of feathers in the diet leads to changes in microbial metabolism (Meyer et al. 2013) supports the first assumption. On the other hand, it was shown that early-life microbiota transplantation can influence behavioral characteristics related to FP, which could support the hypothesis of a causative relationship (van der Eijk et al. 2020). Beyond comparing the microbiota between genotypes or phenotypes, Borda-Molina et al. (2021) estimated the microbiability of FP, i.e., the amount of trait variance explained by gut microbiota composition. While also line differences were detected in this study, the microbiabilities were not significantly different from zero. Thus, the exact role in FP of the gut microbiota remains to be elucidated, but from a breeder's perspective they seem to play a minor role.

### Connections with the Immune System

Similarly to the proposed effect of gut microbiota and possibly linked to it, there might be an influence of the immune system on FP via the serotonergic system. Various studies have reported a relationship between immune response and components of the immune system and FP. Beyond results of mapping studies pointing to cytokine genes (Biscarini et al. 2010), there is convincing evidence from transcriptomic studies (Brunberg et al. 2011; Falker-Gieske et al. 2020a) as will be further explained later in the next section. Interestingly, it also seems that stimulating the humoral immune response in young birds may predispose adult hens to FP (Parmentier et al. 2009), which is in line with a high genetic correlation of 0.79 between severe FP and antibody response (Buitenhuis et al. 2004).

### Transcriptomic Studies

Gene expression studies in FP require a thorough consideration of the study design. Naturally, brain is the target tissue, but based on functional studies (Kops et al. 2017) one could consider sampling particular brain areas. This is challenging as it requires a sophisticated sampling technique, and it is not completely clear, whether the chosen brain areas will deliver meaningful results. Thus, most transcriptome studies have so far considered the whole brain. It is also very important to choose the right time point to sacrifice the animals and to consider the definition of the experimental groups carefully. Most studies have done this based on genotype, i.e., selection lines, so far, but also a phenotypic definition was considered (Brunberg et al. 2011). In our own studies, we chose a genotype-based design comparing hens from selection lines. In addition, we compared baseline samples with triggered samples based on full-sib pairs. The baseline data was generated from one sib of each pair after keeping the animals under dark conditions and thus preventing the expression of actual pecking behavior. Subsequently, the light was turned on and upon clear expression of FP, the second sibling of each pair was sacrificed (Falker-Gieske et al. 2020a).

Unsurprisingly, various genes involved in human psychiatric disorders such as schizophrenia or obsessive-compulsive disorders were identified (Brunberg et al. 2011; Falker-Gieske et al. 2020a; Wysocki et al. 2013) as well as genes involved in neurotransmission, most prominently GABAergic signaling (Falker-Gieske et al. 2020a). Furthermore, clear links with the immune system were found (Brunberg et al. 2011; Falker-Gieske et al. 2020a). Thus, findings from transcriptomic studies are well aligned with the outcome of genomic studies.

A striking observation was made when comparing the transcriptomic response to a light stimulus triggering FP behavior between selection lines. While a strong response occurred in animals selected for low FP, there were only a few differentially expressed genes in high feather peckers (Falker-Gieske et al. 2022). A similar finding had previously been reported as a reduced variance of gene expression in high peckers (Hughes and Buitenhuis 2010). Notably, for genes differentially expressed in both lines, even the direction of response was opposite and among those genes a GABA receptor subunit as well as *Dicer1* was identified. The latter is a key ribonuclease in the biogenesis of regulatory RNAs, which would explain a large number of schizophrenia-associated small RNAs ("psychiatric ncRNAs," Gandal et al. 2018) that are differentially expressed between the FP selection lines (Falker-Gieske et al. 2020a). Furthermore, *Dicer1* might also provide a link to the immune system via interferon signaling (Gurung et al. 2021). A further outcome of this study was the fact that the light-triggered expression patterns of genes related to circadian rhythm differed between the selection lines.

Taking all these results together leads to the hypothesis of a *Dicer1*-dependant dysregulation of miRNA biogenesis resulting in a downregulation of GABA signaling. As GABA is a major inhibitory neurotransmitter, this leads to a constantly high level of excitation in high FP birds (Falker-Gieske et al. 2022). This aligns well with many other findings such as the relationship between activity or fear and FP.

## 5.5 Concluding Remarks

Concerning breeding, the propensity to perform FP is clearly a heritable trait that can be improved by breeding. It is, however, a complex trait influenced by many genes with presumably small effects. It shows a complex genetic relationship with other behavior traits related to foraging, activity, exploration, fear, and aggression, as well as performance traits. A wide variety of functional, genomic, and transcriptomic studies clearly reflect the complex nature of the trait, but also draw a mechanistic picture of the development of FP that gains more and more coherence. It is clear that serotonergic and dopaminergic signaling are involved in FP and that there is an interrelationship with the immune system and possibly also the gut microbiota. But probably these mechanisms are not on top of the regulatory hierarchy orchestrating FP behavior. *Dicer1*-dependent dysregulation of regulatory RNA biogenesis and consequently reduced GABAergic neurotransmission may be crucial to the development of FP. This hypothesis is consistent and could possibly incorporate most existing theories. However, the complexity of FP precludes conclusions that are too simple and general. Especially as much research has been done in different generations of the very same selection lines, there is little knowledge of etiological and genetic heterogeneity. This heterogeneity could be rather striking as shown by de Haas et al. (2014a) who found hens from a brown hybrid more affected by environmental conditions, characterizing FP as maladaptive behavior, while white hybrids were more affected by parental effects pointing to a malfunctioning behavior. Although the selection lines were established from While Leghorn lines, it is unclear whether they still reflect modern layer genetics after several generations of selection in quite small populations. Thus, validation studies in commercial lines are urgently needed in order to make the results useful for practical breeding. It can thus be regarded as crucial to further develop viable phenotyping solutions that can be applied in commercial flocks. Independent from this, evidence for striking similarities between FP and schizophrenia is accumulating. Potentially, results could thus also have implications for other species including humans.

Concerning management, FP can best be prevented by providing a stimulating environment from the very start on. This includes adapted management measures as early as in the rearing phase, providing adequate and enough enrichment material and litter, and a good adaption to humans in order to avoid stress. Even though a lot of research has been carried out regarding FB and the underlying mechanisms and influencing factors, there is still no "one and only solution." As stated by Fijn et al. (2020), more basic research might help to obtain even more comprehensive knowledge of behavioral needs and abilities of laying hens.

Concluding, only by combining both breeding and management aspects, FP as a serious welfare problem can be avoided, thus ensuring healthy and productive laying hen flocks, even with untrimmed beaks.

## References


Åkerfeldt MP, Gunnarsson S, Bernes G, Blanco-Penedo I (2021) Health and welfare in organic livestock production systems—a systematic mapping of current knowledge. Org Agric 11:105–132. https://doi.org/10.1007/s13165-020-00334-y

Bennewitz J, Tetens J (2021) Feather pecking behavior in the Hohenheim selection lines of laying hens-a selective review of genomic, transcriptomic and gut microbiota analyses. Züchtungskunde 93:229–243

Bennewitz J, Bögelein S, Stratz P, Rodehutscord M, Piepho HP, Kjaer JB, Bessei W (2014) Genetic parameters for feather pecking and aggressive behavior in a large F2-cross of laying hens using generalized linear mixed models. Poult Sci 93:810–817

Bessei W (1984) Untersuchungen zur Heritabilität des Federpickverhaltens bei Junghennen. 1. Mitteilung. Archiv für Geflügelkunde 48:224–231

Bessei W, Bauhaus H, Bögelein S (2013) The effect of selection for high and low feather pecking on aggression–related behaviours of laying hens. Archiv für Geflügelkunde 77:10–14

Birkl P, Bharwani A, Kjaer JB, Kunze W, McBride P, Forsythe P, Harlander-Matauschek A (2018) Differences in cecal microbiome of selected high and low feather-pecking laying hens. Poult Sci 97:3009–3014. https://doi.org/10.3382/ps/pey167

Biscarini F, Bovenhuis H, van der Poel JJ, Rodenburg TB, Jungerius AP, van Arendonk JAM (2010) Across-line SNP association study for direct and associative effects on feather damage in laying hens. Behav Genet 40:715–727. https://doi.org/10.1007/s10519-010-9370-0

Blancke B, Klefner W (1908) Unser Hasugeflügel. Das Großgeflügel. Erster Band: Rassenkunde, 2nd edn. Pfennigstorff, Berlin

Blokhuis HJ (1986) Feather-pecking in poultry: its relation with ground-pecking. Appl Anim Behav Sci 16:63–67. https://doi.org/10.1016/0168-1591(86)90040-7

Blokhuis HJ, Arkes JG (1984) Some observations on the development of feather-pecking in poultry. Appl Anim Behav Sci 12:145–157

Blokhuis HJ, van der Haar JW (1992) Effects of pecking incentives during rearing on feather pecking of laying hens. Br Poult Sci 33:17–24. https://doi.org/10.1080/00071669208417440

Blokhuis HJ, van Niekerk TF, Bessei W, Elson A, Guémené D, Kjaer JB, Maria Levrino GA, Nicol CJ, Tauson R, Weeks CA, van de Weerd HA (2007) The LayWel project: welfare implications of changes in production systems for laying hens. Worlds Poult Sci J 63:101–114. https://doi.org/10.1017/S0043933907001328

Borda-Molina D, Iffland H, Schmid M, Müller R, Schad S, Seifert J, Tetens J, Bessei W, Bennewitz J, Camarinha-Silva A (2021) Differences in the gut microbial composition of divergently selected laying hen lines is not associated with feather pecking and agonistic behavior. Life 11:235

Brinker T, Bijma P, Vereijken A, Ellen ED (2018) The genetic architecture of socially-affected traits: a GWAS for direct and indirect genetic effects on survival time in laying hens showing cannibalism. Genet Sel Evol 50:1–11. https://doi.org/10.1186/s12711-018-0409-7

Brunberg E, Jensen P, Isaksson A, Keeling LJ (2011) Feather pecking behavior in laying hens: hypotha-

lamic gene expression in birds performing and receiving pecks. Poult Sci 90:1145–1152
Brunberg EI, Rodenburg TB, Rydhmer L, Kjaer JB, Jensen P, Keeling LJ (2016) Omnivores going astray: a review and new synthesis of abnormal behavior in pigs and laying hens. Front Vet Sci 3:57. https://doi.org/10.3389/fvets.2016.00057
Buitenhuis AJ, Rodenburg TB, Siwek M, Cornelissen SJ, Nieuwland MG, Crooijmans RP, Groenen MA, Koene P, Bovenhuis H, van der Poel JJ (2003) Identification of quantitative trait loci for receiving pecks in young and adult laying hens. Poult Sci 82:1661–1667
Buitenhuis AJ, Rodenburg TB, Wissink PH, Visscher J, Koene P, Bovenhuis H, Ducro BJ, van der Poel JJ (2004) Genetic and phenotypic correlations between feather pecking behavior, stress response, immune response, and egg quality traits in laying hens. Poult Sci 83:1077–1082. https://doi.org/10.1093/ps/83.7.1077
Cloutier S, Newberry RC, Forster CT, Girsberger KM (2000) Does pecking at inanimate stimuli predict cannibalistic behaviour in domestic fowl? Appl Anim Behav Sci 66:119–133. https://doi.org/10.1016/S0168-1591(99)00068-4
Cronin GM, Glatz PC (2021) Causes of feather pecking and subsequent welfare issues for the laying hen: a review. Anim Prod Sci 61:990. https://doi.org/10.1071/AN19628
Cuthbertson GJ (1980) Genetic variation in feather-pecking behaviour. Br Poult Sci 21:447–450
de Haas EN, van der Eijk JAJ (2018) Where in the serotonergic system does it go wrong? Unravelling the route by which the serotonergic system affects feather pecking in chickens. Neurosci Biobehav Rev 95:170–188
de Haas EN, Bolhuis JE, Kemp B, Groothuis TGG, Rodenburg TB (2014a) Parents and early life environment affect behavioral development of laying hen chickens. PLoS One 9:e90577. https://doi.org/10.1371/journal.pone.0090577
de Haas EN, Bolhuis JE, Jong IC d, Kemp B, Janczak AM, Rodenburg TB (2014b) Predicting feather damage in laying hens during the laying period. Is it the past or is it the present? Appl Anim Behav Sci 160:75–85. https://doi.org/10.1016/j.applanim.2014.08.009
Dennis RL, Fahey AG, Cheng HW (2009) Infrared beak treatment method compared with conventional hot-blade trimming in laying hens. Poult Sci 88:38–43. https://doi.org/10.3382/ps.2008-00227
Desbruslais A, Wealleans A, Gonzalez-Sanchez D, Di Benedetto M (2021) Dietary fibre in laying hens: a review of effects on performance, gut health and feather pecking. Worlds Poult Sci J 77:797–823. https://doi.org/10.1080/00439339.2021.1960236
Dinan TG, Stilling RM, Stanton C, Cryan JF (2015) Collective unconscious: how gut microbes shape human behavior. J Psychiatr Res 63:1–9. https://doi.org/10.1016/j.jpsychires.2015.02.021
Drake KA, Donnelly CA, Dawkins MS (2010) Influence of rearing and lay risk factors on propensity for feather damage in laying hens. Br Poult Sci 51:725–733. https://doi.org/10.1080/00071668.2010.528751
Du X, Qin P, Liu Y, Amevor FK, Shu G, Li D, Zhao X (2022) Effects of key farm management practices on pullets welfare-a review. Animals 12:729. https://doi.org/10.3390/ani12060729
Ellen ED, van der Sluis M, Siegford J, Guzhva O, Toscano MJ, Bennewitz J, van der Zande LE, van der Eijk JAJ, de Haas EN, Norton T, Piette D, Tetens J, Klerk B d, Visser B, Rodenburg TB (2019) Review of sensor technologies in animal breeding: phenotyping behaviors of laying hens to select against feather pecking. Animals 9:108. https://doi.org/10.3390/ani9030108
Falker-Gieske C, Mott A, Preuß S, Franzenburg S, Bessei W, Bennewitz J, Tetens J (2020a) Analysis of the brain transcriptome in lines of laying hens divergently selected for feather pecking. BMC Genomics 21:1–14. https://doi.org/10.1186/s12864-020-07002-1
Falker-Gieske C, Iffland H, Preuß S, Bessei W, Drögemüller C, Bennewitz J, Tetens J (2020b) Meta-analyses of genome wide association studies in lines of laying hens divergently selected for feather pecking using imputed sequence level genotypes. BMC Genet 21:114. https://doi.org/10.1186/s12863-020-00920-9
Falker-Gieske C, Bennewitz J, Tetens J (2022) The light response in chickens divergently selected for feather pecking behavior reveals mechanistic insights towards psychiatric disorders. Mol Biol Rep 49:1649–1654. https://doi.org/10.1007/s11033-021-07111-4
Fijn LB, van der Staay FJ, Goerlich-Jansson VC, Arndt SS (2020) Importance of basic research on the causes of feather pecking in relation to welfare. Animals 10:213. https://doi.org/10.3390/ani10020213
Flisikowski K, Schwarzenbacher H, Wysocki M, Weigend S, Preisinger R, Kjaer JB, Fries R (2009) Variation in neighbouring genes of the dopaminergic and serotonergic systems affects feather pecking behaviour of laying hens. Anim Genet 40:192–199. https://doi.org/10.1111/j.1365-2052.2008.01821.x
Gandal MJ, Zhang P, Hadjimichael E, Walker RL, Chen C, Liu S, Won H, van Bakel H, Varghese M, Wang Y, Shieh AW, Haney J, Parhami S, Belmont J, Kim M, Moran Losada P, Khan Z, Mleczko J, Xia Y, Dai R, Wang D, Yang YT, Xu M, Fish K, Hof PR, Warrell J, Fitzgerald D, White K, Jaffe AE, Peters MA, Gerstein M, Liu C, Iakoucheva LM, Pinto D, Geschwind DH (2018) Transcriptome-wide isoform-level dysregulation in ASD, schizophrenia, and bipolar disorder. Science 362:eaat8127. https://doi.org/10.1126/science.aat8127
Gentle MJ, Hughes BO, Fox A, Waddington D (1997) Behavioural and anatomical consequences of two beak trimming methods in 1- and 10-d-old domestic chicks. Br Poult Sci 38:453–463. https://doi.org/10.1080/00071669708418022
Giersberg MF, Spindler B, Kemper N (2017) Assessment of plumage and integument condition in dual-purpose breeds and conventional layers. Animals 7:92. https://doi.org/10.3390/ani7120097

Giersberg MF, Kemper N, Spindler B (2019) Pecking and piling: the behaviour of conventional layer hybrids and dual-purpose hens in the nest. Appl Anim Behav Sci 214:50–56. https://doi.org/10.1016/j.applanim.2019.02.016
Glatz PC (2000) Beak trimming methods - review. Asian Australas J Anim Sci 13:1619–1637. https://doi.org/10.5713/ajas.2000.1619
Glatz PC, Hinch G (2008) Minimise cannibalism using innovative beak trimming methods: Project No. 04-20 Final Report
Grams V, Wellmann R, Preuß S, Grashorn MA, Kjaer JB, Bessei W, Bennewitz J (2015a) Genetic parameters and signatures of selection in two divergent laying hen lines selected for feather pecking behaviour. Genet Sel Evol 47:1–9
Grams V, Bögelein S, Grashorn MA, Bessei W, Bennewitz J (2015b) Quantitative genetic analysis of traits related to fear and feather pecking in laying hens. Behav Genet 45:228–235
Gunnarsson S, Keeling LJ, Svedberg J (1999) Effect of rearing factors on the prevalence of floor eggs, cloacal cannibalism and feather pecking in commercial flocks of loose housed laying hens. Br Poult Sci 40:12–18. https://doi.org/10.1080/00071669987773
Gunnarsson S, Yngvesson J, Keeling LJ, Forkman B (2000) Rearing without early access to perches impairs the spatial skills of laying hens. Appl Anim Behav Sci 67:217–228. https://doi.org/10.1016/S0168-1591(99)00125-2
Gurung C, Fendereski M, Sapkota K, Guo J, Huang F, Guo Y-L (2021) Dicer represses the interferon response and the double-stranded RNA-activated protein kinase pathway in mouse embryonic stem cells. J Biol Chem 296:100264. https://doi.org/10.1016/j.jbc.2021.100264
Habig C, Distl O (2013) Evaluation of bone strength, keel bone status, plumage condition and egg quality of two layer lines kept in small group housing systems. Br Poult Sci 54:413–424. https://doi.org/10.1080/00071668.2013.792405
Harlander-Matauschek A, Häusler K (2009) Understanding feather eating behaviour in laying hens. Appl Anim Behav Sci 117:35–41. https://doi.org/10.1016/j.applanim.2008.11.003
Hartcher KM, Tran MK, Wilkinson SJ, Hemsworth PH, Thomson PC, Cronin GM (2015a) Plumage damage in free-range laying hens: behavioural characteristics in the rearing period and the effects of environmental enrichment and beak-trimming. Appl Anim Behav Sci 164:64–72. https://doi.org/10.1016/j.applanim.2014.12.011
Hartcher KM, Tran KTN, Wilkinson SJ, Hemsworth PH, Thomson PC, Cronin GM (2015b) The effects of environmental enrichment and beak-trimming during the rearing period on subsequent feather damage due to feather-pecking in laying hens. Poult Sci 94:852–859. https://doi.org/10.3382/ps/pev061
Henderson SN, Barton JT, Wolfenden AD, Higgins SE, Higgins JP, Kuenzel WJ, Lester CA, Tellez G, Hargis BM (2009) Comparison of beak-trimming methods on early broiler breeder performance. Poult Sci 88:57–60. https://doi.org/10.3382/ps.2008-00104
Honaker CF, Ruszler PL (2004) The effect of claw and beak reduction on growth parameters and fearfulness of two Leghorn strains. Poult Sci 83:873–881. https://doi.org/10.1093/ps/83.6.873
Huber-Eicher B, Sebö F (2001) The prevalence of feather pecking and development in commercial flocks of laying hens. Appl Anim Behav Sci 74:223–231. https://doi.org/10.1016/S0168-1591(01)00173-3
Huber-Eicher B, Wechsler B (1997) Feather pecking in domestic chicks: its relation to dustbathing and foraging. Anim Behav 54:757–768
Huber-Eicher B, Wechsler B (1998) The effect of quality and availability of foraging materials on feather pecking in laying hen chicks. Anim Behav 55:861–873
Hughes AL, Buitenhuis AJ (2010) Reduced variance of gene expression at numerous loci in a population of chickens selected for high feather pecking. Poult Sci 89:1858–1869. https://doi.org/10.3382/ps.2010-00827
Hughes BO, Duncan IJ (1972) The influence of strain and environmental factors upon feather pecking and cannibalism in fowls. Br Poult Sci 13:525–547. https://doi.org/10.1080/00071667208415981
Iffland H, Wellmann R, Preuß S, Tetens J, Bessei W, Piepho H-P, Bennewitz J (2020a) A novel model to explain extreme feather pecking behavior in laying hens. Behav Genet 50:41–50. https://doi.org/10.1007/s10519-019-09971-w
Iffland H, Wellmann R, Schmid M, Preuß S, Tetens J, Bessei W, Bennewitz J (2020b) Genomewide mapping of selection signatures and genes for extreme feather pecking in two divergently selected laying hen lines. Animals 10:262. https://doi.org/10.3390/ani10020262
Iffland H, Schmid M, Preuß S, Bessei W, Tetens J, Bennewitz J (2021) Phenotypic and genomic analyses of agonistic interactions in laying hen lines divergently selected for feather pecking. Appl Anim Behav Sci 234:105177
Janczak AM, Riber AB (2015) Review of rearing-related factors affecting the welfare of laying hens. Poult Sci 94:1454–1469. https://doi.org/10.3382/ps/pev123
Jensen P, Keeling LJ, Schütz K, Andersson L, Mormède P, Brändström H, Forkman B, Kerje S, Fredriksson R, Ohlsson C, Larsson S, Mallmin H, Kindmark A (2005) Feather pecking in chickens is genetically related to behavioural and developmental traits. Physiol Behav 86:52–60
Jones RB, Blokhuis HJ, Beuving G (1995) Open-field and tonic immobility responses in domestic chicks of two genetic lines differing in their propensity to feather peck. Br Poult Sci 36:525–530
Jung L, Knierim U (2018) Are practice recommendations for the prevention of feather pecking in laying hens in non-cage systems in line with the results of experimental and epidemiological studies? Appl Anim Behav Sci 200:1–12. https://doi.org/10.1016/J.APPLANIM.2017.10.005

Jung L, Knierim U (2019) Differences between feather pecking and non-feather pecking laying hen flocks regarding their compliance with recommendations for the prevention of feather pecking – a matched concurrent case-control design. Appl Anim Behav Sci 219:104839. https://doi.org/10.1016/j.applanim.2019.104839

Kjaer JB (2009) Feather pecking in domestic fowl is genetically related to locomotor activity levels: implications for a hyperactivity disorder model of feather pecking. Behav Genet 39:564–570. https://doi.org/10.1007/s10519-009-9280-1

Kjaer JB, Bessei W (2013) The interrelationships of nutrition and feather pecking in the domestic fowl. Eur Poult Sci 77:1–9

Kjaer JB, Mench JA (2003) Behaviour problems associated with selection for increased production. In: Muir WM, Aggrey SE (eds) Poultry genetics, breeding and biotechnology. CABI, Wallingford, pp 67–82

Kjaer JB, Sørensen P (1997) Feather pecking behaviour in White leghorns, a genetic study. Br Poult Sci 38:333–341

Kjaer JB, Sørensen P, Su G (2001) Divergent selection on feather pecking behaviour in laying hens (Gallus gallus domesticus). Appl Anim Behav Sci 71:229–239. https://doi.org/10.1016/S0168-1591(00)00184-2

Kjaer JB, Würbel H, Schrader L (2015) Perseveration in a guessing task by laying hens selected for high or low levels of feather pecking does not support classification of feather pecking as a stereotypy. Appl Anim Behav Sci 168:56–60. https://doi.org/10.1016/j.applanim.2015.04.014

Kops MS, Kjaer JB, Güntürkün O, Westphal KGC, Korte-Bouws GAH, Olivier B, Bolhuis JE, Korte SM (2014) Serotonin release in the caudal nidopallium of adult laying hens genetically selected for high and low feather pecking behavior: an in vivo microdialysis study. Behav Brain Res 268:81–87. https://doi.org/10.1016/j.bbr.2014.03.050

Kops MS, Kjaer JB, Güntürkün O, Westphal KGC, Korte-Bouws GAH, Olivier B, Korte SM, Bolhuis JE (2017) Brain monoamine levels and behaviour of young and adult chickens genetically selected on feather pecking. Behav Brain Res 327:11–20. https://doi.org/10.1016/j.bbr.2017.03.024

Labouriau R, Kjaer JB, Abreu GCG, Hedegaard J, Buitenhuis AJ (2009) Analysis of severe feather pecking behavior in a high feather pecking selection line. Poult Sci 88:2052–2062

Lambton SL, Knowles TG, Yorke C, Nicol CJ (2010) The risk factors affecting the development of gentle and severe feather pecking in loose housed laying hens. Appl Anim Behav Sci 123:32–42

Lambton SL, Knowles TG, Yorke C, Nicol CJ (2015) The risk factors affecting the development of vent pecking and cannibalism in free-range and organic laying hens. Anim Welf 24:101–111. https://doi.org/10.7120/09627286.24.1.101

Liljedahl LE, Kolstad N, Sørensen P, Maijala K (1979) Scandinavian selection and crossbreeding experiment with laying hens. I. Background and general outline. Acta Agric Scand 29:273–286

Lunam CA, Glatz PC, Hsu YJ (1996) The absence of neuromas in beaks of adult hens after conservative trimming at hatch. Aust Vet J 74:46–49. https://doi.org/10.1111/j.1751-0813.1996.tb13734.x

Lutz V, Kjaer JB, Iffland H, Rodehutscord M, Bessei W, Bennewitz J (2016) Quantitative genetic analysis of causal relationships among feather pecking, feather eating, and general locomotor activity in laying hens using structural equation models. Poult Sci 95:1757–1763

Lutz V, Stratz P, Preuß S, Tetens J, Grashorn MA, Bessei W, Bennewitz J (2017) A genome-wide association study in a large F2-cross of laying hens reveals novel genomic regions associated with feather pecking and aggressive pecking behavior. Genet Sel Evol 49:18. https://doi.org/10.1186/s12711-017-0287-4

Marchant-Forde RM, Fahey AG, Cheng HW (2008) Comparative effects of infrared and one-third hot-blade trimming on beak topography, behavior, and growth. Poult Sci 87:1474–1483. https://doi.org/10.3382/ps.2006-00360

McKeegan DE, Savory C (1999) Feather eating in layer pullets and its possible role in the aetiology of feather pecking damage. Appl Anim Behav Sci 65:73–85

McKeegan DE, Savory C (2001) Feather eating in individually caged hens which differ in their propensity to feather peck. Appl Anim Behav Sci 73:131–140

Mels C, Niebuhr K, Futschik A, Rault J-L, Waiblinger S (2022) Predictors for plumage damage and bloody lesions indicative of feather pecking in pullets reared in aviaries. Appl Anim Behav Sci 250:105607. https://doi.org/10.1016/j.applanim.2022.105607

Mens A, van Krimpen MM, Kwakkel RP (2020) Nutritional approaches to reduce or prevent feather pecking in laying hens: any potential to intervene during rearing? Worlds Poult Sci J 76:591–610. https://doi.org/10.1080/00439339.2020.1772024

Meyer B, Bessei W, Vahjen W, Zentek J, Harlander-Matauschek A (2012) Dietary inclusion of feathers affects intestinal microbiota and microbial metabolites in growing Leghorn-type chickens. Poult Sci 91:1506–1513. https://doi.org/10.3382/ps.2011-01786

Meyer B, Zentek J, Harlander-Matauschek A (2013) Differences in intestinal microbial metabolites in laying hens with high and low levels of repetitive feather-pecking behavior. Physiol Behav 110-111:96–101. https://doi.org/10.1016/j.physbeh.2012.12.017

Mohammed HH, Grashorn MA, Bessei W (2010) The effects of lighting conditions on the behaviour of laying hens. Eur Poult Sci 74:197–202

Newberry RC (2004) Cannibalism. In: Perry G (ed) Welfare of the laying hen: papers from the 27th poultry science symposium of the World's poultry science association (UK branch), Bristol, UK, July 2003. CABI, Wallingford, Cambridge

Newberry RC, Keeling LJ, Estevez I, Bilčík B (2007) Behaviour when young as a predictor of severe feather pecking in adult laying hens: the redirected

foraging hypothesis revisited. Appl Anim Behav Sci 107:262–274
Nicol C, Gregory N, Knowles TG, Parkman I, Wilkins L (1999) Differential effects of increased stocking density, mediated by increased flock size, on feather pecking and aggression in laying hens. Appl Anim Behav Sci 65:137–152. https://doi.org/10.1016/S0168-1591(99)00057-X
Nicol C, Lindberg A, Phillips A, Pope S, Wilkins L, Green L (2001) Influence of prior exposure to wood shavings on feather pecking, dustbathing and foraging in adult laying hens. Appl Anim Behav Sci 73:141–155
Nicol CJ, Bestman M, Gilani A-M, de Haas EN, Jong IC d, Lambton SL, Wagenaar JP, Weeks CA, Rodenburg TB (2013) The prevention and control of feather pecking: application to commercial systems. Worlds Poult Sci J 69:775–788
Parmentier HK, Rodenburg TB, Vries Reilingh G d, Beerda B, Kemp B (2009) Does enhancement of specific immune responses predispose laying hens for feather pecking? Poult Sci 88:536–542. https://doi.org/10.3382/ps.2008-00424
Piepho H-P, Lutz V, Kjaer JB, Grashorn MA, Bennewitz J, Bessei W (2017) The presence of extreme feather peckers in groups of laying hens. Animal 11:500–506
Pötzsch CJ, Lewis K, Nicol CJ, Green LE (2001) A cross-sectional study of the prevalence of vent pecking in laying hens in alternative systems and its associations with feather pecking, management and disease. Appl Anim Behav Sci 74:259–272. https://doi.org/10.1016/S0168-1591(01)00167-8
Preisinger R (2018) Innovative layer genetics to handle global challenges in egg production. Br Poult Sci 59:1–6. https://doi.org/10.1080/00071668.2018.1401828
Prescott NB, Wathes CM, Jarvis JR (2003) Light, vision and the welfare of poultry. Anim Welf 12:269–288
Ramadan SGA, Borell E v (2008) Role of loose feathers on the development of feather pecking in laying hens. Br Poult Sci 49:250–256
Richter F (1954) Experiments to ascertain the causes of feather-eating in the domestic fow. In: Proceedings 10th World's poultry congress, pp 258–262
Riedstra B, Groothuis TG (2002) Early feather pecking as a form of social exploration: the effect of group stability on feather pecking and tonic immobility in domestic chicks. Appl Anim Behav Sci 77:127–138. https://doi.org/10.1016/S0168-1591(02)00031-X
Rodenburg TB, Buitenhuis AJ, Ask B, Uitdehaag KA, Koene P, van der Poel JJ, Bovenhuis H (2003) Heritability of feather pecking and open-field response of laying hens at two different ages. Poult Sci 82:861–867. https://doi.org/10.1093/ps/82.6.861
Rodenburg T, van Hierden YM, Buitenhuis AJ, Riedstra B, Koene P, Korte SM, van der Poel JJ, Groothuis T, Blokhuis HJ (2004a) Feather pecking in laying hens: new insights and directions for research? Appl Anim Behav Sci 86:291–298
Rodenburg TB, Buitenhuis AJ, Ask B, Uitdehaag KA, Koene P, van der Poel JJ, van Arendonk JAM, Bovenhuis H (2004b) Genetic and phenotypic correlations between feather pecking and open-field response in laying hens at two different ages. Behav Genet 34:407–415
Rodenburg TB, Uitdehaag KA, Ellen ED, Komen J (2009) The effects of selection on low mortality and brooding by a mother hen on open-field response, feather pecking and cannibalism in laying hens. Anim Welf 18:427–432
Rodenburg TB, de Haas EN, Nielsen BL, Buitenhuis AJ (2010) Fearfulness and feather damage in laying hens divergently selected for high and low feather pecking. Appl Anim Behav Sci 128:91–96. https://doi.org/10.1016/j.applanim.2010.09.017
Rodenburg TB, Reu K d, Tuyttens FAM (2012) Performance, welfare, health and hygiene of laying hens in non-cage systems in comparison with cage systems. In: Sandilands V (ed) Alternative systems for poultry: health, welfare and productivity. CABI, Wallingford, pp 210–224
Rodenburg TB, van Krimpen MM, Jong IC d, de Haas EN, Kops MS, Riedstra BJ, Nordquist RE, Wagenaar JP, Bestman M, Nicol CJ (2013) The prevention and control of feather pecking in laying hens: identifying the underlying principles. Worlds Poult Sci J 69:361–374. https://doi.org/10.1017/S0043933913000354
Rudkin C (2021) Feather pecking and foraging uncorrelated - the redirection hypothesis revisited. Br Poult Sci 63:1–9
Sandilands V, Savory CJ (2002) Ontogeny of behaviour in intact and beak trimmed layer pullets, with special reference to preening. Br Poult Sci 43:182–189. https://doi.org/10.1080/00071660120121373
Savory CJ (1995) Feather pecking and cannibalism. Worlds Poult Sci J 51:215–219
Savory CJ, Mann JS (1997) Behavioural development in groups of pen-housed pullets in relation to genetic strain, age and food form. Br Poult Sci 38:38–47
Schreiter R, Damme K, Borell E v, Vogt I, Klunker M, Freick M (2019) Effects of litter and additional enrichment elements on the occurrence of feather pecking in pullets and laying hens - a focused review. Vet Med Sci 5:500–507. https://doi.org/10.1002/vms3.184
Sherwin CM, Richards GJ, Nicol CJ (2010) Comparison of the welfare of layer hens in 4 housing systems in the UK. Br Poult Sci 51:488–499. https://doi.org/10.1080/00071668.2010.502518
Shi H, Li B, Tong Q, Zheng W, Zeng D, Feng G (2019) Effects of LED light color and intensity on feather pecking and fear responses of layer breeders in natural mating colony cages. Animals 9:814. https://doi.org/10.3390/ani9100814
Spindler B, Schulze-Hillert M, Hartung J (2013) Praxisbegleitende Untersuchungen zur Prüfung des Praxisbegleitende Untersuchungen zur Prüfung des Verzichts auf Schnabelkürzen bei Legehennen in Praxisbetrieben. TiHo, Abschlussbericht
Spindler B, Giersberg M, Andersson R, Kemper N (2016) Legehennenhaltung mit intaktem Schnabel –

Übersichtsbericht zum aktuellen Stand aus praktisch-wissenschaftlicher Sicht. Züchtungskunde 88:475–493

Su G, Kjaer JB, Sørensen P (2006) Divergent selection on feather pecking behavior in laying hens has caused differences between lines in egg production, egg quality, and feed efficiency. Poult Sci 85:191–197

University of Bristol (2013) Improving Feather Cover: a guide to reducing the risk of injurious pecking occurring in non-cage laying hens

van der Eijk JAJ, Lammers A, Li P, Kjaer JB, Rodenburg TB (2018) Feather pecking genotype and phenotype affect behavioural responses of laying hens. Appl Anim Behav Sci 205:141–150. https://doi.org/10.1016/j.applanim.2018.05.027

van der Eijk JAJ, Vries H d, Kjaer JB, Naguib M, Kemp B, Smidt H, Rodenburg TB, Lammers A (2019) Differences in gut microbiota composition of laying hen lines divergently selected on feather pecking. Poult Sci 98:7009–7021. https://doi.org/10.3382/ps/pez336

van der Eijk JAJ, Rodenburg TB, de Vries H, Kjaer JB, Smidt H, Naguib M, Kemp B, Lammers A (2020) Early-life microbiota transplantation affects behavioural responses, serotonin and immune characteristics in chicken lines divergently selected on feather pecking. Sci Rep 10:1–13. https://doi.org/10.1038/s41598-020-59125-w

van Hierden YM, Boer SF d, Koolhaas JM, Korte SM (2004) The control of feather pecking by serotonin. Behav Neurosci 118:575–583. https://doi.org/10.1037/0735-7044.118.3.575

van Krimpen MM, Kwakkel RP, Reuvekamp B, van der Peet-Schwering C, Hartog LA d, Verstegen M (2005) Impact of feeding management on feather pecking in laying hens. Worlds Poult Sci J 61:663–686. https://doi.org/10.1079/WPS200478

van Niekerk T (2017) Managing laying hen flocks with intact beaks. In: Roberts J (ed) Achieving sustainable production of eggs, vol 2. Burleigh Dodds Science Publishing, pp 163–176

van Niekerk T (2019) Evidence-based management of injurious pecking. In: Olukosi OA, Olori VE, Helmbrecht A, Lambton S, French NA (eds) Poultry feathers and skin: the poultry integument in health and welfare. CABI, Wallingford

van Niekerk T, Jong I d, van Krimpen M, Reuvekamp B, van Tuijl O, Bestman M (2017) Emergency measures against feather pecking. Wageningen Livestock Research

van Staaveren N, Ellis J, Baes CF, Harlander-Matauschek A (2021) A meta-analysis on the effect of environmental enrichment on feather pecking and feather damage in laying hens. Poult Sci 100:397–411. https://doi.org/10.1016/j.psj.2020.11.006

Vestergaard KS, Kruijt JP, Hogan JA (1993) Feather pecking and chronic fear in groups of red junglefowl: their relations to dustbathing, rearing environment and social status. Anim Behav 45:1127–1140. https://doi.org/10.1006/anbe.1993.1137

Wechsler B, Huber-Eicher B, Nash DR (1998) Feather pecking in growers: a study with individually marked birds. Br Poult Sci 39:178–185

Wysocki M, Preuss S, Stratz P, Bennewitz J (2013) Investigating gene expression differences in two chicken groups with variable propensity to feather pecking. Anim Genet 44:773–777. https://doi.org/10.1111/age.12050

Zepp M, Louton H, Erhard M, Schmidt P, Helmer F, Schwarzer A (2018) The influence of stocking density and enrichment on the occurrence of feather pecking and aggressive pecking behavior in laying hen chicks. J Vet Behav 24:9–18. https://doi.org/10.1016/j.jveb.2017.12.005

# Tail Biting in Pigs

Sven König and Kathrin Halli

**Abstract**

Tail biting in pigs is an abnormal behavior with multi-factorial origin, causing wounds, infections, partial or total losses of the tail, and even death in bitten pigs. Tail lesions, caused by "swine inflammation and necrosis syndrome (SINS)," are independent of tail biting. However, such lesions cause pre-damaged tails, resulting in secondary biting. Tail docking, one procedure to restrict tail biting, was banned by the EU Council Directive 2008/120/EC in December 2008, why it is important to automatically detect early warning signs of an upcoming outbreak by use of electronic devices. However, implementation of early warning systems is associated with high costs, stimulating genetic approaches. From a genetic perspective, the multi-factorial effects on tail biting suggest multi-factorial analyses considering breeding, husbandry, and feeding aspects simultaneously, addressing genotype x environment or more specifically genotype x feeding interactions. An indirect breeding approach is selection of pigs with short tails. Generally, due to the quite large direct heritability for tail length in pigs, the possibility of breeding pigs with shorter tails is a promising approach to reduce tail biting without stress, trauma, and pain, but implies small antagonistic effects on weight traits.

**Keywords**

Tail biting · Tail lesions · Tail length · Swine inflammation and necrosis syndrome (SINS) · Genotype x feeding interactions · Actor–victim associations

## 1 General Problem and Background

Tail biting in pigs is a widespread and severe abnormal behavior in pigs indicating a multi-factorial origin, meaning, that it does not have a single cause. It describes a behavior including gentle oral manipulation, chewing, and biting the tail of others, with consequences ranging from skin wounds with possible further infections to partial or total losses of the tail and even death (e.g., Taylor et al. 2010). Furthermore, weight gain is reduced in bitten pigs (Sinisalo et al. 2012) and it results in carcass losses for producers (Kritas and Morrison 2007). Hence, it causes problems addressing animal welfare and economy, mainly in intensive production systems (e.g., Thodberg et al. 2018). Nevertheless, in lower prevalence it has also been observed in out-

S. König (✉) · K. Halli
Institute of Animal Breeding and Genetics, University of Gießen, Giessen, Germany
e-mail: sven.koenig@agrar.uni-giessen.de; kathrin.halli@agrar.uni-giessen.de



J. J. Gross (ed.), *Production Diseases in Farm Animals*,
https://doi.org/10.1007/978-3-031-51788-4_22

door herds (Walker and Bilkei 2006) and under organic conditions (Hansson et al. 2000). One main problem is, that it occurs sporadically and in unpredictable outbreaks (Blackshaw 1981) with a wide range of possibly triggering factors addressing environmental enrichment, social factors, feeding, climate, genetics, and others (D'Eath et al. 2014). Another problem is, that it can spread within the group very quickly (Blackshaw 1981) with rapidly increasing damage (Zonderland et al. 2010), as the bleeding tail causes further biting, either by the principal biter or by other pen-mates. Tail biting can further be categorized into three different types of tail biting. (1) The "two-stage tail biting" is defined by two different stages, the "pre-damage stage" and the "damage stage." During the "pre-damage stage" one pigs' tail is manipulated gently by a pen-mate without any visible damage while during the "damage stage" the manipulation results in a bleeding tail. (2) The "sudden-forceful tail biting" is a forceful biting process without a prior period of gentle manipulation of the tail which can result in a full amputation of the bitten pigs' tail, and (3) The "obsessive tail biting" is defined as forceful bitings which appear repeatedly with a fanatical character and which can also result in a full amputation of the bitten pigs' tail. However, all three different types of tail biting afford the removement of the biting and the bitten individuals (Taylor et al. 2010).

One routine procedure to restrict tail biting on farms is tail docking in piglets. This removal of the distal part of the tail is performed a few days after birth and it usually reduces the likelihood and severity of tail biting (Sutherland and Tucker 2011). Tail docking techniques include (1) the surgical cutting off by use of a sharp knife, (2) severing the tail using a heated cautery, and (3) tail docking by use of a rubber ring which is applied to the tail (Sutherland et al. 2008). However, all techniques cause acute stress response, trauma, and pain, why tail docking is also an issue addressing animal welfare (Sutherland et al. 2008), resulting in growing criticism and concerns from the general public. Not least because it does not eliminate the problem.

In the EU, the "routinely" use of tail docking was banned by the EU Council Directive 2008/120/EC in December 2008, which focuses on restrictions of non-curative invasions. However, there is a huge gap between signals by law and reality, as the directive allows tail docking if "there is evidence that injuries… to other pigs' ears or tails have occurred." Also, there is the permission, that "other measures shall be taken to prevent tail biting… taking into account environment and stocking densities." Therefore, for 95% or more pigs in European countries, such as Germany, Denmark, France, and others, tail docking still continues (D'Eath et al. 2014; EFSA 2007).

## 2 Factors Affecting Tail Biting in Pigs

The complexity of the problem hampers the understanding of the underlying motivation for tail biting. However, some factors are widely accepted to increase the incidence of tail biting. Such factors include enrichment, the floor, air quality, climate, and feed. It can also be social factors such as stocking density or group size, age, and weight. Further factors with impact on the incidence of tail biting are the health status, and characteristics of the pig, including sex and breed (Sonoda et al. 2013; D'Eath et al. 2014).

The most important risk factor for tail biting is the lack of environmental enrichment (EFSA 2007), which relates to the availability of manipulable material attracting pigs' attention (Holm et al. 2008) and which has the potential to reduce tail biting. According to the EU Council Directive (2001/93/EC), suitable material enabling "proper investigation and manipulation activities are straw, hay, wood, sawdust, mushroom compost, peat or a mixture of such, which does not compromise the health of the animals." However, the provision of manipulable materials, such as straw, are problematic for farmers with common indoor production systems, as their main use systems rely on slatted floors and liquid slurry handling via pumps (D'Eath et al. 2014). Objects of non-destructible material seem not sufficient to comply with the EU Council Directive, as pigs lose interest in them very quickly (Van de Perre et al. 2011). The Scientific Veterinary Committee of the European Commission (SVC 1997) also recommended bedding material or earth floors suitable for rooting. A straw–bed floor, in comparison with a part-slatted floor, reduces tail biting, especially during the finishing period (Van de Weerd et al. 2005). Another environmental factor affecting tail biting is a high level of ammonia in the air (>10 ppm) (Sonoda et al. 2013). Also, extreme climate, either low or high temperatures (Geers et al. 1989), rapidly appearing temperature changes and draughts as well as heat stress are likely to cause increasing incidences of tail biting (D'Eath et al. 2014). Climate is especially a problem, as most indoor production systems are limited in their capacity to cool the buildings during summer or heating during winter. A possibility to reduce heat stress during the hot summer months is ventilation. However, according to Hunter et al. (2001), the kind of ventilation used makes a significant difference. They found artificial ventilation to increase the probability of tail biting when compared to natural ventilation.

The aspect of feeding with its impact on the appearance of tail biting is versatile and includes food quality, quantity, space and restriction, and feed type or type of presentation. Inadequate diets in terms of proteins (Fraser et al. 1991) or minerals (Fraser 1987) attract pigs to blood. Furthermore, high energy density and exposure to various pathogens build up endotoxins causing ear or tail necrosis. Such necrosis causes itchiness (Jaeger 2013) why affected pigs may tolerate the investigation and biting by pen-mates. The effect of restriction of feeder space is not clear, as Hansen et al. (1982) found an increasing impact of space restriction on tail biting, while Georgsson and Svendsen (2001, 2002) did not find an effect. In sows, a restriction of feed, instead of feeder space, as well as an impaired foraging behavior are considered as the main points causing stereotypic oral behaviors (Lawrence and Terlouw 1993). Regarding tail biting in weaner pigs, it seems obvious, that tail biting results from the pigs' natural need of rooting and chewing (Sonoda et al. 2013). This behavior is especially important while foraging, but also while exploring their environment, is insufficient when liquid diets are fed (Temple et al. 2012). However, pigs fed pelleted feed showed higher levels of tail injury when compared to pigs fed meal or liquid feed (Hunter et al. 2001).

One social factor with a possible impact on tail biting is stocking density or group size. In Bracke et al. (2013) and Wallgren et al. (2016), farmers indicated overcrowding to be an important risk factor for tail biting. However, Larsen et al. (2018) and Smulders et al. (2008) did not find a significant impact of stocking density on the hazard of tail damage. As tail biting occurs mainly during the finishing period, age and weight are further important influencing factors. Generally, the frequency of tail biting seems to increase when the pigs grow older and heavier (Schrøder-Petersen and Simonsen 2001), which could be connected to a higher level of aggressivity once the pigs have reached a critical age and weight (Hansen et al. 1982). This critical age seems to be at around 90–130 days (Haske-Cornelius et al. 1979; Sambraus 1985; Blackshaw 1981) and at a weight of around 40–50 kg (Schrøder-Petersen and Simonsen 2001).

Furthermore, the health status of the victim and the biter are important factors for tail biting, as a bad health condition can act as a stressor on the animal (Schrøder-Petersen and Simonsen 2001). For example, in herds with higher levels of respiratory problems, the levels of damage at the tails are higher (Edwards 2011). Also, leg disorders and tail damage were highly correlated (Niemi et al. 2012). External parasites, such as mange mites (Colyer 1970) or the bacterium *Streptococcus haemolyticus,* as well as mange and lice are defined as a cause. Mange and lice cause an exudate, which encourages pen-mates to investigate the affected areas. *Streptococcus haemolyticus* causes anemia, predisposing pigs to behaviors such as tail biting (Fritschen and Hogg 1996).

Several studies reported that female pigs are more likely the biters, while castrated males are more likely the victims (Kritas and Morrison

2004, 2007; Wallgren and Lindahl 1996). Nevertheless, there is speculation among scientists that female pigs have a higher motivation for manipulative behavior and a higher interest in ano-genital investigation, once they become sexually mature (Sambraus 1985; Schrøder-Petersen and Simonsen 2001). Both, breed differences in the propensity to become tail bitten, as well as breed differences in the propensity to bite, have been reported (Westin 2003). Nevertheless, the evidence is fairly weak, as there was only a small number of animals which were identified as biters, what impaires the power of genetic anayses. Other studies did not find breed differences for tail biting (Lund and Simonsen 2000; Guy et al. 2002).

## 3 Possibilities of Phenotyping

An alternative to tail docking is to detect an outbreak of tail biting as soon as possible or, even better, before it begins, in order to prevent the outbreak. Once an outbreak has started, it is important to separate the biters from the group. However, it is a lot easier to identify the victims, compared to the biters (Hunter et al. 2001). As leaving the bitten pigs in the group could attract other pen-mates to the blood, it is suggested to separate the bitten pigs as well (Zonderland et al. 2008). Four different signs have been identified as early warning signs, which appear in the days or weeks before an outbreak: (1) increasing general activity (Zonderland et al. 2011), (2) increasing "tail in mouth" behavior (Schrøder-Petersen and Simonsen 2001), (3) down held tails (Zonderland 2010), and (4) changing feeding patterns (Wallenbeck and Keeling 2013). All these signs, however, must be detected, ideally automatically. One possibility to differentiate between the different roles during tail biting is stress measurement, as stress can be a cause for tail biting in biters. Biters are characterized by slightly flattened daytime cortisol patterns and more performed sniffing when compared to victims (Munsterhjelm et al. 2013). The increasing activity before an outbreak could further automatically be detected by "optical flow," a method quantifying overall pixel changes in a video from one moment to the next and which has already been used to detect the reduced movement activity of lame broiler chickens (Dawkins et al. 2009). Furthermore, devices, such as electronic ID-ear tags combined with location detectors could be used, as they also have the potential to detect changes in activity (D'Eath et al. 2014). Differences in heart rate variability between pigs that have performed tail biting, pigs that were victims of tail biting and pigs that have not been involved in tail biting at all were proven by Zupan et al. (2012). In a project called "TailTech," the Scotland's Rural College (SRUC) in Edinburgh is working with industry to develop a 3D-Camera to detect changes in pigtail postures in order to predict an outbreak of tail biting (Scotland's Rural College (SRUC) 2022). The camera is placed above an automatic feeder, controlling pigtail postures several times a day, hence, comparing whether the tail is curled or is held down. Results showed an increasing amount of downheld tails in the time before the first appearance of tail biting (Interessengemeinschaft der Schweinehalter Deutschlands e.V. (ISN) 2022). Nevertheless, electronic devices for activity measurement, location detectors, heart rate belts, or 3D cameras are costly as well, why it remains difficult for farmers to control the problem by tail docking alternatives. However, in regards to address the problem from a breeding perspective (selection against aggressive behavior), it is important to find reliable methods helping to identify biters.

## 4 Genetic Parameters for Tail Biting in Pigs: The Actor and Victim Perspective

The most sustainable selection strategy should focus on the actor perspective, i.e., to eliminate pigs with aggressive behavior. As already mentioned, identification of aggressive pigs implies comprehensive video images, which are very difficult to analyze, or a combination of modern video techniques with complex machine learning algorithms (D'Eath et al. 2014). On the other hand, selection against aggressive individuals might change the social structure in herds, initiating new behavior patterns of formerly calm animals, e.g., stimulating aggressiveness. The group composition reflecting the social component is a crucial effect regarding tail biting in pigs, as identified by Hessel et al. (2006) when comparing the keeping of sows in single compartments or in groups with other sows for farrowing, or by Van der Zande (2017) when applying piglet cross-fostering. An alternative strategy to minimize biting behavior might be group selection, and/or consideration of actors and victims simultaneously in genetic-statistical models with social interaction effects (Canario et al. 2020). Such modelling allows the estimation of correlations between direct genetic effects for biting and social group characteristics, which mostly were close to zero. The close to zero correlations indicate no increasing risk of pigs to receive any bitings when eliminating the aggressive biters with unfavorable breeding values for biting (Canario and Flatres-Grall 2018). The social effect in the group explained up to 93% of the genetic variation for biting, depending on the environmental (feeding and housing) conditions (Canario and Flatres-Grall 2018). Bijma et al. (2007) emphasized the importance of the covariance for residual effects between the individual of interest and the group members in genetic parameter estimations for both victims and actors, reflecting the experienced social interactions. The direct heritability for clinical tail biting in Landrace pigs was 0.05 for a binary trait definition, and 0.27 when applying linear models (Breuer et al. 2005). In a Large White nucleus population, the heritability for biting was very close to zero. Van der Zande (2017) estimated heritability of 0.14 for the binary trait tail damage in models with social group effects. Additional consideration of permanent maternal effects reflecting stress of hormonal imbalances of the sow contributed to an increase of direct genetic variances for tail biting. From a breeding perspective, unfavorable genetic correlations between clinical tail biting and breeding goal traits were estimated, e.g., 0.27 between biting and growth rates (Breuer et al. 2005) indicating improved growth for aggressive pigs.

Regarding genomic studies, 19 potential candidate genes for social behavior and curiosity expressed differentially in pigs characterized as either calm or as aggressive biters (Brunberg et al. 2013a). Brunberg et al. (2013b) studied gene expression patterns, and they assumed susceptibility to abnormal behavior including tail biting only for specific phenotypes.

## 5 Indirect Selection on Tail Necrosis and Tail Lesions

In addition to primary tail biting causing injured tails due to behavioral disturbance or aggressiveness (Taylor et al. 2010), secondary biting of predamaged tails was observed, because the victims perceive the gnawing of the isochemically altered tails as pleasant. Furthermore, Blowey and Done (2003) and Santi et al. (2008) detected tail lesions that were independent of any actions from other pigs. A new syndrome as detected and described by Reiner et al. (2019) and Reiner and Lechner (2019) is defined as "swine inflammation and necrosis syndrome (SINS)." SINS contributes to signs of clinical inflammation and dead tissue in the acral regions but is independent from tail biting. Tail lesions due to SINS can appear in suckling piglets within the first days of life. First SINS indicators are inflammatory changes, continuing with tail necrosis, especially at the base and tip of the tail. In a German research experiment, more than 50% of all litters showed signs of SINS (Reiner and Lechner 2019). The clinical signs for a piglet displaying SINS are illustrated in Fig. 1.

Primary as well as secondary biting was phenotypically assessed via skin lesions and tail necrosis. Respective scores for lesions and necrosis were used in ongoing quantitative genetic analyses. A comprehensive study in this regard was carried out by Aikins-Wilson et al. (2021), additionally investigating the genetic correlations with breeding goal traits. Trait recording for lesions and necrosis focussed on a longitudinal data structure, considering scores from the rearing and the fattening period. The scoring scheme for tail lesions considered three categories: 1 = completely healthy tail without any abnor-

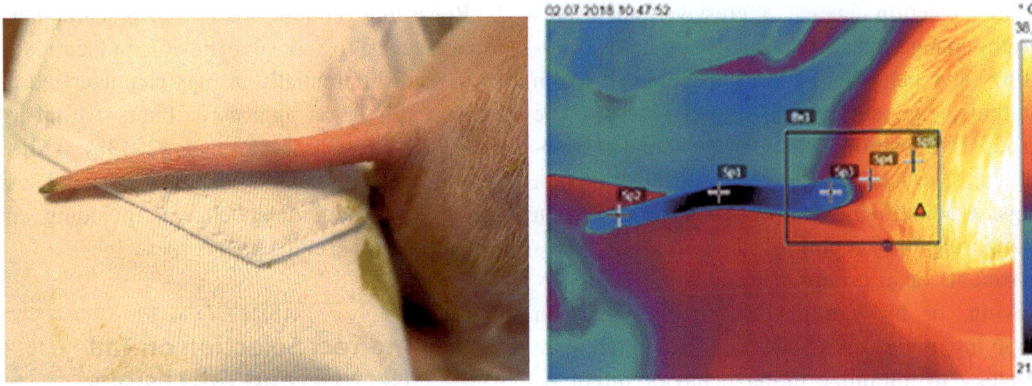

**Fig. 1** (*Left*) SINS piglet from the first day of life with mild inflammatory symptoms (redness and swelling of the base of the tail). (*Right*) The preparation with a FLIR thermographic camera and evaluation of the temperature gradients/breaks proves the connection with significant circulatory (Reiner and Lechner 2019)

malities, 2 = partly tail losses with mild lesions, 3 = complete tail losses with severe lesions. For genetic analyses based on threshold methodology, a binary data structure was created, i.e., with a score = 0 for the completely healthy tails without any abnormalities and a score = 1 for the remaining cases (i.e., combining the animals from categories 2 and 3). Tail scoring was conducted at 30 ± 7 days after entry for rearing, at 50 ± 7 days after entry for rearing (end of rearing period) and at 130 ± 20 days after entry for rearing (end of fattening period). Trait scoring was performed by a single trained person. The 1648 recorded pigs were offspring of crosses from German Landrace sows and Pietrain boars. The genetic-statistical models for the estimation of genetic parameters for tail lesions on the underlying liability scale as well as on the observed scale were (a) repeatability animal models considering lesion scores from all recording dates simultaneously and (b) single trait animal models considering the lesion scores from all recording dates as different traits in consecutive runs. The genetic relationship matrix considered relationships from a comprehensive pedigree traced back to four generations. Further fixed and random model effects included the rearing and/or fattening bay, the piglet supplier, the initial tail assessment at the entry for rearing, the random group effect, and the random permanent environmental effect in the repeatability model. The genetic parameters for tail lesions from this study (Aikins-Wilson et al. 2021) are displayed in Table 1. Heritabilities for tail lesions as presented in Table 1 were in a broad range from 0.01 to 0.39, depending on the recording date and the genetic-statistical modeling approach. Surprisingly, heritabilities from the linear models on the observed scale were throughout larger than from the threshold models on the underlying liability scale. Usually, larger heritabilities from threshold than from linear models are in line with the theory for the analysis of categorical data (Dempster and Lerner 1950). In agreement with the estimates by Aikins-Wilson et al. (2021), the linear model heritabilities for piglet skin lesions were larger than the heritabilities from the threshold model (König von Borstel et al. 2018). The larger heritabilities for lesion scores in pigs on the observed than on the underlying liability scale are very unusual. The additional actor effect on diseases recorded in victims (e.g., tail or skin lesions) might be an explanation in this regard. Nevertheless, several researchers suggested selection strategies based on tail or skin lesions (e.g., Gentz et al. 2019), especially in the context of group selection or when applying genetic evaluation models with social interactions (e.g., Heidaritaba et al. 2019).

Turner et al. (2008) estimated quite large genetic correlations between skin lesions and pig behavior traits, indicating possible favorable indirect selection response in pig aggressiveness. Statham et al. (2009) strongly pointed out the

**Table 1** Genetic parameters ($h^2$ = heritability, $\sigma_a^2$ = additive-genetic variance, $\sigma_{group}^2$ = variance for the group effect, $\sigma_e^2$ = residual variance) for tail lesions from different ages, and for linear (LIN) and threshold (TH) model applications considering single trait or repeated measurements (adopted from Aikins-Wilson et al. 2021)

| Trait[a] | Model | $h^2 \pm SE$ | $\sigma_a^2$ | $\sigma_{group}^2$ | $\sigma_e^2$ |
|---|---|---|---|---|---|
| T-LES-1 | LIN/single trait | 0.18 ± 0.14 | 0.041 | 0.088 | 0.102 |
| T-LES-1 | TH/single trait | 0.01 ± 0.07 | 0.062 | 4.105 | 3.290 |
| T-LES-2 | LIN/single trait | 0.46 ± 0.15 | 0.112 | 0.059 | 0.073 |
| T-LES-2 | TH/single trait | 0.11 ± 0.06 | 0.714 | 2.455 | 3.290 |
| T-LES-3 | LIN/single trait | 0.39 ± 0.19 | 0.084 | 0.048 | 0.083 |
| T-LES-3 | TH /single trait | 0.10 ± 0.09 | 0.587 | 2.257 | 3.290 |
| T-LES | LIN/repeatability | 0.23 ± 0.04 | 0.044 | 0.026 | 0.120 |
| T-LES | TH/repeatability | 0.21 ± 0.06 | 1.646 | 2.898 | 3.290 |

[a]T-LES-1: 30 ± 7 days after entry for rearing; T-LES-2: 50 ± 7 days after entry for rearing (end of rearing period); T-LES-3: 130 ± 20 days after entry for rearing (end of fattening period); T-LES: repeated measurements for tail lesions as used for the repeatability model

effect of tail lesions on the outbreak of tail biting. In conclusion, several effects contribute to tail and skin lesions in pigs, including SINS as well as abnormal behavior. Inferring those effects based on most appropriate phenotyping strategies remains a major challenge. Aikins-Wilson et al. (2021) correlated the estimated breeding values for tail lesions from the different recording dates with the breeding values from the repeatability model. For the same tail lesion recording date, correlations between breeding values from linear and threshold models were throughout larger than 0.90. Scoring of tail lesions is time and labor intensive, suggesting one optimal scoring date. In this regard, the largest correlations with breeding values from the repeatability model were calculated when considering the breeding values from the scoring 50 ± 7 days after entry for rearing. Breeding value correlations between tail lesions and other breeding goal traits (average daily gain, weights at different ages) were close to zero (Aikins-Wilson et al. 2019). Regarding possible genetic-statistical model enhancements, structural equation models have the potential to inferring mutual relationships among traits, i.e., to distinguish between causes (biting) and response variables (tail and skin lesions). König et al. (2008) extended recursive structural equation models by including threshold methodology for categorical traits, in order to understand the genetic background of disease resistance on the underlying liability scale. As a further novelty, Rehbein et al. (2013) enhanced structural equation models via detailed specification of animal-associated environmental effects.

The multi-factorial effects on tail biting, tail necrosis, and tail lesions suggest multi-factorial analyses considering breeding, husbandry, and feeding aspects simultaneously. From a genetics perspective, such a topic addresses possible genotype x environment interactions, or especially for behavior traits in pigs, genotype x feeding (diet) interactions. In such context, nutrition played a major role in tail and ear biting with associated skin lesions, i.e., diet particularities stimulated foraging, exploratory, and social behavior in growing pigs (Holling et al. 2017; Jensen and Pedersen 2010; Beattie et al. 2005). Diets enriched with herbs, or diets with lower protein contents contributed to favorable behavior patterns, less aggressiveness, and fewer tail and skin lesions (Jensen et al. 1993; Kallabis and Kaufmann 2012). With regard to genotype x feeding interactions in pigs, mostly production traits were investigated. Hamilton et al. (2003) identified re-rankings of sires for high heritability carcass and meat quality traits in different production environments, mainly due to differences in nutritional characteristics. Accordingly, Augspurger et al. (2002) highlighted the genotype-specific nutrient requirements in growing pigs and addressed possible genotype x feeding interactions for feed intake and feed efficiency. The aim of an implemented feeding experiment conducted at the research station Oberer Hardthof at the University of Giesen (Aikins-Wilson et al. 2022) was to infer possible genotype x diet interactions for skin lesions in pigs. Cross-classified in this context means an equal distribution of offspring per boar across feeding groups, including a conventional (CON) diet group and a diet enriched with herbs (HD). Lesion scores at age week nine in the post-weaning period were assigned according to Bünger et al. (2015) as follows: Score 1 = no lesion or less than five slight skin lesions; score 2 = mild lesion with more than five mild lesions including hair loss, redness, irritation, scratches, or small abrasions; score 3 = severe lesion, bleeding and loss of tissue. The HD supplement contributed to fewer skin lesions, probably due to the favorable effect on pig behavior. For genetic-statistical analyses via multiple-trait threshold models, a binary data transformation was performed, with score = 0 (no lesions) or score = 1 for any kind of an observed lesion. In genetic analyses, lesion scores (LS) recorded in the two different dietary treatment groups CON or HD were considered as two different traits with $LS_{CON}$ and $LS_{HD}$. The genetic correlation between $LS_{CON}$ and $LS_{HD}$ was 0.35, indicating very obvious genotype x diet interactions for skin lesions in pigs when considering the genetic correlation threshold of 0.80 as suggested by Robertson (1959). Interestingly, the genetic correlations for moderate heritability daily gain and post-weaning weights recorded in both dietary groups CON and HD were close to 0.80. A decline of genetic correlation estimates with decreasing trait heritabilities, i.e., especially small genetic correlations for health or disease indicator traits recorded in different production systems, was proved by Yin and König (2018). Such artifacts or trait-specific genomic mechanisms as observed in pigs as well as in cattle need to be clarified in ongoing studies. From a genetics perspective, increased additive-genetic variances indicated a more pronounced genetic differentiation in the more challenging CON environment. Accordingly, the standard deviation of estimated breeding values of boars was larger for $LS_{CON}$ than for $LS_{HD}$. Generally, the results from the study by Aikins-Wilson et al. (2022) suggest consideration of genotype x feeding interactions in genetic evaluations for behavior and disease indicator traits in swine, and to select specific boars for different feeding environments.

## 6 Selection of Tail Length

Long tails are a major trigger for tail biting, suggesting to evaluate the breeding possibilities on pigs with shorter tails. Short-tailedness occurs in vertebrates across many species, although the genetic causes of this phenomenon are heterogeneous with additive polygenic (e.g., Oltenacu and Boylan 1974) as well as Mendelian mono- and oligogenic factors (e.g., Buckingham et al. 2013). In general, variations in tail length, over generations as well as spontaneous, are not uncommon in vertebrates. Phylogenetic studies indicate that the shortening or loss of the tail evolved independently in different animal species, although the evolutionary "shortening mechanisms" not yet have been clarified (Buckingham et al. 2013). Most of the quantitative-genetic analyses for tail length were carried out in sheep with quite large heritabilities, e.g., 0.77 in Finnsheep (Oltenacu and Boylan 1974). For different breeds, James et al. (1991) reported heritabilities for tail length in the range from 0.39 to 0.77. Scobie and O'Connell (2002) implemented a crossbreeding experiment, and they estimated moderate dominance effects for tail length. Recently, comprehensive genetic analyses for tail length in pigs considering direct as well as maternal genetic effects were carried out by Aikins-Wilson et al. (2019). Further traits in this study for the estimation of covariances and genetic correlations with tail length were birth weight, weaning weight, post-weaning weight, and average daily gain. The direct heritability for tail length was 0.42. The maternal heritability for tail length was 0.05, indicating uterine influence on tail characteristics. Maternal genetic influence through uterine nutrition status and supply levels, uterus capacity, and milk production was detected for piglet weights and weight gains (Kaufmann et al. 2000; Alves et al. 2018). Consequently, Aikins-Wilson et al. (2019) hypothesized uterine impact on piglet conformation traits including vertebrae characteristics, as shown by Haverkamp et al. (2015) for the number of dorsal vertebras and tail length in a Merino sheep population. The maternal permanent environmental effect reflecting the common litter environment contributed to less than 5% of the phenotypic variations. Pronounced negative and antagonistic relationships were estimated between direct and maternal genetic effects for tail length with a correlation of −0.35. The genetic correlation for direct genetic effects between tail length with birth weight was 0.40, and close to zero with the remaining weight traits and with average daily gain. Accordingly, positive but weak correlations were estimated between maternal genetic effects for tail length with the maternal genetic effects of the breeding goal traits.

The quite large heritability for tail length indicates major gene effects. The T-gene is the most prominent gene associated with variations in tail length, but also with deformations of the spine in various vertebrates, e.g., mice, zebrafish, dogs, cats, and cattle (Buckingham et al. 2013; Fujimoto et al. 1995; Haworth et al. 2001; Kispert and Herrmann 1994; Odenthal et al. 1996). The T-gene encodes a transcription factor specifically expressed during the early embryonic stage, influencing the mesoderm development and the differentiation of the chorda dorsalis (Kispert and Herrmann 1994). The phenotypic effects of mutations in the T-gene were discovered a long time ago in mice (Dobrovolskaia-Zavadskaia 1929). Decades later, Herrmann et al. (1990) clarified the molecular genetic background. Initially, a deletion of 200 kb was identified as a causal candidate mutation for extremely short tails, spanning the entire transcription unit of the gene. In a homozygous state, this deletion causes embryonic early death. Wu et al. (2010) associated mutations in the T-Gene with tail length in vital mice populations. In cats, mutations in the T-gene are responsible for short-tailed breeds, such as the Manx cats. In this regard, Buckingham et al. (2013) identified a mutation in exons 8 and 9 of the feline T-gene, causing short tails as well as changes in spine structures. Breeds with genetically shortened tails also occur in dogs (e.g., Pembroke Welsh Corgis), with varying phenotypes, i.e., partially born completely without or with shortened tails (Hytönen et al. 2009). Hytönen et al. (2009) based their analyses on data from 23 different dog breeds. They identified a mutation in exon 1 of the canine T-gene, which was associated with short-tailed phenotypes as well as embryonic mortality (Hytönen et al. 2009). In cattle, Kromik et al. (2015) identified a T-gene influence on the number of cervical and tail vertebrae, as well as on the vertebral shape. A T-gene sequence analysis of the affected animals detected a spontaneous synonymous point mutation (c.196A > G) within the highly conserved T-box, causing spinal anomaly (i.e., vertebral and spinal dysplasia (VSD)). Kromik et al. (2015) gave first evidence for the direct influence of the T-protein (Brachyury) on the number of cervical vertebrae. In addition to the T-gene influence, Greco et al. (1996) and Wilm et al. (1998) identified gender-specific QTL effects on tail length in mice on chromosomes 9 and 15. Several other genes, such as *Pax1* and *Wnt-3a*, influenced tail length in mice (Greco et al. 1996; Wilm et al. 1998). In Asian cats, tail length variations were due to the influence of the *HES7* gene (Xu et al. 2016).

Mono- and oligogenic impact on tail length was associated with pleiotropic effects on tail conformation traits including lethal factors. The T-gene was identified as a major gene influencing tail length, but causing serious pleiotropic effects (Herrmann et al. 1990). Thus, the reported mutations in the T-gene also induced early embryonic death in mice, altered spine structures in cats and cattle (Buckingham et al. 2013; Kromik et al. 2015), and increased embryonic mortality in dogs and sheep (Hytönen et al. 2009; DeForest and Basrur 1979).

## 7 Comparative Genome Analyses Involving Other Species

The identified mutations or major gene effects on tail characteristics and tail length in other species suggest comparative genome analyses including dense genomic data from pigs.

The main paradigm of comparative genomics is that common or similar phenotypic traits in different species are likely to be encoded by evolutionary highly conserved functional DNA variants or motifs (Alföldi and Lindblad-Toh 2013). Highly conserved motifs across multiple (distant) species imply a functional effect (i.e., similarity due to evolutionary pressures on essential biological functions) (Alföldi and Lindblad-Toh 2013). In contrast, non-conserved variants are potentially functional, as novel species- or lineage-specific variants. Hence, comparative genomic methods allow differentiating between old and new, conserved and species-specific functional variants. Comparative genomic studies focusing on livestock species, in particular, pigs and sheep, are scarcely available (e.g., studies by Jann et al. 2009; Rubin et al. 2012). So far, most analyses in this regard have been carried out in humans and model organisms (e.g., in mice and rats) (Alföldi and Lindblad-Toh 2013; Coan et al. 2017). Furthermore, classical comparative studies addressed evolutionary aspects. In this regard, inference of functional gene models based on conservation across species (overview by Alföldi and Lindblad-Toh (2013)). Coan et al. (2017) validated functional gene mechanisms via expression analyses and trait associations in different species. Due to the recent advances in sequencing technologies, novel comparative studies in livestock species based on whole-genome sequencing (WGS) technologies. In this regard, WGS is utilized in the framework of multiple alignment methods, in order to distinguish between variants conserved across closely related (sub)species or breeds in cattle and pigs (Herrero-Medrano et al. 2014; Medugorac et al. 2017). Kim et al. (2012) used an innovative integrated approach combining genome-wide associations with genomic alignments. In this regard, Kim et al. (2012) focused on a large-scale genomic alignment to identify conserved variants in pigs and humans, being significantly associated with obesity traits (i.e., subscapular skinfold thickness and back-fat thickness).

## References


Aikins-Wilson S, Scheper C, Bohlouli M, Yin T, Engel P, König S (2019) Maternal and direct genetic parameters for tail length, tail abnormalities and breeding goal traits in pigs. In: Annual meeting of the German Society of Animal Breeding, Giessen, 11-12.09.2019

Aikins-Wilson S, Bohlouli M, König S (2021) Maternal and direct genetic parameters for tail length, tail lesions, and growth traits in pigs. J Anim Sci 99:1–11

Aikins-Wilson S, Bohlouli M, Engel P, König S (2022) Effects of an herbal diet, diet x boar line and diet x genotype interactions on skin lesions and on growth performance in post-weaning pigs using a cross-classified experiment. Livest Sci 263:105010

Alföldi J, Lindblad-Toh K (2013) Comparative genomics as a tool to understand evolution and disease. Genome Res 23(7):1063–1068

Alves K, Schenkel FS, Brito LF, Robinson A (2018) Estimation of direct and maternal genetic parameters for individual birth weight, weaning weight, and probe weight in Yorkshire and landrace pigs. J Anim Sci 96:2567–2578

Augspurger NR, Ellis M, Hamilton DN, Wolter BF, Beverly JL, Wilson ER (2002) The effect of sire line on the feeding patterns of grow-finish pigs. Appl Anim Behav Sci 75:103–114

Beattie VE, Breuer K, O'Connell NE, Sneddon IA, Mercer JT, Rance KA, Sutcliffe MEM, Edwards SA (2005) Factors identifying pigs predisposed to tail biting. Anim Sci 80:307–312

Bijma P, Muir WM, Ellen ED, Wolf JB, Van Arendonk JAM (2007) Multilevel selection 2: estimating the genetic parameters determining inheritance and response to selection. Genetics 175:289–299

Blackshaw JK (1981) Some behavioural deviations in weaned domestic pigs: persistent inguinal nose thrusting, and tail and ear biting. Anim Prod 33:325–332

Blowey R, Done SH (2003) Tail necrosis in pigs. Pig J 51:155–163

Bracke MBM, De Lauwere CC, Wind SMM, Zonerland JJ (2013) Attitudes of Dutch pig farmers towards tail biting and tail docking. J Agric Environ Ethics 26(4):847–868

Breuer K, Sutcliffe MEM, Mercer JT, Rance KA, O'Connell NE, Sneddon IA, Edwards SA (2005) Heritability of clinical tail-biting and its relation to performance traits. Livest Prod Sci 93:87–94

Brunberg E, Jensen P, Isaksson A, Keeling LJ (2013a) Brain gene expression differences are associated with abnormal tail biting behavior in pigs. Genes Brain Behav 12:275–281

Brunberg E, Jensen P, Isaksson A, Keeling LJ (2013b) Behavioural and brain gene expression profiling in pigs during tail biting outbreaks – evidence of a tail biting resistant phenotype. PLoS One 8(6):e66513

Buckingham KJ, McMillin MJ, Brassil MM, Shively KM, Magnaye KM, Cortes A, Weinmann AS, Lyons LA,

Bamshad MJ (2013) Multiple mutant T alleles cause haploinsufficiency of brachyury and short tails in Manx cats. Mamm Genome 24(9-10):400–408

Bünger B, Schrader L, Schrade H, Zacharias B (2015) Agonistic behaviour, skin lesions and activity pattern of entire male, female and castrated male finishing pigs. Appl Anim Behav Sci 171:64–68

Canario L, Flatres-Grall L (2018) Genetics of tail-biting receipt in gilts from the Tai Zumu line. In: Proceedings of the world congress on genetics applied to livestock production, Vancouver, p 572

Canario L, Bijma P, David I, Camerlink I, Martin A, Rauw WM, Flatres-Grall L, Van der Zande L, Turner SP, Larzul C, Rydhmer L (2020) Prospects for the analysis and reduction of damaging behaviour in group-housed livestock, with application to pig breeding. Front Genet 11:611073

Coan PM, Hummel O, Garcia Diaz A, Barrier M, Alfazema N, Norsworthy PJ, Pravenec M, Petretto E, Hübner N, Aitman TJ (2017) Genetic, physiological and comparative genomic studies of hypertension and insulin resistance in the spontaneously hypertensive rat. Dis Model Mech 10(3):297–306

Colyer RJ (1970) Tail biting in pigs. Agriculture 77:215–218

D'Eath RB, Arnott G, Turner SP, Jensen T, Lahrmann HP, Busch ME, Niemi JK, Lawrence AB, Sandøe P (2014) Injurious tail biting in pigs: how can it be controlled in existing systems without tail docking? Animal 8(9):1479–1497

Dawkins MS, Lee HJ, Waitt CD, Roberts SJ (2009) Optical flow patterns in broiler chicken flocks as automated measures of behaviour and gait. Appl Anim Behav Sci 119:203–209

DeForest ME, Basrur PK (1979) Malformations and the Manx syndrome in cats. Can Vet J 20(11):304–314

Dempster ER, Lerner M (1950) Heritability of threshold characters. Genetics 35:212–286

Dobrovolskaia-Zavadskaia N (1929) The problem of species in view of the origin of some new forms in mice. Biol Rev 4(4):327–351

Edwards S (2011) What do we know about tail biting today? Pig J 66:81–86

EFSA (2007) Scientific report on the risks associated with tail biting in pigs and possible means to reduce the need for tail docking considering the different housing and husbandry systems (question no. EFSA-Q-2006-013). Annex EFSA J 611:1–13

Fraser D (1987) Mineral-deficient diets and the pig's attraction to blood: implications for tail-biting. Can J Anim Sci 67:909–918

Fraser D, Bernon DE, Ball RO (1991) Enhanced attraction to blood by pigs with inadequate dietary-protein supplementation. Can J Anim Sci 71:611–619

Fritschen R, Hogg A (1996) Preventing tail biting in swine (anti-comfort syndrome). Cooperative Extension Institute of Agriculture and Natural Resources, University of Nebraska – Lincoln

Fujimoto A, Wakasugi N, Tomita T (1995) A novel partial t haplotype with a brachyury-independent effect on tail phenotype. Mamm Genome 6(6):396–400

Geers R, Dellaert B, Goedseels V, Hoogerbrugge A, Vranken E, Maes F, Berckmans D (1989) An assessment of optimal air temperatures in pig houses by the quantification of behavioral and health-related problems. Anim Prod 48:571–578

Gentz M, Lange A, Zeidler S, Traulsen I (2019) Classification of pigs with tail lesions from different farrowing and rearing systems during rearing and fattening period. Animals 9:949

Georgsson L, Svendsen J (2001) One or two feeders for groups of 16 growing-finishing pigs: effects on health and production. Acta Agric Scand Sect A Anim Sci 51:257–264

Georgsson L, Svendsen J (2002) Degree of competition at feeding differentially affects behavior and performance of group-housed growing-finishing pigs of different relative weights. J Anim Sci 80:376–383

Greco TL, Takada S, Newhouse MM, McMahon JA, McMahon AP, Camper SA (1996) Analysis of the vestigial tail mutation demonstrates that Wnt-3a gene dosage regulates mouse axial development. Genes Dev 10(3):313–324

Guy JH, Rowlinson P, Chadwick JP, Ellis M (2002) Behaviour of two genotypes of growing/finishing pig in three different housing systems. Appl Anim Behav Sci 75:193–206

Hamilton DN, Ellis M, Wolter BF, McKeith FK, Wilson ER (2003) Carcass and meat quality characteristics of the progeny of two swine sire lines reared under differing environmental conditions. Meat Sci 63:257–263

Hansen LL, Hagelsø AM, Madsen A (1982) Behavioural results and performance of bacon pigs fed ad libitum from one or several self-feeders. Appl Anim Ethol 8:307–333

Hansson I, Hamilton C, Ekman T, Forslund K (2000) Carcass quality in certified organic production compared with conventional livestock production. J Vet Med Ser B Infect Dis Vet Public Health 47:111–120

Haske-Cornelius H, Von Bogner H, Pescheke W (1979) Untersuchungen zum Verhalten von Mastschweinen in verschiedenen Stallsystemen unter besonderer Berücksichtigung des Schwanz- und Ohrenbeissens. Bayerisches landwirtschaftliches Jahrbuch 56:162–200

Haverkamp H, Brandt H, Erhardt G (2015) Variation in der Schwanzlänge beim Merinolandschaf als Basis für züchterische Aktivitäten. In: Vortragstagung der DGfZ/GfT, September 16-17, Berlin, Germany

Haworth K, Putt W, Cattanach B, Breen M, Binns M, Lingaas F, Edwards YH (2001) Canine homolog of the T-box transcription factor T; failure of the protein to bind to its DNA target leads to a short-tail phenotype. Mamm Genome 12(3):212–218

Heidaritaba M, Bijma P, Janss L, Bortuluzzi C, Nielsen HM, Madsen P, Ask B, Christensen OF (2019) Models with indirect genetic effects depending on group sizes: a simulation study assessing the precision of the estimates of the dilution parameter. Genet Sel Evol 51(1):1–10

Herrero-Medrano JM, Megens H-J, Groenen MA, Bosse M, Pérez-Enciso M, Crooijmans RP (2014) Whole-genome sequence analysis reveals differences in population management and selection of European low-input pig breeds. BMC Genomics 15:601

Herrmann BG, Labeit S, Poustka A, King TR, Lehrach H (1990) Cloning of the T gene required in mesoderm formation in the mouse. Nature 343(6259):617–622

Hessel EF, Reiners K, Van den Weghe HF (2006) Socializing piglets before weaning: effects on behavior of lactating sows, pre- and postweaning behavior, and performance of piglets. J Anim Sci 84:2847–2855

Holling C, Beilage EG, Vidondo B, Nathues C (2017) Provision of straw by a foraging tower - effect on tail biting in weaners and fattening pigs. Porcine Health Manag 3:1–14

Holm L, Jensen MB, Pedersen LJ, Ladewig J (2008) The importance of a wood feedback in rooting materials for pigs measured by double demand curves with and without a common scaling factor. Appl Anim Behav Sci 111:68–84

Hunter EJ, Jones TA, Guise HJ, Penny RHC, Hoste S (2001) The relationship between tail biting in pigs, docking procedure and other management practices. Vet J 161:72–79

Hytönen MK, Grall A, Hédan B, Dréano S, Seguin SJ, Delattre D, Thomas A, Galibert F, Paulin L, Lohi H, Sainio K, André C (2009) Ancestral T-box mutation is present in many, but not all, short-tailed dog breeds. J Hered 100(2):236–240

Interessengemeinschaft der Schweinehalter Deutschlands e.V. (ISN) (2022) Schwanzbeissen per 3D-Kamera erkennen. https://www.schweine.net/news/forschung-schwanzbeissen-per-3d-kamera-erkennen.html. Accessed 17 May 2022

Jaeger F (2013) Das Projekt "intakter Ringelschwanz" beim Schwein – stehen wir vor dem Durchbruch? (the project "intact curly tails" in pigs – are we just before a breakthrough?). Tierarztl Umsch 68:3–11

James PJ, Ponzoni RW, Gare DR, Cockrum KS (1991) Inheritance of short tailedness in south Australian merinos. Proc Austral Assoc Anim Breed Genet 9:404–407

Jann OC, King A, Corrales NL, Anderson SI, Jensen K, Ait-Ali T, Tang H, Wu C, Cockett NE, Archibald AL, Glass EJ (2009) Comparative genomics of toll-like receptor signalling in five species. BMC Genomics 10:216

Jensen MB, Pedersen LJ (2010) Effects of feeding level and access to rooting material on behaviour of growing pigs in situations with reduced feeding space and delayed feeding. Appl Anim Behav Sci 123:1–6

Jensen MB, Kyriazakis I, Lawrence AB (1993) The activity and straw directed behaviour of pigs offered foods with different crude protein content. Appl Anim Behav Sci 37:211–221

Kallabis KE, Kaufmann O (2012) Effect of a high-fibre diet on the feeding behaviour of fattening pigs. Arch Anim Breed 55:272–284

Kaufmann D, Hofer A, Bidanel JP, Kunzi N (2000) Genetic parameters for individual birth and weaning weight and for litter size of Large White pigs. J Anim Breed Genet 117:121–128

Kim J, Lee T, Kim T-H, Lee K-T, Kim H (2012) An integrated approach of comparative genomics and heritability analysis of pig and human on obesity trait: evidence for candidate genes on human chromosome 2. BMC Genomics 13:711

Kispert A, Herrmann BG (1994) Immunohistochemical analysis of the Brachyury protein in wild-type and mutant mouse embryos. Dev Biol 161(1):179–193

König von Borstel U, Tönepöhl B, Appel AK, Voß B, Brandt H, Naderi S, Gauly M (2018) Suitability of traits related to aggression and handleability for integration into pig breeding programmes: genetic parameters and comparison between Gaussian and binary trait specifications. PLoS One 13(12):e0204211

König S, Wu X, Gianola D, Heringstad B, Simianer H (2008) Exploration of relationships between claw disorders and milk yield in Holstein cows via recursive linear and threshold models. J Dairy Sci 81:395–406

Kritas SK, Morrison RB (2004) An observational study on tail biting in commercial grower-finisher barns. J Swine Health Prod 12:17–22

Kritas SK, Morrison RB (2007) Relationships between tail biting in pigs and disease lesions and condemnations at slaughter. Vet Rec 160:149–152

Kromik A, Ulrich R, Kusenda M, Tipold A, Stein VM, Hellige M, Dziallas P, Hadlich F, Widmann P, Goldammer T, Baumgärtner W, Rehage J, Segelke D, Weikard R, Kühn C (2015) The mammalian cervical vertebrae blueprint depends on the T (brachyury) gene. Genetics 199(3):873–883

Larsen MLV, Andersen HM-L, Pedersen LJ (2018) Which is the most preventive measure against tail damage in finisher pigs: tail docking, straw provision or lowered stocking density? Animal 12(6):1260–1267

Lawrence AB, Terlouw EMC (1993) A review of behavioural factors involved in the development and continued performance of stereotypic behaviors in pigs. J Anim Sci 71:2815–2825

Lund A, Simonsen HB (2000) Aggression and stimulus-directed activities in two breeds of finishing pig. Pig J 45:123–130

Medugorac I, Graf A, Grohs C, Rothammer S, Zagdsuren Y, Gladyr E, Zinovieva N, Barbieri J, Seichter D, Russ I, Eggen A, Hellenthal G, Brem G, Blum H, Krebs S, Capitan A (2017) Whole-genome analysis of introgressive hybridization and characterization of the bovine legacy of Mongolian yaks. Nat Genet 49(3):470–475

Munsterhjelm C, Brunberg E, Heinonen M, Keeling L, Valros A (2013) Stress measures in tail biters and bitten pigs in a matched case-control study. Anim Welf 22:331–338

Niemi JK, Sinisalo A, Valros A, Heinonen M (2012) Hännänpurenta – syy vai seuraus? (tail biting – cause or effect?). Suomen Maataloustieteellisen Seuran Tiedote 28:1–5

Odenthal J, Haffter P, Vogelsang E, Brand M, van Eeden FJ, Furutani-Seiki M, Granato M, Hammerschmidt M, Heisenberg CP, Jiang YJ, Kane DA, Kelsh RN, Mullins MC, Warga RM, Allende ML, Weinberg ES, Nüsslein-Volhard C (1996) Mutations affecting the formation of the notochord in the zebrafish, Danio rerio. Development 123:103–115

Oltenacu EA, Boylan WJ (1974) Inheritance of tail length in crossbred Finnsheep. J Hered 65(6):331–334

Rehbein P, Brügemann K, Yin T, König V, Borstel U, Wu X-L, König S (2013) Inferring relationships between clinical mastitis, productivity and fertility: a recursive model application including genetics, farm associated herd management, and cow specific antibiotic treatments. Prev Vet Med 112:58–67

Reiner G, Lechner M (2019) Inflammation and necrosis syndrome (SINS) in swine. CAB Rev 2019(040):14

Reiner G, Lechner M, Eisenack A, Kallenbach K, Rau K, Müller S, Fink-Gremmels T (2019) Prevalence of an inflammation and necrosis syndrome in suckling piglets. Animal 13(9):2007–2017

Robertson A (1959) The sampling variance of the genetic correlation coefficient. Biometrics 15:469–485

Rubin C-J, Megens H-J, Martinez Barrio A, Maqbool K, Sayyab S, Schwochow D, Wang C, Carlborg Ö, Jern P, Jørgensen CB, Archibald AL, Fredholm M, Groenen MA, Andersson L (2012) Strong signatures of selection in the domestic pig genome. Proc Natl Acad Sci U S A 109(48):19529–19536

Sambraus HH (1985) Mouth-based anomalous syndromes. In: World animal science, A5, Ethology off arm animals. A comprehensive study of the behavioural features of common farm animals, pp 391–422

Santi M, Gheller NB, Mores TJ, Marques BM, Gonçalves MAD, Gava D (2008) Tail necrosis in piglets – case report. Allen D Leman Swine Conference – Recent Research Reports

Schrøder-Petersen DL, Simonsen HB (2001) Tail biting in pigs. Vet J 162:196–210

Scobie DR, O'Connell D (2002) Genetic reduction of tail length in New Zealand sheep. Proc N Z Soc Anim Prod 62:195–198

Scotsland's Rural College (SRUC) (2022) Using precicion farming tools to improve animal welfare. https://www.sruc.ac.uk/research/research-activities/animal-behaviour-welfare/. Accessed 17 May 2022

Sinisalo A, Niemi JK, Heinonen M, Valros A (2012) Tail biting and production performance in fattening pigs. Livest Sci 143:220–225

Smulders D, Autekiet V, Verbeke G, Geerst R (2008) Tail and ear biting lesions in pigs: an epidemiological study. Anim Welf 17:61–69

Sonoda LT, Fels M, Oczak M, Vranken E, Ismayilova G, Huarino M, Viazzi S, Bahr C, Berckmans D, Hartung J (2013) Tail biting in pigs – causes and management intervention startegies to reduce the behavioural disorder. A review. Berliner und Münchner Tierärztliche Wochenschrift 126(3/4):104–112

Statham P, Green L, Bichard M, Mendl M (2009) Predicting tail-biting from behaviour of pigs prior to outbreaks. Appl Anim Behav Sci 121:157–164

Sutherland MA, Tucker CB (2011) The long and short of it: a review of tail docking in farm animals. Appl Anim Behav Sci 135:179–191

Sutherland MA, Bryer PJ, Krebs N, McGlone JJ (2008) Tail docking in pigs: acute physiological and behavioural responses. Animal 2(2):292–297

SVC (Scientific Veterinary Committee) (1997) The welfare of intensively kept pigs. Report to the directorate general XXIV of the European Commission. Adopted 30 September 1997. Doc. XXIV/ScVc/0005/97. Scientific Veterinary Committee, Animal Welfare Section, Brussels

Taylor NR, Main DCJ, Mendl M, Edwards SA (2010) Tail-biting: a new perspective. Vet J 186:137–147

Temple D, Courboulay V, Velarde A, Dalmau A, Manteca X (2012) The welfare of growing pigs in five different production systems in France and Spain: assessment of health. Anim Welf 21:257–271

The Council of The European Union 2001a. Commission Directive 2001/93/EC of 9 November 2001 amending Directive 91/630/EEC laying down minimum standards of the protection of pigs. Official Journal L316, 01/12/2001

Thodberg K, Herskin MS, Jensen T, Jensen KH (2018) The effect of docking length on the risk of tail biting, tail-directed behaviour, aggression and activity level of growing pigs kept under commercial conditions. Animal 12(12):2609–2618

Turner SP, Roehe R, Mekkawy W, Farnworth MJ, Knap PW, Lawrence A (2008) Bayesian analysis of genetic associations of skin lesions and behavioural traits to identify genetic components of individual aggressiveness in pigs. Behav Genet 38:67–75

Van de Perre V, Driessen B, Van Thielen J, Verbeke G, Geers R (2011) Comparison of pig behaviour when given a sequence of enrichment objects or a chain continuously. Anim Welf 20:641–649

Van de Weerd HA, Docking CM, Day JEL, Edwads SA (2005) The development of harmful social behaviour in pigs with intact tails and different enrichment backgrounds in two housing systems. Anim Sci 80:289–298

Van der Zande L (2017) Advanced statistical analysis of tail bite scores at weaning. A short term scientific mission at GenPhySE INRA

Walker PK, Bilkei G (2006) Tail-biting in outdoor pig production. Vet J 171:367–369

Wallenbeck A, Keeling LJ (2013) Using data from electronic feeders on visit frequency and feed consumption to indicate tail biting outbreaks in commercial pig production. J Anim Sci 91:2879–2884

Wallgren P, Lindahl E (1996) The influence of tail biting on performance of fattening pigs. Acta Vet Scand 37:453–460

Wallgren T, Westin R, Gunnarsson S (2016) A survey of straw use and tail biting in Swedish pig farms rearing undocked pigs. Acta Vet Scand 58:84
Westin R (2003) Svansbitning hos gris relaterat till individuell tillväxt och ras (Tail biting in pigs related to individual growth and breed). Department of Animal Environment and Health, Sveriges Lantbruksuniversitet (SLU), Veterinärmedicinska fakulteten, Veterinärprogrammet
Wilm B, Dahl E, Peters H, Balling R, Imai K (1998) Targeted disruption of Pax1 defines its null phenotype and proves haploinsufficiency. Proc Natl Acad Sci U S A 95(15):8692–8697
Wu B, Shao Y, Chen B, Liu C, Xue Z, Wu P, Li H (2010) Identification of a novel mouse brachyury (T) allele causing a short tail mutation in mice. Cell Biochem Biophys 58(3):129–135
Xu X, Sun X, Hu X-S, Zhuang Y, Liu Y-C, Meng H, Miao L, Yu H, Luo SJ (2016) Whole genome sequencing identifies a missense mutation in HES7 associated with short tails in Asian domestic cats. Sci Rep 6:31583
Yin T, König S (2018) Heritabilities and genetic correlations in the same traits across different strata of herds created according to continuous genomic, genetic, and phenotypic descriptors. J Dairy Sci 101:2171–2186
Zonderland JJ (2010) Talking tails: quantifying the development of tail biting in pigs. PhD, Wageningen UR
Zonderland JJ, Wolthuis-Fillerup M, Van Reenen CG, Bracke MBM, Kemp B, den Hartog LA, Spoolder HAM (2008) Prevention and treatment of tail biting in weaned piglets. Appl Anim Behav Sci 110:269–281
Zonderland JJ, Kemp B, Bracke MBM, den Hartog LA, Spoolder HAM (2010) Individual piglets' contribution to the development of tail biting. Animal 5:601–607
Zonderland JJ, Schepers F, Bracke MBM, den Hartoq LA, Kemp B, Spoolder HAM (2011) Characteristics of biter and victim piglets apparent before a tail-biting outbreak. Animal 5:767–775
Zupan M, Janczak AM, Framstad T, Zanella AJ (2012) The effect of biting tails and having tails bitten in pigs. Physiol Behav 106:638–644

# Breeding for Metabolic Robustness and Longevity in Dairy Cows

Sven König and Hermann H. Swalve

**Abstract**

The relevance of longevity among the traits in dairy cattle for which genetic improvements are desirable is undisputed as welfare and economic considerations underline this importance. Longevity, again as seen in the context of many relevant traits, is a difficult trait since early selection decisions are desired but the precise longevity of an individual animal can only be derived after its death and thus at a later stage in the trajectory of time. Several approaches for modelling of longevity exist, among them are survival analysis and random regression models that by definition are highly suitable to model effects along the trajectory of time. In a broader sense, the concept of longevity has to be extended to robustness and resilience. This latter term has recently gained in popularity as in general definitions are sought that reflect an individual's ability to cope with stressors, be it individual diseases or relevant effects such as hygiene, heat stress, and energy balance. The genetic ability for a long productive life of an individual in a dairy cattle population is only inadequately described by simple measurements of longevity. In this context, cow training sets with precise recordings of diseases and other relevant effects as well as new biomarkers giving more insight into the physiological status of a cow are needed. In future dairy cattle breeding, approaches that jointly combine information on longevity, health, and resilience will be the method of choice.

**Keywords**

Robustness · Resilience · Longevity · Survival analysis · Health traits · Cow training sets

## 1 Genetic Evaluation for Longevity

In general, in livestock breeding, the longevity of animals is an important topic as the productive phase of life never starts with the birth of an animal but rather with a delay in time which usually consists of the rearing period augmented by the time an animal can stay productive. From an economic point of view, the investment made to obtain a productive animal includes the costs of mating of the parents, the costs associated with the pregnancy of the mother, the costs pertaining to the birth of an animal, and the costs of rearing

S. König (✉)
Institute of Animal Breeding and Genetics, University of Giessen, Giessen, Germany
e-mail: sven.koenig@agrar.uni-giessen.de

H. H. Swalve
Institute of Agricultural and Nutritional Sciences, University of Halle, Halle, Germany
e-mail: hermann.swalve@landw.uni-halle.de


J. J. Gross (ed.), *Production Diseases in Farm Animals*,
https://doi.org/10.1007/978-3-031-51788-4_23

of the animal itself. If, as in dairy cattle, the mother itself is a productive animal, usually the main part of the total costs to obtain a new productive animal, i.e., the fresh young cow, are costs of the rearing period. The investments made for rearing then have to be paid back within the early phase of the productive period of the animal. In dairy cattle, it is generally accepted that this payback period spans at least the first lactation of the young cow if not even a total of two lactations. It is also possible to look at the investment costs for rearing as being diluted the longer the cow stays in production. From these economic considerations, two main general conclusions can be derived: (a) It would be desirable to shorten the rearing period as much as possible to limit the costs that have to be paid back and (b) the longer the life of the productive animal, the larger the profit with animals that leave the herd early within the productive phase are most detrimental to profit. However, the two general conclusions can be and have to be challenged by the following arguments: (a) the length of the rearing period is subject to biological constraints. Firstly, age at the time of puberty has to be taken into account and secondly, a desirable age or rather weight of the rearing heifer should be identified in order to determine the optimal time for the first insemination of the rearing heifer. This latter problem not only is a problem to balance economic considerations with biological constraints pertaining to the growing heifer. The impact of the effects of age at first calving on calving ease, production in first—if not also later lactations—and longevity itself have to be considered as well. While impacts of age at first calving on calving ease and dairy production in first lactation are well known to dairy farmers and researchers, the effect of calving ease on longevity often is neglected. Cows that calve young, i.e., around 24 months in Holsteins—but not earlier—exhibit a longer productive life. Conclusion (b) as stated above, i.e., the maximization of profit by a long productive life, can also be challenged as the slaughter value of cows decreases with increasing life, production diseases of older cows are increasing in prevalence as compared to young cows and young heifers bear the most recent genetic progress in all traits that are successfully improved by genetic selection when compared to an older cow that is already productive.

In general, in all kinds of livestock and specifically in dairy cattle, longevity and approaches to improve longevity by genetic selection, are highly complex issues. Problems arise due to the following considerations:

- An exact definition of longevity is needed. It is not advisable to define longevity as the total length of life as this measurement will be highly affected by the length of the rearing period which in turn is strongly influenced by management policies. Rather, the length of productive life (LPL), measured in days or months should be used. In dairy cattle, this measurement is generally defined as the time span from the date of first calving to the date of the cow leaving the herd.
- At any given point in time and for any data originating, e.g., from dairy herd improvement agencies, three types of data exist. Firstly, the LPL of cows is precisely known as her date of first calving and her date of leaving the herd is known. Secondly, a cow may have a known date of leaving herd but not a known date of first calving. Thirdly, a cow may have a known date of first calving but not a date of leaving the herd as she is still alive at the time of analysis. These three types of data are denoted as uncensored, left-censored, and right-censored observations.
- An exception to this classification of censoring is the extraction of a "historic" data set that is defined as a very early date for which dates of first calving are recorded from this point in time onward and which covers a long time span so that all cows "have had the chance" to leave the herd. In dairy cattle, "historic" data sets should at least cover 10 full years if not substantially more years. "Historic" data sets have the intrinsic disadvantage of not referring to the actual dairy population at the time of analysis. However, they may be of some use in studying effects on longevity whose nature does not change severely over time.

- Longevity data is of binary nature as an animal at a specific point in time will be either dead or alive.
- Longevity data, even when aggregated to LPL values for all animals, quite obviously will not follow a normal distribution as some cows will leave the herd early, most cows at some peak in time, but some cows later and even some cows much later resulting in a highly skewed distribution of LPL values.
- Compared to many other trait characteristics commonly analyzed in dairy cattle breeding and also as subjected to estimations of breeding values, an analysis of longevity data besides "common" effects such as herd-year season of calving and age at first calving should account for very specific effects such as "rank in production within herd-year" and "type of herd, growing in size or declining in size." The rank of a cow in production within herd-year is a very important effect on longevity as a low-ranking cow most likely will leave the herd earlier than a high-ranking cow. It is also important to note that the effect of production has to be modelled within herd-year as herds differ in their production average and within a herd the magnitude of production defining a high rank will change over years.
- Cows that leave the herd in general will be cows for which the herd owner or manager decides that they are not "profitable" anymore and/or should be replaced by available young heifers. Often, the decision will be influenced by reasons such as infertility or a predisposition to specific, possibly even recurring disease. When analyzing dairy cattle longevity data, it should be kept in mind that the occurrence of specific diseases in the trajectory of time will be at different points in time. As an example, infertility with the meaning of not being capable of a next pregnancy will lead to a time leaving the herd at the end of the lactation while a severe metabolic disorder lacking a quick recovery most often occurs at the beginning of the lactation and thus may lead to leaving the herd early in lactation.

Longevity can be modelled in a variety of ways. In statistics, the concept of survival to a certain point in time can be modelled using a survivor or also called a survival function. It is beyond the scope of this chapter to cover all scientific aspects of survival analysis, for in-depth studying, the reader should refer to classical textbooks such as Kalbfleisch and Prentice (1980). Survival functions can be either parametric, i.e., defined by parameters, semi-parametric (e.g., Cox's regression), or non-parametric. A well-known example of the latter function is the Kaplan-Meier function (Kaplan and Meier 1958). Empirically, a survival function can simply be plotted by counting the number of individuals that are still alive across a trajectory of time. Under an exponential, parametric, survival function, the probability of failure, i.e., the death of an animal, is constant across time. Using a Weibull function, this concept can be extended to allow for constant, increasing, or decreasing hazard rates. A hazard rate is the probability of death of an individual given that it was still alive at the time interval before. For intervals, any time span can be defined, being it, e.g., days or months, or any other definition of time spans. It is obvious that for the analysis of survival in dairy cattle, a non-constant hazard rate should be assumed as animals age with time and their risk of leaving the herd increases. Largely owing to work done in the animal breeding group at Cornell University in the 1980s, most of the above aspects have been subjected to in-depth scientific analyses. Early literature to cite here is the work by Smith (1983) and Smith and Quaas (1984). This early work was extended by Vincent P. Ducrocq in his PhD dissertation (Ducrocq 1987) and numerous subsequent publications (e.g., Ducrocq et al. 1988; Ducrocq and Casella 1996; Yazdi et al. 2002). Ducrocq showed that the Hazard Model based on the Weibull distribution effectively fitted the problem of longevity in dairy cows as it is capable of accounting for the existence of censored as well as non-censored records and for non-constant hazard rates that are itself modified by time-dependent and time-independent covariates. The general model as proposed by Ducrocq

(1987) also can include random genetic effects as it is needed for genetic evaluation in a dairy cattle population. Ducrocq furthermore developed the concept of true and functional longevity to differentiate between culling for poor production and culling of cows due to reasons such as recurring diseases of an animal. True longevity thus is the ability of a cow to avoid culling no matter the reasons, and functional longevity is the ability of a cow to avoid involuntary culling, i.e., culling for reasons other than the cow's own milk production relative to the herd average in a given year. Ducrocq's approaches later were followed by numerous researchers studying longevity in dairy cattle which was strongly supported by the availability of a statistical software package for genetic evaluation in livestock based on the proportional hazard model and the Weibull function, the Survival Kit (e.g., Ducrocq and Sölkner 1998). The Survival Kit was applied for routine genetic evaluation as early as in the 1990s in countries like France, Austria, Germany, and The Netherlands.

It should be noted that for genetic evaluation of sires in a dairy population based on records of their daughters, even under an application of survival analysis methods, to obtain reliable estimates of sire's breeding values at a point in time when the first daughters of a sire just have calved for the first time will still be difficult. The reason for this is that in this case most observations will be censored. Therefore, genetic evaluation centers in the past have sought to overcome this problem by blending information based on calving dates and dates for leaving the herd with auxiliary data. Auxiliary data in this case will be information on early disease status or, in the case of mastitis, of somatic cell counts, or conformation data of the daughters that will be available early in the life of a sire. The use of conformation data as an indicator of longevity has been demonstrated several times (e.g., Buenger et al. 2001; Caraviello et al. 2004). The problem of early genetic evaluations in the life of a bull used in artificial insemination nowadays has been overcome by Genomic Selection as under this scheme an accurate prediction of a sire's genetic merit can already be made shortly after birth or even before birth while the future bull will still be an embryo. For this statement to hold, however, reference samples of adequate size to estimate genomic effects are needed.

Survival analysis in the form of a proportional hazard model in general should be regarded as a tool to study any systematic effects with a significant impact on longevity. In a setting of genetic evaluation, normally the inclusion of herd effect will account for differing management practices assuming that these will be identical within a herd. For scientific studies of management practices or even any environmental effect, an additional inclusion of such effects into the model will provide insights into the biological nature of longevity of livestock in general and specifically dairy cattle. The prerequisite for such studies obviously is the availability of environmental descriptors beyond the common effects that can be retrieved from milk recording data. An example of this is the inclusion of a type-of-housing effect pertaining to the type of the dairy barn, be it a tie-stall or cubicle-house with differing types of bedding (Buenger et al. 2001). For this study, 43,116 cows were included in a data set from 548 herds for which information on housing systems had been collected. The results are summarized in Fig. 1.

As can be seen, a bedding of some form (straw or sawdust) appears to be advantageous with respect to longevity. Tie stalls were small in numbers but also seemed to be advantageous. This latter finding might be disputed and attributed to the fact that farms with tie stalls were small in herd size and thus the effect of the system for tie stalls might be confounded with better individual care for cows in small herds. However, the results for cubicle housing are in high agreement with recommendations by dairy management extension agencies that recommend some form of bedding in cubicles.

In some countries, for some smaller populations and sometimes in scientific analyses, computationally simpler methods compared to survival analysis have been or are used. Early approaches (e.g., Everett et al. 1976) proposed using the stayability of a cow to a certain limit of lifetime or productive life in the form of a simple binary variable. Other approaches have defined survival as a continuous trait for completed lactations with projecting incomplete to complete lactations or as a binomial trait for having survived consecutive lactations (Visscher and Goddard 1995). Neither of

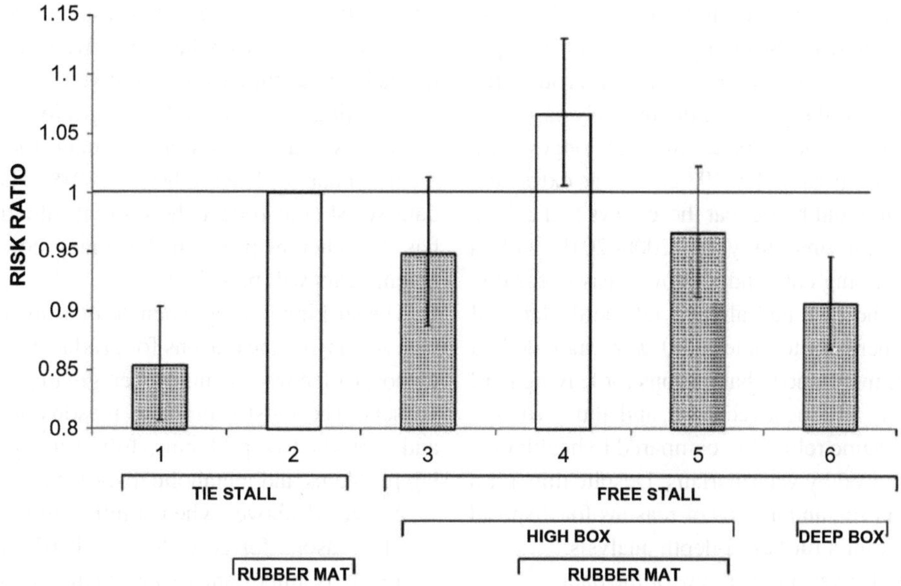


**Fig. 1** Effect of housing system (coding see Table 1) on risk ratio of cows in relation to the most frequent system (code 2; risk ratio = 1.0) for systems with (gray bars) and without (open bars) bedding (straw or sawdust) (Lines on bars indicate 95% confidence intervals)

these approaches can avoid the loss of information that arises from the definition of fixed dates at which survival is measured. Furthermore, with these simpler approaches it is not possible to adequately account for the problem of censoring.

While survival analysis applying a Weibull proportional hazard model, in general, is widely accepted as the method of choice when analyzing dairy cattle longevity data, in recent years a number of studies have raised concerns referring to specific effects found in longevity data and the capability of the proportional hazard model to account for all specificities. Attempts have been made to use alternative methods such as random regression (e.g., Gengler et al. 2005) or other models with a focus on defining specific periods of increasing hazard for cows leaving the herd for different disposal reasons (Holtsmark et al. 2009; Heise et al. 2016). One argument for a slightly sub-optimal performance of the proportional hazard model may be that different reasons for the disposal of a cow occur at different points in time as shown above. Heise et al. (2016) attempted to overcome the loss of information problem arising from defining a few fixed points in time by proposing a model with a total of 9 time periods that were defined reflecting different time spans with differing hazards due to different reasons for disposal and concluded that their model gave an improved fit to the data compared to survival analysis. Accordingly, the official genetic evaluation for longevity has been changed to the model proposed by Heise et al. in 2018.

## 2 Importance of Metabolic Disorders Among All Reasons for Disposal

In Germany, and specifically in the north-eastern state of Mecklenburg-Wester Pomerania as early as 2005 a system of contract herds was established. In these contract herds, health data, i.e., any event comprising the health of a dairy cow is recorded by herd-managers and largely based on the diagnoses of the respective veterinarian in charge of the given herd. The system was extended to more herds and two more states, in 2019, herds from the state of Brandenburg joined and in 2018, also herds from the state of Saxony-Anhalt. Although there has been some fluctuation in the number of herds participating and the extent of recording, the data as of 2022 comprises 119 large commercial herds with average an average herd size of around 700 cows. The data has been shown to be highly suitable for in-depth analysis of various research themes around the health of the dairy cow. For an analysis of the reasons for disposal of a cow and longevity, a data set comprising 279,193 cows was extracted, the requirement being that these cows had calved for the first time in years 2008–2019 with a known calving date and a known reason for disposal if the cow had already left herd, dates of leaving herd up to June 2020 were included. It has to be mentioned, that reasons for leaving herd are purely producer-recorded and thus will not have the same reliability compared to health data as supervised by veterinarians. Despite this, for a large data set, an analysis of reasons for disposal may warrant a further in-depth analysis.

Table 1 gives an overview of the data.

The three subsets correspond to three approaches to analyzing the data. The first data set merely leaves the data set as it is. However, obviously, only cows that had already left the herd can contribute to the analysis. Cows that left the herd as they were sold as breeding stock also were excluded. The second subset forms a "historic" data set of cows that calved for the first time in the years 2008–2010. For this analysis, 0.5% of all cows from the respective years had to be excluded as they were still alive in June 2020. This editing is considered as negligible. For the third subset, all cows were included that had a recorded date of leaving herd in 2019. Thus, this data set should reflect the current situation but has the inherent problem that many cows from recent years will be still alive.

The striking finding when comparing all three data sets is that the reasons for producer-recorded disposal of cows do not differ greatly between subsets. The most prominent reasons are udder and reproductive problems, followed by feet & leg problems and metabolic disorders.

As stated above, when jointly studying disposals, reasons for cows leaving herd, and longevity, it is important to be aware of different points in the life of a cow that will be in focus when looking at different diseases. This can be illustrated by a simple empirical survival function for which the frequency of LPL values is plotted for the time trajectory of a cow's life. In the "historic" data set as explained above, cows lived for a maximum of 4506 days of LPL. The following figures 2, 3, 4, 5 and 6 give these

**Table 1** Reasons for leaving herd (%) based on producer-recorded data from three subsets of data from 279,183 cows in 119 contract herds and years of first calving from 2008 to 2020 in three north-eastern states in Germany. In each subset, only cows that had a recorded date of leaving the herd contributed to the data; cows leaving herd sold as breeding stock were excluded from analysis

| Reason for leaving herd | Subset "All cows that already left herd" N = 194,131 | Subset "All cows that had a first lactation in Years 2008–2010"—Exact date of leaving herd is known—N = 55,883 | Subset "All cows that left herd in 2019" N = 22,487 |
|---|---|---|---|
| Udder problems | 22.1 | 23.0 | 20.4 |
| Reproduction (disorders and infertility) | 21.2 | 21.8 | 20.5 |
| Feet & leg problems | 16.6 | 16.0 | 17.8 |
| Metabolic disorders | 12.5 | 13.1 | 12.4 |
| Other diseases | 7.2 | 6.4 | 7.9 |
| Poor production | 7.8 | 7.0 | 7.4 |
| Milkability problems | 5.0 | 5.2 | 4.3 |
| Other reasons | 7.4 | 7.1 | 8.9 |
| Age | 0.2 | 0.4 | 0.4 |

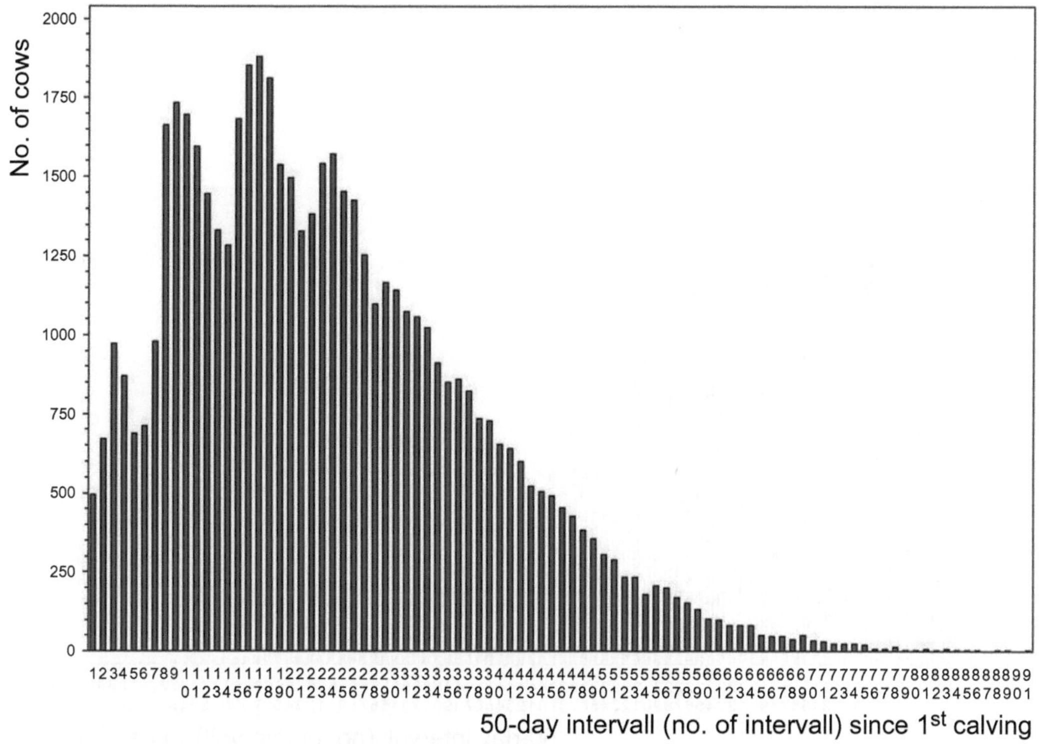


**Fig. 2** Frequency (absolute numbers) of cows leaving herd along time. $N = 55,883$ cows with a known date for leaving herd from years of first calving 2008 to 2010

empirical survival functions for all cows and subsets of this data for cows that left due to udder health, reproduction problems, feet & leg problems, and metabolic reasons.

All cows calving in the years 2008–2010 for the first time had left their herds before 4506 days of productive life, i.e., before or in the 50-day interval 91 of Fig. 2. Precisely 12.62% left herd by interval 8, i.e., before 400 days of LPL. Distinct peaks of time points of leaving herd can be found at the end of each lactation, most likely partly due to cows that failed to be pregnant again and were milked till the end of a sometimes prolonged lactation of around 400 or even more days. Furthermore, peaks after the first lactation partly are due to cows failing to be pregnant in the previous lactation and partly due to acute problems right after calving for later parities. This analysis analogously can be repeated for cows of different categories of disposal.

Figure 3 shows the pattern of time points for disposal of cows for which udder problems had been indicated as the main reason for leaving herd. Already in first lactation, a distinct peak of time for disposal is apparent for the third interval, i.e., between 100 and 150 days. In total 6.34% of all cows leave the herd before day 200 in first lactation. It has to be assumed that these cows at least partly were treated for mastitis early in their first lactation without success. Further peaks point to early and midst of lactation in later parities.

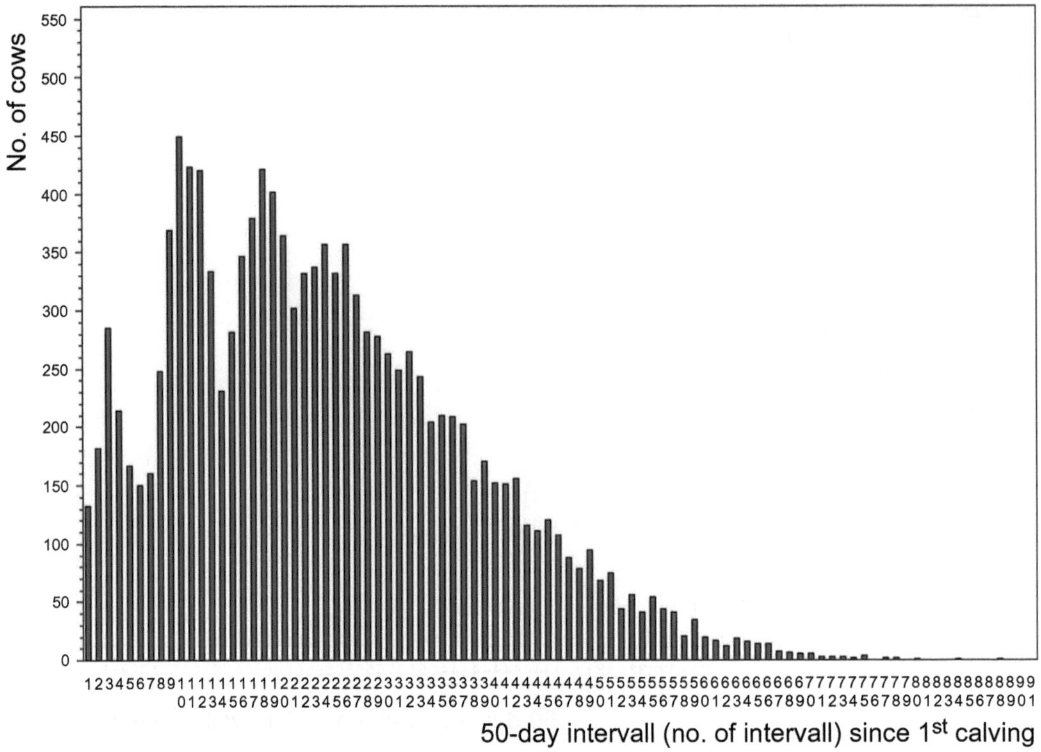


**Fig. 3** Frequency (absolute numbers) of cows leaving herd along time. $N = 12{,}830$ cows with a known date for leaving herd from years of first calving 2008 to 2010 and reason for disposal = udder problems

The disposal reason "reproductive problems" partly may have been indicated for cows not becoming pregnant again and to a lesser extent also to cows having had experienced diseases of the reproductive tract, e.g., endometritis and cystic ovaries. Figure 4 indicates that mostly the reason for disposal will have been not becoming pregnant again as the time points of the peaks are at the end of lactation.

In contrast to the other reasons for leaving herd, time points for cows that leave herd for feet & leg problems are more spread out as displayed in Fig. 5. With the exception of the first lactation, no clear peaks for points in time can be seen. In the first lactation, cows for this category leave herd around 100–200 days in lactation.

Metabolic problems, i.e., ketosis, displaced abomasum, and milk fever do not seem to strongly contribute to reasons for cows leaving herd (Fig. 6). After the first lactation, metabolic problems play a stronger role in the respective beginning of each lactation.

As mentioned before a system of contract herds that records health data of all cows along with calf weights and all standard data (milk recording data, conformation data, insemination data, etc.) has been established in the state of Mecklenburg-Western Pomerania as early as

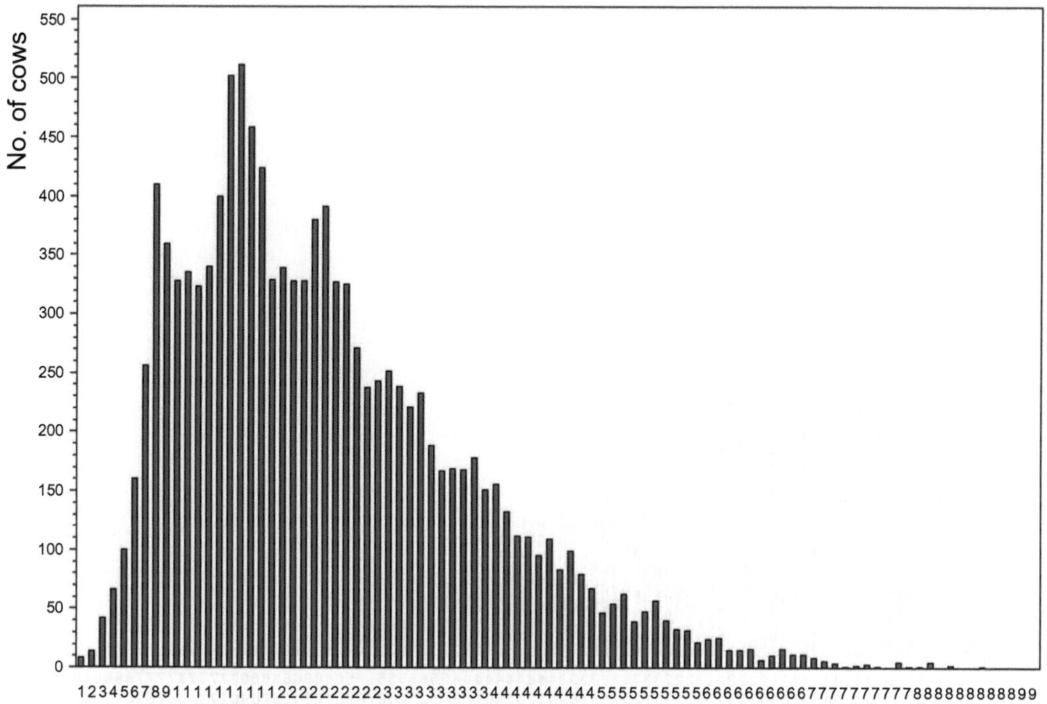


**Fig. 4** Frequency (absolute numbers) of cows leaving herd along time. $N = 12{,}172$ cows with known date for leaving herd from years of first calving 2008 to 2010 and reason for disposal = reproductive problems

2005 in Germany. In 2009, the breeding organization operating in the state of Brandenburg followed, and finally, the state of Saxony-Anhalt was included in 2018. From 2018 onward, the system of health documentation was extended to all Holstein breeding regions in Germany. As the pioneering states for health trait recording, Mecklenburg-Western Pomerania and Brandenburg meanwhile have established more than a decade of health data recording, it is worthwhile to look at the phenotypic trends. The recording of data is supervised as supervision is given by the herd veterinarians and also by the breeding organization and local research institutes for agriculture as these send out specialists to the farms on a regular basis and also are in contact with local veterinarians. All participating farms are relatively large with herd sizes of around 700 cows. Given the time span and also the fact that the participation of individual herds commenced at slightly different times within the region and also differed between regions, for the analysis as presented above, around 203,000 first lactations of individual cows from 95 herds formed the data set. Data was edited quite rigorously, i.e., records were checked for plausible dates and plausibility

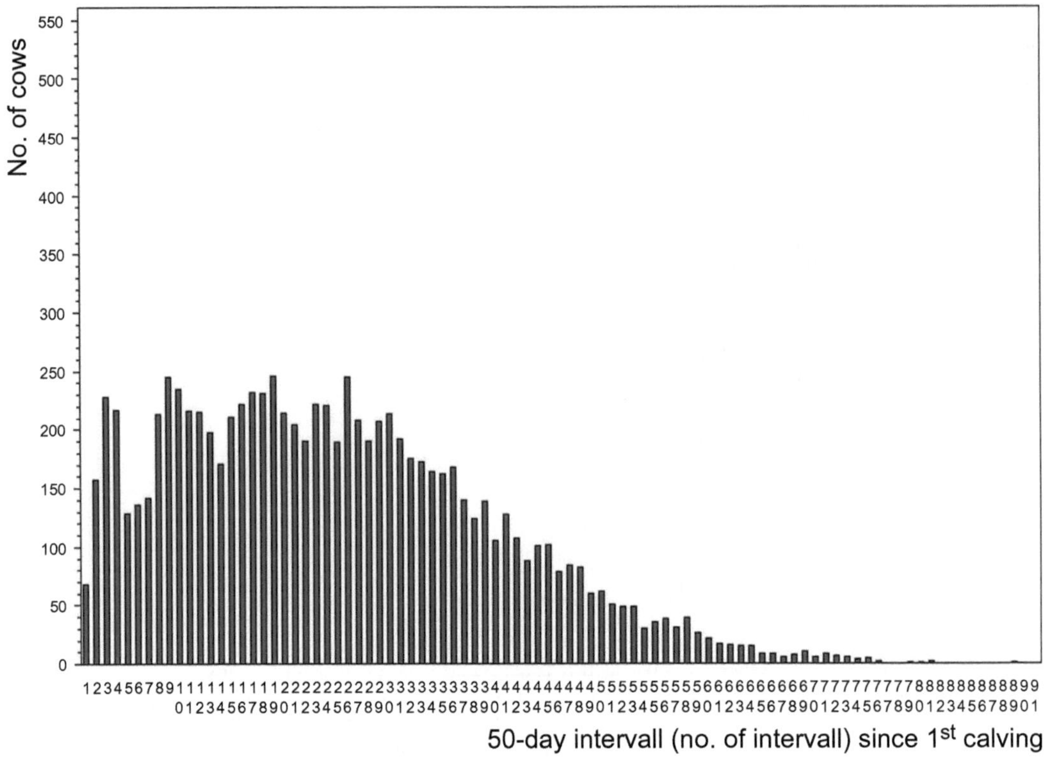


**Fig. 5** Frequency (absolute numbers) of cows leaving herd along time. $N = 8968$ cows with known date for leaving herd from years of first calving 2008 to 2010 and reason for disposal = feet & leg problems

of disease categories as well as incidences for diseases across time. As an example, time periods, i.e., years of recording for an individual herd were completely dropped from the data if the incidences differed drastically from a previous period in the way of being far lower. As a specific example, this might have been the case for recordings of feet & leg problems from observations at the time of hoof trimming if the electronic data transfer from the hoof trimming data had failed and data could not be recovered. Thus, the incidence rates for diseases after this rigorous editing still contain some accuracy issues but will be overestimated rather than underestimated.

Figure 7 presents the phenotypic trends in the 95 commercial dairy farms. While the first 3 years are somewhat affected by merging the two regions, the general trend is very clear: (1) metabolic diseases, at least in the clinical form, are of lesser importance, (2) for reproductive disorders and mastitis, a very favorable trend is observed as the initial values are about halved

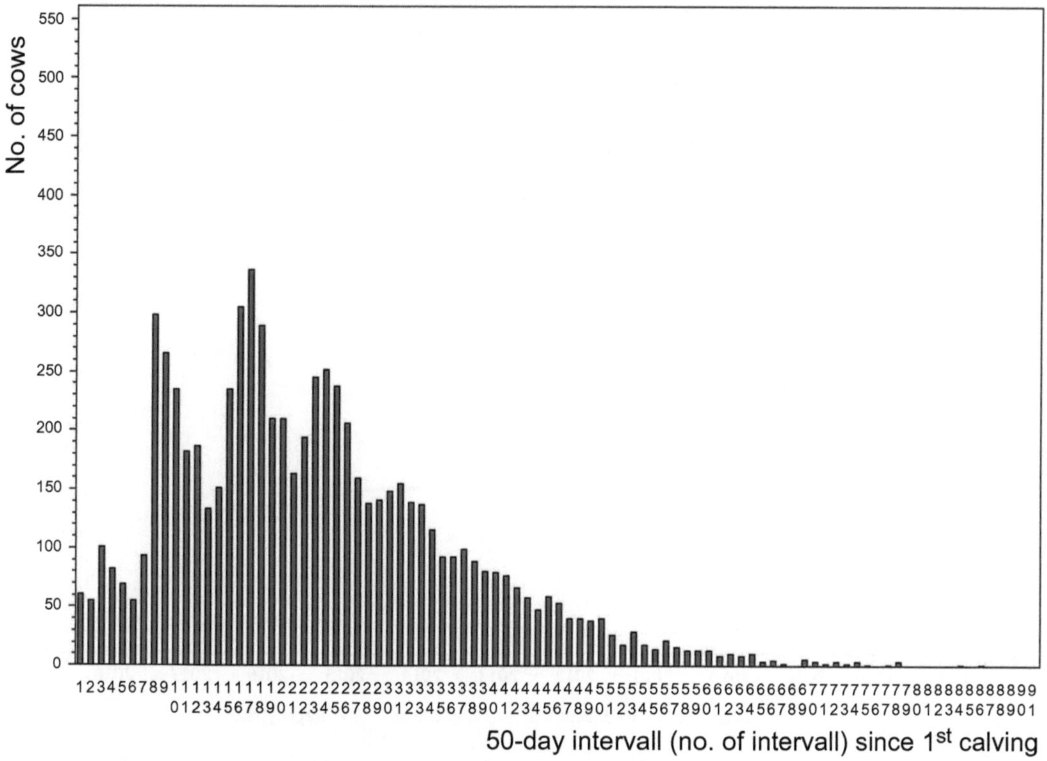


**Fig. 6** Frequency (absolute numbers) of cows leaving herd along time. $N = 7294$ cows with known date for leaving herd from years of first calving 2008 to 2010 and reason for disposal = metabolic problems

during the observation period, and (3) the trend for feet & leg problems and diseases is stable, i.e., no upward or downward trend can be detected.

Concluding from the above analysis, it has to be stated that efforts in the fields of health management as well as genetic selection should focus on improvements of hoof health while not neglecting any measurements in other health categories. The main tools in the field of hoof health are hoof trimming of the entire herd in regular intervals, immediate treatment of acutely sick animals, and data collected at the time of hoof trimming from every cow, being healthy or sick, as a basis for estimating breeding values and genetic selection as suggested by König and Swalve (2006).

Total data: 203,330 first lactations, variable analyzed is "no. of disease events per first lactation." Data edited for plausibility, herd-years with incomplete documentation were excluded.

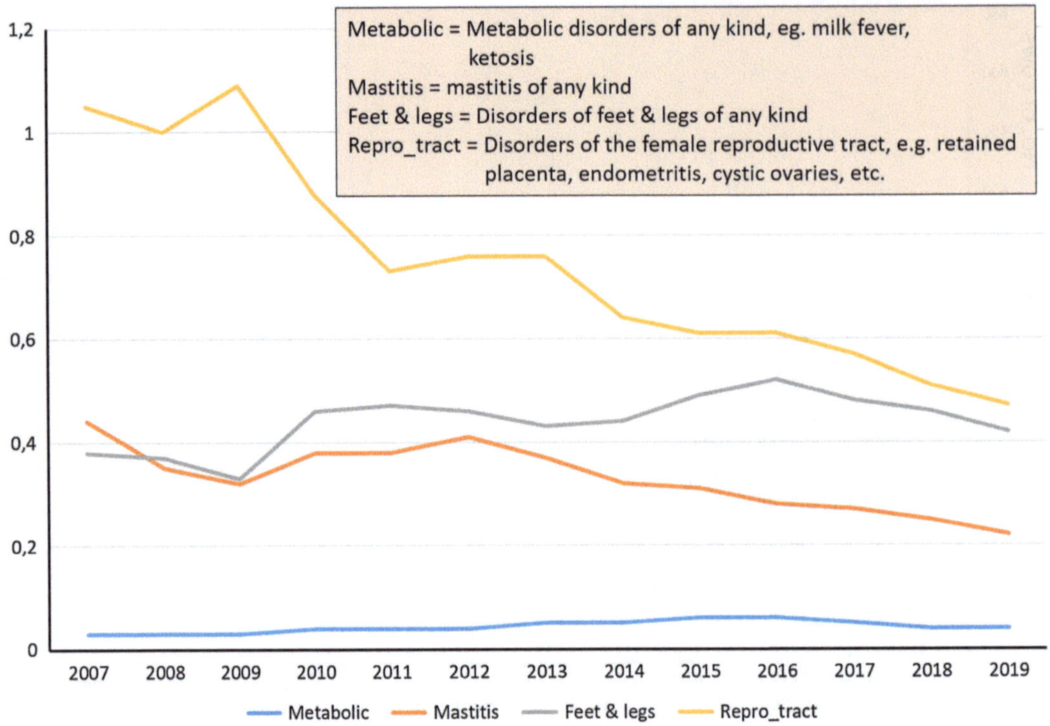


**Fig. 7** Phenotypic trend for health traits in 95 large commercial dairy farms in Mecklenburg-Western Pomerania and Brandenburg for years 2007–2019

## 3 Breeding Approaches to Improve Robustness: Classical Concepts

Breeding approaches on robustness address a currently "hot topic" in dairy cattle, especially focussing on improved animal health and longevity (König and May 2019). In the context of robustness, the term resilience reflects the cow's reactions to external environmental stressors. Classically, resilience is defined as the capability to keep a constant production level or a constant health status, independent from environmental stressors (Rauw and Gomez-Raja 2015). Furthermore, short recovery periods due to external stressors to achieve the former productivity, indicate resilient cows (Berghof et al. 2019). Cow trait reactions to environmental challenges based on repeated measurements can be depicted phenotypically and genetically, i.e., via breeding value alterations on environmental gradients or on time scales. Such strategy implies the identification of proper environmental or herd characteristics, including climatic descriptors, energy, and protein contents in the feeding rations or the herd health and hygiene status. Calus and Veerkamp (2003) used average herd performances for production, functional, and energy indicator traits as continuous gradients in genetic studies. Energy balance indicators are strongly related to disease resistance and resilience (Fikse et al. 2003). For example, impaired defense mechanisms against pathogens were reported for cows with constantly low values for body condition (König and May 2019). Consequently, the average herd body condition score and milk energy content (Calus and Veerkamp 2003) were suggested as explanatory herd variables in random regression or reaction norm models.

Random regression or reaction norm models can be used to detect robust animals and to prove possible genotypes by environmental interactions. Random regression modelling allows the estimation of genetic parameters and breeding values along the environmental or herd gradient. In this regard, Brügemann et al. (2011) identified robust cows and sires, i.e., animals with quite constant and high breeding values for protein yield along a continuous temperature–humidity index. Figure 8 shows breeding values for somatic cell score of three sires A, B, and C along the gradient for the temperature–humidity index. A robust sire in this illustration is sire A, displaying favorable and quite constant breeding values for all climatic levels.

A decline of heritabilities or genetic variances with environmental challenges indicate an environmental effect on gene activities or gene expressions, with indications for optimal test environments. Schierenbeck et al. (2011) analyzed variations of daughter yield deviations and identified a more pronounced genetic differentiation for moderate heritability production traits in superior test environments. In contrast, for low heritability health traits or health indicators such as somatic cell score, a challenging environment might contribute to a clearer separation of genetically favorable breeding sires. Similarly, with regard to feeding characteristics, Hammond (1947) suggested to keep and to select animals in environments reflecting superior nutritional value, with assumed effects of the phenotypic expression of the true genetic potential. Furthermore, random regression models allow the estimation of genetic covariances and genetic correlations between same traits across the whole grid of levels for environmental and time gradients. Classically, a genetic correlation smaller than 0.80 is considered as an indicator for genotype x environment interactions (Robertson 1959). Interestingly, for moderate heritability production traits, the genetic correlations were mostly larger than 0.80, e.g., when correlating protein yield at low with protein yield recorded at high temperatures (Brügemann et al. 2011). However, for low heritability functional traits, a substantial decline in genetic correlation estimates with increasing distances between levels of environmental descriptors were observed, including physiological traits and climate descriptors (Kipp et al. 2021a), male fertility traits of AI sires under different feeding and husbandry conditions (Al-Kanaan et al. 2015) and somatic cell score for the geographical gradients longitude and altitude (Jaeger et al. 2019). In ongoing genetic studies, Yin and König (2018) applied random regression models with phenotypic, genetic, and genomic continuous herd descriptors. Genetic herd descriptors were the percentage of foreign sires within herds, and genomic herd descriptors included levels of linkage disequilibrium and allele frequencies for molecular markers located in close physiological distances to the functional mutation. Again, for low heritability somatic cell score, genetic correlations substantially declined for herd descriptor levels at great distances. Consequently, sire breeding values altered in dependency on phenotypic, genetic, and genomic herd descriptors, suggesting specific sire selec-

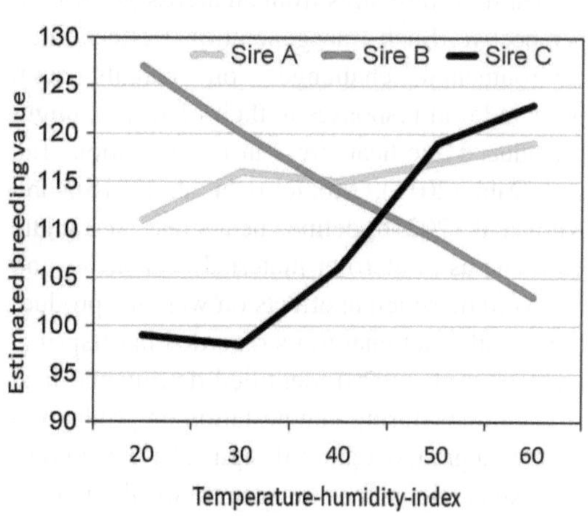


**Fig. 8** Estimated breeding values for somatic cell score of three sires A, B, and C along the continuous environmental gradient "temperature–humidity index"

tion and mating strategies for specific herd architectures.

Similarly, in reaction norm models, solutions for intercept and slope were used as genetic indicators for robustness (e.g., Streit et al. 2012). Specifically, the slope is the first derivative of the reaction norm function, indicating environmental sensitivity. Variations of slope effects indicate possible genotype-by-environment interactions, and specific genotype responses with environmental alterations (Fikse et al. 2003). In genomic studies, 2-step approaches were conducted, implying the utilization of estimated slope effects as dependent traits in ongoing genome-wide association studies (Streit et al. 2013). Hayes et al. (2009) applied reaction norm models to test-day records and modelling temperature–humidity indices as environmental descriptors. In the next step, solutions for intercept and slope were associated with SNP genotypes of sires and cows. Most of the markers significantly contributing to genotype x heat stress interactions were located on chromosomes 9 and 20. Lillehammer et al. (2009) applied reaction norm models to study production trait responses of daughters from genotyped bulls along herd means for test-day records. In this regard, the herd mean for milk yield reflected the herd feeding level. More than 30% of the significantly associated genetic markers affected intercept and slope in opposite directions, indicating that the same marker had opposite effects on environmental sensitivity and on genotype x feeding interactions.

Random regression models also can be applied to identify robust sires from an across-generation perspective. Such transgenerational effects, i.e., environmental challenges on parents with observed trait responses in their offspring, might be induced due heat stress in late gestation. Tao and Dahl (2013), Monteiro et al. (2016), and Kipp et al. (2021b) defined heat stress during late gestation as in utero or maternal heat stress, and observed detrimental effects on weights, productivity, and functional traits recorded in offspring. Laporta et al. (2020) identified detrimental heat stress effects during late gestation on production traits in a granddaughter design. The physiological explanations for transgenerational in utero heat stress effects during late gestation were the stimulation of insulin secretions and alteration of plasma hormone levels, inducing effects on the fetus's development (Bell et al. 1989; Tao and Dahl 2013). Genetically, epigenetic mechanisms inducing genetic modifications in gametes and embryos, which can be persistent across generations, were addressed by Wu and Sirard (2020). With regard to epigenetics, Skibiel et al. (2018) studied DNA patterns and identified modified liver mammary gland DNA methylation due to heat stress effects during late gestation, explaining the differentially expressed genes in offspring.

Kipp et al. (2021c) hypothesized alterations of genetic parameters and breeding values of sires due to epigenetic effects on gene expressions in offspring. Consequently, for the analyses of production and female fertility traits recorded in offspring, they applied a sire-maternal grandsire random regression model with prenatal temperature–humidity index as a continuous environmental descriptor, and additionally considered maternal genetic effects. The temperature–humidity index from late gestation reflecting the prenatal climate conditions contributed to alterations of direct genetic variances and heritabilities of production and female fertility traits in the offspring generation. Furthermore, Kipp et al. (2021c) identified alterations of estimated breeding values of sires in dependency of climatic conditions in the late dry period. Figure 9 displays the breeding values for somatic cell score of 10 influential sires with most daughters in the data set in dependency of the temperature–humidity index from the last week of pregnancy. Accordingly, also from the across-generation perspective, robust sires were identified, i.e., sires with quite constant and favorable values, independent from heat stress challenges in dams during their late dry period.

Moreover, measurements for lactation persistency reflecting the pattern of lactation curves can be used as indicators for metabolic robustness. Impaired persistency, especially high milk yield directly after calving in relation to low milk yield levels at the middle and end of lactation, is strongly related to increased susceptibility to

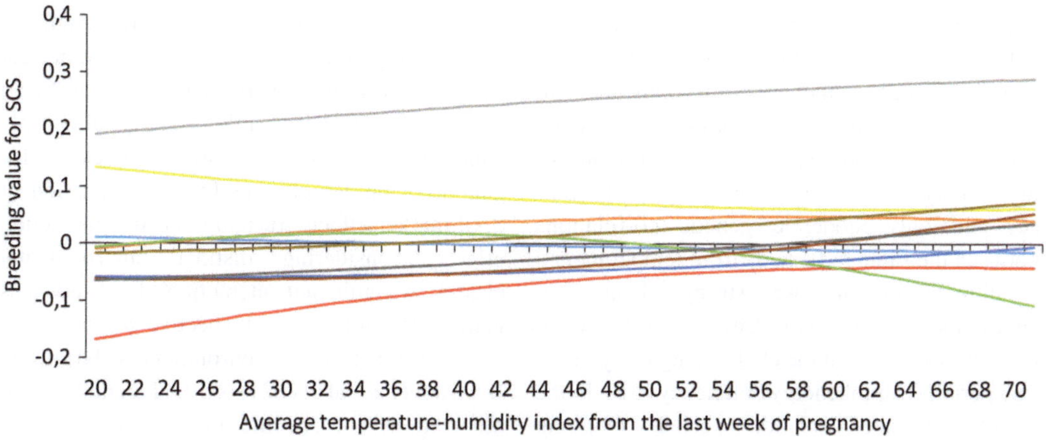


**Fig. 9** Estimated breeding values for somatic cell score of 10 sires along the continuous environmental gradient "temperature-humidity-index" from the last week of pregnancy

stress and diseases (König and May 2019). For different parameters reflecting lactation persistency, Gödert et al. (2020) estimated favorable genetic correlations between lactation persistency and the metabolic disorder ketosis in Holstein-Friesian cows, and they identified the growth hormone receptor gene as a potential candidate gene for both traits persistency and ketosis.

## 4 Breeding Approaches to Improve Robustness: Cow Training Sets and Producer Health Data

The aspects from the previous chapter addressed traits from conventional official recording schemes combined with novel environmental descriptors and enhanced genetic-statistical modelling approaches. The era of genomic selection enabled the implementation of cow training sets, including a large number of genotyped cows with phenotypic records for health traits directly reflecting robustness. Such data sets, e.g., the cow training set considering ~20,000 genotyped Holstein-Friesian cows from large-scale contract herds located in the German federal states of Berlin-Brandenburg and Mecklenburg-West Pomerania, can be used for direct association analyses between cow phenotypes and SNP genotypes (Swalve and König 2007). Originally, in the early period of genomic selection, pure sire training sets have been used (Schaeffer 2006), implying regressions of sire breeding values on sire genotypes. However, the obstacles of sire training sets are possibly biased BLUP breeding values due to pre-selection of sires (Patry and Durcocq 2009) and limited selection response for novel traits in the case of low accuracies for daughter-based breeding values (Buch et al. 2012). The utilization of data from the German cow training set contributed to the implementation of official genomic breeding value estimations for numerous traits directly reflecting robustness, especially for cow health and calf health. Major success in this regard was the implementation of official genetic evaluations for claw disorders, considering a detailed recoding scheme for dermatitis digitalis (Schöpke et al. 2015). Accordingly, a detailed recording scheme for metabolic disorders considering ß-hydroxybutyrate and acetone in milk improved genomic evaluations for ketosis (Klein et al. 2020).

Ongoing genomic studies with a focus on robustness combined the novel cow traits and dense cow genotypes with environmental descriptors. In this regard, especially prompt and time-lagged heat stress studies have been conducted to infer direct genetic, maternal genetic, and maternal sensitivity to heat stress effects (Yin et al.

2022). Bohlouli et al. (2021) indicated the potential of fatty acids combined with temperature–humidity indices as indicators for robustness and environmental sensitivity. Alterations of genetic variances for unsaturated, mono-, and polyunsaturated fatty acids by temperature–humidity indices and only moderate genetic correlations between the same fatty acids from temperate and hot climatic conditions were strong indicators for environmental sensitivity. Environmental sensitivity was stronger in the challenging early lactation period (natural energy deficiency) than in the middle or late lactation period. Ongoing approaches focused on the implementation of cow training sets specifically designed for specific production systems. In this regard, Shabalina et al. (2020) established a specific cow training set for organic herds, only including genotyped cows from organic herds. Interestingly, in genotype x environment interaction analyses considering the two systems organic and conventional, genetic correlations in the same traits across systems were quite small for low heritability udder and claw diseases.

Genomic data contributes to a deeper understanding of genetic mechanisms and gene effects on health and robustness. In this regard, Bohlouli et al. (2022) identified shared potential candidate genes and genomic pathways for different heat stress response traits underlying milk fatty acid composition in dairy cattle. Halli et al. (2021) enhanced a commercial software package for genome-wide associations for the estimation of SNP marker effects for main and for interaction effects, considering distinct environmental classes. A significant interaction effect indicates gene expressions of the respective chromosomal segment in system or environment A, but not in B. Wagner et al. (2021) studied genome-wide associations for specific major pathogens for mastitis considering the two farming systems cubicle and free-walk compost housing. Some significant SNP by housing system interactions indicate opposite effects of the same SNP markers on disease resistance in cubicle and compost, as illustrated in Fig. 10. The SNP BTA-110591-no-rs on BTA6 was significant for the pathogen *Staph. aureus* in both systems, but the SNP ARS-BFGL-NGS-16330 on BTA22 was only significant in the compost group, but not in in cubicle. Interestingly, the annotated potential candidate gene *HEMK1* plays a major role in immune system mechanisms, but depending on the environmental conditions.

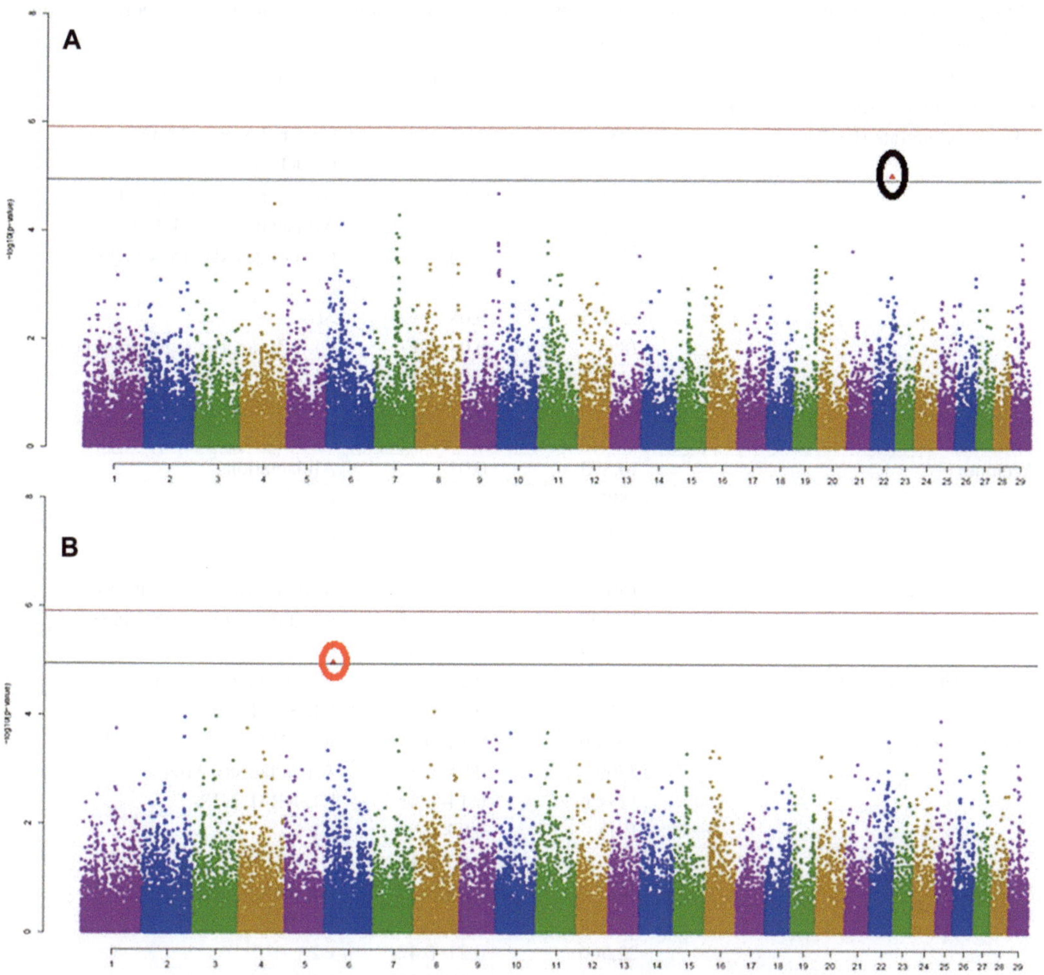


**Fig. 10** Manhattan plot for interaction effects for SNP x housing system effects (**a**) and for main effects (**b**) for *Staph. aureus*. The horizontal line indicates the Bonferroni corrected threshold, and the second horizontal line below indicates the suggestive significance threshold calculated as Psug = 0.05/(number of independent SNP)

## 5 Biomarkers for Metabolic Health Disorders

As stated by König and May (2019), biomarkers are objectively measurable indicators to infer biological and physiological mechanisms and indicate infectious or non-infectious diseases, mainly in response to environmental alterations. As outlined in Table 2, the potential biomarkers as used in genetic analyses (enzymes, serum metabolites, hormone levels, or characteristics of body fluids) have a heritable component, but heritabilities show a very wide range. With regard to metabolic disorders and metabolic stress, Piechotta et al. (2012) showed that a negative energy balance after calving was associated with changing concentrations of the growth hormone IGF-1. The pronounced genetic variation for IGF-1 contributed to a moderate heritability of 0.35 for dairy cows from Poland.

Milk and especially its fine composition, described by mid-infrared (MIR, between 4000 and 400 $cm^{-1}$) absorbance spectra of milk (Gengler et al. 2016), was identified as a very potent tool to detect heat stress (Hammami et al. 2015), and to monitor metabolic diseases such as ketosis (Bastin et al. 2013; Pryce et al. 2016). Unfortunately, acquiring MIR data is, due to price and conditions

**Table 2** Overview of heritability estimates in HF dairy cows for novel biomarkers reflecting metabolic robustness, resistance, and resilience traits (adopted from König and May 2019)

| Biomarker | No. of cows | Heritability | Comment[a] |
|---|---|---|---|
| *Metabolic disease/rumen microbiome* | | | |
| ß-Hydroxybutyrate (BHB) | 175 | 0.25 | Heifers before calving |
| | | 0.08–0.40 | L1, RRM |
| | 1615 | 0.17 | All lactations, plasma BHB |
| | 1565 | 0.16 | All lactations, milk BHB |
| | 61,331 | 0.12 | L1, first test day (5–40 DIM) |
| | 3732 | 0.13 | L1 |
| | 35,575 | 0.08 | >L1 |
| | 7895 | 0.04–0.17 | L1–L3, random regression model |
| | 14,397 | 0.06–0.15 | L1, different lactation stages |
| | 826 | 0.25–0.37 | All lactations, MIR |
| | ~ 18,000 | 0.09–0.14 | L1–4 milk |
| Acetone | 10,375 | 0.01 | All lactations |
| | 7895 | 0.29 | L1 |
| | 7895 | 0.29 | L2 |
| | 7895 | 0.22 | L3 |
| | 1565 | 0.10 | Milk acetone, all lactations |
| | 14,397 | 0.06–0.15 | L1, different lactation stages |
| | ~ 18,000 | 0.23–0.31 | L1–4 |
| Non-esterified free fatty acids | 142 | 0.29 | Heifers before calving |
| | 6015 | 0.08–0.35 | L1, RRM |
| Milk fatty acids | 26,166 | 0.18–0.44 | L1, MIR |
| | 7700 | 0.05–0.38 | All lactations, MIR |
| | 11,626 | 0.14–0.42 | L1, RRM, MIR |
| | 1918 | 0.09–0.54 | L1 |
| | 990 | 0.03–0.19 | L1–7 |
| | 44,000 | 0.26–0.59 | L1, MIR |
| | 37,768 | 0.22–0.46 | L1, MIR |
| | 22,566 | 0.21–0.42 | L2 |
| | 8221 | 0.18–0.39 | L3 |
| Phospholipids (glycerophosphocholine) | 371 | 0.48 | L1–3 |
| Glucose | 192 | 0.21 | L1–3 |
| | 174 | 0.37 | Heifers before calving |
| | 6015 | 0.12–0.39 | L1 |
| Insulin growth factor 1 (IGF-1) | 75 | 0.35 | L1 |
| *Heat stress* | | | |
| Rectal temperature | 1695 | 0.17 | All lactations |
| | 238 | 0.07 | All lactations |
| Vaginal temperature | 238 | 0.04 | All lactations |
| Skin temperature | 238 | 0.02–0.04 | All lactations |
| Respiration rate | 238 | 0.05 | All lactations |
| Pulse rate | 238 | 0.07 | All lactations |
| *Endoparasite infections* | | | |
| Gastrointestinal nematodes | | | |
| Fecal egg counts (FEC) | 1419 | 0.21 | L2–4 |
| | 4053 | 0.07 | L1–4 |
| | 1166 | 0.05–0.06 | All lactations |
| Antibody levels | 10,879 | 0.07 | All lactations |

(continued)

**Table 2** (continued)

| Biomarker | No. of cows | Heritability | Comment[a] |
|---|---|---|---|
| Liver flukes (*F. hepatica*) | | | |
| Fecal egg counts (FEC) | 1166 | 0.33 | All lactations |
| Antibody level | 10,879 | 0.13 | All lactations |

[a]AMS = automatic milking system; DIM = Days in milk; HF = Holstein-Friesian; L = lactation; RRM = random regression model; MIR = assessed by mid-infrared spectrometry

of use of the spectrometers, limited to milk recording. It is, therefore, only done approximately monthly in routine milk recording (i.e., in intervals of 4 (most common), 6 or 8 weeks depending on the sampling plans). However, acute disease and heat stress early warning systems would need near-real-time data acquisition. Research showed the feasibility of transferring the methodology based on MIR technique to a methodology based on to near-infrared (NIR) spectra (Nguyen et al. 2011), allowing data collection at each milking (Diaz-Olivares et al. 2020).

Furthermore, levels of gene expressions can be considered as molecular phenotypic traits or as biomarkers (Nachtomy et al. 2007). During metabolic stress or other stressors such as heat stress, gene expression is altered in various cell types, and a large fraction of these genes are related to immune system mechanisms (Mishra 2021). With regard to transcriptomics, an increase in gene expressions during stress periods is associated with inflammation. Most and Yates (2021) identified relationships between stressors and the number of leukocytes and levels of pro-inflammatory cytokines.

## 6 Outlook

Improving the longevity of dairy cows by genetic selection meanwhile has been the goal in many dairy populations since more than a decade. Later, systems of health data recording as a basis for estimating breeding values have been implemented in many countries. This quite often raises the question if it would be possible to replace genetic selection for longevity by selection for improving health traits, i.e., the resistance against diseases. A definite answer to this question is difficult. However, some approaches do exist that would shed light on this question. Firstly, typical breeding plan calculations considering all genetic correlations between all traits in the breeding goal can be made. The genetic progress for health and longevity then can be studied for various definitions of the breeding goal. Secondly, as a very empirical approach, top lists of sires and their average estimated breeding value for all traits can be studied by comparing success rates for longevity and health. This very empirical approach is well-liked by breeding organizations as it answers their question on 'which bulls are in the top list if we would define the breeding goal in the form of alternative A compared to B?). It is obvious that this empirical approach merely reflects precise breeding plan calculations. Thirdly, again applying the approach of a historic data set for ease of computation, a model to estimate variance components for functional longevity, i.e., accounting for voluntary culling by considering production within herd and also changes in herd size as proposed by Ducrocq et al. (1988) for fixed effects, may be used. The model, for each health trait at a time, is then augmented by sire's breeding values for health traits as well as other traits. It is also conceivable to estimate variance components under a survival model and not being limited to a historic data set.

In the analysis as presented here, a historic data set with cows calving for the first time in 2008–2011 with data available up to and including 2019 was used (Table 3) with a simple sire model. With the base model, a heritability for LPL of 0.109 was estimated which corresponds to most results from the literature. In the sire model, the estimate of heritability is 4 * sire variance/total phenotypic variance with phenotypic variance equals the sum of sire variance and residual variance. Including class levels of EBV for a number of traits, one at a time, it can be seen

**Table 3** Variance components for the sire variance and the residual variance for length of productive life (LPL, in days) for different models ($N = 71{,}098$ cows with known LPL from a "historic data set" from contract herds in Mecklenburg-Western Pomerania and Brandenburg, years of first calving 2008–2011)

| Model | $\sigma_s^2$ | $\sigma_e^2$ | % Red. $\sigma_s^2$ | $h^2$ (* 100) |
|---|---|---|---|---|
| Base model[a] | 11,810 | 419,852 | – | *10.9* |
| + EBV health index[b] | 5671 | 419,903 | 52.0 | 5.3 |
| + EBV mastitis | 7669 | 419,824 | 35.1 | 7.2 |
| + EBV feet & legs[c] | 9355 | 419,822 | 20.8 | 8.7 |
| + EBV metabolic disorder[d] | 9586 | 419,880 | 19.8 | 8.9 |
| + EBV Reproductive_tract[e] | 9107 | 419,887 | 22.9 | 8.5 |
| + EBV somatic cell score | 8655 | 419,690 | 26.7 | 8.1 |
| + EBV Milk production index[f] | 11,759 | 419,707 | 0.4 | 10.9 |
| + EBV conformation index[g] | 11,279 | 419,742 | 4.5 | 10.5 |
| + EBV reproduction index[h] | 8957 | 419,799 | 24.2 | 8.3 |

[a]Model contains herd, year, age at first calving, rank in milk production within herd-year, classes for decreasing/increasing herd size per herd-year as fixed effects and sire as a random effect
[b]Health index comprises mastitis, feet & legs, metabolic disorders, and disorders of the reproductive tract
[c]Feet & legs index comprises a total of six individual hoof diseases (Digital dermatitis, laminitis, sole ulcer, etc.)
[d]Metabolic disorders cover milk fever, ketosis, and displaced abomasum
[e]Disorders of the reproductive tract contain cystic ovaries, endometritis, and retained placenta
[f]Milk production index contains protein yield, fat yield, and protein percentage
[g]Conformation index comprises all linear conformation traits (18), but also subjectively scored composites
[h]Reproduction index covers traits pertaining to success of insemination for heifer and cow inseminations as well as days to first breeding for cows

that the residual variance is hardly affected. However, the sire variance is affected with varying degrees. For the overall health index, the sire variance is reduced by 52%. Within the following four categories of health traits, the EBV for mastitis has the strongest influence on longevity (35.1%), followed by the three other categories with values around 20%. It also can be seen, that an EBV consisting of observations of mastitis cases, as this is the case for "EBV Mastitis," has a much stronger influence on longevity than an EBV for Somatic cell score. Other non-health traits show little influence on longevity with the exception of the reproduction index. Including the production index does not change the variance components. This is expected, as the individual production ranked within herd is in the model. It underlines that the model used truly estimates functional longevity.

The above findings support the conclusion that there is more to longevity than health. Other factors besides health need to be considered if the goal is to replace genetic selection for longevity. Recently, the term resilience, although quite old from its origin has seen increasing popularity. Resilience in essence is the ability of the individual to cope with harmful effects in the specific environment of the animal. One approach in this direction is the work by Poppe et al. (2020, 2021, 2022), who used the variation of daily milk yields and also of a cow's mobility in the form of step counts as indicators for resilience. Other definitions may also be of value and eventually will result in new ways to estimate the genetic merit of dairy cattle jointly for resilience, longevity, and health.

## References

Al-Kanaan A, König S, Brügemann K (2015) Effects of heat stress on semen characteristics of Holstein bulls estimated on a continuous phenotypic and genetic scale. Livest Sci 177:15–24
Bastin C, Soyeurt H, Gengler N (2013) Genetic parameters of milk production traits and fatty acid contents in milk for Holstein cows in parity 1-3. J Anim Breed Genet 130:118–127
Bell AW, McBride BW, Slepetis R, Early RJ, Currie WB (1989) Chronic heat stress and prenatal development in sheep: I. Conceptus growth and maternal plasma hormones and metabolites. J Anim Sci 67:3289–3299
Berghof TVL, Poppe M, Mulder HA (2019) Opportunities to improve resilience in animal breeding programs. Front Genet 9:692

Bohlouli M, Yin T, Hammami H, Gengler N, König S (2021) Climate sensitivity of milk production traits and milk fatty acids in genotyped Holstein dairy cows. J Dairy Sci 104:6847–6860
Bohlouli M, Halli K, Yin T, Gengler N, König S (2022) Genome-wide associations for heat stress response suggest potential candidate genes underlying milk fatty acid composition in dairy cattle. J Dairy Sci 105:3323–3340
Brügemann K, Gernand E, Von Borstel UU, König S (2011) Genetic analyses of protein yield in dairy cows applying random regression models with time-dependent and temperature x humidity-dependent covariates. J Dairy Sci 94:4129–4139
Buch LH, Kargo M, Berg P, Lassen J, Sørensen AC (2012) The value of cows in reference populations for genomic selection of new functional traits. Animal 6:880–886
Buenger A, Ducrocq VP, Swalve HH (2001) An analysis of survival in dairy cows using supplementary data on type scores and housing systems from a region of North West Germany. J Dairy Sci 84:1531–1541
Calus MPL, Veerkamp RF (2003) Estimation of environmental sensitivity of genetic merit for milk production traits using random regression model. J Dairy Sci 86:3756–3764
Caraviello DZ, Weigel KA, Gianola D (2004) Comparison between a Weibull proportional hazard model and a linear model for predicting the genetic merit of US Jersey sires for daughter longevity. J Dairy Sci 87:1469–1476
Diaz-Olivares JA, Adriaens I, Stevens E, Saeys W, Aernouts B (2020) Online milk composition analysis with an on-farm near-infrared sensor. Comput Electron Agric 178:105734
Ducrocq VP (1987) An analysis of length of productive life in dairy cattle. Ph.D. thesis, Cornell University, Ithaca, N.Y, USA; University Microfilms Int., Ann Arbor, MI, USA
Ducrocq V, Casella G (1996) A Bayesian analysis of mixed survival models. Genet Sel Evol 28:505–529
Ducrocq V, Sölkner J (1998) "The SurvivalKit V3.0", a package for large analyses of survival data. In: Proc. 6th world Congress on genetics applied to livestock production, Armidale, vol 27, pp 447–450
Ducrocq VP, Quaas RL, Pollak EJ, Casella G (1988) Length of productive life of dairy cows. 1. Justification of a Weibull model. J Dairy Sci 71:3061–3070
Everett RW, Keown JR, Clapp EE (1976) Relationships among type, production and stayability in Holstein sire evaluation. J Dairy Sci 59:1532–1539
Fikse WF, Rekaya R, Weigel KA (2003) Assessment of environmental descriptors for studying genotype by environment interaction. Livest Prod Sci 82:223–231
Gengler N, Vanderick S, Mayeres P, Gillon A, Croquet C (2005) Genetic evaluation of cow survival using a lactation random regression model. Interbull Bull 33:176–180
Gengler N, Soyeurt H, Dehareng F, Bastin C, Colinet F, Hammami H, Vanrobays ML, Lainé A, Vanderick S, Grelet C, Vanlierde A, Froidmont E, Dardenne P (2016) Capitalizing on fine milk composition for breeding and management of dairy cows. J Dairy Sci 99:4071–4079
Gödert M, Yin T, König S, May K (2020) Maßzahlen der Laktationspersistenz bei Holsteinkühen und deren Beziehung zur Leistung, Klauen- und Stoffwechselgesundheit unter Berücksichtigung von Genotypisierungsdaten. Züchtungskunde 92(6)
Halli K, Vanvanhossou SF, Bohlouli M, König S, Yin T (2021) Identification of candidate genes on the basis of SNP by time-lagged heat stress interactions for milk production traits in German Holstein cattle. PLoS One 16(10):e0258216
Hammami H, Vandenplas J, Vanrobays ML, Rekik B, Bastin C, Gengler N (2015) Genetic analysis of heat stress effects on yield traits, udder health, and fatty acids of Walloon Holstein cows. J Dairy Sci 98:4956–4968
Hammond J (1947) Animal breeding in relation to nutrition and environmental conditions. Biol Rev 22:195–213
Hayes BJ, Bowman PJ, Chamberlain AJ, Savin K, van Tassell CP, Sonstegard TS, Goddard ME (2009) A validated genome wide association study to breed cattle adapted to an environment altered by climate change. PLoS One 4:e6676
Heise J, Liu Z, Stock KF, Rensing S, Reinhardt F, Simianer H (2016) The genetic structure of longevity in dairy cows. J Dairy Sci 99:1253–1265
Heise J, Stock KF, Reinhardt F, Ha N-T, Simianer H (2018) Phenotypic and genetic relationships between age at first calving, its component traits, and survival of heifers up to second calving. J Dairy Sci 101:425–432
Holtsmark M, Heringstad B, Odegard J (2009) Predictive abilities of different statistical models for analysis of survival in dairy cattle. J Dairy Sci 92:5730–5738
Jaeger M, Brügemann K, Naderi S, Brandt H, König S (2019) Variance heterogeneity and genotype by environment interactions in native Black and White dual-purpose cattle for different herd allocation schemes. Animal 13(10):2146–2155
Kalbfleisch JD, Prentice RL (1980) The statistical analysis of failure time data. Wiley, New York
Kaplan EL, Meier P (1958) Nonparametric estimation from incomplete observations. J Am Stat Assoc 53:457–481
Kipp C, Al-Kanaan K, Brügemann K, König S (2021a) Genetic analyses of physiological traits in relation to climatic stresses in German Black Pied cattle. Züchtungskunde 2:81–93
Kipp C, Brügemann K, Zieger P, Mütze K, Möcklinghoff-Wicke S, König S, Halli K (2021b) Across-generation effects of maternal heat stress during late gestation on production, female fertility and longevity traits in dairy cows. J Dairy Res 88:147–153
Kipp C, Brügemann K, Yin T, Halli K, König S (2021c) Genotype by heat stress interactions for production and functional traits in dairy cows from an across-generation perspective. J Dairy Sci 104:10029–10039

Klein SL, Scheper C, May K, König S (2020) Genetic and non-genetic profiling of milk FTIR β-hydroxybutyrate and milk FTIR acetone and their associations with ketosis in Holstein cows. J Dairy Sci 103:10332–10346

König S, May K (2019) Invited review: phenotyping strategies and quantitative-genetic background of resistance, tolerance and resilience associated traits in dairy cattle. Animal 13(5):897–908

König S, Swalve HH (2006) A model calculation on the prospects of an improvement of claw health in dairy cattle via genetic selection. Züchtungskunde 78:345–356

Laporta J, Ferreira FC, Ouellet V, Dado-Senn B, Almeida AK, de Vries A, Dahl GE (2020) Late-gestation heat stress impairs daughter and granddaughter lifetime performance. J Dairy Sci 103:7555–7568

Lillehammer M, Hayes BJ, Meuwissen THE, Goddard ME (2009) Gene by environment interactions for production traits in Australian dairy cattle. J Dairy Sci 92:4008–4017

Mishra SR (2021) Behavioural, physiological, neuroendocrine and molecular responses of cattle against heat stress: an updated review. Trop Anim Health Prod 53:1–40

Monteiro APA, Tao S, Thompson IMT, Dahl GE (2016) In utero heat stress decreases calf survival and performance through the first lactation. J Dairy Sci 99:8443–8450

Most MS, Yates DT (2021) Inflammatory mediation of heat stress-induced growth deficits in livestock and its potential role as a target for nutritional interventions: a review. Animals 11:3539

Nachtomy O, Shavit A, Yakhini Z (2007) Gene expression and the concept of the phenotype. Stud Hist Philos Sci C Stud Hist Philos Biol Biomed Sci 38:238–254

Nguyen HN, Dehareng F, Hammida M, Baeten V, Froidmont E, Soyeurt H, Niemöeller A, Dardenne P (2011) Potential of near infrared spectroscopy for online analysis at the milking parlour using a fibre-optic probe presentation. NIR News 22:11–13

Patry C, Durcocq V (2009) Bias due to genomic selection. Interbull Bull 40:172–177

Piechotta M, Sander AK, Kastelic JP, Wilde R, Heppelmann M, Rudolph B, Schuberth HJ, Bollwein H, Kaske M (2012) Short communication: prepartum plasma insulin-like growth factor-I concentrations based on day of insemination are lower in cows developing postpartum diseases. J Dairy Sci 95:1367–1370

Poppe M, Veerkamp RF, van Pelt ML, Mulder HA (2020) Exploration of variance, autocorrelation, and skewness of deviations from lactation curves as resilience indicators for breeding. J Dairy Sci 103:1667–1684

Poppe M, Mulder HA, Veerkamp RF (2021) Validation of resilience indicators by estimating genetic correlations among daughter groups and with yield responses to a heat wave and disturbances at herd level. J Dairy Sci 104:8094–8106

Poppe M, Mulder HA, van Pelt ML, Mullaart E, Hogeveen H, Veerkamp RF (2022) Development of resilience indicator traits based on daily step count data for dairy cattle breeding. Genet Sel Evol 54:21

Pryce JE, Gaddis KP, Koeck A, Bastin C, Adessayed M, Gengler N, Miglior F, Heringstad B, Egger-Danner C, Stock K (2016) Invited review: opportunities for genetic improvement of metabolic diseases. J Dairy Sci 99:6855–6873

Rauw WM, Gomez-Raja L (2015) Genotype by environment interaction and breeding for robustness in livestock. Front Genet 6:310

Robertson A (1959) The sampling variance of the genetic correlation coefficient. Biometrics 15:469–485

Schaeffer LR (2006) Strategy for applying genome-wide selection in dairy cattle. J Anim Breed Genet 123:218–223

Schierenbeck S, Reinhardt F, Reents R, Simianer H, König S (2011) Identification of informative cooperator herds for progeny testing based on yield deviations. J Dairy Sci 94:6143–6152

Schöpke K, Gomez A, Dunbar KA, Swalve HH, Döpfer D (2015) Investigating the genetic background of bovine digital dermatitis using improved definitions of clinical status. J Dairy Sci 98:164–8174

Shabalina T, Yin T, May K, König S (2020) Proofs for genotype by environment interactions considering pedigree and genomic data from organic and conventional cow reference populations. J Dairy Sci 104:4452–4466

Skibiel AL, Peñagaricano F, Amorín R, Ahmed BM, Dahl GE, Laporta J (2018) In utero heat stress alters the offspring epigenome. Sci Rep 8:14609

Smith SP (1983) The extension of failure time analysis to problems of animal breeding. Ph.D. thesis, Cornell University, Ithaca, N.Y, USA; University Microfilms Int., Ann Arbor, MI, USA

Smith SP, Quaas RL (1984) Productive life span of bull progeny groups: failure time analysis. J Dairy Sci 67:2999–3007

Streit M, Reinhardt F, Thaller G, Bennewitz J (2012) Reaction norms and genotype-by-environment interaction in the German Holstein dairy cattle. J Anim Breed Genet 129:380–389

Streit M, Reinhardt F, Thaller G, Bennewitz J (2013) Genome-wide association analysis to identify genotype x environment interaction for milk protein yield and level of somatic cell score as environmental descriptors in German Holsteins. J Dairy Sci 96:7318–7324

Swalve HH, König S (2007) Test herds in dairy cattle breeding programmes 1st communication: general considerations. Züchtungskunde 79:249–263

Tao S, Dahl GE (2013) Invited review: heat stress effects during late gestation on dry cows and their calves. J Dairy Sci 96:4079–4093

Visscher PM, Goddard ME (1995) Genetic parameters for milk yield, survival, workability and type traits for Australian dairy cattle. J Dairy Sci 78:205–220

Wagner P, Yin T, Brügemann K, Engel P, Weimann C, Schlez K, König S (2021) Genome-wide associations for microscopic differential somatic cell count

and specific mastitis pathogens in Holstein cows in compost-bedded pack and cubicle farming systems. Animals 1(6):1839
Wu C, Sirard MA (2020) Parental effects on epigenetic programming in gametes and embryos of dairy cows. Front Genet 11:557846. https://doi.org/10.3389/fgene.2020.557846
Yazdi MH, Visscher PM, Ducrocq VP, Thompson R (2002) Heritability, reliability of genetic evaluations and response to selection in proportional hazard models. J Dairy Sci 85:1563–1577
Yin T, König S (2018) Heritabilities and genetic correlations in the same traits across different strata of herds created according to continuous genomic, genetic, and phenotypic descriptors. J Dairy Sci 101:2171–2186
Yin T, Halli K, König S (2022) Direct genetic effects, maternal genetic effects, and maternal genetic sensitivity on prenatal heat stress for calf diseases and corresponding genomic loci in German Holsteins. J Dairy Sci 105:6795–6808. https://doi.org/10.3168/jds.2022-21804

# Part VIII

# Strategies and Management Tools for Disease Control

# Milk as Diagnostic Medium for Detection of Health Disorders in Dairy Animals

A. Costa, S. Magro, S. Sterup Moore, G. Visentin, M. Penasa, and M. De Marchi

**Abstract**

Milk is generally analyzed on a monthly basis in dairy animals under official recording schemes. Other than conventional composition, nowadays one sample of milk can provide a variety of interesting phenotypes. This has been mainly possible thanks to infrared spectroscopy, a rapid and low-cost technology implemented worldwide and accurately recognized as a gold standard for some traits (e.g., milk gross composition). Milk is secreted by the mammary gland and mirrors the physiological and metabolic status of the animal. Potentially, milk spectra can be used to predict biomarkers useful for health status assessment and fitness monitoring, such as ketone bodies, urea nitrogen, fatty acid profile, minerals, and protein fractions. In addition, it has been demonstrated that traits not directly measurable in milk can also be predicted with sufficient accuracy, e.g., blood parameters, reproductive traits, body weight, methane emissions, and feed intake. Finally, devices collecting milk spectra are commonly equipped with a flow cytometry machine that can quantify the milk somatic cell count (SCC), widely adopted as an indicator of udder health by farmers, veterinarians, researchers, and breeders. Therefore, milk sampling is a non-invasive procedure that allows for collection of phenotypes even at a population level. This medium can be exploited for diagnostic purposes in dairy species to track difficult-to-be-measured features at individual level, such as health, fertility, and efficiency.

**Keywords**

Indicator · Metabolic disease · Udder health · Phenotype · Indirect response

A. Costa (✉) · G. Visentin
Department of Veterinary Medical Sciences, University of Bologna, Ozzano dell'Emilia, BO, Italy
e-mail: angela.costa2@unibo.it; giulio.visentin@unibo.it

S. Magro · M. Penasa · M. De Marchi
Department of Agronomy, Food, Natural resources, Animals and Environment, University of Padova, Legnaro, PD, Italy
e-mail: silvia.magro.1@studenti.unipd.it; mauro.penasa@unipd.it; massimo.demarchi@unipd.it

S. Sterup Moore
Department of Agronomy, Food, Natural resources, Animals and Environment, University of Padova, Legnaro, PD, Italy

Department of Agroecology, Aarhus University, Tjele, Denmark
e-mail: selina.moore@unipd.it; selinamoore@agro.au.dk


J. J. Gross (ed.), *Production Diseases in Farm Animals*,
https://doi.org/10.1007/978-3-031-51788-4_24

## 1 Introduction

Milk is the secretion of the mammary gland of mammals after parturition and its composition varies according to intrinsic and extrinsic factors, such as species, breed, lactation stage, parity order, farming system, milking practice, and diet. A considerable number of studies have investigated phenotypic and genetic factors affecting milk yield and quality of dairy species of major interest in Europe, i.e., cattle, sheep, goat, and buffalo. The concentration of protein, particularly casein, is important for cheese manufacturers, representing one of the key-features that determine the curd and subsequently cheese yield. However, there are other components in milk which are of interest for the health status monitoring.

Milk production and composition are generally monitored on a monthly basis worldwide in all lactating animals under official milk recording. For this purpose, an accurate, validated, indirect, rapid, and non-destructive analysis is used, the mid-infrared spectroscopy (MIRS). This technique is recognized by the ISO 9622:2013 (IDF 141:2013) as a reference method for the determination of fat, protein, casein, total solids, lactose, and pH. Milk mirrors the physiological and metabolic status of the cow with the milk–blood barrier allowing for a selective permeability of compounds which reflects the circulating components in the blood. Research groups worldwide have therefore focused their activities on the use of milk as a matrix able to provide biomarkers useful for health assessment of cows, such as ketone bodies concentration, urea nitrogen content, fat-to-protein ratio, and fatty acid profile. It is clear that when milk MIRS data is supplemented with additional information, accurate predictive models are developed. Regardless of how they are measured, milk traits collected on a large scale are essential for both the monitoring of health disorders and the genetic improvement of dairy cows, with particular reference to sustainability, animal welfare, health disorders, and efficiency. This is of considerable importance to meet the growing global food demand and increased consumer awareness of product health and animal welfare.

## 2 Milk to Monitor Metabolic Diseases

Early lactation is a critical phase for dairy cows as they often experience a negative energy balance (NEB), a condition caused by a discrepancy in energy intake and requirements needed for the initiation and maintenance of milk production (Berge and Vertenten 2014). This leads to the use of alternative energy sources to compensate for the lack of glucose, a fundamental nutrient for maintenance of regular functions and lactogenesis (Benedet et al. 2019). High-producing cows are often unable to cope with NEB and they thus undergo an excessive mobilization of adipose tissue, in particular triglycerides, which impairs homeorhesis. The physiological capability of the liver to convert such products into usable energy is limited by the low blood glucose availability. As a consequence, the plasma concentration of fatty acids increases rapidly and this is responsible for the abnormal production of non-esterified fatty acids (NEFA) and ketone bodies in the liver, and their subsequent release into the blood (Benedet et al. 2019). Ketone bodies, namely acetone, acetoacetate, and β-hydroxybutyrate (BHB), can also be detected in cows' milk and are indicators of ketosis, one of the most detrimental metabolic diseases for dairy cows. The accumulation of NEFA and ketone bodies in the blood with no clinical signs of ketosis (i.e., sweet smelling breath, loss of appetite, changes in behavior) can be detected in borderline cows suffering from hyperketonemia (HYK). Both ketosis and HYK impair immune function, health, disease resistance, and milk production (McArt et al. 2013). Berge and Vertenten (2014) reported that up to 53% of cows in a dairy herd experienced ketosis within the first 35 days post-calving.

Cows can experience either subclinical (SCK) or clinical ketosis (CK) which vary in severity. While CK refers to a clear disease manifestation with evident symptoms and is thus easily detected, SCK is often overlooked due to the lack of clear physiological manifestations (Berge and Vertenten 2014). In any case, both forms require the use of biochemical analysis for official detection. From an economical point of view, metabolic disorders affect the farmers' income by reducing milk production, impairing reproduction performance and health (McArt et al. 2013), and increasing culling rate and veterinarian costs (Berge and Vertenten 2014; Seifi et al. 2011).

Detection of HYK, SCK, and CK can be performed via milk analysis, exploiting rapid, non-invasive, and accurate techniques. Among the serum biomarkers associated with HYK, BHB is the most abundant and stable circulating ketone body (Duffield et al. 2009). As extensively reviewed by Pralle and White (2020), concentrations of ketone bodies circulating in cows' blood are correlated with milk ketone concentrations. Since milk sampling is a non-invasive procedure routinely conducted on farms, milk is a good alternative matrix to diagnose ketosis and monitor HYK at individual and herd levels. Cows with high BHB concentrations in milk are 14.7 times more likely to encounter HYK (Berge and Vertenten 2014). Milk profile analyses based on photometrical methods, e.g., infrared spectroscopy, or enzymatic assays (Chandler et al. 2018) and strip tests (Keto-Test; Berge and Vertenten 2014) are available. Nevertheless, MIRS remains the most promising tool for the direct determination of milk BHB (Grelet et al. 2016). To a certain extent, MIRS can also be exploited for an indirect prediction of plasma BHB (Benedet et al. 2019); the prediction accuracy in this case can be used for screening purposes at the herd level and not for accurate determination at the individual level. The exact quantification of milk BHB for disease identification in individual cows is not currently feasible due to the inadequate accuracy of developed prediction equations (Pralle and White 2020). It has, nonetheless, proven to be useful in identifying cows with elevated BHB concentrations in milk (Grelet et al. 2016; Santschi et al. 2016). High milk BHB itself is not a direct indicator of ketosis; instead, it must be modelled to validate the correlation with plasma levels and thereby CK and/or SCK events. According to the literature, there is room to improve MIRS accuracy as the correlation coefficients between milk MIRS-predicted ketone bodies and blood BHB are moderate to high (Chandler et al. 2018; Ho et al. 2021). Pralle and White (2020) reported that milk BHB has a sensitivity of 83% and a specificity of 81% for the detection of HYK in dairy cows: though of reasonably promising level, it is still below the diagnostic accuracy of conventional cow-side enzymatic blood tests (90%).

Milk can play a key role in metabolic disease monitoring at the population level, allowing for cost-efficient data collection for improved herd management and genetic strategies. The main advantage of using milk instead of blood-based reference tests is the reduced cost and labor requirements. It is therefore arguable that lower diagnostic accuracy in exchange for cheaper and less demanding methods can be somehow accepted; certainly, the choices must be pondered according to the final purpose. Related to this, it is worth mentioning that a cow suffering from CK or SCK is usually treated with low-cost products such as propylene glycol which requires no milk withdrawal and separation (Pralle and White 2020). In this view, one could argue that the cost of HYK treatment is less than that of mistreating HYK, making the decision to use MIRS on a routine basis justifiable and meaningful.

Attempts to exploit MIRS for the monitoring of other metabolic disorders in dairy cows have been of recent investigation (Luke et al. 2019). For example, low levels of circulating Ca (hypocalcemia) and Mg (hypomagnesemia) can increase the risk of milk fever and tetany and, if not immediately treated, can lead to animals' death. Great interest, therefore, exists in developing rapid and cheap methods to monitor their blood concentrations even through proxies such as their prediction using MIRS. Although this technology is effective in accurately predicting a plethora of mineral concentrations in milk, such as Ca, Mg, P, Na, and K (Malacarne et al. 2018; Soyeurt et al. 2009), the prediction of serum mineral concentration from milk spectra is unfortunately still not sufficient for any purpose.

## 3 Udder Health

### 3.1 Mastitis

Mastitis is one of the most common and impacting diseases that can occur in lactating dairy cows, causing economic losses for the farmer and impairing health in lactating animals. Mastitis costs are attributable to (i) veterinary treatments and culling rates and (ii) reduction of milk yield and deterioration of milk quality, i.e., lower fat and protein concentrations, and higher milk pH (Ruegg 2017). Mammary gland inflammation takes place after an intramammary infection caused by microorganism(s), including pathogenic bacteria, algae, and yeasts. Rarely mastitis shows up due to physical traumatic events (Bradley 2002).

In cattle, mastitis can be either clinical (CM) or subclinical (SCM), with the first being characterized by observable signs of an inflammatory response of udder tissues, e.g., swelling, increased temperature, redness, and pain. The inflammation can affect one or more quarters or even the whole udder, depending on the type of microorganisms involved, intrinsic immune response, and severity. Milk from mastitic quarters/mammary glands is visually abnormal, with anomalous texture, such as clots, and discoloration of the secretum. On the other side, SCM signs are not evident and thus not detectable either by the veterinarian or the farm personnel. For this reason, latent SCM is dangerous for the herd, making accurate and timely identification very important (Heringstad et al. 2000).

### 3.2 Milk Somatic Cells

Milk contains somatic cells, whose concentration is known as SCC (cells/mL). Total SCC is composed of a small amount of epithelial exfoliated cells (<10%) and predominantly of white blood cells. Literature suggests that in physiological/healthy conditions, dairy cows should produce milk with SCC < 200,000 cells/mL. Values of SCC ≥ 200,000 cells/mL are conventionally used as "alert" of inflammation (Sharma et al. 2011).

Currently, the identification of SCM relies on the SCC at individual milk level (i.e., official milk recording). As regards CM, identification could be direct or indirect. The direct identification includes diagnosis of mastitis events recorded in binary format, i.e., presence/absence, evaluated by either experts, veterinarians, or trained personnel in countries with a validated and routine health recording system (Egger-Danner et al. 2015). Scandinavian countries, Canada, and Austria have great experience with cow health data and are good examples of large-scale phenotypes acquisition, integration, and validation. It has been demonstrated that direct binary traits can be successfully used for management and breeding, however, due to their dichotomous nature have some limitations. Research has demonstrated that continuous variables recorded on a monthly basis during official testing, like milk SCC, are able to disclose less severe CM cases as well as cows suffering from SCM or chronic latent inflammation. For this reason, several national selection indexes include milk SCC in addition to, when available, direct diagnoses of mastitis for genetic evaluation of udder health.

Milk SCC is included in various milk payment systems worldwide, especially in countries specializing in cheese manufacturing (Benedet et al. 2018). Penalties are applied by the dairy companies when the bulk milk SCC is above a certain limit. The SCC cut-off can be fixed or dynamic in a dairy company; for instance, one of the most popular criteria used to move/change the SCC maximum level allowed is to consider the monthly average levels of SCC of all milk suppliers. The EU Regulation no. 853/2004 (EU 2004) defines the maximum SCC for bovine bulk tank milk at 400,000 cells/mL. The same Regulation requires that milk is checked for abnormalities by the milking operators or by a method able to achieve similar results. Milk that presents abnormalities is excluded from human consumption.

Another useful trait in mastitis detection is the differential somatic cell count (DSCC, %), which is the amount of polymorphonuclear neutrophils and lymphocytes out of the total SCC (Damm et al. 2017; Schwarz et al. 2020a, b). This trait is

**Table 1** Specific function and proportion of different types of leukocytes, and correlation (r) with SCC (Damm et al. 2017; Halasa and Kirkeby 2020)

| Type of cells | Function | Range | r |
|---|---|---|---|
| Polymorphonuclear neutrophils | Defense against invading bacteria at the beginning of an infection | 13–92% | 0.36 |
| Lymphocytes | Regulation of the induction and suppression of immune responses | 0–19% | −0.17 |
| Macrophages | Ingestion of microorganisms and cellular debris; recognize invading pathogens and trigger an immune reaction, by the rapid recruitment of polymorphonuclear neutrophils | 7–79% | −0.34 |

assessed by FOSS devices, such as the Fossomatic (Damm et al. 2017). Leukocytes present in milk somatic cells are mainly composed of polymorphonuclear neutrophils, macrophages, and lymphocytes, each with a specific function and correlation with total milk SCC (Table 1).

Nevertheless, the composition of SCC fractions changes in presence of intramammary infection with differences before, during, and after the inflammation. In particular, polymorphonuclear neutrophils are the predominant cell type among milk leukocytes in the early stages of mammary infection, while in the chronic stage the lymphocytes increase and neutrophils decrease (Leitner et al. 2003). According to several authors (Bobbo et al. 2020; Schwarz et al. 2011; Zecconi et al. 2019), polymorphonuclear neutrophils increase in milk of infected quarters. For this reason, DSCC has been recently proposed as an additional proxy of udder health. Specifically, DSCC >65% when SCC is >200,000 cells/mL suggests inflammation of the mammary gland (Schwarz et al. 2020a).

A few studies (Costa et al. 2021; Godden et al. 2017) have attempted to investigate detailed SCC fractions determined with instruments different from the Fossomatic, e.g., the QScout® Farm Lab (Advanced Animal Diagnostics, Inc., Morrisville, NC, USA). This portable device is equipped with a software that, through the specific algorithms installed, combines different SCC fractions and performs a discriminant analysis (Costa et al. 2021). This test, known as Qscout Milk Leukocyte Differential test, classifies samples according to mastitis presence ("negative," "borderline," or "positive") and provides concentration of total milk SCC, macrophages, and polymorphonuclear cells, namely neutrophils and lymphocytes. However, still the limitation of the QScout® Farm Lab is that the analysis is time demanding and costly, as it must be carried out at quarter level. In addition, the cost of renting or purchasing the device coupled with the cost of required disposable kits hinder their use for not experimental scopes.

### 3.3 Other Milk Traits

Mastitis can also be directly detected through isolation of pathogen(s) causing the infection and subsequent inflammation. A wide number of studies have evaluated udder health based on bacteriological cultures and polymerase chain reaction (PCR; Kirkeby et al. 2020; Narayana et al. 2017; Nitz et al. 2021). Quantitative and qualitative analyses of milk pathogens are helpful for clinical purposes but are difficult to implement on a routine basis at population level, due to high labor and cost.

On-farm level, indirect identification of mastitis has been traditionally assessed using the California Mastitis Test (CMT), a quick screening procedure to identify potentially infected quarters using a small volume of milk. Using a surfactant, CMT provides a classification of milk based on SCC; despite being old, this method has been extensively used for its cost-effectiveness and ease of interpretation. Nevertheless, the outcome is subjective and is not as accurate as cytometry-based SCC. While official monthly SCC values are measured on a representative pooled sample of milk harvested from all quarters of a cow, CMT is carried out at the quarter level, making identification of problematic quarter(s) easier and straightforward (Sanford et al. 2006).

By exploiting a combination of various milk information, such as gross composition, SCC, DSCC, and electrical conductivity (EC), it is possible to accurately identify udder inflammation (Ebrahimie et al. 2018a, b). In fact, bovine milk is an excellent conductor of electricity and, as soon as a quarter becomes infected, an immune response occurs, leading to a change in the osmotic equilibrium of milk followed by a hampered electrolyte composition. This causes an increase of $Na^+$ and $Cl^-$ with a direct impact on milk EC (Norberg et al. 2004).

Dairy farms under routine herd-testing benefit from milk analysis on a monthly basis and the MIRS devices coupled with a flow cytometric tool are the predominant method for the determination of milk quality traits (Rienesl et al. 2019). Some authors have attempted to predict mastitis through MIRS, but to date the prediction equation is not very accurate. Additional research to further improve the prediction equation for MIRS-predicted mastitis are planned (Rienesl et al. 2019).

Apart from SCC, EC, and DSCC, lactose content has been also recognized as a potential indicator of mastitis in cattle since its concentration decreases when SCC rises. Some studies, mostly intended to valorize and exploit repeated information on milk traits for genetic purposes, observed that lactose content can be used as an indicator trait of mastitis in combination with SCC (Costa et al. 2019b; Ebrahimie et al. 2018a, b). In support of this, lactose content is the most heritable milk constituent, with a heritability of around 0.40 in Holstein-Friesian cows (Costa et al. 2019b). Additionally, negative genetic correlations between lactose and clinical mastitis have been estimated (Miglior et al. 2007); for example, the genetic association between these two traits in Fleckvieh breed was −0.18 in the first 150 DIM (Costa et al. 2019a). The drop in milk lactose during mastitis is due to leakage caused by the compromised alveolar epithelial integrity (Fig. 1). The epithelial permeability is damaged after mastitis due to the agents and the substances released during and after mammary gland tissues inflammation; response-related factors, e.g., bacteriostatic and bactericidal factors such as cytokines and oxidative agents, are responsible for the disruption of tight junctions of the basal membrane of the mammary cells that separate blood and milk (Fig. 1). Part of the lac-

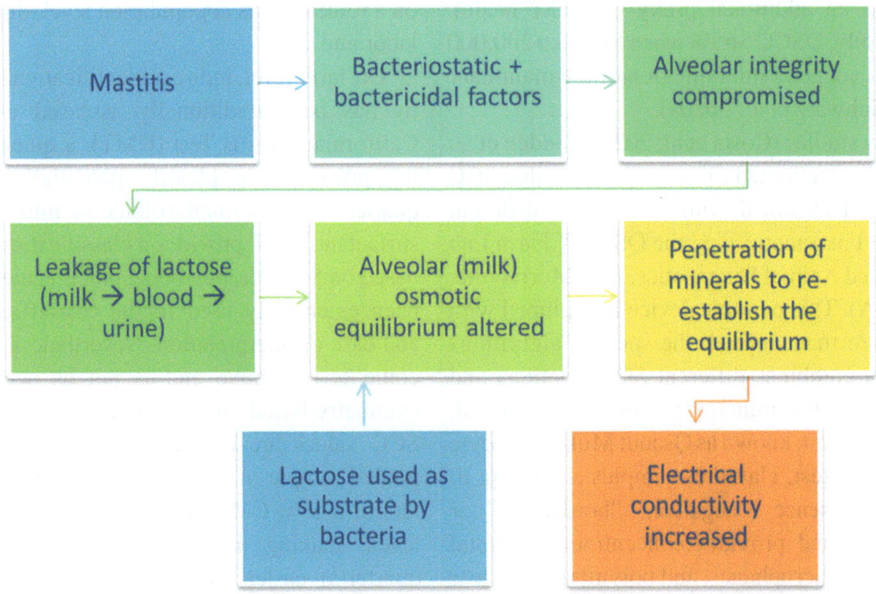


**Fig. 1** Mechanisms responsible for the decrease in milk lactose content when an intramammary inflammation is present

tose is inevitably lost in the bloodstream and is consequently detectable in cow's urine (Herve et al. 2018). It has also been reported that part of the lactose present in the alveoli is used as a substrate by mastitis-causing pathogens, which leads to a decrease in lactose content in milk and to an increase in lactic acid (Costa et al. 2019b).

### 3.4 Milk for Udder Health Genetic Improvement

Data on SCC collected on a large scale are used for genetic purposes, i.e., to improve udder health and mastitis resistance. In particular, the genetic correlation between SCC and mastitis has been estimated at around 0.70 and the heritability of SCS, being approximately 0.17, is greater than that of CM or SCM, which varies from 0.02 to 0.10 (Thompson-Crispi et al. 2014). Selection to improve udder health is often carried out by using SCC which is conventionally log-transformed into SCS. The data is derived from individual test-day records registered on individual cows and generally only the first three lactations are considered (Table 2).

In Scandinavia and countries like Canada and Austria clinical mastitis data in binary form (0/1) are also used for the genetic evaluation of udder health. This phenotype is collected on a large number of animals and farms with validated protocols and recorded by trained personnel (Fuerst et al. 2011). In Denmark, Sweden, and Finland, the udder health index also takes into account fore udder attachment and udder depth linearly evaluated by trained classifiers (Nordic Cattle Genetic Evaluation 2022). In Italy, selection for resistance to mastitis in Holstein is carried out using a composite index that combines several SCC-derived information. The single traits used to create the composite are those that were strongly correlated with the presence of clinical mastitis in a testing data set, but only first parity cows are considered (Finocchiaro et al. 2018). The use of SCC-derived indicator traits for udder health monitoring and breeding started in 2007 in the Netherlands (de Haas et al. 2007). Other traits like EC, lactose content, and DSCC are not yet included in dairy cow selection indexes around the world, although they have the potential to be used as a proxy.

**Table 2** Country-specific traits[a] included in the national official udder health index for the cosmopolitan Holstein breed (source: Interbull, 2022)

| Country | Trait(s) | Lactation(s) |
|---|---|---|
| Australia | SCS + CM | 1–3 |
| Austria, Germany, Luxembourg | CM | All |
| Belgium | SCS | 1–3 |
| Canada | SCS + CM | 1–3 (SCS), all (CM) |
| Denmark, Sweden, Finland | SCS + CM | 1–3 |
| France | SCS + CM | 1–3 |
| Great Britain | SCS + CM | 1–5 (SCS), 1–3 (CM) |
| Ireland | SCS | 1–15 |
| Italy | SCS + Complex Trait | 1–3 (SCS), 1 (Complex Trait) |
| New Zealand | SCS | 1–3 |
| Poland | SCS | 1–3 |
| United States of America | SCS + CM | 1–5 |

[a]CM (clinical mastitis), recorded as 0 = absence or 1 = presence; Complex Trait, combination of different SCC-derived indicators

### 3.5 Automatic Milking Systems Inflammation Alerts

Automatic milking systems (AMS) are able to collect data on milk production and sometimes composition per cow on a daily basis. Also, some AMS can differentiate milk produced from each single quarter and provide quarter-specific information and alerts. Data recorded and stored are of support for farmers, allowing a better understanding of the lactation history of each cow in the herd. Sensors installed in most of AMS inspect milk physical characteristics and quality and thanks to specific algorithms can classify milk, discriminate the normal from the anomalous, and provide alerts to the farmers when needed (Adriaens et al. 2021). Briefly, during milking sensors installed in AMS continuously record EC at quarter level, resulting in time-series data points. Algorithms exclusively based on EC to detect mastitis tend to predict a large number of false positives. This is because EC is not only influenced by mastitis, but also by breed, parity, lactation stage, milking interval, and milk composition (Norberg 2005). Milk color has been shown to be promising for AMS classification of abnormal milk, since it is affected by the presence of blood constituents (white cells) in milk (Hovinen and Pyörälä 2011). In addition to EC and color, AMS algorithms could use other traits derived from milking to obtain a mastitis risk score, such as average milk flow rate, peak milk flow rate, and cow's parity and lactation stage. The models accounting for the combination(s) of these traits are suitable for prediction of mastitis risk at quarter-, cow-, and herd-level (Khatun et al. 2018).

### 3.6 Species Other than Cattle

The identification of mastitis through milk SCC is more challenging in species other than dairy cattle due to different milk synthesis, mammary gland anatomical conformation, and physiological baseline SCC levels. Interpretation and use of milk SCC in buffaloes, ewes, and goats are therefore different compared to cattle and, on top of that, there is still uncertainty on the level of SCC that may be used to identify inflammation in buffaloes, ewes, and goats due to scarce presence of udder health monitoring systems for these species (Costa et al. 2020; Raynal-Ljutovac et al. 2007). In healthy goats, milk is characterized by greater SCC than in healthy cows. This is due to the apocrine secretion of the mammary gland secretum, which leads to greater cytoplasmic particles and epithelial cells into milk. For this reason, SCC of around 1,000,000 cells/mL is considered normal in dairy goats (Granado et al. 2014). The magnitude of the relationship between SCC and mastitis in buffalo, goat, and sheep therefore remains unclear. Thus, there are no specific thresholds of SCC that can be exploited to divide healthy from udder-diseased animals. The EU Regulation no. 853/2004 does not define a legal limit for buffalo, goat, and sheep. Nevertheless, a SCC cut-off is often adopted by dairy companies that process milk of these species.

## 4 Other Monitoring Purposes

### 4.1 Fertility

The benefits of milk as a provider of cow-specific characteristics extend to numerous facets of dairy production, including fertility. Early identification of cows that have been successfully inseminated through milk is important to keep reproductive performance on optimal levels and thereby prevent unnecessary further services. This also allows for a shortening of the interbreed interval for open cows (Giordano et al. 2013). Pregnancy is currently determined either by indirect methods such as the failure to return to estrus, milk progesterone levels or pregnancy-associated glycoproteins or by direct methods such as transrectal palpation and ultrasonography (Tiplady et al. 2022). Several of these methods are carried out by qualified professionals and thereby carry a certain cost (Lainé et al. 2014). It has been well documented that changes in milk yield occur during pregnancy (Lainé et al. 2017; Olori et al. 1997) with effects being small during the early stages after which it starts to increase nonlinearly. In fact, the 305-day milk yield is reduced by 207 kg in pregnant primiparous cows (Lu and Bovenhuis 2020). This reduction has mainly been attributed to a change in the partitioning of energy away from milk synthesis in favor of fetus growth (Penasa et al. 2016). Daily milk yield tends to be lowest in the third month of gestation, i.e., in mid-to-late lactation if the pregnancy started in the optimal time window; this is the point at which the fetus demands the largest amount of nutrients (Moe and Tyrrell 1972; Ho et al. 2019). Furthermore, also the composition of milk is affected by pregnancy, with an increase in fat, protein, and casein content that becomes particularly evident after 6–7 months in gestation (Hu and Bocenhuis, 2020; Penasa et al. 2016; Tiplady et al. 2022).

Milk composition assessed via MIRS is already used worldwide for genetic selection and dairy herd improvement, but spectral data may reveal other features of interest, such as fertility traits. A considerable number of studies have therefore looked at the feasibility and accuracy of MIRS-based prediction models to discriminate between open and pregnant cows. Indeed, by using preadjusted spectral data and discriminant analysis, Lainé et al. (2014) have determined the pregnancy status of a cow with 99.2% accuracy just 30 days after insemination, and without impairment to specificity (89.1%). However, other studies investigating the use of milk as a matrix for pregnancy diagnosis were not as successful in terms of prediction accuracy. While Lainé et al. (2017) identified the spectral region from 968 to 1577 $cm^{-1}$ as being highly influenced by pregnancy status, i.e., the region associated with milk protein content, Toledo-Alvarado et al. (2018) have instead considered the whole spectrum for identification of pregnant cows. Authors concluded that although the use of whole-spectrum profiles increased the accuracy of pregnancy status determination as compared to using single spectral regions, the model is greatly improved by supplementing with data regarding parity and days in milk (Toledo-Alvarado et al. 2018). Ho et al. (2019) similarly concluded that the addition of blood metabolic profiles and milk fatty acid data resulted in better prediction accuracies. Of course, such detailed data is often disconnected from MIRS and is therefore not practical to add. Although there is a correlation between milk composition and pregnancy status of dairy cows, much research remains to develop robust MIRS models that can be applied on a population level.

## 4.2 Animal Welfare

Other than clinical aspects, milk can be informative of the welfare of lactating cows. Bahadi et al. (2021) used movement restrictions as a proxy for cows' welfare and observed that among the four tested configurations, the concentration of certain compounds significantly changed between the most and the least restrictive configurations (tie-rail system; $p = 0.03$). This was observable on MIRS milk spectra, from which the contributing components were attributed to biomarkers related to NEB, such as BHB, citrate, and acetone, probably caused by reduced access to feed (Bahadi et al. 2021). Variations in certain milk composition traits or spectral region can be used to predict lameness as a result of physiological changes in cows, e.g., the onset of inflammation (Bonfatti et al. 2020; Mineur et al. 2017). Lameness is one of the most significant feet and leg disorders and can occur due to physical injury, infection, or physiological changes; moreover, lame cows tend to be more susceptible to mastitis, metabolic disorders, impaired fertility, and reduced daily milk yield (Bonfatti et al. 2020). The onset of lameness symptoms is often delayed with respect to the manifestation of the disorder and with 23–70% of American and European dairy herds showing lameness-related issues, an early detection method would prove to be highly advantageous (Bonfatti et al. 2020). Results based on MIRS have, however, generally been of poor prediction accuracies. Often additional predictors such as age and lactation stage must be supplied to the model to slightly increase the accuracy. Mineur et al. (2017) obtained an overall accuracy of 64.51% which is just 14.51% more than would be determined by chance. However, when subgroups were considered independently instead of all data (e.g., within lactation stages, breeds, or parity classes) higher accuracies were achieved. Predictive models developed by Bonfatti et al. (2020) correctly identified just 57% of lame cows, and like Mineur et al. (2017), higher accuracies were achieved by the authors when subgroups were considered. These models have, nonetheless, higher accuracy than farmers, who were able to identify only 25% of lame cows (Bonfatti et al. 2020). To find strategies to predict lameness in advance in cows, i.e., before the onset of signs, is thus of much interest; the same is valid for issues like mastitis and reproductive problems, that cause the majority of health-related economic losses at farm level (Bruijnis et al. 2010).

## 4.3 Hypocalcemia

Milk fever, also known as Downer cow syndrome, is a metabolic disease that occurs when the homeostatic mechanism of calcium (Ca) fails to adapt in fresh cows during the period in which the most rapid increase in daily milk yield is observed. Incidence rates have been documented to be as high as 25% although with great variation among herds (DeGaris and Lean 2008). Milk synthesis requires huge amounts of nutrients, including minerals and, among the latter, Ca and P stand out. In fact, for each kilogram of colostrum produced, approximately 2.5 g of Ca are extracted from blood, roughly equating to the total amount present in the blood at any given time (Horst 1986). Cows facing a new lactation tend to show a decrease in the plasma Ca concentrations; this mineral, the main mineral responsible for milk fever syndrome, is thus released from body reserves such as bones. As observed for NEB, the use of stored body reserves takes place because a metabolic priority exists in dairy cattle, especially in high-producing specialized breeds (Oltenacu and Broom, 2010). All the nutrients necessary for milk synthesis, including glucose and minerals, are delivered to the mammary gland through plasma (Wellnitz and Bruckmaier 2021), even if this translates into a compromised fitness and health for the cow. Milk fever predisposes the lactating animal to a wide list of disorders and can be diagnosed with the use of rapid on-farm blood test kits as well as through evaluation of symptoms (Aktaş et al. 2010). Indeed, a positive correlation exists between cows suffering from milk fever and cows suffering from mastitis and ketosis. Moreover, high-producing cows are to a larger extent genetically predisposed to certain diseases than cows yielding less milk (Costa et al. 2019a). An excess in dietary cations during the prepartum is known to be the major risk factor exposing cows to milk fever (Goff and Horst 1997), and prepartum transition feeding with an altered macro-mineral concentrate is therefore often practiced (Lean et al. 2006). In fact, for each day a cow is fed with transition feed, she is 1.9% more likely to successfully undergo conception and 4.7% less likely to be removed from the herd due to voluntary culling or death (DeGaris and Lean 2008). Studies published by Rajala-Schultz et al. (1999) demonstrated that milk yield was reduced by 1.1–2.9 kg/d in cows that had contracted milk fever, depending on parity and duration of disease onset. However, since high-producing cows are more susceptible to milk fever, they nonetheless produced 1.1–1.7 kg/d more than healthy cows.

It is plausible that the mineral composition of milk, hereunder particularly Ca, changes upon milk fever manifestation, but this is a topic that is yet to be thoroughly studied. Despite their modest quantities in milk, Visentin et al. (2018) have been able to predict the concentration of minerals in milk using MIRS data. These concentrations, and their variations in milk, were in agreement with previous similar studies, albeit with coefficients of determination ranging from 0.40 to 0.68 depending on the salt considered (Visentin et al. 2018). Therefore, it is possible to assess the mineral content with simple routine techniques. Current diagnostic technologies of milk fever are all blood-based making them tedious and often inaccurate as serum Ca content quickly diminishes upon exposure to air. Also, they carry a high per-sample cost making them too expensive for regular use. Further investigation into possible correlations between milk mineral content, hereunder particularly Ca, serum Ca concentration, hypocalcemia, and milk fever to develop diagnostic models for application in the dairy industry would be of great value as an alternative diagnostic technique.

## 4.4 Conclusion

In conclusion, milk often mirrors the physiology of the cow and can be partly informative of the circulating components, including mineral concentrations and their equilibrium. The secretion of the mammary gland is in equilibrium with blood, in fact, the milk–blood barrier allows for the selective permeability of compounds, ensuring high protein, lactose and potassium concentrations as well as low sodium and chloride concentrations in milk as compared to in blood (Wellnitz and Bruckmaier 2021). Milk can be used as a diagnostic matrix for a wide range of diseases and disorders thanks to MIRS spectra information. Although MIRS has the potential to be used for the prediction of various proxies of health and welfare in dairy cows, there is still room to improve prediction accuracy in order to make models universally implementable and applicable. If the milk mineral content is proven to be genetically rather than phenotypically correlated with blood mineral concentrations and milk fever, MIRS predictions can be successfully adopted for selection scope in addition to management purposes. Milk reflects the metabolic and physiological state of the cow and it therefore represents the most convenient medium for the assessment of wellness, health, fitness, and performance of dairy cows.

## 5 What's Next?

Following trends and EU policies, animal scientists worldwide are currently evaluating if and how milk data can be used to detect and thus monitor cows' resilience, heat stress, methane emissions, and feed efficiency. For instance, research efforts on colostrum composition of dairy animals have dramatically increased in recent years due to its important association with the newborns' health, performance, and survival (Costa et al. 2023). Data of immunoglobulins concentration in both blood and colostrum may drive genetic selection towards more resistant and healthy young stock. Poppe et al. (2020, 2021) found that fluctuations in milk yield are able to disclose more resilient cows; in particular, some parameters obtained from the lactation curve may be heritable and can be used as indicator traits for genetic improvement. Certainly, coupling milk yield records with MIRS data would definitely improve the detection accuracy of best and worst animals in terms of resistance to stressors and disease and ability to restore homeostasis. In addition, in the era of reduction of greenhouse gas emissions, being able to predict the body weight and/or the feed intake of lactating cows from milk records routinely available would be a great achievement allowing indirect estimation of individual methane emissions on a daily basis (Vanlierde et al. 2016, 2018; Denninger et al. 2020).

Last but not least, research is more and more frequently focusing on milk micro RNAs, microbiota, and metabolomic traits which are reported to be highly correlated to certain diseases (Jo et al. 2021; Li et al. 2018; Zheng et al. 2014).

In the "farm to fork" view, milk and its spectra are inexhaustible sources of still unexplored biomarkers and proxies useful to improve animal health and sustainability and efficiency of dairy farming.

## References

Adriaens I, Van Der Brulle I, Geerinckx K, D'Anvers L, De Vliegher S, Aernouts B (2021) Milk losses linked to mastitis treatments at dairy farms with automatic milking systems. Prev Vet Med 194:105420

Aktaş M, Kaynar O, Ozkanlar S, Ozkanlar Y (2010) Diagnosis of milk fever by a water hardness test kit in ewes. Isr J Vet Med 65:108–110

Bahadi M, Ismail A, Vasseur E (2021) Fourier transform infrared spectroscopy as a tool to study milk composition changes in dairy cows attributed to housing modifications to improve animal welfare. Foods 10:450

Benedet A, Manuelian CL, Penasa M, Cassandro M, Righi F, Sternieri M, De Marchi M (2018) Factors associated with herd bulk milk composition and technological traits in the Italian dairy industry. J Dairy Sci 101:934–943

Benedet A, Manuelian CL, Zidi A, Penasa M, De Marchi M (2019) Invited review: β-hydroxybutyrate concentration in blood and milk and its associations with cow performance. Animal 13:1676–1689

Berge AC, Vertenten G (2014) A field study to determine the prevalence, dairy herd management systems, and fresh cow clinical conditions associated with ketosis in western European dairy herds. J Dairy Sci 97:2145–2154

Bobbo T, Penasa M, Cassandro M (2020) Combining total and differential somatic cell count to better assess the association of udder health status with milk yield, composition and coagulation properties in cattle. Ital J Anim Sci 19:697–703

Bonfatti V, Ho PN, Pryce JE (2020) Usefulness of milk mid-infrared spectroscopy for predicting lameness score in dairy cows. J Dairy Sci 103:2534–2544

Bradley AJ (2002) Bovine mastitis: an evolving disease. Vet J 164:116–128

Bruijnis MR, Hogeveen H, Stassen EN (2010) Assessing economic consequences of foot disorders in dairy cattle using a dynamic stochastic simulation model. J Dairy Sci 93:2419–2432
Chandler TL, Pralle RS, Dórea JRR, Poock SE, Oetzel GR, Fourdraine RH, White HM (2018) Predicting hyperketonemia by logistic and linear regression using test-day milk and performance variables in early-lactation Holstein and Jersey cows. J Dairy Sci 101:2476–2491
Costa A, Egger-Danner C, Mészáros G, Fuerst C, Penasa M, Sölkner J, Fuerst-Waltl B (2019a) Genetic associations of lactose and its ratios to other milk solids with health traits in Austrian Fleckvieh cows. J Dairy Sci 102:4238–4248
Costa A, Lopez-Villalobos N, Sneddon NW, Shalloo L, Franzoi M, De Marchi M, Penasa M (2019b) Invited review: milk lactose—Current status and future challenges in dairy cattle. J Dairy Sci 102:5883–5898
Costa A, Neglia G, Campanile G, De Marchi M (2020) Milk somatic cell count and its relationship with milk yield and quality traits in Italian water buffaloes. J Dairy Sci 103:5485–5494
Costa A, De Marchi M, Sagrafoli D, Lanzi H, Amatiste S, Boselli C, Giacinti G (2021) Milk somatic cell count and polymorphonuclear cells in healthy quarters of cows that underwent blanket and selective dry therapy: an Italian case study. Vet Sci 8:298
Damm M, Holm C, Blaabjerg M, Bro MN, Schwarz D (2017) Differential somatic cell count—A novel method for routine mastitis screening in the frame of Dairy Herd Improvement testing programs. J Dairy Sci 100:4926–4940
de Haas Y, Bloemhof S, Ouweltjes W, ten Napel J, de Jong G (2007) Improving selection on udder health by using different trait definitions of somatic cell count. Interbull Bull 37:185–185
DeGaris PJ, Lean IJ (2008) Milk fever in dairy cows: a review of pathophysiology and control principle. Vet J 176:58–69. https://doi.org/10.1016/j.tvjl.2007.12.029
Denninger TM, Schwarm A, Dohme-Meier F, Münger A, Bapst B, Wegmann S, Grandl F, Vanlierde A, Sorg D, Ortmann S, Clauss M, Kreuzer M (2020) Accuracy of methane emissions predicted from milk mid-infrared spectra and measured by laser methane detectors in Brown Swiss dairy cows. J Dairy Sci 103:2024–2039
Duffield TF, Lissemore KD, McBride BW, Leslie KE (2009) Impact of hyperketonemia in early lactation dairy cows on health and production. J Dairy Sci 92:571–580
Ebrahimie E, Ebrahimi F, Ebrahimi M, Tomlinson S, Petrovski KR (2018a) A large-scale study of indicators of sub-clinical mastitis in dairy cattle by attribute weighting analysis of milk composition features: highlighting the predictive power of lactose and electrical conductivity. J Dairy Res 85:193–200
Ebrahimie E, Ebrahimi E, Ebrahimi M, Tomlinson S, Petrovski KR (2018b) Hierarchical pattern recognition in milking parameters predicts mastitis prevalence. Comput Electron Agric 147:6–11
Egger-Danner C, Cole JB, Pryce JE, Gengler N, Heringstad B, Bradley A, Stock KF (2015) Invited review: overview of new traits and phenotyping strategies in dairy cattle with a focus on functional traits. Animal 9:191–207
EU (2004) Regulation (EC) No 853/2004 of the European Parliament and of the Council of 29 April 2004 laying down specific hygiene rules for food of animal origin. https://eur-lex.europa.eu/eli/reg/2004/853/oj
Finocchiaro R, Visentin G, Penasa M, van Kaam JBCHM, Marusi M, Civati G, Cassandro M (2018) Alternative use of somatic cells counts in genetic selection for mastitis resistance: a new estimated breeding value for Italian Holstein breed. Interbull Bull 53:31–33
Fuerst C, Koeck A, Egger-Danner C, Fuerst-Waltl B (2011) Routine genetic evaluation for direct health traits in Austria and Germany. Interbull Bull 44. https://journal.interbull.org/index.php/ib/issue/view/59
Giordano JO, Fricke PM, Cabrera VE (2013) Economics of resynchronization strategies including chemical tests to identify nonpregnant cows. J Dairy Sci 96:949–961
Godden SM, Royster E, Timmerman J, Rapnicki P, Green H (2017) Evaluation of an automated milk leukocyte differential test and the California Mastitis Test for detecting intramammary infection in early- and late-lactation quarters and cows. J Dairy Sci 100:6527–6544
Goff JP, Horst RL (1997) Effects of the addition of potassium or sodium, but not calcium, to prepartum rations on milk fever in dairy cows. J Dairy Sci 80:176–186
Granado RJ, Rodríguez MS, Arce C, Estévez VR (2014) Factors affecting somatic cell count in dairy goats: a review. Span J Agric Res 1:133–150
Grelet C, Bastin C, Gelè M, Davière J-B, Johan M, Werner A, Reding R, Fernandez Pierna JA, Colinet FG, Dardenne P, Gengler N, Soyeurt H, Dehareng F (2016) Development of Fourier transform mid-infrared calibrations to predict acetone, β-hydroxybutyrate, and citrate contents in bovine milk through a European dairy network. J Dairy Sci 99:4816–4825
Halasa T, Kirkeby C (2020) Differential somatic cell count: value for udder health management. Front Vet Sci 7:1153
Heringstad B, Klemetsdal G, Ruane J (2000) Selection for mastitis resistance in dairy cattle: a review with focus on the situation in the Nordic countries. Livest Prod Sci 64:95–106
Herve L, Lollivier V, Quesnel H, Boutinaud M (2018) Oxytocin induces mammary epithelium disruption and could stimulate epithelial cell exfoliation. J Mammary Gland Biol Neoplasia 23:139–147
Ho PN, Bonfatti V, Luke T, Pryce JE (2019) Classifying the fertility of dairy cows using milk mid-infrared spectroscopy. J Dairy Sci 102:10460–10470
Ho PN, Luke T, Pryce JE (2021) Validation of milk mid-infrared spectroscopy for predicting the metabolic status of lactating dairy cows in Australia. J Dairy Sci 104:4467–4477

Horst RL (1986) Regulation of calcium and phosphorus homeostasis in the dairy cow. J Dairy Sci 69:604–616
Hovinen M, Pyörälä S (2011) Invited review: Udder health of dairy cows in automatic milking. J Dairy Sci 94:547–562
Hu H, Bovenhuis H (2020) Phenotypic and genetic effects of pregnancy on milk production traits in Holstein-Friesian cattle. J Dairy Sci 103:11597–11604. https://doi.org/10.3168/jds.2020-18561
Interbull. International Bull Evaluation Service Official Website. https://interbull.org/index. Accessed 25 Mar 2022
Jo J-H, Ghassemi Nejad J, Peng D-Q, Kim H-R, Kim S-H, Lee H-G (2021) Characterization of short-term heat stress in Holstein dairy cows using altered indicators of metabolomics, blood parameters, milk microRNA-216 and characteristics. Animals 11:722
Khatun M, Thomson PC, Kerrisk KL, Lyons NA, Clark CEF, Molfino J, García SC (2018) Development of a new clinical mastitis detection method for automatic milking systems. J Dairy Sci 101:9385–9395
Kirkeby C, Toft N, Schwarz D, Farre M, Nielsen SS, Zervens L, Halasa T (2020) Differential somatic cell count as an additional indicator for intramammary infections in dairy cows. J Dairy Sci 103:1759–1775
Lainé A, Bel Mabrouk H, Dale LM, Bastin C, Gengler N (2014) How to use mid-infrared spectral information from milk recording system to detect the pregnancy status of dairy cows. Commun Agric Appl Biol Sci 79:33–38
Lainé A, Bastin C, Grelet C, Hammami H, Colinet FG, Dale LM, Gillon A, Vandenplas J, Dehareng F, Gengler N (2017) Assessing the effect of pregnancy stage on milk composition of dairy cows using mid-infrared spectra. J Dairy Sci 100:2863–2876
Lean IJ, DeGaris PJ, McNeil DM, Block E (2006) Hypocalcemia in dairy cows: meta-analysis and dietary cation anion difference theory revisited. J Dairy Sci 89(2):669–684. https://doi.org/10.3168/jds.S0022-0302(06)72130-0
Leitner G, Eligulashvily R, Krifucks O, Perl S, Saran A (2003) Immune cell differentiation in mammary gland tissues and milk of cows chronically infected with Staphylococcus aureus. J Veterinary Med Ser B 50:45–52
Li Q, Yang C, Du J, Zhang B, He Y, Hu Q, Li M, Zhang Y, Wang C, Zhong J (2018) Characterization of miRNA profiles in the mammary tissue of dairy cattle in response to heat stress. BMC Genomics 19:975
Lu H, Bovenhuis H (2020) Phenotypic and genetic effects of pregnancy on milk production traits in Holstein-Friesian cattle. J Dairy Sci 103:11597–11604
Luke TDW, Rochfort S, Wales WJ, Bonfatti V, Marett L, Pryce JE (2019) Metabolic profiling of early-lactation dairy cows using mid-infrared spectra. J Dairy Sci 102:1747–1760
Malacarne M, Visentin G, Summer A, Cassandro M, Penasa M, Bolzoni G, Zanardi G, De Marchi M (2018) Investigation on the effectiveness of mid-infrared spectroscopy to predict detailed mineral composition of bulk milk. J Dairy Res 85:83–86
McArt JAA, Nydam DV, Oetzel GR, Overton TR, Ospina PA (2013) Elevated non-esterified fatty acids and β-hydroxybutyrate and their association with transition dairy cow performance. Vet J 198:560–570
Miglior F, Sewalem A, Jamrozik J, Bohmanova J, Lefebvre DM, Moore RK (2007) Genetic analysis of milk urea nitrogen and lactose and their relationships with other production traits in Canadian Holstein cattle. J Dairy Sci 90:2468–2479
Mineur A, Koeck A, Grelet C, Gengler N, Egger-Danner C, Sölkner J (2017) First Results in the use of milk mid-infrared spectra in the detection of lameness in Austrian dairy cows. Agric Conspec Sci 82:163–166
Moe PW, Tyrrell HF (1972) Metabolizable energy requirements of pregnant dairy cows. J Dairy Sci 55:480–483
Narayana SG, Schenkel FS, Fleming A, Koeck A, Malchiodi F, Jamrozik J, Miglior F (2017) Genetic analysis of groups of mid-infrared predicted fatty acids in milk. J Dairy Sci 100:4731–4744
Nitz J, Wente N, Zhang Y, Klocke D, Tho Seeth M, Krömker V (2021) Dry period or early lactation—Time of onset and associated risk factors for intramammary infections in dairy cows. Pathogens 10:224
Norberg E (2005) Electrical conductivity of milk as a phenotypic and genetic indicator of bovine mastitis: A review. Livest Prod Sci 96:129–139
Norberg E, Hogeveen H, Korsgaard IR, Friggens NC, Sloth KHMN, Løvendahl P (2004) Electrical conductivity of milk: ability to predict mastitis status. J Dairy Sci 87:1099–1107
Nordic Cattle Genetic Evaluation. https://nordicebv.info/health. Accessed 7 Apr 2022
Olori VE, Brotherstone S, Hill WG, McGuirk BJ (1997) Effect of gestation stage on milk yield and composition in Holstein Friesian dairy cattle. Livest Prod Sci 52:167–176
Oltenacu P, Broom D (2010) The impact of genetic selection for increased milk yield on the welfare of dairy cows. Anim Welf 19(S1):39–49. https://doi.org/10.1017/S0962728600002220
Penasa M, De Marchi M, Cassandro M (2016) Short communication: effects of pregnancy on milk yield, composition traits, and coagulation properties of Holstein cows. J Dairy Sci 99:4864–4869
Poppe M, Veerkamp RF, van Pelt ML, Mulder HA (2020) Exploration of variance, autocorrelation, and skewness of deviations from lactation curves as resilience indicators for breeding. J Dairy Sci 103:1667–1684
Poppe M, Bonekamp G, van Pelt ML, Mulder HA (2021) Genetic analysis of resilience indicators based on milk yield records in different lactations and at different lactation stages. J Dairy Sci 104:1967–1981
Pralle RS, White HM (2020) Symposium review: Big Data, big predictions: utilizing milk Fourier-transform infrared and genomics to improve hyperketonemia management. J Dairy Sci 103:3867–3873

Rajala-Schultz PJ, Gröhn YT, McCulloch CE (1999) Effects of milk fever, ketosis, and lameness on milk yield in dairy cows. J Dairy Sci 82:288–294
Raynal-Ljutovac K, Pirisi A, De Cremoux R, Gonzalo C (2007) Somatic cells of goat and sheep milk: analytical, sanitary, productive and technological aspects. Small Rumin Res 68:126–144
Rienesl L, Khayatzadeh N, Koeck A, Dale LM, Werner A, Grelet C, Sölkner J (2019) Mastitis detection from milk mid-infrared (MIR) spectroscopy in dairy cows. Mendel University in Brno Acta Universitatis Agriculturae et Silviculturae Mendelianae Brunensis 67:1221–1226
Ruegg PL (2017) A 100-year review: mastitis detection, management, and prevention. J Dairy Sci 100:10381–10397
Sanford CJ, Keefe GP, Sanchez J, Dingwell RT, Barkema HW, Leslie KE, Dohoo IR (2006) Test characteristics from latent-class models of the California Mastitis Test. Prev Vet Med 77:96–108
Santschi DE, Lacroix R, Durocher J, Duplessis M, Moore RK, Lefebvre DM (2016) Prevalence of elevated milk β-hydroxybutyrate concentrations in Holstein cows measured by Fourier-transform infrared analysis in dairy herd improvement milk samples and association with milk yield and components. J Dairy Sci 99:9263–9270
Schwarz D, Diesterbeck US, König S, Brügemann K, Schlez K, Zschöck M, Czerny CP (2011) Flow cytometric differential cell counts in milk for the evaluation of inflammatory reactions in clinically healthy and subclinically infected bovine mammary glands. J Dairy Sci 94:5033–5044
Schwarz D, Kleinhans S, Reimann G, Stückler P, Reith F, Ilves K, Fouz R (2020a) Investigation of dairy cow performance in different udder health groups defined based on a combination of somatic cell count and differential somatic cell count. Prev Vet Med 183:105123
Schwarz D, Santschi DE, Durocher J, Lefebvre DM (2020b) Evaluation of the new differential somatic cell count parameter as a rapid and inexpensive supplementary tool for udder health management through regular milk recording. Prev Vet Med 181:105079
Seifi HA, LeBlanc SJ, Leslie KE, Duffield TF (2011) Metabolic predictors of post-partum disease and culling risk in dairy cattle. Vet J 188:216–220
Sharma N, Singh NK, Bhadwal MS (2011) Relationship of somatic cell count and mastitis: an overview. Asian Australas J Anim Sci 24:429–438
Soyeurt H, Bruwier D, Romnee J-M, Gengler N, Bertozzi C, Veselko D, Dardenne P (2009) Potential estimation of major mineral contents in cow milk using mid-infrared spectrometry. J Dairy Sci 92:2444–2454
Thompson-Crispi K, Atalla H, Miglior F, Mallard BA (2014) Bovine mastitis: frontiers in immunogenetics. Front Immunol 5:493
Tiplady KM, Trinh MH, Davis SR, Sherlock RG, Spelman RJ, Garrick DJ, Harris BL (2022) Pregnancy status predicted using milk mid-infrared spectra from dairy cattle. J Dairy Sci 105:3615–3632
Toledo-Alvarado H, Vazquez AI, de Los CG, Tempelman RJ, Bittante G, Cecchinato A (2018) Diagnosing pregnancy status using infrared spectra and milk composition in dairy cows. J Dairy Sci 101:2496–2505
Vanlierde A, Vanrobays ML, Gengler N, Dardenne P, Froidmont E, Soyeurt H, McParland S, Lewis E, Deighton MH, Mathot M, Dehareng F (2016) Milk mid-infrared spectra enable prediction of lactation-stage-dependent methane emissions of dairy cattle within routine population-scale milk recording schemes. Anim Prod Sci 56:258–264
Vanlierde A, Soyeurt H, Gengler N, Colinet FG, Froidmont E, Kreuzer M, Grandl F, Bell M, Lund P, Olijhoek DW, Eugène M, Martin C, Kuhla B, Dehareng F (2018) Development of an equation for estimating methane emissions of dairy cows from milk Fourier transform mid-infrared spectra by using reference data obtained exclusively from respiration chambers. J Dairy Sci 101:7618–7624
Visentin G, Penasa M, Niero G, Cassandro M, De Marchi M (2018) Phenotypic characterisation of major mineral composition predicted by mid-infrared spectroscopy in cow milk. Ital J Anim Sci 17:549556
Wellnitz O, Bruckmaier RM (2021) Invited review: The role of the blood-milk barrier and its manipulation for the efficacy of the mammary immune response and milk production. J Dairy Sci 104:6376–6388
Zecconi A, Vairani D, Cipolla M, Rizzi N, Zanini L (2019) Assessment of subclinical mastitis diagnostic accuracy by differential cell count in individual cow milk. Ital J Anim Sci 18:460–465
Zheng Y, Chen K, Zheng X, Li H, Wang G (2014) Identification and bioinformatics analysis of microRNAs associated with stress and immune response in serum of heat-stressed and normal Holstein cows. Cell Stress Chaperones 19:973–981

# Consequences of Lactation Length Management for Health and Fertility in Dairy Cows

Ariette T. M. van Knegsel and Akke Kok

**Abstract**

Extending lactation length, as well as shortening or omitting the dry period have interesting perspectives on health and fertility of high-producing dairy cows. By extending lactation length, the frequency of critical calving events for the cow reduces, herewith reducing the frequency of periods with increased risk for health problems. Alternatively, shortening or omitting the dry period reduces the number and magnitude of peripartum transitions for the cow, with less milk and a better energy balance after calving. Potential risks of these management strategies are that milk yield of cows at the end of lactation is too low to extend the lactation period, and that cows fatten in late lactation. Therefore, both extending lactation length and shortening or omitting of the dry period might not fit every cow. Individual cow characteristics like parity, milk yield level, body condition, or health status contribute to the response of the cow to an extended lactation or shortened or omitted dry period. These individual cow characteristics can be used in customized management strategies to optimize lactation length or dry period length for individual cows. Customized lactation or dry period length could limit the impact at herd level of disadvantages concerning milk losses and body fattening and maintain benefits for improved cow health and fertility, reduced number of surplus calves, and increased work satisfaction for the farmer.

**Keywords**

Dry period length · Extended lactation · Voluntary waiting period · Continuous milking · Metabolic status · Metabolic disorders · Reproductive performance · Individual cow variation

## 1 Traditional Lactation Cycle in Dairy Cows

The lactation cycle of dairy cows traditionally consists of a lactation period of about 10 months followed by a non-lactating, or so-called dry period, of 6–8 weeks. This yearly lactation cycle is likely based on the nature of ruminants, with a calf every spring and forage availability matching the energy requirements of the dam and her calf (Balasse et al. 2021). Also, the dry period of dairy cows has a long tradition, built on the experience of dairy farmers and noted as such in handbooks of the early 1900s (Dix Arnold and Becker 1936). Possibly, the dry period in those days was not a

---

A. T. M. van Knegsel (✉) · A. Kok
Adaptation Physiology Group, Wageningen University & Research,
Wageningen, The Netherlands
e-mail: ariette.vanknegsel@wur.nl; akke.kok@wur.nl


J. J. Gross (ed.), *Production Diseases in Farm Animals*,
https://doi.org/10.1007/978-3-031-51788-4_25

farmer's decision or deliberately management strategy, but a consequence of the lower milk production level of dairy cows.

Later, several research groups focused on determining the optimal lactation length (e.g., Dijkhuizen et al. 1985; Schmidt 1989; Strandberg and Oltenacu 1989) or dry period length (Keown and Everett 1986; Kuhn et al. 2006a) for dairy cows. Almost without exception, these studies used a retrospective approach and focused solely on the consequences of milk production. An advantage of these retrospective studies is that they benefit from the power of large datasets and objective monitoring of milk yield of individual cows in a large number of herds. A disadvantage of these retrospective studies is that these large datasets lack information on the reasons why cows had a short or long lactation, or short or long dry period, and that they did not incorporate consequences for health and disease.

More recently, there has been a growing debate concerning the optimal lactation length (Knight 2005) or dry period length (Grummer and Rastani 2004) of high-producing dairy cows. An important driver for this debate is the high production potential of modern dairy cows, with an increased capacity to maintain a high milk yield towards the end of lactation, and the increasingly challenging drying-off process of cows. With an extended lactation or without a dry period, cows could realize more milk in the current lactation and drying-off could be postponed to lower yield levels or avoided (Fig. 1). Moreover, both extending the lactation length as well as shortening the dry period length have interesting perspectives for metabolic status and health of high-producing cows in the peripartum period.

## 1.1 Health Challenges in the Transition Period

Conventionally, the transition period for dairy cows is defined as the period from 3 weeks before calving to 3 weeks after calving (Drackley 1999). This transition period is not only related to a change in physiological status from pregnancy to lactation, but also with regrouping from dry to lactating cows, a change in ration to support lactation and changes in daily routine associated with milking. Moreover, even before the transition period starts, the transition from lactation to the dry period imposes a challenge to health and welfare of dairy cows (Zobel et al. 2015). High milk production at dry-off is not only related to greater udder pressure after dry-off and increased fecal glucocorticoid concentration indicating compromised cow welfare (Bertulat et al. 2013), but also with more milk leakage (Bertulat et al. 2013) and increased risk for new intramammary infections after the next calving moment (Rajala-Schultz et al. 2005).

Each transition period for dairy cows is characterized by an adaptation process to a new physiological state and new housing, ration, and management conditions. Lack of adaptation during a transition period entails an increased risk for compromised health and fertility (The Lancet 2009) or potentially culling and a reduced productive life (Pinedo et al. 2014; Fig. 2). Furthermore, etiology of a negative energy balance and health problems in the peripartum period can be traced back to management and physiology of cows during the transitions from lactation to dry period to calving and start of a new lactation (as reviewed by Ingvartsen 2006; Van Knegsel et al. 2014a).

## 1.2 Aim of This Chapter

With extending the lactation length the frequency of the transitions for the cow reduces, herewith reducing the frequency of periods with increased risk for health problems and reducing the milk yield at dry-off. Alternatively, shortening or omitting the dry period partly avoids these transitions, e.g. when the dry period is completely omitted, and at the same time reduces the magnitude of peripartum transitions with less milk and a better energy balance after calving (Fig. 2).

This chapter focuses on consequences of lactation length and dry period length for milk production, metabolic status, and health and fertility of dairy cows, with a specific focus on production diseases. Moreover, consequences for calves are discussed, as well as customized management

**Fig. 1** Schematic representation of milk production over 3 years following first calving for a conventional lactation cycle, a lactation cycle with an extended lactation, and a lactation cycle without a dry period

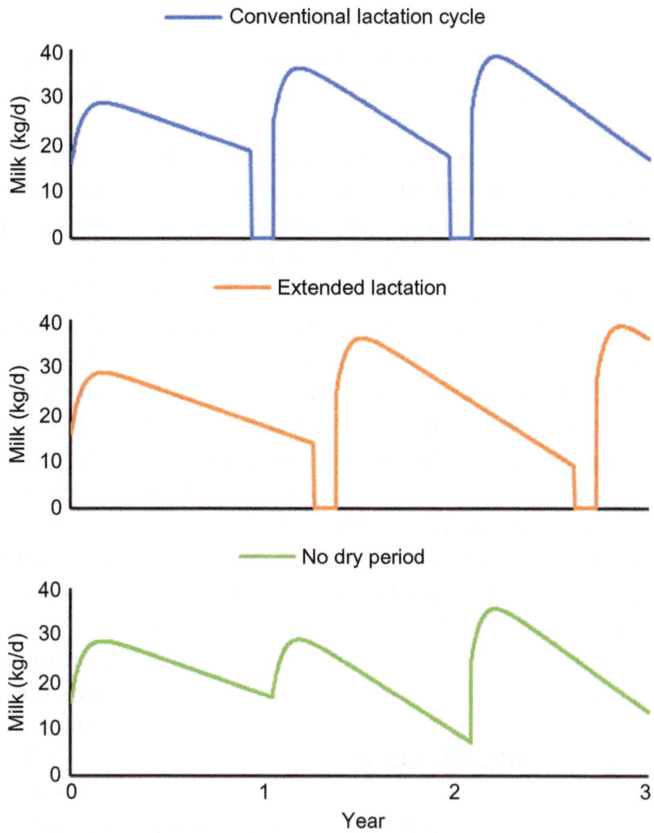


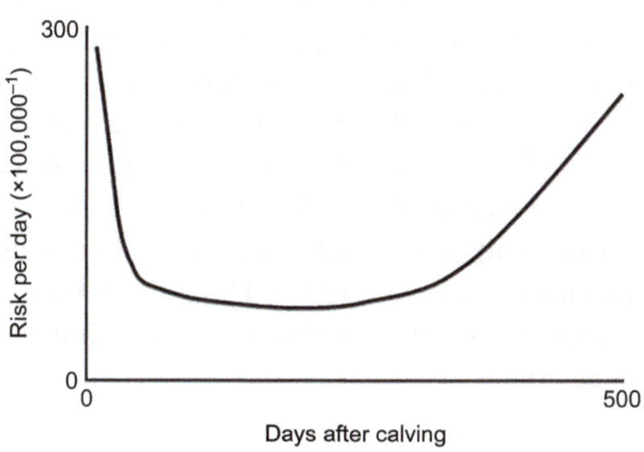


**Fig. 2** Culling risk by days after calving for Holstein dairy cows. Adapted from Pinedo et al. 2014. Culling risk early in lactation was mainly related to culling reasons death, injury or sickness, and mastitis. The increase in culling risk from about 300 days after calving was related to culling reasons "low productivity" and breeding, thus seemed related to subfertility and selection by the farmer

approaches, where performance of individual cows is optimized with a specific lactation or dry period length.

## 2 Lactation Length

In most conventional dairy systems, dairy farmers are advised to aim for a 1-year calving interval (CI), with a yearly calving moment, a yearly peak in milk production and a yearly period with increased risk for disease (Pinedo et al. 2014; Koeck et al. 2015). Originally this yearly calving event might be based on the nature of ruminants, with calves being born in spring when feed availability is abundant (Balasse et al. 2021). Later this advice was further supported by retrospective modelling studies focusing on consequences of lactation length for milk yield (Dijkhuizen et al. 1985; Strandberg and Oltenacu 1989), but lacked impact on cow health, fertility, and farm labor. In the last two decades, the traditional 1-year CI has been increasingly debated because of increased productivity of dairy cows (Knight 2005), health and welfare risks associated with the transition period (Knight 2005), and the more recent debate concerning surplus calves in the dairy sector (Bolton and Von Keyserlingk 2021).

Extending the voluntary waiting period (VWP) can be a strategy to deliberately extend the lactation length and CI of dairy cows. The VWP is the interval after calving during which farmers decide not to inseminate the cow, even if she is seen in estrous. The VWP gives the uterus time to recover from calving, recover from any peripartum infection, and return to its normal involuted state. Traditionally farmers apply a VWP of 40–60 days, aiming at a 1-year CI to maximize milk revenues (Österman and Bertilsson 2003). Deliberately extending the VWP can be a strategy to postpone artificial insemination for cows with high production potential or compromised fertility in early lactation, to reduce the number of calving moments with associated risks and reduce milk yield at dry-off.

## 2.1 Metabolic Status

Longer lactations result specifically in a longer period in late lactation with a relatively lower milk yield, and a smaller proportion of days dry. This implies that consequences of extended lactations for milk yield depend on the capacity of the cow to maintain her milk production, or lactation persistency. For multiparous cows, extending the lactation by extending the VWP resulted in a reduction in milk yield per day of CI (Rehn et al. 2000; Burgers et al. 2021a). Primiparous cows, however, have greater lactation persistency than multiparous cows (Burgers et al. 2021a). As a consequence, extending the VWP for primiparous cows had no effect on milk yield per day of CI (Rehn et al. 2000), or increased milk yield per day of CI in other studies (Arbel et al. 2001; Stangaferro et al. 2018b).

Fattening of dairy cows in late lactation is a risk for cows with a longer lactation, resulting in a greater BCS at dry-off for cows with an extended lactation (Niozas et al. 2019b; Jarman et al. 2020). In another study, specifically multiparous cows, and not primiparous cows, had a greater body condition score (BCS) in the last 3 months before dry-off when the VWP was extended to 200 d, compared with a VWP of 50 d (Burgers et al. 2021a).

Increased BCS at the end of lactation can be a risk factor for a hampered start of the new lactation with a more severe negative energy balance and increased risk for metabolic disorders like ketosis and liver fattening (Morrow 1976). Multiparous cows, but not primiparous cows, with an extended VWP of 200 d and an increased BCS at dry-off had greater plasma non-esterified fatty acid and β-hydroxybutyrate concentrations in the first 6 weeks after the subsequent calving (Burgers et al. 2023). This indicates that when the extended VWP is related to increased BCS at the end of lactation there is a risk for a more compromised metabolic status and more severe negative energy balance after the subsequent calving.

Dietary measures can potentially limit the increase in BCS towards the end of lactation of cows with an extended VWP. Dietary measures to limit fattening of cows in late lactation can either aim to limit energy intake or improve milk production and lactation persistency in late lactation. Feeding a high-energy level in early lactation, aiming at an improved lactation persistency in the complete lactation, however, had a negative carry-over effect on the persistency later in lactation, compared with a standard energy level (Gaillard et al. 2016). On the contrary, a pasture-based diet resulted in a greater proportion of cows milked until the planned time of dry-off, greater milk solid production, and lower body fattening, compared with a TMR diet for cows managed for an extended lactation (Grainger et al. 2009).

## 2.2 Reproduction

Extending the VWP gives cows, and specifically their reproductive system, more time to recover from calving, start of lactation and the negative energy balance, and can therefore be expected to improve reproductive performance. Consequences of an extended VWP for reproduction are, however, not unequivocal (Fig. 3). Cows with an extended VWP in some studies had

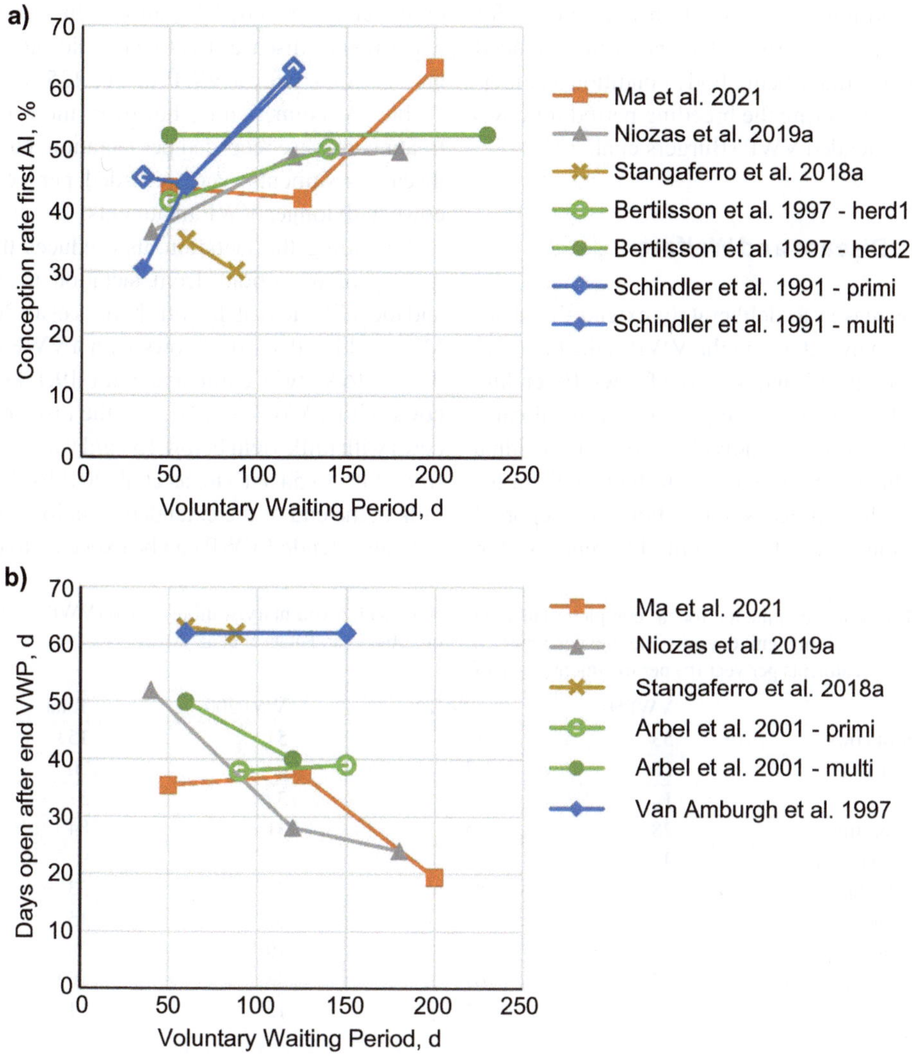


**Fig. 3** Conception rate first AI (**a**) and days open (**b**) after the end of the voluntary waiting period (VWP) for cows that were pregnant per VWP in each experiment (Van Knegsel et al. 2022)

higher conception rates at first AI, and fewer days open after the end of the VWP; though other studies found no effect of VWP on these variables. Differences among studies could be related to (1) Parity of the cows, with no improvement of reproductive performance in primiparous cows (Arbel et al. 2001); (2) Differences in management protocol, i.e., days open after end of the VWP was not reduced when cows were insemination after an estrous synchronization protocol (Stangaferro et al. 2018b) or lactations were managed with use of bovine somatotropin to stimulate milk production (Van Amburgh et al. 1997) or (3) Extent of the VWP contrast and consequences of the VWP extension for body condition.

An improved reproductive performance after extension of the VWP can be hypothesized to be due to more time to resume ovarian cyclicity or a lower milk yield and better metabolic status during the breeding period. Extension of the VWP to 125 or 200 d resulted in a greater percentage of normal ovarian cycles (18–24 d in length) during the 100 d around the end of the VWP, compared with a VWP of 50 d (Ma et al. 2022). In this same experiment, milk yield was lower and plasma insulin and IGF-1 concentration were greater for multiparous cows around the end of the extended VWP, indicating a better body condition and metabolic status during the breeding period for cows with an extended VWP (Burgers et al. 2023).

## 2.3 Disease and Welfare

A major reason to deliberately extend the lactation length by extending the VWP is the potential benefits for health and welfare of cows. Extending the VWP reduces the frequency of critical calving moments for the individual cow and within a herd, which can be expected to be beneficial for cow health. Limited studies, however, reported disease incidence of cows with different VWP or deliberately extended lactations. In our earlier experiment, disease treatments did not differ among cows with a VWP of 50, 125, or 200 d (Table 1). Numerically, however, more disease treatments were recorded per lactation, but fewer disease treatments were recorded per year, for cows with longer VWP treatments.

Extending the lactation also reduces the frequency of transitions from lactation to dry-off and the milk yield at dry-off (Niozas et al. 2019b). Milk yield at dry-off of cows with a VWP of 180 d was 16.9 kg/d compared with 19.1 kg/d for cows with a VWP of 40 d, and the proportion of cows with milk yield below 15 kg/d was increased from 34% to 54% (Niozas et al. 2019b). The low milk at the end of the extended lactation for cows with an extended VWP can be expected to reduce

**Table 1** Disease treatments for a complete lactation (6 weeks after calving until 6 weeks after next calving) (a) and disease treatments per year (b) per treatment group of cows with a voluntary waiting period (VWP) of 50, 125, or 200 d (Van Knegsel et al. 2022)

| (a) | VWP50 | VWP125 | VWP200 | Total |
|---|---|---|---|---|
| Number of cows | 53 | 49 | 51 | 153 |
| Milk fever | 10 | 17 | 9 | 36 |
| Ketosis | 1 | 1 | 3 | 5 |
| Clinical mastitis | 28 | 25 | 31 | 84 |
| Retained placenta | 4 | 1 | 4 | 9 |
| Vaginal discharge | 8 | 8 | 7 | 23 |
| Endometritis | 5 | 4 | 6 | 15 |
| Cystic ovaries | 11 | 8 | 19 | 38 |
| Claw and leg problems | 47 | 67 | 48 | 162 |
| Stomach and intestine problems[a] | 14 | 22 | 15 | 51 |
| Other[b] | 8 | 14 | 14 | 36 |
| Total | 136 | 167 | 156 | 459 |

| (b) | VWP50 | VWP125 | VWP200 | Total |
|---|---|---|---|---|
| Number of cows | 53 | 49 | 51 | 153 |
| Milk fever | 10 | 14 | 7 | 31 |
| Ketosis | 1 | 1 | 2 | 4 |
| Clinical mastitis | 28 | 21 | 24 | 73 |
| Retained placenta | 4 | 1 | 3 | 8 |
| Vaginal discharge | 8 | 7 | 5 | 20 |
| Endometritis | 5 | 3 | 5 | 13 |
| Cystic ovaries | 11 | 7 | 15 | 33 |
| Claw and leg problems | 47 | 56 | 37 | 140 |
| Stomach and intestine problems[a] | 14 | 18 | 12 | 44 |
| Other[b] | 8 | 12 | 11 | 31 |
| Total | 136 | 140 | 121 | 397 |

[a]Main stomach and intestine problems: rotavirus, diarrhea, peritonitis
[b]Main diagnoses in “other”: fever, cobalt deficiency, 3 teats

udder pressure and benefit cow welfare at dry off (Bertulat et al. 2013), and reduce the risk for new intramammary infections at start of the subsequent lactation (Rajala-Schultz et al. 2005).

There are, however, also potential risks of an extended VWP and an extended lactation for cow health. The risk for fattening of cows end of the extended lactation and a compromised metabolic status at the start of the subsequent lactation (Burgers et al. 2023) were already discussed in this chapter. In addition, several studies report a rise in SCC toward the end of an extended lactation (Sorensen et al. 2008; Niozas et al. 2019b; Ma et al. 2021). None of these studies, however, report an effect of an extended lactation for mastitis incidence. This indicates that the rise in SCC at the end of an extended lactation is probably not due to more cases of clinical mastitis, but is likely related to the lower milk yield or reduced epithelial integrity at the end of an extended lactation.

## 2.4 Calves

Extending the lactation length by extending the VWP reduces the number of calving moments and herewith reduces the number of calves born per time unit. Fewer calves born were a major motivation for 13 commercial farmers to extend the VWP in their herd (Van Dooren 2019). First, fewer calving events reduce labor associated with calving and management of neonatal calves and fresh cows. Second, a reduction in number of calves born per year reduces the number of surplus calves that will be mostly transported to the veal industry. These surplus calves are of growing concern not only for farmers but also in science, politics, and society, because of the welfare impact of transport of these calves at a young age and debated management approaches for these young calves (Bolton and Von Keyserlingk 2021).

Extending the VWP not only impacts management of calves, but also hypothetically impacts health and development of the calves born in the herd. Extending the VWP implies that cows are inseminated during a later moment in lactation, where milk production is lower and body condition and metabolic status are improved, compared with insemination earlier in lactation (Burgers et al. 2023). In other species, it is well known that the metabolic status of the dam around conception can have long-term consequences for health and metabolic status of the offspring (as reviewed by Fleming et al. 2018). In dairy cows, milk production level of the dam during the breeding period was negatively related to milk production and survival of the offspring (Berry et al. 2008). Future studies should evaluate the consequences of metabolic status of the cow during the conception period on health and development of her calf during early and later life.

## 2.5 Customized Lactation Length

Although extending the VWP can benefit cow health and fertility, there are also potential trade-offs of an extended VWP with respect to low milk yield and fattening of cows at the end of the lactation. Moreover, the response of cows to an extended VWP depended on individual cow characteristics. In practice, dairy farmers recognized that not all cows are suited for an extended VWP (Burgers et al. 2021b). In this earlier study, the strategy to extend the lactation by extending the VWP varied among farmers. Some farmers (3/13) applied a fixed VWP for the whole herd. Most farmers (10/13) applied a customized strategy based on individual cow characteristics like parity, milk yield level, body condition, or a combination of cow characteristics. As a result, cows with a short calving interval at these farms had a lower peak milk yield and lower 305-d milk yield, while cows with a longer calving interval had a greater 305-d milk yield and a more persistent lactation curve. So farmers seem to select cows with a high milk yield for an extended lactation (Lehmann et al. 2017; Burgers et al. 2021b) to ensure milk yield toward the end of the lactation, limit the risk for body fattening, but possibly also due to the greater benefit of reducing milk yield at dry-off and reducing the frequency of critical calving events specifically for high-producing cows.

The response of individual cows to a VWP was not only evaluated in observational studies (Lehmann et al. 2017; Burgers et al. 2021b), but also in a randomized controlled trial (Burgers et al. 2021a). Parity, milk yield level, and shape of the lactation curve in both the previous and early lactation and body weight predicted the fat-and-protein correct milk yield per day calving interval of cows with different VWP (Burgers et al. 2021a). Moreover, not VWP, but milk yield level, milk composition, and disease treatments determined the net economic result of individual cows with different VWP (Burgers et al. 2022). Until now, especially on-farm cow characteristics were evaluated concerning the prediction of the individual cow response to different VWP. It can be hypothesized that also health characteristics or metabolic status of cows in early lactation is related to the response of cows to an extended VWP. At the moment, studies are ongoing to determine the value of metabolic variables in early lactation to predict the performance of cows in an extended lactation.

## 3 Dry Period Length

The dry period is the non-lactating period before (next) calving, which generally lasts 6 to 8 weeks in modern dairy farming systems. The dry period has multiple functions. First, cows with persistent intramammary infections can be treated with dry cows antibiotics (Bradley and Green 2001), Second, the dry period allows the cow a rest period before the next calving (Kok et al. 2017a). Third, a dry period of 6 to 8 weeks is associated with maximal milk yield in the next lactation (Kuhn et al. 2005, 2006b). During a dry period, renewal of mammary secretory cells is maximal, resulting in a high proportion of new secretory cells at the moment of calving (Capuco et al. 1997), which would clarify the peak in milk yield after calving (Capuco et al. 2001).

Shortening, and omitting, the dry period in dairy cows reduces milk yield after calving, but increases milk yield in the period before calving, when the cow can meet the energy demands for milk production easily (Rastani et al. 2005; Van Knegsel et al. 2014a; Fig. 4a). The reduction in energy partitioned to milk after calving resulted in a better energy balance in early lactation (Rastani et al. 2005; Van Knegsel et al. 2014a; Andrée O'Hara et al. 2018), as earlier reviewed (Kok et al. 2019) (Fig. 4b). Omitting and shortening the dry period resulted in similar or increased dry matter intake in early lactation (as reviewed in Van Knegsel et al. 2013). Reasons for an improved dry matter intake with a short or no dry period are not completely known, but can be speculated to be related to fewer changes in daily routine, ration, and social group, or a less positive energy balance in the precalving period resulting in lower body condition score at calving (Grummer et al. 2004; Rastani et al. 2005). So, shortening and omitting the dry period results in a better energy balance after calving, which is mainly due to the lower milk yield after calving.

### 3.1 Metabolic Status

The better energy balance in early lactation of dairy cows with no or a short dry period is partially reflected in the metabolic status of dairy cows in early lactation. Cows with no dry period had lower plasma non-esterified fatty acids (NEFA) and beta-hydroxybutyrate (BHB) concentrations than cows with a conventional dry period (Rastani et al. 2005; Andersen et al. 2005; Klusmeyer et al. 2009; Chen et al. 2015a), indicating less body fat mobilization and less incomplete hepatic fat oxidation. Moreover, cows with no dry period had greater plasma glucose, insulin, and IGF-1 concentration (Andersen et al. 2005; De Feu et al. 2009; Chen et al. 2015a), mirroring the better body condition in these cows, compared with cows with conventional dry period. Also, a short dry period sometimes reduced plasma NEFA and BHB concentration in early lactation compared with a conventional dry period (Pezeshki et al. 2007; Klusmeyer et al. 2009), but not in all studies (Rastani et al. 2005; Chen et al. 2015a).

Shortening and omitting the dry period not only resulted in a better energy balance in early lactation but also in a better energy balance (Van

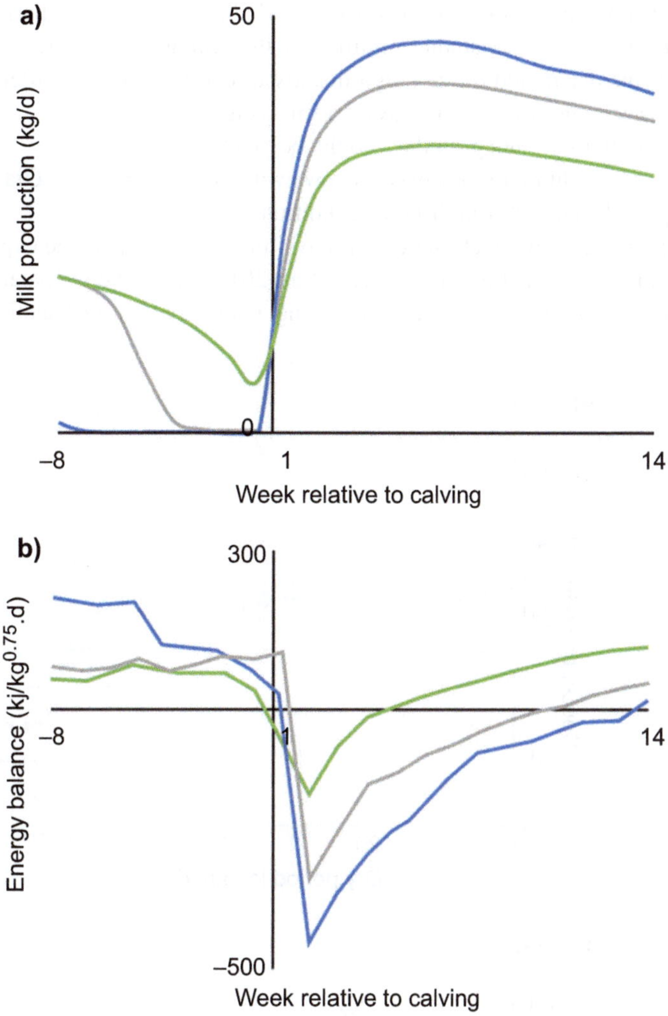


**Fig. 4** Milk production and energy balance from week 8 to 14 relative to calving for cows with a standard (blue), short (grey), or no (green) dry period. Adapted from Van Knegsel et al. (2014b)

Hoeij et al. 2017) and greater body weight (Chen 2016) in mid and late lactation, due to a lower milk output and similar energy intake. The more positive energy balance poses a risk for body fattening in mid and late lactation, with potentially negative consequences for energy balance and metabolic status after the subsequent calving (Chen et al. 2016). To our knowledge, almost all experimental studies evaluated consequences of dry period length while feeding the same energy level during lactation. This made sense from a scientific perspective, avoiding confounding effects of diet and dry period length in the experimental set-up and not limiting the production capacity of cows with a short or no dry period due to feeding a lower energy level in the diet. Under practical circumstances, however, farmers might adjust their diets based on dry period length or expected milk yield. Feeding a lower energy level in early lactation did not affect milk yield or plasma NEFA or BHB concentration, but indeed reduced the positive energy balance in early lactation of cows after no dry period (Van Hoeij et al. 2017).

## 3.2 Reproduction

In some studies, omitting and shortening the dry period reduced the interval between calving and onset of luteal activity (Gümen et al. 2005; Watters et al. 2009) and improved the regularity of subsequent ovarian cycles (Chen et al. 2015b). However, in other studies omitting or shortening of the dry period did not significantly affect the conception rate and days open (Fig. 5). Differences among studies could be related to body condition of the cows, as cows with no dry period that had a high body condition score in early lactation had increased days open and reduced conception rates (Chen et al. 2017; Ma et al. 2020). Management, including voluntary waiting period, estrus detection, and insemination policy, could also affect reproductive results of cows after different dry period lengths.

## 3.3 Disease in Early Lactation

It can be expected that a lower milk yield, better energy balance, and better metabolic status in early lactation due to a short or no dry period

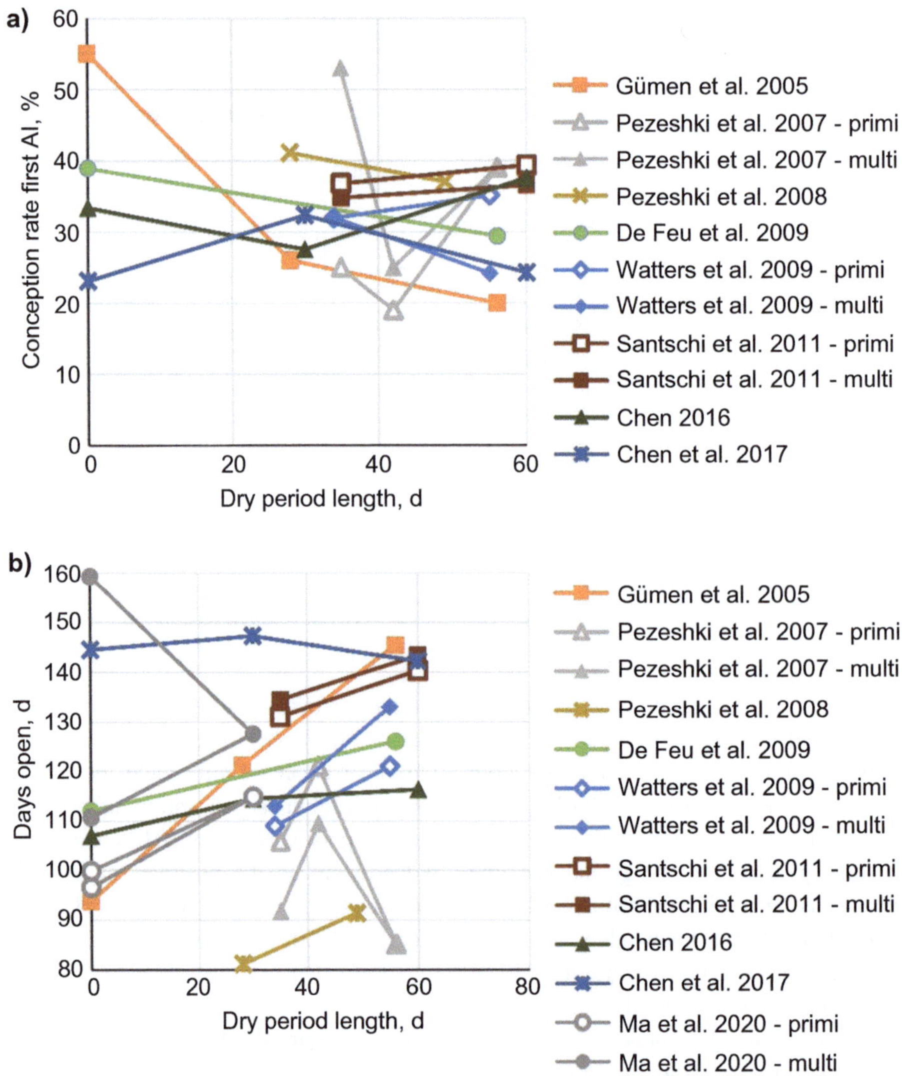


**Fig. 5** Conception rate first AI (**a**) and days open (**b**) after different dry period lengths for cows per dry period length in each experiment. Note Chen et al. 2017: Cows in this study were subjected to a dry period of 0, 30, or 60 days for a second time, where dry period length was confounded with BCS prepartum and in early lactation. Note Ma et al. 2020: Cows with a 0-day dry period length were fed one of two energy levels

reduces the incidence of disease in early lactation. Limited studies, however, reported the disease incidence of cows with different dry period lengths. In a meta-analysis, shortening of the dry period tended to be associated with a lower odds for ketosis (OR = 0.75; P = 0.09), compared with a conventional dry period (Van Knegsel et al. 2013). Three studies reported that ketosis did not occur after no dry period, compared with a conventional dry period (Rastani et al. 2005; Schlamberger et al. 2010; Köpf et al. 2014). In the same meta-analysis, shortening of the dry period was not associated with incidence of metritis, retained placenta, and mastitis, compared with a conventional dry period (Van Knegsel et al. 2013).

## 3.4 Udder Health

Effects of dry period length on udder health are not univocal. Dry period length did not affect the incidence of mastitis in most studies (Santschi et al. 2011; Van Hoeij et al. 2016) nor in a meta-analysis including seven studies (Van Knegsel et al. 2013). In addition, shortening the dry period did not affect SCC after calving compared with a conventional dry period when cows were treated with dry cow antibiotics (Rastani et al. 2005; Pezeshki et al. 2007; Van Knegsel et al. 2014b). Omitting the dry period increased SCC after calving in some (Klusmeyer et al. 2009; Van Knegsel et al. 2014b; Van Hoeij et al. 2016) but not all studies (Rastani et al. 2005; Köpf et al. 2014).

Variation in reported effects of dry period length on udder health might be due to udder health status or immune status of the cows at the end of the lactation, drying off protocol, and other management measures on the farm related to hygiene. A higher SCC after omission of the dry period might be explained by three factors. First, milk yield after calving is lower when the DP is omitted, which reduces the dilution and consequently increases the concentration of the absolute number of cells in milk (Steeneveld et al. 2013). Second, renewal of udder cells during lactation might be increased when the DP is omitted (Annen et al. 2008), whereas renewal of udder cells would normally occur at a high rate during the DP (Capuco et al. 2001). Third, the greater SCC after no dry period might be related to uncured intramammary infections because of the lack of both a dry period and lack of treatment with dry cow antibiotics. To our knowledge, there are two studies that compared udder health of cows without a dry period with udder health of cows with a dry period without use of dry cow antibiotics. Both studies reported an increased risk for mastitis after a short dry period, compared with a conventional dry period (Andrée O'Hara et al. 2019) or an increased risk for mastitis after no dry period, when compared with cows with a dry period without use of dry cow antibiotics (Van Hoeij et al. 2018).

## 3.5 Calves

Calf health could be affected by shortening or omitting the dry period due to differences in physiology and management during late gestation or in colostrum composition during early lactation. Omission of the dry period reduced gestation length by 3 days, and the birth weight of calves by 1.5 kg, compared with a conventional dry period (Mayasari et al. 2015). A short dry period did not affect gestation length or birth weight.

After calving, no dry period resulted in a lower concentration of antibodies, both IgG and IgM, in colostrum compared with a short or conventional dry period (Rastani et al. 2005; Mayasari et al. 2015). The lower concentration of antibodies can be explained by antibodies being secreted in milk, instead of accumulating, in the days before calving (Baumrucker et al. 2014). The quantity of colostrum in case of a short dry period in most studies was sufficient to fulfil the needs of the calf (Rastani et al. 2005; Watters et al. 2008; Shoshani et al. 2014; Mayasari et al. 2015). The high variability in colostrum quality when the DP is omitted (Baumrucker et al. 2014), however, could imply that a larger quantity of colostrum is fed to calves to ensure an adequate intake of antibodies.

## 3.6 Customized Dry Period Length

Although shortening and omitting the dry period benefit energy status and fertility of dairy cows in the subsequent lactation, potential tradeoffs would be a milk yield reduction and loss of opportunity to treat cows with dry cow antibiotics. In addition, the response of dairy cows to a short or no dry period depended on individual cow characteristics. First, primiparous cows gave more milk in the current lactation when the dry period is omitted, but have a greater milk yield reduction after calving, compared with multiparous cows (Pezeshki et al. 2007; Van Knegsel et al. 2014b). Second, the effect of dry period length on SCC depends on SCC elevations and average SCC in the previous lactation and SCC at the last test day before dry-off (Van Hoeij et al. 2016). Third, high-producing cows can be considered to be better suited for shorter dry periods because they produce more additional milk before calving, are more difficult to dry off (Rajala-Schultz et al. 2005; Zobel et al. 2015), benefit more from an improvement in energy balance in the subsequent lactation, and proportionally have the lowest milk yield losses in the subsequent lactation (Kok et al. 2017b).

Customizing dry period length for individual cows might retain the benefits of shorter or no dry periods and limit the negative consequences at herd level. We evaluated two decision trees (Fig. 6) for dry period length and use of dry cow antibiotics based on parity, milk yield level, and SCC in late lactation, and compared these with a

| Decision tree | | 67 days before expected calving | | 37 days before expected calving | | Treatment |
|---|---|---|---|---|---|---|
| C | | | | | | |
| | All | SCC ≥ 150.000 | | | | 60AB |
| | | SCC < 150.000 | | | | 60TS |
| T1 | | | | | | |
| | Par 1 | SCC ≥ 150.000 | | | | 60AB |
| | | SCC < 150.000 | Yield < 12 kg/ d | | | 60TS |
| | | | Yield ≥ 12 kg/ d | | | 30TS |
| | Par >1 | SCC ≥ 50.000 | | | | 60AB |
| | | SCC < 50.000 | Yield < 12 kg/ d | | | 60TS |
| | | | Yield ≥ 12 kg/ d | SCC ≥ 50.000 | | 30TS |
| | | | | SCC < 50.000 | Yield < 12 kg/ d | 30TS |
| | | | | | Yield ≥ 12 kg/ d | 0d |
| T2 | | | | | | |
| | Par 1 | SCC ≥ 200.000 | | | | 60AB |
| | | SCC < 200.000 | Yield < 12 kg/ d | | | 60TS |
| | | | Yield ≥ 12 kg/ d | | | 30TS |
| | Par >1 | SCC ≥ 200.000 | | | | 60AB |
| | | SCC < 200.000 | Yield < 12 kg/ d | | | 60TS |
| | | | Yield ≥ 12 kg/ d | SCC ≥ 200.000 | | 30TS |
| | | | | SCC < 200.000 | Yield < 12 kg/ d | 30TS |
| | | | | | Yield ≥ 12 kg/ d | 0d |

**Fig. 6** Visualization of the dry-period strategies applied under control (CT), treatment 1 (T1), and treatment 2 (T2) decision trees. Combinations of dry-period length of 60, 30, or 0 d, and dry-cow therapy with antibiotics (AB) or teat seal (TS) depended on the SCC in cells/ mL on the last test day before the decision moment at 67 or 37 d before expected calving (adapted from Kok et al. 2021)

fixed dry period length in combination with selective dry cow antibiotics (Kok et al. 2021). In decision tree 1, 36% of the cows qualified for a short dry period of 30 d and 2% of the cows for no dry period. In decision tree 2, 51% of the cows qualified for a short dry period and 30% of the cows for no dry period. Compared with the control group, cows in both decision trees produced more milk in the 8 weeks before calving and less in the 14 weeks after calving. These differences were smaller than when all cows were given a short or no dry period (Van Knegsel et al. 2014a). There was no effect on SCC or mastitis incidence. Due to the larger proportion of short and no dry periods, cows in decision tree 2 had a greater body weight gain in the first 14 weeks of lactation compared with the control group and cows in decision tree 1. Fewer cows tended to be treated for diseases in the first 100 DIM in decision tree 2 (37%) compared with decision tree 1 (54%) and the control treatment (57%).

## 4 Conclusions and Future Perspectives

Both extending the lactation length and shortening or omitting the dry period can improve health and fertility of high-producing dairy cows. With extending the lactation length the frequency of critical calving events for the cow reduces, herewith reducing the frequency of periods with increased risk for health problems. Alternatively, shortening or omitting the dry period partly avoids the transitions for the cow around calving and at the same time reduces the magnitude of peripartum transitions with less milk and a better energy balance after calving. Potential risks of these management strategies are that the milk yield of cows at the end of lactation is too low to extend the lactation period possibly associated with body fattening at end of the lactation. Herewith, both extending the lactation length as well as shortening or omitting of the dry period might not fit every cow. Customized lactation or dry period length could limit the impact at herd level of disadvantages concerning milk losses and body fattening and maintain benefits for improved cow health and fertility, reduced number of surplus calves, and increased work pleasure for the farmer.

The current decision support models to assign individual cows to a VWP or dry period length are based on farmers' experience (Burgers et al. 2021b) or a single dataset from a randomized controlled trial (Kok et al. 2021). To move from prediction models to decision models applicable in dairy practice, these models should be validated under different field conditions. Moreover, fine tuning decision support models based on herd-specific characteristics or including other data sources, like sensor data, can advance the development, predictive value, and applicability of these decision-support models.

Customized management of cows does not end at the right and accurate selection of cows for a specific VWP or dry period length. As described above, dietary measures or selective use of dry cow antibiotics could potentially stimulate lactation persistency or limit body fattening of cows with an extended lactation or no dry period (Grainger et al. 2009; Van Hoeij et al. 2017) or compromised udder health of cows with no dry period (Van Hoeij et al. 2016; Kok et al. 2021). Future studies, including not only dietary measures but possibly also breeding strategies could contribute to further improvement of the health and performance of cows with different VWPs or dry period lengths.

## References

Andersen JB, Madsen TG, Larsen T, Ingvartsen KL, Nielsen MO (2005) The effects of dry period versus continuous lactation on metabolic status and performance in periparturient cows. J Dairy Sci 88:3530–3541

Andrée O'Hara E, Omazic A, Olsson I, Båge R, Emanuelson U, Holtenius K (2018) Effects of dry period length on milk production and energy balance in two cow breeds. Animal 12:508–514

Andrée O'Hara E, Båge R, Emanuelson U, Holtenius K (2019) Effects of dry period length on metabolic status, fertility, udder health, and colostrum production in 2 cow breeds. J Dairy Sci 102:595–606

Annen EL, Stiening CM, Crooker BA, Fitzgerald AC, Collier RJ (2008) Effect of continuous milking and prostaglandin E2 on milk production and mammary epithelial cell turnover, ultrastructure, and gene expression. J Anim Sci 86:1132–1144

Arbel R, Bigun Y, Ezra E, Sturman H, Hojman D (2001) The effect of extended calving intervals in high-yielding lactating cows on milk production and profitability. J Dairy Sci 84:600–608

Balasse M, Gillis R, Živaljević I, Berthon R, Kovačiková L, Fiorillo D, Arbogast RM, Bălăşescu A, Bréhard S, Nyerges É, Dimitrijević V, Bánffy E, Domboróczki L, Marciniak A, Oross K, Vostrovská I, Roffet-Salque M, Stefanović S, Ivanova M (2021) Seasonal calving in European Prehistoric cattle and its impacts on milk availability and cheese-making. Sci Rep 11:1–11

Baumrucker CR, Zbinden RS, van Dorland HA, Remmelink GJ, Kemp B, Van Knegsel ATM, Bruckmaier RM (2014) Continuous milking of dairy cows disrupts timing of peak IgG concentration appearance in mammary secretions. J Dairy Res 81:403–409

Berry DP, Lonergan P, Butler ST, Cromie AR, Fair T, Mossa F, Evans ACO (2008) Negative influence of high maternal milk production before and after conception on offspring survival and milk production in dairy cattle. J Dairy Sci 91:329–337

Bertilsson J, Berglund B, Ratnayake G, Svennersten-Sjaunja K, Wiktorsson H (1997) Optimising lactation cycles for the high-yielding dairy cow. A European perspective. Livest Prod Sci 50:5–13

Bertulat S, Fischer-Tenhagen C, Suthar V, Möstl E, Isaka N, Heuwieser W (2013) Measurement of fecal glucocorticoid metabolites and evaluation of udder characteristics to estimate stress after sudden dry-off in dairy cows with different milk yields. J Dairy Sci 96:3774–3787

Bolton SE, Von Keyserlingk MAG (2021) The dispensable surplus dairy calf: is this issue a "wicked problem" and where do we go from here? Front Vet Sci 8:1–12

Bradley AJ, Green MJ (2001) An investigation of the impact of intramammary antibiotic dry cow therapy on Clinical Coliform Mastitis. J Dairy Sci 84:1632–1639

Burgers EEA, Kok A, Goselink RMA, Hogeveen H, Kemp B, Van Knegsel ATM (2021a) Effects of extended voluntary waiting period from calving until first insemination on body condition, milk yield, and lactation persistency. J Dairy Sci 104:8009–8022

Burgers EEA, Kok A, Goselink RMA, Hogeveen H, Kemp B, Van Knegsel ATM (2021b) Fertility and milk production on commercial dairy farms with customized lactation lengths. J Dairy Sci 104:443–458

Burgers EEA, Kok A, Goselink RMA, Hogeveen H, Kemp B, Van Knegsel ATM (2022) Revenues and costs of dairy cows with different voluntary waiting periods based on data of a randomized control trial. J Dairy Sci 105:4171–4188

Burgers EEA, Goselink RMA, Bruckmaier RM, Gross JJ, Jorritsma R, Kemp B, Kok A, Van Ruitenbeek AH, Van Knegsel ATM (2023) Effect of voluntary waiting period on metabolism of dairy cows during different phases of the lactation. J Anim Sci 101:skad194

Capuco AV, Akers RM, Smith JJ (1997) Mammary growth in Holstein cows during the dry period: quantification of nucleic acids and histology. J Dairy Sci 80:477–487

Capuco A, Wood D, Baldwin R, Mcleod K, Paape M (2001) Mammary cell number, proliferation, and apoptosis during a bovine lactation: relation to milk production and effect of bST. J Dairy Sci 84:2177–2187

Chen J (2016) Shortening or omitting the dry period in dairy cows. PhD thesis. Wageningen University, Wageningen, p 205

Chen J, Gross JJ, van Dorland HA, Remmelink GJ, Bruckmaier RM, Kemp B, Van Knegsel ATM (2015a) Effects of dry period length and dietary energy source on metabolic status and hepatic gene expression of dairy cows in early lactation. J Dairy Sci 98:1033–1045

Chen J, Soede NM, van Dorland HA, Remmelink GJ, Bruckmaier RM, Kemp B, Van Knegsel ATM (2015b) Relationship between metabolism and ovarian activity in dairy cows with different dry period lengths. Theriogenology 84:1387–1396

Chen J, Remmelink GJ, Gross JJ, Bruckmaier RM, Kemp B, Van Knegsel ATM (2016) Effects of dry period length and dietary energy source on milk yield, energy balance, and metabolic status of dairy cows over 2 consecutive years: effects in the second year. J Dairy Sci 99:4862–4838

Chen J, Soede NM, Remmelink GJ, Bruckmaier RM, Kemp B, Van Knegsel ATM (2017) Relationships between uterine health and metabolism in dairy cows with different dry period lengths. Theriogenology 101:8–14

De Feu MA, Evans ACO, Lonergan P, Butler ST (2009) The effect of dry period duration and dietary energy density on milk production, bioenergetic status, and postpartum ovarian function in Holstein-Friesian dairy cows. J Dairy Sci 92:6011–6022

Dijkhuizen AA, Stelwagen J, Renkema JA (1985) Economic aspects of reproductive failure in dairy cattle. I. Financial loss at farm level. Prev Vet Med 3:251–263

Dix Arnold PT, Becker RB (1936) Influence of preceding dry period and of mineral supplement on lactation. J Dairy Sci 19:257–266

Drackley JK (1999) ADSA foundation scholar award. Biology of dairy cows during the transition period: the final frontier? J Dairy Sci 82:2259–2273

Fleming TP, Watkins AJ, Velazquez MA, Mathers JC, Prentice AM, Stephenson J, Barker M, Saffery R, Yajnik CS, Eckert JJ, Hanson MA, Forrester T, Gluckman PD, Godfrey KM (2018) Origins of lifetime health around the time of conception: causes and consequences. Obstet Gynecol Surv 73:555–557

Gaillard C, Friggens NC, Taghipoor M, Weisbjerg MR, Lehmann JO, Sehested J (2016) Effects of an individual weight-adjusted feeding strategy in early lactation on milk production of Holstein cows during extended lactation. J Dairy Sci 99:2221–2236

Grainger C, Auldist MJ, O'Brien G, Macmillan KL, Culley C (2009) Effect of type of diet and energy intake on milk production of Holstein-Friesian cows with extended lactations. J Dairy Sci 92:1479–1492

Grummer RR, Rastani RR (2004) Why reevaluate dry period length? J Dairy Sci 87:E77–E85

Grummer RR, Mashek DG, Hayirli A (2004) Dry matter intake and energy balance in the transition period. Vet Clin North Am Food Anim Pract 20:447–470

Gümen A, Rastani RR, Grummer RR, Wiltbank MC (2005) Reduced dry periods and varying prepartum diets alter postpartum ovulation and reproductive measures. J Dairy Sci 88:2401–2411

Ingvartsen KL (2006) Feeding- and management-related diseases in the transition cow; physiological adaptations around calving and strategies to reduce feeding-related diseases. Anim Feed Sci Technol 126:175–213

Jarman J, Kay J, Neal M, Donaghy D, Tozer P (2020) Implications of using an extended lactation to change from a spring-calving to an autumn- calving farm system in South Taranaki. N Z J Anim Sci Prod 80:143–149

Keown JF, Everett RW (1986) Effect of days carried calf, days dry, and weight of first calf heifers on yield. J Dairy Sci 69:1891–1896

Klusmeyer TH, Fitzgerald AC, Fabellar AC, Ballam JM, Cady RA, Vicini JL (2009) Effect of recombinant bovine somatotropin and a shortened or no dry period on the performance of lactating dairy cows. J Dairy Sci 92:5503–5511

Knight CH (2005) Extended lactation: turning theory into reality. Adv Dairy Technol 17:113–123

Koeck A, Fuerst C, Egger-Danner C (2015) Farmer-observed health data around calving-genetic parameters and association with veterinarian diagnoses in Austrian Fleckvieh cows. J Dairy Sci 98:2753–2758

Kok A, Van Hoeij RJ, Tolkamp BJ, Haskell MJ, Van Knegsel ATM, de Boer IJM, Bokkers EAM (2017a) Behavioural adaptation to a short or no dry period with associated management in dairy cows. Appl Anim Behav Sci 186:7–15

Kok A, Van Middelaar CE, Mostert PF, Van Knegsel ATM, Kemp B, De Boer IJM, Hogeveen H (2017b) Effects of dry period length on production, cash flows and greenhouse gas emissions of the dairy herd: a dynamic stochastic simulation model. PLoS One 12:e0187101

Kok A, Chen J, Kemp B, Van Knegsel ATM (2019) Review: dry period length in dairy cows and consequences for metabolism and welfare and customised management strategies. Animal 13:S42–S51

Kok A, Van Hoeij RJ, Kemp B, Van Knegsel ATM (2021) Evaluation of customized dry-period strategies in dairy cows. J Dairy Sci 104:1887–1899

Köpf M, Gellrich K, Küchenhoff H, Meyer HHD, Kliem H (2014) Effects of continuous milking during a field trial on productivity, milk protein yield and health in dairy cows. Animal 8:1130–1138

Kuhn M, Hutchison JL, Norman HD (2005) Minimum days dry to maximize milk yield in subsequent lactation. Anim Res 54:351–367

Kuhn MT, Hutchison JL, Norman HD (2006a) Dry period length to maximize production across adjacent lactations and lifetime production. J Dairy Sci 89:1713–1722

Kuhn MT, Hutchison JL, Norman HD (2006b) Effects of length of dry period on yields of milk fat and protein, fertility and milk somatic cell score in the subsequent lactation of dairy cows. J Dairy Res 73:154–162

Lancet T (2009) What is health? The ability to adapt. Lancet 373:781

Lehmann JO, Mogensen L, Kristensen T (2017) Early lactation production, health, and welfare characteristics of cows selected for extended lactation. J Dairy Sci 100:1487–1501

Ma J, van Hoeij R, Bruckmaier R, Kok A, Lam TJGM, Kemp B, Van Knegsel ATM (2020) Consequences of transition treatments on fertility and associated metabolic status of dairy cows in early lactation. Animals 10:1100

Ma J, Kok A, Goselink RMA, Lam TJGM, Kemp B, Van Knegsel ATM (2021) Effect of extended voluntary waiting period from calving until first insemination on udder health. In: Abstracts of the 2021 American Dairy Science Association® annual meeting ADSA (J. Dairy Sci. Supplement 1), p 158

Ma J, Burgers EEA, Kok A, Goselink RMA, Lam TJGM, Kemp B, Van Knegsel ATM (2022) Consequences of extending the voluntary waiting period for insemination on reproductive performance in dairy cows. Anim Reprod Sci 244:107046

Mayasari N, de Vries RG, Nieuwland MGB, Remmelink GJ, Parmentier HK, Kemp B, Van Knegsel ATM (2015) Effect of maternal dry period length on colostrum immunoglobulin content and natural and specific antibody titers in calves. J Dairy Sci 98:3969–3979

Morrow DA (1976) Fat cow syndrome. J Dairy Sci 59:1625–1629

Niozas G, Tsousis G, Steinhöfel I, Brozos C, Römer A, Wiedemann S, Bollwein H, Kaske M (2019a) Extended lactation in high-yielding dairy cows. I. Effects on reproductive measurements. J Dairy Sci 102:799–810

Niozas G, Tsousis G, Malesios C, Steinhöfel I, Boscos C, Bollwein H, Kaske M (2019b) Extended lactation in high-yielding dairy cows. II. Effects on milk production, udder health, and body measurements. J Dairy Sci 102:811–823

Österman S, Bertilsson J (2003) Extended calving interval in combination with milking two or three times per day: effects on milk production and milk composition. Livest Prod Sci 82:139–149

Pezeshki A, Mehrzad J, Ghorbani GR, Rahmani HR, Collier RJ, Burvenich C (2007) Effects of short dry periods on performance and metabolic status in Holstein dairy cows. J Dairy Sci 90:5531–5541

Pezeshki A, Mehrzad J, Ghorbani GR, De Spiegeleer B, Collier RJ, Burvenich C (2008) The effect of dry period length reduction to 28 days on the performance of multiparous dairy cows in the subsequent lactation. Can J Anim Sci 88:449–456

Pinedo PJ, Daniels A, Shumaker J, De Vries A (2014) Dynamics of culling for Jersey, Holstein, and Jersey × Holstein crossbred cows in large multibreed dairy herds. J Dairy Sci 97:2886–2895

Rajala-Schultz PJ, Hogan JS, Smith KL (2005) Short communication: association between milk yield at dry-off and probability of intramammary infections at calving. J Dairy Sci 88:577–579

Rastani RR, Grummer RR, Bertics SJ, Gümen A, Wiltbank MC, Mashek DG, Schwab MC (2005) Reducing dry period length to simplify feeding transition cows: milk production, energy balance, and metabolic profiles. J Dairy Sci 88:1004–1014

Rehn H, Berglund B, Emanuelson U, Tengroth G, Philipsson J (2000) Milk production in Swedish dairy cows managed for calving intervals of 12 and 15 months. Acta Agric Scand A Anim Sci 50:263–271

Santschi DE, Lefebvre DM, Cue RI, Girard CL, Pellerin D (2011) Incidence of metabolic disorders and reproductive performance following a short (35-d) or conventional (60-d) dry period management in commercial Holstein herds. J Dairy Sci 94:3322–3330

Schindler H, Eger S, Davidson M, Ochowski D, Schermerhorn E, Foote R (1991) Factors affecting response of groups of dairy cows managed for different calving-conception intervals. Theriogenology 36:495–503

Schlamberger G, Wiedemann S, Viturro E, Meyer HHD, Kaske M (2010) Effects of continuous milking during the dry period or once daily milking in the first 4 weeks of lactation on metabolism and productivity of dairy cows. J Dairy Sci 93:2471–2485

Schmidt GH (1989) Effect of length of calving intervals on income over feed and variable costs. J Dairy Sci 72:1605–1611

Shoshani E, Rozen S, Doekes JJ (2014) Effect of a short dry period on milk yield and content, colostrum quality, fertility, and metabolic status of Holstein cows. J Dairy Sci 97:2909–2922

Sorensen A, Muir DD, Knight CH (2008) Extended lactation in dairy cows: effects of milking frequency, calving season and nutrition on lactation persistency and milk quality. J Dairy Res 75:90–97

Stangaferro ML, Wijma R, Masello M, Thomas MJ, Giordano JO (2018a) Extending the duration of the voluntary waiting period from 60 to 88 days in cows that received timed artificial insemination after the Double-Ovsynch protocol affected the reproductive performance, herd exit dynamics, and lactation performance of dairy. J Dairy Sci 101:717–735

Stangaferro ML, Wijma R, Masello M, Thomas MJ, Giordano JO (2018b) Economic performance of lactating dairy cows submitted for first service timed artificial insemination after a voluntary waiting period of 60 or 88 days. J Dairy Sci 101:7500–7516

Steeneveld W, Schukken YH, van Knegsel ATM, Hogeveen H (2013) Effect of different dry period lengths on milk production and somatic cell count in subsequent lactations in commercial Dutch dairy herds. J Dairy Sci 96:2988–3001

Strandberg E, Oltenacu PA (1989) Economic consequences of different calving intervals. Acta Agric Scand A Anim Sci 39:407–420

Van Amburgh ME, Galton DM, Bauman DE, Everett RW (1997) Management and economics of extended calving intervals with use of bovine somatotropin. Livest Prod Sci 50:15–28

Van Dooren N (2019) The effects of extending the voluntary waiting period on services per conception in dairy cows and the added value of extended lactation for Dutch dairy farmers MSc thesis. Wageningen University, Wageningen, p 105

Van Hoeij RJ, Lam TJGM, de Koning DB, Steeneveld W, Kemp B, Van Knegsel ATM (2016) Cow characteristics and their association with udder health after different dry period lengths. J Dairy Sci 99:8330–8340

Van Hoeij RJ, Dijkstra J, Bruckmaier RM, Gross JJ, Lam TJGM, Remmelink GJ, Kemp B, Van Knegsel ATM (2017) Consequences of dietary energy source and energy level on energy balance, lactogenic hormones, and lactation curve characteristics of cows after a short or omitted dry period. J Dairy Sci 100:8544–8564

Van Hoeij RJ, Lam TJGM, Bruckmaier RM, Dijkstra J, Remmelink GJ, Kemp B, Van Knegsel ATM (2018) Udder health of dairy cows fed different dietary energy levels after a short or no dry period without use of dry cow antibiotics. J Dairy Sci 101:1–16

Van Knegsel ATM, Van der Drift SGA, Cermákovà J, Kemp B (2013) Effects of shortening the dry period of dairy cows on milk production, energy balance, health, and fertility: a systematic review. Vet J 198:707–713

Van Knegsel ATM, Hammon HM, Bernabucci U, Bertoni G, Bruckmaier RM, Goselink RMA, Gross JJ, Kuhla B, Metges CC, Parmentier HK, Trevisi E, Tröscher A, van Vuuren A (2014a) Metabolic adaptation during early lactation: key to cow health, longevity and a sustainable dairy production chain. CAB Rev 9:1–15

Van Knegsel ATM, Remmelink GJ, Jorjong S, Fievez V, Kemp B (2014b) Effect of dry period length and dietary energy source on energy balance, milk yield, and milk composition of dairy cows. J Dairy Sci 97:1499–1512

Van Knegsel ATM, Burgers EEA, Ma J, Goselink RMA, Kok A (2022) Extending lactation length: consequences for cow, calf and farmer. J Anim Sci 100:1–10

Watters RD, Guenther JN, Brickner AE, Rastani RR, Crump PM, Clark PW, Grummer RR (2008) Effects of dry period length on milk production and health of dairy cattle. J Dairy Sci 91:2595–2603

Watters RD, Wiltbank MC, Guenther JN, Brickner AE, Rastani RR, Fricke PM, Grummer RR (2009) Effect of dry period length on reproduction during the subsequent lactation. J Dairy Sci 92:3081–3090

Zobel G, Weary DM, Leslie KE, von Keyserlingk MAG (2015) Invited review: cessation of lactation: effects on animal welfare. J Dairy Sci 98:8263–8277

# Surveillance and Decision-Making in Swine Health Management

Robert Friendship

Abstract

Swine health management requires diligent observation of animals and the collection and analysis of production records to assess the efficacy of on-farm disease control programs and to plan corrective strategies. Diagnostic testing is used to enhance these monitoring efforts and to guide decision-making associated with disease control interventions as well as inform biosecurity and disease eradication programs. There are constantly new diagnostic tests being developed, and it is now possible to use a wide variety of sample materials including serum, oral fluids, feces, semen, and tissues in monitoring programs. The usefulness and reliability of testing is dependent on several factors including the selection of animals to be tested and the attributes of the test that is used. Decision-making to determine intervention strategies, such as mass medication or vaccination, needs to be carefully evaluated to ensure a cost-benefit, in addition to addressing animal welfare and food safety concerns.

Keywords

Swine · Disease control · Biosecurity · Diagnostic testing · Performance data · Partial budgeting

## 1 Introduction

Swine health management programs, in general, focus on three main areas: biosecurity, disease elimination, and control of endemic health issues.

1. An important component of swine health management is implementing programs to keep disease from entering the herd (biosecurity). Modern pig production generally involves large populations of pigs raised indoors, and these herds are very often of a high health status (free of most economically important diseases). However, a consequence of this type of pig rearing is that the introduction of disease into this large naïve population can cause severe losses, and therefore there needs to be a major emphasis placed on

R. Friendship (✉)
Department of Population Medicine, University of Guelph, Guelph, ON, Canada
e-mail: rfriends@ovc.uoguelph.ca


J. J. Gross (ed.), *Production Diseases in Farm Animals*,
https://doi.org/10.1007/978-3-031-51788-4_26

preventing disease entry. This requires active surveillance of incoming replacement breeding stock as well as other measures to reduce the risk of disease introduction.
2. If an infectious disease does enter a herd, in some cases, it is possible to eradicate the disease from the herd without depopulating and repopulating the entire herd. The decision of whether to attempt eradication depends on economic factors, as well as the likelihood of success. In general, disease elimination without complete depopulation and repopulation requires manipulation of the immune status of the breeding herd (i.e., creating sow herd immunity) and control of the pig flow to prevent naïve pigs from becoming exposed to disease-carrying pigs. As pig farms become large and production stages are segregated, there is an increased likelihood that small populations of disease-carrying pigs can persist within the swine operation, making eradication more challenging.
3. Another major emphasis of swine health management programs is to establish intervention strategies to minimize the impact of diseases or health issues that cannot be feasibly eliminated from the herd (e.g., control of endemic infectious diseases and health problems with multiple contributing factors), and these intervention protocols need to be continually re-evaluated to determine their ongoing value and effectiveness. Constant monitoring of animals for clinical signs of disease and evaluating herd productivity are the main methods used to identify health concerns, and when a problem is recognized, mitigating strategies need to be implemented. Further monitoring is needed to verify the effectiveness of interventions, and cost-benefit analyses need to be performed to determine the value of the interventions.

Laboratory testing to detect inapparent carrier animals is an essential component of all three of these swine health management activities. The testing procedure and the diagnostic tests chosen, vary depending on why the testing is being performed. Testing can be expensive, and in the case of an active surveillance program, there needs to be a carefully constructed plan in place to address the possibility that an unexpected disease may be found.

The terms monitoring and surveillance have slightly different meanings but will be used interchangeably in this chapter for sake of simplicity.

## 2 Biosecurity

Preventing diseases from entering a swine population (either a single barn or a network of farms that operate interdependently) has always been important but biosecurity has now become a top priority. This rise in importance has resulted because of two developments in world pork production. Firstly, modern swine operations have greatly increased in herd size, so that today large populations of pigs are housed together, and in many cases, numerous barns are linked by animal or truck movement. Secondly, most modern pork production involves populations of pigs that are free from most of the economically important diseases. The result of raising pigs that have not been exposed to common swine pathogens means that the large population of high-health-status pigs on modern farms have no background immunity to the common disease agents that formerly were endemic on most farms. Therefore, if any of these diseases are introduced to this vulnerable population, a severe outbreak of clinical illness is likely to occur. The impact of disease entry on a naive population is much greater than if the pathogen was introduced to a herd where the animals were carrying at least partial immunity from previous disease exposure.

The term biosecurity is generally used to describe the program created to keep disease from entering a farm, but it also might refer to the strategy designed to limit the spread of a pathogen within a herd (Dewulf and Van Immerseel 2018). The latter form of biosecurity, sometimes called internal biosecurity, will be discussed in the sections describing the ongoing control of endemic diseases in a herd or how prevention of disease spread within a barn is a critical part of a disease eradication program.

Disease can enter a farm by many different routes, with certain diseases more readily transmitted via one route than another (Filippitzi et al. 2017). For example, some respiratory disease agents such as porcine reproductive and respiratory syndrome (PRRS) virus or *Mycoplasma hyopneumoniae* can readily spread by aerosol, while enteric disease agents such as porcine epidemic diarrhea (PED) virus or *Brachyspira hyodysenteriae* are more likely to be spread by fecal contamination. An effective biosecurity program needs to include preventative procedures that recognize the multiple possible sources of risk, but also needs to prioritize the areas that pose the highest probability of disease entry. Although biosecurity programs need to be unique for each farm because circumstances and security risks are different for each farm, there are certain common high-risk practices common to almost all farms.

Generally, the highest risk of disease entry comes from the introduction of live pigs, such as the purchase of replacement breeding stock. Maintaining a closed herd with no live animal introductions can greatly reduce the risk of disease entry but limits genetic improvement and is impractical for many farming operations. The risk can be reduced by minimizing the number of sources of incoming animals, by investigating the health status of the source (or sources) to ensure the purchased pigs are free of specific diseases, and by using an isolation facility to house the incoming animals so that their health status can be verified before exposure to the main herd. The next greatest risk is the entry to the farm of pig "secretions" including semen and feces. There are several important swine diseases that can be spread via semen (Van Rijn et al. 2004), including PRRS so that health monitoring of boar studs and frequent testing of boars and semen is essential to prevent disease spread from a boar stud to large numbers of sow herds.

Contaminated feces can be inadvertently brought into a swine barn on fomites. Footwear is often implicated with the movement of fecal material into the barn. A common requirement for visitors entering a pig barn is to remove footwear at the entrance and wear boots provided by the farm. Possibly the most common source of fecal material that threatens the biosecurity of a farm is a livestock transport vehicle. The movement of a livestock truck from farm to farm or from farm to abattoir and back to a farm can easily spread disease either by infecting pigs being transported or from fecal material on the boots of personnel moving from the trailer and into the barn or loading dock area. The washing and disinfection of livestock trucks is a very important part of a biosecurity program but expensive and difficult. There are certain pathogens such as PED and transmissible gastroenteritis (TGE) viruses that are easily transmitted with very small amounts of fecal matter, and these viruses can survive in cool damp conditions for days or even weeks.

In the case of respiratory diseases, aerosol transmission poses a threat. Pathogens such as *M. hyopneumoniae*, the etiological agent causing enzootic pneumonia, and PRRS virus are examples of microorganisms capable of spreading through the air from farm to farm, prompting some pork producers with barns in very pig dense regions to install filters at the air inlets.

Other routes of disease entry include contaminated feed, water, rodents, and birds. The principles of preventing the introduction of disease onto a swine farm are well described in many papers and book chapters (Neuman and Hall 2019; Pritchard et al. 2005; Amass and Clark 1999). Evaluating and scoring biosecurity is often attempted to rate a farm's biosecurity protocols and possibly point out how they can be improved (for example, Dewulf 2014). The difficulty in rating a farm's biosecurity program is that each farm has unique risks. For example, a farm located far from other farms does not need air filtration, whereas another farm located in a pig dense region may benefit from the sizable investment in air filtration. Likewise, depending on the source of replacement stock, the presence of a gilt isolation barn may be much more important for one farm compared to another. In general, biosecurity can be expensive, and it can add a degree of inconvenience to pig production and needs to be customized to meet the needs of each farm.

It is possible to invest too much time and money into a biosecurity program. Veterinarians and other consultants are very reluctant to suggest that a farm biosecurity program should be reduced to save money because of the fear that as soon as changes are made, a disease outbreak will occur. In reality, despite very strict biosecurity, disease outbreaks can still occur, either because disease spreads in ways we still don't completely understand, or that the biosecurity program was not strict enough or that it was not thoroughly enforced. If biosecurity protocols become too complex and onerous, the likelihood of mistakes increases. Casal et al. (2007) surveyed farm workers and found that many were unsure as to which procedures were most important and what aspects of the farm biosecurity program represented a low risk if not implemented. There needs to be a balance between keeping disease out and ensuring that farm operations can be performed with minimal interference.

## 3 Eradication of Disease

Sometimes the best way to eradicate a disease problem is to depopulate the herd (or to depopulate a specific stage of production) and then repopulate with pigs that are free of the disease. Ideally, before undertaking a depopulation–repopulation exercise, the source of the disease introduction causing the outbreak is discovered, and a plan is put in place to prevent reintroduction of the disease after the herd is re-established. Depopulation–repopulation is an expensive and drastic measure for removing a pathogen from a breeding herd because it generally causes a lengthy disruption in market hog sales (Nathues et al. 2018; Silva et al. 2019). Alternative methods have been established to eliminate certain diseases without the need to depopulate the sow herd.

There are specific examples of medications and vaccines that have been used along with the management of pig movement in the barn to eliminate a specific disease. For example, sarcoptic mange has been eradicated from almost all pig farms and this has been accomplished, for the most part, by the strategic use of ivermectin, usually involving a blanket treatment of the entire population of pigs. It is rare that a medication or vaccine alone, can eliminate a pathogen, and usually there needs to be a combination of vaccination, medication, and sanitation, as well as controlled movement of pigs to segregate non-diseased pigs from possible disease carriers. One of the first examples of using this combination of medication and segregation to create a herd free of a disease has been described as medicated early weaning. The process involves vaccinating and medicating sows, with farrowing occurring in an isolated facility. In addition, piglets are removed from the sows after about 5 days of nursing and moved to a segregated nursery. The weaned piglets are given additional medication. This approach has been used to re-establish high-health breeding nucleus herds, specifically to eradicate *M. hyopneumoniae* (Alexander et al. 1981).

The idea of segregating newly weaned pigs led to the establishment of segregated early weaning and multi-site production. It has become well recognized that the most important aspect of this strategy is not medication or early weaning but the benefit of segregating newly weaned pigs from older animals that might be carrying disease. In general, pigs at weaning have waning passive immunity that was obtained from their dam via colostrum. As passive immunity disappears and piglets become exposed to endemic diseases present in the herd, they will become infected. This phenomenon of naïve pigs becoming infected and then passing disease on to other pigs commonly occurs after weaning. Pigs in the nursery can act as a reservoir for disease for the rest of the herd and preventing this continuous spread of disease in the nursery is one of the biggest challenges of eradicating a disease from a pig farm.

The most common approach to eradicating disease on a pig farm is by means of creating herd immunity in the sows and then controlling the movement of pigs to prevent the naïve weaned pigs from encountering older growing pigs that may be still carrying disease. This is well described elsewhere (for example, Corzo et al. 2010). This approach is commonly used to successfully eliminate PRRS and PED viruses from swine operations. Briefly, when a herd that breaks with PRRS or PED, the first step is to purchase a large number of replacement gilts and then close the herd to any new introductions of animals for about 6 months. Usually, there is an attempt to create widespread immunity among all the sows and gilts, either by actively spreading the disease or by vaccinating the sows. Monitoring using diagnostic tests is performed to determine when all suckling piglets are receiving passive immunity from the sows but none of the piglets are viremic. Usually at this point, if nurseries are not off-site, there is a need to create a gap in production so that the newly weaned pigs are not encountering older viremic pigs. To ensure an eradication scheme has worked, it is important to monitor the weaned pig population to ensure pigs are not becoming infected as they progress through the various production stages.

## 4 Endemic Disease

In modern swine production, pigs are typically free of many of the economically important diseases that historically posed serious problems. The diseases that pigs tend to be free of can also be considered the diseases that are relatively easily eradicated from a herd and diseases that can be prevented from entering a herd with good biosecurity. Harris (2000) stated that if a farm is free of atrophic rhinitis, pleuropneumonia, pseudorabies, swine dysentery, and transmissible gastroenteritis, then it is relatively easy to keep these diseases out. Furthermore, he suggested that it was possible but harder to maintain a pig herd free of PRRS and enzootic pneumonia, and much more difficult to rear pigs free of Glasser's disease and streptococcal meningitis. There are infectious diseases, such as porcine circovirus-associated diseases, that can be found on almost all farms and these diseases appear to be almost impossible to keep out.

In addition to the trend of raising pigs with high health status, modern pig farms have become increasingly large. One consequence of this is that with many pigs on one site, it is easier for a pathogen to continuously find susceptible pigs to infect, and therefore continue to circulate in the herd. Management strategies to minimize spread of disease such as segregating production stages may be useful in minimizing the impact of endemic diseases but can unintentionally make it more difficult to create widespread immunity and thus allow endemic diseases to persist in a herd (Morris et al. 2002).

Although certain pathogens are found on almost all farms, their impact varies greatly from farm to farm. One reason for this variation in severity is that many of these disease agents are considered to require multiple factors for clinical signs to occur. In other words, the presence of a specific pathogen is not necessarily sufficient to cause clinical disease without additional contributing factors, so that the prevalence and severity of disease may depend on the presence of contributing factors rather than attributes of the pathogen. In the case of outbreaks of these diseases, identifying the pathogen does not provide enough information to sufficiently explain why the pigs became sick, and therefore further investigation is needed to determine what other factors contributed to the outbreak (Morris et al. 2002). Endemic health problems also include many conditions which do not appear to have an infectious component. Osteochondrosis, gastric ulceration, tail-biting, and rectal prolapses are all common and economically important swine health problems that do not appear to be caused by an infectious pathogenic agent. Control of these conditions requires the identification of possibly several contributing factors, and these factors may vary from farm to farm, and may include environment, housing, feed, genetics, and husbandry.

In general, the main strategy on pig farms for controlling endemic diseases that have a major infectious component is to attempt to maximize the immunity of the host and at the same time try to minimize the challenge from the pathogen. The most obvious methods for maximizing immunity include, ensuring colostrum intake for neonatal piglets, acclimatizing incoming replacement breeding stock, and instituting comprehensive vaccination programs. Strategies to reduce disease challenges include manipulating pig flow so that different production stages are housed separately, and rooms or entire buildings are managed using an all-in and all-out approach, with thorough washing and disinfection between groups. Housing design has been developed to accommodate all-in and all-out movement by sizing rooms appropriately so that, for example, all the pigs weaned at one time can be accommodated in a nursery room or an off-site nursery barn and when those pigs are moved to a grower-finisher barn, again they enter a facility that has been emptied and cleaned and is the appropriate size to handle the group of pigs until they reach market age. In addition, barns are constructed using materials that facilitate sanitation procedures so that, for example, metal or plastic slats have replaced more traditional porous cement floors in farrowing pens, making it much easier to remove organic matter and reduce the pathogen burden in the pig's environment.

## 5 Monitoring Health

A key component of swine health management is the monitoring of the herd for disease. The definition of disease has come to include a decrease in performance, for example, reduced fertility, slower growth rate, or poorer feed conversion, whether such decreases are caused by an infectious agent or not (Ramirez 2023). Traditionally, health monitoring involved diligent observation of animals for clinical signs of illness or changes in behavior, and this is still an important part of all health programs, although now the monitoring of data including performance records has become equally important. In general, health monitoring deals with endemic diseases and the goal of such programs is to recognize when common health issues change in prevalence and severity. For example, a herdsman may observe a few pigs with diarrhea in the first few weeks post-weaning and accept this as normal or within acceptable limits, but if the prevalence or severity of diarrhea changes, intervention is undertaken to treat these pigs, and plans are made to minimize the likelihood of similar disease levels occurring in subsequent batches of weaned pigs. In addition to observation and the identification of sick animals, there is still a need to verify the cause of the illness (Burrough et al. 2019). This most likely will involve gross post-mortem examinations and gathering tissue samples or obtaining various other biological samples from pen mates for diagnostic testing.

Production records have become an essential part of health monitoring (Stein et al. 1987). Rather than relying solely on observing pigs for clinical signs of disease and using this observation to determine when to treat, a herdsman may use production records as a means of judging when to intervene by setting limits for various production parameters. For example, in the previous case of diarrhea in nursery pigs, the nursery room records might reveal a reduction in growth rate, and this could be used to justify an intervention such as medication. Production records can be used to assess the success of the intervention and help determine the cost-benefit of the treatment or control measure. Polson et al. (1998) listed the following steps in population-based problem solving: (1) use production records to identify a change, (2) quantify the change in biological and financial terms, (3) initiate an investigation to determine the cause, (4) implement an intervention, and (5) continue to monitor and if necessary, change the intervention strategy.

One of the challenges to this approach is to recognize when a decrease in a production parameter is an indication of a problem and when a change merely reflects normal biological variation. Benchmarks are often established with minimum values set that will trigger some type of action. Benchmarks may be based on industry averages but because farms can vary greatly in management, facilities, genetics, health status, etc., often the best source of data to set production targets is to use historical farm records.

Using production records to solve a problem in the breeding herd has been well described elsewhere (for example, Koketsu 2016). Briefly, investigation would typically begin by examining composite parameters such as "pigs weaned per sow per year" which accounts for the reproductive efficiency of the sow herd summarized by "litters per sow per year," as well as the performance in the farrowing rooms measured by "pigs weaned per litter." If this composite parameter is low, then further investigation can determine if the cause is the breeding area or the farrowing room. Litters per sow per year is expected to be 2.3 or better, and if less than this, the cause could be related to preservice issues reflected in an elevated wean-to-service interval or a breeding or post-breeding problem reflected in a lower-than-expected farrowing rate. The farrowing rate (usually expected to be 85% or better) refers to the percentage of sows in a breeding group that farrow at the same time as the rest of the group. If the farrowing rate is low, further analysis of records may reveal whether the problem is a high regular return to service rate indicating a failure in conception, or the problem might be identified as being caused by a high number of irregular returns to service or abortions suggesting a loss of pregnancy. On the other hand, the record analysis might show that reproduction is as expected and the overall poor "pigs per sow per year" appears to be a low number of pigs weaned per litter, which could be further investigated to determine whether the cause of the low weaning numbers is a result of preweaning mortality or caused by too few live born piglets per litter.

Examination of post-weaning production records is possibly more important than using breeding herd performance parameters to solve health problems. Most of the cost of producing a market hog occurs in the grower-finisher stage and is primarily related to feed consumption. Disease during this stage can result in a slow growth rate, poor feed conversion, mortality and culling, and increased production costs associated with medication and extra labor. Sometimes costly disease outbreaks in the grower-finisher barn that negatively impact feed conversion can be overlooked or the importance underestimated because the pigs may display only mild clinical signs. Traditionally, post-weaning records tended to be collected and analyzed when a barn was emptied, providing information that could be used to make decisions regarding the next group of pigs but too late for interventions to be implemented in the current group. However, computerized feeding and recording systems are making it easier for producers to monitor feed use and growth rate continuously and make it possible to identify problems in a timely manner. It is also important to use this information to estimate the cost of losses associated with disease in the grower-finisher stage (Cornelison et al. 2018) so that informed decisions can be made regarding the cost-benefits of intervention programs.

Information from the abattoir can also be used in a health monitoring program. In most countries, the farmer can receive details regarding weight and quality of the carcasses and whether there was any part of the carcass that was removed for health reasons or whether a whole carcass was condemned. Programs have been established to evaluate pluck lesions at slaughter as a health monitoring tool (Scollo et al. 2017). If this type of information is available, it can be used to evaluate the success of health intervention strategies. For example, the efficacy of a vaccination program for enzootic pneumonia could be determined by counting the number of market hogs with lung damage and assessing the proportion of lung damage for each pig.

## 6 Decision-Making

Strategies that are implemented to control endemic diseases need to be periodically examined to determine if they are still effective or still needed, especially if new and possibly better alternative control measures have become available. In certain cases, circumstances change so that an effective control measure is no longer available, and alternatives need to be found. For example, in many countries the routine use of antibiotics or the inclusion of therapeutic levels of zinc oxide in feed has been restricted. Zinc oxide has been widely used at high levels (>2500 ppm) in nursery feeds to control post-weaning diarrhea caused by enterotoxigenic *E. coli*. There have been numerous alternatives promoted to replace zinc oxide and antibiotics for control of post-weaning diarrhea, including vaccines, phytogenic products, organic acids, and probiotics. The usefulness of these alternatives will need to be validated on-farm. This is generally true whenever an endemic disease is identified. In most cases of a disease outbreak, there are various treatment options, including the possibility of not intervening. A sporadic outbreak involving only a small number of animals may be self-limiting. However, if the problem appears to warrant action either because of potential economic loss, or a decrease in animal welfare or a potential zoonotic or food safety concern, then usually some type of cost-benefit analysis is needed to determine the best course of action. One of the most common examples of this type of decision-making happens when a farm manager needs to decide whether to vaccinate for a specific disease.

Vaccination has become an important tool in attempting to maximize immunity, although nutrition, genetics, and good animal husbandry are all important contributors to creating good immune status. In the case of the most common and economically important diseases, vaccines are generally available. The decision as to whether to use a specific vaccine on a specific farm needs to be evaluated using a cost-benefit approach. Sometimes the cost of vaccination will be greater than the cost of the disease. For example, on most farms pigs are exposed to *Glaesserella parasuis* and most young pigs carry the bacteria in their upper respiratory tracts and on rare occasions a pig may develop severe polyserositis (Glasser's disease). The small number of pigs that develop clinical diseases generally costs the farmer far less than a vaccination program for all young pigs. Another consideration is the effectiveness of the vaccine to control the disease. Some vaccines are more effective than others at minimizing clinical signs or reducing pathogen spread. A vaccine with only moderate to poor efficacy is not likely to provide the farmer with a positive economic return. The efficacy of a vaccine can be greatly diminished by vaccine storage and handling but also by the circumstances in the way the vaccine is used. In the case of post-weaning *E. coli* vaccines, immunity must be created soon after weaning because that is when disease challenge occurs but until weaning passive immunity protects the piglet and interferes with efforts to create active gut immunity. Therefore, in the case of an oral live *E. coli* vaccine against post-weaning diarrhea, failure may be the result of logistical difficulties in the timing of vaccination.

Another important consideration regarding whether to use a vaccine is to consider whether there are alternative approaches that might work better than a vaccination program or be easier to institute. For example, the swine industry has long utilized natural exposure to build herd immunity to certain pathogens rather than rely on vaccines directed at those pathogens. This approach has been used in the control of porcine reproductive and respiratory disease (Corzo et al. 2010). Exposure of incoming breeding stock to cull sows is another example of using natural exposure to improve immunity rather than vaccination. There are numerous husbandry practices that have been used to control disease rather than the use of vaccines. For example, the implementation of all-in and all-out pig flow, coupled with good hygiene, has likely contributed to the almost disappearance of several formerly common bacterial diseases of swine such as atrophic rhinitis, pleuropneumonia caused by *Actinobacillus pleuropneumoniae*, and swine dysentery, whereas vaccination strategies to control these diseases had disappointing results.

The negative impact of vaccines also needs to be considered before instituting a vaccine program. There are some obvious issues such as the possibility of a broken needle left in the muscle of a pig following an intramuscular injection, or an abscess resulting from a contaminated injection site. But, in addition, there is the possibility that vaccination during a stressful time such as weaning may produce side effects including fever and inappetence, and therefore cause a growth performance setback. Because the situation as to whether to vaccinate varies from farm to farm and varies from one disease to another, the best way to evaluate whether a vaccination program is appropriate is to conduct on-farm clinical trials, but such trials are quite difficult to do properly and may be impractical. On the other hand, it is quite feasible to periodically reconsider vaccination or other intervention programs and decide whether production performance is as good as expected.

There are numerous studies that have attempted to estimate the cost of important swine diseases. For example, PRRS has been estimated to cost the European and North American swine industries between $6.25 and $15.25 USD per pig annually (Neumann et al. 2005; Holtkamp et al. 2013; Nathues et al. 2018). Knowing roughly what the presence of a disease like PRRS costs, helps to guide the decision as to whether to undertake an intervention to control the disease or even whether to attempt to eliminate the disease from the herd. Partial budgeting is commonly used to create economic models which can help to determine the best course of action. For example, Silva et al. (2019) developed an economic model to estimate the economic value of implementing a herd closure intervention in a 5000-sow herd to eliminate *M. hyopneumoniae* and estimated a gain of $616,121 after 1 year, assuming the program was successful. They estimated that using medication to control the negative growth impact of *M. hyopneumoniae* in the herd would have resulted in a benefit of only about half this amount. However, the results from the medication program may have been more certain and easier for the producer to implement, so the decision as to which strategy to use to control *M. hyopneumoniae* might not be strictly based on the outcome of the financial model. Another study evaluated interventions to control PRRS comparing mass vaccination, herd closure and creation of natural herd immunity, and depopulation/repopulation (Nathues et al. 2018). Although depopulation/repopulation was the most reliable and effective method to eliminate PRRS, it had the lowest expected values for all situations considered in this study, because of the high cost associated with this intervention.

## 7 Diagnostic Tests and Their Use in Surveillance Programs

Diagnostic testing has become a routine part of on-farm swine health management programs. In addition to using laboratory tests to help identify the causative agent in a disease outbreak investigation, it is common for swine farms to use active surveillance programs that incorporate the routine gathering and testing of biological samples to help verify the efficacy of biosecurity and disease control programs. There are important considerations that need to be made regarding the design and implementation of a surveillance program because the labor and expense of gathering samples and laboratory testing can be substantial.

In general, the first consideration should be to outline the program's objectives. The program needs to be focused on one or a small number of economically important diseases that pose a significant risk to the herd to justify the cost. In addition, there needs to be a clear action plan in place to deal with the possibility that the surveillance program will identify positive animals. The objective of diagnostic testing in a biosecurity program is to find all pigs that are carrying a specific disease. If an error in identifying the disease status of a pig is made, it is better to falsely identify a negative pig as being positive rather than admit a positive pig into the herd because the test was misinterpreted as negative. The same objective holds true for testing during a disease eradication exercise. In the case of diagnostic testing for endemic diseases, in general, the pathogenic agent is assumed to be present on the farm so the objective of testing might be to determine disease prevalence or to identify further characteristics of the pathogen such as antibiotic sensitivity or virus serotype because this further information can help determine control options. The consequences of misclassifying a positive animal as negative in the case of endemic disease monitoring are not as important as it is in biosecurity surveillance. Therefore, the testing procedure and the diagnostic tests chosen might vary depending on why the testing is being performed or the objective of the testing program.

The design of the surveillance program needs to consider whether to test for the disease agent directly or to test for the pig's response (i.e., antibodies). This will influence the decision as to what type of diagnostic test to use. If testing is being performed as an investigation into a disease outbreak that happened several weeks earlier, then testing for antibodies in serum is likely the most appropriate approach because the disease agent may have already disappeared. However, if testing is being performed as part of a biosecurity program to determine if newly arrived gilts are carrying a disease, then testing for the disease agent is necessary. Antibody response takes 10–14 days to appear, so if testing is directed at the immune response, then it may miss pigs with an active infection. In addition, antibody response may be a result of vaccination rather than exposure to a disease agent so the presence of antibodies may be difficult to interpret in a vaccinated population.

The two most common assays used for swine disease surveillance purposes are polymerase chain reaction (PCR) and enzyme-linked immunoassay (ELISA). In general, PCR tests can be used on a variety of sample types if the test substance is thought to possibly contain the disease agent. The test is based on the amplification of nucleic acid that can be detected and analyzed. One drawback is that PCR cannot detect whether the genetic material is from a live or dead organism, unlike bacterial or viral culture which can confirm the organism is alive and thus provide more information than PCR. However, culture is much more expensive, time consuming and for many organisms, culture is very difficult and unreliable. Enzyme-linked immunoassays use an enzyme to detect antigen–antibody binding and can be used to either detect the presence of the disease agent or more commonly to detect the presence of antibodies in serum. Although there are numerous other antibody tests, ELISAs are by far the most popular because they tend to be fast, inexpensive, simple, and consistent (Munguia-Ramirez et al. 2023).

Tests vary in their ability to detect the agent or to detect specific antibodies. A test that can correctly identify a high proportion of animals that are either carrying the disease agent or have antibodies to the disease agent, as positive is said to have high sensitivity. Diagnostic tests that result in a high proportion of disease-carrying animals misclassified as negative are considered to have low sensitivity. The other important attribute of a diagnostic test that needs to be considered is specificity. This is the ability of a test to correctly identify pigs that are not carrying the disease or that have not been previously exposed to the disease and classify these animals as negative. A test with low specificity will produce a high proportion of false positives or in other words, pigs that are not infected with a disease may be misclassified as having the disease if the test has low specificity. There are situations where the sensitivity of a test is increased by classifying all suspicious results as positive rather than negative. The result of interpreting the test results in this manner improves the sensitivity but lowers the specificity. As discussed earlier, in the case of a biosecurity program, it is ideal to use a test that has high sensitivity, even if it has a low specificity.

In addition to considering the attributes of the diagnostic test, it is also important to be mindful of which animals to test. In general, it is not practical to test the entire population of pigs to determine if there is a disease present, so often a subset of pigs is chosen for sampling (Arruda and Gauger 2019). In many cases a randomized sampling procedure is appropriate but under certain circumstances, it might be best to sample pigs that appear sick or are in proximity to pigs that have shown clinical signs because they may be the most likely to be positive. To optimize the number of tests to be performed, it is sometimes useful to pool samples. There are many sources for formulas to determine the necessary sample size when calculating the number of pigs to test in a population. These calculations require assumptions regarding the expected prevalence of the disease in the population and they depend on the objective of the testing. If the objective of testing is to determine the prevalence of the disease on the farm, then generally more pigs need to be sampled than if the purpose for testing is to determine if the disease is present or absent. Commonly there is an assumption that the disease will be in the population with a prevalence of at least 10%, and that would require sampling 25 pigs if the population was 100 or about 30 pigs if the population was 1200 (Ramirez 2023). The standard confidence level is generally set at 95%.

A wide choice of sample material can be used for testing, although blood or serum are traditionally used, particularly when testing for immune response. Feces or rectal swabs, especially in the case of enteric disease surveillance are frequently tested. More recently oral fluids have become a popular choice because of the relative ease of collecting these samples by allowing pigs to chew on cotton ropes and then placing the ropes into plastic bags and squeezing the fluid from the ropes (Rotolo et al. 2017). It needs to be noted that the result is an aggregate sample from a group of pigs where not all individuals are represented equally. This is different than a pooled sample and the traditional sample size calculations do not apply to this type of sample collection. In addition, samples can include semen (Van Rijn et al. 2004), processing fluids (Vilalta et al. 2018), laryngeal and tracheal swabs, feed, water, and air samples.

## 8 Summary

A swine health management program involves the prevention of diseases from entering the farm, eliminating diseases that can feasibly be eliminated from the farm, and monitoring the health of the herd to detect outbreaks of clinical signs of disease or decreases in productivity. Decisions to institute control programs require considering the cost-benefit of the interventions, often in the form of a partial budget analysis. The collection of performance data is essential for creating the economic models that guide decisions regarding health management interventions. Active surveillance programs using routine diagnostic testing are helpful in supporting biosecurity and guiding disease control programs.

## References

Alexander TJL, Boon G, Thornton K (1981) The establishment of a nucleus herd by medicated early weaning. Proc Pig Vet Sci 8:74–81

Amass SF, Clark LK (1999) Biosecurity considerations for pork production units. J Swine Health Prod 7:217–228

Arruda HEP, Gauger P (2019) Optimizing sample selection, collection and submission to optimize diagnostic value. In: Zimmerman J, Karriker L, Ramirez A, Schwartz K, Stevenson G (eds) Diseases of swine, 11th edn. Blackwell Publishing, Hoboken, pp 98–111

Burrough ER, Baum DH, Schwartz KJ (2019) Collecting evidence and establishing causality. In: Zimmerman J, Karriker L, Ramirez A, Schwartz K, Stevenson G (eds) Diseases of swine, 11th edn. Blackwell Publishing, Hoboken, pp 112–122

Casal J, De Manuel A, Mateu E, Martin M (2007) Biosecurity measures on swine farms in Spain: perceptions by farmers and their relationship to current on-farm measures. Prev Vet Med 82:138–150

Cornelison AS, Karriker LA, Williams NH, Haberl BJ, Stalder KJ, Schulz LL, Patience JF (2018) Impact of health challenges on pig growth performance, carcass characteristics, and net returns under commercial conditions. Transl Anim Sci 2:50–61

Corzo CA, Mondaca E, Wayne S, Torremorell M, Dee S, Davies P, Morrison RB (2010) Control and elimination of porcine reproductive and respiratory syndrome virus. Virus Res 154:185–192

Dewulf J (2014) An online risk-based biosecurity scoring system for pig farms. Vet Ireland J 4:426–429

Dewulf J, Van Immerseel F (2018) General principles of biosecurity in animal production and veterinary medicine. In: Dewulf J, Van Immerseel F (eds) Biosecurity in animal production and veterinary medicine, from principle to practice. ACCO, pp 295–328

Filippitzi ME, Brinch Kruse A, Postma M, Sarrazin S, Maes D, Alban L, Nielsen LR, Dewulf J (2017) Review of transmission routes of 24 infectious diseases preventable by biosecurity measures and comparison of the implementation of these measures in pig herds in six European countries. Transbound Emerg Dis 65:381–398

Harris DL (2000) Control of common infectious swine diseases. In: Multi-site pig production. Iowa State University Press, pp 97–124

Holtkamp DJ, Kliebenstein JB, Neumann EJ, Zimmerman JJ, Rotto HF, Yoder TK, Wang C, Yeske PE, Mowrer CL, Haley CA (2013) Assessment of the economic impact of porcine reproductive and respiratory syndrome virus on United States pork producers. J Swine Health Prod 21:72–84

Koketsu Y (2016) Factors for high reproductive performance of sows in commercial herds. Proc Int Pig Vet Soc:42–49

Morris RS, Davies PR, Lawton DE (2002) Evolution of diseases in the world's pig industry. In: Proceedings of the 17th international pig veterinary society congress

Munguia-Ramirez B, Armenta-Leyva B, Gimenez-Lirola L, Wang C, Zimmerman J (2023) Surveillance on swine farms using antemortem specimens. In: Maes D, Segales J (eds) Optimizing pig health and production. Burleigh Dodds Science Publishing, Cambridge, pp 97–138

Nathues H, Alarcon P, Rushton J, Jolie R, Fiebi K, Jimenez M, Guerts V, Nathues C (2018) Modelling the economic efficiency of using different strategies to control Porcine Reproductive & Respiratory Syndrome at herd level. Prev Vet Med 152:89–102

Neuman EJ, Hall WF (2019) Disease control, prevention and elimination. In: Zimmerman J, Karriker L, Ramirez A, Schwartz K, Stevenson G (eds) Diseases of swine, 11th edn. Blackwell Publishing, Hoboken, pp 123–157

Neumann EJ, Kliebenstein JB, Johnson CD, Mabry JW, Bush EJ, Seitzinger AH, Green AL, Zimmerman JJ (2005) Assessment of the economic impact of porcine reproductive and respiratory syndrome on swine production in the United States. J Am Vet Med Assoc 227:385–392

Polson DD, Marsh WE, Dial GD (1998) Population-based problem solving in swine herds. J Swine Health Prod 6:267–272

Pritchard G, Dennis I, Waddilove J (2005) Biosecurity: reducing disease risks to pig breeding herds. In Pract 27:230–237

Ramirez A (2023) Improving disease surveillance and monitoring systems in pig herds. In: Maes D, Segales J (eds) Optimizing pig health and production. Burleigh Dodds Science Publishing, Cambridge, pp 67–95

Rotolo ML, Sun Y, Wang C, Gimenez-Lirola L, Baum DH, Gauger PC, Harmon KM, Hoogland M, Main R, Zimmerman JJ (2017) Sampling guidelines for oral fluid-based surveys of group-housed animals. Vet Microbiol 209:20–29

Scollo A, Gottardo F, Conteiro B, Mazzoni C, Leneveu P, Edwards SA (2017) Benchmarking of pluck lesions at slaughter as a health monitoring tool for pigs slaughtered at 170kg (heavy pigs). Prev Vet Med 144:20–28

Silva GS, Yeske P, Morrison RB, Linhares DC (2019) Benefit-cost analysis to estimate the payback time and the economic value of two *Mycoplasma hyopneumoniae* elimination methods in breeding herds. Prev Vet Med 168:95–102

Stein T, Martineau GP, Morris RD, Charette R (1987) A new approach to managing health in swine operations. Can Vet J 28:355–362

Van Rijn PA, Wellenberg GJ, Hakze-van der Hoing R, Jacobs L, Moonen PL, Feitsma H (2004) Detection of economically important viruses in boar semen by quantitative RealTime PCR™ technology. J Virol Methods 120:151–160

Vilalta C, Sanhueza J, Alvarez J, Murray D, Torremorell M, Corzo C, Morrison R (2018) Use of processing fluids and serum samples to characterize porcine reproductive and respiratory syndrome virus dynamics in 3-day-old pigs. Vet Micobiol 225:149–156

# Index




J. J. Gross (ed.), *Production Diseases in Farm Animals*,
https://doi.org/10.1007/978-3-031-51788-4

MIX
Papier aus verantwortungsvollen Quellen
Paper from responsible sources
FSC® C105338

If you have any concerns about our products,
you can contact us on
**ProductSafety@springernature.com**

In case Publisher is established outside the EU,
the EU authorized representative is:
**Springer Nature Customer Service Center GmbH**
**Europaplatz 3, 69115 Heidelberg, Germany**

Printed by Libri Plureos GmbH
in Hamburg, Germany